By Appointment to Her Majesty Queen Elizabeth II
Manufacturers of Motor Cars and Land-Rovers

By Appointment to Her Majesty Queen Elizabeth
the Queen Mother
Suppliers of Motor Cars and Land-Rovers

Rover 3500
Automatic gearbox models
and
Rover 3500S
Synchromesh gearbox models

REPAIR OPERATIONS MANUAL
PUBLICATION PART NUMBER AKM 3621

Issued by the
ROVER TRIUMPH SERVICE DIVISION, LEYLAND CARS
COVENTRY, ENGLAND

CONTENTS

General specification data	04
Engine tuning data	05
Torque wrench settings	06
Recommended lubricants, fuel and fluids—capacities	09
Maintenance	10
Engine	12
Emission control	17
Fuel system	19
Cooling system	26
Manifold and exhaust system	30
Clutch	33
Gearbox—Synchromesh	37
Gearbox—Automatic Type 35 / Type 65	44
Propeller shafts	47
Rear axle and final drive	51
Steering	57
Front suspension	60
Rear suspension	64
Brakes	70
Wheels and tyres	74
Body	76
Heating and ventilation	80
Air conditioning	82
Windscreen wipers and washers	84
Electrical	86
Instruments	88
Service tools	99

Purchasers are advised that the specification details set out in this Manual apply to a range of vehicle and not to any particular vehicle. For the specification of any particular vehicle Purchasers should consult their Distributor or Dealer.

The Manufacturers reserve the right to vary their specifications with or without notice, and at such times and in such manner as they think fit. Major as well as minor changes may be involved in accordance with the Manufacturer's policy of constant product improvement.

Whilst every effort is made to ensure the accuracy of the particulars contained in this Manual, neither the Manufacturer nor the Distributor or Dealer, by whom this Manual is supplied, shall in any circumstances be held liable for any inaccuracy or the consequences thereof.

© **Rover-British Leyland UK Limited 1972**

All rights reserved. No part of this publication may be reproduced, stored in a retrieval system or transmitted, in any form, electronic, mechanical, photocopying, recording or other means without prior written permission of Rover-British Leyland UK Limited.

INTRODUCTION

The purpose of this Manual is to assist skilled mechanics in the efficient repair and maintenance of Rover 3500 and 3500S vehicles.

The procedures detailed, carried out in the sequence given and using the appropriate service tools, will enable the operations to be completed in the time stated in the Repair Operation Times.

Indexing

For convenience, this Manual is divided into a number of divisions. A contents page listing the titles and reference numbers of the various divisions is shown opposite.

A list of the operations within each of the divisions appears in alphabetical order on the contents page preceding each of the divisions.

Operation Numbering

Each operation is followed by the number allocated to it in a master index. The number consists of six digits arranged in three pairs.

The master index of operations has been compiled for universal application to vehicles manufactured by British Leyland Motor Corporation and therefore continuity of the numbering sequence is not maintained throughout the Manual. To assist with locating information, each division of the Manual is preceded by a contents page listing the operations in alphabetical order.

Each instruction within an operation has a sequence number and, to complete the operation in the minimum time it is essential that these instructions are performed in numerical sequence commencing at 1 unless otherwise stated. Where applicable the sequence numbers identify the components in the appropriate illustration.

Where performance of an operation requires the use of a service tool, the tool number is quoted under the operation heading and is repeated in, or following the instruction involving its use.

An illustrated list of all service tools necessary to complete the operations described in the Manual is also included.

References

References to the left or right hand side in the Manual are made when viewing the vehicle from the rear. With the engine and gearbox assembly removed, the water pump end of the engine is referred to as the front.

Repairs and Replacements

When service parts are required it is essential that only genuine Rover or Unipart replacements are used.

Attention is particularly drawn to the following points concerning repairs and the fitting of replacement parts and accessories.

> Safety features embodied in the car may be impaired if other than genuine parts are fitted.
>
> In certain territories, legislation prohibits the fitting of parts not to the vehicle manufacturers' specification.
>
> Torque wrench setting figures given in the Repair Operation Manual must be strictly adhered to. Locking devices, where specified, must be fitted. If the efficiency of a locking device is impaired during removal it must be renewed.
>
> Owners purchasing accessories while travelling abroad should ensure that the accessory and its fitted location on the car conform to mandatory requirements existing in their country of origin.
>
> The car warranty may be invalidated by the fitting of other than genuine Rover or Unipart parts. All Rover or Unipart replacements have the full backing of the factory warranty.
>
> British Leyland Distributors and Dealers are obliged to supply only genuine service parts.

Rover 3500 and 3500S Manual AKM 3621

ABBREVIATIONS AND SYMBOLS IN THIS MANUAL

Across flats (bolt size)	AF
After bottom dead centre	ABDC
After top dead centre	ATDC
Alternating current	a.c.
Ampere	amp
Ampere-hour	amp hr
Atmospheres	Atm
Before bottom dead centre	BBDC
Before top dead centre	BTDC
Bottom dead centre	BDC
Brake mean effective pressure	BMEP
Brake horse power	bhp
British Standards	BS
Carbon monoxide	CO
Centimetre	cm
Centigrade (Celcius)	C
Cubic centimetre	cm³
Cubic inch	in³
Degree (angle)	deg or °
Degree (temperature)	deg or °
Diameter	dia.
Direct current	d.c.
Fahrenheit	F
Feet	ft
Feet per minute	ft/min
Fifth	5th
Figure (illustration)	Fig.
First	1st
Fourth	4th
Gramme (force)	gf
Gramme (mass)	g
Gallons	gal
Gallons (US)	US gal
High compression	h.c.
High tension (electrical)	H.T.
Hundredweight	cwt
Independent front suspension	i.f.s.
Internal diameter	i.dia.
Inches of mercury	in.Hg
Inches	in
Kilogramme (force)	kgf
Kilogramme (mass)	kg
Kilogramme centimetre (torque)	kgf.cm
Kilogramme per square centimetre	kg/cm²
Kilogramme metres (torque)	kgf.m
Kilometres	km
Kilometres per hour	km/h
Kilovolts	kV
King pin inclination	k.p.i.
Left-hand steering	LHStg
Left-hand thread	LHThd
Litres	litre
Low compression	l.c.
Low tension	l.t.
Maximum	max.
Metre	m
Microfarad	mfd
Midget edison screw	MES
Millimetre	mm
Miles per gallon	mpg
Miles per hour	mph
Minimum	min
minute (of angle)	'
Minus (of tolerance)	-
Negative (electrical)	-
Number	No.
Ohms	ohm
Ounces (force)	ozf
Ounces (mass)	oz
Ounce inch (torque)	ozf.in.
Outside diameter	o.dia.
Paragraphs	para.
Part number	Part No.
Percentage	%
Pints	pt
Pints (US)	US pt
Plus (tolerance)	+
Positive (electrical)	+
Pound (force)	lbf
Pounds feet (torque)	lbf.ft.
Pounds inches (torque)	lbf.in.
Pound (mass)	lb
Pounds per square inch	lb/in²
Radius	r
Rate (frequency)	c/min
Ratio	:
Reference	ref.
Revolution per minute	rev/min
Right-hand	RH
Right-hand steering	RHStg
Second (angle)	"
Second (numerical order)	2nd
Single carburetter	SC
Specific gravity	sp.gr.
Square centimetres	cm²
Square inches	in²
Standard	std.
Standard wire gauge	s.w.g.
Synchroniser/synchromesh	synchro.
Third	3rd
Top dead centre	TDC
Twin carburetters	TC
United Kingdom	UK
Volts	V
Watts	W

SCREW THREADS

American Standard Taper Pipe	NPTF
British Association	BA
British Standard Fine	BSF
British Standard Pipe	BSP
British Standard Whitworth	Whit.
Unified Coarse	UNC
Unified Fine	UNF

GENERAL SPECIFICATION DATA

ENGINE
Type — V8
Number of cylinders — Eight, two banks of four
Bore — 88,90 mm (3.500 in)
Stroke — 71,12 mm (2.800 in)
Capacity — 3528 cm³ (215 in³)
Valve operation — Overhead by pushrod

Crankshaft
Main journal diameter — 58,400 to 58,413 mm (2.2992 to 2.2997 in)
Minimum regrind diameter — 57,384 to 57,396 mm (2.2592 to 2.2597 in)
Crankpin journal diameter — 50,800 to 50,812 mm (2.0000 to 2.0005 in)
Minimum regrind diameter — 49,784 to 49,797 mm (1.9600 to 1.9605 in)
Crankshaft end thrust — Taken on thrust faces of centre main bearing
Crankshaft end-float — 0,10 to 0,20 mm (0.004 to 0.008 in)

Main bearings
Number and type — 5 Vandervell shells
Material — Lead indium
Diametrical clearance — 0,023 to 0,065 mm (0.0009 to 0.0025 in)
Undersizes — 0,254 mm, 0,508 mm, 0,762 mm, 1,016 mm
0.010 in, 0.020 in, 0.030 in, 0.040 in

Connecting rods
Type — Horizontally split big end, plain small end
Length between centres — 143,81 to 143,71 mm (5.662 to 5.658 in)

Big end bearings
Type and material — Vandervell VP lead indium
Diametrical clearance — 0,015 to 0,055 mm (0.0006 to 0.0022 in)
End-float on crankpin — 0,15 to 0,37 mm (0.006 to 0.014 in)
Undersizes — 0,254 mm, 0,508 mm, 0,762 mm, 1,016 mm
0.010 in, 0.020 in, 0.030 in, 0.040 in

Gudgeon pins
Length — 72,67 to 72,79 mm (2.861 to 2.866 in)
Diameter — 22,215 to 22,22 mm (0.8746 to 0.8749 in)
Fit-in con rod — Press fit
Clearance in piston — 0,002 to 0,007 mm (0.0001 to 0.0003 in)

Pistons
Early type — Aluminium alloy with apertures below gudgeon pin
  Clearance in bore measured at top of skirt at right angles to gudgeon pin — 0,018 to 0,033 mm (0.0007 to 0.0013 in)

Latest type — Aluminium alloy with 'W' slot skirt
  Clearance in bore measured at bottom of skirt at right angles to gudgeon pin — 0,018 to 0,033 mm (0.0007 to 0.0013 in)

Piston rings
Compression — 2
  Number one compression ring — Chrome faced and marked 'T' or 'TOP'
  Number two compression ring — Stepped 'L' shape and marked 'T' or 'TOP'
  Gap in bore — 0,44 to 0,57 mm (0.017 to 0.022 in)
  Clearance in groove — 0,08 to 0,13 mm (0.003 to 0.005 in)
Oil control — 1
  Type — Perfect circle, type 98
  Gap in bore — 0,38 to 1,40 mm (0.015 to 0.055 in)

Camshaft
Location — Central
Bearings — Non-serviceable
Timing chain — 9,52 mm (0.375 in) pitch x 54 pitches

continued

Rover 3500 and 3500S Manual AKM 3621

04-1

GENERAL SPECIFICATION DATA

Valves
Inlet
Overall length — 116,58 to 117,34 mm (4.590 to 4.620 in)
Head diameter — 37,97 to 38,22 mm (1.495 to 1.505 in)
Angle of face — 45°
Stem diameter — 8,640 to 8,666 mm (0.3402 to 0.3412 in) at the head and increasing to 8,653 to 8,679 mm (0.3407 to 0.3417 in)
Stem to guide clearance: Top — 0,02 to 0,07 mm (0.001 to 0.003 in)
Bottom — 0,013 to 0,063 mm (0.0005 to 0.0025 in)

Exhaust
Overall length — 116,58 to 117,34 mm (4.590 to 4.620 in)
Head diameter — 33,215 to 33,466 mm (1.3075 to 1.3175 in)
Angle of face — 45°
Stem diameter — 8,628 to 8,654 mm (0.3397 to 0.3407 in) at the head and increasing to 8,640 to 8,666 mm (0.3402 to 0.3412 in)
Stem to guide clearance: Top — 0,038 to 0,088 mm (0.0015 to 0.0035 in)
Bottom — 0,05 to 0,10 mm (0.002 to 0.004 in)

Valve lift (both valves) — 9,9 mm (0.39 in)

Valve spring length
Inner — 41,4 mm (1.63 in) under load of 9,75 to 12,02 kg (21.5 to 26.5 lb)
Outer — 40,6 mm (1.6 in) under load of 17,69 to 20,41 kg (39 to 45 lb)

Valve timing
Inlet opens — 30° BTDC
Inlet closes — 75° ABDC
Inlet duration — 285°
Inlet peak — 112.5° ATDC
Exhaust opens — 68° BBDC
Exhaust closes — 37° ATDC
Exhaust duration — 285°
Exhaust peak — 105.5° BTDC

Lubrication
System — Wet sump, pressure fed
System pressure, engine warm at 2400 rev/min — 2,1 to 2,8 kgf/cm² (30 to 40 lbf/in²)
Oil filter — Full-flow, self-contained cartridge
Oil pump type — Gear
Oil pressure relief valve
  Type — Non-adjustable
  Relief valve spring
    Free length — 81,2 mm (3.200 in)
    Compressed length at 4,2 kg (9.3 lb) load — 45,7 mm (1.800 in)
Oil filter by-pass valve
  Type — Non-adjustable
  By-pass valve spring
    Free length — 37,5 mm (1.48 in)
    Compressed length at 0,34 kg (0.75 lb) — 22,6 mm (0.89 in)

CLUTCH
Make/type — Borg & Beck, diaphragm type
Clutch plate diameter — 241,3 mm (9.5 in)
Facing material — Raybestos WR7
Number of damper springs — 6
Damper spring colour — Brown/cream
Clutch release bearing — Ball journal
Clutch fluid — Refer to Division 09

continued

04-2   Rover 3500 and 3500S Manual AKM 3621

GENERAL SPECIFICATION DATA

TRANSMISSION

Gearbox—Synchromesh
Type	Single helical constant mesh
Speeds	4 forward 1 reverse
Synchromesh	All forward speeds
Ratios Fourth (Top)	1.00:1
Third	1.391:1
Second	2.133:1
First	3.625:1
Reverse	3.430:1
Overall ratios (Final drive)	
Fourth (Top)	3.08:1
Third	4.284:1
Second	6.57:1
First	11.165:1
Reverse	10.564:1

Gearbox—Automatic

Type — Hydraulic torque converter, providing torque multiplication between the ratios 2:1 and 1:1. Planetary gear set, comprising two sun gears, two sets of pinions, a pinion carrier and a ring gear. The various mechanical ratios are obtained by the engagement of hydraulically operated multi-disc clutches and brake bands.

Torque converter stall ratio

Gear ratios
- Direct (top)   1:1   2.16
- Intermediate (second)   1.45:1   Torque converter ratio   2.16
- Low (first)   2.39:1   2.16
- Reverse   2.09:1

Overall gear ratios:
1st	7.36
2nd	4.47
3rd	3.08
Reverse	6.45
1,000 engine rpm in 3rd gear	38,5 kph (24 mph)
Maximum automatic upshift speeds (Using kick-down mechanism):	
1st to 2nd	58 to 72 kph (36 to 45 mph)
2nd to 3rd ('P', 'R', 'N', 'D2', 'D1', 'L' selector pattern)	100 to 113 kph (62 to 71 mph)
2nd to 3rd ('P', 'R', 'N', 'D', '2', '1' selector pattern)	109 to 125 kph (68 to 78 mph)
Maximum speed in first gear (at 5,200 rev/min)	77 kph (48 mph)
Maximum speed in second gear (5,200 rev/min)	128 kph (80 mph)
Rear axle gear ratio	3.08:1
Engine stall speed	1,950 to 2,250 rev/min

FINAL DRIVE
Type	Hypoid
Ratio	3.08:1

PROPELLER SHAFT
Type	Hardy Spicer
Universal joints	Needle bearing
Overall free length (face to face)	
Manual gearbox models	1136,65 mm (44.750 in)
Automatic gearbox models	1003,3 mm (39.500 in)

continued

GENERAL SPECIFICATION DATA

COOLING SYSTEM
Type — Pressurized spill return system with thermostat control, pump and fan assisted
Type of pump — Centrifugal
Thermostat — 78 to 84°C (173 to 182°F)
Pressure cap — 1,05 kgf/cm² (15 lbf/in²)

FUEL SYSTEM
Carburetter — Refer to engine tuning data
Fuel pump
Make/type — AC, mechanical
Pressure range — 0,246 to 0,351 kgf/cm² (3.5 to 5.0 lbf/in²)
(Air conditioned models are fitted with a Bendix electrical pump, see Division 82)

SUSPENSION
Coil springs
Front:
Manual steering models:
- Number of working coils — 6 3/8
- Free length — 413,5 mm (16.281 in)
- Rate — 30,35 kg/cm (170 lb/in)
- Identification — 12,5 mm (0.500 in) wide white stripe painted on entire length of spring

Power steering:
- Number of working coils — 7 1/3
- Free length — 434,8 mm (17.120 in)
- Rate — 30,35 kg/cm (170 lb/in)
- Identification — 12,5 mm (0.500 in) wide green stripe painted on entire length of spring

Rear:
Early 3500, up to chassis suffix 'A'
- Number of working coils — 5 1/2
- Free length — 338 mm (13.312 in)
- Rate — 45,43 kg/cm (260 lb in)
- Identification — 12 mm (0.5 in) wide, green and blue stripes painted on entire length of spring.

Latest 3500 from chassis suffix 'B' onwards and all 3500S :
- Number of working coils — 5.4
- Free length — 331,1 mm (13.038 in)
- Rate — 45,43 kg/cm (260 lb in)
- Identification — Two 12 mm (0.5 in) wide white stripes painted on entire length of spring

Hydraulic dampers
Type — Telescopic, double acting
Bore: front — 25,4 mm (1.000 in)
rear — 35,0 mm (1.375 in)

Hubs
Front hub end-float — 0,07 to 0,12 mm (0.003 to 0.005 in)

continued

GENERAL SPECIFICATION DATA

STEERING
Manual steering
Make/type — Burman recirculating ball, worm and nut
Ratio — Variable: Straight ahead 21.5:1
Full lock 26:1

Steering wheel diameter — 431,8 mm (17 in)
Steering wheel turns, lock-to-lock — 4.5

Power steering
Make/type — Adwest Varamatic/linkage
Ratio — Variable: Straight ahead 19.3:1
Full lock 15.4:1

Steering wheel diameter — 406,4 mm (17 in)
Steering wheel turns, lock-to-lock — 3.5

Steering angles and dimensions
Front wheel alignment — 3,0 mm (0.125 in) toe in ± 1,5 mm (0.625 in)
Camber angle — 0° ± 1° — Check with vehicle in static unladen condition, that is, vehicle with water, oil and five gallons of fuel. Rock the vehicle up and down at the front to allow it to take up a static position
Castor angle — 1½° positive ± ½°
Swivel pin inclination — 8°

BRAKES
Foot brake
Type — Disc
Operation — Hydraulic, servo assisted. Self adjusting

Front brake
Type — Outboard discs with three pistons
Disc diameter — 274 mm (10.82 in)
Disc thickness (new) — 12,7 mm (0.505 in)
Disc minimum worn thickness — 11,43 mm (0.455 in)

Rear brake
Type — Inboard discs with one piston
Disc diameter — 272 mm (10.690 in)
Disc thickness (new) — 9,65 mm (0.380 in)
Disc minimum worn thickness — 8,38 mm (0.330 in)

Total pad area — 264 cm² (41 in²)
Total swept area — 2400 cm² (372 in²)
Pad material — Mintex NNTX M114 GG or Mintex M. 108 or DON 230
Hydraulic fluid — Refer to Division 09

Handbrake
Type — Mechanical, operating rear calipers
Handbrake lever ratio — 80:1 overall

WHEELS
Size/type — 5½JSL x 14 in

Tyres
Size/type — 185 HR 14

continued

Rover 3500 and 3500S Manual AKM 3621

GENERAL SPECIFICATION DATA

ELECTRICAL EQUIPMENT

System	12 volt, negative earth
Fuses	35 amp, blow rating — Horns, cigar lighter, interior lights, windscreen washer, stop lights, flasher lights, heater and wiper motor
	8 amp, blow rating cigar, clock and panel illumination

Battery
Make/type	Lucas RCAZ 11/7 or Exide 6-XAZ 11R or Exide 6-XAZ 13R
Capacity at 20-hr. rate	60 amp-hour

Alternator
Early models:

Make/type	Lucas 11AC
Nominal output	45 amps at 6000 alternator rev/min
Field resistance	3.8 ohms ± 5%
Brush spring pressure	212 to 241 gm (7.5 to 8.5 oz)
Brush minimum length	4 mm (0.15 in)

Control unit, 11AC alternator

Make/type	Lucas 4TR
Regulating voltage (preset)	13.9 to 14.3 volts

Latest models

Make/type	Lucas 18ACR
Nominal output	45 amps at 6000 alternator rev/min
Field resistance	3.2 ohms
Brush spring pressure	255 to 368 gf (9 to 13 ozf.)
Brush spring minimum length	8 mm (0.300 in)
Regulating voltage	13.6 to 14.4 volts

Starter motor

Make/type	
Early models	Lucas M45G, pre-engaged
Latest models	Lucas 3M100, pre-engaged

Wiper motor

Early 3500 models	Lucas DL3A or 15W
Latest 3500 and 3500S models	Lucas 16W

Horns
Make/type	Lucas 6H

Distributor
Refer to engine tuning data

GENERAL DIMENSIONS

Overall length	4560 mm (179.75 in)
Overall width	1680 mm (66.0 in)
Overall height	1420 mm (56.25 in)
Wheelbase	2630 mm (103.375 in)
Track, front	1350 mm (53.375 in)
Track, rear	1310 mm (51.75 in)
Ground clearance under differential	216 mm (8.5 in)
Turning circle	9600 mm (31.5 ft)

WEIGHTS

	3500	3500S
Kerbside (includes oil, water and 22,5 litres—5 gallons—of fuel)	1303 kg (2872 lb)	1306 kg (2878 lb)
Maximum towing weight, car fully laden and using Rover approved towing attachment	1250 kg (25 cwt)	1250 kg (25 cwt)
Maximum roof rack load	50 kg (112 lb)	50 kg (112 lb)

ENGINE TUNING DATA

ENGINE

Type	V8
Capacity	3528 cm³ (215 in³)
Compression ratio:	
Engines numbered in the range commencing 425, 427 and, 451, 453, 455 with a suffix 'A' 'B' or 'C'	10.5:1 (8.5:1 to special order)
Engines numbered in the range commencing 451, 453, and 455 from suffix 'D' onwards	9.25:1
Firing order	1—8—4—3—6—5—7—2
Cylinder numbering system, front to rear	
Left bank	1—3—5—7
Right bank	2—4—6—8
Compression pressure (minimum)	
10.5:1 compression ratio	12,5 kgf/cm² (175 lbf/in²)
9.25:1 compression ratio	9,5 kgf/cm² (135 lbf/in²)
8.5:1 compression ratio	10,9 kgf/cm² (155 lbf/in²)
Idling speeds:	
10.5:1 compression ratio	600 to 650 rev/min
9.25:1 compression ratio	700 to 750 rev/min
8.5:1 compression ratio	600 to 650 rev/min
Air conditioned models	700 to 750 rev/min with the compressor disengaged
Fast idle speed	1100 to 1200 rev/min
Ignition timing, static and dynamic at 600 engine rev/min	
10.5:1 compression ratio	6° BTDC for use with fuel of 100 minimum research octane number
	TDC for use with fuel of 96 to 99 research octane number
9.25:1 compression ratio	6° BTDC for use with fuel of 96 minimum research octane number
8.5:1 compression ratio	6° BTDC for use with fuel of 90 minimum research octane number
Timing marks	On crankshaft pulley
Valve clearance	Not adjustable

DISTRIBUTOR

Make/type	Lucas 35 D8
Rotation of rotor	Clockwise
Dwell angle	26° to 28°
Contact breaker gap	0,36 mm to 0,40 mm (0.014 in to 0.016 in)
Condenser capacity	0.18 to 0.25 microfarad
Serial number	1569

continued

Rover 3500 and 3500S Manual AKM 3621

ENGINE TUNING DATA

Centrifugal advance—Decelerating check with vacuum unit disconnected

Early models with Lucas No. 41176 or 4278 stamped on the distributor body
   Ignition timing

Crankshaft angle	Engine rev/min
6° BTDC	
30° to 34°	4800
26° to 30°	3800
18° to 22°	1800
14° to 18°	1400
8° to 12°	1000
6°	Below 600

Vacuum advance
  starts   101 mm (4 in) Hg
  finishes   508 mm (20 in) Hg

Models with Lucas No. 41317 stamped on the distributor body
   Ignition timing

Crankshaft angle	TDC	Engine rev/min
6° BTDC		
30° to 34°	24° to 28°	4800
26° to 30°	20° to 24°	3800
20° to 24°	14° to 18°	2400
16° to 20°	10° to 14°	1800
9° to 12°	3° to 6°	1200
6° to 9°	0° to 3°	900
6°	0°	Below 700

Vacuum advance
  starts   101 mm (4 in) Hg
  finishes   508 mm (20 in) Hg

Models with Lucas No. 41392 or 41393 stamped on the distributor body
   Ignition timing

Crankshaft angle	Engine rev/min
6° BTDC	
25° to 29°	4400
20° to 24°	2600
15° to 20°	1800
8° to 12°	1200
6°	Below 600

Vacuum advance
  starts   101 mm (4 in) Hg   101 mm (4 in) Hg
  finishes   381 mm (15 in) Hg } 41392   508 mm (20 in) Hg } 41393

Models with 41394 stamped on the distributor body
   Ignition timing

Crankshaft angle	Engine rev/min
TDC	
26° to 30°	4800
22° to 26°	4000
15° to 19°	2400
9° to 14°	1800
2° to 6°	1200
0°	Below 700

Vacuum advance
  starts   101 mm (4 in) Hg
  finishes   508 mm (20 in) Hg

continued

ENGINE TUNING DATA

Models with Lucas No. 41573 stamped on the distributor body
   Ignition timing

Crankshaft angle	Engine rev/min
6° BTDC	
26° to 30°	4200
23° to 27°	3500
18° to 22°	2600
12° to 16°	1500
8° to 12°	1200
6° to 7°	800
6°	Below 600

Vacuum advance
   starts   90 mm (3.500 in) Hg
   finishes   381 mm (15 in) Hg

SPARKING PLUGS
Make   Champion
Type   L92Y
Gap   0,60 mm (0.025 in)

IGNITION COIL
Make/type   Lucas 16 C 6 with ballast resistor
Primary resistance at 20°C (68°F)   1.2 to 1.4 ohms
Consumption—ignition on at 2000 rev/min   1 amp

CARBURETTERS
Engines numbered in the range commencing 425 and 427
Make/type   SU HS6
Bore   44,45 mm (1.75 in)
Needle: Early 3500   KO
        New look 3500   BAK
Jet size   2,54 mm (0.100 in)
Float level   3,0 to 4,5 mm (0.062 to 0.187 in)

Engines numbered in the range commencing 451, 453 and 455 with a suffix 'A' 'B' or 'C'
Make/type   SU HIF6
Bore   44,45 mm (1.75 in)
Needle   BBG
Starter valve spindle   1 mm orifice
Piston spring colour   Yellow
Float level   1,0 mm ± 0,5 mm (0.040 in ± 0.020 in)
Damper oil   SAE 20

Engines numbered in the range commencing 451, 453 and 455 from suffix 'D' onwards
Make/type   SU HIF6
Bore   44,45 mm (1.75 in)
Needle   BBV
Starter valve spindle   1 mm orifice
Piston spring colour   Yellow
Float level   1,0 mm ± 0,5 mm (0.040 in ± 0.020 in)
Damper oil   SAE 20

NOTE: Refer to Division 17 for emission controlled models.

TORQUE WRENCH SETTINGS

ENGINE

	kgf m	lbf ft	
Connecting rod cap nuts	6,2	40	
Main bearing cap bolts, numbers one to four	7,6	55	
Rear main bearing cap bolts	9,6	70	
Cylinder head bolts	9,6	70	see 12.29.10
	and 6,2	45	
Rocker shaft bolts	4,0	30	
Flywheel bolts	8,5	60	
Oil pump cover bolts	1,2	9	
Oil pressure relief valve	4,9	35	
Timing chain cover bolts	3,5	25	
Crankshaft pulley	22,3	160	
Distributor drive gear to camshaft bolt	6,2	45	

COOLING SYSTEM

	kgf m	lbf ft
Water pump housing bolts 7/16 in AF	1,0	8
Water pump housing bolts 1/2 in AF	3,5	25

MANIFOLDS AND EXHAUST SYSTEM

	kgf m	lbf ft
Induction manifold bolts	4,0	30
Induction manifold gasket clamp bolts	2,0	15
Exhaust manifold bolts	2,3	16

CLUTCH

	kgf m	lbf ft
Clutch cover bolts	2,8	20
Clutch pedal lever nut	4,1	30
Clutch withdrawal unit bolts	2,0	15
Cross-shaft end cover bolts	1,0	8
External clutch lever nut	2,0	15
Slave cylinder bolts	3,5	25

GEARBOX—AUTOMATIC, BORG-WARNER TYPE 35

	kgf m	lbf ft
Torque converter to drive plate	4,0	30
Drive plate to crankshaft	7,0	50
Transmission case to bell housing	1,7	13
Extension housing to transmission case	1,7	13
Oil pan to transmission case	1,7	13
Front servo to transmission case	1,7	13
Rear servo to transmission case	3,6	27
Pump adaptor to front pump body (locating screw)	0,4	3 (35 lbf in)
Pump adaptor to front pump body	3,0	22
Pump adaptor to transmission case	2,5	18
Rear adaptor to transmission case	0,7	5
Centre support to transmission case	1,7	13
Outer lever to manual valve shaft	2,0	15
Pressure take-off plug	0,7	5 (60 lbf in)
Oil pan drain plug	1,6	12
Oil tube collector to lower valve body	0,35	2.5 (30 lbf in)
'D1–D2' control valve to lower valve body	0,35	2.5 (30 lbf in)

continued

TORQUE WRENCH SETTINGS

	kgf m	lbf ft
Lower body end plate to lower valve body	0,35	2.5 (30 lbf in)
Upper body end plate, front or rear, to upper valve body	0,35	2.5 (30 lbf in)
Upper valve body to lower valve body	0,35	2.5 (30 lbf in)
Valve bodies assembly to transmission case	0,7	5 (66 lbf in)
Pump strainers to lower valve body	0,35	2.5 (30 lbf in)
Downshift valve cam bracket to valve body	0,35	2.5 (30 lbf in)
Body to governor sleeve	Tighten fully with large screwdriver	
Cover plate to governor body	0,35	2.5 (30 lbf in)
Adjusting screw, front servo	0,11	10 lbf in
Locknut, front servo lever	2,8	20
Adjusting screw, rear servo	1,4	10
Locknut, rear servo to case	4,0	30
Starter inhibitor switch (self-adjusting type)	1,10	8
Starter inhibitor switch locknut	0,8	6 (72 lbf in)
Downshift valve cable adaptor to transmission case	1,2	9 (108 lbf in)
Filler tube connector adaptor to transmission case	4,0	30
Filler tube to connector sleeve nut	2,4	18
Stone guards to converter housing	0,22	1.6 (19 lbf in)
Coupling flange to driven shaft	3,5	25
Oil cooler to transmission case connector	1,4	10 (120 lbf in)
Extension housing to mounting bracket on sub-frame	3,5	25
Torque converter housing to flexible drive plate housing	4,0	30
Downshift cable clevis to link	0,3	2 (25 lb in)
Range control valve to lower valve body	0,35	2.5 (30 lbf in)
Sighting and filler tube to adaptor and adaptor to rear of oil pan	1,4	10

GEARBOX—AUTOMATIC, BORG-WARNER TYPE 65

	kgf m	lbf ft
Torque converter to drive plate	4,0	30
Drive plate to crankshaft	7,0	50
Gearbox case to converter housing, upper	4,1	30
Gearbox case to converter housing, lower	6,9	50
Fluid pan to gearbox case	0,9	7.0
Front servo cover	3,4	25
Rear servo cover	3,4	25
Pump adaptor to pump housing	0,42	36 lbf in
Pump adaptor to gearbox case	3,4	25
Pressure take-off plug	1,2	8
Upper valve body to lower valve body	0,35	30 lbf in
Lower valve body to upper valve body	0,35	30 lbf in
Lower valve body (200) to lower valve body (202)	0,35	30 lbf in
Pump oil strainer to lower valve body	0,35	30 lbf in
Oil tube collector to lower valve body	0,35	30 lbf in
End plate to lower valve body	0,35	30 lbf in
End plate to upper valve body	0,35	30 lbf in
Lower valve body to case	0,35	30 lbf in
Tube location plate	0,35	30 lbf in
Detent spring to lower valve body	0,35	30 lbf in
Front servo adjusting screw locknut	5,5	40
Rear servo adjusting screw locknut	5,5	40
Connector, oil tube	3,0	22
Extension housing to gearbox case	7,6	55
Inhibitor switch to gearbox case	0,69	60 lbf in
Parking brake cam plate to gearbox case	0,83	70 lbf in
Governor retaining bolt	2,5	18
Coupling flange bolt	6,9	50
Valve bodies assembly to transmission case	0,7	5

continued

Rover 3500 and 3500S Manual AKM 3621

TORQUE WRENCH SETTINGS

GEARBOX—SYNCHROMESH	kgf m	lbf ft
Bell housing cover plate bolts	1,0	8
Bell housing to engine bolts	3,5	25
Bell housing to main case nuts	7,0	50
Drain plug	2,8	20
External clutch lever nut	2,0	15
Filler plug	2,8	20
Gearchange shaft housing nuts	1,0	8
Rear output driving flange nut	10,5	75
Reverse gear shaft retaining plate bolt	1,0	8
Reverse light switch nuts	0,5	4
Selector fork pinch bolt, 1st—2nd and 3rd—4th	2,8	20
Selector fork pinch bolt, reverse	1,6	12
Selector fork stop bolt locknuts	1,0	8
Selector shaft detent ball retaining plate bolts	1,0	8
Selector shaft seal retaining plate bolts	1,0	8
Speedometer drive housing:		
$\frac{1}{4}$ in UNF bolts	1,0	8
$\frac{3}{8}$ in UNF bolts	3,5	25
Top cover bolts	2,0	15

PROPELLER SHAFT		
Coupling flange bolts	4,0	30

STEERING		
Steering box sector shaft cover	6,0	42
Steering pump pulley bolt	1,6	12
Steering pump valve cap	4,9	35

SUSPENSION		
Front		
Brake caliper mounting bolts	8,5	60
Top ball swivel nut	7,5 to 11,5	55 to 85
Bottom ball swivel nut	8,5 to 10,0	60 to 75
Bottom link strut to bottom link	8,5 to 10,0	60 to 75
Bottom link to base unit	7,5	55
Bottom link strut to base unit	7,5	55
Bottom link to swivel pillar	8,5 to 10,0	60 to 75
Top link securing bolts	4,0	30
Anti-roll bar cap bolts	4,0	30
Rear		
Bottom link bolts	7,5	55
de Dion tube elbows	1,0	8
Top link bolts	7,5	55
Hub to de-Dion tube	2,8	20
Axle flange to brake disc	11,7	85
Shock absorber lower mounting nut	2,0	15
Shock absorber lower mounting locknut	3,2	23

continued

TORQUE WRENCH SETTINGS

	kgf m	lbf ft
REAR AXLE AND FINAL DRIVE		
Crown wheel bolts		
Early type (ten $\tfrac{5}{16}$ in diameter and two fitted bolts)	3,5	25
Intermediate type—(twelve $\tfrac{5}{16}$ in diameter bolts)	4,8	35
Latest type—(twelve $\tfrac{3}{8}$ in diameter bolts)	5,5 to 6,2	40 to 45
Pinion nut	10,0	75
Drive shaft flanges to disc bolts	11,7	85
Cover to pinion housing		
$\tfrac{5}{16}$ in diameter bolts	2,0	15
$\tfrac{3}{8}$ in diameter bolts	4,0	30
Bearing housing to pinion housing bolts	4,0	30
Extension housing to pinion housing bolts	4,0	30
Caliper hinge pin sealing nut	4,9	35
Caliper hinge pin plug	3,5 to 4,9	25 to 35
Coupling flange to extension shaft bolt: $\tfrac{3}{8}$ in UNF	4,9	35
$\tfrac{7}{16}$ in UNF	10,0	75
Hub to de-Dion tube bolts	2,8	20
Panhard rod to final drive fixing	4,0	30
Extension housing to flexible mounting	4,9	35
Flexible mounting to front mounting bracket	2,3	17
Final drive rear mounting bracket to flexible mounting	6,2	45
Differential case fixings	6,2	45
BRAKES		
Brake caliper to front suspension member	8,5	60
Brake caliper to final drive bearing housing	8,5	60
Disc to front hub	6,0	44
Disc to rear drive flange	11,7	85
AIR CONDITIONING		
Compressor rear bearing cover plate	2,3	17
Compressor seal cover plate	1,7	13
Compressor connecting rod bolts	2,2	16
Compressor base plate	3,0	22
Compressor valve plate	3,0	22
Compressor "Rotalock" valve hexagon	4,9	35
Compressor crankcase filler plug	0,8	6
ELECTRICAL		
Alternator—Lucas 11AC and 18ACR		
Brush box fixing screws	0,1	10 lbf in
Diode heat sink fixings	0,3	2 (25 lbf in)
Through bolts	0,50 to 0,57	4 (45 to 50 lbf in)
Starter motor—Lucas 3M100PE		
Through bolts	1,1	8
Solenoid stud fixing nuts	0,6	4.5
Solenoid upper terminal nuts	0,4	3
Starter motor—Lucas M45G		
Securing bolts	4,9	35
Eccentric pivot pin locknut	2,2	16

Rover 3500 and 3500S Manual AKM 3621

RECOMMENDED LUBRICANTS, FLUIDS AND FUEL—CAPACITIES

RECOMMENDED LUBRICANTS AND FLUIDS

These recommendations apply to temperate climates where operational temperatures may vary between 10°C (14°F) and 32°C (90°F).

Lubricants marked with an asterisk (*) are multigrade oils suitable for all temperature ranges.

Information on oil recommendations for use under extreme winter or tropical conditions can be obtained from your local Rover Distributor or Dealer or

The Rover-British Leyland UK Limited, Technical Service Department.

Dia. No.	Components	SAE	BP	Castrol	Duckhams	Esso	Mobil	Texaco	Shell
8, 3, 15	Engine, 3500S Gearbox—Synchromesh, de Dion tube	20W	*BP Super Visco-Static 10-40W or 20-50	*Castrol GTX	Duckhams Q20-50 Motor Oil	Uniflow or Esso Extra Motor Oil 20W-30	Mobiloil Super or Mobiloil Special 20W-50	Havoline 20/20W or 20W-50	*Shell Super Oil
11, 2	3500 Gearbox—Automatic, Power steering fluid reservoir		BP Autran B	Castrol TQF	Duckhams Q-Matic	Esso-Glide	Mobil ATF 210	Texamatic Type F	Shell Donax T7
13, 12	Final drive unit, Steering box—manual steering	90EP	BP Gear Oil SAE 90 EP	Castrol Hypoy	Duckhams Hypoid 90	Esso Gear Oil GX90/140	Mobilube GX 90	Multigear Lubricant EP90	Spirax 90 EP
1, 9, 5, 14, 4	Front hub, L.H., Front hub, R.H., Rear hub, L.H., Rear hub, R.H., Propeller shaft	—	BP Energrease L2	Castrol LM Grease	Duckhams LB 10 Grease	Esso Multi-purpose Grease H	Mobilgrease MP or Mobilgrease Super	Marfak Allpurpose	Retinax A or Darina AX
6	Fuel tank	100 Research Octane Fuel 5-Star rating in the United Kingdom with standard ignition timing (6° BTDC static and dynamic) 96-99 Research Octane fuel with reset ignition timing. (TDC static and dynamic)
10	Brake and clutch fluid reservoir	Castrol Girling Brake and Clutch Fluid, Green † Specification, current SAE J1703 and US Federal Standard 116 or Unipart Brake and Clutch Fluid type 550 (Coloured green) † The latest Green fluid is miscible with the earlier Crimson specification
7	Radiator	Anti-freeze solution (use summer and winter)	Bluecol A A, coloured green for summer and winter use, or Anti-freeze conforming to British Standard No. 3150, or Prestone, or Anti-freeze to MIL-E-5559 formulation, providing mixed anti-freezes are miscible.
		Inhibitor solution (use instead of Anti-freeze if frost precautions are NOT required)	Marston Lubricants SQ36—Coolant inhibitor concentrate

AIR CONDITIONED CARS

Air conditioning compressor	BP Energol LPT 100	Texaco Capella E	Shell Clavus 33

Refrigerant: Refrigerant 12. This includes Freon 12 and Arcton 12.

CAPACITIES

The following capacity figures are approximate and are provided as a guide only. All oil levels must be set using the dipstick or level plug, as applicable.

	Litres	Imperial unit	US unit
Engine sump oil	4,5	8 pints	9.5 pints
Extra when refilling after fitting new filter	0,56	1 pint	1.2 pints
3500 Gearbox-Automatic	8,0	14 pints	17 pints
3500S Gearbox-Synchromesh	1,75	3.25 pints	4.0 pints
Final drive unit oil	1,3	2.25 pints	2.5 pints
de Dion tube	0,19	0.33 pints	0.4 pints
Steering box—manual steering	0,4	0.75 pints	0.9 pints
Hydraulic fluid—power steering	2,25	4 pints	4.75 pints
Cooling system	8,5	15.25 pints	18.5 pints
Fuel tank	68,0	15 gallons	18 gallons

AIR CONDITIONING

	Litres	Imperial unit	US unit
Air conditioning compressor			
Initial charge	0,28	0.5 pints	0.6 pints
Normal running	0,23	0.4 pints	0.5 pints
Refrigerant	0,9 kg	2 lb	

Rover 3500 and 3500S Manual AKM 3621

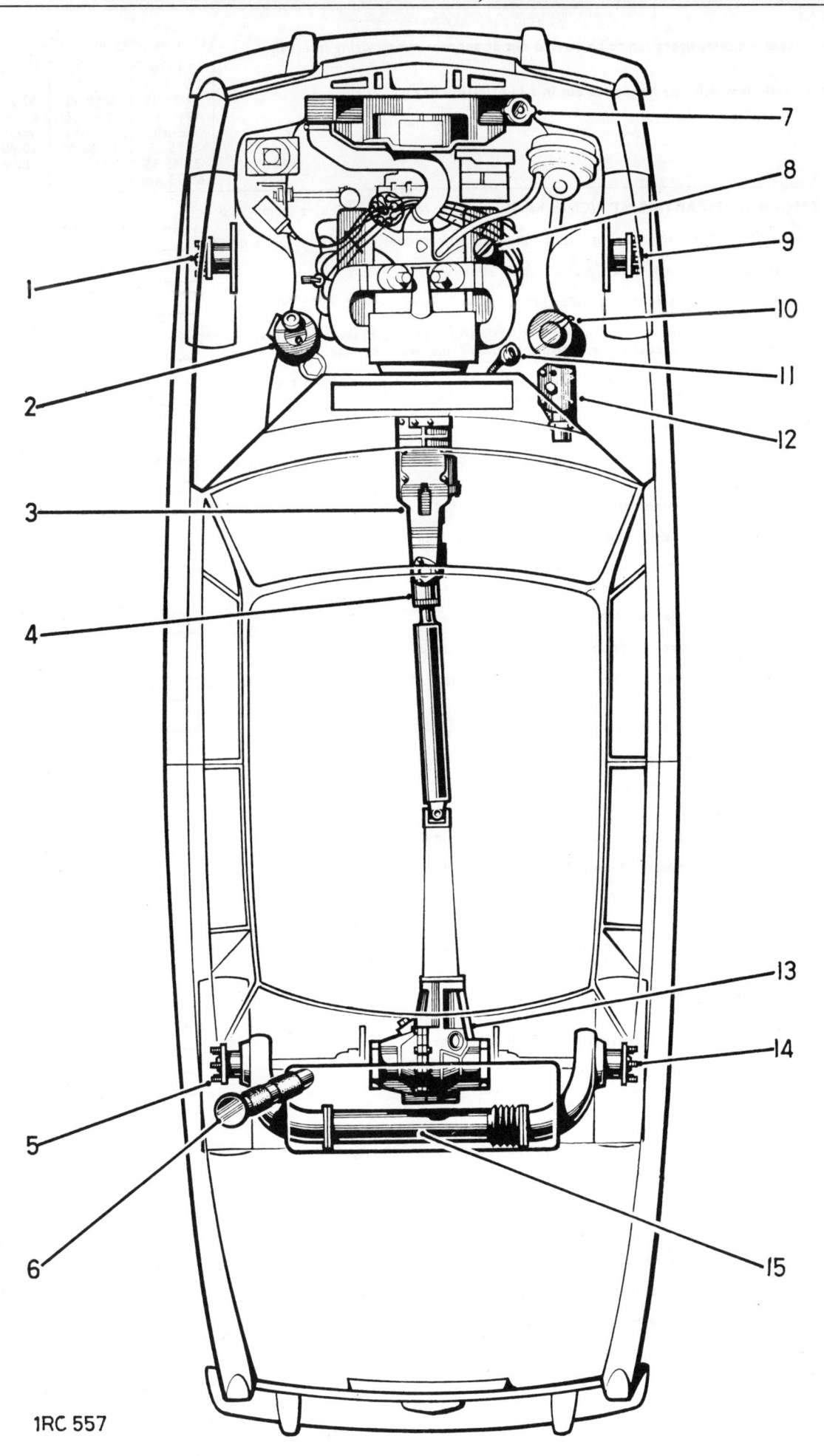

RECOMMENDED LUBRICANTS, FLUIDS AND FUEL—CAPACITIES

1RC 557

Rover 3500 and 3500S Manual AKM 3621

09-2

MAINTENANCE

☐ Box symbol indicates operation to be carried out at appropriate servicing interval.

LUB Defines operations which may be carried out in a Lubrication Bay.

	Servicing interval X 1000		
	10.10.06 3 9 15 21 27 33 39 45 miles/ months 5 15 25 35 45 55 65 75 km	10.10.12 6 18 30 42 miles/ months 10 30 50 70 km	10.10.24 12 24 36 48 miles/ months 20 40 60 80 km

PASSENGER COMPARTMENT (Clean hands or fit gloves when checking items 1 to 8 inclusive)

#	Operation	06	12	24	
1	Fit seat covers. Place floor covers in position. Fit steering wheel cover if necessary. Drive on lift	☐	☐	☐	1
2	Check steering wheel backlash	☐	☐	☐	2
3	Check handbrake to Manufacturer's specification	☐	☐	☐	3
4	Check foot brake operation to Manufacturer's specification. Ensure brake pedal travel is not excessive and maintains a satisfactory pressure under normal working load	☐	☐	☐	4
5	Check/adjust operation of windscreen washers. Ensure jets are not blocked and that liquid is confined to area covered by windscreen wiper blades	☐	☐	☐	5
6	Check function of original equipment, i.e. interior and exterior lamps, horns and indicators	☐	☐	☐	6
7	Check condition and security of seats and seat belts. Seat belts which are frayed or cut must be reported to the Owner	☐	☐	☐	7
8	Check rear view mirror for looseness, cracks and crazing	☐	☐	☐	8
9	Check/top-up gearbox oil 3500S. Top-up to bottom of filler plug hole		☐	☐	9

EXTERIOR AND BOOT COMPARTMENT

10	Lubricate all locks and hinges. Include bonnet release		☐	☐	10
11	Check operation of all window controls		☐	☐	11
12	Check/top-up battery electrolyte. Follow Manufacturer's instructions	☐	☐	☐	12
13	Clean/grease battery connections. Grease with petroleum jelly. Check spare wheel, see items 18, 19 and 20		☐	☐	13
14	Check headlamp alignment. Use authorised equipment		☐	☐	14
15	Check, if necessary renew windscreen wiper blades		☐	☐	15

Raise lift to a convenient height and wheel free
Remove front road wheels

#	Operation				
16	Inspect front brake pads for wear. Minimum thickness 3,0 mm (0.215 in.)	☐	☐	☐	16
17	Inspect front discs for condition. Check for wear, cracks, pitting, corrosion or damage	☐	☐	☐	17
18	Check that tyres comply with Manufacturer's specifications	☐	☐	☐	18
19	Check tyres for tread depth, cuts in fabric, exposure of ply or cord structure, lumps or bulges. Tread depth must be at least 1 mm throughout at least three-quarters of the breadth of tread and round the entire outer circumference	☐	☐	☐	19
20	Check/adjust tyre pressures: Front: 1,9 kg cm² (28 lb/sq in) 1.93 bars Rear: 2,1 kg cm² (30 lb/sq in) 2.07 bars	☐	☐	☐	20
21	Replace road wheels. Check tightness of road wheel fastenings	☐	☐	☐	21

UNDERBODY Raise lift to full height

#	Operation				
22	Check for oil fluid leaks	☐	☐	☐	22
23	Drain engine oil	☐	☐	☐	23
24	Check exhaust system for leakage and security	☐	☐	☐	24
25	Check visually fuel, clutch and hydraulic pipes, unions for chafing, leaks and corrosion	☐	☐	☐	25
26	Check condition and security of steering joints and gaiters. Check that gaiters have not become dislodged or damaged	☐	☐	☐	26
27	Lubricate propeller shaft sliding joint ..LUB	☐	☐	☐	27
28	Lubricate handbrake mechanical linkage and cable ..LUB	☐	☐	☐	28
29	Inspect rear brake pads for wear. Minimum thickness 1,5 mm (0.062 in)	☐	☐	☐	29
30	Inspect rear discs for condition. Check for wear damage, pitting and corrosion	☐	☐	☐	30
31	Check/top-up oil level of final drive/rear axle oil. Top-up to bottom of filler plug hole LUB		☐	☐	31

Rover 3500 and 3500S Manual AKM 3621

MAINTENANCE

#	Task	
32	Check/top-up de Dion tube oil. Top-up to bottom of filler plug hole .. .. .. LUB	
33	Check that rubber boot on de Dion tube is not dislodged or damaged	
34	Renew engine oil filter. Refit engine sump drain plug and lower lift	
35	Fit exhaust extractor pipe. Open bonnet and fit wing covers	

ENGINE COMPARTMENT

#	Task	
36	Check/top-up cooling system: 25 mm (1 in) below bottom of filler neck (engine cold) ..	
37	Refill engine with oil. Capacity 4,5 litres (8 Imperial pints) Start and run engine for a short period to prime oil pump ..	
38	Check for oil leaks in engine compartment	
39	Check/top-up steering box oil. Top-up to bottom of filler plug hole in cover plate .. LUB	
40	Check condition of steering unit and relay. Ensure fixings are secure	
41	Check condition and security of steering track rod ball joints	
42	Check/top-up clutch fluid reservoir. 3500. Castrol Girling Brake and Clutch Fluid 'Green' Specification SAE J1703 .. .. LUB	
43	Check/top-up brake fluid reservoir. Castrol Girling Brake and Clutch Fluid 'Green' Specification SAE J1703 .. .. LUB	
44	Check operation of brake fluid level warning light switch: Ignition 'on' handbrake 'off'. Lift filler cap 25 mm (1 in), warning light should be illuminated	
45	Clean/adjust spark plugs. Gap: 0,60 mm (0.025 in)	
46	Renew spark plugs. Champion L92Y	
47	Check distributor points. Adjust or renew. Dwell angle 26° to 28°..	
48	Lubricate distributor	
49	Check/adjust ignition timing and distributor characteristics using electronic equipment: 6° BTDC at 600 revs/min.	
50	Check driving belts. Adjust or renew. Engine belt: 11 to 14 mm (0.437 to 0.562 in) free movement between alternator and crankshaft pulley. Power steering, as applicable: 6 to 9 mm (0.250 to 0.375 in) free movement between crankshaft and pump pulley	
51	Check/report cooling and heater system for leaks	
52	Check/top-up windscreen washer reservoir: 25 mm (1 in) below bottom of filler neck ..	
53	Renew cartridge type fuel filter	
54	Renew engine flame traps	
55	Renew air cleaner elements	
56	Clean engine breather filter. Renew every 40.000 km (24,000 miles)	
57	Check/top-up engine oil (See item 23 for oil change intervals) .. .. .. LUB	
58	Top-up carburetter piston dampers: 25 mm (1 in) below top of tube	
59	Lubricate accelerator control linkage and check operation	
60	Check/adjust carburetter idle speed and mixture settings: 600 revs/min	
61	Check/top-up automatic gearbox fluid. 3500 .. .. LUB	
62	Check downshift cable. 3500	
63	Check/top-up fluid in power steering reservoir, as applicable	
64	Check for fluid leaks from power steering	
65	Drive car off lift Check front wheel alignment: 3,0 mm (0.125 in) toe-in plus or minus 1,5 mm (0.062 in) .. .. Remove wing covers, lower and close bonnet	

ROAD OR DYNAMOMETER TEST

#	Task	
66	Carry out a short but careful test. Prior to commencing test, check function of ignition, oil pressure, brakes, direction indicator and headlamp main beam. During the test particular attention should be paid to the behaviour of the engine, the transmission, brakes, steering, also any body noises	
67	Check function of all instrumentation	
68	Check heater and heater controls	
69	Check operation of safety harness inertia reel locking mechanism	
70	Check for oil leaks	
71	Remove covers from seat, steering wheel and floor	
72	Ensure cleanliness of controls, door handles, steering wheel	

Rover 3500 and 3500S Manual AKM 3621

MAINTENANCE

This Division of the Manual includes a maintenance summary chart and the procedures for carrying out routine maintenance on Rover 3500 and 3500S models. The procedures are generally applicable but apply specifically to current models, for specific details of carburetter tuning, ignition timing and downshift cable setting for earlier models, refer to the applicable Operation in the relative Division of this Manual.

MAINTENANCE

Fuel recommendations

1. The engine is designed to run on 100 research octane fuel, five star grade in the United Kingdom. See 86.35.20 for alternative fuel and ignition timing recommendations.

Engine

2. Under adverse conditions such as driving over dusty roads or where short stop-start runs are made, oil changes, attention to the engine flame traps and breather filter replacements must be more frequent.

Air cleaner

3. When the car is driven over dusty roads, the elements should be changed more frequently.

Propeller shaft

4. Under tropical or sandy and dusty conditions, the sliding joint must be lubricated frequently to prevent ingress of abrasive materials.

Lubricants

5. The recommended lubricants have been found pre-eminently suitable for Rover cars, and should be used whenever possible in the grades specified.
   When ordering oil, the correct grade, as well as the make should be clearly stated.
   The oils recommended by Rover British Leyland UK Limited are complete in themselves and additives should not be used.
   Rover British Leyland UK Limited attaches very great importance to the nature of the lubricants used in its products and therefore, gives specific recommendations. See Division 09.

Summary chart

IMPORTANT

1. Every 1.000 km (750 miles) check engine oil level, power steering reservoir fluid level, when fitted, and water level in radiator and screen washer reservoir.

2. Drain and refill engine sump every 10.000 km (6,000 miles) or six months, whichever comes first.

3. Every week and every maintenance inspection check tyre pressures and inspect tyre treads; when high speed touring, the tyre pressures should be checked much more frequently, even to the extent of a daily check. If front wheel tread wear is uneven, check wheel alignment.

4. Every month and every maintenance inspection check fluid level in brake fluid reservoir, and battery acid level.

5. Brakes. Change brake fluid every 30.000 km (18,000 miles) or eighteen months. The fluid should also be changed before touring in mountainous areas if not done in the previous nine months. Use only Castrol Girling Brake and Clutch Fluid Green Specification, current SAE J1703 and US Federal Standard 116 from sealed tins, or Unipart brake and clutch fluid type 550 (coloured green) from sealed tins.

6. Owners are under a legal obligation to maintain all exterior lights in good working order, this also applies to headlamp beam setting, which should be checked at regular intervals by your Rover Distributor or Dealer.

   NOTE: The sequence of operations under the headings of Passenger Compartment, Boot Compartment, Exterior, Engine Compartment, Under Body and Road Test will enable the work to be carried out in the most efficient manner.

MAINTENANCE

PASSENGER COMPARTMENT

Foot and handbrake—Every 5.000 km (3,000 miles) or 3 months.

Check operation of foot and handbrake, ensure that the brake pedal travel is not excessive and maintains a satisfactory pressure under normal working load. Excessive pedal travel indicates worn brake pads.

If the brakes feel spongy, this may be caused by air in the hydraulic system and this must be removed by bleeding the system at the disc cylinders.

Prior to this operation all hydraulic hoses, pipes and connections should be checked for leaks and any leaks rectified.

Check operation of handbrake, ensure that it holds the car satisfactorily. If the handbrake lever can be pulled up to the stop, check the rear brake pads, adjust the handbrake mechanism or overhaul the handbrake mechanism and the rear calipers as necessary.

Electrical equipment—Every maintenance inspection.

Check operation of all lamps, direction indicators, warning lights, horns, instruments and other equipment. See Division 86 for replacement bulbs and units.

Rear view mirror, seats and safety belts—Every 5.000 km (3,000 miles) or 3 months.

Check rear view mirrors for security and examine mirror face for signs of cracking or crazing.

Check all seat fixings for security and examine condition of safety harness. Safety harness which have been used in an accident or are frayed or cut must be replaced.

Inertia reel mechanism check—Every 10.000 km (6,000 miles) or 6 months.

To ensure that the safety harness inertia reel mechanism is in satisfactory operating condition, it should be checked at the above intervals.

Gearbox oil level, 3500S. Every 10.000 km (6,000 miles) or 6 months.

1. Gearbox and clutch withdrawal are lubricated as one unit on 3500S models. Check oil level and top-up if necessary to the bottom of the filler plug hole.
2. Do not add anti-friction additives to the gearbox oil.
3. Lift the carpets, on the left-hand side of the gearbox cover, to one side and remove the large rubber plug from the cover.
4. The oil filler and level plug will now be accessible through the inspection hole.
5. If significant topping-up is required, check for oil leaks at the drain and filler plugs, all joint faces and through the drain hole in the bell housing.

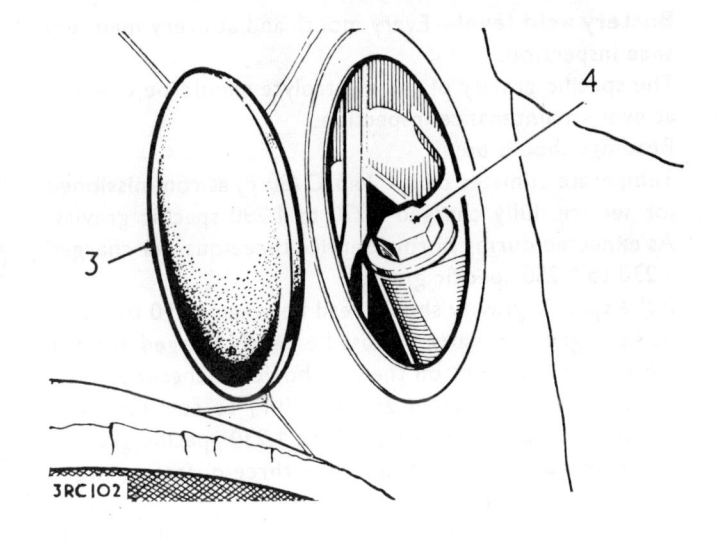

Rover 3500 and 3500S Manual AKM 3621

BOOT COMPARTMENT

Battery acid level—Every month and at every maintenance inspection.

The specific gravity of the electrolyte should be checked at every maintenance inspection.

Readings should be:

Temperate climates below 26.5°C (80°F) as commissioned for service, fully charged 1.270 to 1.290 specific gravity. As expected during normal service three-quarter charged 1.230 to 1.250 specific gravity.

If the specific gravity should read between 1.190 to 1.210, half-charged, the battery must be bench charged and the electrical equipment on the car should be checked.

Tropical climate, above 26.5°C (80°F) as commissioned for service, fully charged 1.210 to 1.230 specific gravity. As expected during normal service three-quarter charged 1.170 to 1.190 specific gravity.

If the specific gravity should read between 1.130 to 1.150, half-charged, the battery must be bench charged and the electrical equipment on the car should be checked.

The battery is located in the boot at the right-hand side.

Check acid level as follows:

1. Place car on a level surface.
2. Remove knurled nuts securing cover and withdraw the cover.
3. Wipe all dirt and moisture from the battery top.
4. Remove the vent cover. Top-up only if the acid levels are below the bottoms of the filling tubes. If necessary add sufficient distilled water into the trough until all the tubes are filled.
5. Immediately replace the cover to allow the distilled water in the trough and tubes to flow into the cells. Each cell will automatically receive the correct amount of water.
6. Reverse removal procedure. Avoid the use of a naked light when examining the cells.
7. In hot climates it will be necessary to top-up the battery at more frequent intervals.
8. In very cold weather it is essential that the car is used immediately after topping-up, to ensure that the distilled water is thoroughly mixed with the electrolyte. Neglect of this precaution may result in the distilled water freezing and causing damage to the battery.

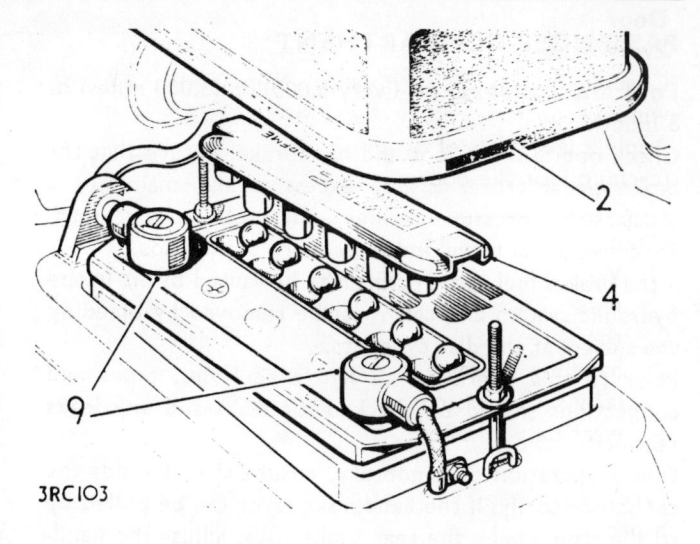

Battery terminals—Every 10.000 km (6,000 miles) or 6 months.

9. Remove battery terminals, clean, grease with petroleum jelly and refit.
10. Replace terminal screw—do not overtighten. Do not use the screw for pulling down the terminal.
    On some batteries, 'clamp' type terminals may be fitted.
11. Do not disconnect the battery cables whilst the engine is running or irreparable damage to the control unit may occur. It is also inadvisable to break or make any connection in the alternator charging and control circuits whilst the engine is running.
12. It is essential to observe the polarity of connections to the battery, alternator and regulator, as any incorrect connections when reconnecting cables may cause irreparable damage to semi-conductor devices in the alternator and other transistorised equipment fitted to the car.

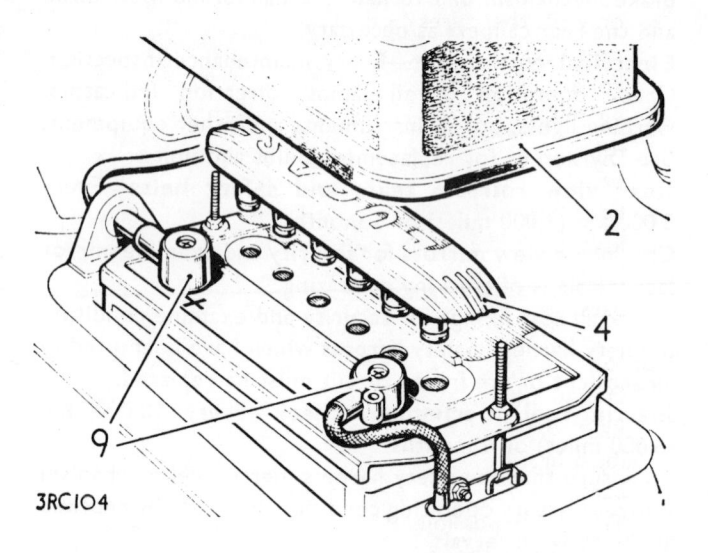

EXTERIOR

Door locks, bonnet release and window controls—
Every 10.000 km (6,000 miles) or 6 months
Check operation of door locks, bonnet release control and window controls, rectify any faults as necessary.
Apply a few spots of oil as required, using thin machine (cycle) oil for the door key locks.

Changing wheel position
1. With the type of tyre used on Rover 3500 and 3500S models, it is not considered advantageous to change the wheel positions, this in fact can give unpleasant handling characteristics when carried out, particularly if there is considerable difference between the wear pattern of one tyre and the other.

Tyre pressures—Every week and at every maintenance inspection.
2. Maximum tyre life and performance will be obtained only if the tyres are maintained at the correct pressures.

For all speeds and normal loads:	Front	Rear
kg/cm²	1,9	2,1
lb/sq in	28	30
bars	1,93	2,07
For all speeds when fully laden:	Front	Rear
kg/cm²	2,1	2,4
lb/sq in	30	34
bars	2,07	2,36

Dunlop Denovo tyres

For all speeds and loads	Front	Rear
kg/cm²	2,1	2,2
lb/sq in	30	32
bars	2,07	2,21

3. Whenever possible check with the tyres cold, as the pressure is about 0,2 kg/cm² (3 lb/sq in, 0,21 bars) higher at running temperature.
4. Always replace the valve caps, as they form a positive seal on the valves.

Road wheels—Every maintenance inspection.

Remove the wheel trim and check tightness of road wheels nuts. Replace wheel trim carefully ensuring that valves are in the centre of the wheel trim aperture.

Wheel alignment—Every 10.000 km (6,000 miles) or 6 months.
Special equipment is required to check wheel alignment, the alignment should be 3,0 mm (0.125 in) toe-in plus or minus 1,5 mm (0.062 in).
To adjust:
1. Remove air cleaner, as detailed under 'Air cleaner'.
2. Slacken the locknuts at each end of steering track rod at the rear of engine.
3. Turn rod to obtain the correct alignment.
4. Tighten locknuts and re-check. Ensure that the track rod ball joints are correctly aligned; that is, the top of the ball joint should be horizontal in the fore and aft direction.

5. When high-speed touring, the tyre pressures should be checked much more frequently, even to the extent of a daily check. If front wheel tread wear is uneven check wheel alignment.
6. Any unusual pressure loss in excess of 0,05 kg/cm² (1 lb/sq in, 0,07 bars) per week should be investigated and corrected.
7. Always check the spare wheel, so that it is ready for use at any time.
8. At the same time, remove embedded flints, etc. from the tyre treads with the aid of a penknife or similar tool. Clean off any oil or grease on the tyres, using white spirit sparingly.
9. Minimum tread depth 1 mm in the United Kingdom but may be subject to local safety regulations in other countries.
10. Check that there are no lumps or bulges in the tyres or exposure of the ply or cord structure.
11. Wheel and tyre units are accurately balanced on initial assembly with the aid of clip-on weights secured to the wheel rims.
12. Wheel balance should always be checked whenever new tyres are fitted to ensure that the dynamic balance of the wheel and tyre is correct.
13. When tyres are changed, road wheels should be carefully checked for possible damage.
14. When replacements are required, the tyres should be as currently specified by the Company. They should be of the same type and make as those previously fitted.
15. In the case of tubeless tyres a new Schrader snap-in valve must be fitted whenever a tubeless tyre is replaced.
16. It is advisable to run-in new tyres by driving at reasonable speeds for the first 400 kilometres (250 miles) or so before driving at high speeds.

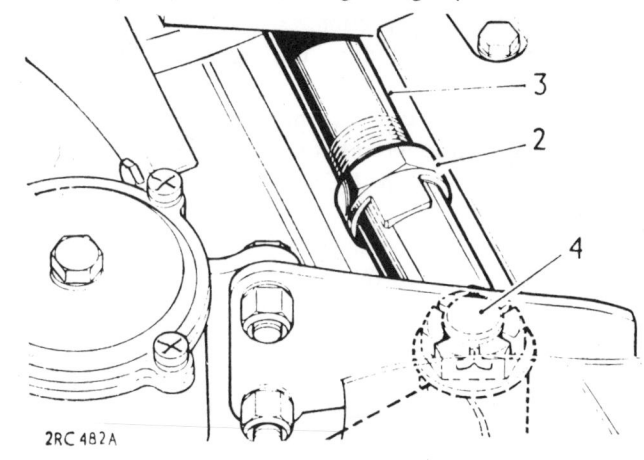

2RC 482A

5. After each adjustment of the track rod, move the car forward for a short distance before rechecking, it will be necessary to repeat this check three times to ensure accuracy of readings.
6. Refit air cleaner.
7. It should be noted that the camber of the front wheels on this car is 0°, with a tolerance of plus or minus 1°, that is, the bottom of the wheels can splay out slightly.

Rover 3500 and 3500S Manual AKM 3621

EXTERIOR

Headlamp beam setting—Every maintenance inspection.

This operation requires special equipment.
1. In an emergency each headlamp unit can be adjusted by means of the headlamp horizontal adjusting screw.
2. The headlamp vertical adjusting screw.

Wiper blades—Check, if necessary replace every 5.000 km (3,000 miles) or 3 months.

To replace wiper blades:
1. Pull wiper arm forward.
2. Lift spring clip and withdraw blade from wiper arm.
3. To fit new blade reverse removal procedure.

Oil leaks—Every maintenance inspection.
Check for oil leaks in engine compartment, rectify as necessary.

Steering box lubrication, manual type—Every 20.000 km (12,000 miles) or 12 months.
1. Remove the rubber plug from the top of the steering box cover plate.
2. Check oil level and top-up if necessary to the bottom of the filler plug hole.
3. If significant topping-up is required, check for oil leaks at rocker shaft oil seal and joint faces.

Steering box—Every 10.000 km (6,000 miles) or 6 months.
Check security of steering box mountings and backlash at steering wheel. Rectify as necessary.

Brake fluid reservoir. Combined brake and clutch fluid reservoir on 3500S models—Every month and at every maintenance inspection.
The reservoir is located on the right-hand front wing valance.
1. The reservoir cap incorporates a float and level switch which operates the red brake warning light, should the level in the reservoir fall below the safe limit.
2. Check the fluid level in the reservoir. If necessary, replenish to the level mark.
3. Ensure joint washer is in good condition.
Use Castrol Girling Brake and Clutch Fluid Green Specification, current SAE J1703 and US Federal Standard 116 from sealed tins, or Unipart brake and clutch fluid type 550 (coloured green) from sealed tins.
4. If significant topping-up is required check master cylinders, clutch, slave cylinder and connecting pipe as applicable, brake disc cylinders and brake pipes for leakage; any leakage must be rectified immediately.
5. When removing the reservoir cap, do not disconnect the wires; hold centre and rotate cap. Care should be taken when withdrawing the float unit to ensure that the brake fluid does not drip on to the car.

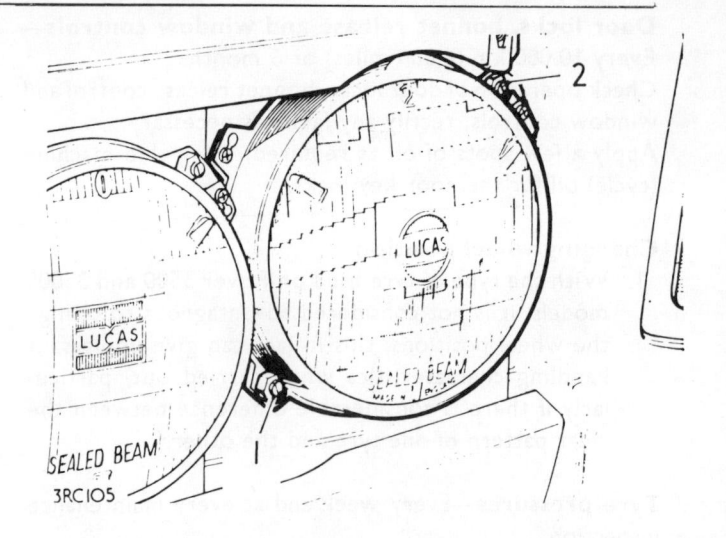

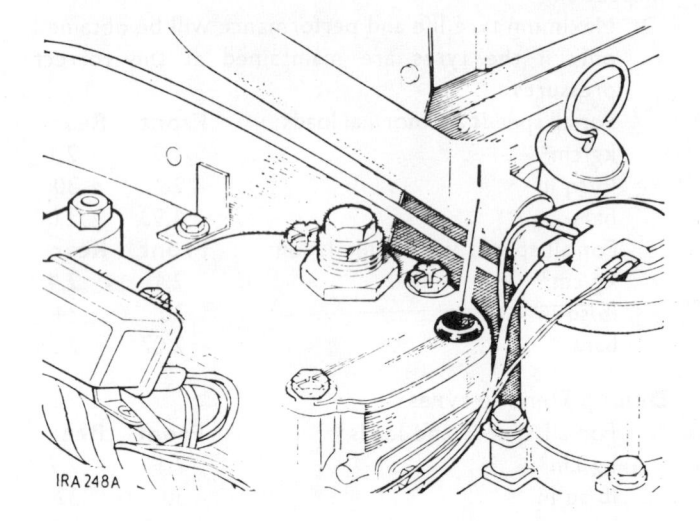

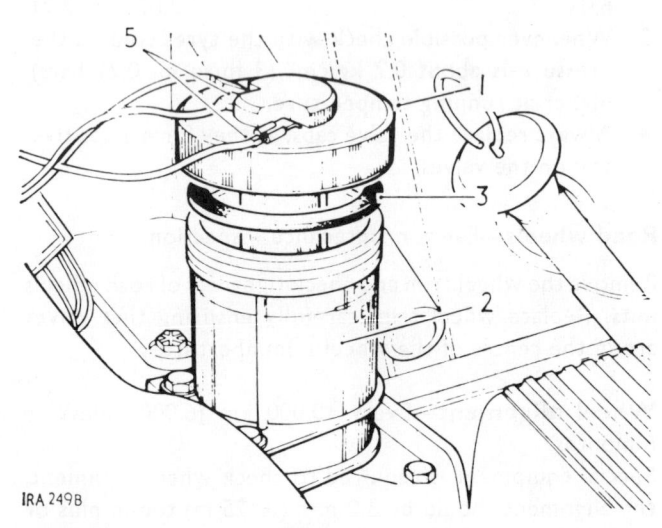

6. Check operation of reservoir level safety switch as follows:
   (a) Ignition 'on', handbrake 'off', unscrew and lift filler cap 25 mm (1 in), brake warning light should be illuminated.
   (b) If the warning light is not illuminated, the operation of the float unit and the wiring connections must be investigated.

ENGINE COMPARTMENT

Automatic gearbox fluid level—Every 10.000 km (6,000 miles) or 6 months.

The torque converter and automatic gearbox are lubricated as one unit. As the fluid for operating the torque converter is fed from the gearbox casing, it is essential when checking the level or topping-up the automatic gearbox that the engine is run at idling speed for about 2 minutes to transfer fluid from gearbox casing to torque converter, otherwise a false level reading will be obtained. Do not add anti-friction additives to the automatic gearbox fluid. Check level as follows:

1. Absolute cleanliness is essential. Use only nylon rag for cleaning and checking fluid level.
2. Stand car on level ground. With the foot brake firmly applied and engine idling, pass the gear selector through the complete range of gears to ensure that the transmission system is fully primed. Place the selector in the 'P' position and switch off the engine.
3. Lift the bonnet to expose the dipstick which is adjacent to the brake fluid reservoir.
4. Clean area round dipstick and oil filler hole.
5. Remove dipstick, wipe dry and dip immediately to check fluid level.
6. Top up if necessary to the 'H' mark'. Do not overfill. The difference between high and low marks on the dipstick represents approximately 0,5 litre (one pint). After topping up repeat items 2 and 5.
7. If significant topping-up is required, check for leakage at oil seals and sump joint; rectify immediately.

Spark plugs. Check every 10.000 km (6,000 miles) or 6 months; replace every 20.000 km (12,000 miles).

Use the special spark plug spanner and tommy bar supplied in the tool kit when removing or refitting spark plugs.

IMPORTANT: Take great care when fitting spark plugs not to cross-thread the plug otherwise costly damage to the cylinder head will result.

1. Check or replace the spark plugs as applicable; if the plugs are in good condition clean and reset the electrode gaps to 0,60 mm (0.025 in), at the same time file the end of the central electrode until bright metal can be seen.
2. It is important that only Champion L92Y spark plugs are used for replacements.
   Incorrect grades of plug may lead to piston overheating and engine failure.

Proceed as follows:

1. Remove the leads from the spark plugs.
2. **Using the special spark plug spanner and tommy bar** supplied in the vehicle tool kit, remove the plugs and washers. This is best done by leaning over the radiator and reaching along each cylinder bank, rather than the usual practice of leaning over the wing valance. Clean the spark plugs as follows:

continued

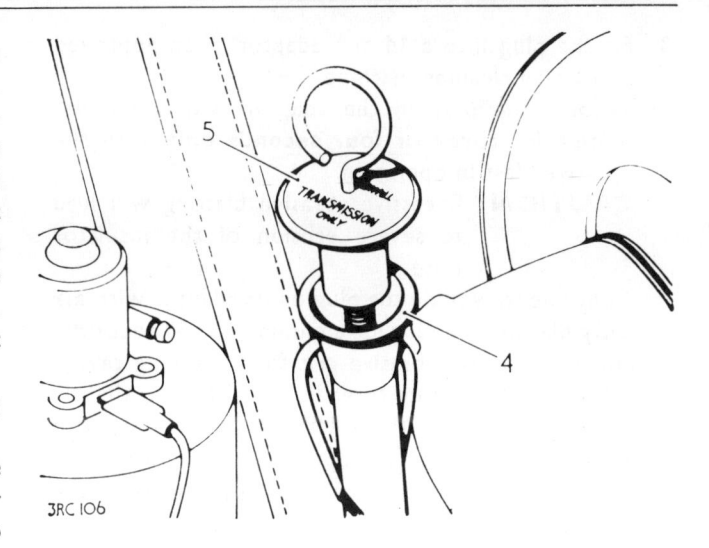

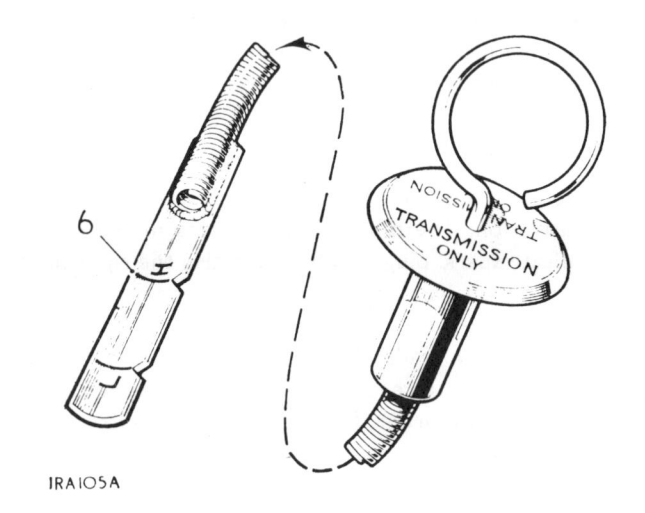

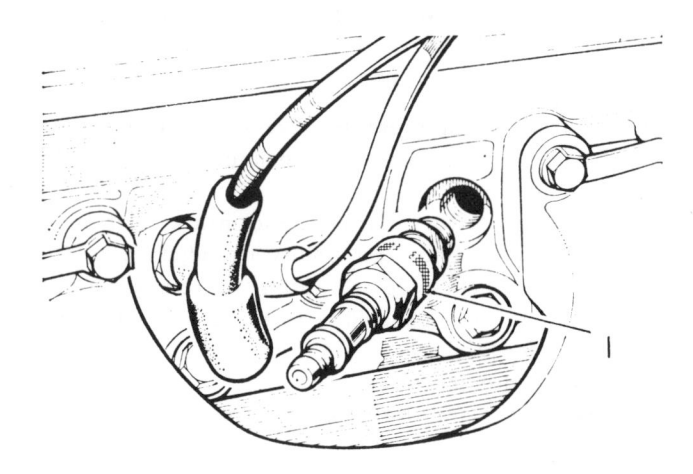

Rover 3500 and 3500S Manual AKM 3621

10-8

ENGINE COMPARTMENT

3. Fit the plug into a 14 mm adaptor of an approved spark plug cleaning machine.
4. Wobble the plug in the adaptor with a circular motion for **three or four seconds only** with the abrasive blast in operation.
   CAUTION: Excessive abrasive blasting will lead to severe erosion of the insulator nose.
   Continue to wobble the plug on its adaptor with **air only** blasting the plug for a minimum of **30 seconds**; this will remove abrasive grit from the plug cavity.
5. Wire brush the plug threads, open the gap slightly and vigorously file the electrode sparking surfaces using a point file. This operation is important to ensure correct plug operation by squaring the electrode sparking surfaces.
6. Set the electrode gap to the recommended clearance of 0,60 mm (0.025 in).
   Test the plugs in accordance with the plug cleaning machine manufacturer's recommendations.
   If satisfactory, the plugs may be replaced in the engine.

Distributor leads

1. The correct sequence of plug leads is shown in this illustration. The numbers and letters in the circles indicate spark plug numbers and also the right-hand (RH) or left-hand (LH) bank of the engine to which the leads go.
2. High tension leads must be replaced in the correct relationship to each other, as well as ensuring correct firing order. Failure to do this may result in cross firing.
3. Loose clips locating leads.
4. Locating clips fixed to rocker cover.
   NOTE: The electrical leads to the ignition coil are fitted with male and female connectors; ensure that they are fitted to the correct blade on the coil.

Distributor contact points—Every 10.000 km (6,000 miles) or 6 months. If necessary replace every 20.000 km (12,000 miles) or 12 months.

To obtain satisfactory engine performance it is most important that the contact points are adjusted to the dwell angle which is 26°–28° using suitable workshop equipment. This work should be carried out by your local Distributor or Dealer.

1. Remove distributor cap.
2. Remove the nut on the terminal block.
3. Lift off the spring and moving contact.
4. Remove adjustable contact secured with a screw.
5. Add a smear of grease to the contact pivot before fitting new contact points. Then carry out distributor maintenance followed by setting the ignition timing and dwell angle.

However, when it becomes necessary to change the contact points, and specialised checking equipment is not available, they may be adjusted either by the feeler gauge or alternatively the timing lamp method. Proceed as follows:

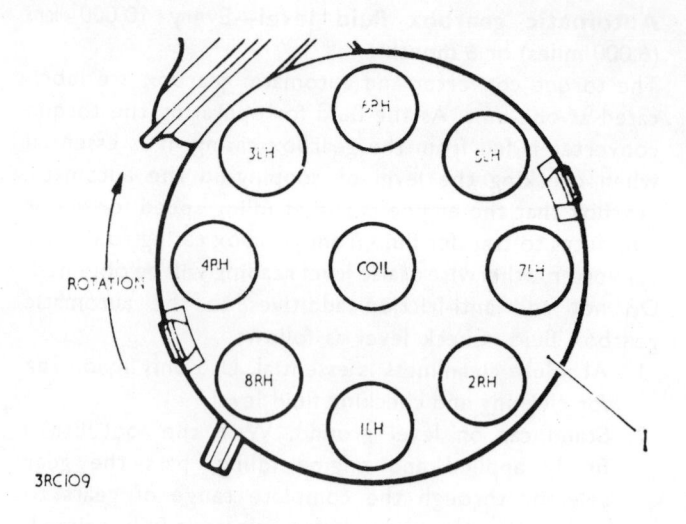

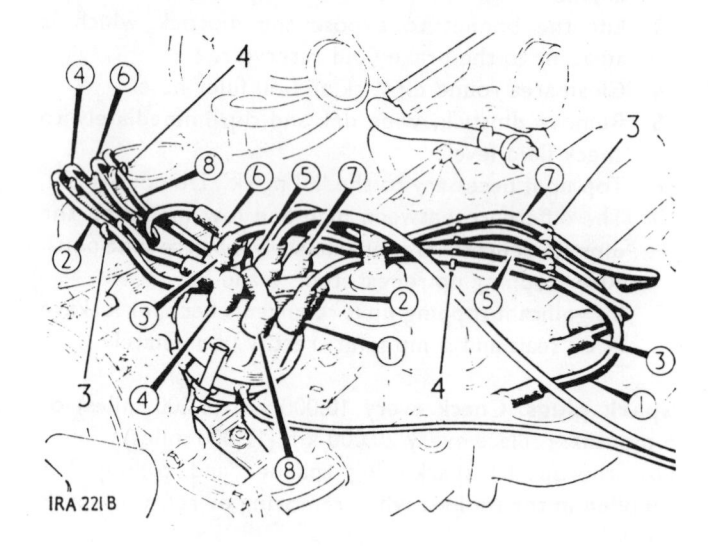

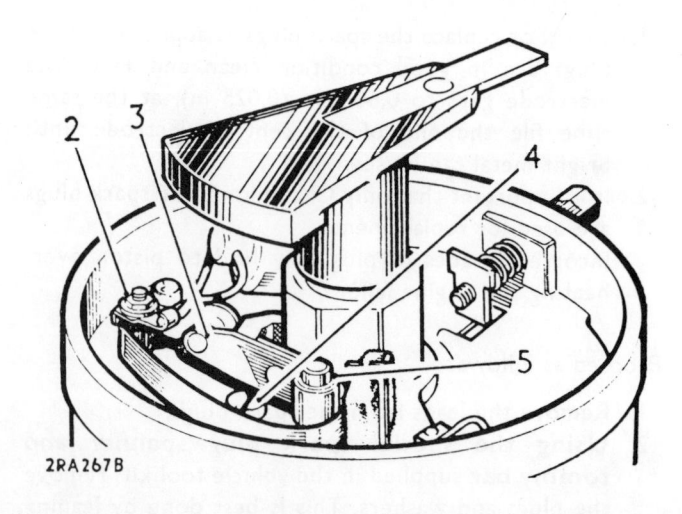

continued

Rover 3500 and 3500S Manual AKM 3621

ENGINE COMPARTMENT

Checking contact points—*feeler gauge method*

6. Turn the engine using a 0.937 mm ($\frac{15}{16}$ in) AF socket spanner on the front pulley retaining bolt, until the contacts are fully open.
7. The clearance should be 0,35 to 0,40 mm (0.014 to 0.016 in) with the feeler gauge a sliding fit between the contacts.
8. Adjust by turning the adjusting nut clockwise to increase gap and anti-clockwise to reduce gap.
9. Replace the distributor cap.
   At the first available opportunity after the contact points have been adjusted as detailed above, they must be finally set to the dwell angle using specialised equipment.
10. At the same time check the ignition timing which should be dynamically set to 6° BTDC at 600 revs/min. (100 Research Octane fuel). When new contact points have been fitted, the dwell angle must be checked after a further 1.500 km (1,000 miles) running.

Checking contact points—*timing lamp method*

11. Remove distributor cap.
12. Turn the engine using a 0.937 mm ($\frac{15}{16}$ in) AF socket spanner on the front pulley retaining bolt until the contact breaker heel is on the peak of number one cylinder cam. Points should be fully open.
13. Connect a 12 volt timing lamp, or suitable voltmeter, across the contact breaker lead terminal and a suitable earth point.
14. Switch on the ignition.
15. Turn the distributor adjusting nut **anti-clockwise** until the timing lamp goes out, or there is no reading on the voltmeter.
16. Continue a further two turns of the adjuster in an anti-clockwise direction.
    During this operation the adjusting nut should be pressed inwards with the thumb to assist the helical return spring. 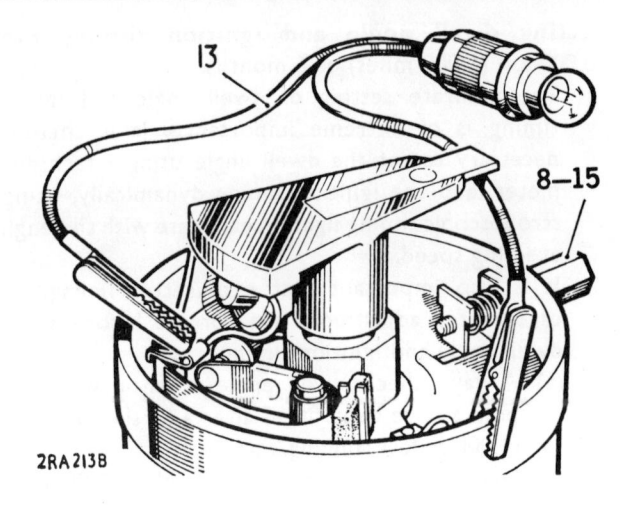
17. Slowly turn the adjusting nut **clockwise** until the timing lamp just comes on, or there is a voltage shown on the voltmeter.
18. Noting the position of the flats on the adjusting nut, continue in a clockwise rotation for a further **five** flats.
19. Remove timing lamp or voltmeter and switch off ignition.
20. Replace the distributor cap.
    At the first available opportunity after the contact points have been adjusted as detailed above, they must be finally set to the dwell angle using specialised equipment.
21. At the same time, check the ignition timing, which should be dynamically set to 6° BTDC at 600 revs/min. (100 Research octane fuel.) When new contact points have been fitted, the dwell angle must be checked after a further 1.500 km (1,000 miles) running.

Distributor maintenance—Every 10.000 km (6,000 miles) or 6 months.

Lubricate as follows:

1. Remove the distributor cap.
2. Remove the rotor arm.
3. Lightly smear the cam with clean engine oil.
4. Add a few drops of thin machine oil to lubricate the cam bearing and distributor shaft.
5. Wipe the inside and outside of the distributor cap with a soft dry cloth.
6. Ensure that the carbon brush works freely in its holder.
7. Replace rotor arm and distributor cap.

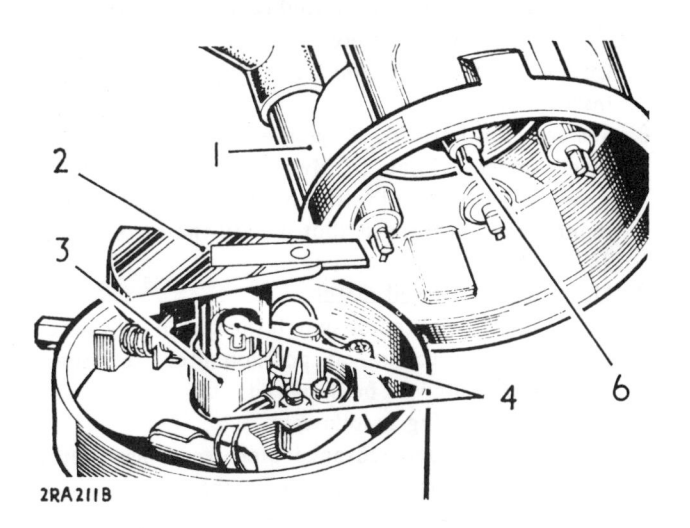

ENGINE COMPARTMENT

Setting dwell angle and ignition timing—Every 10.000 km (6,000 miles) or 6 months.

1. The accurate setting of dwell angle and ignition timing is of extreme importance. It is therefore necessary to set the dwell angle using a tach/dwell meter, and the ignition timing dynamically, using a stroboscopic timing light. In each case with the engine at idling speed.
2. It is also important that carburetter linkage and carburetter adjustment are correct before setting dwell angle and ignition timing.
    (a) Always check dwell angle when new contact points have been fitted after a further 1.500 km (1,000 miles) running.
    (b) Ignition timing when using fuel of 100 octane rating; five-star rating in the United Kingdom.
    Static ignition timing: 6° BTDC.
    Dynamic ignition timing: 6° BTDC at 600 revs/min.
    Dwell angle 26° to 28°.
    (c) Ignition timing when using fuel of 96 to 98 octane rating.
    Static ignition timing: TDC.
    Dynamic ignition timing: TDC at 600 revs/min.
    Dwell angle: 26° to 28°.

Carry out item 3 only if distributor has been disturbed.

3. Set ignition timing statically to 6° BTDC prior to engine being run, by the basic timing lamp method. (This sequence is to give only an approximation in order that the engine may be run. The engine must not be started after distributor replacement until this check has been carried out.)
4. Set dwell angle as follows:
5. Start engine and set to an idling speed of 600 revs/min. (700 to 750 revs/min. on models fitted with air conditioning, compressor isolating switch disengaged), using an accurate tachometer.
6. Set selector knob to 'calibrate' position on the tach/dwell meter. Adjust calibration knob to give a zero reading on the meter.
7. Couple meter to engine following manufacturer's instructions.
8. Set selector knob to 8 cylinder position and tach/dwell selector knob to 'dwell'. Adjust distributor dwell angle by turning the hexagon headed adjustment screw on the distributor until the meter reads 26° to 28°.
9. Uncouple tach-dwell meter.
    Care should be taken to switch the tach/dwell meter selector switch to the 'off' position after use, otherwise battery life will be impaired.
10. Set ignition timing as follows:
11. Disconnect vacuum advance pipe from distributor and block the vacuum pipe by suitable means.
12. Couple stroboscopic timing lamp to engine following the manufacturer's instructions with the high tension lead attached into No. 1 cylinder plug lead (Front cylinder on left-hand bank).
13. Ensure engine is still idling at the correct speed.
14. Slacken distributor clamping bolt.
15. Turn distributor until stroboscopic lamp synchronises the timing pointer and the timing mark at 6° BTDC on the vibration damper rim.
16. Arrow (R) indicates direction to retard ignition. Arrow (A) indicates direction to advance ignition.
17. Re-tighten the distributor clamping bolt securely.
18. Switch off engine and disconnect stroboscopic timing lamp.
19. Refit vacuum advance pipe.

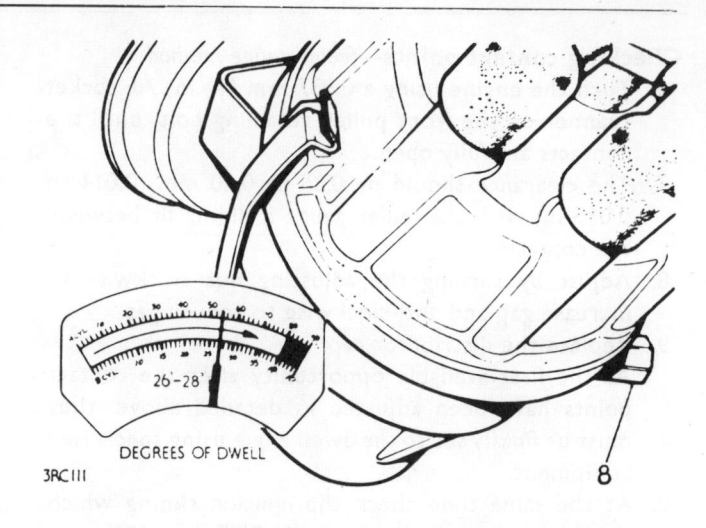

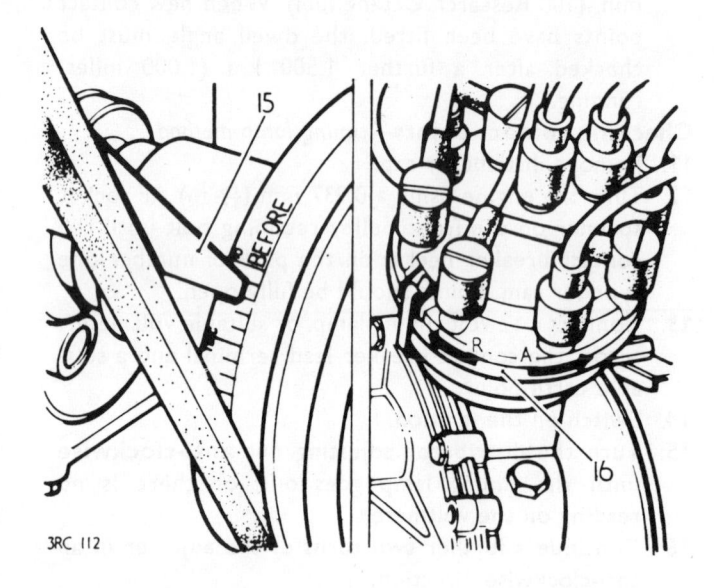

    NOTE: Engine speed accuracy during ignition timing is of paramount importance. Any variation from the required idling speed, particularly in an upward direction, will lead to wrongly set ignition timing.

ENGINE COMPARTMENT

Fan belt adjustment—Every maintenance inspection.
1. Check by thumb pressure between alternator and crankshaft pulleys, movement should be: 11 to 14 mm (0.437 to 0.562 in); if necessary adjust as follows:
2. Slacken the bolts securing the alternator to the front cover.
3. Slacken the fixing at the adjustment link.
4. Pivot the alternator inwards or outwards as necessary and adjust until the correct tension is obtained.
5. Tighten alternator adjusting bolts.

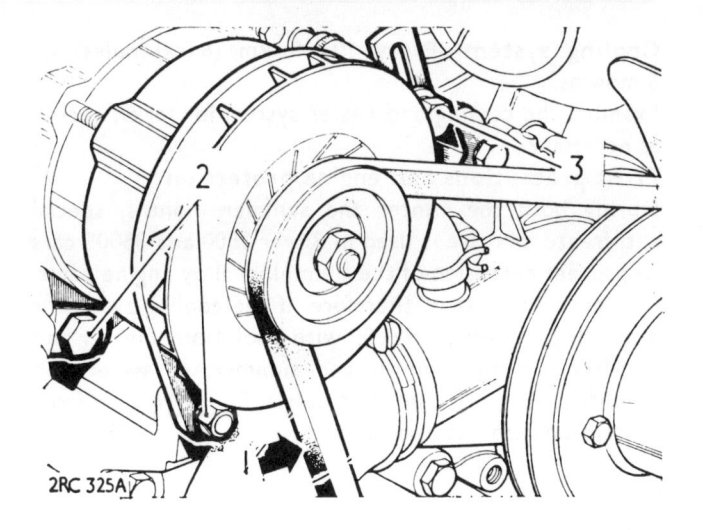

Power steering pump belt adjustment, as applicable—Every maintenance inspection.
Whenever a new belt is fitted, check adjustment again after approximately 1.500 km (1,000 miles) running.
1. Check by thumb pressure between the crankshaft and pump pulley. Movement should be 6 to 9 mm (0.250 to 0.375 in.)

If adjustment is necessary:
2. Slacken the pivot bolt securing the pump to the timing cover bracket.
3. Slacken the fixing at the adjustment link.
4. Pivot the pump as necessary and adjust until the correct belt tension is obtained.
5. Tighten the pump adjusting and pivot bolts.

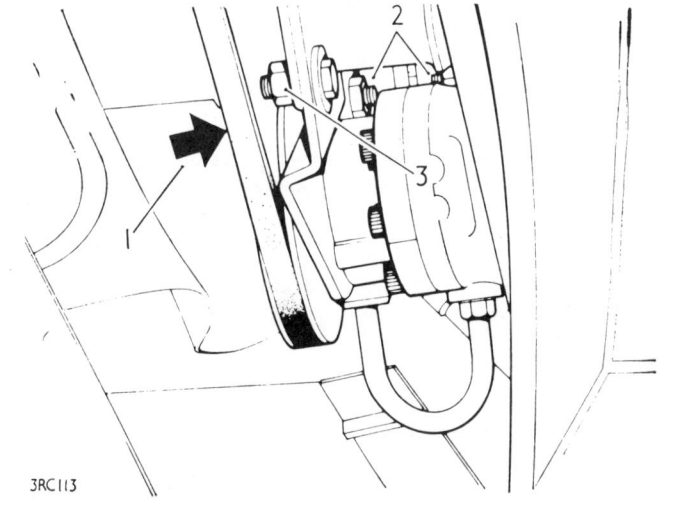

Radiator water level—Every 1.000 km (750 miles) and at every maintenance inspection.
1. To prevent corrosion of the aluminium alloy engine parts it is imperative that the cooling system is filled with a solution of water and anti-freeze, winter and summer, or water and inhibitor during the summer only. Never fill or top-up with plain water.
2. The radiator filler cap is under the bonnet panel.
3. With a cold engine the correct water level is 25 mm (1 in) below the bottom of the filler neck.

   WARNING: Do NOT remove the radiator filler cap when the engine is hot because the cooling system is pressurised and personal scalding could result.

4. When the engine has cooled down, first turn it anti-clockwise a quarter of a turn and allow all pressure to escape, before turning further in the same direction to lift it off.
5. When replacing the filler cap it is important it is tightened down fully, not just to the first stop. Failure to tighten the filler cap properly may result in water loss, with possible damage to the engine through overheating.
   If the cooling system is being refilled after draining or a large quantity of water needs to be added proceed as follows:
6. Fill radiator with a solution of either water and anti-freeze or water and inhibitor. See next item for details of anti-freeze and inhibitor solutions and quantity to be used. 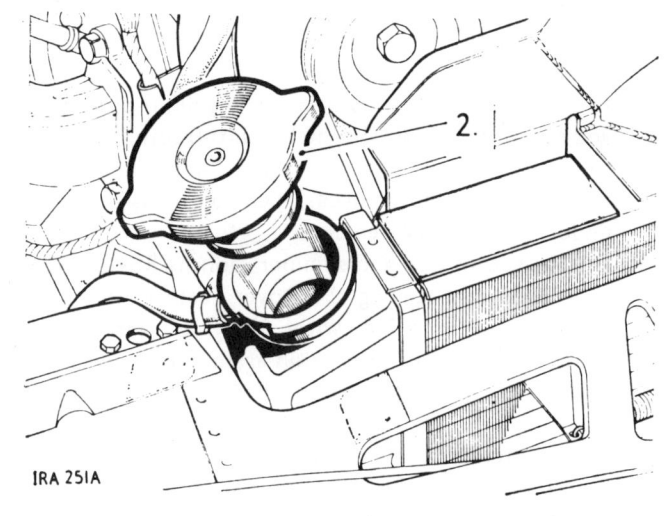
7. Run engine at a fast idle until top radiator hose is warm, that is thermostat open.
8. With the engine still running at a fast idle, fill radiator to the bottom of the neck of the filler tube.
9. The water level will fall to 25 mm (1 in) below the bottom of the filler neck when engine is stationary and cold.
10. Replace filler cap with the engine still running at a fast idle.

Use soft water wherever possible; if the local water supply is hard, rainwater should be used.

ENGINE COMPARTMENT

Cooling system—Every 10.000 km (6,000 miles) or 6 months.
Examine the cooling and heater system for leaks; rectify as necessary.

Frost precautions and engine protection
During both the winter and summer months, special anti-freeze mixture is used in Rover 3500 and 3500S cars to prevent corrosion of the aluminium alloy engine parts. It is most important, therefore, if the cooling system is drained at any time, to refill with a solution of water and anti-freeze during winter and summer, or water and inhibitor during the summer only if for any reason frost precautions are not necessary.

Recommended solutions are:
Anti-freeze—Bluecol AA coloured green or anti-freeze to BS 3150 or Prestone, or anti-freeze to MIL-E 5559. See note on following page.
Inhibitor—Marston Lubricants SQ36. Coolant inhibitor concentrate.
Use one part of anti-freeze to two parts of water.
Use 3 fluid ounces of inhibitor per 4.5 litres (one gallon) of water.
To ensure that the solution is fully effective at all times the cooling system should be drained and refilled every 12 months.

Proceed as follows:
1. Ensure that the cooling system is leak-proof, anti-freeze solutions are far more 'searching' at joints than water.
2. Drain and flush the system. Radiator drain plug under front valance on left-hand side.
3. Drain tap on cylinder block, two fitted – one each side of engine.
4. Pour approximately 4,5 litres (one gallon) of water, add solution 3 litres (5 pints) of anti-freeze, summer or winter, or 6 fluid ounces of inhibitor if frost protection is not required. Then top up as detailed under 'Radiator Water Level'.
5. During the winter and summer months Rover 3500 and 3500S cars leaving the Rover Factory have the cooling system filled with $33\frac{1}{3}\%$ of anti-freeze mixture. This gives protection against frost down to minus 32°C (-25°F). Cars so filled can be identified by the green label affixed to the right-hand side of the windscreen and a green label tied to the engine.
6. Bluecol AA is a British Standard 3150 type anti-freeze. When this type is not available in service, use either 'Prestone' or anti-freeze to specification MIL-E 5559.
7. Where 'Prestone' is to be used after B.S. 3150, SQ36 or MIL-E 5559, empty the coolant, re-fill with water, run the engine to circulate the coolant throughout the system, stop the engine, empty the coolant, repeat filling with water running and emptying once more, finally swill out by use of a hose and running water into the top header tank for a few minutes with the exit taps open, then close the taps and fill with the appropriate amount of water and 'Prestone' as detailed above.

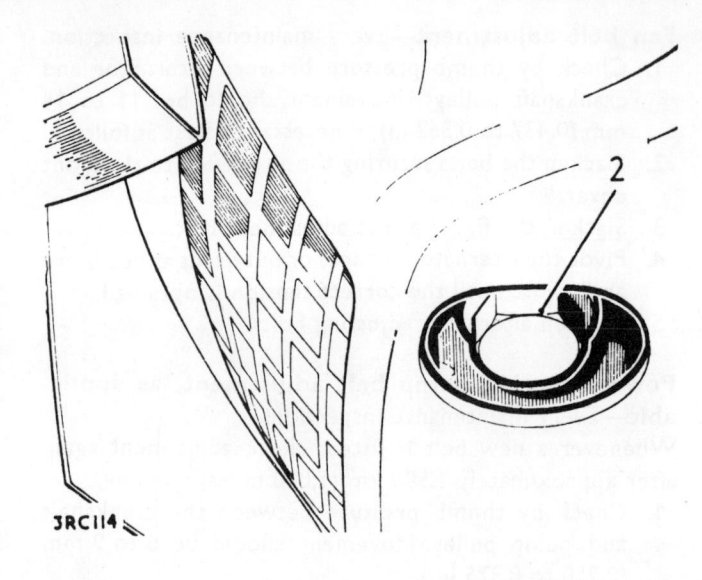

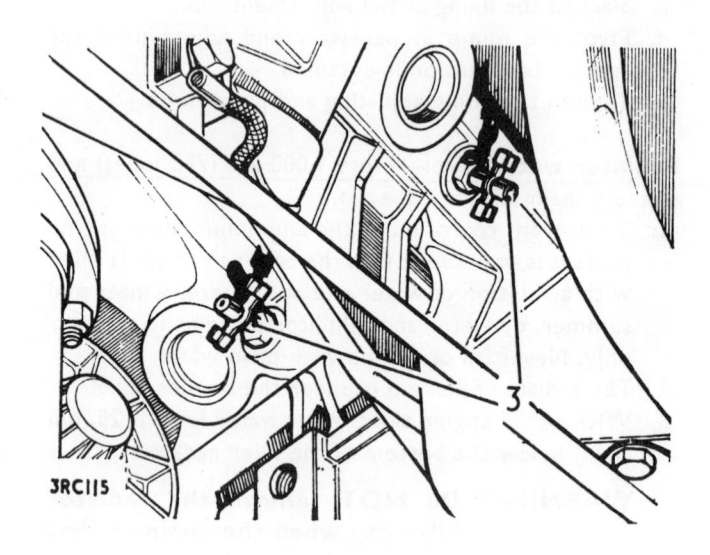

ENGINE COMPARTMENT

Water level—windscreen washer—Every 1.000 km (750 miles) and at every maintenance inspection.
The windscreen washer reservoir is located on the left-hand wing valance on right-hand steering models and on the opposite side on left-hand steering models.
1. Top up reservoir to within approximately 25 mm (1 in) below bottom of filler neck.
2. Use a windscreen washer solvent in the bottle, this will assist in removing mud, flies and road film.
3. In cold weather, to prevent freezing of the water, the reservoir should have Isopropyl Alcohol added. Do not use methylated spirits, which has a detrimental effect on the screenwasher impeller.

Fuel filter, cartridge type—Every 20.000 km (12,000 miles) or 12 months.
The cartridge provides an additional filter between pump and carburetter.
Replace as follows:
1. Disconnect fuel pipes from each end of filter.
2. Slacken clip securing filter and withdraw unit.
3. Fit new filter with end marked 'IN' downwards. Alternatively if the filter is marked with arrows they must point upwards. Tighten securing clip and refit fuel pipes.

Engine flame traps—Every 20.000 km (12,000 miles) or 12 months.
Replace as follows:
1. Pull off flame trap hoses.
2. Remove flame traps, one on top of each rocker cover.
3. Replace with new flame traps, which are located in position by the hoses.

Air cleaner—Every 20.000 km (12,000 miles) or 12 months.
Attention to the air cleaner is extremely important. Replace elements every 10.000 km (6,000 miles) or 6 months under severe dusty conditions, as performance will be seriously affected if the engine is run with an excessive amount of dust or industrial deposits in the element.
Proceed as follows:
1. Slacken off the hose clips on each side of the air cleaner.
2. Release the two elbows.
3. Disconnect the hose from the engine breather filter.
4. The air cleaner can now be removed by easing it up from the domed retaining studs.

Air cleaner element replacement.
1. Release the three clips at the side of the air cleaner casing and withdraw the frames and elements.
2. To replace the elements, remove the screw and washer on the frame.
3. Withdraw frame.
4. Remove end cap.
5. Remove sealing washers.
6. Discard old elements and replace with new units.

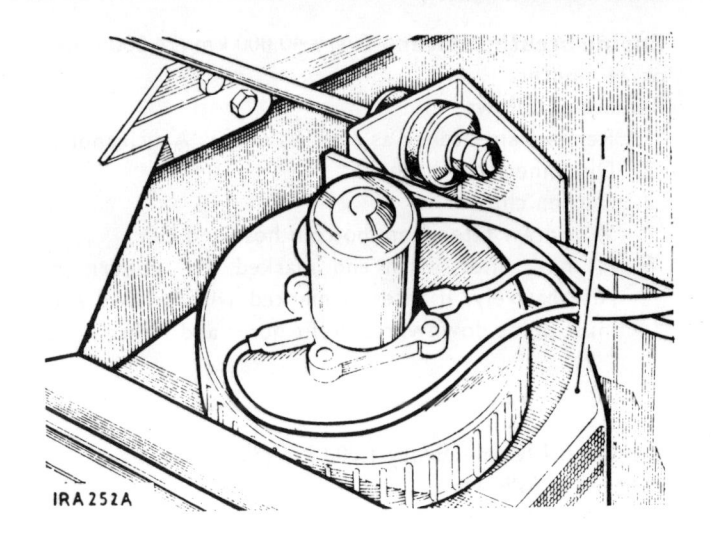

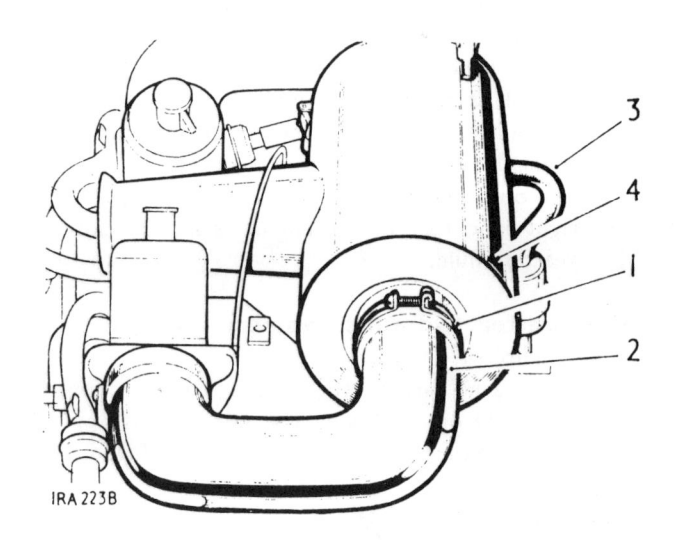

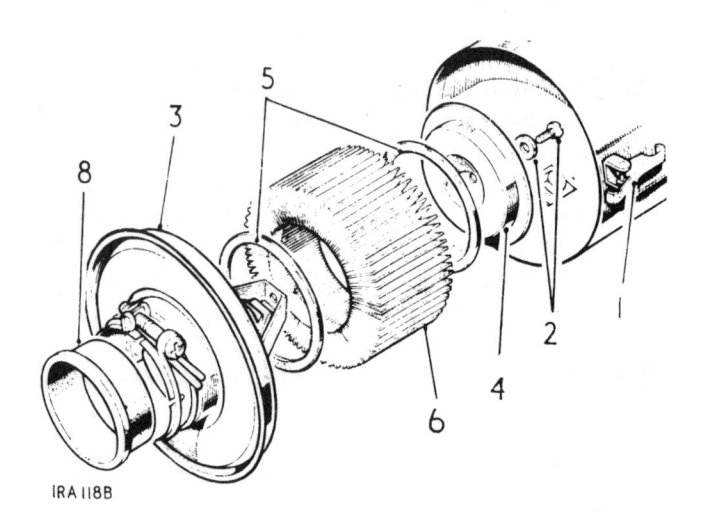

7. Ensure that sealing washers are in good condition, and correctly located.
8. Check condition of rubber seals on end of air cleaner frame – replace if necessary.
9. Reassemble elements to air cleaner and air cleaner to engine by reversing the removal procedure.

Rover 3500 and 3500S Manual AKM 3621

ENGINE COMPARTMENT

Engine breather filter—Every 40.000 km (24,000 miles) or 24 months.
Replace as follows:
1. Remove air cleaner as detailed under 'Air cleaner'.
2. Disconnect top hose.
3. Slacken clip.
4. Withdraw filter from bottom hose and clip.
5. Fit new filter, with end marked 'IN' uppermost. Alternatively, if filter is marked with arrows they must point downwards. Refit hoses and tighten clip.

Engine oil level—Every 1.000 km (750 miles).
Proceed as follows:
1. Stand the car on level ground and allow the oil to drain back into the sump.
2. Withdraw the dipstick at left-hand side of engine; wipe it clean, re-insert to its full depth and remove a second time to take the reading.
3. Add oil as necessary through the screw-on filler cap marked 'engine oil' on the right-hand front rocker cover. Never fill above the 'High' mark.
The oil filler cap with the dipstick attached at the right-hand rear of the engine is for the automatic gearbox fluid.

Carburetter hydraulic dampers—Every 10.000 km (6,000 miles) or 6 months.
Top up each carburetter piston damper reservoir using SAE 20 oil, following the instructions applicable to the type of carburetter suction chamber fitted.

Plain suction chamber
1. Unscrew the oil cap and withdraw the damper.
2. Top up with oil to within about 25 mm (1 in) of the top of the tube.
3. Then replace the cap and hydraulic damper.

Ball bearing lined suction chamber
4. Unscrew the cap and lift the piston and damper to the top of their travel.
5. Fill the retainer recess with oil and push the damper down until the cap contacts the top of the suction chamber. Repeat this filling procedure until oil just remains visible at the bottom of the retainer recess with the piston down. Screw the cap up firmly**.

Accelerator linkage—Every 10.000 km (6,000 miles) or 6 months.
1. Prior to carburetter adjustments, lubricate the accelerator linkage using clean engine oil paying particular attention to the accelerator coupling shaft and control rod ball joint.
2. Check the linkage for correct operation and ensure that there is no tendency to stick.
Badly worn parts should be replaced.

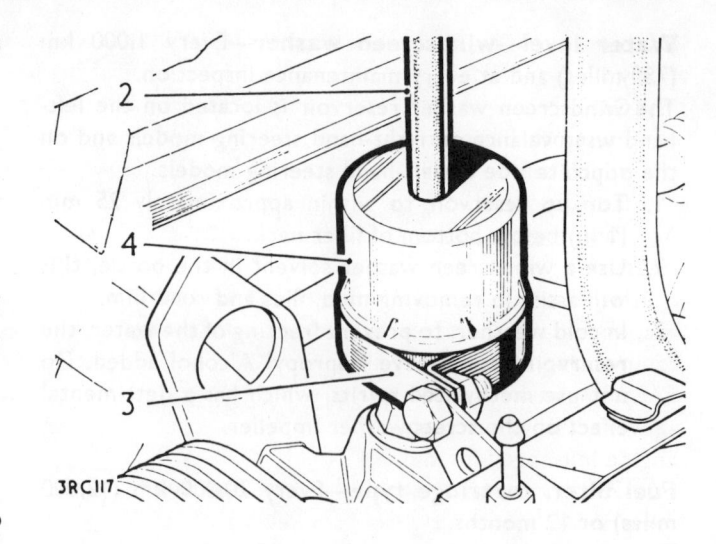

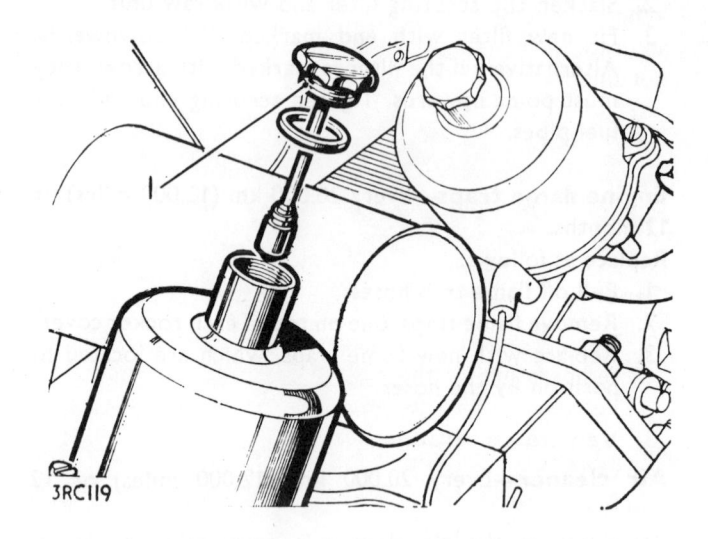

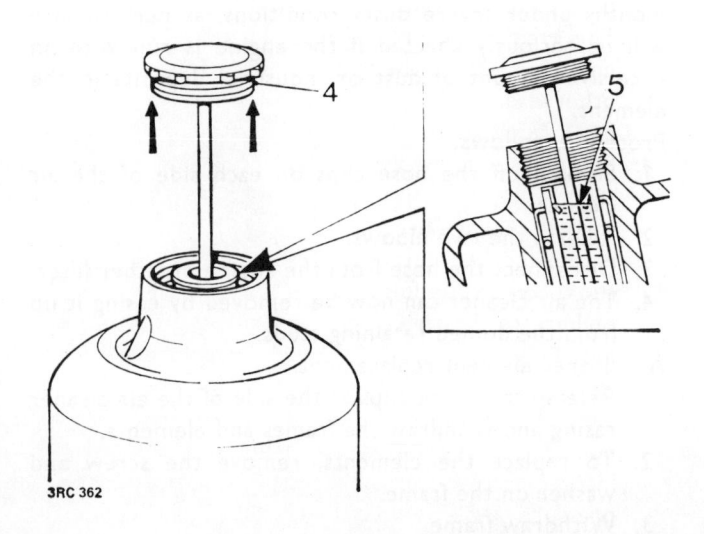

10–15

Rover 3500 and 3500S Manual AKM 3621

ENGINE COMPARTMENT

Carburetter linkages and carburetter adjustments–
Every 10.000 km (6,000 miles) or 6 months.
1. Carburetter mixture ratio is pre-set and sealed and must not be interfered with. The only adjustments which can be carried out are engine idle speed and fast idle speed.

Accurate engine speed is essential during carburetter adjustments, therefore, the contact breaker dwell angle, ignition timing and automatic ignition advance mechanism, should all be checked and reset if necessary before commencing carburetter adjustments.

Check engine idle speed at the service intervals mentioned above, adjust if necessary.

Engine idle speed for standard vehicles 600–650 revs/min. (700–750 revs/min. on models fitted with air conditioning, compressor isolation switch disengaged.)

Engine idle speed for emission controlled vehicles 700–750 revs/min.
Fast idle speed for all vehicles 1100–1200 revs/min.
When checking engine speed, use an independent and accurate tachometer. The tachometer fitted to the car is not suitable.

General requirements when setting carburetters

2. **Temperature:** Whenever possible the ambient air temperature of the setting environment should be between 15.5 to 26.5°C (60° to 80°F).
3. **Vehicle conditions.** Idling adjustments should be carried out on a fully warmed up engine, that is, at least 5 minutes after the thermostat has opened. This should be followed by a run of one minute duration at an engine speed of approximately 2,500 revs/min. before further adjustments or checks are carried out. This cycle may be repeated as often as required. It is important that the above cycle is adhered to, otherwise, overheating may result and settings may be incorrect.
4. Before any attempt is made to check settings a thorough check should be carried out to see that the throttle linkage between the pedal and carburetters is free and has no tendency to stick. Ensure that the choke control is fully pushed in. If at any time the carburetter linkage is adjusted on automatic gearbox models, it must be followed by a check of the automatic gearbox downshift cable, see following page. This is to ensure that the downshift cable has not been disturbed during carburetter adjustments. Incorrect downshift cable setting will give wrong oil pressure and can cause automatic gearbox clutch failure.

To adjust engine speed, proceed as follows:
1. Move the gear selector to the 'P' (Park) position on automatic gearbox models.
2. Run the engine until warm. See note concerning general requirements when setting carburetters.
3. Switch off engine and remove air cleaner as described previously.
4. Slacken the screws securing the throttle lever to the carburetter lever on each carburetter, thus allowing individual adjustment of carburetters.
5. Start the engine.

Engine idle speed adjustment

6. Use special spanner and slacken off carburetter idle adjusting screw lock nuts.
7. Adjust idle screws by equal amounts to give an idle speed of 600–650 revs/min. for standard vehicles and 700–750 revs/min. for emission controlled vehicles and models fitted with air conditioning equipment, compressor isolating switch disengaged.
8. When both carburetters have been adjusted, tighten the idle screw lock nuts.

Fast idle speed adjustment

9. Pull out the mixture control until the mark on the fast idle cam is opposite the centre line of the fast idle screw.
10. Using the special spanner, slacken the lock nut.

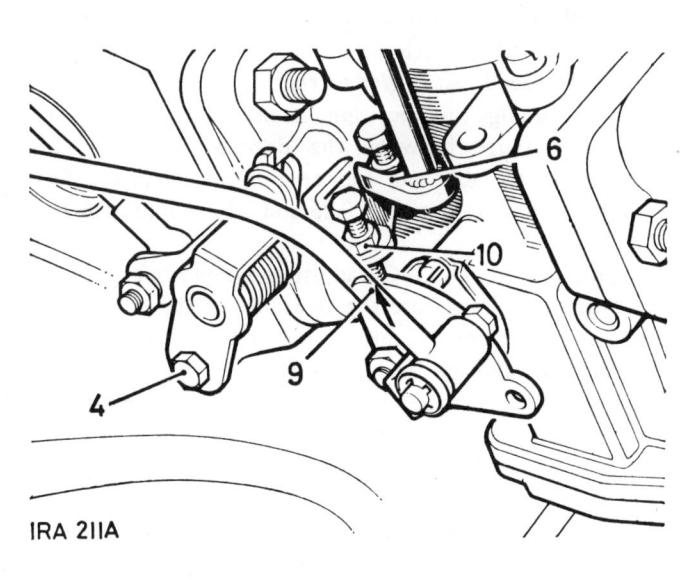

continued

IRA 211A

Rover 3500 and 3500S Manual AKM 3621

ENGINE COMPARTMENT

11. Adjust the fast idle screw to give an engine speed of between 1100 and 1200 revs/min.
12. Balance carburetter air flow as follows:
    Check and if necessary zero the gauge, by means of the adjustment screw on the carburetter balancing device. Place the balancer on to the carburetter adaptors, ensuring that there are no air leaks.
    Note the reading on the gauge; if the pointer is in the zero sector of the gauge, no adjustment is required. If the needle moves to the right, decrease the air flow through the left-hand carburetter by unscrewing the idle adjusting screw or increase the air flow through the right-hand carburetter by screwing in the idle adjusting screw. Reverse the procedure if the needle moves to the left.
13. Should the idling speed rise too high or drop too low during balancing, adjust to the correct idle speed maintaining the gauge needle in the zero sector. With the carburetter balance correctly adjusted, the difference in engine speed with the balancer on or off will be negligible, approximately plus or minus 25 revs/min.
    If there is a considerable change of engine speed this indicates incorrect mixture setting and specialised workshop attention to the carburetter will be necessary.
14. On the right-hand carburetter place a 0,15 mm (0.006 in) feeler between the right leg of the fork on the adjusting lever and the pin on the throttle lever.
15. Apply light pressure to the linkage to hold the feeler, then tighten the throttle lever securing screws.

Downshift cable check, automatic gearbox—Every 10.000 km (6,000 miles) or 6 months as part of carburetter adjustment procedure.

16. With the engine running at the correct idling speed, operate the accelerator coupling shaft until the idling speed just begins to rise.
17. Maintaining this position, check the gap with a feeler gauge between the crimped stop on the downshift cable and the end of the adjuster. The gap should be between 0,25 and 0,50 mm (0.010: and 0.020 in).
    Important: DO NOT adjust the downshift cable to increase or decrease the clearance. If the gap is not within the above limits this indicates a fault in the throttle linkage and/or automatic gearbox unit. Special workshop procedure and equipment is required to carry out checks to these items and this work should be carried out as described in Division 44.
18. Switch off the engine and replace the air cleaner.

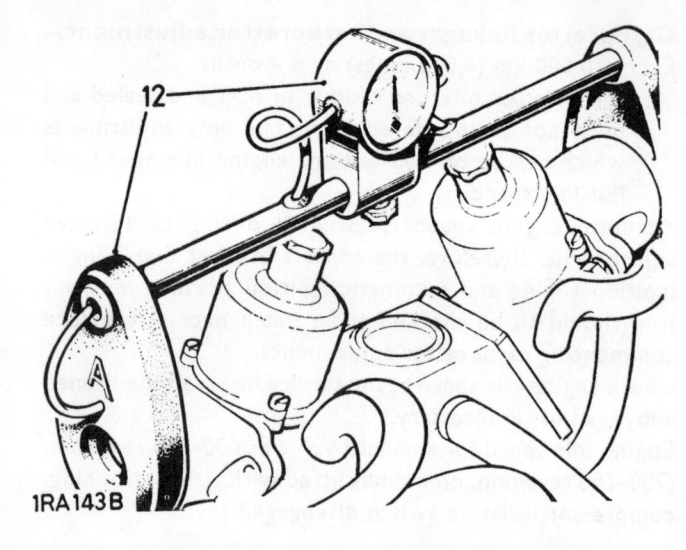

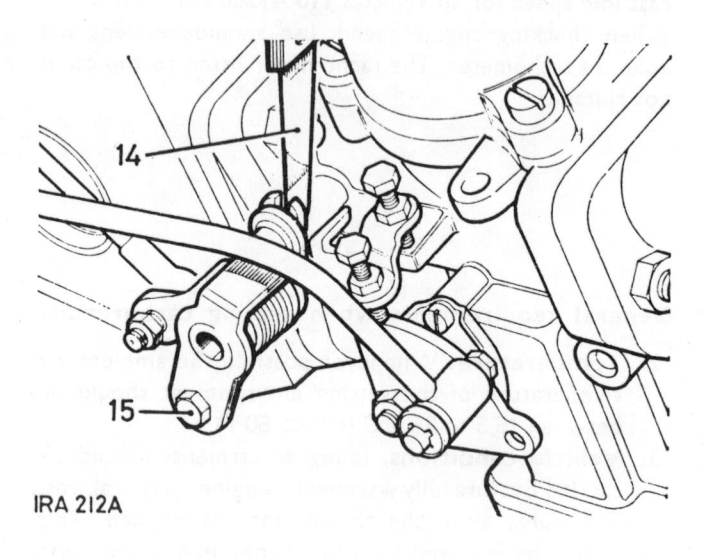

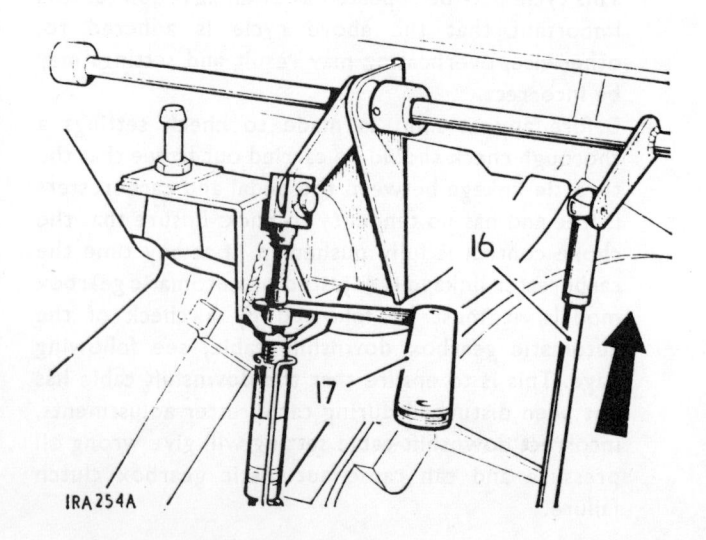

10-17

Rover 3500 and 3500S Manual AKM 3621

ENGINE COMPARTMENT

Power steering fluid reservoir, as applicable—At 1.000 km (750 miles) and at every maintenance inspection. The power steering units are lubricated by the operating fluid. The only lubrication required is to check the reservoir level as follows:

1. Unscrew the fluid reservoir cap.
2. Check that the fluid is up to the mark on the dipstick.
3. If necessary, top up using one of the recommended grades of fluid.

Power steering, as applicable—Every 5.000 km (3,000 miles) or 3 months.
Check power steering for oil leaks at hose connections, oil seals and joint faces on steering unit and power steering pump. Rectify as necessary.

UNDERBODY

Oil leaks—Every maintenance inspection.
Check for oil leaks; rectify as necessary.

Engine oil changes—After every 10.000 km (6,000 miles) or 6 months.
To change the engine oil:

1. Run the engine to warm up the oil, switch off the ignition.
2. Remove the drain plug in the bottom of the sump at left-hand side. Allow oil to drain away completely and replace the plug.
3. Replenish the sump as follows.

Filter replacement—Every 10.000 km (6,000 miles) or 6 months.

To change filter:

1. Place oil tray under engine.
2. Unscrew the filter anti-clockwise and discard. It may be necessary to use a 'strap spanner' or similar tool to release the filter.
3. Smear a little clean engine oil on the rubber washer of the new filter, then screw the filter on clockwise until the rubber sealing ring touches the oil pump cover face, then tighten a further half turn by hand only. Do not overtighten.
   Refill with oil of the correct grade through the screw-on filler cap on the right-hand front rocker cover; the capacity is 5,0 litres (9 Imperial pints), 10.5 US pints. This includes 0,5 litres (1 Imperial pint), 1 US pint for the filter.
   Run engine and check for oil leaks at filter and drain plug.

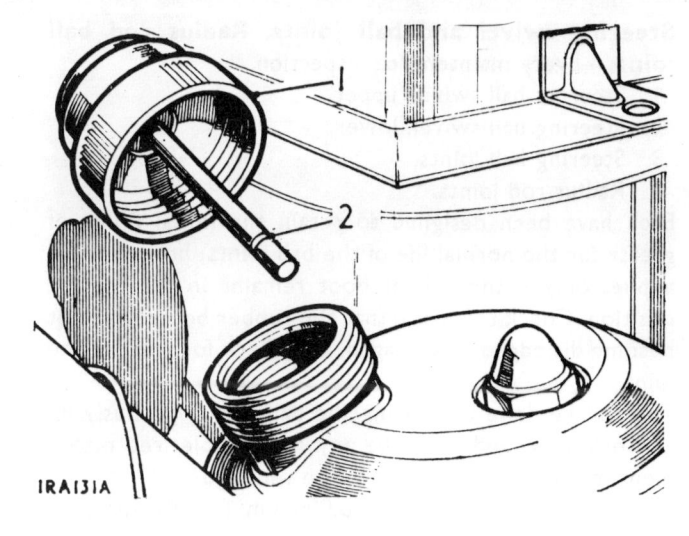

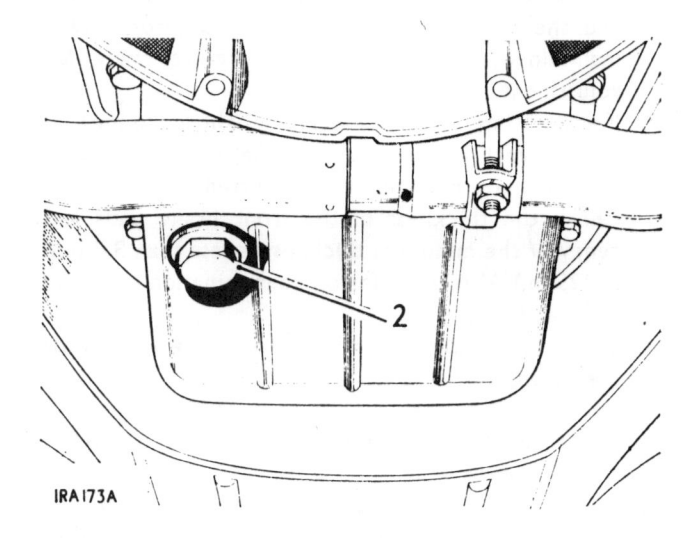

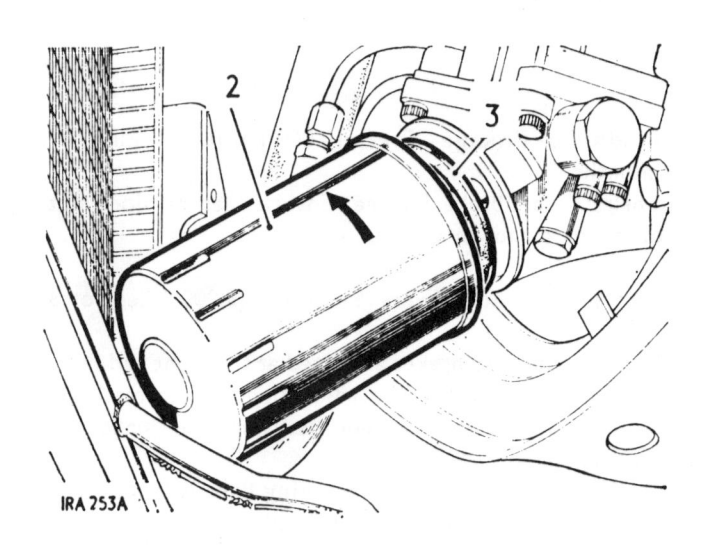

Rover 3500 and 3500S Manual AKM 3621

UNDERBODY

Steering swivel and ball joints. Radius rod ball joints—Every maintenance inspection.
1. Steering ball swivel, upper
2. Steering ball swivel, lower.
3. Steering ball joints.
4. Radius rod joints.

Each have been designed to retain the initial filling of grease for the normal life of the ball joints; however this applies only if the rubber boot remains in the correct position. Check to ensure that the rubber boots have not become dislodged or damaged, and check for wear in the joint.

This can be done by moving the ball joint vigorously up and down. Should there be any appreciable free movement, the complete joint must be replaced.

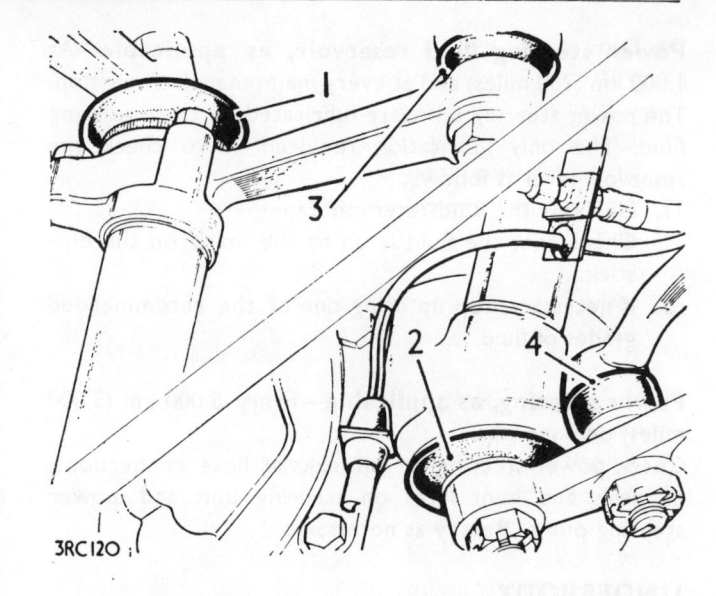

Front brake pads—Every 5.000 km (3,000 miles) or 3 months.
1. Hydraulic disc brakes are fitted at the front and rear and the correct brake adjustment is automatically maintained; no provision is therefore made for adjustment.
2. If at any time the brake warning light is illuminated due to worn brake pads, immediate action must be taken to have replacement pads fitted.
3. Check the thickness of the front brake pads and renew if the minimum thickness is less than 3,0 mm (0.125 in.) Also check for oil contamination on brake pads, and discs.
4. Check condition of brake discs for wear and/or corrosion.

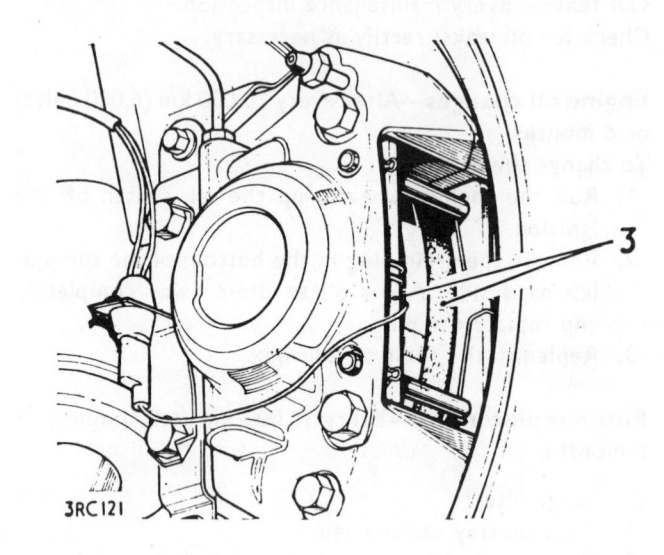

Propeller shaft lubrication—Every 10.000 km (6,000 miles) or 6 months.
1. Apply one of the recommended greases at the lubrication nipple on the sliding portion of propeller shaft.
2. Fully sealed journals are fitted and these require no lubrication.

Handbrake linkage—Every 10.000 km (6,000 miles) or 6 months.
Lubricate handbrake linkage. Ensure that oil does not contaminate pads or discs.

Rear brake pads—Every 5.000 km (3,000 miles) or 3 months.
1. Hydraulic disc brakes are fitted at the front and rear, and the correct brake adjustment is automatically maintained; no provision is therefore made for adjustment.
2. If at any time the brake warning light is illuminated due to worn brake pads, immediate action must be taken to have replacement pads fitted.
3. Check the thickness of the rear brake pads and renew if the minimum thickness is less than: 1,5 mm (0.062 in). Also check for oil contamination on brake pads, and discs.
4. When new, the brake pads on the rear calipers are tapered; however, as wear takes place the friction faces become parallel with the backing plates.
5. Check condition of brake discs for wear and/or corrosion. Fit new parts, as required, see Division 70.

10–19

Rover 3500 and 3500S Manual AKM 3621

UNDERBODY

De Dion tube, rear suspension—Check rubber boot every 10.000 km (6,000 miles) or 6 months. Check oil level every 20.000 km (12,000 miles) or 12 months.
1. Check oil level and top-up if necessary to the bottom of the filler plug hole.
2. At the same time check that the rubber boot on the de Dion tube is not dislodged or damaged.
3. If significant topping-up is required, check for oil leaks at drain hole in elbows and underneath rubber boot.

Final drive oil level—Every 10.000 km (6,000 miles) or 6 months.
1. Check oil level and top-up if necessary to the bottom of the filler plug hole.
2. If significant topping-up is required check for oil leaks at plugs, joint faces and oil seals adjacent to disc driving flanges and propeller shaft driving flange.

Exhaust system—Every 5.000 km (3,000 miles) or 3 months.
Check exhaust system fixings for security, paying particular attention to flexible mounting plates and clamps etc. Examine the system for signs of leakage and blowing. Any silencers or pipes found to be leaking or badly corroded should be replaced.

Fuel and brake pipes—Every maintenance inspection. Check all fuel and brake pipes, unions and hoses for signs of leakage, corrosion, chafing or damage.

3500 and 3500S models fitted with air conditioning
Fuel pump filter—Clean every 80.000 km (48,000 miles) or 48 months.
The electric fuel pump is located underneath the right-hand wheel arch.
To clean the filter proceed as follows:

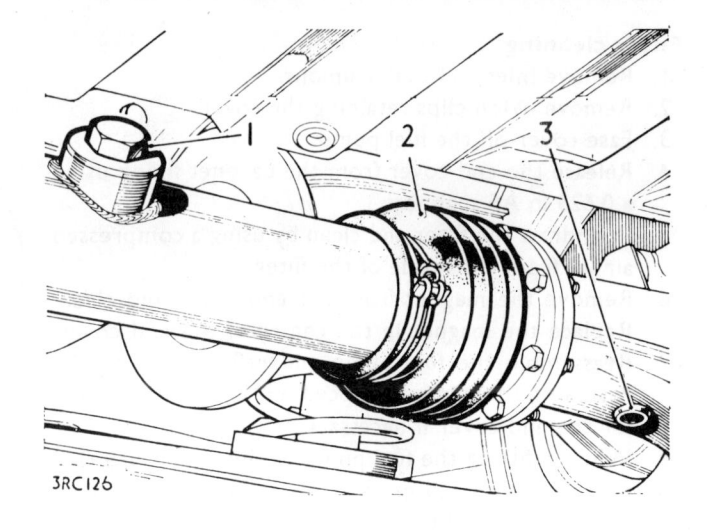

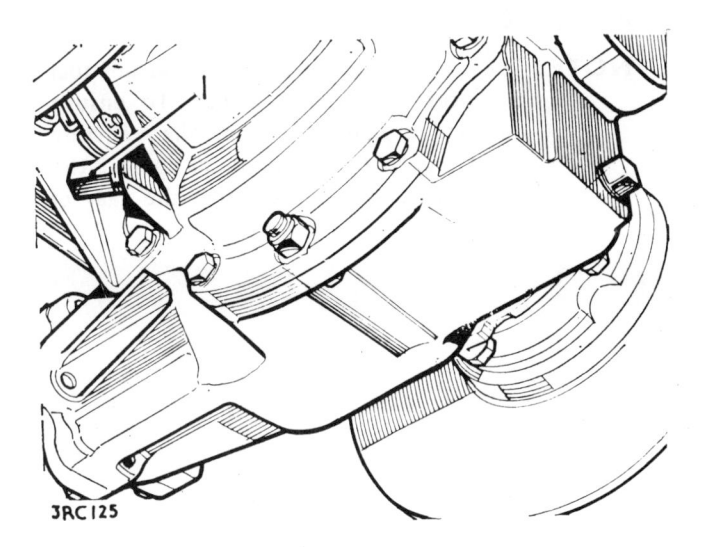

Pump removal
1. Open the bonnet.
2. Disconnect the fuel return pipe at the left-hand carburetter, to prevent draining the fuel tank by syphoning.
3. Open the boot and withdraw the rear trim panel.
4. Disconnect the fuel pump wire at the snap connector.
5. From underneath the car push in the knob for the fuel pipe cut-off valve.
6. Disconnect the inlet pipe together with the fuel pipe cut-off valve.
7. The closed cut-off valve will prevent the fuel tank from draining through the inlet pipe.
8. Disconnect the outlet pipe.
9. Remove the two bolts securing the fuel pump to the support bracket.
10. When the fuel pump is released the feed wire can be withdrawn through the grommet in the boot floor.

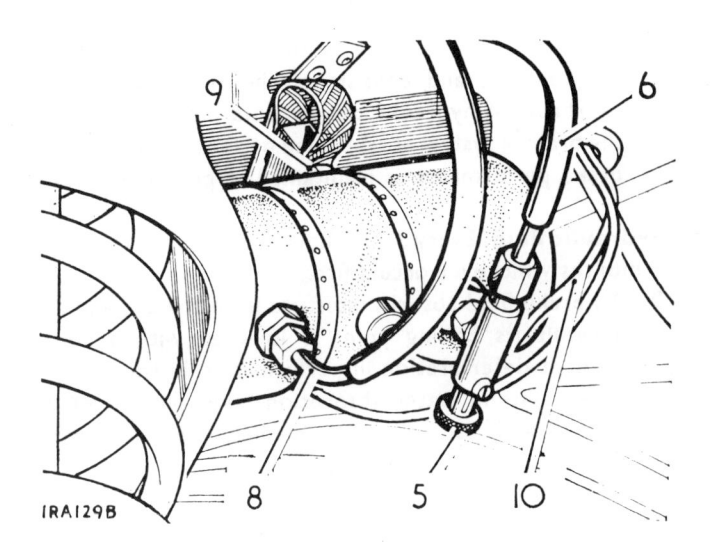

Rover 3500 and 3500S Manual AKM 3621

MAINTENANCE

Filter cleaning
1. Remove inlet and outlet unions.
2. Remove nylon clips retaining the cover.
3. Ease cover off the fuel pump.
4. Release the end cover from the bayonet fixing using a 0.625 in AF spanner.
5. Withdraw the filter and clean by using a compressed air jet from the inside of the filter.
6. Remove the magnet from the end cover and clean. Replace the magnet in the centre of the end cover.
7. Reassemble the fuel pump and refit to the car by reversing the removal procedure. Use a new gasket for the end cover if necessary.
8. When refitting the fuel pump to the car ensure that the earth wire and earth strip are fitted to the bolts securing the fuel pump to the bracket.
9. Pull out the cut-off valve knob.

   NOTE: The inlet and outlet unions are at unequal distances from the feed lead. To ensure correct fuel pipe connections when refitting the pump, position the union nearest the feed lead connection towards the rear of the vehicle.

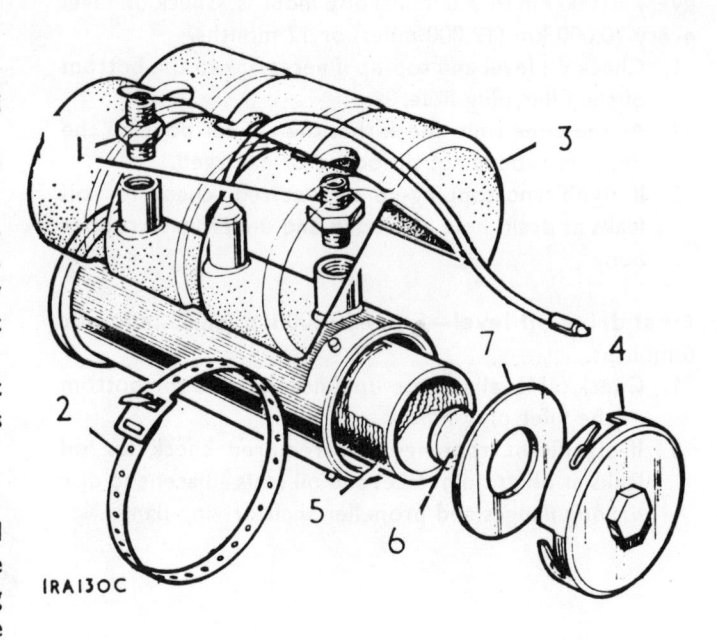

Road test—Every 10.000 km (6,000 miles) or 6 months. Give the car a thorough road test and carry out any further adjustments required.
Check operation of all instruments and warning lights. After test, check for oil, fluid or grease leaks at all plugs, flanges, joints and unions.

Preventive maintenance, all models
1. Preventive maintenance is in addition to routine maintenance, and consists of the replacement, or overhaul, of hydraulic components incorporated in the braking system at scheduled periods, in order that brake performance is maintained at peak efficiency.

Hydraulic fluid. Every 30.000 km (18,000 miles) or every 18 months whichever occurs first.
2. All brake fluid absorbs moisture from the air, and as a result its boiling point is lowered with a consequent deterioration in performance. In the sealed brake system, water absorption takes place over a period and can, if not remedied, reduce brake performance to a dangerous level.

All the fluid in the brake system should be changed every 30.000 km (18,000 miles) or eighteen months. It should also be changed before touring in mountainous areas if not done in the previous nine months. Use only Castrol Girling Brake and Clutch Fluid Green Specification, current SAE J1703 and US Federal Standard 116, from sealed tins, or Unipart brake and clutch fluid type 550 (coloured green) from sealed tins.
Never use fluid which has been left in an unsealed tin, nor re-use fluid already drained.

Rubber seals in brake system—Every 60.000 km (36,000 miles) or every three years, whichever occurs first.
3. Renew all rubber seals in complete brake system and all brake hydraulic hoses. Drain the brake fluid reservoir and flush the system. Refill with the correct fluid, that is, Castrol Girling Brake and Clutch Fluid Green Specification, current SAE J1703 and US Federal Standard 116, from sealed tins, or Unipart brake and clutch fluid type 550 (coloured green) from sealed tins.

ENGINE

LIST OF OPERATIONS

Camshaft—remove and refit..	12.13.01
Connecting rods and pistons	
—remove and refit	12.17.01
—overhaul	12.17.10
Crankshaft	
—remove and refit	12.21.33
—overhaul	12.21.46
—rear oil seal—remove and refit	12.21.20
—spigot bearing—remove and refit	12.53.20
Cylinder	
—heads—remove and refit	12.29.10
—heads—overhaul	12.29.18
—pressures—check	12.25.01
Drive plate—remove and refit	12.53.13
Engine assembly—remove and refit	12.41.01
Engine and gearbox assembly—remove and refit	12.37.01
Flywheel	
—remove and refit	12.53.07
—overhaul	12.53.10
—starter ring gear—remove and refit	12.53.19
Oil	
—filter assembly, external—remove and refit	12.60.01
—pump—remove and refit	12.60.26
—pump—overhaul	12.60.32
—sump—remove and refit	12.60.44
Rocker shafts	
—remove and refit	12.29.54
—overhaul	12.29.55
Timing	
—chain and gears—remove and refit	12.65.12
—gear cover—remove and refit	12.65.01
—gear cover oil seal—remove and refit	12.65.05
Valve gear—remove and refit	12.29.34

Rover 3500 and 3500S Manual AKM 3621

ENGINE

CAMSHAFT

—Remove and refit   12.13.01

Removing

1. Drain the cooling system. 26.10.01.
2. Remove the radiator block. 26.40.04.

Early style grille: 3 to 6

3. Remove the badge from the centre of the radiator grille by tapping the knurled studs at the rear.
4. Prise the plastic caps from the screw heads.
5. Remove both sections of the radiator grille.
6. Remove the badge mounting bracket.

Late style grille: 7 and 8

7. Remove the fixings from both headlamp bezels and the radiator grille.
8. Withdraw the headlamp bezels and the radiator grille.
9. Remove the alternator. 86.10.02. (AA1-6)
10. Remove the air cleaner. 19.10.01. (AA1-3)
11. Remove the induction manifold. 30.15.02. (AA1-9)
12. Remove the valve gear. 12.29.34.
13. Remove the timing chain cover. 12.65.01.
14. Remove the timing chain. 12.65.12.

> **CAUTION:** Do not damage the bearings when withdrawing the camshaft as the camshaft bearings are not serviceable.

15. Withdraw the camshaft.

Inspecting

16. Check all bearing surfaces for excessive wear and score marks. Also check cam lobes for excessive wear. Check key and keyway.

Refitting

17. Reverse 1 to 12.

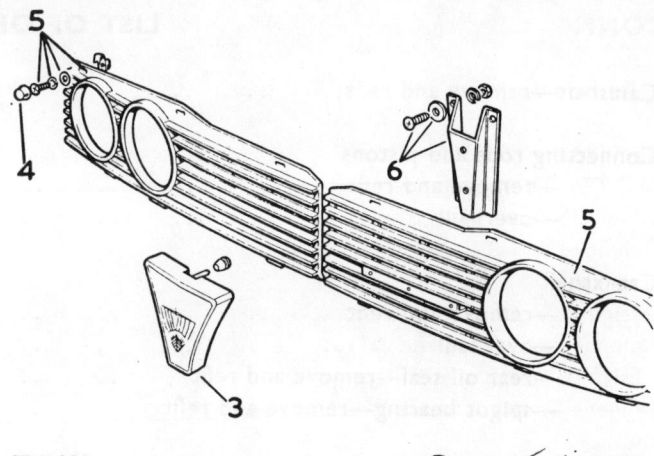

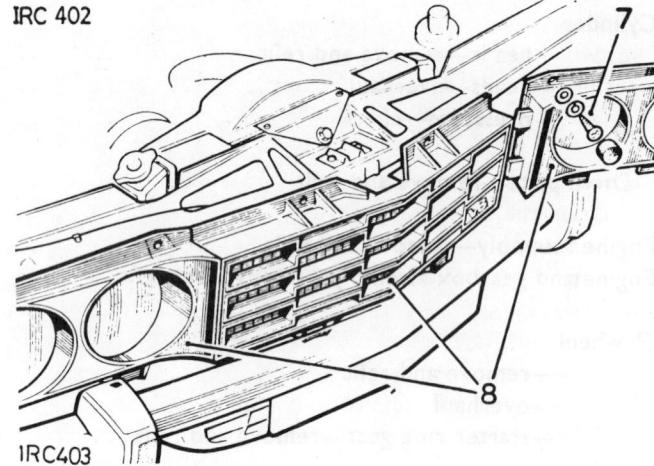

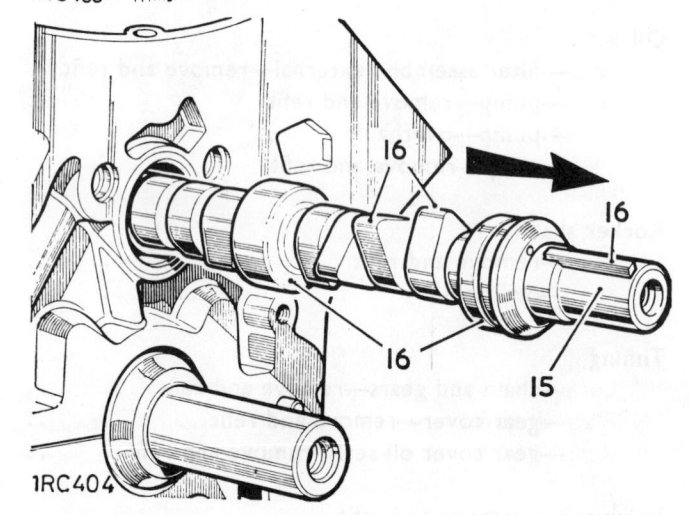

DATA

Camshaft

Position	Central
Material	Cast iron
Drive	Chain
Bearings Type	Steel backed babbit lined
Number	5

Valve timing

	Inlet	Exhaust
Opens	30 degrees BTDC	68 degrees BBDC
Closes	75 degrees ABDC	37 degrees ATDC
Duration	285 degrees	285 degrees
Valve peak	112.5 degrees ATDC	105.5 degrees BTDC

Rover 3500 and 3500S Manual AKM 3621

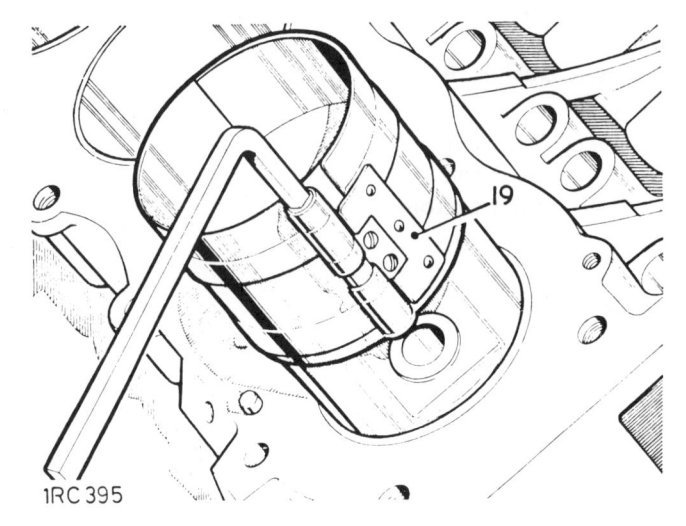

ENGINE

CONNECTING RODS AND PISTONS

—Remove and refit 12.17.01

Service tool 605351—Guide bolts for connecting rods
NOTE: There are two designs of pistons in use on Rover 3500 engines, the two designs differ slightly in weight and this is compensated for by two standards of crankshaft balance.

Design 'A' pistons are fitted with a crankshaft which has a plain face at the starter dog end.

Design 'B' pistons are fitted with a crankshaft which has an identification groove 'C' in the face at the starter dog end.

The two designs are interchangeable providing that the eight pistons and applicable crankshaft are fitted initially as a set.

Removing

1. Drain the cooling system. 26.10.01.
2. Remove the air cleaner. 19.10.01. (AA1-3)
3. Remove the alternator. 86.10.02. (AA1-6)
4. Remove the induction manifold. 30.15.02. (AA1-9)
5. Remove the valve gear. 12.29.34.
6. Remove the cylinder heads. 12.29.10.
7. Remove the oil sump. 12.60.44.
8. Remove the baffle plate and sump oil strainer.
9. Remove the connecting rod caps and retain them in sequence for reassembly.
10. Screw the guide bolts 605351, on to the connecting rods.
11. Push the connecting rod and piston assembly up the cylinder bore and withdraw it from the top. Retain the connecting rod and piston assemblies in sequence with their respective caps.
12. Remove the guide bolts 605351, from the connecting rod.

Refitting

13. Locate the applicable crankshaft journal at BDC.
14. Place the bearing upper shell in the connecting rod.
15. Retain the upper shell by screwing the guide bolts 605351, on to the connecting rods.
16. Insert the connecting rod and piston assembly into its respective bore, noting that the domed shape boss on the connecting rod must face towards the front of the engine on the right-hand bank of cylinders and towards the rear on the left-hand bank. When both connecting rods are fitted, the bosses will face inwards towards each other.
17. Position the oil control piston rings so that the ring gaps are all at one side, between the gudgeon pin and piston thrust face. Space the gaps in the ring rails approximately 25 mm (1 in) each side of the expander ring joint.
18. Position the compression rings so that their gaps are on opposite sides of the piston between the gudgeon pin and piston thrust face.

continued

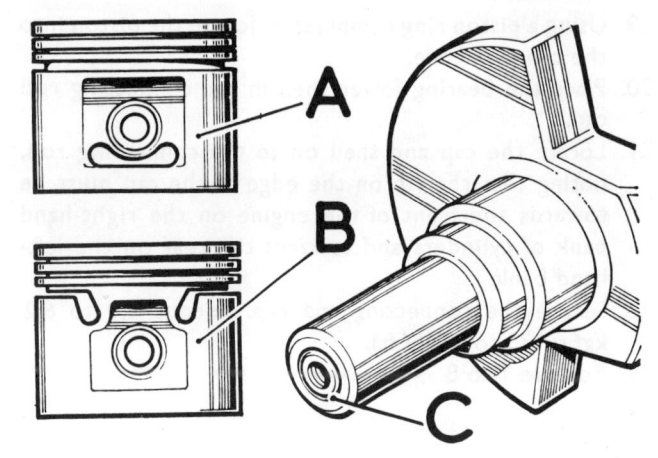

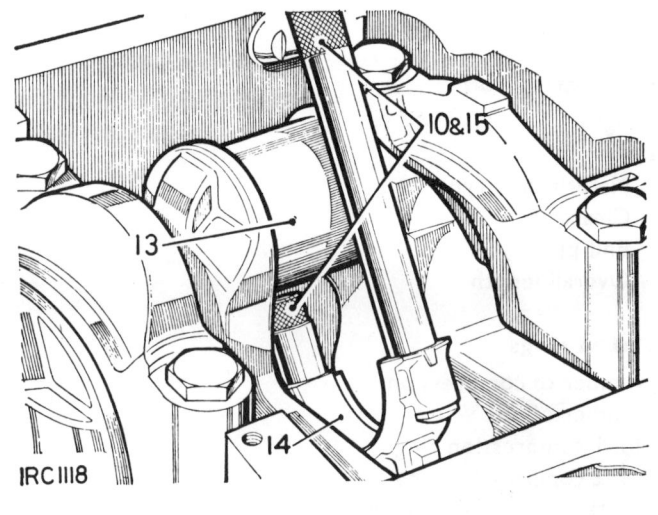

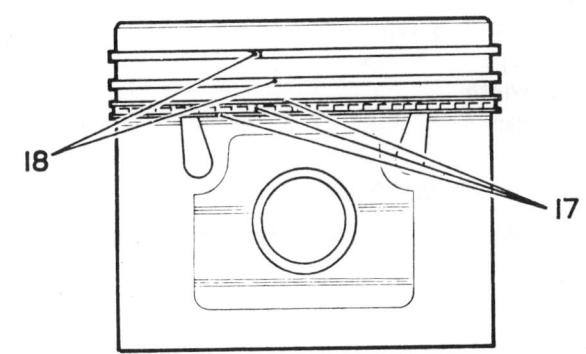

Rover 3500 and 3500S Manual AKM 3621

12.17.01
Sheet 1

ENGINE

19. Using a piston ring compressor, locate the piston into the cylinder bore.
20. Place the bearing lower shell in the connecting rod cap.
21. Locate the cap and shell on to the connecting rod, noting that the rib on the edge of the cap must be towards the front of the engine on the right-hand bank of cylinders and towards the rear on the left-hand bank.
22. Secure the connecting rod cap. Torque 4,9 to 6,2 kgf.m (35 to 40 lbf ft).
23. Reverse 1 to 8.

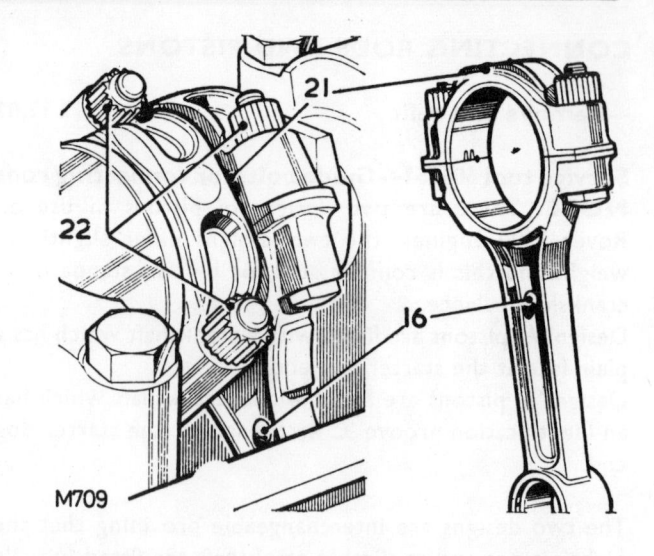

DATA

Connecting rod

Length	Centres 143,81 mm to 143,71 mm (5.662 to 5.658 in)
Bearings:	
Material and type	Vandervell VP lead indium
Clearance	0,015 to 0,055 mm (0.0006 to 0.0022 in)
End play	0,15 to 0,37 mm (0.006 to 0.014 in)
Overall length	18,60 to 18,85 mm (0.732 to 0.742 in)

Piston rings

Number of compression	2
Number of oil	1
No. 1 compression ring	Chrome faced marked 'T' or 'TOP'
No. 2 compression ring	Stepped to 'L' shape and marked 'T' or 'TOP'
Width of compression rings	1,98 to 2,01 mm (0.078 to 0.079 in)
Compression ring gap	0,44 to 0,57 mm (0.017 to 0.022 in)
Oil ring type	Perfect circle, type 98
Oil ring width	4,811 mm (0.1894 in) max.
Oil ring gap	0,38 to 1,40 mm (0.015 to 0.055 in)

Pistons and gudgeon pins

Pistons—Design 'A' (see illustration at beginning of this operation)

Type	Aluminium alloy
Clearance: Top land	0,65 to 0,81 mm (0.0255 to 0.0320 in)
Skirt top	0,018 to 0,033 mm (0.0007 to 0.0013 in)
Skirt bottom	0,008 to 0,043 mm (0.0003 to 0.0017 in)

Pistons—Design 'B'

Type	Aluminium alloy—'W' slot skirt
Clearance: Top land	0,73 to 0,88 mm (0.0296 to 0.0350 in)
Skirt top	0,040 to 0,071 mm (0.0016 to 0.0028 in)
Skirt bottom	0,018 to 0,033 mm (0.0007 to 0.0013 in)

Gudgeon pins

Length	72,67 to 72,79 mm (2.861 to 2.866 in)
Diameter	22,215 to 22,220 mm (0.8746 to 0.8749 in)
Fit in rod	Press fit
Clearance in piston	0,002 to 0,007 mm (0.0001 to 0.0003 in)

12.17.01
Sheet 2

Rover 3500 and 3500S Manual AKM 3621

ENGINE

CONNECTING RODS AND PISTONS

—Overhaul   12.17.10

Service tools:
- 605350—Tool for removing and refitting gudgeon pin
- 605238—Plastigauge

NOTE: There are two designs of pistons in use on Rover 3500 engines, the two designs differ slightly in weight and this is compensated for by two standards of crankshaft balance.

Design 'A' pistons are fitted with a crankshaft which has a plain face at the starter dog end.

Design 'B' pistons are fitted with a crankshaft which has an identification groove 'C' in the face at the starter dog end. The two designs are interchangeable providing that the eight pistons and applicable crankshaft are fitted initially as a set.

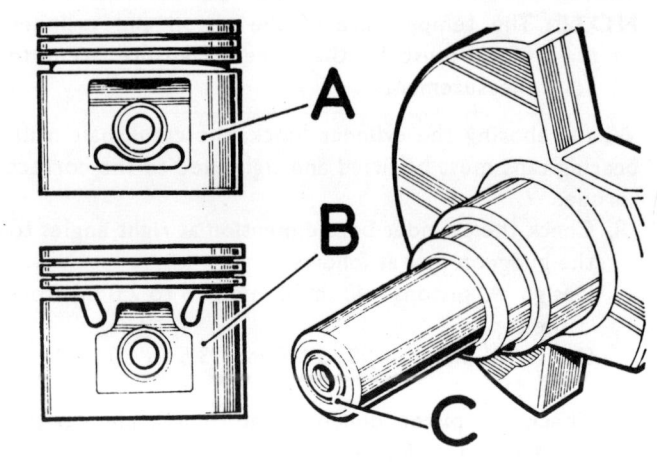

1RC393

Dismantling

1. Remove the connecting rods and pistons. 12.17.01.
2. If the same piston is to be refitted, add location marks to ensure reassembling in the same relative position.
3. Locate the piston and connecting rod assembly on tool 605350.
4. Locate the drift, part of tool 605350, on to the gudgeon pin.
5. Using a hydraulic press—8 tonne (8 ton) capacity—press out the gudgeon pin.

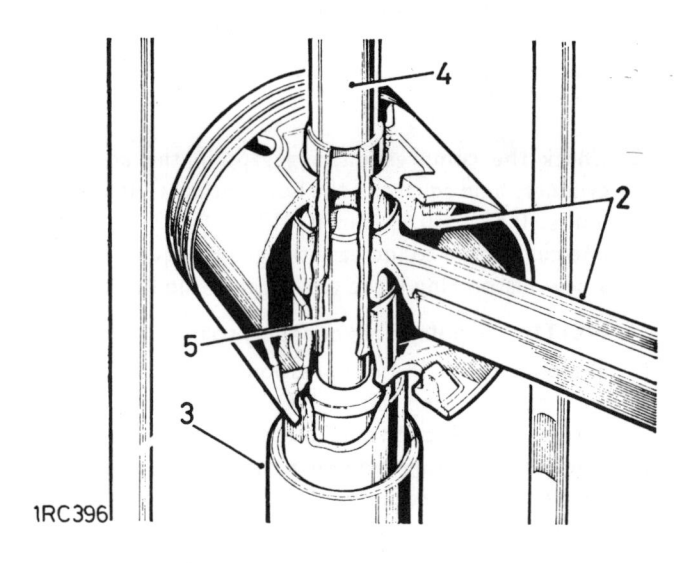

Overhauling pistons

Original pistons

6. Remove carbon and deposits, particularly from the ring grooves.
7. Examine the pistons for damage or excess wear—see under 'New pistons' for clearance dimensions—fit new replacements as necessary.

1RC396

New pistons

Pistons are available in graded standard sizes and in ungraded oversizes of 0,25 mm (0.010 in) and 0,50 mm (0.020 in).

Standard pistons are graded in diameter, and the grade letter is stamped on the crown of the piston and on the cylinder block.

Grade letter 'Z' cylinder bore diameter nominal to +0,0075 mm (0.0003 in).

'A' 0,0075 mm (0.0003 in) to 0,015 mm (0.0006 in) above nominal.

'B' 0,015 mm (0.0006 in) to 0,0225 mm (0.0009 in) above nominal.

'C' 0,0225 mm (0.0009 in) to 0,03 mm (0.0012 in) above nominal.

'D' 0,03 mm (0.0012 in) to 0,0375 mm (0.0015 in) above nominal.

'S' 0,081 mm (0.0032 in) to 0,088 mm (0.0035 in) above nominal.

If new pistons are required for a standard size bore, check the bore size and fit the grade of piston that provides the correct clearance.

The clearance limits with new pistons and a new or rebored cylinder are 0,018 to 0,033 mm (0.0007 to 0.0013 in).

continued

Rover 3500 and 3500S Manual AKM 3621

ENGINE

NOTE: The temperature of the piston and cylinder block must be the same to ensure accurate measurements.

When reboring the cylinder block, the crankshaft main bearing caps must be fitted and tightened to the correct torque.

8. Check the cylinder bore dimension at right angles to the gudgeon pin, as follows:
   Design 'A' pistons—40 to 50 mm (1.5 to 2.0 in) from the top.
   Design 'B' pistons—90 to 100 mm (3.5 to 4.0 in) from the top.
9. Check the piston dimension at right angles to the gudgeon pin, as follows:
   Design 'A' pistons—at the top of the skirt.
   Design 'B' pistons—at the bottom of the skirt.
10. The piston dimension must be 0,018 to 0,033 mm (0.0007 to 0.0013 in) smaller than the cylinder.
11. If new piston rings are to be fitted without reboring, deglaze the cylinder walls with a hone, without increasing the bore diameter.

    NOTE: A deglazed bore must have a cross-hatch finish.

12. Check the compression ring gaps in the applicable cylinder, held square to the bore with the piston. Gap limits: 0,46 to 0,60 mm (0.017 to 0.022 in). Use a fine-cut flat file to increase the gap if required. Select a new piston ring if the gap exceeds the limit.

    NOTE: Gapping does not apply to oil control rings.

13. Temporarily fit the compression rings to the piston with the marking 'T' or 'TOP' uppermost and the chrome compression ring in the top groove.
14. Check the compression ring clearance in the piston groove. Clearance limits: 0,08 to 0,13 mm (0.003 to 0.005 in).

Fitting piston rings

15. Fit the expander ring into the bottom groove making sure that the ends abut and do not overlap.
16. Fit two ring rails to the bottom groove, one above and one below the expander ring.
17. Fit the compression rings with the marking 'T' or 'TOP' uppermost and the chrome compression ring in the top groove.

Connecting rods

18. Check the alignment of the connecting rod.
19. Check the connecting rod small end, the gudgeon pin must be an interference fit.

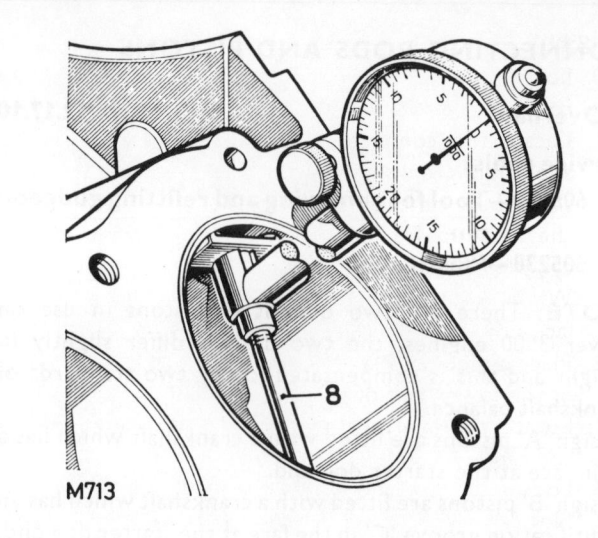

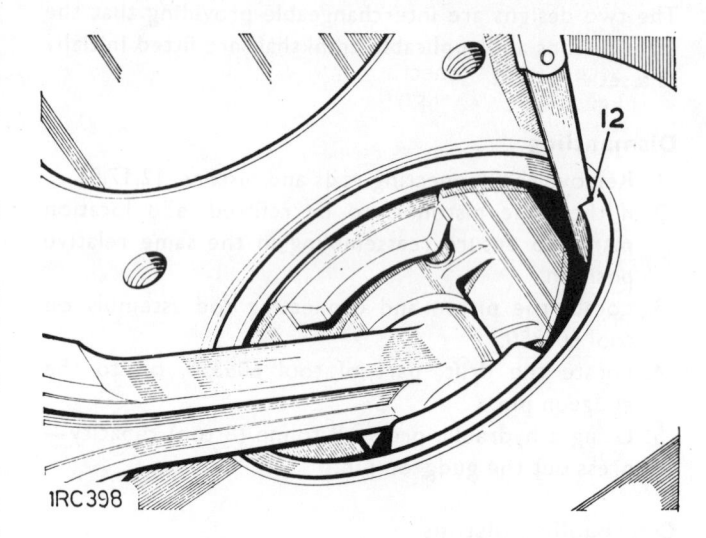

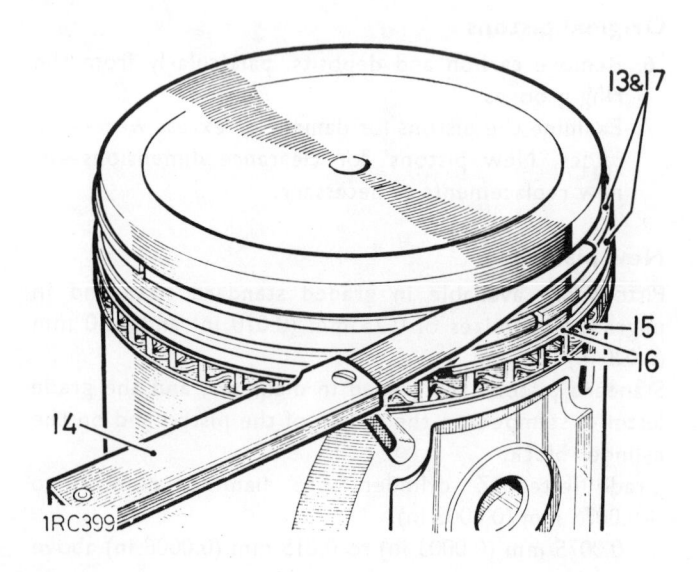

continued

12.17.10
Sheet 2

Rover 3500 and 3500S Manual AKM 3621

ENGINE

Big-end bearings

20. Locate the bearing upper shell into the connecting rod.
21. Locate the connecting rod and bearing on to the applicable crankshaft journal, noting that the domed shape boss on the connecting rod must face towards the front of the engine on the right-hand bank of cylinders and towards the rear on the left-hand bank. When both connecting rods are fitted, the bosses will face inwards towards each other.
22. Place a piece of Plastigauge 605238, across the centre of the lower half of the crankshaft journal.
23. Locate the bearing lower shell into the connecting rod cap.
24. Locate the cap and shell on to the connecting rod. Note that the rib on the edge of the cap must be the same side as the domed shape boss on the connecting rod.
25. Secure the connecting rod cap. Torque 4,9 to 6,2 kgf.m (35 to 40 lbf ft).

    NOTE: Do not rotate the crankshaft while the Plastigauge is fitted.

26. Remove the connecting rod cap and shell.
27. Using the scale printed on the Plastigauge packet, measure the flattened Plastigauge at its widest point. The graduation that most closely corresponds to the width of the Plastigauge indicates the bearing clearance.
28. The correct bearing clearance with new or overhauled components is 0,015 to 0,055 mm (0.0006 to 0.0022 in).
29. If a bearing has been in service, it is advisable to fit a new bearing if the clearance exceeds 0,08 mm (0.003 in).
30. If a new bearing is being fitted, use selective assembly to obtain the correct clearance.
31. Wipe off the Plastigauge with an oily rag. DO NOT scrape it off.

    NOTE: The connecting rods, caps and bearing shells must be retained in sets, and in the correct sequence.

Reassembling

32. Locate the guide for the gudgeon pin on tool 605350.
33. Locate the piston and connecting rod on tool 605350.
34. Insert the gudgeon pin into the piston and locate it over the guide.
35. Locate the drift, part of 605350, on to the gudgeon pin.
36. Using a hydraulic press—8 tonne (8 ton) capacity—press in the gudgeon pin until it abuts the shoulder of the guide.
37. Check that the piston moves freely on the gudgeon pin and that no damage has occurred during pressing.
38. Fit the connecting rods and pistons, 12.17.01, carrying out the following checks during fitting. 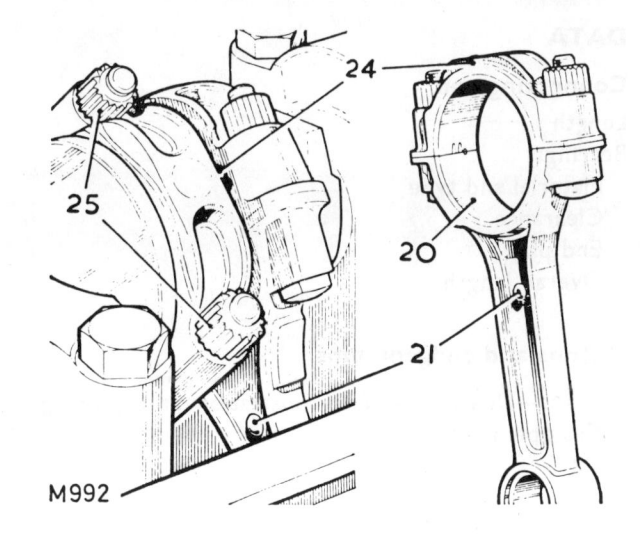 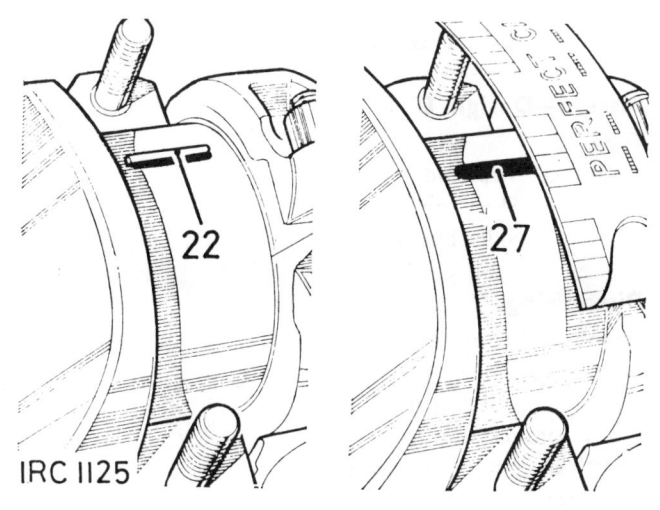 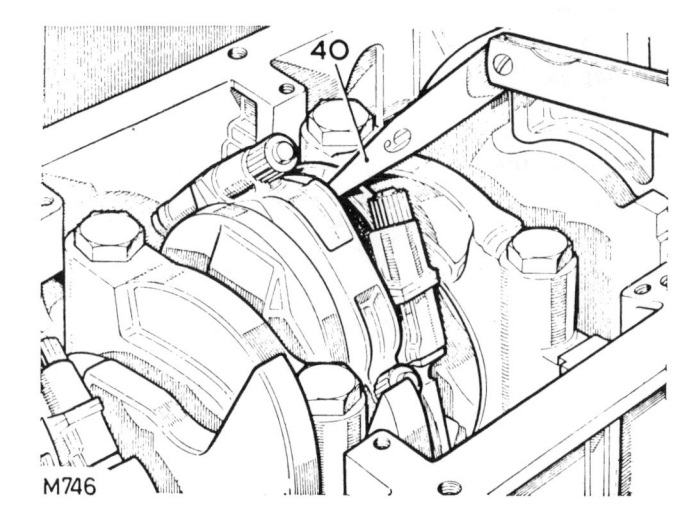
39. Check that the connecting rods move freely sideways on the crankshaft. Tightness indicates insufficient bearing clearance or a mis-aligned connecting rod.
40. Check the end-float between the connecting rods on each crankshaft journal. Clearance limits: 0,15 to 0,37 mm (0.006 to 0.014 in).

continued

Rover 3500 and 3500S Manual AKM 3621

12.17.10
Sheet 3

ENGINE

DATA

Connecting rod

Length	Centres 143,81 mm to 143,71 mm (5.662 to 5.658 in)
Bearings:	
Material and type	Vandervell VP lead indium
Clearance	0,015 to 0,055 mm (0.0006 to 0.0022 in)
End play	0,15 to 0,37 mm (0.006 to 0.014 in)
Overall length	18,60 to 18,85 mm (0.732 to 0.742 in)

Pistons and gudgeon pins

Pistons—Design 'A' (see illustration at beginning of this operation)

Type	Aluminium alloy
Clearance: Top land	0,65 to 0,81 mm (0.0255 to 0.0320 in)
Skirt top	0,018 to 0,033 mm (0.0007 to 0.0013 in)
Skirt bottom	0,008 to 0,043 mm (0.0003 to 0.0017 in)

Pistons—Design 'B'

Type	Aluminium alloy—'W' slot skirt
Clearance: Top land	0,73 to 0,88 mm (0.0296 to 0.0350 in)
Skirt top	0,040 to 0,071 mm (0.0016 to 0.0028 in)
Skirt bottom	0,018 to 0,033 mm (0.0007 to 0.0013 in)

Piston rings

Number of compression	2
Number of oil	1
No. 1 compression ring	Chrome faced marked 'T' or 'TOP'
No. 2 compression ring	Stepped to 'L' shape and marked 'T' or 'TOP'
Width of compression rings	1,98 to 2,01 mm (0.078 to 0.079 in)
Compression ring gap	0,44 to 0,57 mm (0.017 to 0.022 in)
Oil ring type	Perfect circle, type 98
Oil ring width	4,811 mm (0.1894 in) max.
Oil ring gap	0,38 to 1,40 mm (0.015 to 0.055 in)

Gudgeon pins

Length	72,67 to 72,79 mm (2.861 to 2.866 in)
Diameter	22,215 to 22,220 mm (0.8746 to 0.8749 in)
Fit in rod	Press fit
Clearance in piston	0,002 to 0,007 mm (0.0001 to 0.0003 in)

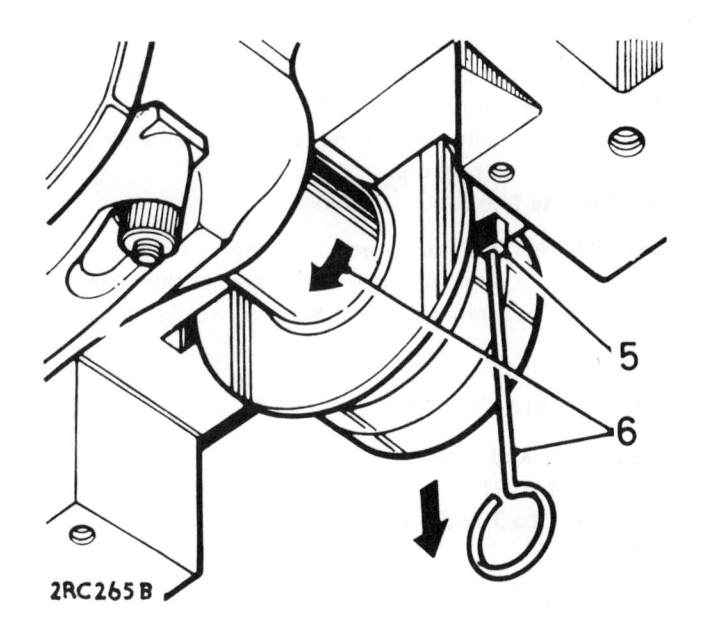

ENGINE

CRANKSHAFT REAR OIL SEAL—Braided type seal

—Remove and refit   12.21.20
3500   1 to 27
3500S 2 to 27

Service tools: RO 1009 Oil seal, upper, remover and replacer

NOTE: The braided type crankshaft rear oil seal is fitted to engines numbered suffix 'A'. For engines numbered suffix 'B' onwards, refer to separate Operation with same number on sheet 2.

Removing

1. Remove the front exhaust pipe.
2. Remove the oil sump. 12.60.44.
3. Remove the cover plate from the gearbox bell housing.
4. Remove the rear main bearing cap.
5. Insert the screwed end of the remover into the centre of the upper seal.
6. Rotate the crankshaft in the direction illustrated and extract the old seal.

Refitting

7. Prepare and shape the new seal, utilising the seal groove in the rear main bearing cap, as follows:
   a. Place the new seal in the groove with both ends projecting above the face of the cap.
   b. Force the seal into the groove by rubbing down with a hammer handle until the seal projects above the groove not more than 1,5 mm (0.030 in).
   c. Gently remove the seal.
8. Spread the 'Chinese Finger' with the tapered end of the installer handle to receive the end of the seal.
9. Insert 6 mm (0.250 in) of one end of the seal into the 'Chinese Finger', compress the 'Chinese Finger' against the seal by rolling between the fingers.
10. Soak the seal with heavy engine oil.
11. Select the funnel with the larger internal diameter and slip it over the installer with the countersunk end towards the seal.
12. Fit the installer into the seal groove.
13. Attach the handle to the installer and position the flat of the funnel against the crankshaft.
14. Carefully pull the seal into the groove, while simultaneously turning the flywheel or drive plate as applicable, stopping when an even amount of seal shows each side. DO NOT pull the seal too far, otherwise a new seal may have to be fitted.
15. Remove the installer.
16. Using the pusher, pack the protruding seal material up into the groove.
17. Using a sharp, flat blade, trim off any excess seal to provide a perfectly smooth mating surface for the bearing cap.
18. Place a new seal in the groove in the rear main bearing cap with both ends projecting above the face of the cap.

continued

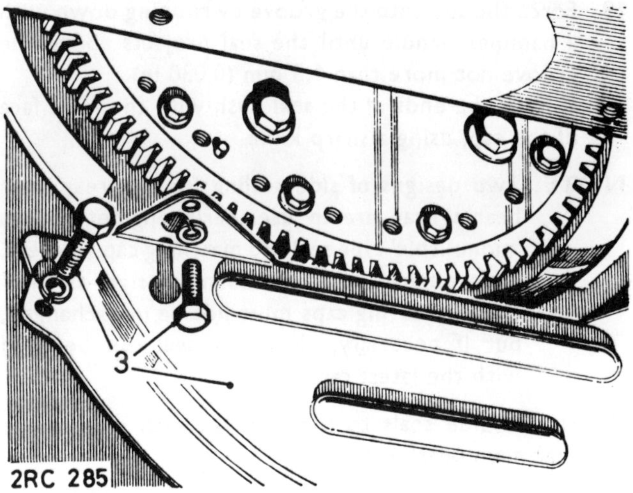

Rover 3500 and 3500S Manual AKM 3621

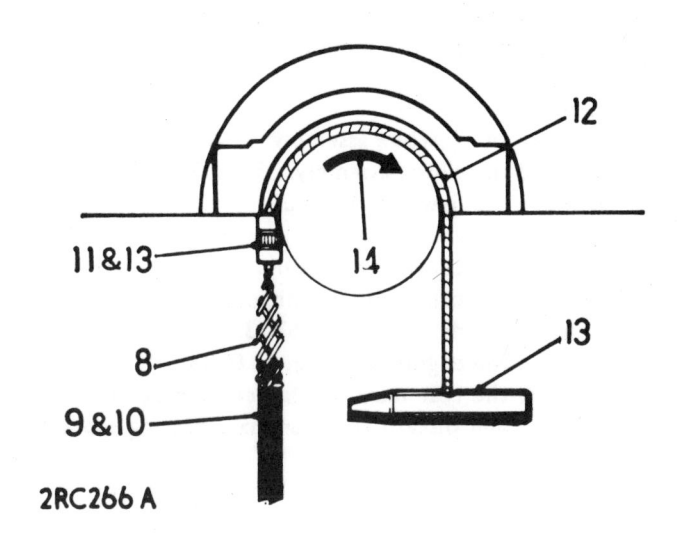

12.21.20
Sheet 1

ENGINE

19. Force the seal into the groove by rubbing down with a hammer handle until the seal projects above the groove not more than 1,5 mm (0.030 in).
20. Cut off the ends of the seal flush with the joint face of the cap, using a sharp knife.

NOTE: Two designs of side sealing for the rear main bearing cap are in use. Early engines have a 'monolithic' type seal and matching cap whereas, later engines use 'cruciform' design as illustrated. Bearing caps must not be interchanged, but if necessary, the early seal can be used with the latest cap.

21. Fit the side seals to the grooves each side of the rear main bearing cap.
22. Do not cut the side seals to length; they must protrude approximately 1,5 mm (0.062 in) above the bearing cap parting face.
23. Lubricate the rear oil seal with heavy engine oil.
24. Lubricate the lower bearing shell and the side oil seals with light engine oil.
25. Fit the rear main bearing cap and shell. Torque 9,0 to 9,6 kgf. m (65 to 70 lbf ft).
26. Using a blunt instrument, push the side oil seals into the bearing cap.
27. Reverse 1 to 3.

CAUTION: Do not exceed 1000 engine rev/min when first starting the engine, otherwise the crankshaft rear oil seal will be damaged.

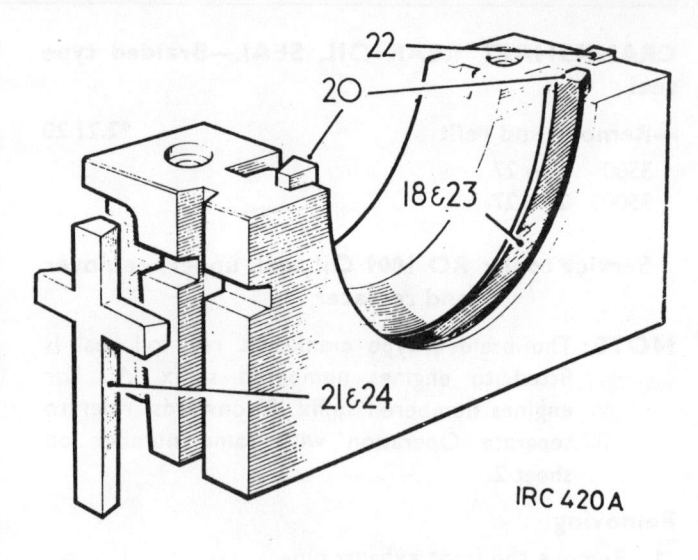

CRANKSHAFT REAR OIL SEAL—Lip type seal

—Remove and refit   12.21.20
3500   4 to 23
3500S   1 to 3 and 8 to 23

Service tool: RO. 1014 Seal guide

NOTE: The lip type crankshaft rear oil seal is fitted to engines numbered suffix 'B' onwards. For engines numbered suffix 'A', refer to separate Operation with same number on sheet one.

Removing

1. Remove the gearbox assembly. 37.20.01. ⎫
2. Remove the clutch assembly. 33.10.01. ⎬ *manual gearbox only.*
3. Remove the flywheel. 12.53.07. ⎭
4. Remove the gearbox assembly. 44.20.01. ⎫
5. Remove the converter housing. 44.17.01. ⎪ *automatic transmission only.*
6. Remove the converter assembly. 44.17.07. ⎬
7. Remove the drive plate. 12.53.13. ⎭
8. Remove the engine oil sump. 12.60.44.
9. Remove the rear main bearing cap and bearing half.
10. Remove and discard the rear main bearing cap side seals.
11. Remove and discard the crankshaft rear oil seal.

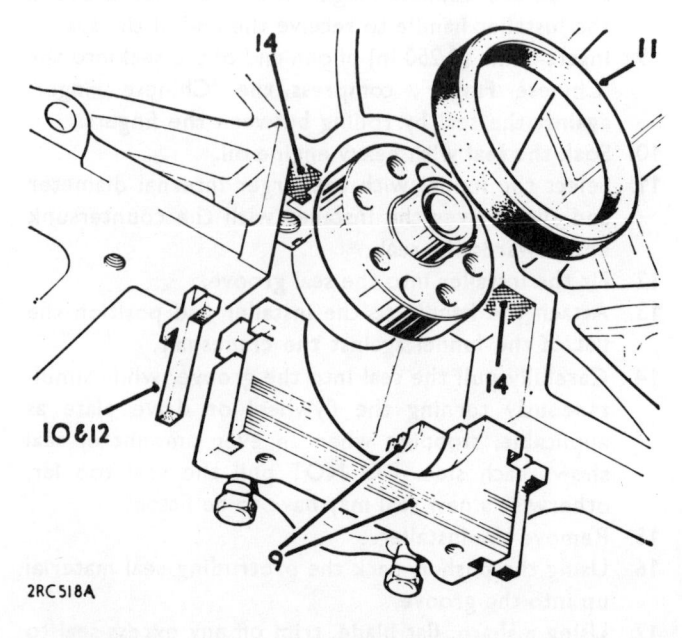

continued

12.21.20
Sheet 2

Rover 3500 and 3500S Manual AKM 3621

ENGINE

Refitting

CAUTION: Do not handle the seal lip at any time, visually check that it is not damaged and ensure that the outside diameter remains clean and dry.

12. Fit the side seals to the grooves each side of the rear main bearing cap.
13. Do not cut the side seals to length; they must protrude approximately 1,5 mm (0.062 in) above the bearing cap parting face.
14. Apply Hylomar PL 32M jointing compound, Part No. 534244 (or Unipart No. GGC 102) to the rearmost half of the rear main bearing cap parting face or, if preferred, to the equivalent area on the cylinder block, as illustrated.
15. Lubricate the bearing half and bearing cap side seals with clean engine oil.
16. Fit the bearing cap assembly to the engine. Do not tighten the fixings at this stage but ensure that the cap is fully home and squarely seated on the cylinder block.
17. Tension the cap bolts equally by one-quarter turn approximately, then back off one complete turn on each fixing bolt.
18. Position the seal guide RO.1014 on the crankshaft flange.
19. Ensure that the oil seal guide and the crankshaft journal are scrupulously clean then coat the seal guide and oil seal journal with clean engine oil.
    NOTE: The lubricant coating must cover the seal guide outer surface completely to ensure that the oil seal lip is not turned back during assembly.
20. Position the oil seal, lipped side towards the engine, on to the seal guide. The seal outside diameter must be clean and dry.
21. Push home the oil seal fully and squarely by hand into the recess formed in the cap and block until it abuts against the machined step in the recess. Withdraw the seal guide.
22. Tighten the rear main bearing cap fixings fully and evenly. Torque: 9,0 to 9,6 kgf. m (65 to 70 lbf ft).
23. Reverse 1 to 8, as applicable.

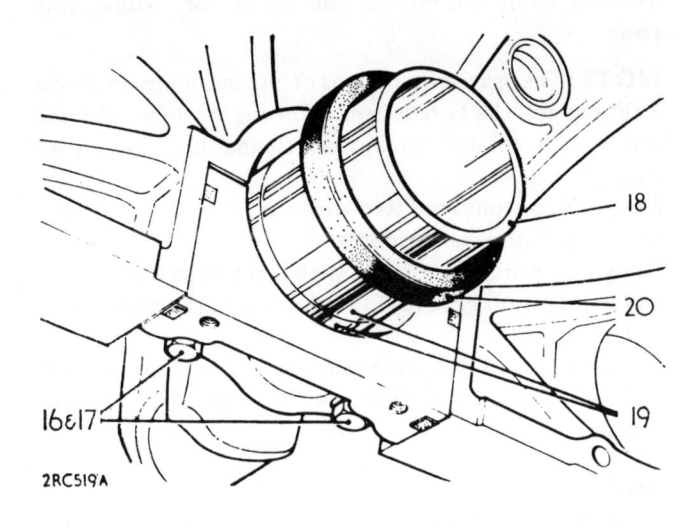

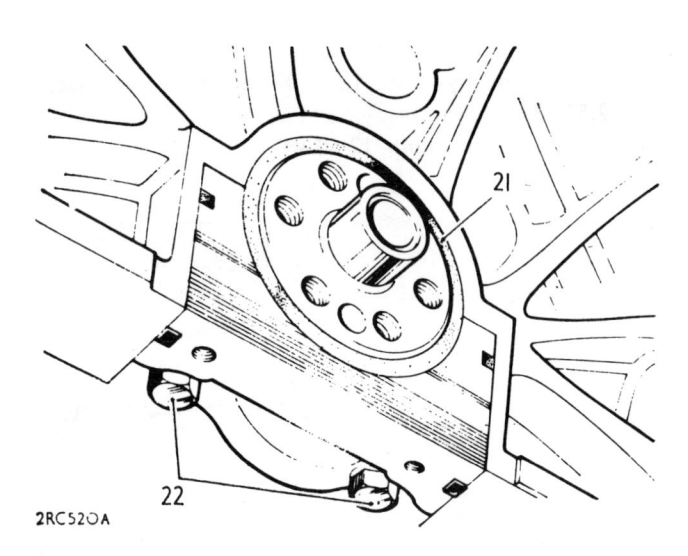

Rover 3500 and 3500S Manual AKM 3621

ENGINE

CRANKSHAFT

—Remove and refit   12.21.33

Service tools 605351—Guide bolts for connecting rods

NOTE: There are two designs of pistons in use on Rover 3500 engines, the two designs differ slightly in weight and this is compensated for by two standards of crankshaft balance.

Design 'A' pistons are fitted with a crankshaft which has a plain face at the starter dog end.

Design 'B' pistons are fitted with a crankshaft which has an identification groove 'C' in the face at the starter dog end.

The two designs are interchangeable providing that the eight pistons and applicable crankshaft are fitted initially as a set.

Removing

1. Remove the engine assembly. 12.41.01.
2. Remove the timing gear cover. 12.65.01.
3. Remove the timing chain and gears. 12.65.12.
4. Remove the clutch. 33.10.01.
5. Remove the flywheel. 12.53.07. or the drive plate. 12.53.13.
6. Remove the oil sump. 12.60.44.

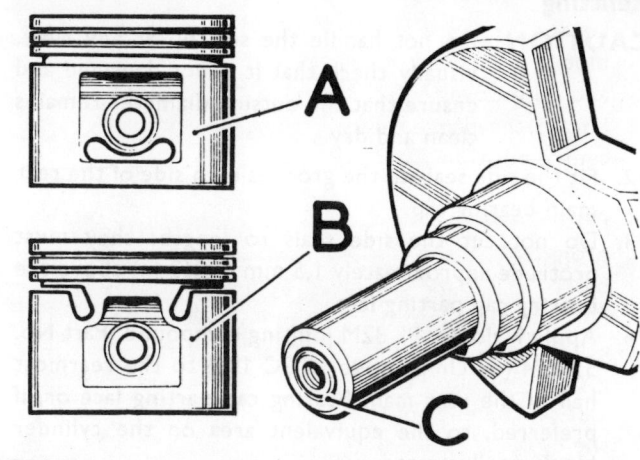

1RC393

7. Remove the baffle plate and sump oil strainer.
8. Remove the connecting rod caps and lower bearing shells and retain in sequence.
9. Remove the main bearing caps and lower bearing shells and retain in sequence.

NOTE: If the same bearing shells are to be refitted, retain them in pairs and mark them with the number of the respective journal.

10. Withdraw the crankshaft.

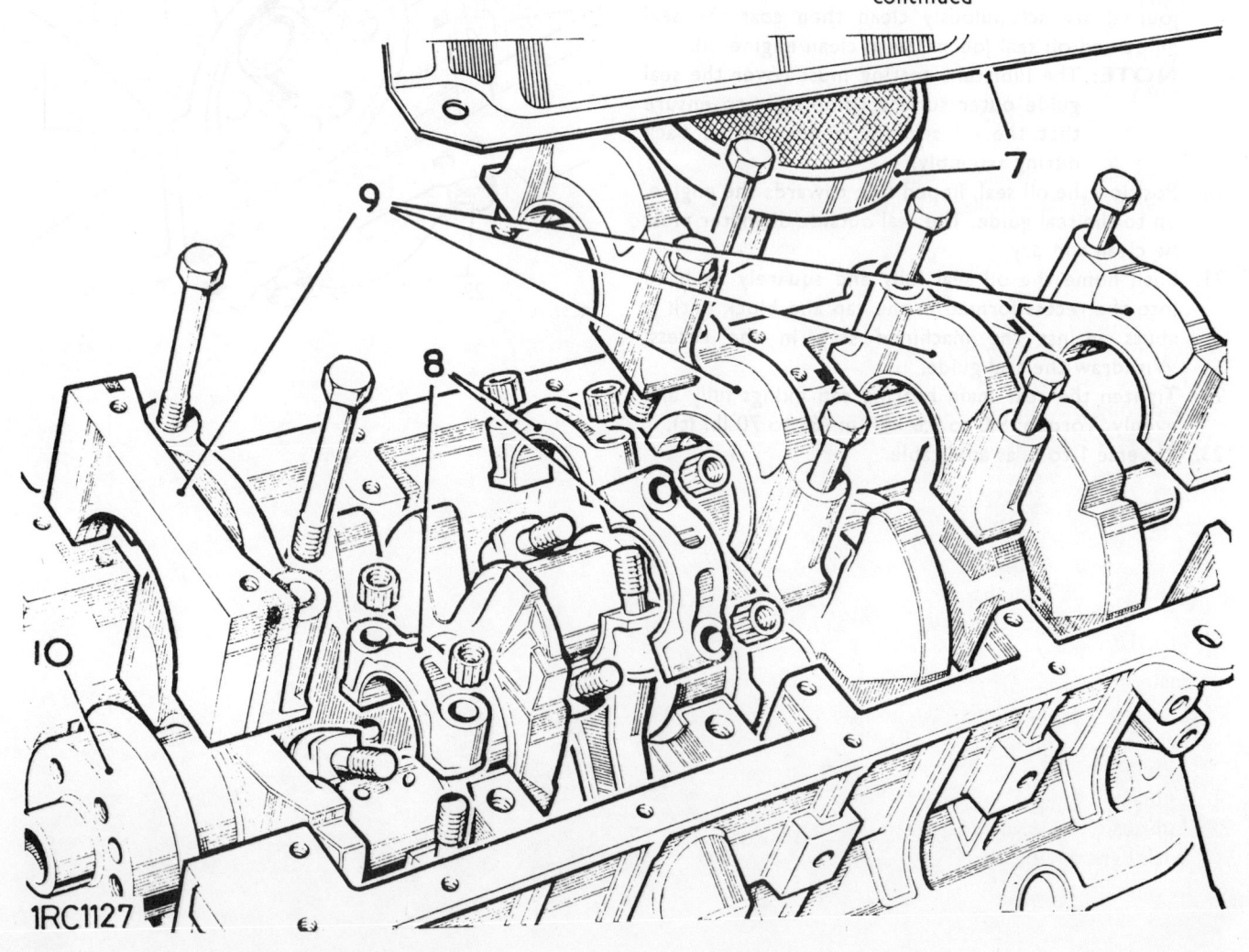

continued

1RC1127

12.21.33
Sheet 1

Rover 3500 and 3500S Manual AKM 3621

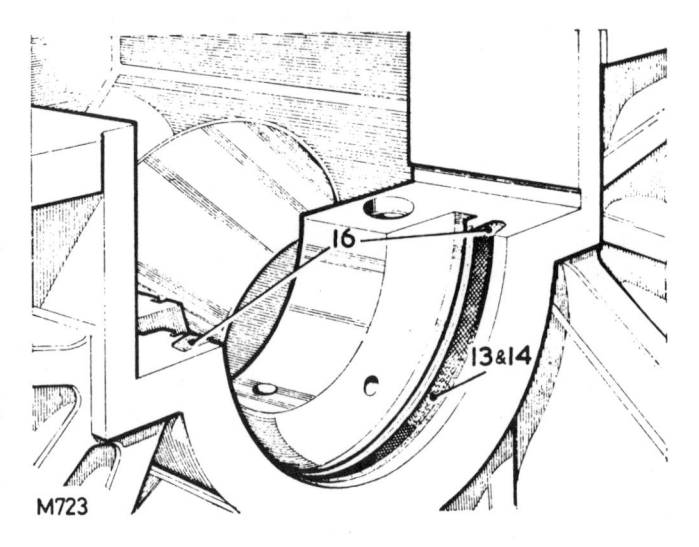

ENGINE

Refitting

11. Locate the upper main bearing shells into the cylinder block; these must be the shells with the oil drilling and oil grooves.
12. Locate the flanged upper main bearing shell in the centre position.
13. Remove the rear oil seal from the cylinder block.
14. Place a new oil seal in the groove with both ends projecting above the parting face of the cylinder block.
15. Force the seal into the groove by rubbing down with a hammer handle until the seal projects above the groove not more than 1,5 mm (0.031 in).
16. Cut off the ends of the seal flush with the surface of the cylinder block, using a sharp knife.
17. Place suitable blocks, approximately 12,5 mm (0.500 in) thick, on to each end of the cylinder block so that they cover the front and rear upper main bearing shells.
18. Lift the crankshaft into position with the ends supported on the blocks.
19. Lubricate the crankshaft journals and bearing shells with engine oil.
20. Lubricate the rear oil seal with heavy engine oil.
21. Holding the connecting rods in position, **remove one** of the blocks and lower the crankshaft on to the connecting rod bearings. Repeat for the opposite end.
22. Where necessary, use the guide bolt 605351 to draw the connecting rods up to the crankshaft journal.
23. Locate the bearing caps and lower shells on to the connecting rods, noting that the rib on the edge of the cap must be towards the front of the engine on the right-hand bank of cylinders, and towards the rear on the left-hand bank.
24. Secure the connecting rod caps. Torque: 4,9 to 6,2 kgf.m (30 to 35 lbf.ft).
25. Lubricate the lower main bearing shells with engine oil.
26. Fit numbers one to four main bearing caps and shells, leaving the fixing bolts finger tight at this stage.
27. Fit the rear oil seal to the rear main bearing cap following the same procedure as items 13 to 16.

    NOTE: Two designs of side sealing for the rear main bearing cap are in use. Early engines have a 'monolithic' type seal and matching cap whereas, later engines use a 'cruciform' design as illustrated. Bearing caps must not be interchanged, but if necessary, the early seal can be used with the latest cap.

28. Fit the side seals to the grooves each side of the rear main bearing cap.
29. Do not cut the side seals to length; they must protrude approximately 1,5 mm (0.062 in) above the bearing cap parting face.
30. Lubricate the rear oil seal with heavy engine oil.
31. Lubricate the lower bearing shell and the side oil seal; with light engine oil.

continued

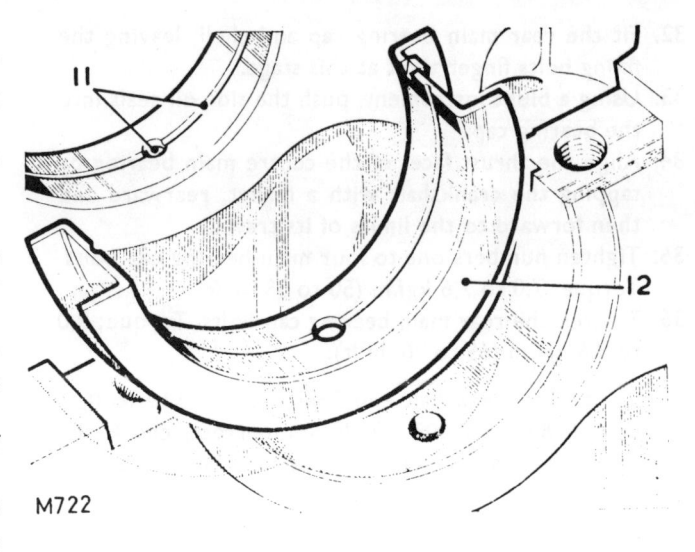

Rover 3500 and 3500S Manual AKM 3621

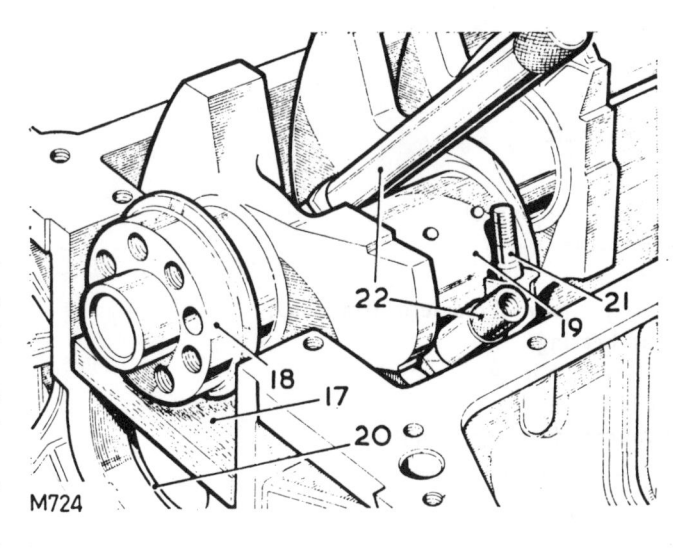

12.21.33
Sheet 2

ENGINE

32. Fit the rear main bearing cap and shell, leaving the fixing bolts finger tight at this stage.
33. Using a blunt instrument, push the side oil seals into the bearing cap.
34. Align the thrust faces of the centre main bearing by tapping the crankshaft with a mallet, rearward and then forward to the limits of its travel.
35. Tighten numbers one to four main bearing cap bolts. Torque: 7,0 to 7,6 kgf.m (50 to 55 lbf.ft).
36. Tighten the rear main bearing cap bolts. Torque: 9,0 to 9,6 kgf.m (65 to 70 lbf.ft).
37. Check the crankshaft end-float. Limits: 0,10 to 0,20 mm (0.004 to 0.008 in). If not correct, check the components and assembly procedure for faults.
38. Reverse 1 to 7.

CAUTION: Do not exceed 1,000 engine rev/min when first starting the engine, otherwise the crankshaft rear oil seal will be damaged.

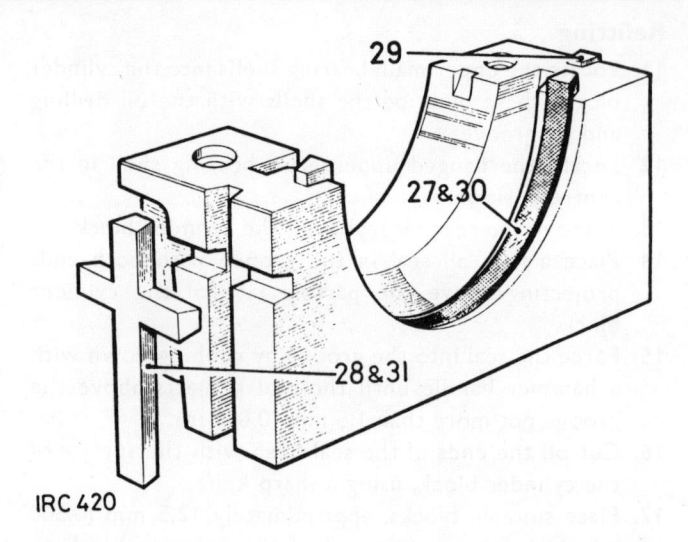

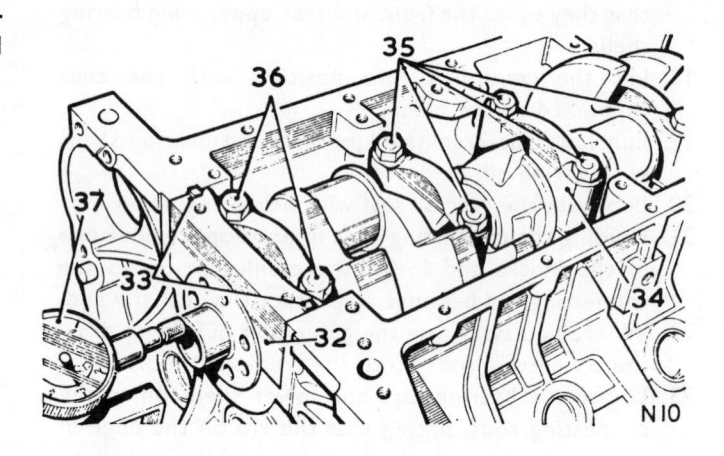

DATA

Crankshaft

Material	Iron, spheroidal graphite
No. of main journals	5
End thrust	Taken on Number 3
End-play	0,10 to 0,20 mm (0.004 to 0.008 in)
Crankpin journal diameter (standard)	50,800 to 50,813 mm (2.0000 to 2.0005 in)
Main bearing:	
Material and type	Vandervell lead indium
Clearance	0,023 to 0,065 mm (0.0009 to 0.0025 in)
Journal diameter (standard)	58,400 to 58,413 mm (2.2992 to 2.2997 in)
Bearing overall length	20,24 to 20,49 mm (0.797 to 0.807 in). Nos. 1, 2, 4 and 5
	26,82 to 26,87 mm (1.056 to 1.058 in) Number 3
Crankshaft vibration damper type	Torsional

Rover 3500 and 3500S Manual AKM 3621

ENGINE

CRANKSHAFT

—Overhaul 12.21.46

Service tools 605238—Plastigauge

NOTE: There are two designs of pistons in use on Rover 3500 engines, the two designs differ slightly in weight and this is compensated for by two standards of crankshaft balance.

Design 'A' pistons are fitted with a crankshaft which has a plain face at the starter dog end.

Design 'B' pistons are fitted with a crankshaft which has an identification groove 'C' in the face at the starter dog end.

The two designs are interchangeable providing that the eight pistons and applicable crankshaft are fitted initially as a set.

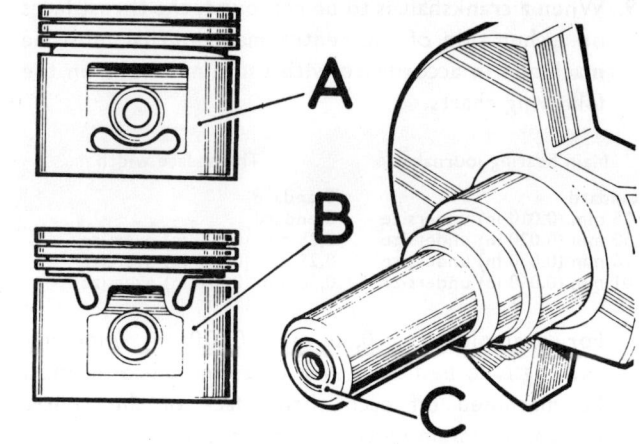

Procedure

1. Remove the crankshaft. 12.21.33.

Inspecting

2. Rest the crankshaft on vee-blocks at numbers one and five main bearing journals.
3. Using a dial test indicator, check the run-out at numbers two, three and four main bearing journals. The total indicator readings at each journal should not exceed 0,08 mm (0.003 in).
4. While checking the run-out at each journal, note the relation of maximum eccentricity on each journal to the others. The maximum on all journals should come at very near the same angular location.
5. If the crankshaft fails to meet the foregoing checks it is bent and is unsatisfactory for service.
6. Check each crankshaft journal for ovality. If ovality exceeds 0,040 mm (0.0015 in), a reground or new crankshaft should be fitted.
7. Bearings for the crankshaft main journals and the connecting rod journals are available in the following undersizes:
   0,25 mm (0.010 in)
   0,50 mm (0.020 in)
   0,76 mm (0.030 in)
   1,01 mm (0.040 in)
8. The centre main bearing shell, which controls crankshaft thrust, has the thrust faces increased in thickness when more than 0,25 mm (0.010 in) undersize, as shown on the following chart.

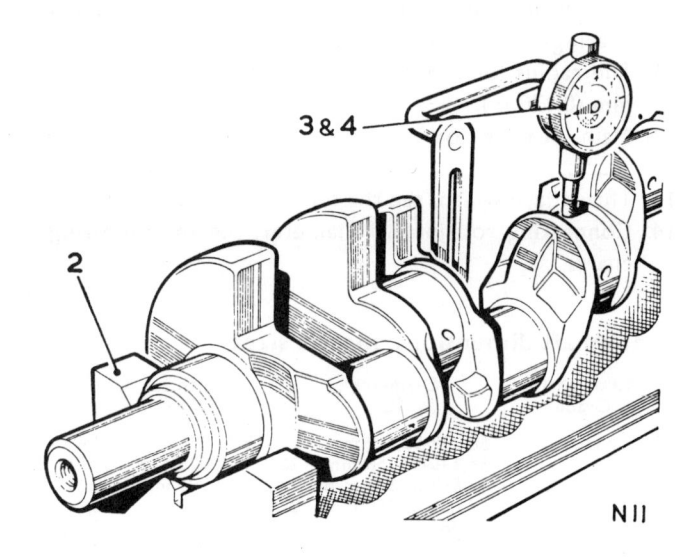

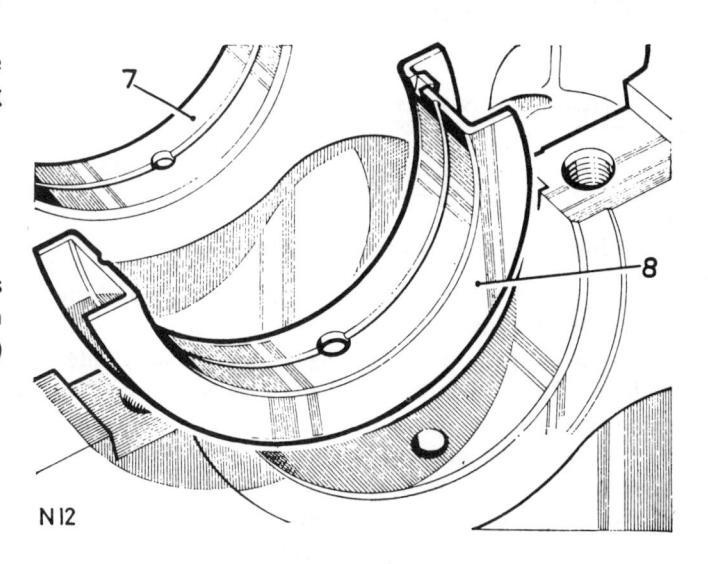

continued

Rover 3500 and 3500S Manual AKM 3621

ENGINE

9. When a crankshaft is to be reground, the thrust faces on either side of the centre main journal must be machined in accordance with the dimensions on the following charts.

Main bearing journal size	Thrust face width
Standard	Standard
0,25 mm (0.010 in) undersize	Standard
0,50 mm (0.020 in) undersize	0,25 mm (0.010 in) oversize
0,76 mm (0.030 in) undersize	0,25 mm (0.010 in) oversize
1,01 mm (0.040 in) undersize	0,50 mm (0.020 in) oversize

For example: If a 0,50 mm (0.020 in) undersize bearing is to be fitted, then 0,12 mm (0.005 in) must be machined off each thrust face of the centre journal, maintaining the correct radius.

Crankshaft dimensions 10 to 14

10. The radius for all journals except the rear main bearings is 1,90 to 2,28 mm (0.075 to 0.090 in).
11. The radius for the rear main bearing journal is 3,04 mm (0.120 in).
12. Main bearing journal diameter, see the following charts.
13. Thrust face width, see the following charts.
14. Connecting rod journal diameter, see the following charts.

Crankshaft dimensions, metric sizes

Crankshaft Grade	Diameter '12'	Width '13'	Diameter '14'
Standard	58,399-58,412 mm	26,975-27,026 mm	50,800-50,812 mm
0,254 mm U/S	58,146-58,158 mm	26,975-27,026 mm	50,546-50,559 mm
0,508 mm U/S	57,892-57,904 mm	27,229-27,280 mm	50,292-50,305 mm
0,762 mm U/S	57,638-57,650 mm	27,229-27,280 mm	50,038-50,051 mm
1,016 mm U/S	57,384-57,396 mm	27,483-27,534 mm	49,784-49,797 mm

Crankshaft dimensions, British sizes

Crankshaft Grade	Diameter '12'	Width '13'	Diameter '14'
Standard	2.2992-2.2997 in	1.062-1.064 in	2.0000-2.0005 in
0.010 in U/S	2.2892-2.2897 in	1.062-1.064 in	1.9900-1.9905 in
0.020 in U/S	2.2792-2.2797 in	1.072-1.074 in	1.9800-1.9805 in
0.030 in U/S	2.2692-2.2697 in	1.072-1.074 in	1.9700-1.9705 in
0.040 in U/S	2.2592-2.2597 in	1.082-1.084 in	1.9600-1.9605 in

Checking the main bearing clearance

15. Remove the oil seals from the cylinder block and the rear main bearing cap.
16. Locate the upper main bearing shells into the cylinder block. These must be the shells with the oil drilling and oil grooves.
17. Locate the flanged upper main bearing shell in the centre position.
18. Place the crankshaft in position on the bearings.
19. Place a piece of Plastigauge 605238 across the centre of the crankshaft main bearing journals.
20. Locate the bearing lower shell into the main bearing cap.
21. Fit numbers one to four main bearing caps and shells. Torque: 7,0 to 7,6 kgf m (50 to 55 lbf ft).

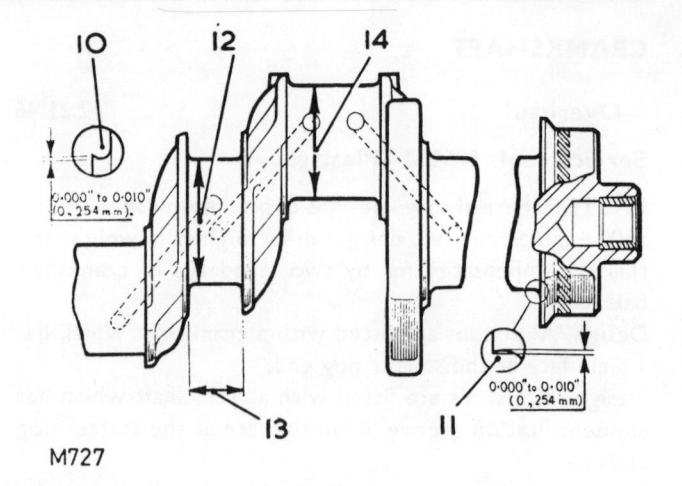

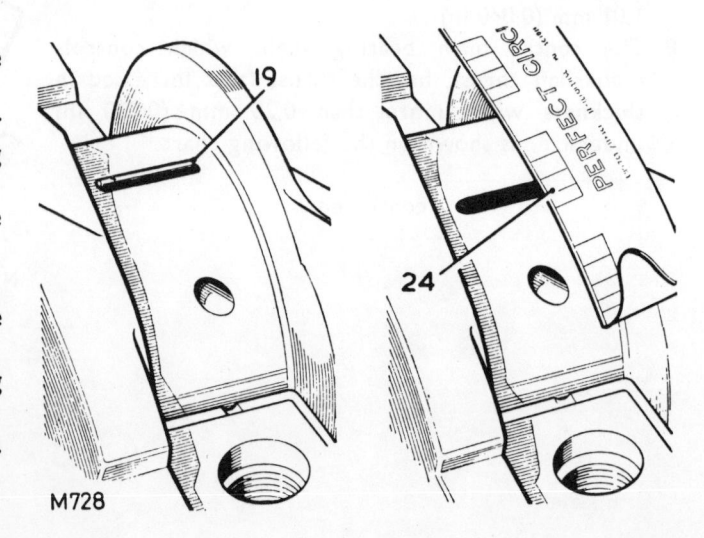

continued

ENGINE

22. Fit the rear main bearing cap and shell. Torque: 9,0 to 9,6 kgf m (65 to 70 lbf ft).

    NOTE: Do not rotate the crankshaft while the Plastigauge is fitted.

23. Remove the main bearing caps and shells.
24. Using the scale printed on the Plastigauge packet, measure the flattened Plastigauge at its widest point. The graduation that most closely corresponds to the width of the Plastigauge indicates the bearing clearance.
25. The correct bearing clearance with new or overhauled components is 0,023 to 0,065 mm (0.0009 to 0.0025 in).
26. If the correct clearance is not obtained initially, use selective bearing assembly.
27. Wipe off the Plastigauge with an oily rag. Do NOT scrape it off.

    NOTE: The bearing shells must be retained in sets and in the correct sequence.

28. If required, check the connecting rod big-end bearing clearance. 12.17.10.
29. Refit the crankshaft. 12.21.33.

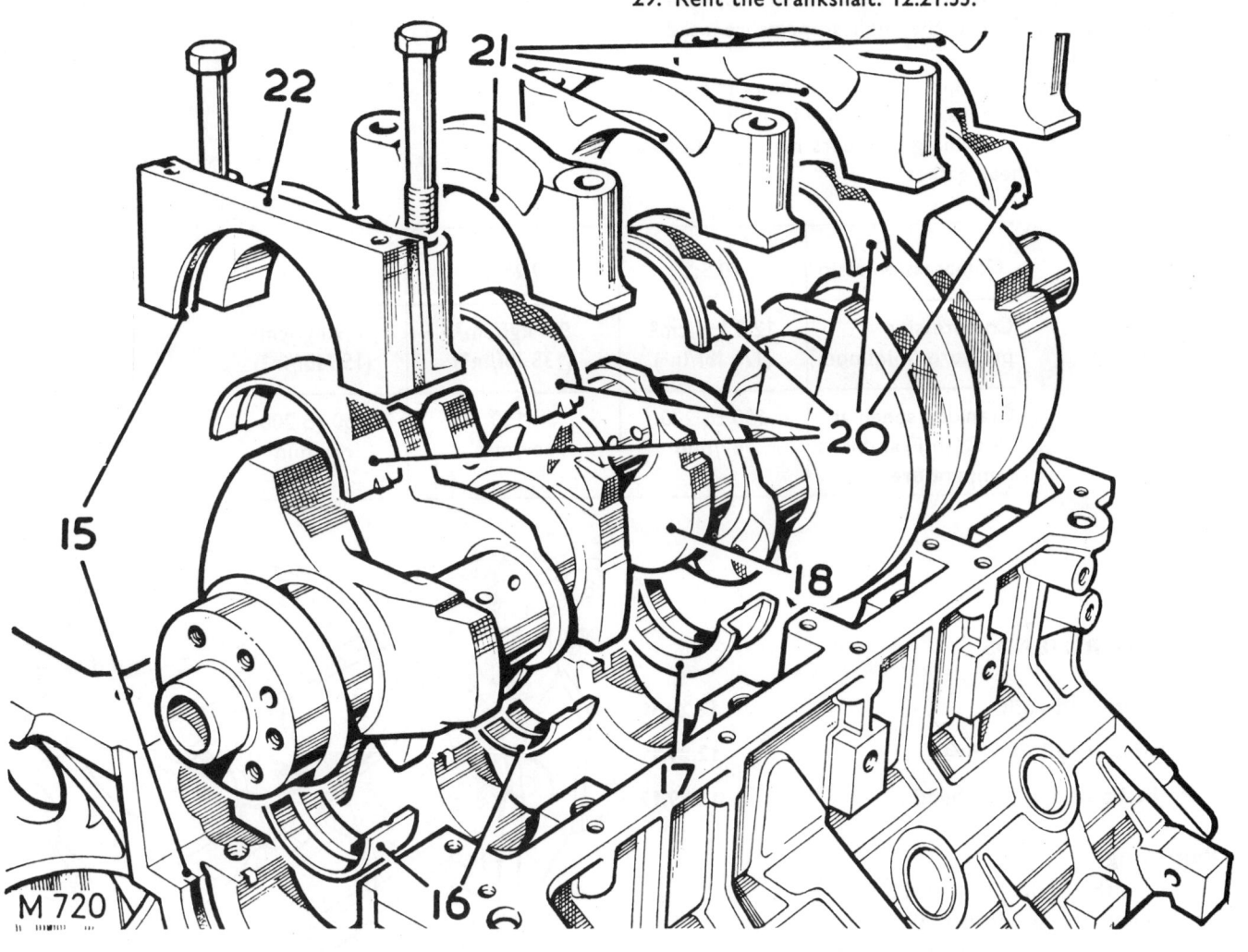

DATA

Crankshaft

Material	Iron, spheroidal graphite
No. of main journals	5
End-thrust	Taken on Number 3
End-play	0,10 to 0,20 mm (0.004 to 0.008 in)
Crankpin journal diameter (standard)	50,800 to 50,812 mm (2.0000 to 2.0005 in)
Main bearing:	
Material and type	Vandervell lead indium
Clearance	0,023 to 0,065 mm (0.0009 to 0.0025 in)
Journal diameter (standard)	58,399 to 58,412 mm (2.2992 to 2.2997 in)
Bearing overall length	20,24 to 20,49 mm (0.797 to 0.807 in). Nos. 1, 2, 4 and 5
	26,82 to 26,87 mm (1.056 to 1.058 in) Number 3
Crankshaft vibration damper type	Torsional

Rover 3500 and 3500S Manual AKM 3621

ENGINE

CYLINDER PRESSURES

— Check   12.25.01

Checking

1. Run the engine until it attains normal operating temperature.
2. Remove all the sparking plugs.
3. Secure the throttle in the fully open position.
4. Check each cylinder in turn as follows:
5. Insert a suitable pressure gauge into the sparking plug hole.
6. Crank the engine with the starter motor for several revolutions and note the highest pressure reading obtainable.
7. If the compression is appreciably less than the correct figure, the piston rings or valves may be faulty.
8. Low pressure in adjoining cylinders may be due to a faulty cylinder head gasket.

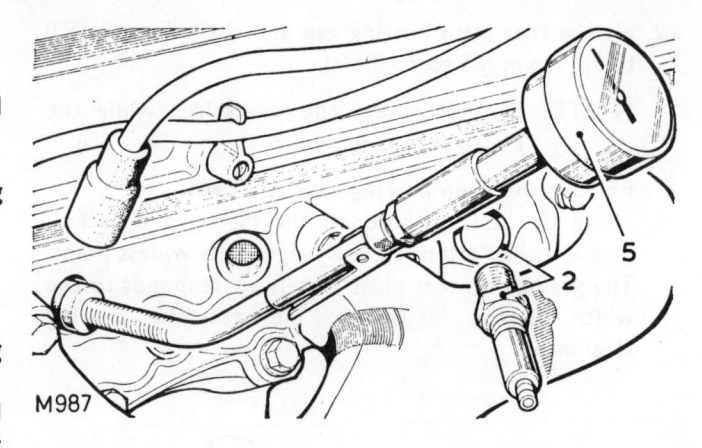

Compression ratio	10.5:1	9.25:1	8.5:1
Compression pressure (minimum)	12,5 kgf/cm² (175 lbf/in²)	9,5 kgf/cm² (135 lbf/in²)	10,9 kgf/cm² (155 lbf/in²)
Cranking speed at 15°C (60°F) ambient temperature	150 to 200 rev/min	150 rev/min	150 to 200 rev/min

CYLINDER HEADS

— Remove and refit   12.29.10

    Left-hand cylinder head   12.29.11

    Right-hand cylinder head   12.29.12

Removing

1. Drain the cooling system. 26.10.01.
2. Remove the air cleaner. 19.10.01. (AA1-3)
3. Remove the induction manifold. 30.15.02. (AA1-9)
4. Remove the valve gear. 12.29.34.
5. Disconnect the front exhaust pipes from the manifolds.
6. LH cylinder head—Disconnect the fuel line filter bracket from the engine lifting hook.
7. RH cylinder head—Remove the alternator. 86.10.02. (AA1-6)
8. Slacken the cylinder head bolts evenly.
9. If both cylinder heads are being removed, mark them relative to LH and RH sides of the engine.

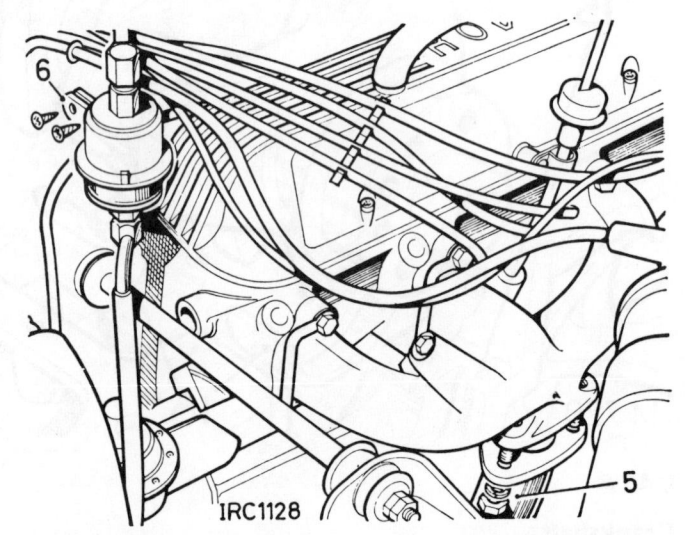

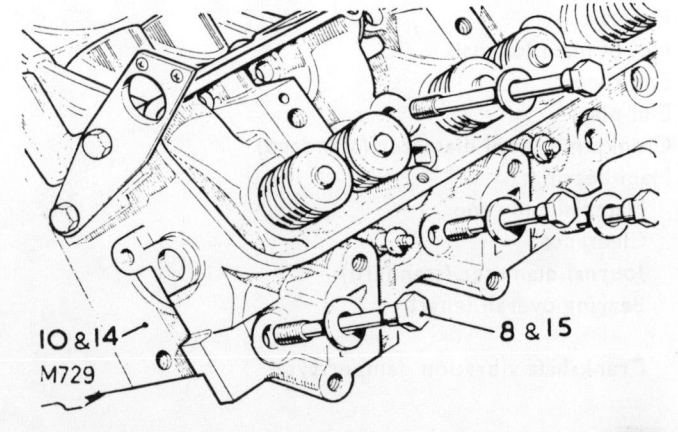

continued

Rover 3500 and 3500S Manual AKM 3621

ENGINE

10. Remove the cylinder heads and discard the gaskets.
11. If required, remove the exhaust manifolds. 30.15.10, 30.15.11.

    NOTE: On stripping an engine, the cylinder head bolts should be immediately wire brush washed in 3M Solvent No. 2, manufactured by The 3M Company Ltd., 3M House, Wigmore Street, London W1A 1ET and available through normal trade channels in the British Isles and overseas. It is not available from Rover Parts Dept.

    If the bolts cannot be cleaned immediately, it is essential that they be stored in a bath of trichlorethylene, petrol or paraffin etc., which may also be used as a cleaner if 3M Solvent No. 2 is not available, otherwise the sealant used on previous assembly will tend to air harden, making subsequent removal very difficult.

    After four re-assembly operations renew all bolts. When re-assembling at any time, renew all bolts if more than two bolts exhibit evidence of elongation. If one or two bolts are elongated, they must be replaced.

Refitting

12. If removed, fit the exhaust manifolds. 30.15.10, 30.15.11.
13. Fit new cylinder head gaskets with the word 'TOP' uppermost. Do NOT use any sealant.
14. Locate the cylinder heads on the block dowel pins.
15. Clean the threads of the cylinder head bolts then coat them with Thread Lubricant-Sealant 3M EC776, Rover Part No. 605764.
16. Locate the cylinder head bolts in position:
    Long bolts—1, 2 and 4.
    Medium bolts—3, 5, 6, 7, 8, 9 and 10.
    Short bolts—11, 12, 13 and 14.
17. Tighten the cylinder head bolts a little at a time in the sequence shown. Final torque: 9,0 to 9,6 kgf.m (65 to 70 lbf.ft) for bolts 1 to 10; 5,5 to 6,2 kgf.m (40 to 45 lbf. ft) for bolts 11, 12, 13 and 14.
18. Reverse 1 to 7.

Cylinder head thread insert salvage instructions

19. These three holes may be drilled 0.3906 in dia x 0.937 + 0.040 in deep. Tapped with Helicoil Tap No. 6 CPB or 6CS x 0.875 in (min.) deep ($\frac{3}{8}$ UNC 1$\frac{1}{2}$D insert).
20. These eight holes may be drilled 0.3906 in dia x 0.812 + 0.040 in deep. Tapped with Helicoil Tap No. 6 CBB 0.749 in (min.) deep ($\frac{3}{8}$ UNC 1$\frac{1}{2}$D insert).
21. These four holes may be drilled 0.3906 in dia x 0.937 + 0.040 in deep. Tapped with Helicoil Tap No. 6 CPB or 6CS x 0.875 in (min) deep ($\frac{3}{8}$ UNC 1$\frac{1}{2}$D insert).

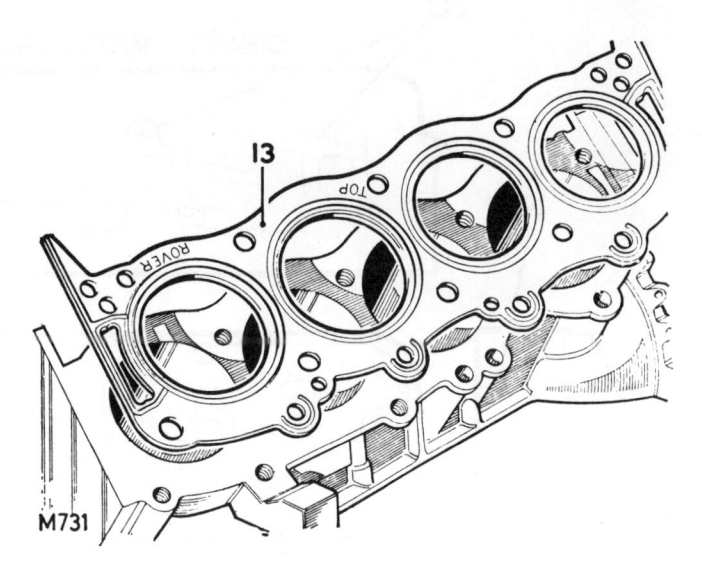

L.H. HEAD ILLUSTRATED

22. These four holes may be drilled 0.261 in dia x 0.675 + 0.040 in deep. Tapped with Helicoil Tap No. 4CPB or 4CS x 0.625 in (min) deep ($\frac{1}{4}$ UNC 1$\frac{1}{2}$D insert).
23. These six holes may be drilled 0.3906 in dia x 0.937 + 0.040 in deep. Tapped with Helicoil Tap No. 6 CPB or 6CS x 0.875 in (min) deep ($\frac{3}{8}$ UNC 1$\frac{1}{2}$D insert).

    CAUTION: Any attempt to salvage the sparking plug threads in the cylinder head may result in breaking into the water jacket, rendering the head scrap.

continued

Rover 3500 and 3500S Manual AKM 3621

ENGINE

EXHAUST MANIFOLD FACE

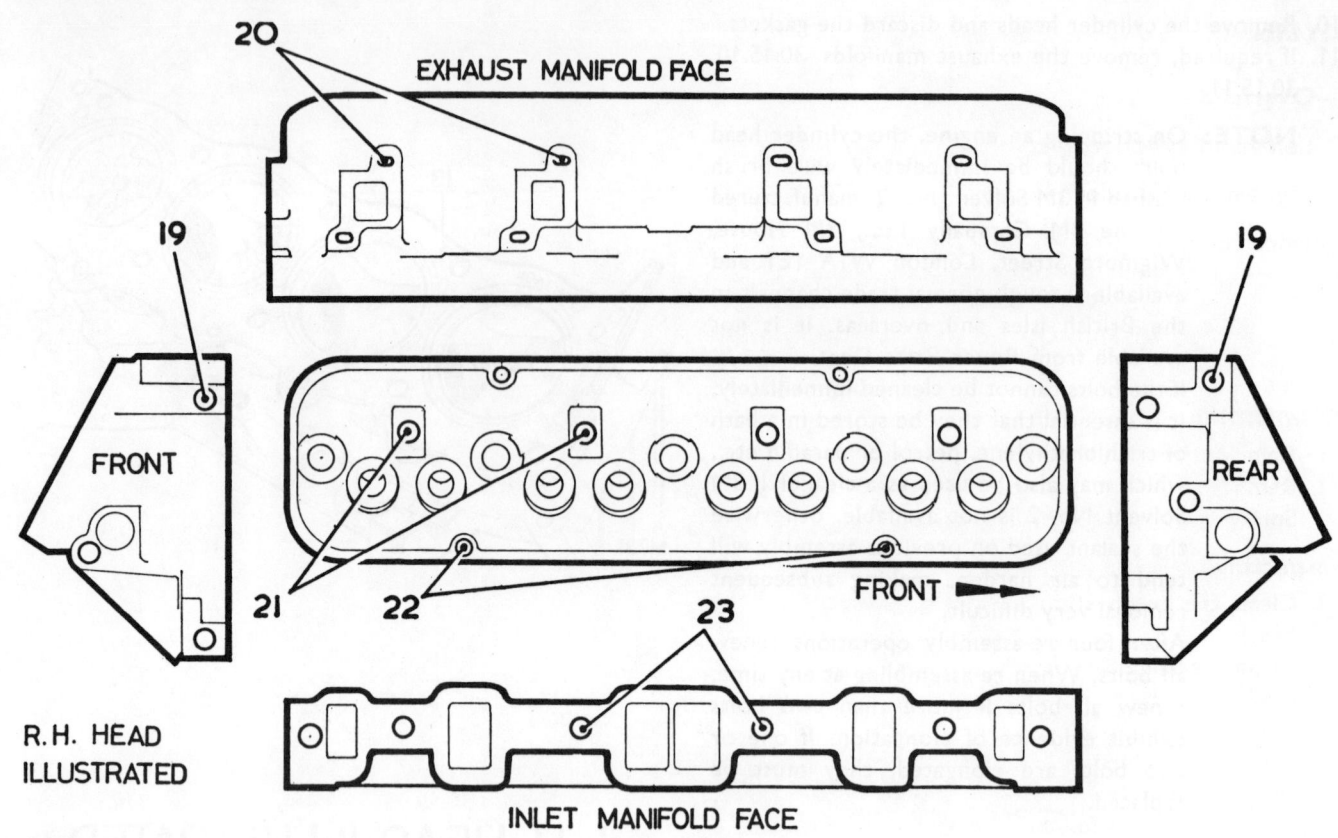

R.H. HEAD ILLUSTRATED

M732

DATA

Cylinder heads
Material	Aluminium alloy
Type	Two heads with separate alloy inlet manifold
Inlet and exhaust valve seat material	Piston ring iron
Inlet and exhaust valve seat angle	$46 + \frac{1}{4}$ degrees

Valves

Valves, inlet

Overall length	116,58 to 117,34 mm (4.590 to 4.620 in)
Actual overall head diameter	37,97 to 38,22 mm (1.495 to 1.505 in)
Angle of face	45 degrees
Stem diameter	8,640 to 8,666 mm (0.3402 to 0.3412 in) at the head and increasing to 8,653 to 8,679 mm (0.3407 to 0.3417 in)
Stem clearance in guide	Top 0,02 to 0,07 mm (0.001 to 0.003 in)
	Bottom 0,013 to 0,063 mm (0.0005 to 0.0025 in)

Valves, exhaust

Overall length	116,58 to 117,34 mm (4.590 to 4.620 in)
Actual overall head diameter	33,215 to 33,466 mm (1.3075 to 1.3175 in)
Angle of face	45 degrees
Stem diameter	8,628 to 8,654 mm (0.3397 to 0.3407 in) at the head and increasing to 8,640 to 8,666 mm (0.3402 to 0.3412 in)
Stem clearance in guide	Top 0,038 to 0,088 mm (0.0015 to 0.0035 in)
	Bottom 0,05 to 0,10 mm (0.002 to 0.004 in)
Valve lift	9,9 mm (0.39 in) both valves
Valve spring length:	
Outer	40,6 mm (1.6 in) at pressure of 17,69 to 20,41 kg (39 to 45 lb)
Inner	41,2 mm (1.63 in) at pressure of 9,75 to 12,02 kg (21.5 to 26.5 lb)

Rover 3500 and 3500S Manual AKM 3621

ENGINE

CYLINDER HEADS

—Overhaul	12.29.18
Left-hand cylinder head	12.29.19
Right-hand cylinder head	12.29.30

Service tools: 276102—Valve spring compressor
274401—Valve guide remover
600959—Valve guide drift
605774—Distance piece for valve guide drift

Dismantling

1. Remove the cylinder heads. 12.29.10.
2. Remove the valves and retain in sequence for refitting. Spring compressor 276102.

Inspecting

3. Clean the combustion chambers with a soft wire brush.
4. Clean the valves.
5. Clean the valve guide bores.
6. Regrind or fit new valves as necessary.
7. If a valve must be ground to a knife-edge to obtain a true seat, fit a new valve.
8. The correct angle for the valve face is 45 degrees.
9. The correct angle for the seat is $46 + \frac{1}{4}$ degrees, and the seat witness should be towards the outer edge.
10. Check the valve guides and fit replacements as necessary. 11 to 15.
11. Using the valve guide remover 274401, drive out the old guides from the combustion chamber side.
12. Locate the distance piece for the valve guide drift 605774 on the valve spring seat in the top of the cylinder head.
13. Lubricate the new valve guide and insert it into the distance piece.
14. Using the valve guide drift 600959, drive the valve guide into the cylinder head until the drift bottoms on the distance piece.
15. The fitted guide should stand 19 mm (0.750 in) above the step surrounding the valve guide boss in the cylinder head.

    NOTE: Service valve guides are 0,02 mm (0.001 in) larger on the outside diameter than the original equipment to ensure interference fit.

16. Check the valve seats and fit replacements as necessary. 17 to 19.
17. Remove the old seat inserts by grinding them away until they are thin enough to be cracked and prised out.
18. Heat the cylinder head evenly to approximately 65 degrees C (150 degrees F).

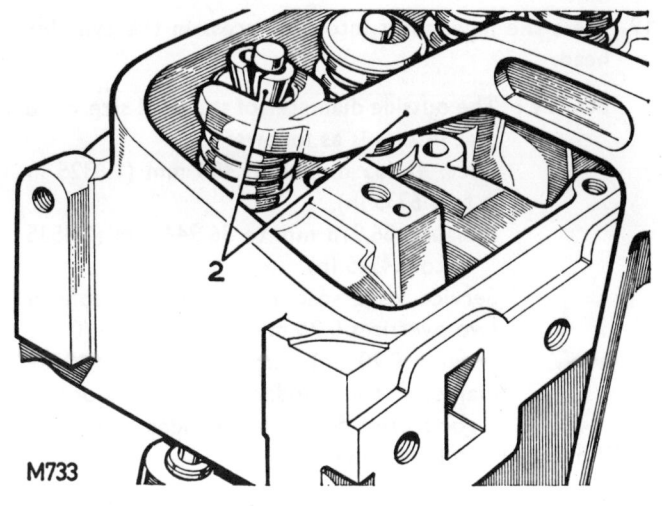

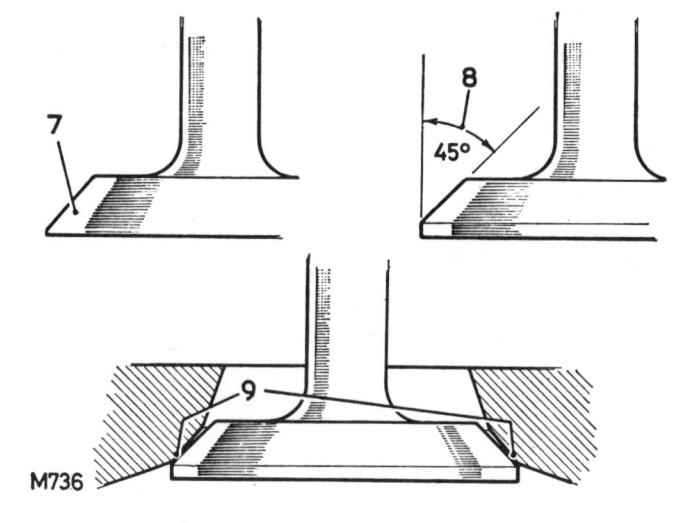

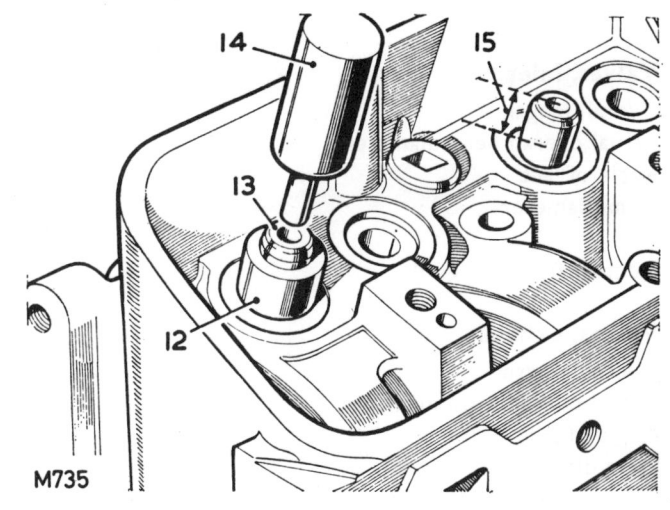

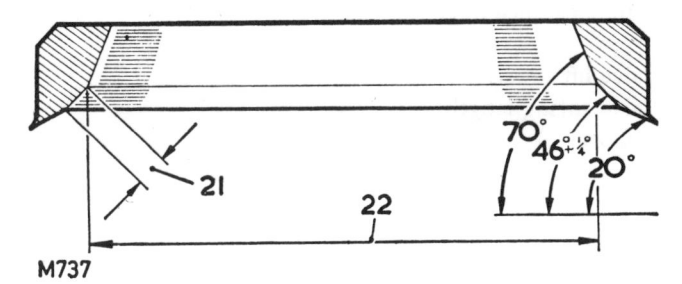

continued

Rover 3500 and 3500S Manual AKM 3621

12.29.18
12.29.30
Sheet 1

ENGINE

19. Press the new insert into the recess in the cylinder head.

    NOTE: The outside diameter of standard size valve seat inserts is as follows:
    Inlet: 41,465 mm to 41,490 mm (1.6325 in to 1.6335 in).
    Exhaust: 36,918 mm to 36,944 mm (1.4535 in to 1.4545 in).
    Service valve seat inserts are available in two oversizes 0,25 mm and 0,50 mm (0.010 in and 0.20 in) larger on the outside diameter than standard in order to obtain a good press fit in the cylinder head.

20. If necessary, cut the valve seats to 46 + ¼ degrees.

21. The nominal seat width is 1,5 mm (0.031 in). If the seat exceeds 2,0 mm (0.078 in) it should be reduced to the specified width by the use of 20 and 70 degree stones.

22. The inlet valve seat is 35,25 mm (1.388 in) diameter and the exhaust seat is 30,48 mm (1.200 in) diameter.

23. Check the height of the valve stems above the outer valve spring seat surface of the cylinder head. This MUST NOT exceed 47,63 mm (1.875 in). If necessary grind the end of the valve stem or fit new parts.

Reassembling

24. Lubricate the valve stems and guides with engine oil and fit each valve as follows:
25. Insert the valve into its guide.
26. Place the valve springs in position.
27. Locate the cap on the springs.
28. Using the valve spring compressor 276102, fit the valve collets.
29. Refit the cylinder heads. 12.29.10.

DATA

Cylinder heads

Material	Aluminium alloy
Type	Two heads with separate alloy inlet manifold
Inlet and exhaust valve seat material	Piston ring iron
Inlet and exhaust valve seat angle	46 + ¼ degrees

Valves

	Inlet	Exhaust
Opens	30 degrees BTDC	68 degrees BBDC
Closes	75 degrees ABDC	37 degrees ATDC
Duration	285 degrees	285 degrees
Valve peak	112.5 degrees ATDC	105.5 degrees BTDC

Valves, inlet

Overall length	116,58 to 117,34 mm (4.590 to 4.620 in)
Actual overall head diameter	37,97 to 38,22 mm (1.495 to 1.505 in)
Angle of face	45 degrees
Stem diameter	8,640 to 8,666 mm (0.3402 to 0.3412 in) at the head and increasing to 8,653 to 8,679 mm (0.3407 to 0.3417 in)
Stem clearance in guide	Top 0,02 to 0,07 mm (0.001 to 0.003 in)
	Bottom 0,013 to 0,0635 mm (0.0005 to 0.0025 in)

Valves, exhaust

Overall length	116,58 to 117,34 mm (4.590 to 4.620 in)
Actual overall head diameter	33,215 to 33,466 mm (1.3075 to 1.3175 in)
Angle of face	45 degrees
Stem diameter	8,628 to 8,654 mm (0.3397 to 0.3407 in) at the head and increasing to 8,640 to 8,666 mm (0.3402 to 0.3412 in)
Stem clearance in guide	Top 0,038 to 0,088 mm (0.0015 to 0.0035 in)
	Bottom 0,05 to 0,10 mm (0.002 to 0.004 in)
Valve lift	9,9 mm (0.39 in) both valves
Valve spring length:	
Outer	40,6 mm (1.6 in) at pressure of 17,69 to 20,41 kg (39 to 45 lb)
Inner	41,4 mm (1.63 in) at pressure of 9,75 to 12,02 kg (21.5 to 26.5 lb)

12.29.18
12.29.30
Sheet 2

Rover 3500 and 3500S Manual AKM 3621

ENGINE

VALVE GEAR

—Remove and refit 1 to 7, 15 to 21 and 29 to 37
 12.29.34

ROCKER SHAFTS

—Remove and refit 2, 4, 5 and 30 to 37 12.29.54

—Overhaul 2, 4, 5, 8 to 13, 22 to 28 and 30 to 37
 12.29.55

Removing

1. Drain the cooling system. 26.10.01.
2. Remove the air cleaner. 19.10.01. (AA1-3)
3. Remove the induction manifold. 30.15.02. (AA1-9)
4. Remove the rocker covers.
5. Remove the rocker shaft assemblies.
6. Withdraw the pushrods and retain in the sequence removed.
7. Withdraw the tappets and retain with respective pushrods.

   NOTE: If a tappet cannot be withdrawn, remove the camshaft and withdraw the tappet from the bottom.

Dismantling rocker shafts

8. Remove the split pin from one end of the rocker shaft.
   Withdraw the following components and retain them in the correct sequence for reassembly:
9. A plain washer.
10. A wave washer.
11. Rocker arms.
12. Brackets.
13. Springs.

Inspection of hydraulic tappets and pushrods

14. Hydraulic tappet; inspect inner and outer surfaces of body for blow holes and scoring. Replace hydraulic tappet if body is roughly scored or grooved, or has a blow hole extending through the wall in a position to permit oil leakage from lower chamber.
15. The prominent wear pattern just above lower end of body should not be considered a defect unless it is definitely grooved or scored; it is caused by side thrust of cam against body while the tappet is moving vertically in its guide.

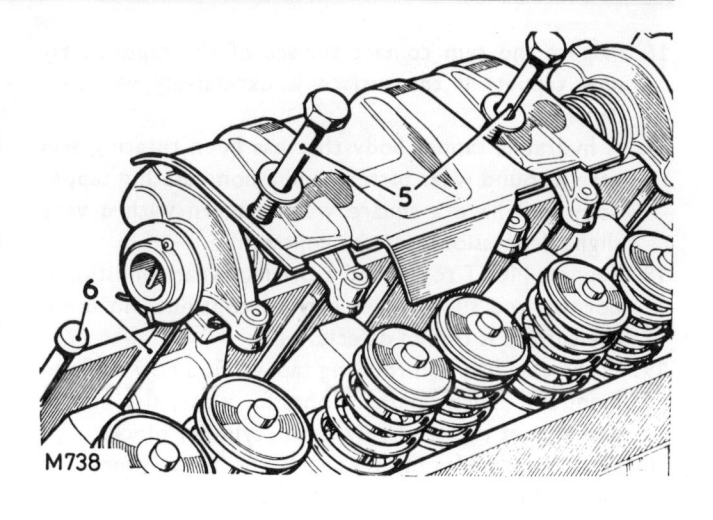

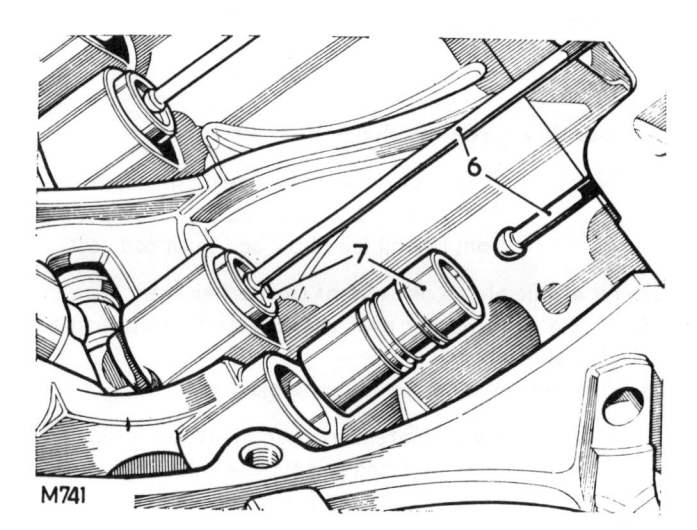

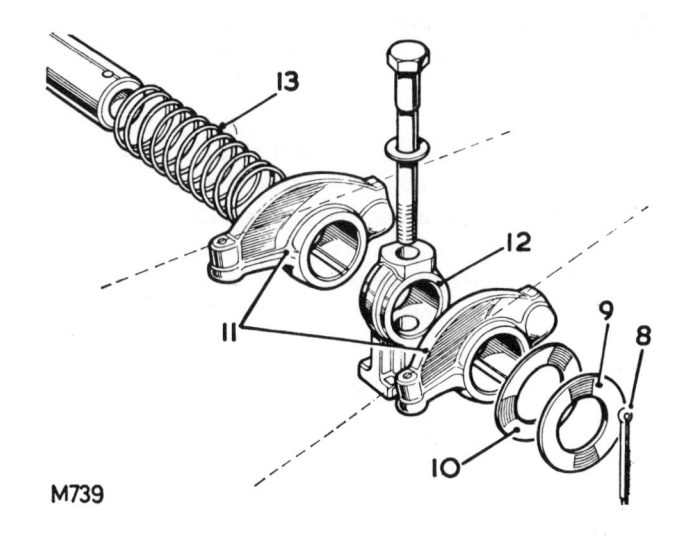

continued

Rover 3500 and 3500S Manual AKM 3621

12.29.34
Sheet 1

ENGINE

16. Inspect the cam contact surface of the tappets. Fit new tappets if the surface is excessively worn or damaged.
17. A hydraulic tappet body that has been rotating will have a round wear pattern and a non-rotating tappet body will have a square wear pattern with a very slight depression near the centre.
18. Tappets MUST rotate and a circular wear condition is normal, and such bodies may be continued in use if the surface is free of defects.
19. In the case of a non-rotating tappet, fit a new replacement and check camshaft lobes for wear; also ensure new tappet rotates freely in the cylinder block.
20. Fit a new hydraulic tappet if the area where the pushrod contacts is rough or otherwise damaged.
21. Pushrod. Replace with new, any pushrod having a rough or damaged ball end or seat.

Refitting

Reassembling rocker shafts 22 to 28

NOTE: If new rocker arms are being fitted, ensure that the protective coating is removed from the oil feed hole and push rod seat.

22. Fit a split pin to one end of the rocker shaft.
23. Slide a plain washer over the long end of the shaft to abut the split pin.
24. Fit a wave washer to abut the plain washer.

NOTE: Two different rocker arms are used and must be fitted so that the valve ends of the arms slope away from the brackets.

25. Assemble the rocker arms, brackets and springs to the rocker shaft.
26. Compress the springs, brackets and rockers, and fit a wave washer, plain washer and split pin to the end of the rocker shaft.
27. Locate the oil baffle plates in place over the rockers furthest from the notched end of the rocker shaft.
28. Fit the bolts through the brackets and shaft so that the notch on the one end of the shaft is uppermost and towards the front of the engine on the right-hand side, and towards the rear on the left-hand side.
29. Fit the tappets and pushrods in the original sequence.

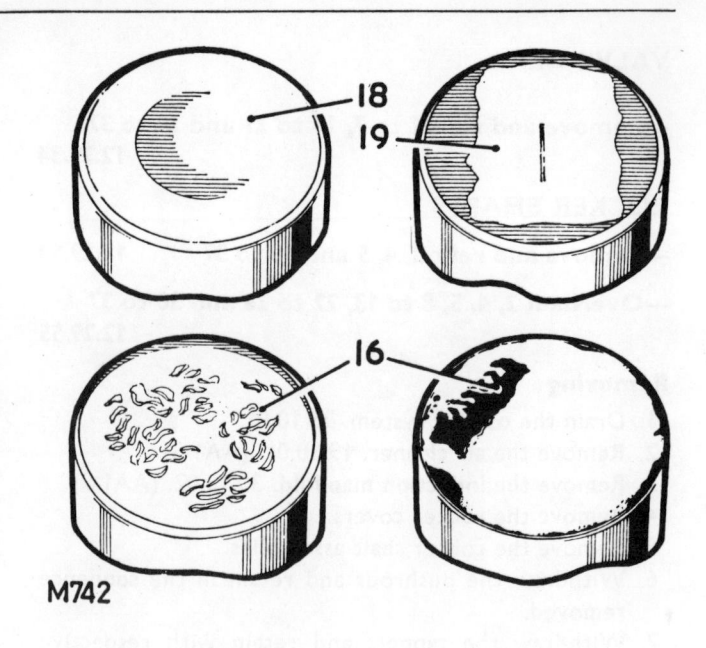

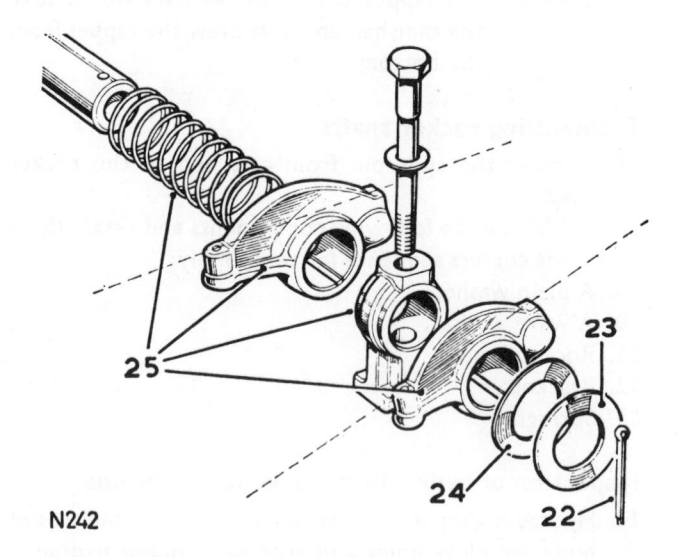

continued

12.29.34
Sheet 2

Rover 3500 and 3500S Manual AKM 3621

ENGINE

NOTE: The rocker shafts are handed and must be fitted correctly to align the oilways.

30. Each rocker shaft is notched at one end and on one side only. The notch must be uppermost and towards the front of the engine on the right-hand side, and towards the rear on the left-hand side.
31. Fit the rocker shaft assemblies. Ensure that the pushrods engage the rocker cups and that the baffle plates are fitted to the front on the left-hand side, and to the rear on the right-hand side. Tighten the bolts evenly. Torque: 3,5 to 4,0 kgf.m (25 to 30 lbf. ft).

If it is necessary to fit a new rocker cover gasket, proceed as follows, 32 to 36.

32. Clean and dry the gasket mounting surface, using Bostik cleaner 6001.
33. Apply Bostik 1775 impact adhesive, Rover Part No. 601736 to the seal face and the gasket, using a brush to ensure an even film.
34. Allow the adhesive to become touch-dry, approximately fifteen minutes.

    NOTE: The gasket fits one way round only and must be fitted accurately first time, any subsequent movement would destroy the bond.

35. Place one end of the gasket into the cover recess with the edge firmly against the recess wall; at the same time hold the remainder of the gasket clear; then work around the cover, pressing the gasket into place ensuring that the outer edge firmly abuts the recess wall.
36. Allow the cover to stand for thirty minutes before fitting it to the engine.
37. Reverse 1 to 4.

NOTE: Tappet noise

It should be noted that tappet noise can be expected on initial starting up after an overhaul due to oil drainage from the tappet assemblies or indeed if the vehicle has been standing over a very long period. If excessive noise should be apparent after an overhaul, the engine should be run at approximately 2,500 rev/min for a few minutes, when the noise should be eliminated.

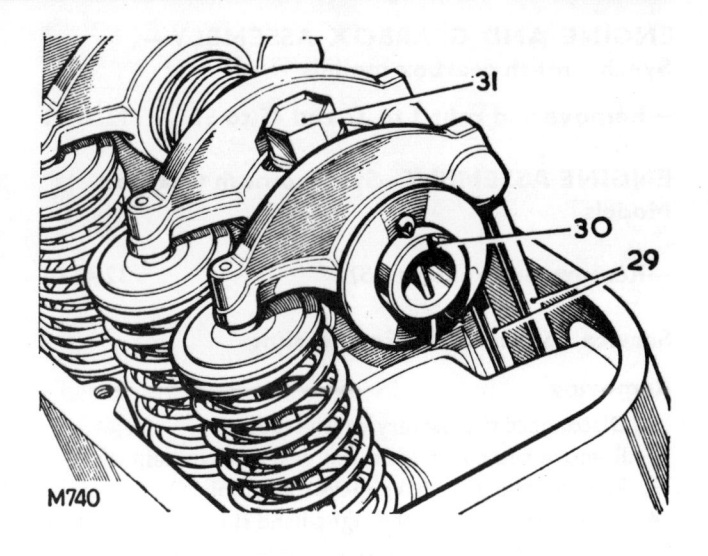

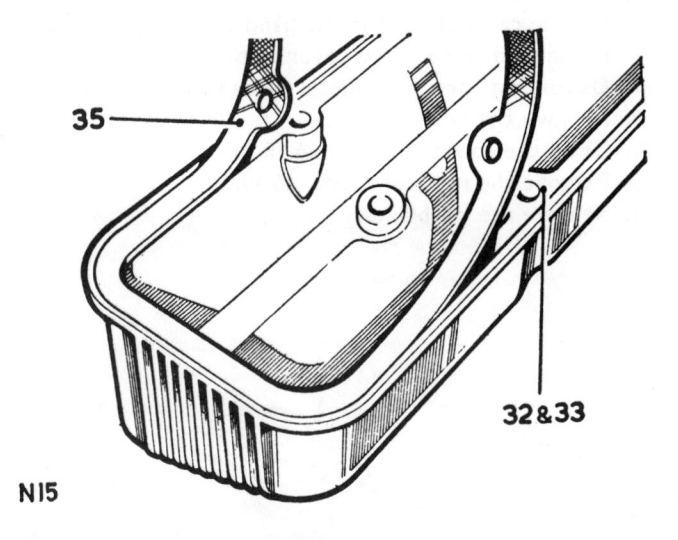

ENGINE

ENGINE AND GEARBOX ASSEMBLY—
Synchromesh gearbox models

—Remove and refit 1 to 37 and 45 to 57   12.37.01

ENGINE ASSEMBLY—Synchromesh Gearbox Models

—Remove and refit 1 to 57   12.41.01

Service tools: 600963 Engine sling

Removing

1. Disconnect the battery earth lead.
2. Remove the cover from the gearbox tunnel.
3. Remove the gearchange lever assembly.
4. Lift the carpet from the right-hand side of the gearbox tunnel and remove the large rubber grommet.
5. Disconnect the leads from the reverse lamp switch.
6. Lift the carpet from the left-hand side of the gearbox tunnel and remove the cover plate.
7. Disconnect the speedometer cable from the gearbox.
8. Remove the bonnet.
9. Drain the cooling system. 26.10.01.
10. Remove the radiator block. 26.40.04.
11. Remove the fan blades.

continued

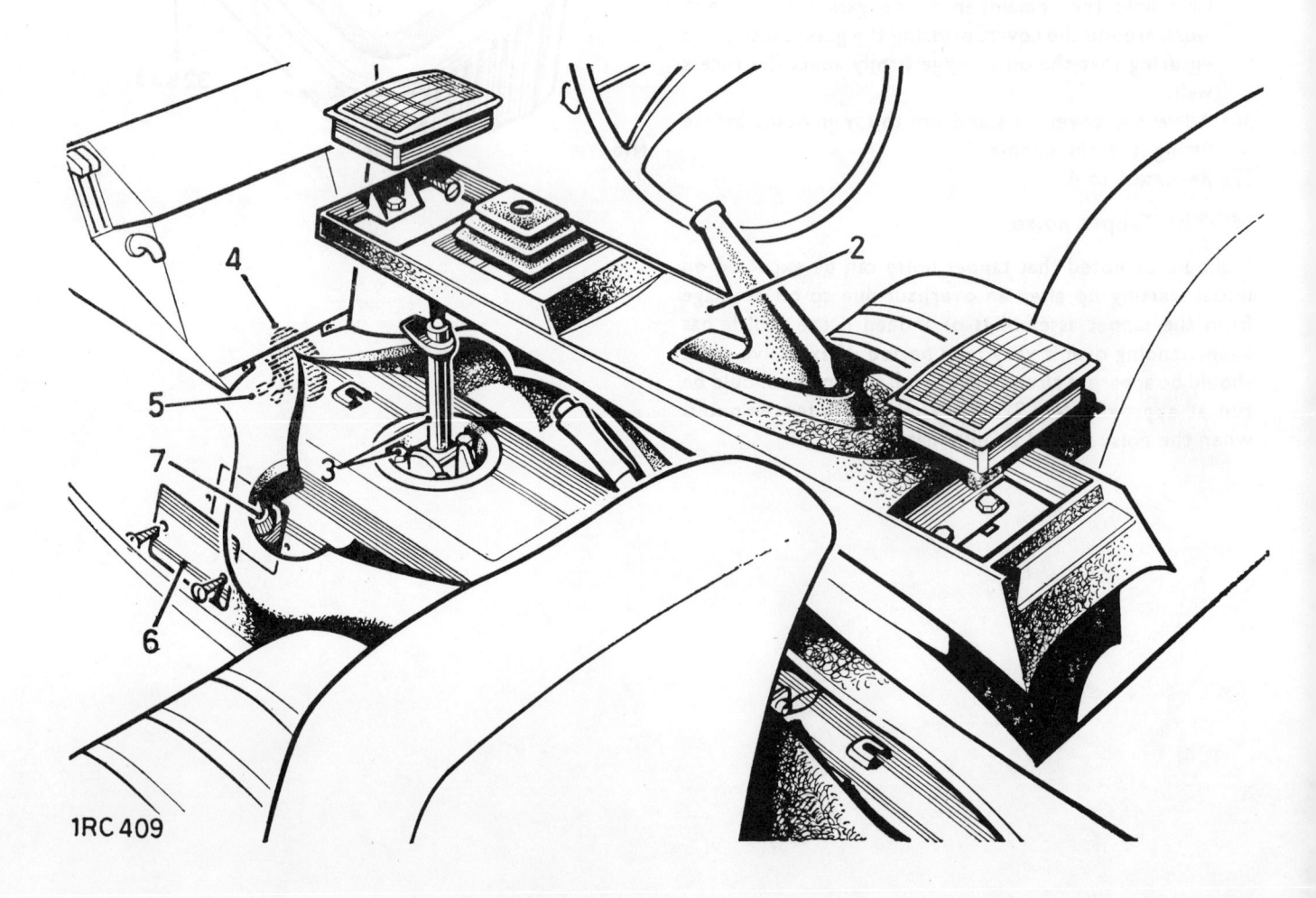

1RC 409

Rover 3500 and 3500S Manual AKM 3621

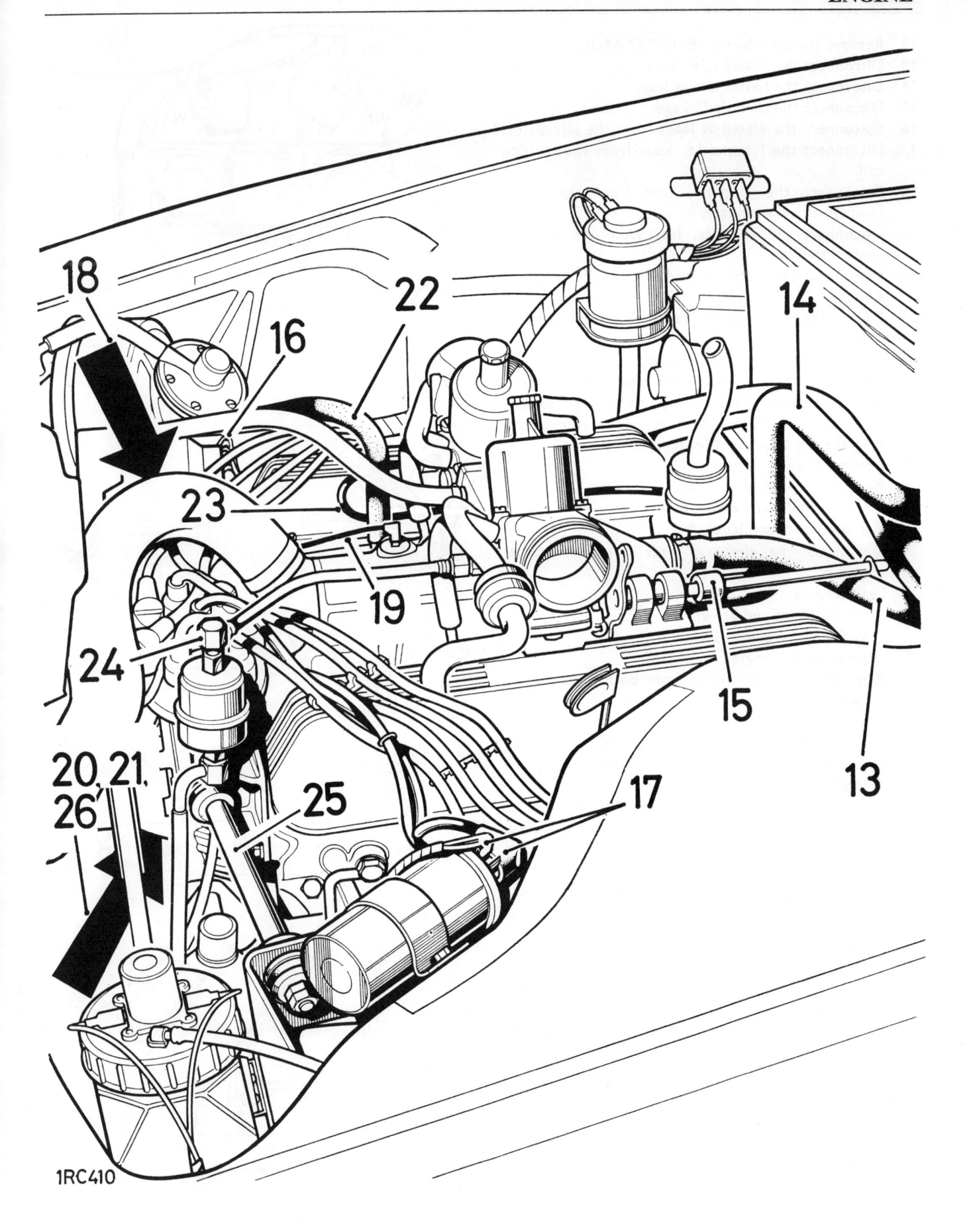

ENGINE

1RC410

Rover 3500 and 3500S Manual AKM 3621

12.37.01
12.41.01
Sheet 2

ENGINE

12. Remove the air cleaner. 19.10.01 (AA1-3).
13. Disconnect the heater inlet hose.
14. Disconnect the heater outlet hose.
15. Disconnect the throttle linkage.
16. Disconnect the electrical leads from the alternator.
17. Disconnect the h.t. and l.t. leads from the ignition coil.
18. Disconnect the electrical lead from the water temperature transmitter.
19. Disconnect the electrical lead from the mixture control switch.
20. Disconnect the electrical lead from the oil pressure switch.
21. Disconnect the electrical lead from the oil pressure transmitter.
22. Disconnect the brake servo pipe.
23. Disconnect the fuel spill return pipe from the right-hand carburetter. Plug the pipe to prevent leakage.
24. Disconnect the outlet pipe from the fuel reserve tap. Plug the tap to prevent leakage.
25. Remove the engine tie rod.
26. Remove the engine oil filter.
27. Remove the front exhaust pipes.
28. Remove the propeller shaft.
29. Disconnect the electrical leads from the starter motor.
    Lead colours:
       WN—White and brown
       WY—White and yellow
30. Disconnect the engine earth lead.
31. Remove the lower fixings from both front engine mountings.
32. Remove the fixings from the clutch slave cylinder, withdraw the slave cylinder from the bell housing and move it to one side without disconnecting the fluid pipe.
33. Attach the engine lifting sling and tension the hoist sufficient to support the weight of the engine. 600963.
34. Support the gearbox with a jack.
35. Remove the fixings from the gearbox rear mounting.
36. Lower the rear end of the gearbox and remove the jack.
37. Hoist out the engine and gearbox assembly.

If required, separate the engine and gearbox 38 to 40.

38. Remove the cover plate from the bell housing.
39. Remove the fixings securing the bell housing to the engine.
40. Withdraw the gearbox from the engine.

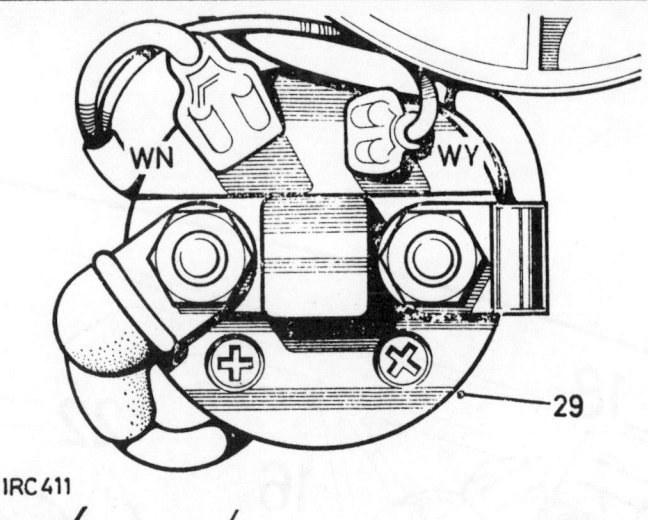

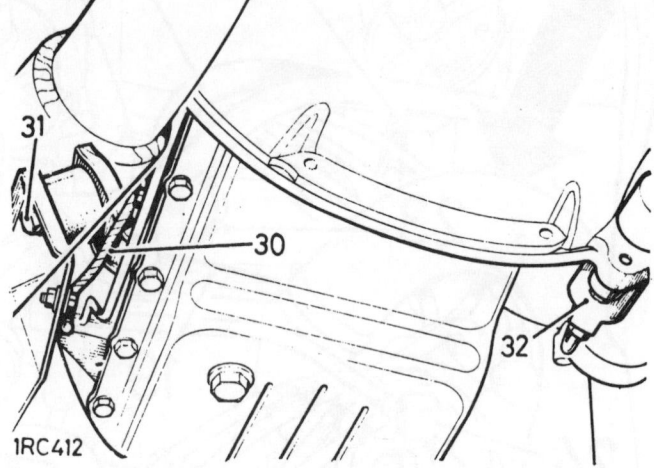

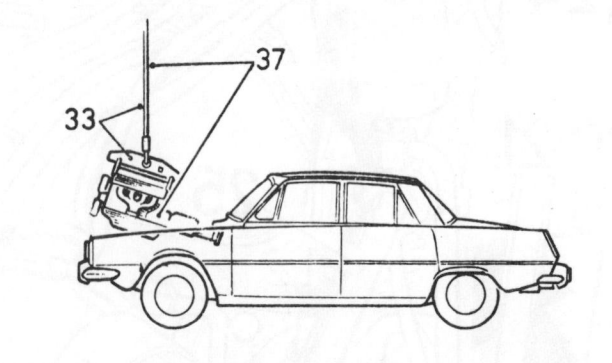

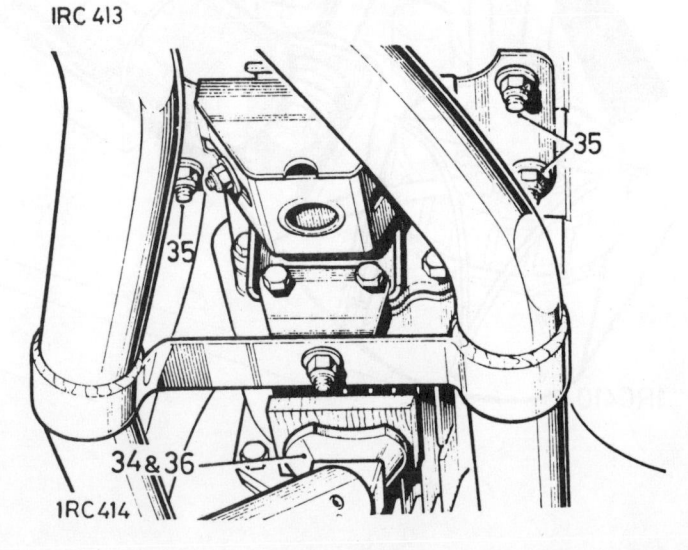

continued

12.37.01
12.41.01
Sheet 3

Rover 3500 and 3500S Manual AKM 3621

ENGINE

Refitting

If separated, assemble the gearbox to the engine 41 to 44.

41. Smear the splines of the primary pinion, the clutch centre and the withdrawal unit abutment faces with PBC (Poly Butyl Cuprysil) grease.
42. Offer the gearbox to the engine, locating the primary pinion into the clutch and engage the bell housing dowels.
43. Secure the bell housing to the engine. Torque 3,5 kgf.m (25 lbf. ft).
44. Fit the bell housing cover plate. Torque 1,0 kgf.m (8 lbf. ft).
45. Rotate the clutch withdrawal shaft until the withdrawal sleeve contacts the clutch, then fit the external clutch lever vertically downwards. If an exact vertical position cannot be obtained, choose the nearest serration either side of the vertical. Secure the lever. Torque 2,0 kgf.m (15 lbf. ft).

    NOTE: Prior to commencing the refitting, the engine and gearbox assembly should be complete, including rubber mountings and external clutch lever, but minus the engine oil filter and the driver's gearchange lever.

46. Reverse 13 to 37. Locate the speedometer cable through the hole in the gearchange shaft housing before securing the gearbox rear mounting.
47. Check, and if necessary adjust, the throttle linkage. 19.15.02.
48. Fit the air cleaner. 19.10.01 (AA1-3).
49. Fit the fan blades. An off-set dowel location ensures that the fixing bolt holes only align when the blades are the correct way round.
50. Reverse 1 to 10.
51. Check, and if necessary replenish, the gearbox lubricating oil.
52. Check, and if necessary replenish, the engine lubricating oil.
53. Start the engine and check that the oil pressure warning light goes out. If the light remains on, the engine must be stopped and the oil pump dismantled and primed. 12.60.26.
54. Check the cooling system for leaks.
55. Check, and if necessary adjust, the engine idle speed.
56. Check, and if necessary adjust, the distributor dwell angle and ignition timing. 86.35.20.
57. When the engine is cold, check the coolant level in the radiator and top up if necessary.

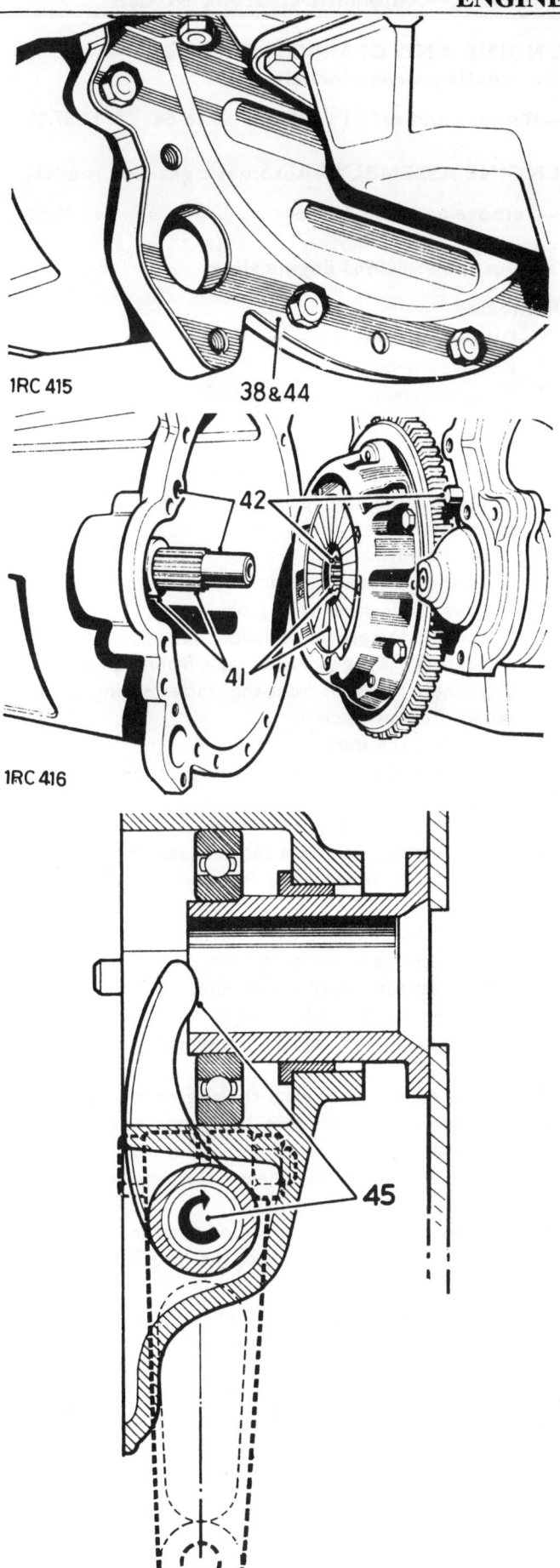

Rover 3500 and 3500S Manual AKM 3621

ENGINE—Automatic Gearbox Models

ENGINE AND GEARBOX ASSEMBLY—
Automatic gearbox models

—Remove and refit 1 to 42 and 50 to 64 12.37.01

ENGINE ASSEMBLY—Automatic gearbox models

—Remove and refit 1 to 64 12.41.01

Service tools: 600963 Engine sling

Removing

1. Disconnect the battery.
2. Remove the bonnet.
3. Drain the cooling system. 26.10.01
4. Remove the radiator assembly. 26.40.04.
5. If applicable, remove the power steering hose clip from the engine sump. Remove both hoses from the power steering pump and drain into a suitable container.
6. Drain the engine sump.
7. Drain the gearbox fluid pan.
8. Remove the front exhaust pipe.
9. Disconnect the propeller shaft.
10. Disconnect the speedometer cable from the gearbox.
11. Disconnect the leads from the starter solenoid, and release from cable cleat.
12. Disconnect the starter cable.
13. Disconnect the leads from the gearbox inhibitor switch.
14. Disconnect the engine earth lead.
15. Disconnect the rods from the gear selector lever.
16. Remove the lower fixings from both front engine mountings.
17. Remove the air cleaner. 19.10.01.
18. Disconnect the heater inlet hose.
19. Disconnect the heater outlet hose.
20. Disconnect the throttle linkage.
21. Disconnect the h.t. and l.t. leads from the ignition coil.
22. Disconnect the leads from the alternator.
23. Disconnect the lead from the water temperature transmitter.
24. Where applicable, disconnect the lead from the choke thermostat switch.
25. Disconnect the lead from the oil pressure switch.
26. Where applicable, disconnect the lead from the oil pressure transmitter.
27. Disconnect the brake servo pipe.
28. Disconnect the fuel spill return pipe from the R.H. carburetter. Plug the pipe to prevent leakage.
29. Where applicable, disconnect the choke control cable from the carburetters and unclip the cable from the gearbox fluid filler tube.
30. Disconnect the inlet pipe from the fuel filter. Plug the pipe to prevent leakage.
31. Remove the fan blades assembly.
32. Release the fuel supply and spill return pipes from the clipping to the engine.
33. Remove the engine tie rod.
34. Where applicable, release the power steering hoses from the clipping to the engine.
35. Where applicable, remove the power steering pump, leaving the driving belt on the crankshaft pulley.
36. Remove the engine oil filter.
37. To allow for any variations or additional equipment, check that all necessary pipes, leads etc. are disconnected. Refer to Division 82 if air conditioning equipment is fitted.

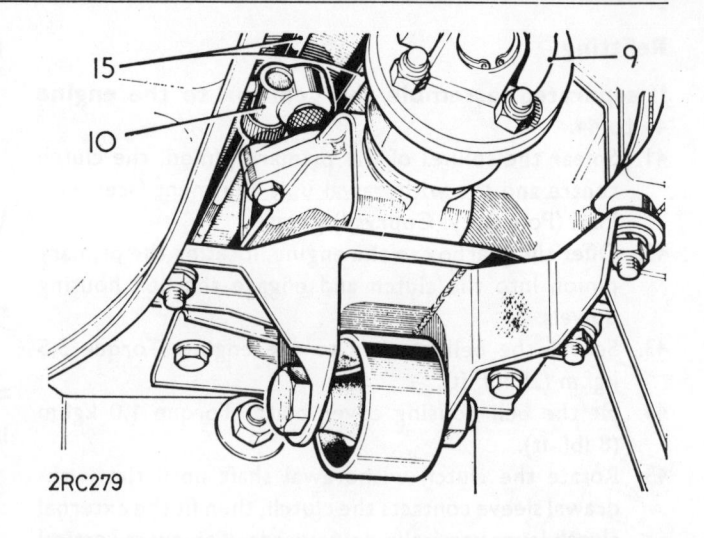

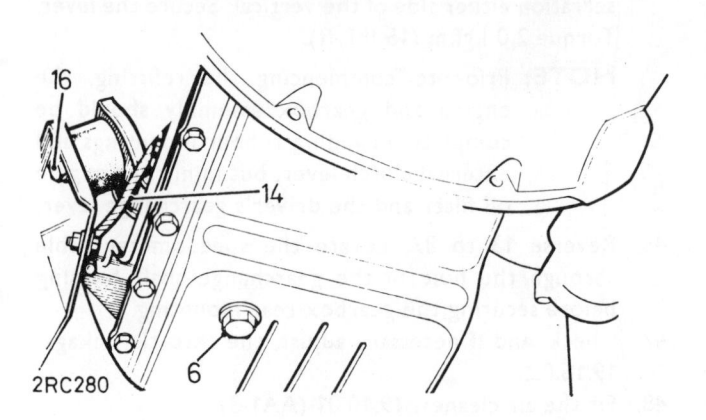

continued

ENGINE—Automatic Gearbox Models

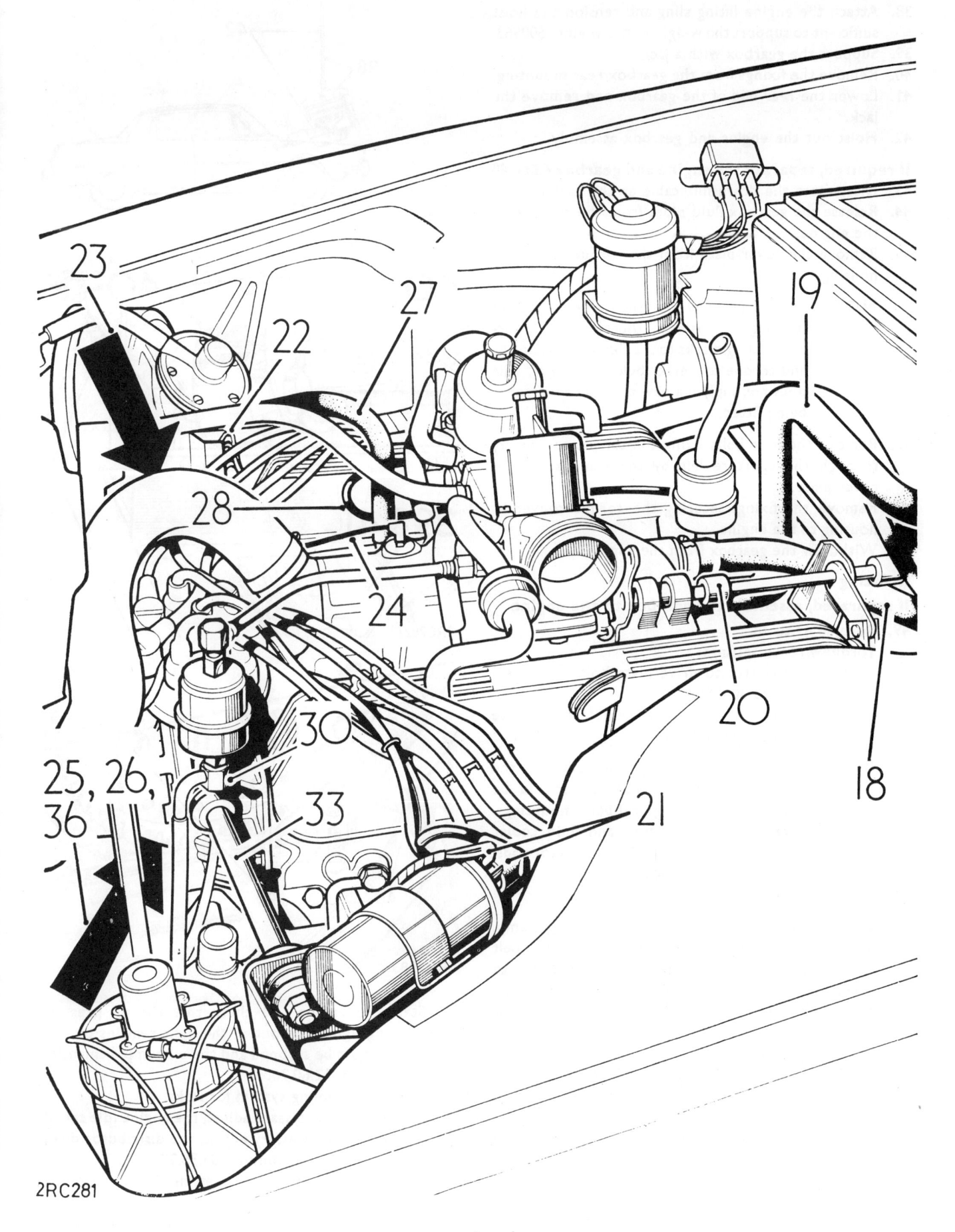

continued

Rover 3500 and 3500S Manual AKM 3621

12.37.01
12.41.01
Sheet 6

ENGINE—Automatic Gearbox Models

38. Attach the engine lifting sling and tension the hoist sufficient to support the weight of the engine. 600963.
39. Support the gearbox with a jack.
40. Remove the fixings from the gearbox rear mounting.
41. Lower the rear end of the gearbox and remove the jack.
42. Hoist out the engine and gearbox assembly.

If required, separate the engine and gearbox 43 to 48

43. Disconnect the downshift cable at the carburetter.
44. Release the gearbox fluid pipes from the clipping to the engine.
45. Remove the cover plate from the torque converter housing.

   NOTE: The torque converter should be refitted in its original position, therefore, add corresponding marks to the flexible drive plate and torque converter before removal. This does not apply if a new torque converter is being fitted.

46. Remove the four bolts fixing the torque converter, through the aperture left by the removal of the cover plate.
47. Remove the fixings securing the torque converter housing to the engine.
48. Withdraw the gearbox from the engine.

Refitting

If separated, assemble the gearbox to the engine

49. Reverse 43 to 48.
    Torque wrench settings:
    Torque converter to drive plate bolts 3,5 to 4,0 kgf.m (25 to 30 lbf ft).
    Torque converter housing to flexible drive plate housing 4,0 kgf.m (30 lbf ft).

    NOTE: Prior to commencing the refitting, the engine and gearbox assembly should be complete, including rubber mountings, gearbox fluid filler tube and fluid cooler pipes, starter motor and gear selector rods, but minus the oil filter.

50. Reverse 36 to 42.
51. Where applicable, refit the power steering pump, locating the driving belt in position. Adjust the belt tension to give a deflection of 11 to 14 mm (0.437 to 0.562 in) when checked midway between the crankshaft and power steering pump pulleys, then tighten the pump securing bolts.
52. Reverse 32 to 34.
53. Fit the fan blades. An off-set dowel location ensures that the fixing bolt holes only align when the blades are the correct way round.
54. Reverse 1 to 30, using the correct mixture for the cooling system. 26.10.01.
55. Replenish the gearbox fluid system. 44.24.02.
56. Replenish the engine oil sump.
57. Where applicable, replenish the power steering fluid system.
58. Start the engine and check that the oil pressure warning light goes out. If the light remains on, the engine must be stopped and the oil pump dismantled and primed. 12.60.26.
59. Check the cooling system for leaks.
60. Check, and if necessary adjust, the engine idle speed.
61. Check, and if necessary adjust, the distributor dwell angle and ignition timing. 86.35.20.
62. Check the downshift cable setting. Division 44.
63. Stop the engine.
64. When the engine is cold, check the coolant level in the radiator and top up if necessary.

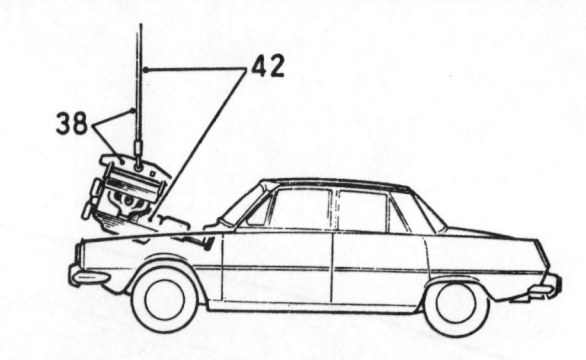

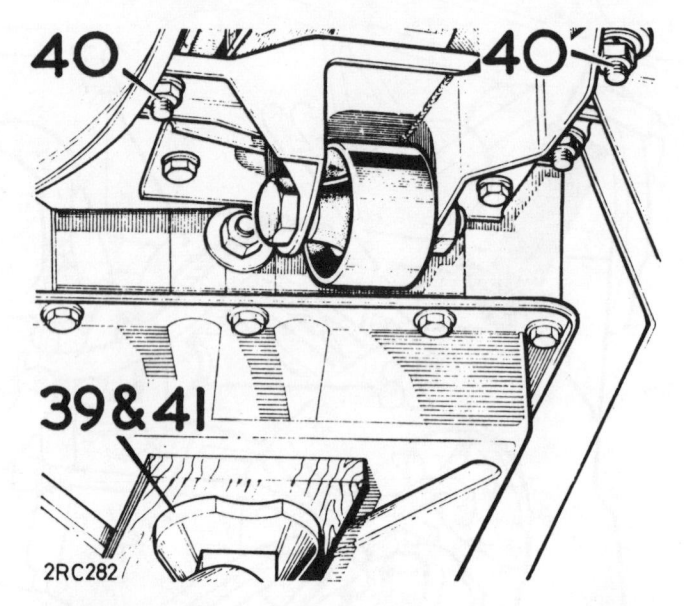

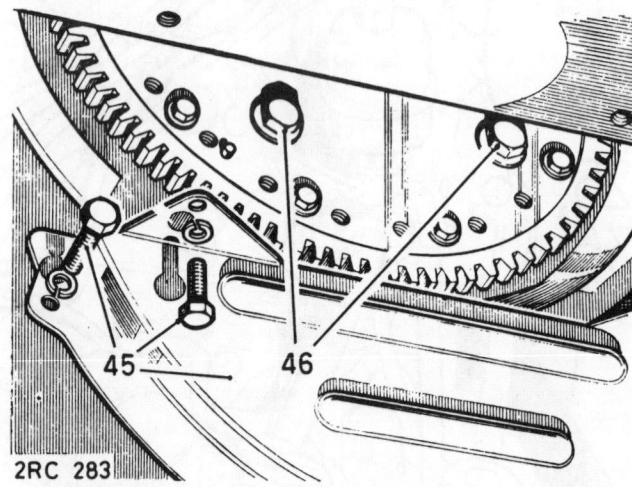

ENGINE

FLYWHEEL—Synchromesh Gearbox Models

—Remove and refit   12.53.07

Removing

1. Remove the gearbox assembly. 37.20.01.
2. Remove the clutch assembly. 33.10.01.
3. Remove the flywheel.

Refitting

4. Locate the flywheel in position on the crankshaft spigot, with the ring gear towards the engine.
5. Align the flywheel fixing bolt holes which are off-set to prevent incorrect assembly.
6. Fit the flywheel fixing bolts and before finally tightening, take up any clearance by rotating the flywheel against the direction of engine rotation. Torque 7,0 to 8,5 kgf.m (50 to 60 lbf. ft).
7. Reverse 1 and 2.

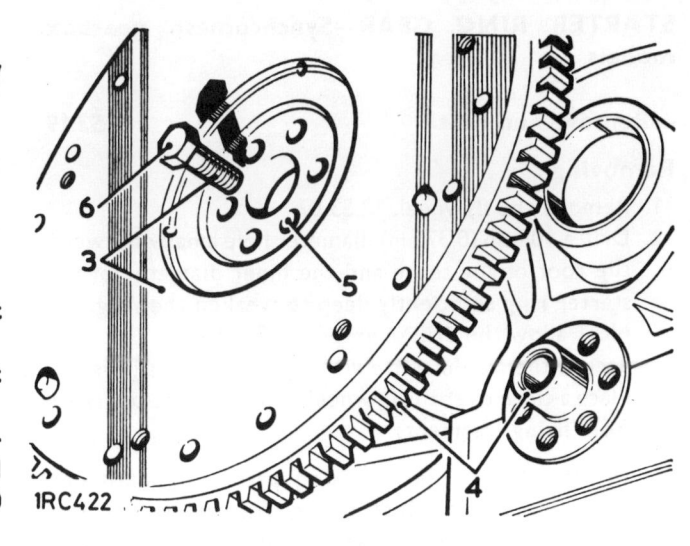

FLYWHEEL—Synchromesh Gearbox Models

—Overhaul   12.53.10

Procedure

1. Remove the flywheel. 12.53.07.
2. Measure the overall thickness of the flywheel. Fit a new flywheel if it is less than 29,33 mm (1.155 in).
3. If the flywheel is above the minimum thickness, the clutch face can be refaced as follows.
4. Remove the dowels.
5. Reface the flywheel over the complete surface.
6. Check the overall thickness of the flywheel to ensure that it is still above the minimum thickness.
7. Refit the flywheel. 12.53.07.

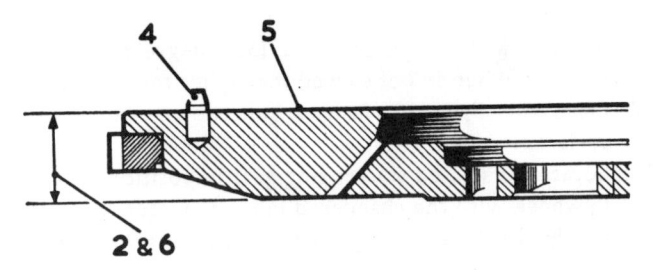

DRIVE PLATE—Automatic gearbox models

—Remove and refit   12.53.13

Removing

1. Remove the gearbox assembly. 44.20.01.
2. Remove the converter housing. 44.17.01.
3. Remove the converter assembly. 44.17.07.
4. Remove the six bolts securing the drive plate to the crankshaft.
5. Withdraw the reinforcing plate, drive plate and spacer ring.

Refitting

6. Reverse 1 to 5 noting the following:
7. The spacer ring and the reinforcing plate each have a chamfer on one outer edge and this chamfer must be fitted towards the drive plate.
8. The bolt holes are unevenly spaced so that the drive plate can only be fitted in one position, with the starter ring bolt heads towards the engine. 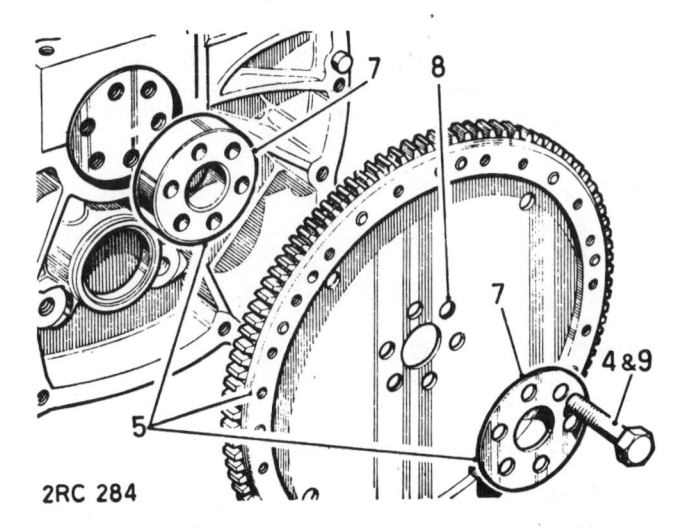
9. Before tightening the drive plate securing bolts, take up any clearance in the drive plate by rotating it in the direction of engine rotation, then secure the bolts. Torque 7,0 kgf.m (50 lbf ft).
10. Replenish the gearbox fluid system. 44.24.02.
11. Check the downshift cable setting. Division 44.

Rover 3500 and 3500S Manual AKM 3621

ENGINE

STARTER RING GEAR—Synchromesh gearbox models

—Remove and refit 12.53.19

Removing

1. Remove the flywheel. 12.53.07.
2. Drill a 10 mm (0.375 in) diameter hole axially between the root of any tooth and the inner diameter of the starter ring sufficiently deep to weaken the ring. Do NOT allow the drill to enter the flywheel.
3. Secure the flywheel in a vice fitted with soft jaws.
4. Place a cloth over the flywheel to protect the operator from flying fragments.

   WARNING: Take adequate precautions against flying fragments as the starter ring gear may fly asunder when being split.

5. Place a chisel immediately above the drilled hole and strike it sharply to split the starter ring gear.

Refitting

6. Heat the starter ring gear uniformly to between 170 degrees and 175 degrees C (338 degrees to 347 degrees F) but do not exceed the higher temperature.
7. Place the flywheel, flanged side down, on a flat surface.
8. Locate the heated starter ring gear in position on the flywheel, with the chamfered inner diameter towards the flywheel flange. If the starter ring gear is chamfered both sides, it can be fitted either way round.
9. Press the starter ring gear firmly against the flange until the ring contracts sufficiently to grip the flywheel.
10. Allow the flywheel to cool gradually. Do NOT hasten cooling in any way and thereby avoid the setting up of internal stresses in the ring gear which may cause fracture or failure in some respect.
11. Fit the flywheel. 12.53.07.

SPIGOT BEARING

—Remove and refit 12.53.20

Synchromesh gearbox 1 and 2, and 6 to 9

Automatic gearbox 3 to 9

Removing

1. Remove the gearbox assembly. 37.20.01. ⎫ manual gear-
2. Remove the clutch assembly. 33.10.01. ⎭ box only.
3. Remove the gearbox assembly. 44.20.01. ⎫ automatic
4. Remove the converter housing. 44.17.01. ⎬ transmission
5. Remove the converter assembly. 44.17.07. ⎭ only.
6. Remove the spigot bearing.

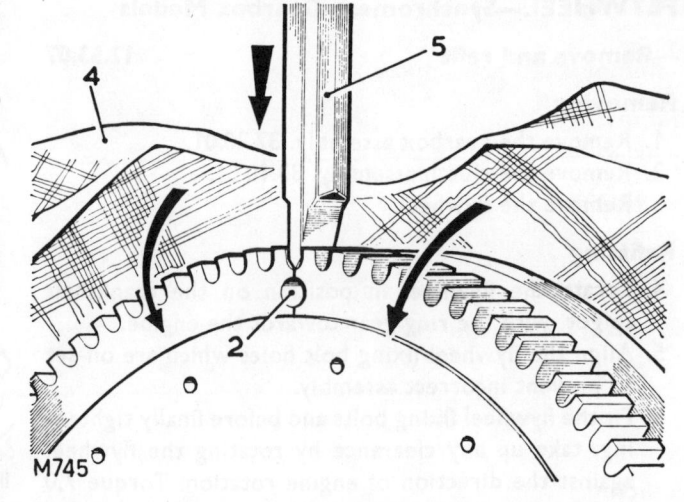

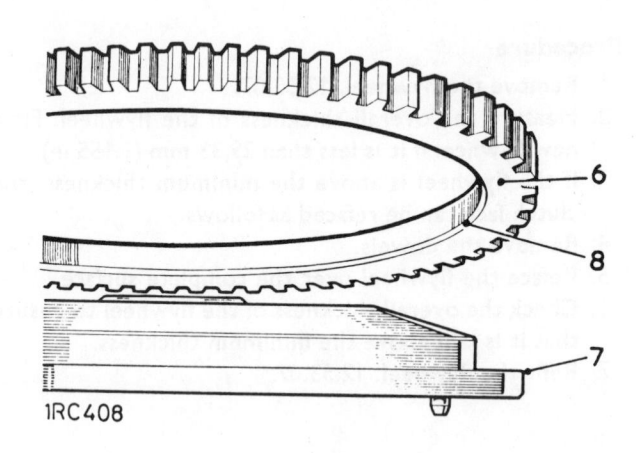

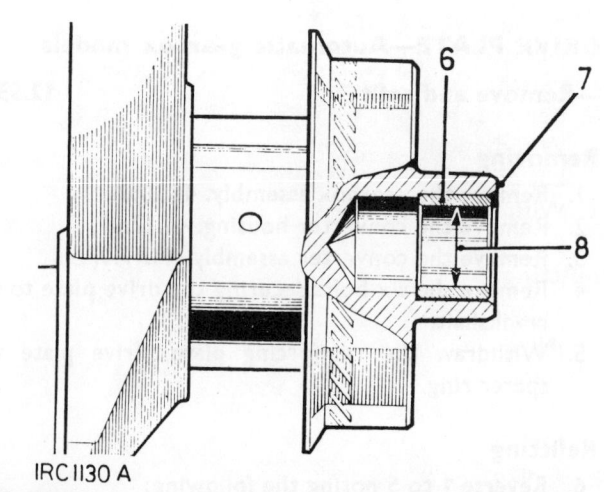

Refitting

7. Fit the spigot bearing flush with, or to a maximum of 1,6 mm (0.063 in) below the end face of the crankshaft.
8. Check and, if necessary, reamer the spigot bearing to 19,177+0,025 mm (0.7504+0.001 in) inside diameter.
9. Reverse 1 to 5.

Rover 3500 and 3500S Manual AKM 3621

ENGINE

OIL FILTER ASSEMBLY—EXTERNAL

—Remove and refit   12.60.01

Removing

1. Unscrew the filter anti-clockwise and discard.

   NOTE: If the filter is difficult to remove, use a strap spanner.

2. Withdraw the sealing washer and discard.

   CAUTION: Do NOT delay fitting a new filter, otherwise the oil pump may drain and require priming (12.60.26) before running the engine.

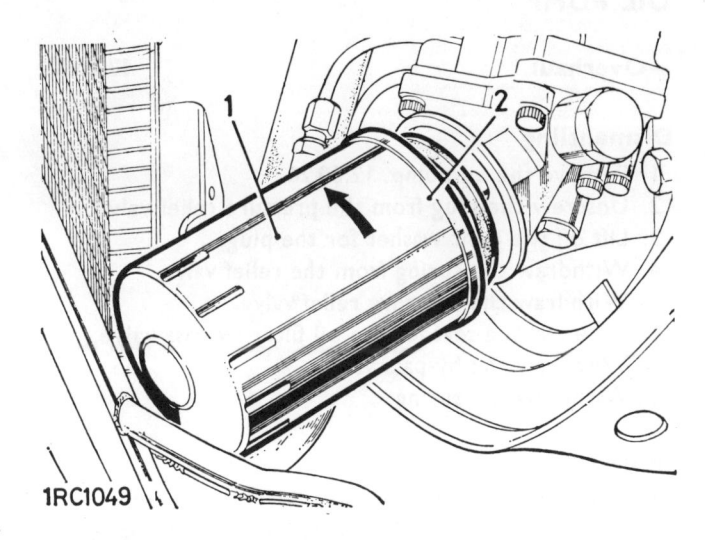

Refitting

3. Smear a little clean engine oil on the rubber washer of the new filter.
4. Screw the filter on clockwise until the rubber sealing ring touches the oil pump cover face, then tighten a further half turn by hand only. Do not overtighten.
5. Refill with oil of the correct grade through the screw on filler cap on the R.H. front rocker cover; the capacity is 0,5 litre (1 imperial pint or 1.2 US pints) if the oil filter only has been drained.
6. Start the engine and check that the oil pressure warning light goes out. If the light remains on, the engine must be stopped and the oil pump dismantled and primed. 12.60.26. (AA1-19)
7. Run the engine and check the filter joint for leaks.
8. Check the oil sump level after the engine has been stopped for a few minutes and replenish if necessary.

OIL PUMP

—Remove and refit   12.60.26

Removing

1. Remove the oil filter assembly. 12.60.01.
2. Disconnect the electrical lead from the oil pressure switch.
3. Remove the bolts from the oil pump cover.
4. Withdraw the oil pump cover.
5. Lift off the cover gasket.
6. Withdraw the oil pump gears.

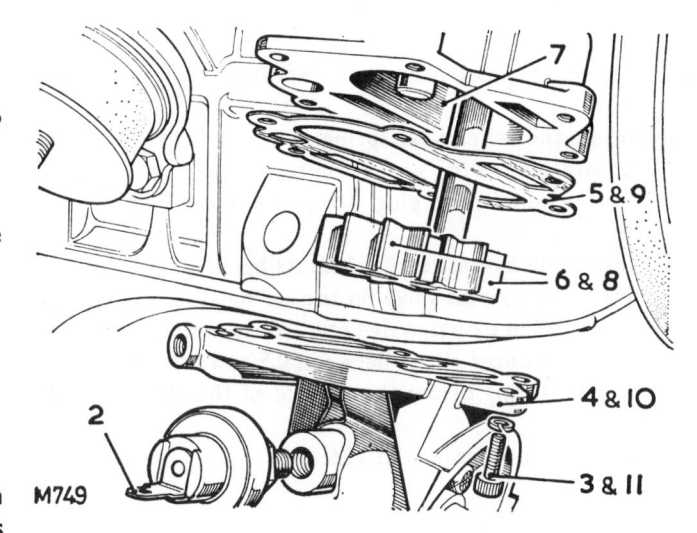

Refitting

7. Fully pack the oil pump gear housing with petroleum jelly. Use only petroleum jelly; no other grease is suitable.
8. Fit the oil pump gears so that the petroleum jelly is forced into every cavity between the teeth of the gears.

   NOTE: Unless the pump is fully packed with petroleum jelly it may not prime itself when the engine is started.

9. Place a new gasket on the oil pump cover.
10. Locate the oil pump cover in position.
11. Fit the special fixing bolts and tighten alternately and evenly. Torque: 1,2 kgf.m (9 lbf ft).
12. Reverse 1 and 2.
13. Check the oil level in the engine sump and replenish as necessary.

Rover 3500 and 3500S Manual AKM 3621

ENGINE

OIL PUMP

—Overhaul 12.60.32

Dismantling

1. Remove the oil pump. 12.60.26.
2. Unscrew the plug from the pressure relief valve.
3. Lift off the joint washer for the plug.
4. Withdraw the spring from the relief valve.
5. Withdraw the pressure relief valve.
6. Prise out the seat for the oil filter by-pass valve.
7. Withdraw the by-pass valve.
8. Withdraw the spring.

Inspecting

9. Check the oil pump gears for wear or scores.
10. Fit the oil pump gears and shaft into the front cover.
11. Place a straight edge across the gears.
12. Check the clearance between the straight edge and the front cover. If less than 0,05 mm (0.0018 in) check the front cover gear pocket for wear.
13. Check the oil pressure relief valve for wear or scores.
14. Check the relief valve spring for wear at the sides or signs of collapse.
15. Clean the gauze filter for the relief valve.
16. Check the fit of the relief valve in its bore. The valve must be an easy slide fit with no perceptible side movement.
17. Check the oil filter by-pass valve for cracks, nicks or scoring.

Reassembling

18. Lubricate the oil pressure relief valve and fit it into its bore.
19. Insert the relief valve spring.
20. Locate the sealing washer on to the relief valve plug.
21. Fit the relief valve plug. Torque: 4,0 to 4,9 kgf.m (30 to 35 lbf. ft).
22. Insert the by-pass spring into its bore.
23. Place the by-pass valve on the spring.
24. Press in the by-pass valve seat, concave side outward, until the outer rim is between 0,5 and 1,0 mm (0.020 and 0.040 in) below the surface of the surrounding casting.
25. Refit the oil pump. 12.60.26.

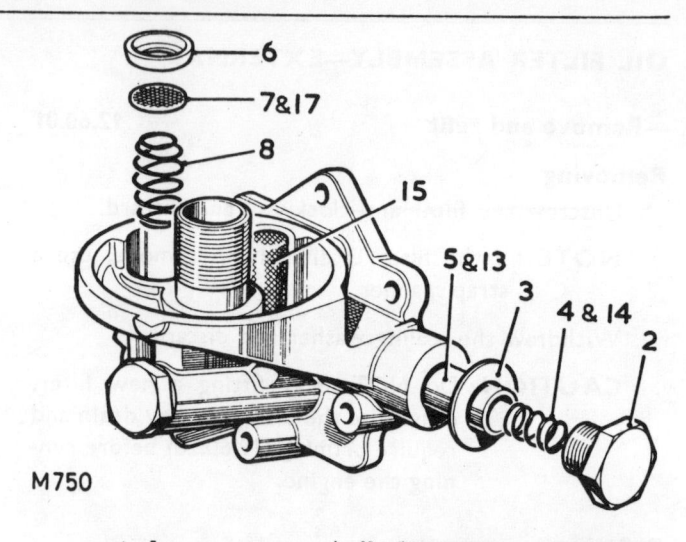

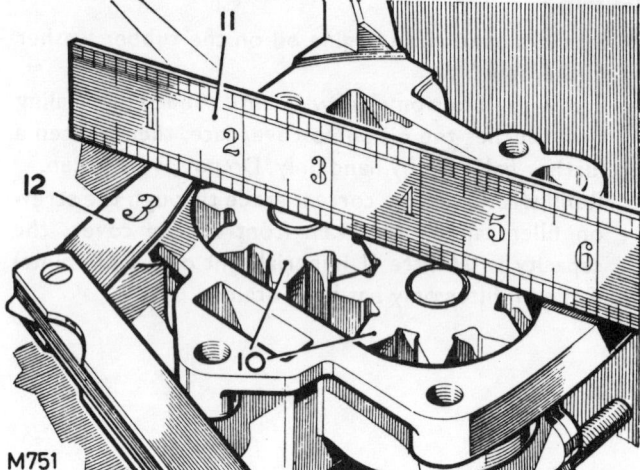

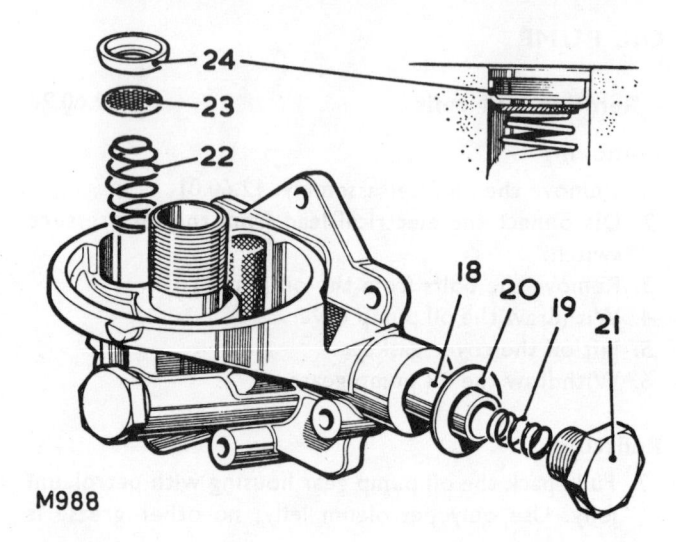

Rover 3500 and 3500S Manual AKM 3621

ENGINE

OIL SUMP

—Remove and refit 12.60.44

Removing
1. Remove the sump drain plug.
2. Allow all the oil to drain, then refit the plug and sealing washer.
3. Remove the sump.

Refitting
4. Clean the sump mating surfaces at the join between the timing chain cover and the cylinder block.
5. Apply a coating of 'Hylomar PL 32/M sealing compound, Rover Part No. 534244, across the join.
6. Place the sump gasket in position.
7. Fit the sump, locating the reinforcing strip under the rear fixings.
8. Unscrew the oil filler cap from the R.H. rocker cover.
9. Using the correct grade oil, see division 09, replenish the sump through the rocker cover.
10. Use the sump dipstick to set the final level. DO NOT fill above the 'HIGH' mark.

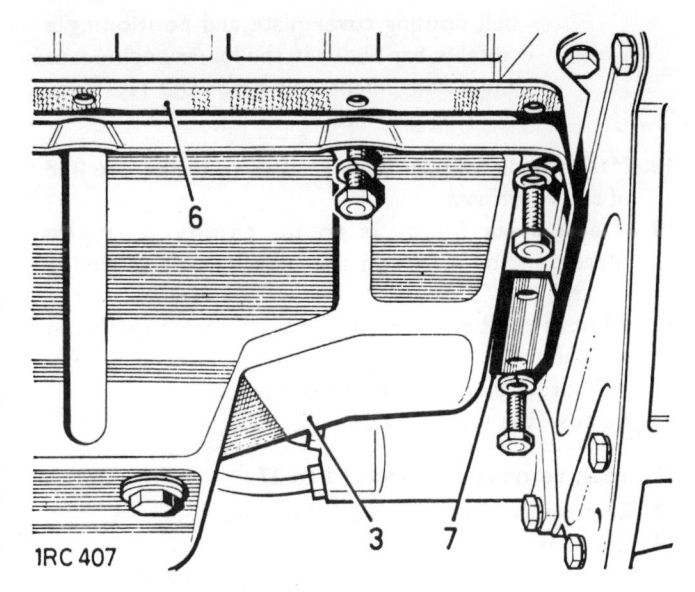

TIMING GEAR COVER AND OIL SEAL

—Remove and refit

Gear cover 1 to 21 and 27 to 42   12.65.01

Oil seal 1 to 42   12.65.05

Removing
1. Disconnect the battery earth lead.
2. Drain the cooling system. 26.10.01.
3. Remove the radiator block. 26.40.04.
4. Remove the fan blades
5. Remove the fan blade pulley.
6. Release the alternator adjusting link from the water pump.
7. Disconnect the by-pass hose from the thermostat.
8. Disconnect the heater return hose from the water pump.
9. Disconnect the inlet hose from the water pump.
10. Disconnect the vacuum pipe from the distributor.
11. Release the distributor cap, unclip the leads and move the cap to one side.
12. Disconnect the low tension lead from the ignition coil.
13. Disconnect the electrical lead from the oil pressure switch.
14. Disconnect the electrical lead from the oil pressure transmitter.
15. Release the fuel filter retaining clip from the engine lifting bracket.
16. Remove the fixings from the fuel pump and move the pump to one side. DO NOT disconnect the fuel pipes.

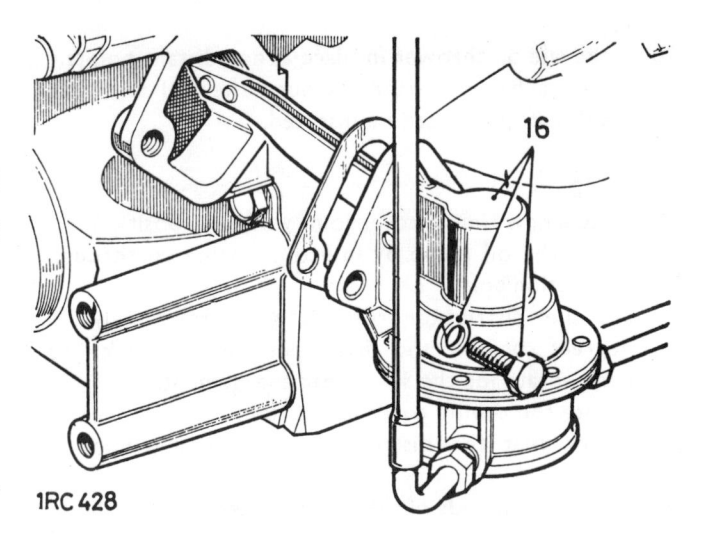

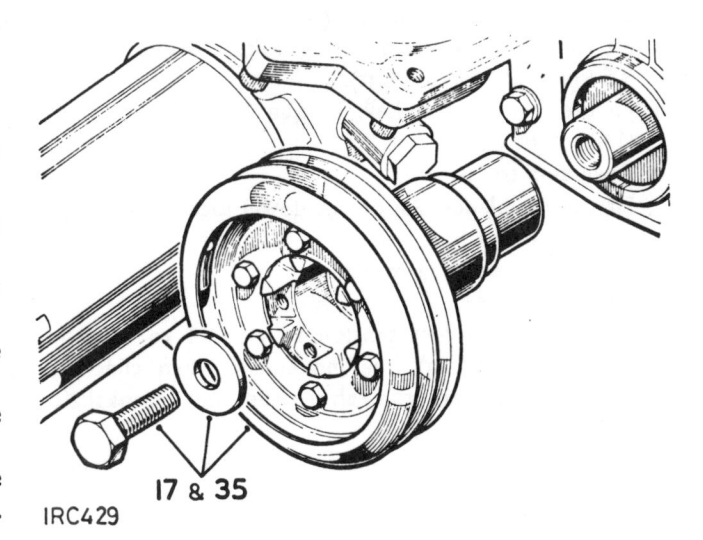

continued

Rover 3500 and 3500S Manual AKM 3621

12.60.44
12.65.01/05
Sheet 1

ENGINE

17. Remove the crankshaft pulley.

    NOTE: Automatic transmission models: To enable the crankshaft pulley fixings to be removed, the engine can be locked by removing the bell housing cover plate and positioning a suitable bar between the lower engine face and a convenient bolt head. DO NOT use the ring gear teeth.

18. Mark the distributor body relative to the centre line of the rotor arm.
19. If the distributor is to be removed, make corresponding marks on the distributor and timing cover.
20. Remove the timing cover fixings, including two from the sump.
21. Withdraw the timing cover complete.
22. Remove the joint washer.

Oil seal, remove and refit 23 to 27

23. Remove the oil seal and oil thrower.
24. Coil a new seal into a new oil thrower.
25. Fit the oil seal and thrower assembly into the front cover, ensuring that the ends of the seal are at the top.
26. Stake the oil thrower in place at equidistant points.
27. Rotate a hammer handle around the seal until the crankshaft pulley can be inserted.

Refitting

28. Place a new timing cover joint washer in position.
29. Prime the oil pump by injecting engine oil through the suction port.
30. Set the distributor rotor arm approximately 30 degrees before the final positioning mark, to compensate for the skew gear engagement.
31. Locate the timing cover in position.
32. Check that the distributor marking alignment is correct.
33. Clean the threads of the timing cover securing bolts then coat them with Thread Lubricant-Sealant 3M EC776, Rover Part No. 605764.
34. Fit the timing cover securing bolts. Torque: 2,8 to 3,5 kgf.m (20 to 25 lbf. ft).
35. Fit the crankshaft pulley. Torque: 19,3 to 22,3 kgf.m (140 to 160 lbf. ft).
36. Reverse 5 to 16.
37. Fit the fan blades. An off-set dowel location ensures that the fixing bolt holes only align when the blades are the correct way round.
38. Reverse 1 to 4.
39. Adjust the fan belt tension to give 11 to 14 mm (0.437 to 0.562 in) free movement when checked midway between the alternator and crankshaft pulleys, by hand.
40. Start the engine and check that the oil pressure warning light goes out. If the light remains on, the engine must be stopped and the oil pump dismantled and primed. 12.60.26.
41. Check the cooling system for leaks.
42. Check, and if necessary adjust, the distributor dwell angle and ignition timing. 86.35.20
43. When the engine is cold, check the coolant level in the radiator and top up if necessary.

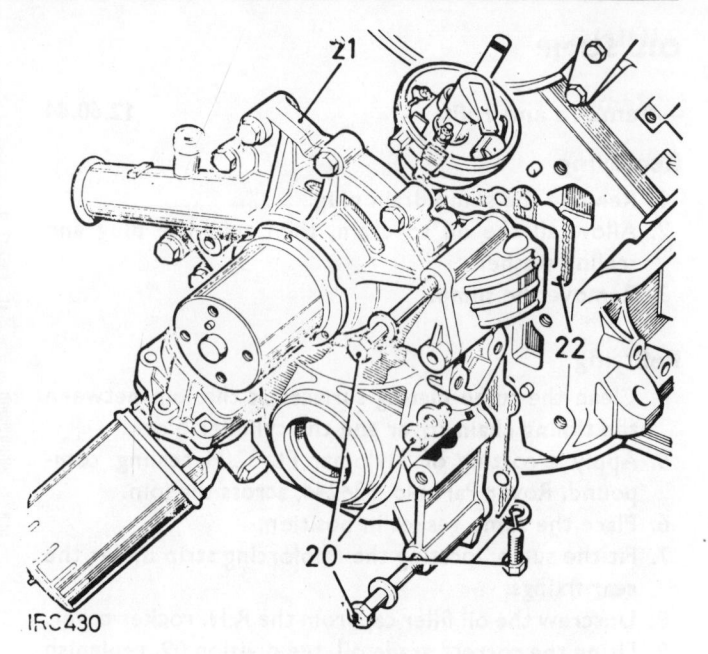

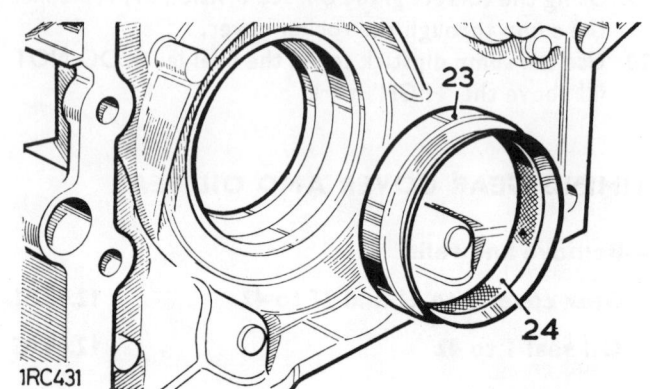

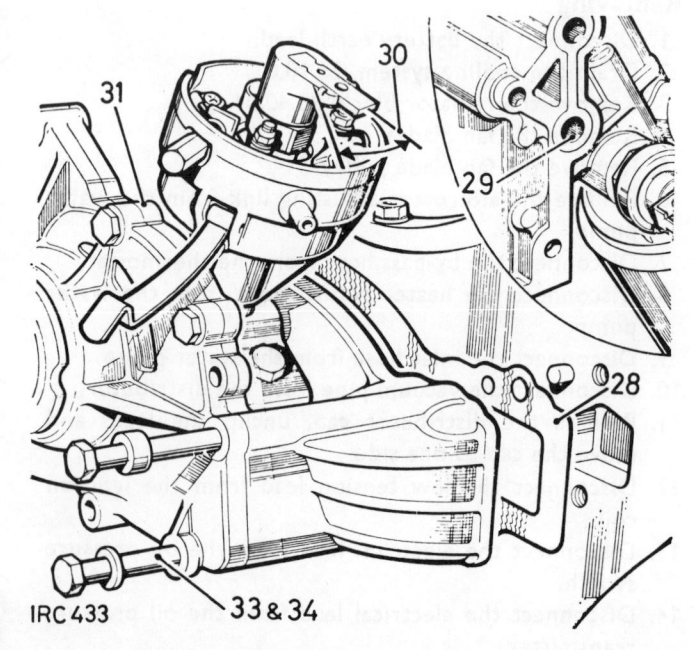

ENGINE

TIMING CHAIN AND GEARS

—Remove and refit   12.65.12

Removing

1. Set the engine—number one piston at TDC.
2. Remove the timing chain cover. 12.65.01.
3. Check that number one piston is still at TDC.
4. Withdraw the oil thrower.
5. Remove the distributor drive gear.
6. Withdraw the fuel pump cam.
7. Withdraw the chain complete with the chainwheels.

CAUTION: Do NOT rotate the engine if the rocker shafts are fitted, otherwise the valve gear and pistons will be damaged.

Refitting

   NOTE: If the crankshaft and/or camshaft have not been rotated, commence at item 12. If they have been rotated, commence at item 8.

8. Remove the rocker shafts. 12.29.54.
9. Set the engine—number one piston at TDC.
10. Temporarily fit the camshaft chainwheel with the marking 'FRONT' outward.
11. Turn the camshaft until the mark on the chainwheel is at the six o'clock position, then remove the chainwheel without disturbing the camshaft.
12. Locate the chainwheels to the chain with the timing marks aligned.
13. Engage the chainwheel assembly on to the camshaft and crankshaft key locations.
14. Check that the timing marks are in line.
15. Fit the fuel pump cam with the marking 'F' outward.
16. Fit the distributor drive gear, washer and bolt. Torque: 5,5 to 6,2 kgf.m (40 to 45 lbf. ft).
17. Fit the oil thrower, concave side outward.
18. Fit the timing chain cover. 12.65.01.

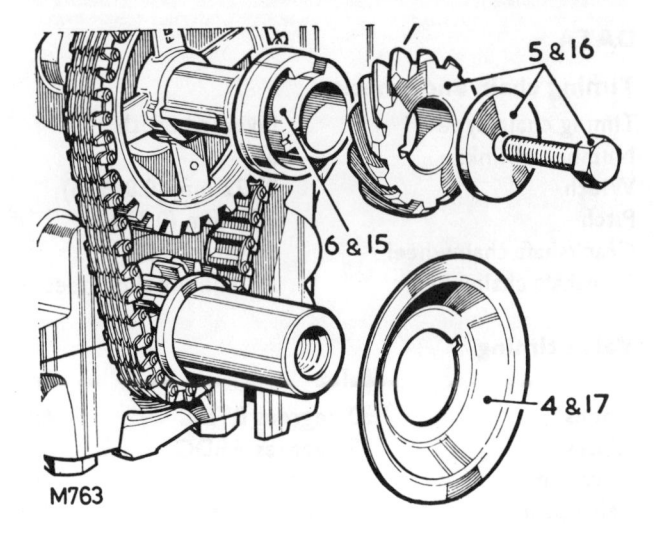

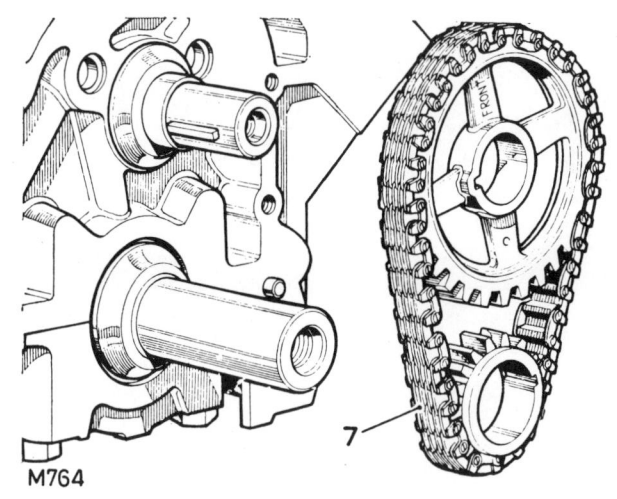

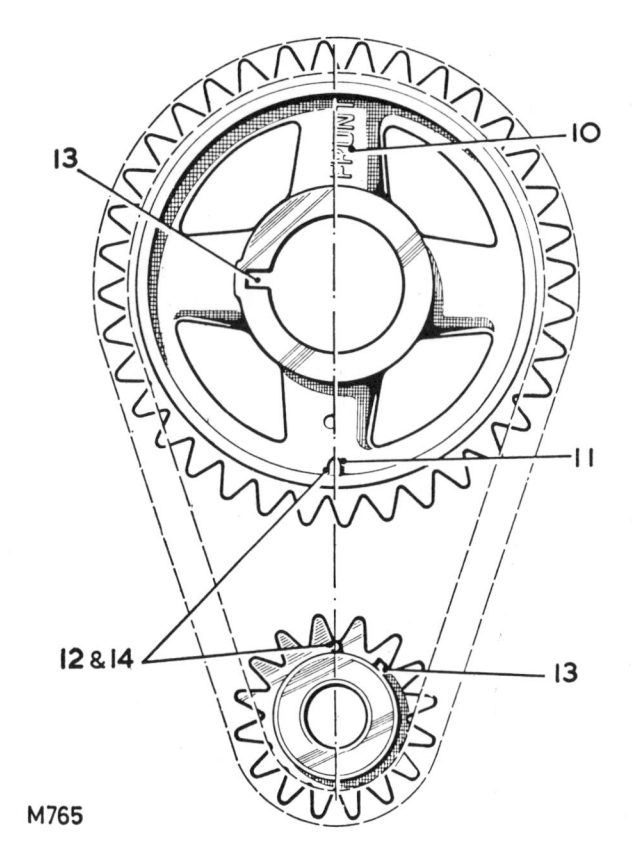

continued

Rover·3500 and 3500S Manual AKM 3621

12.65.12
Sheet 1

ENGINE

DATA

Timing chain and wheels

Timing chain type	Inverted tooth
Number of links	54
Width	22,22 mm (0.875 in)
Pitch	9,52 mm (0.375 in)
Crankshaft chainwheel	Sintered iron
Camshaft chainwheel	Aluminium alloy, teeth covered with nylon

Valve timing

	Inlet	Exhaust
Opens	30 degrees BTDC	68 degrees BBDC
Closes	75 degrees ABDC	37 degrees ATDC
Duration	285 degrees	285 degrees
Valve peak	112.5 degrees ATDC	105.5 degrees BTDC

EMISSION CONTROL

EMISSION CONTROL OPERATIONS

Adsorption canister—remove and refit	17.15.13
Adsorption canister air filter—remove and refit	17.15.07
Breather pipe, expansion tank to filter—remove and refit	17.15.32
Engine breather filter—remove and refit	17.10.02
Engine flame traps—remove and refit	17.10.03
Expansion tank—remove and refit	17.15.19

(Refer to Divisions 19 and 86 for special instructions covering carburation and ignition.)

EMISSION CONTROL

EMISSION CONTROL

Rover 3500 and 3500S models supplied to countries that have emission control regulations, conform to all applicable regulations and amendments in effect on the date of manufacture of the vehicle.

The following details provide a general description of the emission control features, but due to varying regulations in particular countries, all these features may not be applicable, or incorporated.

Crankcase emission control

All Rover 3500 and 3500S models are equipped with crankcase emission control breathing, in which the crankcase fumes are recirculated through the intake manifold to be eventually burned in the normal combustion process, instead of emerging into the atmosphere.

The breathing cycle is performed by tapping clean air from the rear of the air cleaner and feeding it via a hose and filter to the crankcase, the crankcase fumes rise via the push rod tubes to the rocker covers where they are then tapped off via hoses and flame traps connected to the carburetter intake adaptors, finally the fumes are drawn into the engine and are burned in the combustion chamber.

The following maintenance must be carried out to ensure the continued efficiency of the crankcase emission control system.

Fit new engine flame traps every 20 000 km (12 000 miles) or 12 months, whichever is first.

Fit new engine breather filter every 40 000 km (24 000 miles) or 24 months, whichever is first.

Instructions covering the replacement of the engine breather filter and flame traps are contained in 17.10.02–03.

Exhaust emission control

All Rover 3500 and 3500S models supplied to countries where emission control regulations apply, are specially equipped to control the emission of hydrocarbons and carbon monoxide from the exhaust system. Conformity to the regulations is achieved by alterations to the carburation and combustion characteristics.

Each carburetter needle is spring loaded and the needle is biased by the spring against the retainer, this maintains the needle in its correct relationship with the carburetter jet, thus improving the control of emission. Each throttle butterfly has a spring loaded poppet valve. With low manifold depression, the poppet valve remains closed. At high manifold depression, that is, over-run at closed throttle, the valve opens and prevents incorrect combustion of the fuel by supplementing the volume of fuel/air mixture. This, together with the ignition timing set at TDC or 6° BTDC, as applicable (see Data), maintains correct combustion.

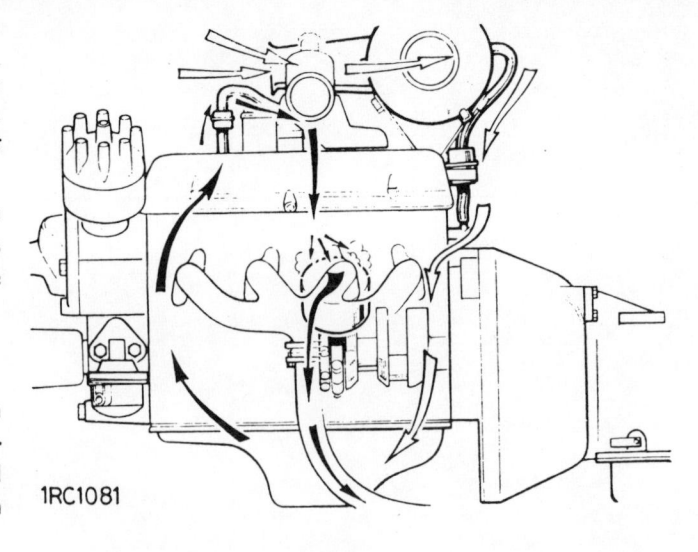

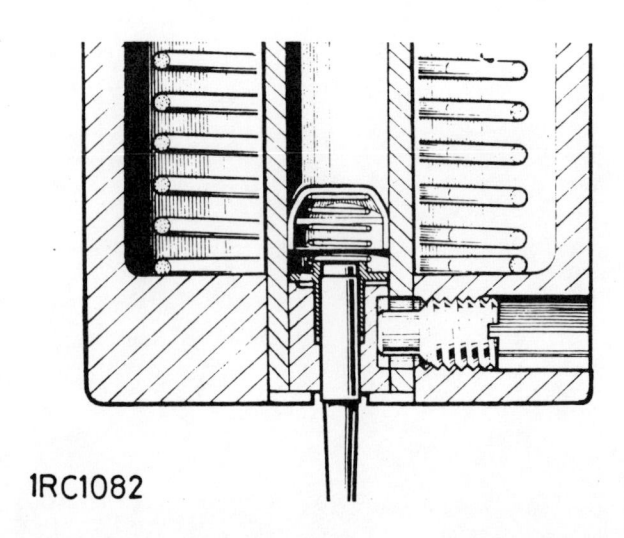

continued

EMISSION CONTROL

The correct ignition timing and carburetter settings must be maintained to ensure continued compliance with the exhaust emission control regulations. Therefore, all adjustments and repairs to the ignition and carburation systems must be carried out strictly in accordance with the special Operations and Instructions included in Divisions 19—Fuel System and 86—Electrical equipment. Failure to comply with the foregoing instructions would almost certainly result in the car failing to meet the legal requirements in respect of air pollution.

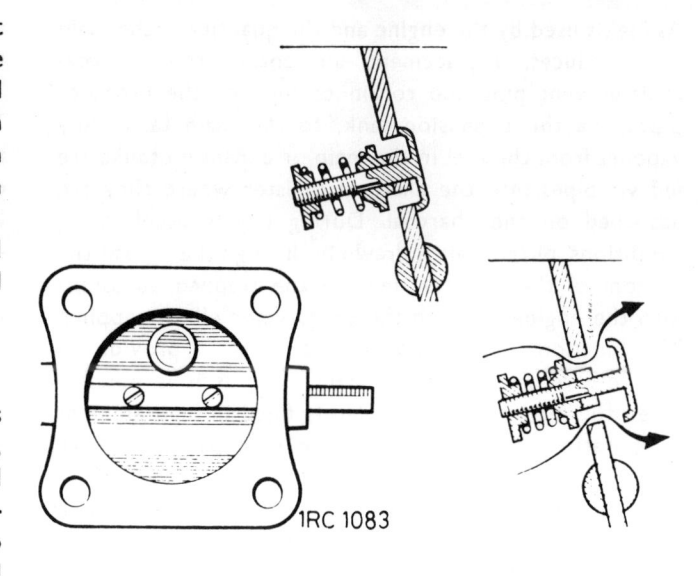

Evaporative emission control

All Rover 3500 and 3500S models supplied to countries where evaporative emission control regulations apply, are specially equipped to control the emission of fuel vapours from the carburetters and the fuel tank. Conformity to the regulations is achieved by alterations to the venting systems for the carburetters and the fuel tank.

Control of evaporative emissions from the carburetters

A charcoal filled canister is situated on the left-hand wing valance to deal with evaporative emissions from the carburetters.

During hot soak conditions, emissions from the carburetters are fed via a pipe to the top of the charcoal canister and are adsorbed on the charcoal. Then, under engine accelerating conditions, a small negative pressure in the carburetter elbow adaptor draws purge air through the atmospheric vent in the bottom of the charcoal canister, thus feeding the emissions trapped in the canister back into the engine through the pipe to the carburetter elbow tapping. An additional advantage given by this system is the prevention of an accumulation of fuel vapour in the intake system, which could cause hot starting difficulties.

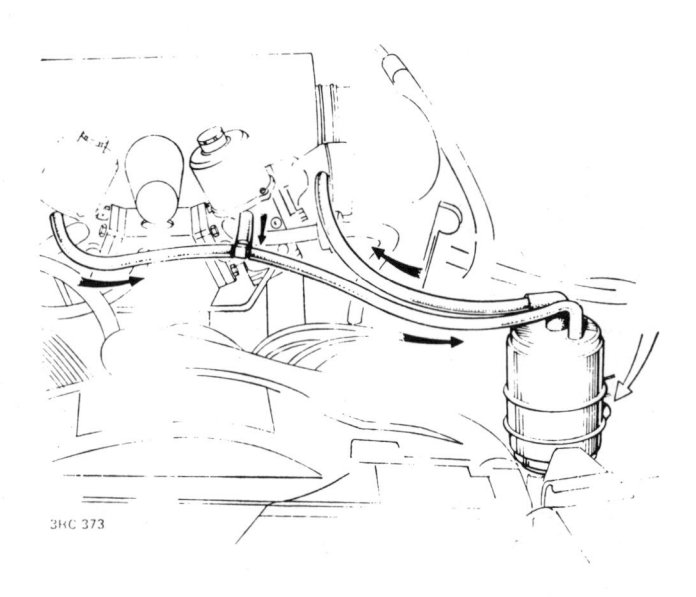

Control of evaporative emissions from the fuel tank

A charcoal filled canister is situated on the right-hand wing valance to deal with evaporative emissions from the fuel tank.

Adjacent to the main fuel tank is a separate expansion tank into which is fed the main tank breather pipe. From the expansion tank a further breather pipe leads to the charcoal canister. At the bottom of the canister a vent pipe is open to atmosphere and, from the top of the canister a pipe leads to the carburetter elbow adaptor on the right-hand side of the engine.

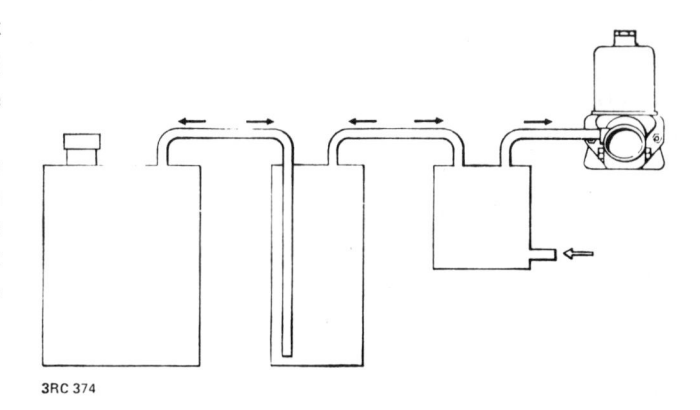

Rover 3500 and 3500S Manual AKM 3621

17.00.00
Sheet 2

EMISSION CONTROL

As fuel is used by the engine and the quantity in the main tank reduces, replacement air enters the charcoal canister vent pipe and continues through the breather pipes, via the expansion tank, to the main tank. Any vapours from the fuel in the main or expansion tanks are fed via pipes into the charcoal canister where they are adsorbed on the charcoal. During engine accelerating conditions, purging air is drawn in through the pipe at the bottom of the canister, feeding the trapped emissions into the engine through the carburetter elbow tapping. The function of the expansion tank is to provide an overflow reservoir for the main tank, as it is possible when the main tank is completely filled in high ambient temperature conditions for the fuel to expand and force a large quantity along the breather pipe. The size of the expansion tank allows for maximum expansion conditions, under such conditions evaporative emissions are still controlled by the charcoal canister and, due to the location of the breather pipe at the bottom of the expansion tank, the overflow fuel will eventually be drawn back into the main tank as fuel is used.

The following maintenance must be carried out to ensure the continued efficiency of the evaporative emission control system. Fit new charcoal canisters every 64.000 km (24,000 miles). If however, severe flooding of the canister takes place, which may result in fuel weeping from the tank filler cap or the emission system or, erratic engine running due to enriched mixture, check that all pipes are connected correctly, and that there are no obstructions. If flooding persists, change the filter pad in the base of the canister. If flooding still persists, replace the canister immediately regardless of mileage.

Instructions covering the replacement of the charcoal canisters and air filters are contained in 17.15.13 and 17.15.07.

DATA

Engines numbered in the range commencing 425 and 427

Compression ratio	10.5:1
Engine idle speed	700 to 750 rev/min
Fast idle speed	1100 to 1200 rev/min
Ignition timing, static and dynamic at 600 rev/min	TDC for use with fuel of not less than 96 research octane number
Sparking plugs	Champion L87Y
Carburetters:	
Type	SU HS6
Needle	BAK
Jet size	0.100 in (2,54 mm)
CO emission	4% maximum

Engines numbered in the range commencing 451, 453 and 455 with a suffix 'A', 'B' or 'C'

Compression ratio	10.5:1
Engine idle speed	700 to 750 rev/min
Fast idle speed	1100 to 1200 rev/min
Ignition timing static and dynamic at 600 rev/min	TDC for use with fuel of not less than 96 research octane number
Sparking plugs	Champion L92Y
Carburetters:	
Type	SU HIF6
Needle	BBG
CO emission	4% maximum

Engines numbered in the range commencing 451, 453 and 455 from suffix 'D' onwards

Compression ratio	9.25:1
Engine idle speed	700 to 750 rev/min
Fast idle speed	1100 to 1200 rev/min
Ignition timing, static and dynamic at 600 rev/min	6° BTDC for use with fuel of not less than 96 research octane number
Sparking plugs	L92Y
Carburetters:	
Type	SU HIF6
Needle	BBV
CO emission	4% maximum

EMISSION CONTROL

ENGINE BREATHER FILTER

—Remove and refit   17.10.02

Removing

1. Remove the air cleaner. 19.10.01 (AA1-3).
2. Disconnect the top hose.
3. Slacken the clip.
4. Withdraw the filter from the bottom hose and clip.

Refitting

5. Reverse 1 to 4 noting the following. Fit the filter with the end marked 'IN' uppermost. Alternatively, if the filter is marked with arrows they must point downwards.

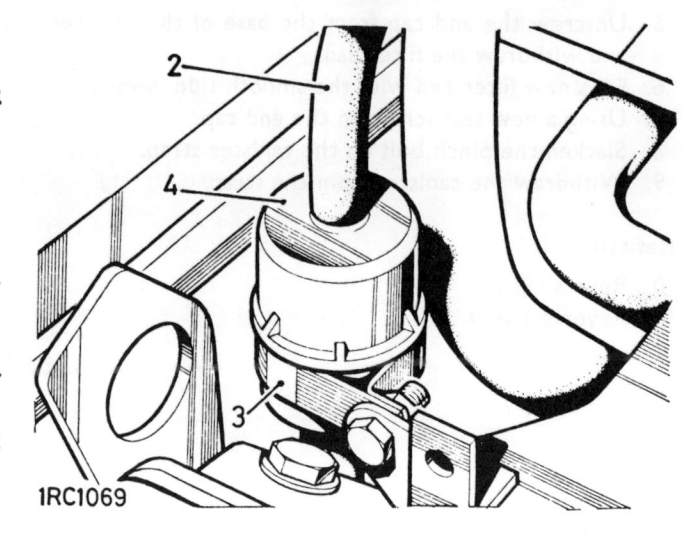

1RC1069

ENGINE FLAME TRAPS

—Remove and refit   17.10.03

Removing

1. Pull off the flame trap hoses.
2. Remove the flame traps, one on top of each rocker cover.

Refitting

3. Replace with new flame traps, which are located in position by the hoses.

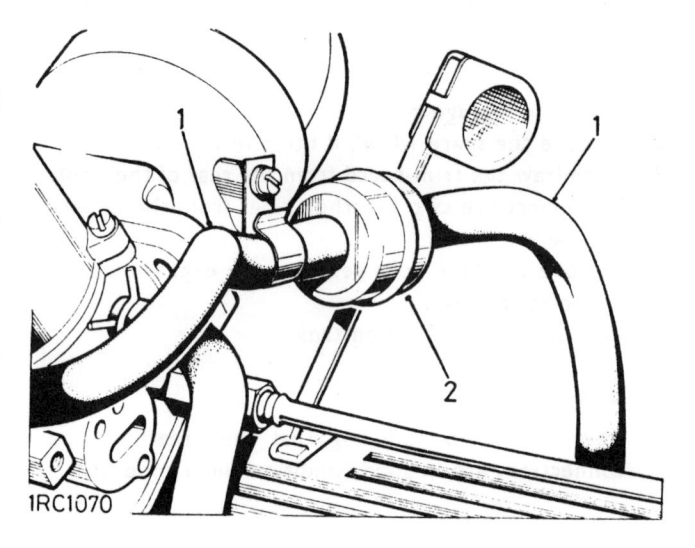

1RC1070

ADSORPTION CANISTER AIR FILTER

—Remove and refit 1 to 7 and 11   17.15.07

ADSORPTION CANISTER

—Remove and refit 1 to 4 and 8 to 11   17.15.13

WARNING: No attempt should be made to cleanse the canister. The use of compressed air could cause the activated charcoal filling to ignite.

1. Prop open the bonnet.
2. Disconnect the hoses from the canister.
3. Remove the fixings, container strap to mounting bracket.
4. Withdraw the canister.

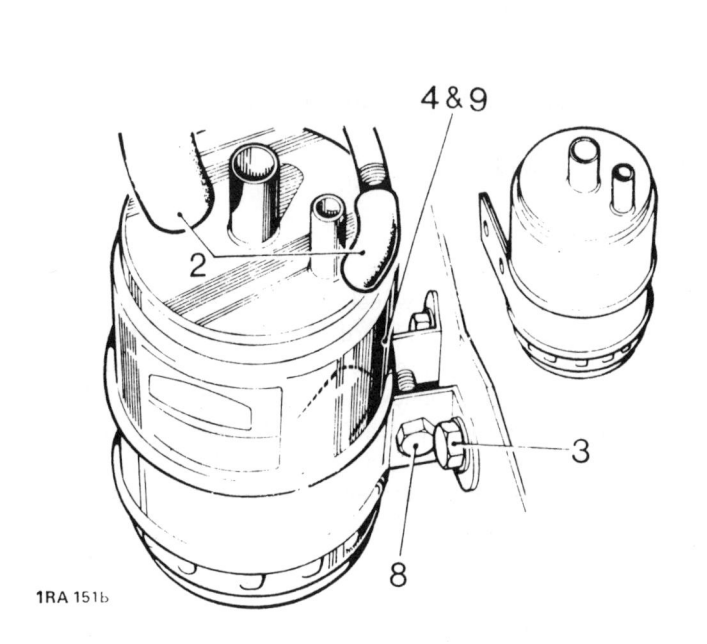

1RA 151b

continued

Rover 3500 and 3500S Manual AKM 3621

17.10.02
17.15.13
Sheet 1

EMISSION CONTROL

5. Unscrew the end cap from the base of the canister and withdraw the filter pad.
6. Fit a new filter pad with the smooth side inward.
7. Using a new seal screw on the end cap.
8. Slacken the pinch bolt on the canister strap.
9. Withdraw the canister from the strap.

Refitting

10. Reverse 8, 9.
11. Reverse 1 to 4.

EXPANSION TANK

—Remove and refit   17.15.19

Removing

1. Open the luggage boot.
2. Remove the spare wheel, if fitted in the boot.
3. Withdraw the trim board from the rear of the boot.
4. Disconnect the two breather pipes from the expansion tank.
5. Remove the four bolts securing the expansion tank to the main tank.
6. Withdraw the expansion tank.

Refitting

7. Reverse 1 to 6, ensuring that the breather pipe connections are in good condition and a good fit.

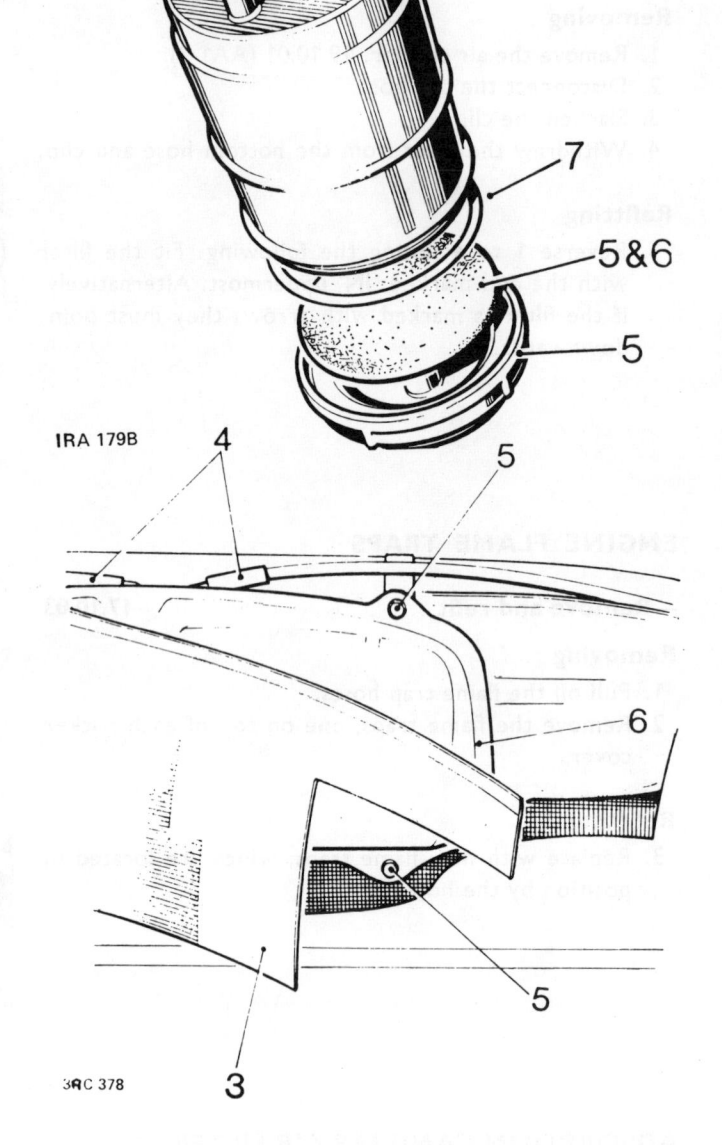

BREATHER PIPE—Expansion tank to filter

—Remove and refit   17.15.32

Removing

1. Open the luggage boot.
2. Remove the spare wheel, if fitted in the boot.
3. Withdraw the trim board from the rear of the boot.
4. To remove the breather pipe clipped around the main fuel tank, remove the main fuel tank. 19.55.01.
5. To remove the pipe leading to the charcoal canister:
   a. Remove the head lining. 76.64.01.
   b. Release the breather pipe from the clips and connections and withdraw.

Refitting

6. Reverse 1 to 5, using MS4 silicone grease on the outside of the pipes to facilitate passing through grommets.

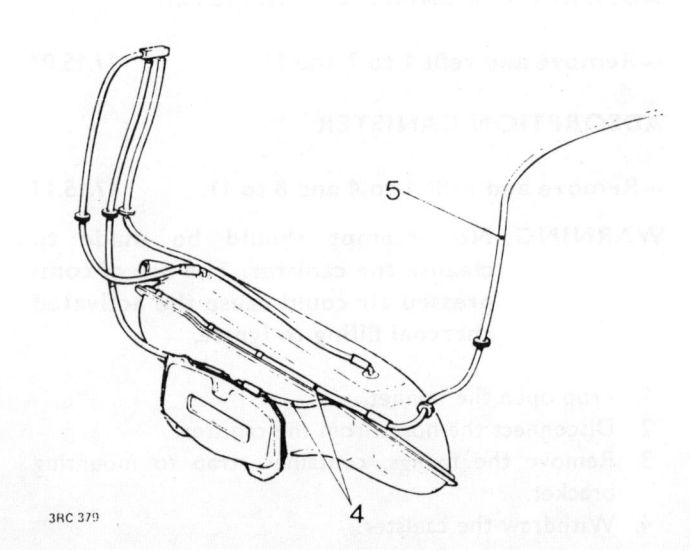

17.15.13
Sheet 2
17.15.32

Rover 3500 and 3500S Manual AKM 3621

FUEL SYSTEM

LIST OF OPERATIONS

Adsorption canister—remove and refit	19.25.10
Air cleaner	
—remove and refit	19.10.01
—renew elements	19.10.08
—overhaul	19.10.14
Automatic enrichment device (A.E.D.)—remove and refit	19.15.38
A.E.D. fuel and air system—remove and refit	19.15.39
Carburetters	
—tune and adjust	19.15.02
—remove and refit	19.15.11
—overhaul and adjust	19.15.18
Choke control cable—remove and refit	19.20.13
Fuel filler lock—remove and refit	19.55.09
Fuel line filter—remove and refit	19.25.01
Fuel pipes	
—main line complete—remove and refit	19.40.01
—spill return—remove and refit	19.40.08
—reserve line—remove and refit	19.40.07
Fuel pump	
—remove and refit	19.45.08
—clean filter	19.45.05
—overhaul	19.45.15
Fuel reserve control cable—remove and refit	19.20.20
Fuel tank—remove and refit	19.55.01
Pipe—fuel tank vent—remove and refit	19.40.35
Reserve tap—remove and refit	19.55.06
Thermostat switch—remove and refit	19.15.50
Throttle linkage—remove and refit	19.20.07
Throttle pedal—remove and refit	19.20.01

Rover 3500 and 3500S Manual AKM 3621

FUEL SYSTEM

AIR CLEANER

—Remove and refit   19.10.01

  Basic type 1, 5 to 9, 14 and 15

  Thermostatic type 1 to 9, 14 and 15

—Renew element 10 to 13   19.10.08

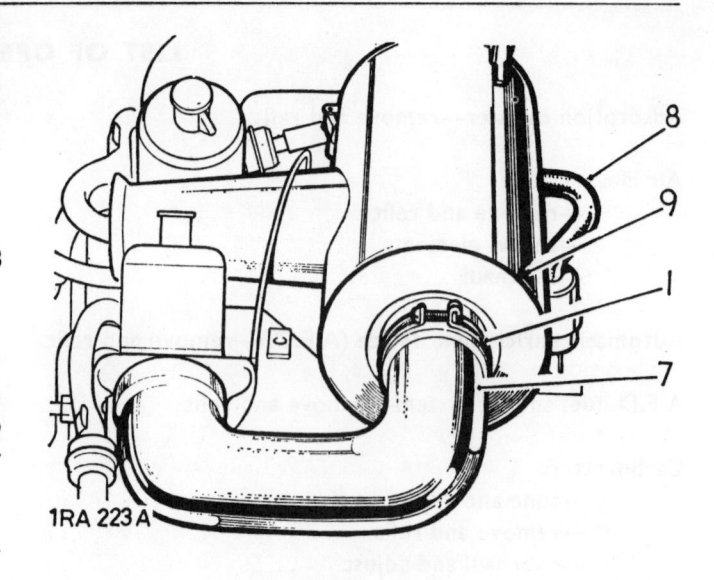

Removing

1. Release the hose clips each side of the air cleaner.
2. Disconnect and blank off the vacuum pipe between the inlet manifold and the thermostatic switch. Also release the flame traps from the front of the air cleaner.
3. Disconnect the hose from the exhaust hot box.
4. Disconnect the hose(s), adsorption canister(s) to air cleaner.
5. Release the choke cable from the clip on the air cleaner, or if an automatic enrichment device is fitted, disconnect the hose from the left hand elbow.
6. Automatic gearbox models. Release the fluid filler tube from the clip on the air cleaner.
7. Withdraw the air cleaner elbows.
8. Disconnect the hose from the engine breather filter.
9. Withdraw the air cleaner body from the mounting pegs on the engine.
10. Air cleaners with detachable plate at both ends. Release the three clips at both ends of the air cleaner, withdraw the end plates and remove the old elements.
11. Air cleaners with detachable plate at one end only. Release the three clips and withdraw the end plate. Remove the old element from the end plate and, using a long screwdriver, release the fixings from the inner element and withdraw it from the end plate side.
12. Fit new sealing washers, as necessary.
13. Fit new paper elements.

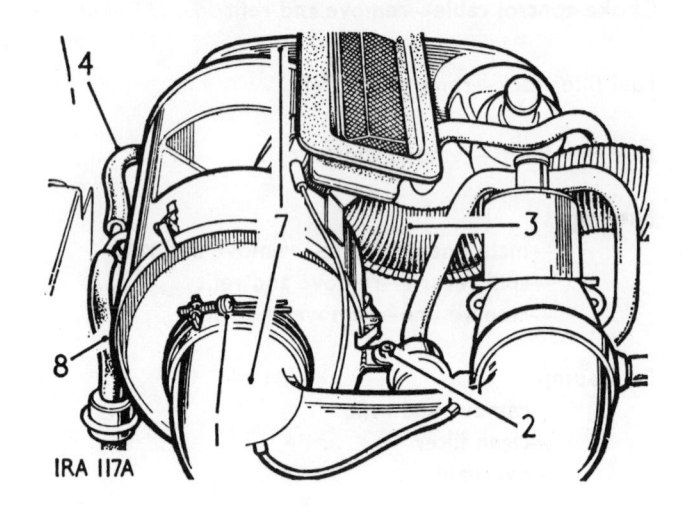

Refitting

14. Smear the 'O'-rings at the carburetter intakes with Silicon MS4 grease.
15. Reverse 1 to 13 as appropriate.

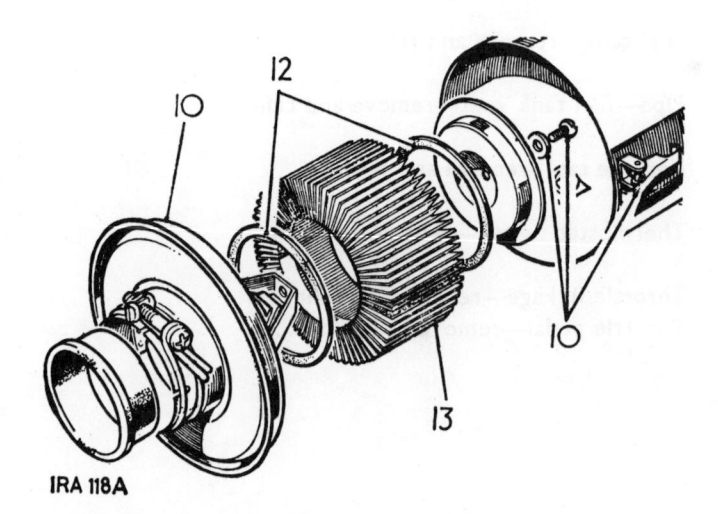

Rover 3500 and 3500S Manual AKM 3621

FUEL SYSTEM—HS6 Carburetter with A.E.D.

AIR CLEANER—Thermostatic type

—Overhaul 19.10.14

Testing in position

1. Check operation of mixing flap valve in air cleaner by starting up engine from cold and observing the flap valve as the engine temperature rises. The valve should start to open slowly within a few minutes of starting and continue to open until a stabilised position is achieved. This position and the speed of operation will be entirely dependent on prevailing ambient conditions. Failure to operate indicates failure of the operating vacuum capsule or the thermostatically controlled vacuum switch or both.
2. Check by connecting a pipe shown in dotted line direct from manifold tapping to operating vacuum capsule. If movement is present with the engine idling the thermostat control vacuum switch is faulty, if no movement is detected the operating capsule is faulty. Replace faulty parts.

Dismantling

3. Remove the air cleaner. 19.10.01.
4. Remove the RH end plate.
5. Withdraw the thermostatic vacuum switch.
6. Remove the air cleaner elements.
7. Drill out the rivets securing the vacuum motor.

Reassembly

8. Fit the vacuum motor and secure with drive screws.
9. Smear the pipe connection adaptors for the thermostatic vacuum switch with Silicon MS4 grease.
10. Reverse 3 to 7. The arrangement of the pipe and adaptor connections for the thermostatic vacuum switch is not of particular importance.

CARBURETTER SU Type HS6 with automatic enrichment device

—Tune and adjust 19.15.02

Service tools: 605330 Carburetter balancer
605927 Carburetter adjustment spanner

The service tool 605330 carburetter balancer must be used to adjust the carburetters. Primarily this instrument is for balancing the air flow through the carburetters, but it also gives a good indication of the mixture setting. Investigation has shown that incorrect mixture setting causes either stalling of the engine or a considerable drop in engine rev/min. If the balancer is fitted when the mixture is too rich or a considerable increase in rev/min when used with the mixture setting too weak. Before balancing the carburetters it is most important therefore that the following procedure be carried out:

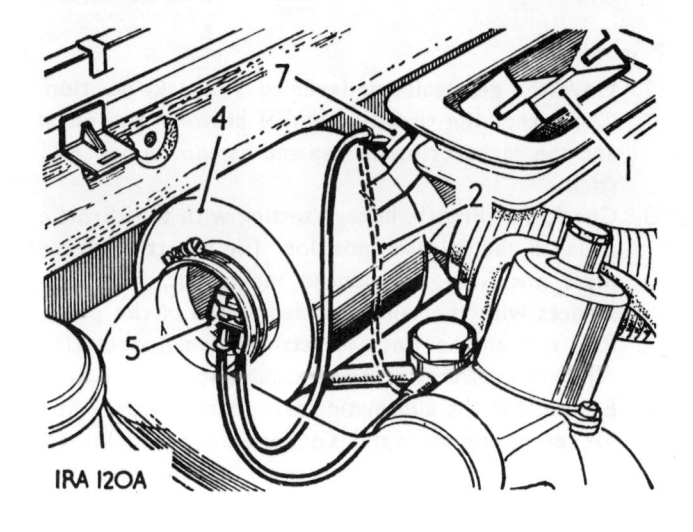

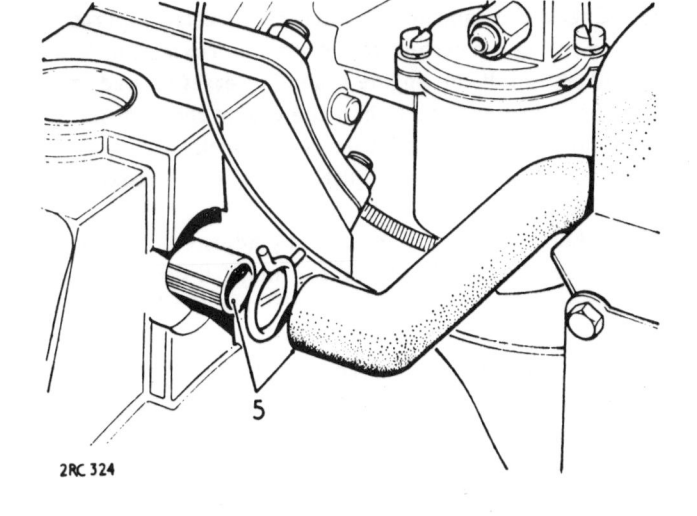

continued

Rover 3500 and 3500S Manual AKM 3621

FUEL SYSTEM—HS6 Carburetter with AED

Procedure

1. Move the gear selector lever to 'P' (park) position.
2. Check that the throttle control between the pedal and the carburetters is free and has no tendency to stick.
3. Check the throttle linkage setting with the throttle pedal in the released position. The throttle linkage must not have commenced movement, but commences with the minimum depression of the pedal.
4. Run the engine until it attains normal operating temperature, that is, thermostat open.
5. Ensure that the automatic enrichment device (AED) has ceased operation and is not enriching the mixture. Check by first noting the engine idling speed, then stop the engine, withdraw the delivery hose from the induction manifold and blank off the manifold pipe. Start the engine, the idling speed should be slightly lower than that previously noted, as there is an air bleed through the AED when connected. If the engine speed drops more than approximately 50 rev/min this indicates a leak in the AED and a new AED should be fitted. Finally, reconnect the delivery hose.

   NOTE: The air cleaner can be pivoted upward temporarily to give access to the rear of the carburetters.

6. Slacken the screws securing the throttle adjusting levers on both carburetters.
7. Start the engine and check the idle speed. If necessary, adjust the throttle stop screws to give an idle speed of 600 to 650 rev/min. 605927.

   NOTE: An independent and accurate tachometer must be used to set the engine idle speed. The vehicle rev/min indicator is not suitable.

8. Check the mixture on each carburetter separately, by lifting the carburetter piston approximately 1 mm (0.040 in). If the engine speed increases immediately, the mixture is too rich. If the engine speed decreases immediately, the mixture is too weak.
9. Screw the mixture adjusting nut in to weaken the mixture, or out to enrich.
10. When the mixture is correctly adjusted, the engine speed will remain constant or may fall slowly a slight amount as the carburetter piston is lifted.
11. Remove the air cleaner. 19.10.01.
12. Check, and if necessary, zero the gauge on tool 605330.
13. Place tool 605330 on to the carburetter adaptors, ensuring that there are no air leaks. If the engine stalls or decreases considerably in speed, the mixture is too rich. If the engine speed increases, the mixture is too weak.
14. If necessary, remove tool 605330 and readjust the mixture, then refit the tool.

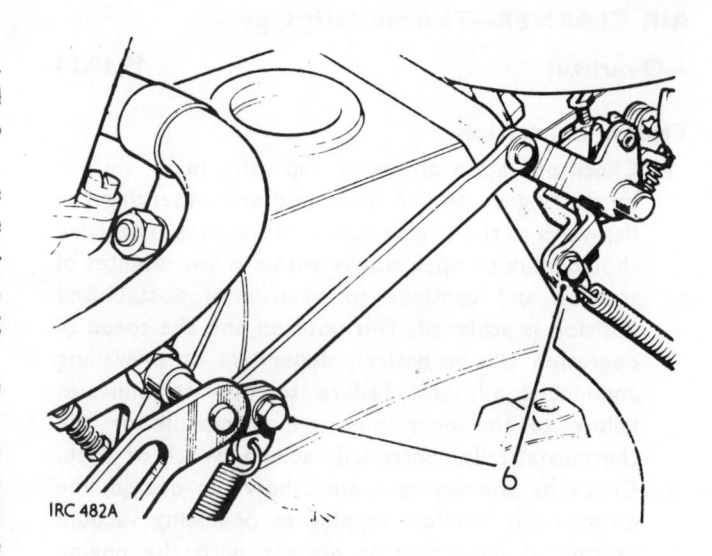

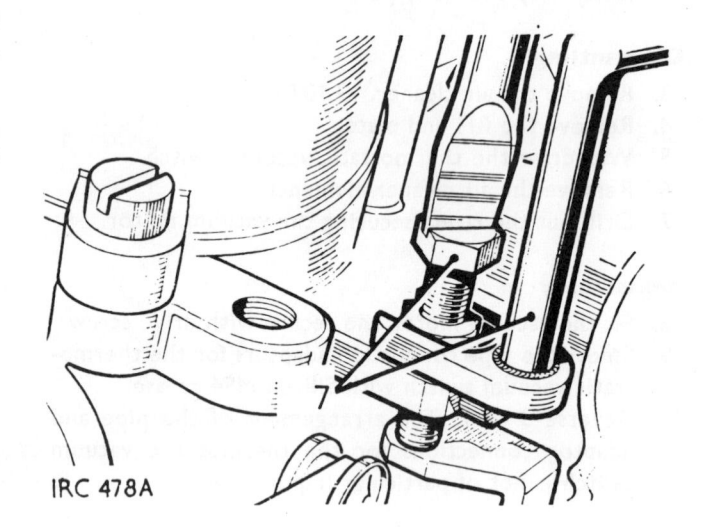

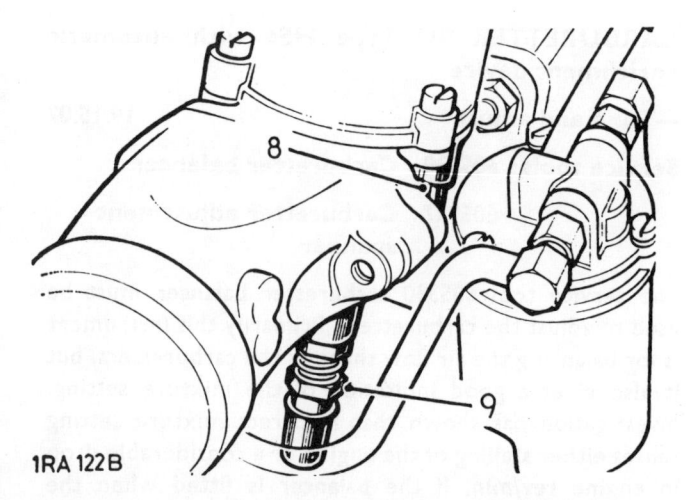

19.15.02
Sheet 2

Rover 3500 and 3500S Manual AKM 3621

FUEL SYSTEM—HS6 Carburetter with Manual Choke

15. Check tool 605330 gauge reading.
16. If the gauge pointer is in the 'ZERO' sector at the correct engine idle speed of 600 to 650 rev/min, no adjustment is required.
17. If the engine idle speed rises too high or drops too low during balancing, adjust to the correct idle speed of 600 to 650 rev/min, maintaining the gauge pointer in the 'ZERO' sector.
18. If the gauge pointer moves to the right, decrease the air flow through the left hand carburetter by unscrewing the throttle stop screw or increase the air flow through the right hand carburetter by screwing in the throttle stop screw. Reverse the procedure if the pointer moves to the left.
19. Remove tool 605330. With the mixture setting and carburetter balance correctly adjusted the difference in engine rev/min with the tool 605330 on or off will be negligible, approximately plus or minus 25 rev/min.
20. On the left hand carburetter, place a 0,15 mm (0.006 in) feeler between the roller on the counter shaft lever and the throttle lever.
21. Apply pressure to the throttle lever to hold the feeler.
22. Tighten the screw to secure the throttle adjusting lever, then withdraw the feeler.
23. On the right hand carburetter, place a 0,15 mm (0.006 in) feeler between the right leg of the fork on the adjusting lever and the pin on the throttle lever.
24. Apply pressure to the linkage to hold the feeler.
25. Tighten the screw to secure the throttle adjusting lever, then withdraw the feeler.
26. Check the downshift cable setting. 44.30.02.
27. Fit the air cleaner. 19.10.01.
28. Road test the car. During test it may be necessary to slightly readjust the mixture setting.

DATA
Engine idle speed   600 to 650 rev/min.

CARBURETTER—SU Type HS 6 with manual choke

—Tune and adjust   19.15.02

Service tools 605330 Carburetter balancer

605927 Carburetter adjustment spanner

The service tool 605330 carburetter balancer must be used to adjust the carburetters. Primarily this instrument is for balancing the air flow through the carburetters, but it also gives a good indication of the mixture setting. Investigation has shown that incorrect mixture setting causes either stalling of the engine or a considerable drop in engine rev/min. if the balancer is fitted when the mixture is too rich or a considerable increase in rev/min. when used with the mixture setting too weak. Before balancing the carburetters it is most important therefore that the following procedure be carried out:

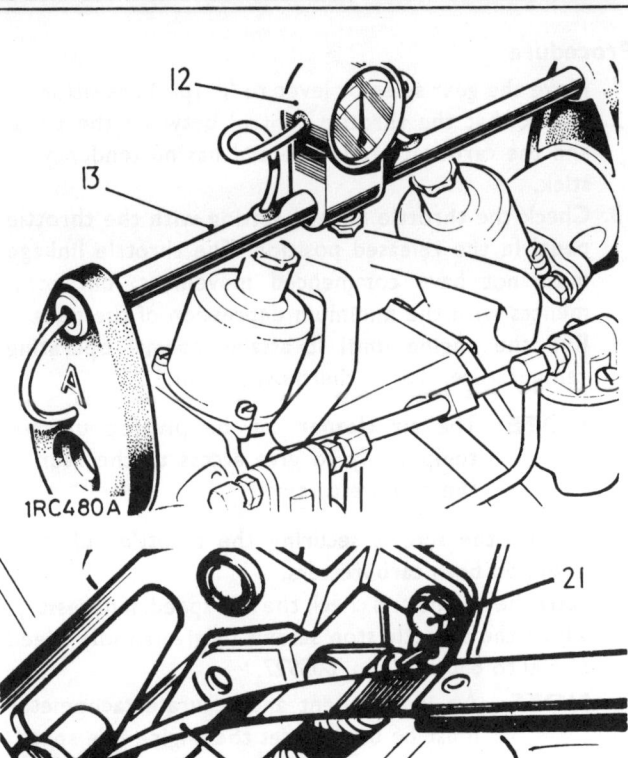

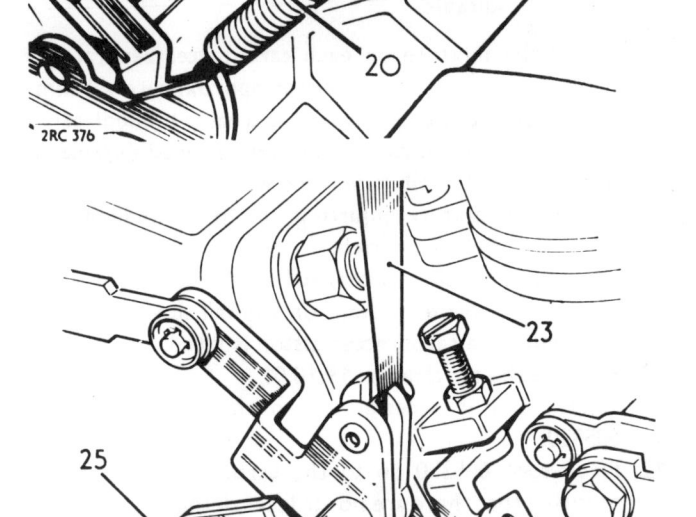

continued

Rover 3500 and 3500S Manual AKM 3621

FUEL SYSTEM—HS6 Carburetter with Manual Choke

Procedure

1. Move the gear selector lever to 'P' (park) position.
2. Check that the throttle control between the pedal and the carburetters is free and has no tendency to stick.
3. Check the throttle linkage setting with the throttle pedal in the released position. The throttle linkage must not have commenced movement, but commences with the minimum depression of the pedal.
4. Run the engine until it attains normal operating temperature, that is, thermostat open.

   NOTE: The air cleaner can be pivoted upward temporarily to give access to the rear of the carburetters.

5. Slacken the screws securing the throttle adjusting levers on both carburetters.
6. Start the engine and check the idle speed. If necessary, adjust the throttle stop screws to give an idle speed of 600 to 650 rev/min. 605927.

   NOTE: An independent and accurate tachometer must be used to set the engine idle speed. The vehicle rev/min indicator is not suitable.

7. Check the mixture on each carburetter separately, by lifting the carburetter piston approximately 1 mm (0.040 in). If the engine speed increases immediately, the mixture is too rich. If the engine speed decreases immediately, the mixture is too weak.
8. Screw the mixture adjusting nut in to weaken the mixture, or out to enrich.
9. When the mixture is correctly adjusted, the engine speed will remain constant or may fall slowly a slight amount as the carburetter piston is lifted.
10. Remove the air cleaner. 19.10.01.
11. Check, and if necessary, zero the gauge on tool 605330.
12. Place tool 605330 on to the carburetter adaptors, ensuring that there are no air leaks. If the engine stalls or decreases considerably in speed, the mixture is too rich. If the engine speed increases, the mixture is too weak.
13. If necessary, remove tool 605330 and readjust the mixture, then refit the tool.
14. Check tool 605330 gauge reading.
15. If the gauge pointer is in the 'ZERO' sector at the correct engine idle speed of 600 to 650 rev/min, no adjustment is required.
16. If the engine idle speed rises too high or drops too low during balancing, adjust to the correct idle speed of 600 to 650 rev/min., maintaining the gauge pointer in the 'ZERO' sector.
17. If the gauge pointer moves to the right, decrease the air flow through the left hand carburetter by unscrewing the throttle stop screw or increase the air flow through the right hand carburetter by screwing in the throttle stop screw. Reverse the procedure if the pointer moves to the left.

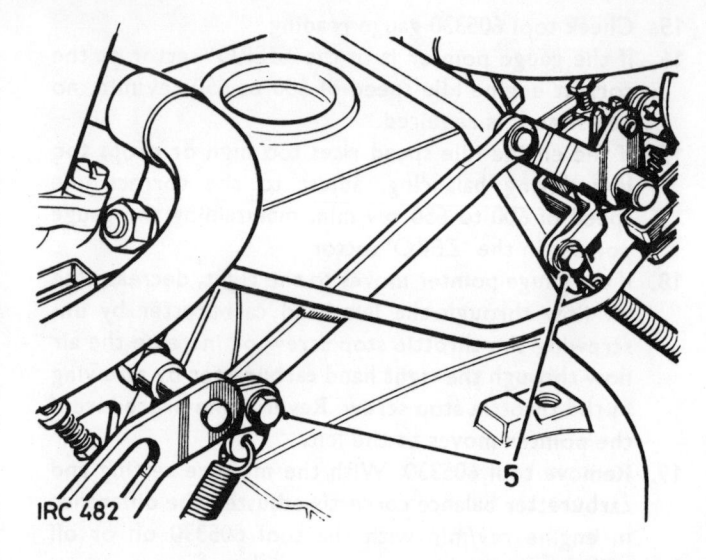

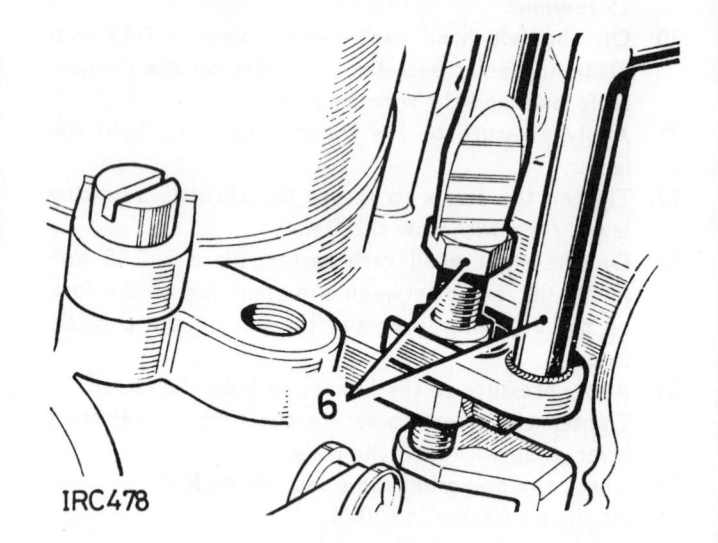

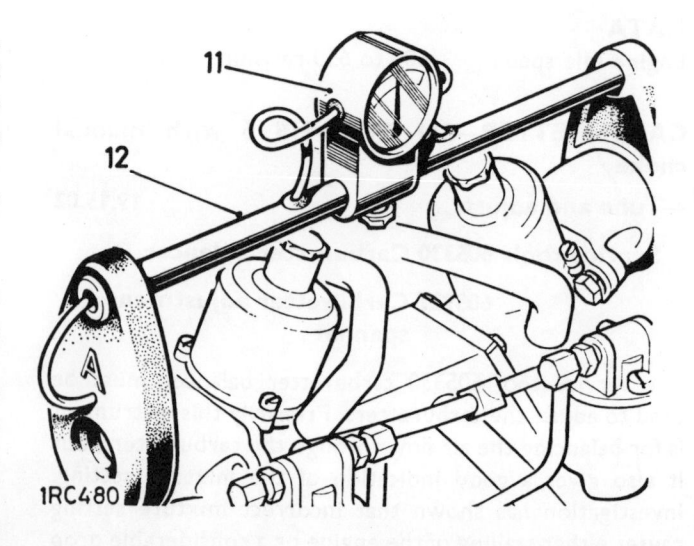

continued

19.15.02
Sheet 4

Rover 3500 and 3500S Manual AKM 3621

FUEL SYSTEM—HS6 Carburetter with Manual Choke

18. Remove tool 605330. With the mixture setting and carburetter balance correctly adjusted the difference in engine rev/min. with the tool 605330 on or off will be negligible, approximately plus or minus 25 rev/min.
19. Stop the engine.
20. On the right hand carburetter, place a 0,15 mm (0.006 in) feeler between the right leg of the fork on the adjusting lever and the pin on the throttle lever.
21. Apply light pressure to the linkage to hold the feeler.
22. Tighten the screw to secure the throttle adjusting lever, then withdraw the feeler.
23. On the left hand carburetter, apply pressure to the throttle lever to hold it lightly against the roller on the countershaft lever.
24. Tighten the screw to secure the throttle adjusting lever, then release the pressure on the throttle lever.
25. **IMPORTANT:** Check and if necessary, adjust the downshift cable. 44.30.02.

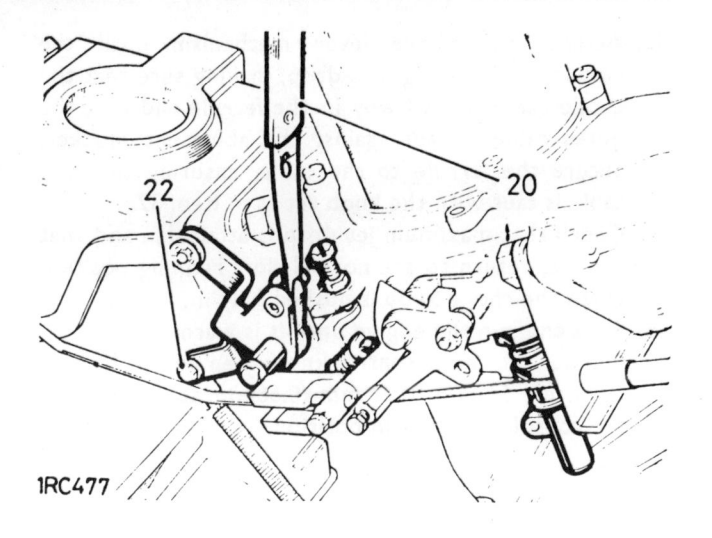

Manual choke linkage, check and adjust

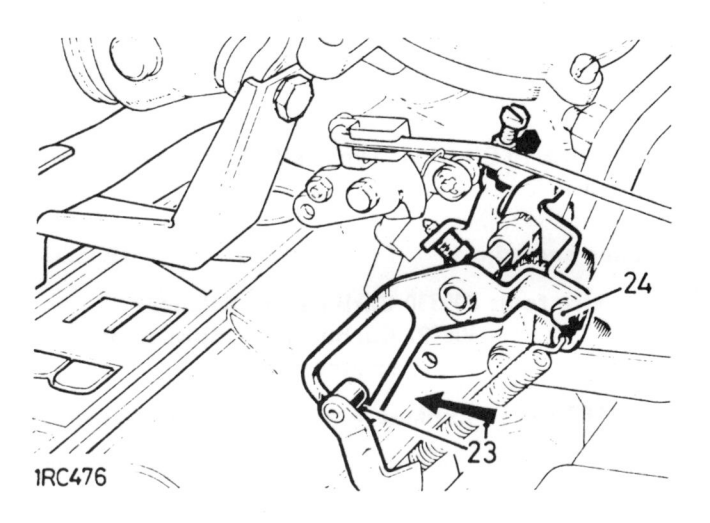

26. Slacken the ferrule screw securing the choke inner cable.
27. Check that the interconnecting rod operates the cam levers so that the carburetter jets will drop to their full extent.
28. Slacken the trunnion screw.
29. Operate the left hand choke lever by hand, noting that the right hand lever still operates with the trunnion screw loose.
30. Adjust the position of the interconnecting rod relative to the right hand lever so that both carburetter jets begin to drop at the same time when the mechanism is operated.
31. Secure the interconnecting rod to the right hand lever with the trunnion screw.
32. Pull out the choke control knob 14 mm to 16 mm (0.562 in to 0.625 in) from the console.

continued

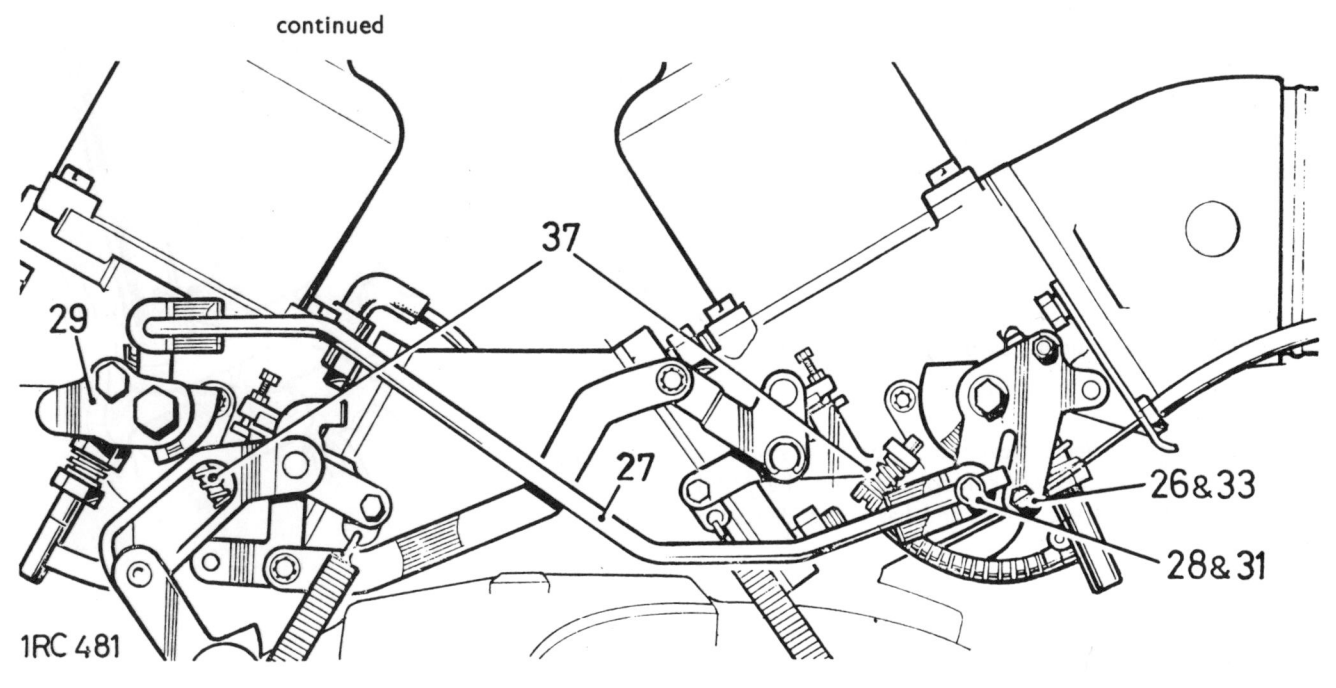

Rover 3500 and 3500S Manual AKM 3621

19.15.02
Sheet 5

FUEL SYSTEM—HS6 Carburetter with Emission Control

33. Actuate the choke lever mechanism until the carburetter jets begin to drop, making sure that the choke cable slides freely in the ferrule and that the outer cable is taut against the abutment bracket. Secure the ferrule to the cable, ensuring that the cable is taut with the knob set as in item 32.
34. Check that maximum jet drop is attainable and that the jets return to the normal slow running position when the choke knob is pushed fully in.
35. Start and run the engine until it is warm.
36. Pull out the choke control knob, then push it back in to within 14 mm to 16 mm (0.562 in. to 0.625 in.) of the console and lock in position.
37. Unscrew both fast idle adjustment screws until they are just clear of their cams, then screw them in evenly until the engine runs at 1200 to 1400 rev/min.
38. Lubricate the trunnion bearing and cam spindles with an oil based graphite or molybdenum disulphide mixture.
39. Fit the air cleaner. 19.10.01.
40. Road test the car. During test it may be necessary to slightly readjust the mixture setting.

DATA

Engine idle speed — 600 to 650 rev/min
Fast idle engine speed — 1200 to 1400 rev/min

CARBURETTER—SU Type HS 6 with emission control

—Tune and adjust   19.15.02

Service tools 605330 Carburetter balancer
              605927 Carburetter adjustment spanner

IMPORTANT. Exhaust emission control system.

The carburetters are fitted with locking devices on the mixture adjusting jet nuts. These locking devices permit approximately 3 flats total movement of the jet nut, and are so set at the factory as to be on the maximum richness which will comply with the regulations. This means that the mixture may only be weakened. Should the amount of adjustment available be insufficient to obtain reasonable idling conditions, the only permissible course of action is to overhaul or fit replacement carburetters. Carburetter overhaul is restricted to the instructions given in 19.15.18.

Under no circumstances must the mixture locking devices be disturbed, as this would almost certainly result in the vehicle failing to meet with legal requirements in respect of air pollution.

The idle adjuster screws also have locknuts fitted to prevent ready unauthorised adjustment or movement due to vibration. It is permissible to unlock and adjust these screws. It is important, however, that following adjustments the locknuts be securely re-tightened.

Procedure

1. Move the gear selector lever to 'P' (park) position.
2. Check that the throttle control between the pedal and the carburetters is free and has no tendency to stick.

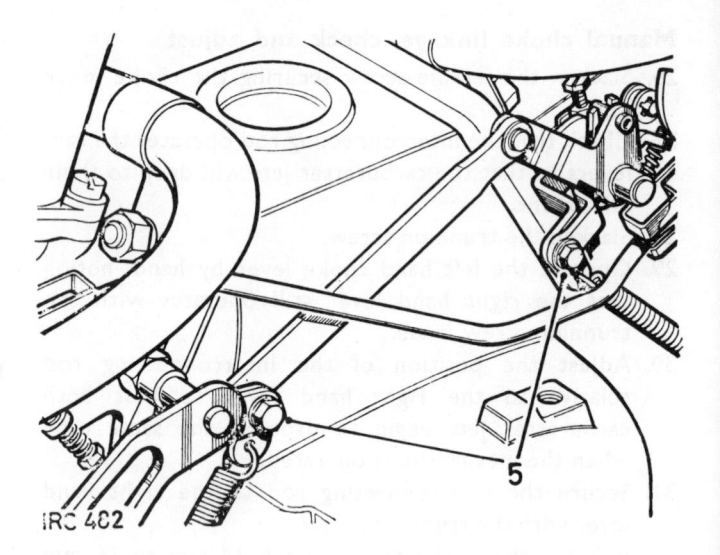

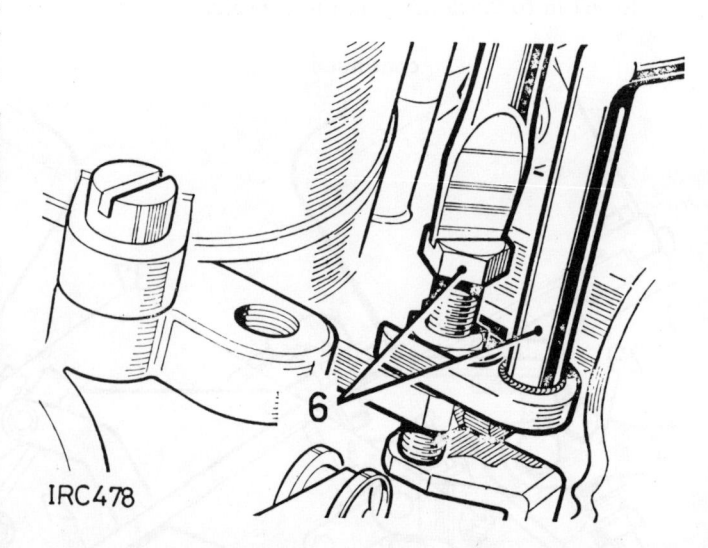

continued

19.15.02
Sheet 6

Rover 3500 and 3500S Manual AKM 3621

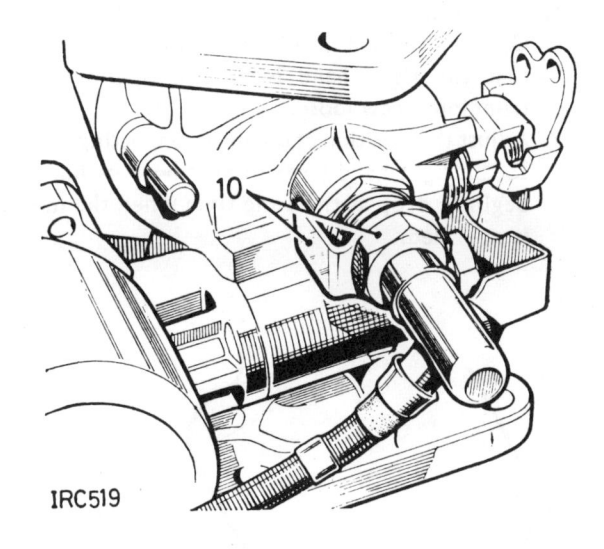

FUEL SYSTEM—HS6 Carburetter with Emission Control

3. Check the throttle linkage setting with the throttle pedal in the released position. The throttle linkage must not have commenced movement, but commences with the minimum depression of the pedal.
4. Run the engine until it attains normal operating temperature, that is, thermostat open.

    NOTE: The air cleaner can be pivoted upward temporarily to give access to the carburetters.

5. Slacken the screws securing the throttle adjusting levers on both carburetters.
6. Start the engine and check the idle speed. If necessary, adjust the throttle stop screws to give an idle speed of 700 to 750 rev/min. 605927.

    NOTE: An independent and accurate tachometer must be used to set the engine idle speed. The vehicle rev/min indicator is not suitable.

7. Remove the air cleaner. 19.10.01
8. Check, and if necessary, zero the gauge on tool 605330.
9. Place tool 605330 on to the carburetter adaptors, ensuring that there are no air leaks. If the engine stalls or decreases considerably in speed, the mixture is too rich. If the engine speed increases, the mixture is too weak.
10. If necessary, remove tool 605330 and readjust the mixture within the bounds of the locking devices on the mixture adjustment nuts, then refit the tool.
11. Check tool 605330 gauge reading.
12. If the gauge pointer is in the 'ZERO' sector at the correct engine idle speed of 700 to 750 rev/min, no adjustment is required.
13. If the engine idle speed rises too high or drops too low during balancing, adjust to the correct idle speed of 700 to 750 rev/min, maintaining the gauge pointer in the 'ZERO' sector.
14. If the gauge pointer moves to the right, decrease the air flow through the left hand carburetter by unscrewing the throttle stop screw or increase the air flow through the right hand carburetter by screwing in the throttle stop screw. Reverse the procedure if the pointer moves to the left.
15. Remove tool 605330. With the mixture setting and carburetter balance correctly adjusted the difference in engine rev/min with the tool 605330 on or off will be negligible, approximately plus or minus 25 rev/min.
16. Stop the engine.
17. On the right hand carburetter, place a 0,15 mm (0.006 in) feeler between the right leg of the fork on the adjusting lever and the pin on the throttle lever.
18. Apply light pressure to the linkage to hold the feeler.

continued

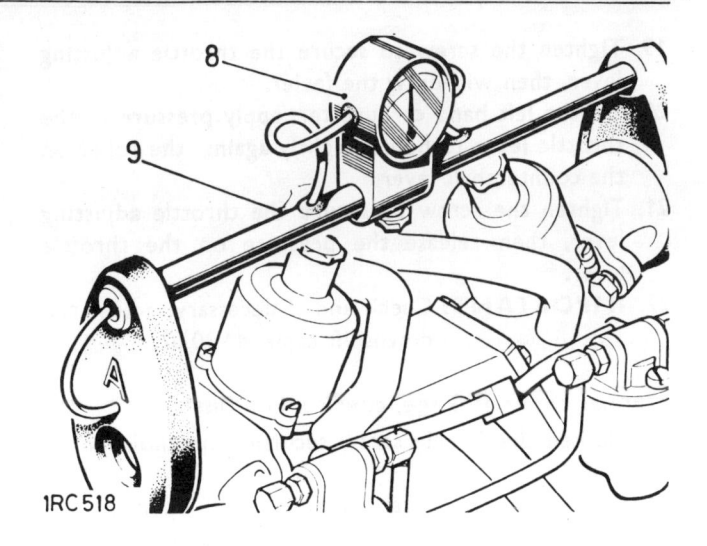

Rover 3500 and 3500S Manual AKM 3621

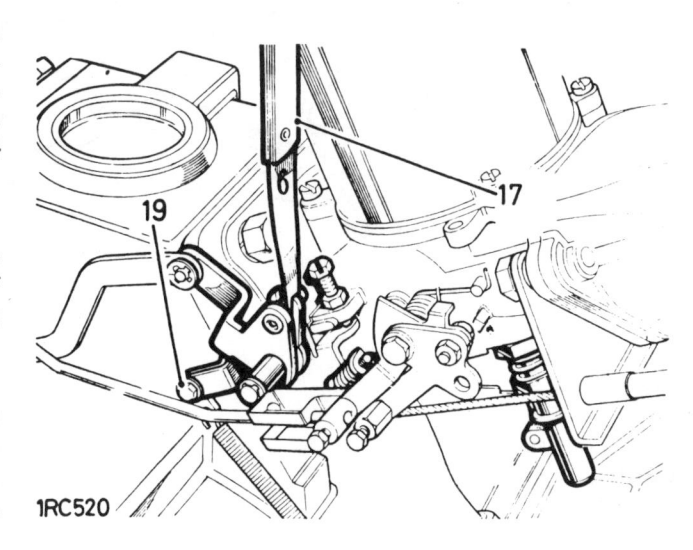

19.15.02
Sheet 7

FUEL SYSTEM—HS6 Carburetter with Emission Control

19. Tighten the screw to secure the throttle adjusting lever, then withdraw the feeler.
20. On the left hand carburetter, apply pressure to the throttle lever to hold it lightly against the roller on the countershaft lever.
21. Tighten the screw to secure the throttle adjusting lever, then release the pressure on the throttle lever.
22. **IMPORTANT:** Check and if necessary, adjust the downshift cable. 44.30.02.

Manual choke linkage, check and adjust

23. Slacken the ferrule screw securing the choke inner cable.
24. Check that the interconnecting rod operates the cam levers so that the carburetter jets drop to their full extent.
25. Slacken the trunnion screw.
26. Operate the left hand choke lever by hand, noting that the right hand lever still operates with the trunnion screw loose.
27. Adjust the position of the interconnecting rod relative to the right hand lever so that both carburetter jets begin to drop at the same time when the mechanism is operated.
28. Secure the interconnecting rod to the right hand lever with the trunnion screw.
29. Pull out the choke control knob 14 mm to 16 mm (0.562 in to 0.625 in) from the console.
30. Actuate the choke lever mechanism until the carburetter jets begin to drop, making sure that the choke cable slides freely in the ferrule and that the outer cable is taut against the abutment bracket. Secure the ferrule to the cable, ensuring that the cable is taut with the knob set as in item 29.
31. Check that maximum jet drop is attainable and that the jets return to the normal slow running position when the choke knob is pushed fully in.
32. Start and run the engine until it is warm.
33. Pull out the choke control knob, then push it back in to within 14 mm to 16 mm (0.562 in to 0.625 in) of the console and lock in position.
34. Unscrew both fast idle adjustment screws until they are just clear of their cams, then screw them in evenly until the engine runs at 1200 to 1400 rev/min.
35. Lubricate the trunnion bearing and cam spindles with an oil based graphite or molybdenum disulphide mixture.
36. Fit the air cleaner. 19.10.01
37. Road test the car. During test it may be necessary to slightly readjust the mixture setting.

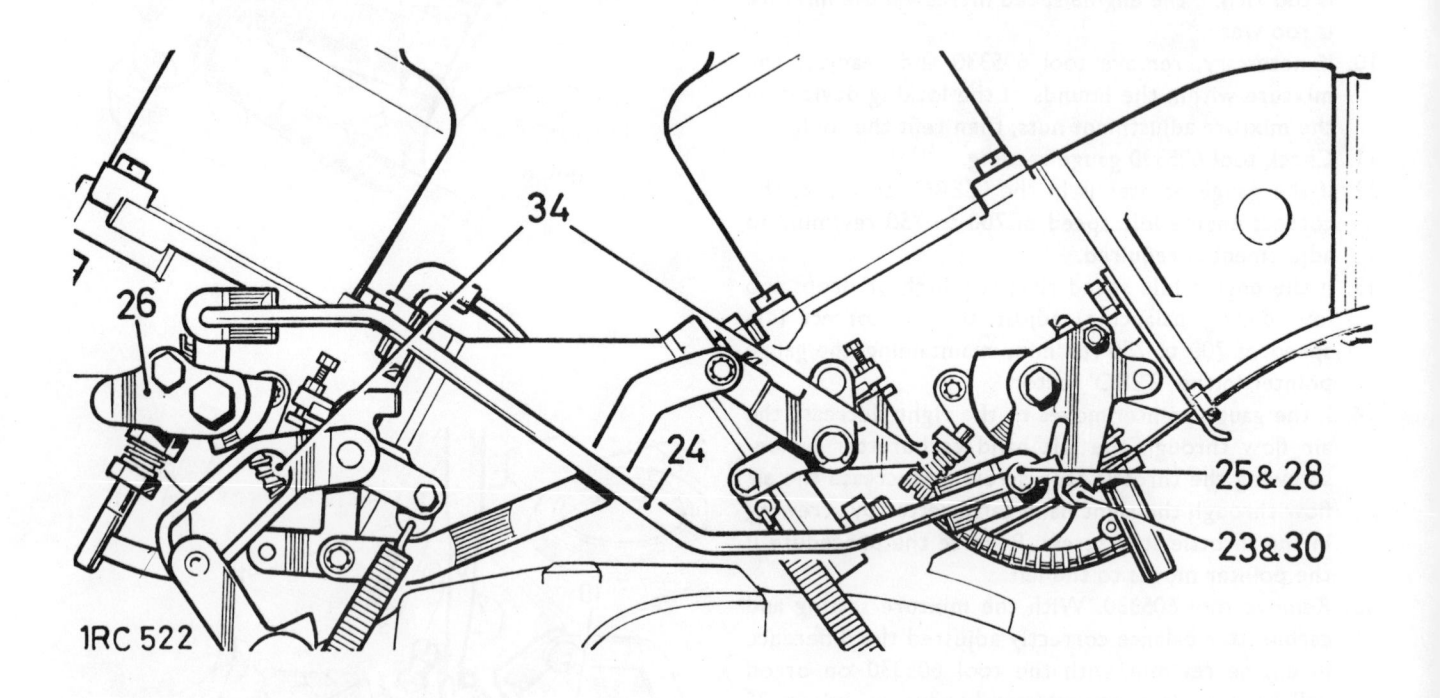

DATA

Engine idle speed   700 to 750 rev/min
Fast idle engine speed   1200 to 1400 rev/min

FUEL SYSTEM—HIF6 Carburetter

CARBURETTER—SU Type HIF 6

(Standard and emission controlled)

—Tune and adjust   19.15.02

Service tools 605330 Carburetter balancer
                605927 Carburetter adjustment spanner

NOTE:
Carburetter mixture ratio is pre-set and sealed and should not normally be interfered with. The only adjustments that should normally be carried out are engine idle speed and fast idle speed. In circumstances where satisfactory carburation cannot be obtained or where the emission control does not meet the regulations, the mixture can be checked and reset as necessary in accordance with the following procedure, using an exhaust gas analyser.

Accurate engine speed is essential during carburetter adjustments, therefore, the contact breaker dwell angle, ignition timing and automatic ignition advance mechanism, should all be checked and reset if necessary before commencing carburetter adjustments.

Engine idle speed:

10.5:1	Compression ratio	600 to 650 rev/min
	Emission controlled	700 to 750 rev/min
8.5:1	Compression ratio	600 to 650 rev/min
	Emission controlled	700 to 750 rev/min
9.25:1	Compression ratio	700 to 750 rev/min

(All are emission controlled)
Air conditioned models   700 to 750 rev/min
Fast idle speed for all models 1100 to 1200 rev/min

When checking engine speed, use an independent and accurate tachometer. The rev/min indicator fitted to the car is not suitable.

General requirements when setting carburetters

Temperature: Whenever possible the ambient air temperature of the setting environment should be between 15° to 26°C (60° to 80°F).

Vehicle conditions. Idling adjustments should be carried out on a fully warmed up engine, that is, at least 5 minutes after the thermostat has opened. This should be followed by a run of one minute duration at an engine speed of approximately 2,500 rev/min in neutral, after which three minutes may be taken in which to check and carry out adjustments; a further one minute run at 2500 rev/min must be made before further checks or adjustments are carried out. This cycle may be repeated as often as required. It is important that the above cycle is adhered to, otherwise, overheating may result and settings may be incorrect.

Before any attempt is made to check settings a thorough check should be carried out to see that the throttle linkage between the pedal and carburetters is free and has no tendency to stick. Ensure that the choke control knob is pushed fully in. If at any time the carburetter linkage is adjusted on automatic gearbox models, it must be followed by a check of the downshift cable. 44.30.02. This will enable any disturbance of downshift cable adjustment to be corrected. Incorrect downshift cable adjustment will give wrong fluid pressure and can cause automatic gearbox clutch failure. Any faults must be corrected before proceeding with the checks or settings.

continued

Rover 3500 and 3500S Manual AKM 3621

FUEL SYSTEM—HIF6 Carburetter

Procedure

1. Move the gear selector to the 'P' (Park) position on automatic gearbox models.
2. Run the engine until warm. See note concerning general requirements when setting carburetters.
3. Switch off engine and remove air cleaner.
4. Slacken the screws securing the throttle lever to the carburetter lever on each carburetter, thus allowing individual adjustment of carburetters.
5. Start the engine.

Engine idle speed adjustment

NOTE: Emission controlled models—Just before the readings of the tachometer and exhaust gas analyser are taken, gently tap the neck of the carburetter suction chamber with a light non-metalic instrument, (e.g. a screwdriver handle).

6. Use special spanner and slacken off carburetter idle adjusting screw lock nuts 605927.
7. Adjust idle screws by equal amounts to give a speed of 600 to 650 rev/min or 700 to 750 rev/min as applicable.
8. When both carburetters have been adjusted, tighten the idle screw lock nuts.

Fast idle speed adjustment

9. Pull the mixture control until the mark on the fast idle cam is opposite the centre line of the fast idle screw.
10. Using the special spanner, slacken the lock nut 605927
11. Adjust the fast idle screw to give an engine speed of between 1100 and 1200 rev/min for both standard and emission controlled vehicles
12. Balance carburetter air flow as follows:
    Check and if necessary zero the gauge, by means of the adjustment screw on the carburetter balancing device. Place the balancer on to the carburetter adaptors, ensuring that there are no air leaks.
    Note the reading on the gauge; if the pointer is in the zero sector of the gauge, no adjustment is required. If the needle moves to the right, decrease the air flow through the left-hand carburetter by unscrewing the idle adjusting screw or increase the air flow through the right-hand carburetter by screwing in the idle adjusting screw. Reverse the procedure if the needle moves to the left.
13. Should the idling speed rise too high or drop too low during balancing, adjust to the correct idle speed maintaining the gauge needle in the zero sector. With the carburetter balance correctly adjusted, the difference in engine speed with the balancer on or off will be negligible, approximately plus or minus 25 rev/min.
    If there is a considerable change of engine speed this indicates incorrect carburation; in which case, the mixture can be checked and if necessary readjusted, providing that an exhaust gas analyser is used and, that the regulations are strictly adhered to.

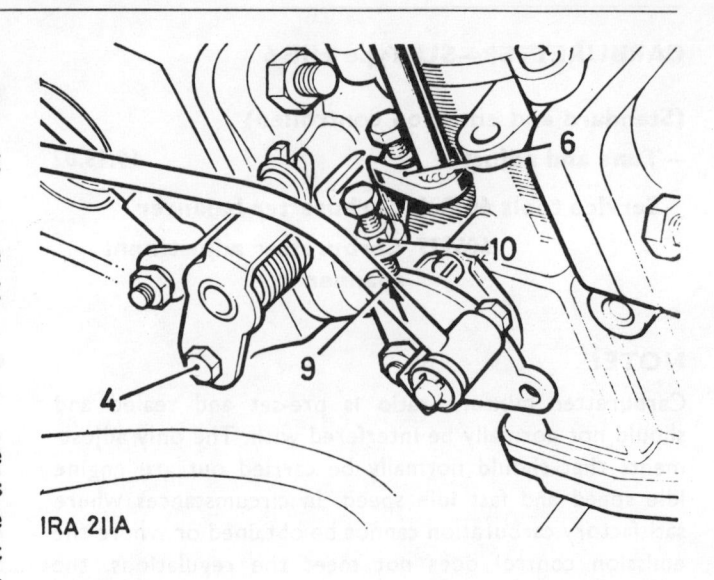

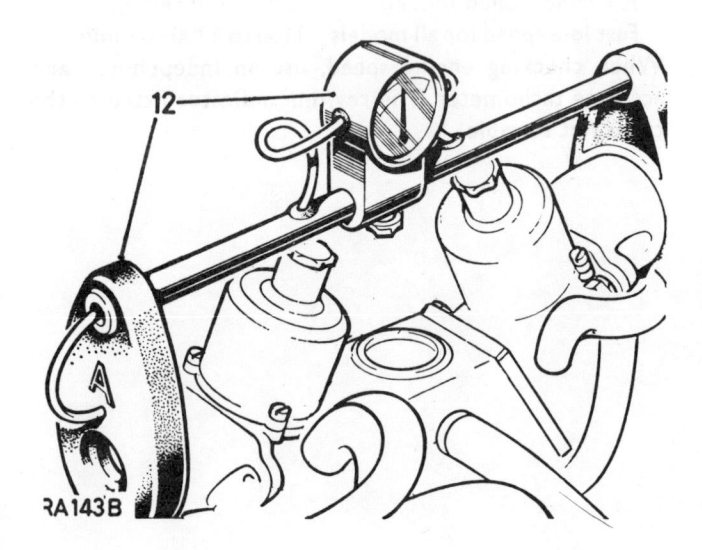

continued

FUEL SYSTEM—HIF6 Carburetter

14. On the right-hand carburetter place a 0,15 mm (0.006 in) feeler between the right leg of the fork on the adjusting lever and the pin on the throttle lever.
15. Apply light pressure to the linkage to hold the feeler, then tighten the throttle lever securing screws.
16. Switch off engine and replace the air cleaner.
17. 3500 models—Check the downshift cable setting. 44.30.02.

Carburetter mixture setting

Do not adjust the carburetter mixture unnecessarily, as this is pre-set and sealed for emission control purposes. If faulty carburation is suspected, mixture adjustment should only be attempted after all other possible factors have been eliminated, then proceed as follows:

18. Set the jet datum position on each carburetter, as follows:
    a. Remove the air cleaner elbow.
    b. Remove the sealing plug from the carburetter jet adjustment screw.
    c. Turn the jet adjusting screw until the jet is below the level of the carburetter bridge.
    d. Lift the carburetter piston by hand and insert a steel rule, approximately 13 mm (0.500 in) wide, alongside the needle in a vertical plane.
    e. Turn the jet adjusting screw until the jet just contacts the steel rule. This will accurately position the jet level with the carburetter bridge, then remove the steel rule and screw the jet adjusting screw in 2½ turns.
19. Refit the air cleaner elbows.
20. Start the engine and allow it to warm up, as described under 'General requirements'.
21. Check, and if necessary adjust, the engine idle speed to the previously stated figure.

NOTE: Just before the readings of the tachometer and exhaust gas analyser are taken, gently tap the neck of the carburetter suction chamber with a light, non-metalic instrument (e.g. a screwdriver handle).

22. Connect a suitable exhaust gas analyser to the vehicle exhaust and allow to stabilise for a minimum of one minute, then check the CO emission.
23. If necessary, turn the jet adjustment screws, out to weaken, in to enrich, to obtain a suitable mixture within the CO emission regulations.
24. Fit new sealing plugs to the jet adjustment screws.
25. Repeat 1 to 8 and 14 to 17.

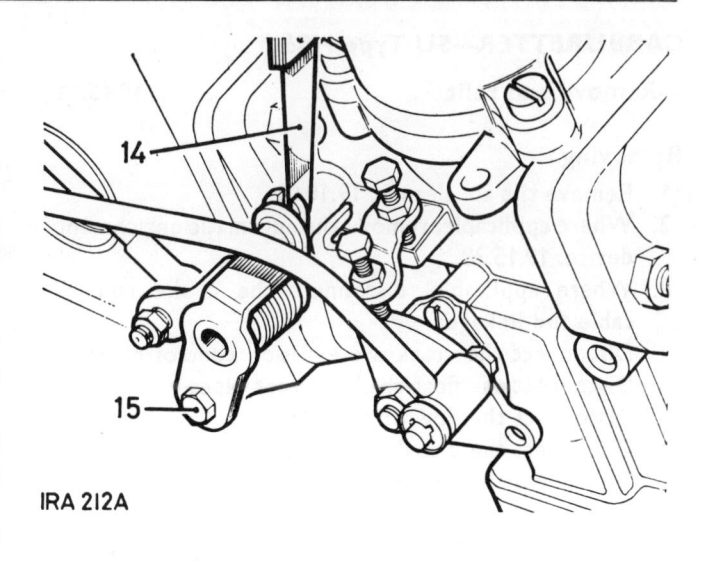

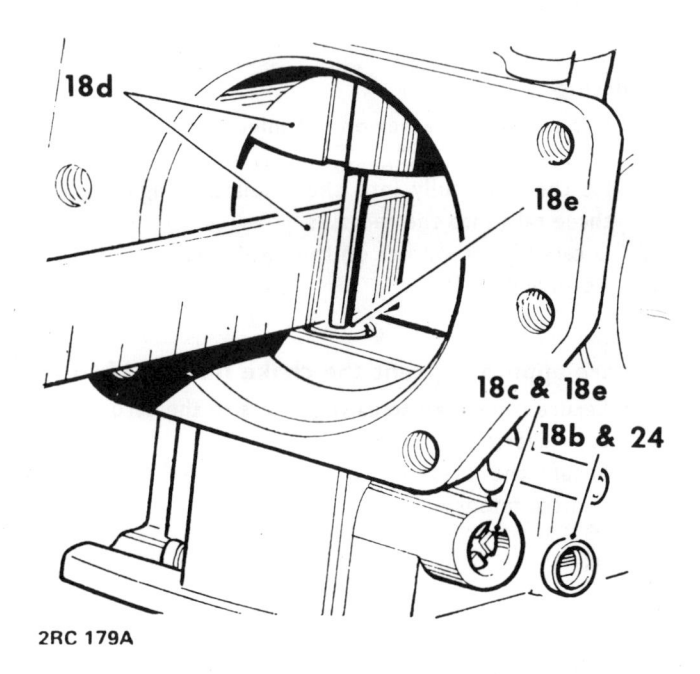

IRA 212A

2RC 179A

Rover 3500 and 3500S Manual AKM 3621

19.15.02
Sheet 11

FUEL SYSTEM—HS6 Carburetter

CARBURETTER—SU Type HS6

—Remove and refit 19.15.11

Removing

1. Remove the air cleaner. 19.10.01.
2. Where applicable, remove the automatic enrichment device. 19.15.38.
3. Where applicable, disconnect the choke control cable and linkage.
4. Disconnect the crankcase emission control pipes.
5. Disconnect the float chamber vent pipes.
6. Disconnect the downshift cable.
7. Disconnect the distributor vacuum pipe.
8. Disconnect the throttle linkage and return springs.
9. Disconnect the fuel pipes.
10. Remove the carburetters.
11. If required, withdraw the joint washers, insulator and liner.

Refitting

12. Locate a joint washer on the inlet manifold.
13. Fit the insulator, aligning the arrows.
14. Fit the liner fully into the insulator, engaging the three tabs into the recesses.
15. Locate a joint washer on the insulator.
16. Reverse 4 to 10.

Where applicable, refit the choke linkage 17 to 20.

17. Secure the choke actuating levers to the carburetter cam levers.
18. Secure the interconnecting rod, using the two clips and trunnions.
19. Check that the assembled rod will operate the cam levers so that the carburetter jets will drop to their full extent.
20. Lubricate the trunnion bearing and cam spindles with an oil based graphite or molybdenum-disulphide mixture.
21. Reverse 2 and 3.
22. Check and, if necessary, replenish the carburetter damper reservoirs with SAE 20 oil to within approximately 12 mm (0.500 in) of the top of the hollow piston rod.
23. Tune and adjust the carburetters. 19.15.02.
24. Refit the air cleaner. 19.10.01.
25. Check the downshift cable setting. 44.30.02.

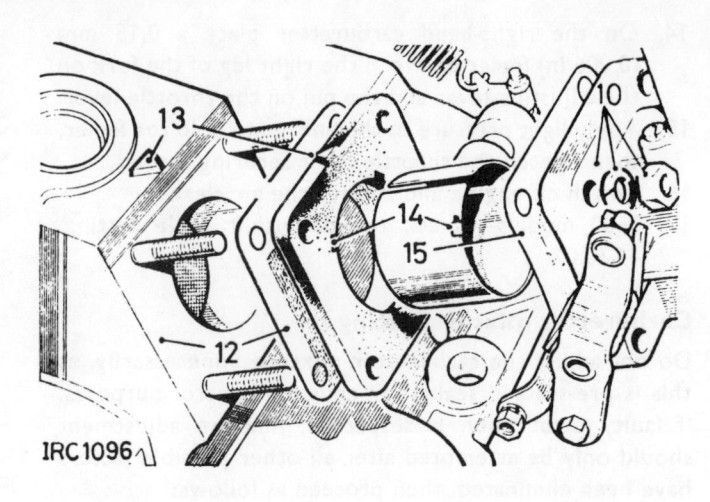

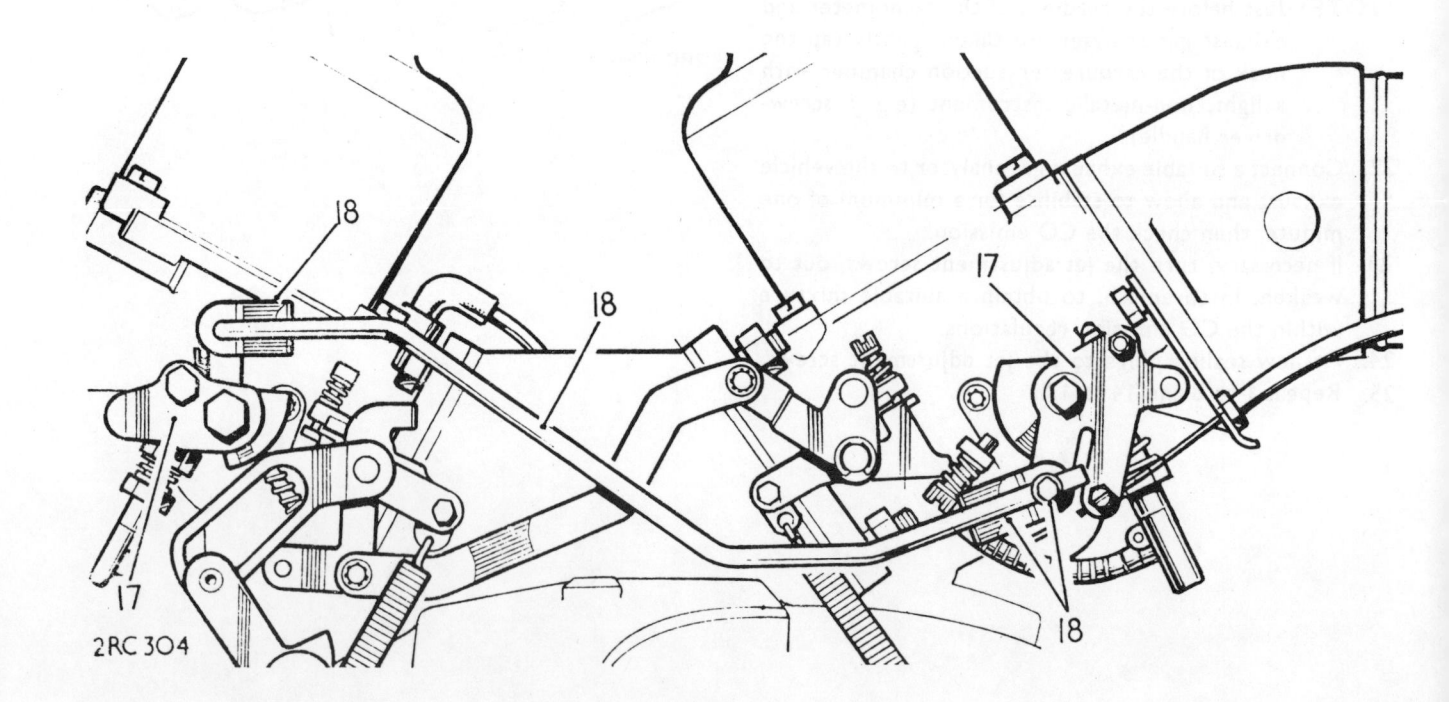

19.15.11
Sheet 1

Rover 3500 and 3500S Manual AKM 3621

FUEL SYSTEM—HIF6 Carburetter

CARBURETTERS—SU Type HIF 6

—Remove and refit 19.15.11

Removing

1. Remove the air cleaner. 19.10.01.
2. Disconnect the crankcase emission control pipes.
3. Disconnect the float chamber vent pipes.
4. Disconnect the fuel pipes.
5. Disconnect the distributor vacuum pipe.
6. Disconnect the choke control cable.
7. Disconnect the choke link rod.
8. Disconnect the throttle vertical link rod.
9. Disconnect the throttle link rod.
10. Remove the carburetters.
11. If required, withdraw the joint washers, insulator and liner.

Refitting

12. Locate a joint washer on the inlet manifold.
13. Fit the insulator, aligning the arrows.
14. Fit the liner fully into the insulator, engaging the three tabs into the recesses.
15. Locate a joint washer on the insulator.
16. Reverse 8 to 10.
17. Assemble the choke link rod to both carburetters leaving the trunnion bolt loose.
18. Ensure that both choke cams are in the closed stop position, then secure the link rod trunnion bolt.
19. Fit the choke cable, outer casing to the carburetter body with the spring clip, and inner cable through the trunnion on the left-hand choke cam, secure the trunnion bolt with the choke in the closed position.
20. Reverse 2 to 5.
21. Check and, if necessary, replenish the carburetter damper reservoirs with SAE 20 oil to within approximately 25 mm (1 in) of the top of the tube, then replace the hydraulic dampers and caps.
22. Tune and adjust the carburetters. 19.15.02.

   NOTE: If difficulty is experienced in obtaining the correct throttle linkage setting, check the linkage as detailed in 19.20.07.

23. Refit the air cleaner. 19.10.01.
24. 3500 models—Check the downshift cable setting. 44.30.02.

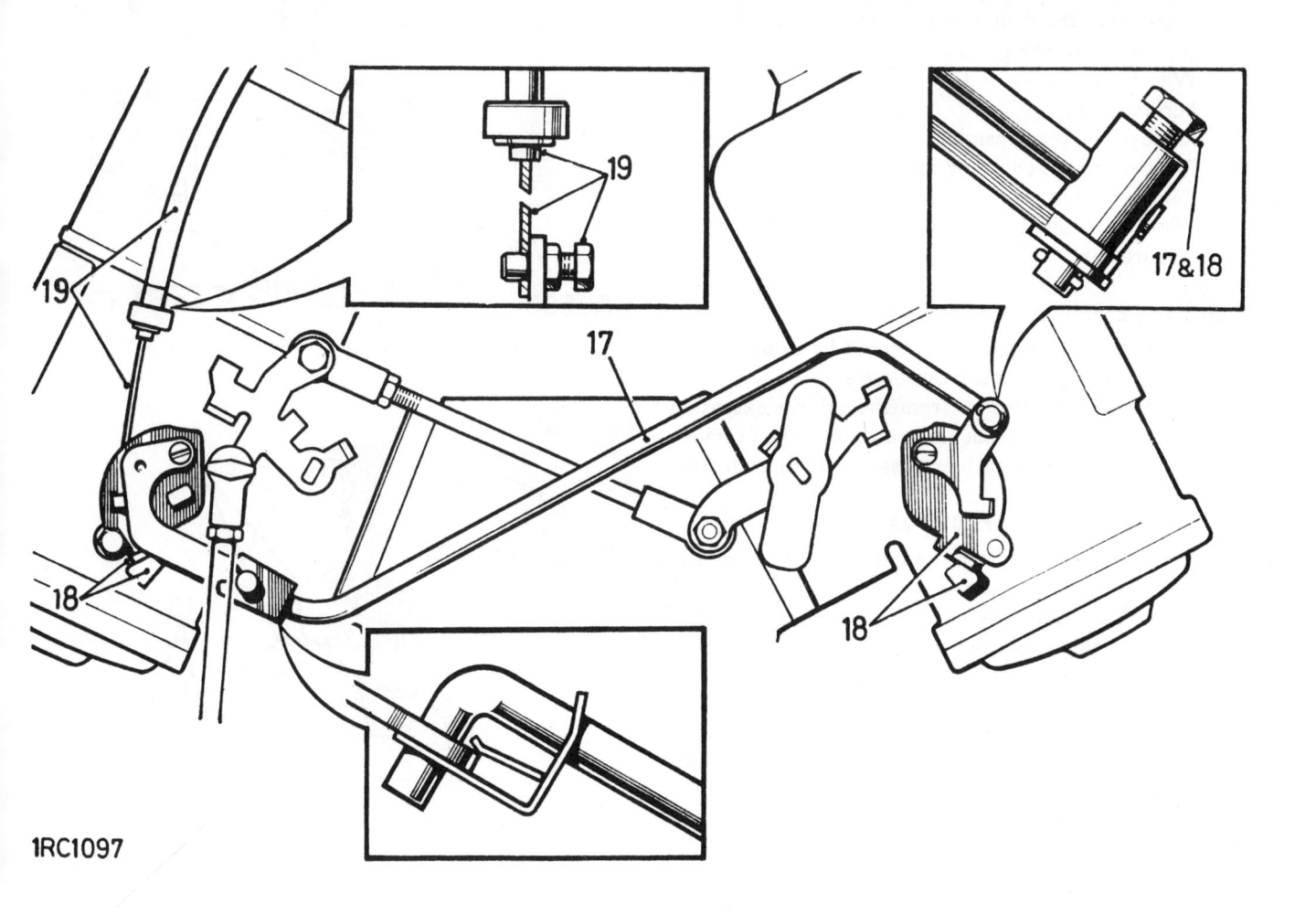

Rover 3500 and 3500S Manual AKM 3621

FUEL SYSTEM—HS6 Carburetter

CARBURETTER—SU Type HS6

—Overhaul and adjust   19.15.18

NOTE: For instructions covering carburetters with emission control, see separate Operation of same number, sheet 4.

Dismantling

1. Remove the carburetters. 19.15.11.
2. Remove the oil cap and damper.
3. Remove the suction chamber and spring.
4. Withdraw the piston and needle assembly.
5. If required, slacken the needle retaining screw at the side of the piston and withdraw the needle.
6. Remove the float chamber lid and gasket.
7. Withdraw the hinge pin and remove the float.
8. Remove the needle valve and seat.
9. Relieve the load on the pick-up lever return spring by applying pressure to the cam lever and remove the Phillips drive screw from the base of the jet head assembly and release the pick-up link.
10. Remove the bolt and washer securing the pick-up link assembly, at the same time releasing the return spring.
11. Withdraw the pick-up link assembly.
12. Disconnect the flexible pipe from the float chamber and withdraw the gland washer and metal washer.
13. Withdraw the jet and flexible pipe from the carburetter body.
14. Remove the float chamber.
15. Remove the jet adjusting nut and spring.
16. Remove the jet bearing nut, bearing and washer.
17. Disconnect the throttle link from the right hand carburetter.

Left hand carburetter 18 and 19

18. Remove the throttle lever secured by a bolt, spring and brass washer to the carburetter lever.
19. Release the tab washer from the spindle extension nut and remove the nut and tab washer, then remove the carburetter lever, brass washer and throttle link.

Right hand carburetter 20 to 23

20. Remove the circlip from the spindle and draw off the right hand throttle lever.
21. Remove the adjusting bolt, spring and plain washer from the adjusting lever and draw off.
22. Release the tab washer from the spindle extension nut and remove the nut and tab washer.
23. Remove the carburetter lever and brass washer.
24. Close the ends and remove the two special screws retaining the throttle butterfly.
25. Withdraw the butterfly.
26. Withdraw the butterfly spindle.
27. If required, remove the piston lift pin.

continued

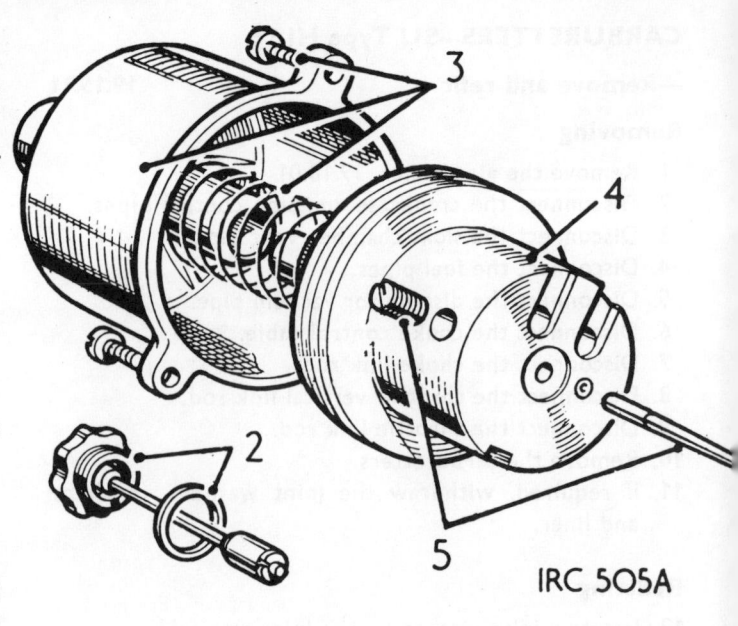

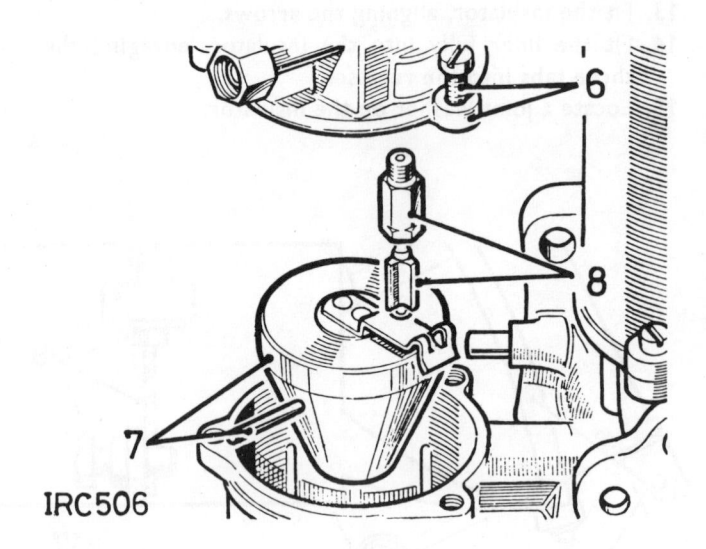

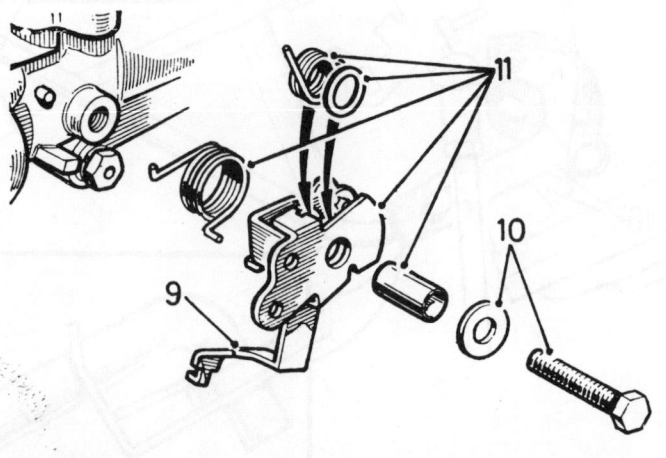

19.15.18
Sheet 1

Rover 3500 and 3500S Manual AKM 3621

FUEL SYSTEM—HS6 Carburetter

Inspecting

28. Check the throttle spindle and carburetter body for wear.
29. Check all levers and linkage for excessive free movement, indicating wear.
30. Clean the inside of the suction chamber and the outside diameters of the piston with a clean cloth moistened with petrol.
31. Inspect the jet and jet bearing for signs of wear.
32. Obtain new replacements as necessary and ensure that all components are thoroughly clean in preparation for assembly.

Reassembling

33. If previously removed, position the spring over the lift pin and locate into the carburetter body. Fit the neoprene washer, brass washer and secure with the circlip.
34. Insert the spindle into the carburetter body with the plain end on the float chamber side.
35. Slide the butterfly through the spindle so that the top of the butterfly will move away from the engine when opening the throttle and the chamfered edge locates with the bore when closed.
36. Fit two new screws, but do not fully tighten at this stage.
37. Place the brass washer for throttle spindle on to the threaded end of the spindle ensuring that it is located on the larger diameter of the spindle which will just be protruding from the carburetter body.
38. Fit the carburetter lever and, on the left hand carburetter, the throttle link, locating it correctly on the threaded portion of the spindle, then fit the tab washer and extension nut.
39. Tighten the extension nut to secure the lever to the spindle and lock with the tab washer.
40. With the throttle butterfly held in the fully closed position, apply light pressure to the spindle at the end with the extension nut to bring the lever into contact with the brass washer. Hold this position and fully tighten the two screws fixing the throttle butterfly to the spindle and lock them by opening the slots.
41. **Left hand carburetter.** Refit the throttle lever to carburetter lever with bolt, spring and plain brass washer. Do not tighten at this stage.

Right hand carburetter 42 to 44

42. Refit adjusting lever, and secure with adjusting bolt, spring and plain washer. Do not tighten at this stage.
43. Refit the right hand throttle lever to extension, and secure to spindle with circlip.
44. Refit throttle link to throttle lever, and secure with the circlip.
45. Locate the jet bearing, short end first, into the carburetter body followed by the brass washer and bearing nut. Do not tighten at this stage.
46. Insert the jet into the jet bearing.

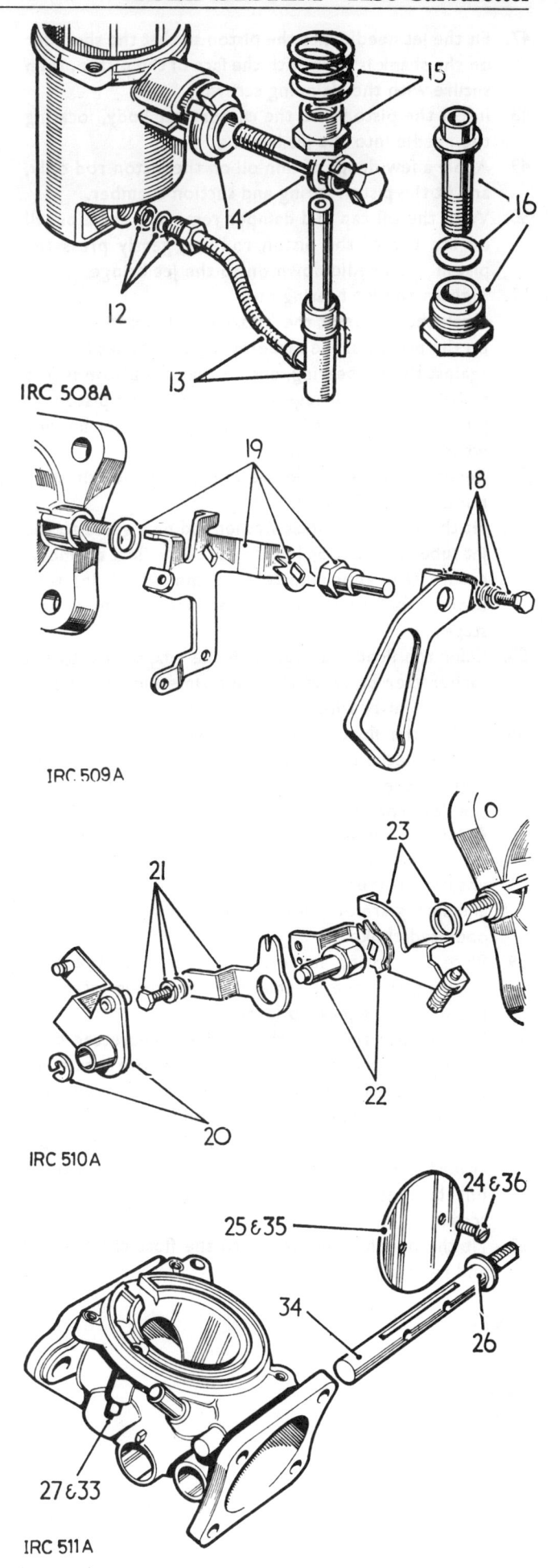

Rover 3500 and 3500S Manual AKM 3621

19.15.18
Sheet 2

FUEL SYSTEM—HS6 Carburetter

47. Fit the jet needle into the piston so that the shoulder on the shank is flush with the face of the piston, then secure with the clamping screw.
48. Insert the piston into the carburetter body, locating the needle into the jet bearing.
49. Apply a few drops of thin oil on the piston rod only, and fit the piston spring and suction chamber.
50. With the oil cap and damper removed, use a pencil on the top of the piston rod and gently press the piston and needle down on to the jet bridge.
51. Tighten the jet bearing nut.
52. With the piston at the bottom of its travel (that is, on the bridge of the carburetter) and the jet hard up against the jet bearing, ensure that the piston is free to fall on to the bridge when lifted by the piston lift pin and that the jet is not binding in the jet bearing when moved in and out.
53. Remove the jet and locate the spring and adjusting nut on to the bearing.
54. Fit the gland nut, brass washer and new gland to the jet tube and position in the base of the float chamber, ensure that approximately 4,7 mm (0.187 in) tube projects before fitting, but do not tighten at this stage.
55. Offer the float chamber with the adaptor on to the carburetter body, at the same time locating the jet into the jet bearing.
56. Secure the float chamber to the carburetter body with a special bolt, plain washer and spring washer.
57. Tighten the gland nut on the jet pipe at the base of the float chamber.
58. Assemble the cam lever, pick-up lever and link, cam lever spring, distance tube and shim washer, ensuring correct sequence. Note that the cam lever and spring are only fitted to models with a manually operated choke.
59. Place the spring for pick-up lever over the bolt, ensuring that the outer end of the spring locates in the slot in the pick-up lever.
60. Now secure this assembly to the carburetter body and using a suitable piece of hooked wire locate the inner end of the pick-up lever spring over the projection on the carburetter body.
61. Turn the cam lever in a clockwise direction and hold against spring pressure while securing the pick-up link bracket to the jet head, with the self-tapping screw.
62. Fit the needle valve seat into the float chamber lid and replace the valve.

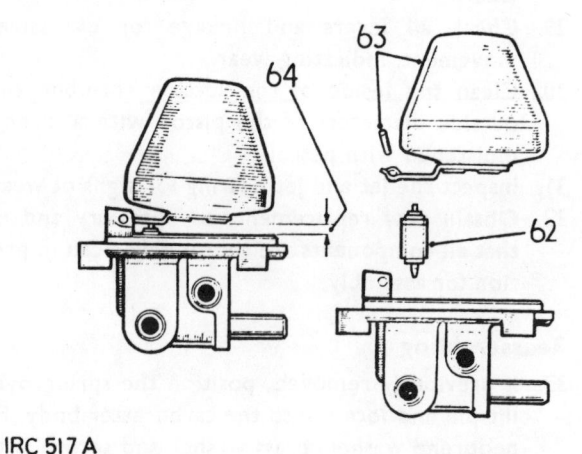

IRC 517 A

63. Fit the float to the float chamber lid and secure with the hinge pin.
64. A check should be made to ensure that the float level will be correct. With the float chamber lid inverted and the needle valve in the shut off position there should be a 3 mm to 4,5 mm (0.125 in. to 0.187 in) gap between the end of the float lever and the rim of the float chamber lid. If necessary, adjust by bending the float lever.

   NOTE: Where the float and lever are integral and not riveted (some later models), measure between the lid register and the nearest point on the float. This type of float and lever is not adjustable for height setting. A float setting out of limits indicates a damaged or faulty part which must be identified and replaced.

65. Fit the lid complete with a new gasket to the float chamber and secure with the three screws and spring washers.
66. Fill the damper reservoir in the suction piston with SAE 20 engine oil to within 12 mm (0.500 in) of the top of the hollow piston rod.
    Fit the damper and washer.
    Refit the carburetters. 19.15.11.
    Tune and adjust the carburetters. 19.15.02.

DATA

Carburetter

Type	SU HS6
Number	Two
Bore	44,45 mm (1.75 in)
Needle	K.O.
Jet size	2,54 mm (0.100 in)
Float level	3,0 mm to 4,5 mm (0.125 to 0.187 in)
Damper oil	SAE 20

19.15.18
Sheet 3

Rover 3500 and 3500S Manual AKM 3621

FUEL SYSTEM—HS6 Carburetter with Emission Control

CARBURETTER—SU Type HS 6 with emission control

—Overhaul and adjust 19.15.18

IMPORTANT

The carburetter jet height setting is determined using a flow test rig and unless similar facilities are available and approved by the carburetter manufacturer, the jet height must not be reset in service. For this reason, the jet adjustment nut is fitted with a locking device which limits the amount of mixture adjustment available. The height setting is locked at a point giving the maximum richness which will comply with the emission control regulations, and the small amount of adjustment permitted by the locking device can only weaken the mixture.

The carburetter may be dismantled for cleaning and replacement of components with the exception of the jet bearing, jet adjustment nut and locking device.

If satisfactory carburation is not achieved after overhaul and tuning, fit a new carburetter.

Dismantling

1. Remove the carburetters. 19.15.11.
2. Remove the oil cap and damper.
3. Remove the suction chamber and spring.
4. Withdraw the piston and needle assembly.
5. If required, slacken the needle retaining screw at the side of the piston and withdraw the needle, bias sleeve and spring.
6. Remove the float chamber lid and gasket.
7. Withdraw the hinge pin and remove the float.
8. Remove the needle valve and seat.
9. Relieve the load on the pick-up lever return spring by applying pressure to the cam lever and remove the Phillips drive screw from the base of the jet head assembly and release the pick-up link.
10. Remove the bolt and washer securing the pick-up link assembly, at the same time releasing the return spring.
11. Withdraw the pick-up link assembly.
12. Disconnect the flexible pipe from the float chamber and withdraw the gland washer and metal washer.
13. Withdraw the jet and flexible pipe from the carburetter body.

    CAUTION: Do not remove the jet bearing assembly.

14. Remove the float chamber.
15. Disconnect the throttle link from the right hand carburetter.

Left hand carburetter 16 and 17

16. Remove the throttle lever secured by a bolt, spring and brass washer to the carburetter lever.
17. Release the tab washer from the spindle extension nut and remove the nut and tab washer, then remove the carburetter lever, brass washer and throttle link.

continued

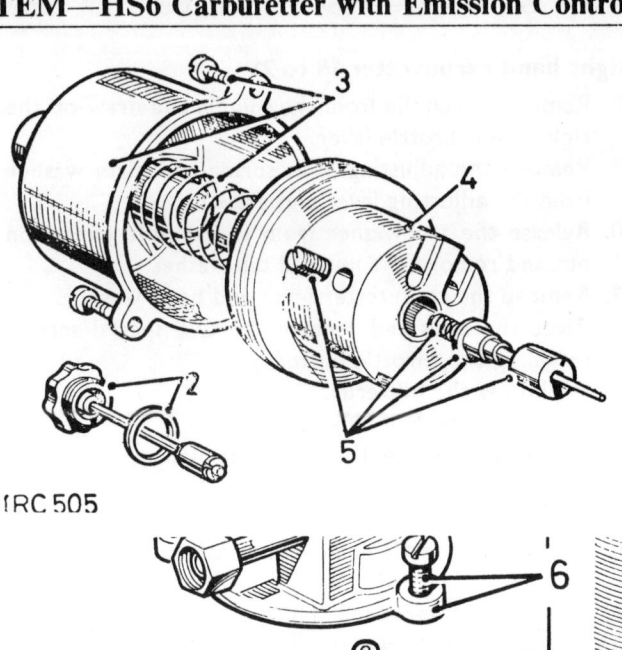

IRC 505

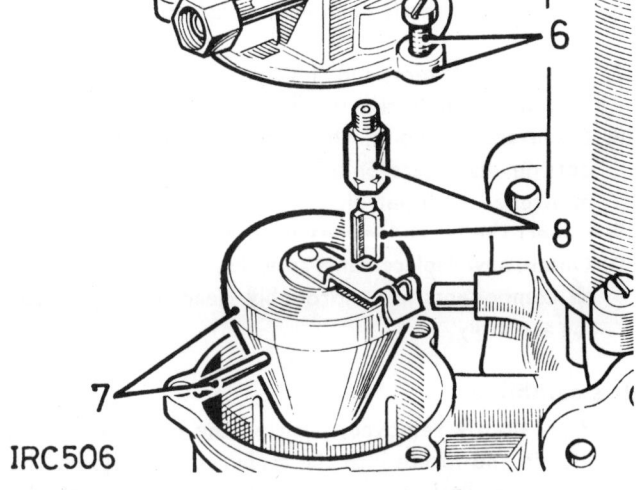

IRC 506

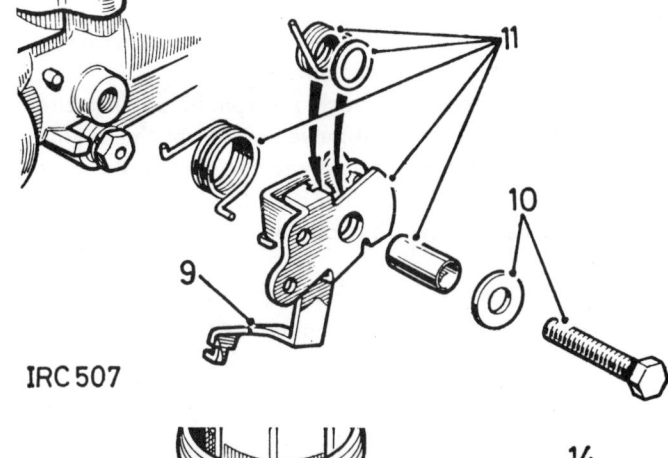

IRC 507

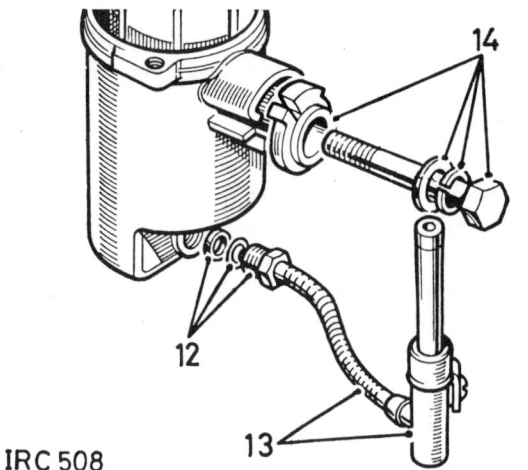

IRC 508

Rover 3500 and 3500S Manual AKM 3621

FUEL SYSTEM—HS6 Carburetter with Emission Control

Right hand carburetter 18 to 21

18. Remove the circlip from the spindle and draw off the right hand throttle lever.
19. Remove the adjusting bolt, spring and plain washer from the adjusting lever and draw off.
20. Release the tab washer from the spindle extension nut and remove the nut and tab washer.
21. Remove the carburetter lever and brass washer.
22. Close the ends and remove the two special screws retaining the throttle butterfly.
23. Withdraw the butterfly.
24. Withdraw the butterfly spindle.
25. If required, remove the piston lift pin.

Inspecting

26. Check the throttle spindle and carburetter body for wear.
27. Check all levers and linkage for excessive free movement, indicating wear.
28. Clean the inside of the suction chamber and the outside diameters of the piston with a clean cloth moistened with petrol.
29. Inspect the jet and jet bearing for signs of wear.
30. Obtain new replacements as necessary and ensure that all components are thoroughly clean in preparation for assembly.

Reassembling

31. If previously removed, position the spring over the lift pin and locate into the carburetter body. Fit the neoprene washer, brass washer and secure with the circlip.
32. Insert the spindle into the carburetter body with the plain end on the float chamber side.
33. Slide the butterfly through the spindle so that the top of the butterfly will move away from the engine when opening the throttle and the chamfered edge locates with the bore when closed.
34. Fit two new screws, but do not fully tighten at this stage.
35. Place the brass washer for throttle spindle on to the threaded end of the spindle ensuring that it is located on the larger diameter of the spindle which will just be protruding from the carburetter body.
36. Fit the carburetter lever and, on the left hand carburetter, the throttle link locating it correctly on the threaded portion of the spindle, then fit the tab washer and extension nut.
37. Tighten the extension nut to secure the lever to the spindle and lock with the tab washer.
38. With the throttle butterfly held in the fully closed position, apply light pressure to the spindle at the end with the extension nut to bring the lever into contact with the brass washer. Hold this position and fully tighten the two screws fixing the throttle butterfly to the spindle and lock them by opening the slots.

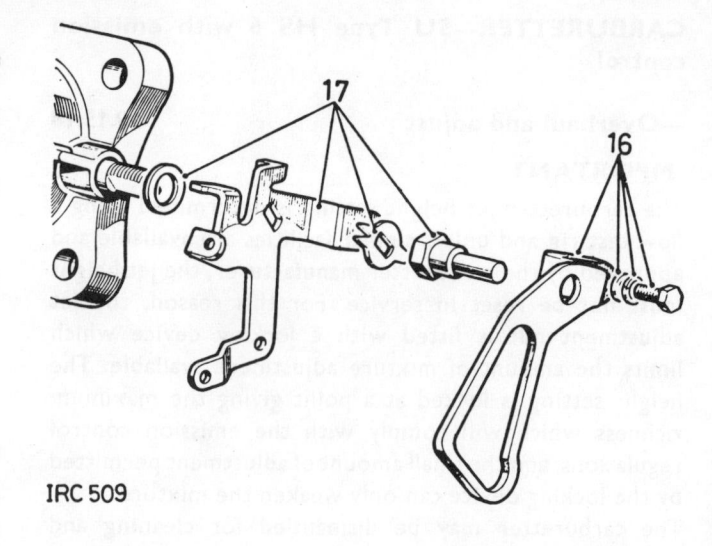

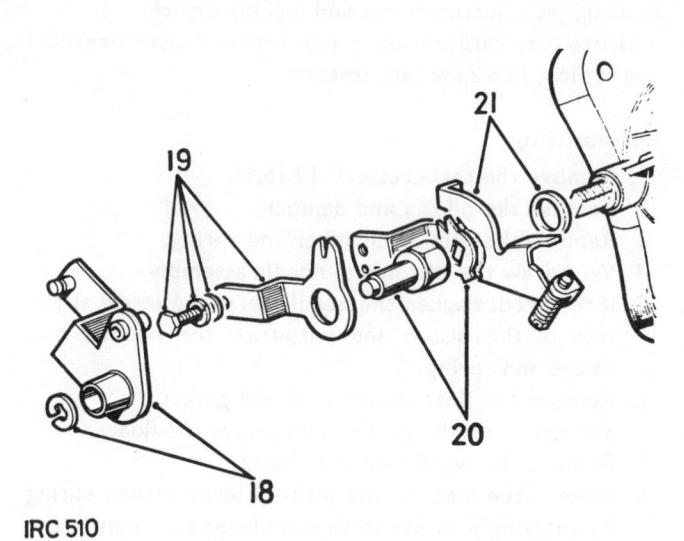

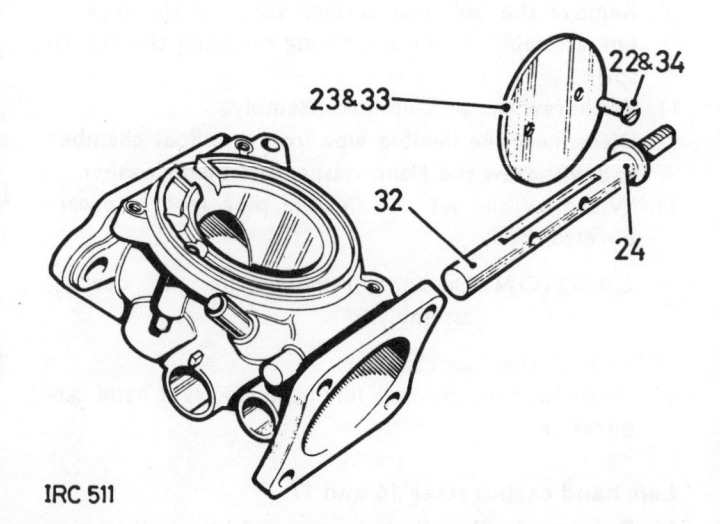

continued

IRC 509

IRC 510

IRC 511

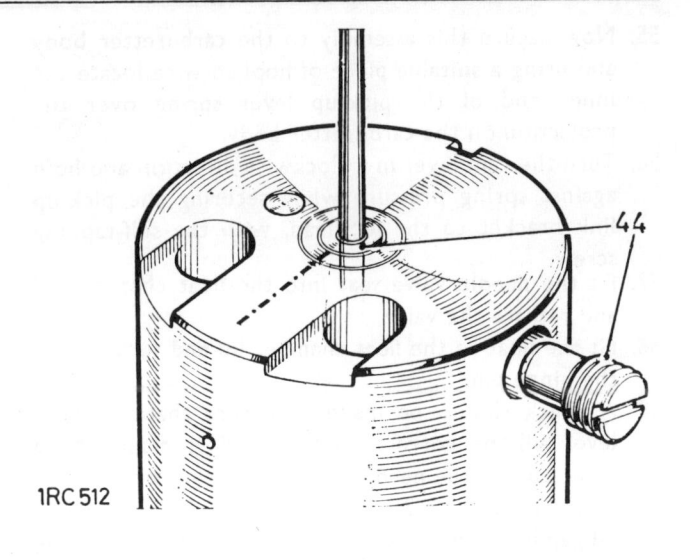

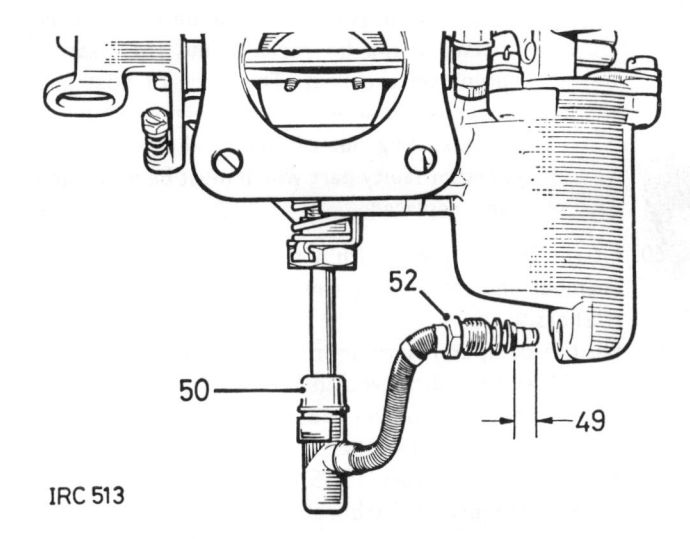

FUEL SYSTEM—HS6 Carburetter with Emission Control

39. **Left hand carburetter.** Refit the throttle lever to carburetter lever with bolt, spring and plain brass washer. Do not tighten at this stage.

Right hand carburetter 40 to 42

40. Refit adjusting lever, and secure with adjusting bolt, spring and plain washer. Do not tighten at this stage.
41. Refit the right hand throttle lever to extension, and secure to spindle with circlip.
42. Refit throttle link to throttle lever, and secure with the circlip.
43. Insert the jet into the jet bearing.
44. Fit the jet needle, spring and bias sleeve into the piston so that the shoulder on the shank is flush with the face of the piston and secure with the clamping screw.

   NOTE: Where an alignment mark is provided on the bias sleeve, the mark must be aligned as illustrated. Later models provide a full length locating flat for engaging the needle retaining screw.

45. Insert the piston into the carburetter body locating the needle into the jet bearing.
46. With a few drops of thin oil on the piston rod only, position the piston spring and suction chamber on to the carburetter body and secure with the three screws.
47. Place the carburetter upright with the air cleaner flange face downwards.
48. With the piston at the bottom of its travel (that is, on the bridge of the carburetter) and the jet hard up against the jet bearing, ensure that the piston is free to fall on to the bridge when lifted by the piston lift pin and that the jet is not binding in the jet bearing when moved in and out.
49. Fit the gland nut, brass washer and new gland to the jet tube and position in the base of the float chamber, ensure that approximately 4,7 mm (0.187 in) tube projects before fitting, but do not tighten at this stage.
50. Offer the float chamber with the adaptor on to the carburetter body, at the same time locating the jet into the jet bearing.
51. Secure the float chamber to the carburetter body with a special bolt, plain washer and spring washer.
52. Tighten the gland nut on the jet pipe at the base of the float chamber.
53. Assemble the cam lever, pick-up lever and link, cam lever spring, distance tube and shim washer, ensuring correct sequence.
54. Place the spring for pick-up lever over the bolt, ensuring that the outer end of the spring locates in the slot in the pick-up lever.

continued

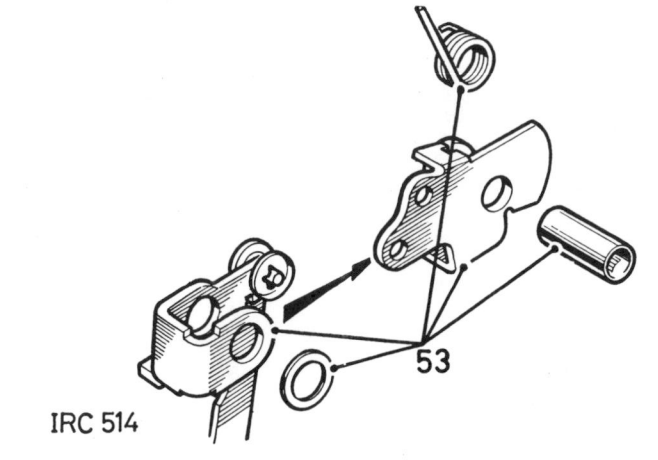

FUEL SYSTEM—HS6 Carburetter with Emission Control

55. Now secure this assembly to the carburetter body and using a suitable piece of hooked wire locate the inner end of the pick-up lever spring over the projection on the carburetter body.
56. Turn the cam lever in a clockwise direction and hold against spring pressure while securing the pick-up link bracket to the jet head, with the self-tapping screw.
57. Fit the needle valve seat into the float chamber lid and replace the valve.
58. Fit the float to the float chamber lid and secure with the hinge pin.
59. A check should be made to ensure that the float level will be correct. With the float chamber lid inverted and the needle valve in the shut off position there should be a 3 mm to 4,5 mm (0.125 in to 0.187 in) gap between the end of the float lever and the rim of the float chamber lid.

   NOTE: Where the float and lever are integral and not riveted (some later models), measure between the lid register and the nearest point on the float. This type of float and lever is not adjustable for height setting. A float setting out of limits indicates a damaged or faulty part which must be identified and replaced.

60. Fit the lid complete with a new gasket to the float chamber and secure with the three screws and spring washers.
61. Fill the damper reservoir in the suction piston with SAE 20 engine oil to within 12 mm (0.500 in) of the top of the hollow piston rod.
62. Fit the damper and washer.
63. Refit the carburetters. 19.15.11.
64. Tune and adjust the carburetters. 19.15.02.

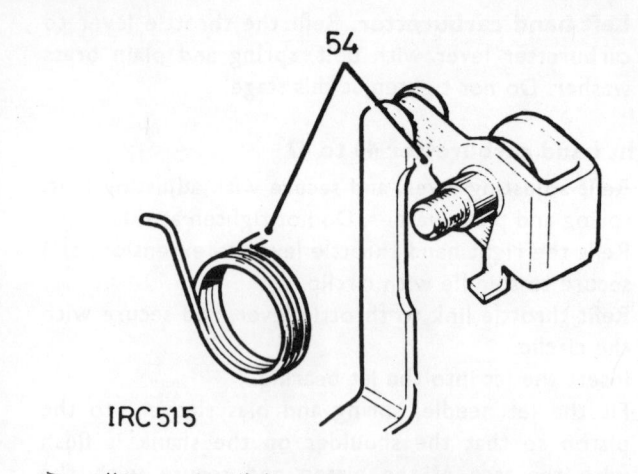

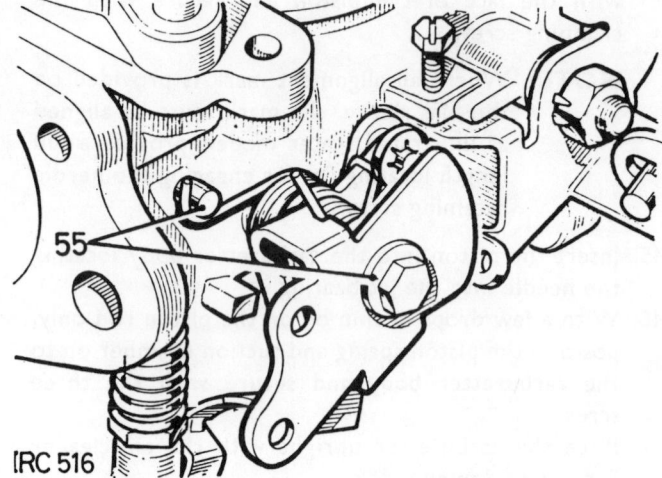

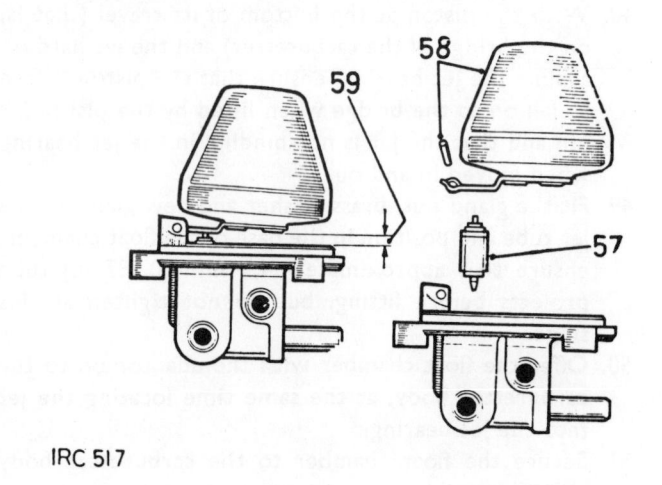

DATA

Carburetter

Type	SU HS6
Number	Two
Bore	44,45 mm (1.75 in)
Needle	BAK
Jet size	2,54 mm (0.100 in)
Float level	3,0 to 4,5 mm (0.125 to 0.187 in)
Damper oil	SAE 20

FUEL SYSTEM—HIF6 Carburetter

CARBURETTER—SU Type HIF6 (standard and emission controlled)

—Overhaul and adjust   19.15.18

IMPORTANT

The carburetter jet height setting is determined using a CO meter and, unless a suitable meter or approved alternative equipment is available, the jet height must not be reset in service. For this reason, the access hole to the jet adjustment screw is plugged and sealed. The carburetters may be dismantled for cleaning and replacement of component parts, but the final jet height can only be set when the carburetters have been reassembled and refitted to the engine. See Operation 19.15.02

Dismantling

1. Remove the carburetters. 19.15.11.
2. Remove the oil cap and damper complete with sealing washer.

NOTE: If the carburetter is of the ball bearing lined suction chamber type, the damper will require a sharp pull to release the retaining clip.

3. Remove the fixings and the identity tag from the suction chamber.
4. Withdraw the suction chamber without tilting it.
5. Withdraw the spring.
6. Withdraw the piston assembly and empty the oil from the piston rod.
7. Remove the needle locking screw and withdraw the needle complete with guide and spring.
8. Remove the circlip and spring, and withdraw the piston lift pin.
9. Remove the bottom cover plate complete with sealing ring.
10. Remove the screw and spring securing the bi-metal and adjusting lever.
11. Withdraw the bi-metal blade and jet assembly. Disengage the jet.
12. Remove the float pivot spindle and sealing washer.
13. Withdraw the float and needle valve.
14. Unscrew the needle valve seat.
15. Unscrew the jet bearing locking nut and withdraw the jet bearing and sealing washer.
16. Remove the cam lever and spring.
17. Remove the cold start assembly.
18. Remove the throttle levers.
19. Remove the throttle butterfly.
20. Remove the throttle spindle complete with seals.

Inspecting

21. Examine all joint faces for deep scores which would lead to leakage taking place when assembled.
22. New seals should be used throughout the carburetter rebuild.
23. Inspect the metering needle for signs of wear, bend and twist, renew if necessary.
24. Examine the throttle spindle, its bearings in the carburetter body and the rubber end seals. Renew as necessary.
25. Examine the suction chamber.
    a. Plain suction chamber:
    Clean inside the suction chamber and piston rod guide using petrol or methylated spirit. Refit the damper assembly and washer. Seal the transfer holes in the piston assembly with rubber plugs, and refit the assembly to the suction chamber. Invert the complete assembly and allow the suction chamber to fall away from the piston. Check the time this takes, which should be from five to seven seconds for the HIF6 carburetters of 44,5 mm (1.750 in) bore. If the time taken is in excess of that quoted, the cause will be thick oil on the piston rod, or an oil film on the piston or inside the suction chamber. Remove the oil from the points indicated and re-check.
    b. Ball bearing lined suction chamber:
    Check that all the balls are in the piston ball race (2 rows of 6). Fit the piston into the suction chamber, without the damper and spring, hold the assembly horizontally and spin the piston. The piston should spin freely in the suction chamber without any tendency to stick.

Reassembling

26. Fit the throttle spindle into the body with the threaded end protruding from the cold start side.
27. Fit the throttle butterfly so that the bottom edge will move away from the engine when opening the throttle and the chamfered edge locates with the bore when closed. Use new retaining screws, but do not fully tighten at this stage.

    NOTE: Standard carburetters have a plain butterfly, emission controlled carburetters have a poppet valve in the butterfly.

continued

Rover 3500 and 3500S Manual AKM 3621

FUEL SYSTEM—HIF6 Carburetter

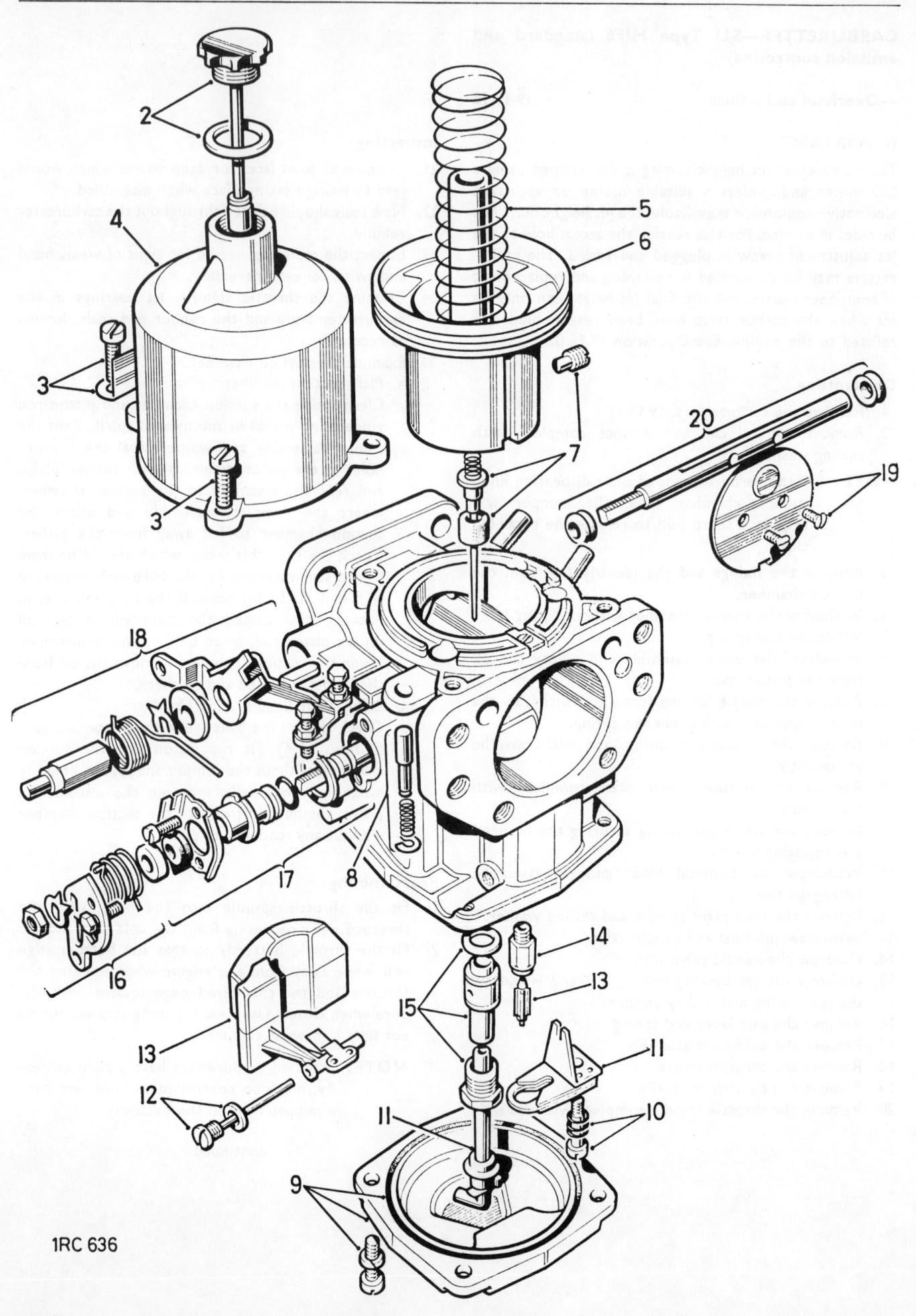

1RC 636

19.15.18
Sheet 9

Rover 3500 and 3500S Manual AKM 3621

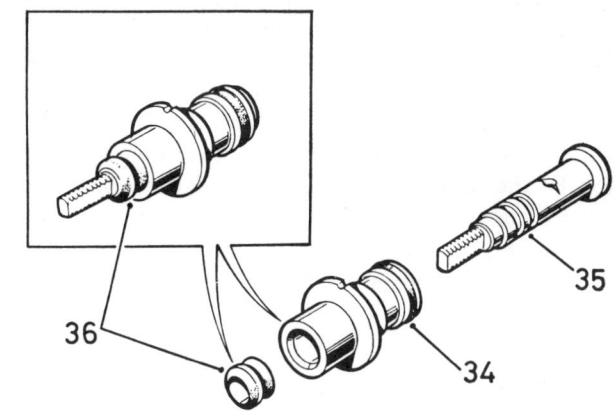

FUEL SYSTEM—HIF6 Carburetter

28. Actuate the throttle several times to centralise the butterfly, then tighten the retaining screws and lock by peening the ends.
29. Fit the end seals over the throttle spindle with the dished face towards the throttle. Position the seals 0,9 mm (0.035 in) below the spindle housing flange.
30. Close the throttle and fit the throttle return lever.
31. Fit the throttle lever.
32. Fit the bush with its spigot located through the throttle lever.
33. Fit the tab washer and nut. Tighten the nut and engage the tab washer.
34. Lubricate the 'O' ring for the cold start valve with oil and fit it to the valve body.
35. Insert the spindle into the valve body from the 'O' ring end.
36. Fit the end seal, dished side first, and locate it completely on to the spindle shoulder.
37. Place the carburetter body on its side, cold start valve bore uppermost, and place the valve gasket in position with the cut-out towards the top screw hole.
38. Insert the cold start assembly into the carburetter body, aligning the cut-out in the valve flange with that in the paper gasket.
39. Fit the cover over the end seal.
40. Fit the cold start retaining plate with the slotted flange towards the throttle spindle. Evenly tighten the two retaining screws.
41. Position the cam return spring over the cold start valve body with the straight end located in the top slot of the retaining bracket.
42. Fit the cam lever to the cold start spindle with the right-angled extension towards the carburetter body and in front of the cast stop.
43. Fit the tab washer and nut. Tighten the nut and engage the tab washer.
44. Using a suitable wire hook, locate the right-angled end of the cam lever return spring around the right angled extension on the cam lever.
45. Fit the throttle return spring, long extension side first.
46. Locate the short extension of the throttle return spring into the slot in the right-angled extension of the throttle lever.
47. Locate the long extension of the throttle return spring in the bottom slot of the cold start retaining bracket.
48. Fit the fibre washer over the jet bearing and assemble complete with the locking nut into the carburetter body.
49. Fit the float chamber needle valve seating.
50. Position the carburetter body with the float chamber uppermost.
51. Place the needle valve, conical end first, into the seating.
52. Insert the float with the moulded Part No. uppermost and the hinge tab abutting the needle valve.
53. Fit the float pivot spindle complete with sealing washer.

continued

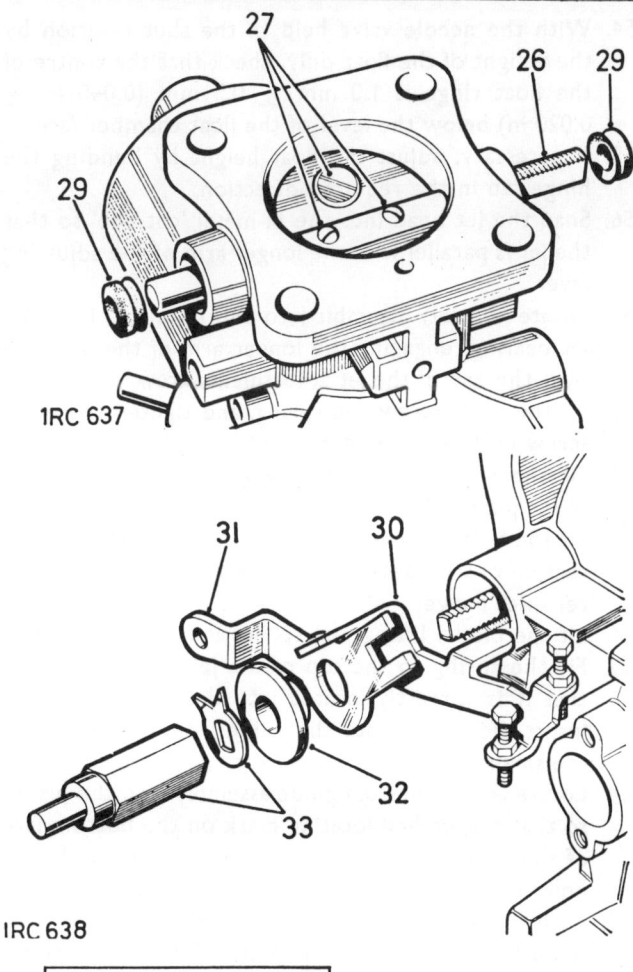

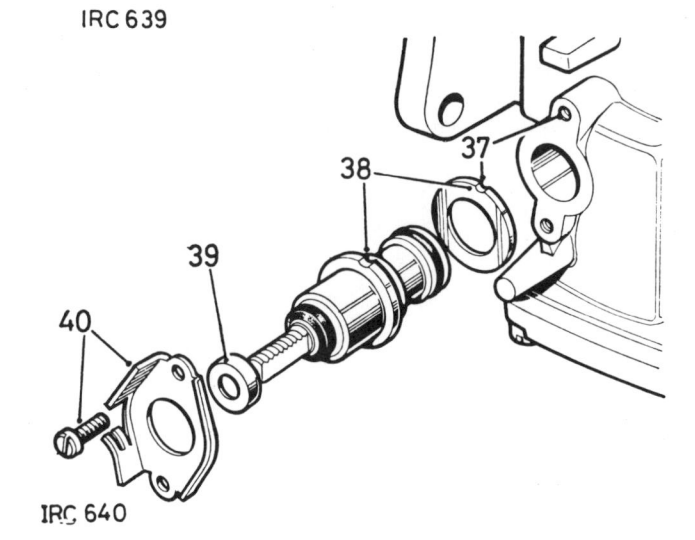

FUEL SYSTEM—HIF6 Carburetter

54. With the needle valve held in the shut position by the weight of the float only, check that the centre of the float ridge is 1,0 mm ± 0,5 mm (0.040 in ± 0.020 in) below the level of the float chamber face.
55. If necessary, adjust the float height by bending the hinge tab in the required direction.
56. Snap the jet head into the bi-metal 'cut-out' so that the jet is parallel with the longer arm of the adjusting lever.
57. Locate the jet, assembled to the bi-metal, into the jet bearing, engaging the longer arm of the bi-metal over the end of the jet adjustment screw.
58. Fit the pivot screw and spring, and tighten the pivot screw up to its shoulder.
59. Fit the sealing ring into the groove in the bottom cover plate.
60. Fit the bottom cover plate, ensuring that the cut-out section of the cover plate contour faces the carburetter inlet flange.
61. Fit the piston lifting pin, spring and circlip.
62. Fit the spring to the top of the jet needle ensuring that it is correctly located into the annular groove.
63. Insert the needle into the guide from the side with the small protrusion.
64. Locate the needle and guide assembly into the piston so that the etched locating mark on the bottom face of the guide is central to the slot between the air holes in the piston.
65. Push the guide into the piston until its bottom face is flush with the bottom of the slot in the piston, then secure with the locking screw.
66. Fit the piston assembly to the carburetter body. Lightly oil the outside of the piston rod.
67. Place the piston spring over the piston rod.
68. Fit the suction chamber without rotating it.
69. Fit the identification tag to the rear screw.
70. Fit the oil cap and damper complete with sealing washer.
71. Refit the carburetters. 19.15.11.

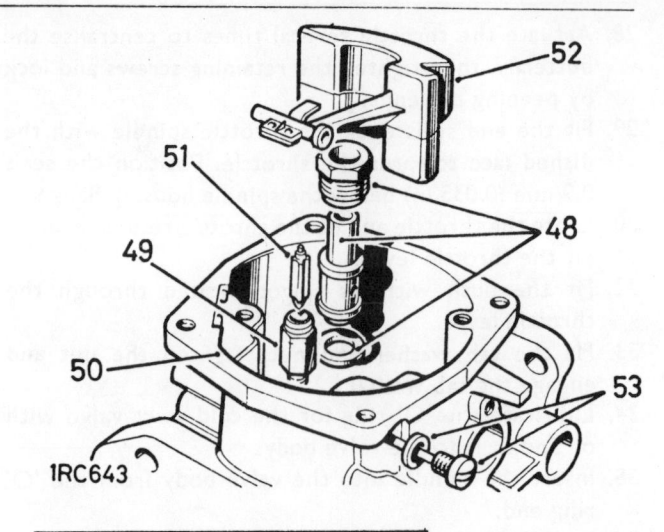

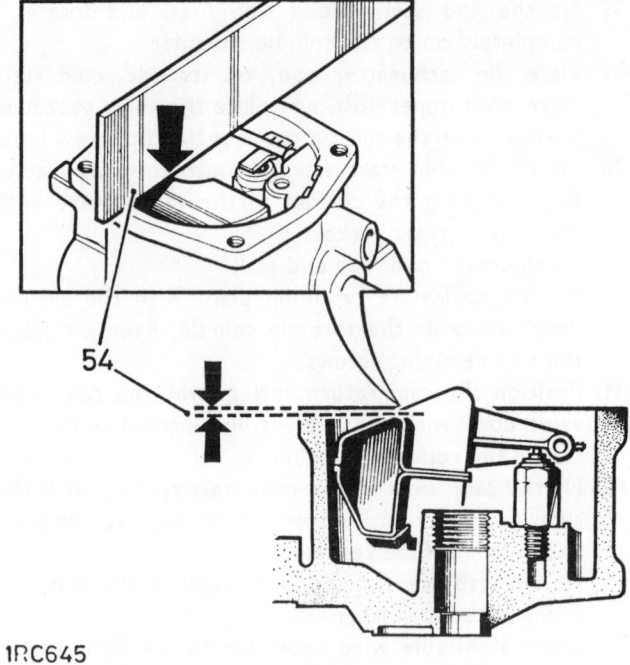

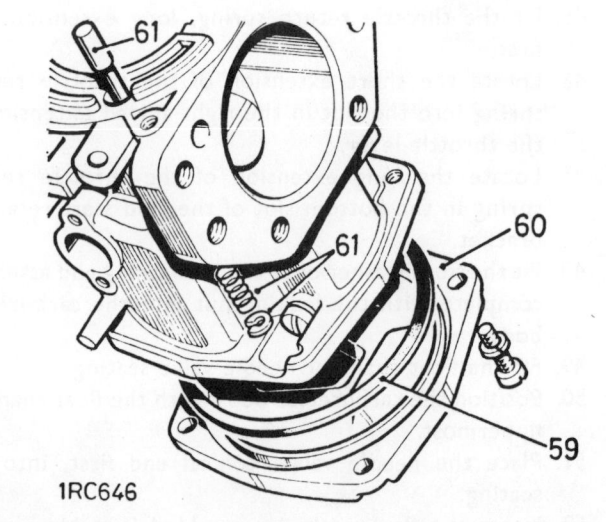

FUEL SYSTEM

AUTOMATIC ENRICHMENT DEVICE (A.E.D.)

—Remove and refit 19.15.38

Removing

1. Prop open the bonnet.
2. Disconnect the fuel inlet pipe.
3. Disconnect the air inlet hose.
4. Disconnect the vent hose from the float chamber.
5. Disconnect the delivery hose to the inlet manifold.
6. Remove the A.E.D. from the support bracket.

Refitting

7. Reverse 1 to 6.

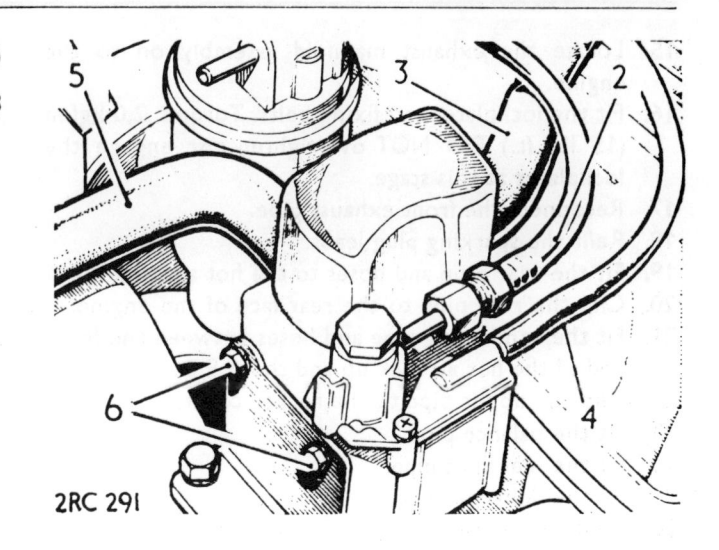

A.E.D. FUEL AND AIR SYSTEM

—Remove and refit 19.15.39

Removing

1. Prop open the bonnet.
2. Remove the air cleaner.
3. Depress the spring clips and withdraw the delivery hose from the A.E.D. and inlet manifold.
4. Remove the fuel pipe from the A.E.D. and left-hand float chamber.
5. Disconnect the fuel pipe from the top of the fuel filter.
6. Withdraw the balance pipe from the A.E.D. and left-hand carburetter adaptor.
7. Release the clip and withdraw the upper pipe from the hot air pick up.
8. Release the clip and withdraw the rear pipe from the hot air pick up.
9. Disconnect the front exhaust pipe from the left-hand exhaust manifold.
10. Withdraw the leads from the sparking plugs.
11. Remove the left-hand exhaust manifold.
12. Remove the hot air pick up, together with gasket, from the left-hand exhaust manifold.

Refitting

13. Locate the hot air pick up and gasket on to the left-hand exhaust manifold.
14. Ensure that the front pipe faces upward when the cover plate is fitted, then secure the fixings, engaging the tab washers.

continued

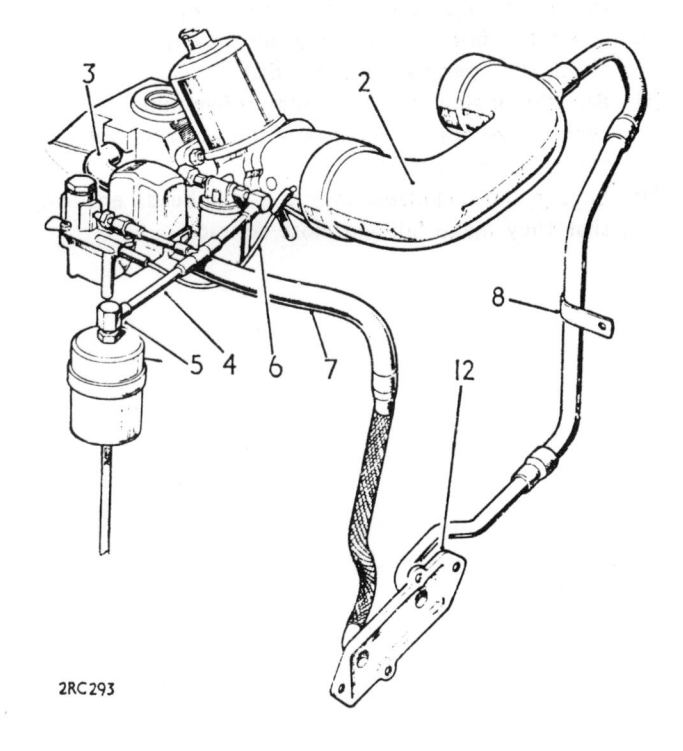

Rover 3500 and 3500S Manual AKM 3621

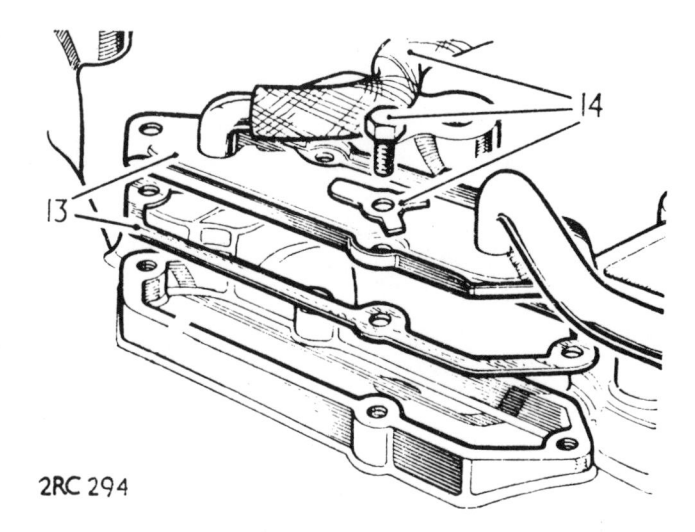

FUEL SYSTEM

15. Locate the exhaust manifold assembly on to the engine.
16. Fit the lockplates and fixing bolts. Torque: 2,0 kgf.m (15 lbf ft.) DO NOT overtighten nor engage the lockplates at this stage.
17. Reconnect the front exhaust pipe.
18. Refit the sparking plug leads.
19. Fit the rear pipe and hoses to the hot air pick up.
20. Clip the rear pipe to the rear face of the engine.
21. Fit the connecting pipe and hoses between the front end of the hot air pick up and the A.E.D.
22. Clip the upper pipe to the rocker cover.
23. Fit the balance pipe.
24. Fit the fuel inlet pipe.
25. Fit the delivery hose.
26. Refit the air cleaner.
27. Connect the rear pipe hose to the adaptor at the rear of the left-hand air cleaner elbow.
28. Close the bonnet and carry out an engine start to check the operation of the A.E.D.
29. Run the engine for a minimum of five minutes, then recheck the torque settings of the securing bolts for the exhaust manifolds.
30. Engage the lockplates over the bolt heads ensuring that they make full and close contact.

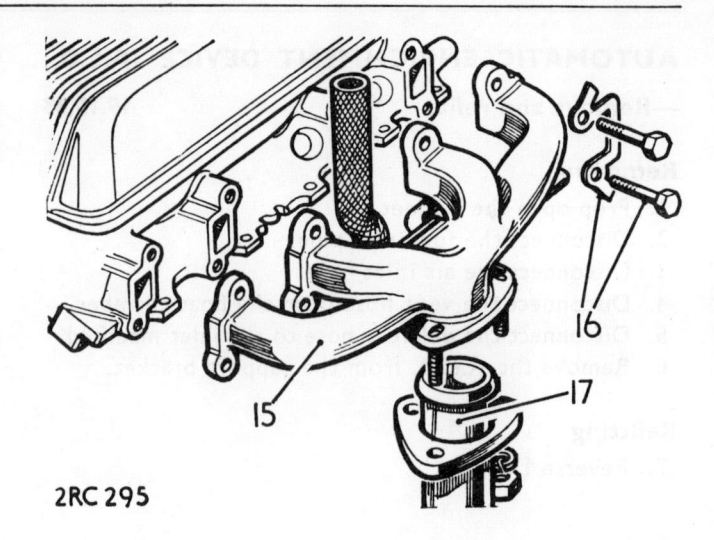

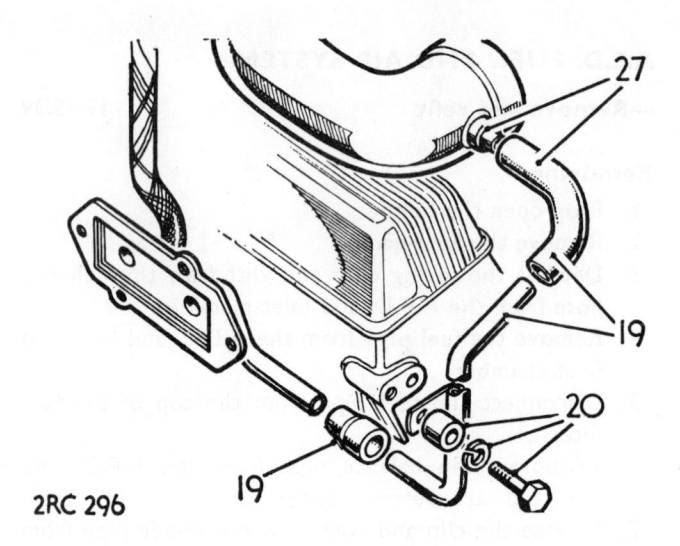

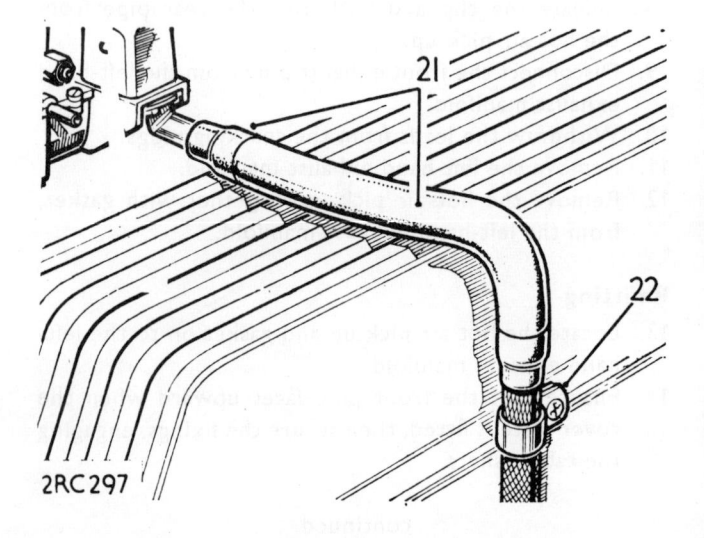

2RC 295

2RC 296

2RC 297

FUEL SYSTEM

THERMOSTAT SWITCH

—Remove and refit   19.15.50

Removing

1. Disconnect the battery earth lead.
2. Disconnect the lead from the switch.
3. Remove the thermostat switch.

Refitting

4. Using a new joint washer, reverse 1 to 3.

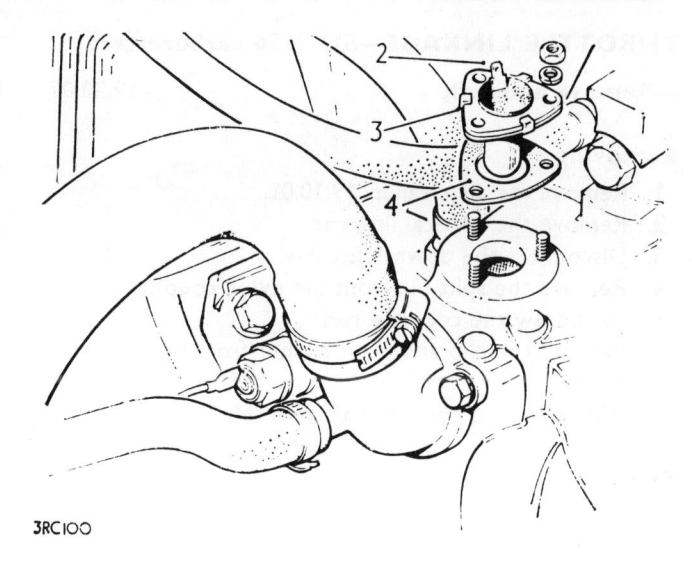

THROTTLE PEDAL

—Remove and refit   19.20.01

Removing

1. Prop open the bonnet.
2. Disconnect the vertical rod at the lower ball joint.
3. Disconnect the return spring.
4. From under the front wing, remove the two nuts and bolts securing the pedal pivot.
5. Early models, remove the lever from the cross-shaft.
6. Remove the cross-shaft mounting bracket.
7. Withdraw the throttle pedal and cross-shaft assembly from inside the car.

Refitting

8. Ensure that the grommet for the cross-shaft is correctly located in the bulkhead.
9. Reverse 1 to 7.
10. 3500 models. Check the downshift cable setting. 44.30.02.

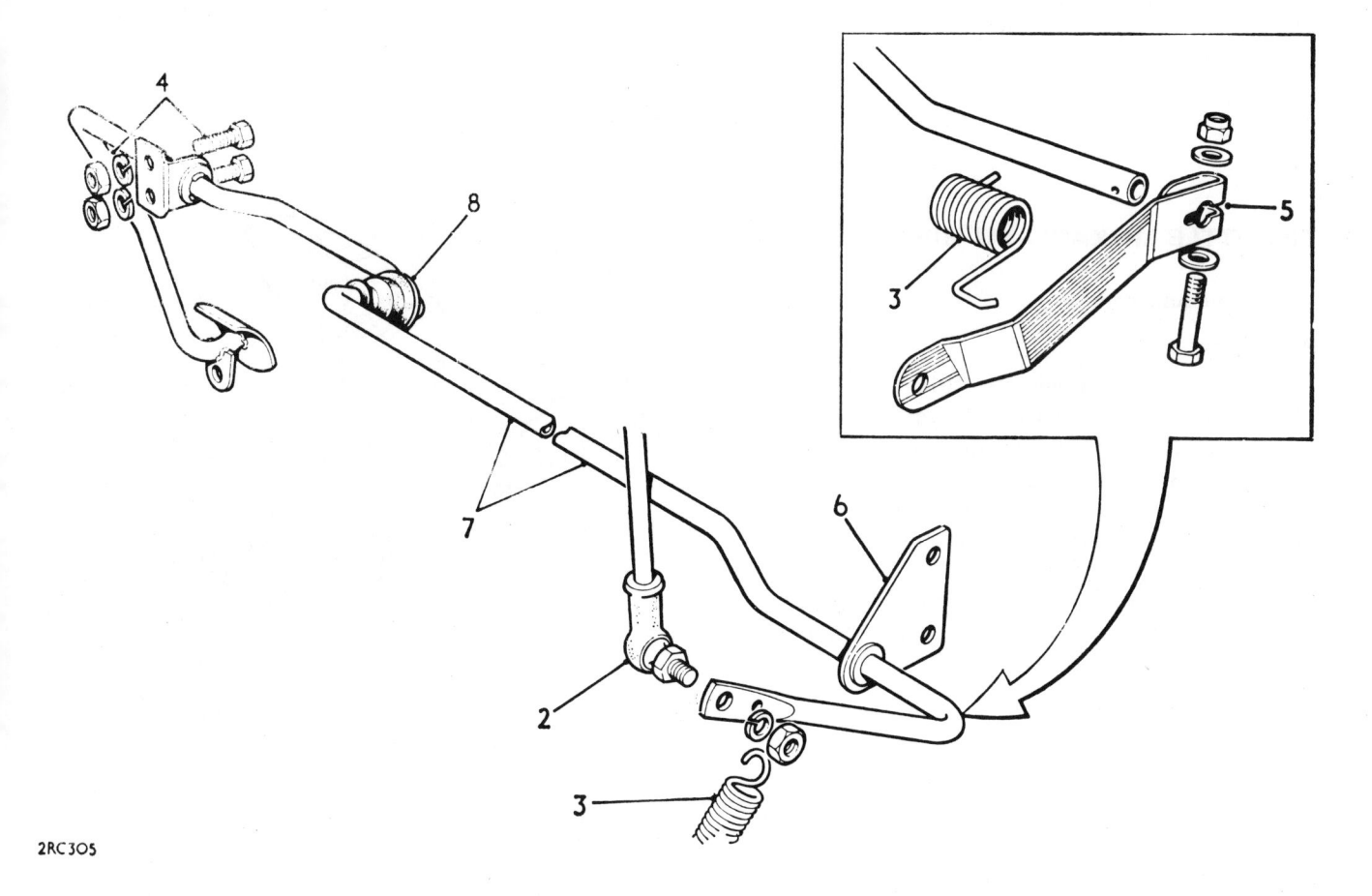

Rover 3500 and 3500S Manual AKM 3621

19.15.50
19.20.01

FUEL SYSTEM

THROTTLE LINKAGE—SU HS6 carburetters

—Remove and refit   19.20.07

Removing

1. Remove the air cleaner. 19.10.01.
2. Remove the vertical link rod.
3. Disconnect the downshift cable.
4. Remove the split pin from the nylon coupling.
5. Withdraw the coupling rod.
6. Remove the clip and plain washer from the countershaft.
7. Withdraw the countershaft assembly.

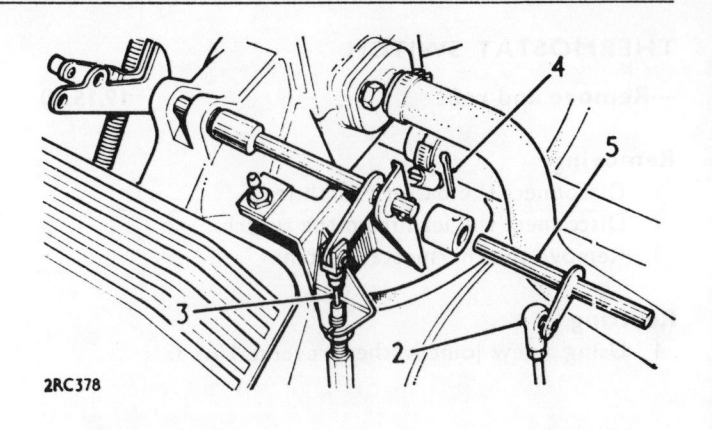

Refitting

8. If the vertical link rod ball joints have been disturbed, reset initially, as follows:
   RHStg   193,6 mm (7.625 in).
   LHStg   225,4 mm (8.875 in).
9. Reverse 3 to 7.
10. Secure the lower end of the vertical link rod to the throttle pedal linkage.
11. Hold the throttle pedal hard against the carpet, in the full throttle position.
12. Hold the coupling shaft in the full throttle position.
13. Adjust the upper ball joint as necessary, by turning it on the link rod until it will just slip over the ball on the coupling shaft.
14. Assemble the ball joint and secure the locknut.
15. Release the pedal and check that the pedal and linkage is free throughout its travel and full movement of the carburetter butterfly is obtained.
16. Check the downshift cable setting. 44.30.02.
17. Refit the air cleaner. 19.10.01.

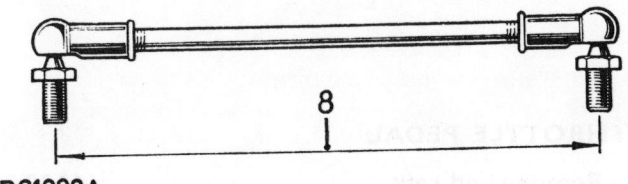

THROTTLE LINKAGE—SU HIF 6 carburetters

—Remove and refit   19.20.07

Removing

1. Remove the air cleaner. 19.10.01.
2. Remove the long vertical link rod.
3. Remove the split pin from the nylon coupling.
4. Withdraw the coupling rod.
5. Remove the short vertical link rod.
6. Remove the clip and plain washer from the countershaft.
7. Withdraw the countershaft assembly.
8. Remove the throttle link rod from between the carburetters.

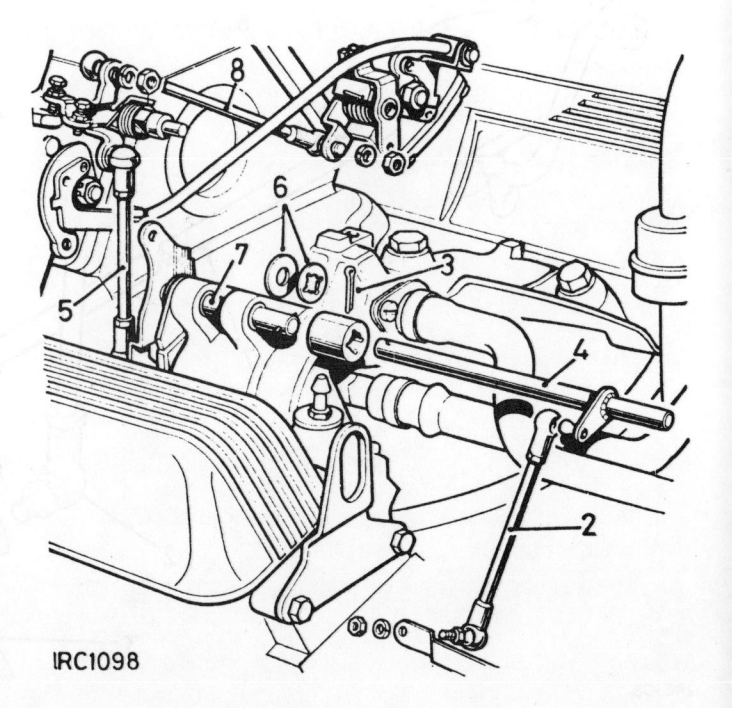

19.20.07
Sheet 1

Rover 3500 and 3500S Manual AKM 3621

FUEL SYSTEM

Refitting

9. If any of the link rod ball joints have been disturbed, reset as follows, between centres.
   Long vertical rod:
   RH Stg 193,6 mm (7.625 in)
   LH Stg 202,3 mm (7.968 in)
   Short vertical rod: 99,6 ± 0,7 mm (3.922 ± 0.030 in)
   Carburetter link rod: 128,7 ± 0,7 mm (5.070 ± 0.030 in)
10. Reverse 6 to 8.
11. Hold the countershaft in the closed stop position and offer up the short vertical link rod, the ball sockets must fit the ball ends without interference or bias. If necessary readjust the length of the rod and refit it.
12. Reverse 3 and 4.
13. Hold the throttle pedal hard against the carpet in the full throttle position.
14. Hold the coupling rod in the full throttle position.
15. Locate the long vertical link rod in position and tighten the locknut on the lower ball joint.
16. Adjust the upper ball socket by turning it on the link rod until it will just slip over the ball on the coupling rod.
17. Assemble the ball joint and secure the locknut.
18. Release the throttle pedal and check that the pedal and linkage is free throughout its travel and that full movement of the carburetter butterflies is obtained.
19. Tune and adjust the carburetters. 19.15.02.

    NOTE: If any difficulty is experienced in setting the 0,15 mm (0.006 in) clearance on the RH carburetter linkage, readjust the short vertical rod as necessary.

20. Refit the air cleaner. 19.10.01.
21. 3500 models—Check the downshift cable setting. 44.30.02.

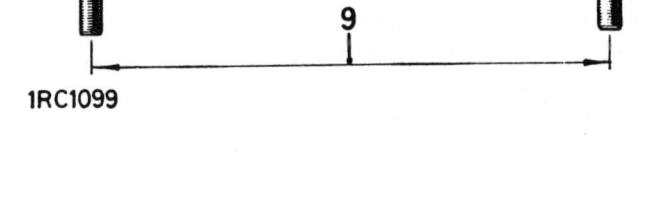

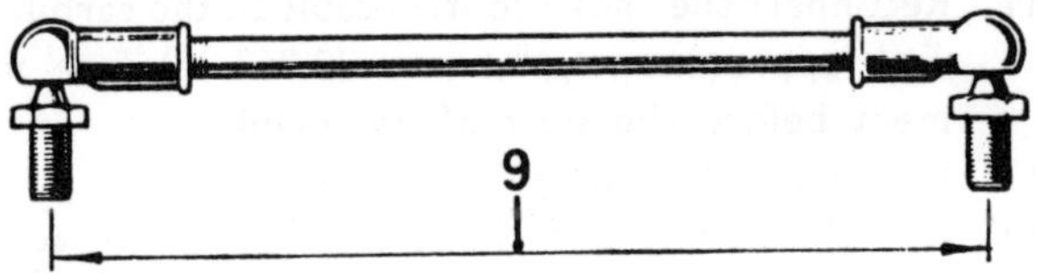

1RC1099

CHOKE CONTROL CABLE—SU HS6 carburetters

—Remove and refit 19.20.13

Service tool: 600967 Cable nut spanner

Removing

1. Disconnect the battery earth lead.
2. Disconnect the choke control cable from the carburetter.
3. Remove the radio speaker panel.
4. Disconnect the electrical leads from the cold start switch, behind the panel.
5. Release the cable clamping bolt and withdraw the switch.
6. Release the nut securing the outer cable. 600967.
7. Withdraw the cable assembly.

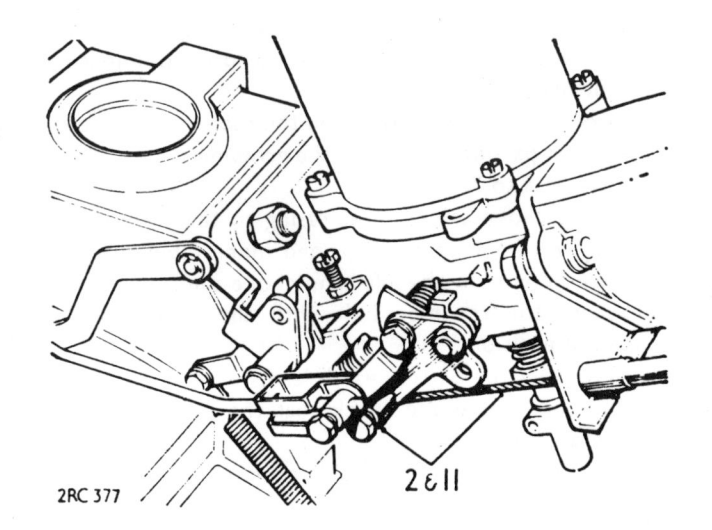

2RC 377

continued

Rover 3500 and 3500S Manual AKM 3621

19.20.07
Sheet 2
19.20.13
Sheet 1

FUEL SYSTEM

Refitting

8. Ensure that the securing nut and spring washer are in place on the outer cable before commencing refitting.
9. Smear MS4 silicone grease on the cable outer case to assist in passing it through the grommets.
10. Reverse 3 to 7.
11. Reconnect the choke control cable to the carburetter with approximately 1,5 mm (0.062 in) free movement before the start of its travel.
12. Reconnect the battery earth lead.
13. Run the engine and check that the choke controls operate correctly.
14. Cut off any excess inner cable protruding beyond the trunnion bolt.

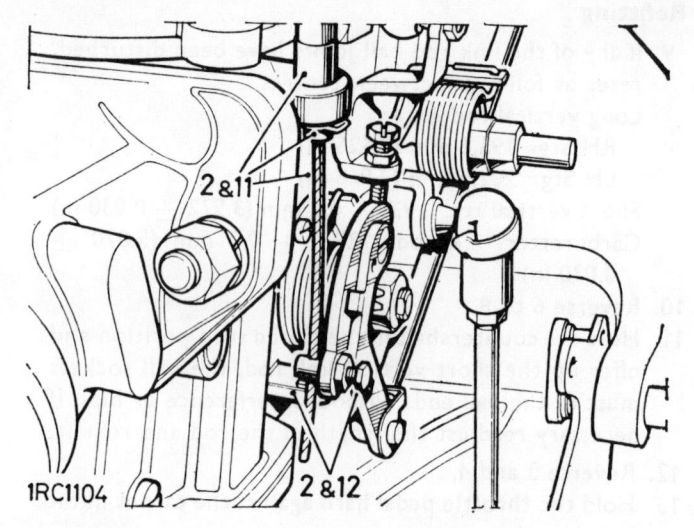

CHOKE CONTROL CABLE—SU HIF 6 carburetters

—Remove and refit   19.20.13

Service tool: 600967 Cable nut spanner

Removing

1. Disconnect the battery earth lead.
2. Disconnect the choke control cable from the carburetter.
3. Remove the radio speaker panel.
4. Disconnect the electrical leads from the cold start switch, behind the panel.
5. Release the cable clamping bolt and withdraw the switch.
6. Release the nut securing the outer cable. 600967.
7. Withdraw the cable assembly.

Refitting

8. Ensure that the securing nut and spring washer are in place on the outer cable before commencing refitting.
9. Smear MS4 silicone grease on the cable outer case to assist in passing it through the grommets.
10. Reverse 3 to 7.
11. Fit the choke cable, outer casing to the carburetter body with the spring clip, and inner cable through the trunnion on the left-hand choke cam.
12. Secure the trunnion bolt with the choke in the closed position.
13. Reconnect the battery earth lead.
14. Run the engine and check that the choke controls operate correctly.
15. Cut off any excess inner cable protruding beyond the trunnion bolt.

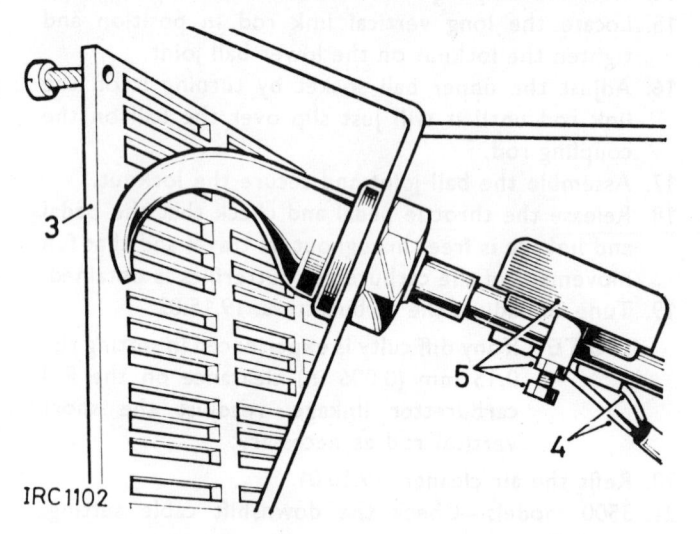

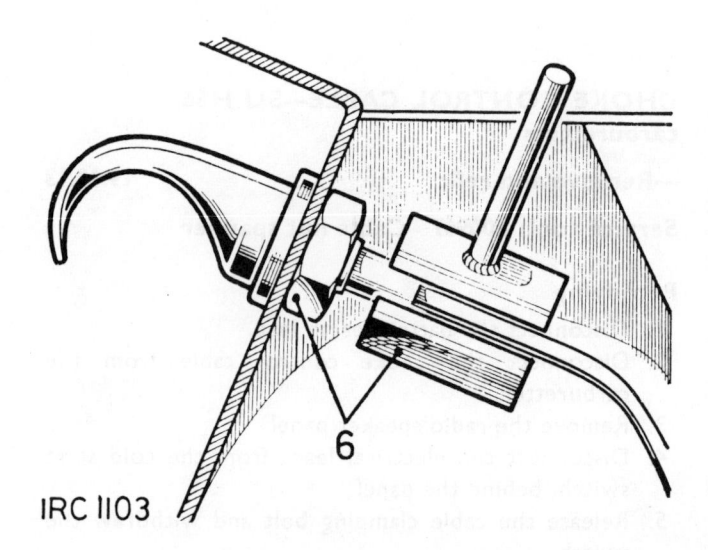

19.20.13
Sheet 2

Rover 3500 and 3500S Manual AKM 3621

FUEL SYSTEM

FUEL RESERVE CONTROL CABLE

—Remove and refit   19.20.20

Service tool: 600967   Cable nut spanner

Removing

1. Disconnect the cable from the reserve tap.

   NOTE: Some models, the reserve tap is fitted beneath the fuel tank, as illustrated. Other models, the reserve tap is fitted in the engine compartment.

2. Remove the radio speaker panel.
3. Release the nut securing the outer cable. 600967.
4. Withdraw the cable assembly.

Refitting

5. Ensure that the securing nut and spring washer are in place on the outer cable before commencing refitting.
6. Smear MS4 silicone grease on the cable outer case to assist in passing it through the grommets.
7. Reverse 1 to 4.
8. Check the reserve tap control for correct operation.

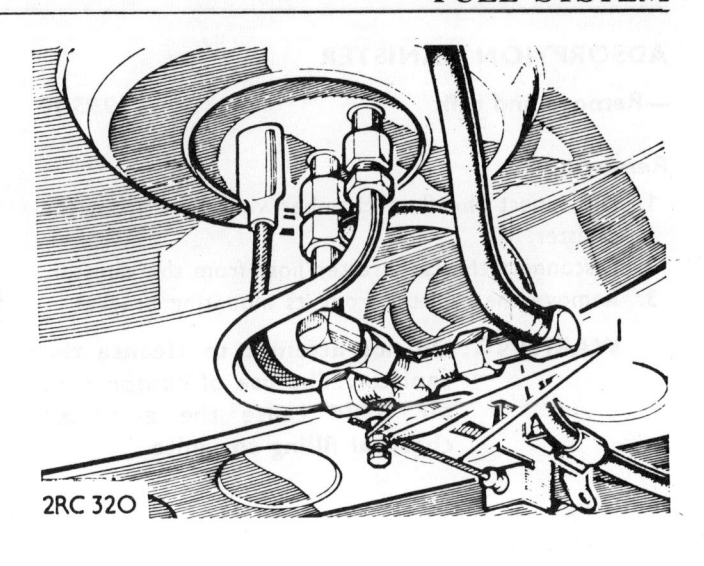

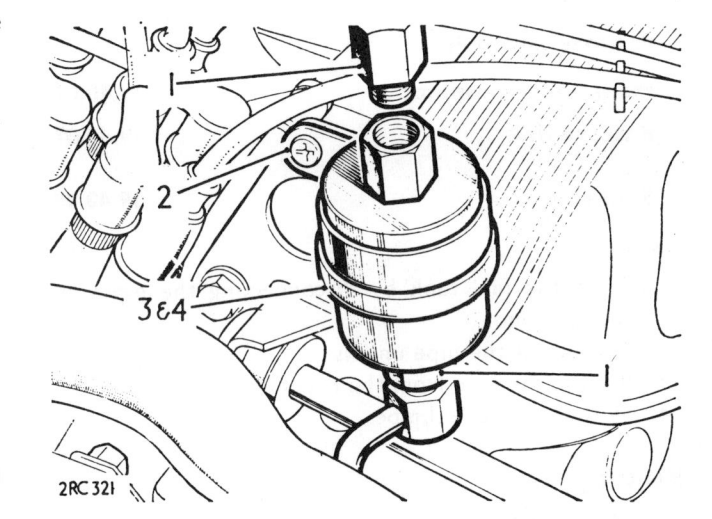

2RC 320

FUEL LINE FILTER

—Remove and refit   19.25.01

Removing

1. Disconnect the fuel pipes.
2. Slacken the clip securing the filter.
3. Withdraw the filter.

Refitting

4. Vertically mounted filter—Reverse 1 to 3, fitting the filter with the end marked 'IN' downwards. Alternatively, if the filter is marked with arrows, they must point upwards.
5. Horizontally mounted filter—Reverse 1 to 3, fitting the filter with the end marked 'IN' to the rear. Alternatively, if the filter is marked with arrows, they must point to the front.

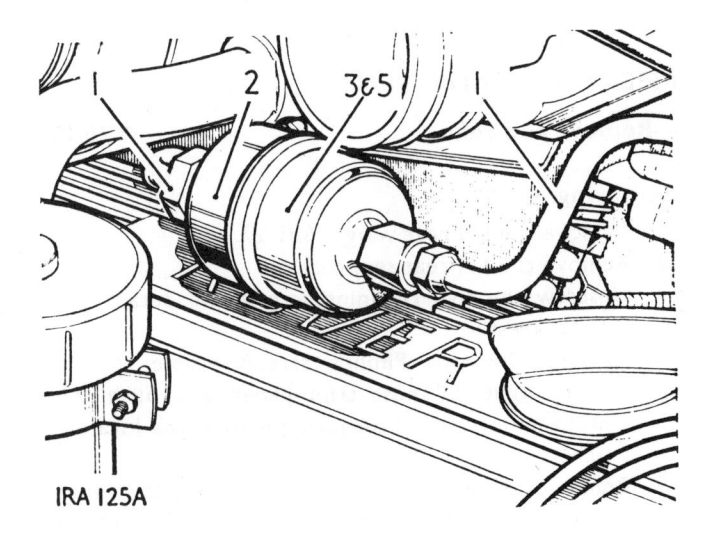

2RC 321

IRA 125A

Rover 3500 and 3500S Manual AKM 3621

FUEL SYSTEM

ADSORPTION CANISTER

—Remove and refit 19.25.10

Removing

1. Disconnect the float chamber vent hose from the canister.
2. Disconnect the carburetter hose from the canister.
3. Remove the canister from its mounting bracket.

   WARNING: Do not attempt to cleanse the canister. The use of compressed air could cause the activated charcoal filling to ignite.

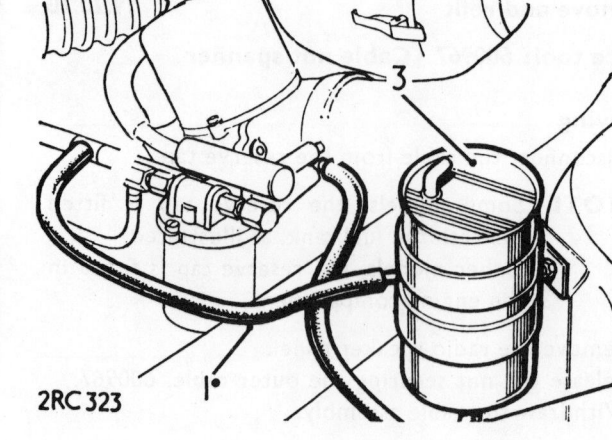

Refitting

4. Reverse 1 to 3.

FUEL PIPES

—Remove and refit

Main line—complete	19.40.01
Reserve line	19.40.07
Spill return	19.40.08

Removing

1. Drain the fuel tank by disconnecting the pipes at the tank end.
2. Disconnect the pipe unions.
3. Release the pipe from the retaining clips.
4. Withdraw the fuel pipe.

Refitting

5. Reverse 1 to 4.

PIPE—FUEL TANK VENT

—Remove and refit 19.40.35

Removing

1. Remove the LH rear wing. 76.10.27.
2. Remove the LH rear seat cushion.
3. Unscrew the two retaining screws and detach the seat backrest.
4. Remove the two Phillips screws visible in the quarter light 'D' post trim panel, and push the trim panel forward, disengaging it from the locating brackets.
5. Remove the three retaining screws from the inner face of the 'D' post panel and remove the screws visible along the door edge of the outer panel.
6. Lift off the rear quarter panel, carefully disengaging the rear edge from the rear window surround.

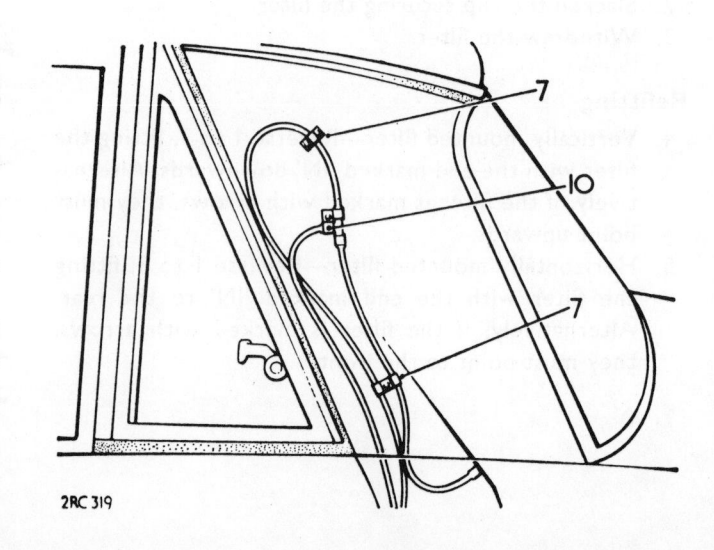

continued

19.25.10
19.40.35
Sheet 1

Rover 3500 and 3500S Manual AKM 3621

7. Slacken the six drive screws securing the vent and drain pipe clips to the base unit.
8. Remove the fuel tank trim panel in boot and disconnect the vent pipe at the fuel tank union.
9. Remove nut and olive from pipe and withdraw pipe through grommet in base unit.

Refitting

10. Locate the vent and drain pipe 'tee' junction in the appropriate clip on the base unit, as illustrated, and secure with drive screw.
11. Thread a length of 3,0 mm (0.125 in) welding wire through the two grommets in the base unit.
12. Lubricate the vent pipe with MS4 silicone grease and feed the vent pipe over the welding wire from the outside. Remove welding wire.
13. Pull the vent pipe through until the union can be secured to the tank, clip the pipe as illustrated.
14. Reverse 1 to 8. Before replacing the outer body quarter panel, apply sealing compound to drive screws which retain the vent pipe.
    Ensure that the outer edge of the rear quarter panel is sealed satisfactorily.

FUEL PUMP—Electrical

| —Clean filter 1 to 19 | 19.45.05 |
| —Remove and refit 1 to 11 and 19 | 19.45.08 |

Removing

1. Prop open the bonnet.
2. Disconnect the fuel spill return pipe from the carburetter, to prevent draining the fuel tank by syphoning.
3. Open the luggage boot.
4. Disconnect the battery earth lead.
5. Withdraw the trim board from the rear of the boot.
6. Disconnect the electrical lead from the fuel pump.
7. From underneath the vehicle, push in the knob for the fuel pipe cut-off valve.
8. Disconnect the fuel inlet pipe together with the cut-off valve.
9. Disconnect the fuel outlet pipe.
10. Remove the two bolts securing the pump to the support bracket.
11. When the fuel pump is released, the feed lead can be withdrawn through the grommet in the boot floor.

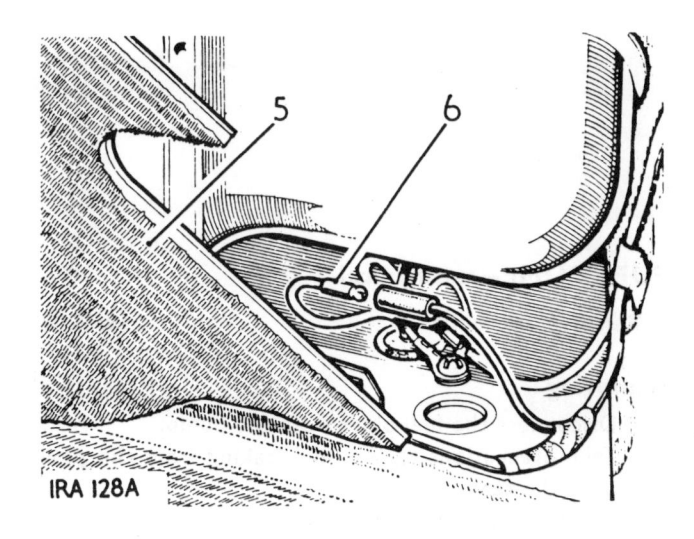

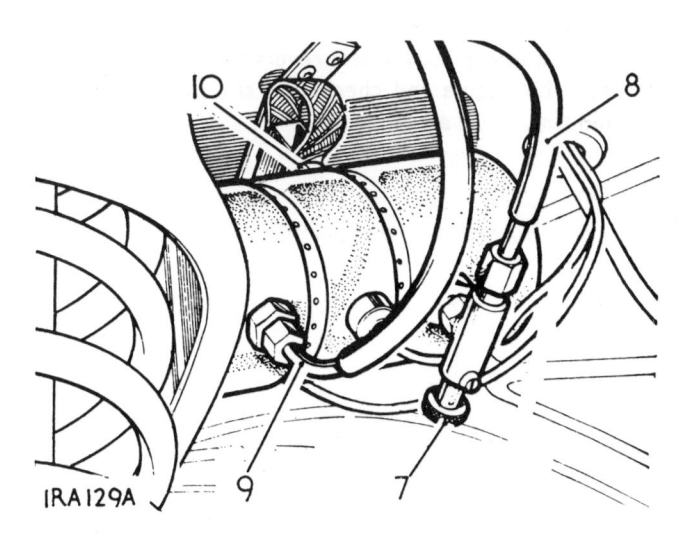

continued

Rover 3500 and 3500S Manual AKM 3621

FUEL SYSTEM

Cleaning filter

12. Remove the inlet and outlet unions.
13. Remove the clips retaining the sponge rubber cover.
14. Ease the sponge rubber casing off the fuel pump.
15. Using a spanner on the hexagon provided, release the end cover from the bayonet fixing.
16. Withdraw the filter, and clean by using a compressed air jet from the inside of the filter.
17. Remove the magnet from the end cover and clean both components, then refit the magnet in the centre of the end cover.
18. Reverse 12 to 16.

Refitting

19. Reverse 1 to 11, ensuring that the earth wire and earth strip are fitted to the bolts securing the fuel pump to the bracket.

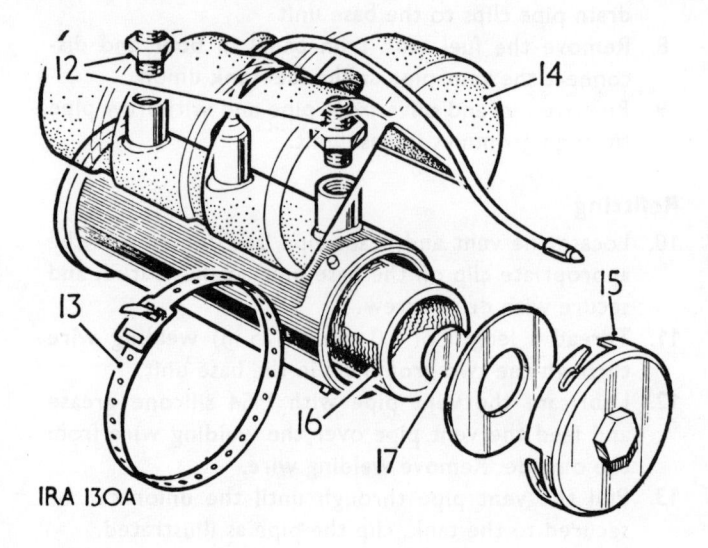

FUEL PUMP—mechanical

—Remove and refit   19.45.08

Removing

1. Disconnect the fuel pipes from the pump.
2. Take precautions against fuel leaking from the pipe from the tank.
3. Remove the fuel pump.

Refitting

4. Using a new joint washer, reverse 1 and 3.
5. Run the engine and check that the pump operates and that there are no leaks.

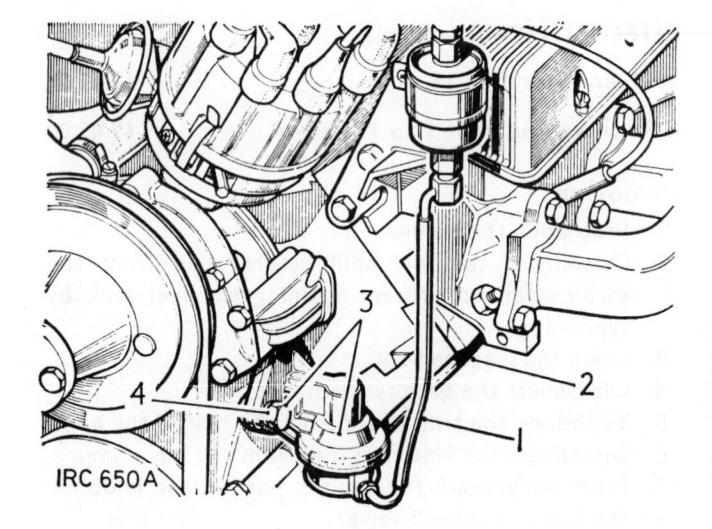

FUEL SYSTEM

FUEL PUMP—mechanical

—Overhaul 19.45.15

Dismantling

1. Remove the fuel pump. 19.45.08.
2. Add alignment marks between the pump body and the fuel cover.
3. Remove the fuel cover.
4. Remove the pulsator cover.
5. Withdraw the pulsator diaphragm.
6. Scrape out the burs produced by staking the valves and remove both valves and gaskets.
7. Remove the return spring from the rocker arm.
8. Hold the link depressed.
9. Depress the diaphragm and tilt the lower end away from the link to unhook the pull rod.
10. Withdraw the diaphragm.
11. Withdraw the spring.
12. Withdraw the oil seal.
13. Drive out the rocker arm pivot pin.
14. Withdraw the rocker arm and link.

continued

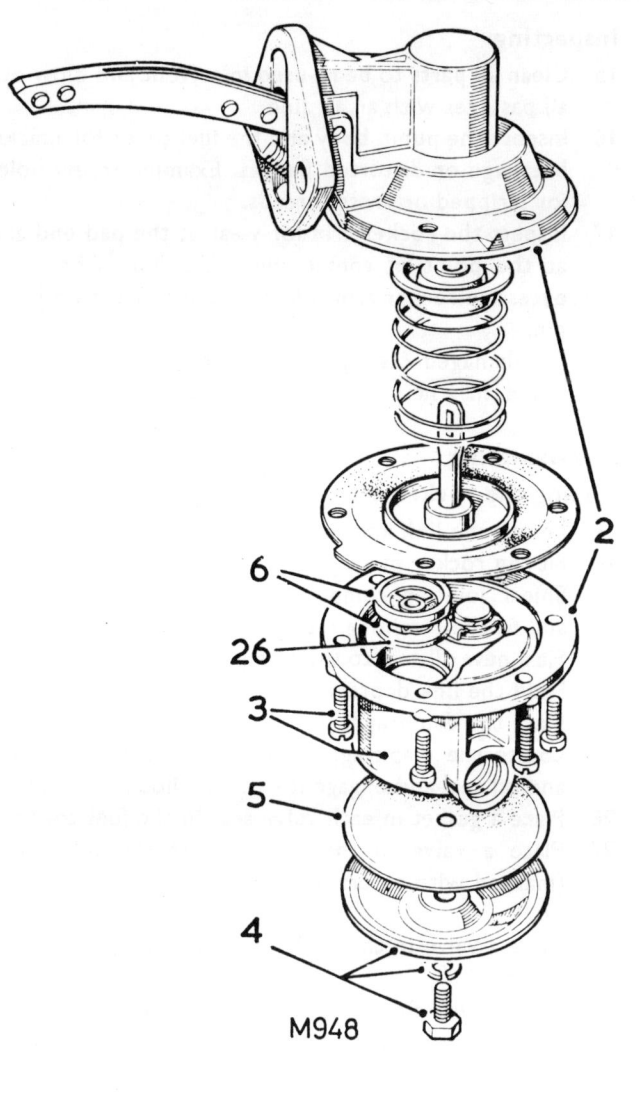

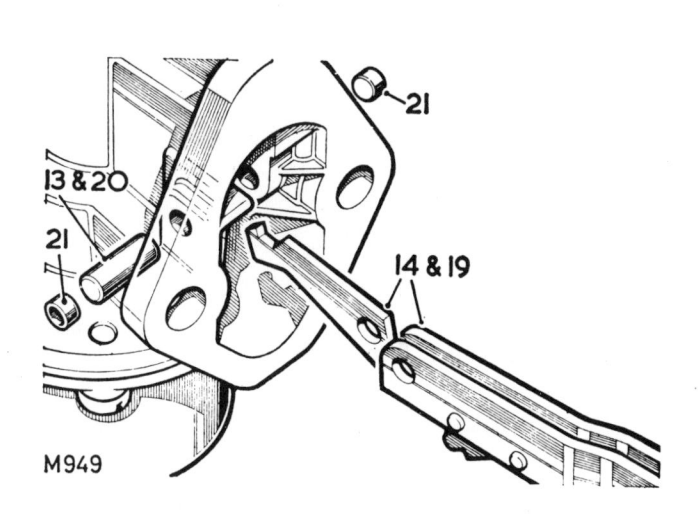

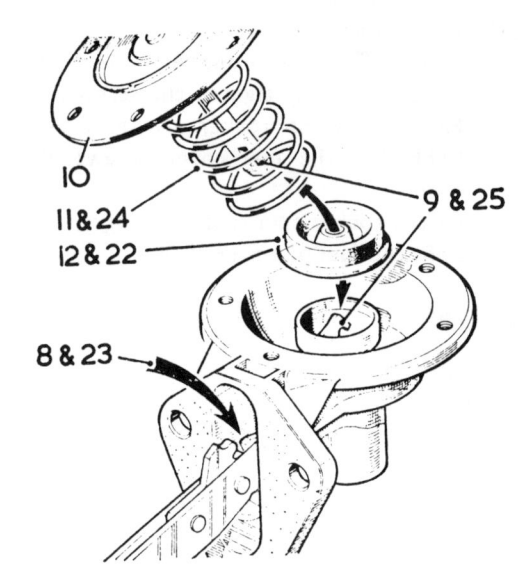

Rover 3500 and 3500S Manual AKM 3621

FUEL SYSTEM

Inspecting

15. Clean all parts to be re-used in solvent and blow out all passages with an air line.
16. Inspect the pump body and the fuel cover for cracks, breakage or distorted flanges. Examine screw holes for stripped or cross threads.
17. Inspect the rocker arm for wear at the pad end and at the point of contact with the link. Check for excessive rocker arm side play due to wear on pivot pin.
18. If a damaged casting or loose rocker pivot pin are found, fit a new fuel pump complete.

Reassembling

Fitting the rocker arm 19 to 21

19. Locate the link and rocker arm into the pump body.
20. Fit the rocker pivot pin.
21. Smear jointing compound on each end of the pin and fit the retaining caps.
22. Fit a new oil seal to the pump body.
23. Hold the link depressed.
24. Place the diaphragm spring in position.
25. Locate the diaphragm pull rod through the spring and oil seal and engage it over the hook on the link.
26. Place a gasket in each valve seat in the fuel cover.
27. Place a valve in the seat nearest the inlet port (marked with an arrow) with the spring cage facing up.
28. Place a valve in the outlet seat with the spring cage down.
29. Seat valves firmly against gaskets and secure by staking the cover in four places around the edge of each valve.
30. Fit a new pulsator diaphragm.
31. Fit the pulsator cover.
32. Fit the fuel cover, maintaining the original alignment, but do not fully tighten the screws.
33. Hold the pump rocker arm fully depressed while tightening the fuel cover screws alternately and evenly.
34. Fit the rocker arm return spring.
35. Refit the fuel pump 19.45.08.

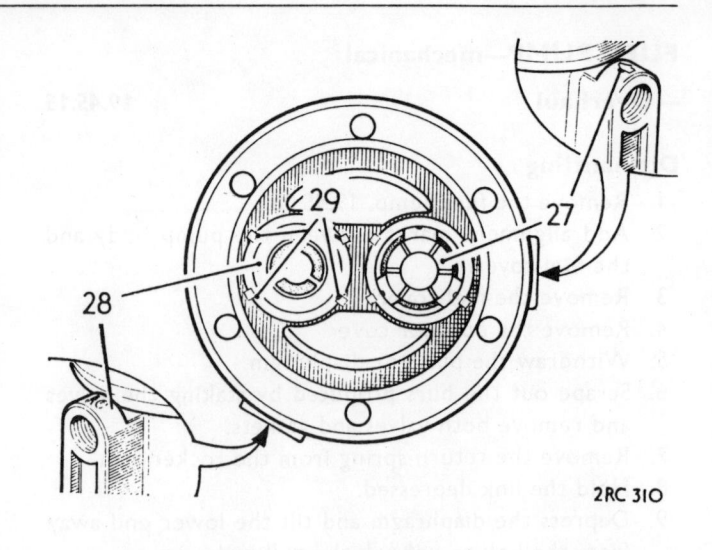

DATA

Fuel pump

Type   A C mechanical
Delivery pressure   0,25 to 0,35 kgf/cm² (3.5 to 5.0 lbf/in²).

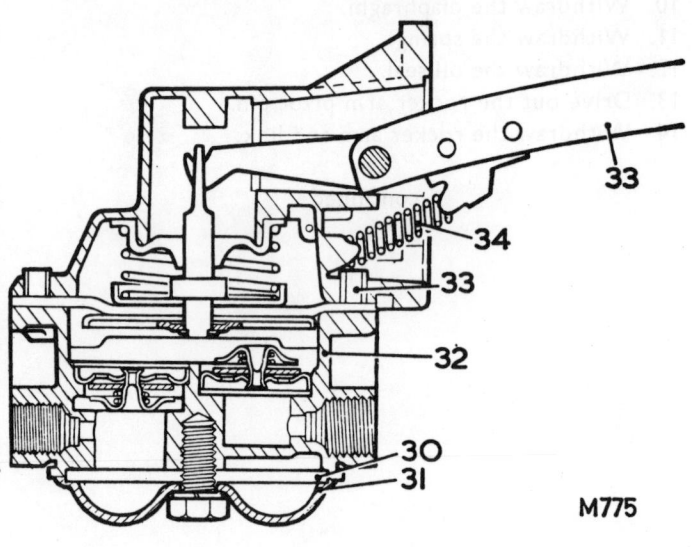

FUEL SYSTEM

FUEL TANK

—Remove and refit 19.55.01

Removing

1. Disconnect the battery earth lead.
2. Disconnect the electrical leads from the tank gauge unit.
3. Drain the fuel by disconnecting the fuel pipes from the tank unit.
4. Remove the spare wheel, if fitted in the boot.
5. Withdraw the trim board from the rear of the boot.
6. Slacken the hose clips securing the upper hose and cut hose off.
7. Remove the fuel filler unit.
8. Disconnect the breather pipe.
9. Remove the lower hose.
10. Release the clamping straps and withdraw the fuel tank.

Refitting

11. Reverse 1 to 10, using Bostik 1753 to secure filler hoses.
12. Refill the fuel tank and check for leaks.

RESERVE TAP

—Remove and refit 19.55.06

Models with tap in engine compartment 1 to 4.
Models with tap under boot floor 3 and 5.

Removing

1. Drain the fuel tank by disconnecting the fuel pipes from the tank unit.
2. Disconnect the three pipe connections from the reserve tap.
3. Disconnect the operating cable.
4. Remove the reserve tap.
5. Drain the fuel tank by disconnecting the three pipe connections from the reserve tap, and lift the tap clear.

Refitting

6. Reverse 1 to 5.
7. Refill the fuel tank and check for leaks, and correct operation of tap.

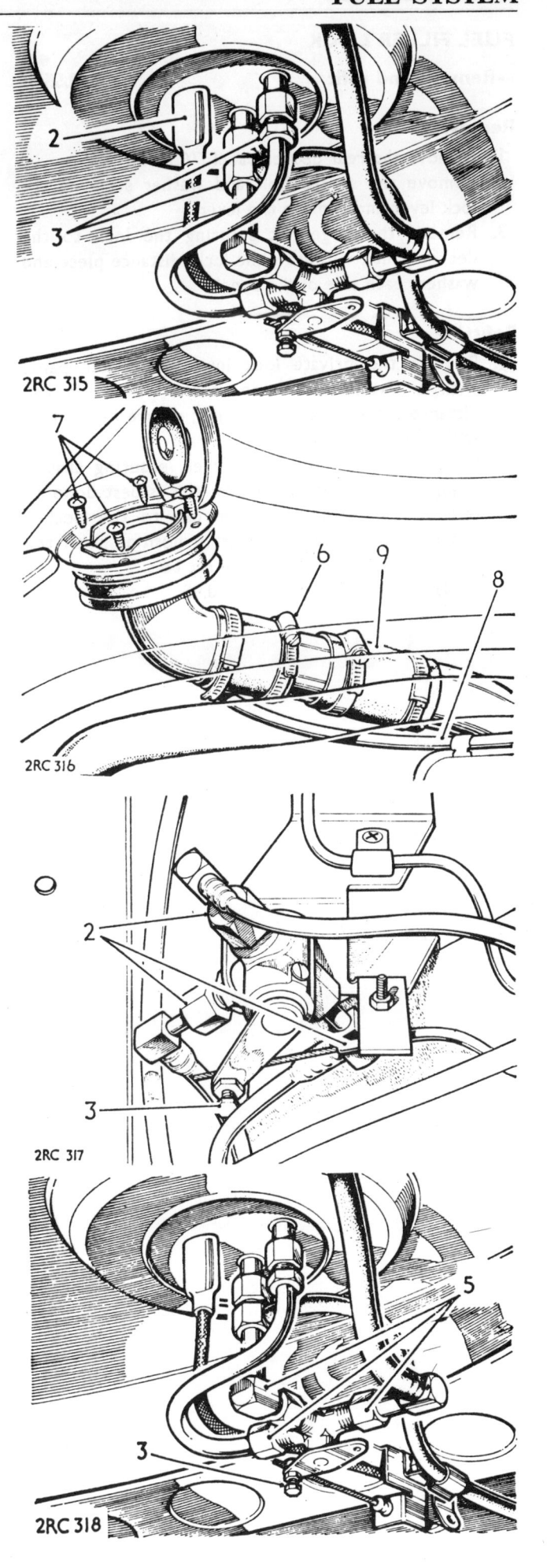

Rover 3500 and 3500S Manual AKM 3621

FUEL SYSTEM

FUEL FILLER LOCK

—Remove and refit   19.55.09

Removing

1. Remove the rear decker panel. 76.10.45.
2. Remove the special bolt and washer securing the lock lever and remove the lever.
3. Remove the large nut securing the lock to the decker panel, withdraw the lock, distance piece and washers each side of the panel.

Refitting

4. Position the private lock into the decker panel with a washer each side of the panel and the distance piece on the underside; secure in position with the large nut.
5. Fit the special washer and lever, ensuring that it engages in the filler cap catch; this is best done by temporarily fitting the filler cap to the decker panel. Finally, lock the lever in position with the shakeproof washer and special set bolt.
6. Replace the decker panel. 76.10.45.

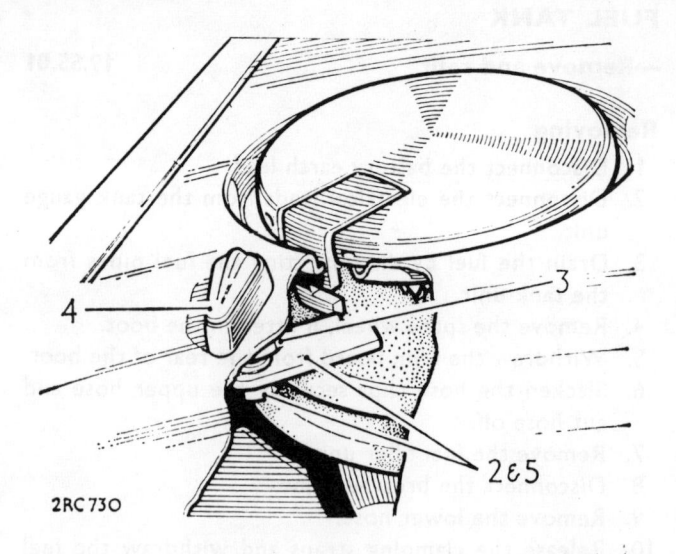

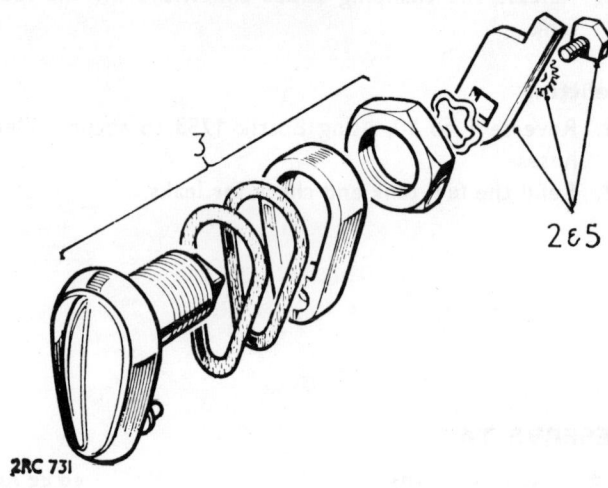

19.55.09

Rover 3500 and 3500S Manual AKM 3621

COOLING SYSTEM

LIST OF OPERATIONS

Coolant—drain and refill	26.10.01
Fan belt	
—check and adjust	26.20.01
—remove and refit	26.20.07
Fan blades and pulley—remove and refit	26.25.01
Fan blades—remove and refit	26.25.06
Radiator block—remove and refit	26.40.04
Thermostat—remove and refit	26.45.01
Water pump—remove and refit	26.50.01

Rover 3500 and 3500S Manual AKM 3621

26-1

COOLING SYSTEM

COOLANT

—Drain and refill   **26.10.01**

WARNING: Do not remove the radiator filler cap when the engine is hot because the cooling system is pressurised and personal scalding could result.

Draining

1. Remove the radiator filler cap.
2. Remove the plug and drain the radiator. As the system is filled with a solution of anti-freeze or inhibitor, use a clean container if the coolant is to be re-used.
3. Refit the drain plug and washer.
4. Open the tap(s) on the side of the cylinder block, and drain the engine.
5. Close the tap(s).

Coolant requirements

Frost precautions and engine protection

The engine cooling system MUST ALWAYS be filled or topped up with a solution of water and anti-freeze, winter and summer, or water and inhibitor all year if frost precautions are not required. NEVER use water alone as this would corrode the aluminium alloy engine.

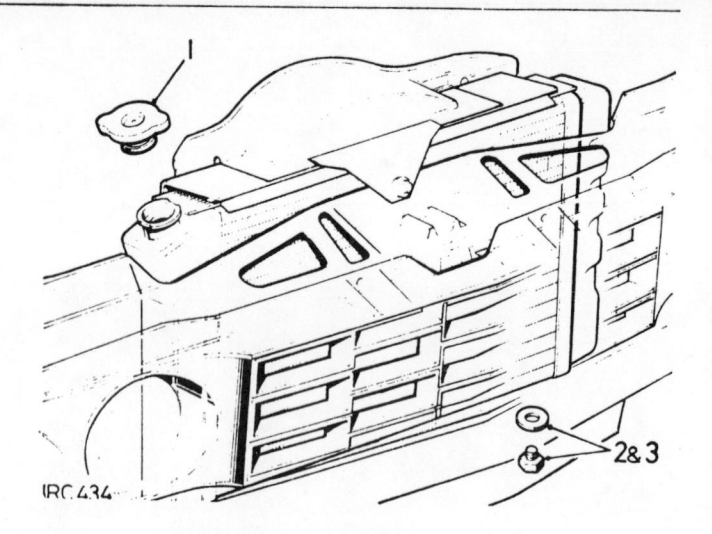

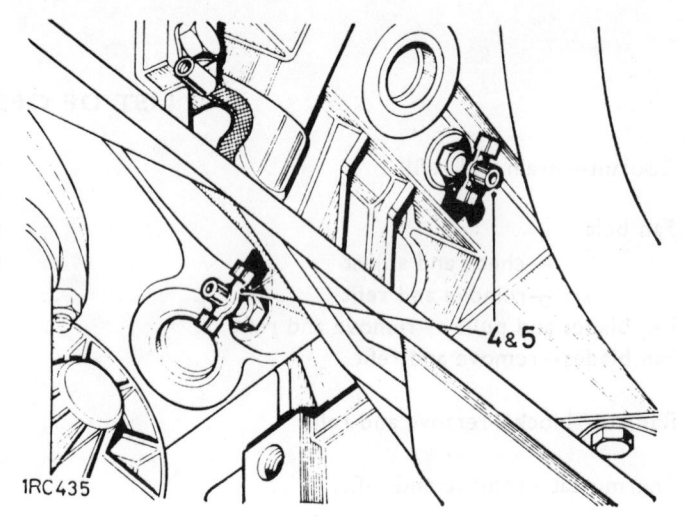

Recommended solutions

Anti-freeze—

Bluecol AA, coloured green, or Anti-freeze solution conforming to British Standard No. 3150, or
Prestone, or
Anti-freeze solution to MIL-E-5559 formulation.

Inhibitor—

Marston Lubricants SQ35 Coolant Inhibitor Concentrate. Rover Part No. 605765.
Use soft water wherever possible. If the local water supply is hard, use rain water.
To ensure that the solution is effective, the coolant should be changed annually.

IMPORTANT

Points to observe with 'Prestone' anti-freeze

Where 'Prestone' is to be used after BS 3150, SQ36 or MIL-E-5559, empty the coolant; refill with water. Run the engine to circulate the coolant throughout the system, stop the engine, empty the coolant.
Repeat filling with water running and emptying once again; finally swill out by use of a hose and running water into the top header tank for a few minutes with the exit taps open, then close taps and fill with the appropriate amount of water and 'Prestone'.

COOLING SYSTEM

Use the correct anti-freeze mixture according to local climatic conditions as follows:

Cooling system capacity	Frost precaution	Proportion of anti-freeze	Anti-freeze required to raise 33⅓% solution to 50%
15.25 UK pints 8,5 litres 18.5 US pints	−25°F (−32°C)	33⅓% 5.25 UK pints 3.0 litres 6.3 US pints	3.75 UK pints 2,1 litres 4.5 US pints
	−33°F (−36°C)	50% 7.75 UK pints 4,2 litres 9.25 US pints	

To raise a 33⅓% solution to 50% drain off the appropriate amount of the original 33⅓% solution and add the corresponding quantity of neat anti-freeze, this amount compensates for the anti-freeze lost when the radiator is partially drained.

If frost precautions are not required, use 19 cc of inhibitor per litre of water (3 fluid ounces of inhibitor per gallon of water).

Rover cars leaving the Rover factory (except to the U.S.A. and Canada) have the cooling system filled with 33⅓% of anti-freeze mixture.

Rover cars despatched to the United States and Canada have the cooling system filled with 50% anti-freeze mixture.

At the same time a 25% solution of Isopropyl Alcohol is added to the windscreen washer reservoir, during winter on home market models and all year on export models.

Refilling

6. Remove the radiator filler cap.
7. Pour 4,5 litres (1 gallon) of water into the radiator.
8. Add the recommended quantity of anti-freeze or inhibitor.
9. Top up the radiator with water until the coolant is level with the bottom of the filler neck.
10. Start and run the engine at approximately 1500 rev/min for one minute to purge all air from the heater system.
11. With the engine idling, top up the radiator to bottom of the filler neck.
12. Run the engine until it attains normal running temperature, i.e. thermostat open. Check the level of the coolant and top up if necessary.
13. With the engine cold, the final level of the coolant will be 25 mm (1 in) below the lowest point of the filler neck.

Rover 3500 and 3500S Manual AKM 3621

COOLING SYSTEM

FAN BELT

—Check and adjust 1 and 5 to 6   26.20.01

—Remove and refit 1 to 6   26.20.07

Removing

1. Slacken the alternator fixings, three bolts and nuts.
2. Pivot the alternator inwards.
3. Lift off the fan belt.

Refitting

4. Locate the fan belt on the pulleys.
5. Using the alternator slotted fixing, adjust the fan belt tension to give 11 to 14 mm (0.437 to 0.562 in) free movement when checked midway between the alternator and crankshaft pulleys, by hand.
6. Secure the alternator fixings.

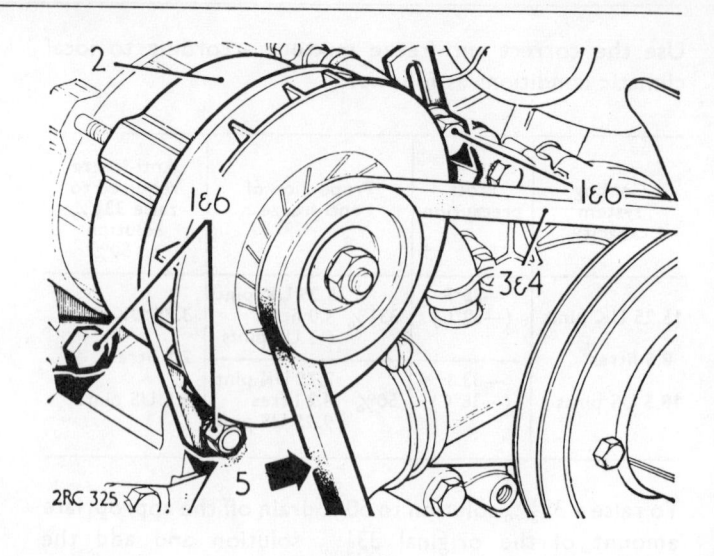

FAN BLADES AND PULLEY

—Remove and refit 1 to 9   26.25.01

FAN BLADES

—Remove and refit 1, 2 and 6 to 8   26.25.06

Removing

1. Remove the fan guard, four bolts.
2. Remove the fan blades.
3. Standard type pulley. Remove the pulley fixings.
4. Viscous coupling type pulley. Remove the coupling fixings.
5. Slacken the alternator fixings, remove the fan belt and lift off the viscous coupling and/or the fan pulley.

Refitting

6. Reverse 1 to 5, noting the following.
7. Standard type pulley. An off-set dowel location ensures that the fixing bolt holes only align when the blades are the correct way round.
8. Viscous coupling type pulley. Fit the fan blades with the larger diameter fixing bosses to the front.
9. Adjust the fan belt. 26.20.01.

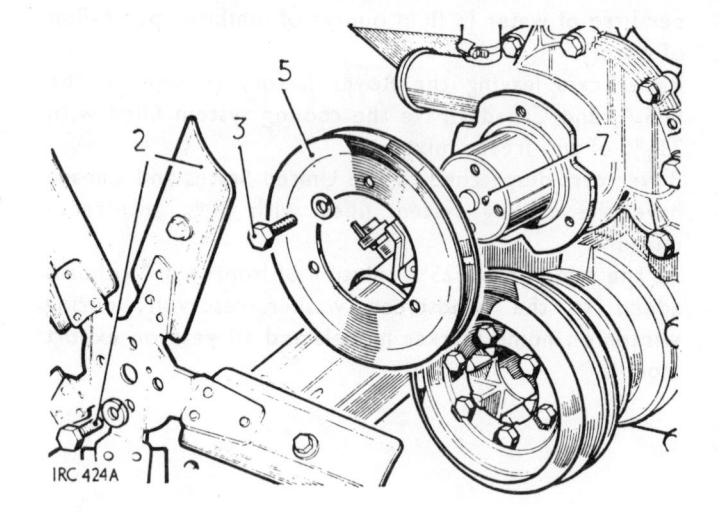

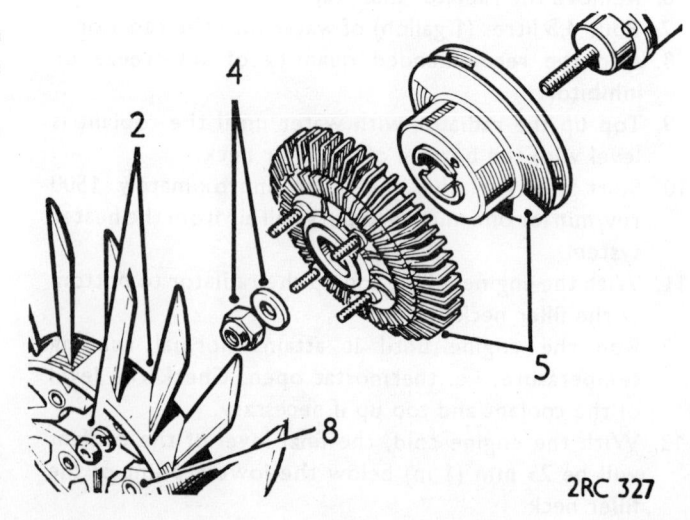

26.20.01
26.25.06

Rover 3500 and 3500S Manual AKM 3621

COOLING SYSTEM

RADIATOR BLOCK

—Remove and refit   **26.40.04**

Removing
1. Drain the cooling system. 26.10.01.
2. Disconnect the top hose from the radiator.
3. Disconnect the hose to the induction manifold, and the overflow hose (later models).
4. Disconnect the bottom hose from the radiator.
5. **Automatic gearbox models:** Disconnect the two oil cooler pipes from the radiator.
6. Remove the fixings from the top of the radiator.
7. Remove the fixings from each side of the radiator.
8. Lift out the radiator.
9. If required, remove the fan guard.

Refitting
10. Reverse 1 to 9. Use the correct mixture to refill the cooling system. 26.10.01.
11. Run the engine and check the cooling system for leaks.

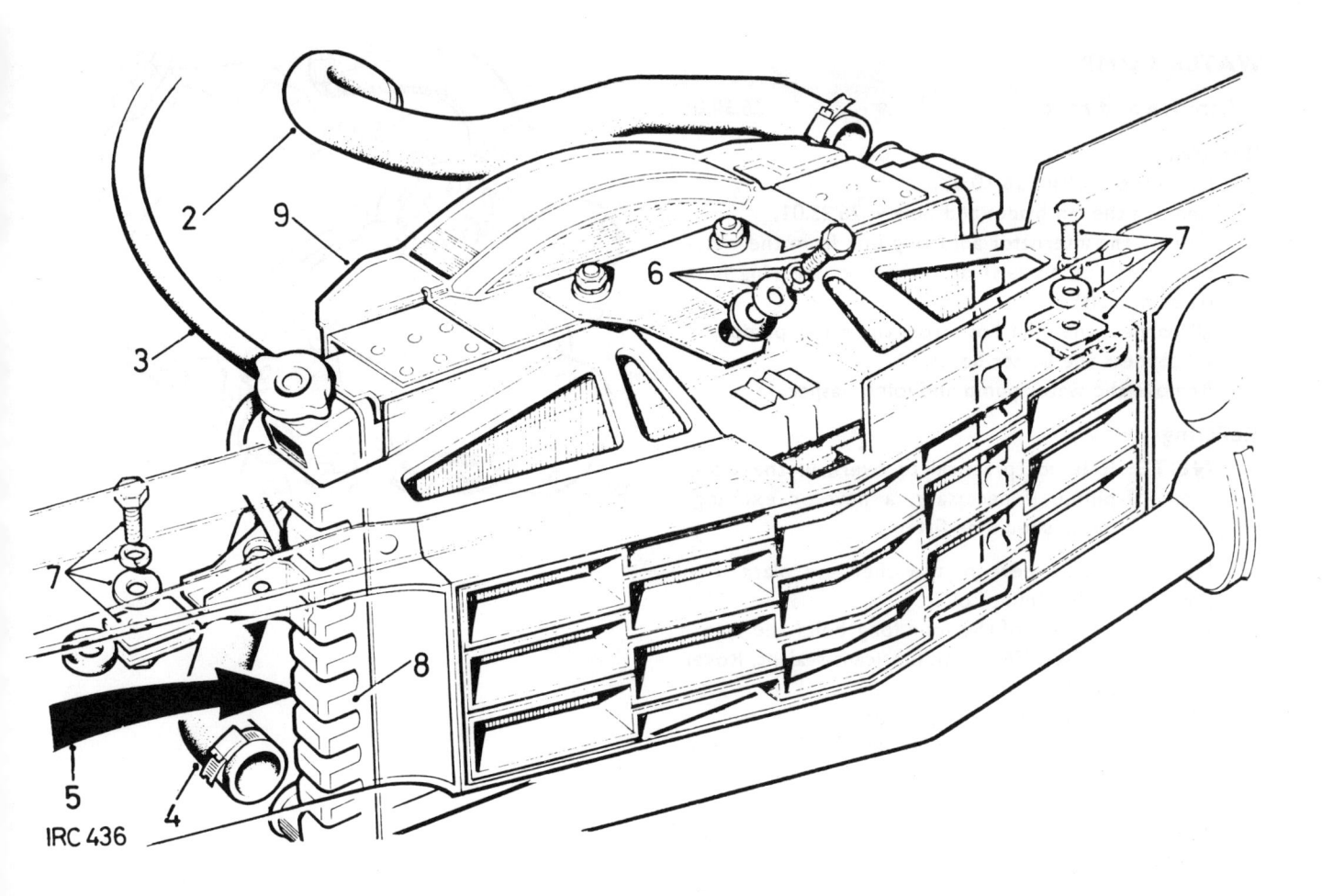

Rover 3500 and 3500S Manual AKM 3621

COOLING SYSTEM

THERMOSTAT

—Remove and refit 26.45.01

Removing

1. Drain the cooling system. 26.10.01, sufficient to drain the induction manifold.
2. Disconnect the hose to the radiator.
3. Disconnect the thermostat by-pass hose (earlier models).
4. Remove the outlet elbow.
5. Withdraw the thermostat.

Testing

6. When immersed in hot water, the thermostat should commence expansion between 80 to 84 degrees centigrade (176 to 183 degrees fahrenheit).

Refitting

7. Insert the thermostat with the jiggle pin uppermost (12 o'clock).
8. Using a new joint washer, fit the outlet elbow.
9. Reverse 1 to 3.

WATER PUMP

—Remove and refit 26.50.01

Removing

1. Drain the coolant. 26.10.01.
2. Remove the fan blades and pulley. 26.25.01.
3. Release the alternator adjusting link from the water pump (earlier models) or the power steering hose (later models).
4. Disconnect the inlet hose from the water pump.
5. Disconnect the heater hoses.
6. Remove the water pump and joint washer.

Refitting

   NOTE: It is not practical to overhaul the water pump, if necessary, a new or exchange pump should be fitted.

7. Lightly grease a new joint washer and place it in position on the timing cover.
8. Clean the threads of the four long bolts and smear them with 3M-EC776 thread lubricant-sealant, Rover Part No. 605764.
9. Locate the water pump in position.
10. Locate the alternator adjusting link (earlier models) or the power steering hose (later models) on the water pump.
11. Leave the alternator adjusting link loose and tighten the remaining water pump fixings gradually.
    Torque: Large bolts 2,8 to 3,5 kgf.m (20 to 25 lbf ft)
    Small bolts 0,8 to 1,0 kgf.m (6 to 8 lbf ft)
12. Connect the heater hoses to the water pump.
13. Connect the inlet hose to the water pump.
14. Fit the fan pulley.
15. Fit and adjust the fan belt. 26.20.07.
16. Fit the fan blades. 26.25.06.
17. Refill the cooling system. 26.10.01.

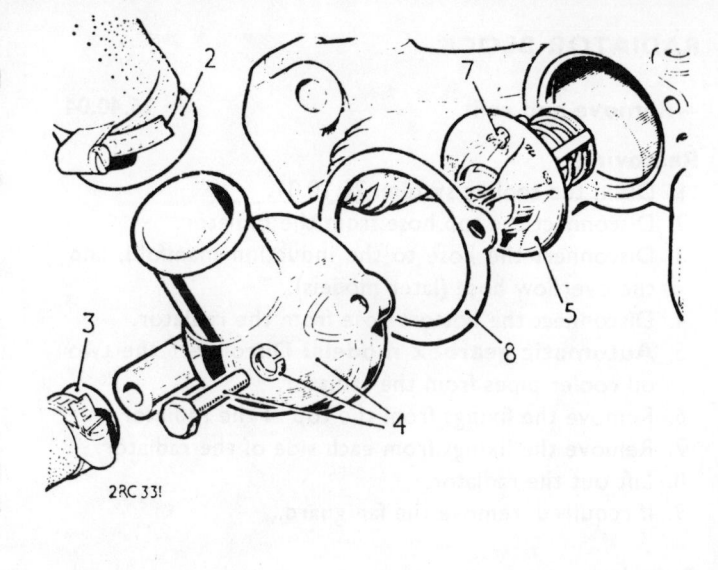

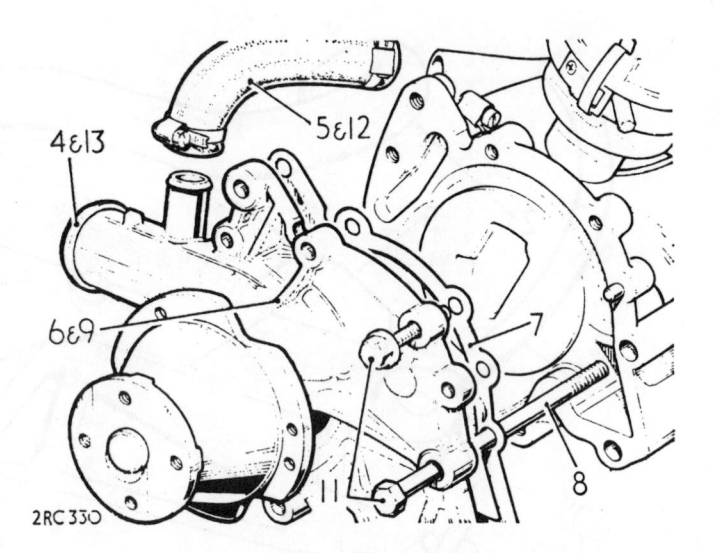

MANIFOLD AND EXHAUST SYSTEM

LIST OF OPERATIONS

Exhaust manifold	
—left hand—remove and refit	30.15.10
—right hand—remove and refit	30.15.11
Exhaust system—remove and refit	30.10.01
—front pipe—remove and refit	30.10.05
—intermediate pipe—remove and refit	30.10.18
—tail pipe—remove and refit	30.10.22
Induction manifold—remove and refit	30.15.02

MANIFOLD AND EXHAUST SYSTEM

EXHAUST SYSTEM COMPLETE—Automatic Gearbox Models

—Remove and refit	30.10.01
Front pipe 1 to 4 and 11	30.10.05
Intermediate pipe 5 to 7 and 12	30.10.18
Tail pipe 8, 9 and 13	30.10.22

Removing

1. Disconnect the front pipe from the manifolds.
2. Disconnect the front pipe from the intermediate pipe.
3. Release the front pipe from the gearbox mounting.
4. If required, separate the left hand and right hand sections of the front pipe.
5. Disconnect the intermediate pipe from the front pipe.
6. Disconnect the intermediate pipe from the tail pipe.
7. Disconnect the rubber mounting rings from the silencer.
8. Disconnect the tail pipe from the intermediate pipe.
9. Disconnect the silencer from the rear mounting bracket.

Refitting

10. Complete system, reverse 1 to 9.
11. Front pipe, reverse 1 to 4.
12. Intermediate pipe, reverse 5 to 7.
13. Tail pipe, reverse 8 and 9.

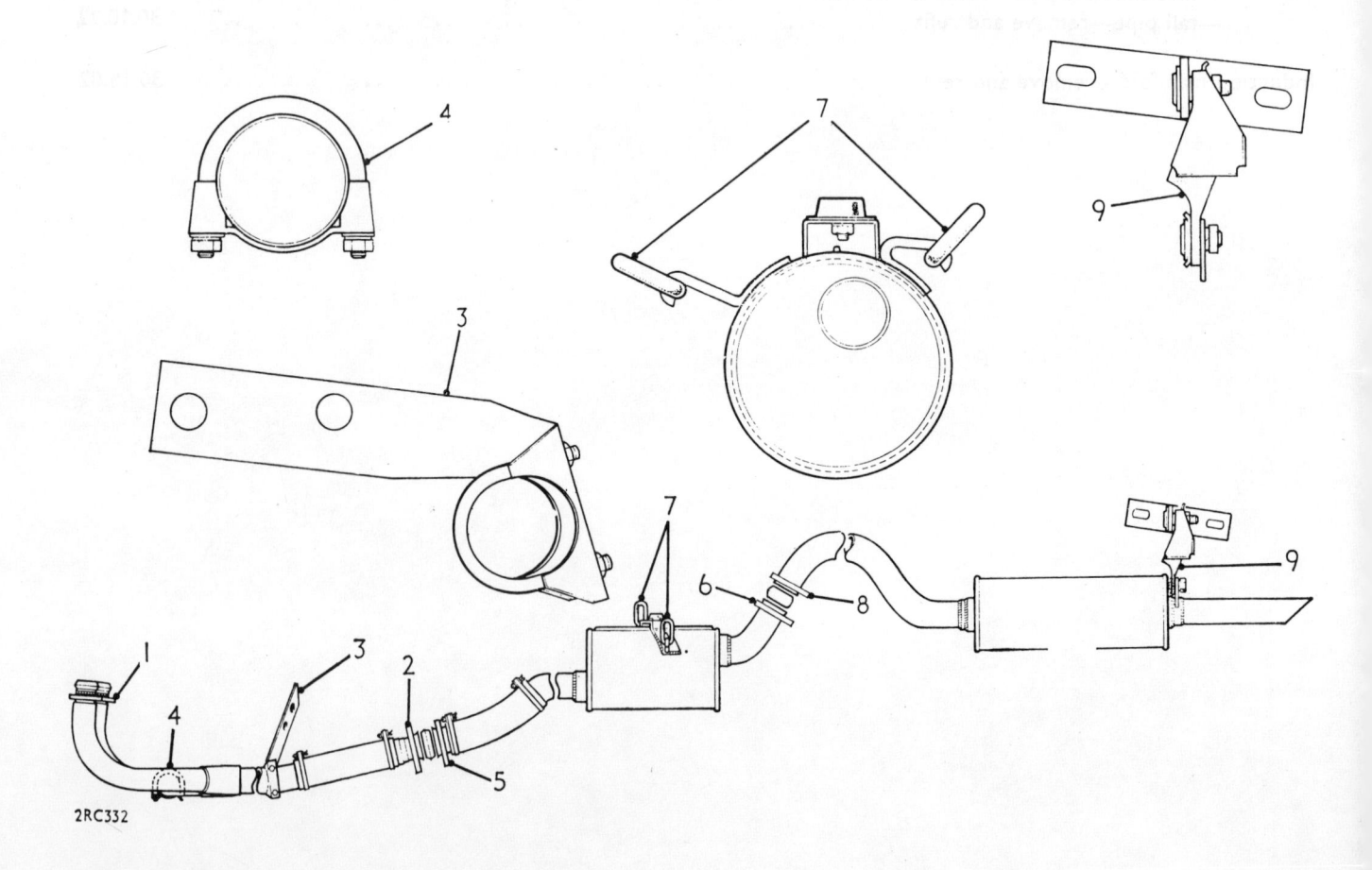

Rover 3500 and 3500S Manual AKM 3621

MANIFOLD AND EXHAUST SYSTEM

EXHAUST SYSTEM COMPLETE—Synchromesh Gearbox Models

—Remove and refit	30.10.01
Front pipe 1 to 4 and 11	30.10.05
Intermediate pipe 5 to 7 and 12	30.10.18
Tail pipe 8, 9 and 13	30.10.22

Removing

1. Disconnect the front pipe from the manifolds.
2. Disconnect the front pipe from the intermediate pipe.
3. Release the front pipe from the gearbox mounting.
4. If required, separate the left hand and right hand sections of the front pipe.
5. Disconnect the intermediate pipe from the front pipe.
6. Disconnect the intermediate pipe from the tail pipe.
7. Disconnect the rubber mounting rings from the silencer.
8. Disconnect the tail pipe from the intermediate pipe.
9. Disconnect the silencer from the rear mounting bracket.

Refitting

10. Complete system, reverse 1 to 9.
11. Front pipe, reverse 1 to 4.
12. Intermediate pipe, reverse 5 to 7.
13. Tail pipe, reverse 8 and 9.

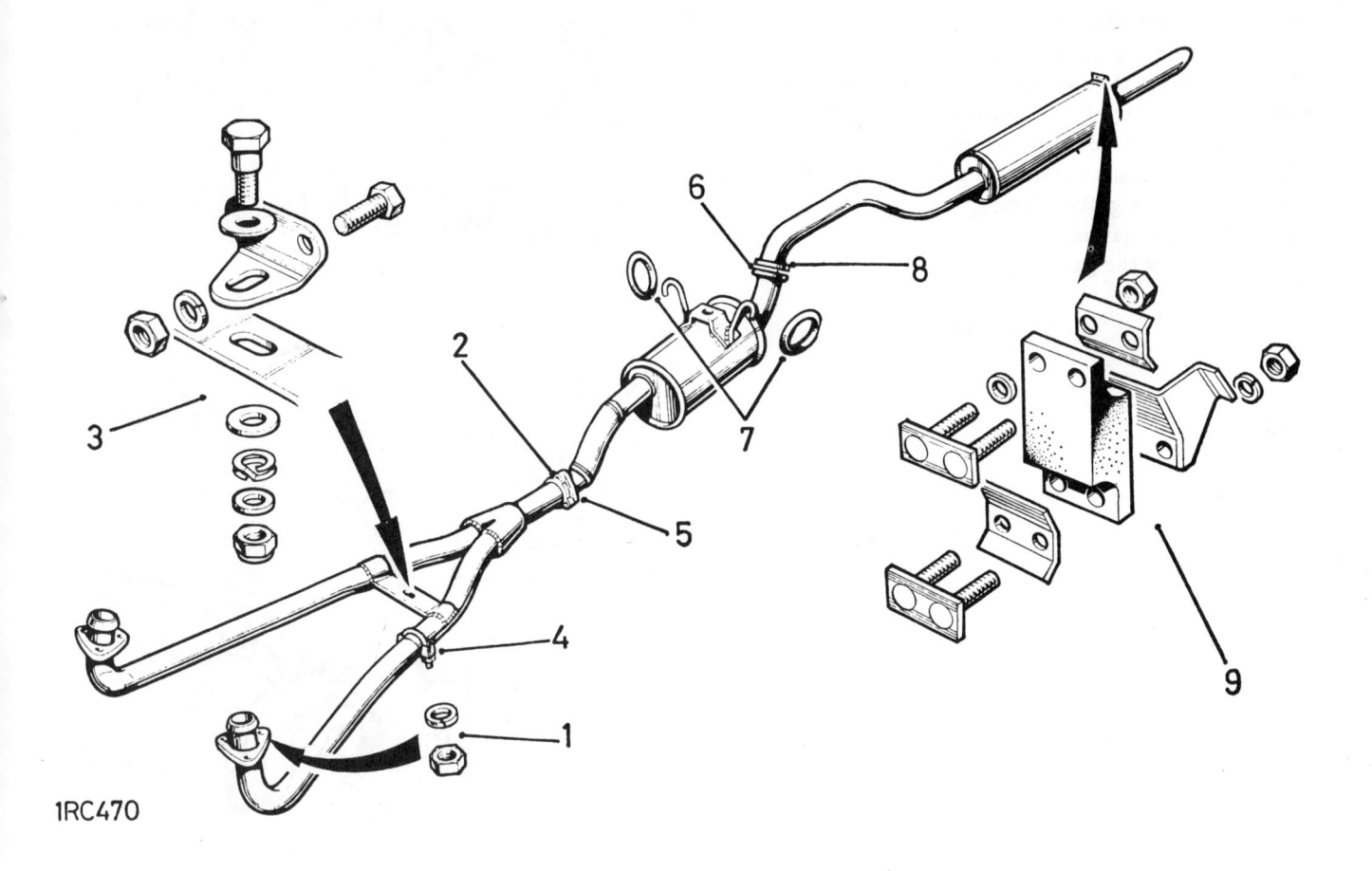

1RC470

Rover 3500 and 3500S Manual AKM 3621

30.10.01
30.10.22
Sheet 2

MANIFOLD AND EXHAUST SYSTEM

INDUCTION MANIFOLD

—Remove and refit 30.15.02

Removing

1. Drain the cooling system. 26.10.01.
2. Remove the air cleaner. 19.10.01.
3. Remove the engine breather filter. 17.10.02.
4. Disconnect the throttle linkage.
5. Disconnect the choke control cable from the carburetter or, if an automatic enrichment device is fitted, remove the A.E.D. 19.15.38.
6. Automatic gearbox models. Disconnect the downshift cable from the throttle linkage.
7. Disconnect the fuel spill return pipe from the RH carburetter.
8. Remove the fuel supply pipe from the carburetters.
9. Disconnect the lead from the choke thermostat switch.
10. Disconnect the lead from the water temperature transmitter.
11. Disconnect the flame trap hoses from the carburetters.
12. Disconnect the vacuum pipe for the brake servo.
13. Disconnect the vacuum pipe from the distributor.
14. Disconnect the h.t. lead from the distributor and release the distributor cap.
15. Disconnect the outlet hose from the heater.
16. Disconnect the inlet hose to the heater.
17. Disconnect the return hose to the radiator.
18. Disconnect the return hose from the top of the induction manifold.
19. Disconnect the by-pass hose from the thermostat.
20. Disconnect the heater return pipe from the manifold.
21. Remove the induction manifold.
22. Wipe away any coolant lying on the manifold gasket.
23. Remove the gasket clamps.
24. Lift off the gasket.
25. Withdraw the gasket seals.

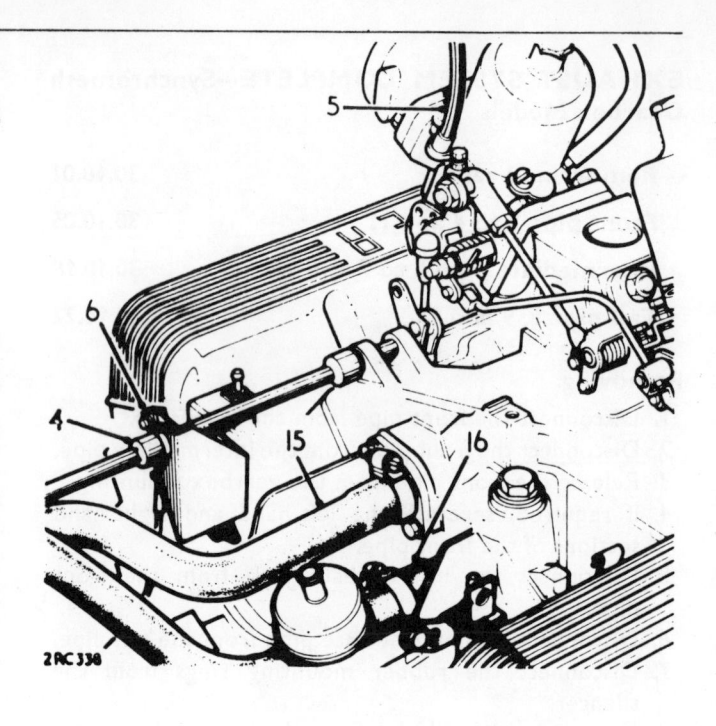

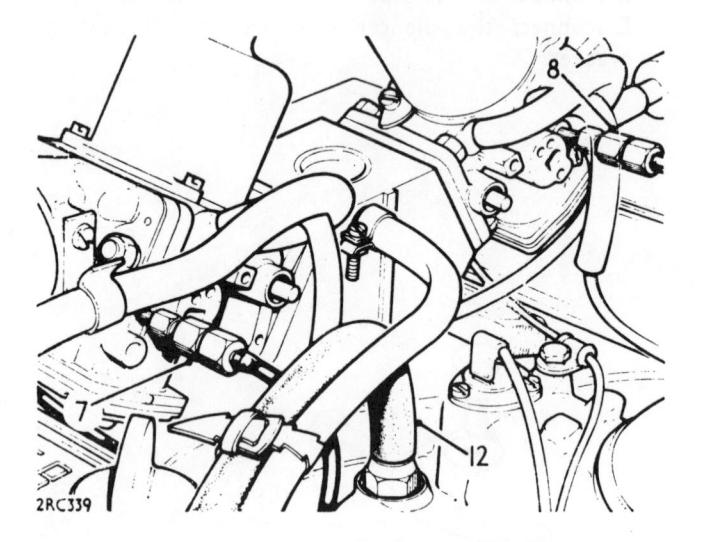

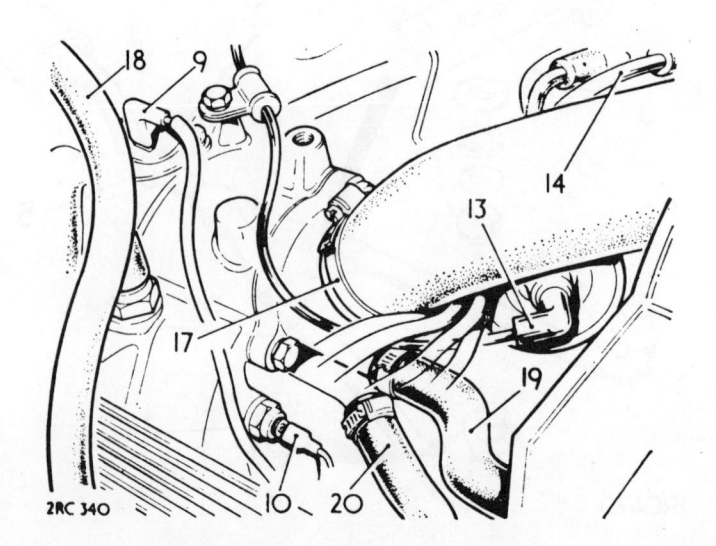

continued

30.15.02
Sheet 1

Rover 3500 and 3500S Manual AKM 3621

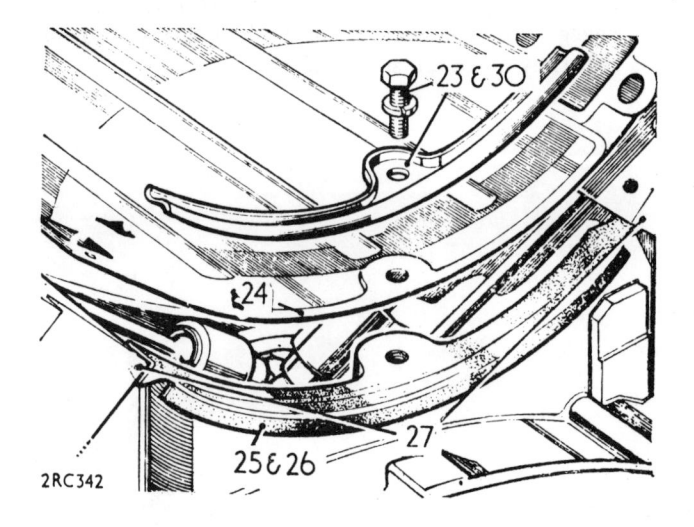

MANIFOLD AND EXHAUST SYSTEM

Refitting

26. Using new seals, smear them on both sides with silicon grease.
27. Locate the seals in position with their ends engaged in the notches formed between the cylinder head and block.
28. Apply 'Hylomar' sealing compound PL 32M, Part No. 534244 on the corners of the cylinder head, manifold gasket and manifold, around the water passage joints.
29. Fit the manifold gasket with the word 'FRONT' to the front and the open bolt hole at the front RH side.
30. Fit the gasket clamps but do not fully tighten the bolts at this stage.
31. Locate the manifold on to the cylinder head.
32. Clean the threads of the manifold securing bolts and then coat them with Thread Lubricant-sealant 3M EC776, Rover Part Number 605764.
33. Fit all the manifold bolts and tighten them a little at a time, evenly, alternate sides working from the centre to each end. Torque: 3,5 to 4,0 kgf. m (25 to 30 lbf ft).
34. Tighten the gasket clamp bolts. Torque: 1,4 to 2,0 kgf. m (10 to 15 lbf ft).
35. Reverse 1 to 20.
36. Run the engine and check for water leaks.

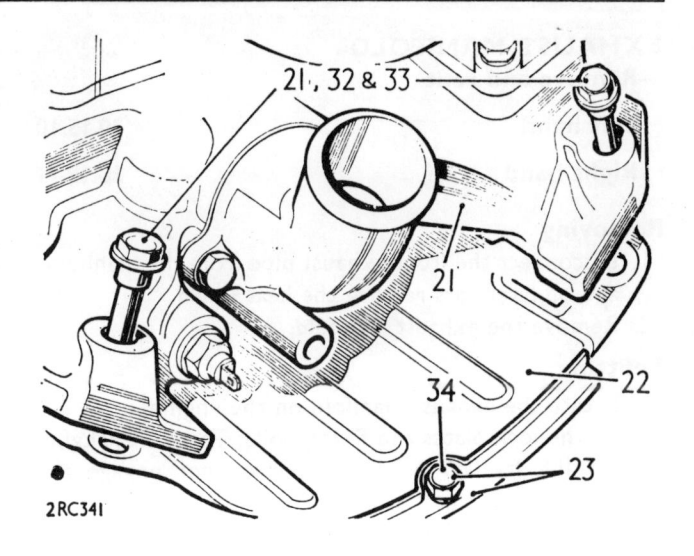

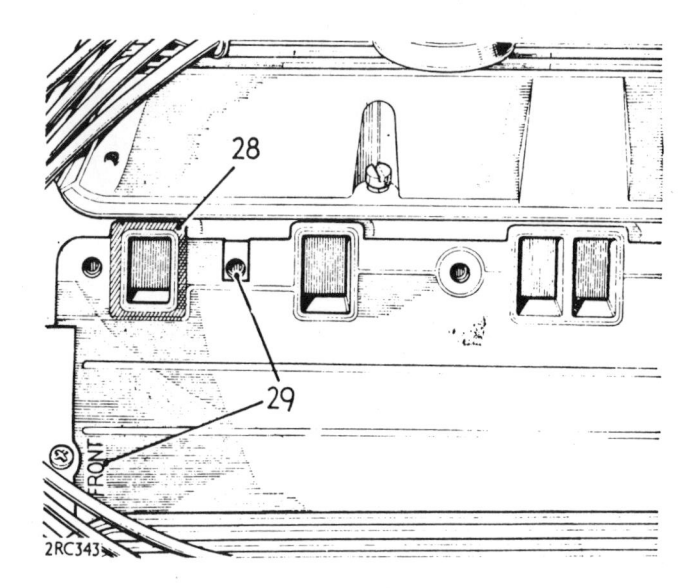

Rover 3500 and 3500S Manual AKM 3621

30.15.02
Sheet 2

MANIFOLD AND EXHAUST SYSTEM

EXHAUST MANIFOLD
—Remove and refit

Left hand	30.15.10
Right hand	30.15.11

Removing

1. Disconnect the front exhaust pipe from the manifold, and on later cars remove the heat shield.
2. Remove the exhaust manifold.

Refitting

3. Locate the exhaust manifold on the engine.
4. Fit the lock plates and fixing bolts. Torque: 2,0 kgf.m (15 lbf. ft) DO NOT overtighten nor engage the lockplates at this stage.
5. Reconnect the front exhaust pipe.
6. Start and run the engine for a minimum of five minutes, then recheck the torque settings.
7. Engage the lock plates over the bolt heads ensuring that they make full and close contact.

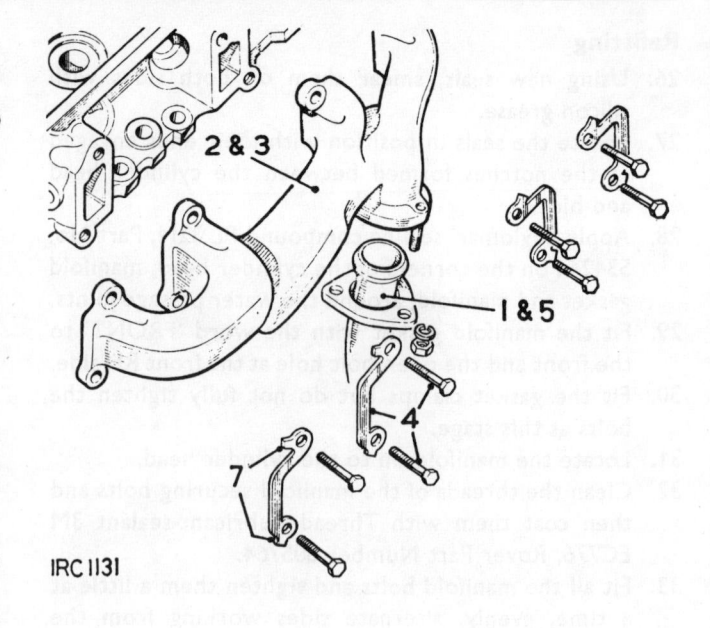

IRC 1131

30.15.10
30.15.11

Rover 3500 and 3500S Manual AKM 3621

CLUTCH

CLUTCH OPERATIONS

Clutch assembly
    —remove and refit 33.10.01
    —overhaul 33.10.08
Clutch and brake pedal—remove and refit 33.30.01
Clutch pedal and linkage—adjust 33.25.01
Hydraulic system—bleed 33.15.01
Master cylinder
    —remove and refit 33.20.01
    —overhaul 33.20.07
Slave cylinder
    —remove and refit 33.35.01
    —overhaul 33.35.07
Withdrawal unit
    —remove and refit 33.25.12
    —overhaul 33.25.17

CLUTCH

CLUTCH ASSEMBLY

—Remove and refit 33.10.01

Service tool 18G79 Clutch centralising tool

Removing

1. Remove the gearbox assembly. 37.20.01.
2. Mark the clutch cover fitted position relative to the flywheel.
3. Do not disturb the three bolts located in the apertures in the clutch cover.
4. Remove the clutch assembly.
5. Withdraw the clutch driven plate.

Refitting

6. Reverse 4 and 5, aligning the assembly marks. Centralising tool 18G79.
7. Secure the cover fixings evenly, using diagonal selection. Torque: 2,8 kgf.m (20 lbf. ft).
8. Fit the gearbox. 37.20.01.

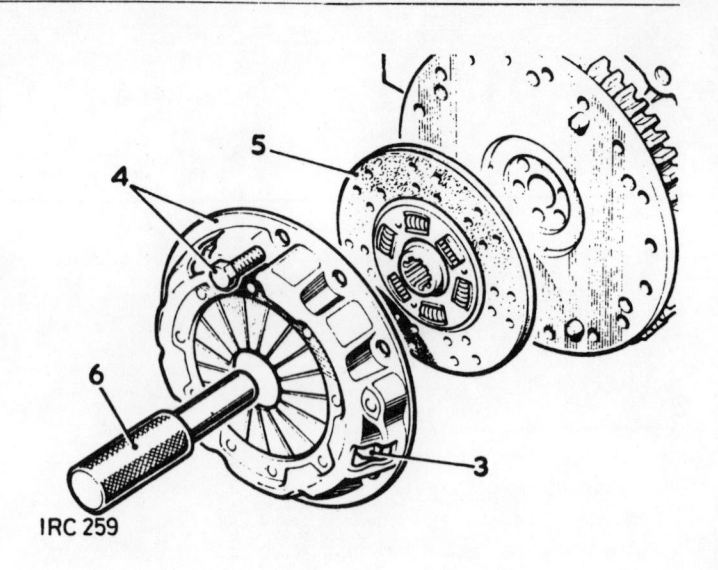

DATA

Clutch driven plate diameter	241,3 mm (9.5 in)
Damper springs colour identification	Brown/cream

CLUTCH ASSEMBLY

—Overhaul 33.10.08

Clutch assembly

The clutch assembly is of the diaphragm spring type and no overhaul procedures are applicable. Repair is by replacement only.

Clutch driven plate

Examine clutch driven plate for wear and signs of oil contamination. Examine all rivets for pulling and distortion, rivets must be below the friction surface. If oil contamination is present on the friction linings or if they are appreciably worn, replace the clutch driven plate assembly complete or alternatively, replace the friction linings following standard workshop practices.

DATA

Clutch driven plate diameter	241,3 mm (9.5 in)
Damper springs colour identification	Brown/cream

CLUTCH

HYDRAULIC SYSTEM

—Bleed   33.15.01

Procedure

NOTE: During the procedure, keep the fluid reservoir topped up to avoid introducing further air into the system. Use only the recommended type of hydraulic fluid. Division 09 refers.

1. Attach a length of suitable tubing to the slave cylinder bleed screw.
2. Place the free end of the tube in a glass jar containing clutch fluid.
3. Slacken the bleed screw.
4. Pump the clutch pedal, pausing at the end of each stroke, until the fluid issuing from the tubing is free of air with the tube free end below the surface of the fluid in the container.
5. Hold the tube free end immersed and tighten the bleed screw when commencing a pedal down stroke.

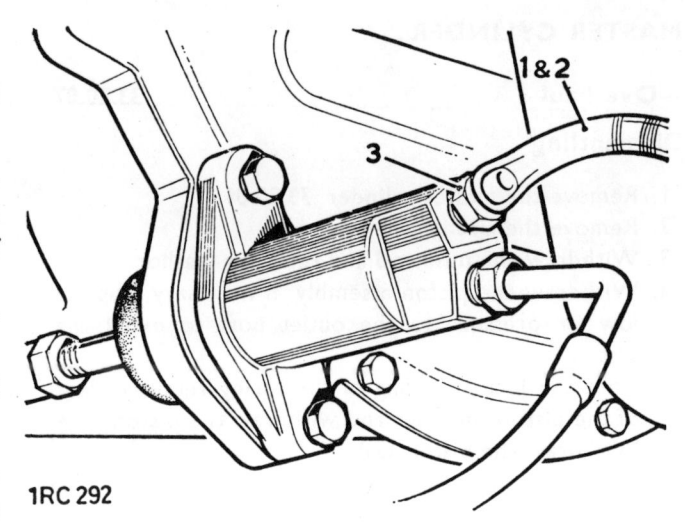

MASTER CYLINDER

—Remove and refit   33.20.01

Removing

1. Remove the air cleaner 19.10.01.
2. Clamp off the rubber inlet hose to the master cylinder, using a Girling Hose Clamp, or drain the reservoir.
3. Disconnect the inlet hose.
4. Disconnect the outlet pipe.
5. Disconnect the push rod.

NOTE: RH Stg push rod connection is illustrated. LH Stg push rod must be released from the pedal trunnion in the driving compartment.

6. Remove the master cylinder.
7. RH Stg if required, remove the clevis jaw from the push rod.

Refitting

8. RH Stg. If removed, fit the clevis jaw and secure in position with the push rod fully withdrawn and a dimension of 120 mm (4.750 in) between the centre of the hole in the jaw and the master cylinder mounting flange.
9. Reverse 1 to 6.
10. Bleed and replenish the hydraulic system. 33.15.01.
11. Check, and if necessary adjust, the clutch pedal height, 33.25.01, items 1 to 5.

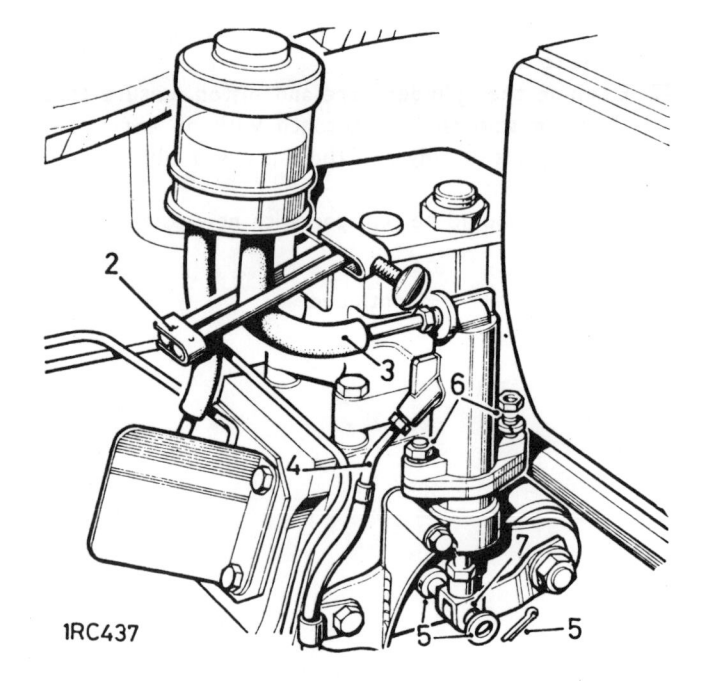

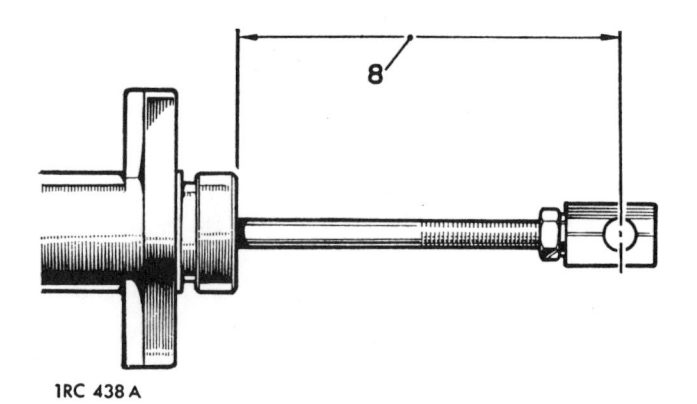

Rover 3500 and 3500S Manual AKM 3621

CLUTCH

MASTER CYLINDER

—Overhaul 33.20.07

Dismantling

1. Remove the master cylinder. 33.20.01.
2. Remove the circlip.
3. Withdraw the push rod and retaining washer.
4. Withdraw the piston assembly. If necessary, apply a low air pressure to the outlet port to expel the piston.
5. Prise the locking prong of the spring retainer clear of the piston shoulder and withdraw the piston.
6. Withdraw the piston seal.
7. Compress the spring and position the valve stem to align with the larger hole in the spring retainer.
8. Withdraw the spring and retainer.
9. Withdraw the valve spacer and spring washer from the valve stem.
10. Remove the valve seal.

Inspecting

11. Clean all components in Girling cleaning fluid and allow to dry.
12. Examine the cylinder bore and piston, ensure that they are smooth to the touch with no corrosion, score marks or ridges. If there is any doubt, fit new replacements.
13. The seals should be replaced with new components.

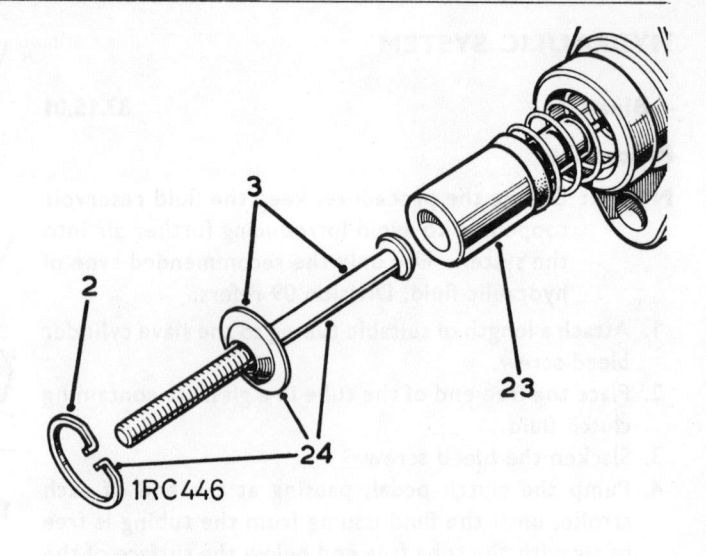

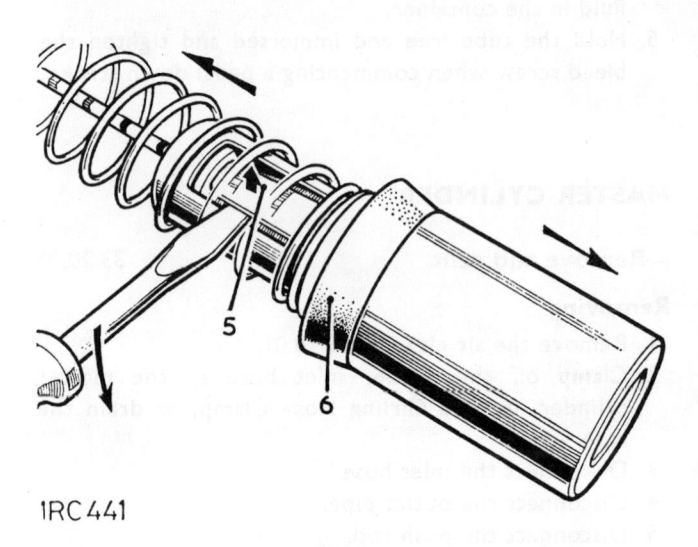

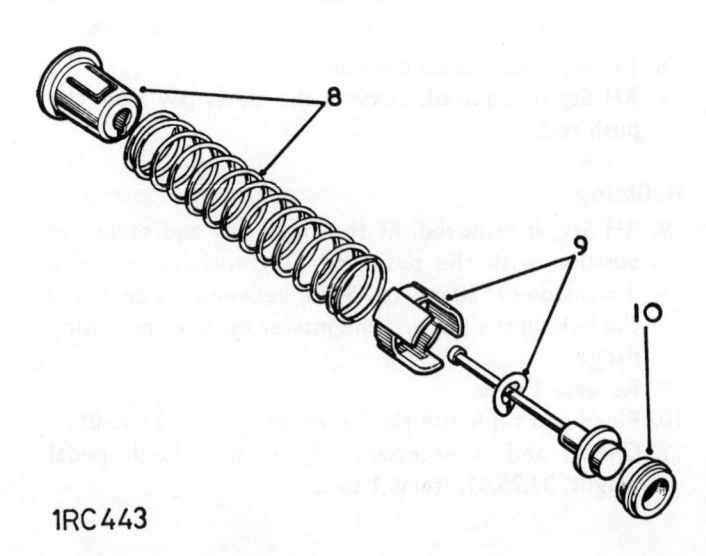

continued

Rover 3500 and 3500S Manual AKM 3621

CLUTCH

Reassembling

14. Smear the seals with Castrol-Girling rubber grease and the remaining internal items with Castrol-Girling Brake and Clutch Fluid.
15. Fit the valve seal, flat side first, on to the end of the valve stem.
16. Place the spring washer, domed side first, over the small end of the valve stem.
17. Fit the spacer, legs first.
18. Place the coil spring over the valve stem.
19. Insert the retainer into the spring.
20. Compress the spring and engage the valve stem in the keyhole slot in the retainer.
21. Fit the seal, large diameter last, to the piston.
22. Insert the piston into the spring retainer and engage the locking prong.
23. Smear the piston with Castrol-Girling rubber grease and insert the assembly, valve end first, into the cylinder.
24. Fit the push rod, retaining washer and circlip.
25. Refit the master cylinder. 33.20.01.

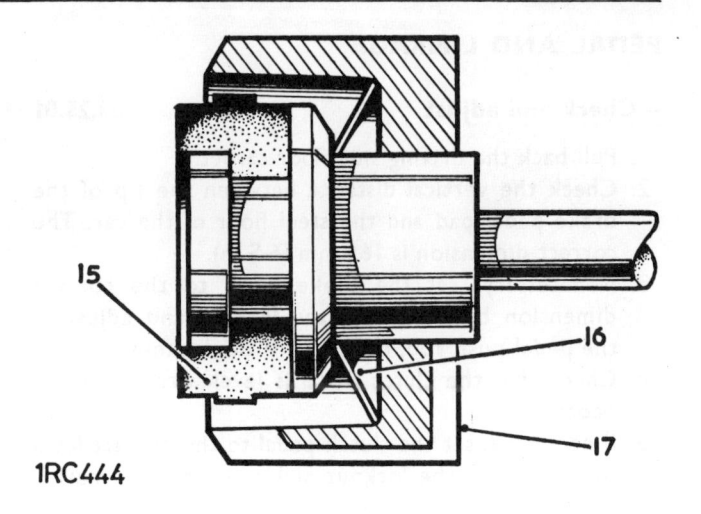

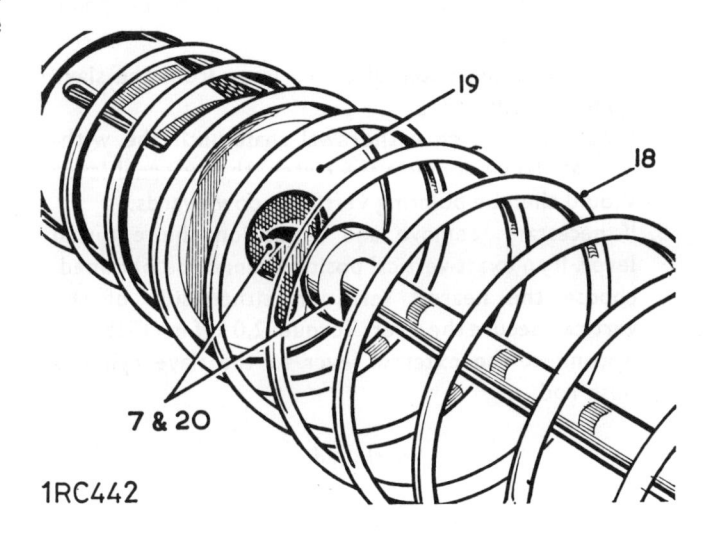

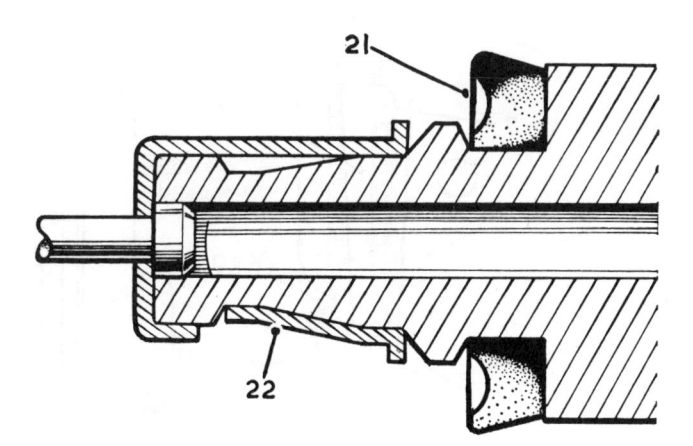

1RC444

1RC442

1RC445

Rover 3500 and 3500S Manual AKM 3621

33.20.07
Sheet 2

CLUTCH

PEDAL AND LINKAGE

—Check and adjust 33.25.01

1. Pull back the driving side floor carpet.
2. Check the vertical distance between the tip of the brake pedal pad and the steel floor of the car. The correct dimension is 165 mm (6.5 in).
3. If necessary, set the brake pedal to the correct dimension by slackening the locknut and adjusting the pedal push rod, then secure the locknut.
4. Check that the clutch pedal is level with the brake pedal.
5. If necessary, set the clutch pedal to the correct level by slackening the locknut and adjusting the master cylinder push rod, then secure the locknut.

NOTE: RH Stg clutch pedal adjustment is illustrated. LH Stg is as illustrated for the brake pedal (item 3).

6. Disconnect the external clutch lever from the slave cylinder push rod.
7. Rotate the clutch withdrawal shaft until the withdrawal sleeve contacts the clutch, the external lever should then be pointing vertically downwards.
8. If necessary, remove and reposition the external lever. If an exact vertical position cannot be obtained, choose the nearest serration either side of the vertical. Secure the lever. Torque: 2,0 kgf.m (15 lbf. ft).
9. Reconnect the external lever to the slave cylinder push rod.

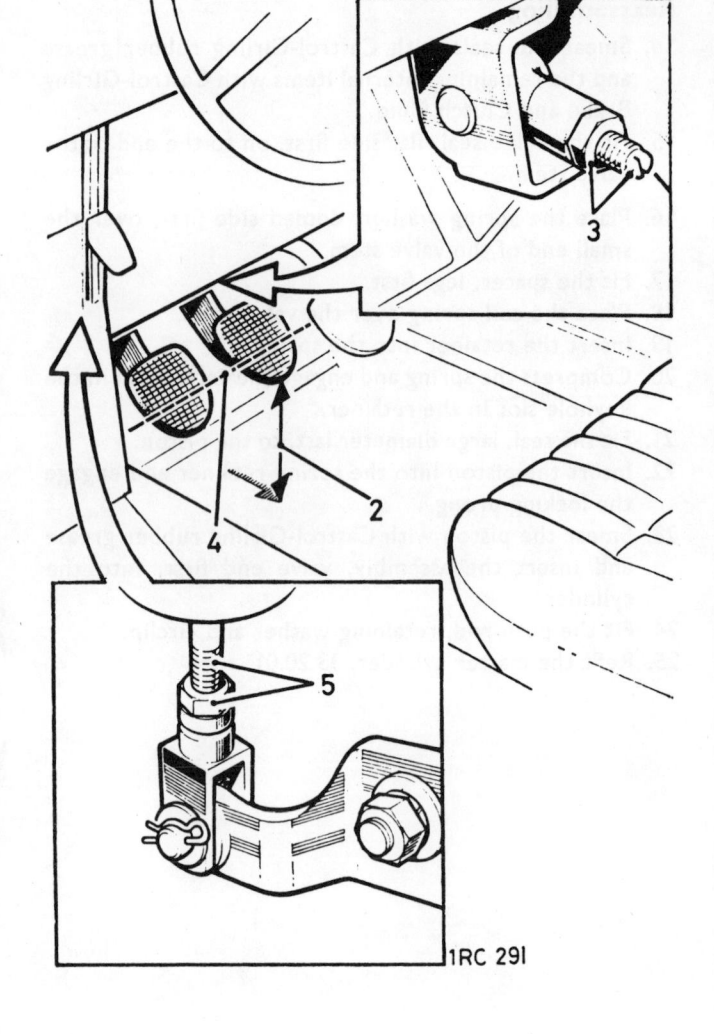

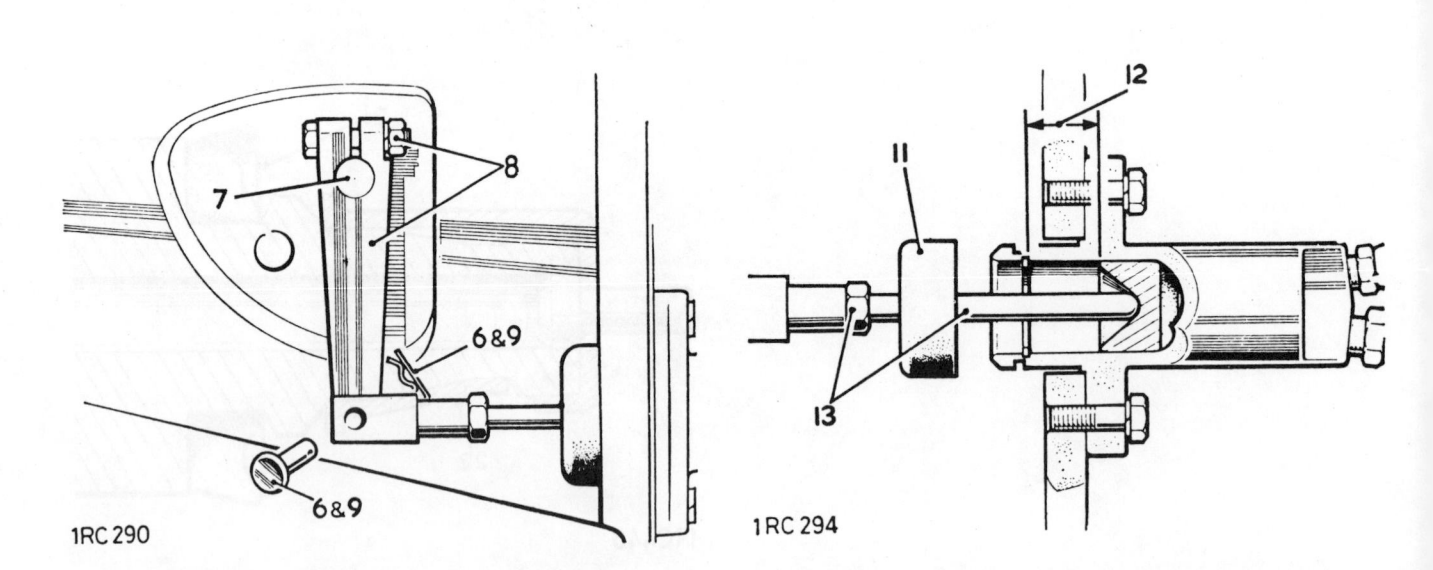

continued

1RC 291

1RC 290

1RC 294

33.25.01
Sheet 1

Rover 3500 and 3500S Manual AKM 3621

CLUTCH

10. Slacken the locknut on the clutch pedal stop, and screw the stop fully down.
11. Pull back the rubber dust cover from the slave cylinder.
12. Check the dimension between the rear face of the piston and the inner face of the circlip, this should be 20,0 mm (0.790 in).
13. If necessary, slacken the locknut and adjust the slave cylinder push rod to obtain the correct dimension, then secure the locknut.
14. Depress the clutch pedal until the slave cylinder piston just touches the circlip, and hold in this position.
15. Screw the clutch pedal stop up until it abuts the pedal release the pedal and screw the stop up a further full turn, then secure the locknut.
16. Replace the floor carpet.
17. Check, and if necessary top up, the fluid level in the reservoir.

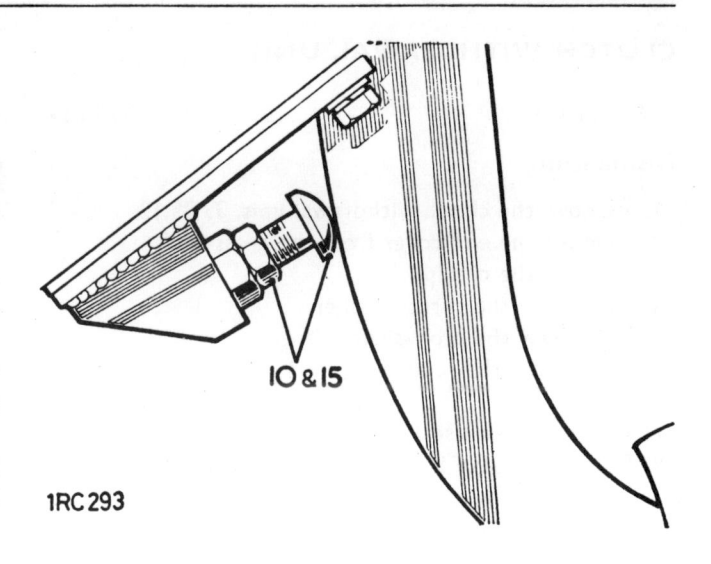

DATA

Clutch and brake pedal height — 165 mm (6.5 in)
External clutch lever static setting — Parallel to the bell housing to engine joint faces
Slave cylinder piston to circlip static setting — 20,0 mm (0.790 in).

CLUTCH WITHDRAWAL UNIT

—Remove and refit 33.25.12

Removing
1. Remove the gearbox assembly 37.20.01.
2. Drain the oil from the gearbox.
3. Remove the external clutch lever.
4. Remove the clutch withdrawal housing.
5. Withdraw the joint washer.

Refitting
6. Remove all traces of old jointing compound.
7. Smear a new joint washer with Hylomar PL 32/M jointing compound, Rover Part No. 534244.
8. Fit the joint washer and clutch withdrawal unit to the bell housing. Torque: 2,0 kgf.m (15 lbf. ft).
9. Fit the gearbox assembly to the engine.
10. Turn the cross-shaft in a clockwise direction and hold the withdrawal sleeve in contact with the clutch thrust pad.
11. Fit the external lever pointing vertically downwards. If an exact vertical position cannot be obtained, choose the nearest serration either side of the vertical. Secure the lever. Torque: 2,0 kgf.m (15 lbf. ft).
12. Refit the gearbox. 37.20.01.
13. Replenish the gearbox lubricating oil.

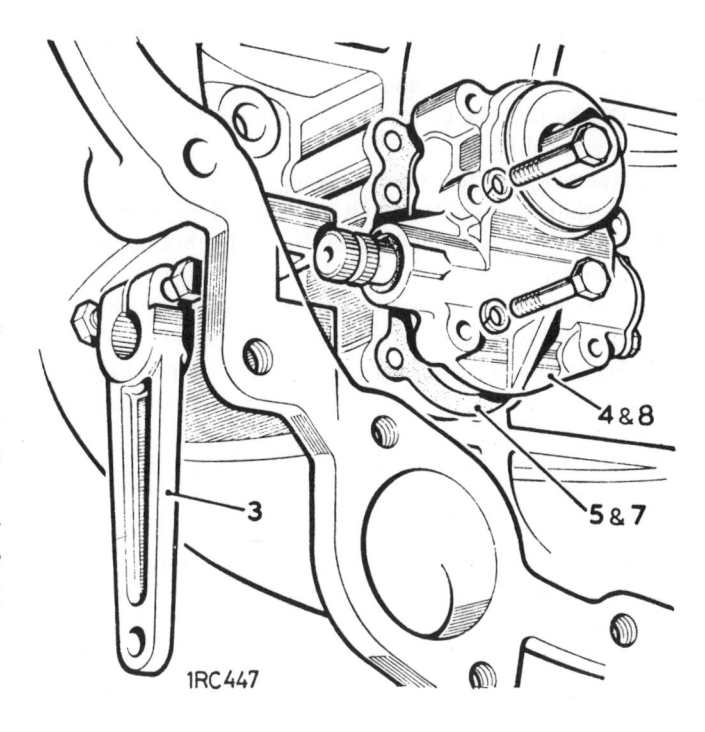

Rover 3500 and 3500S Manual AKM 3621

CLUTCH

CLUTCH WITHDRAWAL UNIT

—Overhaul 33.25.17

Dismantling

1. Remove the clutch withdrawal unit. 33.25.12.
2. Remove the end cover from the cross-shaft.
3. Remove the circlip.
4. Withdraw the thrust washer.
5. Withdraw the cross-shaft.
6. Withdraw the operating fork, spring and thrust washer.
7. Remove the oil seal from the housing.
8. Press the sleeve from the withdrawal race.
9. Withdraw the race.
10. Withdraw the bush.

Inspecting

11. Fit a new oil seal for the cross-shaft.
12. Fit a new joint washer for the cross-shaft end cover.
13. Clean and examine the remaining components and fit new replacements as necessary.

Reassembling

14. Reverse 1 to 10. Ensure that the withdrawal race abuts the shoulder on the sleeve. Tighten the bolts securing the cross-shaft end cover to 1,0 kgf.m (8 lbf. ft) torque.

CLUTCH AND BRAKE PEDAL ASSEMBLY

—Remove and refit 33.30.01

RH Stg. 1 to 7, 10 to 12 and 15 to 26.
LH Stg. 1, 5, 6, 8 to 16 and 18 to 26.

Removing

1. Remove the air cleaner. 19.10.01
2. Disconnect the clutch master cylinder push rod from the pedal lever.
3. Remove the clutch pedal lever.
4. Withdraw the felt washer and steel shim washers.
5. Disconnect the fluid pipes from the brake master cylinder.
6. Remove the two bolts and spring washers securing the brake master cylinder to the bulkhead.
7. From under the right-hand front wing valance, release the fixings securing the mounting bracket for the accelerator cross-shaft.
8. Disconnect the fluid pipes from the clutch master cylinder.
9. Remove the two bolts and spring washers securing the clutch master cylinder to the bulkhead.
10. From inside the car, pull back the carpet from the pedal box.
11. Remove the locknut from the brake pedal push rod.

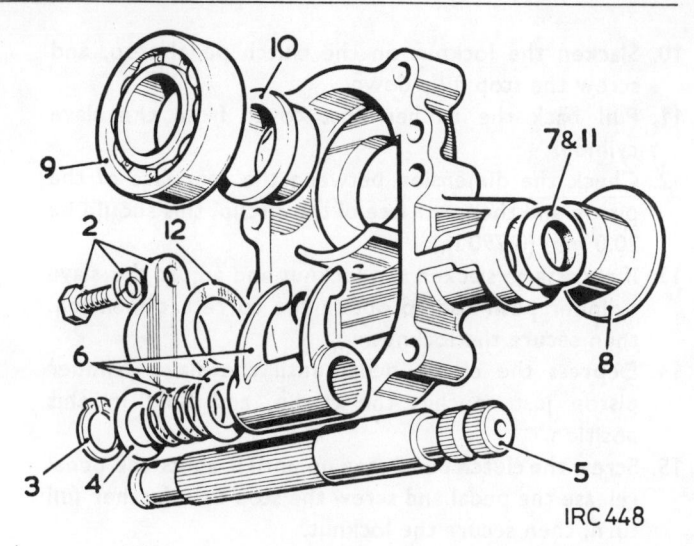

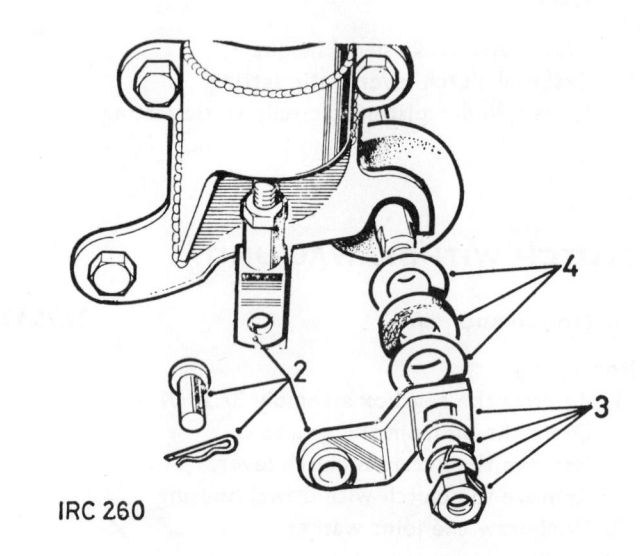

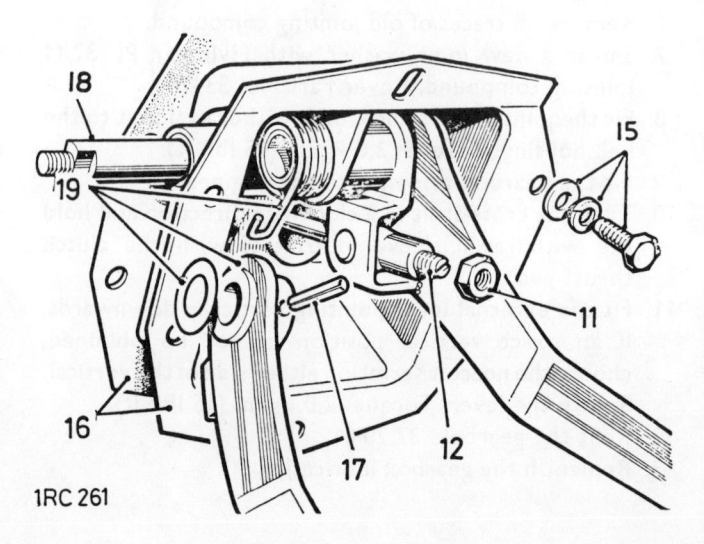

continued

33.25.17
33.30.01
Sheet 1

Rover 3500 and 3500S Manual AKM 3621

CLUTCH

12. Screw the brake pedal push rod forward clear of the threads in the trunnion and withdraw the master cylinder from the engine compartment.
13. Remove the locknut from the clutch pedal push rod.
14. Screw the clutch pedal push rod forward clear of the threads in the trunnion and withdraw the master cylinder from the engine compartment.
15. Remove the fixings from the pedal box.
16. Withdraw the pedal box and gasket.
17. Release the wire-locking and drive out the roll-pin from the clutch pedal.
18. Withdraw the clutch pedal shaft.
19. Withdraw the clutch pedal, shim and spring.
20. If required, remove the brake pedal shaft, shim and 'O'-ring, and withdraw the brake pedal and spring.
21. If required, remove the bushes from the brake pedal and pedal box.

Refitting

22. If removed, fit new bushes to the brake pedal. Ensure that the internal diameter of the fitted bushes is 15,87 to 15,92mm (0.625 to 0.627 in.)
23. If removed, fit new bushes to the pedal box. The bushes must be spaced 21,7 mm—1 mm (0.855 in—0.040 in) apart. Ensure that the internal diameter of the fitted bushes is 15,87 to 15,92 mm (0.625 to 0.627 in) and that they are in line.
24. Reverse 1 to 20 noting the following:
    a. Lightly oil the bushes before fitting the pedal shafts.
    b. Grease the 'O'-ring for the brake pedal shaft.
    c. RH Stg. Use a new roll-pin to retain the clutch pedal, and wire-lock in position.
    d. Use a suitable adhesive, such as Bostik or Holdite to secure the gasket to the pedal box.
    e. Fit the clutch pedal stop at the bottom left-hand corner of the pedal box.
    f. RH Stg. Soak the felt washer for the clutch pedal shaft with SAE 30 oil.
    g. RH Stg. Grease the exposed end of the clutch pedal shaft.
    h. RH Stg. Tighten the pedal lever nut to 4,1 kgf.m (30 lbf. ft) torque.
25. Check, and if necessary adjust, the clutch pedal and linkage. 33.25.01.
26. Bleed the brakes. 70.25.02.

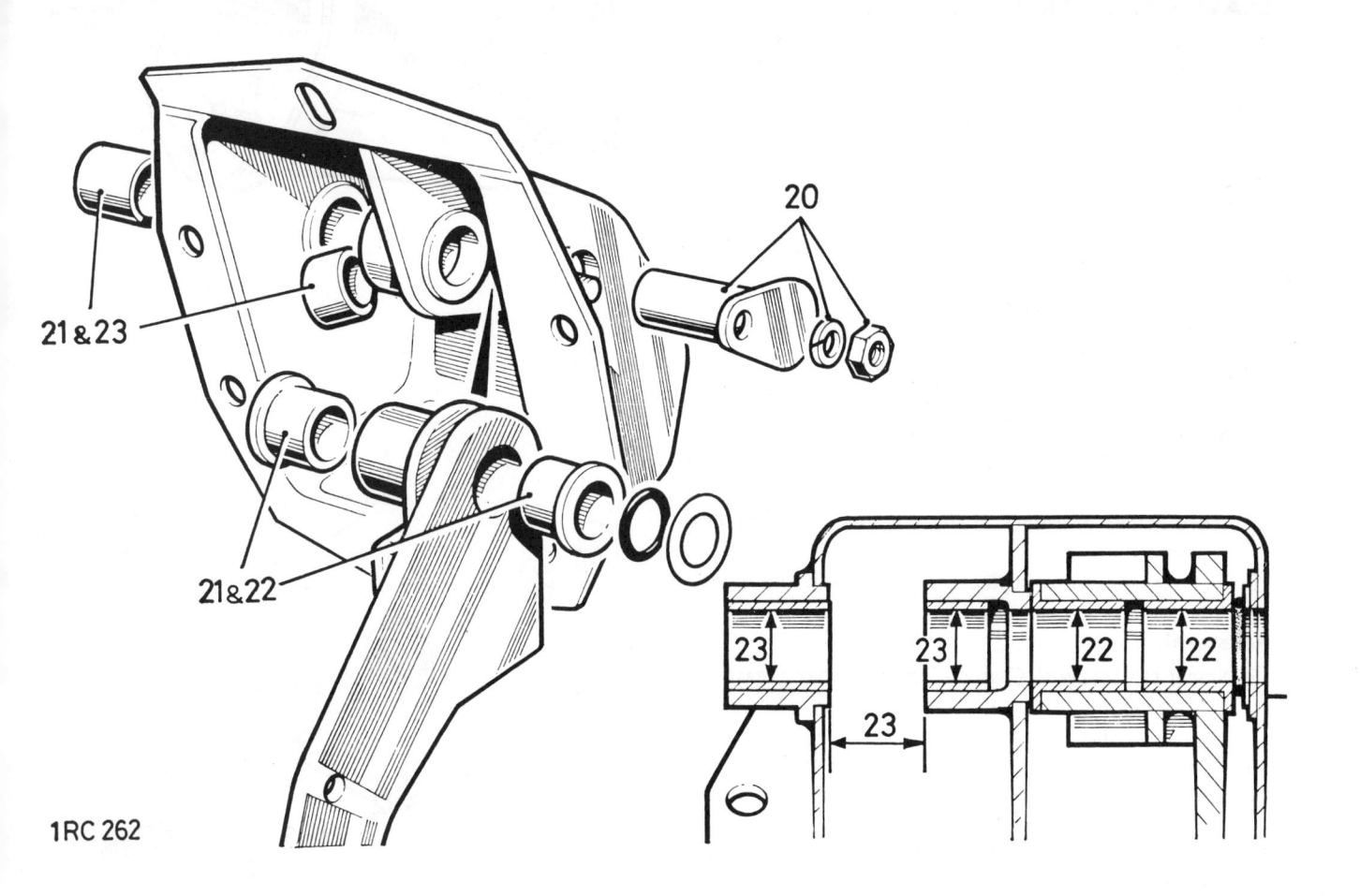

1RC 262

DATA

Pedal shaft bush internal diameter	15,87 to 15,92 mm (0.625 to 0.627 in)
Distance between pedal box bushes	21,7 — 1 mm (0.855 — 0.040 in)

Rover 3500 and 3500S Manual AKM 3621

CLUTCH

SLAVE CYLINDER

—Remove and refit 33.35.01

Removing
1. Prop open the bonnet.
2. Clamp off the rubber inlet hose to the master cylinder, using a Girling hose clamp, or drain the reservoir.
3. Evacuate the clutch system fluid at the slave cylinder bleed valve.
4. Disconnect the fluid pipe.
5. Remove the slave cylinder.

Refitting
6. Secure the slave cylinder to the bell housing, locating the push-rod. Torque: 3,5 kgf.m (25 lbf. ft).
7. Reverse 2 and 4.
8. Bleed and replenish the hydraulic system. 33.15.01.
9. Check, and if necessary adjust, the clutch pedal and linkage. 33.25.01.
10. Close the bonnet.

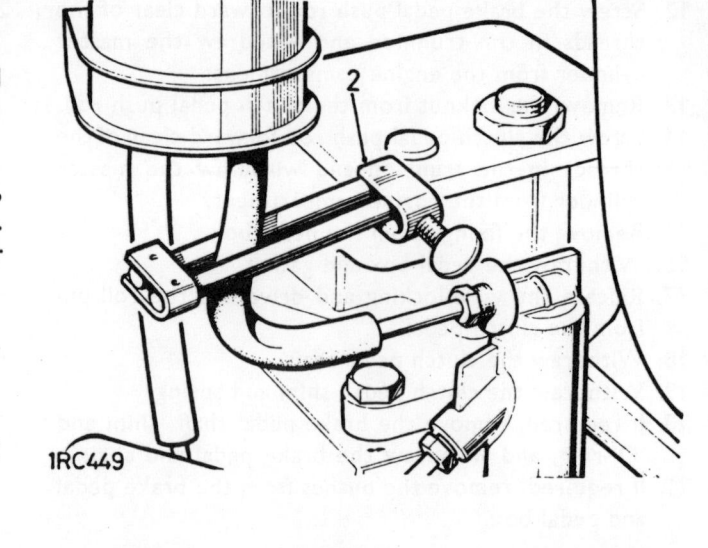

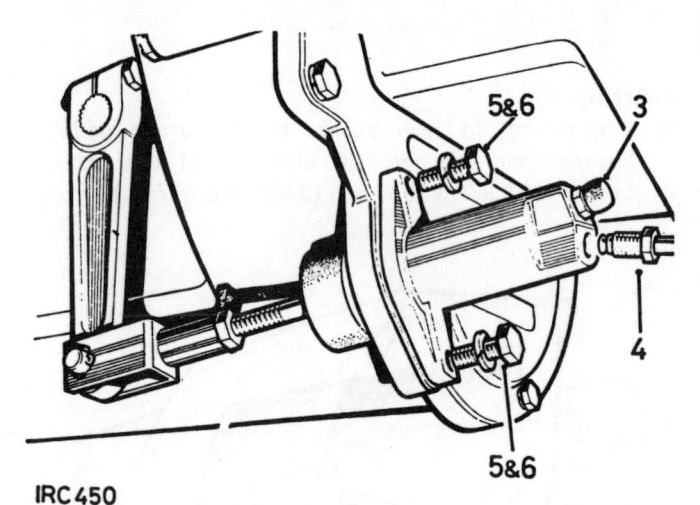

SLAVE CYLINDER

—Overhaul 33.35.07

Dismantling
1. Remove the slave cylinder. 33.35.01.
2. Withdraw the dust cover.
3. Remove the circlip.
4. Expel the piston assembly, applying low pressure air to the fluid inlet.
5. Withdraw the spring.

Inspecting
6. Clean all components in Girling cleaning fluid and allow to dry.
7. Examine the cylinder bore and piston, ensure that they are smooth to the touch with no corrosion, score marks or ridges. If there is any doubt, fit new replacements.
8. The seal should be replaced with a new component.

Reassembling
9. Smear the seal with Castrol-Girling rubber grease and the remaining internal items with Castrol-Girling brake and clutch fluid.
10. Fit the seal, large diameter last, to the piston.
11. Locate the conical spring, small diameter first, over the front end of the piston.
12. Smear the piston with Castrol-Girling rubber grease and insert the assembly, spring end first, into the cylinder.
13. Fit the circlip.
14. Fill the dust cover with Castrol-Girling rubber grease and fit the cover to the cylinder.
15. Refit the slave cylinder. 33.35.01.

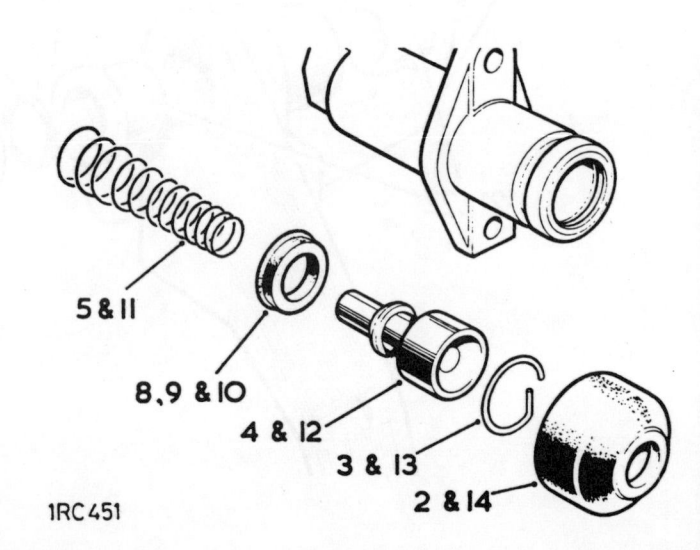

33.35.01
33.35.07

Rover 3500 and 3500S Manual AKM 3621

GEARBOX—SYNCHROMESH

SYNCHROMESH GEARBOX OPERATIONS

Gearbox assembly—remove and refit	37.20.01
Gearchange selectors—remove and refit	37.16.31
Layshaft	
—remove and refit	37.20.29
—bearings—remove and refit	37.20.22
Maincase—remove and refit	37.12.40
Mainshaft	
—remove and refit	37.20.25
—overhaul	37.20.31
—bearings—remove and refit	37.20.26
Oil pump—remove and refit	37.12.34
Primary pinion	
—remove and refit	37.20.29
—bearings—remove and refit	37.20.17
Speedometer drive housing	
—remove and refit	37.25.09
—overhaul	37.25.13
Synchromesh assemblies—overhaul	37.20.08

Rover 3500 and 3500S Manual AKM 3621

GEARBOX—SYNCHROMESH

GEARBOX OIL PUMP

—Remove and refit   37.12.34

Removing

1. Remove the speedometer drive housing. 37.25.09.
2. Withdraw the oil pump complete.
3. Withdraw the joint washer.
4. Retain the shim for the layshaft bearing.

Refitting

5. Prime the pump with clean gearbox oil.
6. Smear a new joint washer with Hylomar PL 32/M jointing compound, Rover Part No. 534244.
7. Reverse 1 to 4 ensuring that the oil pump drive is aligned.

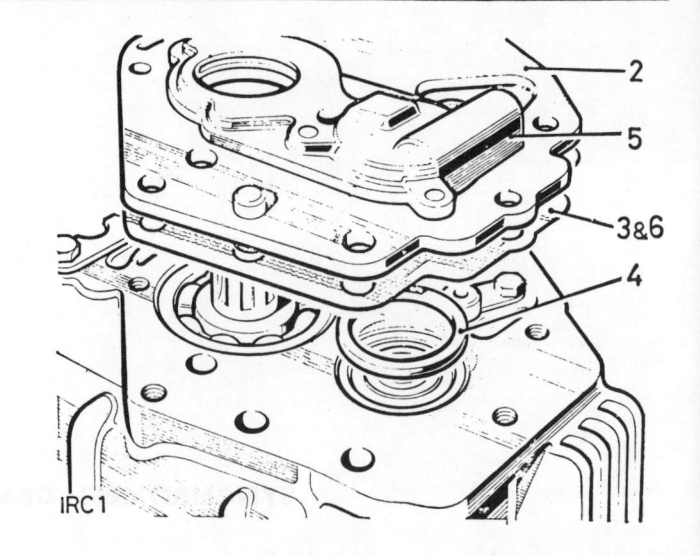

GEARBOX MAIN CASE

—Remove and refit   37.12.40

Removing

1. Remove the gearbox. 37.20.01.
2. Drain the gearbox oil.
3. Remove the speedometer drive housing. 37.25.09.
4. Remove the gearbox oil pump. 37.12.34.
5. Remove the selector shafts. 37.16.31.
6. Remove the bolt and retaining plate from the reverse gear idler shaft.
7. Withdraw the shaft.
8. Remove the fixings securing the bell housing to the gearbox main case.
9. Position the gearbox upright on the bell housing.
10. Push the reverse gear to one side clear of the mainshaft first gear.
11. Withdraw the gearbox main case.
12. Withdraw the bell housing joint washer.

Refitting

13. Smear the joint faces of the bell housing and the gearbox main case with Hylomar PL 32/M jointing compound, Rover Part No. 534244.
14. Place a new joint washer in position on the bell housing.
15. Place the reverse gear in position in the main case. The lead on the gear teeth must be towards the front of the gearbox.
16. Lower the gearbox main case on to the bell housing while keeping the reverse gear pushed to one side.
17. Locate the main case to the bell housing with two bolts, but do not fit the nuts at this stage.
18. Fit the reverse idler shaft.
19. Fit the retaining plate, engaging the slot in the reverse idle shaft and secure with a bolt only (no washer). Torque: 1,0 kgf.m (8 lbf ft).
20. Fit the main case to bell housing fixings. Torque: 7,0 kgf.m (50 lbf ft).
21. Reverse 1 to 5.

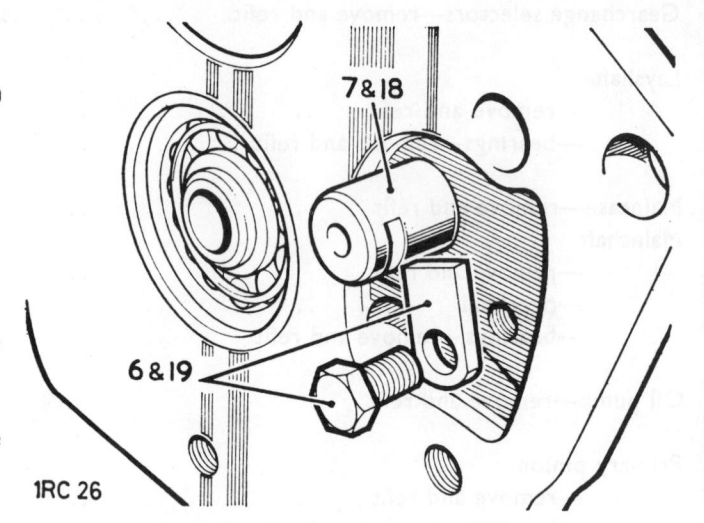

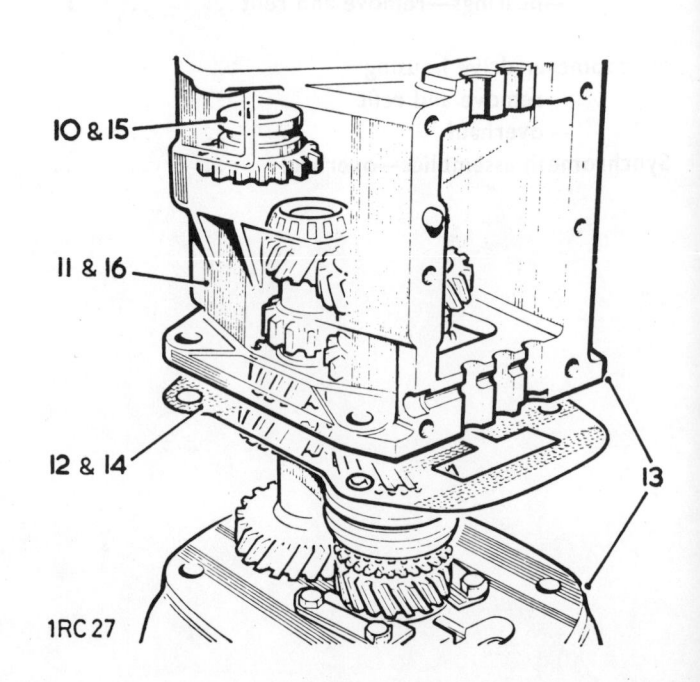

GEARBOX—SYNCHROMESH

GEAR CHANGE SELECTORS

—Remove and refit 37.16.31

Removing

1. Remove the gearbox assembly. 37.20.01.
2. Drain the gearbox oil.
3. Remove the speedometer drive housing. 37.25.09.
4. Remove the gearbox oil pump. 37.12.34.
5. Remove the gearchange shaft housing, noting the position of any packing washers fitted.
6. Remove the reverse light switch complete with its mounting bracket.
7. Remove the seal retaining plates.
8. Remove the retaining plate, joint washer and detent springs, noting position for reassembly.
9. Withdraw the rubber plug.
10. Pack grease into the three selector spring location holes to retain the three balls.
11. Remove the gearbox top cover complete with the selector balls held by the grease.
12. Remove the balls and grease from the detent spring location holes.
13. Withdraw the interlock plungers.

continued

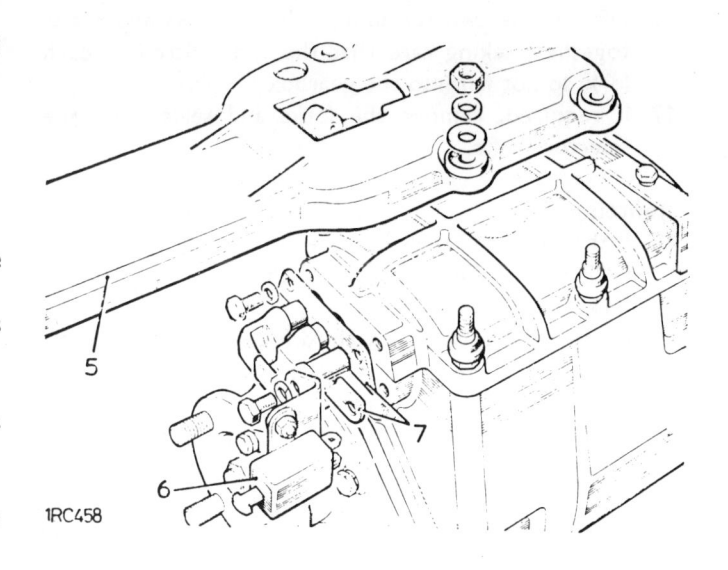

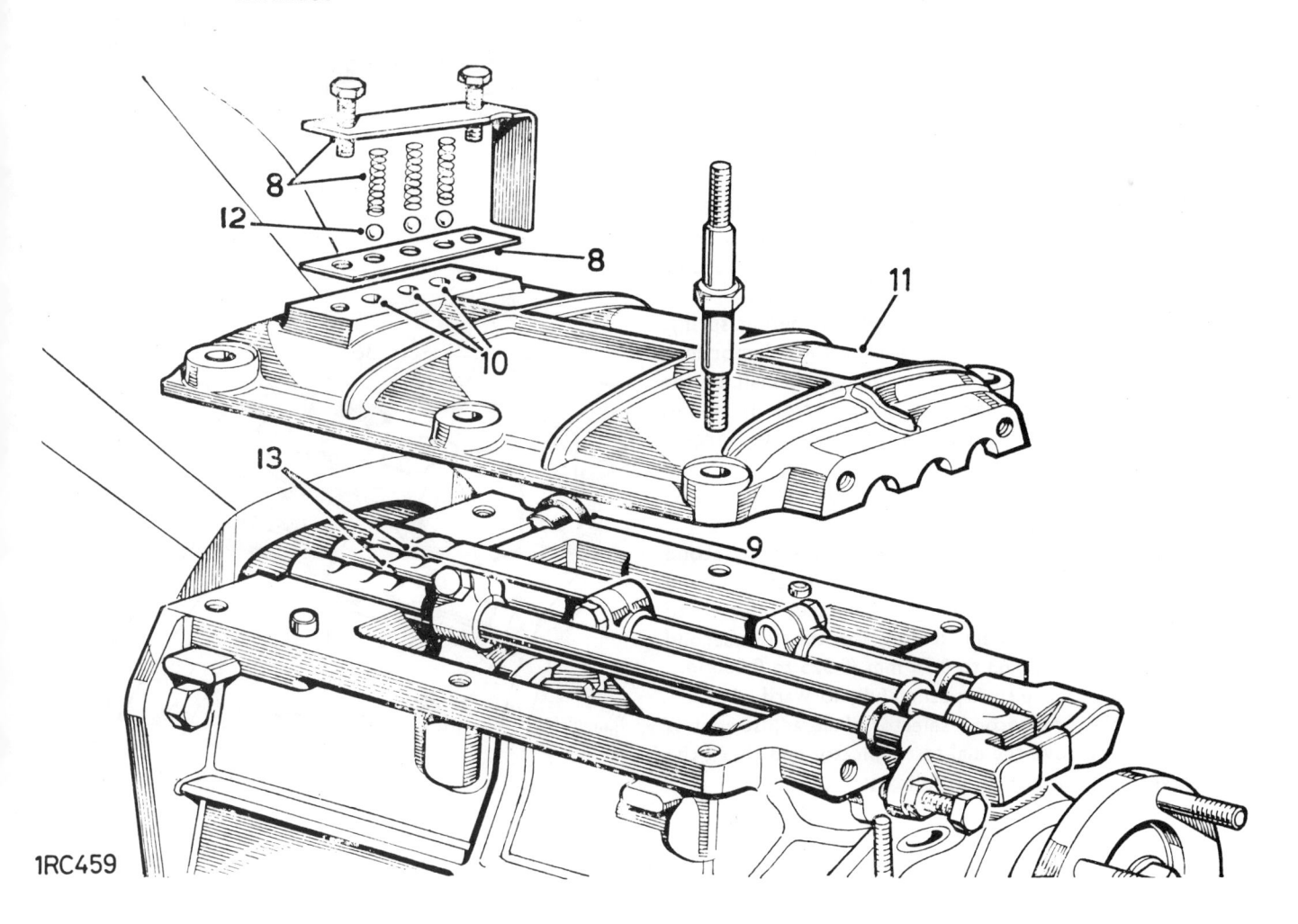

Rover 3500 and 3500S Manual AKM 3621

37.16.31
Sheet 1

GEARBOX—SYNCHROMESH

14. Screw a slave bolt (10 UNF) into the reverse selector guide rod and withdraw the rod.
15. Lift out the reverse gear selector fork and shaft.
16. Lift out the two remaining selector forks and shafts together, taking care that the shoes fitted in each fork do not fall into the gearbox.
17. If required, remove the forks and seals from the selector shafts.

Refitting

18. If the forks and seals have been removed, refit them with the smaller diameter of the seals towards the forks.
19. Fit the first/second and third/fourth forks and shaft assemblies. Ensure that the shoes locate their respective synchromesh unit.
20. Fit the reverse fork and shaft assembly, locating the reverse gear.
21. Fit the reverse selector guide rod.
22. Move the seals along the selector shafts to clear their grooves.
23. Fit the interlock plungers.

    CAUTION: Do not damage the bell housing gasket when fitting and removing the gearbox top cover.

24. Locate the top cover in position on the dowels and secure with the two centre fixings.
25. Select third gear, selector to rear.
26. Mount a dial test indicator to read of the rear of the third/fourth selector shaft.
27. Insert the third/fourth detent ball and spring.
28. Hold the spring compressed to correctly locate the selector shaft then zero the dial test indicator.
29. Release the pressure on the detent spring.
30. Pull the third/fourth selector shaft fully rearward and note the maximum reading on the dial test indicator. With the selector fork correctly adjusted, the reading must be 0,63 to 0,76 mm (0.025 to 0.030 in.).
31. If adjustment is required, remove the detent spring, top cover and detent ball. Slacken the selector fork pinch bolt and move the fork along the shaft by the amount of error, forward if the error is minus, rearward if it is plus. Tighten the pinch bolt. Torque: 2,8 kgf.m (20 lbf ft).
32. Recheck the adjustment. 24 to 31. If necessary, readjust until correct.
33. Repeat the foregoing procedure for the first/second selector shaft, with the first/second detent ball and spring fitted and second gear selected, selector to front. The adjustment setting for the first/second selector shaft is 0,63 to 0,76 mm (0.025 to 0.030 in.), checked with the shaft pushed fully forward.
34. When the first/second selector shaft adjustment is correct, remove the dial test indicator and check the reverse selector shaft, as follows:

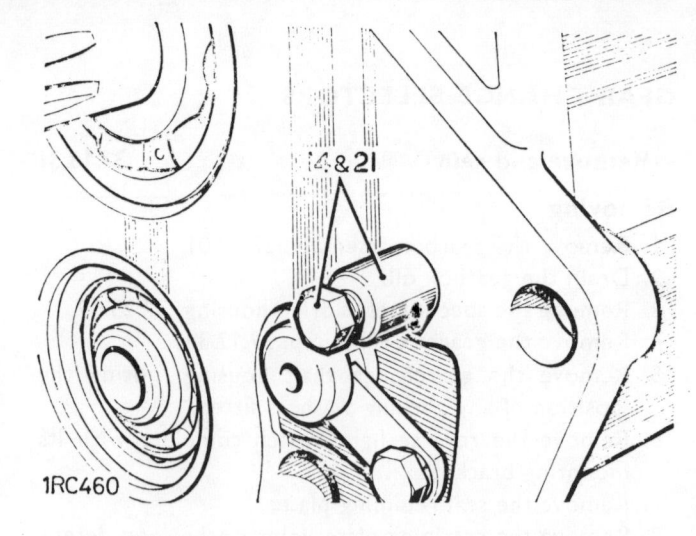

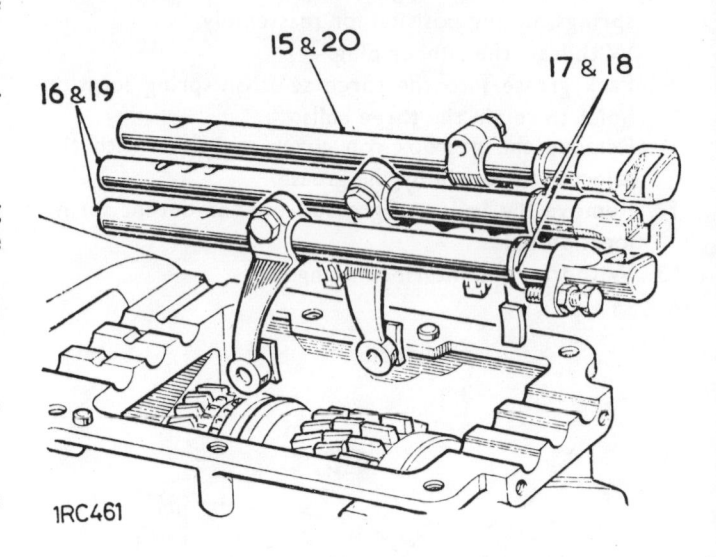

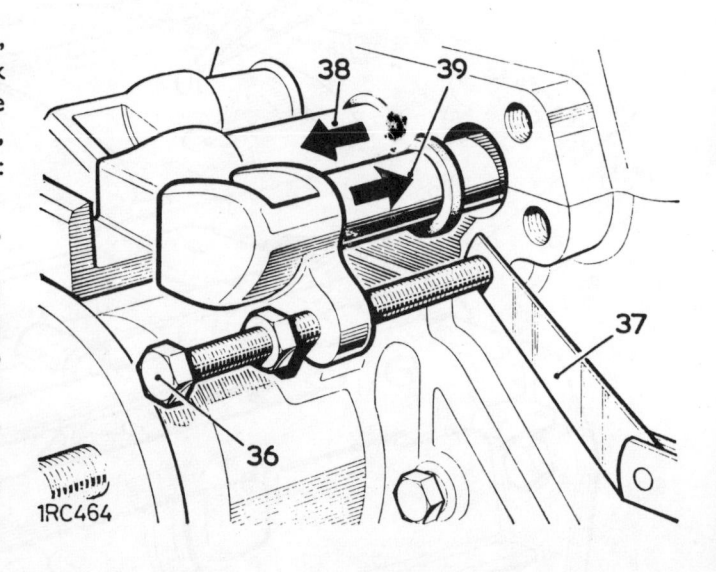

continued

GEARBOX—SYNCHROMESH

35. Remove the top cover, detent springs and balls.
36. Fit the reverse stop bolt, minus the distance piece and switch striker arm.
37. Fit the interlocking plunger to locate and hold the reverse selector shaft in neutral position, then using feeler gauges, adjust the stop bolt to give 0,9 mm (0.035 in.) clearance between the end of the bolt and the gearbox case.
38. Select first gear and hold in the over-travel position, that is fully rearward.
39. Push the reverse shaft forward until the stop bolt contacts the gearbox case.
40. With the first/second and reverse selector shafts held as described, the reverse idler gear should just contact the gear teeth on the outer member of the low gear synchromesh, but not be rotated with it when it is turned.
41. If necessary, adjust the position of the fork on the reverse selector shaft to obtain the above condition. Tighten the pinch bolt. Torque: 1,6 kgf.m (12 lbf ft).

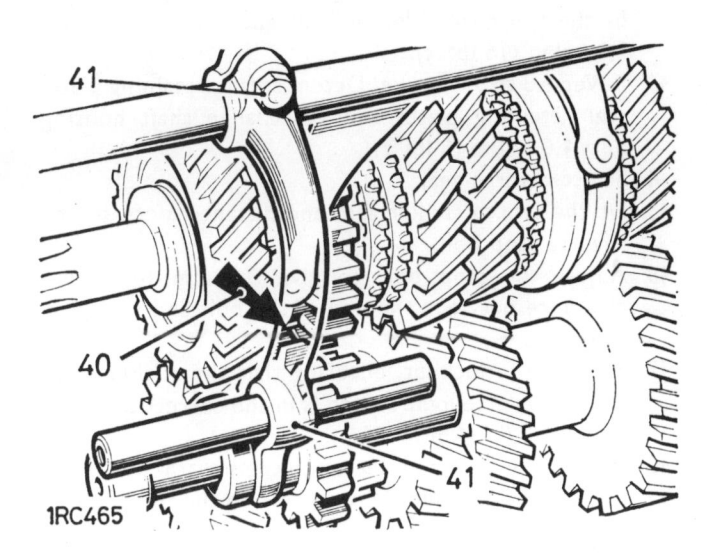

continued

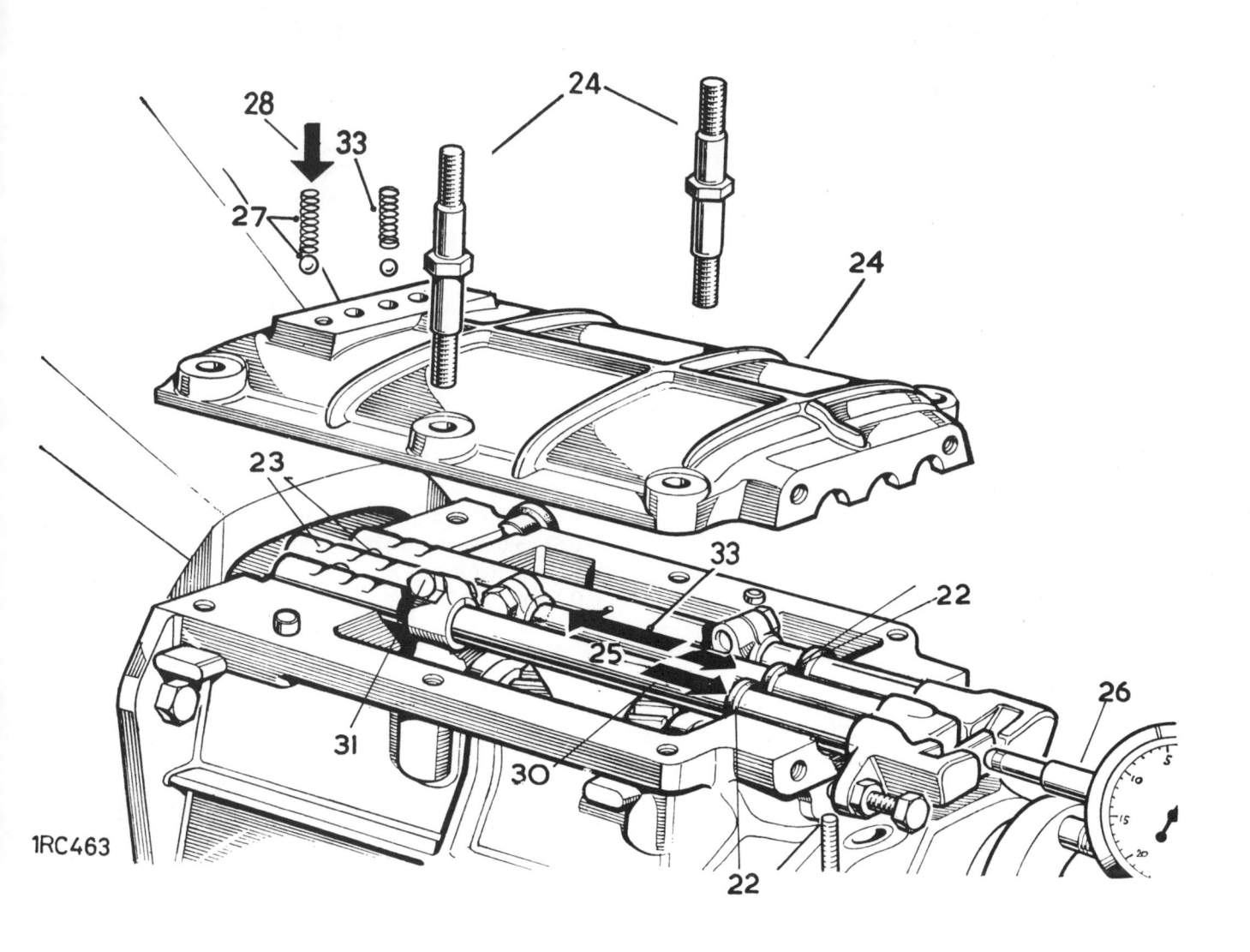

Rover 3500 and 3500S Manual AKM 3621

37.16.31
Sheet 3

GEARBOX—SYNCHROMESH

42. When the selector shafts are correctly adjusted, fit the top cover, detent balls and springs. Torque: 2,0 kgf.m (15 lbf ft).
43. Reverse 3 to 9. Torque: Detent spring retaining plate, seal retaining plates and gearchange shaft housing nuts 1,0 kgf.m (8 lbf ft).
44. Select reverse gear.
45. Fit the switch striker arm and distance piece to the reverse stop bolt.
46. Adjust the stop bolt to give 0,4 mm (0.015 in.) clearance between the end of the bolt and the gearbox case.
47. Tighten the locknut, ensuring that the striker plate for the reverse light switch is in the correct position. Torque: 1,0 kgf.m (8 lbf ft).
48. Return the reverse selector to the neutral position.
49. Slacken the locknut and stop bolt on the third/fourth selector shaft.
50. Select fourth gear.
51. Adjust the fourth gear selector stop bolt to give 0,4 mm (0.015 in.) clearance between the end of the bolt and the gearbox case.
52. Secure the locknut and return the selector to the neutral position. Torque: 1,0 kgf.m (8 lbf ft).
53. Reverse 1 and 2.

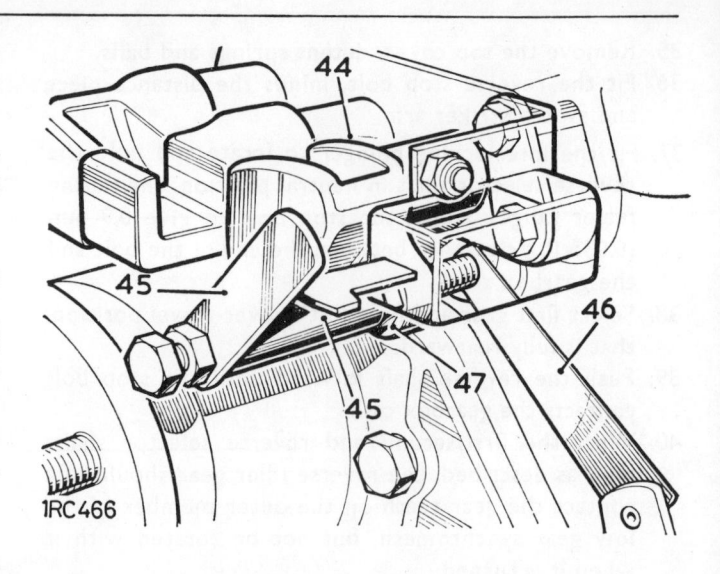

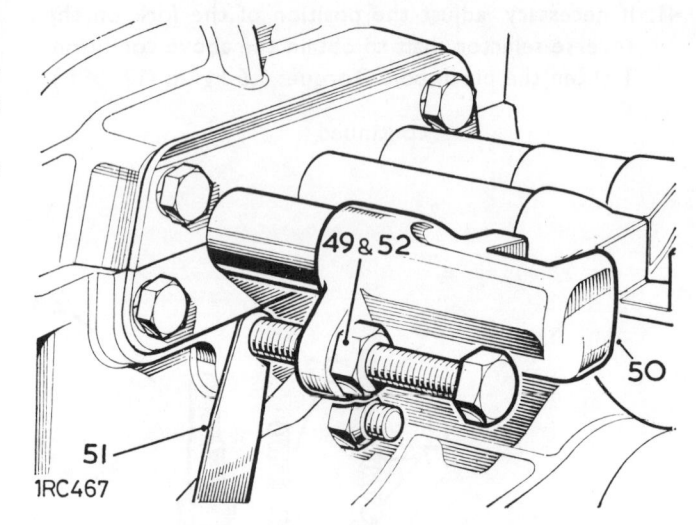

DATA

Selector settings:

1st/2nd speed	0.63 to 0.76 mm (0.025 to 0.030 in.)
3rd/4th speed	0,63 to 0,76 mm (0.025 to 0.030 in.)
Reverse	0,9 mm (0.035 in.)
Overtravel stops	0,4 mm (0.015 in.)

GEARBOX—SYNCHROMESH

GEARBOX ASSEMBLY

—Remove and refit 37.20.01

Removing

1. Disconnect the battery earth lead.
2. Remove the bonnet.
3. Remove the air cleaner. 19.10.01
4. Remove the fan blades.
5. Remove the cover from the gearbox tunnel.
6. Remove the gearchange lever assembly.
7. Lift the carpet from the left hand side of the gearbox tunnel and remove the cover plate.
8. Disconnect the speedometer cable from the gearbox.
9. Lift the carpet from the right hand side of the gearbox tunnel and remove the large rubber grommet.
10. Disconnect the electrical leads from the reverse light switch.
11. Remove the three bolts that are accessible from the engine compartment, securing the bell housing to the engine.
12. Drain the gearbox lubricating oil.
13. Remove the front exhaust pipe.
14. Remove the propeller shaft.
15. Remove the fixings from the clutch slave cylinder, withdraw the slave cylinder from the bell housing and move it to one side without disconnecting the fluid pipe.

continued

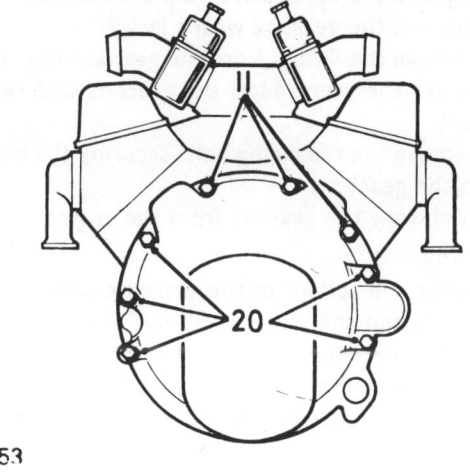

IRC 453

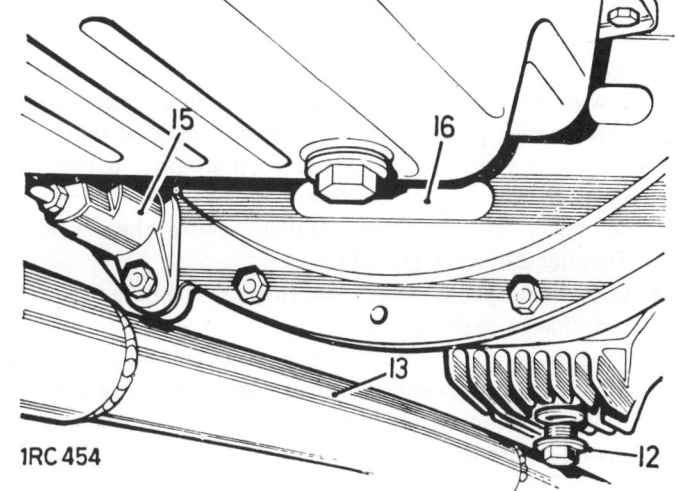

1RC 454

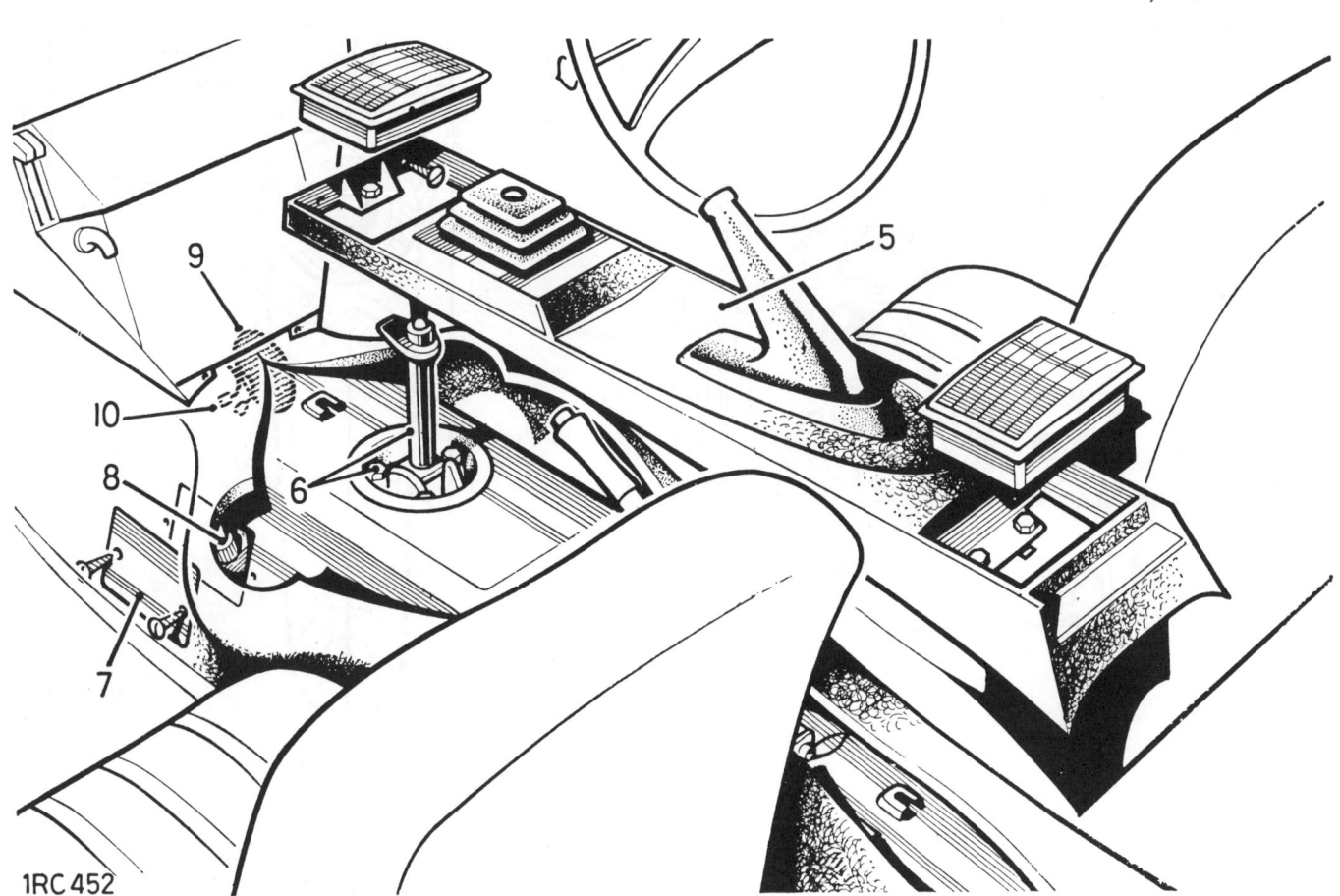

1RC 452

Rover 3500 and 3500S Manual AKM 3621

GEARBOX—SYNCHROMESH

16. Remove the cover plate from the bell housing.
17. Support the gearbox with a jack.
18. Remove the fixings from the gearbox rear mounting.
19. Lower the rear end of the gearbox and remove the jack.
20. Remove the remaining bolts securing the bell housing to the gearbox.
21. Withdraw the gearbox from the engine.

Refitting

22. Smear the splines of the primary pinion, the clutch centre and the withdrawal unit abutment faces with PBC (Poly Butyl Cuprysil) grease.
23. Offer the gearbox to the engine, locating the primary pinion into the clutch and engage the bell housing dowels.
24. Secure the bell housing to the engine with the five bolts that are accessible from beneath the car. Torque: 3,5 kgf.m (25 lbf ft).
25. Rotate the clutch withdrawal shaft until the withdrawal sleeve contacts the clutch, then fit the external clutch lever pointing downwards and parallel to the bell housing to engine joint faces. If an exact parallel position cannot be obtained, choose the nearest serration either side of parallel. Secure the lever. Torque: 2,0 kgf.m (15 lbf ft).
26. Locate the speedometer cable through the hole in the gearchange shaft housing.
27. Reverse 1 to 19. When fitting the fan blades, an off-set dowel location ensures that the fixing bolt holes only align when the blades are the correct way round. Torque bell housing cover plate bolts 1,0 kgf.m (8 lbf ft) slave cylinder bolts 3,5 kgf.m (25 lbf ft).
28. Check, and if necessary adjust, the clutch pedal and linkage. 33.25.01.

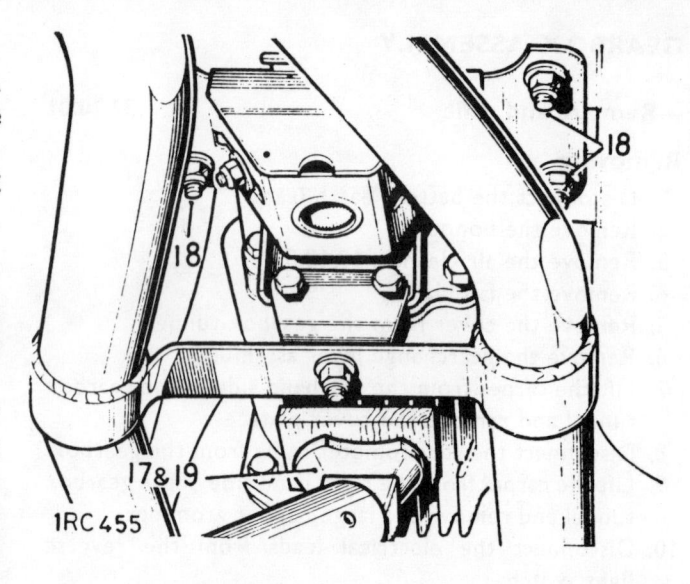

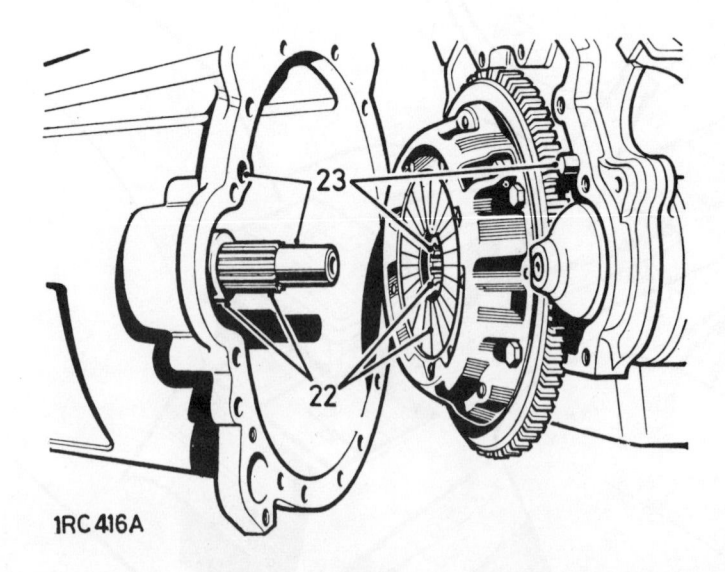

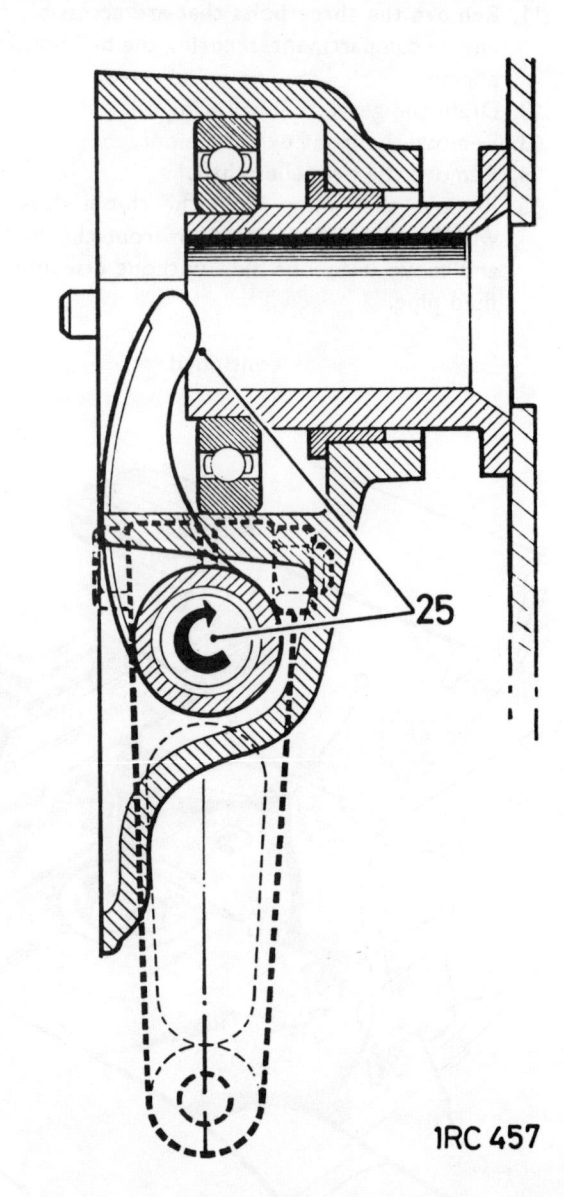

Rover 3500 and 3500S Manual AKM 3621

GEARBOX—SYNCHROMESH

SYNCHROMESH ASSEMBLIES

—Overhaul 37.20.08

Dismantling

1. Remove the synchromesh assemblies. 37.20.31.
2. Place a cloth around the synchromesh unit to prevent losing the balls and springs when the unit is parted.
3. Push the synchromesh inner and outer members apart.
4. Remove the cloth and collect the three balls, springs and sliding blocks.

Inspecting

5. Synchromesh inner and outer members are only supplied as matched sets, and inner and outer members must not be changed individually.
6. Examine all components and fit new replacements as necessary.

Reassembling

7. During manufacture synchromesh inner and outer members are checked for optimum location, then alignment marks are added to enable the unit to be reassembled in the optimum position.
8. Place the low gear synchromesh outer member, gear side downward, on a clean surface.
9. Fit the inner member, internal splined side first, maintaining the alignment marking.
10. Fit the sliding blocks, radiused face outward.
11. Locate the springs through the sliding blocks into the housing bores in the inner member.
12. Position the balls on the spring ends, press home in sequence and retain by hand.
13. Lift the outer member to retain the balls. Continue lifting until the balls spring home into the grooves in the outer member.
14. Assemble the high gear synchromesh unit as described for the low gear unit, except that the outer member may be fitted either way, the alignment marks being the only important location.
15. Refit the synchromesh assemblies. 37.20.31.

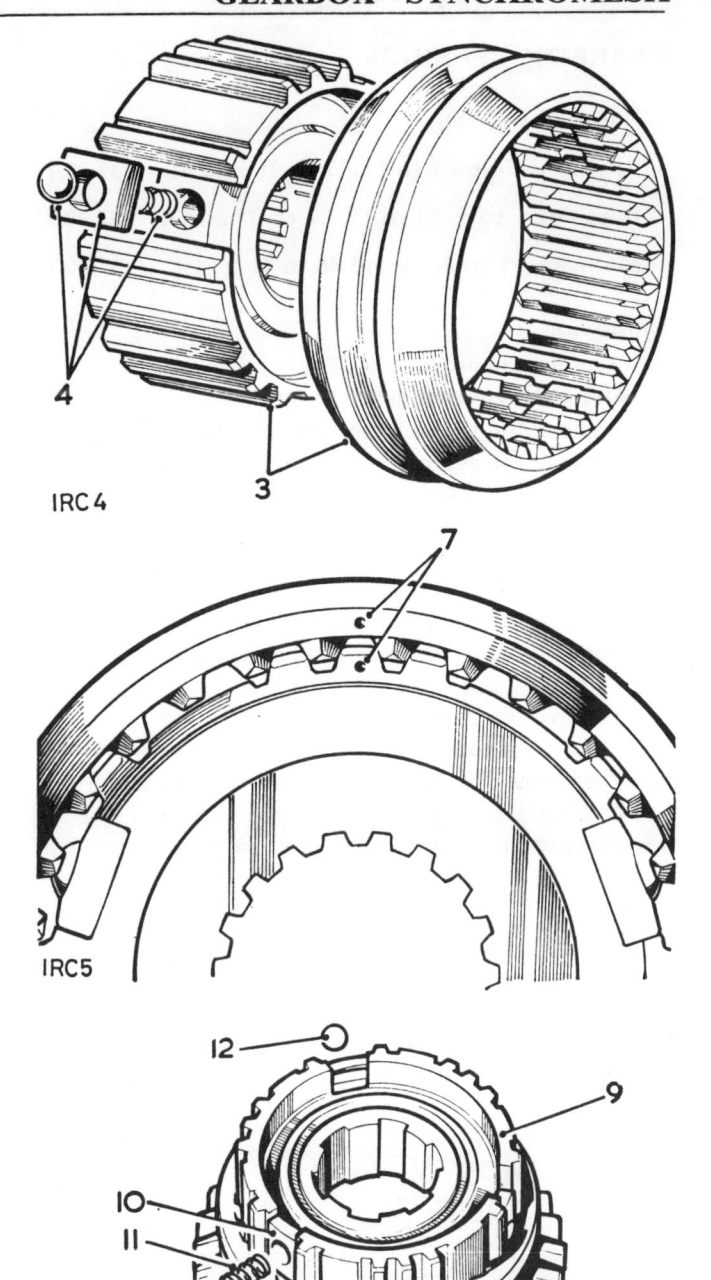

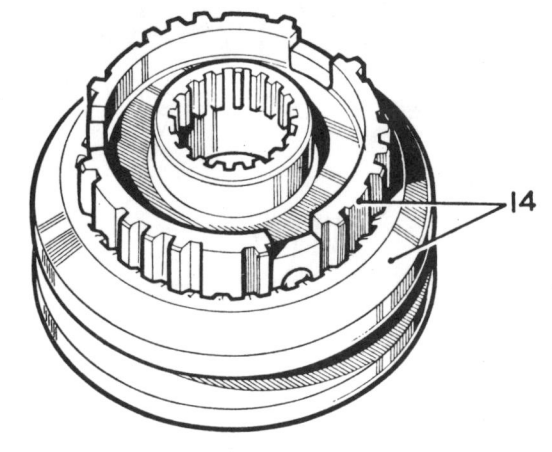

IRC 4

IRC 5

IRC 3

IRC 2

Rover 3500 and 3500S Manual AKM 3621

37.20.08

GEARBOX—SYNCHROMESH

GEARBOX BEARINGS

—Remove and refit

Primary pinion 1 to 12, 17 to 21 and 41	37.20.17
Layshaft 1 to 9, 13 to 15, 22 to 37 and 41	37.20.22
Mainshaft 1 to 8, 16 and 38 to 41	37.20.26

Removing

1. Remove the gearbox. 37.20.01.
2. Drain the gearbox oil.
3. Remove the speedometer drive housing. 37.25.09.
4. Remove the gearbox oil pump. 37.12.34.
5. Remove the selector shafts. 37.16.31.
6. Remove the clutch withdrawal unit. 33.25.12.
7. Remove the gearbox main case. 37.12.40
8. Remove the mainshaft. 37.20.25.
9. Remove the primary pinion and layshaft. 37.20.29.
10. Remove the circlip from the primary pinion shaft.
11. Withdraw the shim.
12. Remove the bearing.
13. Extract the taper roller bearings from the layshaft.
14. Press the layshaft bearing outer race from the bell housing.
15. Press the layshaft bearing outer race from the main case.
16. Press the mainshaft rear bearing from the main case.

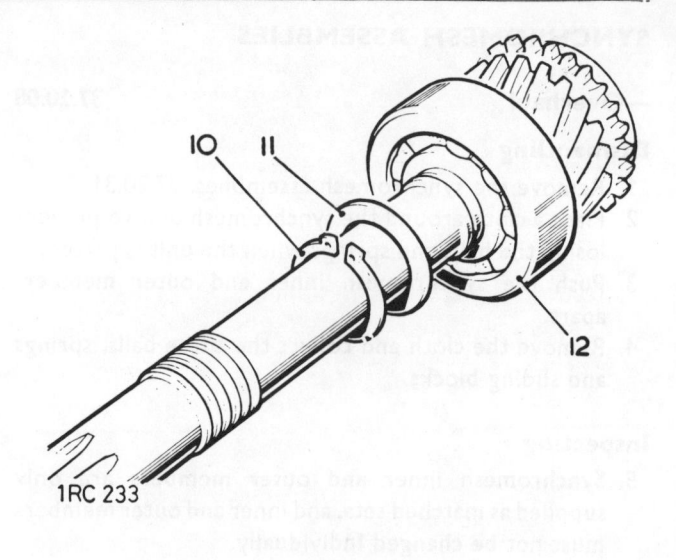

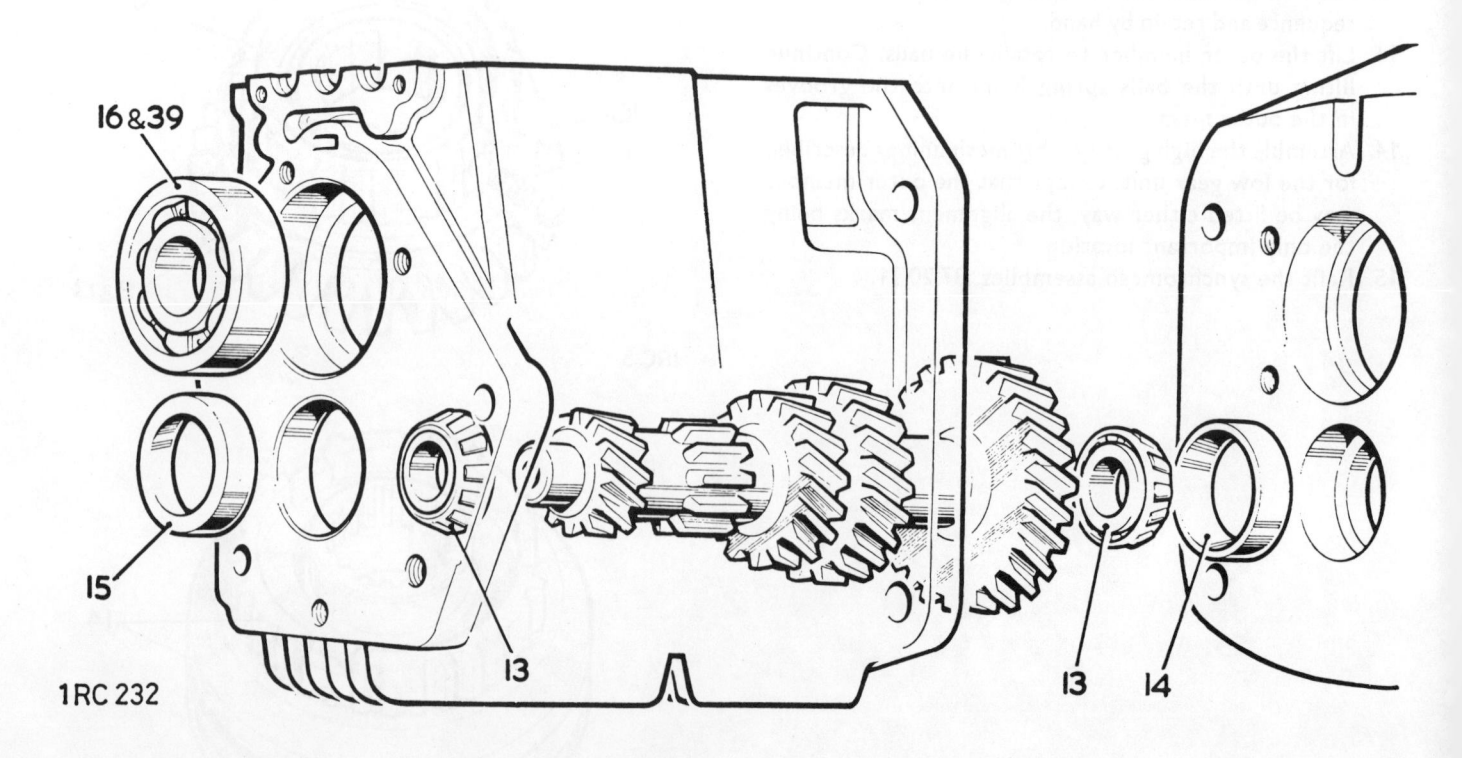

continued

GEARBOX—SYNCHROMESH

Refitting

17. Fit a new bearing to the primary pinion.
18. Fit the original shim.
19. Fit the circlip.
20. Using a feeler gauge, check the clearance between the circlip and the shim. The clearance must not exceed 0,05 mm (0.002 in.). If necessary, fit a thicker shim to reduce the clearance. Shims are available in 0,05 mm (0.002 in.) steps, ranging from 2,23 mm (0.088 in.) to 2,38 mm (0.094 in.).
21. When the clearance is correct, fit a new circlip.
22. Press a new taper roller bearing on to each end of the layshaft.
23. Press a new layshaft bearing outer race into the bell housing.
24. Press the layshaft rear bearing outer race into the main case until it is flush with the rear face.

    NOTE: The following assembly procedure is temporary in order to check and adjust the layshaft bearing preload.

25. Place a new joint washer on to the bell housing.
26. Place the layshaft in position on the bell housing.
27. Fit the main case to the bell housing and secure the fixings. Torque: 7,0 kgf.m (50 lbf ft).
28. Lightly oil both layshaft bearings.
29. Place the original shim on to the layshaft rear bearing.
30. Place a new joint washer on the main case.
31. Fit the oil pump, aligning the drive.
32. Place a new joint washer on the oil pump.
33. Fit the speedometer drive housing. Torque:
    $\frac{7}{16}$ in AF fixings 1,0 kgf.m (8 lbf ft).
    $\frac{9}{16}$ in AF fixings 3,5 kgf.m (25 lbf ft).
34. Measure the rolling resistance of the layshaft, using a spring balance and a cord coiled around the layshaft large diameter groove.
35. The rolling resistance must be 0,9 to 2,7 kg (2 to 6 lb) after having overcome inertia.
36. To adjust the bearing pre-load, fit a replacement shim of suitable thickness between the oil pump and the layshaft bearing outer race. Shim range is from 2,36 mm (0.093 in.) to 3,12 mm (0.123 in.) in 0,02 mm (0.001 in.) steps—earlier models, and 2,64 mm (0.104 in.) to 3,51 mm (0.138 in.) in 0,02 mm (0.001 in.) steps—later models with thicker joint washer.
37. When the layshaft bearing pre-load is correct, remove the speedometer housing, oil pump, main case and layshaft. Retain the selected shim.
38. Coat the outside diameter of the mainshaft rear bearing with Loctite grade 'Bearing Fit'.
39. Press a new mainshaft rear bearing into the main case.
40. The mainshaft front bearing fits inside the primary pinion and a new bearing can be fitted during re-assembly of the gearbox. In the event of excessive wear, it may be necessary to fit a new primary pinion.
41. Reverse 1 to 9.

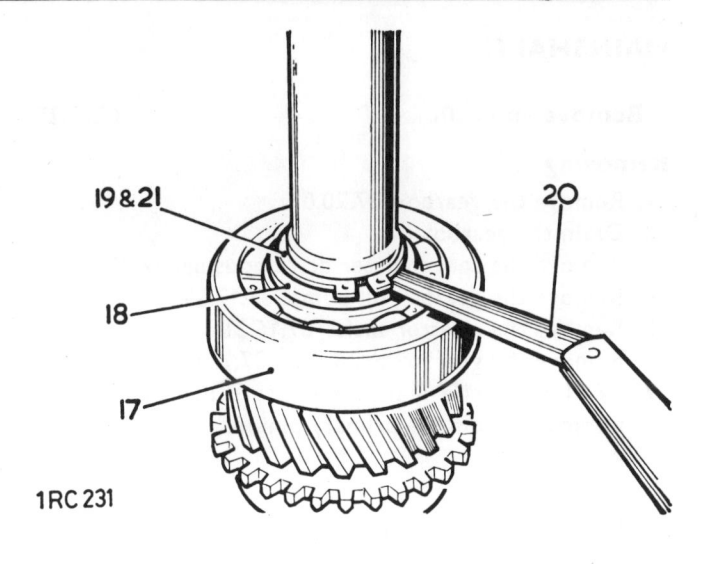

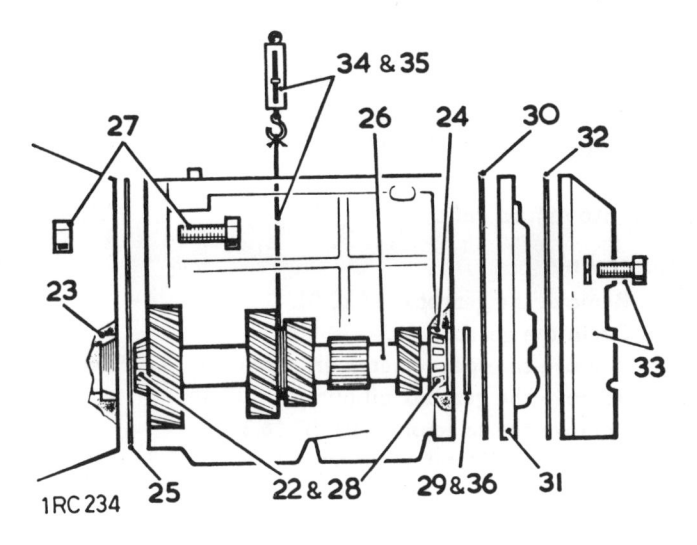

Rover 3500 and 3500S Manual AKM 3621

GEARBOX—SYNCHROMESH

MAINSHAFT

—Remove and refit 37.20.25

Removing

1. Remove the gearbox. 37.20.01.
2. Drain the gearbox oil.
3. Remove the speedometer drive housing. 37.25.09.
4. Remove the gearbox oil pump. 37.12.34.
5. Remove the selector shafts. 37.16.31.
6. Remove the gearbox main case. 37.12.40.
7. Withdraw the mainshaft complete with gear assemblies from the primary pinion.

Refitting

8. Ensure that the synchromesh cone and needle roller bearing are fitted to the primary pinion.
9. Reverse 1 to 7.

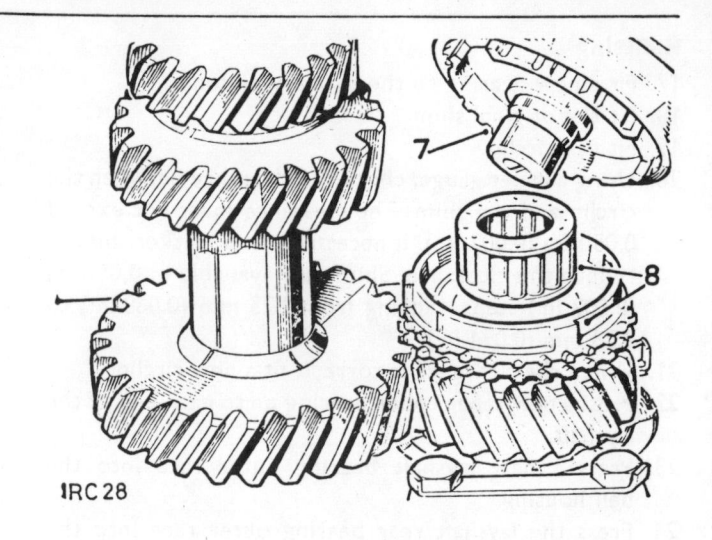

PRIMARY PINION AND LAYSHAFT

—Remove and refit 37.20.29

Removing

1. Remove the gearbox. 37.20.01.
2. Drain the gearbox oil.
3. Remove the speedometer drive housing. 37.25.09.
4. Remove the gearbox oil pump. 37.12.34.
5. Remove the selector shafts. 37.16.31.
6. Remove the clutch withdrawal unit. 33.25.12.
7. Remove the gearbox main case. 37.12.40.
8. Withdraw the mainshaft complete with gear assemblies from the primary pinion.
9. Withdraw the synchromesh cone and needle roller bearing from the primary pinion.
10. Remove the bearing retaining plates from the rear of the bell housing.
11. Hold the layshaft and, using a soft faced mallet, tap the primary pinion shaft rearward until the layshaft can be withdrawn.
12. Continue to tap the primary pinion from the bell housing.
13. Withdraw the bearing baffle plate.

Refitting

   NOTE: If new bearings are being fitted, carry out the checks described in 37.20.17 and 37.20.22.

14. Heat the bell housing by immersing in boiling water or a degreasing plant for a few minutes.
15. Place the bearing baffle plate into the primary pinion bore in the bell housing.
16. Locate the primary pinion and layshaft together and fit them to the bell housing.
17. Reverse 1 to 10. Ensure that the lock plates for the bearing retaining plate bolts are fully engaged.

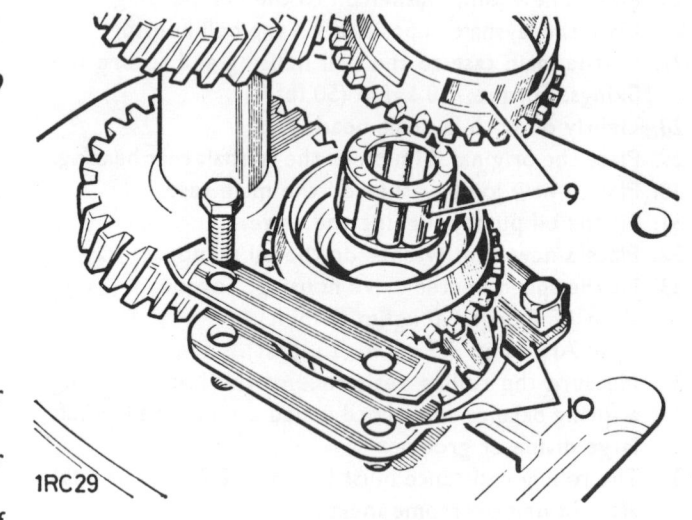

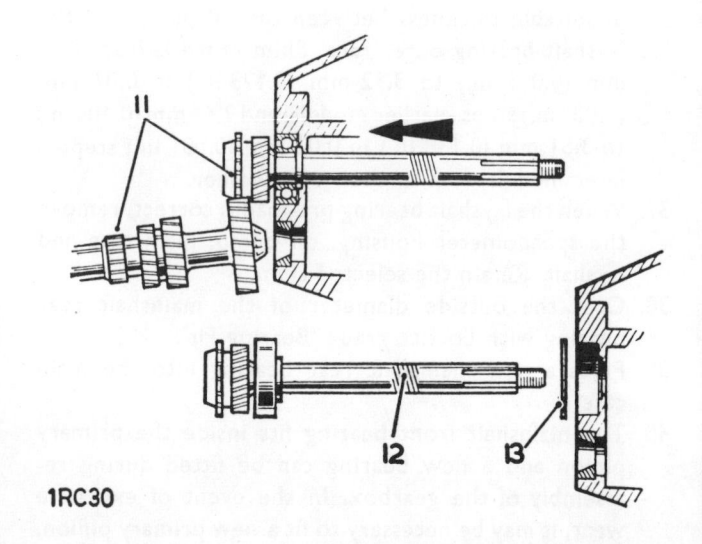

Rover 3500 and 3500S Manual AKM 3621

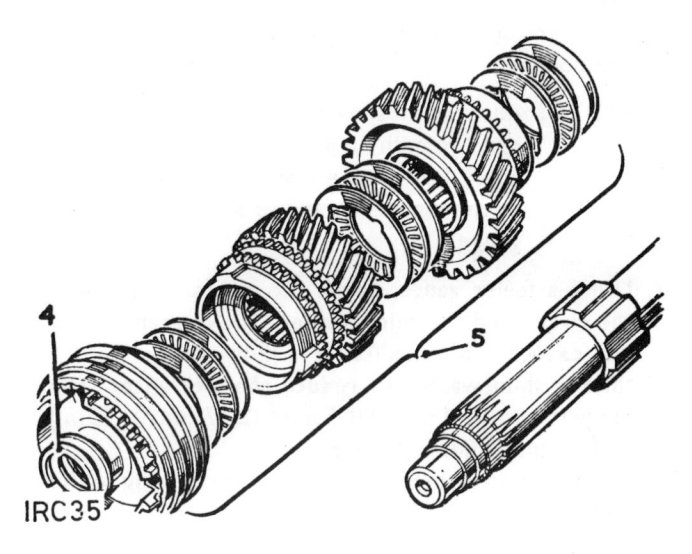

GEARBOX—SYNCHROMESH

MAINSHAFT (with needle roller thrust washers)

—Overhaul   37.20.31

NOTE: Two designs of mainshaft thrust washers are in use, either needle roller or bronze type. The two designs are interchangeable in complete sets providing that the relative assembly procedure is followed. For details of the bronze thrust washer design, refer to the accompanying operation of the same number.

Dismantling

1. Remove the mainshaft. 37.20.25.
2. Support the first speed gear, then drift the mainshaft to free the thick thrust washer.
3. Withdraw the components fitted to the rear end of the mainshaft.
4. Remove the spring ring from the front end of the mainshaft.
5. Withdraw the remaining components from the mainshaft.

Inspecting

6. Obtain a new spring ring for the mainshaft front end. Retain the original spring ring until the mainshaft checks have been completed.
7. Examine all components and fit new replacements as necessary. For details of synchromesh units see 37.20.08.

Reassembling

NOTE: Do not degrease new needle roller thrust bearings and thrust washers.

8. Assemble the following components to the front end of the mainshaft. 9 to 25.
9. Thrust washer for low gear synchromesh inner member, front, with chamfer towards the rear of the mainshaft.
10. Needle roller thrust bearing.
11. Scalloped thrust washer.
12. Needle roller bearing.

continued

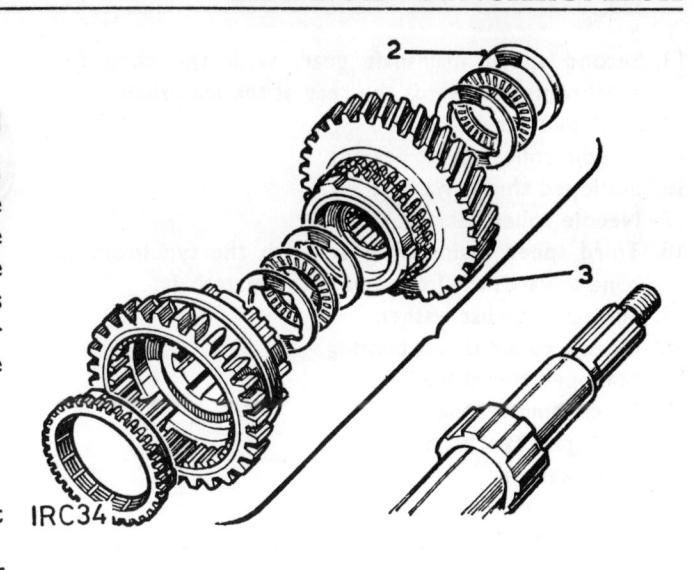

Rover 3500 and 3500S Manual AKM 3621

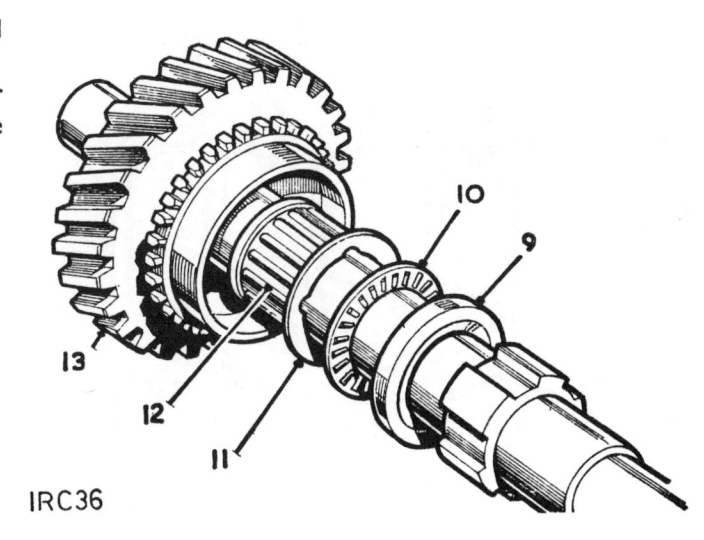

37.20.31
Sheet 1

GEARBOX—SYNCHROMESH

13. Second speed mainshaft gear, with the cone for synchromesh towards the rear of the mainshaft.
14. Scalloped thrust washer.
15. Needle roller thrust bearing.
16. Scalloped thrust washer.
17. Needle roller bearing.
18. Third speed mainshaft gear, with the synchromesh cone towards the front end of the mainshaft.
19. Scalloped thrust washer.
20. Needle roller thrust bearing.
21. Scalloped thrust washer.
22. Synchromesh cone.
23. High gear synchromesh unit.
24. Shim washer.
25. Original spring ring.
26. Apply a load of approximately 28 kg (60 lb) to the inner member of the high gear synchromesh unit, to take up all clearance between the mainshaft components.

    NOTE: Illustration shows part of a Land-Rover synchromesh unit which can be used to apply load to the mainshaft assembly, and at the same time allows access for a feeler gauge.

27. Using a feeler gauge, check the clearance between the circlip and the shim washer. The clearance must not exceed 0,07 mm (0.003 in.). If necessary, fit a thicker shim washer to reduce the clearance. Shim range is from 2,41 mm (0.095 in.) to 2,87 mm (0.113 in.) in 0,07 mm (0.003 in.) steps.
28. When clearance is correct, fit a new spring ring to the mainshaft.

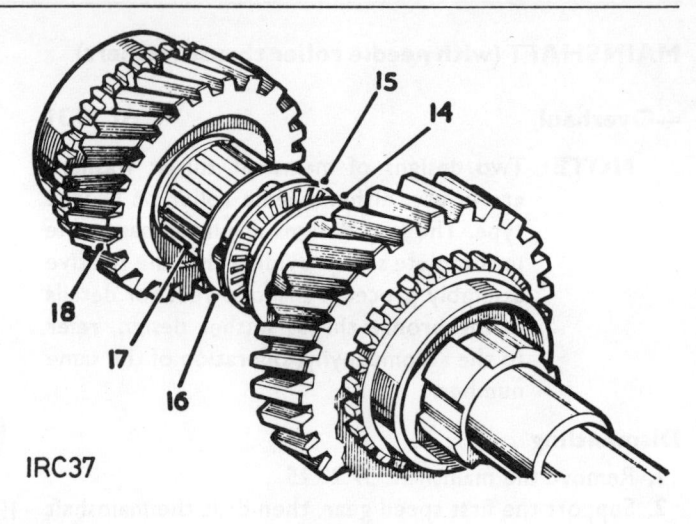

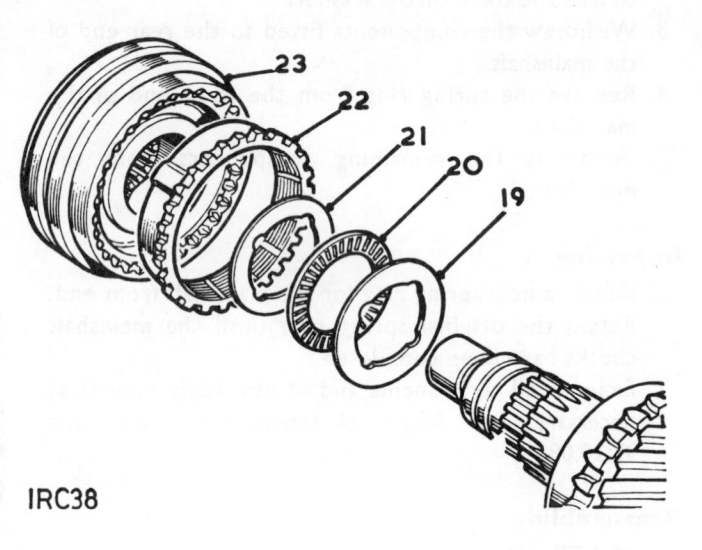

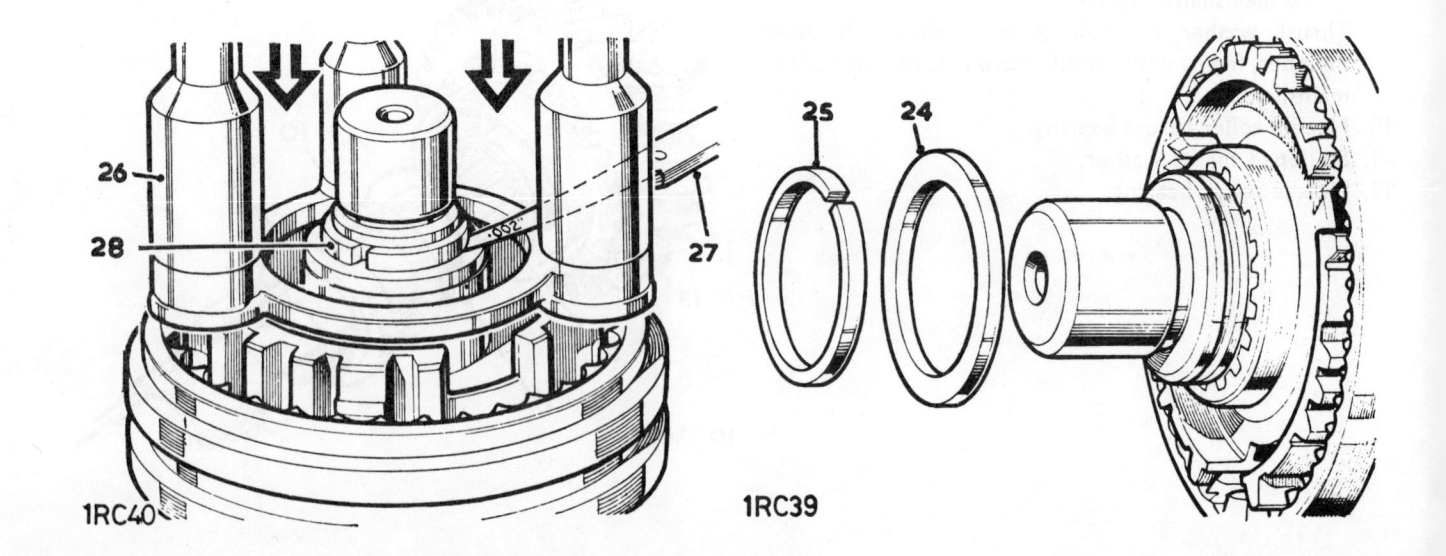

continued

37.20.31
Sheet 2

Rover 3500 and 3500S Manual AKM 3621

GEARBOX—SYNCHROMESH

29. Assemble the following components to the rear end of the mainshaft. 30 to 40.
30. Synchromesh cone.
31. Low gear synchromesh unit.
32. Plain thrust washer.
33. Needle roller thrust bearing.
34. Scalloped thrust washer.
35. Synchromesh cone.
36. First speed mainshaft gear.
37. Needle roller bearing.
38. Scalloped thrust washer.
39. Needle roller thrust bearing.
40. The thrust washer for the first speed mainshaft gear must be pressed on to the mainshaft, with the inner chamfer towards the gears, so that the rear face is flush with the mainshaft shoulder.
41. Refit the mainshaft. 37.20.25.

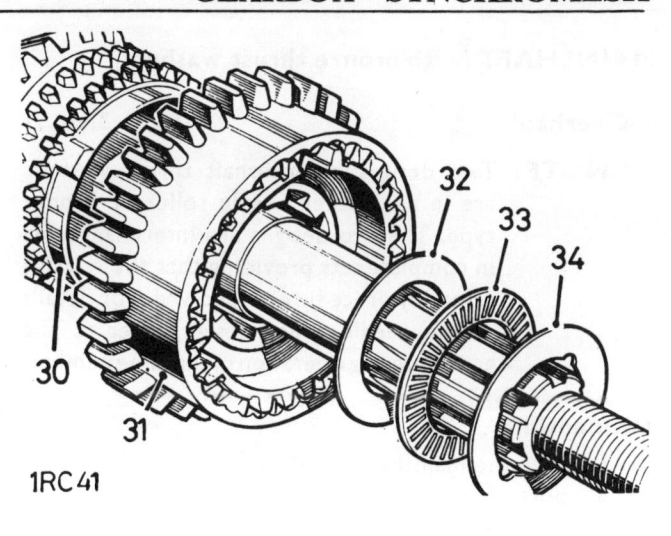

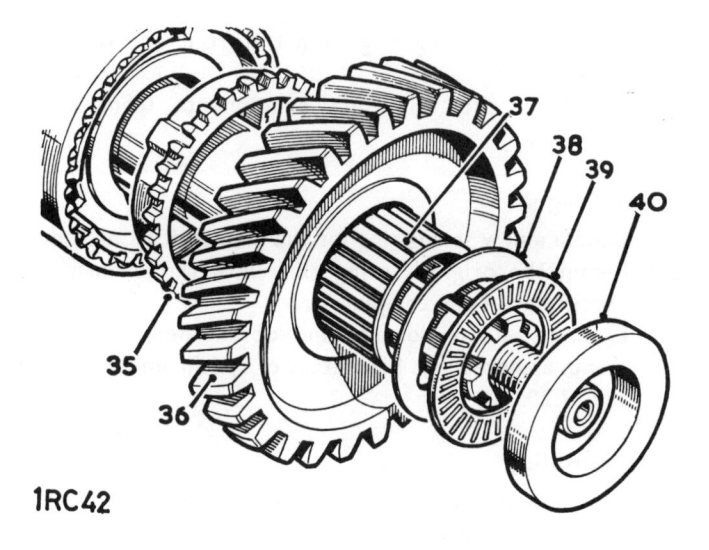

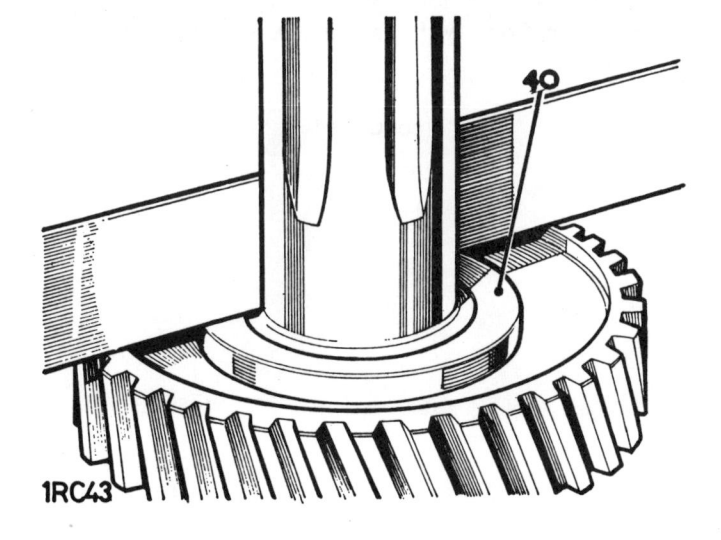

DATA

Shim clearance, mainshaft front end   0,07 mm (0.003 in.)

Rover 3500 and 3500S Manual AKM 3621

37.20.31
Sheet 3

GEARBOX—SYNCHROMESH

MAINSHAFT (with bronze thrust washers)

—Overhaul   37.20.31

NOTE: Two designs of mainshaft thrust washers are in use, either needle roller or bronze type. The two designs are interchangeable in complete sets providing that the relative assembly procedure is followed. For details of the needle roller design, refer to the accompanying operation of the same number.

Dismantling

1. Remove the mainshaft. 37.20.25.
2. Support the first speed gear, then drift the mainshaft to free the thick thrust washer.
3. Withdraw the components fitted to the rear end of the mainshaft.
4. Remove the spring ring from the front end of the mainshaft.
5. Withdraw the remaining components from the mainshaft.

Inspecting

6. Obtain a new spring ring for the mainshaft front end. Retain the original spring ring until the mainshaft checks have been completed.
7. Examine all components and obtain new replacements as necessary. For details of synchromesh units see 37.20.08.

Reassembling

8. Assemble the following components to the front end of the mainshaft. 9 to 20.
9. Thrust washer for low gear synchromesh inner member, front, with chamfer towards the rear of the mainshaft.

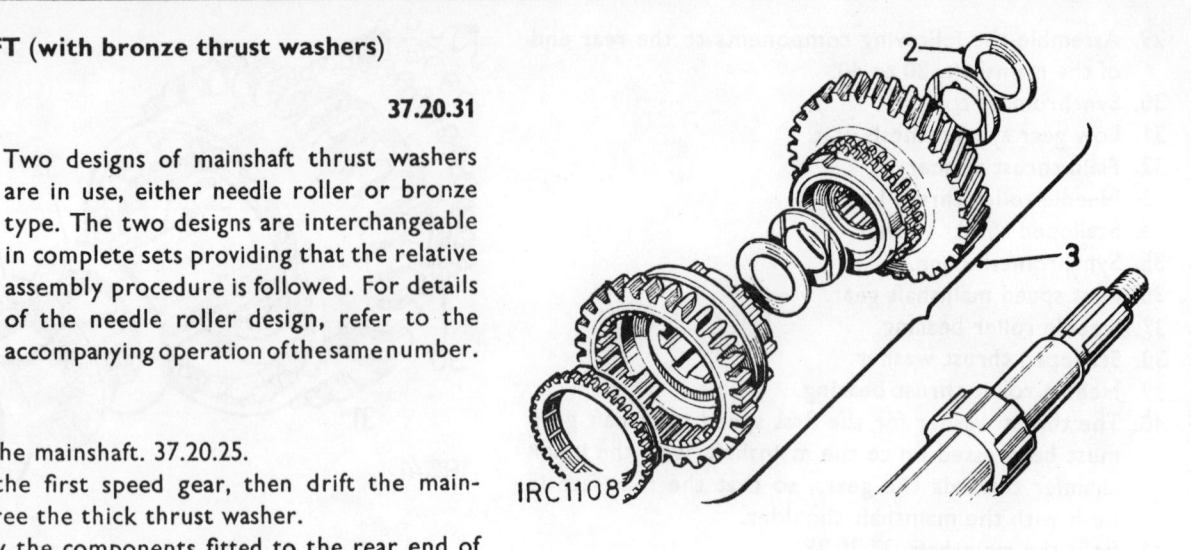

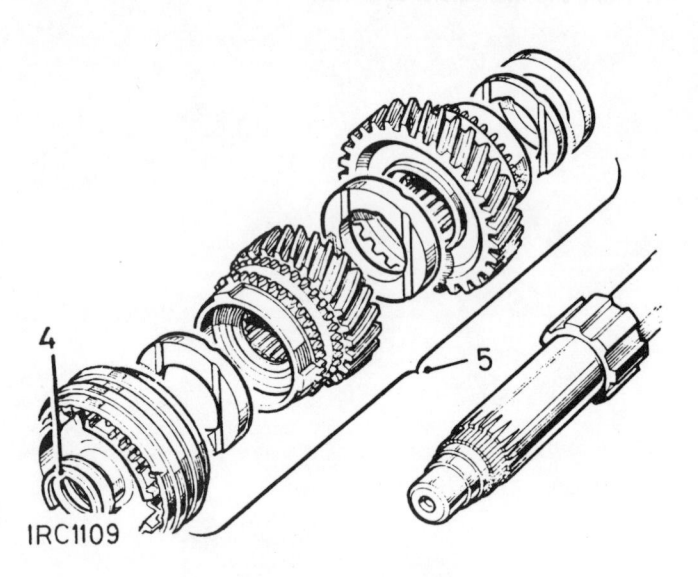

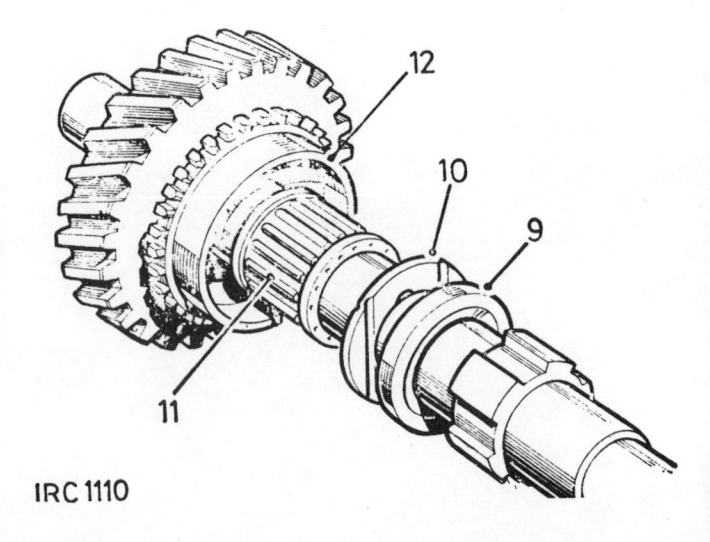

continued

37.20.31
Sheet 4

Rover 3500 and 3500S Manual AKM 3621

GEARBOX—SYNCHROMESH

10. Thin bronze thrust washer.
11. Needle roller bearing.
12. Second speed mainshaft gear, with the cone for the syncromesh towards the rear of the mainshaft.
13. Thick bronze thrust washer.
14. Needle roller bearing.
15. Third speed mainshaft gear, with the syncromesh cone towards the front end of the mainshaft.
16. Thick bronze thrust washer.
17. Syncromesh cone.
18. High gear synchromesh unit.
19. Shim washer.
20. Original spring ring.
21. Using a feeler gauge, check the clearance between the circlip and the shim washer. The clearance must be 0,1 to 0,2 mm (0.004 to 0.008 in.). If necessary, adjust the clearance by fitting an alternative size shim washer. Shim range is from 2,23 to 2,69 mm (0.088 to 0.106 in.) in 0,07 mm (0.003 in.) steps.
22. When the clearance is correct, fit a new spring ring to the mainshaft.
23. Assemble the following components to the rear end of the mainshaft. 24 to 32.

continued

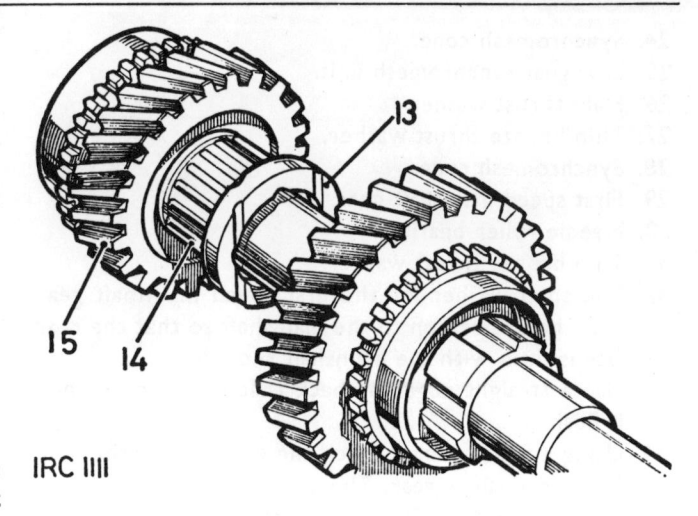

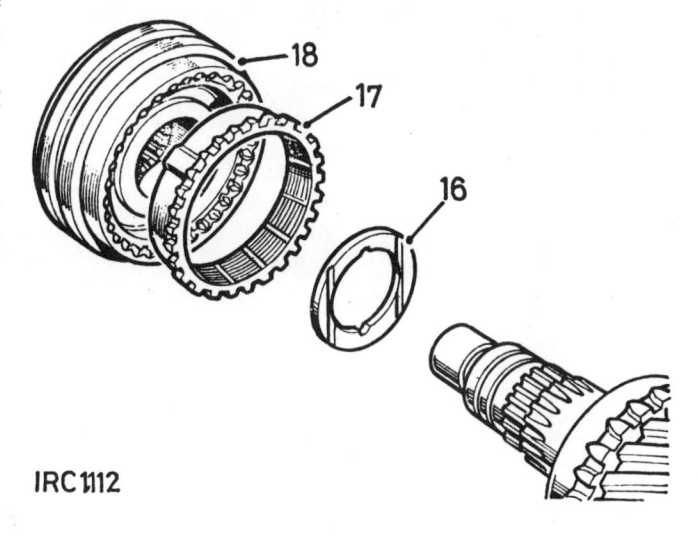

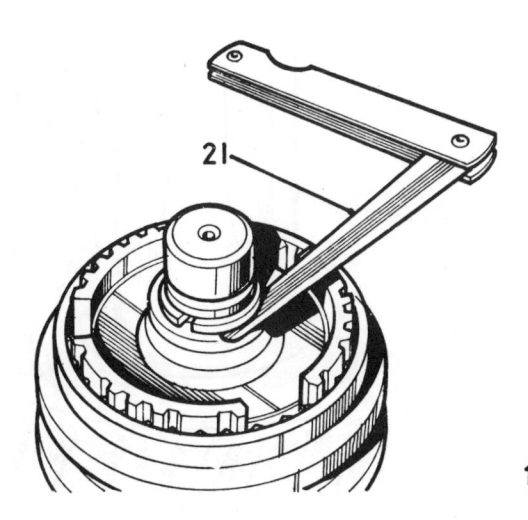

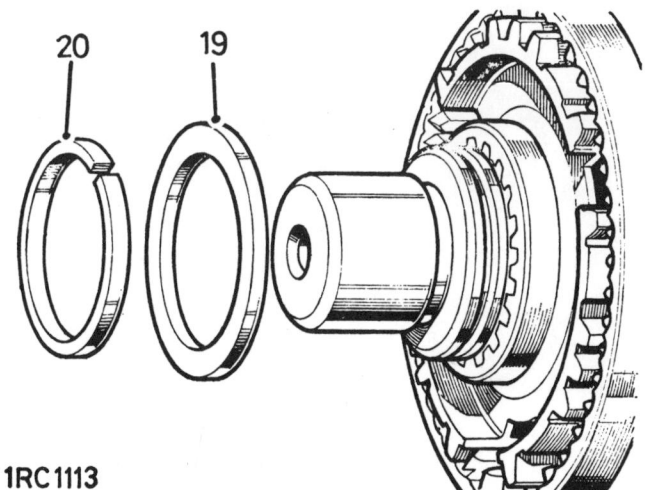

IRC 1111

IRC 1112

IRC 1114

1RC 1113

Rover 3500 and 3500S Manual AKM 3621

GEARBOX—SYNCHROMESH

24. Synchromesh cone.
25. Low gear synchromesh unit.
26. Plain thrust washer.
27. Thin bronze thrust washer.
28. Synchromesh cone.
29. First speed mainshaft gear.
30. Needle roller bearing.
31. Thin bronze thrust washer.
32. The shim washer for the first speed mainshaft gear must be pressed on to the mainshaft so that the rear face is flush with the mainshaft shoulder.
33. Use a straight edge to check that the shim washer is flush.
34. Using a feeler gauge, check the end float of the first speed mainshaft gear. The end float must be 0,1 to 0,2 mm (0.004 to 0.008 in.). If necessary, adjust the end float by fitting an alternative size shim washer. Shim range is 5,38 to 5,58 mm (0.212 to 0.220 in.) in 0,05 mm (0.002 in.) steps.
35. Refit the mainshaft. 37.20.25.

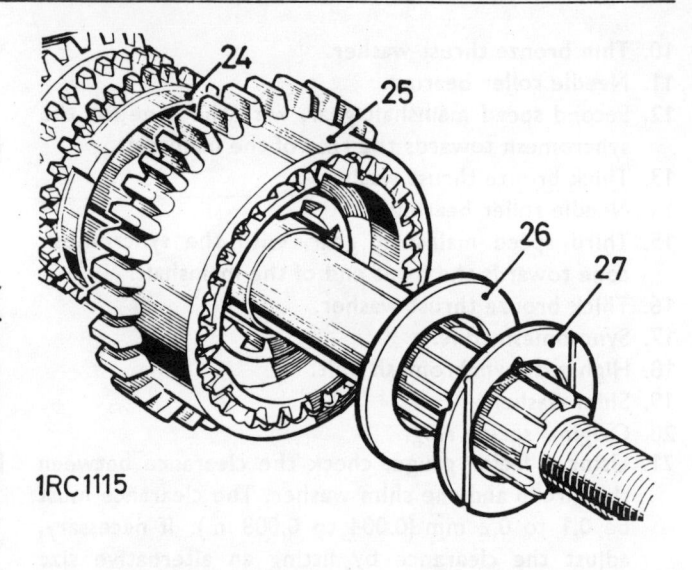

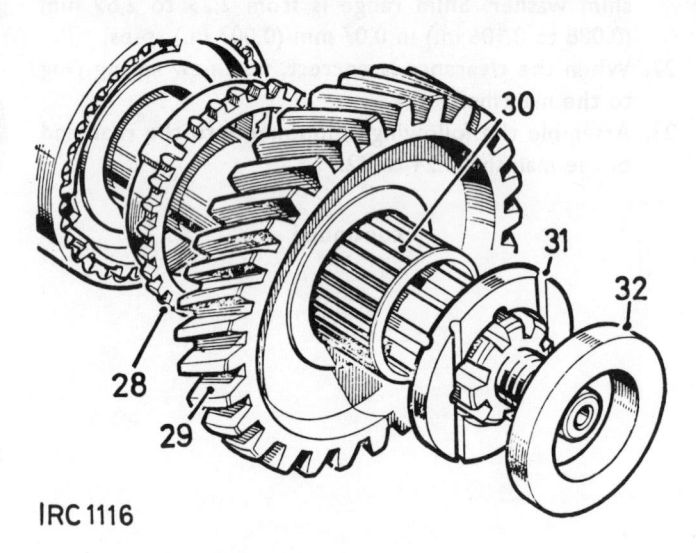

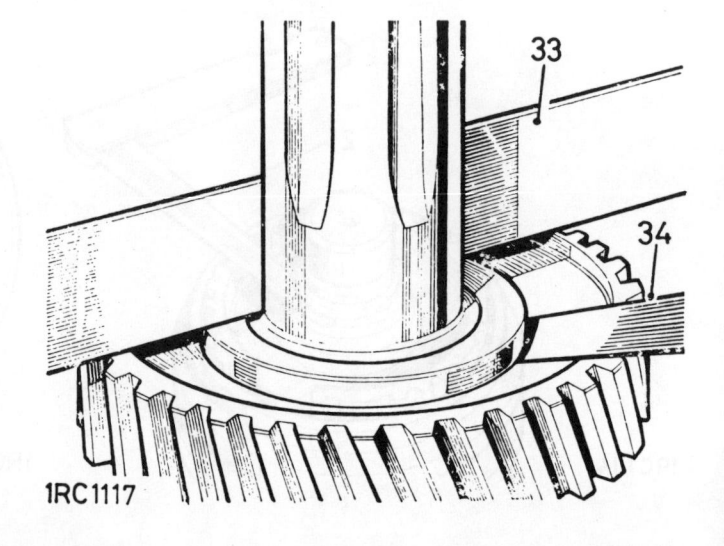

DATA

Shim clearance, mainshaft front end	0,1 to 0,2 mm (0.004 to 0.008 in)
End float, mainshaft first speed gear	0,1 to 0,2 mm (0.004 to 0.008 in)

GEARBOX—SYNCHROMESH

SPEEDOMETER DRIVE HOUSING

—Remove and refit 37.25.09

Removing

1. Disconnect the battery earth lead.
2. Remove the air cleaner. 19.10.01.
3. Remove the fan blades.
4. Remove the cover from the gearbox tunnel.
5. Remove the gearchange lever assembly.
6. Lift the carpet from the left hand side of the gearbox tunnel and remove the cover plate.
7. Disconnect the speedometer cable from the gearbox.
8. Lift the carpet from the right hand side of the gearbox tunnel and remove the large rubber grommet.
9. Disconnect the electrical leads from the reverse light switch.
10. Drain the gearbox lubricating oil.
11. Remove the front exhaust pipe.
12. Remove the propeller shaft.
13. Remove the fixings from the clutch slave cylinder, withdraw the slave cylinder from the bell housing and move it to one side without disconnecting the fluid pipe.

continued

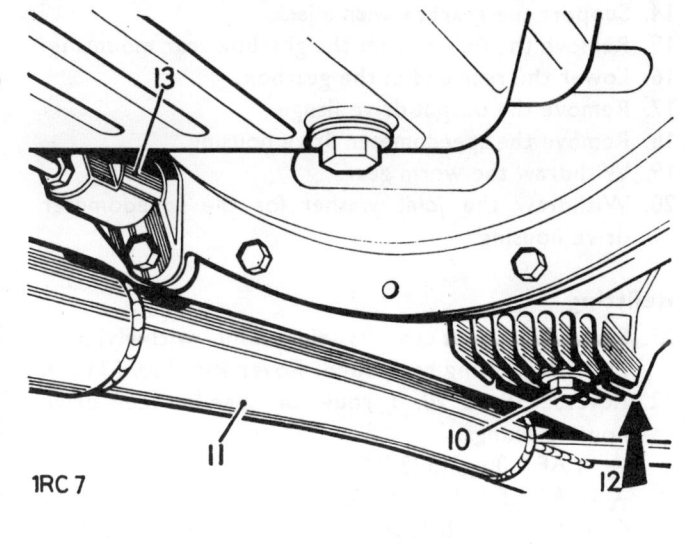

Rover 3500 and 3500S Manual AKM 3621

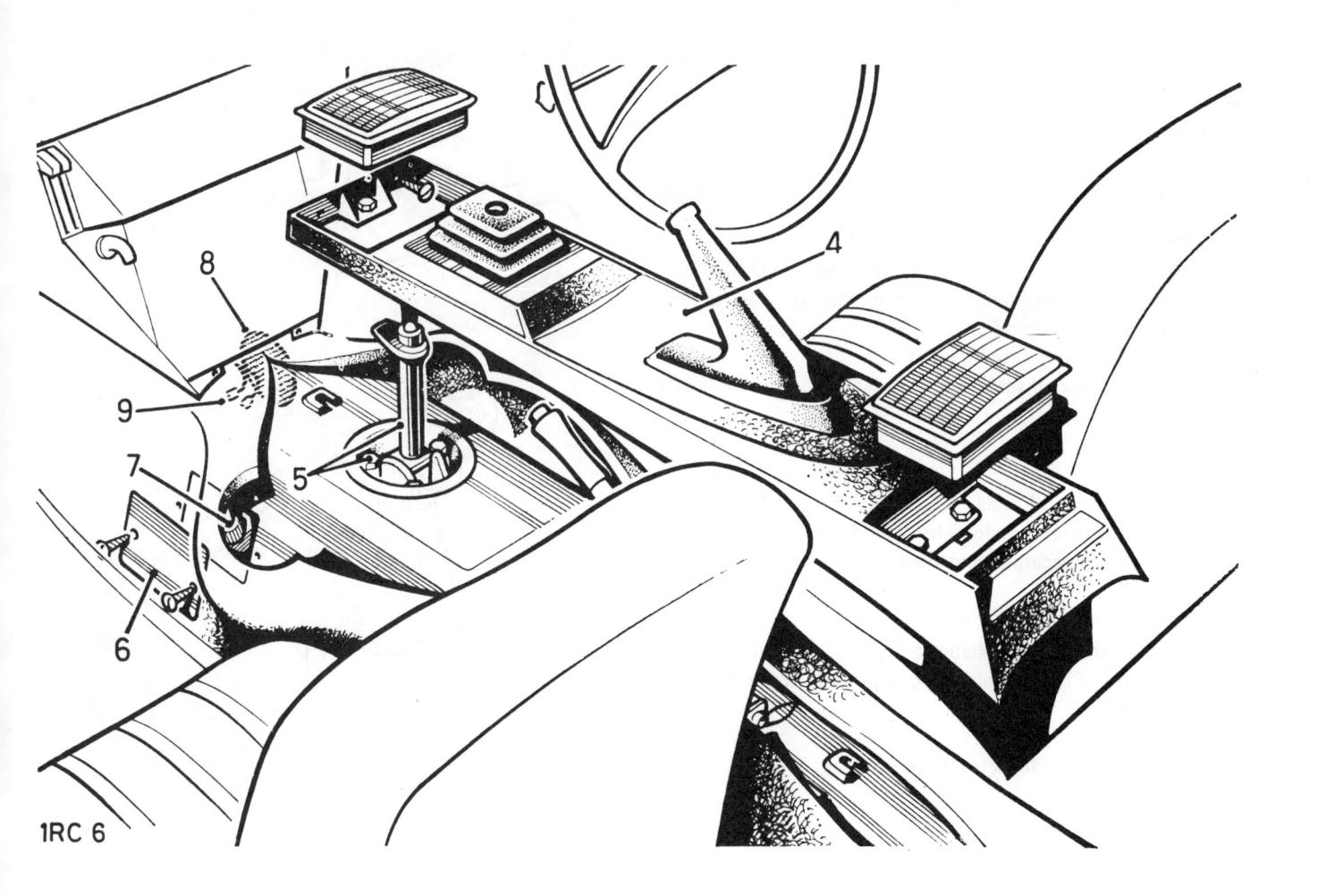

37.25.09
Sheet 1

GEARBOX—SYNCHROMESH

14. Support the gearbox with a jack.
15. Remove the fixings from the gearbox rear mounting.
16. Lower the rear end of the gearbox.
17. Remove the output drive flange.
18. Remove the speedometer drive housing.
19. Withdraw the worm gear.
20. Withdraw the joint washer for the speedometer drive housing.

Refitting

21. Smear both sides of a new joint washer with Hylomar PL 32/M jointing compound, Rover Part No. 534244.
22. Reverse 18 to 20. Torque for speedometer drive housing fixings:
    $\frac{7}{16}$ in AF 1,0 kgf.m (8 lbf ft).
    $\frac{9}{16}$ in AF 3,5 kgf.m (25 lbf ft).
23. Coat the mainshaft splines and thread with Loctite Sealant, Grade AVV, Rover Part No. 600303.
24. Fit the output driving flange and secure with the special washer and self-locking nut. Torque: 10,5 kgf.m (75 lbf ft).
25. Reverse 1 to 16.

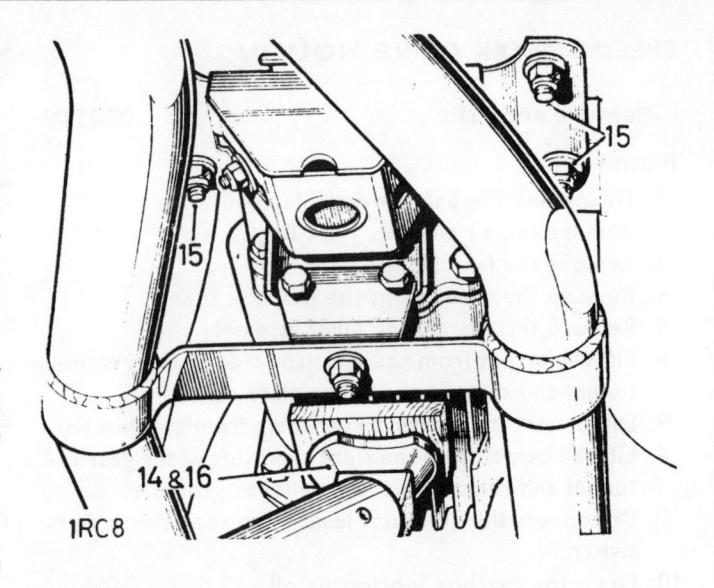

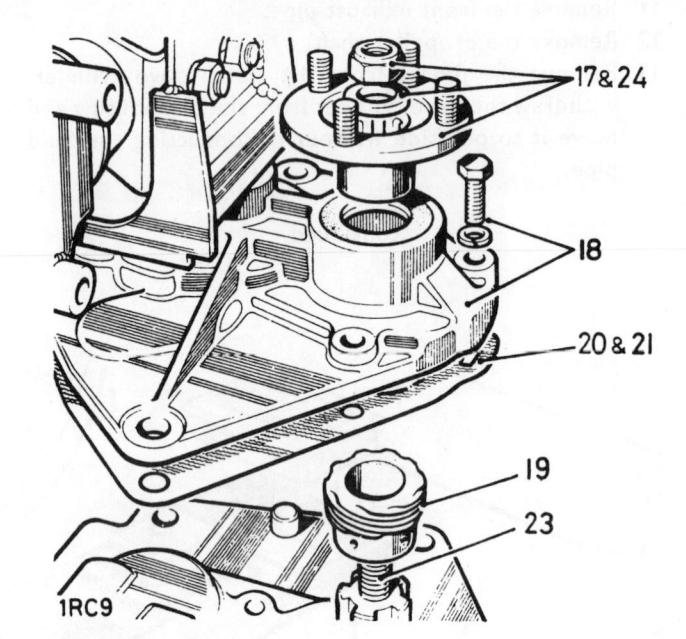

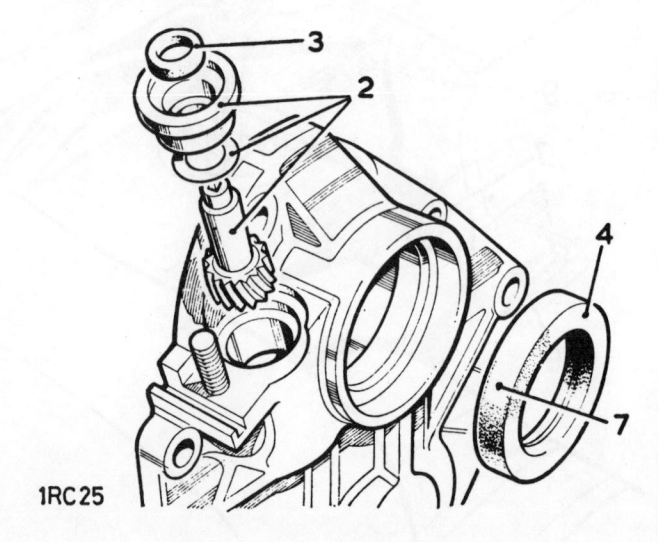

SPEEDOMETER DRIVE HOUSING

—Overhaul   37.25.13

Dismantling

1. Remove the speedometer drive housing. 37.25.09.
2. Withdraw the nylon housing, thrust washer and speedometer spindle.
3. Withdraw the spindle oil seal.
4. Remove the oil seal from the rear of the speedometer drive housing.

Inspecting

5. Obtain new oil seals.
6. Examine the speedometer spindle and worm gear, fit new replacements as necessary.

Reassembling

7. Smear the outside diameter of the rear oil seal with Hylomar PL 32/M jointing compound, Rover Part No. 534244.
8. Reverse 2 to 4.
9. Refit the speedometer drive housing. 37.25.09.

37.25.09
Sheet 2
37.25.13

Rover 3500 and 3500S Manual AKM 3621

GEARBOX—AUTOMATIC—Borg-Warner type 35

LIST OF OPERATIONS

Brake bands
    —front—remove and refit 44.10.01
    —rear—remove and refit 44.10.09

Clutches
    —front and rear—remove and refit 44.12.01
    —front—overhaul 44.12.10
    —rear—overhaul 44.12.13

Converter
    —assembly—remove and refit 44.17.07
    —assembly—balance 44.17.08
    —housing—remove and refit 44.17.01

Downshift cable
    —adjustment check 44.30.02
    —pressure check 44.30.03

Fluid system—drain and refill 44.24.02

Front pump
    —remove and refit 44.32.01
    —overhaul 44.32.04

Gearbox assembly—remove and refit 44.20.01

Governor assembly
    —remove and refit 44.22.01
    —overhaul 44.22.04

Hand selector lever assembly
    —remove and refit 44.15.04
    —overhaul 44.15.05

Input shaft—end float check 44.36.08

Manual valve lever shaft—remove and refit 44.26.04

Mechanical operation—air pressure checks 44.30.16

Output shaft and ring gear—remove and refit 44.36.01

Parking brake pawl assembly
    —remove and refit 44.28.07
    —overhaul 44.28.08

Planet gear and centre support
    —remove and refit 44.36.04
    —overhaul 44.36.05

Rover 3500 and 3500S Manual AKM 3621

GEARBOX—AUTOMATIC—Borg-Warner type 35

Rear extension housing—remove and refit 44.20.15

Road test 44.30.17

Selector rod assembly—remove and refit 44.15.08

Servos
    —front—remove and refit 44.34.07
    —rear—remove and refit 44.34.13
    —front—overhaul 44.34.10
    —rear—overhaul 44.34.16

Speedometer drive pinion—remove and refit 44.38.04

Stall test 44.30.13

Starter and reverse light switch—remove and refit 44.15.15

Valve body assembly
    —remove and refit 44.40.01
    —overhaul 44.40.04

GEARBOX—AUTOMATIC—Borg-Warner type 35

Borg-Warner automatic gearbox, model 35

Description and operation

The model 35 three-speed automatic gearbox is coupled to the engine by a three-element torque converter. Engine power is converted into hydro-kinetic energy and this provides smooth application or driving torque.

Torque multiplication in all gear ratios is provided by the converter, which is infinitely variable between the ratio of 2:1 and 1:1. Extreme low-speed flexibility in third gear is the result. The speed range during which torque multiplication can be achieved is also variable, depending upon the position of the accelerator.

The torque converter

The torque converter consists of an impeller, connected to the engine crankshaft; a turbine connected to the gearbox input shaft and a stator incorporating a sprag-type one-way clutch. This assembly is supported by a tube attached to the gearbox; it is co-axial with the input shaft.

The impeller, driven by the engine, causes fluid to flow from its vanes to the turbine vanes and to return to the impeller through the stator vanes.

The vanes of the components are designed and curved to affect the angle of fluid flow when a speed differential exists between impeller and turbine. The angle of the fluid flow from the turbine is changed by the stator vanes in such a way that the fluid from the stator assists in driving the impeller; torque multiplication taking place. This varies from 2:1 when the turbine is stalled* to 1:1 when the turbine speed reaches approximately 90 per cent of the impeller.

With the speed differential between impeller and turbine achieved the angle of fluid flow from the turbine drives the stator in the same direction as the turbine and impeller. In this state the converter acts as a fluid coupling and no torque multiplication takes place.

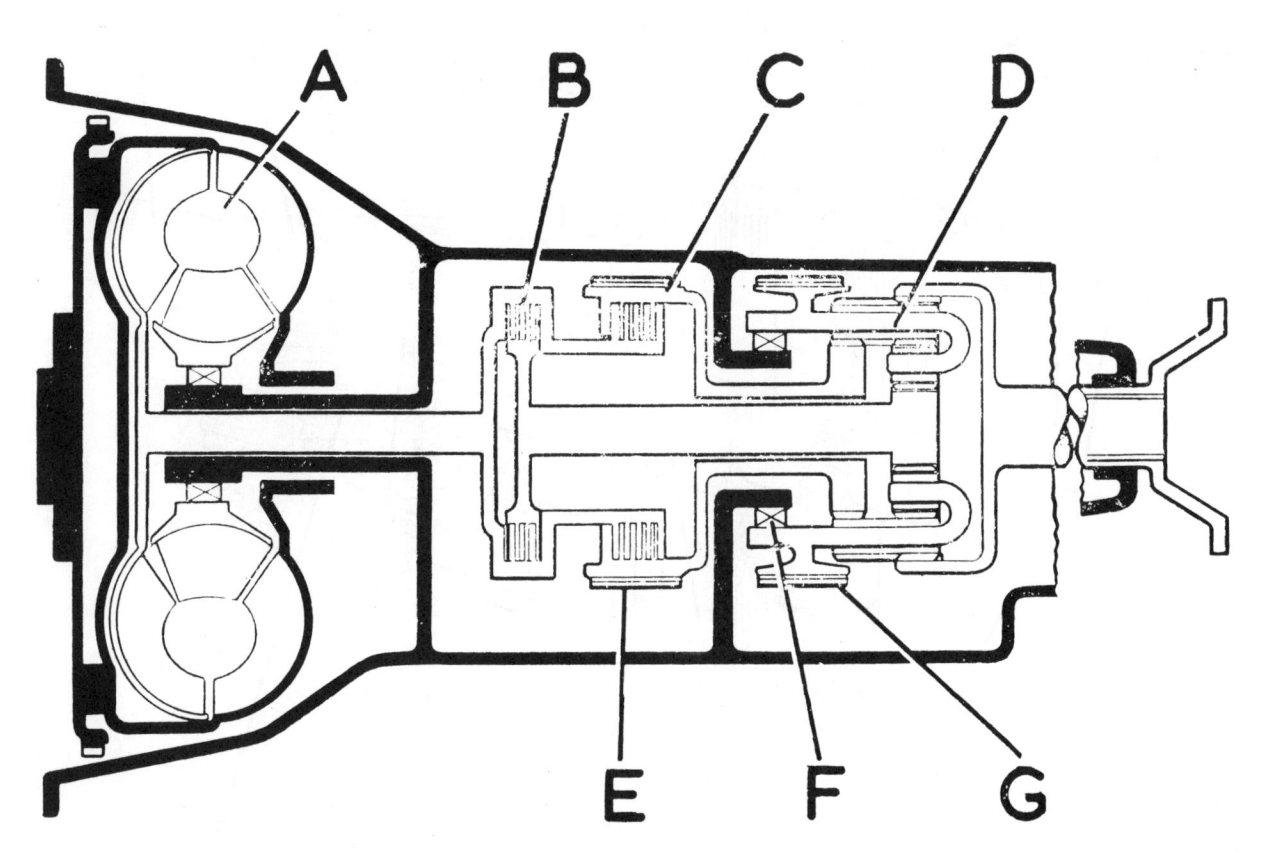

Mechanical layout

A—Torque converter
B—Front clutch
C—Rear clutch
D—Planetary gear set
E—Front band
F—One-way clutch
G—Rear band

Rover 3500 and 3500S Manual AKM 3621

GEARBOX—AUTOMATIC—Borg-Warner type 35

The mechanical system

A planetary gear set, having helical involute tooth form throughout, provides three forward speeds and reverse. The planetary gear set comprises of two sun gears (forward and rear); two sets of pinions located in a pinion carrier; and ring gear which is attached to the output shaft.

Power enters through one of the sun gears and leaves the gear set through the ring gear. In all forward gears, power enters through the forward sun gear and in reverse through the rear sun gear.

A single set of pinions is engaged when reverse is selected, causing the ring gear to rotate in the opposite direction to the sun gear. A double set of pinions, engaged when forward gears are selected, cause the ring gear to rotate in the same direction as the sun gear.

The pinions are housed in a carrier which locates them in their respective position relative to the two sun gears and the ring gear. The carrier is in the form of a drum which may rotate or be held stationary by a brake band or the one-way clutch.

Gear ratios are engaged by hydraulically-operated clutches and brake bands, of which there are two of each.

The clutches

These are operated by hydraulic pistons and connect the torque converter to the gear set.

In forward gears the front clutch connects the converter to the forward sun gear and in reverse the rear clutch connects the torque converter to the reverse sun gear. Both clutches are of the multi-disc type.

The one-way clutch

This functions in place of the rear brake band when the selector is at the 'D'—Drive position. It prevents anti-clockwise rotation of the planetary gear carrier.

In first gear the gear set therefore freewheels, providing smooth changes first to second and vice versa.

The brake bands

There are two bands which hold elements of the gearing stationary to effect lower ratio output and increased torque. Both bands are hydraulically operated by servos. The front band holds the reverse sun gear stationary; the rear band holds the planetary gear carrier stationary.

In 'lock-up' on early selector pattern, or '1' on latest selector pattern, the pinion or planetary gear carrier is held stationary by the rear band and provides the first gear ratio of 2.39:1.

A double set of pinions are engaged so that the output shaft rotates in the same direction as the input shaft.

In reverse gear the planetary gear carrier is again held stationary, but a single set of pinions causes the driven shaft to rotate in the opposite direction.

For second gear ratio the front band holds the reverse sun gear stationary and the gearing produces a ratio of 1.45:1.

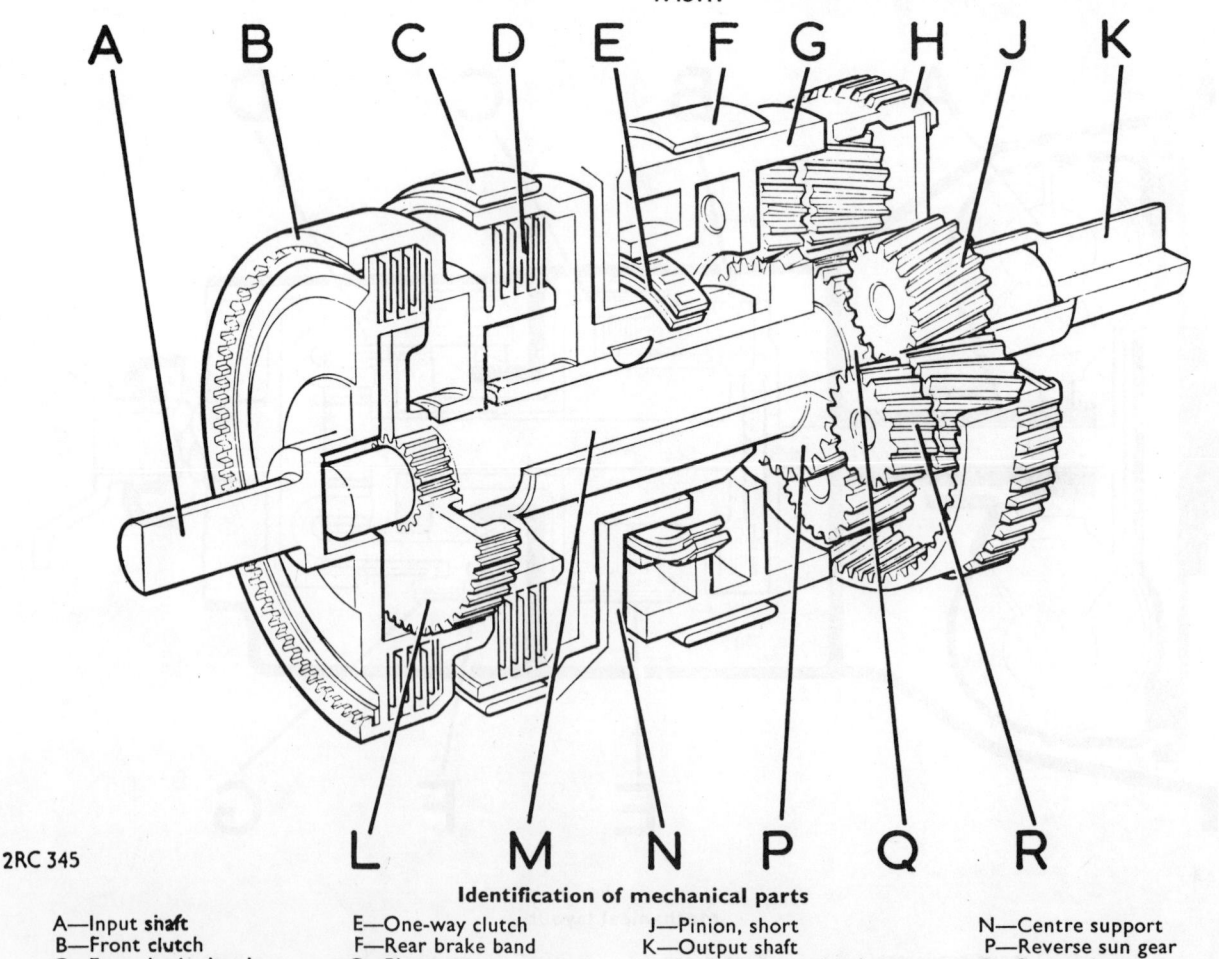

Identification of mechanical parts

A—Input shaft
B—Front clutch
C—Front brake band
D—Rear clutch
E—One-way clutch
F—Rear brake band
G—Planetary gear carrier
H—Ring gear
J—Pinion, short
K—Output shaft
L—Hub for front clutch
M—Shaft for forward sun gear
N—Centre support
P—Reverse sun gear
Q—Forward sun gear
R—Pinion, long

* When the car is held stationary with the engine operating at maximum throttle opening and any one of the driving ranges selected.

Rover 3500 and 3500S Manual AKM 3621

GEARBOX—AUTOMATIC—Borg-Warner type 35

Mechanical power flow—first gear; selector at the 'L' position on early selector pattern or '1' on latest selector pattern

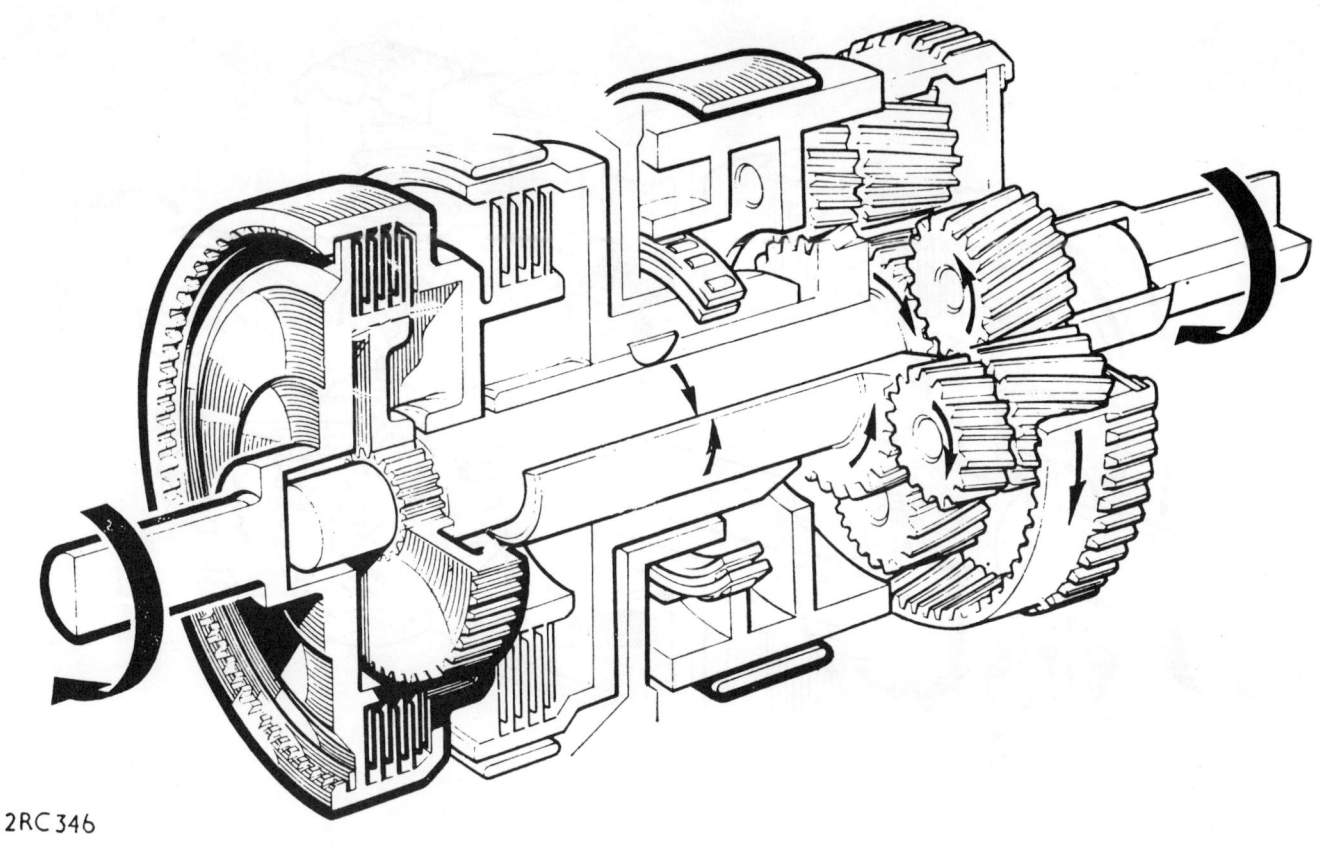

Application of bands and clutches—front clutch and rear band

P	R	N	D2	D1	▼L▼
P	R	N	D	2	1

Power from the turbine of the torque converter is transmitted through the front clutch to the forward sun gear, which is in mesh with a double set of pinions in the planetary gear carrier.

Power leaves the planetary gear set by the ring gear, which is attached to the driven shaft. The carrier itself is held stationary by the rear brake band.

The reverse sun gear, which is also in mesh with a single set of pinions in the planetary set, rotates freely in the opposite direction to the forward sun gear.

GEARBOX—AUTOMATIC—Borg-Warner type 35

Mechanical power flow—first gear; selector at the 'D1' position on early selector pattern or 'D' on latest selector pattern

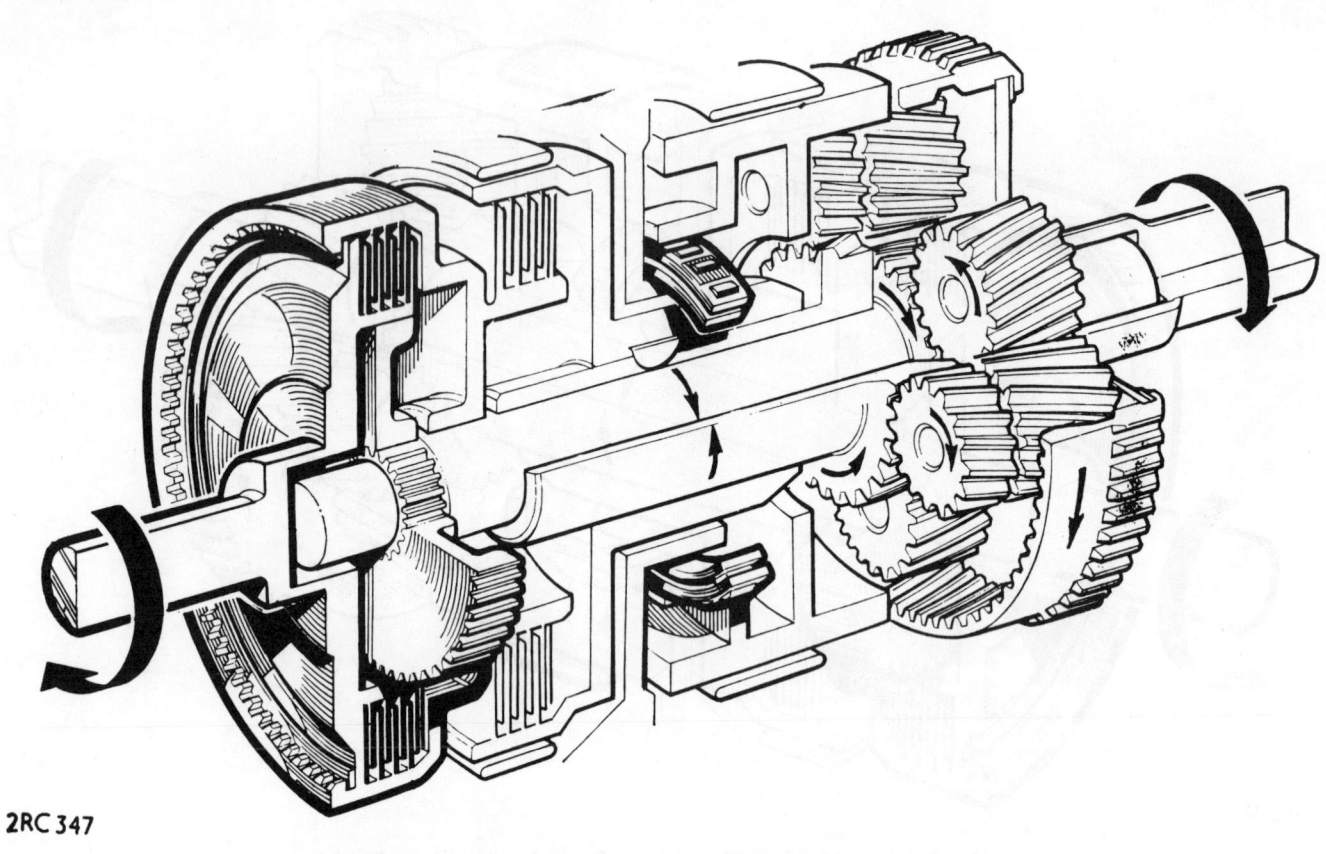

Application of bands and clutches—front clutch and one-way clutch

```
            ▼
P  R  N  D2 D1 L
         ▼
P  R  N  D  2  1
```

Power is again transmitted from the turbine of the torque converter to the forward sun gear, through the front clutch. The forward sun gear is in mesh with a double set of pinions in the planetary gear set.

A one-way clutch, incorporated in the carrier of the planetary gears, prevents anti-clockwise rotation of the carrier and allows the gearbox to freewheel when on the overrun.

Power leaves the planet gear set by the ring gear and driven shaft.

GEARBOX—AUTOMATIC—Borg-Warner type 35

Mechanical power flow—second gear; selector at the 'L', 'D1' or 'D2' position on early selector pattern, or '2' or 'D' on latest selector pattern

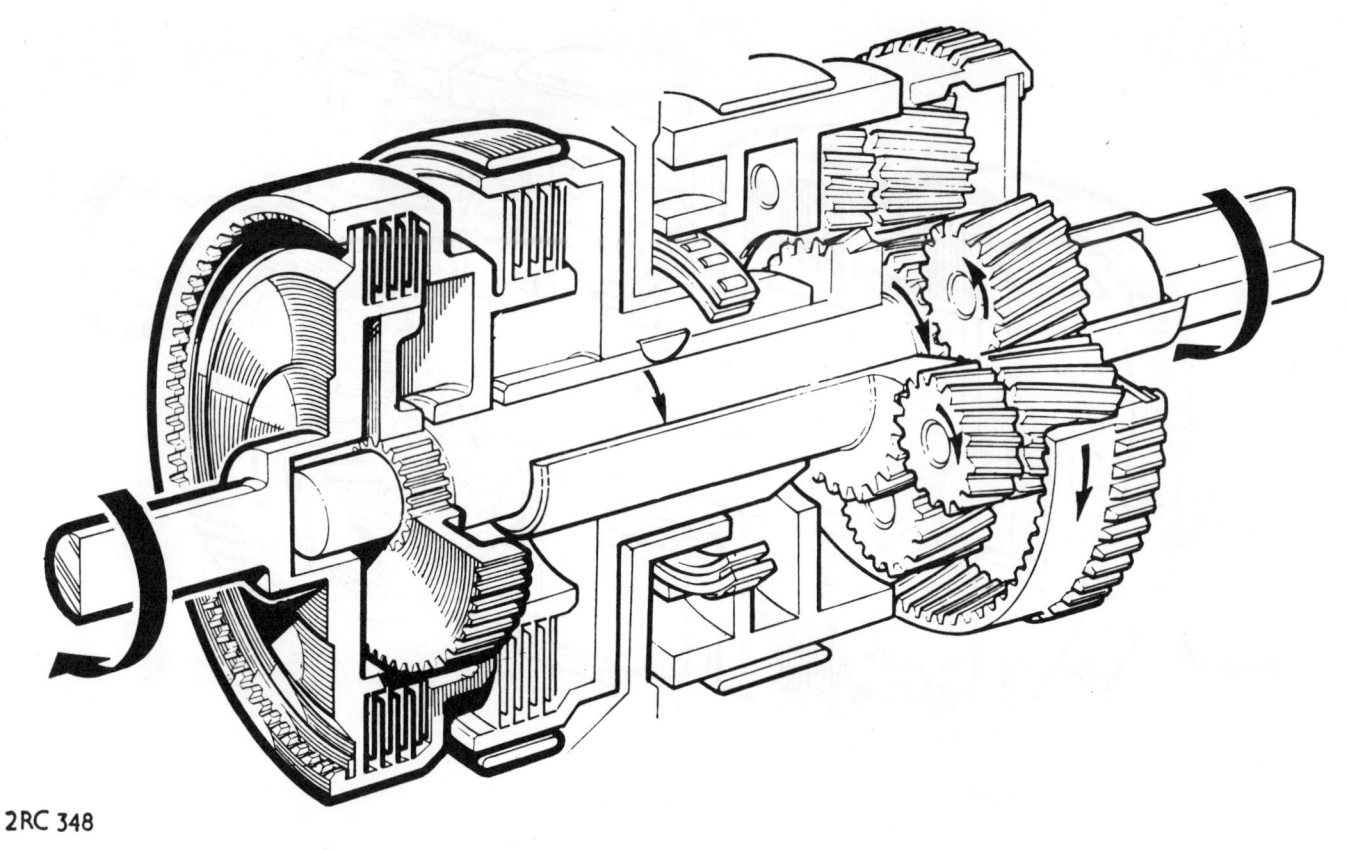

Application of bands and clutches—front clutch and front brake band

P	R	N	▼D2	▼D1	▼L
P	R	N	▼D	▼2	1

The front clutch is applied, connecting the power from the converter to the forward sun gear and the planetary gear set. The front band is applied holding the reverse sun gear stationary; this allows the planetary gear carrier to be driven around the stationary reverse sun gear and provides the reduction of 1.45:1.

Power again leaves the gearbox by the ring gear and driven shaft.

GEARBOX—AUTOMATIC—Borg-Warner type 35

Mechanical power flow—top gear; selector at the 'D1' or 'D2' position on early selector pattern, or 'D' on latest selector pattern

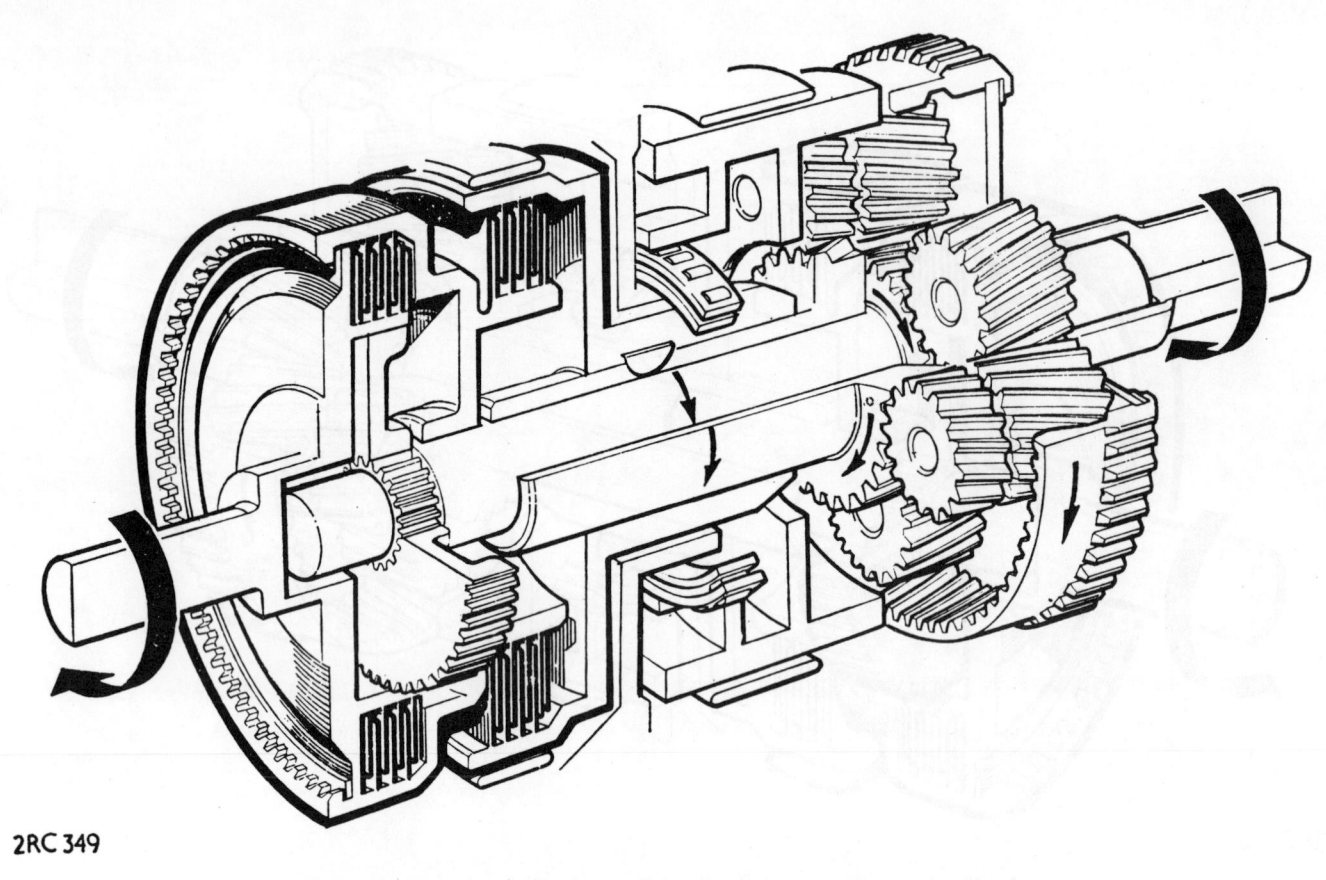

2RC 349

Application of bands and clutches—front clutch and rear clutch

| P | R | N | ▼D2 | ▼D1 | L |
| P | R | N | ▼D | 2 | 1 |

Power enters the gearbox from the torque converter through the front clutch, to the forward sun gear.
The rear clutch is applied which connects the power from the converter to the reverse sun gear; both sun gears being locked together, the gear set rotates as a unit providing a 1:1 ratio.

44—8

Rover 3500 and 3500S Manual AKM 3621

GEARBOX—AUTOMATIC—Borg-Warner type 35

Mechanical power flow—selector at the 'R' position

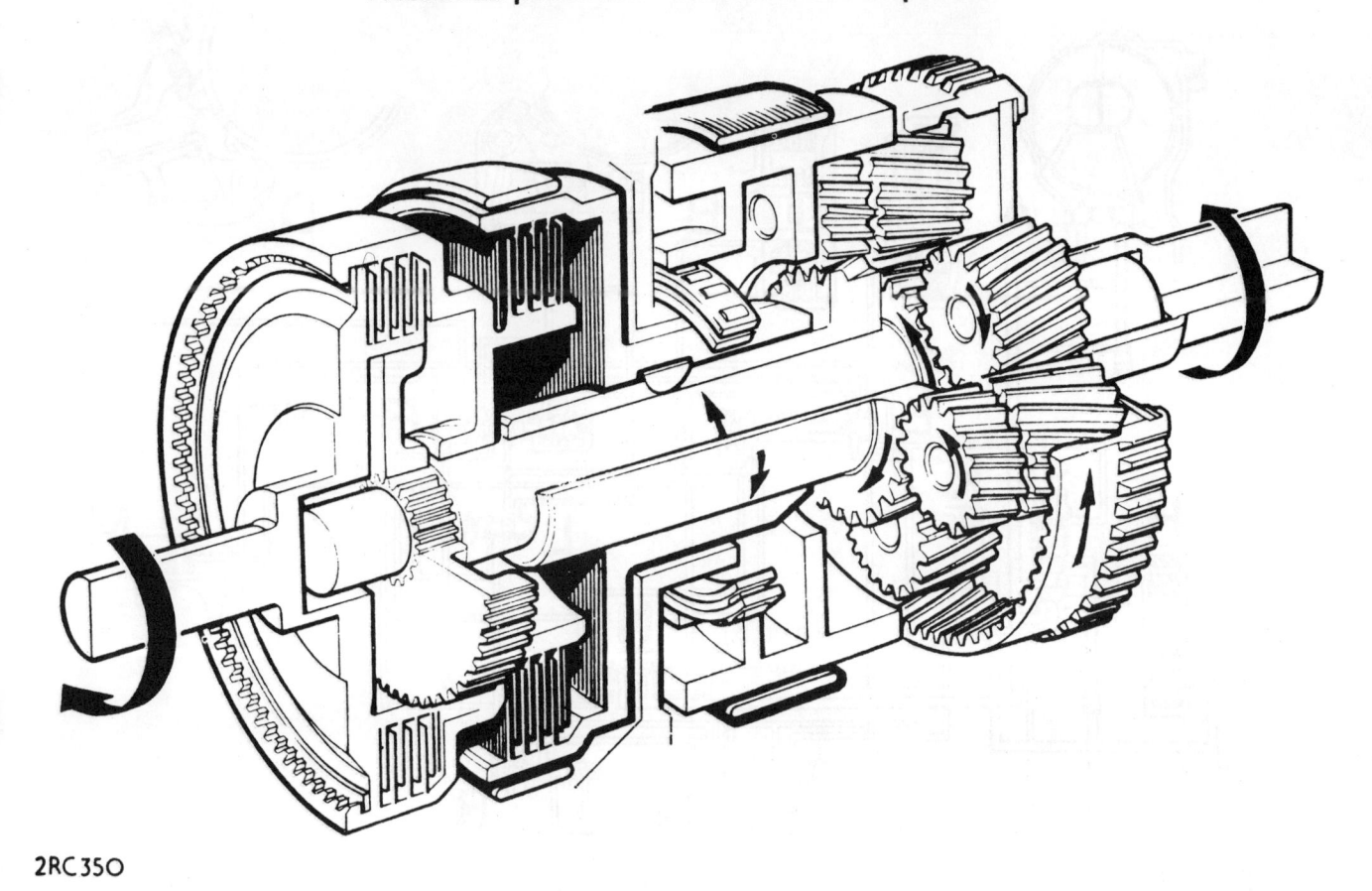

2RC 350

Application of bands and clutches—rear clutch and rear band

```
   ▼
P  R  N  D2  D1  L
   ▼
P  R  N  D   2   1
```

Power enters the gearbox from the converter and through the rear clutch to the reverse sun gear.

The rear band is applied which holds the planetary gear carrier stationary. A single set of pinions in the planetary set between the reverse sun gear and the ring gear of the driven shaft provides a reduction of 2.09:1 in the reverse direction to the input shaft.

Neutral and Park

The front and rear clutches are off and no power is transmitted from the converter to the gear set. The front and rear bands are also released, except in 'P', where for constructional reasons the rear band is applied as long as the engine is running.

GEARBOX—AUTOMATIC—Borg-Warner type 35

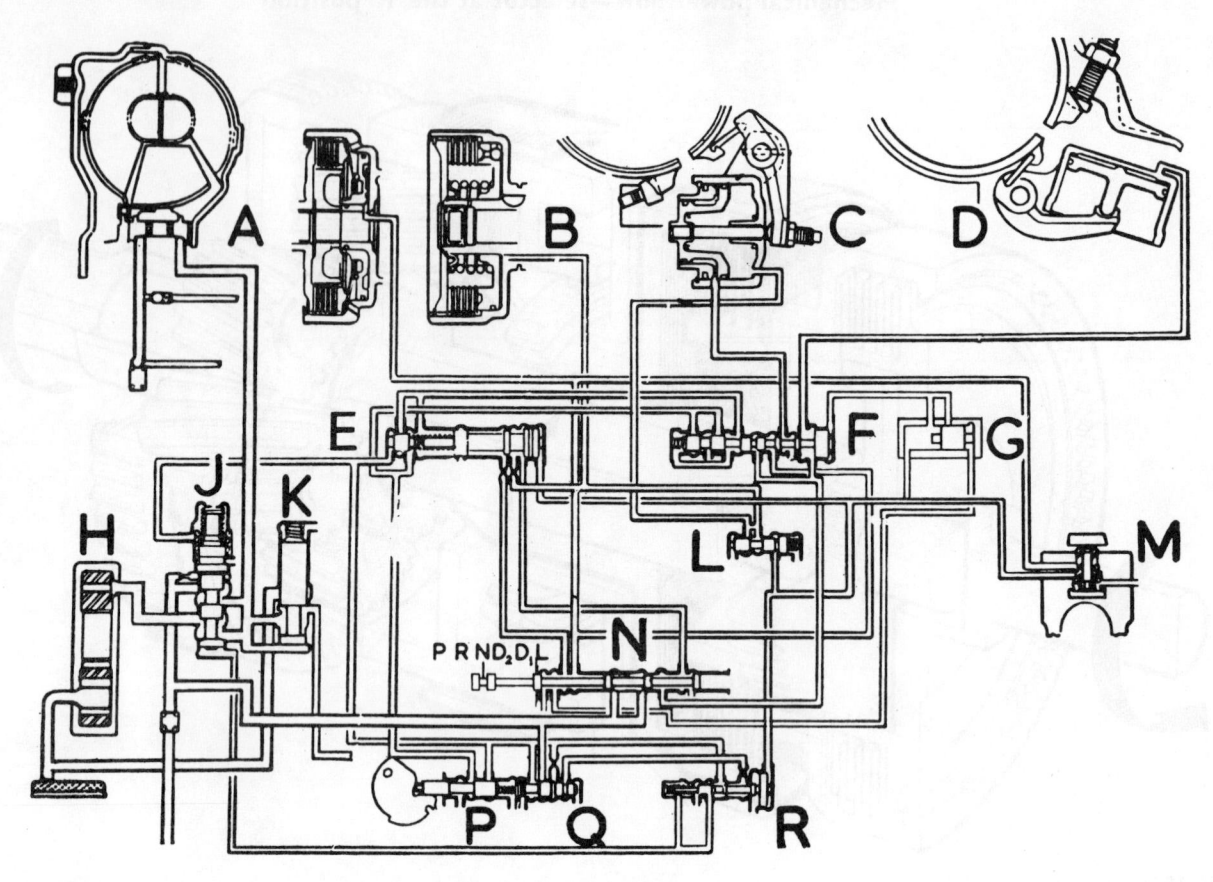

Identification of hydraulic parts 'P' 'R' 'N' 'D2' 'D1' 'L' selector pattern

A—Front clutch
B—Rear clutch
C—Front servo
D—Rear servo

E—2 to 3 shift valve
F—1 to 2 shift valve
G—D1–D2 control valve
H—Front pump

J—Primary regulator valve
K—Secondary regulator valve
L—Servo orifice control valve
M—Governor valve

N—Manual control valve
P—Downshift valve
Q—Throttle valve
R—Modulator valve

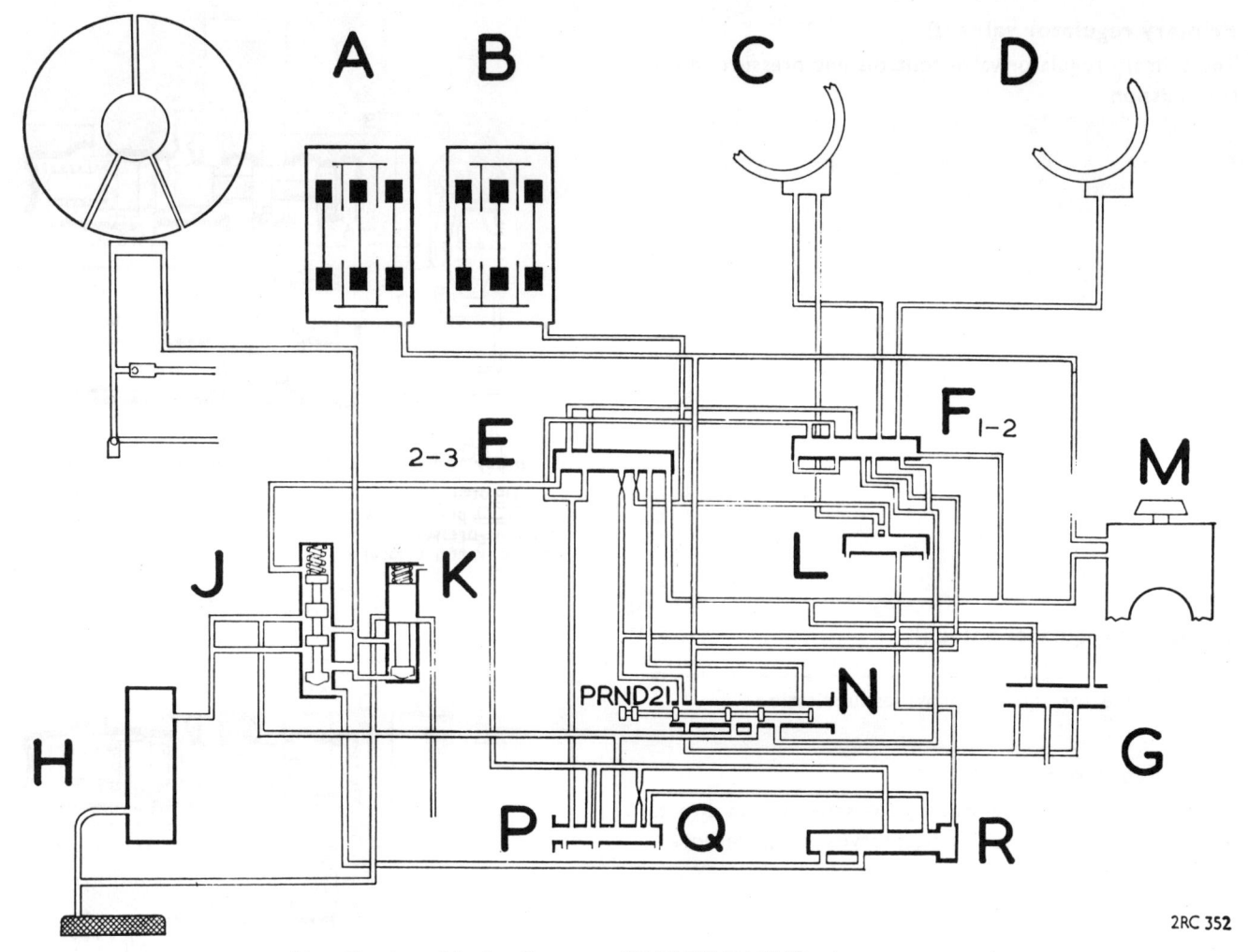

GEARBOX—AUTOMATIC—Borg-Warner type 35

Identification of hydraulic parts—'P' 'R' 'N' 'D' '2' '1' selector pattern

- A—Front clutch
- B—Rear clutch
- C—Front servo
- D—Rear servo
- E—2 to 3 shift valve
- F—1 to 2 shift valve
- G—Range control valve
- H—Front pump
- J—Primary regulator valve
- K—Secondary regulator valve
- L—Servo orifice control valve
- M—Governor valve
- N—Manual control valve
- P—Downshift valve
- Q—Throttle valve
- R—Modulator valve

The hydraulic system consists of a front pump and a centrifugally-operated hydraulic governor. Pressure and direction of flow of fluid to the various mechanical components is regulated by a series of valves.

Front pump (H)

The front pump is driven by the converter impeller and is in operation whenever the engine is running. This pump supplies fluid to the hydraulic control system and to the torque converter, and also controls lubrication.

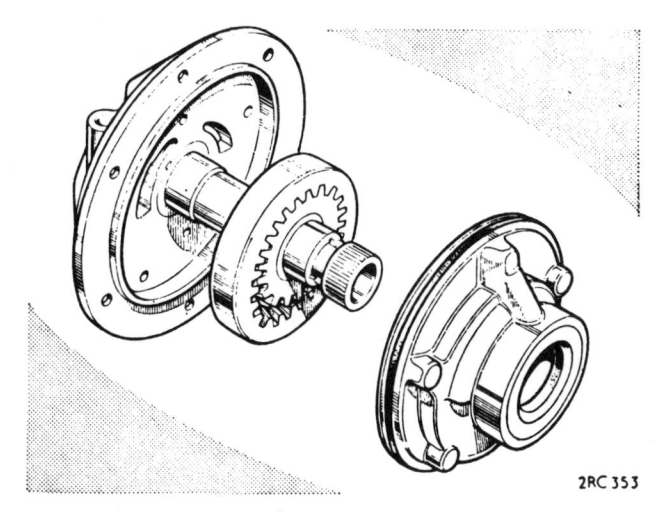

Front pump

Rover 3500 and 3500S Manual AKM 3621

44–11

GEARBOX—AUTOMATIC—Borg-Warner type 35

Primary regulator valve (J)

The primary regulator valve controls line pressure in the transmission.

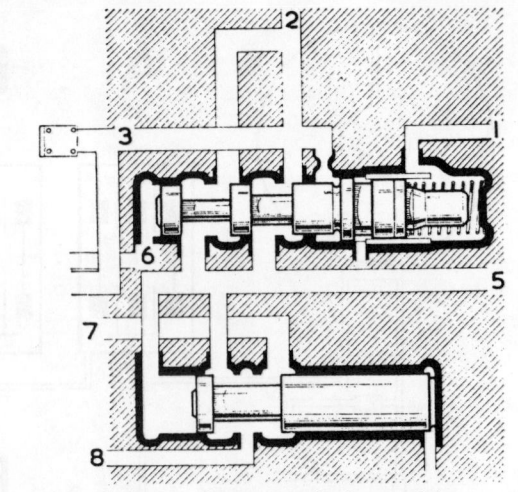

Primary and secondary valve at regulating position
1—Throttle pressure
2—Front pump pressure
3—Line pressure
5—Converter pressure
6—Modulator pressure
7—Return to front pump
8—Lubrication

Secondary regulator valve (K)

The secondary regulator valve controls pressure in the torque converter and also provides lubrication.

Manual control valve (N)

The manual valve is operated by the selector and directs fluid to the shift valves, clutches, servos and governor.

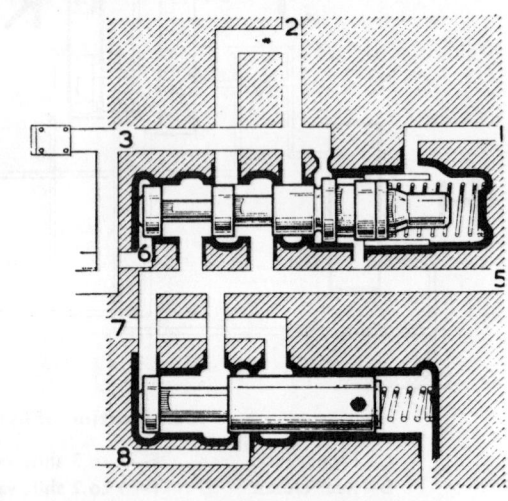

Primary and secondary valve at closed position
1—Throttle pressure
2—Front pump pressure
3—Line pressure
5—Converter pressure
6—Modulator pressure
7—Return to front pump
8—Lubrication

Governor valve (M)

The governor, driven by the output shaft, regulates the fluid pressure applied to the ends of the shift valves, according to the vehicle road speed.

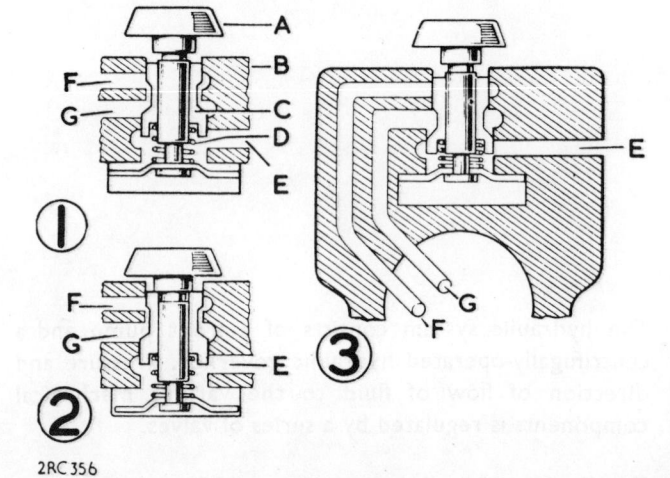

Governor valve
1 Low speed   2 At rest   3 High speed
A—Weight for governor
B—Body for governor
C—Valve for governor
D—Spring for governor
E—Exhaust port
F—Main line pressure
G—Governor pressure

Rover 3500 and 3500S Manual AKM 3621

GEARBOX—AUTOMATIC—Borg-Warner type 35

Throttle valve (Q)

The throttle valve is operated by the downshift valve via a compression spring and directs fluid to the shift valves to delay upshifts and effect downshifts.

Downshift valve (P)

The downshift valve is operated by the downshift cable and cam and directs fluid to the shift valves, further delaying upshifts and effecting downshifts.

Downshift and throttle valve
A—Kickdown position
B—Part throttle position
1—Throttle pressure
1a—Throttle pressure, modulated
3—Line pressure
9—Kickdown throttle pressure
10—Exhaust ports

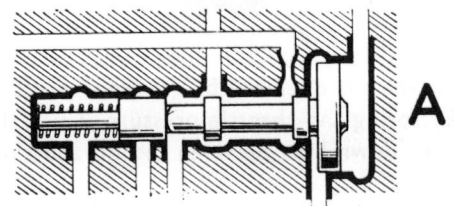

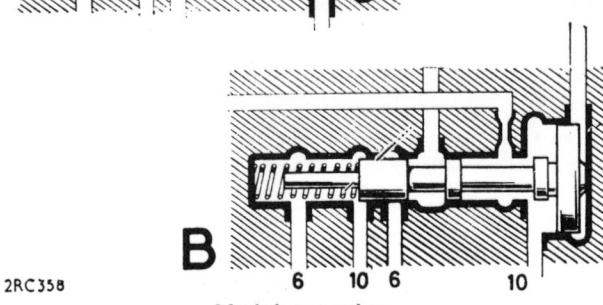

Modulator valve (R)

The modulator valve provides high line pressure during stall or part throttle conditions. This valve varies the line pressure relative to throttle and governor pressure.

Modulator valve
A—Governor pressure applied
B—Closed position
1—Throttle pressure
1a—Throttle pressure, modulated
6—Modulator pressure
10—Exhaust ports
11—Governor pressure

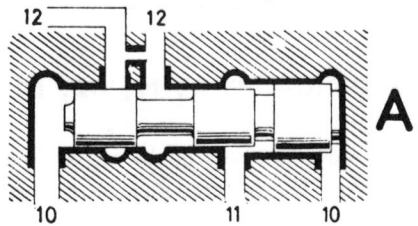

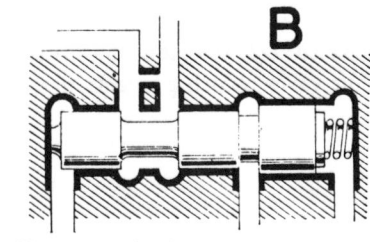

Servo orifice control valve (L)

The servo orifice control valve is operated by governor pressure, and opposed by spring pressure. This valve controls the quality of 2 to 3 and 3 to 2 shifts.

Servo orifice control valve
A—Operation at high speed, orifice in circuit
B—Operation at low speed, orifice by-passed
10—Exhaust ports
11—Governor pressure
12—Line pressure to front servo release

GEARBOX—AUTOMATIC—Borg-Warner type 35

1 to 2 shift valve (F)

The 1 to 2 shift valve is operated by governor pressure and opposed by spring and throttle pressure. This valve effects upshifts or downshifts between first and second gears.

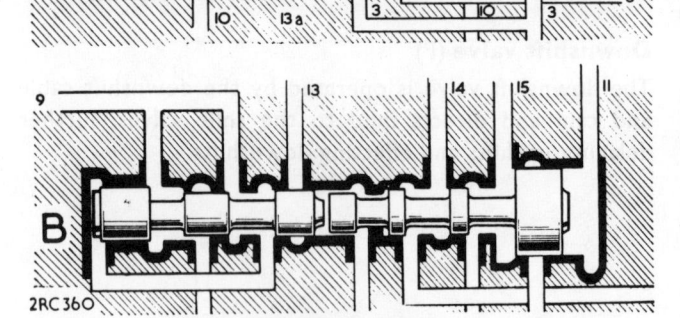

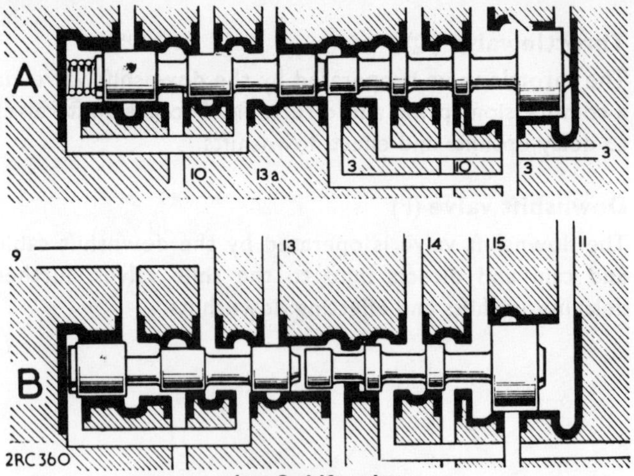

1 to 2 shift valve
A—First gear position
B—Second gear position
3—Line pressure
9—Kickdown throttle pressure
10—Exhaust ports
11—Governor pressure
13—Reduced throttle pressure
13a—Reduced throttle pressure
14—Line pressure to front servo apply
15—Line pressure to rear servo

2 to 3 shift valve (E)

The 2 to 3 shift valve is operated by governor pressure and opposed by spring and throttle pressure. This valve effects upshifts or downshifts between second and third gears.

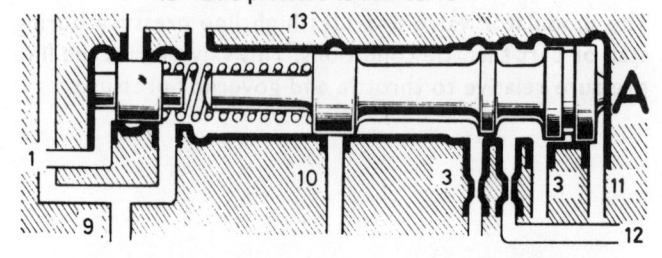

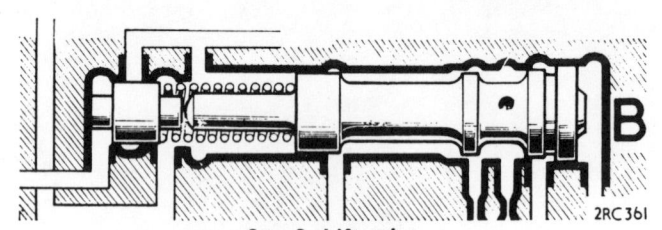

2 to 3 shift valve
A—Second gear position
B—Third gear position
1—Throttle pressure
3—Line pressure
9—Kickdown throttle pressure
10—Exhaust port
11—Governor pressure
12—Line pressure to front servo release
13—Reduced throttle pressure

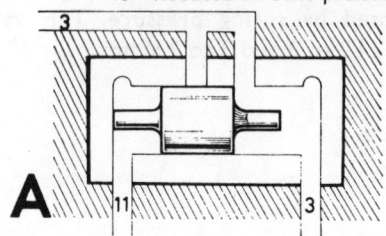

Control valve 'D1-D2' (G)—'P', 'R', 'N', 'D2', 'D1' 'L' selector pattern only

The 'D1-D2' control valve is operated by either line or governor pressure. With 'D1' selected governor pressure is applied to the 1–2 shift valve, allowing the three speed ranges to be used.

With 'D2' selected line pressure is applied to the 1–2 shift valve, which prevents first gear engagement.

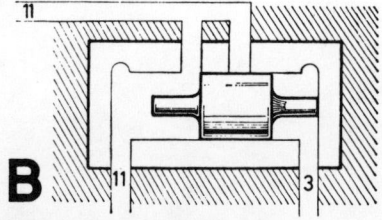

Control valve, 'D1–D2'—'P', 'R', 'N', 'D2', 'D1', 'L' selector pattern only
A—'D2' selected
B—'D1' selected
3—Line pressure
11—Governor pressure

44-14

Rover 3500 and 3500S Manual AKM 3621

GEARBOX—AUTOMATIC—Borg-Warner type 35

Range control valve (G)—'P', 'R', 'N', 'D', '2', '1' selector pattern

The range control valve is operated by line and spring pressure. With 'D' selected, the range control valve is ineffective allowing the three speed ranges to be used. With either '1' or '2' selected, the valve directs either governor or line pressure to the 1-2 shift valve to hold first or second gear accordingly.

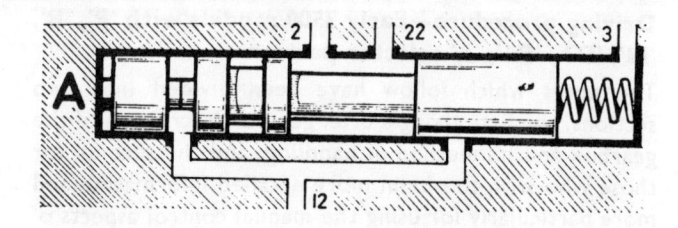

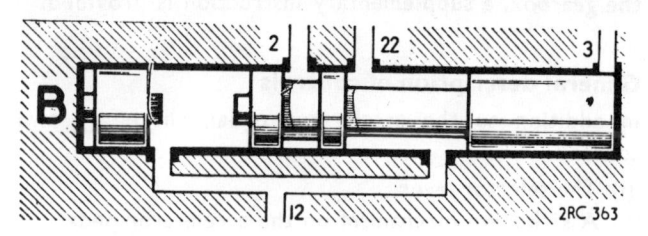

Range control valve—'P', 'R', 'N', 'D', '2', '1' selector pattern

A—'D' or '1' selected
B—'2' selected
2—Governor pressure
3—Line pressure
12—Line pressure
22—To 1-2 shift valve

Front servo (C)

The front servo is a double-acting unit, hydraulically operated by line pressure to engage or release the front band (second gear).

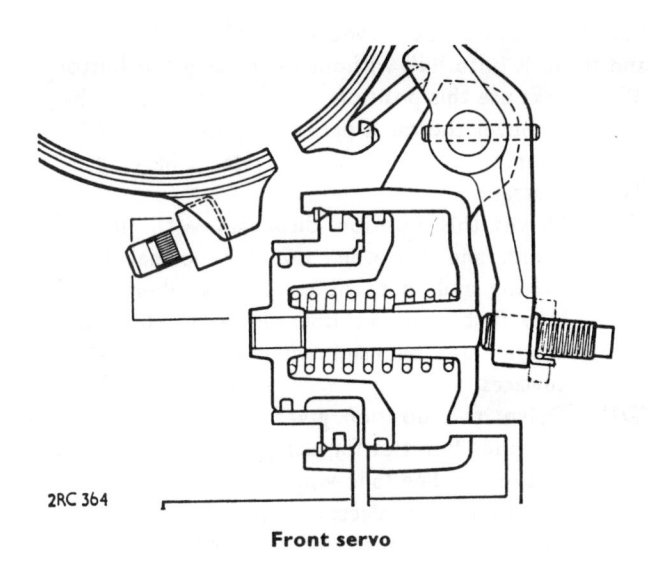

Front servo

Rear servo (D)

The rear servo is a single-acting unit, hydraulically operated by line pressure to apply the rear band (lock-up and reverse gear).

Front and rear clutches (A and B)

Both front and rear clutches are of the wet multi-plate type operated by line pressure.

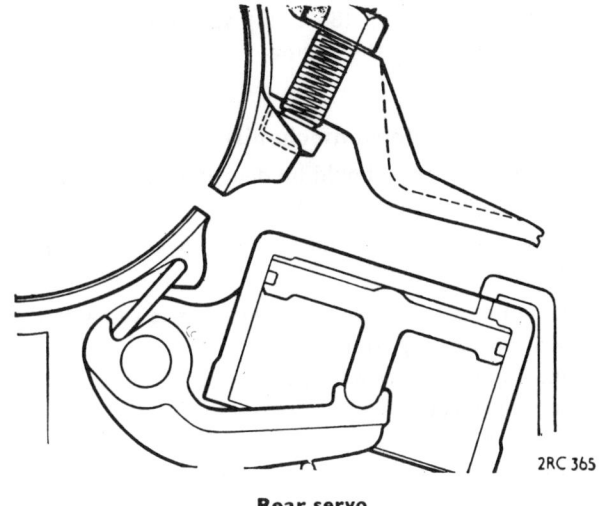

Rear servo

Rover 3500 and 3500S Manual AKM 3621

44-15

GEARBOX—AUTOMATIC—Borg-Warner type 35

Driving procedure—Early 3500 models with 'P', 'R', 'N', 'D2', 'D1', 'L' selector pattern

The notes which follow have been divided into two sections, the first giving a brief general description of the gearbox and its method of handling, and the second, for those who wish to obtain more detailed information and more particularly for using the manual control aspects of the gearbox, a supplementary instruction is provided.

General description of controls

In addition to the accelerator pedal, there are two controls:

(i) The selector lever.
(ii) A kickdown mechanism on the accelerator pedal.

Selector lever

The selector lever moves through a quadrant and by means of a pointer indicates the various positions, which are marked 'P', 'R', 'N', 'D2', 'D1' and 'L'. Depress button in centre of selector lever, then move to the position required. Depression of the button is necessary to engage all positions, except between 'D2' and 'D1'; the lever can also be moved from 'L' into 'D1' or 'D2' and from 'R' into 'N', without depressing the button.

'P' – Park: use this position whenever the car is parked ⎫
'R' – Reverse ⎬ Never select these positions when car is moving
'N' – Neutral. ⎭

'D2' – Drive: this position dispenses with first gear. Select and leave in 'D2' for normal leisurely driving and in town conditions where changes into first gear are not required. This position should also be used when driving on slippery surfaces.

'D1' – Drive: this position gives a first gear start and provides facilities for using the full performance of the car. See following detailed description for the use of this selector position.

'L' – Lock-up: not normally required. See following detailed description for the use of this selector position. Never select 'L' when road speed is above 128 km/h (80 mph) as there is a risk of over speeding the engine.

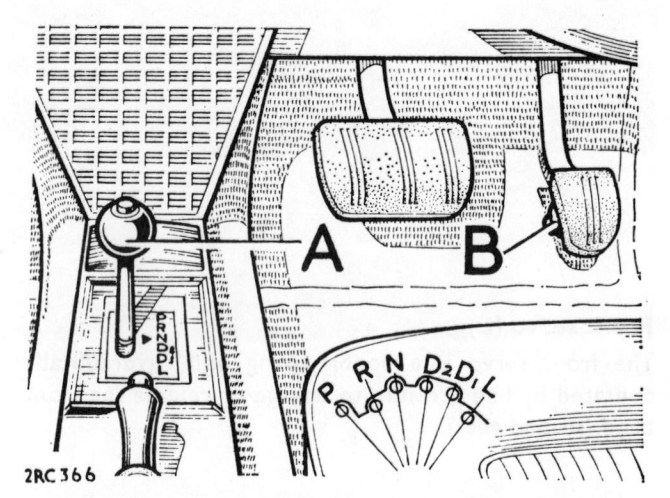

Selector lever and accelerator kickdown mechanism
A—Gear selector lever   B—Kickdown mechanism

Accelerator kickdown mechanism

The accelerator pedal should be depressed beyond its full throttle travel to operate the kickdown mechanism when:

(i) A lower gear is required for rapid overtaking or hill climbing.

(ii) Maximum acceleration is required, and can safely be used, on starting from rest.

The maximum kickdown speeds are pre-set to give optimum performance without overspeeding the engine. If the accelerator pedal is eased back from the kickdown position during acceleration, up-changes will be made at lower road speeds.

44-16   Rover 3500 and 3500S Manual AKM 3621

GEARBOX—AUTOMATIC—Borg-Warner type 35

Driving the car

Ensure that the handbrake is firmly applied and that the selector is in either the 'P' or 'N' position. As a safety measure an isolation switch prevents the starter being operated in any other position. Operate the combined ignition and starter switch. With the engine started and idling easily, depress the button and move the selector lever to either the 'D2', 'D1' or 'R' position as required, release handbrake and gently depress the accelerator pedal. Use footbrake to control car when manoeuvring in confined spaces. With the selector at 'D2' or 'D1', the car will move forward controlled by the accelerator and gear changes will be made automatically, depending on road speed and accelerator position; minimum accelerator pedal pressure will result in low up-change speeds. The further the accelerator pedal is depressed the faster the car will go before changing into a higher gear. The kickdown position referred to previously can be used for changing down into a lower gear or to obtain maximum acceleration as required.

Manual control aspects of the gearbox

The following paragraphs provide a more detailed description of the use of manual control aspects of the gearbox.

'D2'—Drive

The drive position 'D2' covers an automatic range of two ratios with second gear start, operating in conjunction with a torque converter, the ratios being engaged progressively according to car speed and accelerator position.

Top gear will be retained, unless using the kickdown, until the car speed has fallen to about 16 km/h (10 mph), when second gear will be automatically engaged.

Provided road speed is below 88 km/h (55 mph), second gear can be obtained by use of the kickdown mechanism and can be held up to a maximum speed of 112 km/h (70 mph), at which it will automatically change into top gear.

With the engine running and the selector at the 'D2' position the car will not roll back on an incline. This feature can be used when car is stopped on an upward slope; slip selector into 'D1' as car accelerates away.

First gear cannot be engaged with the selector at the 'D2' position.

'D1'—Drive

The drive position 'D1' covers a fully-automatic range of three ratios, with first gear start, operating in conjunction with a torque converter, the ratios being engaged progressively according to car speed and accelerator position.

When the road speed has fallen to about 8 km/h (5 mph), first gear will be engaged.

Provided road speed is below 88 km/h (55 mph), second gear can be obtained by use of the kickdown mechanism and can be held up to a maximum speed of 112 km/h (70 mph), at which it will automatically change into top gear. Similarly, first gear can be engaged below 48 km/h (30 mph) and can be held up to 64 km/h (40 mph), at which speed it will automatically change into second.

'L'—Lock-up

The 'L' position (lock-up) provides the ability to hold the car in either first gear or second gear and therefore, in effect, to drive the car in manual control, the torque converter, however, replacing the conventional clutch.

At any time, when first or second gear is in operation in lock-up, full overrun braking is obtained from the gear in question. The lock-up should never be engaged at speeds above 128 km/h (80 mph).

If the selector lever is placed in 'L' when starting from rest, the transmission starts and will remain locked in first gear regardless of road speed or throttle position. This gear gives maximum engine braking, but a speed of 77 km/h (48 mph) must not be exceeded in first gear. To move from first gear lock-up to second gear lock-up, the lever must be moved to 'D1' or 'D2' and then returned to 'L', which will put the transmission into second gear, where it will remain locked.

Maximum speed in this gear should not exceed 128 km/h (80 mph).

The change to top gear is made simply by moving the lever to 'D1' or 'D2' and leaving it there.

The maximum recommended speeds in first and second gear, referred to previously, are indicated by yellow figures on the speedometer dial.

When the transmission is in top gear, selector position 'D2' and 'D1', movement of the lever to 'L' will immediately give second gear, but first gear will be automatically selected if the road speed drops below 20 km/h (12 mph).

With 'D2' engaged, second gear in operation, first gear can be engaged at speeds below 66 km/h (35 mph) by operating the kickdown mechanism, but to change up into second gear the lever must be moved to 'D1' or 'D2' and returned to 'L', as referred to above.

The 'L' position is particularly useful in the following conditions:

(i) For unusually long and steep up-gradients or descents where a low gear is required, either for pulling or overrun braking.

(ii) Where hard pulling may be encountered, as in deep snow or heavy mud.

(iii) Driving slowly but where it is desirable to maintain reasonable engine speed so that the alternator can balance the heavy electrical load placed upon it by lights, windscreen wiper, heater, etc.; that is, in foggy conditions at night.

(iv) To select and hold a lower gear when overtaking, within the permissible maximum speed limits stated previously.

GEARBOX—AUTOMATIC—Borg-Warner type 35

Driving in mountainous areas

If the car is heavily laden, or towing a trailer, then 'D1' selector position should always be used when climbing any long hill of alpine type. This will make available first gear, by use of the kickdown, when negotiating hairpin corners and also allow a restart in first gear should this become necessary.

When descending a hill of this type, under similar load conditions, use the 'L' selector position. This will provide engine braking in second and first gears, so reducing the load on the wheel brakes.

When approaching a corner, if the speed drops below 20 km/h (12 mph), first gear will be obtained. Should this not be required during acceleration between corners, second gear may be obtained by moving the selector lever to 'D1' or 'D2', and then back into 'L' after the transmission has changed gear.

It will then be locked in second gear until the next occasion on which the car speed falls below 20 km/h (12 mph).

Emergency and manual starting

The design of the automatic gearbox prevents a tow start being made.

In the unlikely event of the car having to be towed for more than approximately 1,0 km (½ mile) the propeller shaft must be disconnected, or the car towed with the rear wheels off the road.

This precaution must be taken to prevent possible damage to the transmission due to lack of lubrication, as the front pump is not operating unless the engine is running.

Driving procedure—Latest 3500 models, with 'P', 'R', 'N', 'D', '2', '1' selector lever positions

The automatic gearbox on latest Rover 3500 models gives fully automatic gear changes for all normal driving conditions, but can, when circumstances demand, be used to give manual control over both up-change and down-change requirements.

Automatic gearbox controls

In addition to the accelerator pedal there are two controls:
(i) The selector lever.
(ii) A kickdown mechanism under the accelerator pedal.

Selector lever

The selector lever moves through a quadrant, and by means of a pointer indicates the various positions which are marked 'P', 'R', 'N', 'D', '2' and '1'.

Depress button in centre of selector lever, then move to the position required. Depression of the button is necessary to engage the gear positions, except between 'D' and '2'; from '1' into '2' or 'D' and from 'R' into 'N'.

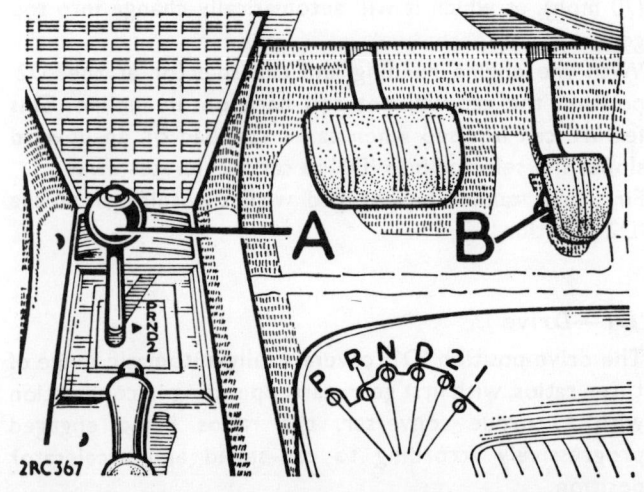

Selector lever and accelerator kick-down mechanism
A—Gear selector lever
B—Kick-down mechanism

44-18

Rover 3500 and 3500S Manual AKM 3621

GEARBOX—AUTOMATIC—Borg-Warner type 35

'P' – Park: Use this position whenever the car is parked
'R' – Reverse
'N' – Neutral

} Never select these positions when the car is moving

'D' – Drive: All normal driving should be done with the lever at the 'D' position. This gives a fully automatic range of three ratios, with first gear start, the ratios being engaged progressively, according to car speed and accelerator position.

With the selector at the 'D' position and the car in top gear, second gear can be engaged by simply moving the lever to the '2' position. Second gear will be immediately engaged and held, until the gear lever is moved to either 'D' or '1'. By depressing the knob, and moving the lever to the '1' position, first gear will be engaged when the road speed falls to or is below approximately 35 km/h (22 mph), again first gear will be held until the gear lever is moved to '2' or 'D'.

Never attempt to engage a lower gear, that is, from 'D' to '2' or 'D' to '1' when the road speed is above 120 km/h (75 mph), as there is a grave risk of over speeding the engine.

To engage a higher gear, merely push the lever from '1' to '2' or from '2' to 'D' as required; this can be done without depressing the button in the centre of the gear selector lever.

'2' – Drive: This position gives a second speed start and the transmission will remain in second gear regardless of road speed and throttle position, until the gear lever is moved to the 'D' or '1' position.

Maximum speed in second gear is 124 km/h (80 mph).

To engage first gear depress knob in gear selector lever and move lever to '1' position, as detailed above.

'1' – Drive: With the gear lever at position '1', the transmission will start in first gear and remain in that gear regardless of road speed and throttle position, until the gear lever is moved to '2' or 'D'.

Maximum speed in first gear is 72 km/h (48 mph).

It will be seen that using the gear lever as described above, complete manual control of the gear required to suit any circumstances, can be obtained.

The maximum recommended speeds in first and second gear, referred to above, are indicated by yellow figures on the speedometer dial.

The two lower gear positions are particularly useful under the following conditions:

(i) For unusually long and steep up-gradients or descents, where a low gear is required, either for pulling or over-run braking.

(ii) Where hard pulling may be encountered, as in deep snow or heavy mud.

(iii) Where the car is being driven slowly, but it is desirable to maintain a reasonable engine speed, so that the alternator can balance the heavy electrical load placed upon it by lights, windscreen wipers, heater, etc., that is, in foggy conditions at night.

(iv) To select and hold a lower gear when overtaking within the permissible maximum speed limits stated previously.

(v) Use the '2' position when driving on slippery surfaces.

Accelerator kick-down mechanism

The kick-down mechanism is only effective with the selector lever at the 'D' position.

The accelerator pedal should be depressed beyond its full throttle travel to operate the kick-down mechanism when:

(i) A lower gear is required for rapid overtaking or hill climbing.

(ii) Maximum acceleration is required, which can safely be used when starting from rest.

The maximum 'kick-down' speeds are pre-set to give optimum performance without over speeding the engine. If the accelerator pedal is eased back from the kick-down position during acceleration, up-changes will be made at lower road speeds.

With the gear selector lever at the 'D' position, the kick-down will operate as follows:

(i) Road speed below 88 km/h (55 mph) and above 48 km/h (30 mph). Second gear will be obtained and held up to a maximum speed of 112 km/h (70 mph) at which it will automatically change into top gear.

(ii) Road speed below 48 km/h (30 mph). First gear will be engaged and can be held up to 64 km/h (40 mph) at which speed it will automatically change into second. If the kick-down position is maintained on the accelerator pedal, it will hold second gear up to a maximum speed of 112 km/h (70 mph), when a change into top gear will be made.

Driving the car

Ensure that the handbrake is firmly applied, and the selector is at either the 'P' or 'N' position. As a safety measure, an isolation switch prevents the starter being operated in any other position. Operate the combined ignition and starter switch. With the engine started and idling easily, depress the button and move the selector lever to either the 'D', '2', '1' or 'R' position as required. Release handbrake and gently depress the accelerator pedal. Use footbrake to control car when manoeuvring in confined spaces.

Rover 3500 and 3500S Manual AKM 3621

GEARBOX—AUTOMATIC—Borg-Warner type 35

With the selector at 'D', the car will move forward, controlled by the accelerator, and gear changes will be made automatically, depending on road speed and accelerator position; minimum accelerator pedal pressure will result in low up-change speeds. The further the accelerator pedal is depressed the faster the car will go before changing into a higher gear. The kick-down position or the use of the gear lever to obtain second or first gear referred to previously, can be used for changing down into a lower gear or to obtain maximum acceleration as required.

With the engine running and the selector at the '2' position, the car will not roll back on an incline. This feature can be used when the car is stopped on an upward slope; slip selector into 'D' as the car accelerates away.

Driving in mountainous areas

If the car is heavily laden or towing a trailer, then the '2' or '1' gear position should be used as circumstances dictate, when climbing or descending any long hill of alpine type.

When descending, the use of a lower gear will provide engine braking, so reducing the load on the wheel brakes.

Emergency and manual starting

The design of the automatic gearbox prevents a tow start being made.

In the unlikely event of the car having to be towed for more than approximately 1,0 km (½ mile) the propeller shaft must be disconnected or the car towed with the rear wheels off the road. This precaution must be taken to prevent possible damage to the transmission due to lack of lubrication, as the front pump is not operating unless the engine is running.

General service information 44.00.00

The following information is provided to assist in servicing Rover 3500 Automatic gearbox models, and gives details of important changes, service modifications and fault diagnosis.

Corrective maintenance

A large proportion of the faults encountered with automatic gearboxes can be directly attributed to deterioration from the initial settings causing clutch failure as a result of:

a. Low fluid level.
b. Incorrect pressure setting.
c. Brake band setting, front and rear (debris ingress).
d. Manual linkage maladjustment, covered under pressure setting.

In view of the foregoing, it is important to carry out the following checks and corrective maintenance before action is taken to remove and dismantle the automatic gearbox. When it is necessary to dismantle the gearbox after any failure which can deposit debris in the fluid, that is brake band slipping etc, the torque converter and fluid cooler should be flushed to prevent any contamination being pumped from one unit to the other. In cases of excessive contamination, consideration should be given to fitting a new torque converter or a reconditioned gearbox complete.

1. Fluid level

When checking the fluid level, follow the instructions in 44.24.02.

2. Fluid loss

Fluid loss can result from ejection through the gearbox breather pipe. This only occurs during or immediately after braking with a hot transmission. The most common causes, in order of probability, are:

i. Overfilling of gearbox. Check that correct dipstick is being used. See item 3.
ii. Misplacement or omission of fluid cooler transfer pipe between valve block and case.
iii. Damage to the front pump suction tube or 'O' ring, causing heavy fluid aeration.

continued

iv. Faulty non-return valve in forward sun gear shaft. This cause is an exception to the occurrence detailed above, as fluid ejection will take place immediately after a cold start, when the car has been left standing for a prolonged period, at least over-night, thus allowing fluid to drain from torque converter into fluid pan. This condition may be aggravated if the car is parked on a steeply sloping drive with the front of the car downwards.
v. Water in the transmission fluid.
vi. Overheating due to a defective cooler or blocked fluid cooler circuit.
vii. If fluid is ejected from the throttle cable, then in addition to one of the above faults, it will be found that the breather tube is blocked.
viii. Fluid ejection from the breather tube not caused by one of the above reasons may sometimes occur. To rectify this condition, re-locate the breather tube as follows:
   a. Remove the fluid filler and breather tube—'A' Fig. 1. Withdraw dipstick.
   b. Cut off the breather tube flush with the filler tube.
   c. Fill this hole by brazing.
   d. Drill a new hole of the same diameter, 75 mm (3 in.), from the top of the filler tube.
   e. Cut off surplus length of breather tube.
   f. Bend the shortened breather tube to suit the new hole, and braze in position, as shown at 'B' Fig. 1.
   g. Refit the filler tube and clip breather pipe to casing.
   h. Insert dipstick.
   NOTE: This is a Service modification only, and will not be incorporated on Production vehicles.

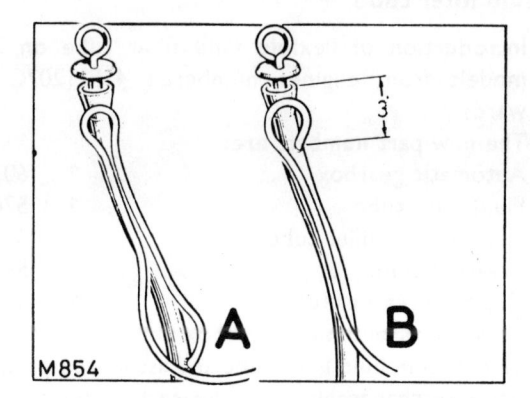

Fig. 1. Automatic gearbox breather tube modifications.

3. Fluid level dipstick

Varying lengths of fluid level dipstick are in use on the range of Rover Automatic gearbox cars, and it is essential that only the correct length dipstick is used for the particular model.

The dipsticks currently in use on Rover 3500 Automatic Gearbox models are identified as shown at 'A' in Fig. 2.
A=646 mm (25.437 in.) up to engine serial number 42521206C
A=681 mm (26.812 in.) from engine serial number 42521207C onwards for use with flexible oil filler tube.
Failure to use the correct dipstick will result in either a low or excessively high fluid level.
An incorrect fluid level can cause any or a combination of the following faults:
i. No engagement or bumpy engagement of 'R', 'D', '2', '1', 'D2', 'D1' or 'L'.
ii. Slip during upshift change from '1' to '2' and '2' to '3'.
iii. Involuntary high downshift speed from '3' to '2'.
iv. Low line pressure at idling speed.
v. Low line pressure at stall speed.
vi. Overheating.

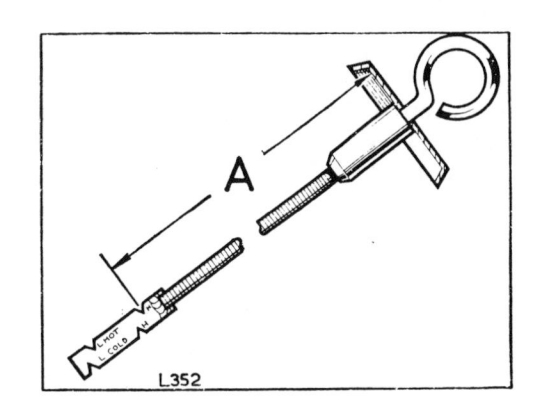

continued

Fig. 2. Dipstick identification.

GEARBOX—AUTOMATIC—Borg-Warner type 35

4. Fluid filler tube

i. Introduction of flexible fluid filler tube on 3500 models from engines numbered 42521207C onwards.

The new part numbers are:

Automatic gearbox	1	607081
Fluid filler tube	1	576979
Clip for fluid filler tube	1	508035
Breather tube	1	591208
Clip for breather tube	1	576185
Dipstick for automatic gearbox	1	591207

The flexible fluid filler tube and associated parts are not interchangeable with the earlier items except as a complete kit, which can be obtained under Part No. 607158.

The correct length dipstick for use with the flexible fluid filler tube is 681 mm (26.812 in.), See Item 3.

ii. The later automatic gearbox, which can be identified by a grey plate with '267' in the lower R.H. corner on initial supplies and a French blue plate on later supplies, can be used as a replacement on all earlier 3500 models.

Note the following:

a. As the later automatic gearbox has the 'D', '2', '1' selector pattern, fit indicator plate Part No. 586159 when the unit replaces an automatic gearbox with the 'D1', 'D2', 'L' type selector pattern.

b. Where the original long downshift cable is fitted the later short downshift cable, supplied as a conversion kit under Part No. 606704, must be used. See item 7.

5. Fluid leaks

Fluid leaks in the vicinity of the sump pan should be carefully examined to determine the exact source, before removing the pan. Fluid leaking from the manual valve lever control shaft seal will frequently run down on to the sump pan, and then around it, giving the impression that the sump pan is leaking.

It is not necessary to remove the sump pan, to replace the manual lever control shaft seals.

To replace the seals, proceed as follows:

i. Remove lever 'A' Fig. 3 for manual valve detent.
ii. Prise old seal 'B' Fig. 3 out of gearbox casing, using a small screwdriver or suitable tool.
iii. Fit new seal and replace lever.

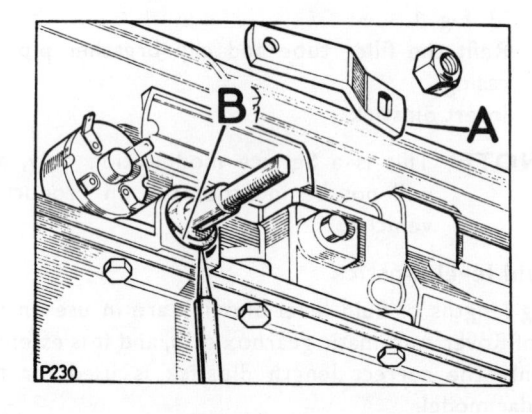

continued

Fig. 3. Shaft and oil seal for manual valve lever control shaft.

44.00.00
Sheet 3

Rover 3500 and 3500S Manual AKM 3621

GEARBOX—AUTOMATIC—Borg-Warner type 35

6. Pressure setting—Series 3FU, 7FU, 9FU and 267 gearboxes.

i. The only method currently recommended of checking downshift cable adjustment is to ensure that the fluid pressure in the automatic gearbox is within certain limits when checked at engine idle speed; at 800 rev/min and also at 1,200 rev/min. See Operation 44.30.03.

ii. Incorrect fluid pressure and particularly no increase or an incorrect rise in fluid pressure as engine speed increases when the accelerator pedal is operated can lead to clutch failure or rough and bumpy gear changes.
The fluid pressure should be checked at the 1500 km (1,000 miles) Free Service Inspection and subsequently every 8000 km (5,000 miles) or whenever the accelerator linkage has been adjusted.
NOTE: Series 303 gearboxes do not normally require pressure checks, it is sufficient to maintain the downshift cable in its original setting. See Operation 44.30.02.

7. Front brake band

i. Correct front brake band adjustment, where applicable, is most important and should be carefully carried out. See Operation DDD2-9. Later models are fitted with a self-adjusting front brake band servo. See Operation 44.34.07.

ii. Automatic gearboxes with the latest self-adjusting front brake band servo are numbered from 7 FU 10164 onwards.
When stocks of the earlier automatic gearbox unit and the front servo are exhausted, only the latest type will be supplied for all replacements.

iii. Varying designs of front and rear brake bands are in use on the range of automatic gearboxes fitted to Rover 3500 models, and it is essential to fit the correct brake band to the applicable gearbox. The brake band type can be identified by the colour of the lining material and the finish on the inside diameter. Also, some designs have a one-piece lining, while others have three pieces composed of a long centre piece and two short end pieces. The range of automatic gearbox types can be identified by the serial number prefix, stamped on the plate that is attached to the gearbox case. The following list provides details of brake band identification and gearbox applicability.
Part No. 601104. Front brake band.
Two types of brake band are supplied under this Part No., they are interchangeable with each other and may be either:
One-piece red coloured lining with a smooth internal diameter or
One-piece brown coloured lining with a grooved internal diameter.
These are applicable to gearbox serial number prefix 3FU, 7FU and 9FU.
Part No. 605377. Rear brake band.

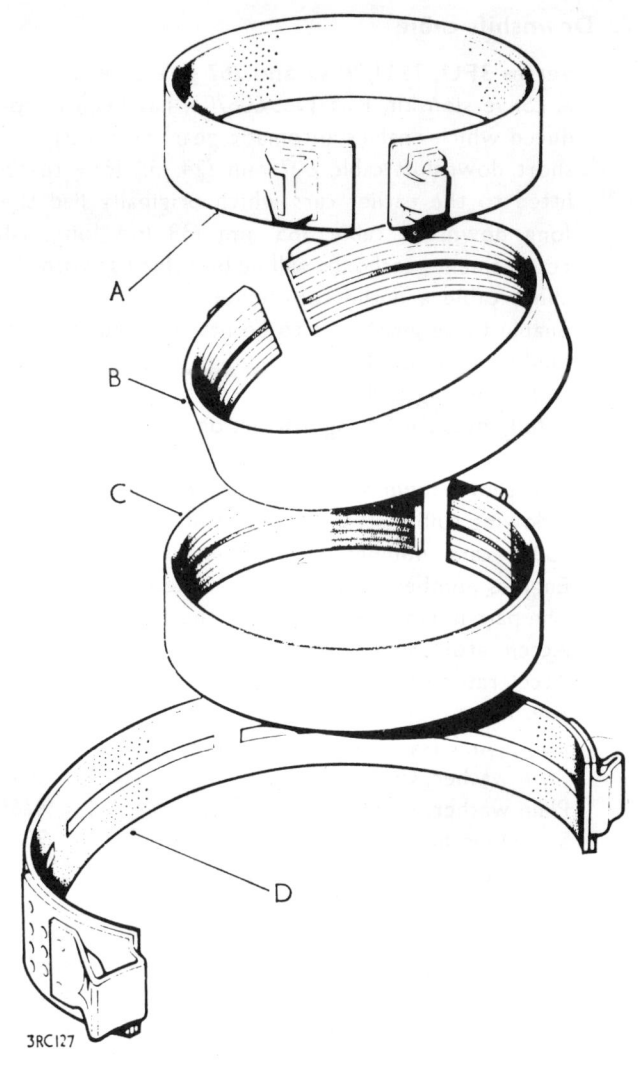

Brake band identification
A—One-piece lining with smooth internal diameter
B—One-piece lining with grooved internal diameter
C—Three-piece lining with grooved internal diameter
D—One-piece lining with slotted internal diameter and flexible outer case

Three-piece lining coloured brown with grey end pieces and with a grooved internal diameter.
Applicable to gearbox serial number prefix 3FU, 7FU and 9FU.
Part No. 607671. Front brake band (flexible).
One-piece brown coloured lining with a slotted internal diameter and flexible outer case.
Applicable to gearbox serial number prefix 303.
Part No. 608387. Rear brake band.
One-piece red lining with diagonal black stripes on the internal diameter.
Applicable to gearbox serial number prefix 303.

continued

GEARBOX—AUTOMATIC—Borg-Warner type 35

8. Downshift cable

i. **Series 3FU, 7FU, 9FU and 267 gearboxes**
A conversion kit, Part No. 606704, has been introduced which enables automatic gearboxes with the short downshift cable 607 mm (24 in.) long to be fitted to the earlier cars which originally had the long downshift cable 965 mm (38 in.) long. All replacement gearboxes will be built for use with the short cable and the conversion kit is essential to enable these gearboxes to be fitted to earlier cars. Under no circumstances should an attempt be made to fit the long cable.
A fully detailed Fitting Instruction is included in the kit.
The short downshift cable was introduced at the following suffix letter change points.
3500. cars numbered suffix letter 'F' onwards.
Engines numbered suffix letter 'C' onwards.
The part numbers applicable to the change are:

Accelerator coupling shaft, front	1 610997
Accelerator coupling shaft, rear	1 611007
Abutment bracket	1 610973
Bolt (⅜ in. UNC x 1⅜ in. long) Fixing	1 253048
Plain washer, thin ⎫bracket to	1 4094
Plain washer, thick ⎬cylinder	1 4266
Spring washer ⎭head	1 3076
Plastic bearing for coupling shaft	1 553851
Tubular pin, fixing coupling shaft to cross shaft	1 611021
Downshift, cable short	1 606514
Clip for downshift cable	1 576788
Bracket for breather hose	1 611032

The downshift cable on these later models is clipped to the bell housing and is supported vertically at a cylinder head bracket, the engine breather outlet hose is secured at a clip retained at a carburetter fixing.

ii. **Series 303 gearboxes**
A new, shorter downshift cable is introduced on Series 303 gearboxes, and must not be interchanged with previous cables.

iii. **Downshift cable identification**
'84'—Downshift cable for Series 3FU, 7FU, 9FU and 267 gearboxes.
'148'—Downshift cable for Series 303 gearboxes.
It is essential to fit the correct downshift cable to the applicable Series gearbox; cables can be identified by the type number which is stamped on the hexagon nut where the cable enters the gearbox. See Fig. 4.

9. Flushing automatic gearbox fluid cooler

It is possible that after automatic gearbox failure, debris may have been deposited in the fluid. It is therefore important that the fluid cooler should be flushed, otherwise this debris will contaminate the new or repaired gearbox and may cause another failure. See Operation 44.24.02. Under no circumstances should old fluid be used in a new gearbox.

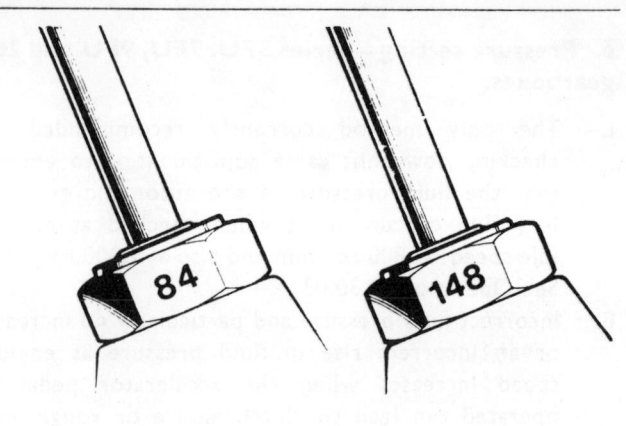

1RC1232

Fig. 4. Downshift cable identification

The torque converter should also be flushed as far as is practical. In circumstances where the debris is excessive, consideration should be given to fitting a new torque converter or a reconditioned gearbox complete.

10. Starter inhibitor switch

A new starter inhibitor switch incorporating automatic setting has been introduced. The new switch is fully interchangeable with the earlier type. See Operation 44.15.15.

11. Gear selector buzz

Introduction of a new gearchange housing assembly, in which the gear selector lever slot is brush masked, in place of the earlier steel shield.
A modified gear selector lever with a smaller diameter on the lever ball is also introduced to facilitate gear selection. The brush assembly incorporated in the housing provides smoother operation of the gearchange selector and obviates vibrations sometimes experienced with the earlier steel shield. To replace the brushes, locate the assemblies in the housing slots provided and turn the brush carrier ends over the housing to secure.
The new indicator plate is opaque, except at the gear position indicator panel, to improve light masking.
Cars fitted with the new gear selector housing can easily be identified by the brush assemblies. See Fig. 5.
When refitting a gear selector lever, apply a small quantity of grease, MS4 or suitable equivalent, to ball and seat.
The latest gearchange housing is fully interchangeable with the earlier type, but when fitting it to early cars, indicator plate, Part No. 586159 must also be ordered and fitted.
The part numbers applicable to the change are:

Gearchange housing assembly	1 591388
Brush assembly (Part of assembly 591388)	2 591386
Gearchange indicator plate	1 586159
Gear selector lever assembly, part number unchanged	1 591304

continued

GEARBOX—AUTOMATIC—Borg-Warner type 35

12. Gear change lever knob

A few instances have been reported where the fibre washer, Part No. 267721, which fits underneath the new shaped gear lever knob, Part No. 576413, has been omitted during assembly. This could lead to a situation where it is possible to freely move the selector lever through all the gear positions without resistance, even with the vehicle in motion.

If the gear lever is dismantled for any reason, always **ensure that the fibre washer is fitted on assembly.**

13. Vibration

Before replacing the torque converter on cars which vibrate because of an out-of-balance condition on the rear of the engine, the torque convertor should be rebalanced. See Operation 44.17.08.

14. Automatic gearbox series numbers

a. **Series 3FU**

Automatic gearboxes with the serial number prefix 3FU, were fitted at the commencement of the Rover 3500 model.

b. **Series 7FU**

A season change modification introduced automatic gearboxes with the serial number prefix 7FU, and these featured a new gear change pattern with positions 'P', 'R', 'N', 'D', '2', '1'. Later versions of this series incorporated a shorter downshift cable and a self-adjusting front servo. The 7FU series gearbox is interchangeable with the earlier series, providing that the later gear change indicator plate and short downshift cable are also fitted.

c. **Series 9FU**

Automatic gearboxes with the serial number prefix 9FU were first introduced on the Rover 3500 S special export model. Later, a commonised version was introduced for the Rover 3500, 3500 S and 3½ litre models, and featured a flexible fluid filler tube. The 9FU series gearbox can be fitted to earlier models.

d. **Series 267**

The introduction of automatic gearboxes with the serial number prefix 267, incorporated a number of features to facilitate interchangeability between Rover 3500 and 3½ litre models. All previous automatic gearboxes and torque convertor assemblies that are returned to Borg Warners for reconditioning will be built to the 267 series specification, and these can be identified by a new serial number in the range commencing 04 35 000 387, stamped on the identification plate. It is therfore important to change the indicator plate and use the conversion kit for the short downshift cable, as applicable, when fitting the 267 specification gearbox to earlier cars. The owner of the car should be made aware of the conversion and the method of operation with the later gear change pattern.

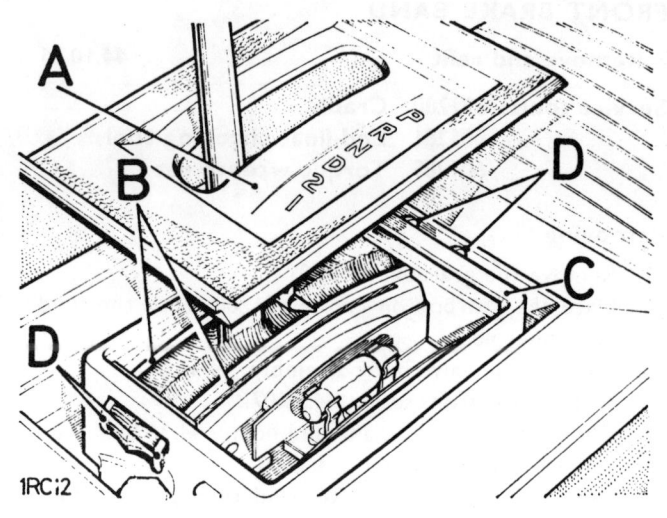

Fig. 5. Modified gearchange housing with brush light masking
A—Opaque cover C—Gearchange housing
B—Brush assemblies D—Brush carrier ends turned down to secure

Fig. 6. Gear lever knob and fibre washer

e. **Series 303**

The introduction of automatic gearboxes with the serial number prefix 303 incorporated a number of modified components, including the torque converter, bell housing, front brake band, front drum, downshift cam, downshift cable and fluid filter. The components are not generally interchangeable with earlier types, but the gearbox complete is interchangeable. A revised method of checking the downshift cable is introduced for this series gearbox, the procedure is covered in a new Operation 44.30.02, and where necessary, followed by the applicable version of Operation 44.30.03.

Rover 3500 and 3500S Manual AKM 3621

GEARBOX—AUTOMATIC—Borg-Warner type 35

FRONT BRAKE BAND

—Remove and refit   44.10.01

Service tools: 601280   Cradle
601281   End float checking tool
601285   Torque wrench set

Removing

1. Remove the gearbox. 44.20.01
2. Place the gearbox on cradle 601280, with the fluid pan uppermost.
3. Remove the valve body assembly. 44.40.01.
4. Remove the front servo. 44.34.07.
5. Check the input shaft end-float, 44.36.08, then remove the front pump. 44.32.01.
6. Withdraw the front brake band through the front of the main casing without disturbing the front clutch and input shaft.

Refitting

7. Reverse 1 to 6, ensuring that the brake band is fitted into the gearbox with the larger external lug located on the support pin in the casing.
8. Refill the fluid system. 44.24.02.

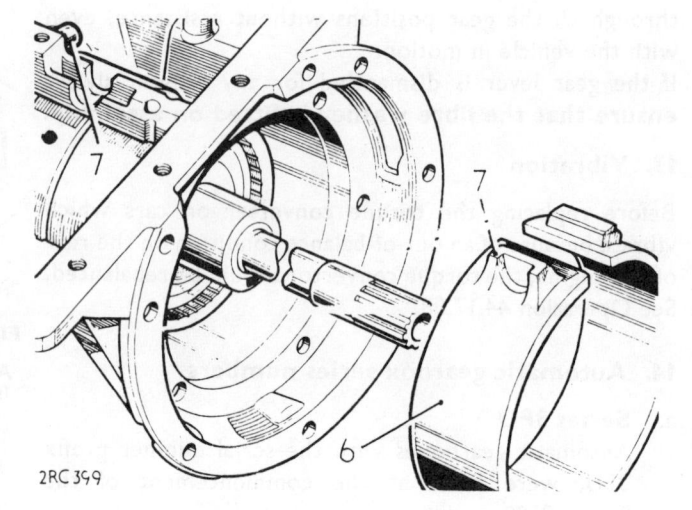

REAR BRAKE BAND

—Remove and refit   44.10.09

Service tools: 601280   Cradle
601281   End-float checking tool
601285   Torque wrench set
601286   Screwdriver set, spintorq

Removing

1. Remove the gearbox. 44.20.01.
2. Place the gearbox on cradle 601280, with the fluid pan uppermost.
3. Remove the valve body assembly. 44.40.01.
4. Remove the front servo. 44.34.07.
5. Remove the rear servo. 44.34.13.
6. Check the input shaft end-float. 44.36.08, then remove the front pump. 44.32.01.
7. Remove the front brake band. 44.10.01.
8. Remove the front and rear clutch and forward sun gear assembly. 44.12.01.
9. Remove the planet gears and centre support. 44.36.04.
10. Withdraw the rear brake band by tilting it to clear the casing.

Refitting

11. Reverse 1 to 10. Ensure that the rear brake band is fitted with the slotted lug for the servo strut towards the valve bodies, and the larger lug located on the head of the adjuster bolt.
12. Refill the fluid system. 44.24.02.

44.10.01
44.10.09

Rover 3500 and 3500S Manual AKM 3621

GEARBOX—AUTOMATIC—Borg-Warner type 35

FRONT AND REAR CLUTCH AND FORWARD SUN GEAR

—Remove and refit 44.12.01

Service tools: 601280 Cradle
601281 End-float checking tool
601285 Torque wrench set

Removing

1. Remove the gearbox. 44.20.01.
2. Place the gearbox on cradle 601280, with the fluid pan uppermost.
3. Remove the valve body assembly. 44.40.01.
4. Remove the front servo. 44.34.07.
5. Check the input shaft end-float. 44.36.08, then remove the front pump. 44.32.01.
6. Withdraw the front brake band.
7. Ease the rear clutch forward with a screwdriver.
8. Withdraw the front and rear clutch and forward sun gear as an assembly.
9. Ensure that the needle thrust bearing and bearing plate, located on the forward sun gear shaft, have not dropped into the planetary gear set.
   NOTE: Take care not to withdraw the front clutch from the rear clutch, unless some repair work is to be carried out on these units. If the units are inadvertently disengaged, refer to 44.12.13 for assembly procedure.

Refitting

10. Fit a new seal on the forward sun gear shaft.
11. Fit the needle thrust bearing.
12. Fit the bearing plate, lip towards the seal.
13. Ensure that the oil sealing rings are correctly interlocked.
14. Install the assembly of the front and rear clutch. Turn the output shaft to engage the gears of the planetary set.
15. Fit the front brake band into the gearbox, with the larger external lug located on the support pin in the casing.
16. Reverse 1 to 5.
17. Refill the fluid system. 44.24.02.

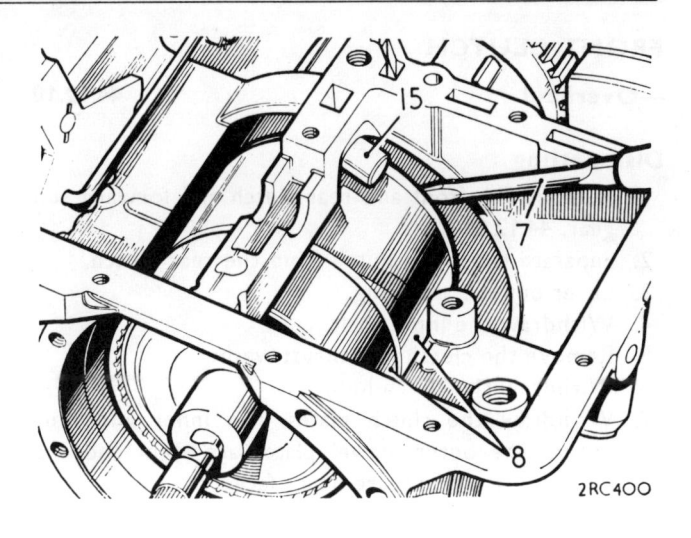

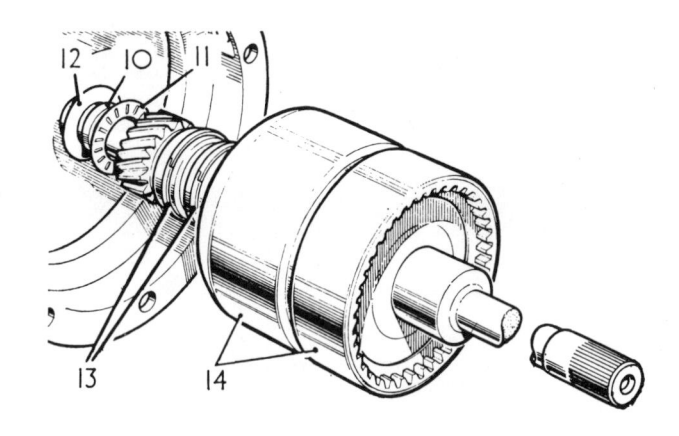

GEARBOX—AUTOMATIC—Borg-Warner type 35

FRONT CLUTCH

—Overhaul　　　　　　　　　　　　　44.12.10

Dismantling

1. Remove the front and rear clutch and forward sun gear. 44.12.01.
2. Separate the front clutch from the rear clutch.
3. Lever out the snap ring
4. Withdraw the input shaft.
5. Remove the clutch hub thrust washer.
6. Withdraw the clutch hub.
7. Withdraw the clutch plates, five inner and four outer. The outer 'drive' plates are steel and the inner friction plates are paper-faced.

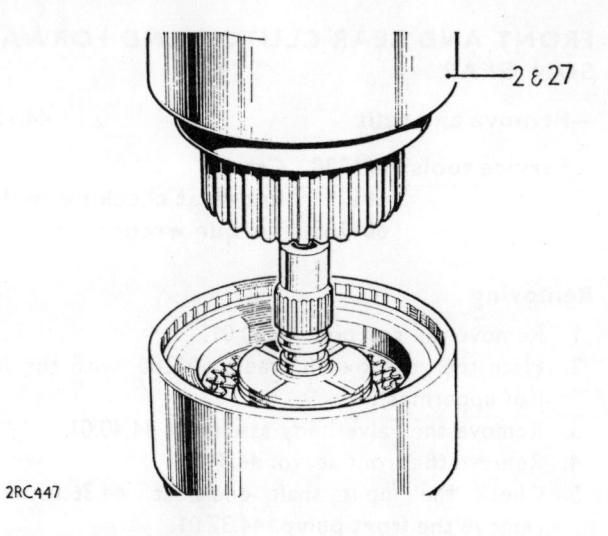

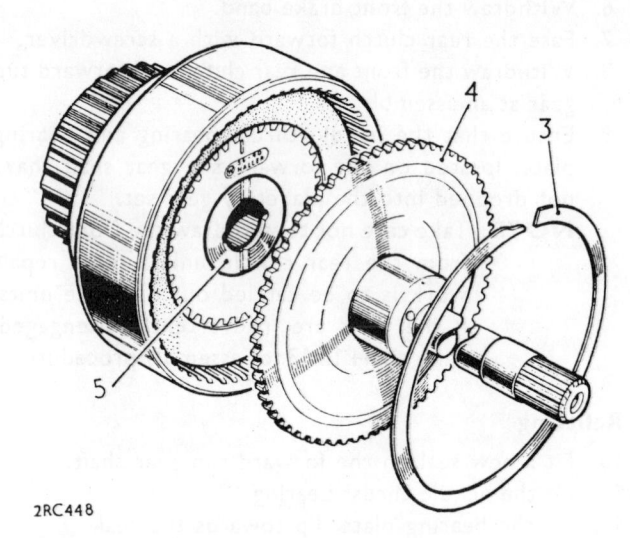

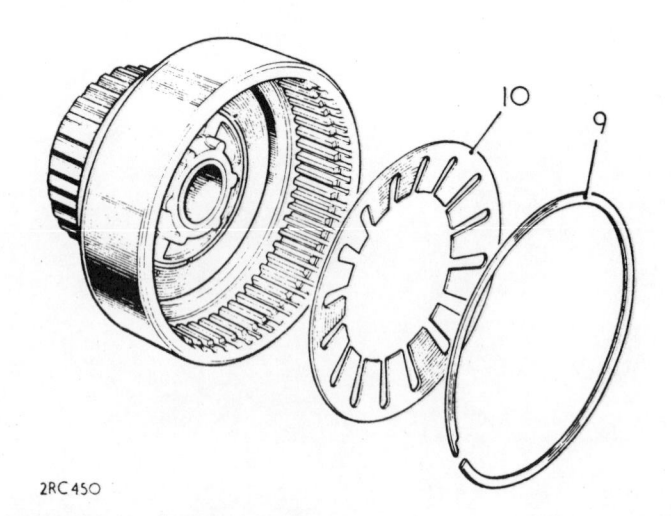

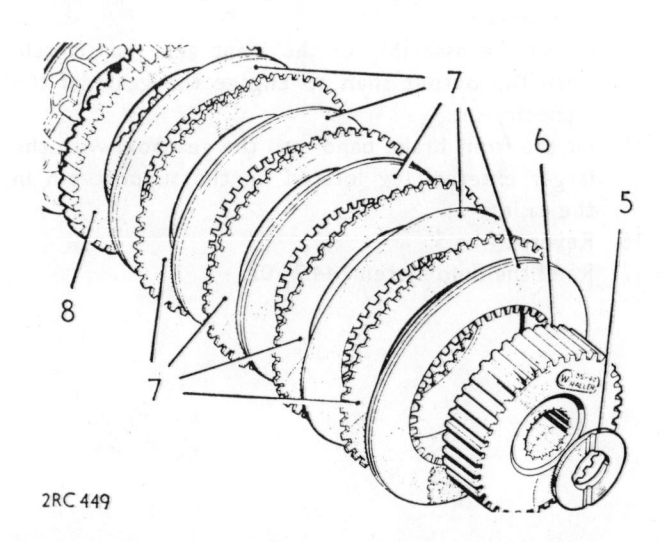

continued

Rover 3500 and 3500S Manual AKM 3621

GEARBOX—AUTOMATIC—Borg-Warner type 35

8. Lift out the pressure plate.
9. Lever out the snap ring retaining the diaphragm spring.
10. Remove the diaphragm spring, noting that the inner diameter is dished inwards.
11. Withdraw the piston by applying air pressure to the hole in the internal bore whilst the second hole is blanked off.
12. Withdraw the bearing ring from the top of the piston.
13. Remove the sealing ring.
14. Remove the 'O' ring from the clutch housing.

Inspecting

15. Obtain new seals.
16. Check that the piston one-way valve will operate.

    NOTE: If the rear clutch is not being overhauled, check the sealing rings on the forward sun gear shaft for wear, and renew as necessary.

17. Check the friction plates for wear and burning. Renew as a set.
18. Check the steel plates for distortion; if the distortion exceeds 0,12 mm (0.005 in) the plates must be renewed as a set.

Reassembling

19. Lubricate the cylinder with clean transmission fluid. Smear a new small 'O' ring with clean fluid and fit the ring into the cylinder.
20. Fit a new sealing ring on to the clutch piston, after lubricating both parts with clean transmission fluid. Place the bearing ring into position in its groove on top of the piston.
21. Replace the diaphragm spring, dished edge inwards, and secure in position with the snap ring.
22. Install the clutch pressure plate with the plain side uppermost.
23. Replace the clutch plates; an inner splined plate first, followed by an outer splined plate, alternately.
24. Line up the internal teeth of the inner plates, then fit the clutch hub, with the flat face uppermost. If necessary, turn the hub slightly to pick up the splines of the clutch plates.
25. Place the bronze thrust washer into the recess of the hub.
26. Fit the input shaft to the cylinder and secure with the large snap ring.
27. Support the rear clutch and forward sun gear assembly in a vertical position, centralise the two sealing rings on the shaft of the forward sun gear assembly. Lubricate freely with petroleum jelly and gently lower the front clutch cylinder into position.
28. Refit the front and rear clutch and forward sun gear. 44.12.01.

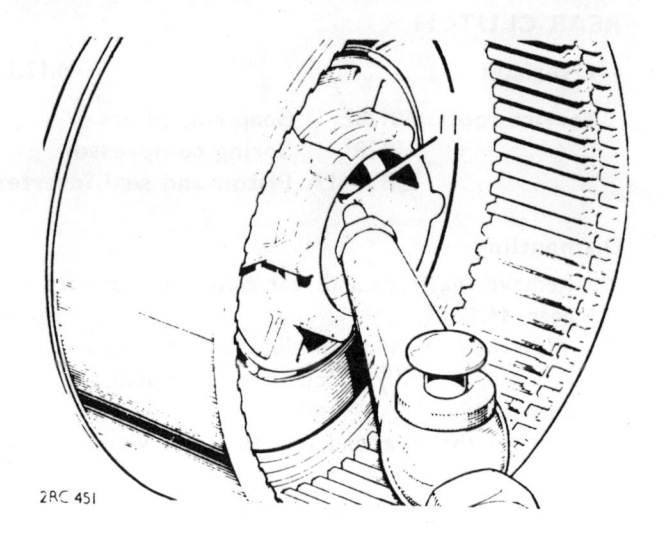

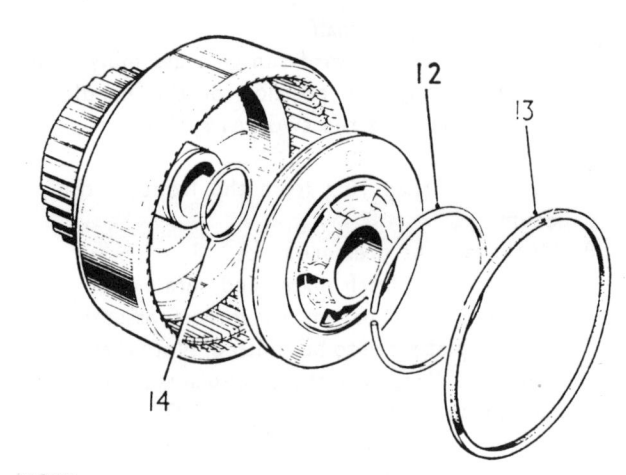

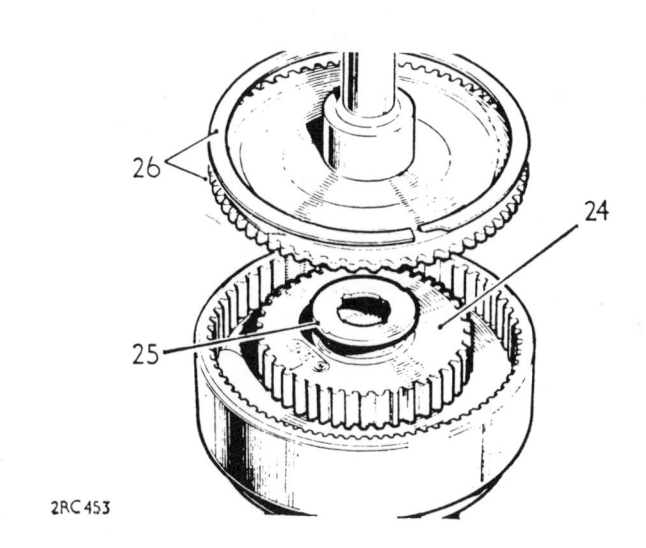

Rover 3500 and 3500S Manual AKM 3621

GEARBOX—AUTOMATIC—Borg-Warner type 35

REAR CLUTCH

—Overhaul   44.12.13

Service tools: 601282   Snap ring pliers
601283   Spring compressor
CBW41A Piston and seal inserter

Dismantling

1. Remove the front and rear clutch and forward sun gear. 44.12.01.
2. With the front and rear clutch assembly in a vertical position, carefully lift off the front clutch and place on one side.
3. Remove the steel and bronze thrust washers from the rear clutch housing.
4. Remove the needle bearing and thrust plate from the rear end of the shaft.
5. Remove the sealing rings from the front end of the forward sun gear shaft.
6. Withdraw the forward sun gear assembly from the clutch, noting the position of the needle thrust bearing between the two gears. The sun gear cannot be removed from its shaft.
7. Lever the snap ring from the front of the clutch drum.
8. Withdraw the pressure plate.
9. Remove the clutch plates and retain them in their removal order to ensure correct assembly if the original plates are to be refitted. Note that the outer plates are dished and the inner plates are paper faced.
10. Compress the piston spring. 601283.
11. Remove the spring retaining circlip. 601282.
12. Withdraw the seat and spring.
13. Remove the piston by applying air pressure to the hole between the two inner oil rings at the rear of the housing.
14. Remove the piston seal.
15. Remove the 'O' ring from the clutch housing.

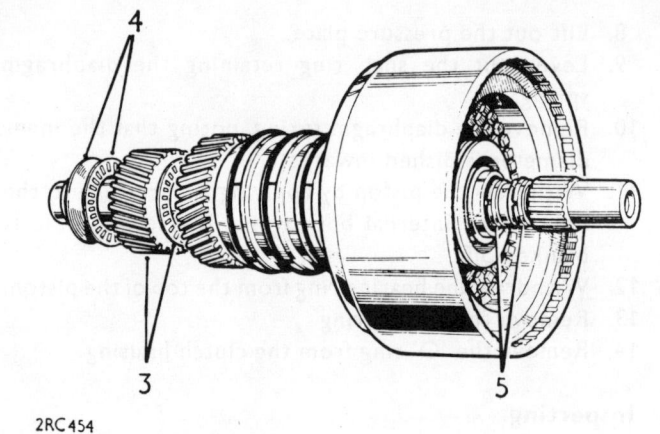

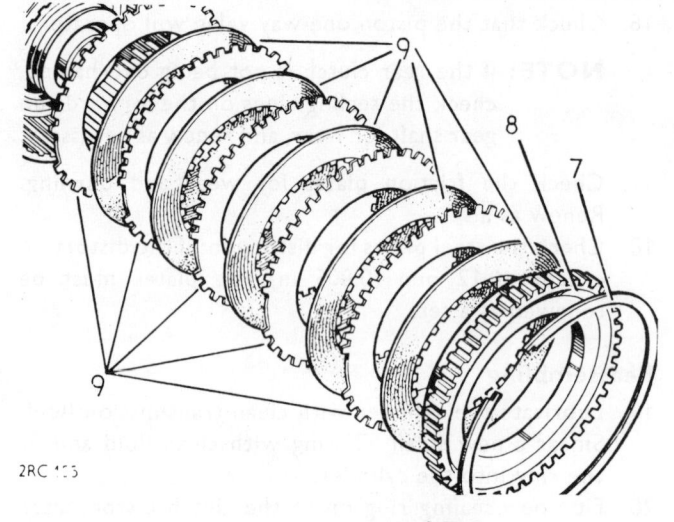

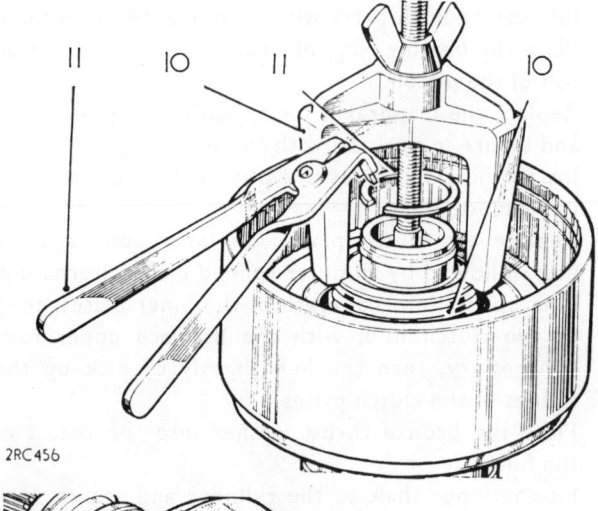

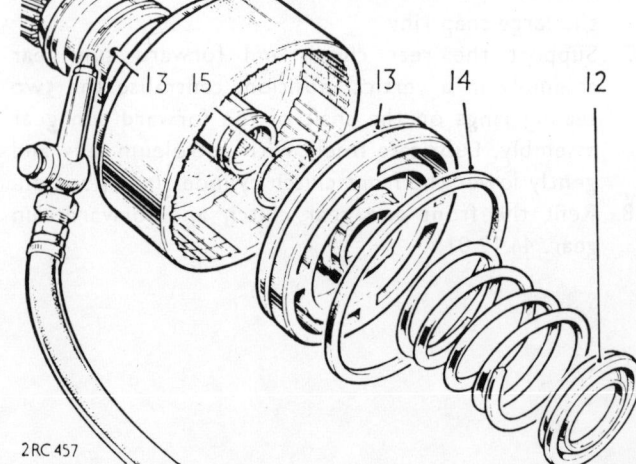

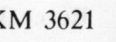

continued

44.12.13
Sheet 1

Rover 3500 and 3500S Manual AKM 3621

GEARBOX—AUTOMATIC—Borg-Warner type 35

Inspecting

16. Obtain new seals.
17. Check that the piston one-way valve will operate.
18. Check the friction plates for wear; the plates are coned 0.03 to 0.05 mm (0.010 to 0.020 in). Renew as a set.
19. Check the ring seals and the drum bearing for wear or damage.
20. Forward sun gear shaft. Check the needle thrust washers and ring seals for wear or damage.

Reassembling

21. Fit a new 'O' ring to the boss in the clutch housing.
22. Fit a new sealing ring to the piston.
23. Locate inserter CBW41A into the clutch housing, then install the piston.
24. Remove the inserter.
25. Locate the piston spring and spring seat on to the piston.
26. Position the circlip on the spring seat, then use 601283 to compress the spring. Fit the circlip, using 601282.
27. Fit the inner and outer clutch plates in alternate sequence and with the coning in the same direction.
28. Place a needle thrust washer over the shaft against the front face of the forward sun gear, then insert shaft into the clutch.
29. With the assembly in a vertical position, insert the front clutch in order to line up the internal teeth of the front clutch plates.
30. Remove the front clutch and fit two new oil sealing rings to the front end of the forward sun gear shaft.
31. Locate the steel thrust washer on to the two flats of the centre hub, followed by the bronze thrust washer.
32. Centralise the two small oil sealing rings on the forward sun gear shaft and gently lower the front clutch over the shaft.
33. If difficulty is experienced when carrying out this operation, the front clutch should be checked to ensure that the internal thrust washer has not been displaced.
34. Fit new nylon sealing ring to rear end of forward sun gear shaft and place needle thrust washer and thrust plate in position and retain with petroleum jelly.
35. Refit the front and rear clutch and forward sun gear. 44.12.01.

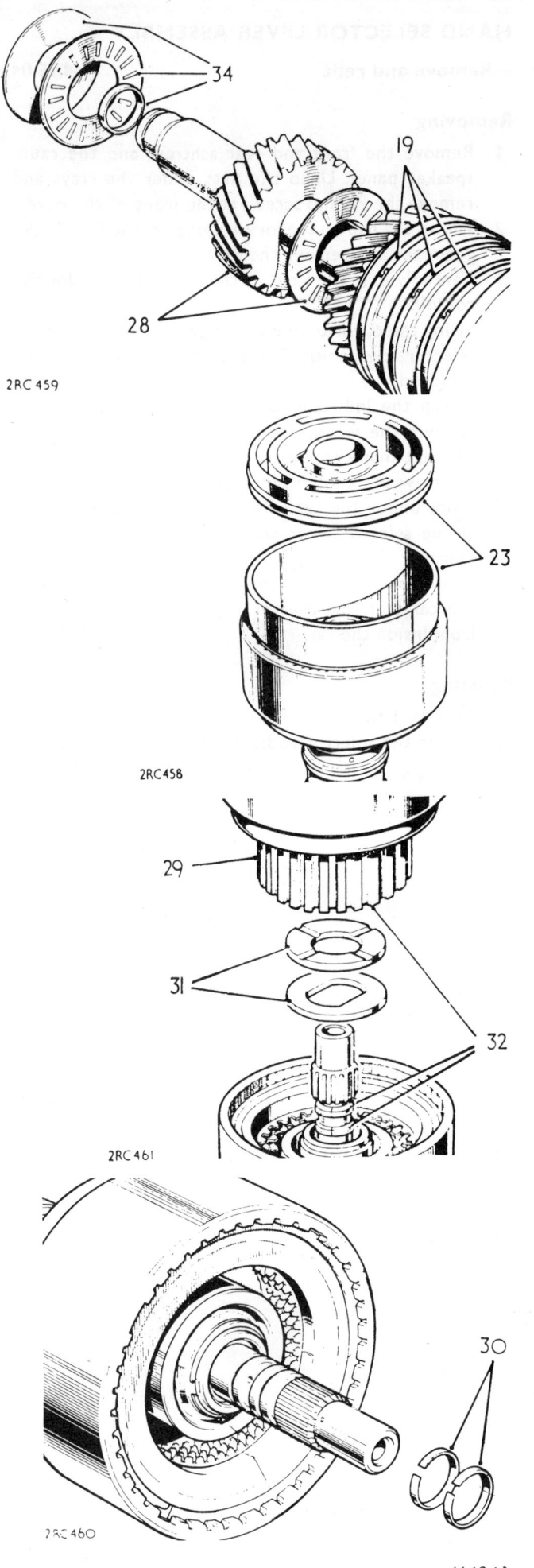

Rover 3500 and 3500S Manual AKM 3621

44.12.13
Sheet 2

GEARBOX—AUTOMATIC—Borg-Warner type 35

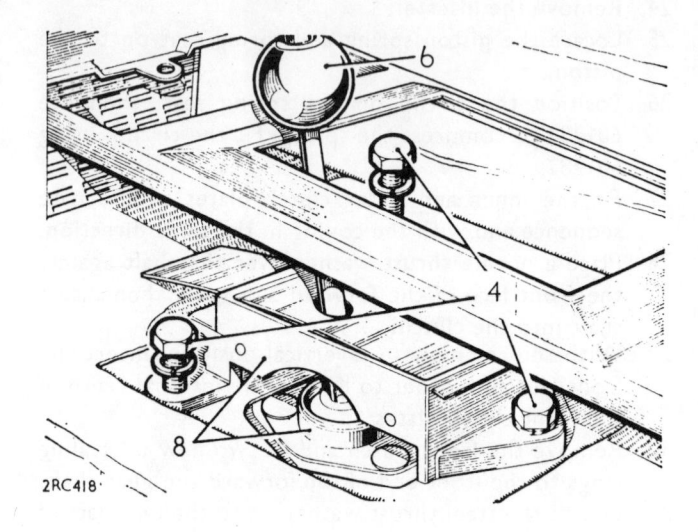

HAND SELECTOR LEVER ASSEMBLY

—Remove and refit 44.15.04

Removing

1. Remove the front and rear ashtrays and the radio speaker panel. Undo the nuts under the trays and remove the Phillips screws at the front of the cover.
2. Lift up the flap on the forward edge of the handbrake grommet and remove the drive screw.
3. Carefully ease the front of the cover from under the console.
4. Swing the cover to one side to gain access to the hand selector mechanism fixing bolts, which should be removed.
5. Unclip the indicator for the gear selector and disconnect the wiring from the illumination lamp, at adjacent snap connectors.
6. Unscrew the knob from the selector lever.
7. From under the car, disconnect the operating rod at the selector lever and the tie rod at the pivot housing.
8. The pivot housing can now be removed from under the car and the selector lever housing withdrawn from inside the car.

Refitting

9. Reverse 1 to 8.
10. Adjust the selector rods, if necessary. 44.15.08.

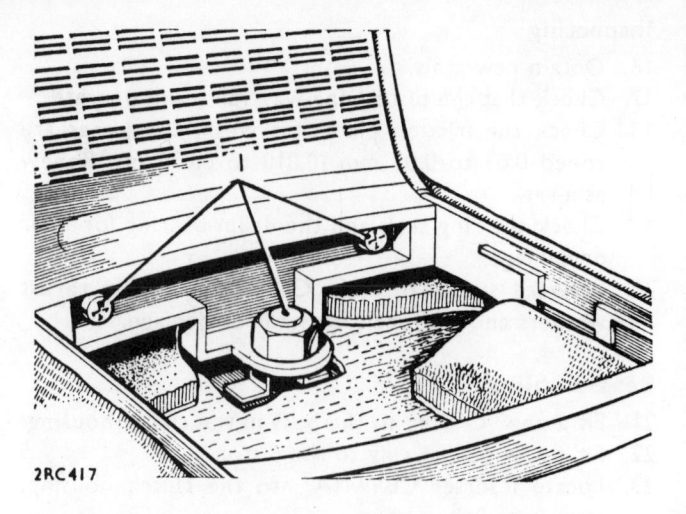

44.15.04

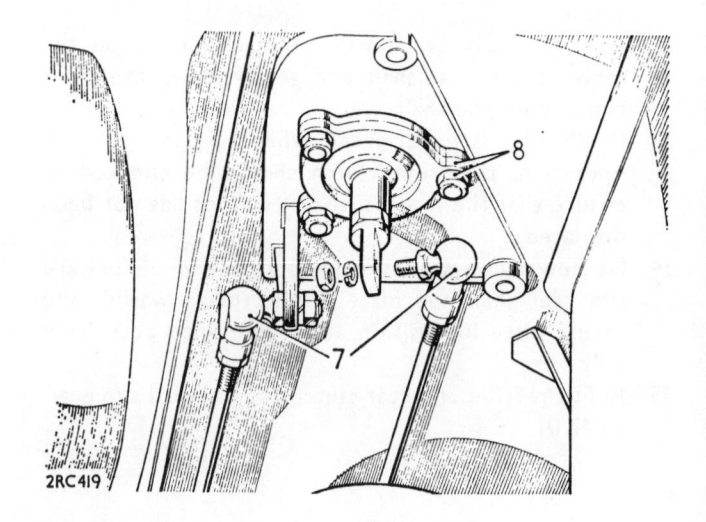

Rover 3500 and 3500S Manual AKM 3621

GEARBOX—AUTOMATIC—Borg-Warner type 35

HAND SELECTOR LEVER ASSEMBLY

—Overhaul 44.15.05

Dismantling

1. Remove the hand selector lever assembly. 44.15.04.
2. If required, remove the rubber insert, 'O' ring and button.
3. Remove the shield and the illumination lamp from the housing.
4. Withdraw the control rod.
5. Remove the retaining cap for the spherical seat.
6. Remove the spherical seat by forcing it off the lever by hand. Note the locating pin.
7. Remove the eye bolt, locknut and spring.
8. Remove the detent pin and plug from the lever.

Inspecting

9. Obtain new parts as necessary.
10. If new control rod is being fitted, it should be cut to length on assembly so that 12,5 mm (0.500 in) protrudes from the top of the gear selector lever. This dimension is important to ensure correct engagement of the detent pin in the selector gate.

Reassembling

11. Reverse 1 to 8.

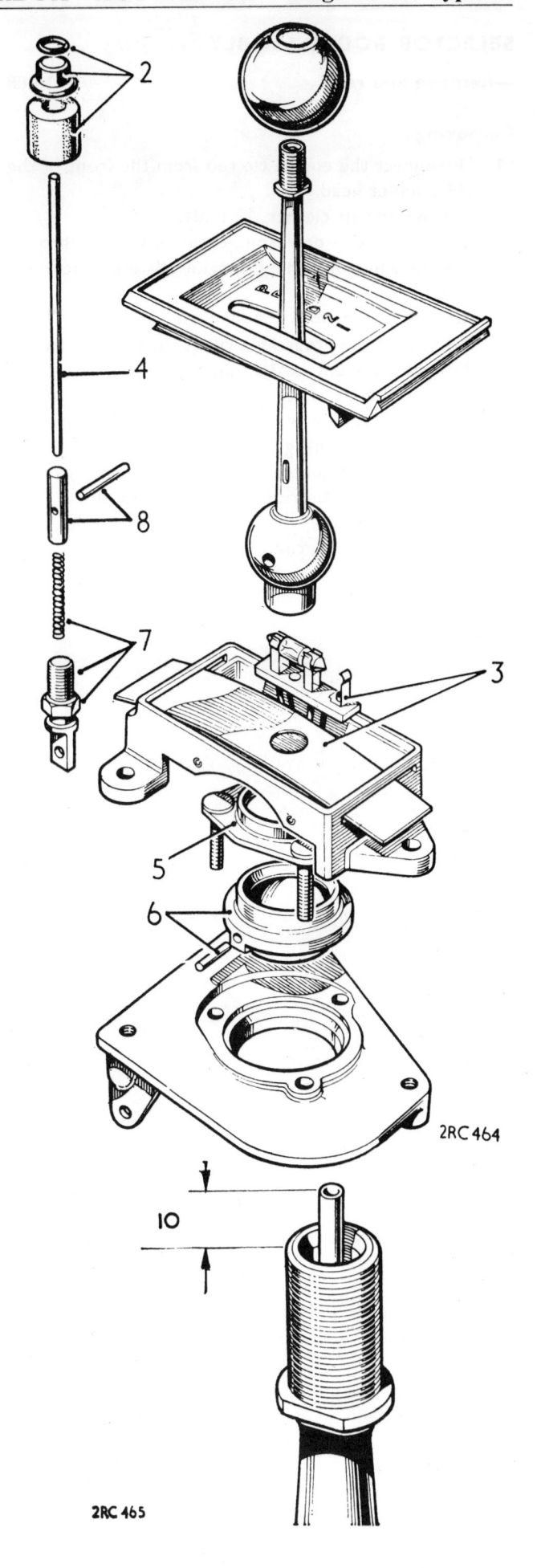

Rover 3500 and 3500S Manual AKM 3621

GEARBOX—AUTOMATIC—Borg-Warner type 35

SELECTOR ROD ASSEMBLY

—Remove and refit 44.15.08

Removing

1. Disconnect the engine tie rod from the front of the LH cylinder head.
2. Remove the air cleaner. 19.10.01.
3. Disconnect the downshift cable from the throttle linkage and adjustment bracket (latest type illustrated).
4. Remove the throttle coupling rod.
5. Disconnect the propeller shaft at the gearbox flange.
6. Remove the front exhaust pipe.
7. Support the engine then release the rear mounting.
8. Lower the engine rear end sufficient to give access to the selector linkage.
9. Disconnect the control rod at the selector lever.
10. Disconnect the tie rod at the pivot housing.

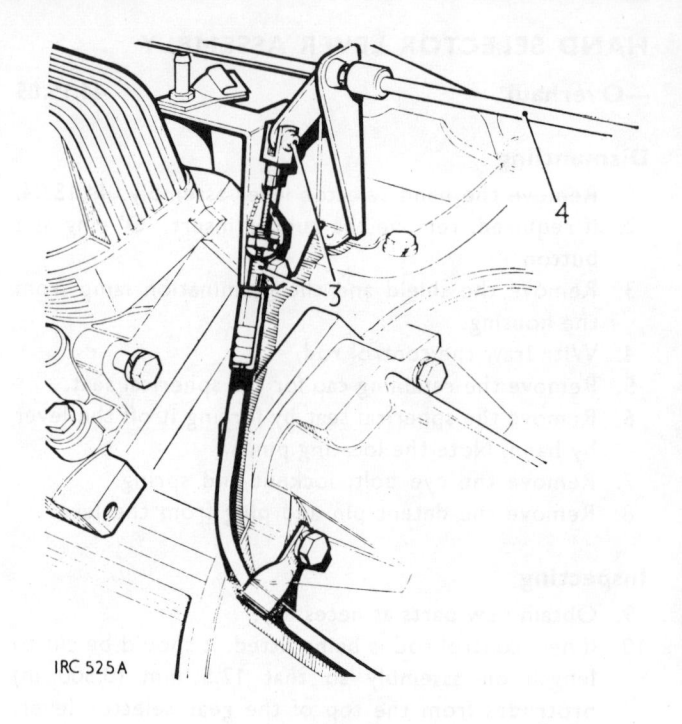

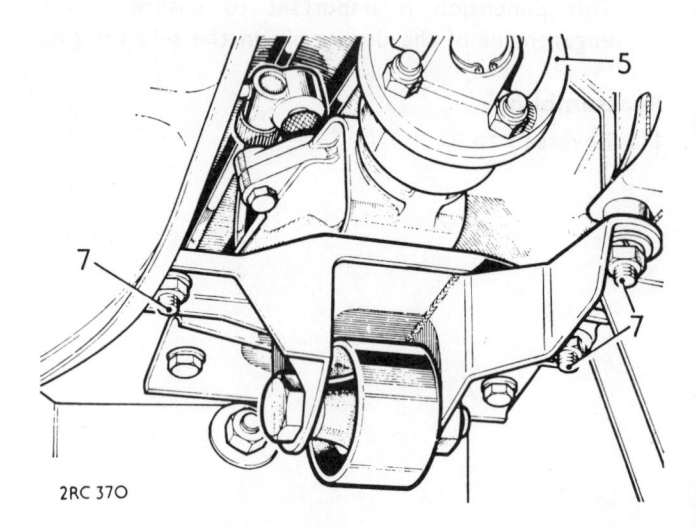

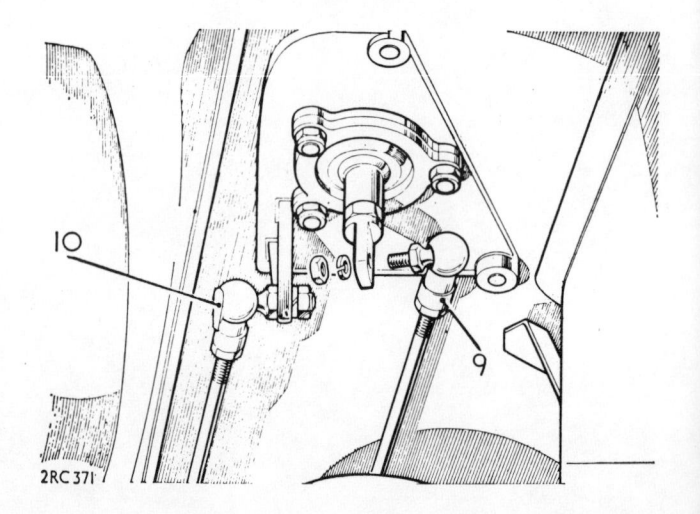

continued

44.15.08
Sheet 1

Rover 3500 and 3500S Manual AKM 3621

GEARBOX—AUTOMATIC—Borg-Warner type 35

11. Disconnect the vertical rod at the gearbox.
12. Release the compensator bracket from the converter housing.
13. Withdraw the compensator and rod assembly.
14. If required, remove the upper selector lever, control rod and tie rod from the compensator.
15. Remove the two drive screws to release the compensator block from the bracket.

Refitting

16. Reverse 1 to 15, noting the following.
17. Fit the upper selector lever facing forwards and at right angles to the other lever.
18. Ensure that the compensator bracket is in line with the axis of the transmission.
19. Connect the control rod and tie rod at both ends. Connect the vertical rod at the top end only.
    NOTE: The gear selector linkage must be correctly adjusted to give satisfactory operation of the gearbox. Ensure that the linkage is never allowed to over-ride the gearbox detent; a definite 'click' must be felt in each selector position.
20. Check, and if necessary adjust, the length of the control rods to the following dimensions, which represent the length between the centres of the ball joints. The latest selector pattern is illustrated, but the dimensions are the same for the earlier 'P' 'R' 'N' 'D2' 'D1' 'L' selector pattern.
21. Vertical rod:
    Models with adjustable rod 190,0 mm (7.500 in).
    Models with non-adjustable rod 190,5 mm (7.700 in).
22. Tie rod: 401,6 mm (15.812 in).
23. Control rod: 528,6 mm (20.812 in).
24. Check and if necessary, adjust, the downshift cable. 44.30.02/03.
25. Select 'N' at the selector lever on the gearbox.
26. Place the gear selector lever in the neutral ('N') position.
27. Adjust as necessary at lower control rod, ensuring the ball joint can be connected to the lower selector lever without strain. Check quadrant gating in all selector positions, ensuring that linkage is not over-riding the transmission detent.

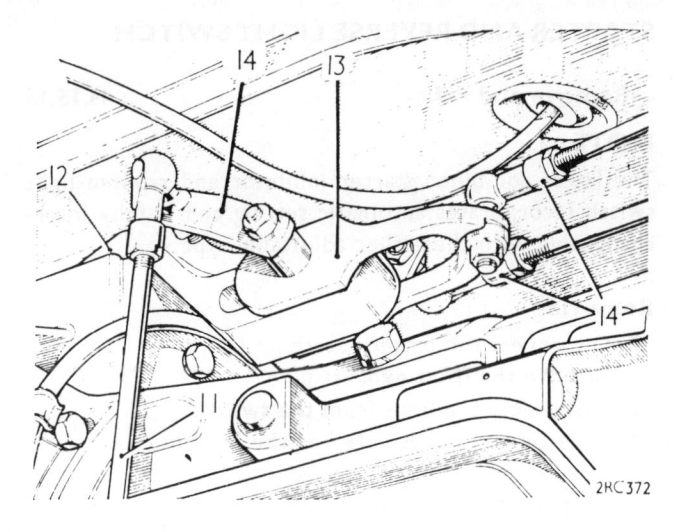

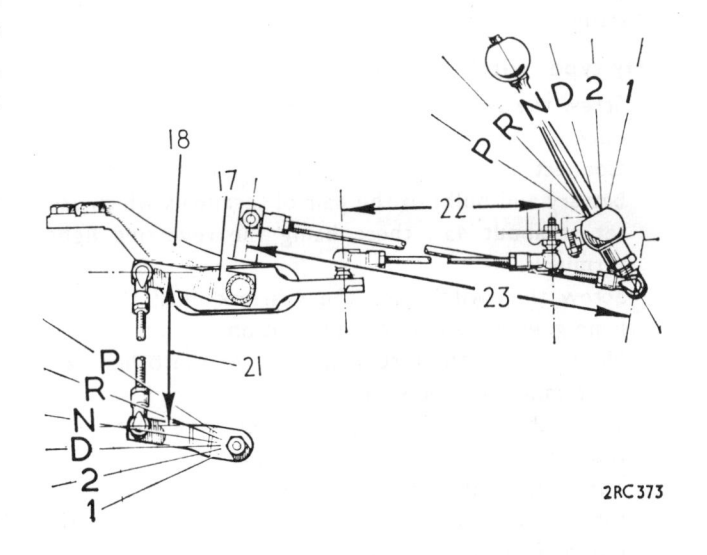

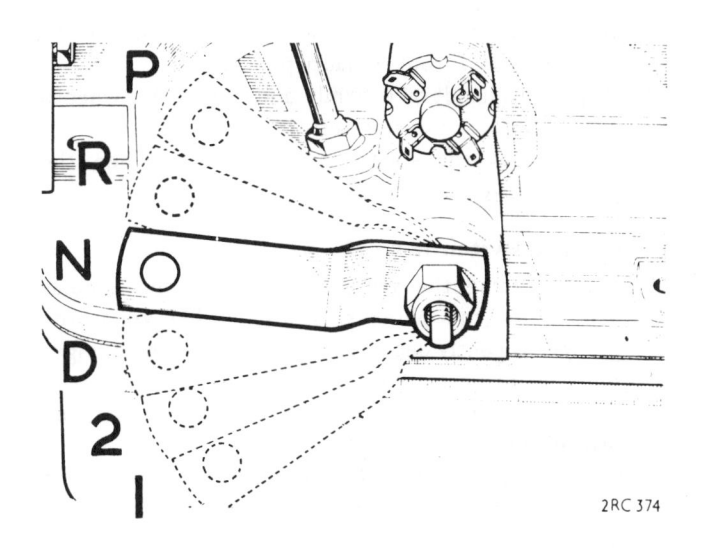

Rover 3500 and 3500S Manual AKM 3621

44.15.08
Sheet 2

GEARBOX—AUTOMATIC—Borg-Warner type 35

STARTER AND REVERSE LIGHT SWITCH

—Remove and refit   **44.15.15**

NOTE.
The latest combined starter inhibitor and reverse light switch incorporates automatic setting and is fully interchangeable with the earlier adjustable type.

Removing

1. Disconnect the battery earth lead.
2. Remove the front exhaust pipe.
3. Disconnect the leads from the switch.
4. Early type switches—slacken the locknut.
5. Latest type switches—slacken the switch, using a spanner on the square section.
6. Unscrew the switch by hand.

Refitting

Early type switch 7 to 15

7. Screw the switch into the gearbox by a few threads only.
8. Connect a test lamp (consisting of a small dry battery and bulb) to the pair of terminals which are set at about 45°; these being the reversing light terminals.
9. Screw the switch into the gearbox until the test lamp goes out and note the position.
10. Now connect the test lamp leads to the other pair of terminals on the switch.
11. Screw the switch in approximately one turn until the light comes on and note the position.
12. Turn the switch back approximately half a turn; that is, midway between the two points previously noted. Then tighten the locknut, taking care not to strip the thread of the light alloy switch body.
13. Disconnect the test lamp wiring and reconnect the car wiring and front exhaust pipe.
14. Apply the handbrake firmly and chock the wheels. Apply footbrake during test.
15. Check that the starter operates only when the selector lever is at 'P' or 'N'.

Latest type switch 16 to 25

16. Before fitting the switch, check that the annular groove on the plunger is just visible when the plunger is gently depressed to its stop. If the plunger does not project sufficiently to expose the groove in the fully depressed position carry out the following procedure:
    a. Screw a 7/16 in UNC nut on to the end of the switch thread flush with the end face.
    b. Clamp the plunger in a vice so that the nut on the switch end face is up against the vice jaws.
    c. Turn the nut against the jaws of the vice, thus gradually jacking out the switch body as necessary. Take care not to over draw.

   NOTE: This type of inhibitor switch may be re-set in the above manner up to six times before its normal operation is affected.

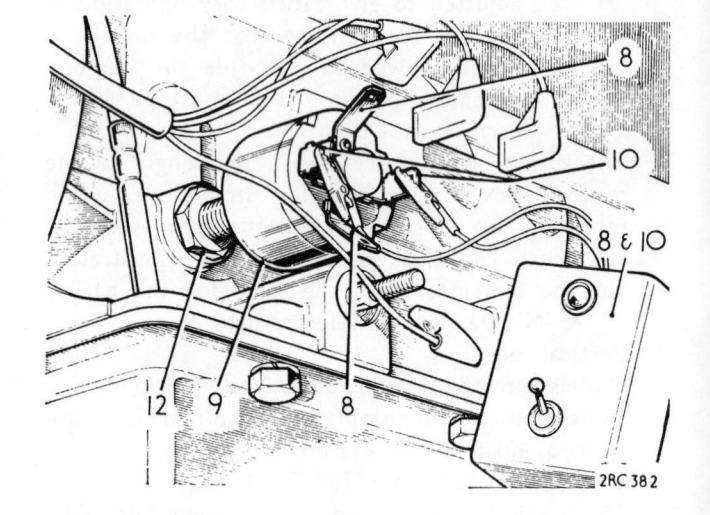

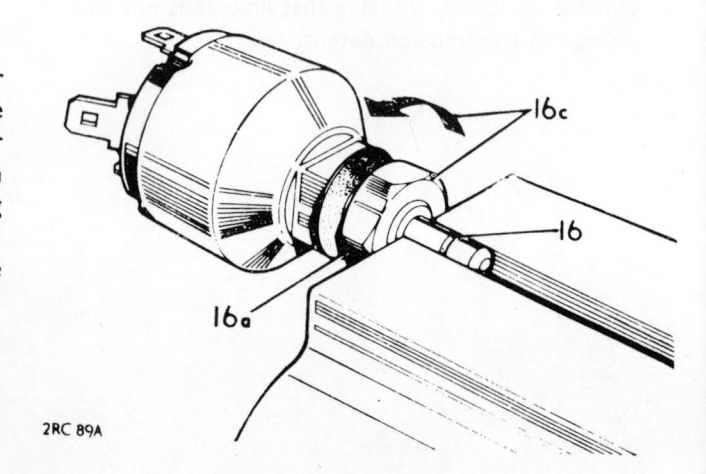

continued

Rover 3500 and 3500S Manual AKM 3621

GEARBOX—AUTOMATIC—Borg-Warner type 35

17. Move the gear selector to 'P' park position.
18. Place the sealing washer on to the switch.
19. Screw the switch into the gearbox case by hand, then tighten to a torque of 0,83 to 1,10 kgf.m (6 to 8 lbf ft).
20. Connect the starter inhibitor leads (white and red) to the straight switch terminals.
21. Connect the reverse light leads (green and green/brown) to the angled switch terminals.
22. Reverse 1 and 2.
23. Apply the hand brake firmly, chock the wheels and apply the foot brake during the following tests.
24. Check that the starter operates only when the selector lever is at 'P' or 'N'.
25. Check that the reverse lights only operate when the selector lever is at 'R'.

CONVERTER HOUSING

—Remove and refit 1 to 3, 6 and 8   44.17.01

CONVERTER ASSEMBLY

—Remove and refit 1 to 8   44.17.07

Removing

1. Remove the gearbox assembly. 44.20.01.
2. Remove the cover plate from the lower front face of the converter housing.
3. Remove the converter housing.
   NOTE: The converter should be refitted in its original position, and to enable this to be done, corresponding marks should be made on the flexible drive plate and the converter before removal. This does not apply if a new converter is being fitted.
4. Support the converter and remove the four securing bolts, turning the engine as necessary.
5. Withdraw the converter taking precautions against spilling any residual fluid.

Refitting

6. Reverse 1 to 5, noting the following.
7. Tightening torque for converter securing bolts: 3,5 to 4,0 kgf.m (25 to 30 lbf ft).
8. Tightening torque for converter housing securing bolts: 4,0 kgf.m (30 lbf ft).

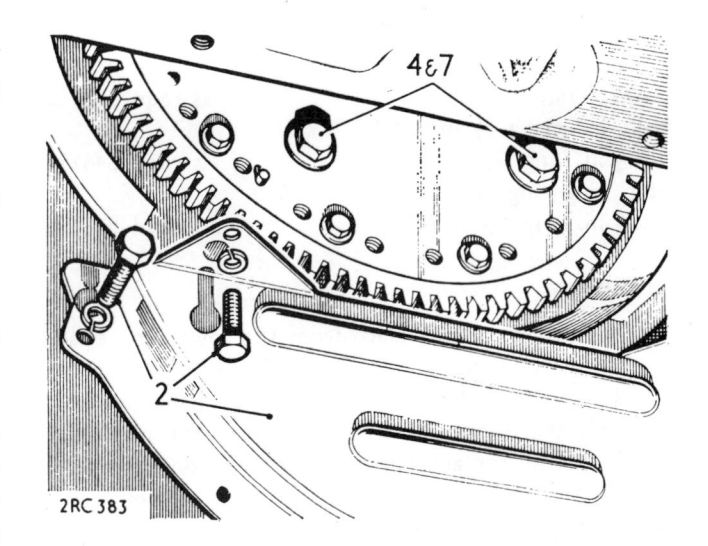

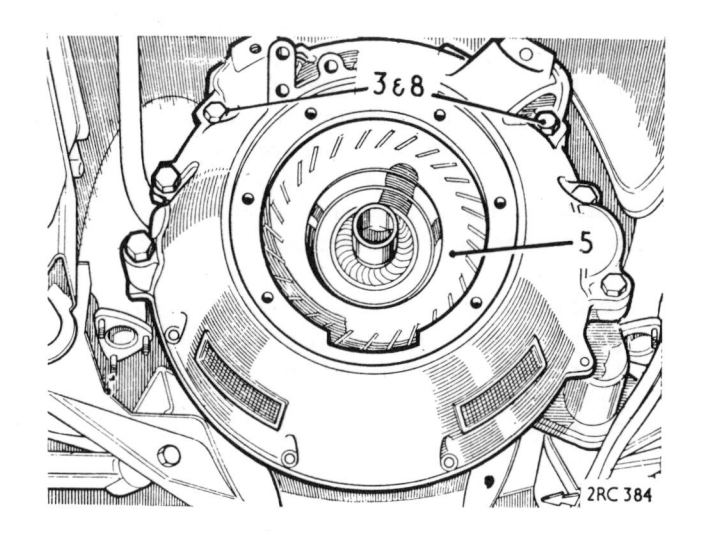

GEARBOX—AUTOMATIC—Borg-Warner type 35

CONVERTER

—Balance 44.17.08

NOTE:

Final balance of the engine and gearbox assembly is achieved by selective fitting of special balance bolts in the forward facing side of the starter ring gear. If a car vibrates because of an out-of-balance condition on the rear of the engine, the following rebalancing procedure should be carried out. If rebalancing fails to eliminate the vibrations, it may be necessary to fit a new torque converter.

To balance a torque converter in position, normal engine balancing equipment may be used or it may be successfully carried out manually.

Parts which may be required are as follows:

Bolt, balancing (¼ in. UNF × ⅜ in. long, 7/16 in. AF) Head thickness 4 mm (0.160 in.)	For balancing torque converter	As reqd 535781
Bolt, balancing (¼ in. UNF × ⅜ in. long, 7/16 in. AF) Head thickness, 6,3 mm (0.250 in.)		As reqd 535782
Bolt, balancing (¼ in. UNF × ⅜ in. long, 9/16 in. AF)		As reqd 546194
Bolt, balancing (¼ in. UNF × ⅜ in. long, 5/16 in. AF)		As reqd 546198

Procedure

1. Remove the cover plate from the lower front face of the bell housing.
2. Remove the balance bolt(s) from the original position.
3. Refit the balance bolt(s) 180° around the ring gear.
4. Sit in the car and run the engine at 2000 rev/min; feel the vibration through the steering wheel.
5. If the vibration is reduced with the bolt(s) in this position, substitute bolt(s) for heavier or lighter bolt(s) until vibration is acceptable.

   However, if the vibration is worse after moving the bolt(s) through 180°, return the bolt(s) to the original position, substituting heavier or lighter bolt(s) and checking again until vibration is acceptable.
6. To achieve balance it may be necessary to place two bolts close together, or move a bolt only a short distance round the circumference of the ring gear, up to 90° from the original position.
7. Alternatively, if position 3 gives a better balance condition than position 2, use two bolts close together, or move a bolt only a short distance round the circumference of the ring gear, up to 90° from the 180° position.
8. Refit the bell housing cover plate.

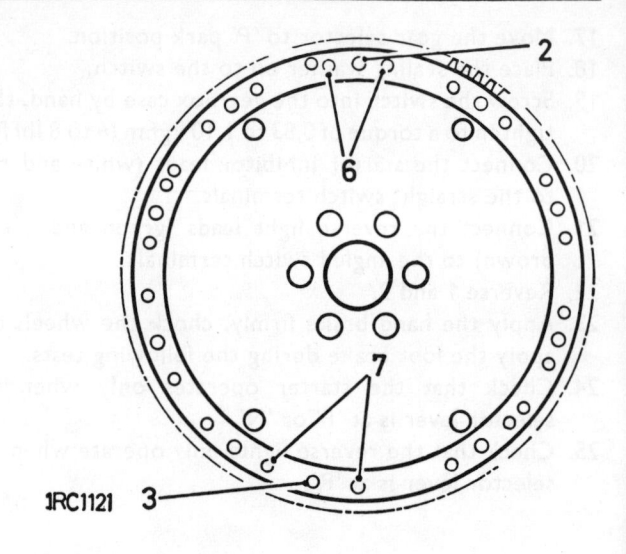

GEARBOX—AUTOMATIC—Borg-Warner type 35

GEARBOX ASSEMBLY

—Remove and refit   44.20.01

Removing

1. Disconnect the battery earth lead.
2. Remove the air cleaner. 19.10.01.
3. Disconnect the downshift cable from the throttle linkage, engine and bell housing.
4. Disconnect the throttle linkage.
5. Remove the fan guard from the top of the radiator.
6. Disconnect the breather pipe between the radiator header tank and the induction manifold.
7. Release the gearbox fluid filler and breather tubes from the rear of the engine.
8. **Air conditioned models**—Drain the radiator and disconnect the water hose between the inlet manifold and the water valve.
9. Place the car on a suitable ramp.
10. Remove the front exhaust pipe.
11. Remove the propeller shaft.
12. Drain the gearbox fluid pan.
13. Disconnect the speedometer cable from the gearbox.
14. Disconnect the leads from the starter inhibitor switch, noting the colour and position of the leads to facilitate re-assembly.
15. Disconnect the gear selector rods from the underside of the driver's control lever, and disconnect the lower control rod from the selector lever on the gearbox.
16. Support the engine then release the rear mounting.
17. Lower the rear end of the gearbox sufficient only to give access from the rear, to the upper hexagon securing the fluid filler tube. Do not lower the gearbox excessively, otherwise water hoses and other components may be overstressed.
18. Place a drip tray beneath the gearbox to catch residual fluid which will be released.
19. Disconnect the two oil cooler pipes from the gearbox.
20. Unscrew the upper hexagon securing the fluid filler tube to the gearbox. Withdraw the fluid level dipstick followed by the filler tube.
21. Remove the three bolts and spring washers securing the gear selector rods to the converter housing, and lift the rod assembly clear.

    NOTE: During the next item (22), the gearbox **must** be held in position on the bell housing, to prevent it falling as the securing bolts are removed.

22. Using a socket spanner with suitable extensions, remove the six bolts and spring washers securing the gearbox to the converter housing, then carefully draw the gearbox rearward and clear. **Do not** allow the weight of the gearbox to be supported by the input shaft, otherwise damage to the shaft and torque converter may ensue.

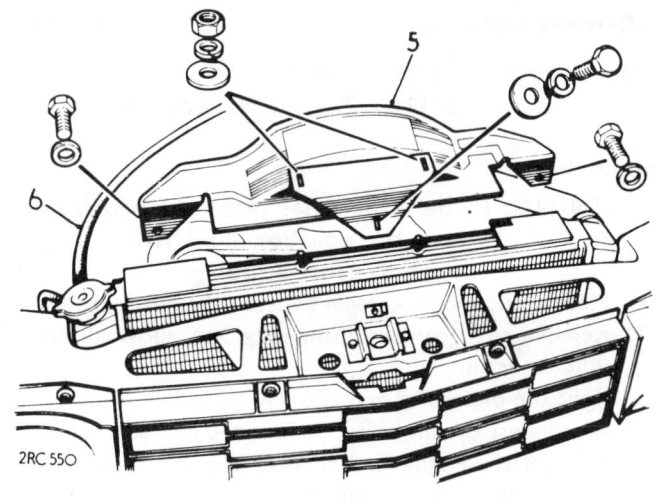

continued

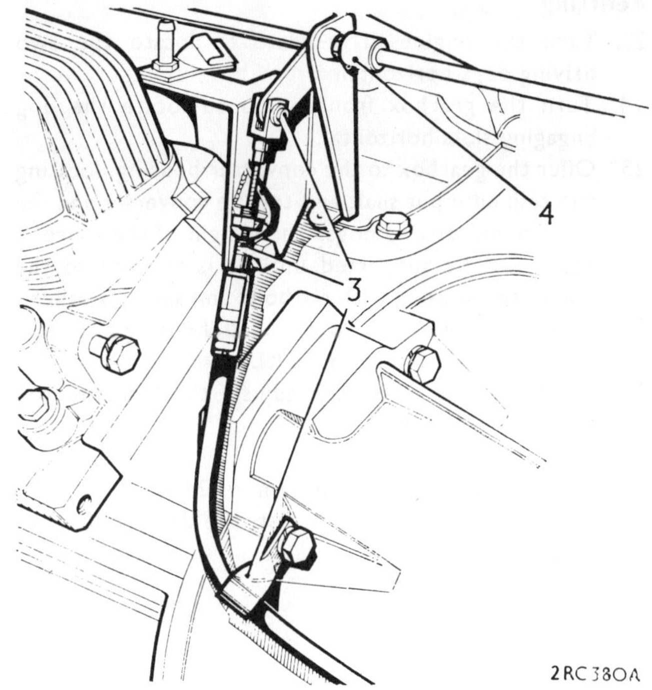

Rover 3500 and 3500S Manual AKM 3621

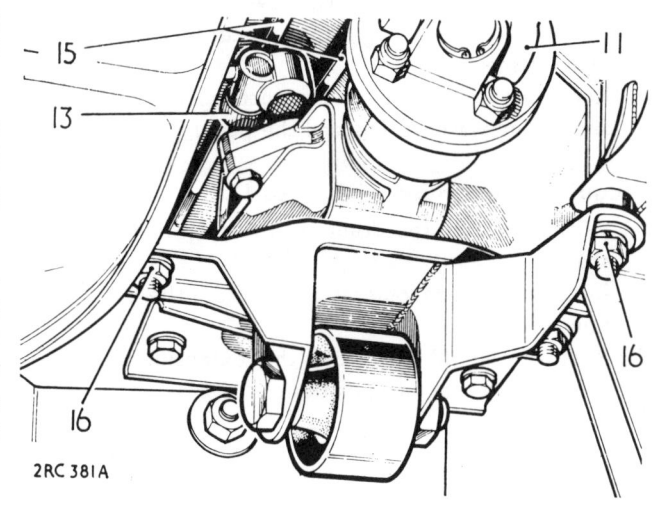

44.20.01
Sheet 1

GEARBOX—AUTOMATIC—Borg-Warner type 35

Refitting

23. Turn the engine drive plate to locate the two driving dogs horizontal.
24. Turn the gearbox front pump to locate the dog engaging slots horizontal.
25. Offer the gearbox to the converter housing, locating the splined input shaft and torque converter and the front pump driving dogs. The weight of the gearbox case must be supported until it is secured to the converter housing with six bolts and spring washers.
26. Secure the gearbox to the bell housing. Torque: 1,0 to 1,7 kgf.m (8 to 13 lbf ft).
27. Check, and if necessary adjust, the selector rod assembly. 44.15.08.
28. Reverse 1 to 21.
29. Refill the gearbox fluid system. 44.24.02.
30. Check, and if necessary adjust, the downshift cable. 44.30.02/03.

REAR EXTENSION HOUSING

—Remove and refit 44.20.15

Service tools: 601280 Cradle
601285 Torque wrench set

Removing

WARNING: If the vehicle has just returned from a run, the gearbox fluid will be hot enough to cause scalding.

1. Place a suitable receptacle beneath the gearbox.
2. Remove the gearbox drain plug and allow all the fluid to drain.
3. Refit the drain plug and washer.
4. Disconnect the propeller shaft at the gearbox flange.
5. Support the engine and remove the rear mounting.
6. Disconnect the speedometer cable.
7. Lock the gearbox in the park position. If the unit is out of the vehicle, turn the lever clockwise.
8. Withdraw the driving flange, secured by bolt and plain washer.
9. Remove the extension case, secured by four bolts and spring washers. Discard the paper gasket.

If required, continued with 10 to 12

10. Withdraw the speedometer drive gear from the shaft.
11. Remove the speedometer pinion from the extension housing.
12. Remove the oil seal from the extension housing. Press a new seal into position.

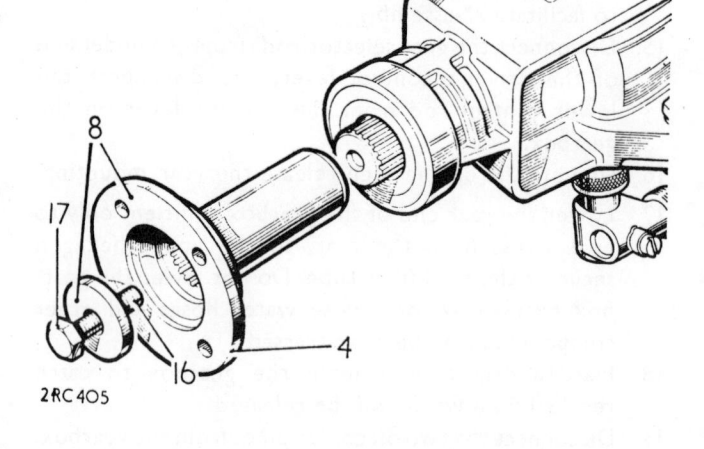

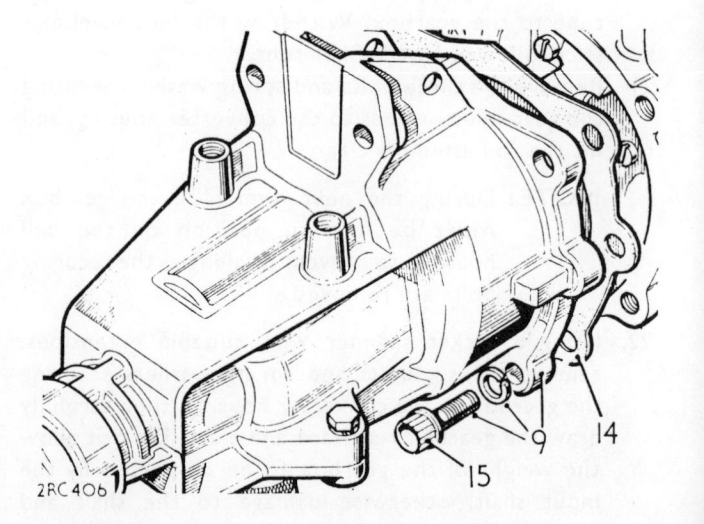

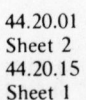

continued

Rover 3500 and 3500S Manual AKM 3621

GEARBOX—AUTOMATIC—Borg-Warner type 35

Refitting

13. Reverse 1 to 12, noting the following.
14. Use a new paper gasket for the extension housing.
15. Tightening torque for extension housing securing bolts: 1,0 to 1,7 kgf.m (8 to 13 lbf ft).
16. Smear the threads of the coupling flange securing bolt with 'Loctite' sealant grade 'CV', Rover Part No. 601168.
17. Tightening torque for coupling flange securing bolt: 2,8 to 3,5 kgf.m (20 to 25 lbf ft).
18. Release the parking pawl engagement.
19. If the gearbox is completely dismantled, hold the ring gear and driven shaft firmly whilst fitting the extension housing and the coupling flange.
20. Refill the fluid system. 44.24.02.

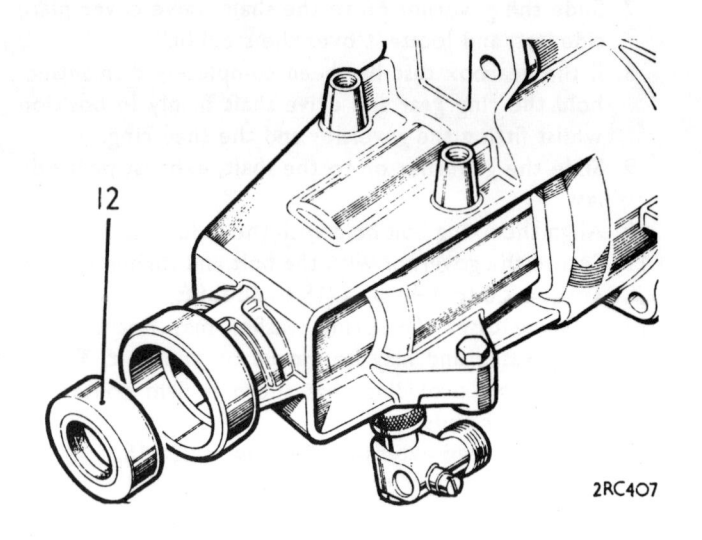

GOVERNOR ASSEMBLY

—Remove and refit   44.22.01

Early, two piece, type governor 1 to 4, 7, 8 and 13
Latest, one piece, type governor 1, 5, 6 and 9 to 13

Service tools—601280 Cradle, gearbox
—601282 Pliers, snap rings
—601285 Torque wrench set

Removing

1. Remove the rear extension housing. 44.20.15 (DDD2-17).
2. Release park position, then rotate the driven shaft so that the governor centrifugal weight is at the 6 o'clock position, that is, downwards as illustrated. This is a precaution to prevent the steel retaining ball from being lost when the governor is withdrawn from the shaft.
3. Remove the snap ring locating the governor on to the driven shaft, using the special snap ring pliers, 601282.
4. Slide the governor off the shaft, taking care not to lose the steel ball which keys the governor to the shaft.
5. Remove the special bolt and spring washer securing the governor.
6. Slide the governor off the shaft.

    NOTE: The latest one piece governor is interchangeable with the earlier type as a complete unit. The latest type governor is not serviceable and should not be dismantled, if a fault is suspected, fit a new governor complete.

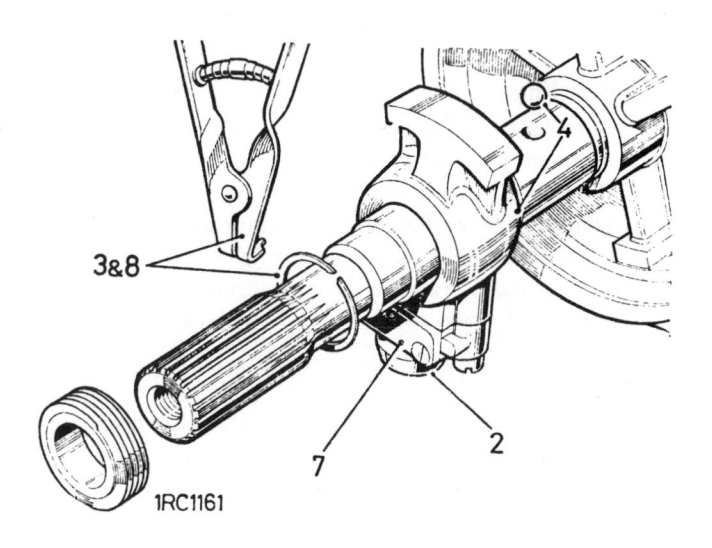

continued

GEARBOX—AUTOMATIC—Borg-Warner type 35

Refitting

7. Slide the governor on to the shaft, valve cover plate side last, and locate it over the steel ball.
8. If the gearbox case has been completely dismantled, hold the ring gear and drive shaft firmly in position whilst fitting the governor and the snap ring.
9. Slide the governor on to the shaft, exhaust port side last.
10. Align the fixing bolt hole with the indent in the shaft.
11. Secure the governor with the bolt and spring washer. Torque 2,0 to 2,4 kgf.m (15 to 18 lbf ft).
12. In some cases the spring washer may not be fully compressed and may be moved on the bolt. This is permissible providing that there is firm contact at the washer ends.
13. Refit the rear extension housing. 44.20.15.

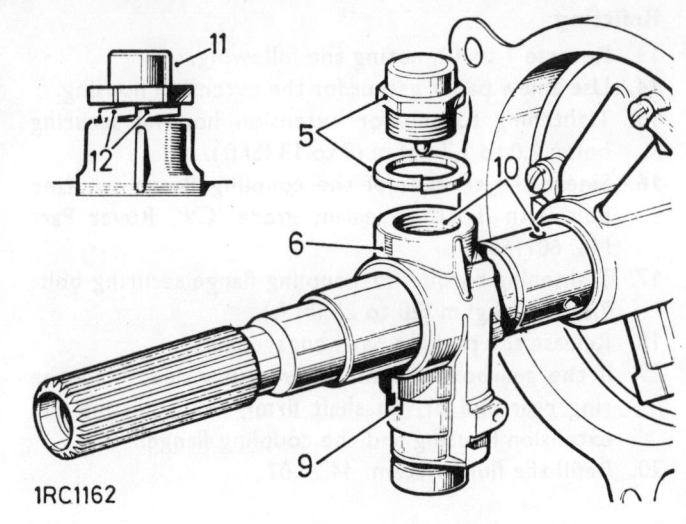

GOVERNOR ASSEMBLY

—Overhaul   44.22.04

Service tool: 601286   Screwdriver set, spintorq

NOTE: The following procedure applies to the early design governor only, the later type is not serviceable, but can be fitted in place of the earlier type.

Dismantling

1. Remove the governor assembly. 44.22.01.
2. Remove the two screws and spring washers retaining the body to the governor sleeve.
3. Depress and withdraw the retainer and remove the governor weight, valve and spring.
4. If necessary, remove the governor cover plate, retained by two countersunk-head lock screws.

Inspecting

5. Clean and examine all components. Use nylon rag and transmission fluid for cleaning.

Reassembling

6. Reverse 1 to 3, noting the following.
7. Align the fluid holes between the body and the sleeve.
8. Tighten the valve body screws using a large screwdriver.
9. Tighten the cover plate screws to 0,23 to 0,35 kgf.m (20 to 30 lbf in).
10. Check that the valve and governor weight can move freely.
11. Refit the governor assembly. 44.22.01.

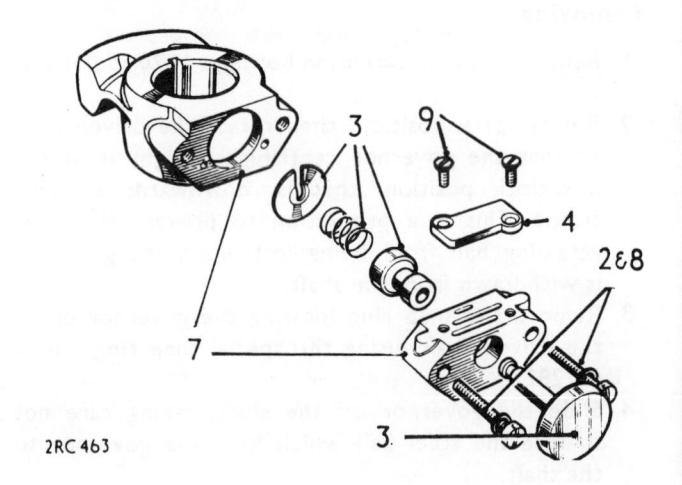

44.20.01
Sheet 2
44.22.04

Rover 3500 and 3500S Manual AKM 3621

GEARBOX—AUTOMATIC—Borg-Warner type 35

FLUID SYSTEM

—Drain and refill 44.24.02

Draining

WARNING: If the vehicle has just returned from a run the fluid will be hot enough to cause scalding.

1. Place a suitable receptacle beneath the gearbox fluid pan.
2. Remove the fluid drain plug and allow all the fluid to drain.
3. Refit the drain plug and washer.

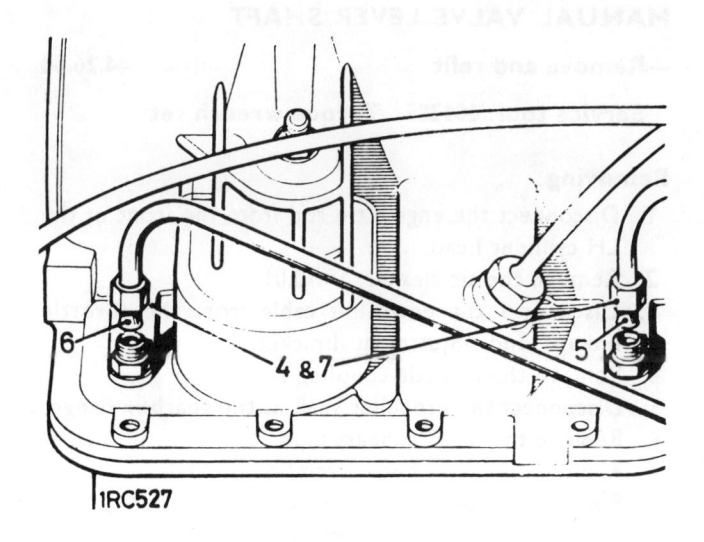

Flushing fluid cooler

Whenever a new or overhauled automatic gearbox is fitted, the fluid cooler should be flushed, otherwise any debris or foreign matter from the original gearbox deposited in the fluid, will contaminate the new or overhauled gearbox and may result in a failure.

Procedure

4. Disconnect both pipes from the gearbox to the fluid cooler, at the gearbox end.
5. Position a water-free air line into the input pipe of the fluid cooler and blow all the fluid out through the return pipe.
6. Using a syringe, pump fresh automatic gearbox fluid through the fluid cooler until at least 0,6 litre (1 pint) has been pumped out of the return pipe.
7. Reconnect the fluid cooler pipes to the automatic gearbox.

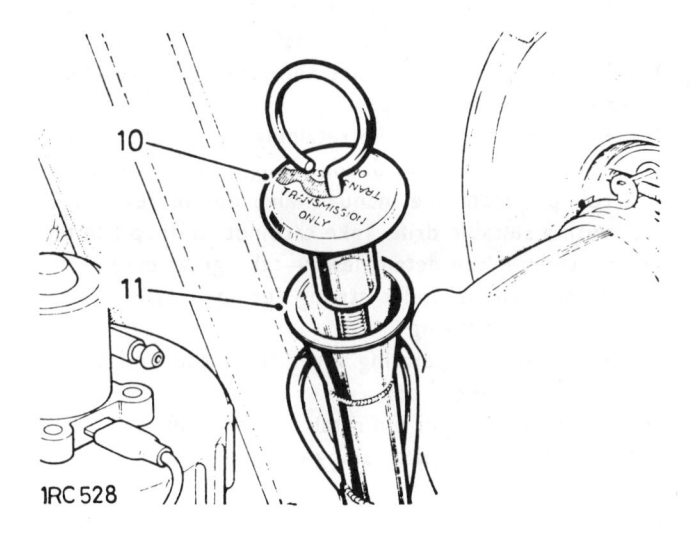

Refilling

8. Stand the car on level ground.
9. Prop open the bonnet.
10. Withdraw the fluid level dipstick.
11. Using the correct grade fluid, see Division 09, fill the automatic gearbox fluid system via the dipstick filler tube to within 6 mm (0.250 in.) of the low mark on the dipstick. The final level must be set with the gearbox at normal operating temperature.
12. Run the car for a minimum of 4 km (3 miles) to ensure that the automatic gearbox is at normal operating temperature.
13. Stand the car on level ground, engine idling and selector at 'P' (park) position.
14. Prop open the bonnet.
15. Remove the dipstick, wipe dry and check the fluid level. Take the dipstick reading immediately after the dipstick has been fully inserted to avoid misreadings by splashing.

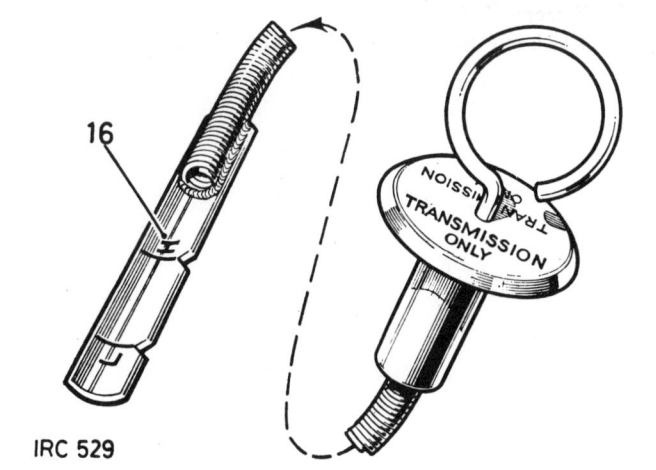

16. If necessary, top up to the 'H' mark. Do not overfill.

NOTE: The latest type dipstick is illustrated, on earlier models top up to the 'H' mark on the side marked 'HOT'.

17. Close the bonnet.

Rover 3500 and 3500S Manual AKM 3621

GEARBOX—AUTOMATIC—Borg-Warner type 35

MANUAL VALVE LEVER SHAFT

—Remove and refit 44.26.04

Service tool: 601285 Torque wrench set

Removing

1. Disconnect the engine tie rod from the front of the LH cylinder head.
2. Remove the air cleaner. 19.10.01.
3. Disconnect the downshift cable from the throttle linkage and adjustment bracket.
4. Remove the throttle coupling rod.
5. Disconnect the propeller shaft at the gearbox flange.
6. Remove the front exhaust pipe.
7. Remove the valve body assembly. 44.40.01.
8. Remove the starter and reverse light switch. 44.15.15.
9. Support the engine then release the rear mounting.
10. Lower the engine rear end sufficient to give access to the manual valve lever shaft.
11. Remove the clip retaining the parking brake link at the torsion lever end.
12. Remove the spring clip retaining the detent lever.
13. Push the lever towards the casing, then drive out the roll pin from the manual valve detent lever shaft with a suitable drift. Take care not to drop the pin.
14. Withdraw the detent lever—take great care as the detent ball is under strong spring pressure—then remove the spring.
15. Remove the operating lever from the end of the shaft outside the box.
16. Remove the taper pin retaining the shaft.
17. Withdraw the shaft, taking out the oil seal.

Refitting

18. Fit a new oil seal in the gearbox.
19. Reverse 1 to 17.
20. Refill the fluid system. 44.24.02.

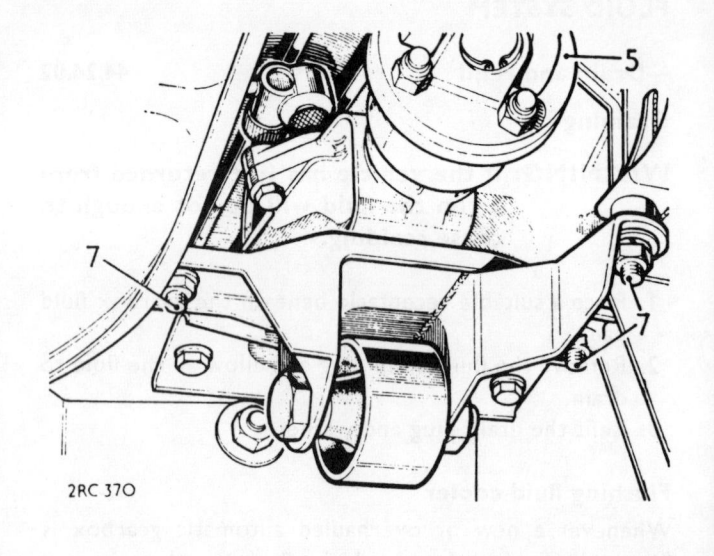

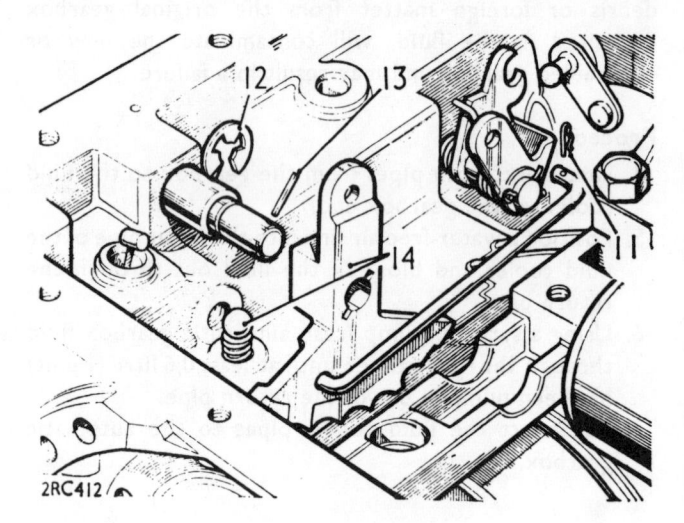

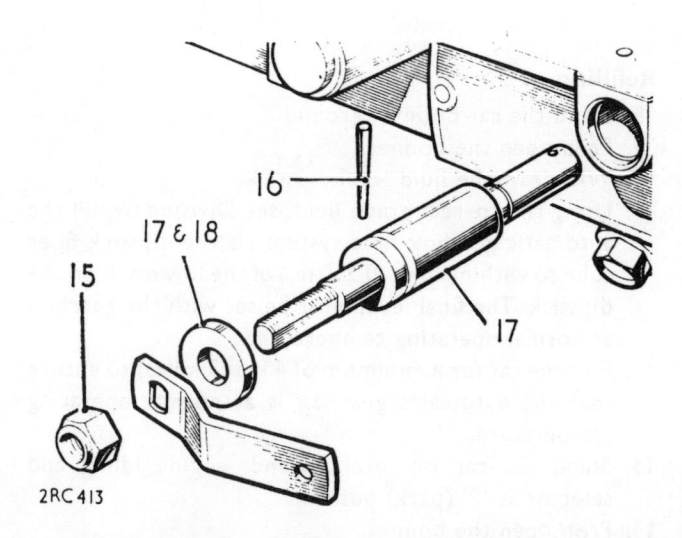

GEARBOX—AUTOMATIC—Borg-Warner type 35

PARKING BRAKE PAWL ASSEMBLY

—Remove and refit 1 to 5, 11 and 12   44.28.07
—Overhaul 1 to 13   44.28.08

Service tool: 601285   Torque wrench set

Removing

1. Remove the valve body assembly. 44.40.01.
2. Remove the rear extension housing. 44.20.15.
3. Tap out the roll pin and extract the toggle pin, using a magnet.
4. Ease out the anchor pin.
5. Withdraw the parking brake pawl assembly.
6. Remove the spring clip from the toggle lever pin.
7. Withdraw the spring retainer and disconnect the release spring.
8. Separate the components by withdrawing the toggle lever pin and toggle link pin; note plain washer on link pin.
9. Remove the rear servo and slide the brake band fully rearward.
10. Remove the spring clip retaining the link rod, also the spring clip retaining the torsion lever assembly to the pin in the casing and withdraw the assembly.

Refitting

11. Fit a new 'O' ring seal to the toggle pin, if necessary.
12. Reverse 1 to 10, ensuring that the toggle pin and toggle lift lever are aligned.
13. The torsion lever pin is a press fit in the case and should protrude 3.2 mm (0.125 in) approximately.

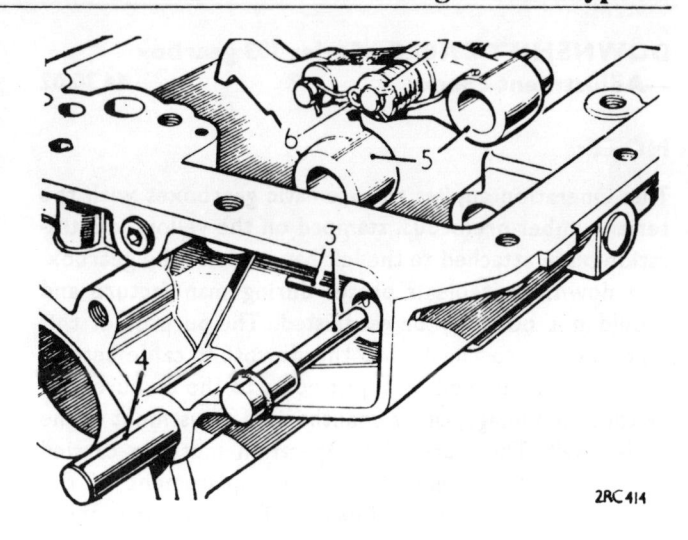

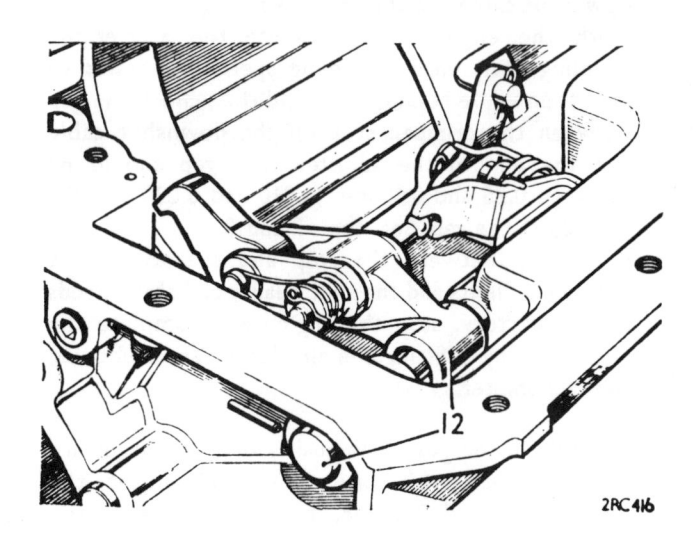

Rover 3500 and 3500S Manual AKM 3621

GEARBOX—AUTOMATIC—Borg-Warner type 35

DOWNSHIFT CABLE—Series 303 gearbox
—Adjustment check   44.30.02

NOTE:

This Operation applies to automatic gearboxes with the serial number prefix 303, stamped on the yellow identification plate attached to the left hand side of the gearbox. The downshift cable is preset during manufacture and should not normally be readjusted. The purpose of this Operation is to check that the downshift cable setting has not been affected by adjustments to the carburetters or throttle linkage, or by unauthorised adjustment of the cable itself. Therefore, this Operation must be carried out following any adjustments or replacement of the carburetter or throttle linkage. The pressure check Operation 44.30.03, should not normally be required and should only be carried out as indicated in the following instructions.

Checking

1. Start and run the engine until it attains normal operating temperature, that is, five minutes after the thermostat opens.
2. Check the distributor dwell angle and ignition timing. 86.35.20.
3. Using an independent and accurate tachometer, check the engine idle speed, this must be 600 to 650 rev/min (700 to 750 rev/min for emission controlled vehicles). If the idle speed is not correct, tune and adjust the carburetters. 19.15.02.
4. Release the fluid filler tube from the clip at the rear of the air cleaner, move the choke cable aside and pivot the air cleaner forward to give access to the downshift cable and throttle linkage.
5. With the engine idling, actuate the accelerator coupling shaft until the idling speed just starts to rise, then while holding this condition, check the gap between the crimped stop on the downshift cable and the end of the adjuster. The gap should be between 0,25 and 0,50 mm (0.010 and 0.020 in.). Do not adjust the gap. See items 7 and 8.
6. If the gap between the stop and the adjuster is within the limits and the shift quality and shift speeds are satisfactory, stop the engine and complete the Operation by refitting the air cleaner and removing the tachometer.

    NOTE: If the gear shift quality or shift speeds are not satisfactory, DO NOT adjust the downshift cable, but proceed with the pressure check Operation 44.30.03, applicable to the Series 303 gearbox.

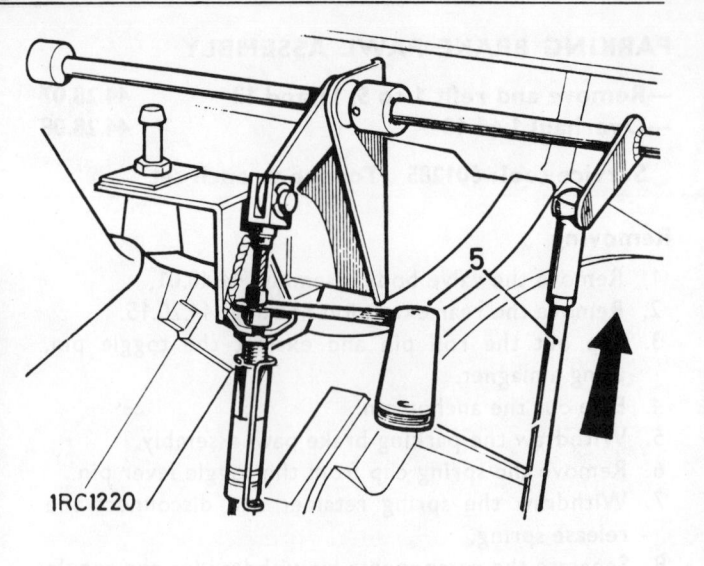

7. If the gap is not within the limits, check all the throttle linkage and if necessary reset the rods and linkage to the correct dimensions, 19.20.07.
8. If all the throttle linkage settings are correct and the downshift cable gap is still not within the limits, reset the gap to between 0,25 to 0,50 mm (0.010 to 0.020 in.) using the screwed adjuster, then proceed with the pressure check Operation 44.30.03, applicable to the 303 series gearbox.

Rover 3500 and 3500S Manual AKM 3621

GEARBOX—AUTOMATIC—Borg-Warner type 35

DOWNSHIFT CABLE—Series 3FU, 7FU, 9FU and 267 gearboxes

—Pressure check   44.30.03

Service tools. 601284. Pressure gauge and tachometer

NOTE:

This Operation applies to automatic gearboxes with the serial number prefix 3FU, 7FU, 9FU or 267, stamped on the identification plate attached to the left-hand side of the gearbox. For Series 303 gearboxes, see the accompanying Operation of the same number, also Operation 44.30.02.

Procedure

1. Run the car for a minimum of 4 km (3 miles) to ensure that the engine and gearbox are at normal running temperature.
2. Chock the front wheels of the car and apply the handbrake firmly. Apply the footbrake during the test.
3. Remove the line pressure take-off plug from the rear of the gearbox case.
4. Connect the pressure gauge to the pressure take-off point. 601284.
5. Connect the tachometer 'CB' lead to the negative terminal on the ignition coil, and the earth lead to an earthing point. 601284.

   NOTE: If the tachometer in use is for 4 and 6 cylinder models, use the 4 cylinder position and note that the reading on the tachometer will be double the actual engine rev/min, that is 1000 will show as 2000 rev/min.

6. Check that the downshift cable is correctly clipped to the bell housing and that the cable adjuster is fully abutted on the outer cable.
7. Check that the fluid level in the automatic gearbox is correct on the dipstick with the engine idling in 'P'.

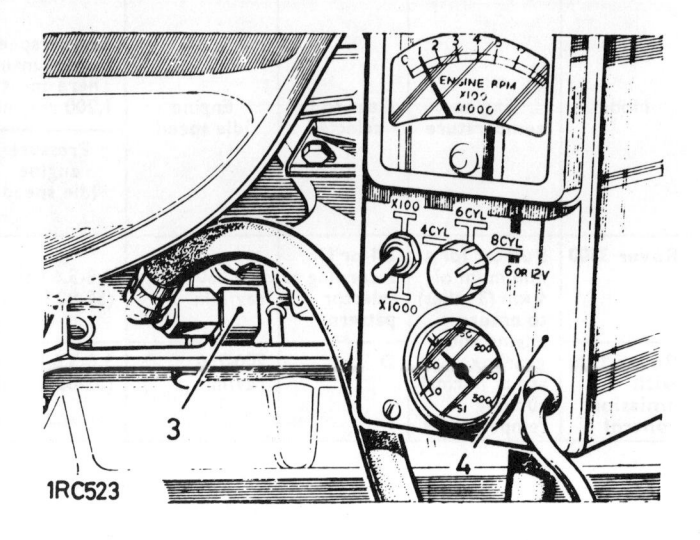

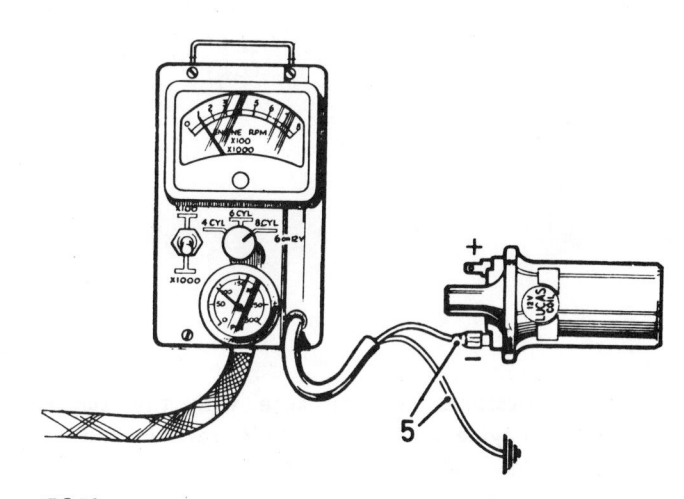

continued

Rover 3500 and 3500S Manual AKM 3621

GEARBOX—AUTOMATIC—Borg-Warner type 35

Pressure check speeds:
Engine speed must be increased by the use of the accelerator pedal, not by manual manipulation of the carburetter linkage.
There must be a progressive rise in pressure between engine idle speed and 1,200 rev/min corresponding to accelerator pedal movement.

Model	Engine temperature	Gear to be selected	Engine idle speed	Pressure engine idle speed	Increase engine speed to:	Pressure at 800 rev/min	Further increase engine speed to:	Pressure at 1200 rev/min
Rover 3500	Run car for a minimum of 4 km (3 miles) to ensure engine and gearbox are at normal running temperatures	D1 or D according to selector pattern	600-650 rev/min	3,8-5,6 kgf/cm² (55-80 lbf/in²)	800 rev/min	4,9-7,7 kgf/cm² (70-110 lbf/in²)	1,200 rev/min	7,0 kgf/cm² (100 lbf/in²) minimum
Rover 3500 with emission control		D	700-750 rev/min	3,8-5,6 kgf/cm² (55-80 lbf/in²)	800 rev/min	4,9-7,7 kgf/cm² (70-110 lbf/in²)	1,200 rev/min	7,0 kgf/cm² (100 lbf/in²) minimum

8. Check pressures in accordance with the above chart at engine idle speed; at 800 rev/min and at 1,200 rev/min. **There must be a progressive rise in pressure between engine idle speed and 1,200 rev/min corresponding to accelerator pedal movement.**
9. If the pressure is less than that indicated for the model and rev/min concerned then it should be increased by screwing out the cable adjuster. If the pressure is more than that indicated for the model and rev/min concerned then it should be decreased by screwing in the cable adjuster.
10. If the correct pressures cannot be achieved, check:
    (a) That the downshift cable is not sticking and that the inner cable is not frayed.
    (b) That the fixing screws on the valve body are tight.
11. When the pressure settings are correct, stop the engine.
12. Remove the pressure gauge and tachometer.
13. Refit the line pressure take-off plug.

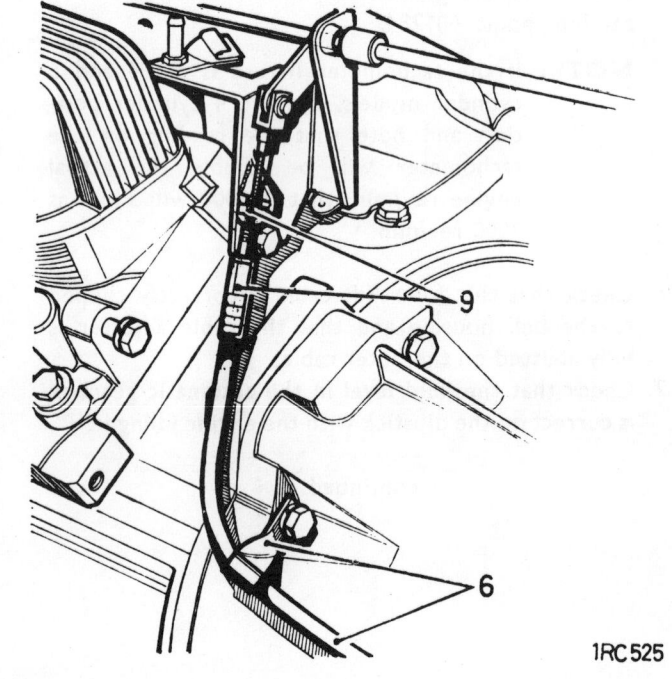

GEARBOX—AUTOMATIC—Borg-Warner type 35

DOWNSHIFT CABLE—Series 303 gearbox
—Pressure check   **44.30.03**

   Service tool—**601284 Pressure gauge and tachometer**

NOTE:

This Operation applies to automatic gearboxes with the serial number prefix 303, stamped on the yellow identification plate attached to the left-hand side of the gearbox. The gearbox operating pressure is preset during manufacture and is prevented from falling below the minimum requirement by a crimped stop on the downshift cable. Therefore, the downshift cable setting, which also represents the gearbox operating pressure setting, must be maintained in the original preset position, as described in Operation 44.30.02. The purpose of this Operation is to check the gearbox operating pressure relative to the correct downshift cable setting in order to diagnose a maladjusted downshift cable or internal gearbox fault, which would be suspected if the results of the downshift cable adjustment check, Operation 44.30.02, were not satisfactory.

Checking

1. Start and run the engine until it attains normal operating temperature, that is, five minutes after the thermostat opens.
2. Check the distributor dwell angle and ignition timing. 86.35.20.
3. Using an independent and accurate tachometer, check the engine idle speed; this must be 600 to 650 rev/min (700 to 750 rev/min for emission controlled vehicles). If the idle speed is not correct, tune and adjust the carburetters. 19.15.02.
4. Chock the front wheels of the car and apply the handbrake firmly. Apply the footbrake during the test.
5. Remove the line pressure take-off plug from the rear of the gearbox case.
6. Connect the pressure gauge to the pressure take-off point. 601284.
7. Connect the tachometer 'CB' lead to the negative terminal on the ignition coil, and the earth lead to an earthing point. 601284.

   NOTE: If the tachometer in use is for 4 and 6 cylinder models, use the 4 cylinder position and note that the reading on the tachometer will be double the actual engine rev/min, that is 1000 will show as 2000 rev/min.

8. Check that the fluid level in the automatic gearbox is correct on the dipstick with the engine idling in 'P'.
9. Move the gear selector to 'D' and check the gearbox operating pressures at the following engine speeds:
   (a) At engine idle speed the pressure must be between 3,86 to 5,62 kgf/cm² (55 to 80 lbf/in²).
   (b) At 1,200 rev/min the pressure must be 5,62 kgf/cm² (80 lbf/in²) minimum.
   (c) There must be a rise in pressure of at least 1,05 kgf/cm² (15 lbf/in²) between the check at engine idle speed and the check at 1200 rev/min.
10. If the downshift cable has been readjusted in accordance with instruction 8 in Operation 44.30.02, and the operating pressures are correct, it must be accepted that the downshift cable was previously maladjusted and is now correct.
11. If the downshift cable setting is as specified in Operation 44.30.02 and the operating pressures are not correct, a gearbox internal fault must be suspected. Do not adjust the downshift cable in an attempt to correct the pressure.

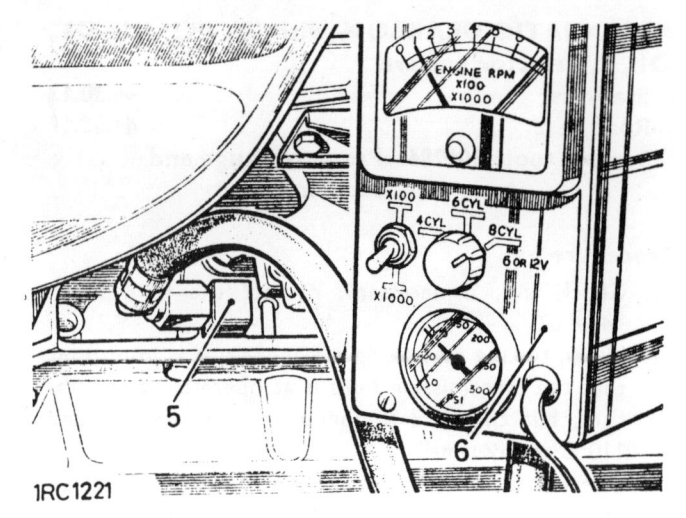

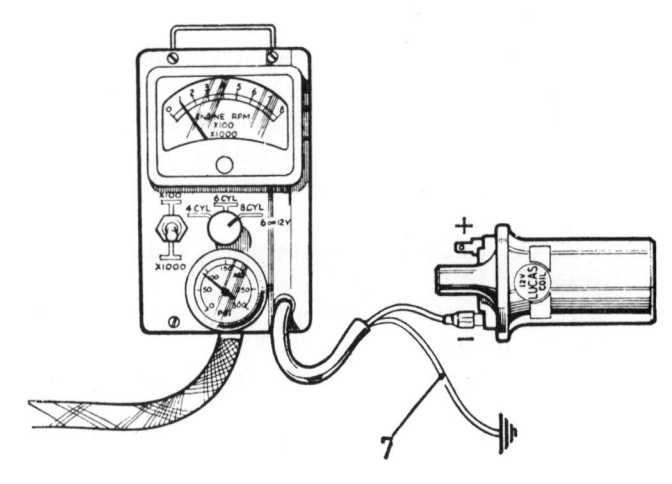

   NOTE: Fitting an incorrect downshift cable to the gearbox would give the above symptoms. If a new cable has been fitted, check that it is the correct type with number '148' stamped on the hexagon nut where the cable enters the gearbox.

12. Stop the engine.
13. Remove the pressure gauge and tachometer.
14. Refit the line pressure take-off plug. Torque 0,55 to 0,7 kgf.m (4 to 5 lbf ft).

GEARBOX—AUTOMATIC—Borg-Warner type 35

AUTOMATIC GEARBOX with 'P', 'R', 'N', 'D2', 'D1', 'L' selector pattern
—Stall test   44.30.13
—Road test   44.30.17
Service tool: 601284   Pressure gauge and tachometer

Procedure

1. Check and, if necessary, correct the fluid level. 44.24.02.
2. When testing or making a diagnosis, it is important that the transmission fluid is at operating temperature, but the fluid temperature must **not** exceed 110°C at any time.
3. Check that starter will operate only with the selector in 'P' and 'N' positions—but **not** in 'D1', 'D2', or 'L' or 'R'.
4. Check that reverse light comes on when 'R' is selected.
5. Check the freedom of the transmission with the selector lever at 'N'. There should be no tendency for the engine to drive the car, nor should there be any engine braking effect.
6. Apply the hand and foot brakes firmly and chock the wheels.
   With the engine at normal idling speed, select 'N–D1', 'N–D2', 'N–L' and 'N–R'.
   Gearbox engagement should be felt in each position selected.
   Check for slip or clutch squawk.

Stall speeds

'Stall speed' is the maximum speed at which the engine can drive the torque converter impeller whilst the turbine is held stationary. It is dependent on both engine and torque converter characteristics and will vary according to the condition of the engine.

An engine is poor condition will give a lower stall speed.

7. Due to the high output of the 3500 engine, it is necessary to position the car with the front bumper against a wall, not a work bench, with the bumper and overriders suitable protected.
8. Also firmly apply hand and foot brakes and chock the rear wheels.
9. Connect tachometer, 601284 to the engine, start engine, select 'L' and allow to idle for approximately one minute to ensure circulation of fluid.
10. Depress accelerator to full throttle position and note the revolution counter reading.
    DO NOT STALL FOR LONGER THAN TEN SECONDS OR TRANSMISSION WILL OVERHEAT.
11. The stall speed of an engine in good condition should be 1,950 to 2,250 rev/min.
12. If the stall speed is below 1,950 rev/min, carry out compression test on engine to determine engine condition.

continued

44.30.13
44.30.17
Sheet 1

Rover 3500 and 3500S Manual AKM 3621

GEARBOX—AUTOMATIC—Borg-Warner type 35

13. If after engine rectification stall speed is still below 1,950 rev/min renew torque converter.
14. Stall speeds above 2,250 rev/min indicate band or clutch slip.

Road test

15. Check and, if necessary, correct the fluid level. 44.24.02.
16. All normal forward running is done in 'D2' or 'D1' selector position. The following charts of change speeds apply with the selector linkage correctly adjusted and selector at the 'D1' position.
    The speeds in the following charts are 'indicated'. Maximum permissible tolerances in both speedometer and transmission have been taken into account, together with tyre make and expansion due to speed.

Road test procedure and change-speed charts
'P', 'R', 'N', 'D2', 'D1', 'L' selector pattern
Tests 1, 2 and 3—Upchange speed checks

The first three tests which must be carried out from a standing start are the light throttle (LT), full throttle (FT) and kick-down (KD) or maximum throttle. Note that the 'full throttle' test is the least important of the three, since if the 'light throttle' and the 'kick-down' speeds are correct, the 'full throttle' will usually fall within the tolerances.

The arrow heads in the chart below indicate the following:
→ acceleration, ← deceleration, ↑ upchange point, ↓ downchange point.

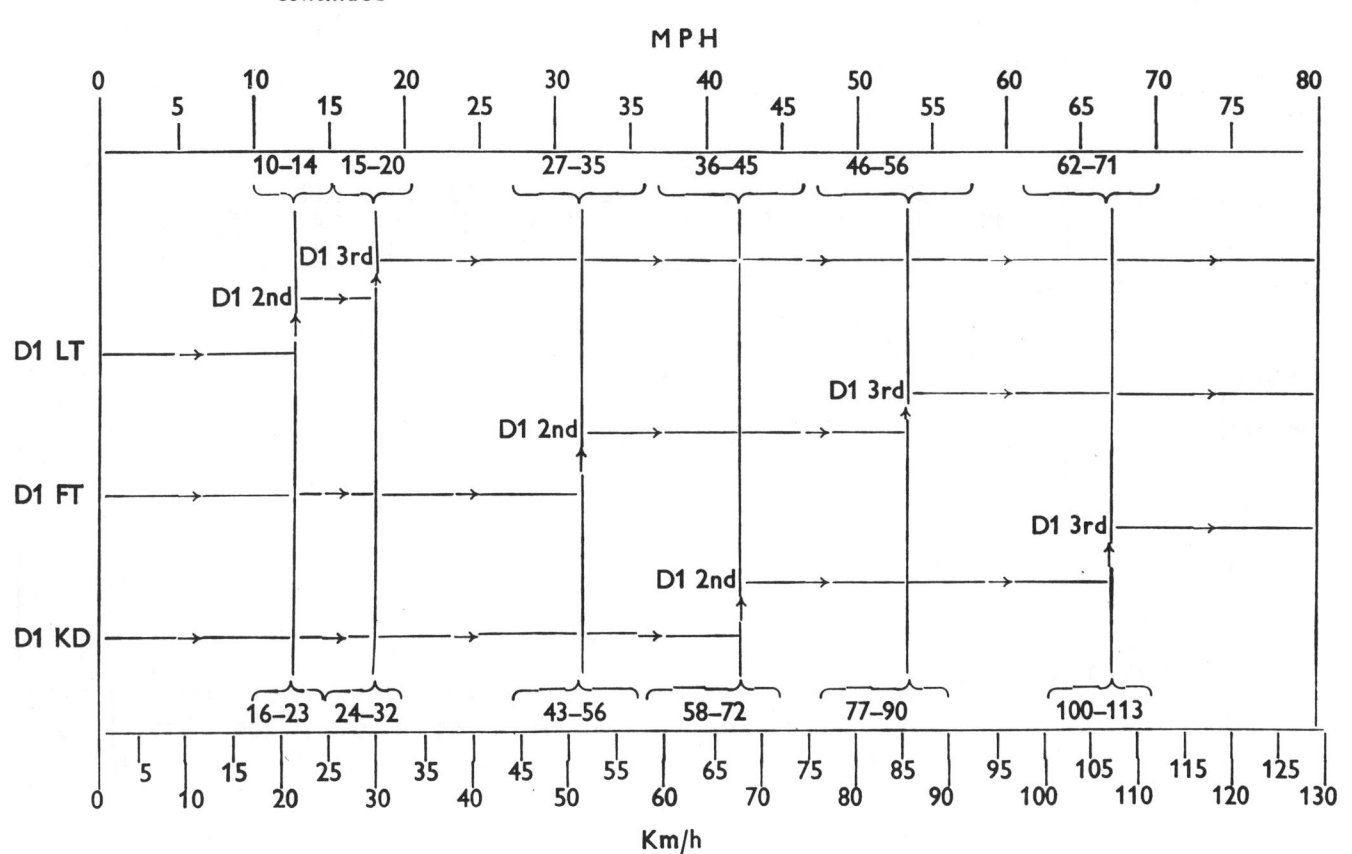

continued

Rover 3500 and 3500S Manual AKM 3621.

44.30.13
44.30.17
Sheet 2

GEARBOX—AUTOMATIC—Borg-Warner type 35

Note: If the 1–2 or 2–3 upchanges have not occurred by the time the specified maximums are reached, do not continue to accelerate otherwise serious damage to the engine may result.

During the 'full throttle' and 'kick-down' upchange tests, a check should be made for clutch squawk or slip when moving from rest and as the upchanges occur. A momentary slip between the 2–3 upchange followed by a thump in the transmission may be due to the servo orifice valve sticking. This condition is called 'run-up'. If it is suspected that 'run-up' is present then the transmission will also be suffering from a condition called 'tie up', which has the effect of slowing down the engine just before the 2–3 change down occurs.

Test 4—'D1' roll out change down check followed by reverse check

At 65 km/h (40 mph), 'D1' selected and 3rd engaged, release throttle and allow car to slow at 8 km/h (5 mph) or below. The transmission should change from 3rd to 1st. To check that a change down has occurred, stop and select reverse and ensure that car does 'reverse'. If car fails to reverse, this indicates 1–2 shift valve sticking.

Test 5—'L' roll out change down check

At 65 km/h (40 mph), 'D1' selected and 3rd engaged, release throttle and select 'L'. Gearbox will immediately change down into 'L2' giving engine braking. Allow car to slow. At between 11 km/h and 21 km/h (7 mph and 13 mph) the gearbox should change down to 'L1', giving increased engine braking. If gearbox does not change down, repeat reverse test above.

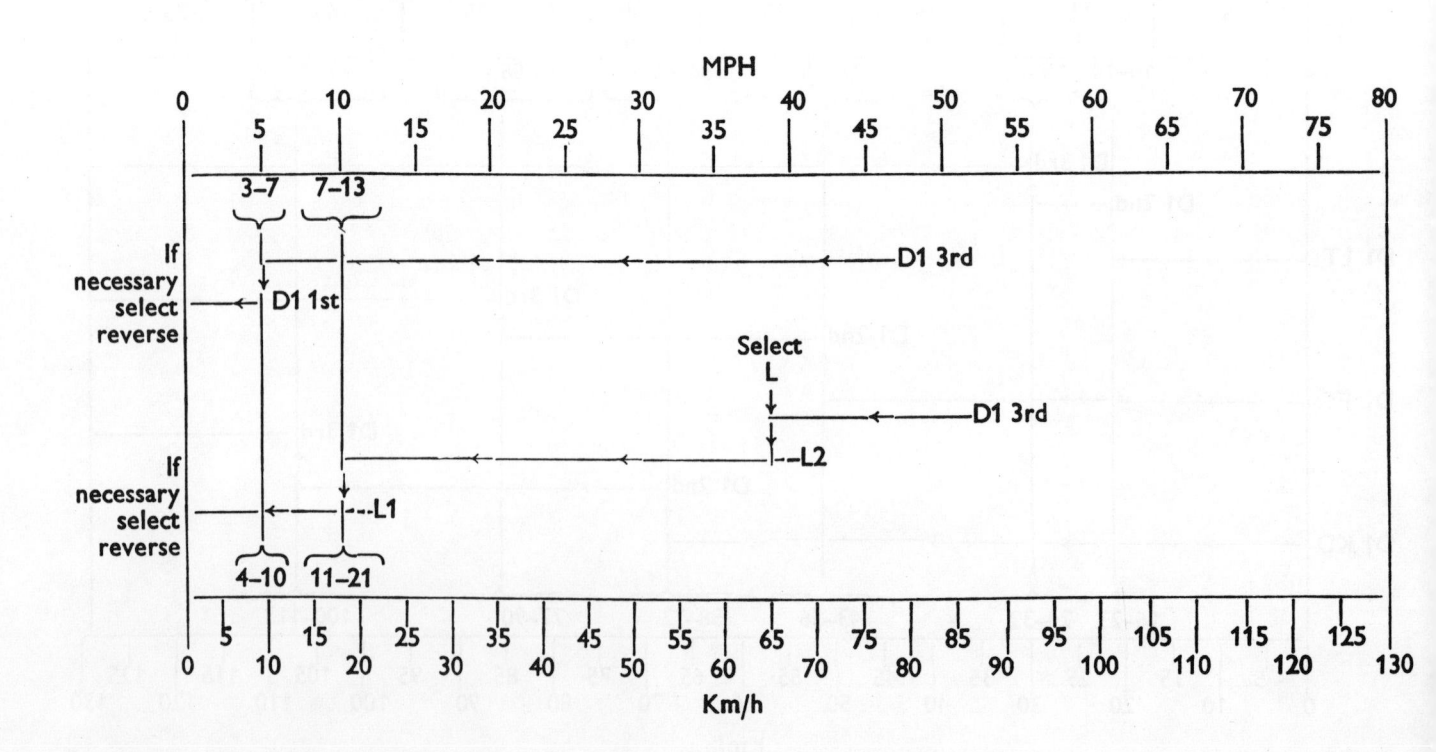

continued

44.30.13
44.30.17
Sheet 3

Rover 3500 and 3500S Manual AKM 3621

GEARBOX—AUTOMATIC—Borg-Warner type 35

Test 6—Kick-down check, 3-2 change down

At 83 km/h (50 mph), 'D1' selected, 3rd engaged, decelerate to between 64-72 km/h (40-45 mph), depress throttle to maximum, i.e. 'KD'. Gearbox should immediately change down into 2nd gear. If the throttle is now released, the gearbox should revert to 3rd gear (shown in dotted on chart).

Test 7—Kick-down check, 3-1 change down

At 48 km/h (30 mph), 'D1' selected, 3rd engaged, decelerate to between 32-40 km/h (20-25 mph), depress throttle to maximum, i.e. 'KD'. Gearbox should immediately change down into 1st gear. If the throttle is now released, the gearbox should revert to 3rd gear (shown in dotted on chart).

CAUTION: THIS GEARBOX DOES **NOT** INCORPORATE ANY SAFETY INTERLOCKS ON 'PARK' OR 'REVERSE'. 'P' or 'R' must never be selected unless the vehicle is stationary.

17. Stop and select 'P'. Release the brakes when facing up or down hill, the car should remain stationary. On steep slopes some force may be needed to disengage the 'Park' position.

18. Stop and select 'R'. Release brakes and reverse, using full throttle if possible. Check for slip or clutch squawk.

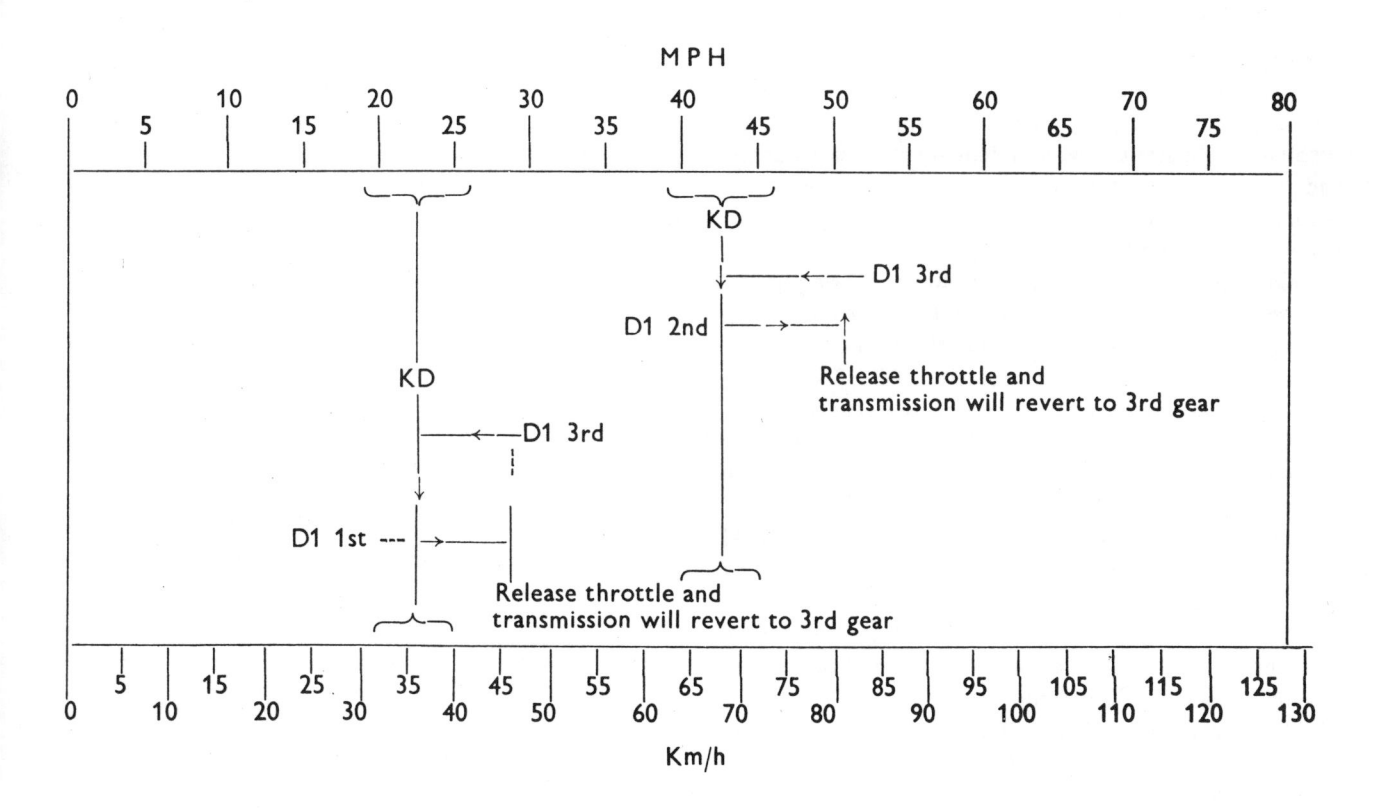

Rover 3500 and 3500S Manual AKM 3621

44.30.13
44.30.17
Sheet 4

GEARBOX—AUTOMATIC—Borg-Warner type 35

AUTOMATIC GEARBOX with 'P', 'R', 'N', 'D', '2', '1' selector pattern

—Stall test   44.30.13
—Road test   44.30.17

Service tool: 601284 Pressure gauge and tachometer

Procedure

1. Check and, if necessary, correct the fluid level. 44.24.02.
2. When testing or making a diagnosis, it is important that the transmission fluid is at operating temperature but the fluid temperature must **not** exceed 110°C at any time.
3. Check that starter will operate only with the selector in 'P' and 'N' positions—but **not** in 'D', '2', '1' or 'R'.
4. Check that reverse light comes on when 'R' is selected.
5. Check the freedom of the transmission with the selector lever at 'N'. There should be no tendency for the engine to drive the car, nor should there be any engine braking effect.
6. Apply the hand and foot brakes firmly and chock the wheels.
   With the engine at normal idling speed, select 'N–D', 'N–2', 'N–1' and 'N–R'.
   Gearbox engagement should be felt in each position selected.
   Check for slip or clutch squawk.

Stall speeds

'Stall speed' is the maximum speed at which the engine can drive the torque converter impeller whilst the turbine is held stationary. It is dependent on both engine and torque converter characteristics and will vary according to the condition of the engine.

An engine in poor condition will give a lower stall speed.

7. Due to the high output of the Rover V8 engine it is necessary to position the car with the front bumper against a wall, not a work bench, with the bumper and overriders suitably protected. Also firmly apply hand and foot brakes and chock the rear wheels.
8. Connect tachometer, 601284, to the engine, start engine, select 'D' and allow to idle for approximately one minute to ensure circulation of fluid.
9. Depress accelerator to full throttle position, **NOT** kick-down, and note the revolution counter reading. DO NOT STALL FOR LONGER THAN TEN SECONDS OR TRANSMISSION WILL OVERHEAT.
10. The stall speed of an engine in good condition should be 1,950 to 2,250 rev/min.
11. If the stall speed is below 1,950 rev/min, carry out compression test on engine to determine engine condition.
12. If after engine rectification stall speed is still below 1,950 rev/min renew torque converter.

continued

44.30.13
44.30.17
Sheet 5

Rover 3500 and 3500S Manual AKM 3621

GEARBOX—AUTOMATIC—Borg-Warner type 35

13. Stall speeds above 2,250 rev/min indicate band or clutch slip.

Road test

14. Check and, if necessary, correct the fluid level. 44.24.02.
15. All normal forward running is done in 'D' selector position. The following charts of change speeds apply with the selector linkage correctly adjusted and selector at the 'D' position.

   The speeds in the following charts are 'indicated'. Maximum permissible tolerances in both speedometer and transmission have been taken into account, together with tyre make and expansion due to speed.

**Road test procedure and change-speed charts—
'P', 'R', 'N', 'D', '2', '1' selector pattern
Tests 1, 2 and 3—Upchange speed checks**

The first three tests which must be carried out from a standing start are the light throttle (LT), full throttle (FT) and kick-down (KD) or maximum throttle. Note that the 'full throttle' test is the least important of the three, since if the 'light throttle' and the 'kick-down' speeds are correct, the 'full throttle' will usually fall within the tolerances.

The arrow heads in the chart below indicate the following:
→ acceleration, ← deceleration, ↑ upchange point,
↑ downchange point.

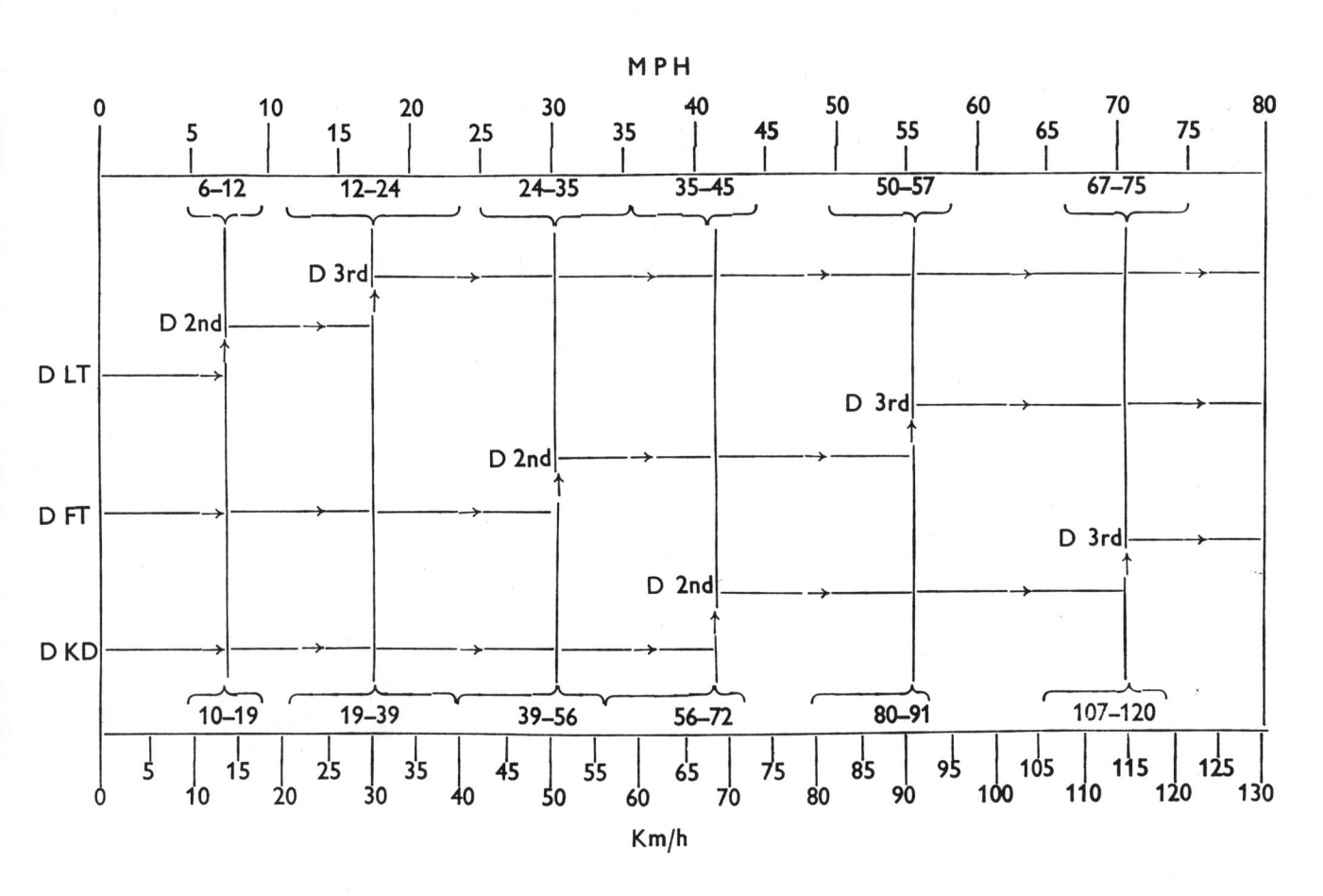

continued

Rover 3500 and 3500S Manual AKM 3621

44.30.13
44.30.17
Sheet 6

GEARBOX—AUTOMATIC—Borg-Warner type 35

Note: If the 1–2 or 2–3 upchanges have not occurred by the time the specified maximums are reached, do not continue to accelerate otherwise serious damage to the engine may result.

During the 'full throttle' and 'kick-down' upchange tests, a check should be made for clutch squawk or slip when moving from rest and as the upchanges occur. A momentary slip between the 2–3 upchange followed by a thump in the transmission may be due to the servo orifice valve sticking. This condition is called 'run-up'. If it is suspected that 'run-up' is present then the transmission will also be suffering from a condition called 'tie-up', which has the effect of slowing down the engine just before the 2–3 change down occurs.

Test 4—'D' roll out change down check followed by reverse check

At 65 km/h (40 mph), 'D' selected and 3rd engaged, release throttle and allow car to slow to 8 km/h (5 mph) or below. The transmission should change from 3rd to 1st. To check that a change down has occurred, stop and select reverse and ensure that car does 'reverse'. If car fails to reverse, this indicates 1–2 shift valve sticking.

Test 5—'1' roll out change down check

At 48 km/h (30 mph), 'D' selected and 3rd engaged, release throttle and select '1'. Gearbox will immediately change down into 2nd, giving engine braking. Allow car to slow. At about 35 km/h (22 mph) the gearbox should change down to 1st, giving increased engine braking. If gearbox does not change down, repeat reverse test as before.

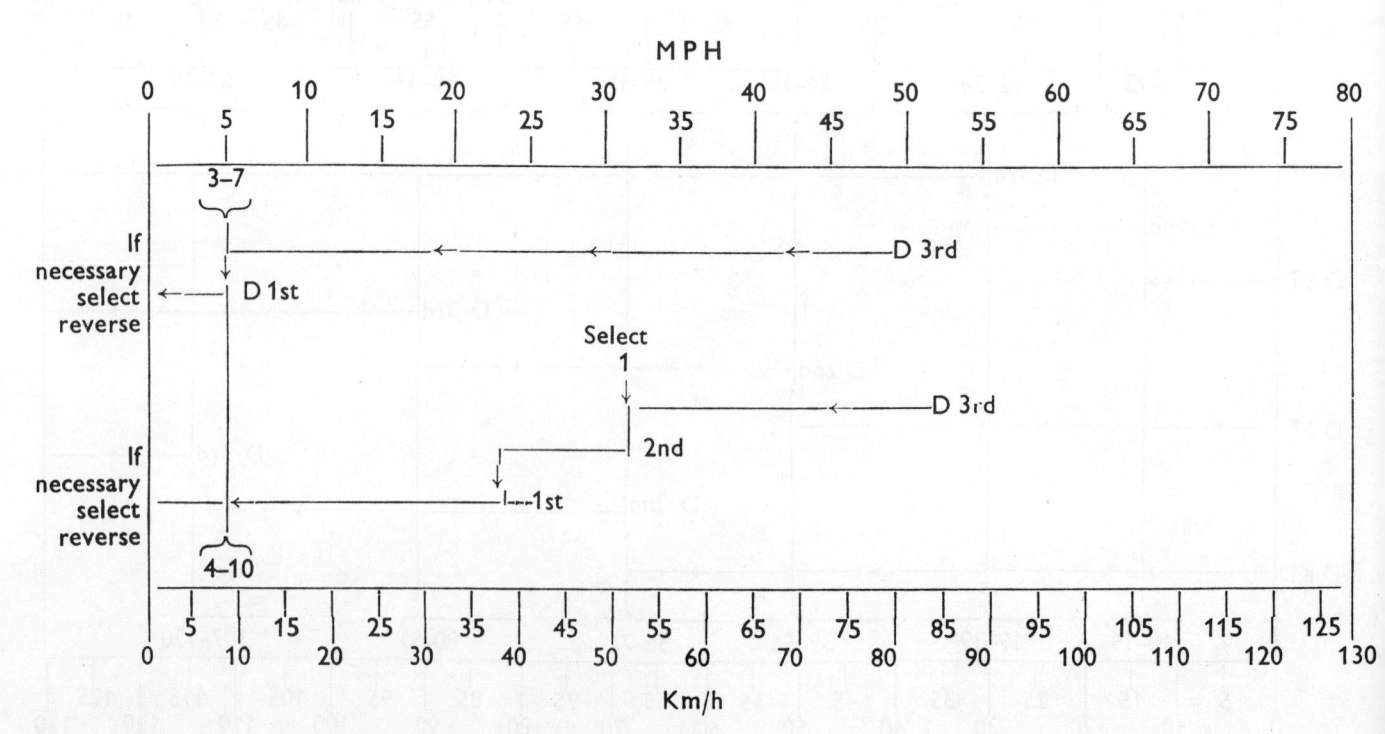

continued

44.30.13
44.30.17
Sheet 7

Rover 3500 and 3500S Manual AKM 3621

GEARBOX—AUTOMATIC—Borg-Warner type 35

Test 6—Kick-down check, 3-2 change down

At 83 km/h (50 mph), 'D' selected, 3rd engaged, decelerate to between 64–72 km/h (40–45 mph), depress throttle to maximum, i.e. 'KD'. Gearbox should immediately change down into 2nd gear. If the throttle is now released, the gearbox should revert to 3rd gear (shown in dotted on chart).

Test 7—Kick-down check, 3-1 change down

At 48 km/h (30 mph), 'D' selected, 3rd engaged, decelerate to between 32–40 km/h (20–25 mph), depress throttle to maximum, i.e. 'KD'. Gearbox should immediately change down into 1st gear. If the throttle is now released, the gearbox should revert to 3rd gear (shown in dotted on chart).

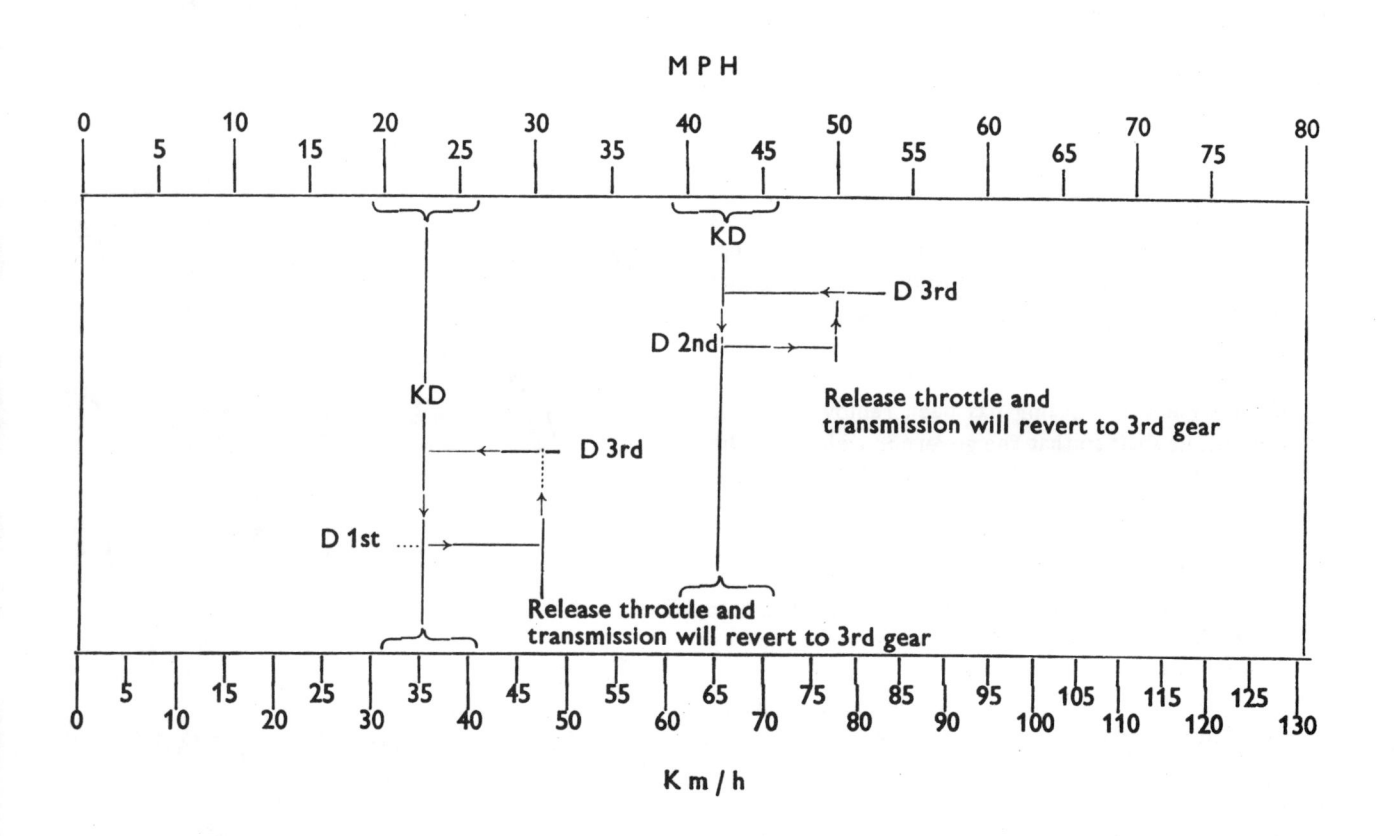

CAUTION: THIS TRANSMISSION DOES **NOT** INCORPORATE ANY SAFETY INTERLOCKS ON 'PARK' OR 'REVERSE'. 'P' or 'R' must never be selected unless the vehicle is stationary.

16. Stop and select 'P'. Release the brakes when facing up or down hill, the car should remain stationary. On steep slopes some force may be needed to disengage the 'Park' position.
17. Stop and select 'R'. Release brakes and reverse, using full throttle if possible. Check for slip or clutch squawk.

Rover 3500 and 3500S Manual AKM 3621

44.30.13
44.30.17
Sheet 8

GEARBOX—AUTOMATIC—Borg-Warner type 35

MECHANICAL OPERATION

—Air pressure checks   44.30.16

Service tools: 601285 Torque wrench set

Air pressure checks to determine whether the clutch and brake bands are operating can be made with the gearbox in the car, or on the bench.

If the clutch and bands operate satisfactorily with air pressure, faulty operation of the gearbox must be due to malfunction of the hydraulic control system.

The valve bodies assembly must then be dismantled, cleaned, inspected and reassembled. 44.40.04.

Procedure

1. Remove the valve body assembly. 44.40.01.
2. Remove the rear extension housing. 44.20.15.
3. **Front clutch and governor feed**

    Apply air pressure to the passage (3). Listen for a thump, indicating that the clutch is functioning. With the unit on a bench, verify by rotating the input shaft with air pressure applied. Keep air pressure applied for several seconds to check for leaks in the circuit.

    If the extension housing has been removed, rotate the output shaft so that the governor weight will be at the bottom of the assembly. Verify that the weight moves inwards with air pressure applied.

4. **Rear clutch**

    Apply air pressure to the passage (4). With the unit on the bench, verify that the clutch is functioning by turning the input shaft. Keep air pressure applied for several seconds to check for leaks; then listen for a thump indicating that the clutch is releasing.

5. **Front servo**

    Apply air pressure to the apply tube location (5) immediately adjacent to the rear retaining bolt. Observe the movement of the piston pin.

6. **Rear servo**

    Apply air pressure to the tube location (6). Observe the movement of the servo lever.

Conclusions

If the clutch and bands operate satisfactorily with air pressure, faulty operation of the transmission indicates malfunction of the hydraulic control system which will necessitate removing and overhauling the valve bodies.

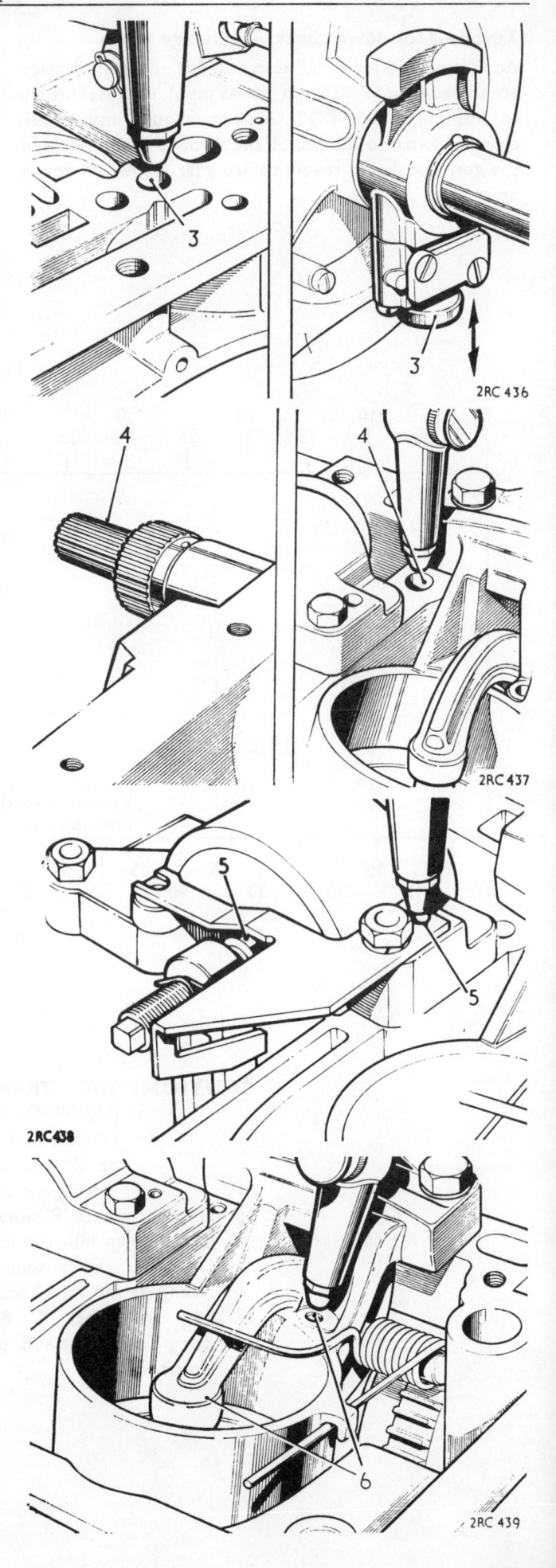

Rover 3500 and 3500S Manual AKM 3621

GEARBOX—AUTOMATIC—Borg-Warner type 35

FRONT PUMP ASSEMBLY

—Remove and refit 44.32.01

Service tool: 601280 Cradle

Removing

1. Remove the gearbox. 44.20.01.
2. Place the gearbox on cradle 601280, with the fluid pan uppermost.
3. Remove the valve body assembly. 44.40.01.
4. Pull out the following fluid tubes:
   a. Converter outlet.
   b. Front pump inlet.
   c. 'O' ring, pump inlet.
   d. Converter inlet.
   e. Front pump outlet.
5. Check the input shaft end float. 44.36.08.

   NOTE: The existing input shaft end-float must be checked before the front pump is removed so that, subsequently, compensation can be made for thrust washer wear.

6. Remove the six bolts and spring washers.
7. Remove the front pump, joint washer and the input shaft thrust washer.

Refitting

8. Reverse 1 to 9, noting the following.
9. Using the correct thrust washer, determined in item 5, retain it in position on the front pump assembly using petroleum jelly.
10. Locate a new joint washer in position on the front pump, using petroleum jelly.
11. Fit the front pump with the fluid tube locations facing the valve body. Torque: 1,0 to 2,5 kgf.m (8 to 18 lbf ft).
12. Fit a new 'O' ring to the pump inlet tube.
13. Refill the fluid system. 44.24.02.

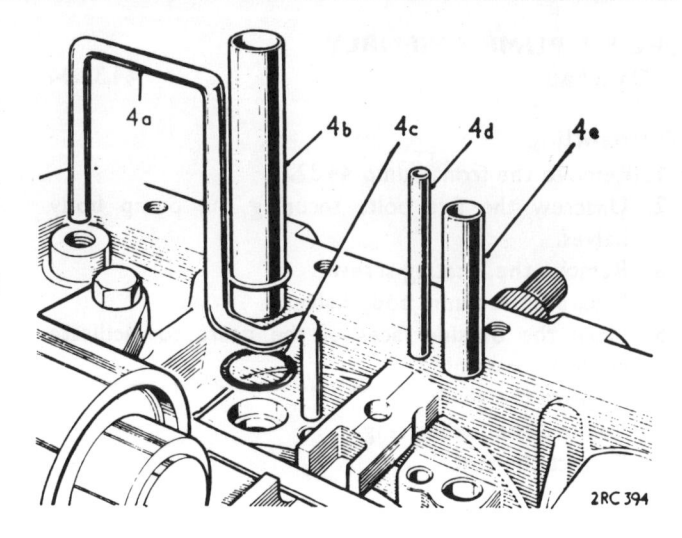

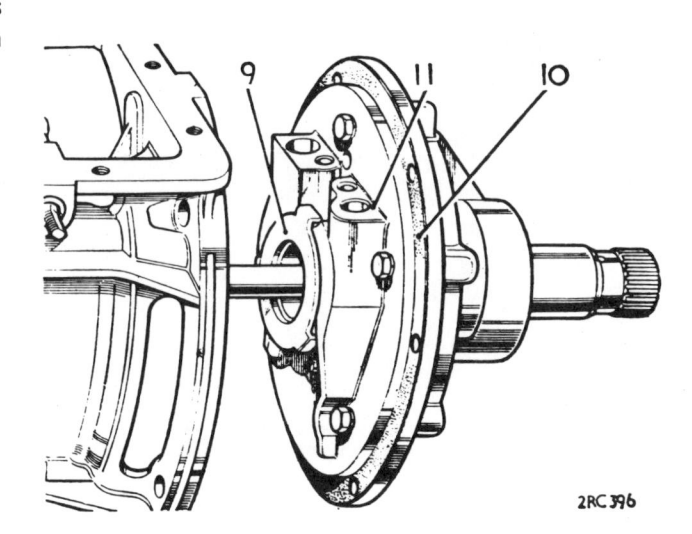

Rover 3500 and 3500S Manual AKM 3621

44.32.01

GEARBOX—AUTOMATIC—Borg-Warner type 35

FRONT PUMP ASSEMBLY
—Overhaul 44.32.04

Dismantling

1. Remove the front pump 44.32.01
2. Unscrew the five bolts securing the pump body halves.
3. Remove the locating screw.
4. Separate the pump body halves.
5. Mark the outside faces of the gears to facilitate correct reassembly.
6. Remove the gears.
7. Remove the sealing ring.
8. Extract the seal.

Inspecting

9. Check the pump body and gear teeth for scores and excessive wear.

Reassembling

10. Fit a new seal.
11. Fit a new sealing ring.
12. Assemble the pump gears; locate in position with petroleum jelly the correct way round, as previously marked.
13. Lubricate the seals with clean transmission fluid.
14. Locate the pump body halves together.
15. Fit and tighten the locating screw with the lock washer to 0,3 to 0,4 kgf.m (2 to 3 lbf ft).
16. Fit and tighten the five bolts and with spring washers to 2,3 to 3,0 kgf.m (17 to 22 lbf ft).
17. Refit the front pump. 44.32.01.

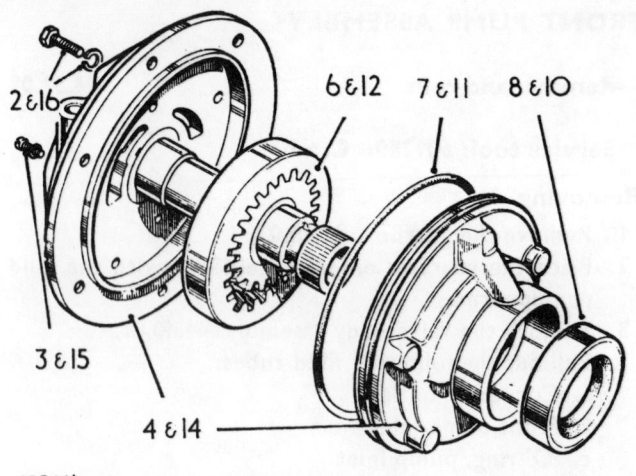

GEARBOX—AUTOMATIC—Borg-Warner type 35

FRONT SERVO—Manually adjusted

—Remove and refit 44.34.07

Service tool: 606328 Adjusting tool

Removing

WARNING: If the vehicle has just returned from a run, the gearbox fluid will be hot enough to cause scalding.

1. Place a suitable receptacle beneath the gearbox.
2. Remove the gearbox drain plug and allow all the fluid to drain.
3. Refit the drain plug and washer.
4. Remove the fluid pan and gasket.
5. Withdraw the two fluid tubes from the front servo.
6. Remove the front servo.
7. Withdraw the strut.

Refitting

8. To facilitate assembly, attach the strut to the servo lever with petroleum jelly.
9. Fit the servo, ensuring that the strut engages with the slot in the brake band.
10. Tighten the servo securing bolts. Torque: 1,0 to 1,7 kgf.m (8 to 13 lbf ft).
11. Fit the two fluid tubes between the servo and the valve body assembly.

Adjusting front servo

12. Slacken the locknut and unscrew the adjuster.
13. Pull the servo lever outward and insert a 6,3 mm (0.250 in) thick spacer between the adjusting screw and the servo piston pin.
14. Fit the adjusting tool to the servo lever adjusting screw, 606328.
15. Attach a spring balance to the small hole in the end of the adjusting tool.
16. Tighten the adjusting screw by tensioning the spring balance to a reading of 2,25 kg (5 lb). It is important that the spring balance reading is obtained with the adjusting tool and spring balance at 90° to each other.
17. Remove the spring balance and adjusting tool.
18. Using a screwdriver between the servo body and the adjusting lever, release the pressure from the spacer and withdraw it.
19. Using a new fluid pan gasket, locate it on the gearbox casing with petroleum jelly.
20. Fit the fluid pan. Tighten the bolts to a torque of 1,0 to 1,7 kgf.m (8 to 13 lbf ft).
21. Refill the fluid system. 44.24.02.

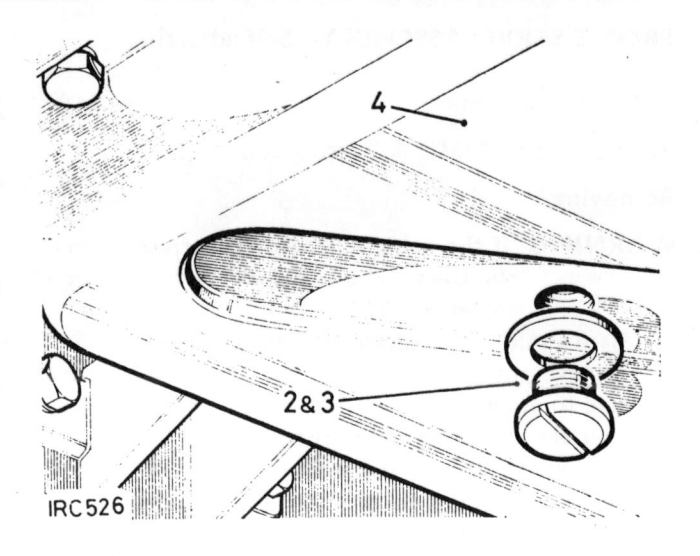

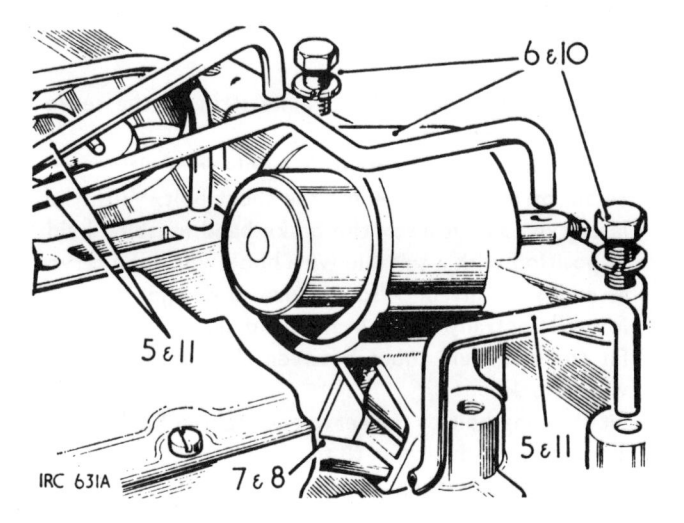

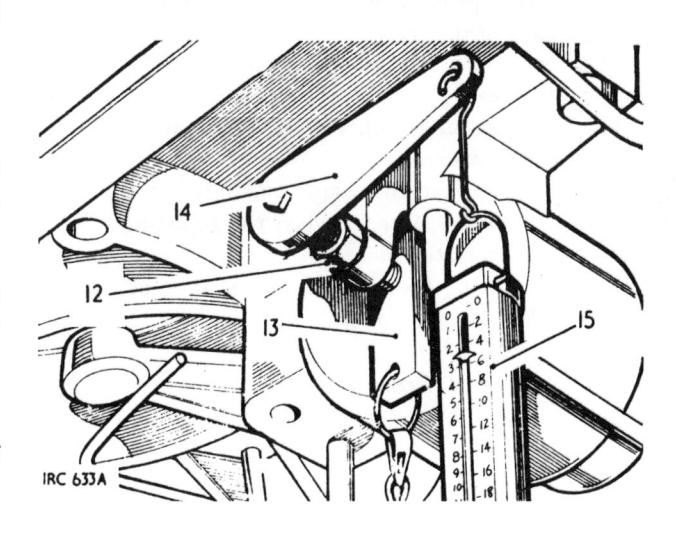

Rover 3500 and 3500S Manual AKM 3621

GEARBOX—AUTOMATIC—Borg-Warner type 35

FRONT SERVO ASSEMBLY—Self adjusting

—Remove and refit 44.34.07

Service tools—606328 Adjusting tool

Removing

WARNING: If the vehicle has just returned from a run the gearbox fluid will be hot enough to cause scalding.

1. Place a suitable receptacle beneath the gearbox.
2. Remove the gearbox drain plug and allow all the fluid to drain.
3. Refit the drain plug and washer.
4. Remove the fluid pan and gasket.
5. Withdraw the two fluid tubes from the servo.
6. Remove the servo assembly.
7. Withdraw the strut.

Refitting

8. To facilitate assembly, attach the strut to the servo lever with petroleum jelly.
9. Locate the servo and strut assembly in position and retain by loosely tightening the short bolt. With the later type 303 gearbox the brake band must be held in position with a suitable wire hook while the servo assembly is fitted. Also, ensure that the strut is located centrally in the brake band slot.
10. Fit the two fluid tubes between the servo and the valve block.
11. Fit the long bolt to the servo and tighten both bolts. Torque 1,0 to 1,7 kgf.m (8 to 13 lbf ft).
12. Screw the adjusting screw in until it protrudes approximately 1,5 mm (0.062 in.) through the lever.
13. Position the spring between one and two turns clear of the back of the lever.
14. Using feeler gauges, check the clearance between the spring and lever. The correct clearance is 1,5 to 3,0 mm (0.062 to 0.124 in.). Adjust the position of the spring as necessary.
15. Pull the servo lever outwards and insert a 6,3 mm (0.250 in.) thick spacer between the adjusting screw and the servo piston pin.
16. Fit the adjusting tool to the servo lever adjusting screw. 606328.

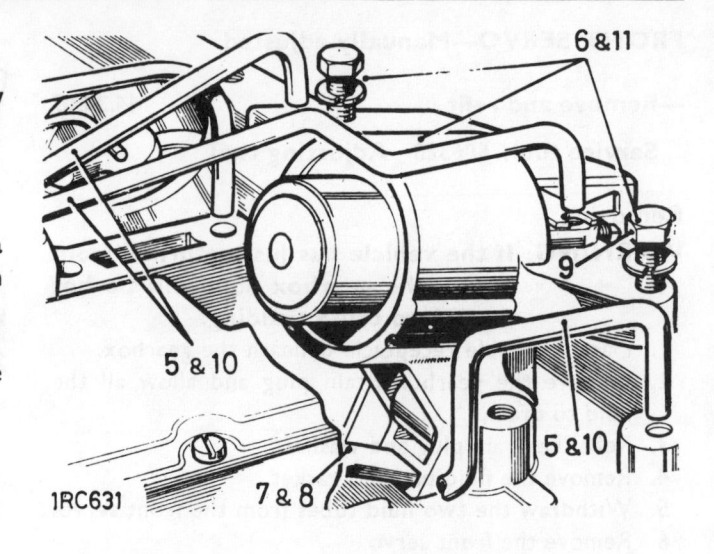

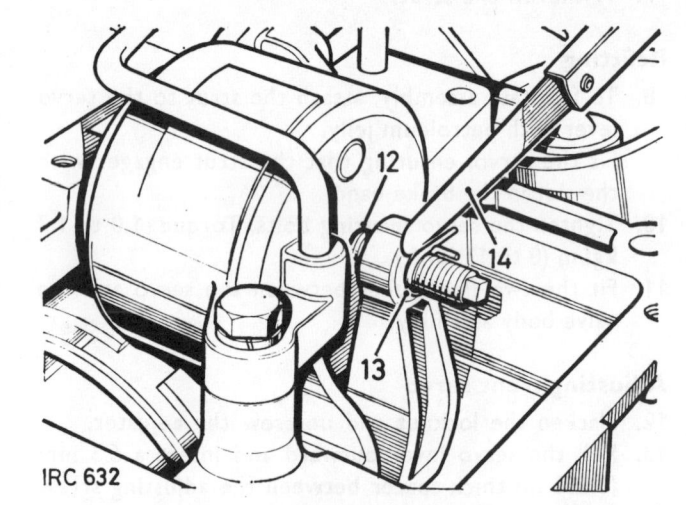

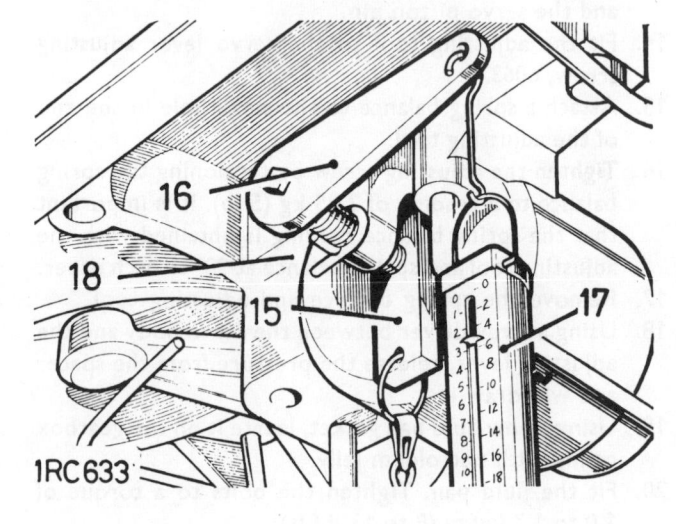

continued

44.34.07
Sheet 2

Rover 3500 and 3500S Manual AKM 3621

GEARBOX—AUTOMATIC—Borg-Warner type 35

17. Attach a spring balance to the small hole in the end of the adjusting tool.
18. Hold the long leg of the adjusting spring to prevent it rotating and tighten the adjusting screw by tensioning the spring balance to a reading of 2,25 kg (5 lb). It is important that the spring balance reading is obtained with the adjusting tool and spring balance at 90° to each other.
19. Remove the spring balance and adjusting tool.
20. Using a screwdriver between the servo body and the adjusting lever, release the pressure from the spacer and withdraw it.
21. Remove the rear (long) bolt and the adjacent fluid tube from the servo assembly.
22. Fit the adjusting plate to the servo, locating the long leg of the spring into the slot.
23. Refit the fluid tube and the rear bolt to the servo assembly. Torque 1,0 to 1,7 kgf.m (8 to 13 lbf ft).
24. Using a new fluid pan gasket, locate it on the gearbox casing with petroleum jelly.
25. Fit the fluid pan. Tighten the bolts to a torque of 1,0 to 1,7 kgf.m (8 to 13 lbf ft).
26. Replenish the gearbox fluid. 44.24.02.

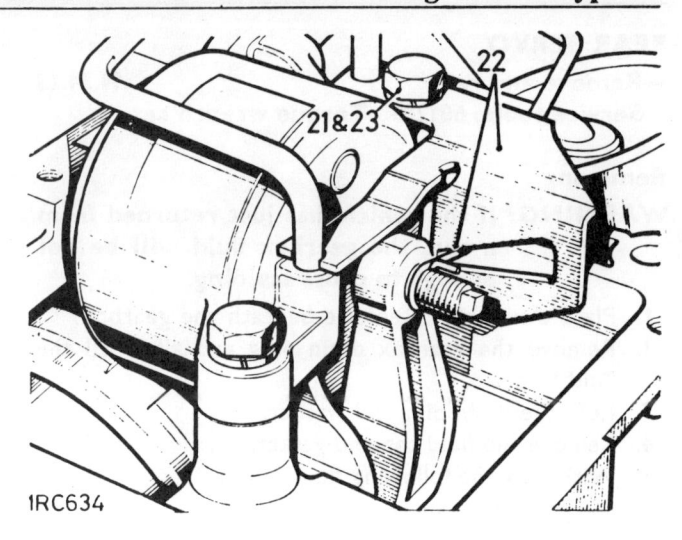

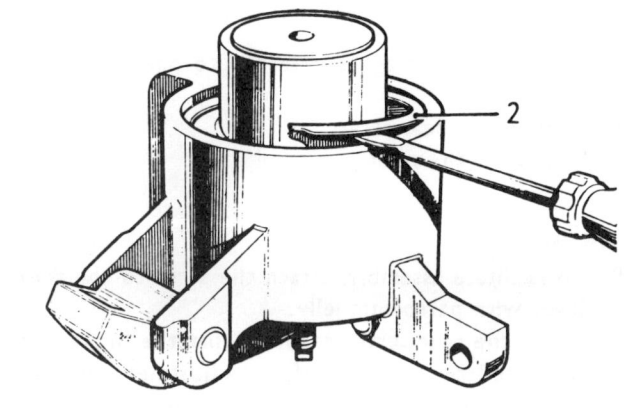

FRONT SERVO ASSEMBLY

—Overhaul   44.34.10

Dismantling

1. Remove the front servo assembly. 44.34.07.
2. Remove the snap ring.
3. Remove the piston sleeve, piston and spring.
4. Remove the piston from the sleeve.
5. Press out the lever pivot pin from the body and remove the lever.

Inspecting

6. Check the 'O' rings and oil sealing rings for signs of deterioration or damage; renew the rings as necessary.
7. Examine the piston, sleeve and body for cracks, scratches and wear.

Reassembling

8. Prior to assembly, lubricate the components with clean transmission fluid.
9. Assemble the components and secure with the snap ring. Note that the collar of the sleeve fits inwards.
10. Refit the front servo assembly. 44.34.07.

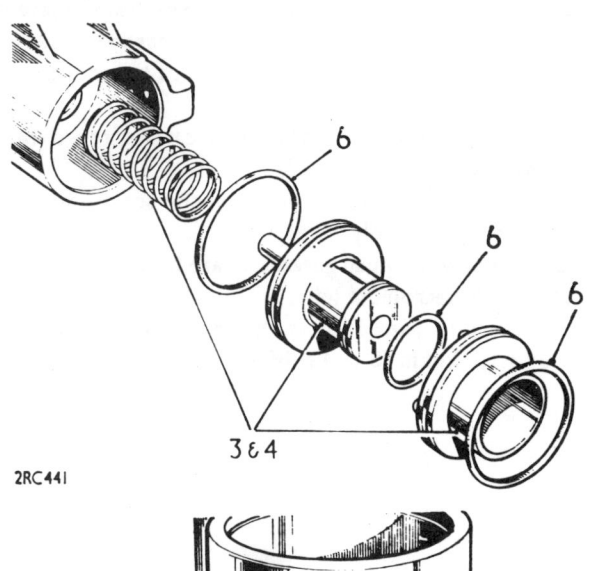

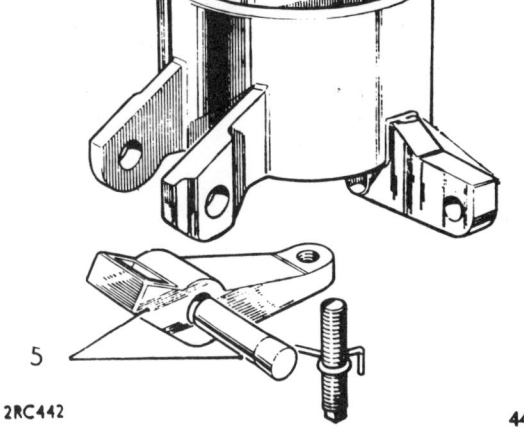

Rover 3500 and 3500S Manual AKM 3621

GEARBOX—AUTOMATIC—Borg-Warner type 35

REAR SERVO
—Remove and refit 44.34.13
Service tools: 601285 Torque wrench set

Removing
WARNING: If the vehicle has just returned from a run, the gearbox fluid will be hot enough to cause scalding.

1. Place a suitable receptacle beneath the gearbox.
2. Remove the gearbox drain plug and allow all the fluid to drain.
3. Refit the drain plug and washer.
4. Remove the fluid pan and gasket.
5. Withdraw the following fluid tubes:
   a. Front servo apply.
   b. Rear clutch.
   c. Rear servo.
6. Withdraw the magnet, attached to one of the bolt heads.
7. Remove the rear servo.
8. Withdraw the strut.

Refitting
9. To facilitate assembly, attach the strut to the servo lever with petroleum jelly.
10. Install the rear servo, ensuring that the brake band is located on the head of the adjustment bolt and that the strut engages with the slot in the brake band.
11. Locate the special servo bolt into the hole adjacent to the centre cross-support of the gearbox case. Fit the rear bolt and tighten both fixings to a torque of 1,4 to 3,6 kgf.m (13 to 27 lbf ft).
12. Locate the magnet on the head of the servo rear fixing bolt.
13. Fit the three fluid tubes between the servos and the valve body assembly.
14. Using a new fluid pan gasket, locate it on the gearbox case with petroleum jelly.
15. Fit the fluid pan. Tighten the bolts to a torque of 1,0 to 1,7 kgf.m (8 to 13 lbf ft).

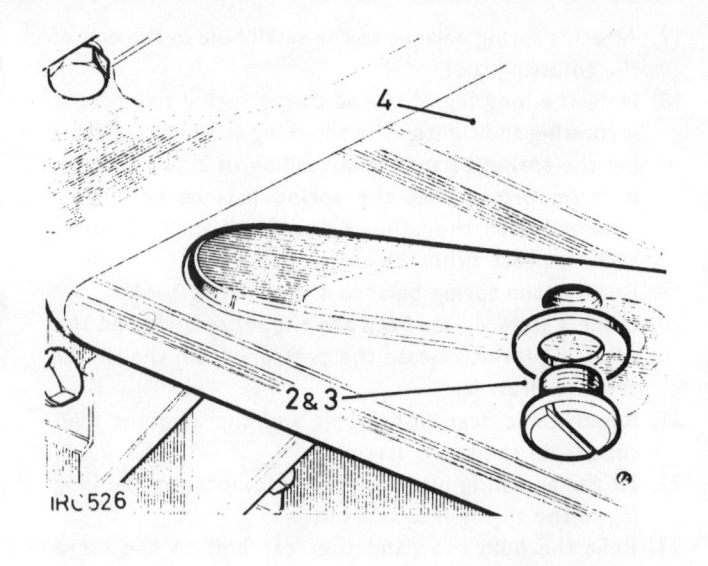

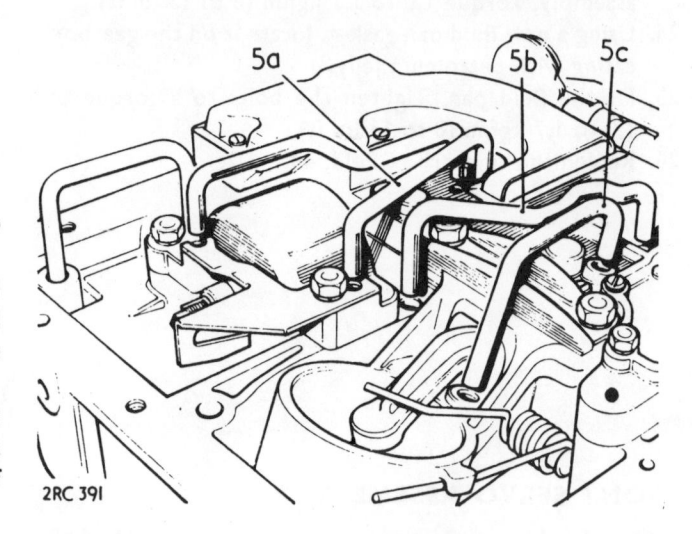

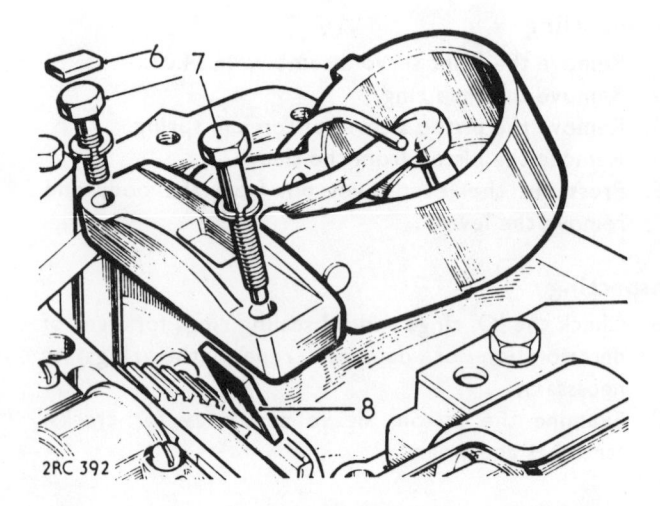

continued

GEARBOX—AUTOMATIC—Borg-Warner type 35

Adjusting rear servo

NOTE: The adjuster for the rear band servo is located on the outside of the transmission case. When the unit is fitted in the car, the adjuster is accessible from inside the car by removing the tunnel carpet and rubber grommet.

16. Slacken the locknut of the adjusting screw.
17. Using the torque wrench set 601285, tighten the servo adjusting screw to 1,4 kgf.m (10 lbf ft), then turn it back one complete turn and tighten the locknut to 3,5 to 4,0 kgf.m (25 to 30 lbf ft).
18. Refill the fluid system. 44.24.02.

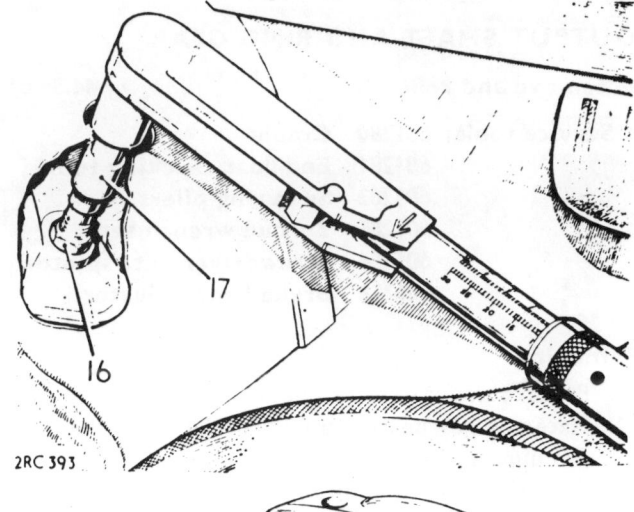

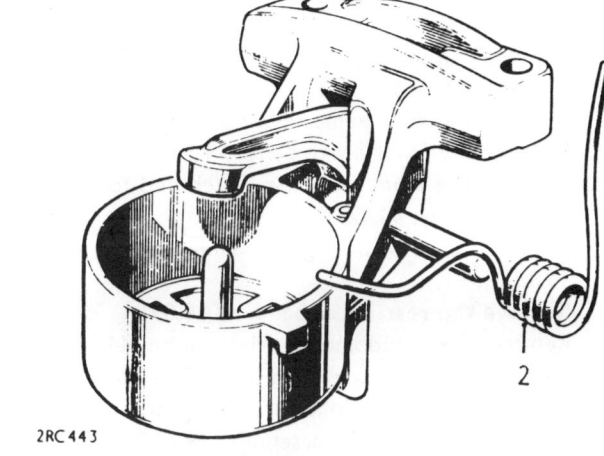

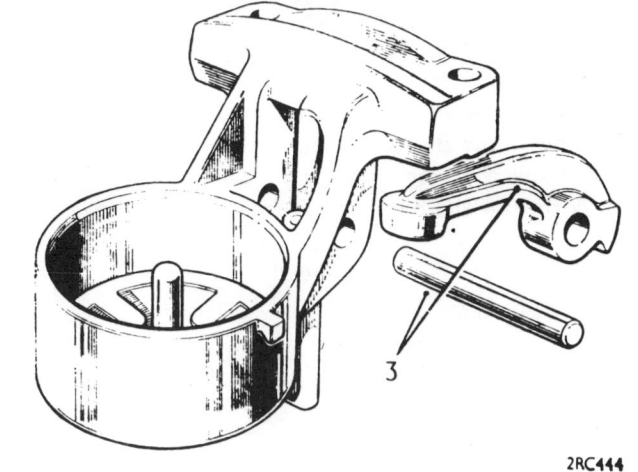

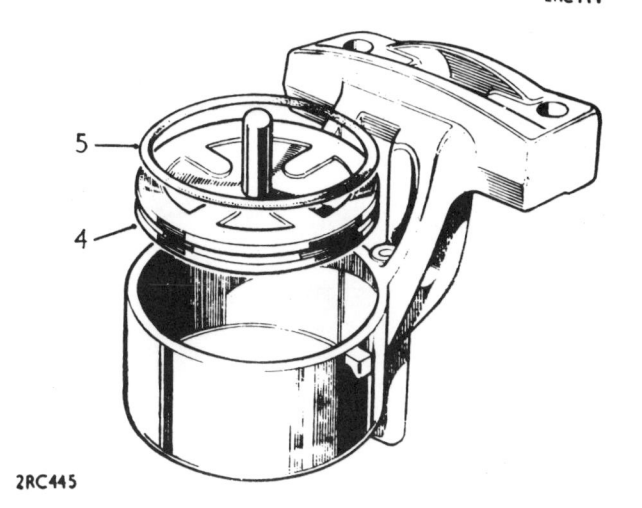

REAR SERVO ASSEMBLY

—Overhaul   44.34.16

Dismantling

1. Remove the rear servo assembly. 44.34.16.
2. Release the return spring.
3. Press the lever pivot pin from the body and remove the lever.
4. Withdraw the piston.

Inspecting

5. Check the 'O' ring for signs of deterioration or damage; renew the ring if necessary.
6. Check the piston and bore for cracks, scratches and wear.

Reassembling

7. Prior to assembly of the piston into the body, lubricate all components with transmission fluid.
8. Reverse 1 to 4.

Rover 3500 and 3500S Manual AKM 3621

44.34.13
Sheet 2
44.34.16

GEARBOX—AUTOMATIC—Borg-Warner type 35

OUTPUT SHAFT AND RING GEAR

—Remove and refit 44.36.01

Service tools:
- 601280 Cradle
- 601281 End-float checking tool
- 601282 Snap ring pliers
- 601285 Torque wrench set
- 601286 Screwdriver set, spintorq
- 606328 Brake band adjuster

Removing

1. Remove the gearbox. 44.20.01.
2. Place the gearbox on cradle 601280, with the fluid pan uppermost.
3. Remove the valve body assembly. 44.40.01.
4. Remove the front servo. 44.34.07.
5. Remove the rear servo. 44.34.13.
6. Check the input shaft end-float. 44.36.08, then remove the front pump. 44.32.01.
7. Remove the front brake band. 44.10.01.
8. Remove the front and rear clutch and forward sun gear assembly. 44.12.01.
9. Remove the planet gears and centre support. 44.36.04.
10. Remove the rear brake band. 44.10.09.
11. Remove the speedometer drive pinion. 44.38.04.
12. Remove the rear extension housing. 44.20.15.
13. Withdraw the speedometer drive gear.
14. Remove the governor assembly. 44.22.01.
15. Remove the adaptor plate.
16. Withdraw the output shaft and ring gear assembly, taking care not to damage the surface of the rear support bearing in the casing.
17. Remove the thrust washer.
18. Remove the snap ring.
19. Withdraw the output shaft from the ring gear.

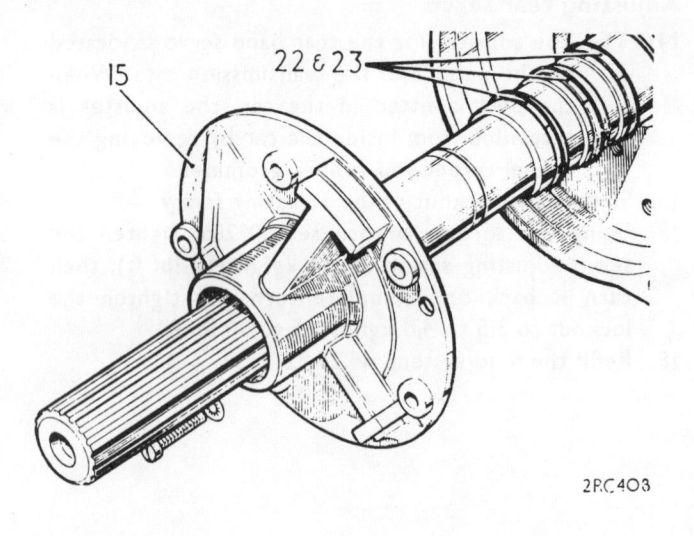

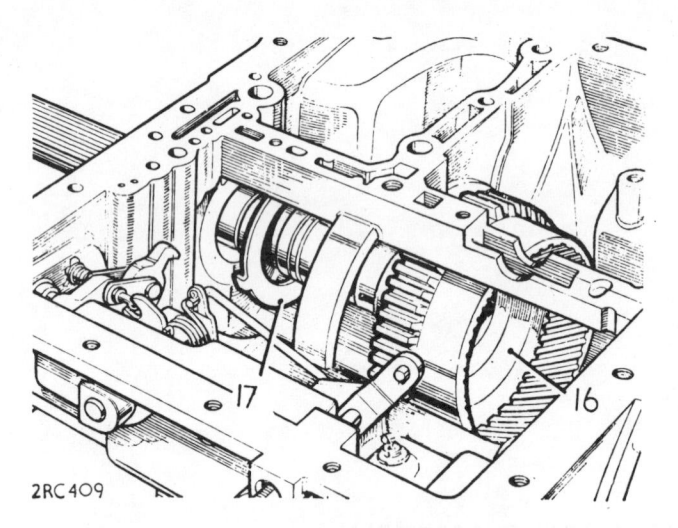

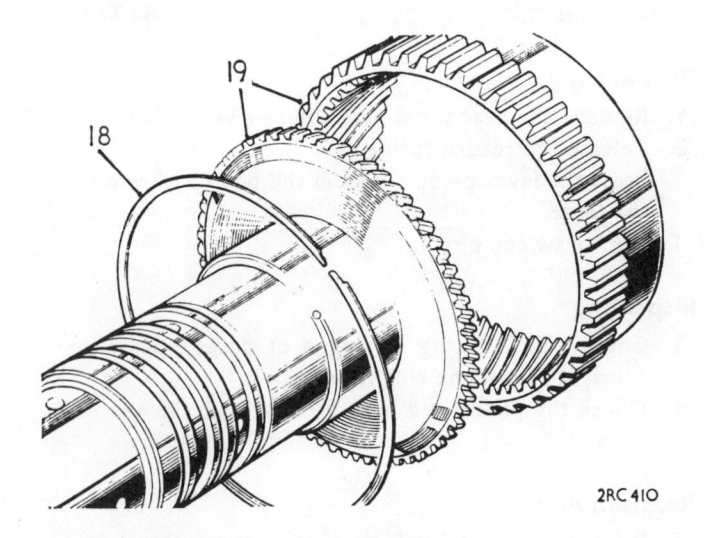

continued

GEARBOX—AUTOMATIC—Borg-Warner type 35

Refitting

NOTE: If the oil cooler pipe unions have been removed from the gearbox, before refitting, use 'Loctite' sealant grade 'AVV', Rover Part No. 600303, on the threads.

20. Fit the output shaft to the ring gear and secure with the snap ring.
21. Check and ensure that there is no end-float between the output shaft and the ring gear. To enable any end-float to be eliminated, snap rings are available in the following thicknesses: 1,39, 1,44 and 1,49 mm (0.055, 0.057 and 0.059 in).
22. Fit new oil sealing rings if necessary.
23. Align the oil sealing rings with their gaps uppermost as viewed with the box inverted, centralise the rings in their grooves, retaining them in position with a little petroleum jelly.
24. Locate the tabbed thrust washer on to the boss of the gearbox case with petroleum jelly. so that one tab is at the top with the gearbox inverted.
25. Reverse 1 to 16.
26. Refill the fluid system. 44.24.02

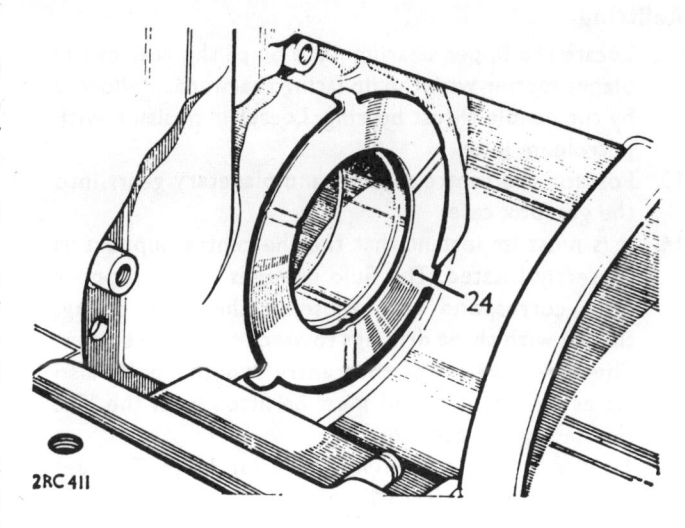

PLANET GEARS AND CENTRE SUPPORT

—Remove and refit   44.36.04

Service tools:
- 601280 Cradle
- 601281 End-float checking tool
- 601285 Torque wrench set
- 601286 Screwdriver set, spintorq

Removing

1. Remove the gearbox. 44.20.01.
2. Place the gearbox on cradle 601280, with the fluid pan uppermost.
3. Remove the valve body assembly. 44.40.01.
4. Remove the front servo. 44.34.07.
5. Remove the rear servo. 44.34.13.
6. Check the input shaft end-float. 44.36.08, then remove the front pump. 44.32.01.
7. Remove the front brake band. 44.10.01.
8. Remove the front and rear clutch and forward sun gear assembly. 44.12.01.
9. Remove the two special bolts and lock washers retaining the centre support. These fixings are located on the outside of the gearbox case.
10. Withdraw the centre support and planet gears through the front of the gearbox case.
11. Note the needle thrust bearing and plate, located on the rear end of the planetary gear assembly.

continued

GEARBOX—AUTOMATIC—Borg-Warner type 35

Refitting

12. Locate the lipped bearing plate on to the rear of the planet carrier with the lip facing rearwards, followed by the needle thrust bearing. Locate in position with petroleum jelly.
13. Position the centre support and planetary gears into the gearbox case.
14. It is most important that the the centre support is correctly located. The fluid passages in the support must correspond with those in the main casing, that is, with three drillings towards the valve bodies.
15. The lock washers of the centre support bolts also act as oil retainers and must be fitted with the 'flat rim' facing the bolt head.
16. Tighten the centre support securing bolts. Torque: 1,4 to 1,7 kgf.m (10 to 13 lbf ft).
17. Reverse 1 to 8.
18. Refill the fluid system. 44.24.02.

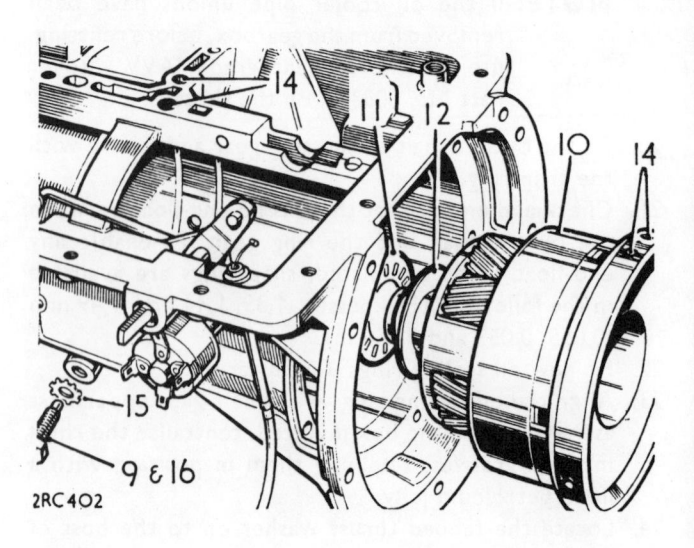

PLANET GEARS AND CENTRE SUPPORT

—Overhaul 44.36.05

Dismantling

1. Remove the planet gears and centre support. 44.36.04.
2. Remove the needle thrust bearing and plate from the rear end of the drum.
3. Separate the centre support from the planet gear assembly.
4. Withdraw the one-way clutch from its outer race.
5. Support the drum in a soft jawed vice and prise out the snap ring.
6. Remove the one-way clutch outer race by levering alternately in the three slots.

Inspecting

7. Clean all the components, using transmission fluid and nylon rag. Examine all components for defects.

Reassembling

8. Lubricate all the components with clean transmission fluid.
9. Lightly tap the one-way clutch outer race into position, using a soft-faced mallet, and ensure correct alignment. Secure with the snap ring.
10. Position the one-way clutch in the outer race with the lip of the one-way clutch outwards.
11. Place the centre support into position on the drum.
12. First place the bearing plate lip outwards, followed by the needle thrust washer in position at the rear of the assembly and retain, using petroleum jelly.
13. Refit the planet gears and centre support. 44.36.04.

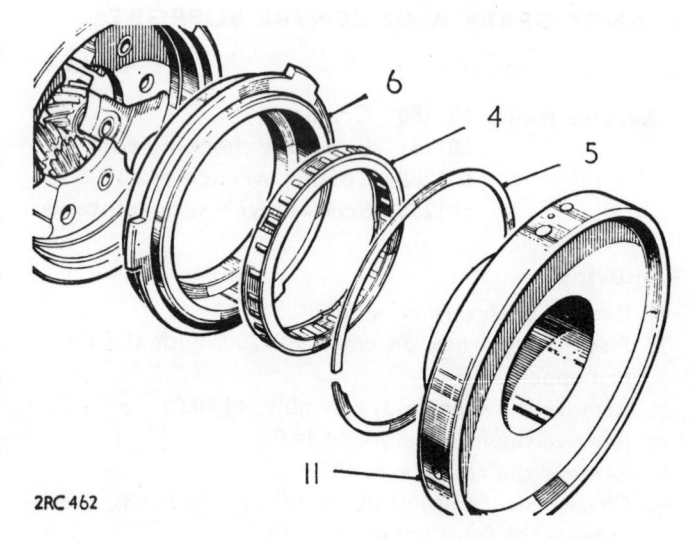

Rover 3500 and 3500S Manual AKM 3621

GEARBOX—AUTOMATIC—Borg-Warner type 35

INPUT SHAFT

—End-float check   44.36.08

Service tools: 601280 Cradle
601281 End-float checking tool

Checking

1. Remove the gearbox. 44.20.01.
2. Place the gearbox on cradle 601280, with the fluid pan uppermost.
3. Remove the valve body assembly. 44.40.01.
4. Clamp 601281 to the converter support shaft.
5. Gently lever the gear train forward and adjust the screw of the tool until it just contacts the end of the input shaft.
6. Lever the clutch back, using light pressure, and measure the gap produced between the tool and the end of the shaft, see 'Data'. Remove the tool.
7. If the end-float is excessive, remove the front pump and fit new thrust washers, as required.

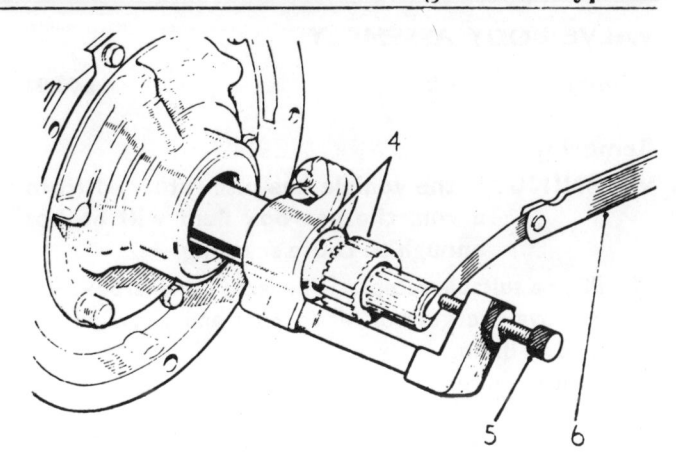

Refitting

8. Reverse 1 to 7.
9. Refill the fluid system. 44.24.02.

Data

Input shaft:	End-float	0,25 to 0,75 mm (0.010 to 0.030 in).
	Thrust washer thickness	1,54 to 1,6 mm (0.061 to 0.063 in).
	(two ranges)	1,97 to 2,3 mm (0.078 to 0.080 in).

SPEEDOMETER DRIVE PINION

—Remove and refit   44.38.04

Removing

1. Disconnect the speedometer cable at the gearbox end.
2. Remove the speedometer drive housing.
3. Withdraw the 'O' ring seal from the housing.
4. Withdraw the speedometer drive pinion.
5. Unscrew the right angled drive.
6. Unscrew the adaptor from the drive housing.
7. Withdraw the seal from the drive housing.

Refitting

8. Fit a new seal into the drive housing.
9. Smear the larger diameter threads of the adaptor with 'Loctite' sealant grade 'AVV', Rover Part No. 600303.
10. Reverse 1 to 6.

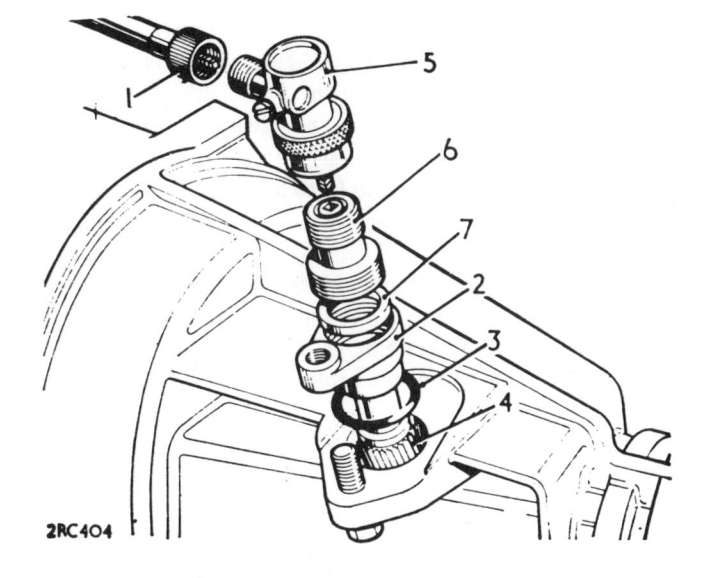

GEARBOX—AUTOMATIC—Borg-Warner type 35

VALVE BODY ASSEMBLY

—Remove and refit   44.40.01

Removing

WARNING: If the vehicle has just returned from a run, the gearbox fluid will be hot enough to cause scalding.

1. Place a suitable receptacle beneath the gearbox.
2. Remove the gearbox drain plug and allow all the fluid to drain.
3. Refit the drain plug and washer.
4. Remove the fluid pan.
5. Withdraw the magnet attached to one of the bolt heads.
6. Pull out the following fluid tubes:
   a. Front servo release.
   b. Front servo apply.
   c. Rear clutch.
   d. Rear servo.
7. If the gearbox is installed in the car, disconnect the downshift cable at the throttle end.
8. Disconnect the downshift cable from the cam.
9. Remove the three securing bolts and withdraw the valve body assembly.

Refitting

10. Reverse 1 to 9, noting the following.
11. When placing the valve body assembly into position, ensure that the oil tubes between valve body and front pump casing are correctly located, flange on front pump inlet tube must be towards front pump, and that the peg on the manual valve detent lever engages in the groove of the manual control valve.
12. If the oil tubes have been inadvertently disturbed, it is important that the 'O' ring, which is located at the pump end of the front pump inlet tube, is not omitted or dropped into the gearbox casing.
13. Tightening torque for valve body assembly securing bolts: 0,55 to 0,7 kgf.m (4 to 5 lbf ft).
14. Ensure that the four fluid tubes are pushed fully into place.
15. Use a new gasket and tighten the fluid pan bolts to a torque of 1,0 to 1,7 kgf.m (8 to 13 lbf ft).
16. Refill the fluid system. 44.24.02.
17. Check the downshift cable adjustment. 44.30.02/03.

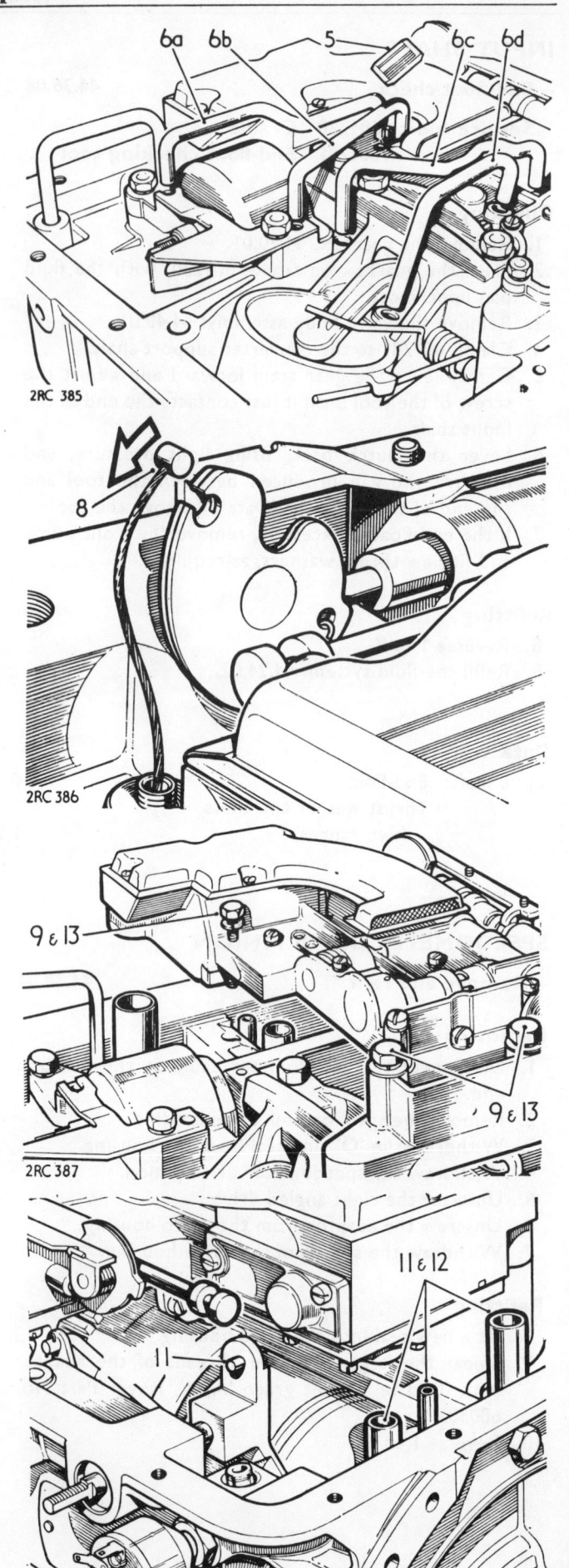

Rover 3500 and 3500S Manual AKM 3621

GEARBOX—AUTOMATIC—Borg-Warner type 35

VALVE BODY ASSEMBLY

—Overhaul 44.40.04

Service tool: 601286 Screwdriver set, spintorq

Dismantling

1. Remove the valve body assembly. 44.40.01.
2. Withdraw the manual control valve.
3. Remove the front pump oil strainer, secured by screws, plain and spring washers.
4. 'P' 'R' 'N' 'D2' 'D1' 'L' selector pattern—Remove eight screws and spring washers securing the lower valve body to the upper valve body; two screws and washers are located in the oil strainer housing, these retain the 'D1-D2' control valve.
5. 'P' 'R' 'N' 'D' '2' '1' selector pattern—Withdraw the fluid tube between the range control valve and the valve body, then remove the eight screws and spring washers securing the lower valve body to the upper valve body; two screws and washers are located in the oil strainer housing, these retain the range control valve.
6. If required, withdraw the components of the 'D1-D2' valve or range control valve, as applicable.
7. Turn the unit over and remove the oil tube collector plate, secured by six short and two long screws and spring washers to the lower valve body.

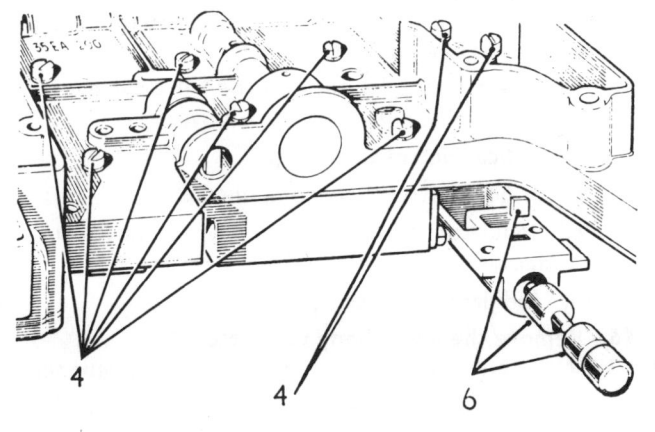

continued

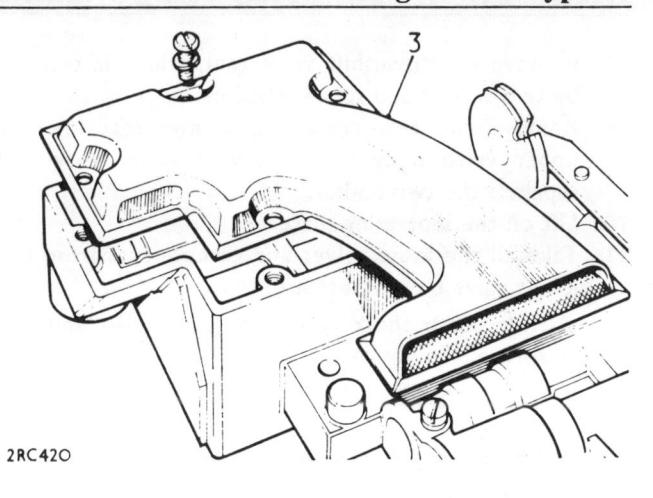

Rover 3500 and 3500S Manual AKM 3621

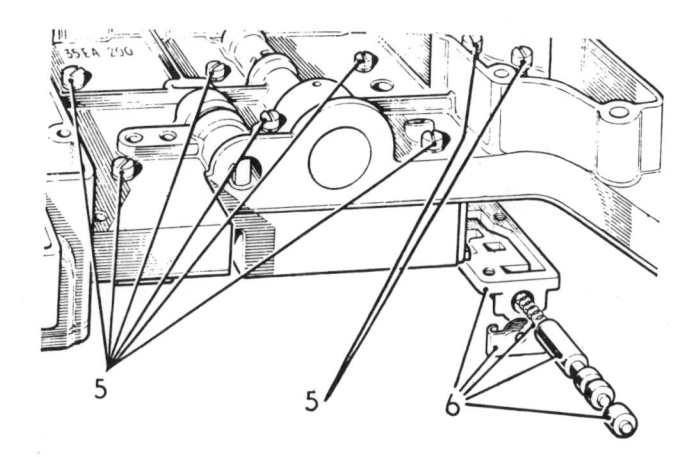

44.40.04
Sheet 1

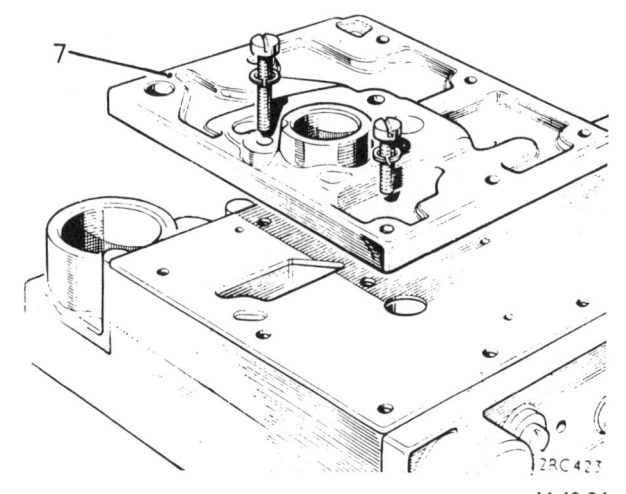

GEARBOX—AUTOMATIC—Borg-Warner type 35

8. Remove the downshift valve cam, which is secured by two screws and spring washers.
9. Remove the two remaining screws securing the upper valve body to the lower valve body and separate the two bodies.
10. Lift off the separating plate.
11. Take all the loose valves and components from the lower valve body, these are:
    a. Rear pump check valve and spring, disc and coil spring.
    b. Converter check valve and spring, ball and coil spring.
12. Withdraw the downshift valve and the retainer for the throttle valve.
13. Tilt the valve body to allow the throttle valve and spring to slide out.
14. Remove the stop for the throttle valve spring.
15. Turn the valve body over to remove:
    a. Orifice control valve stop.
    b. Orifice control valve components.
    c. Modulator valve dowel. It may be necessary to depress the valve plug to release the dowel.
    d. Modulator valve components.
16. Remove the lower body end plate.
17. Withdraw the primary regulator valve components.
18. Withdraw the secondary regulator valve components.

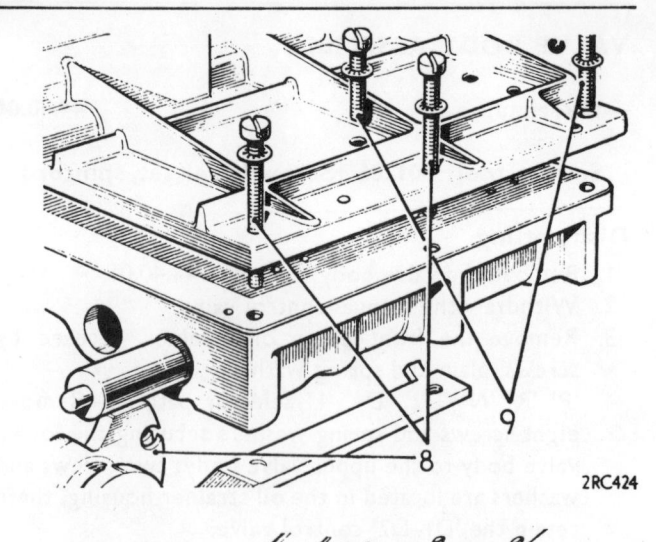

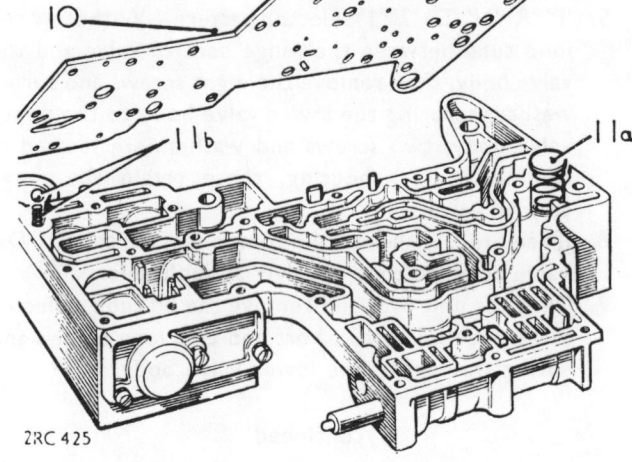

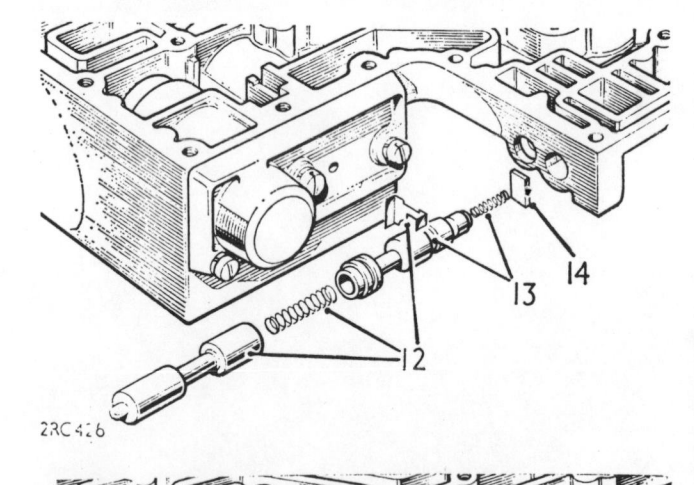

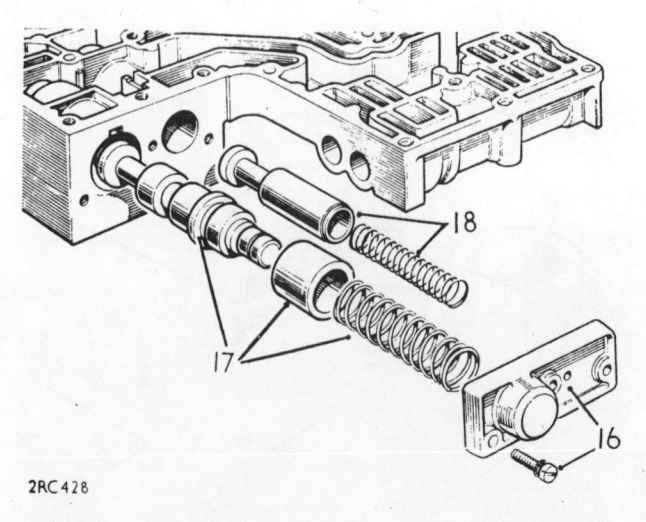

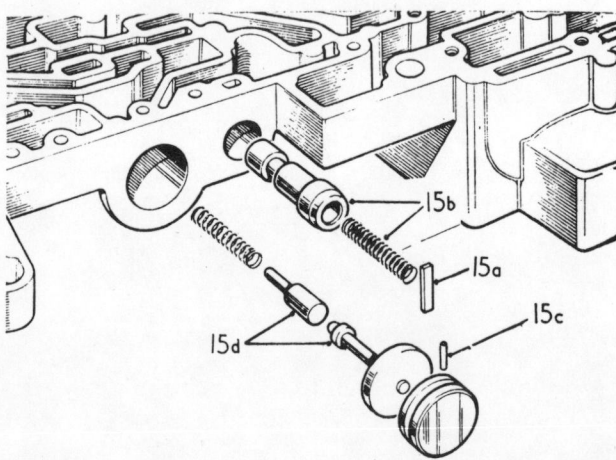

continued

44.40.04
Sheet 2

Rover 3500 and 3500S Manual AKM 3621

GEARBOX—AUTOMATIC—Borg-Warner type 35

19. Remove the end plates from the upper valve body.
20. Withdraw the 1st-2nd shift valve components.
21. Withdraw the 2nd-3rd shift valve components.

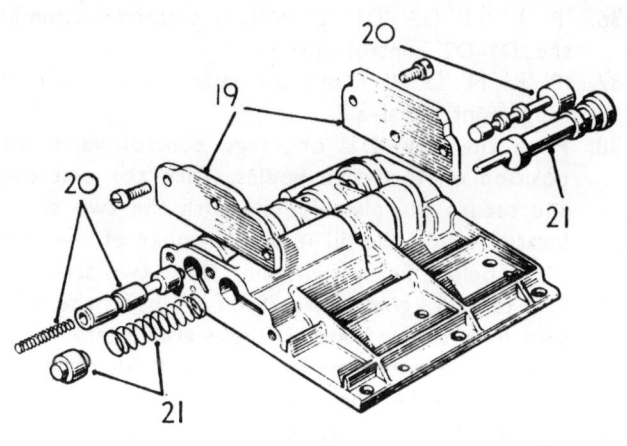

Inspecting

22. Carefully clean all the components, using petrol as the cleaning medium. Use only nylon or non-fluffy rag. Clear fluid passages with the aid of an air line. Assemble parts dry to check fit in bores.
23. Examine all components for wear and verify the free movement of all the valves in their bores.

Reassembling

NOTE: No gaskets are used throughout the valve body assembly. It is essential that a torque screwdriver is used exclusively on reassembly.

24. Assemble the upper valve body in the order as illustrated (items 19 to 21). Tighten the six end-plate fixing screws to 0,23 to 0,35 kgf.m (20 to 30 lbf in).
25. Place the upper valve body on one side until required for final assembly.
26. Assemble, in accordance with the illustration (items 15a to 15d), the servo orifice control valve and secure with the stop.
27. Fit the modulator valve to the lower valve body, and secure the assembly with the dowel pin.
28. Insert the primary and secondary regulator valves into the lower valve body and fit the cover plate. Tighten the three screws securing the cover plate to the body to a torque figure of 0,23 to 0,35 kgf.m (20 to 30 lbf in).
29. Insert the rear pump check valve.
30. Insert the converter 'out' check valve.
31. Place the stop for the throttle valve spring into position.
32. Fit the throttle valve and spring and locate valve retainer, as illustrated.
33. Carefully place the separating plate into position, ensuring that the valves beneath it are locating correctly.
34. Position the upper valve body assembly on to the separating plate and locate with the two screws as shown, but do not fully tighten the screws at this stage.
35. Fit the oil tube collector plate to the valve body. Tighten the eight fixing screws, six short and two long, to 0.23 to 0,35 kgf.m (20 to 30 lbf in).

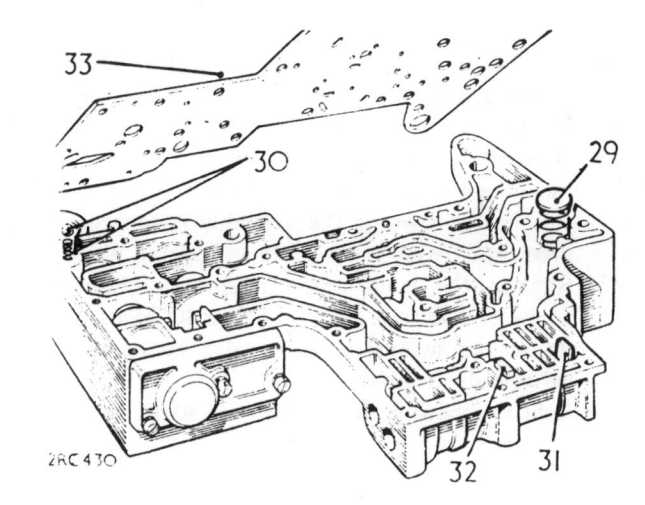

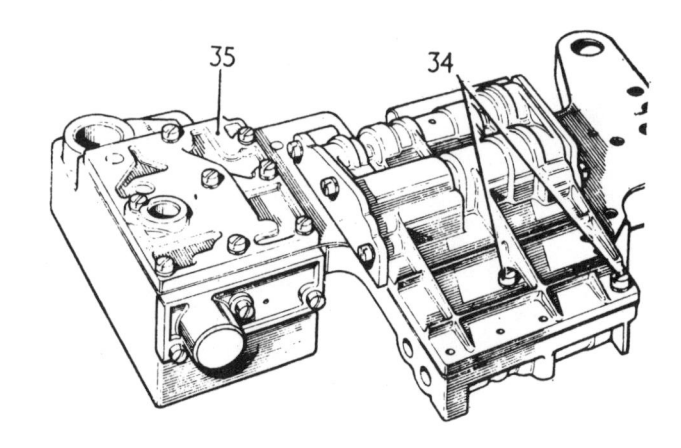

continued

Rover 3500 and 3500S Manual AKM 3621

GEARBOX—AUTOMATIC—Borg-Warner type 35

36. 'P' 'R' 'N' 'D2' 'D1' 'L' selector pattern—Assemble the 'D1–D2' control valve.
37. 'P' 'R' 'N' 'D' '2' '1' selector pattern—Assemble the range control valve.
38. Place the 'D1'–'D2' or range control valve into position on the valve bodies. Turn the unit over and secure the plate loosely with the two screws located inside the oil intake chamber of the valve body. Before finally tightening these two screws to 0,23 to 0,35 kgf.m (20 to 30 lbf in) insert the other two long screws loosely (these also locate the oil strainer) to ensure correct alignment. After tightening, remove the long screws.
39. Fit the six remaining screws securing the upper valve body to the lower valve body. Tighten to a torque figure of 0,23 to 0,35 kgf.m (20 to 30 lbf in).
40. If a range control valve is fitted, replace the oil tube between the valve and the main valve body. The tube is a push fit in its locations.
41. Fit the downshift valve and throttle valve in the order previously illustrated (items 12 to 14).
42. Fit the downshift valve cam. To ensure that the return spring is correctly tensioned, the cam should be turned in the direction of the arrow and held in this position whilst the assembly is secured to the valve body with two screws and spring washers.
43. Replace the front pump strainer, tightening the screw to 0,23 to 0,35 kgf.m (20 to 30 lbf in).
44. Insert the manual control valve to complete the valve body assembly.
45. Refit the valve body assembly. 44.40.01.

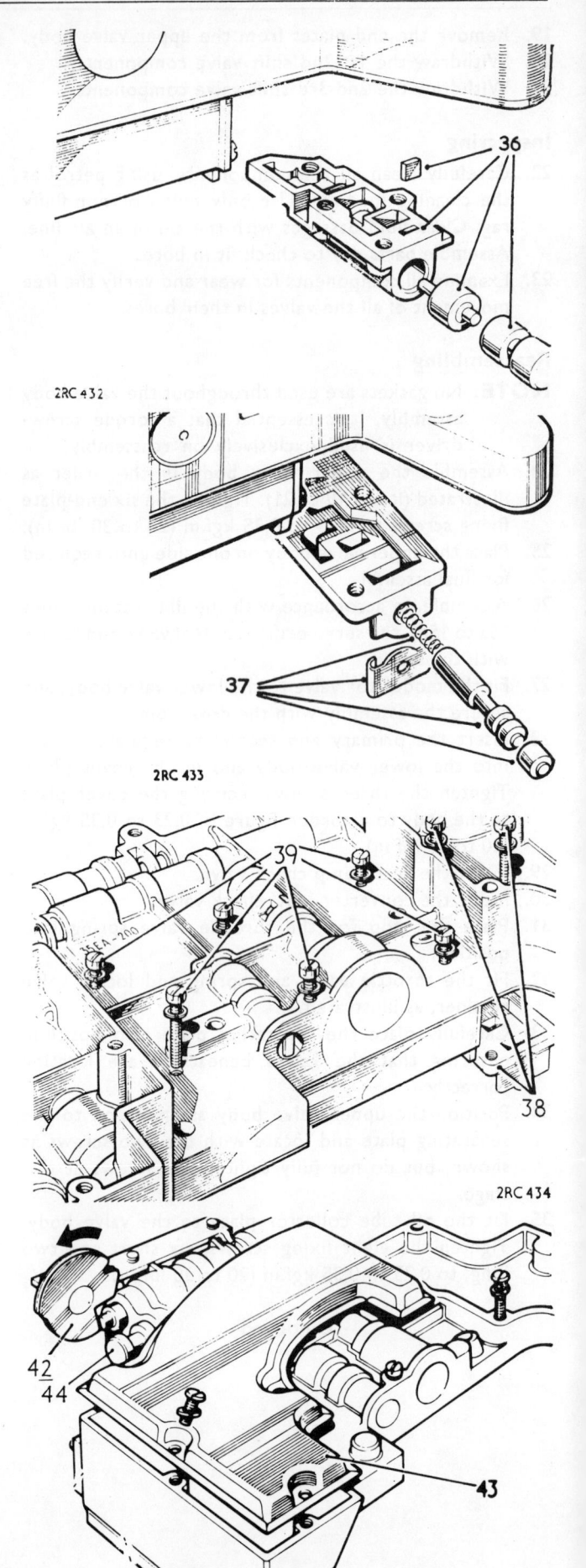

44.40.04
Sheet 4

Rover 3500 and 3500S Manual AKM 3621

GEARBOX—AUTOMATIC—Borg-Warner type 65

LIST OF OPERATIONS

Brake bands
- front—check and adjust .. 44.30.07
- front—remove and refit .. 44.10.01
- rear—check and adjust .. 44.30.10
- rear—remove and refit .. 44.10.09

Clutches
- front—remove and refit .. 44.12.04
- front—overhaul .. 44.12.10
- rear—remove and refit .. 44.12.07
- rear—overhaul .. 44.12.13

Converter
- assembly—remove and refit .. 44.17.07
- housing—remove and refit .. 44.17.01

Dipstick and fluid filler tube—remove and refit .. 44.24.01

Downshift cable
- check and adjust .. 44.30.02
- pressure check .. 44.30.03
- remove and refit .. 44.15.01

Fluid pan—remove and refit .. 44.24.04

Fluid system—drain and refill .. 44.24.02

Front pump
- remove and refit .. 44.32.01
- overhaul .. 44.32.04

Gearbox assembly
- remove and refit .. 44.20.01
- overhaul .. 44.20.06

Governor assembly
- remove and refit .. 44.22.01
- overhaul .. 44.22.04

Hand selector lever assembly
- remove and refit .. 44.15.04
- overhaul .. 44.15.05

Mechanical operation—air pressure checks .. 44.30.16

One way clutch—remove and refit .. 44.12.16

Output shaft and ring gear—remove and refit .. 44.36.01

Planet gears and centre support—remove and refit .. 44.36.04

Rear extension housing
- remove and refit .. 44.20.15
- oil seal—remove and refit .. 44.20.18

Restrictor valve—remove and refit .. 44.24.22

Road test .. 44.30.17

Selector cable assembly—remove and refit .. 44.15.08

Rover 3500 and 3500S Manual AKM 3621

44-1

GEARBOX—AUTOMATIC—Borg-Warner type 65

Servo assemblies
    —front—remove and refit 44.34.07
    —front—overhaul 44.34.10
    —rear—remove and refit 44.34.13
    —rear—overhaul 44.34.16

Speedometer
    —drive gear—remove and refit 44.38.07
    —drive pinion—remove and refit 44.38.04

Stall test 44.30.13

Starter and reverse light switch—remove and refit 44.15.15

Valve body assemly
    —remove and refit 44.40.01
    —overhaul 44.40.04

44-2   Rover 3500 and 3500S Manual AKM 3621

GEARBOX—AUTOMATIC—Borg-Warner type 65

EXAMINATION OF COMPONENTS

Gearbox Case and Servo Covers	Check for cracks and obstructions in passages
Front Pump	Check for scoring and excessive wear
Shafts	Check bearing and thrust faces for scoring
Output shaft and ring gear	Check that there is no end float between the shaft and ring gear
Clutch Plates	Check for warping, scoring, overheating and excessive wear
Bands	Check for scoring, overheating and excessive wear
Drums	Check for overheating and scoring
Gears	Check teeth for chipping, scoring, wear and condition of thrust faces
One way Clutch and Races	Check for scoring, overheating and wear
Valve body and Governor	Check for burrs, crossed or stripped threads, and scored sealing faces
Impeller Hub abd Front Pump Drive Gear	Check for pitting and wear. Ensure good contact
Thrust washers	Check for burrs, scoring and wear
White Metal Bushes	Check for scoring and loss of white metal
Lip Seals	Check for cuts, hardening of rubber, leakage past outer diameter
Rubber 'O' Rings and Seals	Check for hardening, cracking, cuts or damage
Cast Iron Sealing Rings	Check fit in groove and wear (evident by lip overhanging the groove)
Teflon Sealing Rings	Check for cracking, cuts or damage

SERVICING REQUIREMENTS

1. For all operations high standards of cleanliness are essential.
2. Rags and cloths must be clean and free from lint; nylon cloths are preferable.
3. Prior to assembly all components must be cleaned thoroughly with petrol, paraffin or an industrial solvent and dried immediately.
4. All defective items must be renewed.
5. Components should be lubricated with transmission fluid before assembly.
6. New joint washers should be fitted where applicable.
7. Where jointing compound is required, the use of Hylomar PL32M, Hermetite or Wellseal is approved. Apply jointing compound sparingly.
8. All screws, bolts and nuts must be tightened to the recommended torque figure.
9. Thrust washers and bearings should be coated with petroleum jelly to facilitate retaining them in position during assembly operations. Grease should not be used as it may be insoluble in the transmission fluid and could subsequently cause blockage of fluid passages and contamination of brake band and clutch facings.

IMPORTANT: Metric threads are used throughout most of the gearbox unit and it is therefore essential that fastenings, and especially lock washers, are segregated into sets and not intermixed with those from other parts of the vehicle.

GEARBOX—AUTOMATIC—Borg-Warner type 65

FRONT BRAKE BAND

—Remove and refit 44.10.01

Service tools: CBW547A–50 or 601285 torque wrench
CBW 60 Bench cradle

Removing

1. Remove the gearbox assembly. 44.20.01.
2. Wash the exterior of the unit in clean petrol or paraffin, invert it and place on a bench cradle CBW 60. Remove the starter and reverse switch. 44.15.15.
3. Unscrew the bolts securing the torque converter housing.
4. Remove the torque converter housing.
5. Unscrew 12 bolts.
6. Remove the fluid pan, joint washer and magnet.
7. Pull out the five fluid tubes.
8. Release the downshift inner cable from the downshift cam.
9. Take out three bolts and washers.
10. Lift off the valve body assembly.
11. Unscrew two bolts.
12. Remove the fluid tube locating plate.
13. Pull out the fluid tubes. (Note the 'O' ring on the pump suction tube.)
14. Take out five bolts.
15. Remove the pump and joint washer.
16. Remove the thrust washer.
17. Withdraw the front clutch.
18. Remove the thrust washers.
19. Withdraw the rear clutch, forward sun gear and bearing plate.
20. Squeeze together the ends of the front brake band and remove it together with the strut.

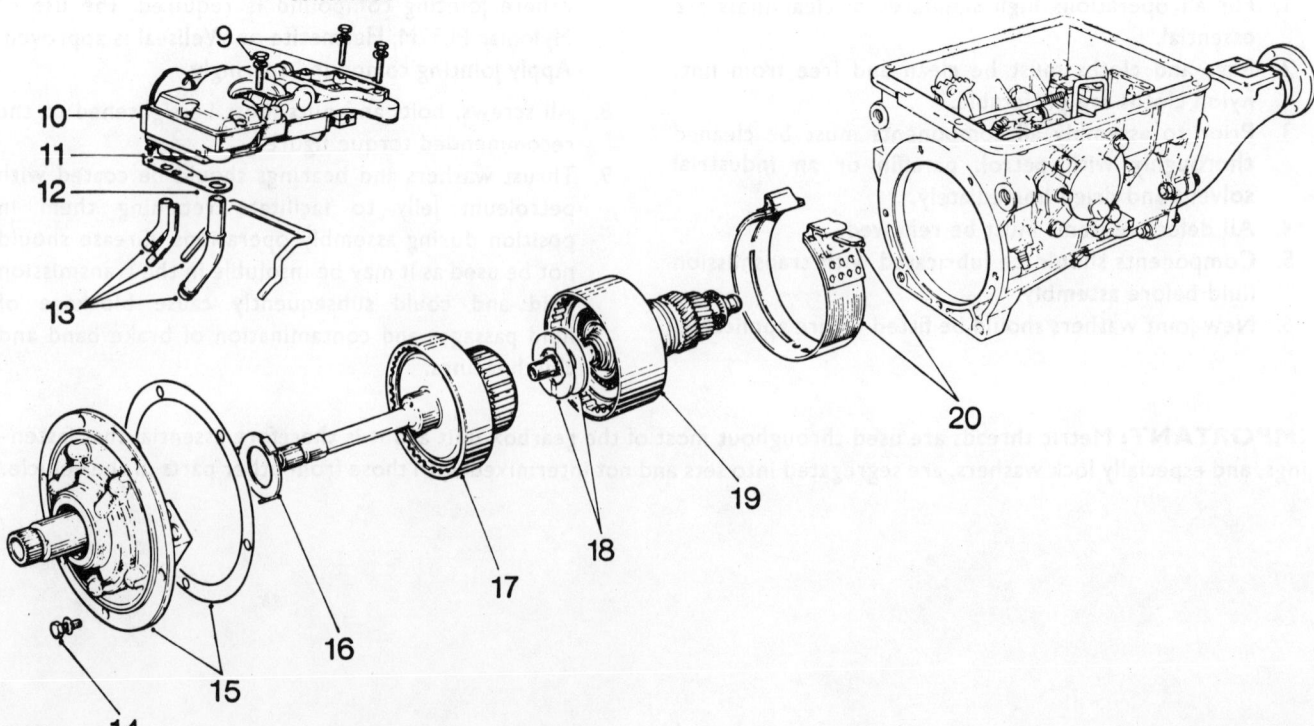

continued

44.10.01
Sheet 1

Rover 3500 and 3500S Manual AKM 3621

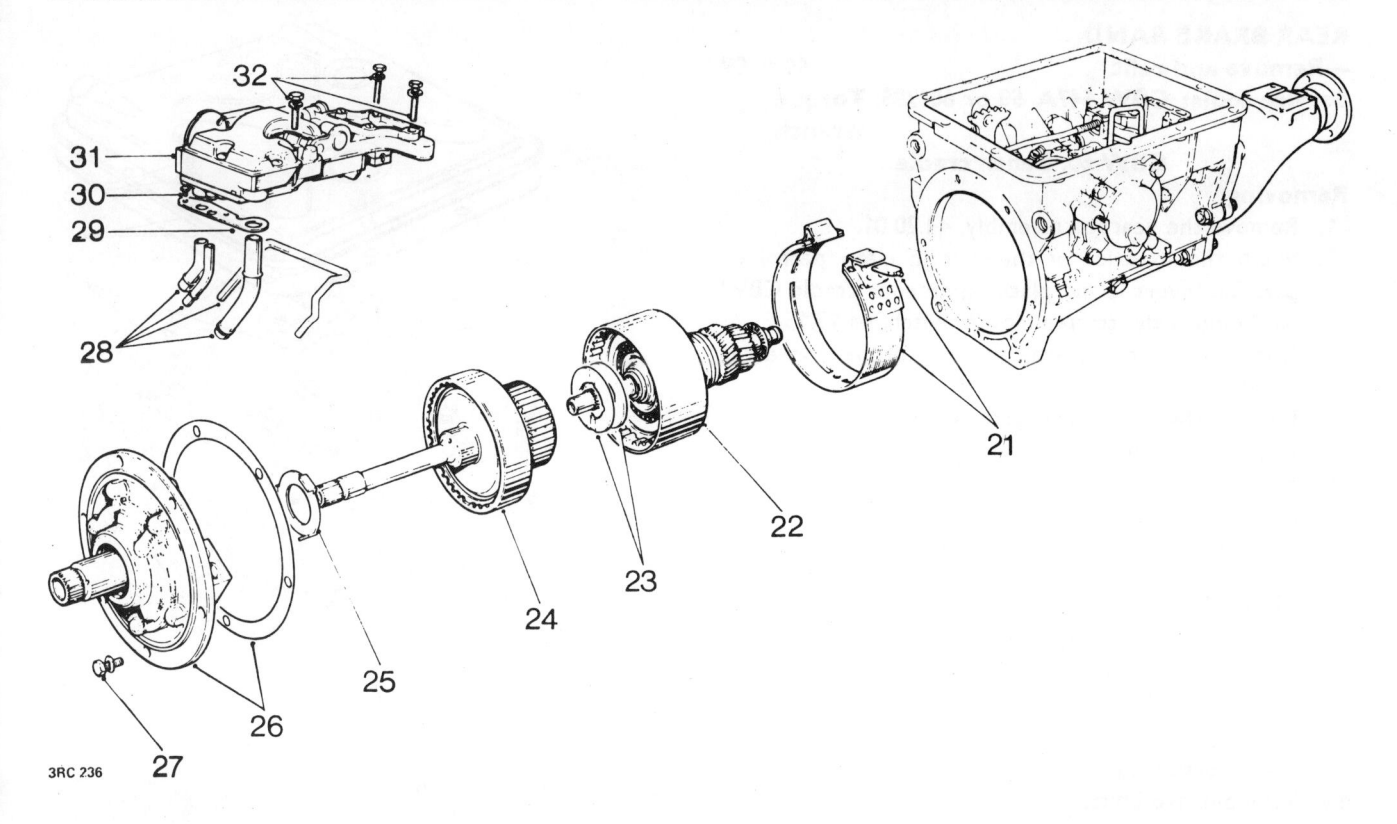

GEARBOX—AUTOMATIC—Borg-Warner type 65

Refitting

21. Squeeze together the ends of the front brake band and fit it in position together with the strut.
22. Refit the bearing plate, rear clutch and forward sun gear assembly.
23. Using petroleum jelly, stick the thrust washers to the rear clutch assembly (phosphor bronze towards the front clutch).
24. Refit the front clutch assembly.
25. Using petroleum jelly, stick the thrust washer to the pump assembly.
26. Refit the pump assembly and joint washer.
27. Fit and tighten the bolts, 1,8 to 3,4 kgf.m (13 to 25 lbf ft).
28. Refit the fluid tubes. (Note the 'O' ring on the pump suction tube.)
29. Refit the fluid tube locating plate.
30. Fit and tighten the two bolts, 0,24 to 0,35 kgf.m (20 to 30 lbf in).
31. Carefully refit the valve body assembly, ensuring that the tubes are not distorted.
32. Fit and tighten the three bolts and washers, 0,24 to 0,35 kgf.m (20 to 30 lbf in).
33. Connect the downshift inner cable to the downshift cam.
34. Refit the five fluid tubes.
35. Replace the magnet and refit the fluid pan and joint washer.
36. Fit and tighten 12 bolts, 0,6 to 0,9 kgf.m (4.5 to 7.0 lbf ft).
37. Locate the torque converter housing in place.
38. Fit and tighten four bolts securing the torque converter housing, 10 mm diameter—2,7 to 4,1 kgf.m (20 to 30 lbf ft) 12 mm diameter 4,1 to 6,9 kgf.m (30 to 50 lbf ft). 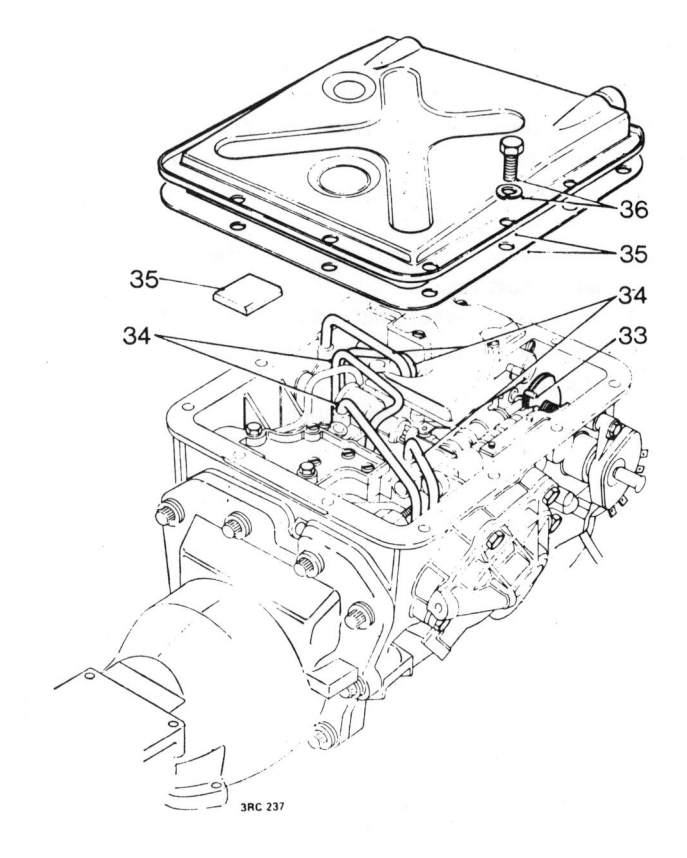
39. Refit the starter and reverse switch 44.15.15. 0,42 to 0,69 kgf.m (36 to 60 lbf in).
40. Check, and if necessary adjust, the front brake band. 44.30.07.
41. Refit the gearbox assembly.

Rover 3500 and 3500S Manual AKM 3621

44.10.01
Sheet 2

GEARBOX—AUTOMATIC—Borg-Warner type 65

REAR BRAKE BAND
—Remove and refit 44.10.09
Service tools: CBW 547A–50 or 601285, Torque wrench
CBW 60, Bench cradle

Removing
1. Remove the gearbox assembly. 44.20.01.
2. Wash the exterior of the unit in clean petrol or paraffin, invert it and place on a bench cradle CBW 60. Remove the starter motor switch. 44.15.15.
3. Unscrew the bolts securing the torque converter housing.
4. Remove the torque converter housing.
5. Unscrew 12 bolts.
6. Remove the fluid pan, joint washer and magnet.
7. Pull out the fluid tubes.
8. Release the downshift inner cable from the downshift cam.
9. Take out three bolts and washers.
10. Lift off the valve body assembly
11. Unscrew two bolts.
12. Remove the fluid tube locating plate.
13. Pull out the fluid tubes. Note the 'O', ring on the pump suction tube.)
14. Take out five bolts.
15. Remove the pump and joint washer.
16. Remove the thrust washer.
17. Withdraw the front clutch.
18. Remove the thrust washers.
19. Withdraw the rear clutch and forward sun gear.
20. Squeeze together the ends of the front brake band and remove it together with the strut.
21. Unscrew the bolts.
22. Withdraw the centre support/planet gear assembly and thrust race.
23. Squeeze together the ends of the rear brake band, tilt and withdraw it from the casing together with the strut.

continued

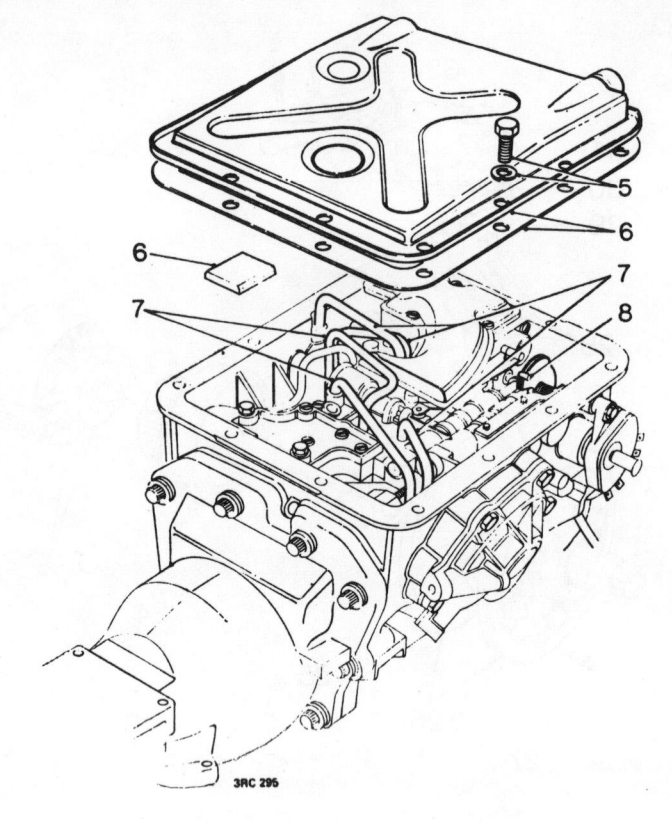

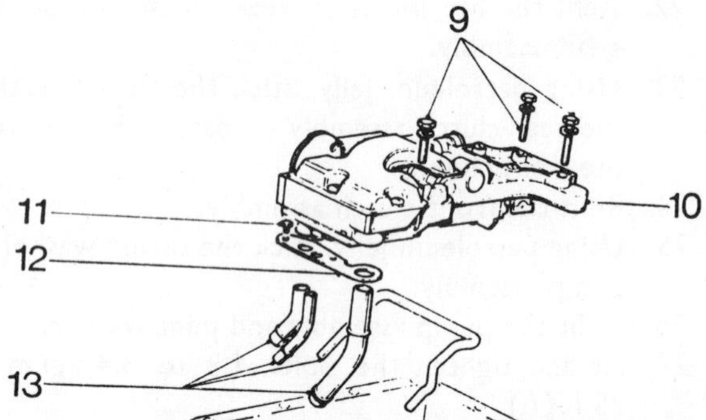

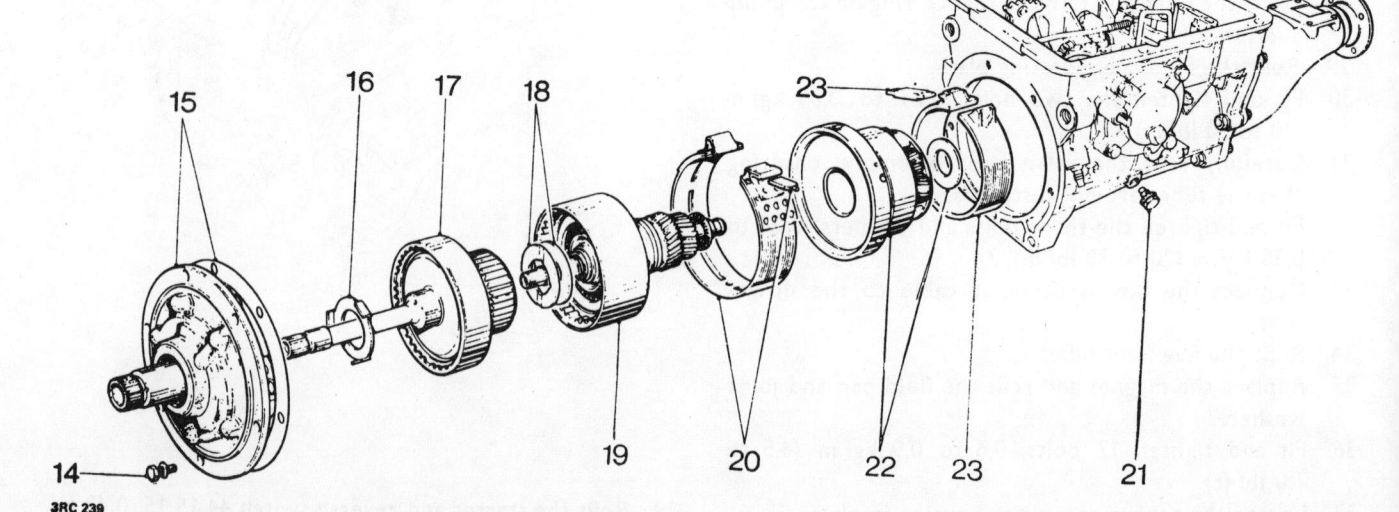

44.10.09
Sheet 1

Rover 3500 and 3500S Manual AKM 3621

GEARBOX—AUTOMATIC—Borg-Warner type 65

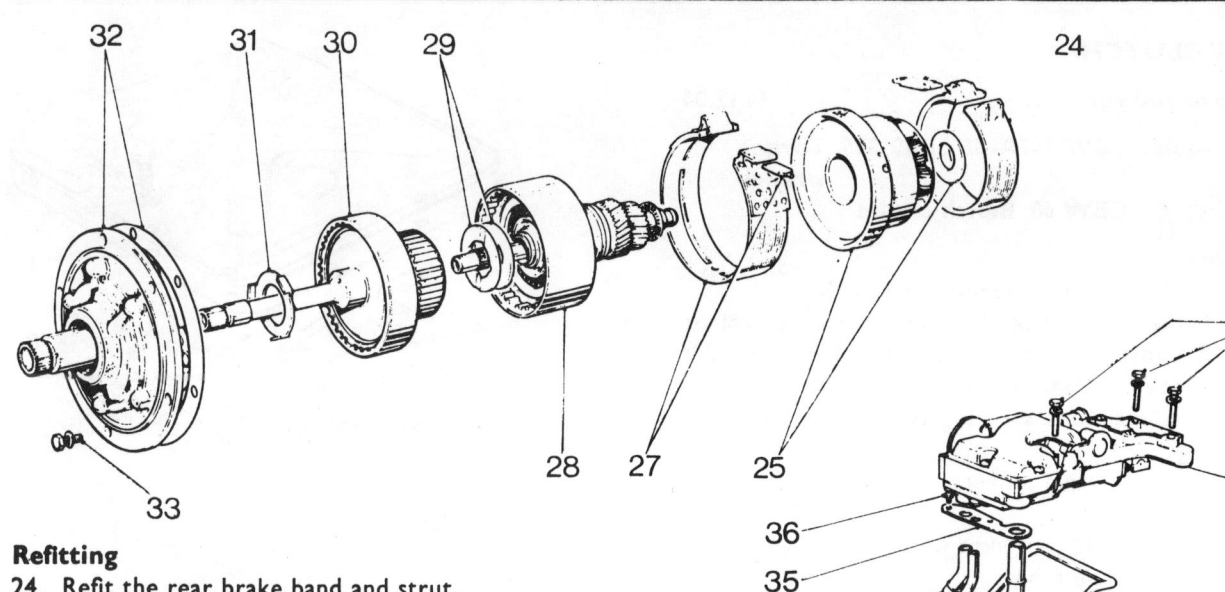

Refitting

24. Refit the rear brake band and strut.
25. Refit the centre support and planet gear assembly, ensuring that the fluid holes in the centre support are aligned with those in the casing.
26. Fit and tighten the bolts.
27. Squeeze together the ends of the front brake band and fit it in position together with the strut.
28. Refit the rear clutch and forward sun gear assembly.
29. Using petroleum jelly, stick the thrust washers to the rear clutch assembly (phosphor bronze towards the front clutch).
30. Refit the front clutch assembly.
31. Using petroleum jelly, stick the thrust washer to the pump assembly.
32. Refit the pump assembly and joint washer.
33. Fit and tighten the bolts, 1,8 to 3,4 kgf.m (13 to 25 lbf ft).
34. Refit the fluid tubes. (Note the 'O' ring on the pump suction tube).
35. Refit the fluid tube locating plate.
36. Fit and tighten the two bolts, 0,24 to 0,35 kgf.m (20 to 30 lbf in).
37. Carefully refit the valve body assembly, ensuring that the fluid tubes are not distorted.
38. Fit and tighten the three bolts and washers, 0,24 to 0,35 kgf m (20 to 30 lbf in).
39. Connect the downshift inner cable to the downshift cam.
40. Refit the fluid tubes.
41. Replace the magnet and refit the fluid pan and joint washer.
42. Fit and tighten 12 bolts, 0,6 to 0,9 kgf.m (4.5 to 7.0 lbf ft).
43. Locate the torque converter housing in place.
44. Fit and tighten four bolts securing the torque converter housing. 10 mm diameter—2,7 to 4,1 kgf.m (20 to 30 lbf ft) 12 mm diameter—4,1 to 6,9 kgf.m (30 to 50 lbf ft).
45. Refit the starter and reverse switch. 44.15.15. 0,42 to 0,69 kgf.m (36 to 60 lbf in).
46. Check, and if necessary adjust, the front and rear brake bands. 44.30.07/10.
47. Refit the gearbox assembly. 44.20.01.

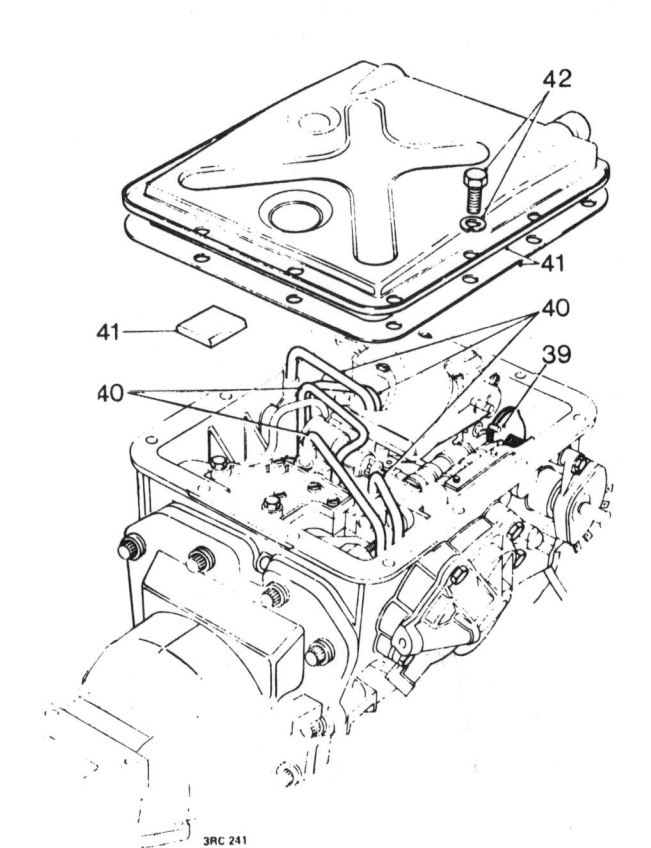

Rover 3500 and 3500S Manual AKM 3621

44.10.09
Sheet 2

GEARBOX—AUTOMATIC—Borg-Warner type 65

FRONT CLUTCH

—Remove and refit   44.12.04

Service tools: CBW 547A–50 or 601285, Torque wrench
           CBW 60, Bench cradle

Removing

1. Remove the gearbox assembly. 44.20.01.
2. Wash the exterior of the unit in clean petrol or paraffin, invert it and place on a bench cradle CBW 60. Remove the starter and reverse switch. 44.15.15.
3. Unscrew the bolts securing the torque converter housing.
4. Remove the torque converter housing.
5. Unscrew 12 bolts.
6. Remove the fluid pan, joint washer and magnet.
7. Pull out the fluid tubes.
8. Release the downshift inner cable from the downshift cam.
9. Take out three bolts and washers.
10. Lift off the valve body assembly.
11. Unscrew two bolts.
12. Remove the fluid tube locating plate.
13. Pull out the fluid tubes. (Note the 'O' ring on the pump suction tube).
14. Take out five bolts.
15. Remove the pump joint washer.
16. Remove the thrust washer.
17. Withdraw the front clutch.
18. Remove the thrust washers.

                 continued

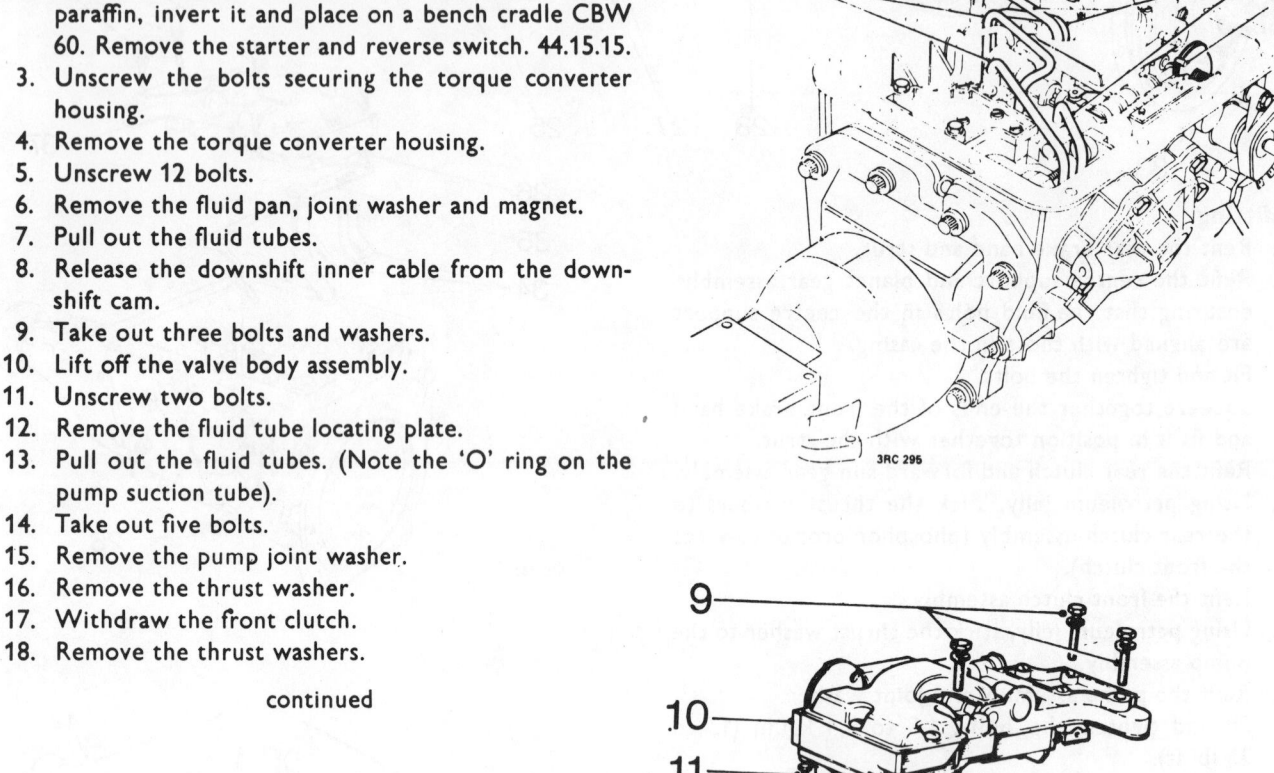

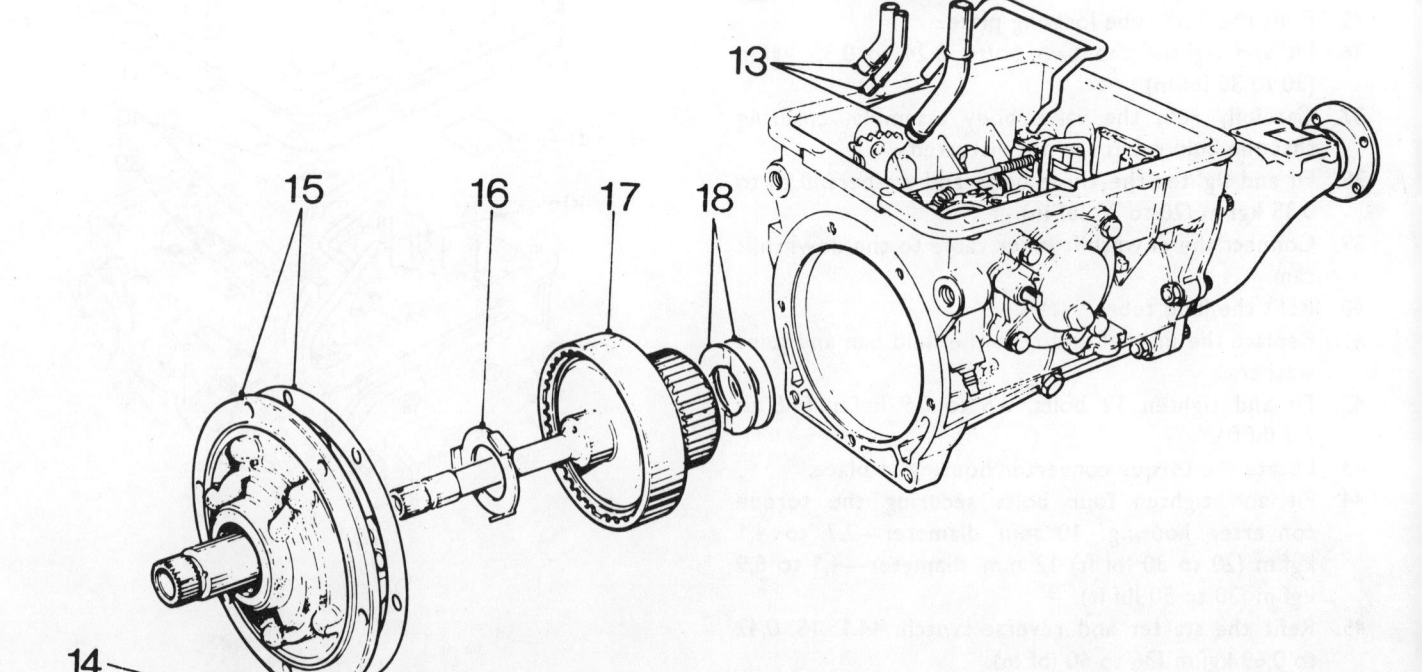

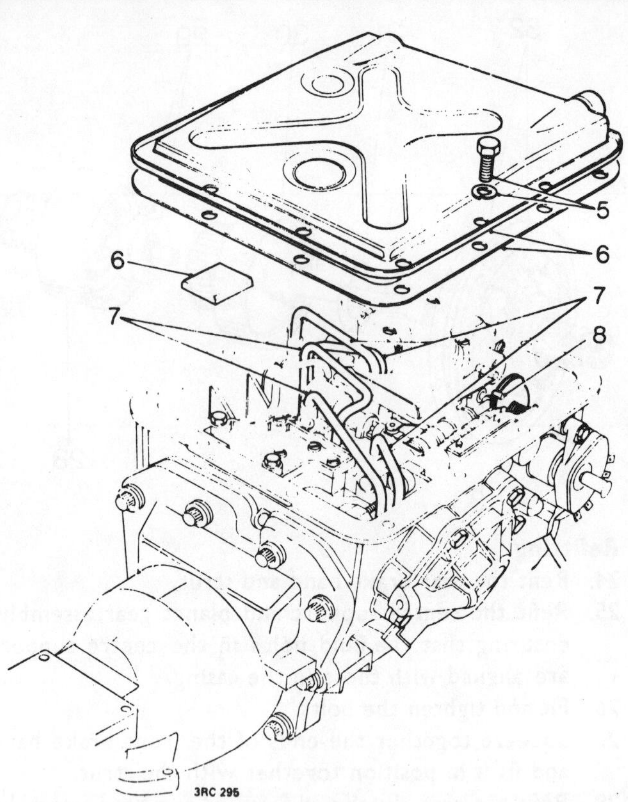

GEARBOX—AUTOMATIC—Borg-Warner type 65

Refitting

19. Using petroleum jelly, stick the thrust washers to the rear clutch assembly (phosphor bronze towards the front clutch).
20. Refit the front clutch assembly.
21. Using petroleum jelly, stick the thrust washer to the pump assembly.
22. Refit the pump assembly and joint washer.
23. Fit and tighten the bolts, 1,8 to 3,4 kgf.m (13 to 25 lbf ft).
24. Refit the fluid tubes. (Note the 'O' ring on the pump suction tube).
25. Refit the fluid tube locating plate.
26. Fit and tighten the two bolts, 0,24 to 0,35 kgf.m (20 to 30 lbf in).
27. Carefully refit the valve body assembly, ensuring that the fluid tubes are not distorted.
28. Fit and tighten the three bolts and washers, 0,24 to 0,35 kgf.m (20 to 30 lbf in).
29. Connect the downshift inner cable to the downshift cam.
30. Refit the fluid tubes.
31. Replace the magnet and refit the fluid pan and joint washer.
32. Fit and tighten the 12 bolts, 0,6 to 0,9 kgf.m (4.5 to 7.0 lbf ft).
33. Locate the torque converter housing in place.
34. Fit and tighten four bolts securing the torque converter housing. 10 mm diameter—2,7 to 4,1 kgf.m (20 to 30 lbf ft) 12 mm diameter—4,1 to 6,9 kgf.m (30 to 50 lbf ft).
35. Refit the starter and reverse switch. 44.15.15. 0,42 to 0,69 kgf.m (36 to 60 lbf in).
36. Refit the gearbox assembly. 44.20.01.

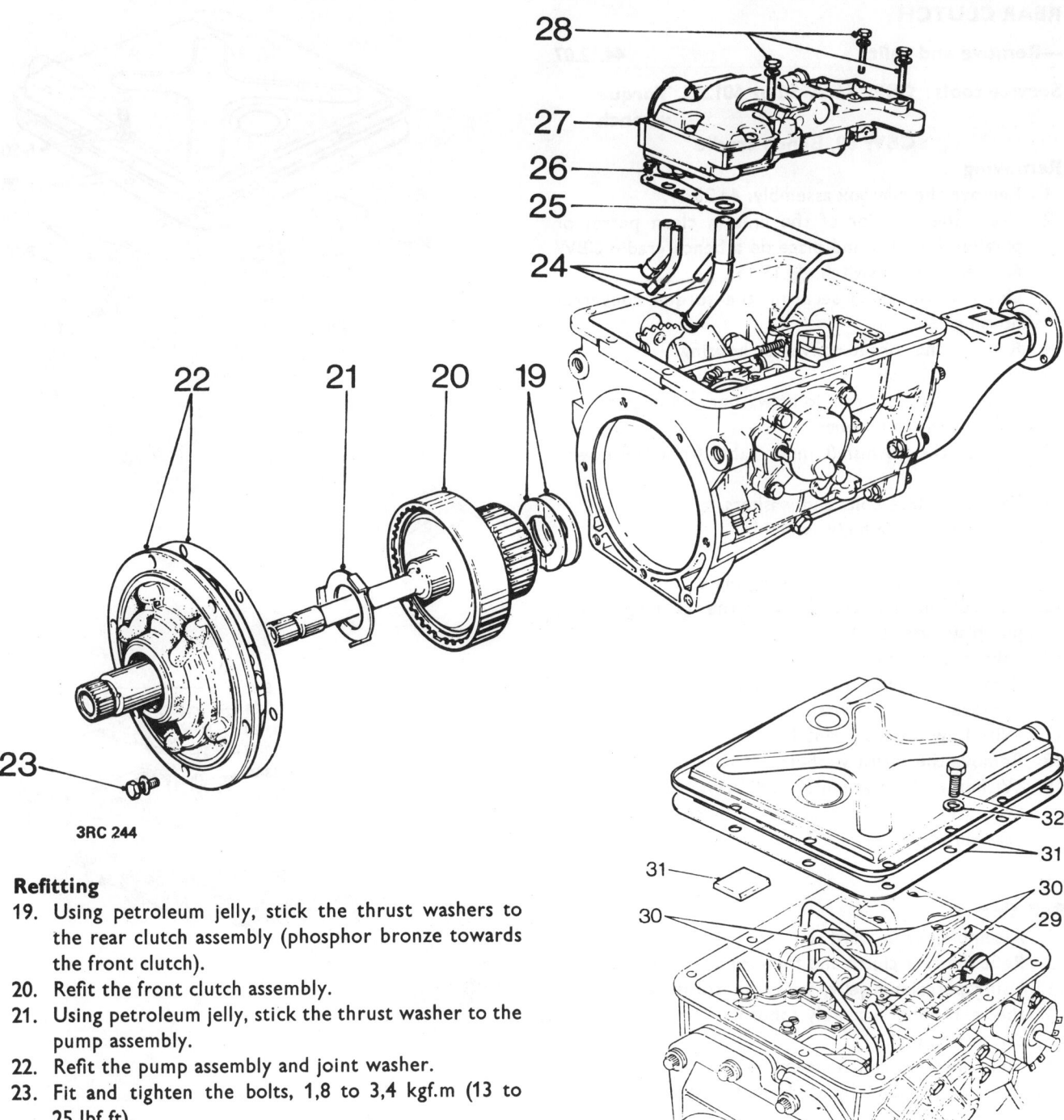

Rover 3500 and 3500S Manual AKM 3621

GEARBOX—AUTOMATIC—Borg-Warner type 65

REAR CLUTCH

—Remove and refit 44.12.07

Service tools: CBW 547A–50 or 601285, Torque wrench
CBW 60, Bench cradle

Removing

1. Remove the gearbox assembly. 44.20.01.
2. Wash the exterior of the unit in clean petrol or paraffin, invert it and place on a bench cradle CBW 60. Remove the switch. 44.15.15.
3. Unscrew the bolts securing the torque converter housing.
4. Remove the torque converter housing.
5. Unscrew 12 bolts.
6. Remove the fluid pan, joint washer and magnet.
7. Pull out the fluid tubes.
8. Release the downshift inner cable from the downshift cam.
9. Take out three bolts and washers.
10. Lift off the valve body assembly.
11. Unscrew two bolts.
12. Remove the fluid tube locating plate.
13. Pull out the fluid tubes. (Note the 'O' ring on the pump suction tube).
14. Take out five bolts.
15. Remove the pump and joint washer.
16. Remove the thrust washer.
17. Withdraw the front clutch.
18. Remove the thrust washers.
19. Withdraw the rear clutch and forward sun gear.
20. Separate the forward sun gear assembly from the rear clutch.

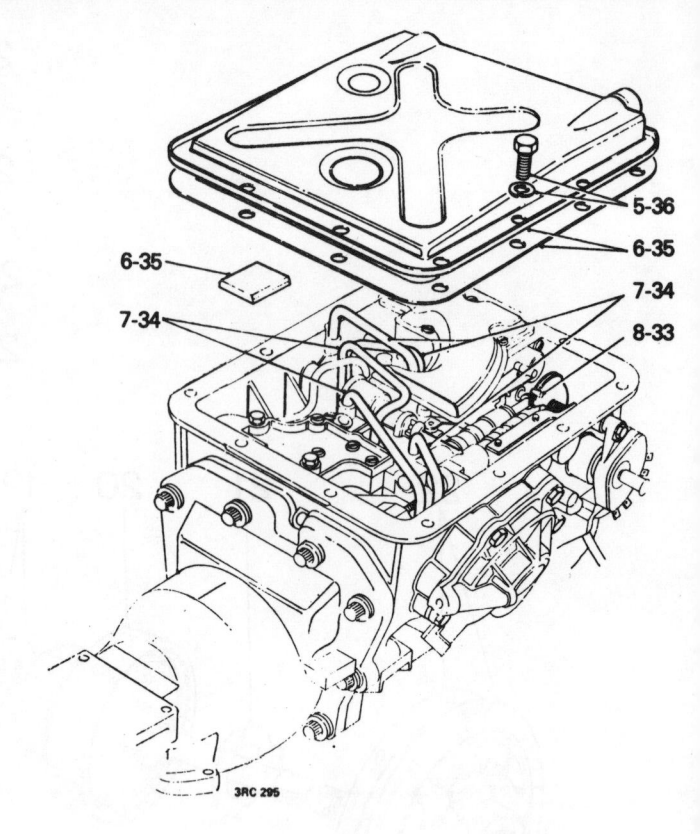

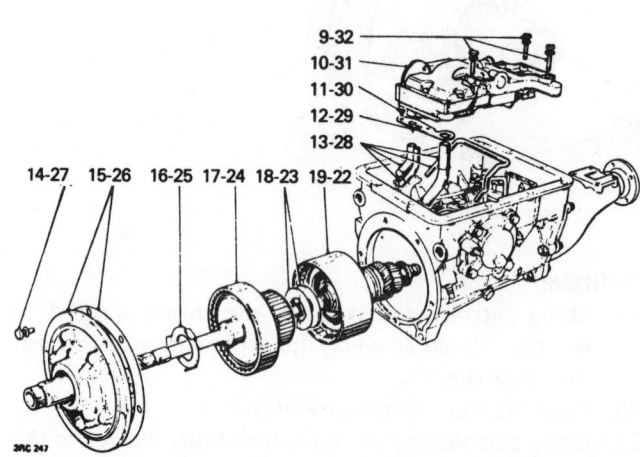

Refitting

21. Assemble the forward sun gear to the rear clutch.
22. Refit the rear clutch and forward sun gear assembly.
23. Using petroleum jelly, stick the thrust washers to the rear clutch assembly (phosphor bronze towards the front clutch).
24. Refit the front clutch assembly.
25. Using petroleum jelly, stick the thrust washer to the pump assembly.
26. Refit the pump assembly and joint washer.
27. Fit and tighten the bolts, 1,8 to 3,4 kgf.m (13 to 25 lbf ft).
28. Refit the fluid tubes. (Note the 'O' ring on the pump suction tube).
29. Refit the fluid tube locating plate.
30. Fit and tighten the two bolts. 0,24 to 0,35 kgf.m (20 to 30 lbf in).
31. Carefully refit the valve body assembly, ensuring that the fluid tubes are not distorted.
32. Fit and tighten the three bolts and washers, 0,24 to 0,35 kgf.m (20 to 30 lbf in).
33. Connect the downshift inner cable to the downshift cam.
34. Refit the fluid tubes.
35. Replace the magnet and refit the fluid pan and joint washer.
36. Fit and tighten 12 bolts, 0,6 to 0,9 kgf.m (4.5 to 7.0 lbf ft).
37. Locate the torque converter housing in place.
38. Fit and tighten four bolts securing the torque converter housing. 10 mm diameter—2,7 to 4,1 kgf.m (20 to 30 lbf ft) 12 mm diameter—4,1 to 6,9 kgf.m (30 to 50 lbf ft).
39. Refit the starter and reverse switch. 44.15.15, 0,42 to 0,69 kgf.m (36 to 60 lbf in).
40. Refit the gearbox assembly. 44.20.01.

GEARBOX—AUTOMATIC—Borg-Warner type 65

FRONT CLUTCH

—Overhaul 44.12.10

Service tool: CBW 42, Front clutch piston replacer

Dismantling

1. Remove the front clutch. 44.12.04.
2. Remove the circlip.
3. Withdraw the input shaft.
4. Remove the thrust washer.
5. Remove the hub.
6. Take out the inner and outer friction plates.
7. Remove the pressure plate.
8. Remove the circlip.
9. Take out the spring.
10. Remove the spring bearing.
11. Withdraw the piston. (If necessary, blank off the bores of the clutch drum and apply a compressed air line to the piston valve hole).
12. Withdraw the two bellvile washers and the flat washer.
13. Remove the oil seal from the drum.
14. Remove the seal from the piston.
15. Withdraw the collar from the drum.
16. Remove the 'O' ring from the drum.

Inspecting

17. Obtain new seals and 'O' rings.
18. Examine all components for wear or damage.

Reassembling

19. Fit the 'O' ring in the drum.
20. Fit the collar into the drum.
21. Fit the seal to the piston.
22. Fit the oil seal to the drum.
23. Fit the flat washer and two bellvile washers
24. Fit the piston into tool no. CBW 42 and place the tool in the drum. Push the piston into the drum and remove the tool.
25. Locate the spring bearing in position.
26. Refit the spring.
27. Fit the circlip.
28. Refit the pressure plate.
29. Refit the hub.
30. Fit the inner and outer friction plates in alternate sequence.
31. Using petroleum jelly, stick the thrust washer to the hub.
32. Locate the input shaft in position.
33. Refit the circlip.
34. Refit the front clutch. 44.12.04.

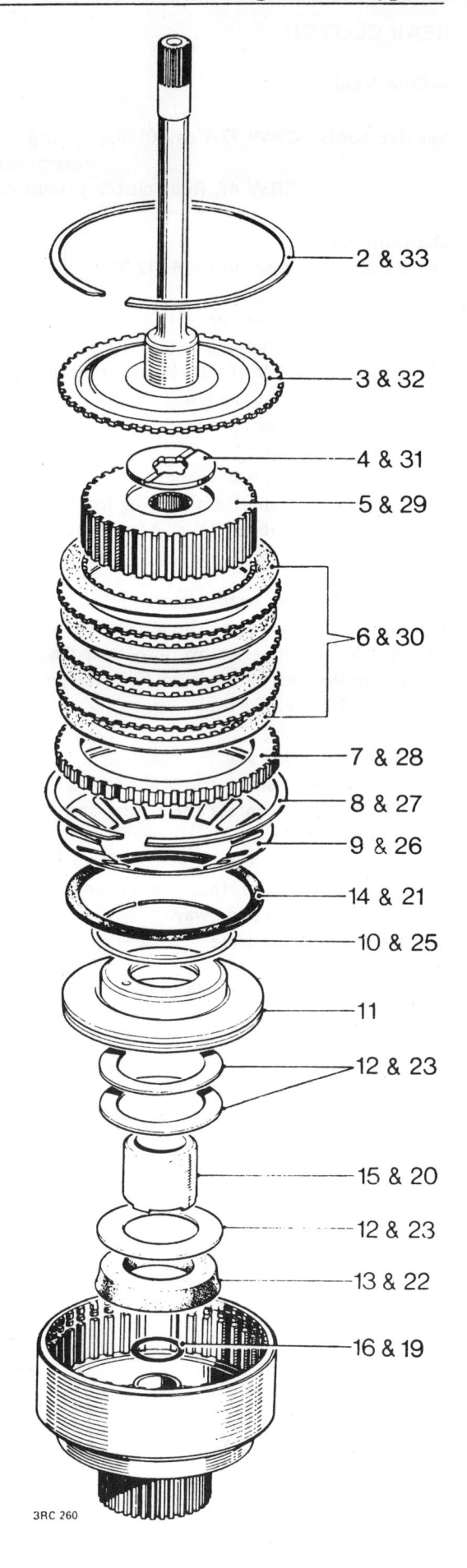

Rover 3500 and 3500S Manual AKM 3621

GEARBOX—AUTOMATIC—Borg-Warner type 65

REAR CLUTCH

—Overhaul   44.12.13

Service tools: CBW 37A or 601283, Spring compressor
CBW 41, Rear clutch piston replacer

Dismantling
1. Remove the rear clutch. 44.12.07.
2. Remove the circlip.
3. Take out the pressure plate.
4. Remove the inner and outer friction plates.
5. Using tool CBW 37A as shown, compress the spring and remove the spring seat circlip. Remove the tool.
6. Take out the spring seat.
7. Remove the spring.
8. Withdraw the piston.
9. Remove the rubber sealing ring from the piston.
10. Remove the rubber 'O' ring from the drum.

Reassembling
11. Fit the 'O' ring to the drum.
12. Fit the sealing ring to the piston drum.
13. Fit the piston assembly into tool CBW 41A and locate the tool in the drum. Push the piston into the drum. Remove the tool.
14. Refit the spring.
15. Refit the spring seat.
16. Locate the circlip on top of the spring seat and, using CBW 37A, compress the spring and fit the circlip. Remove the tool.
17. Refit the inner and outer clutch plates in alternate sequence. Ensure that the dishing on the metal plates faces the same way.
18. Fit the pressure plate.
19. Refit the circlip.
20. Refit the rear clutch. 44.12.07.

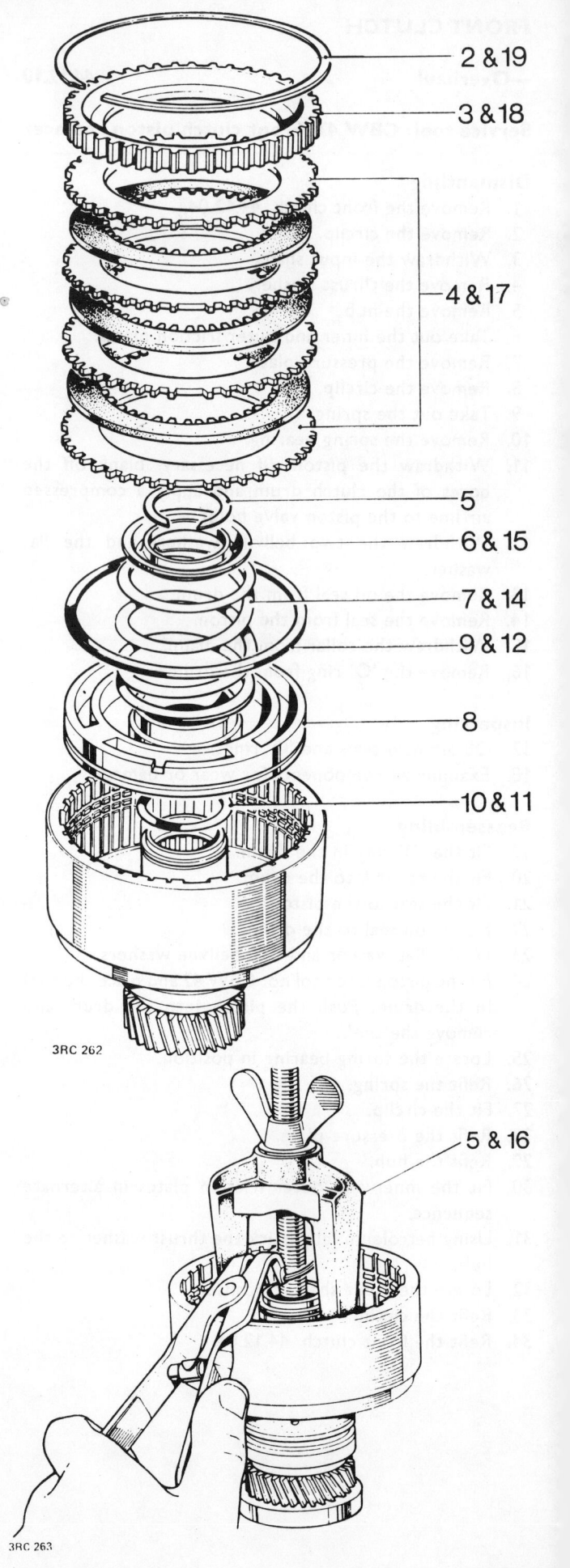

Rover 3500 and 3500S Manual AKM 3621

GEARBOX—AUTOMATIC—Borg-Warner type 65

ONE-WAY CLUTCH

—Remove and refit 44.12.16

Service tools: CBW 547A–50 or 601285, Torque wrench
CBW 60 Bench cradle

Removing

1. Remove the gearbox assembly. 44.20.01.
2. Wash the exterior of the unit in clean petrol or paraffin, invert it and place on a bench cradle CBW 60. Remove the starter and reverse switch. 44.15.15.
3. Unscrew the bolts securing the torque converter housing.
4. Remove the torque converter housing.
5. Unscrew 12 bolts.
6. Remove the fluid pan, joint washer and magnet.
7. Pull out the fluid tubes.
8. Release the downshift inner cable from the downshift cam.
9. Take out three bolts and washers.
10. Lift off the valve body assembly.
11. Unscrew two bolts.
12. Remove the fluid tube locating plate.
13. Pull out the fluid tubes. (Note the 'O' ring on the pump suction tube).
14. Take out five bolts.
15. Remove the pump and joint washer.
16. Remove the thrust washer.
17. Withdraw the front clutch.
18. Remove the thrust washer.
19. Withdraw the rear clutch and forward sun gear.
20. Squeeze together the ends of the front brake band and remove it together with the strut.
21. Unscrew the bolts.

continued

Rover 3500 and 3500S Manual AKM 3621

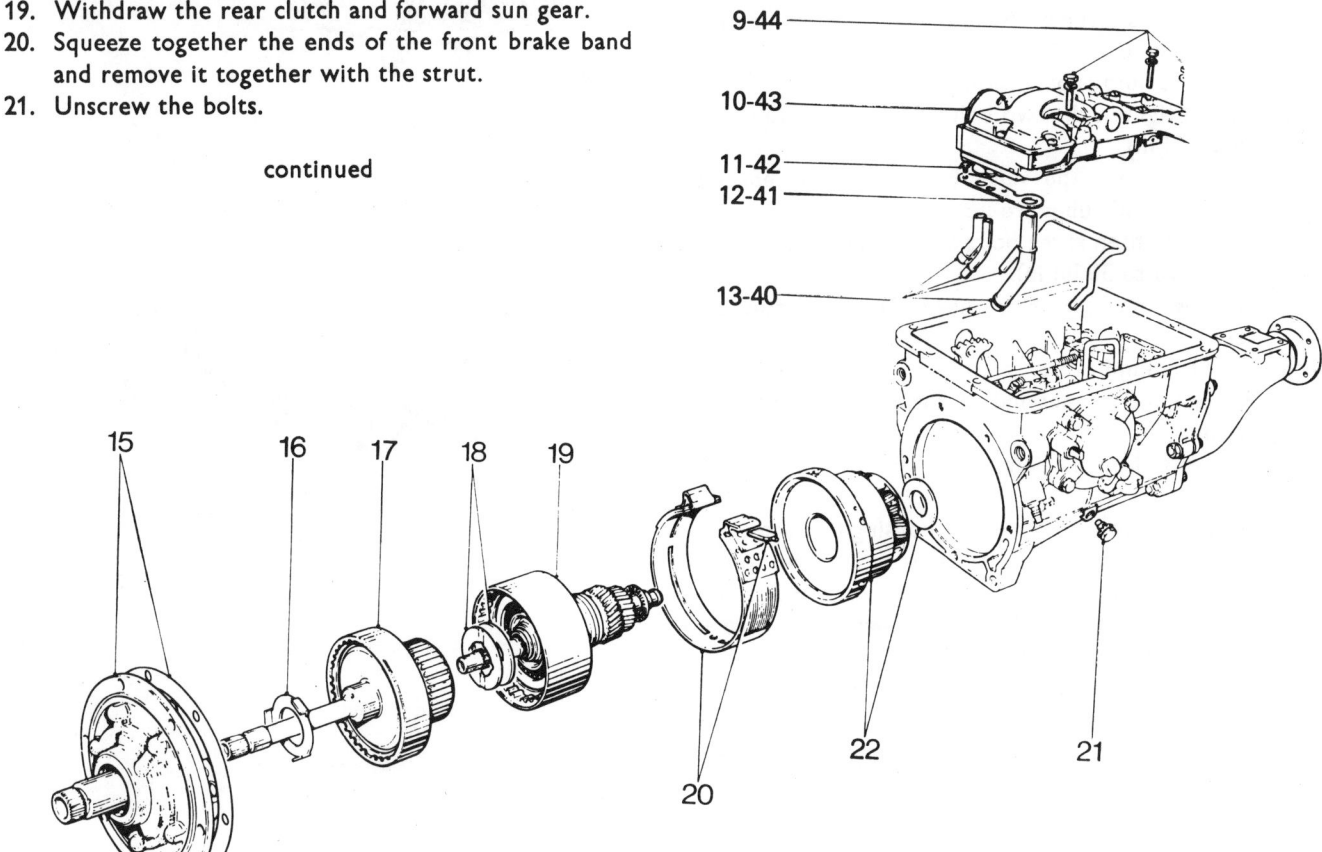

44.12.16
Sheet 1

GEARBOX—AUTOMATIC—Borg-Warner type 65

22. Withdraw the centre support/planet gear assembly.
23. Separate the centre support from the planet gear assembly.
24. Withdraw the one-way clutch.
25. Remove the circlip.
26. Remove the one-way clutch outer race.

Refitting

27. Refit the one-way clutch outer race to the rear drum.
28. Refit the circlip.
29. Refit the one-way clutch.
30. Assemble the centre support and planet gear assembly.
31. Refit the assembly, ensuring that the fluid and locating holes in the centre support align with those in the casing.
32. Fit and tighten the bolts.
33. Squeeze together the ends of the front brake band and fit it in position together with the strut.
34. Refit the rear clutch and forward sun gear assembly.
35. Using petroleum jelly, stick the thrust washer to the rear clutch assembly (phosplor bronze towards the front clutch).
36. Refit the front clutch assembly.
37. Using petroleum jelly, stick the thrust washer to the pump assembly.
38. Refit the pump assembly and joint washer.
39. Fit and tighten the bolts, 1,8 to 3,4 kgf.m (13 to 25 lbf ft).
40. Refit the fluid tubes. (Note the 'O' ring on the pump suction tube.)
41. Refit the fluid tube locating plate.
42. Fit and tighten the two bolts, 0,24 to 0,35 Kgf.m (20 to 30 lbf in).
43. Carefully refit the valve body assembly, ensuring that the fluid tubes are not distorted.
44. Fit and tighten three bolts and washers, 0,24 to 0,35 kgf.m (20 to 30 lbf in).
45. Connect the downshift inner cable to the downshift cam.
46. Refit the fluid tubes.
47. Replace the magnet and refit the fluid pan and joint washer.
48. Fit and tighten 12 bolts, 0,6 to 0,9 kgf.m (4.5 to 7.0 lbf ft).
49. Locate the torque converter housing in place.
50. Fit and tighten four bolts securing the torque converter housing. 10 mm diameter—2,7 to 4,1 kgf.m (20 to 30 lbf ft) 12 mm diameter—4,1 to 6,9 kgf.m (30 to 50 lbf ft).
51. Refit the starter and reverse switch. 44.15.15. 0,42 to 0,69 kgf.m (36 to 60 lbf in).
52. Check, and if necessary adjust, front brake band. 44.30.07.
53. Refit the gearbox assembly. 44.20.01.

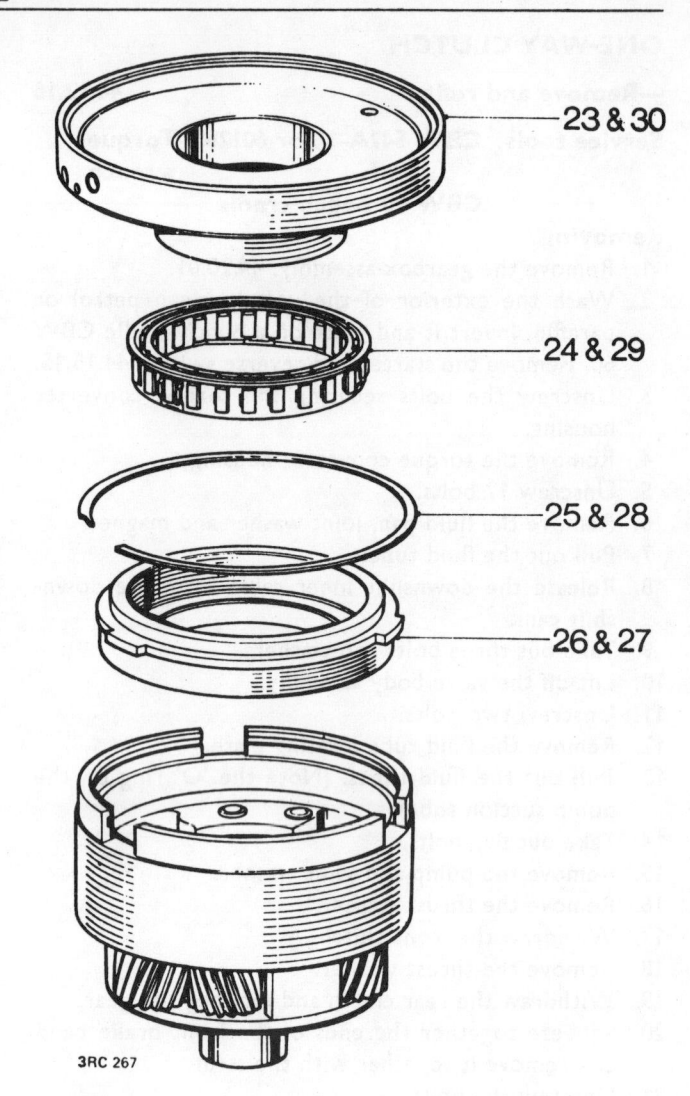

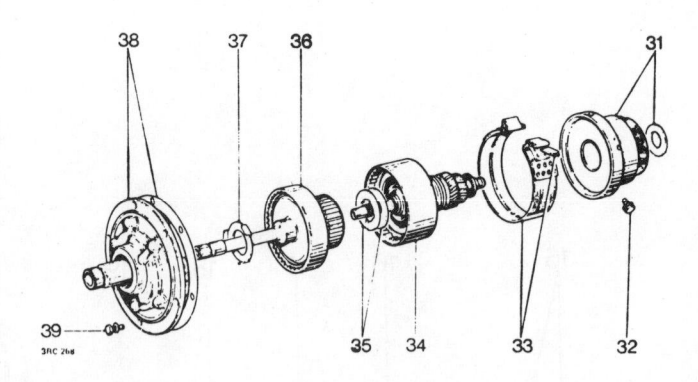

GEARBOX—AUTOMATIC—Borg-Warner type 65

DOWNSHIFT CABLE

—Remove and refit 44.15.01

Service tool: CBW 62 Extractor

Removing

1. Drive the vehicle on to a ramp, select 'N' and apply the hand brake.
2. Prop open the bonnet.
3. Remove the air cleaner. 19.10.01.
4. Disconnect the downshift cable from the throttle linkage and unscrew it from the adjuster.
5. Close the bonnet and raise the ramp.
6. Release the downshift cable from the clip at the converter housing flange.
7. Place a tray beneath the fluid pan.
8. Drain the gearbox fluid by disconnecting the filler pipe from the fluid pan.
9. Remove the fluid pan. 44.24.04.
10. Disconnect the downshift inner cable from the cam.
11. Using CBW 62, push the downshift outer cable from the gearbox casing.
12. Withdraw the downshift cable assembly.

Refitting

13. Reverse 4 to 12.
14. Set the downshift cable adjuster to give a gap of 0,25 to 0,50 mm (0.010 to 0.020 in) between the crimped stop on the downshift inner cable and the end of the adjuster.
15. Reverse 1 to 3.

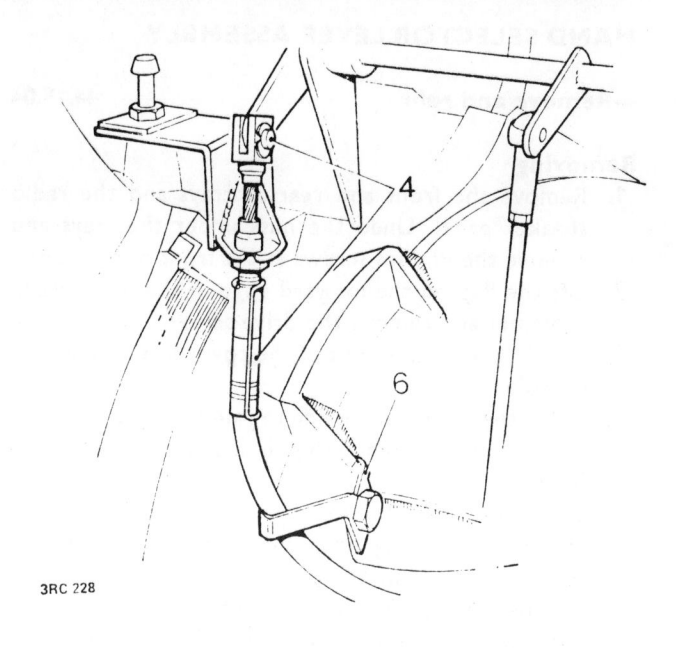

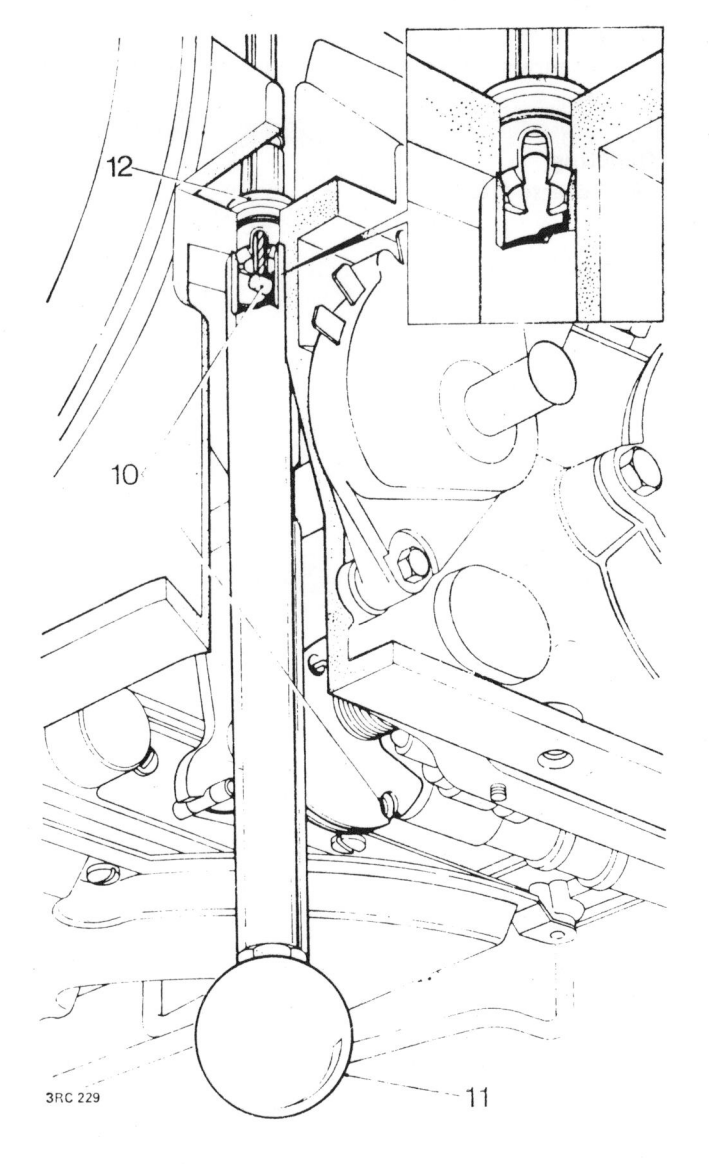

Rover 3500 and 3500S Manual AKM 3621

44.15.01

GEARBOX—AUTOMATIC—Borg-Warner type 65

HAND SELECTOR LEVER ASSEMBLY

—Remove and refit   44.15.04

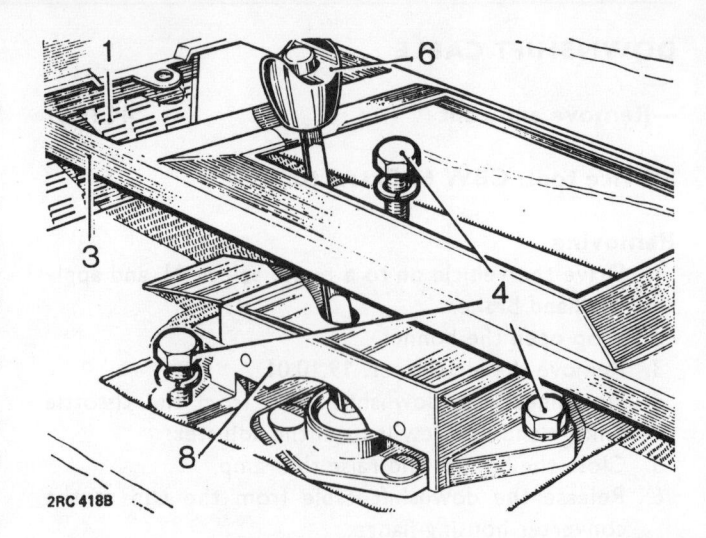

Removing

1. Remove the front and rear ashtrays and the radio speaker panel. Undo the nuts under the trays and remove the Phillips screws at the front of the cover.
2. Lift the flap on the forward edge of the hand brake grommet and remove the drive screw.
3. Carefully ease the front of the cover from under the console.
4. Swing the cover to one side to gain access to the hand selector mechanism fixing bolts, and remove the bolts.
5. Unclip the indicator for the gear selector and disconnect the wiring from the illumination lamp, at adjacent snap connectors.
6. Unscrew the knob from the selector lever.
7. From under the car, disconnect the selector cable from the lever.
8. Remove the pivot housing from under the car, and withdraw the selector lever housing from inside the car.

Refitting

9. Reverse 1 to 8.

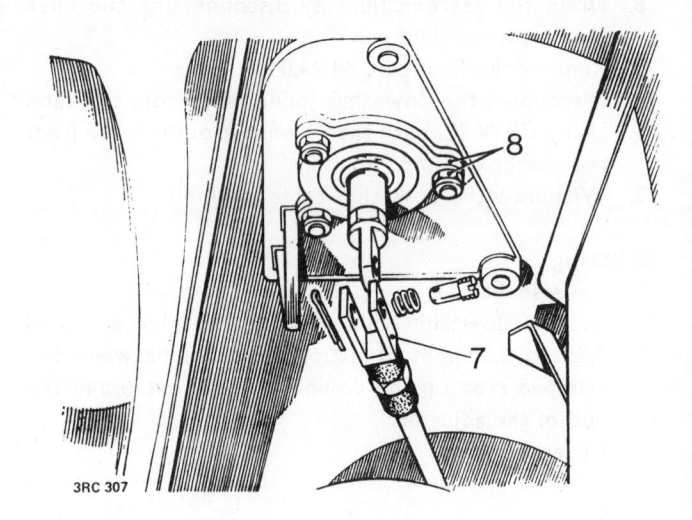

44.15.04

Rover 3500 and 3500S Manual AKM 3621

GEARBOX—AUTOMATIC—Borg-Warner type 65

HAND SELECTOR LEVER ASSEMBLY

—Overhaul 44.15.05

Dismantling

1. Remove the hand selector lever assembly. 44.15.04.
2. If required, remove the rubber insert, 'O' ring and button.
3. Remove the shield and the illumination lamp from the housing.
4. Withdraw the control rod.
5. Remove the retaining cap for the spherical seat.
6. Remove the spherical seat by forcing it off the lever by hand. Note the locating pin.
7. Remove the eye bolt, locknut and spring.
8. Remove the detent pin and plug from the lever.

Inspecting

9. Obtain new parts as necessary.
10. If new control rod is being fitted, it should be cut to length on assembly so that 12,5 mm (0.500 in) protrudes from the top of the gear selector lever. This dimension is important to ensure correct engagement of the detent pin in the selector gate.

Reassembling

11. Reverse 1 to 8.

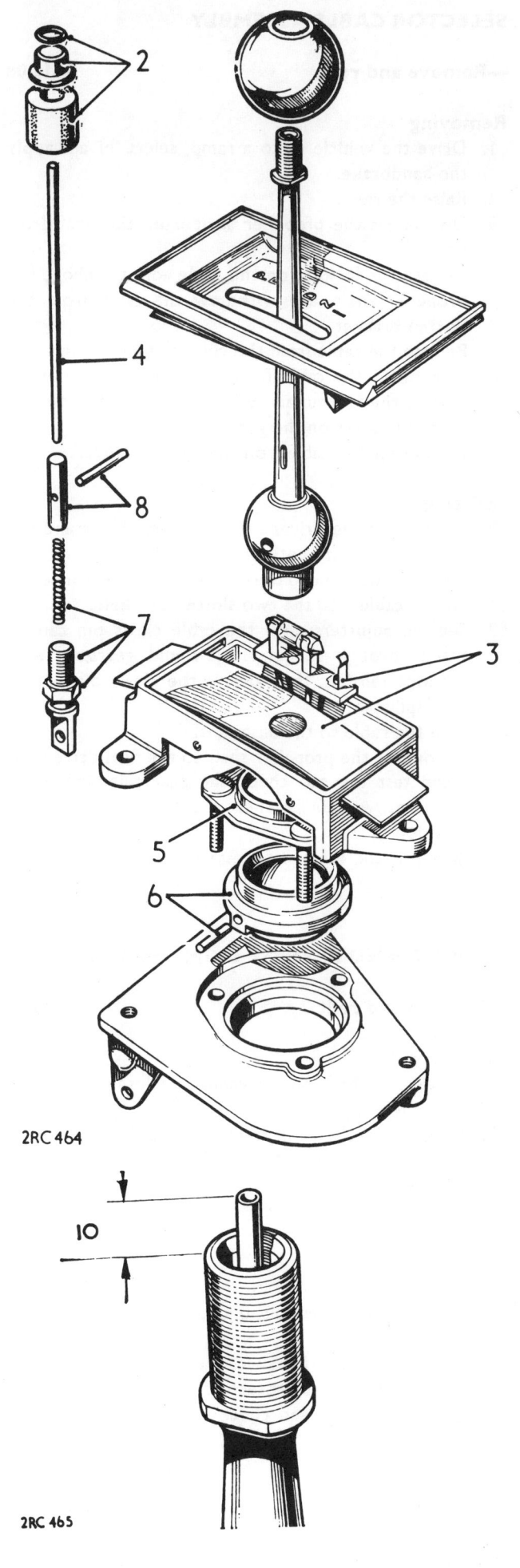

Rover 3500 and 3500S Manual AKM 3621

GEARBOX—AUTOMATIC—Borg-Warner type 65

SELECTOR CABLE ASSEMBLY

—Remove and refit 44.15.08

Removing

1. Drive the vehicle on to a ramp, select 'N' and apply the handbrake.
2. Raise the ramp.
3. Disconnect the propeller shaft from the final drive unit.
4. Release the locknut on the cable adjuster above the propeller shaft, and withdraw the cable from the slotted bracket.
5. Release the cable from the clip on the base unit.
6. Disconnect the cable from the gear selector lever.
7. Release the locknut and withdraw the cable from the slotted bracket on the gearbox.
8. Disconnect the cable from the gearbox lever.

Refitting

9. Check that the driver's gear selector lever and the gearbox lever are both in 'N' position.
10. Connect the cable to the driver's gear selector lever.
11. Fit the cable into the two slotted brackets.
12. Set the adjusters until the cable clevis pin can be fitted through the clevis jaw and gearbox lever without tension, then secure the clevis pin with a new split pin.
13. Clip the cable to the base unit.
14. Reconnect the propeller shaft to the final drive unit.
15. Road test the car, check the gear selection in all positions. A definite click should be felt in each position.
16. Readjust the cable as necessary.

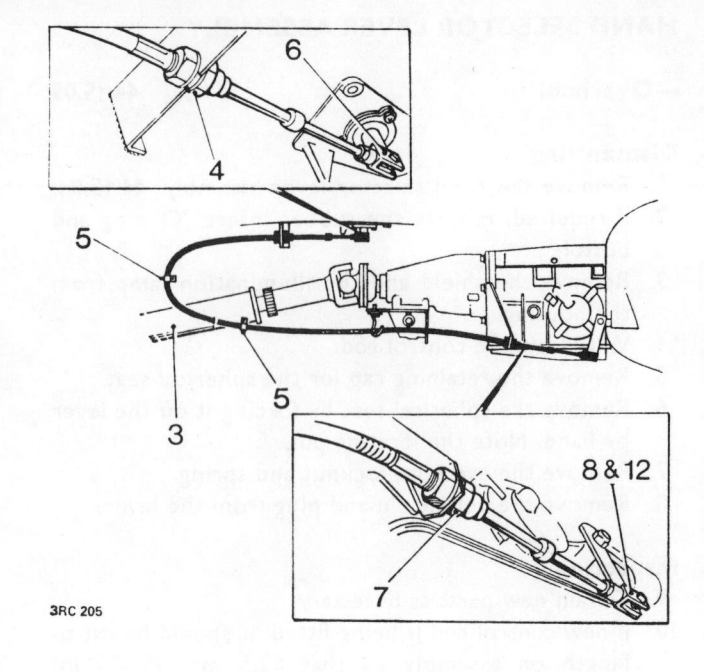

STARTER AND REVERSE LIGHT SWITCH

—Remove and refit 44.15.15

Removing

1. Drive the vehicle on to a ramp, chock the wheels and raise the ramp.
2. Remove the thread protector (if fitted).
3. Disconnect the leads from the switch.
4. Unscrew the bolt.
5. Remove the switch.

Refitting

6. Refit the switch.
7. Fit and tighten the bolt, 0,42 to 0,69 kgf.m (36 to 60 lbf in).
8. Refit the leads to the switch terminals.
9. Replace the thread protector.
10. Lower the ramp and check the switch for correct operation.

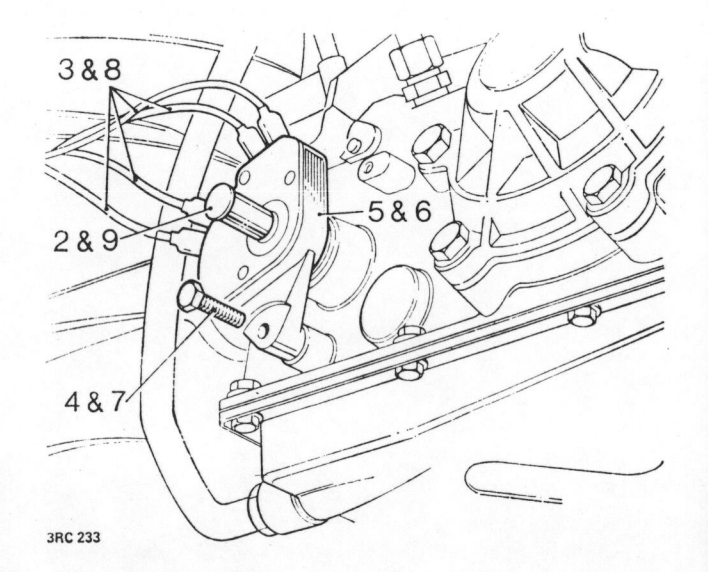

GEARBOX—AUTOMATIC—Borg-Warner type 65

CONVERTER HOUSING
—Remove and refit 44.17.01

Service tool: CBW 547A-50 or 601285, Torque wrench

Removing
1. Remove the gearbox assembly. 44.20.01.
2. Unscrew the four bolts securing the converter housing to the gearbox.
3. Remove the converter housing.

Refitting
4. Place the converter housing in position.
5. Fit and tighten the four bolts.
   10 mm diameter: 2,7 to 4,1 kgf.m (20 to 30 lbf ft).
   12 mm diameter: 4,1 to 6,9 kgf.m (30 to 50 lbf ft).
6. Refit the gearbox assembly. 44.20.01.

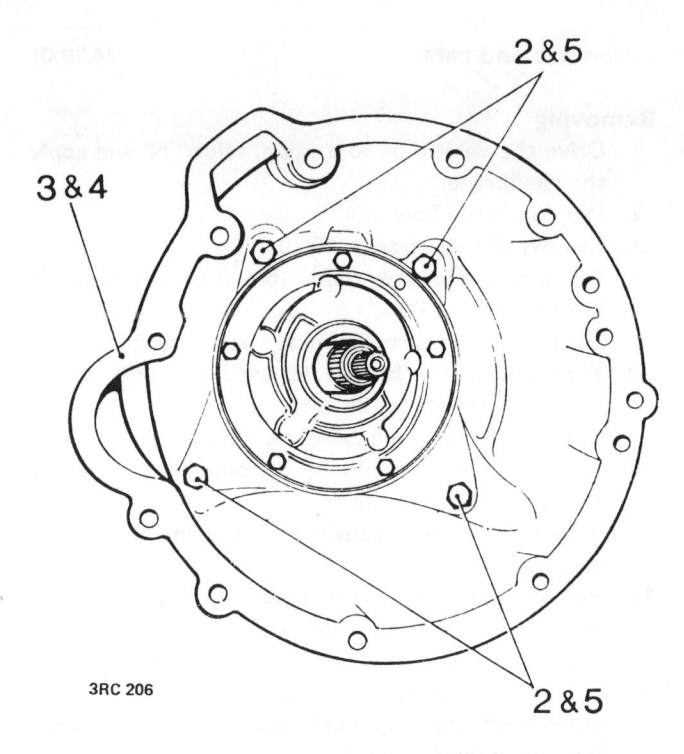

CONVERTER ASSEMBLY

—Remove and refit 44.17.07

Service tool: CBW 547A or 601285, Torque wrench

Removing
1. Remove the fan blades. 26.25.06.
2. Remove the gearbox assembly. 44.20.01.
3. Add alignment marks to the converter and drive plate.
4. Support the converter and remove the four securing bolts, turning the engine as necessary.
5. Withdraw the converter taking precautions against spilling any residual fluid.

Refitting
6. Reverse 1 to 5, including the following.
7. Tightening torque for converter securing bolts: 3,5 to 4,0 kgf.m (25 to 30 lbf ft).

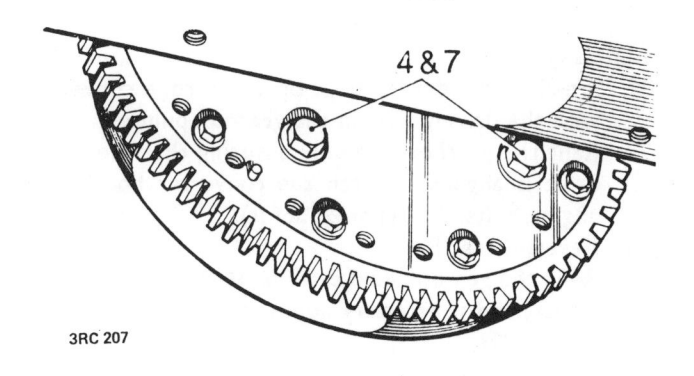

Rover 3500 and 3500S Manual AKM 3621

GEARBOX—AUTOMATIC—Borg-Warner type 65

GEARBOX ASSEMBLY

—Remove and refit 44.20.01

Removing

1. Drive the vehicle on to a ramp, select 'N' and apply the handbrake.
2. Prop open the bonnet.
3. Remove the air cleaner. 19.10.01.
4. Disconnect the downshift cable from the throttle linkage.
5. Disconnect the throttle linkage at the coupling.
6. Turn the engine fan blades until the two widest spaced blades are at the top.
7. Close the bonnet and raise the ramp.
8. Drain the gearbox fluid by disconnecting the filler pipe from the fluid pan.
9. Remove the cover plate from the front of the converter housing.
10. Remove the front exhaust pipe.
11. Remove the propeller shaft.
12. Disconnect the speedometer cable from the gearbox.
13. Disconnect the gearchange cable from the gearbox.
14. Disconnect the electrical leads from the starter inhibitor/reverse switch.
15. Remove the fixings from the gearbox rear mounting, and lower the assembly until the exhaust manifolds are resting on the base unit.
16. Release the gearbox breather tube from the retaining clip.
17. Disconnect the oil cooler pipes from the gearbox.
18. Support the gearbox.
19. Remove the bolts securing the converter housing to the engine.
20. Withdraw the gearbox assembly.

Refitting

21. Pre-align the driving dogs on the torque converter with the driving slots in the gearbox pump.
22. Refit the gearbox assembly ensuring that the pump drive is aligned. Tighten the converter housing to engine bolts: 3,5 kgf.m (25 lbf ft).
23. Reverse 3 to 18.
24. Refill the gearbox fluid system. 44.24.02.
25. Check, and if necessary adjust, the downshift cable. 44.30.02/03.

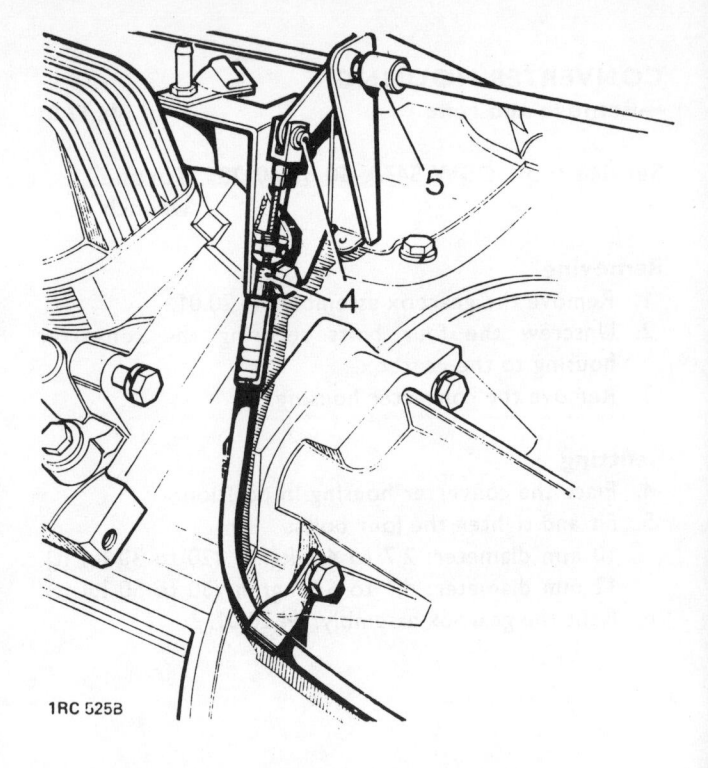

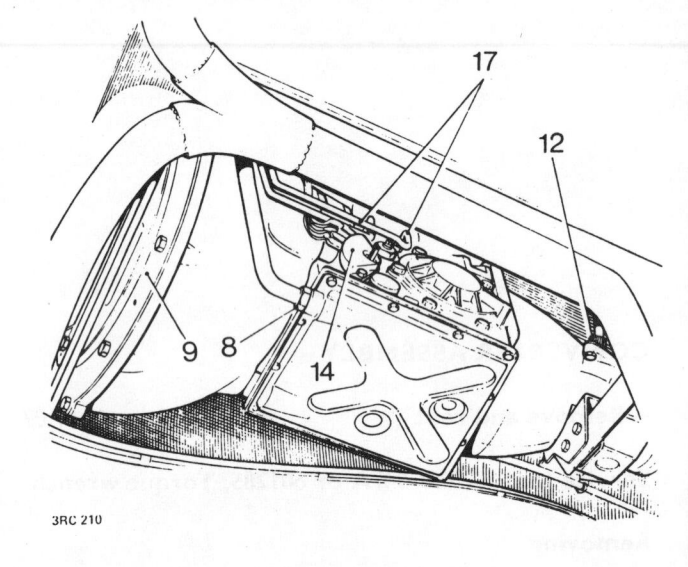

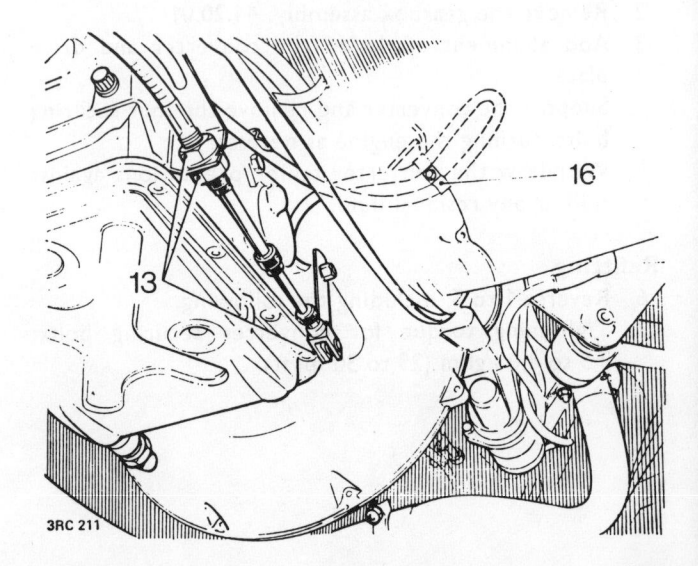

Rover 3500 and 3500S Manual AKM 3621

GEARBOX—AUTOMATIC—Borg-Warner type 65

GEARBOX ASSEMBLY

—Overhaul   44.20.06

Service tools: CBW 60, Bench cradle
                 CBW 87, Mainshaft end float gauge
                 CBW 62, Extractor, downshift cable
                 CBW 547A–50 or 601285, Torque wrench
                 CBW 61, Spanners, servo adjusters
                 18G 1205, Retainer, coupling flange

Dismantling

1. Remove the gearbox assembly. 44.20.01.
2. Wash the exterior of the unit in clean petrol or paraffin, invert it and place on a bench cradle CBW 60.
3. Unscrew the bolts securing the torque converter housing.
4. Remove the torque converter housing.
5. Unscrew 12 bolts.
6. Remove the fluid pan, joint washer and magnet.
7. Pull out the fluid tubes.
8. Release the downshift inner cable from the downshift cam.
9. Take out three bolts and washers.
10. Lift off the valve body assembly.
11. Unscrew two bolts.
12. Remove the fluid tube locating plate.
13. Pull out the fluid tubes. (Note the 'O' ring on the pump suction tube).
14. Take out five bolts.
15. Remove the pump and joint washer.
16. Remove the thrust washer.
17. Withdraw the front clutch.
18. Remove the thrust washers.
19. Slacken the adjusters for the front and rear servos.
20. Withdraw the rear clutch and forward sun gear.
21. Squeeze together the ends of the front brake band and remove it together with the strut.

continued

Rover 3500 and 3500S Manual AKM 3621

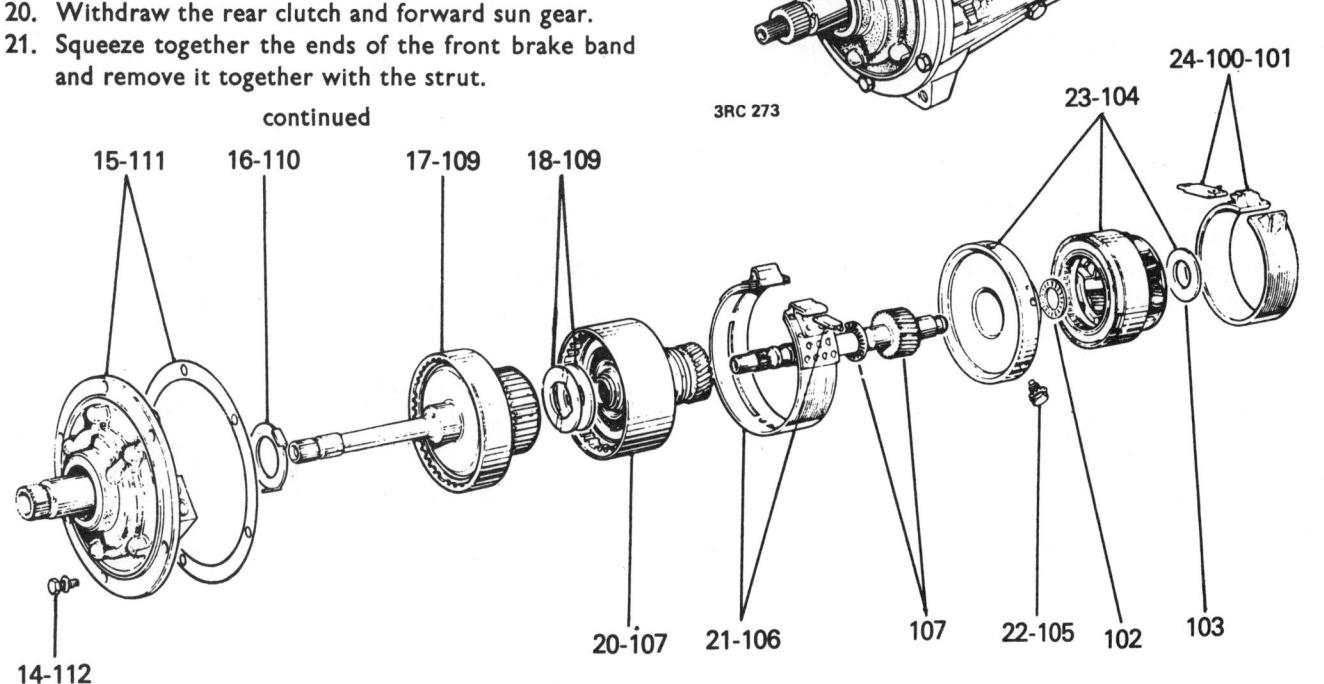

44.20.06
Sheet 1

GEARBOX—AUTOMATIC—Borg-Warner type 65

22. Unscrew the three bolts.
23. Withdraw the centre support/planet gear assembly.
24. Squeeze together the ends of the rear brake band, tilt and withdraw it together with the strut.
25. Using tool no. 18G 1205 to hold the flange, unscrew the bolt.
26. Withdraw the flange.
27. Unscrew the bolts.
28. Remove the rear extension and joint washer.
29. Withdraw the speedometer drive gear.
30. Unscrew the counterweight.
31. Withdraw the governor assembly.
32. Withdraw the output shaft assembly and thrust washer.
33. Remove the fluid tubes.
34. Unscrew the bolts.
35. Remove the rear servo assembly, joint washer and 'O' rings.
36. Unscrew the nut and remove the selector lever.
37. Unscrew the bolt and remove the starter and reverse switch.
38. Unscrew the bolts.
39. Remove the front servo and joint washer.
40. Remove the spring clip.
41. Withdraw the pin.
42. Withdraw the cross-shaft and remove the 'O' ring.
43. Remove the detent lever, collar and washers.
44. Remove the oil seal.
45. Unscrew two screws and remove the cam plate.
46. Remove the parking brake rod assembly.
47. Withdraw the parking brake pawl pivot pin.
48. Remove the parking brake pawl.
49. Remove the spring.
50. Remove the relay lever pivot pin.
51. Remove the relay lever.
52. Remove the torsion spring.
53. Using tool no. CBW 62, remove the downshift cable assembly.
54. Using tool no. CBW 62, remove the breather adaptor.
55. Unscrew the adaptor.
56. Unscrew the return valve.
57. Withdraw the rear servo lever pivot pin.
58. Remove the rear servo lever.
59. Unscrew the locknuts.
60. Unscrew the adjusting screws.
61. Unscrew the pressure take-off plug.

Reassembling

62. Fit the pressure take-off plug.
63. Refit the adjusting screws.
64. Loosely refit the locknuts.
65. Replace the rear servo lever.
66. Refit the rear servo lever pivot pin.
67. Fit the fluid return valve.

continued

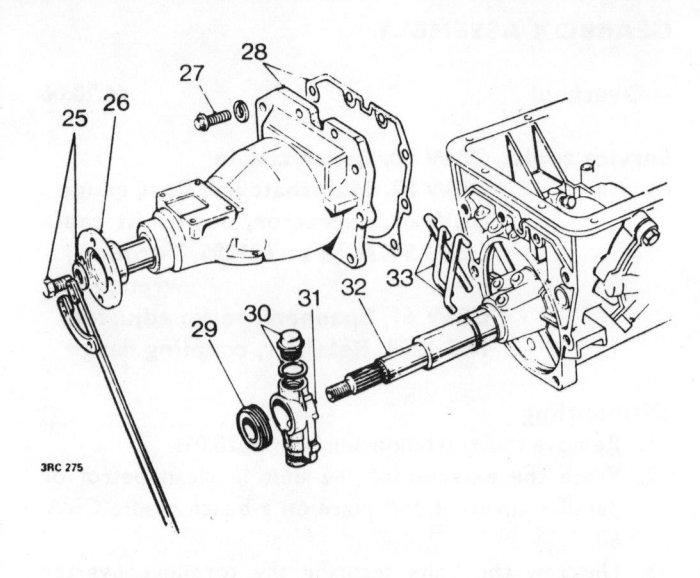

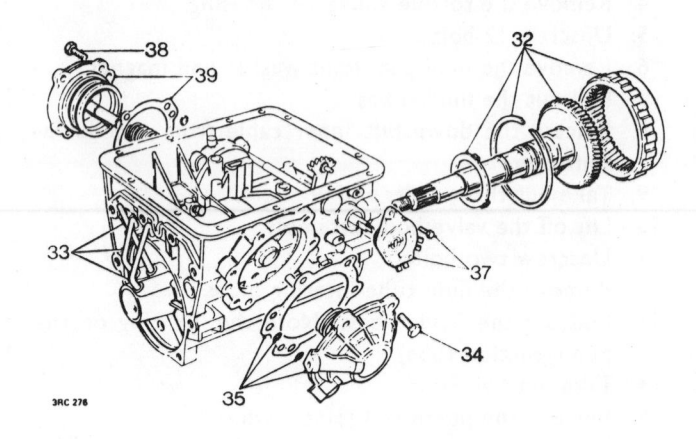

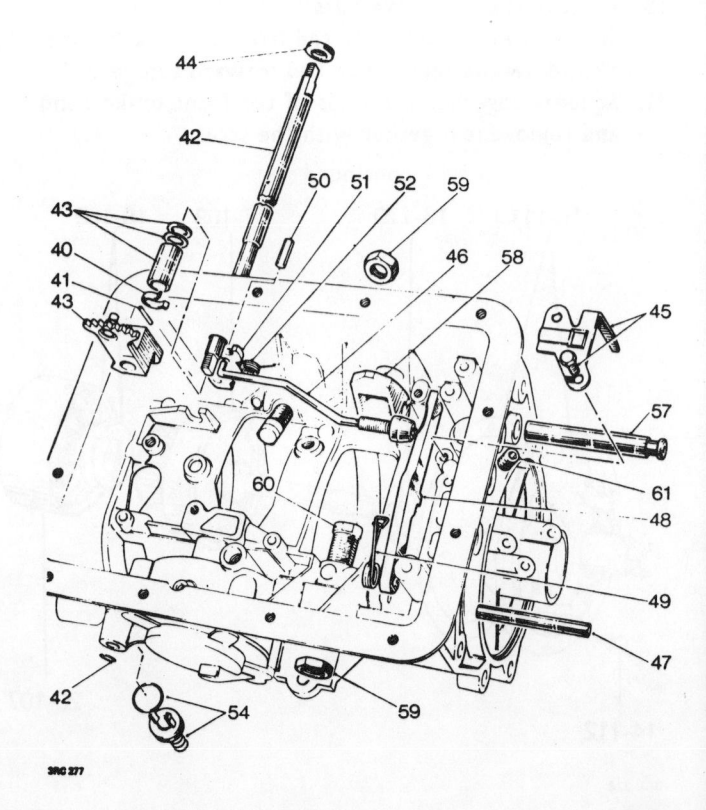

44.20.06
Sheet 2

Rover 3500 and 3500S Manual AKM 3621

GEARBOX—AUTOMATIC—Borg-Warner type 65

68. Fit the adaptor.
69. Refit the breather adaptor.
70. Refit the downshift cable.
71. Replace the relay lever and torsion spring.
72. Refit the relay lever pivot pin.
73. Replace the parking brake pawl and spring.
74. Refit the parking brake pawl pivot pin.
75. Refit the parking brake rod.
76. Refit the cam plate, ensuring that the tag end locates in the groove in the rear servo lever pivot pin.
77. Fit and tighten the bolts, 0,62 to 0,83 kgf.m (54 to 70 lbf in).
78. Fit a new cross-shaft oil seal.
79. Locate the cross-shaft through the oil seal and fit the washers.
80. Fit the collar and detent lever and push the cross-shaft fully home.
81. Refit the pin.
82. Refit the clip.
83. Refit the 'O' ring.
84. Refit the front servo and joint washer.
85. Fit and tighten the bolts, 1,8 to 3,4 kgf.m (13 to 25 lbf ft).
86. Refit the selector lever and secure with the nut.
87. Refit the switch and secure with the bolt, 0,42 to 0,69 kgf.m (36 to 60 lbf in).
88. Refit the rear servo assembly, joint washer and 'O' rings, retaining them in position using petroleum jelly.
89. Fit and tighten the bolts. 1,8 to 3,4 kgf.m (13 to 25 lbf ft).
90. Refit the fluid tubes, ensuring that they are correctly located.
91. Locate the thrust washer on the end wall of the casing, using petroleum jelly.
92. Carefully refit the output shaft assembly.
93. Refit the governor assembly, correct way round, tighten the bolt to 2,5 kgf.m (18 lbf ft).
94. Fit and tighten the counterweight.
95. Refit the speedometer drive gear.
96. Refit the rear extension housing and joint washer.
97. Fit and tighten the bolts. 4,1 to 7,6 kgf.m (30 to 55 lbf ft).
98. Tap the drive flange into position.
99. Coat the threads of the bolt with Loctite grade 'CV', and fit the bolt and washer, using tool 18G 1205 to hold the flange. 4,8 to 6,9 kgf.m (35 to 50 lbf ft).
100. Squeeze together the ends of the rear brake band, tilt and locate it in position.
101. Refit the rear brake band strut.
102. Check that the bearing track is located inside the planet carrier.
103. Using petroleum jelly, locate the thrust race on the planet carrier spigot.
104. Refit the centre support/planet gear assembly ensuring that the fluid and locating holes align with those in the casing.

continued

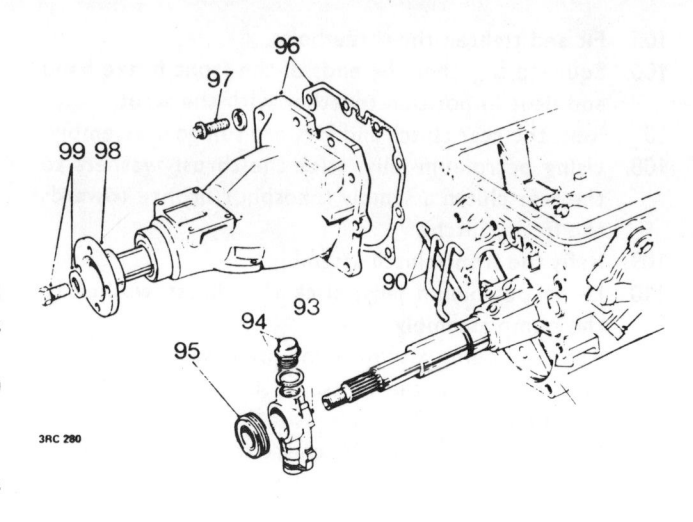

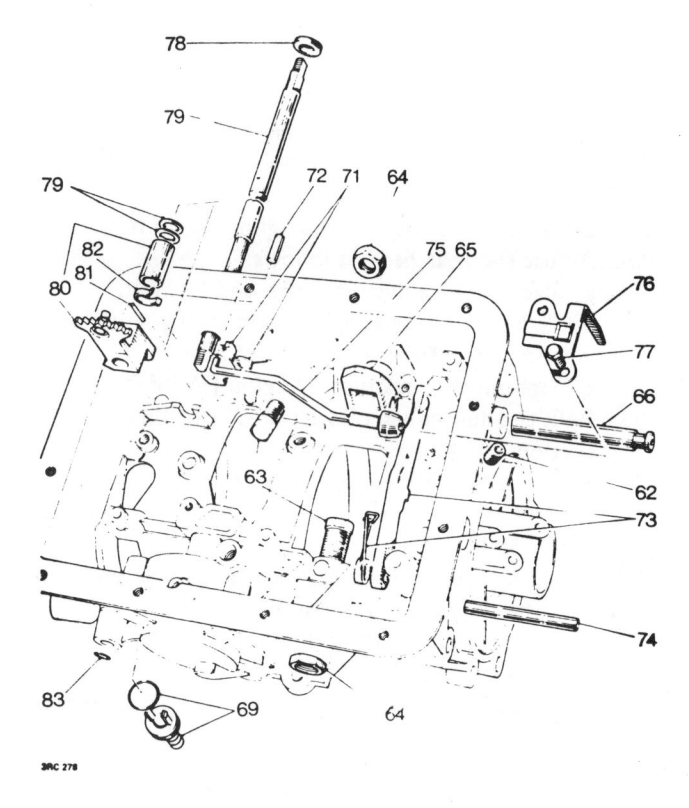

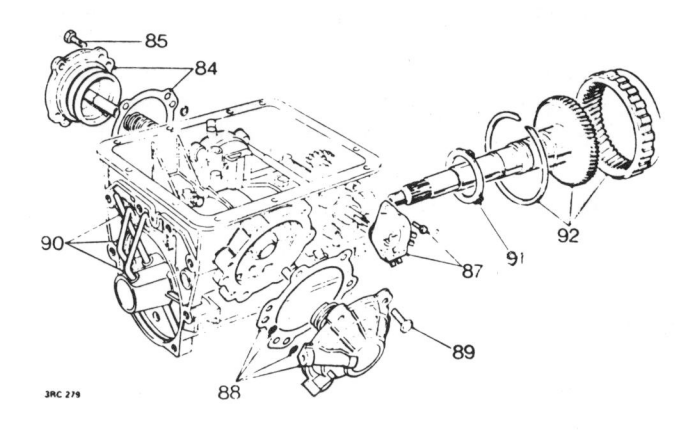

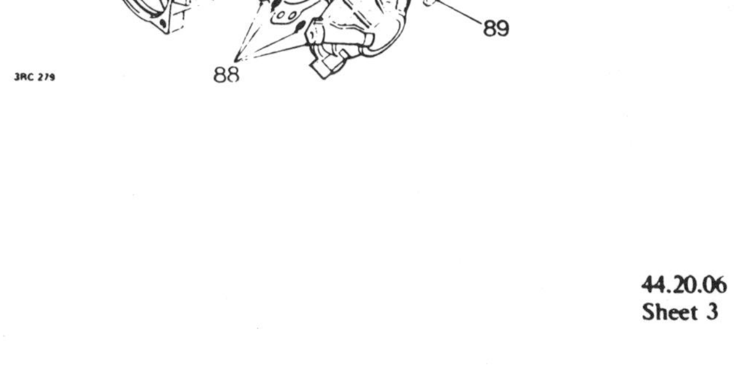

GEARBOX—AUTOMATIC—Borg-Warner type 65

105. Fit and tighten the three bolts.
106. Squeeze together the ends of the front brake band and fit it in position together with the strut.
107. Refit the rear clutch and forward sun gear assembly.
108. Using petroleum jelly, stick the thrust washers to the rear clutch assembly (phosphor bronze towards the front clutch).
109. Refit the front clutch assembly.
110. Using petroleum jelly, stick the thrust washer to the pump assembly.
111. Refit the pump assembly and joint washer.
112. Fit and tighten the bolts. 1,8 to 3,4 kgf.m (13 to 25 lbf ft).
113. Using tool no. CBW 87, check the gear train end-float, and if necessary, adjust by the selective use of the thrust washer fitted between the pump and the front clutch.
    Recommended end-float 0.25 mm to 0.75 mm (0.010 to 0.030 in).
114. Adjust the front band as follows:
    a. Slacken the adjusting screw and locknut.
    b. Tighten the adjusting screw to 0.7 kgf.m (5 lbf ft) and back off three-quarters of a turn.
    c. Tighten the locknut. 4,8 kgf.m (35 lbf ft).
115. Adjust the rear band as follows:
    a. Slacken the adjusting screw and locknut.
    b. Tighten the adjusting screw to 0.7 kgf.m (5 lbf ft) and back off three-quarters of a turn.
    c. Tighten the locknut. 4,8 kgf.m (35 lbf ft).
116. Refit the fluid tubes. (Note the 'O' ring on the pump suction tube).
117. Refit the fluid tube locating plate.
118. Fit and tighten the two bolts. 0,24 to 0,35 kgf.m (20 to 30 lbf in).
119. Carefully refit the valve body assembly, ensuring that the fluid tubes are not distorted.
120. Fit and tighten the three bolts and washers. 0,24 to 0,35 kgf.m (20 to 30 lbf in).
121. Connect the downshift inner cable to the downshift cam.
122. Refit the fluid tubes.
123. Replace the magnet and refit the fluid pan and joint washer.
124. Fit and tighten 12 bolts. 0,6 to 0,9 kgf.m (4.5 to 7.0 lbf ft).
125. Locate the torque converter housing in place.
126. Fit and tighten four bolts securing the torque converter housing. 10 mm diameter—2,7 to 4,1 kgf.m (20 to lbf ft) 12 mm diameter—4,1 to 6,9 kgf.m (30 to 50 lbf ft).
127. Refit the gearbox assembly.

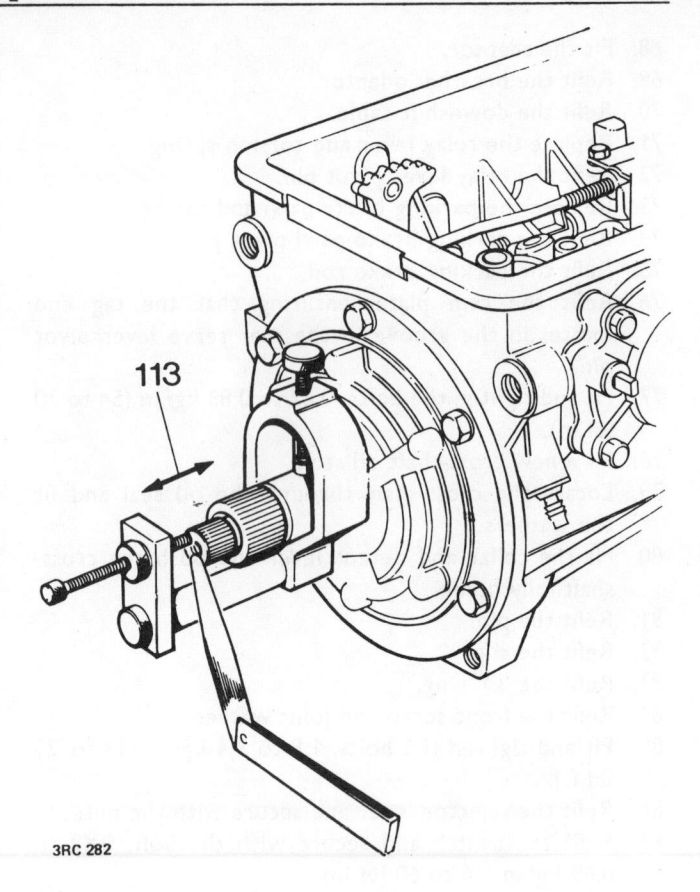

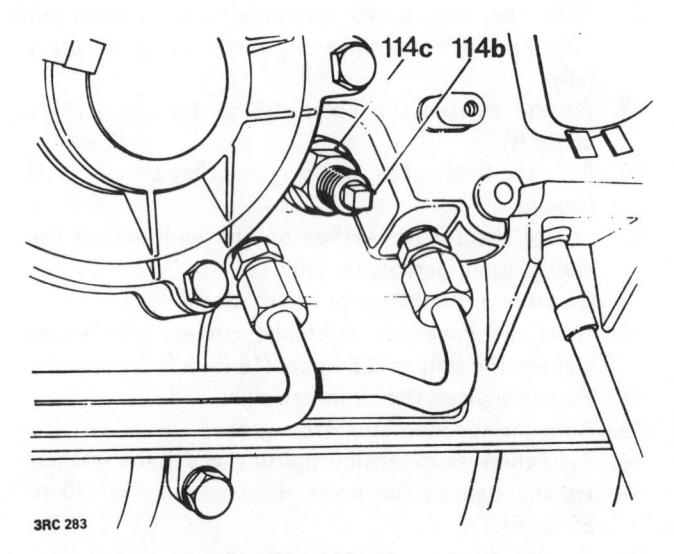

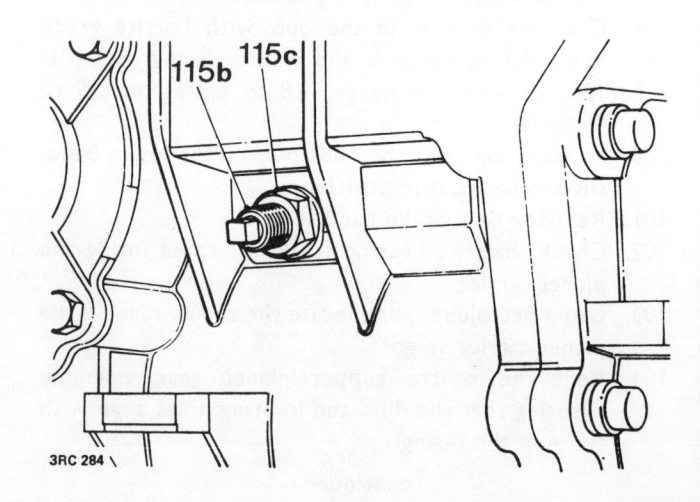

44.20.06
Sheet 4

Rover 3500 and 3500S Manual AKM 3621

GEARBOX—AUTOMATIC—Borg-Warner type 65

REAR EXTENSION HOUSING

—Remove and refit 44.20.15

Service tools: CBW 547A–50 or 601285, Torque wrench
18G 1205 Retainer, coupling flange

Removing

1. Drive the vehicle on to a ramp, select 'N' and chock the wheels.
2. Drain the gearbox fluid by disconnecting the filler pipe from the fluid pan.
3. Remove the front exhaust pipe.
4. Remove the propeller shaft.
5. Turn the engine fan blades until the two widest spaced blades are at the top.
6. Disconnect the speedometer cable from the gearbox.
7. Disconnect the electrical leads from the starter inhibitor/reverse switch.
8. Release the gear change cable from the clip on the base unit.
9. Remove the fixings from the gearbox rear mounting, and lower the assembly until the exhaust manifolds are resting on the base unit.
10. Using 18G 1205 to retain the coupling flange, unscrew the bolt.
11. Withdraw the coupling flange.
12. Remove eight bolts.
13. Withdraw the rear extension housing and joint washer.
14. If required, remove the mounting bracket from the rear extension housing.

Refitting

15. If removed, refit the mounting bracket to the rear extension housing.
16. Refit the rear extension housing and a new joint washer.
17. Fit and tighten the bolts. 4,1 to 7,6 kgf.m (30 to 55 lbf ft).
18. Place the coupling flange in position.
19. Coat the threads of the flange retaining bolt with Loctite Grade 'CV'.
20. Using 18G 1205 to retain the coupling flange, fit and tighten the bolt and washer. 4,8 to 6,9 kgf.m (35 to 50 lb ft).
21. Reverse 1 to 9.
22. Refill the gearbox fluid system. 44.24.02.

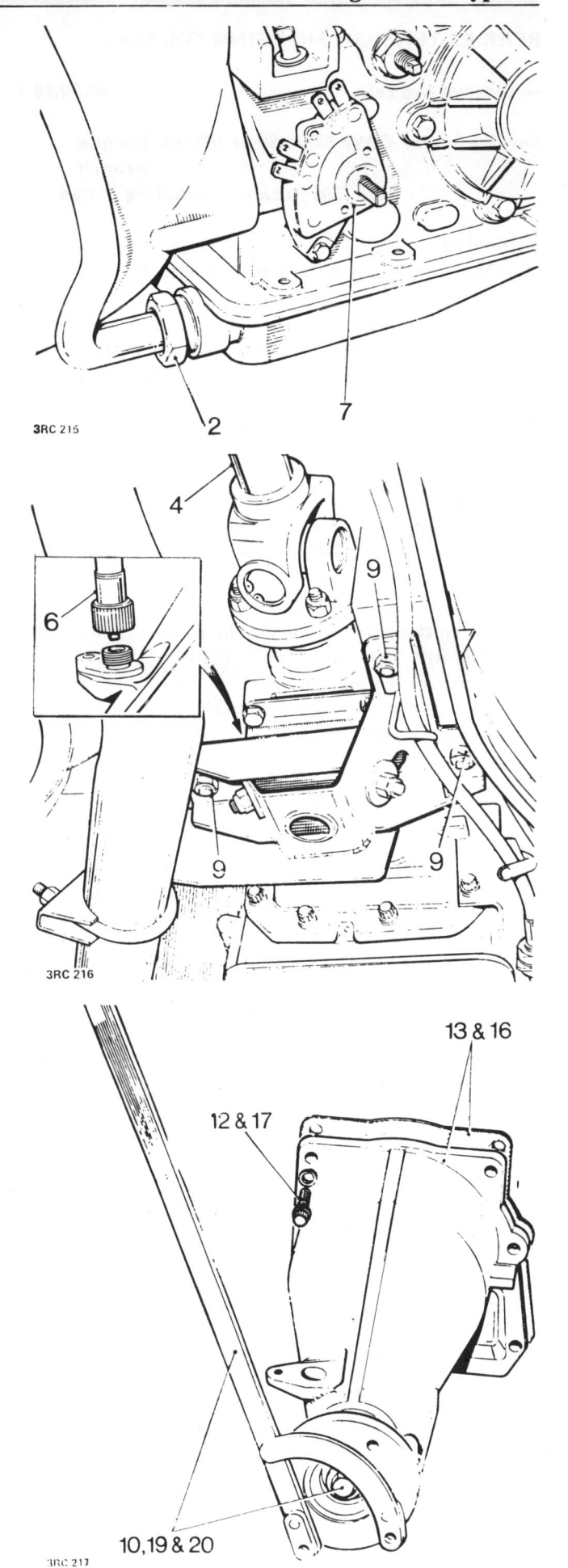

Rover 3500 and 3500S Manual AKM 3621

44.20.15

GEARBOX—AUTOMATIC—Borg-Warner type 65

REAR EXTENSION HOUSING OIL SEAL

—Remove and refit　　　　　　　　　44.20.18

Service tools: CBW 547A–50 or 601285, Torque wrench
18G 1205 Retainer, coupling flange

Removing

1. Drive the vehicle on to a ramp, select 'N' and chock the wheels.
2. Raise the ramp.
3. Disconnect the propeller shaft from the gearbox.
4. Using 18G 1205 to retain the coupling flange, unscrew the bolt.
5. Withdraw the coupling flange.
6. Prise out the oil seal.

Refitting

7. Fit a new oil seal to the extension housing.
8. Coat the threads of the flange retaining bolt with Loctite Grade 'CV'.
9. Using 18G 1205, to retain the coupling flange, fit and tighten the bolt and washer. 4,8 to 6,9 kgf.m (35 to 50 lbf ft).
10. Reconnect the propeller shaft.
11. Check and if necessary top up the gearbox fluid to the dipstick level mark.

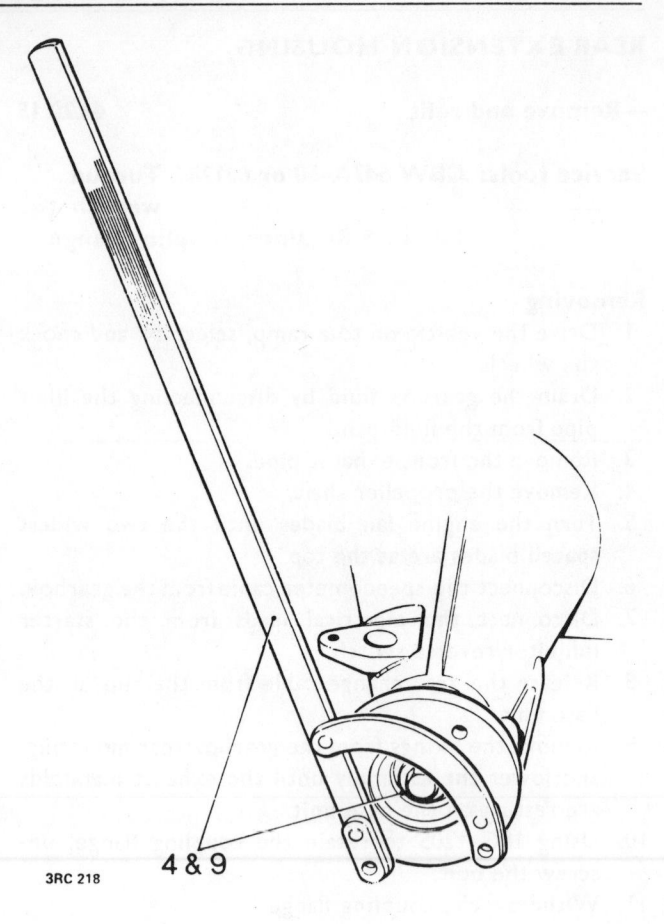

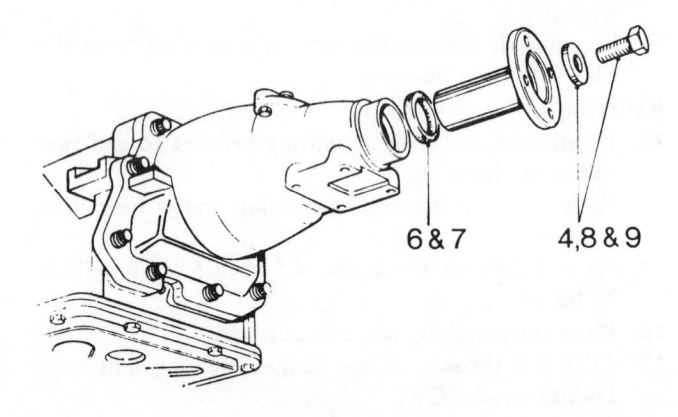

Rover 3500 and 3500S Manual AKM 3621

GEARBOX—AUTOMATIC—Borg-Warner type 65

GOVERNOR ASSEMBLY

—Remove and refit   44.22.01

Removing
1. Remove the rear extension housing. 44.20.15.
2. Withdraw the speedometer drive gear.
3. Unscrew the counterweight.
4. Withdraw the governor assembly.

Refitting
5. Slide the governor assembly into position, correct way round.
6. Locate and secure the counterweight. 2,0 to 2,4 kgf.m (15 to 18 lbf ft).
7. Refit the speedometer drive gear, rear extension housing and coupling flange. 44.20.15.

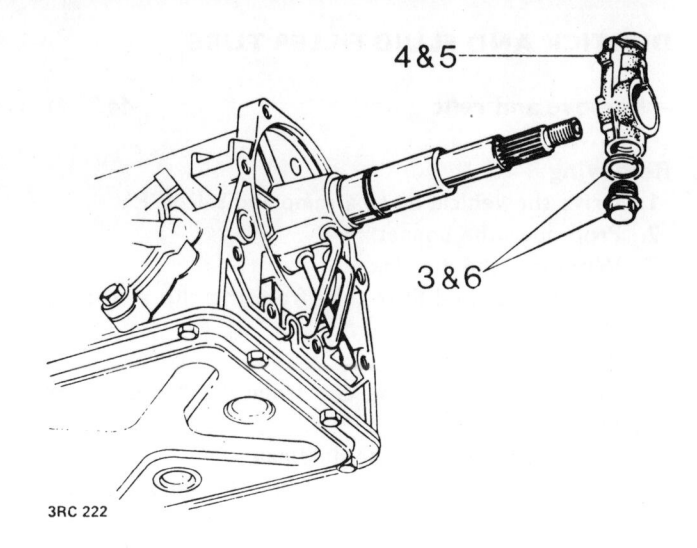

3RC 222

GOVERNOR ASSEMBLY

—Overhaul   44.22.04

Dismantling
1. Remove the governor assembly. 44.22.01.
2. Pull off the retainer.
3. Withdraw the weight.
4. Withdraw the stem.
5. Remove the spring.
6. Withdraw the valve.

Reassembling
7. Insert the valve.
8. Refit the spring on to the stem.
9. Refit the stem and spring.
10. Refit the weight.
11. Refit the retainer.
12. Refit the governor assembly. 44.22.01.

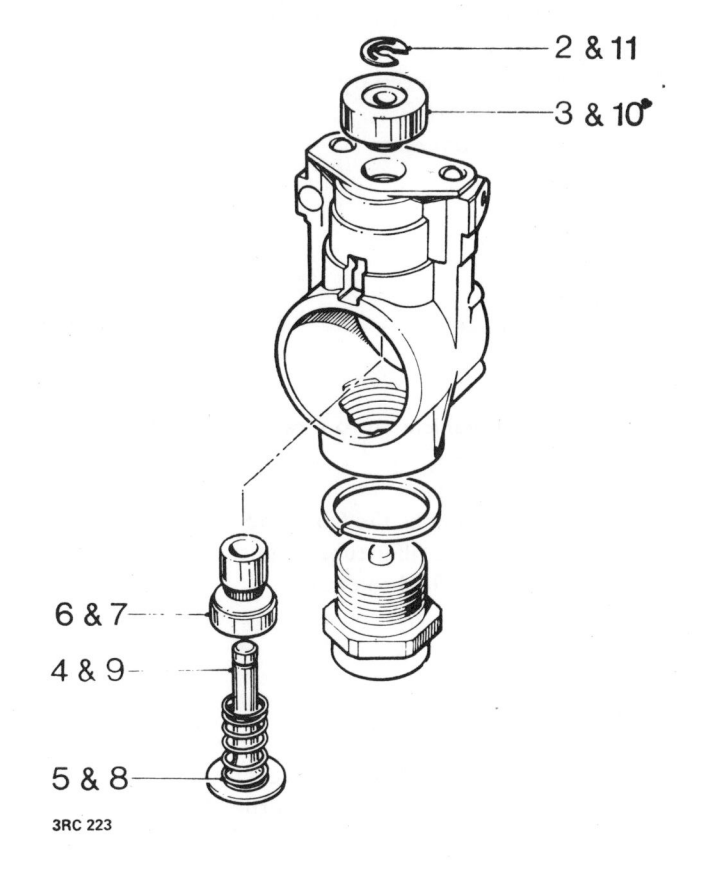

3RC 223

Rover 3500 and 3500S Manual AKM 3621

44.22.01
44.22.04

GEARBOX—AUTOMATIC—Borg-Warner type 65

DIPSTICK AND FLUID FILLER TUBE

—Remove and refit 44.24.01

Removing

1. Drive the vehicle on to a ramp and select 'P'.
2. Prop open the bonnet.
3. Withdraw the dipstick.
4. Release the fluid filler tube from the clip at the air cleaner.
5. Raise the ramp.
6. Release the filler tube clip from the flange of the converter housing.
7. Place a tray beneath the fluid pan.
8. Disconnect the filler pipe from the fluid pan and allow all the fluid to drain.
9. Withdraw the fluid filler tube from beneath the vehicle.

Refitting

10. Reverse 3 to 9.
11. Using the correct fluid, see Division 09, fill the gearbox to the level mark on the dipstick.
12. Run the engine at idle speed for a few minutes, selecting each gear position for a short period to distribute the fluid through the gearbox.
13. Stop the engine.
14. Check, and if necessary top up the fluid to the dipstick level mark.
15. Close the bonnet.

FLUID SYSTEM

—Drain and refill 44.24.02

Draining

1. Drive the vehicle on to a ramp, select 'P' and apply the handbrake.
2. Raise the ramp.
3. Place a tray under the front end of the fluid pan.
4. Disconnect the filler tube from the fluid pan.
5. Allow all the fluid to drain.
   NOTE: It is not possible to drain the torque converter.

Refilling

6. If applicable, reconnect the filler tube to the fluid pan.
7. Using the correct fluid, see Division 09, fill the gearbox to the level mark on the dipstick.
8. Run the engine at idle speed for a few minutes, selecting each gear position for a short period to distribute the fluid through the gearbox.
9. Stop the engine.
10. Check, and if necessary top up the fluid to the dipstick level mark.

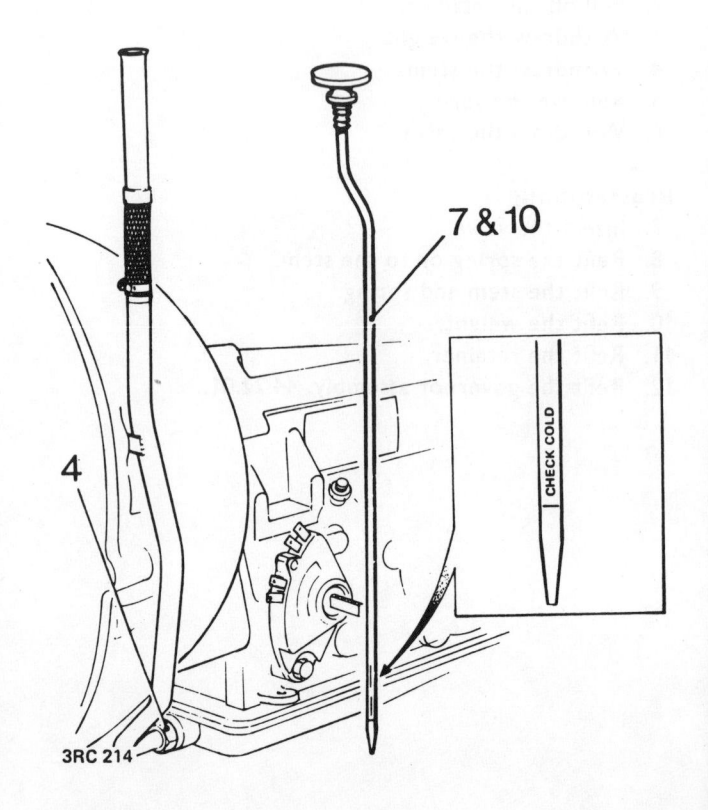

GEARBOX—AUTOMATIC—Borg-Warner type 65

FLUID PAN

—Remove and refit 44.24.04

Removing
1. Drive the vehicle on to a ramp, select 'N' and apply the handbrake.
2. Raise the ramp.
3. Place a tray beneath the fluid pan.
4. Drain the gearbox fluid by disconnecting the filler pipe from the fluid pan.
5. Remove the bolts and withdraw the fluid pan and gasket.

Refitting
6. Using a new gasket, refit the fluid pan and tighten the bolts, 0,6 to 0,9 kgf.m (4.5 to 7.0 lbf ft).
7. Reverse 2 to 4.
8. Using the correct fluid, see Division 09, fill the gearbox to the level mark on the dipstick.
9. Run the engine at idle speed for a few minutes, selecting each gear position for a short period to distribute the fluid through the gearbox.
10. Stop the engine.
11. Check, and if necessary top up the fluid to the dipstick level mark.

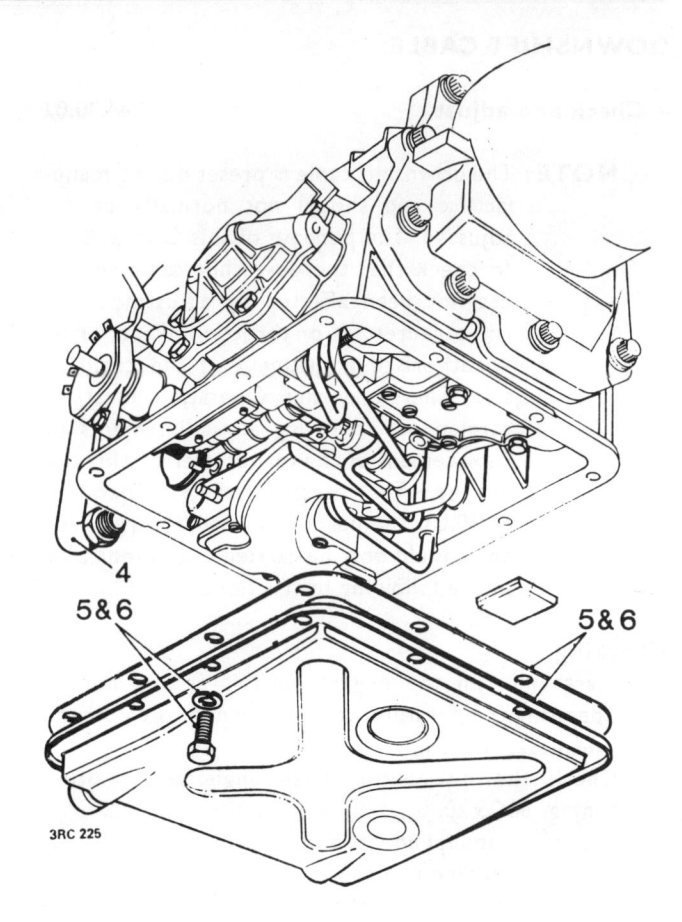

RESTRICTOR VALVE—FLUID RETURN

—Remove and refit 44.24.22

Removing
1. Drive the vehicle on to a ramp, select 'P' and apply the handbrake.
2. Disconnect the oil cooler pipe from the rearmost connection at the gearbox.
3. Unscrew the restrictor valve.

Refitting
4. Reverse 1 to 3.

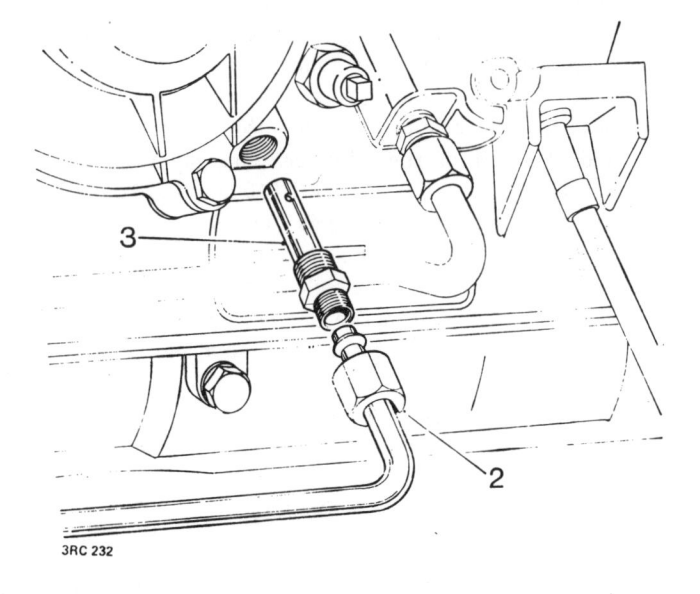

Rover 3500 and 3500S Manual AKM 3621

44.24.04
44.24.22

GEARBOX—AUTOMATIC—Borg-Warner type 65

DOWNSHIFT CABLE

—Check and adjust 44.30.02

> **NOTE:** The downshift cable is preset during manufacture and should not normally be re-adjusted. The purpose of this Operation is to check that the downshift cable setting has not been affected by adjustments to the carburetters or throttle linkage, or by unauthorised adjustment of the cable itself. Therefore, this Operation must be carried out following any adjustments or replacement of the carburetter or throttle linkage. The pressure check Operation 44.30.03, should not normally be required and should only be carried out as indicated in the following instructions.

Checking

1. Start and run the engine until it attains normal operating temperature, that is, five minutes after the thermostat opens.
2. Check the distributor dwell angle and ignition timing. 86.35.20.
3. Using an independent and accurate tachometer, check the engine idle speed, this must be 600 to 650 rev/min (700 to 750 rev/min for emission controlled vehicles). If the idle speed is not correct, tune and adjust the carburetters. 19.15.02.
4. Release the fluid filler tube from the clip at the rear of the air cleaner, move the choke cable aside and pivot the air cleaner forward to give access to the downshift cable and throttle linkage.
5. With the engine idling, actuate the accelerator coupling shaft until the idling speed just starts to rise, then while holding this condition, check the gap between the crimped stop on the downshift cable and the end of the adjuster. The gap should be between 0,25 and 0,50 mm (0.010 and 0.020 in). Do not adjust the gap. See items 7 and 8.
6. If the gap between the stop and the adjuster is within the limits and the shift quality and shift speeds are satisfactory, stop the engine and complete the Operation by refitting the air cleaner and removing the tachometer.
   > **NOTE:** If the gear shift quality or shift speeds are not satisfactory, DO NOT adjust the downshift cable, but proceed with the pressure check Operation 44.30.03.
7. If the gap is not within the limits, check all the throttle linkage and if necessary reset the rods and linkage to the correct dimensions, 19.20.07.
8. If all the throttle linkage settings are correct and the downshift cable gap is still not within the limits, reset the gap to between 0,25 to 0,50 mm (0.010 to 0.020 in.) using the screwed adjuster, then proceed with the pressure check Operation 44.30.03.

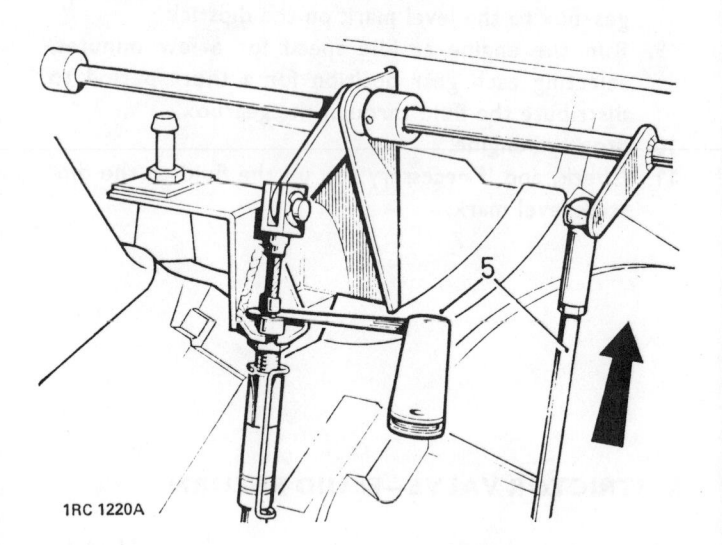

44.30.02

Rover 3500 and 3500S Manual AKM 3621

GEARBOX—AUTOMATIC—Borg-Warner type 65

DOWNSHIFT CABLE

—Pressure check   44.30.03

Service tools: 601284, Pressure gauge and tachometer
CBW 547A–50–4 Take-off plug adaptor
CBW 1C–2 Pressure test adaptor

NOTE: The gearbox operating pressure is preset during manufacture and is prevented from falling below the minimum requirement by a crimped stop on the downshift cable. Therefore, the downshift cable setting, which also represents the gearbox operating pressure setting, must be maintained in the original preset position, as described in Operation 44.30.02. The purpose of this Operation is to check the gearbox operating pressure relative to the correct downshift cable setting in order to diagnose a maladjusted downshift cable or internal gearbox fault, which would be suspected if the results of the downshift cable adjustment check, Operation 44.30.02, were not satisfactory.

Checking

1. Start and run the engine until it attains normal operating temperature, that is, five minutes after the thermostat opens.
2. Check the distributor dwell angle and ignition timing. 86.35.20.
3. Using an independent and accurate tachometer, check the engine idle speed; this must be 600 to 650 rev/min (700 to 750 rev/min for emission controlled vehicles). If the idle speed is not correct, tune and adjust the carburetters. 19.15.02.
4. Chock the front wheels of the car and apply the handbrake firmly. Apply the footbrake during the test.
5. Using CBW 547A–50–4, remove the line pressure take-off plug from the rear of the gearbox case.
6. Using CBW 1C–2, connect the pressure gauge, 601284, to the pressure take-off point.
7. Connect the tachometer 'CB' lead to the negative terminal on the ignition coil, and the earth lead to an earthing point. 601284.
   NOTE: If the tachometer in use is for 4 and 6 cylinder models, use the 4 cylinder position and note that the reading on the tachometer will be double the actual engine rev/min, that is 1000 will show as 2000 rev/min.
8. Check that the fluid level in the automatic gearbox is correct on the dipstick with the engine idling in 'P'.
9. Move the gear selector to 'D' and check the gearbox operating pressures at the following engine speeds:
   a. At engine idle speed the pressure must be between 3,86 to 4,96 kgf/cm² (55 to 70 lbf/in²).
   b. At 1200 rev/min the pressure must be 5,62 kgf/cm² (80 lbf/in²) minimum.
   c. There must be a rise in pressure of at least 1,05 kgf/cm² (15 lbf/in²) between the check at engine idle speed and the check at 1200 rev/min. For example, if the pressure at engine idle speed is 4,9 kgf/cm² (70 lbf/in²), then the minimum pressure at 1200 rev/min must be 5,9 kgf/cm² (85 lbf/in²)
10. If the downshift cable has been readjusted in accordance with instruction 8 in Operation 44.30.02, and the operating pressures are correct, it must be accepted that the downshift cable was previously maladjusted and is now correct.
11. If the downshift cable setting is as specified in Operation 44.30.02 and the operating pressures are not correct, a gearbox internal fault must be suspected. Do not adjust the downshift cable in an attempt to correct the pressure.
12. Stop the engine.
13. Remove the pressure gauge and tachometer.
14. Refit the line pressure take-off plug. Torque 0,8 to 1,2 kgf.m (6 to 8 lbf ft).

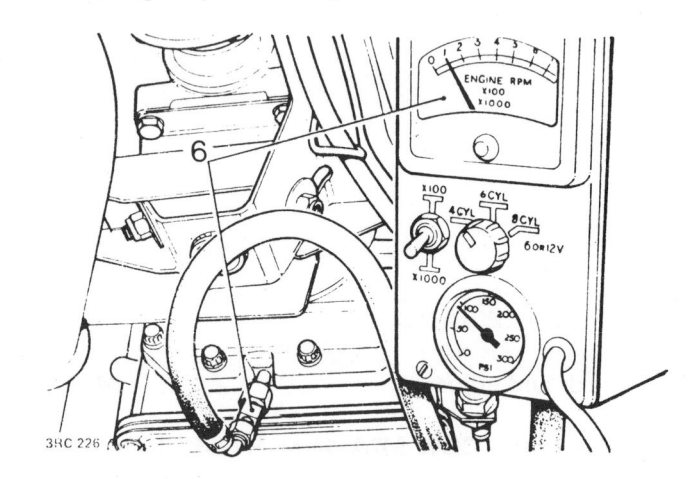

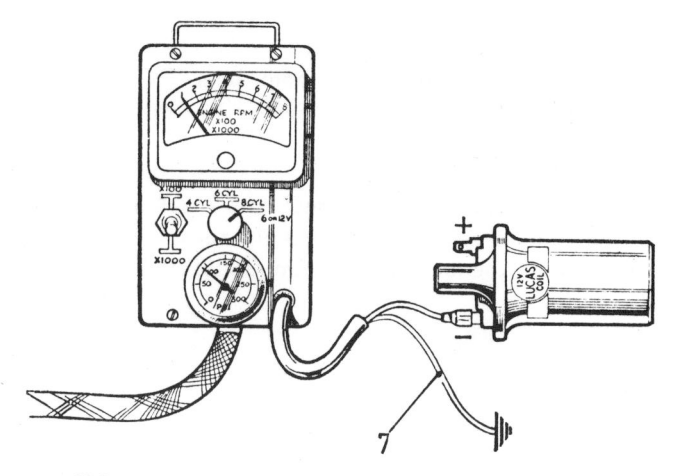

Rover 3500 and 3500S Manual AKM 3621

GEARBOX—AUTOMATIC—Borg-Warner type 65

FRONT BRAKE BAND

—Check and adjust 44.30.07

Service tools: CBW 547A–50, or 601285, Torque wrench
CBW 61, Spanners, servo adjusters

Procedure
1. Drive the vehicle on to a ramp, select 'N' and apply the handbrake.
2. Raise the ramp.
3. Slacken the locknut.
4. Tighten the adjusting screw to 0,7 kgf.m (5 lbf ft) then unscrew three-quarters of a turn.
5. Tighten the locknut 4,8 kgf.m (35 lbf ft).

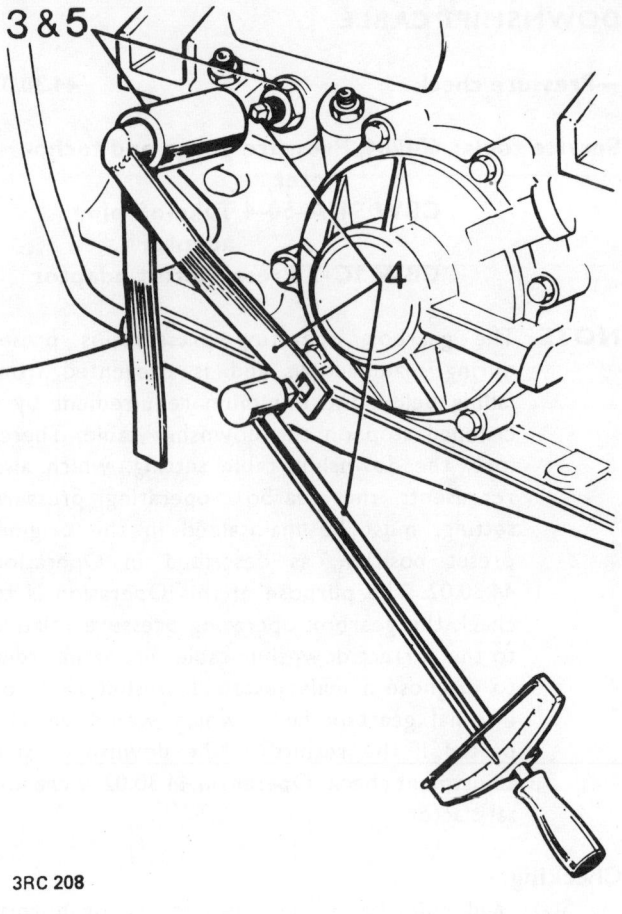

3RC 208

REAR BRAKE BAND

—Check and adjust 44.30.10

Service tools: CBW 547A–50 or 601285, Torque wrench
CBW 61, Spanners, servo adjusters

Procedure
1. Drive the vehicle on to a ramp, select 'N' and apply the handbrake.
2. Raise the ramp.
3. Slacken the locknut.
4. Tighten the adjusting screw to 0,7 kgf.m (5 lbf ft) then unscrew three-quarters of a turn.
5. Tighten the locknut 4,8 kgf.m (35 lbf ft).

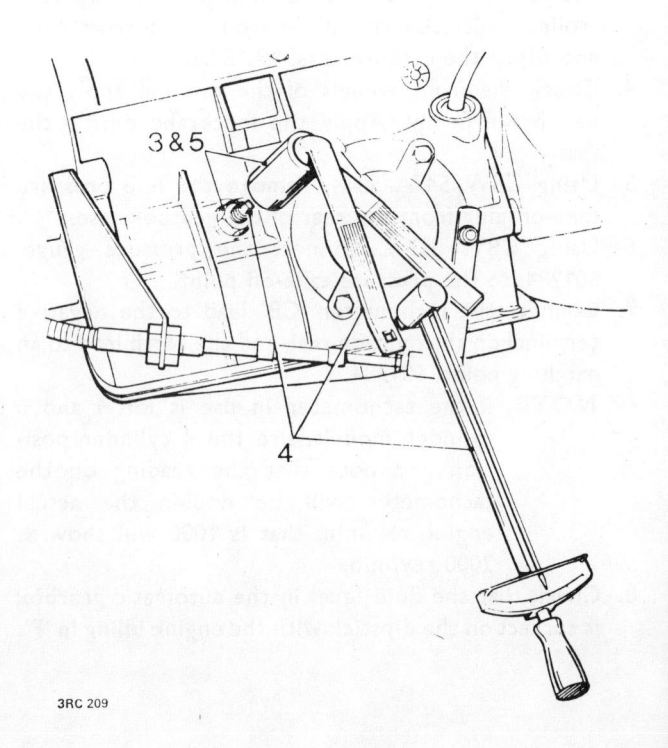

3RC 209

Rover 3500 and 3500S Manual AKM 3621

GEARBOX—AUTOMATIC—Borg-Warner type 65

STALL TEST   44.30.13

Service tool: 601284, Pressure gauge and tachometer

NOTE: 'Stall speed' is the maximum speed at which the engine can drive the torque converter impeller whilst the turbine is held stationary. It is dependent on both engine and torque converter characteristics and will vary according to the condition of the engine.
An engine in poor condition will give a lower stall speed.

Procedure

1. Check and, if necessary, correct the fluid level. 44.24.02.
2. When testing or making a diagnosis, it is important that the transmission fluid is at operating temperature but the fluid temperature must **not** exceed 110°C at any time.
3. Due to the high output of the Rover V8 engine it is necessary to position the car with the front bumper against a wall, not a work bench, with the bumper and overiders suitably protected. Also firmly apply hand and foot brakes and chock the rear wheels.
4. Connect tachometer, 601284, to the engine, start engine, select 'D' and allow to idle for approximately one minute to ensure circulation of fluid.
5. Depress accelerator to full throttle position, **NOT** kick-down, and note the revolution counter reading. DO NOT STALL FOR LONGER THAN TEN SECONDS OR TRANSMISSION WILL OVERHEAT.
6. The stall speed of an engine in good condition should be 1,950 to 2,250 rev/min.
7. If the stall speed is below 1.950 rev/min, carry out compression test on engine to determine engine condition.
8. If after engine rectification stall speed is below 1,200 rev/min renew torque converter.
9. If the stall speed is in excess of 2,400 rev min, suspect the gearbox for brake band or clutch slip.

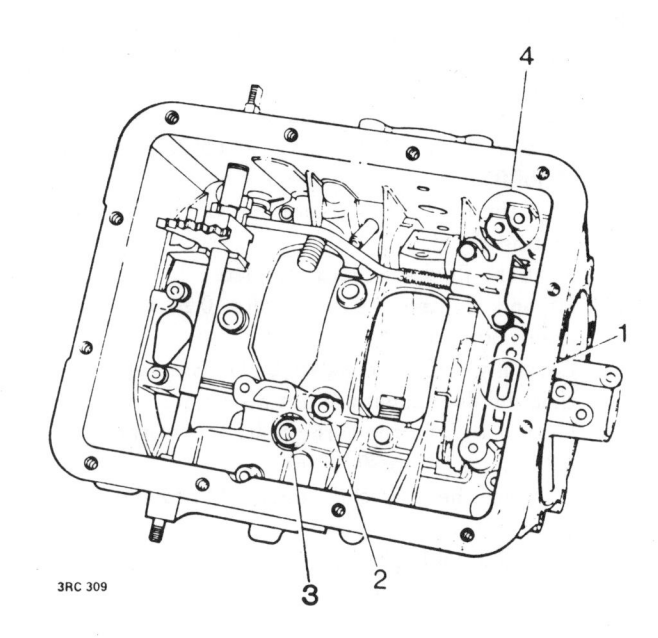

MECHANICAL OPERATION

—Air pressure checks   44.30.16

Air pressure checks can be made on the gearbox assembly to determine whether the clutches and brake bands are operating. These checks can be made with the gearbox in the car or on the bench, using a high pressure air-line. Remove the fluid pan, the valve body and fluid tubes.

1. **Front Clutch and Governor Feed**
   Apply air pressure to the passage (1). Listen for a thump, indicating that the clutch is functioning. With the unit on a bench, verify by rotating the input shaft with air pressure applied. Keep air pressure applied for several seconds to check for leaks in the circuit.
   If the extension housing has been removed, rotate the output shaft so that the governor weight will be at the bottom of the assembly. Verify that the weight moves inwards with air pressure applied.

2. **Rear Clutch**
   Apply air pressure to the passage (2). With the unit on the bench, verify that the clutch is functioning by turning the input shaft. Keep air pressure applied for several seconds to check for leaks; then listen for a thump indicating that the clutch is releasing.

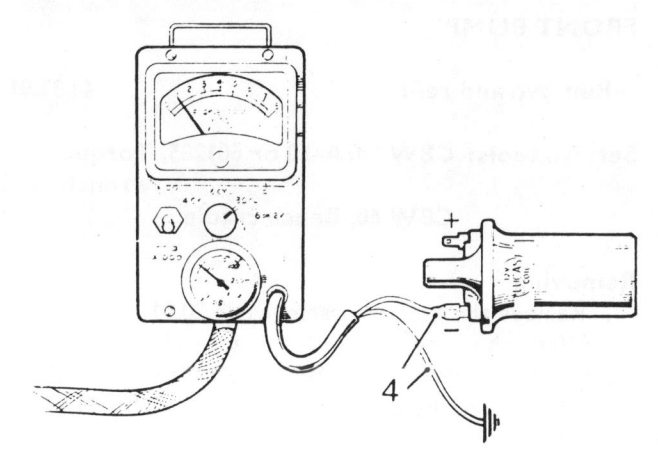

3. **Front Servo**
   Apply air pressure to the apply tube location (3). Observe the movement of the piston pin.

4. **Rear Servo**
   Apply air pressure to the tube location (4). Observe the movement of the servo lever.

Rover 3500 and 3500S Manual AKM 3621

GEARBOX—AUTOMATIC—Borg-Warner type 65

FRONT PUMP

—Remove and refit 44.32.01

Service tools: CBW 547A–50 or 601285, Torque wrench
CBW 60, Bench cradle

Removing

1. Remove the gearbox assembly. 44.20.01.
2. Wash the exterior of the unit in clean petrol or paraffin, invert it and place on a bench cradle CBW 60. Remove the starter and reverse switch. 44.15.15.
3. Unscrew the bolts.
4. Remove the torque converter housing.
5. Unscrew 12 bolts.
6. Remove the fluid pan, joint washer and magnet.
7. Pull out the fluid tubes.
8. Release the inner downshift cable from downshift cam.
9. Take out three bolts and washers.
10. Lift off the valve body assembly.
11. Unscrew two bolts.
12. Remove the fluid tube locating plate.
13. Pull out the fluid tubes. (Note the 'O' ring on the pump suction tube).
14. Take out five bolts.
15. Remove the pump and joint washer.
16. Remove the thrust washer.

Refitting

17. Using petroleum jelly, stick the thrust washer to the pump assembly.
18. Refit the pump assembly and joint washer.
19. Fit and tighten the bolts. 1,8 to 3,4 kgf.m (13 to 25 lbf ft).
20. Refit the fluid tubes. (Note the 'O' ring on the pump suction tube).
21. Refit the fluid tube locating plate.
22. Fit and tighten the two bolts. 0,24 to 0,35 kgf.m (20 to 30 lbf in).
23. Carefully refit the valve body assembly, ensuring that the fluid tubes are not distorted.
24. Fit and tighten the three bolts and washers. 0,24 to 0,35 kgf.m (20 to 30 lbf in).
25. Connect the downshift inner cable to the downshift cam.
26. Refit the fluid tubes.
27. Replace the magnet and refit the fluid pan and joint washer.
28. Fit and tighten 12 bolts. 0,6 to 0,9 kgf.m (4.5 to 7.0 lbf ft).
29. Locate the torque converter housing in place.
30. Fit and tighten four bolts. 10 mm diameter—2,7 to 4,1 kgf.m (20 to 30 lbf ft) 12 mm diameter—4,1 to 6,9 kgf.m (30 to 50 lbf ft).
31. Refit the starter and reverse switch. 44.15.15. 0,42 to 0,69 kgf.m (30 to 60 lbf in).
32. Refit the gearbox assembly. 44.20.01.

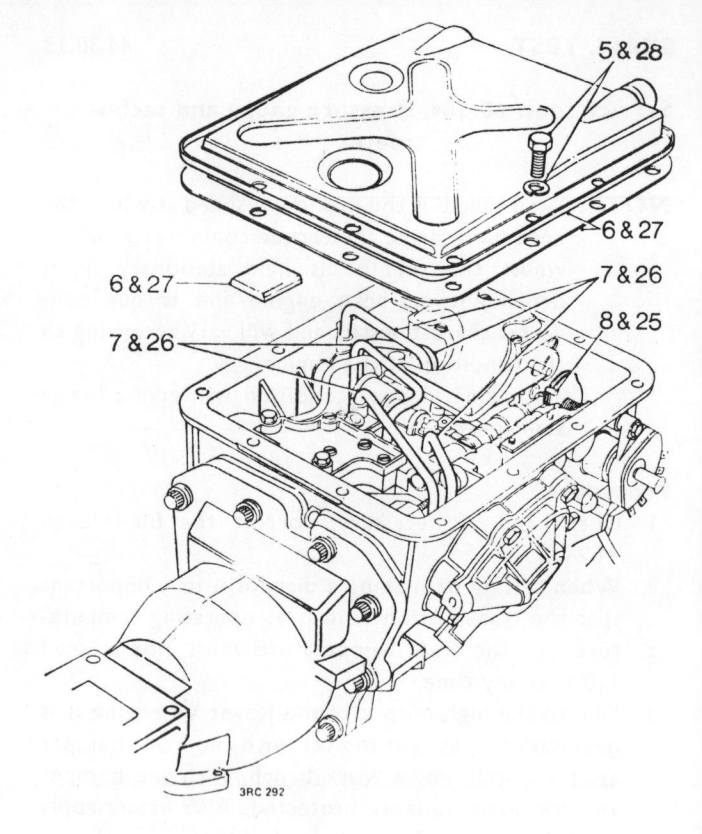

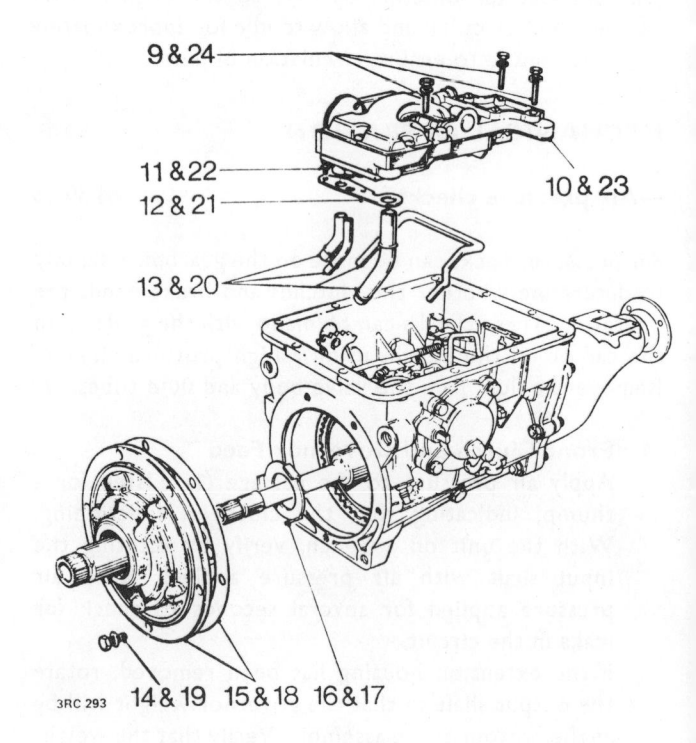

44.32.01

Rover 3500 and 3500S Manual AKM 3621

GEARBOX—AUTOMATIC—Borg-Warner type 65

ROAD TEST 44.30.17

Service tools: 601284 Pressure gauge and tachometer
CBW 547A–50–4 Adaptor, pressure take-off
CBW 1C–2 Adaptor, pressure gauge

Procedure

1. Check, and if necessary correct, the fluid level. 44.24.02.
2. Check, and if necessary adjust, the front and rear brake bands. 44.30.07/10.
3. When testing or making a diagnosis, it is important that the transmission fluid is at operating temperature.
4. Check that the starter will operate only with the selector in 'P' and 'N' positions—but **not** in 'D', '2', '1' or 'R'.
5. Check that reverse light comes on when 'R' is selected.
6. Check the freedom of the transmission with the selector lever at 'N'. There should be no tendency for the engine to drive the car, nor should there be any engine braking effect.
7. Apply the hand and foot brakes firmly and chock the wheels.
   With the engine at normal idling speed, select 'N–D', 'N–2', 'N–1' and 'N–R'.
   Gearbox engagement should be felt in each position selected.
   Check for slip or clutch squawk.

Test procedures

The following test procedures, diagram of test patterns, gear change speeds and fluid pressure figures are for an average gearbox in good condition. Therefore, as some allowance for tolerances must be taken into consideration, the tester must use discretion.

Two fault diagnosis charts are included at the end of the road test procedure, the first chart is related directly to the road test procedure and covers hydraulic rectifications which can be carried out with the gearbox installed. The second chart is more general, covering removal and dismantling the gearbox.

continued

GEARBOX—AUTOMATIC—Borg-Warner type 65

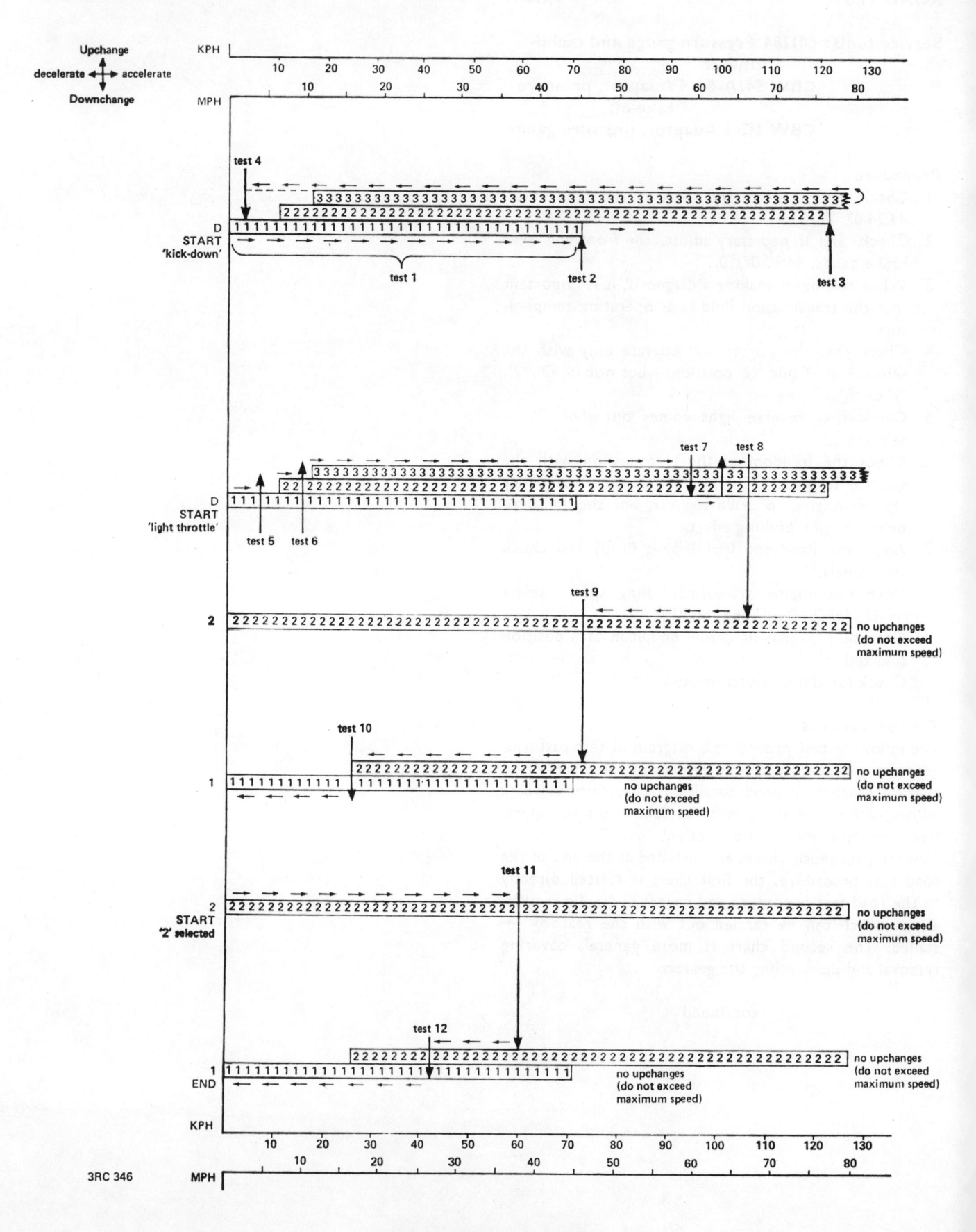

Diagram of Road test patterns

44.30.17
Sheet 2

Rover 3500 and 3500S Manual AKM 3621

GEARBOX—AUTOMATIC—Borg-Warner type 65

Test 1. Cut-back pressure check
Using CBW 547A-50-4, remove the line pressure take-off plug from the rear of the gearbox.
Place the pressure gauge 601284, on the floor in the passengers side of the car, feed the hose through a grommet hole in the gearbox tunnel and connect it to the pressure take-off point, using CBW 1C-2.
Commencing from a standing start, select 'D', 'kick-down' and hold, and check the gearbox pressure readings which must rise to approximately 14,0 to 17,5 kgf/cm^2 (200 to 250 lbf/in^2) and then cut-back to approximately 7,0 kgf/cm^2 (100 lbf/in^2) before second gear engagement. If the pressure does not cut-back, the modulator or the governor valves are sticking. Check the modulator first.
If Test 1 is satisfactory, continue with tests 2 and 3.

Tests 2 and 3—Maximum automatic upchange speeds and run-up.
Commencing from a standing start, select 'D' and, using and holding 'kick-down', check the 1–2 and 2–3 upchange speeds.
If the 1–2 or 2–3 upchanges have not occurred by the time the specified maximums are reached, do not continue to acclerate otherwise serious damage to the engine may result.
During the 'kick-down' upchange tests, a check should be made for clutch squawk or slip when moving from rest and as the upchanges occur. A momentary slip between the 2–3 upchange followed by a thump in the transmission may be due to the servo orifice valve sticking. This condition is called 'run-up'. If it is suspected that 'run-up' is present then the transmission will also be suffering from a condition called 'tie-up', which has the effect of slowing down the engine just before the 3–2 change down occurs. Note that tie-up is difficult to detect.

Test 4—Roll out change down, 3–1
With 'D' selected and third speed engaged, release the throttle pedal and check the 3–1 roll out downchange speed.
To check that a downchange has occured, stop and select reverse and ensure that the car does reverse without rear brake band slip. If band slip is evident, this indicates that the 1–2 shift valve is sticking. However, this condition can be intermittent and is affected by the fluid temperature.

Tests 5 and 6—Light throttle upchange speeds
Commencing from a standing start, select 'D' and applying 'light throttle', check the 1–2 and 2–3 upchange speeds.

Test 7—Maximum 3–2 'kick-down' speed
At 95 km/h (60 mph) 'D' selected, third speed engaged, depress the throttle to 'kick-down'. The gearbox should immediately change down into second speed. If the throttle is then released, the gearbox should revert to third speed.

Test 8—'D' selected (third speed engaged) to '2' selected downchange
With 'D' selected, third speed engaged, at 105 km/h (65 mph), simultaneously select '2' and release the throttle. The gearbox should immediately change down.

Test 9—'2' selected to '1' selected
With '2' selected, release the throttle and at 70 km/h (30 mph) select '1'. The gearbox should not change down immediately.

Test 10—Roll out downchange 2–1
Continuing from Test 9 ('1' selected, second speed engaged) continue to decelerate when at 26 to 42 km/h (16 to 26 mph) an automatic roll-out downchange to first speed should occur.
To check that a downchange has occured, stop and select reverse and ensure that the car does reverse. If the car fails to reverse, this indicates that the 1–2 shift valve is sticking. However, this condition can be intermittent and be affected by the fluid temperature, being more obvious when the fluid is cold.

Test 11—'2' start to '1' selected
Select '2' and at 60 km/h (37 mph) release the throttle and select '1'. The gearbox should not change down immediately.

Test 12—Maximum 2–1 kickdown speed from '2' selected to '1' selected
NOTE: This test is for maximum engine braking deceleration.
Continuing from test 11 ('1' selected, second speed engaged) continue to decelerate and, at 40 km/h (25 mph) rapidly 'kick-down' and release the throttle pedal. The gearbox should immediately change down to first speed, giving maximum engine braking.
NOTE: The maximum 'kick-down' downshift speeds, 3–2 and 2–1, are approximately 24 km/h (15 mph) slower than the 'kick-down' upshift speeds.

continued

Rover 3500 and 3500S Manual AKM 3621

44.30.17
Sheet 3

GEARBOX—AUTOMATIC—Borg-Warner type 65

ROAD TEST RELATED DIAGNOSIS CHART

NOTE: This chart assumes that the stall test and static pressure checks have been carried out and are correct

NOTE: The numbers indicate the recommended sequence of investigation.

Test Number	Fault	1–2 Shift Valve	2–3 Shift Valve	Modulator Valve	Servo Orifice Valve	P.R.V. Control Valve	Front Servo/Band	Rear Servo/Band	Governor Valve
1	Pressure does not fall			1		3			2
2a	Test 1 correct but no upchange 1–2	1					2		
2b	Test 1 correct but high upchange 1–2	1							
2c	Test 1 correct but slip 1–2 shift	1					2		
3a	Slip 2–3 shift run up		2		1				
3b	High upchange 2–3		1						2
3c	No 2–3 change		1						2
4	No change down. Carry out reverse test, below, (see also tests 10 and 12)		1						2
5	Delayed or no upchange 1–2	1							2
6	Delayed or no upchange 2–3		1						2
7	No change down 3–2		1						
8	Change down to first gear (see tests 9 and 11)					1			
9	Change down gear immediately (see tests 8 and 11)					1			
10	No change down. Carry out reverse test, below, see also tests 4 and 12)	1							2
11	Change down to first gear immediately. See tests 8 and 9					1			
12	No change down. Carry out reverse test, below. See also test 4 and 10	1							
Reverse test	Band slip in reverse **NOTE:** If rear brake band slip is evident, this indicates a sticking 1–2 shift valve	1						2	

44.30.17
Sheet 4

Rover 3500 and 3500S Manual AKM 3621

GEARBOX—AUTOMATIC—Borg-Warner type 65

General fault diagnosis chart

Note: The numbers indicate the recommended sequence of fault investigation
See opposite page for Action Key

	1	2	3	4	5	6	7	8	9	10	11	12	13
ENGAGEMENT OF 'R', 'D', '1', or '2'													
Bumpy	D	B	d	f	c	O	Q	—	—	—	—	—	—
Delayed	A	C	D	a	d	c	b	s	N	P	S	X	q
None	A	C	a	b	c	d	V	W	X	Y	—	—	—
TAKE-OFF													
None forward	C	c	b	N	T	—	—	—	—	—	—	—	—
None reverse	C	F	m	n	c	b	a	S	P	—	—	—	—
Seizure reverse	E	O	—	—	—	—	—	—	—	—	—	—	—
No neutral	C	O	c	—	—	—	—	—	—	—	—	—	—
UPSHIFTS													
No. 1 to 2	C	E	h	m	R	f	g	a	b	c	t	—	—
No. 2 to 3	C	h	n	p	P	f	g	a	b	c	—	—	—
Above normal speeds	B	f	h	m	n	p	g	b	c	d	—	—	—
Below normal speeds	B	f	h	p	b	c	—	—	—	—	—	—	—
UPSHIFT QUALITY													
Slip 1 to 2	A	B	C	E	R	d	f	a	b	c	—	—	—
Slip 2 to 3	A	B	C	E	P	R	d	f	a	b	c	—	—
Rough 1 to 2	B	E	d	f	g	h	T	U	N	c	—	—	—
Rough 2 to 3	B	E	d	f	Q	c	—	—	—	—	—	—	—
Seizure 1 to 2	F	Q	T	U	b	c	—	—	—	—	—	—	—
Seizure 2 to 3	E	a	b	c	—	—	—	—	—	—	—	—	—
DOWNSHIFTS													
No. 2 to 1	B	m	h	S	t	—	—	—	—	—	—	—	—
No. 3 to 2	B	m	h	Q	R	—	—	—	—	—	—	—	—
Involuntary high speed 3 to 2	A	a	P	—	—	—	—	—	—	—	—	—	—
Above normal shift speeds	B	h	p	f	b	c	—	—	—	—	—	—	—
Below normal shift speeds	B	h	p	f	b	c	m	n	—	—	—	—	—
DOWNSHIFT QUALITY													
Slip 2 to 1	T	—	—	—	—	—	—	—	—	—	—	—	—
Slip 3 to 2	E	R	l	d	f	a	b	c	P	—	—	—	—
Rough 2 to 1	T	N	b	—	—	—	—	—	—	—	—	—	—
Rough 3 to 2	E	I	d	f	c	P	O	R	—	—	—	—	—
LINE PRESSURE													
Low, idling	A	C	D	d	c	a	s	b	W	—	—	—	—
High, idling	B	D	d	f	e	—	—	—	—	—	—	—	—
Low at stall	A	B	d	g	f	a	c	b	h	X	—	—	—
High at stall	d	f	g	c	—	—	—	—	—	—	—	—	—
STALL SPEED													
Below 1,000	Y	—	—	—	—	—	—	—	—	—	—	—	—
Over 2,250	A	C	F	a	b	c	d	N	P	S	T	V	Y
OVERHEATING	A	E	F	Y	—	—	—	—	—	—	—	—	—

Rover 3500 and 3500S Manual AKM 3621

44.30.17
Sheet 5

GEARBOX—AUTOMATIC—Borg-Warner type 65

PRELIMINARY ADJUSTMENT FAULTS

A Fluid level incorrect
B Downshift valve cable incorrectly assembled or adjusted
C Manual linkage incorrectly assembled or adjusted
D Incorrect engine idling speed
E Incorrect front band adjustment
F Incorrect rear band adjustment

MECHANICAL FAULTS

N Front clutch slipping due to worn plates or faulty parts
O Front clutch seized or plates distorted
P Rear clutch slipping due to worn plates or faulty check valve in piston
Q Rear clutch seized or plates distorted
R Front band slipping due to faulty servo, broken or worn band
S Rear band slipping due to faulty servo, broken or worn band
T One-way clutch slipping or incorrectly installed
U One-way clutch seized
V Input shaft broken
W Front pump drive tangs on converter hub broken
X Front pump worn
Y Converter blading and/or one-way clutch failed

HYDRAULIC CONTROL FAULTS

a Fluid tubes missing or not installed correctly
b Sealing rings missing or broken
c Valve body assembly screws missing or not correctly tightened
d Primary regulator valve sticking
e Secondary regulator valve sticking
f Throttle valve sticking
g Modulator valve sticking
h Governor valve sticking, leaking or incorrectly assembled
l Orifice control valve sticking
m 1 to 2 shift valve sticking
n 2 to 3 shift valve sticking
p 2 to 3 shift valve plunger sticking
q Converter 'out' check valve missing or sticking
s Pump check valve missing or sticking
t 'D1–D2' control valve or range control valve sticking

44.30.17
Sheet 6

Rover 3500 and 3500S Manual AKM 3621

GEARBOX—AUTOMATIC—Borg-Warner type 65

FRONT PUMP

—Overhaul 44.32.04

Service tool: CBW 547A-50 or 601285, Torque wrench

Dismantling

1. Remove the front pump. 44.32.01.
2. Unscrew the bolts.
3. Take out the locating screw.
4. Separate the stator support from the pump body assembly.
5. Mark the outside faces of the gears to facilitate correct assembly.
6. Remove the gears.
7. Remove the 'O' ring.
8. Extract the seal.

Reassembling

9. Fit a new seal.
10. Fit a new 'O' ring.
11. Fit the gears into the pump body.
12. Lightly lubricate the gears and the 'O' ring.
13. Refit the stator support.
14. Fit and tighten the locating screw and lock washer.
15. Fit and tighten the bolts and lock washers.
16. Refit the front pump. 44.32.01.

FRONT SERVO ASSEMBLY

—Remove and refit 44.34.07

Service tool: CBW 547A-50 or 601285, Torque wrench

Removing

1. Drive the vehicle on to a ramp, select 'N', apply the handbrake and raise the ramp.
2. Remove the gearbox selector lever.
3. Take precautions against fluid spillage.
4. Take out the four bolts.
5. Withdraw the front servo assembly, spring and joint washer, taking care that strut does not drop.

Refitting

6. Locate the joint washer on to the servo body flange.
7. Refit the servo and spring.
8. Fit and tighten the bolts. 1,8 to 3,4 kgf.m (13 to 25 lbf ft).
9. Check, and if necessary replenish, the gearbox fluid level.
10. Check the gearbox for fluid leaks.

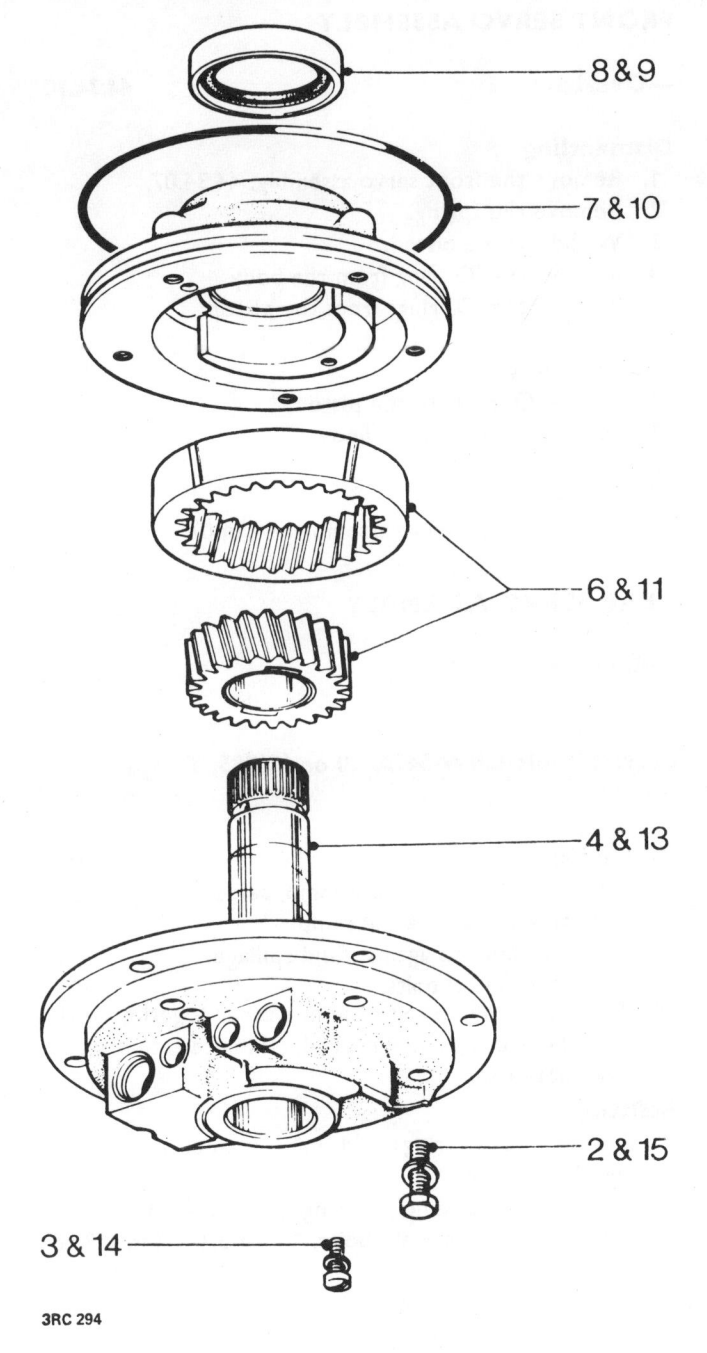

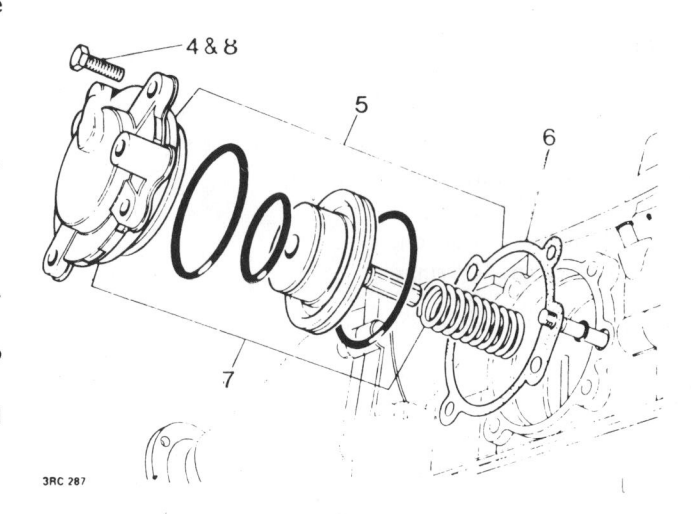

Rover 3500 and 3500S Manual AKM 3621

44.32.04
44.34.07

GEARBOX—AUTOMATIC—Borg-Warner type 65

FRONT SERVO ASSEMBLY

—Overhaul   44.34.10

Dismantling

1. Remove the front servo assembly. 44.34.07.
2. Remove the spring.
3. Withdraw the piston.
4. Remove the 'O' ring from the body.
5. Remove the 'O' rings from the piston.

Reassembling

6. Fit the 'O' rings to the piston.
7. Fit the 'O' ring to the body.
8. Refit the piston.
9. Fit the spring.
10. Refit the front servo assembly. 44.34.07.

REAR SERVO ASSEMBLY

—Remove and refit   44.34.13

Service tool: CBW 547A–50 or 601285, Torque wrench

Removing

1. Drive the vehicle on to a ramp, select 'N', apply the handbrake and raise the ramp.
2. Take precautions against fluid spillage.
3. Unscrew the six bolts.
4. Withdraw the servo and joint washer together with 'O' rings, spring and push-rod, taking care that strut does not drop.

Refitting

5. Locate the 'O' rings and joint washer on to the gearbox casing.
6. Fit the servo assembly, spring and push-rod.
7. Fit and tighten the six bolts. 1,8 to 3,4 kgf.m (13 to 25 lbf ft).
8. Replenish the gearbox fluid.
9. Check the gearbox for fluid leaks.

REAR SERVO ASSEMBLY

—Overhaul   44.34.16

Dismantling

1. Remove the rear servo assembly. 44.34.13.
2. Remove the push-rod.
3. Remove the spring.
4. Withdraw the piston.
5. Remove the 'O' rings.

Reassembling

6. Fit the 'O' rings to the piston.
7. Refit the piston.
8. Refit the spring.
9. Refit the push-rod.
10. Refit the rear servo assembly. 44.34.13.

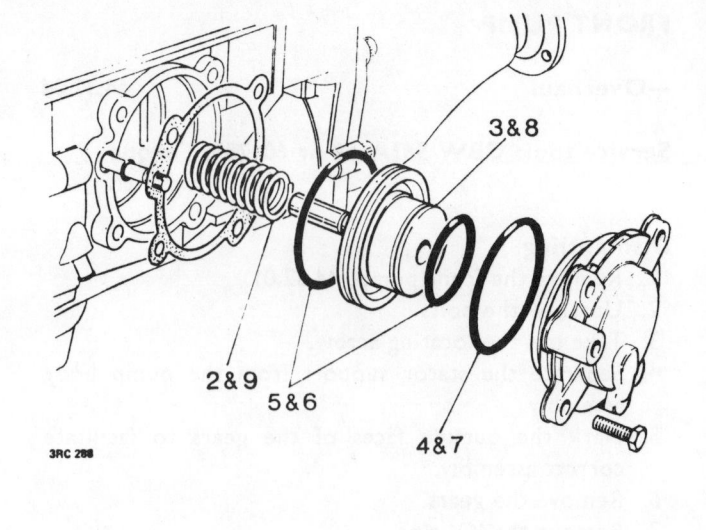

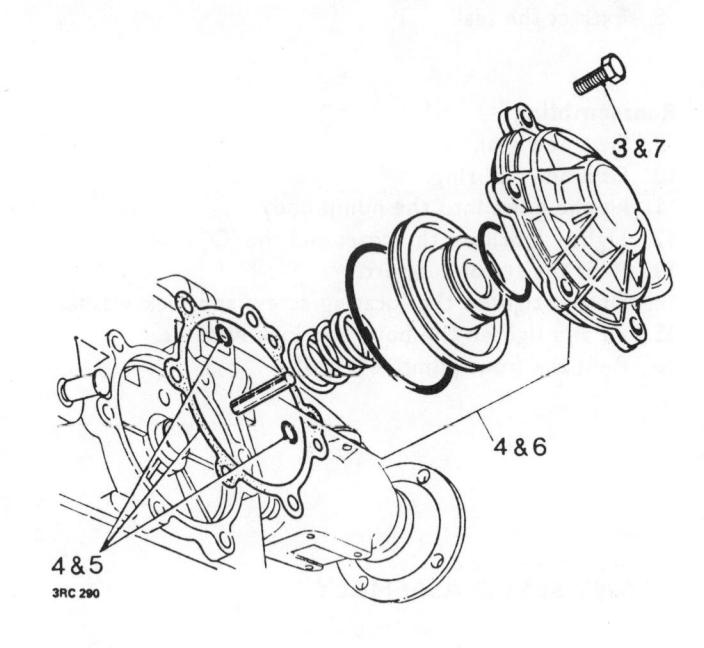

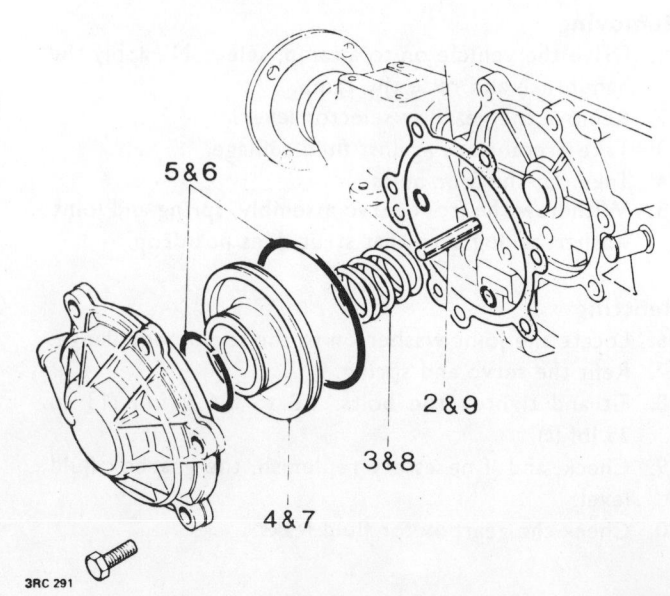

44.34.10
44.34.16

Rover 3500 and 3500S Manual AKM 3621

GEARBOX—AUTOMATIC—Borg-Warner type 65

OUTPUT SHAFT AND RING GEAR

—Remove and refit 44.36.01

Service tool: CBW 547A–50 or 601285, Torque wrench
CBW 60, Bench cradle
18G 1205, Retainer, coupling flange

Removing

1. Remove the gearbox assembly. 44.20.01.
2. Wash the exterior of the unit in clean petrol or paraffin, invert it and place on a bench cradle CBW 60. Remove the switch. 44.15.15.
3. Unscrew the bolts securing the torque converter housing.
4. Remove the torque converter housing.
5. Unscrew 12 bolts.
6. Remove the fluid pan, joint washer and magnet.
7. Pull out the fluid tubes.
8. Release the downshift inner cable from the downshift cam.
9. Take out three bolts and washers.
10. Lift off the valve body assembly.
11. Unscrew two bolts.
12. Remove the fluid tube locating plate.
13. Pull out the fluid tubes. (Note the 'O' ring on the pump suction tube).
14. Take out five bolts.
15. Remove the pump and joint washer.
16. Remove the thrust washer.
17. Withdraw the front clutch.
18. Remove the thrust washers.
19. Withdraw the rear clutch and forward sun gear.
20. Squeeze together the ends of the front brake band and remove it together with the strut.
21. Unscrew the three bolts.
22. Withdraw the centre support/planet gear assembly and needle thrust assembly.
23. Squeeze together the ends of the rear brake band, tilt and withdraw together with the strut.
24. Using tool no. 18G 1205 to retain the flange, unscrew the bolt.
25. Withdraw the flange.
26. Unscrew the bolts.
27. Withdraw the rear extension and joint washer.
28. Withdraw the speedometer drive gear.
29. Unscrew the counterweight and remove the governor.

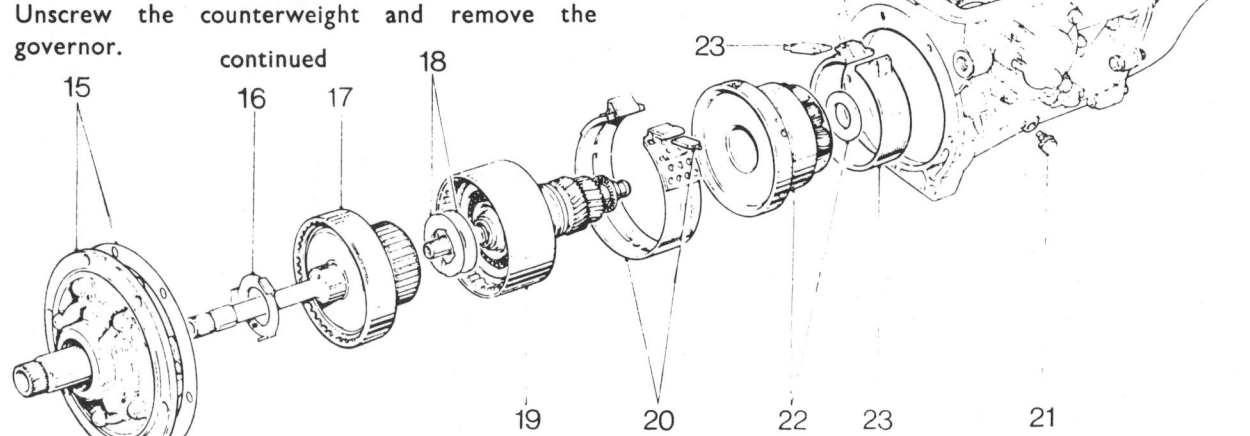

continued

Rover 3500 and 3500S Manual AKM 3621

44.36.01
Sheet 1

GEARBOX—AUTOMATIC—Borg-Warner type 65

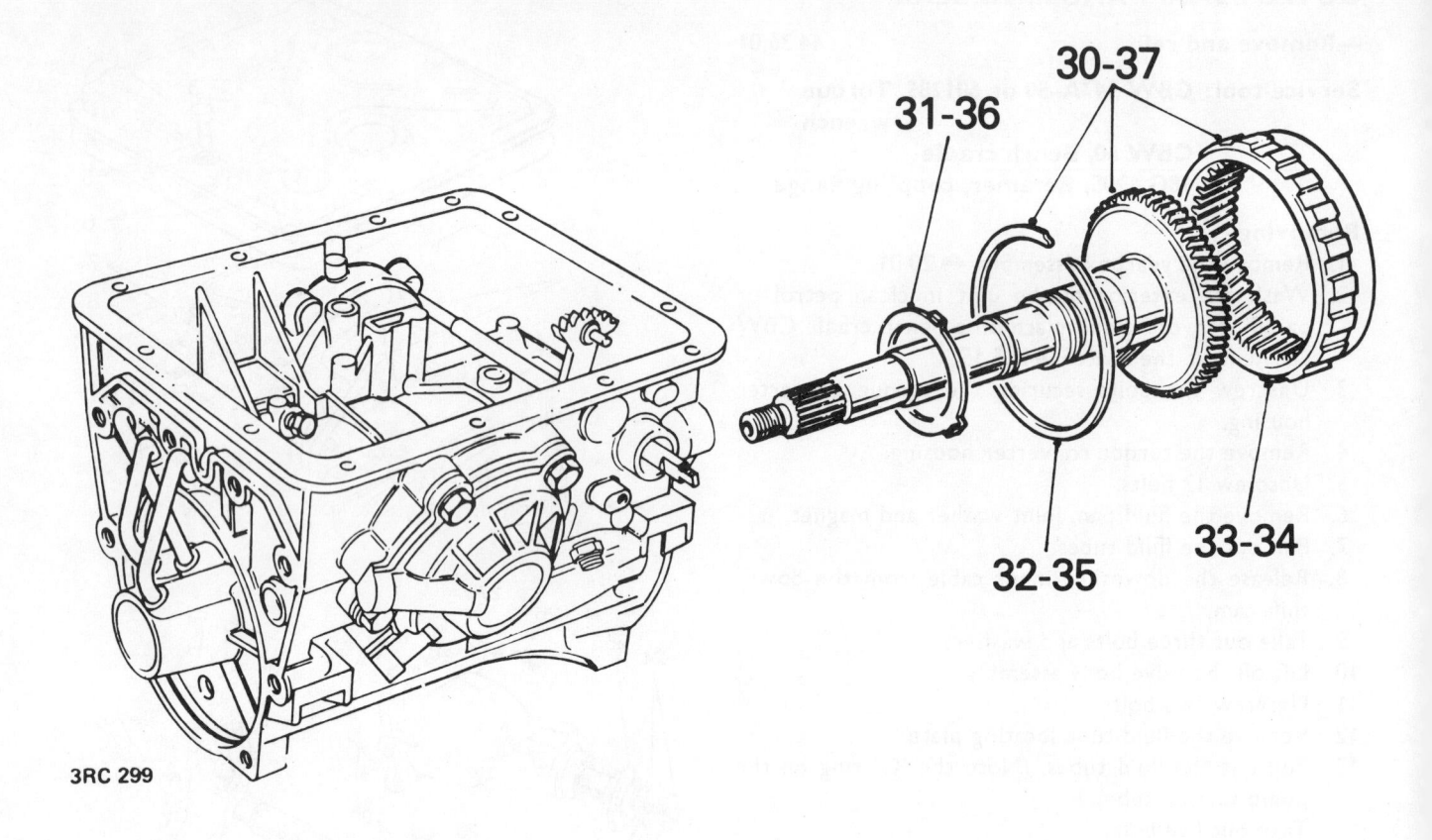

30. Withdraw the output shaft assembly.
31. Remove the thrust washer.
32. Remove the circlip.
33. Detach the outer annulus from the output shaft.

Refitting

34. Assemble the outer annulus and the output shaft.
35. Fit the circlip.
36. Using petroleum jelly, stick the thrust washer to the casing.
37. Refit the output shaft assembly.
38. Refit the governor and secure it with the counter-weight.
39. Refit the speedometer drive gear.
40. Refit the rear extension, using a new joint washer if necessary.
41. Fit and tighten the bolts. 4,1 to 7,6 kgf.m (30 to 55 lbf ft).
42. Refit the flange.
43. Coat the threads of the securing bolt with Loctite Grade 'CV', then holding the flange with tool 18G 1205, fit and tighten the bolt. 4,8 to 6,9 kgf.m 35 to 50 lbf ft).
44. Using petroleum jelly, stick the needle thrust bearing on to the planet gear case (rear drum).
45. Refit the rear brake band and strut.
46. Refit the centre support/planet gear assembly, ensuring that the fluid and locating holes align with those in the casing.

continued

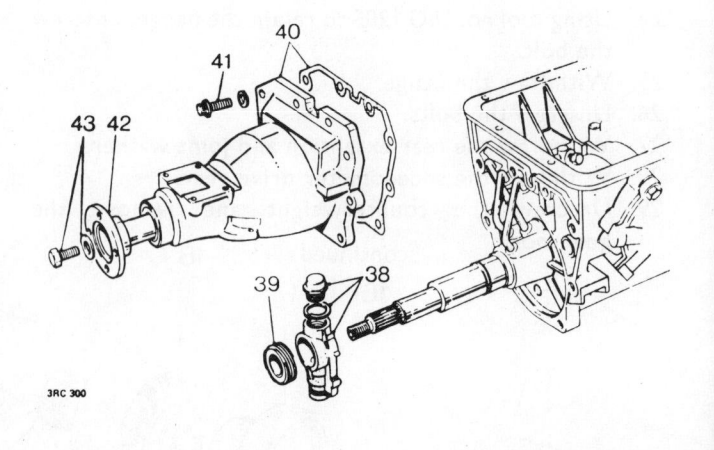

44.36.01
Sheet 2

Rover 3500 and 3500S Manual AKM 3621

GEARBOX—AUTOMATIC—Borg-Warner type 65

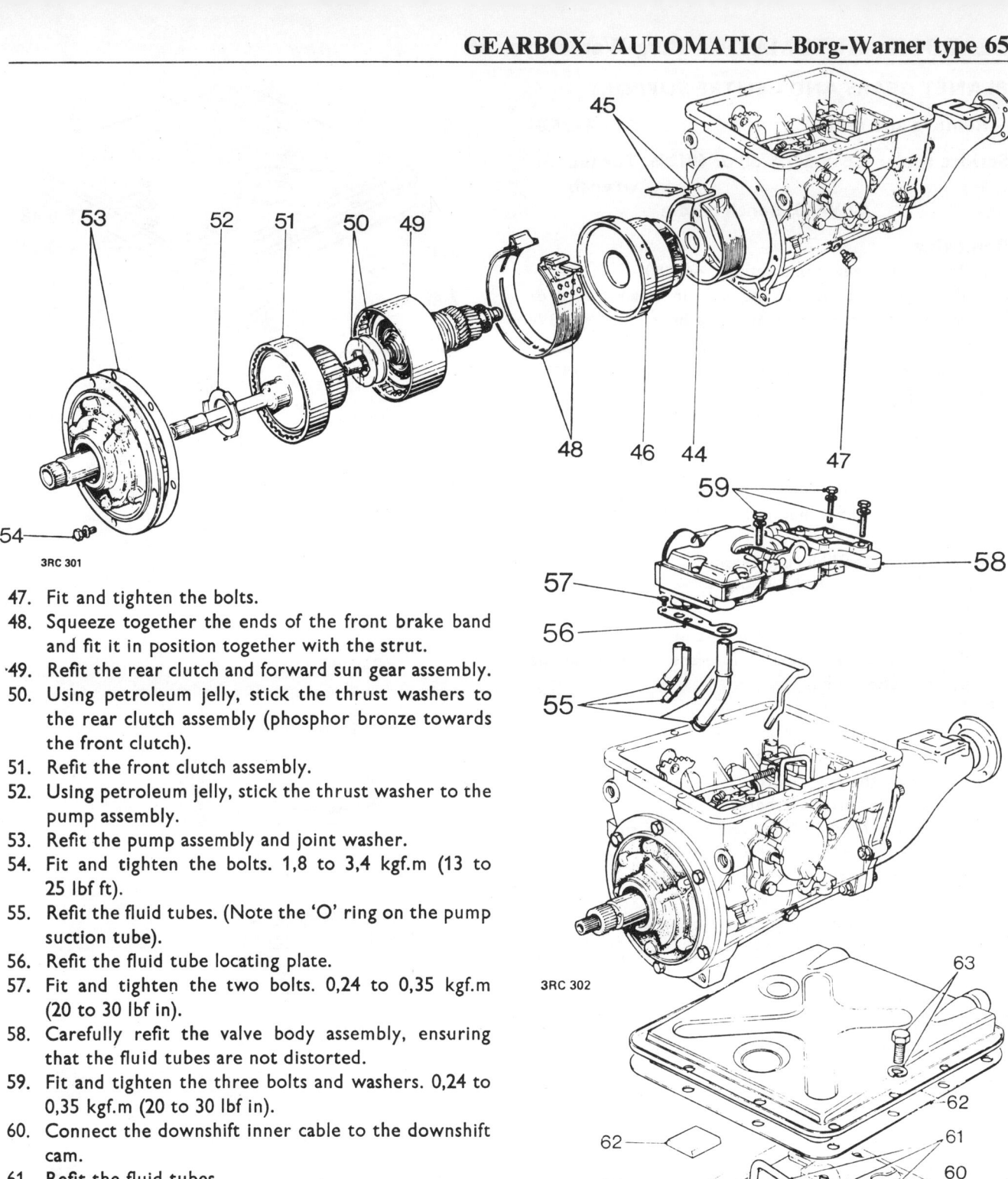

47. Fit and tighten the bolts.
48. Squeeze together the ends of the front brake band and fit it in position together with the strut.
49. Refit the rear clutch and forward sun gear assembly.
50. Using petroleum jelly, stick the thrust washers to the rear clutch assembly (phosphor bronze towards the front clutch).
51. Refit the front clutch assembly.
52. Using petroleum jelly, stick the thrust washer to the pump assembly.
53. Refit the pump assembly and joint washer.
54. Fit and tighten the bolts. 1,8 to 3,4 kgf.m (13 to 25 lbf ft).
55. Refit the fluid tubes. (Note the 'O' ring on the pump suction tube).
56. Refit the fluid tube locating plate.
57. Fit and tighten the two bolts. 0,24 to 0,35 kgf.m (20 to 30 lbf in).
58. Carefully refit the valve body assembly, ensuring that the fluid tubes are not distorted.
59. Fit and tighten the three bolts and washers. 0,24 to 0,35 kgf.m (20 to 30 lbf in).
60. Connect the downshift inner cable to the downshift cam.
61. Refit the fluid tubes.
62. Replace the magnet and refit the fluid pan and joint washer.
63. Fit and tighten 12 bolts. 0,6 to 0,9 kgf.m (4.5 to 7.0 lbf ft).
64. Locate the torque converter housing in place.
65. Fit and tighten four bolts securing the torque converter housing. 10 mm diameter—2,7 to 4,4 kgf.m (20 to 30 lbf ft) 12 mm diameter— 4,1 to 6,9 kgf.m (30 to 50 lbf ft).
66. Refit the starter and reverse switch. 44.15.15. 0,42 to 0,69 kgf.m (36 to 60 lbf in).
67. Check, and if necessary adjust, the front and rear brake bands. 44.30.07 and 44.30.10.
68. Refit the gearbox assembly. 44.15.15.

Rover 3500 and 3500S Manual AKM 3621

44.36.01
Sheet 3

GEARBOX—AUTOMATIC—Borg-Warner type 65

PLANET GEARS AND CENTRE SUPPORT
—Remove and refit 44.36.04

Service tools: CBW 547A–50 or 601285, Torque wrench
CBW 60, Bench cradle

Removing
1. Remove the gearbox assembly. 44.20.01.
2. Wash the exterior of the unit in clean petrol or paraffin, invert it and place on a bench cradle CBW 60. Remove the switch. 44.15.15.
3. Unscrew the bolts securing the torque converter housing.
4. Remove the torque converter housing.
5. Unscrew 12 bolts.
6. Remove the fluid pan, joint washer and magnet.
7. Pull out the fluid tubes.
8. Release the downshift inner cable from the downshift cam.
9. Take out three bolts and washers.
10. Lift off the valve body assembly.
11. Unscrew two bolts.
12. Remove the fluid tube locating plate.
13. Pull out the fluid tubes. (Note the 'O' ring on the pump suction tube).
14. Take out five bolts.
15. Remove the pump and joint washer.
16. Remove the thrust washer.
17. Withdraw the front clutch.
18. Remove the thrust washers.
19. Withdraw the rear clutch and forward sun gear.
20. Squeeze together the ends of the front brake band and remove it together with the strut.
21. Take out three bolts.
22. Withdraw the centre support/planet gear assembly.
23. Separate the centre support from the planet gear assembly.
24. Withdraw the one-way clutch.
25. Remove the circlip.
26. Detach the one-way clutch outer race.

Refitting
27. Fit the one-way clutch outer race to the rear drum assembly.
28. Fit the circlip.
29. Refit the one-way clutch.
30. Assemble the centre support and planet gear assembly.
31. Refit the centre support/planet gear assembly, ensuring that the fluid and locating holes align with those in the casing.
32. Fit and tighten the bolts.
33. Squeeze together the ends of the front brake band and fit it in position together with the strut.

continued

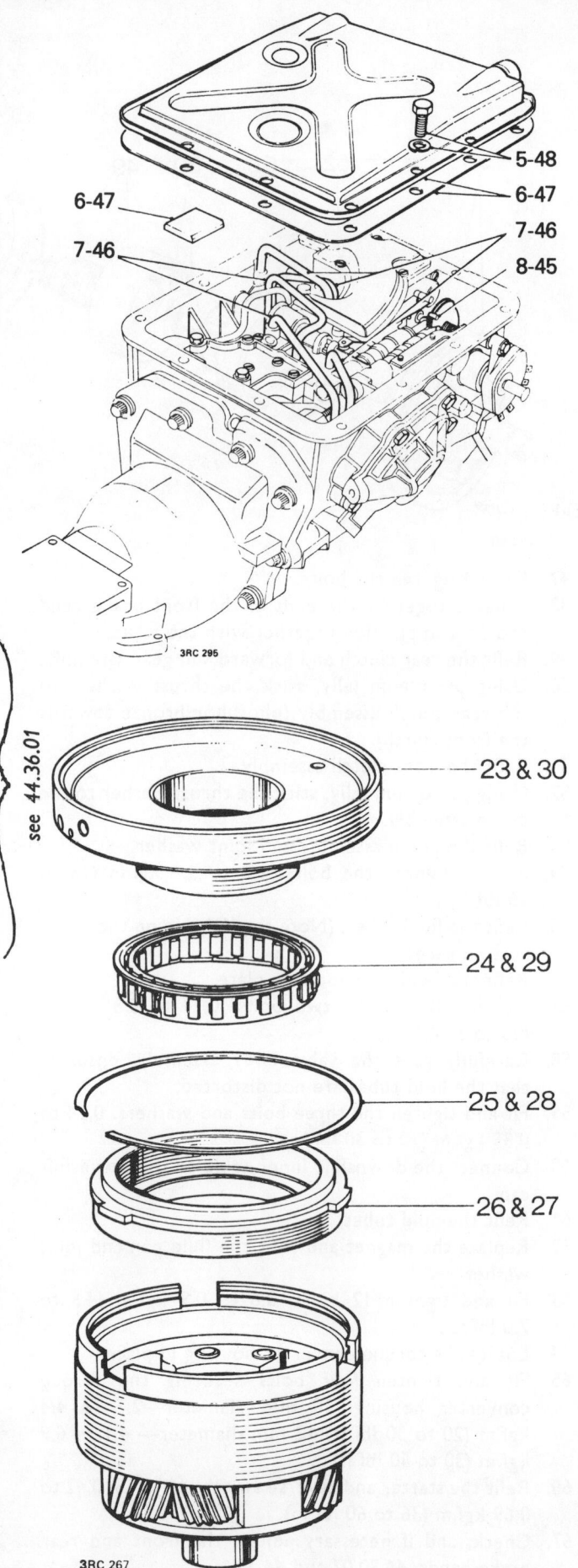

44.36.04
Sheet 1

Rover 3500 and 3500S Manual AKM 3621

GEARBOX—AUTOMATIC—Borg-Warner type 65

34. Refit the rear clutch and forward sun gear assembly.
35. Using petroleum jelly, stick the thrust washers to the rear clutch assembly (phosphor bronze towards the front clutch).
36. Refit the front clutch assembly.
37. Using petroleum jelly, stick the thrust washer to the pump assembly.
38. Refit the pump assembly and joint washer.
39. Fit and tighten the bolts. 1,8 to 3,4 kgf.m (13 to 25 lbf ft).
40. Refit the fluid tubes. (Note the 'O' ring on the pump suction tube).
41. Refit the fluid tube locating plate.
42. Fit and tighten the two bolts. 0,24 to 0,35 kgf.m (20 to 30 lbf in).
43. Carefully refit the valve body assembly, ensuring that the fluid tubes are not distorted.
44. Fit and tighten the three bolts and washers. 0,24 to 0,35 kgf.m (20 to 30 lbf in).
45. Connect the downshift inner cable to the downshift cam.
46. Refit the fluid tubes.
47. Replace the magnet and refit the fluid pan and joint washer.
48. Fit and tighten 12 bolts. 0,6 to 0,9 kgf.m (4.5 to 7.0 lbf ft).
49. Locate the torque converter housing in place.
50. Fit and tighten four bolts securing the torque converter housing. 10 mm diameter—2,7 to 4,5 kgf.m (20 to 30 lbf ft) 12 mm diameter— 4,1 to 6,9 kgf.m (30 to 50 lbf ft).
51. Refit the starter and reverse switch. 44.15.15. 0,42 to 0,69 kgf.m (36 to 60 lbf in).
52. Check, and if necessary adjust, the front brake band. 44.30.07.
53. Refit the gearbox assembly. 44.20.01.

SPEEDOMETER DRIVE PINION

—Remove and refit   44.38.04

Removing

1. Drive the vehicle on to a ramp, apply the handbrake and raise the ramp.
2. Disconnect the speedometer cable from the gearbox.
3. Remove the bolt and prise the speedometer pinion housing out of the extension.
4. Withdraw the speedometer drive pinion.
5. Remove the 'O' ring.
6. Extract the seal.

Refitting

7. Press a new seal into the housing.
8. Fit a new 'O' ring to the housing.
9. Fit the drive pinion into the housing.
10. Press the housing into the rear extension and secure with the bolt.
11. Refit the speedometer cable.

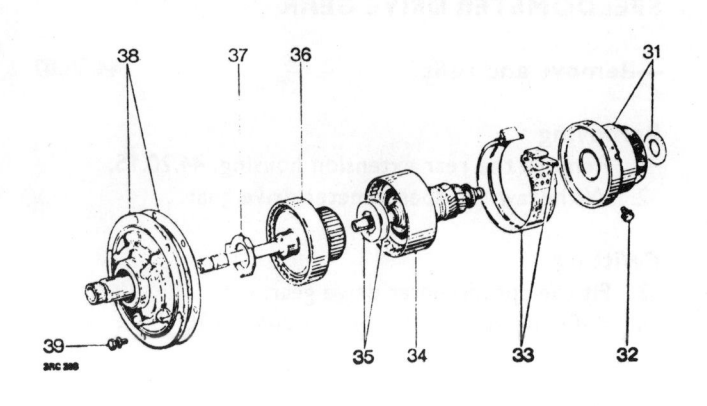

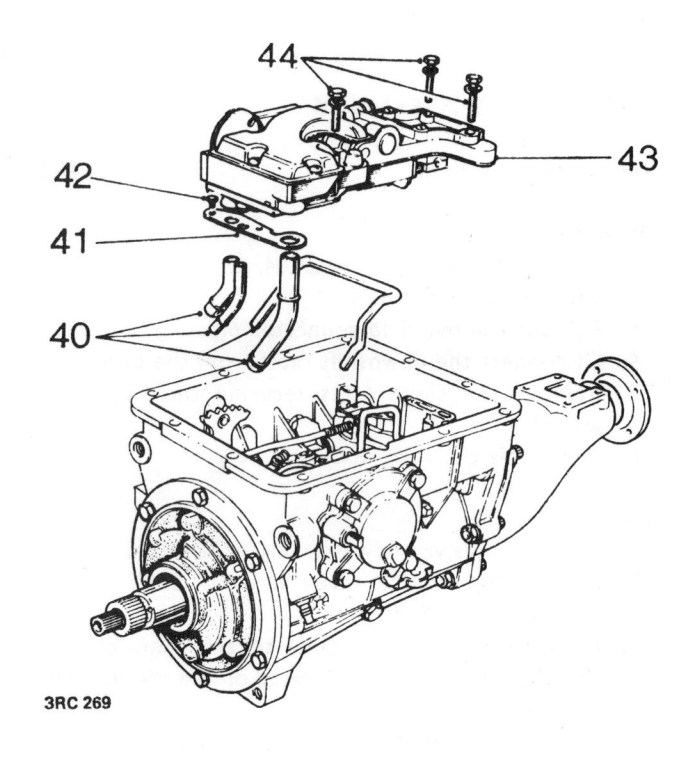

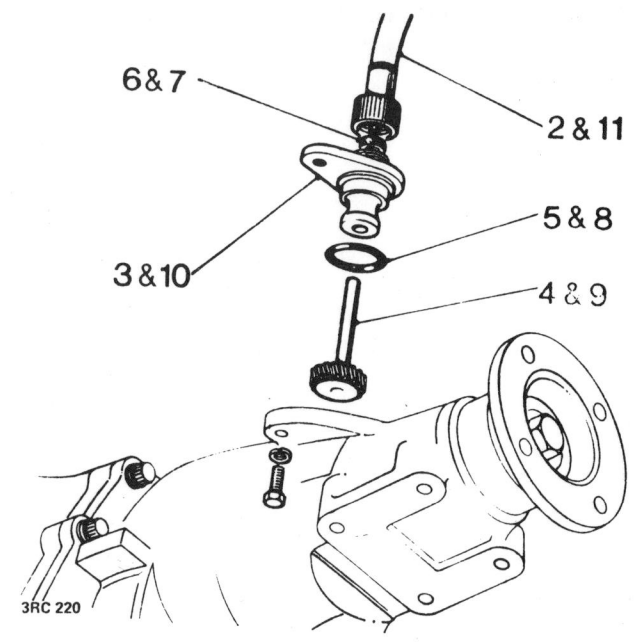

Rover 3500 and 3500S Manual AKM 3621

GEARBOX—AUTOMATIC—Borg-Warner type 65

SPEEDOMETER DRIVE GEAR

—Remove and refit   44.38.07

Removing
1. Remove the rear extension housing. 44.20.15.
2. Withdraw the speedometer drive gear.

Refitting
3. Fit the speedometer drive gear.
4. Refit the rear extension housing. 44.20.15.

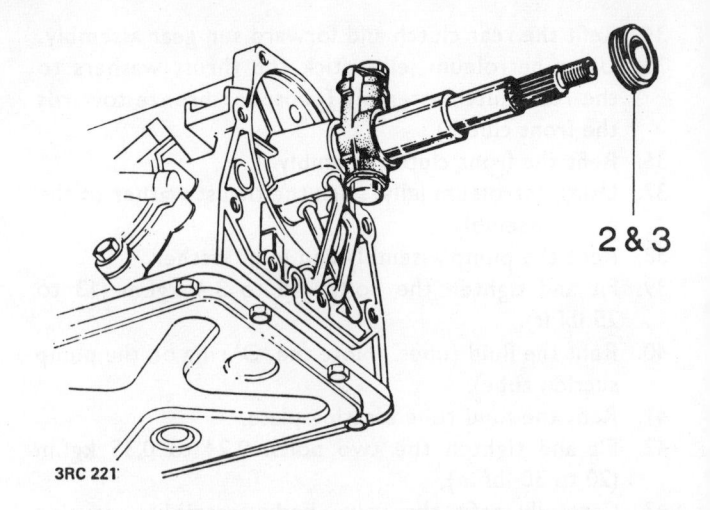

VALVE BODY ASSEMBLY

—Remove and refit   44.40.01

Removing
1. Remove the fluid pan. 44.24.04.
2. Withdraw the magnet.
3. Pull out the five fluid connector pipes.
4. Disconnect the downshift cable from the cam.
5. Remove the three bolts securing the valve body assembly.
6. Withdraw the valve body assembly.

Refitting
7. Locate the valve body assembly in position.
8. Fit and tighten the three securing bolts. 0,55 to 0,7 kgf.m (4 to 5 lbf ft).
9. Attach the downshift cable to the cam, ensuring that the cam is correctly located on the manual valve.
10. Refit the fluid connector pipes.
11. Attach the magnet to one of the bolt heads.
12. Refit the fluid pan. 44.24.04.

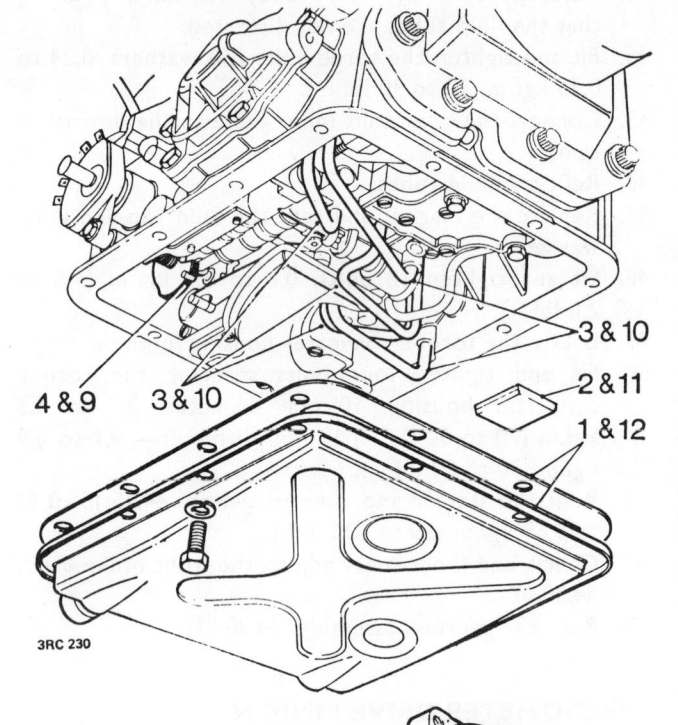

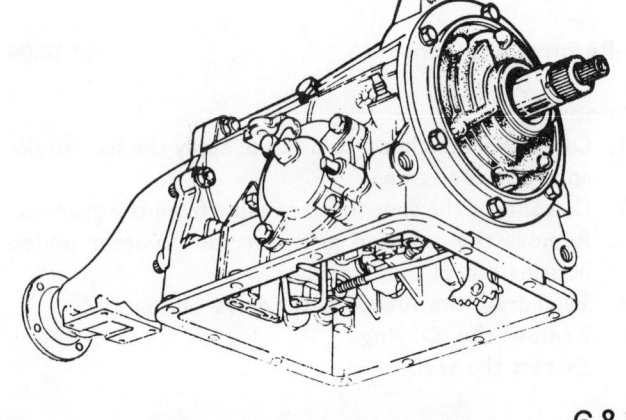

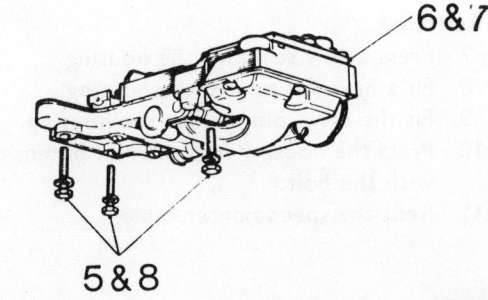

GEARBOX—AUTOMATIC—Borg-Warner type 65

VALVE BODY ASSEMBLY

—Overhaul 44.40.04
Service tool: CBW 548 or 601286 Torque screwdriver

Dismantling

1. Remove the valve body assembly. 44.40.01.
2. Withdraw the manual control valve.
3. Take out two screws.
4. Remove the downshift cam assembly.
5. Take out four screws.
6. Remove the oil strainer and gasket.
7. Take out screw.
8. Remove detent spring and spacer.
9. Withdraw the fluid pipe.
10. Remove four screws.
11. Withdraw the range control valve.
12. Take out eight screws.
13. Remove upper valve body.
14. Take out eight screws.
15. Remove the fluid tube collector.
16. Remove the separating plate.
17. Remove the check valve ball and spring.
18. Remove the servo orifice control valve spring and stop.
19. Remove the throttle valve stop and return spring.
20. Remove the throttle valve plate.
21. Withdraw the downshift valve.
22. Remove the throttle valve spring.
23. Withdraw the throttle valve.
24. Tap out the dowel pin, applying light pressure to the plug.
25. Withdraw the modulator plug.
26. Withdraw the modulator valve.
27. Withdraw the modulator valve spacer.
28. Withdraw the modulator valve spring.
29. Withdraw the servo orifice control valve.
30. Slacken progressively the three screws.
31. Carefully remove the end plate.
32. Remove the spring.
33. Withdraw the sleeve.
34. Take out the primary regulator valve.
35. Remove the spring.
36. Withdraw the secondary regulator valve.
37. Remove the screws from the upper valve body.
38. Remove the front end plate.
39. Take out three screws.
40. Remove the rear end plate.
41. Withdraw the 2–3 shift valve from the rear.
42. Remove the spring.
43. Withdraw the plunger.
44. Withdraw the 1–2 shift valve from the rear.
45. Remove the spring.
46. Withdraw the plunger.
47. Apply light pressure to the plug and remove the retainer from the range control valve.
48. Withdraw the plug, valve and spring.

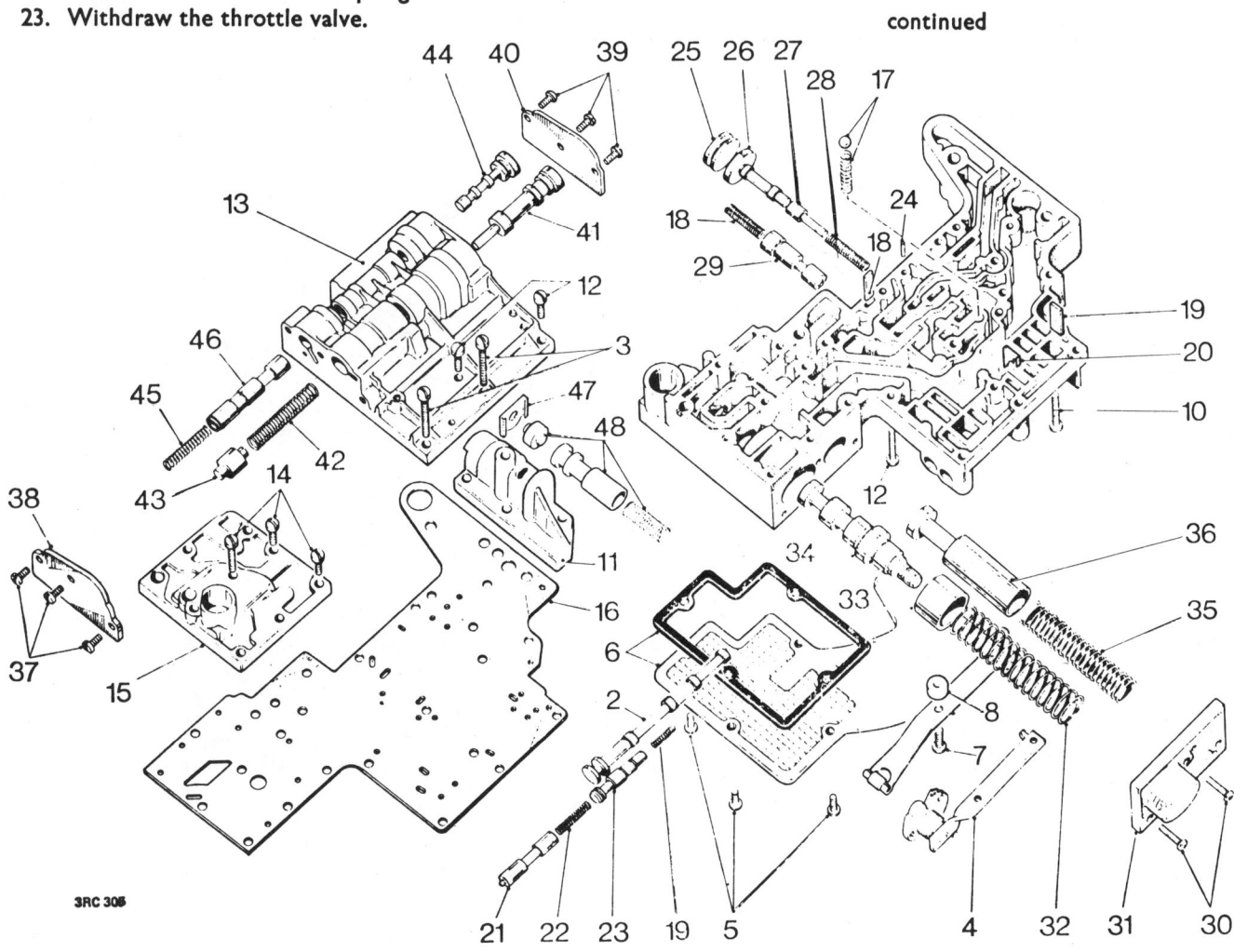

continued

Rover 3500 and 3500S Manual AKM 3621

44.40.04
Sheet 1

GEARBOX—AUTOMATIC—Borg-Warner type 65

Reassembling

49. Insert the spring, valve and plug into the range control valve body.
50. Apply light pressure to the plug and fit the retainer.
51. Insert the 1–2 shift valve.
52. Insert the 2–3 shift valve.
53. Replace the rear end plate.
54. Fit and tighten the three screws. 0,24 to 0,35 kgf.m (20 to 30 lbf in).
55. Insert the 1–2 shift valve plunger.
56. Insert the 2–3 shift valve spring.
57. Insert the 1–2 shift valve spring.
58. Insert the 2–3 shift valve plunger.
59. Locate the front end plate in position.
60. Fit and tighten three screws. 0,24 to 0,35 kgf.m (20 to 30 lbf in).
61. Insert the secondary regulator valve into the lower valve body.
62. Refit the spring.
63. Insert the primary regulator valve.
64. Insert the sleeve.
65. Insert the spring.
66. Hold the end plate in position.
67. Fit and tighten the three screws. 0,24 to 0,35 kgf.m (20 to 30 lbf in).
68. Insert the servo orifice control valve.
69. Insert the spring.
70. Depress the spring and fit the stop.
71. Insert the modulator control valve spring.
72. Insert the spacer.
73. Insert the modulator control valve.
74. Insert the plug.
75. Fit the dowel pin.
76. Refit the valve stop plate and spring.
77. Insert the throttle valve.
78. Insert the throttle valve return spring.
79. Insert the downshift valve.
80. Refit the throttle valve plate.
81. Refit the check valve ball and spring.
82. Place the separating plate in position.
83. Replace the oil tube collector.
84. Fit and loosely tighten the eight screws.
85. Replace the upper valve body.
86. Fit and loosely tighten the eight screws.
87. Tighten the sixteen screws, previously left loose. 0,24 to 0,35 kgf.m (20 to 30 lbf in).
88. Refit the range control valve.
89. Fit and tighten four screws.
90. Refit the detent spring and spacer.
91. Fit and tighten the screw. 0,24 to 0,35 kgf.m (20 to 30 lbf in).
92. Tension the downshift cam and refit the assembly.
93. Fit and tighten two screws. 0,24 to 0,35 kgf.m (20 to 30 lbf in).
94. Refit the oil strainer and gasket.
95. Fit and tighten four screws. 0,24 to 0,35 kgf.m (20 to 30 lbf in).
96. Insert the manual control valve.
97. Refit the valve body assembly.

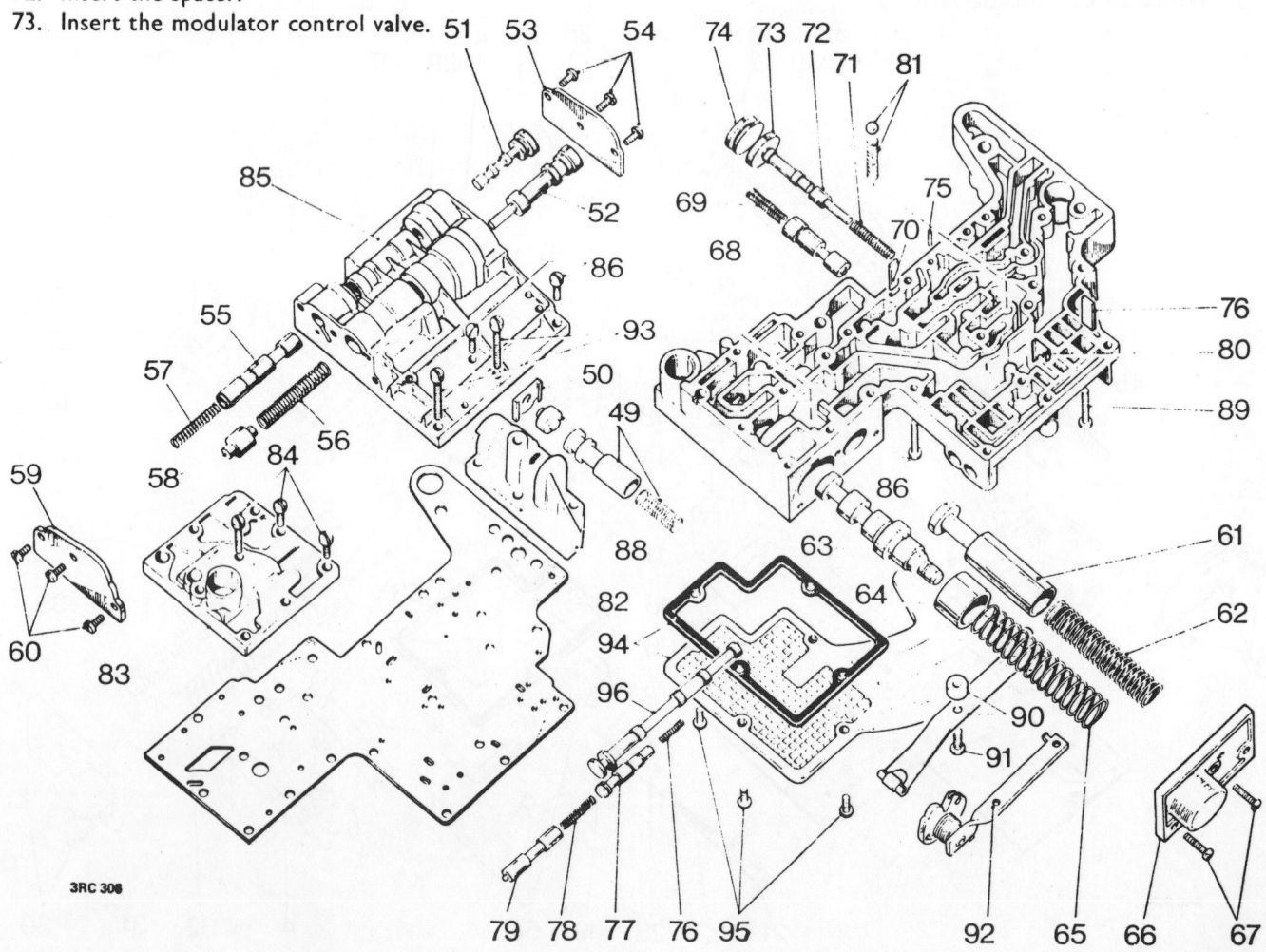

3RC 308

44.40.04
Sheet 2

Rover 3500 and 3500S Manual AKM 3621

PROPELLER SHAFTS

LIST OF OPERATIONS

Propeller shaft
    —remove and refit 47.15.01
    —overhaul 47.15.10

PROPELLER SHAFTS

PROPELLER SHAFT

—Remove and refit 47.15.01

Removing
1. Disconnect the coupling flanges.
2. Withdraw the propeller shaft.

Refitting
3. Ensure that the arrows marked on the splined sleeve yoke and shaft are in line.
4. Locate the propeller shaft in position with the sleeve end towards the gearbox.
5. Ensure that the registers on the coupling flanges engage.
6. Secure the coupling flange fixings. Torque 4,0 kgf.m (30 lbf ft).

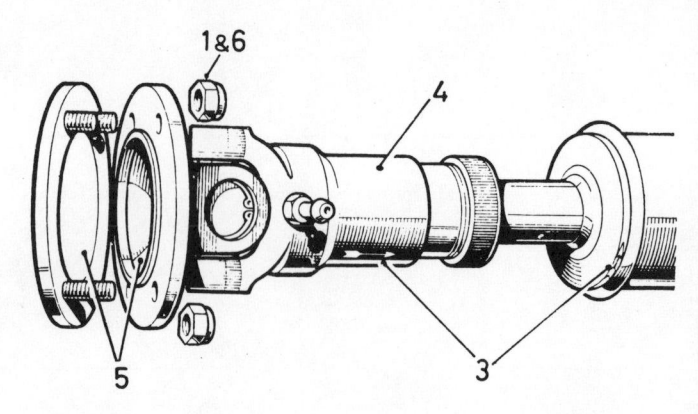

DATA

Manual gearbox models

Propeller shaft type	Hardy-Spicer needle bearing
Tubular shaft	63,5 mm (2.500 in.) diameter
Overall free length (face to face)	1136,65 mm (44.750 in.)

Automatic transmission models

Propeller shaft type	Hardy Spicer needle bearing
Tubular shaft	63,5 mm (2.500 in.) diameter
Overall free length (face to face)	1003,3 mm (39.5 in.)
Yoke angular displacement (see illustration)	
Up to final drive suffix 'A'	107°
From final drive suffix 'B' onwards	144°

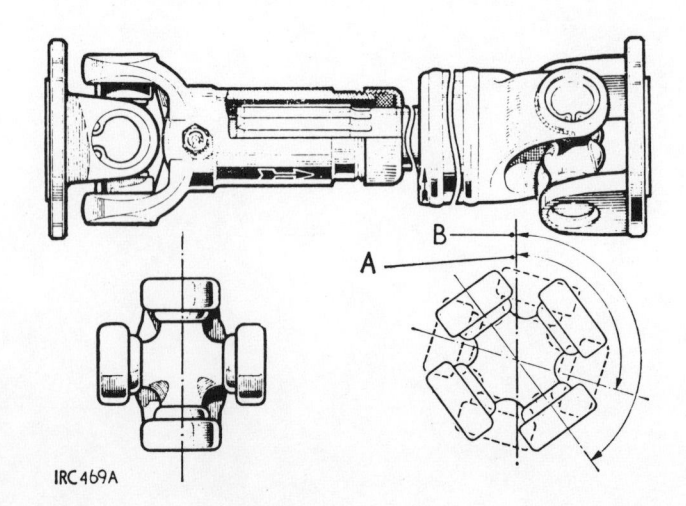

Rover 3500 and 3500S Manual AKM 3621

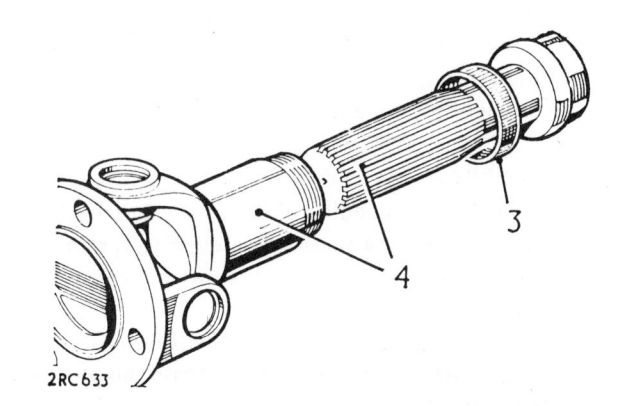

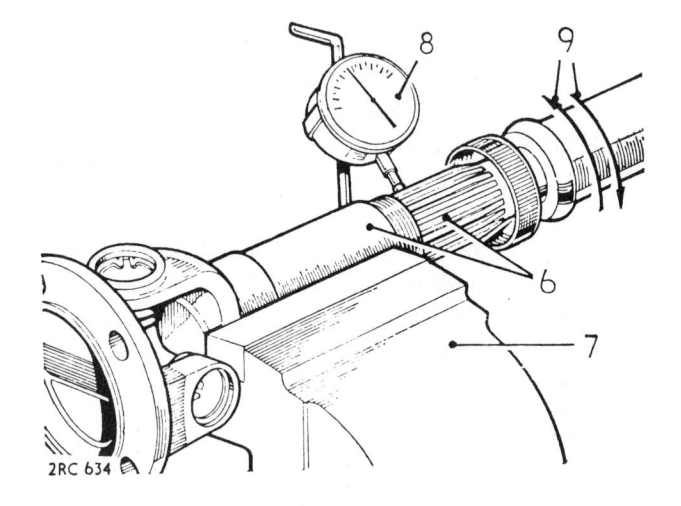

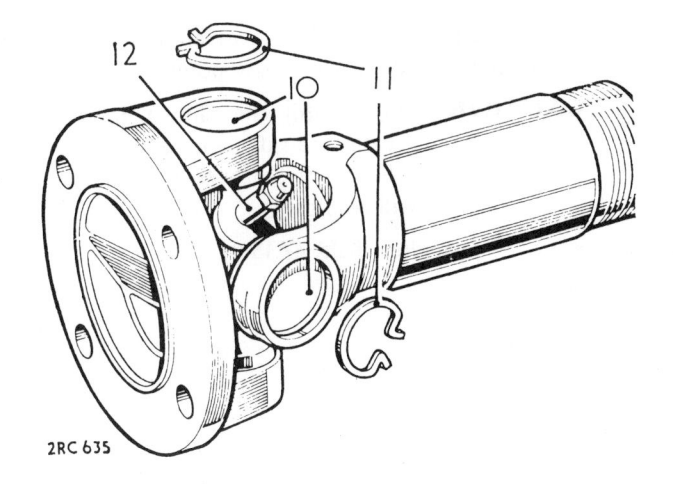

PROPELLER SHAFTS

PROPELLER SHAFT
—Overhaul 47.15.10

Dismantling

1. Remove the propeller shaft. 47.15.01.
2. Check that the alignment marks on the splined sleeve and the splined shaft are clearly visible. If necessary, make new alignment marks.
3. Unscrew the dust cap.
4. Withdraw the sliding joint.
5. Clean the splined shaft and the splined sleeve.
6. Temporarily locate the splined shaft into the sleeve, maintaining the marked alignment.
7. Secure the shaft in a vice.
8. Mount a dial test indicator to read off the outside diameter of the shaft splines.
9. Check the circumferential movement between the sleeve and shaft. If the movement exceeds 0,1 mm (.004 in) fit a new propeller shaft complete.
10. Clean any dirt and enamel from the circlips and the tops of the bearing races.
11. Remove the circlips.
12. Remove the grease nipple from the universal joint.
13. Locate the yoke of the splined sleeve on to a suitable piece of tube which has a slightly larger internal diameter than the journal bearing.
14. Using a brass drift, drive the universal joint downward until it is just clear of the lower yoke.
15. Lift the sleeve clear of the tube and withdraw the bearing downward to avoid dropping the needle rollers.

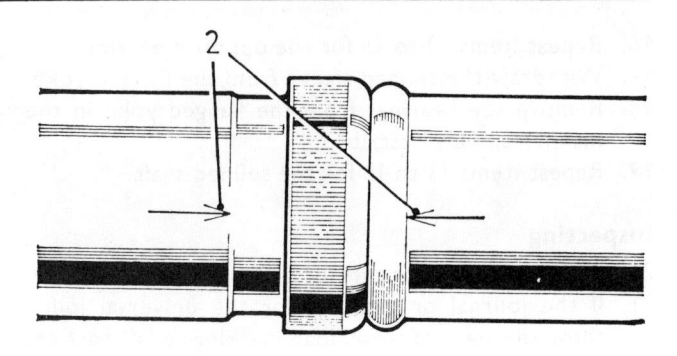

continued

Rover 3500 and 3500S Manual AKM 3621

47.15.10
Sheet 1

PROPELLER SHAFTS

16. Repeat items 13 to 15 for the opposite bearing.
17. Withdraw the splined sleeve from the flanged yoke.
18. Remove the bearings from the flanged yoke in the manner already described.
19. Repeat items 13 to 18 for the splined shaft.

Inspecting

20. Examine all components for obvious wear or damage.
21. If the journal or bearings for the universal joints show any signs of wear, load markings or distortion, they must be replaced complete. Replacement journal assemblies comprise a spider complete with oil seals and bearings.
22. In the event of wear in any of the eight yoke cross holes, rendering them oval, a new propeller shaft complete must be fitted.

Reassembling

23. Assemble the needle rollers in the bearing races, if necessary using a smear of vaseline to retain them in place. About half fill the races with a recommended grease.
24. Insert the journal, complete with seals, into the flange yoke holes with the grease nipple tapping pointing away from the flange.
25. Place the flanged yoke on a suitable flat support.
26. Place the first bearing in position.
27. Using a brass drift, slightly smaller in diameter than the hole in the yoke, tap the bearing into position.
28. Fit the circlip to retain the bearing.

    NOTE: The bearing outer races must be a drive fit, otherwise fit a new propeller shaft complete.

29. Repeat 25 to 28 for the other three bearings comprising the universal joint.
30. Ensure that all four circlips are firmly located in their grooves. If the joint appears to bind, tap the yoke ears lightly with a soft mallet.
31. Repeat 23 to 30 for the other universal joint.
32. Liberally smear the splines of the shaft and sleeve with the recommended grease.
33. Assemble the splined shaft and sleeve maintaining the marked alignment.
34. Fit the grease nipple to the universal joint. Lubricate the propeller shaft at the grease points.

    CAUTION: Do not fill the sliding joint with grease, use sufficient to lubricate the splines only, otherwise hydraulicing will result.

35. Refit the propeller shaft. 47.15.01.

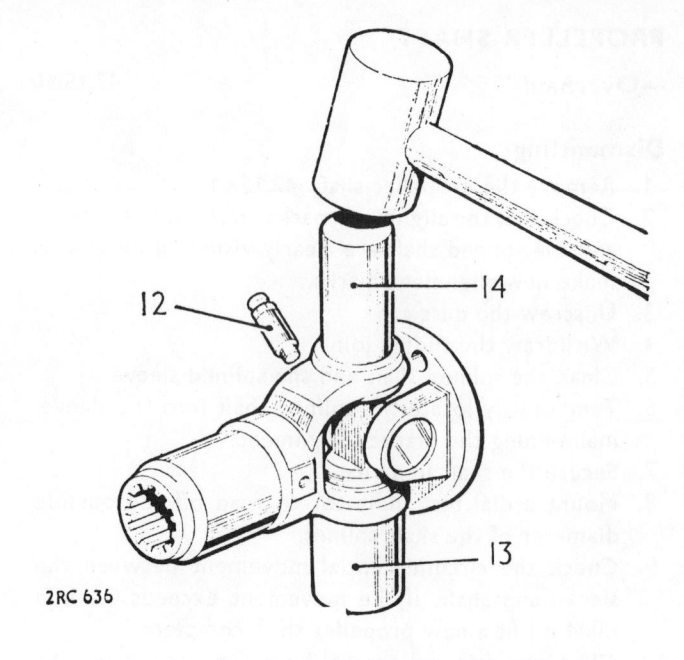

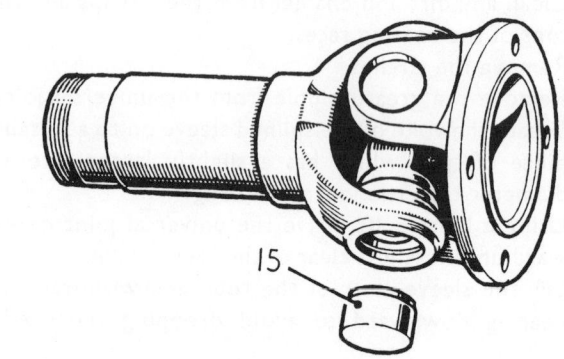

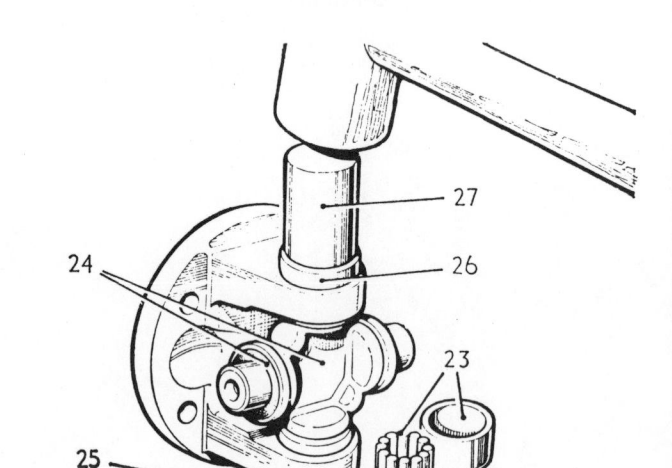

47.15.10
Sheet 2

Rover 3500 and 3500S Manual AKM 3621

REAR AXLE AND FINAL DRIVE

LIST OF OPERATIONS

Differential drive shafts
    —remove and refit 51.10.07
    —overhaul 51.10.14

Final drive unit
    —remove and refit 51.25.13
    —overhaul 51.25.19
    —mountings—remove and refit 51.25.31

Pinion case extension
    —remove and refit 51.25.37
    —overhaul 51.25.49
    —mounting—remove and refit 51.25.55

Rover 3500 and 3500S Manual AKM 3621

REAR AXLE AND FINAL DRIVE

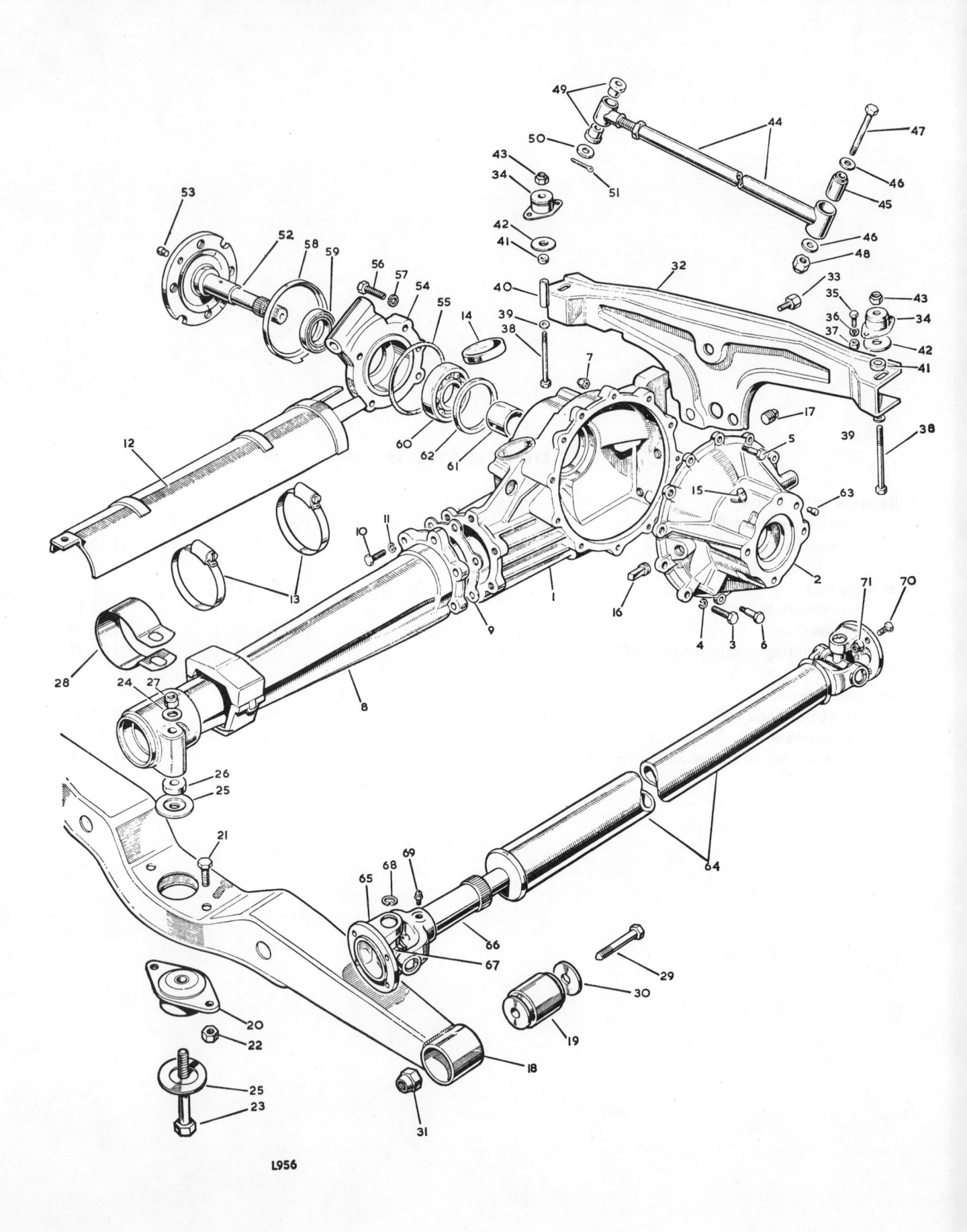

51–2

Rover 3500 and 3500S Manual AKM 3621

REAR AXLE AND FINAL DRIVE

Key to final drive casings and propeller shaft

1. Housing for pinion and drive shaft
2. Cover for pinion and drive shaft housing
3. Set bolt ($\frac{5}{16}$ in UNF x 1 in long) ⎫
4. Spring washer ⎬ Fixing cover to housing
5. Fitting bolt, top ($\frac{3}{8}$ in UNF special) ⎬
6. Fitting bolt, bottom ($\frac{3}{8}$ in UNF special) ⎬
7. Self-locking nut ($\frac{3}{8}$ in UNF) ⎭
8. Extension case for final drive
9. Joint washer for extension case
10. Bolt ($\frac{3}{8}$ in UNF x $1\frac{1}{8}$ in long) ⎫ Fixing extension case
11. Spring washer ⎭ to pinion housing
12. Damper plate for extension case
13. Clip fixing damper plate to extension case
14. Cup plug for inspection hole
15. Breather for pinion housing
16. Filler plug ⎫ For
17. Drain plug ⎭ final drive
18. Crossmember, front, for final drive
19. Bush for mounting bracket
20. Flexible mounting, front, for final drive
21. Set bolt ($\frac{5}{16}$ in UNF x $\frac{3}{4}$ in long) ⎫ Fixing flexible
22. Self-locking nut ($\frac{5}{16}$ in UNF) ⎬ mounting to front mounting bracket
23. Bolt ($\frac{7}{16}$ in UNF x $5\frac{1}{2}$ in long) ⎫ Fixing
24. Plain washer ⎬ flexible
25. Dished washer ⎬ mounting
26. Distance piece ⎬ to
27. Self-locking nut ($\frac{7}{16}$ in UNF) ⎬ extension
28. Support strap ⎭ casing
29. Set bolt ($\frac{7}{16}$ in special) ⎫ Fixing front
30. Plain washer ⎬ mounting bracket
31. Nut plate ⎭ to base unit
32. Mounting bracket, rear, for final drive
33. Special countersunk bolt fixing rear mounting bracket to final drive
34. Flexible mounting, rear, for final drive
35. Bolt ($\frac{5}{16}$ in UNF x $\frac{3}{4}$ in long) ⎫ Fixing rear flexible
36. Nut ($\frac{5}{16}$ in UNF) ⎬ mounting to
37. Spring washer ⎭ base unit
38. Bolt ($\frac{7}{16}$ in UNF x $5\frac{3}{4}$ in long) ⎫
39. Plain washer, small ⎬ Fixing final drive
40. Distance tube ⎬ to rear
41. Distance washer ⎬ flexible
42. Plain washer, large ⎬ mountings
43. Self-locking nut ($\frac{7}{16}$ in UNF) ⎭
44. Rear stabiliser rod
45. Bush for rear stabiliser rod
46. Thrust washer for rear stabiliser rod
47. Bolt ($\frac{3}{8}$ in UNF x $2\frac{1}{4}$ in long) ⎫ Fixing rod to final
48. Self-locking nut ($\frac{3}{8}$ in UNF) ⎭ drive rear bracket
49. Rubber bush ⎫
50. Plain washer ⎬ Fixing rod to base unit
51. Split pin ⎭
52. Differential drive shaft
53. Dowel locating brake disc
54. Bearing housing
55. 'O' ring for bearing housing
56. Set bolt ($\frac{3}{8}$ in UNF x $1\frac{1}{4}$ in long) ⎫ Fixing bearing
57. Spring washer ⎬ housing to pinion housing and cover
58. Oil catcher
59. Oil seal for drive shaft
60. Bearing for drive shaft
61. Thrust collar for bearing
62. Spacer
63. Dowel, bearing housings to pinion housing and cover
64. Propeller shaft assembly
65. Flange yoke
66. Sleeve yoke
67. Journal complete
68. Circlip for journal
69. Grease nipple for sleeve yoke
70. Special bolt ⎫ Fixing propeller shaft
71. Self-locking nut ($\frac{3}{8}$ in UNF) ⎭ to final drive flange

REAR AXLE AND FINAL DRIVE

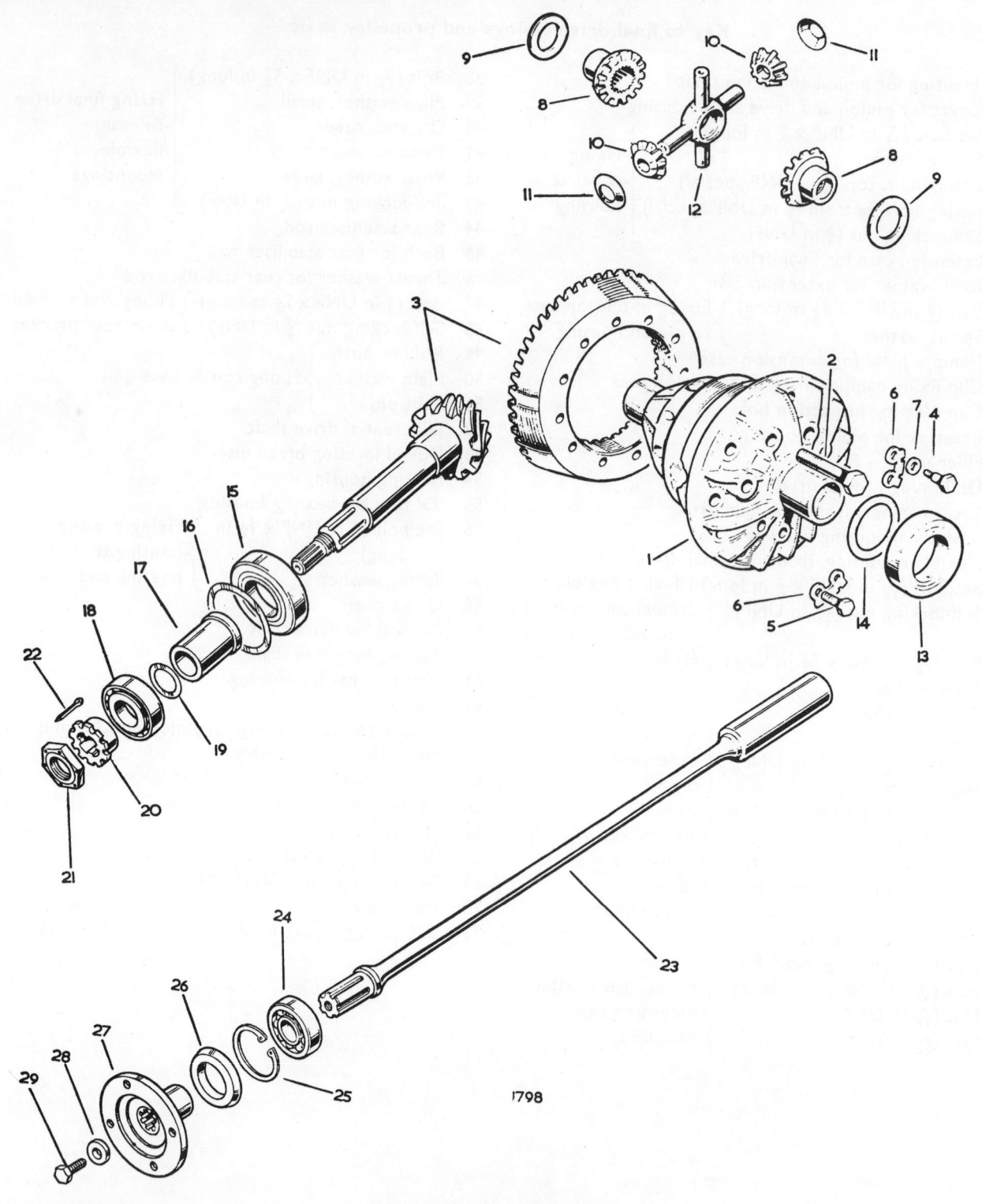

Differential and extension shaft

1. Differential case
2. Set bolt (⅜ in UNF x 2¼ in long) for differential case
3. Crownwheel and hypoid pinion
4. Special bolt (1.125 in long)
5. Fitting bolt (1.125 in long) } Fixing crownwheel to differential case See 51.25.19
6. Double locker
7. Plain washer
8. Differential wheel
9. Thrust washer for differential wheel
10. Differential pinion
11. Thrust washer, dished, for differential pinion
12. Cross-pin for pinion, early type
    Cross-pin for pinion, latest type
13. Bearing for differential
14. Shim for differential bearing
15. Bearing for hypoid pinion, pinion end
16. Shim for bearing adjustment, pinion end
17. Spacer for pinion bearings
18. Bearing for hypoid pinion, front end
19. Shim for bearing adjustment, front end
20. Serrated locking collar for pinion front bearing
21. Special locking nut for pinion
22. Split pin for locking nut
23. Extension shaft for final drive
24. Bearing for extension shaft
25. Circlip retaining bearing to extension case
26. Oil seal for extension shaft
27. Coupling flange
28. Plain washer
29. 'Wedglock' bolt (⅜ in UNF x 1 in long) } Fixing flange to extension shaft

51-4

Rover 3500 and 3500S Manual AKM 3621

REAR AXLE AND FINAL DRIVE

General service information 51.00.00

IMPORTANT: Do not jack under the de Dion tube, use the rear jacking point as illustrated.

The serial numbering of the final drive units for the Rover 3500 commences at 42500001A, however 314 units are stamped with the serial number beginning 401, and are in the range 40190066 to 40199919 inclusive. Therefore, as the 401 range of numbers normally applies to the Rover 2000 final drive units, care must be taken to correctly identify the car model when ordering replacement parts or carrying out work on the final drive. The oil filler/level plug provides an easy means of identification, being in the front of the 3500 unit and in the rear of the 2000 unit.

3500 models from final drive unit suffix 'B' onwards and all 3500S models—Fitted to the pinion housing extension case is a harmonic damper which effectively reduces transmission resonance. After fitting, this damper must not be disturbed, as special equipment is necessary for correct tuning of the component.

The final drive unit and drive shafts fitted to early 3500 models are superseded by modified components, which are fitted to latest 3500 and all 3500S models. See beginning of Division 64 for full description of modification.

When topping up or refilling the final drive unit, it is important to ensure that the car is standing on a level surface; do not jack up the rear.

Great care must be taken not to overfill the final drive housing. Allow all surplus oil to drain away before replacing the oil level filler plug. The oil level must never exceed the bottom of the filler plug hole.

Rear jacking point
A—Jacking bracket

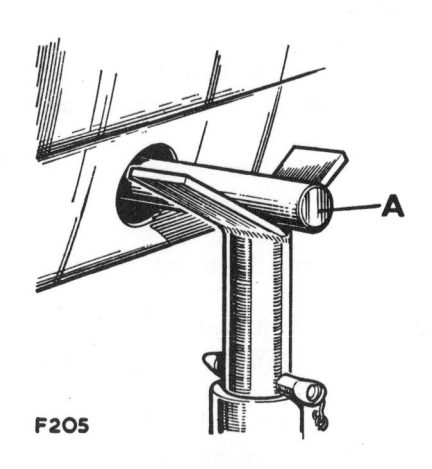

Position of support stands
A—Suitable length of 22 mm ($\frac{7}{8}$ in) diameter rod inserted in jacking tube

DIFFERENTIAL DRIVE SHAFT

—Remove and refit 51.10.07

Removing

1. Remove the rear brake disc. 70.10.11.
2. Disconnect the connecting pipe from between the calipers and plug to prevent leakage. If removing the right-hand side drive shaft, disconnect the main supply flexible brake hose and plug to prevent leakage.
3. Remove the brake caliper(s). 70.55.03.
4. Place a drip tray under the final drive assembly.
5. Remove the four bolts and spring washers securing the bearing housing to the pinion housing. Rotate flange to gain access to bolts.
6. Withdraw the differential drive shaft and bearing housing complete with spacer and, where fitted, the 'O' ring.
7. Ensure that the spacer for drive shaft bearing is carefully retained for reassembly.

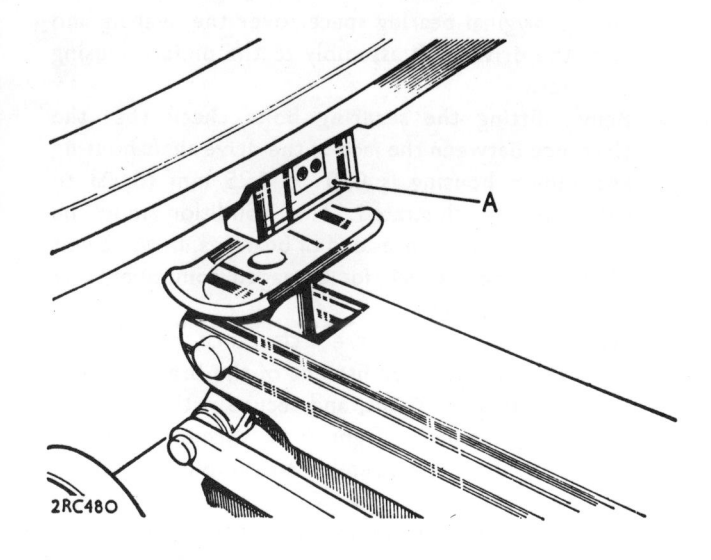

continued

Rover 3500 and 3500S Manual AKM 3621

REAR AXLE AND FINAL DRIVE

Refitting

8. Fit the original bearing spacer over the bearing and offer the drive shaft assembly to the pinion housing or cover.

9. Before fitting the securing bolts check that the clearance between the face of the drive shaft housing and pinion housing is 0,07 to 0,25 mm (0.003 to 0.010 in), as illustrated. This condition must be achieved to ensure pre-load of bearings. In the event of discrepancy, check for damaged components or incorrect assembly.

10. Having obtained the correct clearance, remove the drive shaft housing for fitment of the brake caliper, using the original shims, and secure with two bolts and spring washers. Tighten to torque 8,5 kgf.m (60 lbf ft). Refit the assembly to the pinion housing and, where fitted, a new 'O' ring, using Hylomar PL 32M sealing compound on both joint faces.

11. Tighten to a torque of 4,0 kgf.m (30 lbf ft).

12. Refit the caliper to the housing. 70.55.03.

13. Replace the brake disc for checking alignment and secure with four bolts, packing washers being necessary at this stage.
    It will be necessary to rotate the disc and drive shaft until the dowel holes are parallel with the caliper, then ease the disc between the caliper pads and on to the dowels.

14. Check that the disc runs true to within 0,17 mm (0.007 in) at 254 mm (10 in).

15. Refit the outer brake pad(s).

16. Reconnect all hydraulic fluid pipes that have been previously disconnected.

17. Reconnect the handbrake linkage.

18. Remove the four bolts and packing washers securing the brake disc.

19. Position the external drive shaft assembly on to the brake disc.

20. Secure drive shaft flange yoke to the disc and differential drive assembly with four bolts and new lock plates and tighten to 11,7 kgf.m (85 lbf ft). Ensure that when fitted there is a clearance of at least 0,2 mm (0.010 in) between the set bolts and the oil catcher.
    The bolts and washers are natural finish and must not be rustproofed until after assembly, when they are to be brush painted.

21. Bleed the brakes. 70.25.02.

22. Fit the road wheel.

23. Adjust the handbrake linkage, as necessary.

24. Remove the axle stands and jack.

25. Check the final drive oil level, and replenish as necessary.

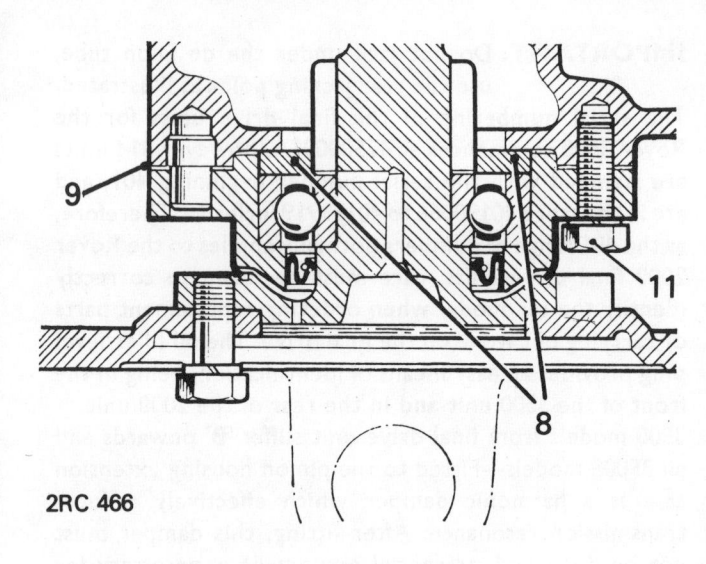

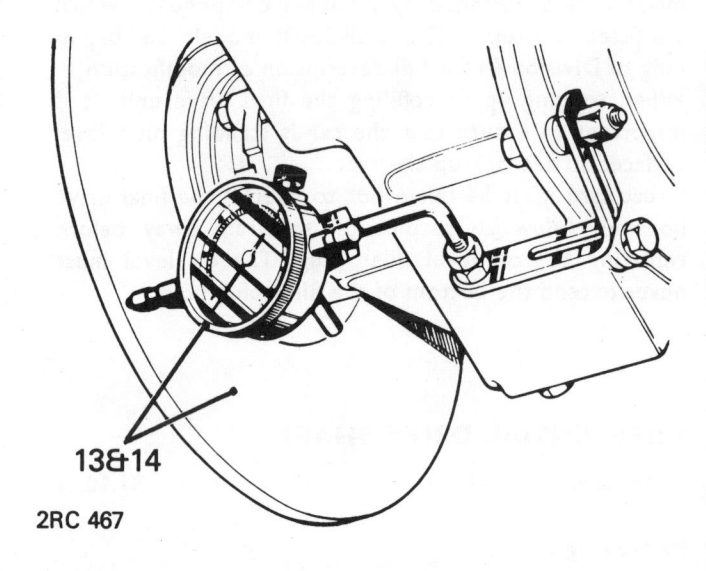

2RC 466

2RC 467

REAR AXLE AND FINAL DRIVE

DIFFERENTIAL DRIVE SHAFT

—Overhaul   51.10.14

Service tools: 275870   Axle shaft tool
605641   Collar adaptor
605008   Plunger tool

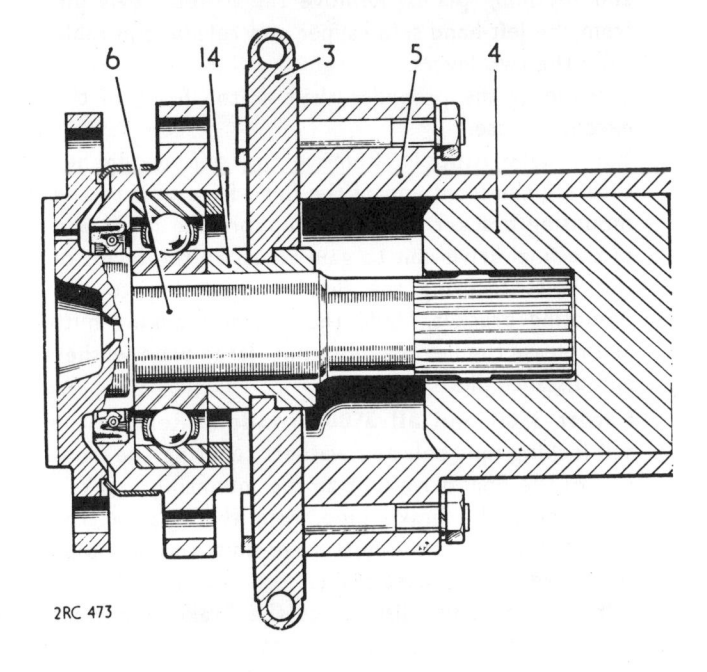

Dismantling

1. Remove the differential drive shaft. 51.10.07.
2. Remove the bearing spacer from the housing and ensure that it is retained for refitment to the appropriate housing.
3. Attach 605641 to the thrust collar machined groove.
4. Locate 605008 over the differential drive shaft.
5. Bolt 275870 to the adaptor 605641.
6. Press out the drive shaft.
7. Remove the oil seal.
8. Drift the bearing from the drive shaft housing.
9. Do not remove the oil catcher unless necessary.

Reassembling

10. If the oil catcher has been removed, a new one should be replaced by carefully tapping into position, using a wooden block. Ensure that the lip is adjacent to the oil drain hole at the underside of the housing, and that it is fully home to avoid fouling the shaft flange.
11. Smear the running diameter of the drive shaft with Silicone Compound MS4, apply Bostik 1776 to the outside diameter of the seal, then press the seal into the housing.
12. Press the new bearing into the housing.
13. Position the bearing and housing over the drive shaft and, using a suitable tube, press into position.
14. Ensure that the breather slots are clean and position a new thrust collar over the shaft with the small slots to the bearing face, and press home. The minimum pressure required to press the thrust collar to the shaft is two tons.
15. Refit the differential drive shaft. 51.10.07.

FINAL DRIVE UNIT

—Remove and refit   51.25.13

Service tool: RO 1007   Final drive cradle

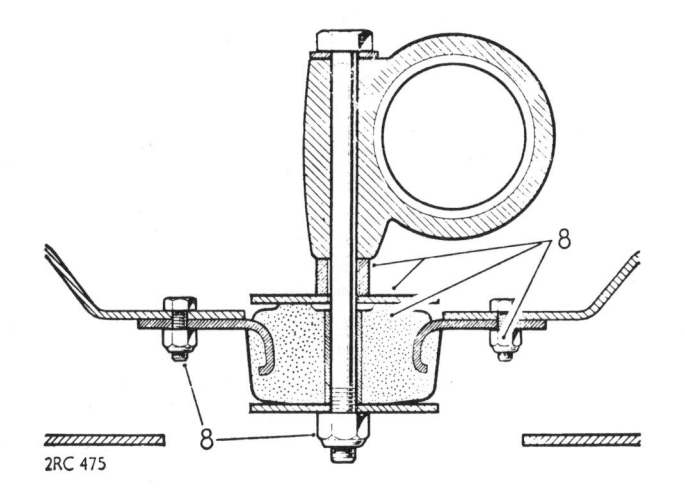

Removing

1. Slacken the road wheel nuts.
2. Jack up the car and support at jacking points at each side of the car.
3. Remove the road wheels.
4. Fit RO 1007 to a trolley jack and position it under the final drive unit.

continued

Rover 3500 and 3500S Manual AKM 3621

51.10.14
51.25.13
Sheet 1

REAR AXLE AND FINAL DRIVE

5. Remove the four special bolts and two lock-washers from each side securing left-hand and right-hand hub drive shafts to the brake discs.
6. Disconnect the handbrake cable from the rear brake calipers by slackening the two Phillips head screws and retaining plates. Remove the lower clevis pin from the left-hand side caliper and release the cable from the two levers.
7. Disconnect the propeller shaft at the front of the extension case.
8. **Early 3500 models.** Remove the self-locking nut securing the front end of the final drive unit to the mounting bracket; raise and support the front end of the final drive unit to gain access to the heads of the bolts securing the flexible mounting to the mounting bracket. Undo the two self-locking nuts and release the flexible mounting, large plain washer and distance piece.
9. **Latest 3500 and all 3500S models.** Remove the self-locking nut, bolt and washers securing the front end of the final drive unit to the mounting bracket.
10. Disconnect the brake pipe from the hose at the right-hand side of the final drive unit, and seal pipe with bleed nipple dust cover.
11. Where applicable, disconnect the brake pad wear warning leads at the plug connectors.
12. Remove the three bolts securing the final drive unit to the rear mounting bracket.
13. Carefully lower the rear end of the final drive unit, while at the same time manoeuvring the front end free of the mounting bracket, and remove the unit from under the car.
14. **Latest 3500 and all 3500S models.** Take care not to lose the dished washer and distance piece located above the flexible mounting.

Refitting

NOTE: If a new replacement final drive unit is being fitted to an early 3500 model with original final drive unit serial number up to suffix 'A' and chassis serial number up to suffix 'B' inclusive, the new unit will be of a modified type which is supplied for all Service replacements. However, as the mounting on the new unit has been moved forward by 9,5 mm (0.375 in), it is necessary to fit the rear mounting bracket, Part No. 557333.

15. Place the final drive unit on a jack and position under the car.

Early 3500 models—Items 16 to 19

16. Insert the front securing bolt with a plain washer through the lug on the front end of the final drive extension case. Ensure that the four bolts are located in the propeller shaft coupling flange at the front of the extension case.

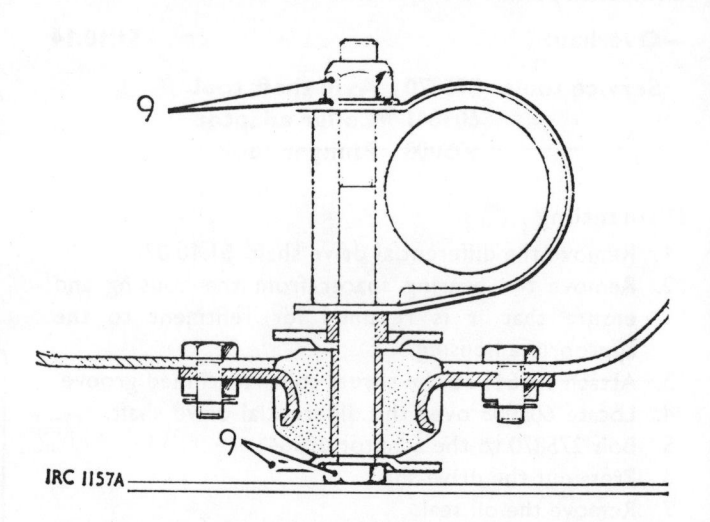

17. Raise the front end only of the final drive unit and position it above the mounting bracket. Locate the distance piece, large plain washer and flexible mounting through the aperture in the mounting bracket and on to the bolt in the extension case. Fit a large plain washer and self-locking nut to the bolt; tighten the nut only a few turns sufficient to retain the assembly.
18. Raise the rear end of the final drive unit, locate the rear mounting bracket, and secure with the three special bolts.
19. Support the front end of the final drive unit clear of the mounting bracket and secure the flexible mounting, thick side downward, with two bolts and self-locking nuts. Tighten to a torque of 2,3 kgf.m (17 lbf ft). Tighten the nut and bolt securing the extension case to the flexible mounting to a torque of 4,9 kgf.m (35 lbf ft).

Latest 3500 and all 3500S models—Items 20 to 23

20. Locate the support bracket in position around the lug on the front end of the final drive extension case. Ensure that the four bolts are located in the propeller shaft coupling flange at the front end of the extension case.
21. Place the dished washer, washer and distance piece in position above the flexible mounting in the front support bracket.
22. Raise the front end only of the final drive unit and position it above the distance piece on the flexible mounting.
23. Insert the front securing bolt with dished washer through the flexible mounting, upper dished washer, distance piece, support bracket and lug on the front end of the final drive extension case. Fit a plain washer and self-locking nut to the bolt; tighten the nut only a few turns sufficient to retain the assembly.

continued

REAR AXLE AND FINAL DRIVE

24. Raise the rear end of the final drive unit, locate the rear mounting bracket, and secure with the three special bolts.
25. Tighten the nut and bolt securing the extension case to the flexible mounting to a torque of 4,9 kgf.m (35 lbf ft).
26. Refit the propeller shaft.
27. Reconnect the brake pipes and handbrake cable. Where applicable, reconnect the brake pad wear warning leads.
28. Fit the two drive-shafts using new lock plates, and ensure that when fitted there is a minimum clearance of 0,254 mm (0.010 in) between the threaded end of the bolts and the oil catcher.
29. Bleed the brakes. 70.25.02.
30. Ensure that final drive is filled with correct grade oil to level plug, and that all fixings are secure.

NOTE: With a new or overhauled final drive unit, it is important to use one of the following especially suitable lubricants for the initial fill. Then, after completing 1.600 km (1,000 miles) running, the initial fill oil should be drained and the final drive refilled with the standard recommended lubricant. Division 09.

Component	SAE	BP	Castrol	Duckhams	Esso	Mobil	Shell	Texaco
Final drive	90EP	BP Gear Oil SAE 90 EP	Castrol Hypoy 'B'	Duckhams Hypoid 90 'S'	Esso Gear Oil GX 90/140	Mobilube HD 90	Spirax HD 90	Multigear Lubricant EP 90

31. Fit road wheels and fully tighten nuts when car is standing on floor.

FINAL DRIVE UNIT

—Overhaul   51.25.19

Service tools:
- 18G47 Press
- 262757 Pinion head outer race fitting tool
- 530102 Pinion nut spanner
- 530106 Dial gauge bracket
- 541885 Adaptor for pinion head bearing
- 600192 Adaptor for differential bearings
- 600447 Adaptor for pinion
- 600968 Pre-load checking tool
- 605004 Pinion height gauge

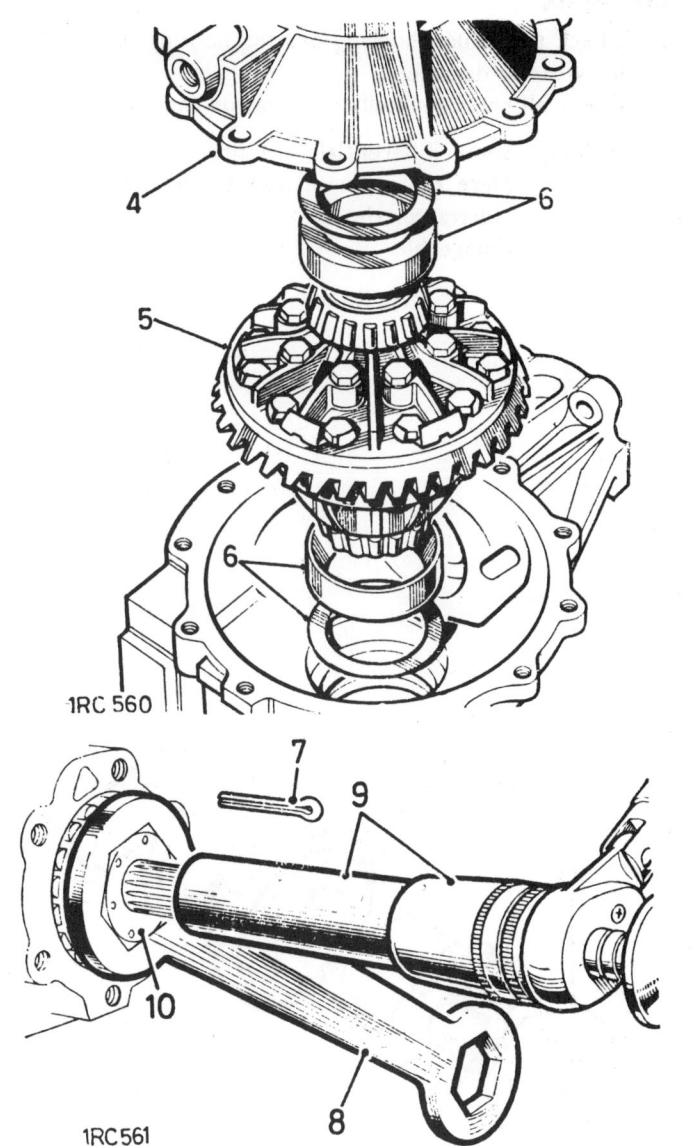

Dismantling

1. Remove the final drive unit. 51.25.13.
2. Remove the differential drive shafts. 51.10.07.
3. Remove the extension case.
4. Remove the pinion housing cover.
5. Withdraw the crownwheel and differential case assembly.
6. Remove the differential case bearing races and shims from the pinion housing and cover.
7. Withdraw the split pin from the pinion nut.
8. Fit the pinion nut spanner. 530102.
9. Fit the adaptor 600447, together with a wrench to the pinion shaft.
10. Remove the pinion nut.

continued

Rover 3500 and 3500S Manual AKM 3621

REAR AXLE AND FINAL DRIVE

11. Withdraw the serrated collar and taper roller bearing.
12. Withdraw the pinion from the housing and remove the shims and distance sleeve.
13. Using a suitable brass drift, remove the pinion head bearing race. Do not damage the shims fitted under the race.
14. Remove the drain plug from the housing.
15. Insert a long brass drift through the drain plug hole and remove the pinion front bearing race.
16. Extract the roller bearing and shims from the pinion head. 18G47 press and 541885 adaptor.
17. Extract the roller bearings from the differential case. 18G47 press and 600192 adaptor.
18. Remove the crownwheel from the differential case.
19. Add alignment marks between the two halves of the differential case to ensure correct re-assembly.
20. Remove the inner ring of bolts from the differential case and separate the two halves.
21. Withdraw the differential cross pin, pinions and thrust washers.
22. Withdraw the differential wheels and thrust washers.

Inspection

23. Check the pinion housing bearing housing locations for wear or damage.
24. Check all parts generally for wear or damage.

   NOTE: Some differentials are fitted with a two piece cross pin instead of the one piece illustrated, the two types are fully interchangeable.

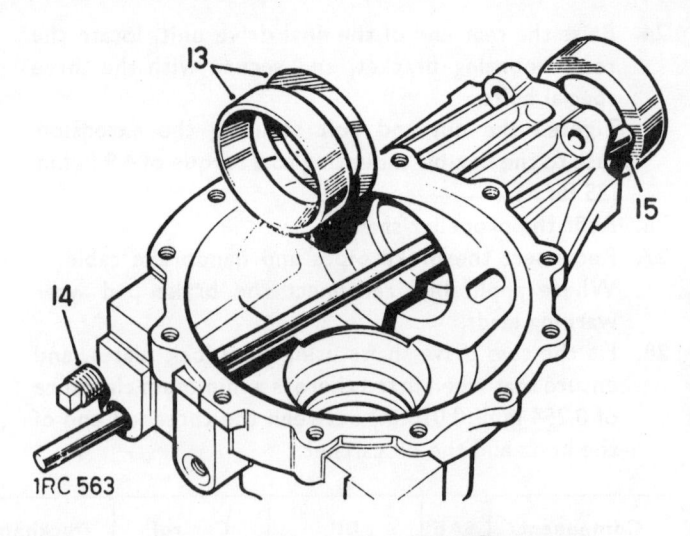

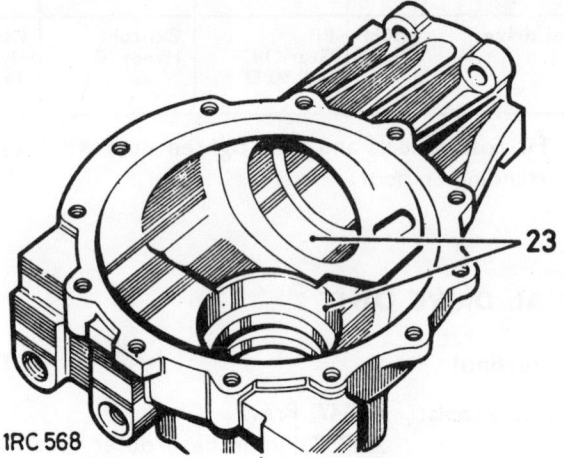

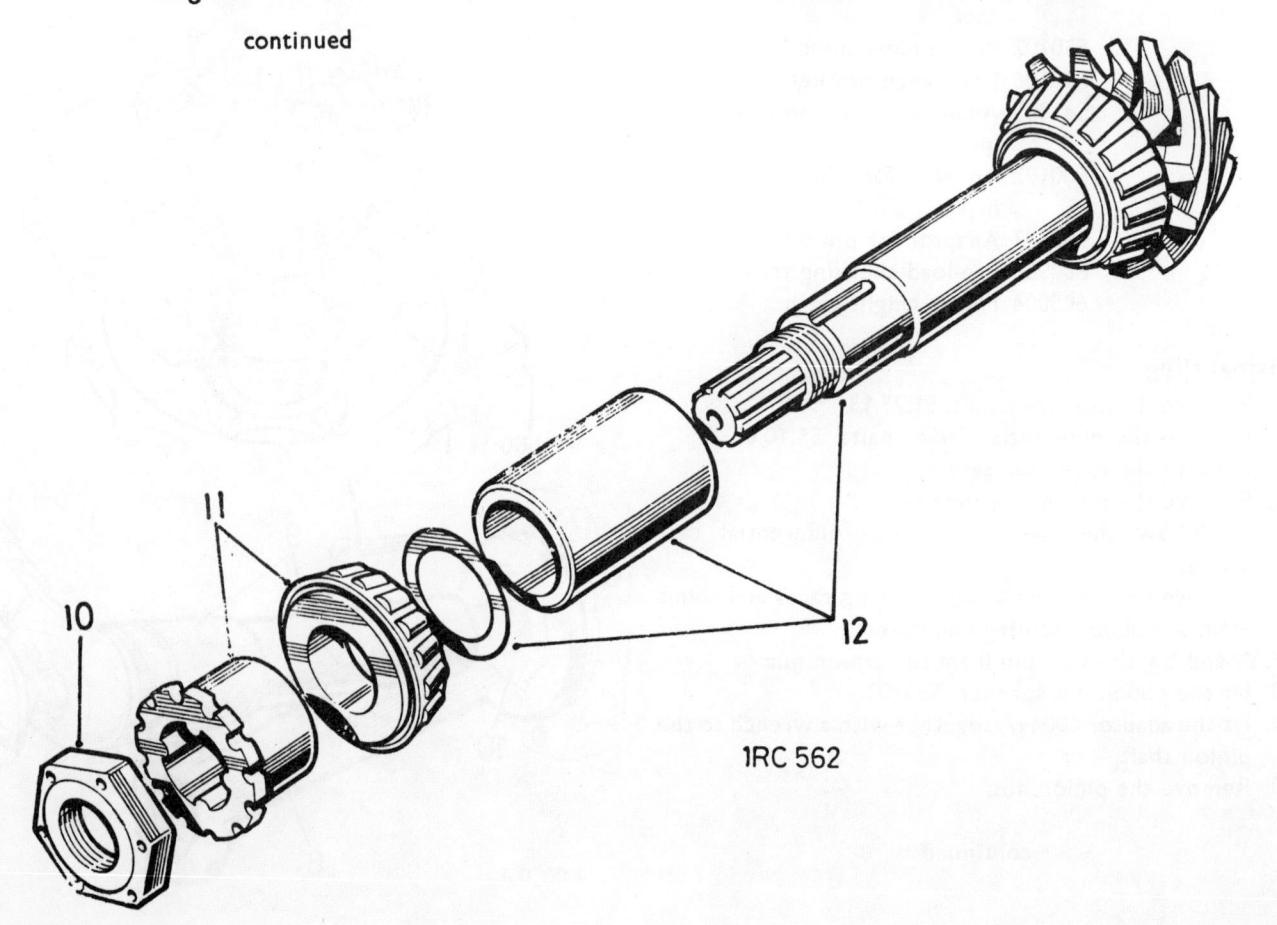

   continued

REAR AXLE AND FINAL DRIVE

Reassembling

25. Reverse 20 to 22 ensuring that the alignment marks on the differential case halves match. Torque for differential case fixings 5,3 to 6,2 kgf.m (40 to 45 lbf ft).
26. Fit the crown wheel to the differential case as follows:
Early models with ten $\frac{5}{16}$ in. diameter bolts and two fitting bolts—Fit the twelve securing bolts complete with plain washers and lockplates, locating the two fitting bolts diagonally opposite. Torque 3,5 kgf.m (25 lbf ft).

Intermediate models with twelve $\frac{5}{16}$ in. diameter bolts—Fit the twelve securing bolts complete with 'Kantlink' spring washers. Torque 4,8 kgf.m (35 lbf ft).

Latest models with twelve $\frac{3}{8}$ in. diameter special bolts—Coat the threads of the twelve securing bolts with 'Loctite Stud Lock Retaining Compound', and fit them complete with special plain washers. Torque 5,5 to 6,2 kgf.m (40 to 45 lbf ft).

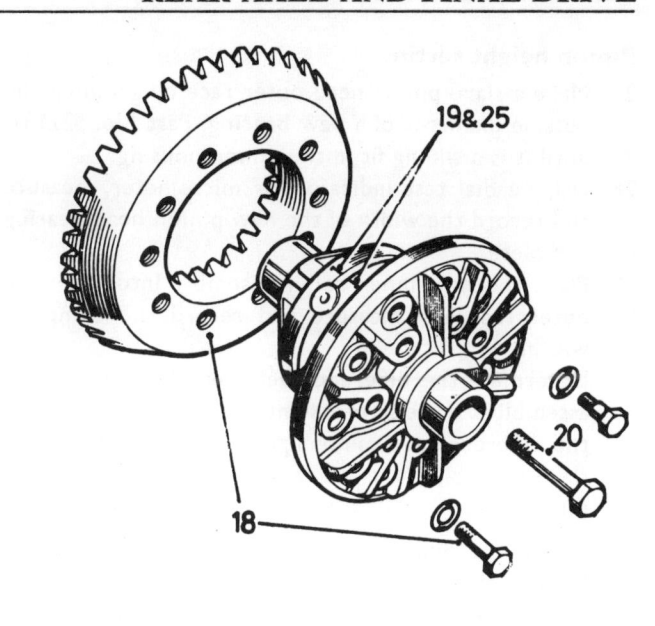

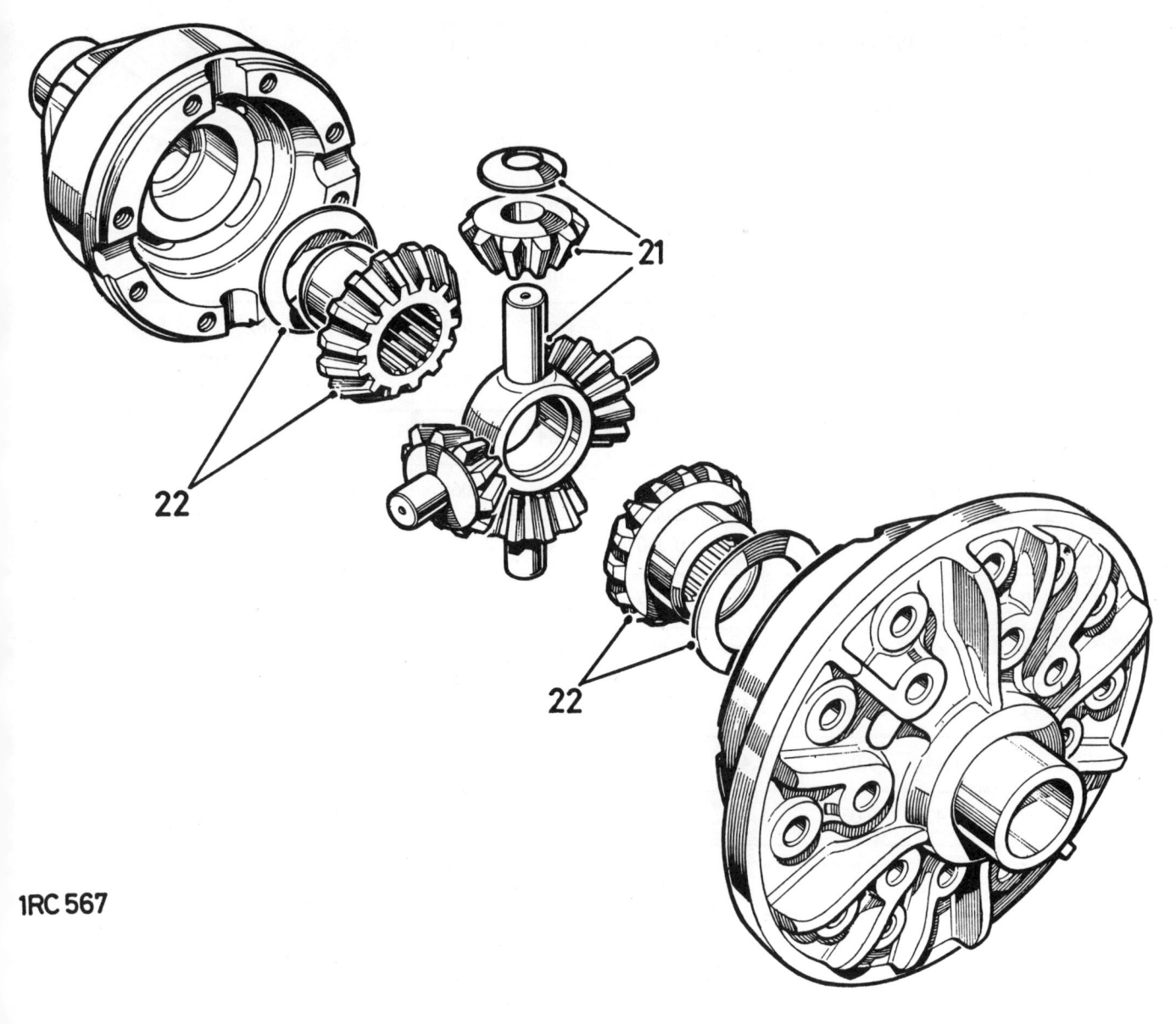

continued

1RC 567

Rover 3500 and 3500S Manual AKM 3621

REAR AXLE AND FINAL DRIVE

Pinion height setting

27. Make a slave pinion head outer race by grinding the outside diameter of a new bearing, Part No. 522181, until it is a sliding fit in the pinion housing.
28. Using a dial test indicator, or micrometer, measure and record the width of the new pinion head bearing complete.
29. Place the new inner roller assembly into the slave outer race and measure and record the complete width.
30. Determine the difference between the two bearing assemblies. If the slave assembly is the smaller, record the difference as a minus figure. Alternatively, if the slave assembly is the larger, record the difference as a plus figure.
31. Check the pinion face for dimensional markings as follows, then refer to the following chart to ascertain the final feeler gauge clearance required between the pinion height gauge and the slip gauge.
32. Pinion height dimension. This may be —5 to +5.
33. Head depth. This may be zero to —4.
34. The production batch number. This is for identification purposes only.

> **NOTE:** Where a pinion has no dimensional markings, use the pinion height 'O' and the head depth 'O' on the chart for final feeler gauge clearance.

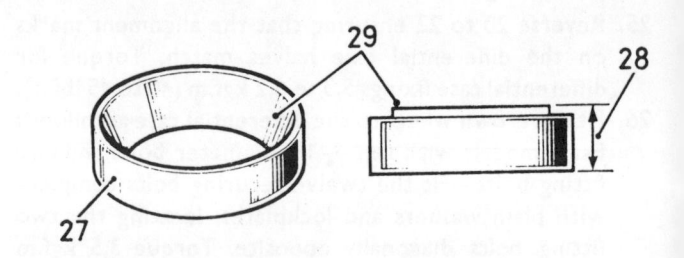

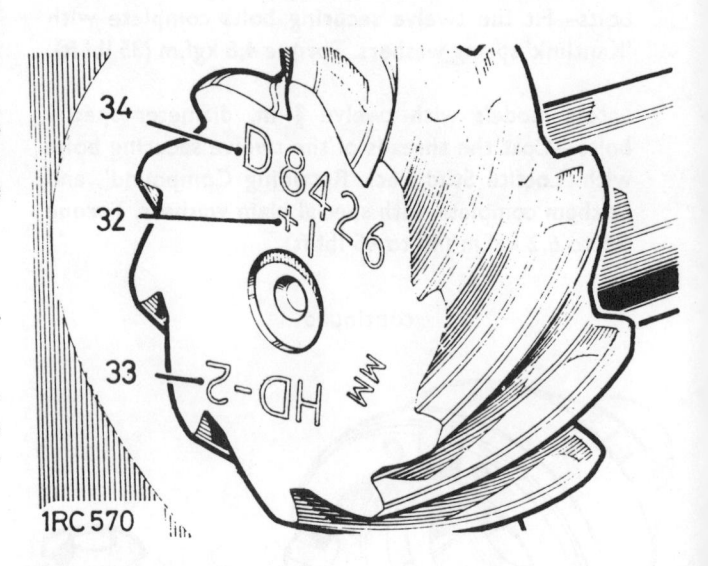

continued

Final drive feeler gauge clearance

	mm	in	mm	in	mm	in	mm	in	mm	in
+5	1,65	0.065	1,62	0.064	1,60	0.063	1,57	0.062	1,54	0.061
+4	1,62	0.064	1,60	0.063	1,57	0.062	1,54	0.061	1,52	0.060
+3	1,60	0.063	1,57	0.062	1,54	0.061	1,52	0.060	1,49	0.059
+2	1,57	0.062	1,54	0.061	1,52	0.060	1,49	0.059	1,47	0.058
+1	1,54	0.061	1,52	0.060	1,49	0.059	1,47	0.058	1,44	0.057
0	1,52	0.060	1,49	0.059	1,47	0.058	1,44	0.057	1,42	0.056
—1	1,49	0.059	1,47	0.058	1,44	0.057	1,42	0.056	1,39	0.055
—2	1,47	0.058	1,44	0.057	1,42	0.056	1,39	0.055	1,37	0.054
—3	1,44	0.057	1,42	0.056	1,39	0.055	1,37	0.054	1,34	0.053
—4	1,42	0.056	1,39	0.055	1,37	0.054	1,34	0.053	1,32	0.052
—5	1,39	0.055	1,37	0.054	1,34	0.053	1,32	0.052	1,29	0.051

Pinion height markings (rows) / Head depth H or HD markings: —4, —3, —2, —1, 0 (columns)

Rover 3500 and 3500S Manual AKM 3621

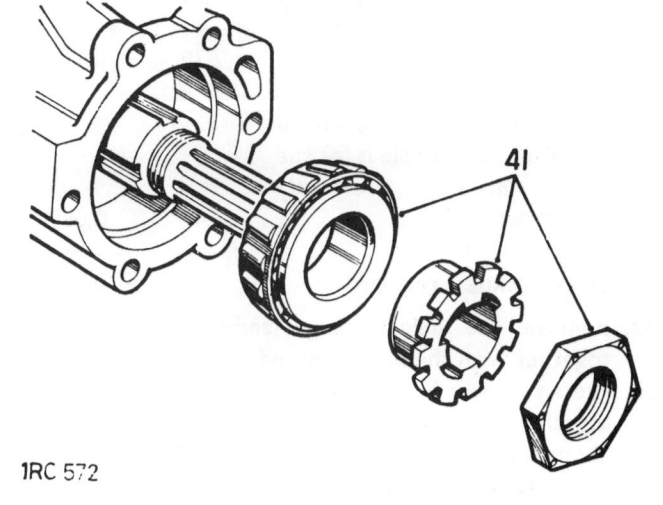

REAR AXLE AND FINAL DRIVE

35. Measure and record the thickness of the original shim fitted under the pinion head bearing race.
36. Press the new roller bearing on to the pinion head.
37. Press the pinion front bearing race into the pinion housing.
38. Place the original shim and the pinion head slave race into the pinion housing.

    NOTE: If the outer edge of the shim is chamfered, the chamfer must face away from the bearing.

39. Place the distance piece on to the pinion shaft with the internal chamfer towards the pinion.
40. Insert the pinion assembly into the housing.
41. Locate the pinion front roller bearing, serrated collar and nut in position.
42. Fit the pinion nut spanner. 530102.
43. Fit the adaptor 600447, together with a wrench to the pinion shaft.
44. Tighten the pinion nut sufficient to give a pinion bearing pre-load of 9 to 20 kg cm (8 to 18 lb in) using pre-load tool 600968 to check the setting.
45. Secure the pinion height gauge 605004, in the pinion housing.
46. Place the slip gauge on to the pinion face and hold firmly in position.
47. Using feeler gauges, measure and record the clearance between the height gauge and the slip gauge.

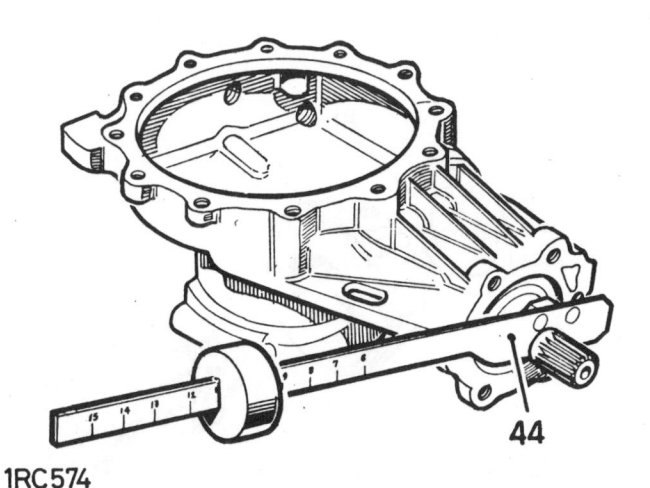

continued

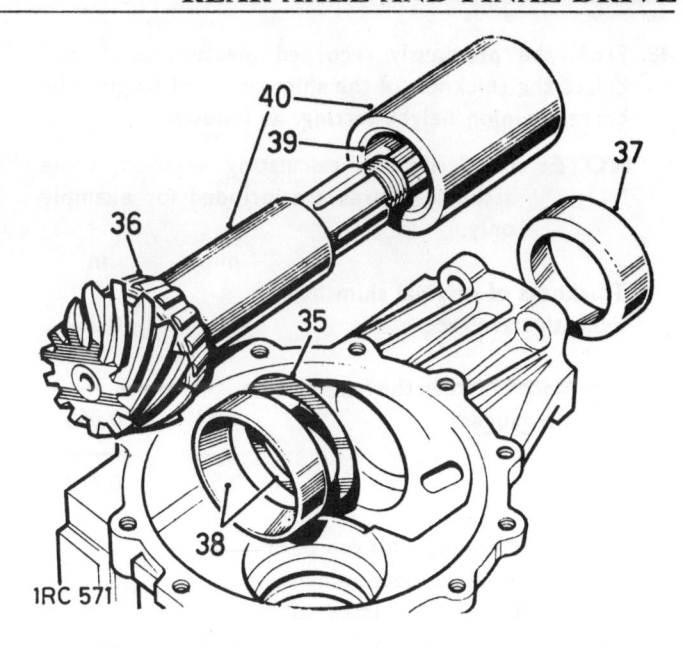

Rover 3500 and 3500S Manual AKM 3621

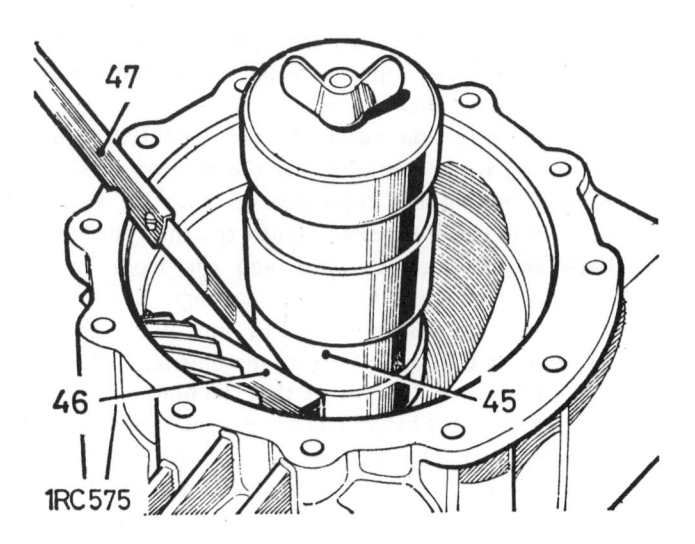

51.25.19
Sheet 5

REAR AXLE AND FINAL DRIVE

48. From the previously recorded measurements, calculate the thickness of the shim required to give the correct pinion height setting, as follows.

    NOTE: To clarify the calculating method, some assumed figures are included for example only.

	mm	in
Thickness of original shim fitted beneath slave race	1,57	0.062
Clearance between the height and slip gauges	1,27	0.050
	2,84	0.112
Bearing difference. This can be plus or minus as determined in item 30 and is assumed to be minus for this example	0,05	0.002
	2,79	0.110
—The chart dimension for the pinion fitted. The assumed figure in the example is for the pinion marking previously illustrated	1,49	0.059
=Shim thickness required	1,30	0.051

49. Remove the height gauge and pinion assembly together with the slave race and original shim.
50. Press the new bearing race into the pinion housing together with the correct thickness shim, 262757. Shims are available from 1,29 mm to 1,72 mm (0.051 in to 0.068 in) in 0,02 mm (0.001 in) increments.

    NOTE: If the outer edge of the shim is chamfered, the chamfer must face away from the bearing.

51. Refit the pinion assembly into the housing together with the original shims between the distance piece and the front roller bearing.
52. Tighten the pinion nut to a torque of 10,0 kgf.m (75 lbf ft) while checking that the bearing pre-load remains within the tolerance of 9 to 20 kg cm (8 to 18 lb in.). If necessary adjust the shim thickness between the distance piece and the front roller bearing to obtain the above settings. Shims are available from 1,37 mm to 1,82 mm (0.054 in. to 0.072 in.) in 0,02 mm (0.001 in.) increments.
53. Make a final check to ensure that the pinion height is correct. Refit the height and slip gauges, the clearance between the gauges must agree with the chart dimension for the pinion fitted ±0.02 mm (0.001 in.). If necessary, the shim thickness must be readjusted. Remove the gauges.

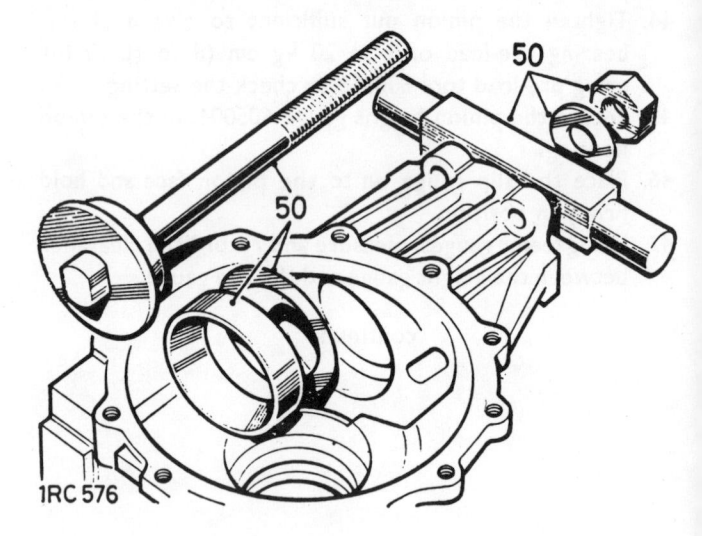

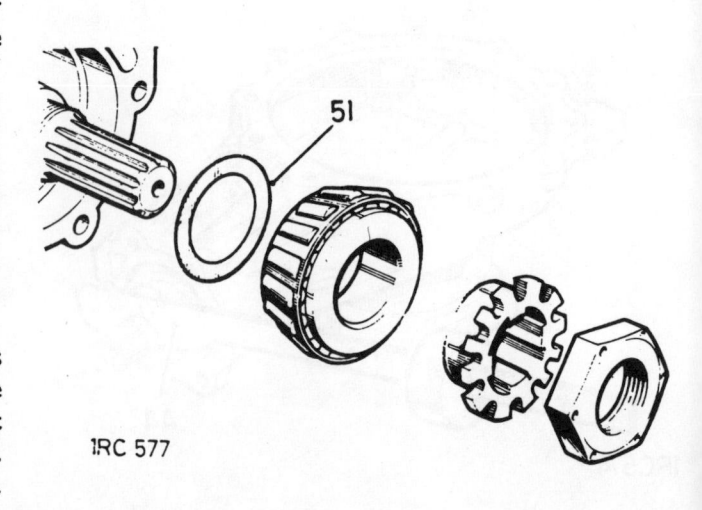

continued

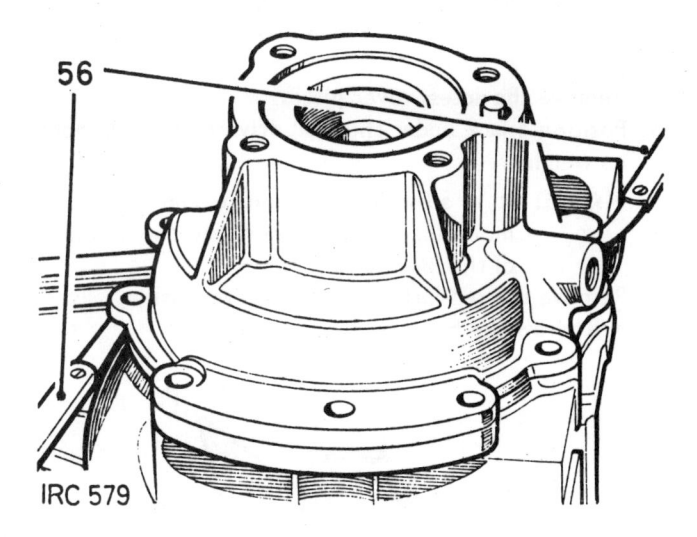

REAR AXLE AND FINAL DRIVE

Differential case pre-load setting

54. Initially, insert both the original shims or a minimum of 2,6 mm (0.100 in.) into the differential race location in the pinion housing and no shims in the cover.
55. Install the differential assembly complete with bearings into the pinion housing and fit the cover without bolts.
56. Apply a steady pressure to the pinion housing cover and insert two feeler gauges of equal thickness and diagonally opposite to each other to determine the clearance between the pinion housing and cover. The clearance must be 0,07 mm to 0,12 mm (0.003 in. to 0.005 in.) and must be obtained by adjusting the thickness of the shims.
57. When the pre-load clearance is correct, withdraw the differential assembly, bearings and shims, and measure the thickness of the shims.
58. Place the thicker shim into the race location in the pinion housing and refit the bearing race and differential assembly.
59. Check that there is some backlash between the crown wheel and pinion. If necessary, use a thicker shim in the pinion housing, but any increase in the thickness of the pinion housing shim must be compensated by an equivalent decrease in the cover shim to maintain the correct overall shim thickness.
60. Place the selected shim into the race location in the pinion housing cover and refit the bearing race.
61. Fit the cover to the pinion housing and secure with four bolts, including the two fitting bolts, spaced evenly around the cover.
62. Insert a suitable steel rod through the hole in the pinion shaft so that it protrudes 25 mm (1,0 in.) from the shaft.
63. Mount a dial test indicator to read off the steel rod 25 mm (1.0 in.) from the centre of the pinion shaft. 530106.
64. Insert a screwdriver in the drain or filler plug hole to secure the crown wheel.
65. Check the backlash between the crown wheel and pinion. The correct backlash is 0,12 mm to 0,20 mm (0.005 in. to 0.008 in.). If adjustment is necessary, change the shim fitted behind each differential bearing race for shims of alternative thickness without altering the overall shim thickness. Shims are available from 1,06 mm to 1,87 mm (0.042 in. to 0.074 in.) in 0,02 mm (0.001 in.) increments.
66. When the crown wheel and pinion backlash is correct, remove the pinion housing cover.
67. Apply 'Hylomar PL 32/M' jointing compound to the pinion housing and cover joint faces.
68. Secure the cover to the pinion housing ensuring that the two 'fitted' bolts are correctly located. Torque 4,0 kgf.m (30 lbf ft).
69. Secure the pinion nut with a new split pin.
70. Refit the oil drain and filler plugs.
71. Reverse 1 to 3.

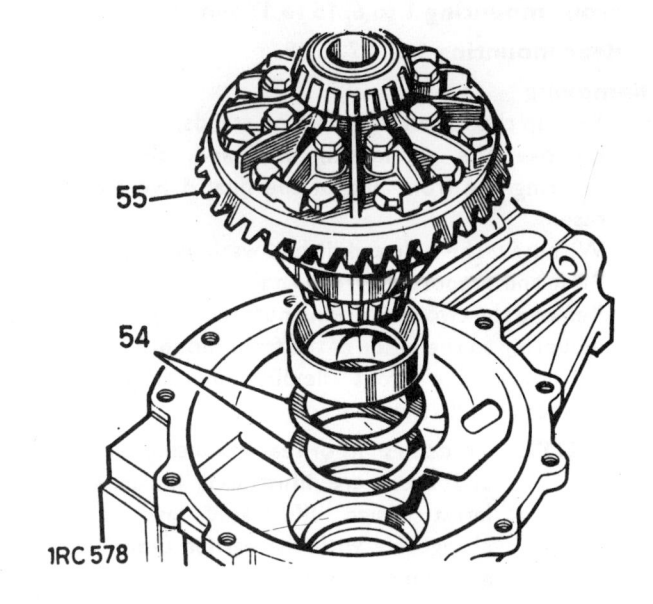

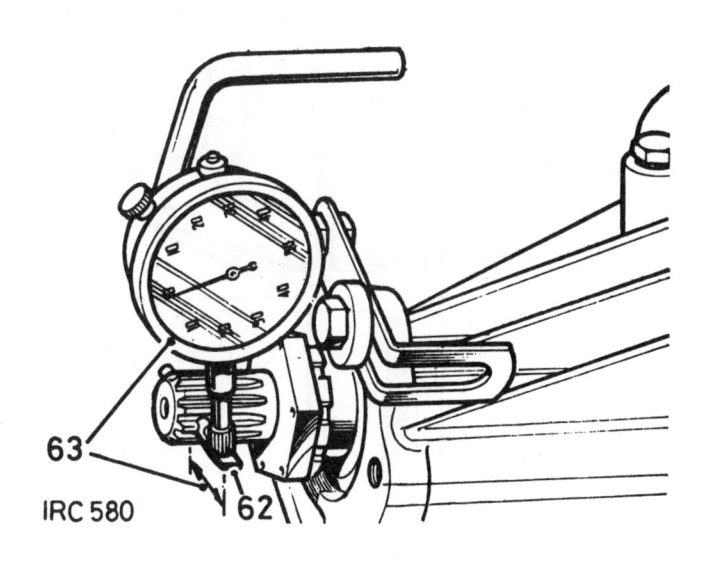

Rover 3500 and 3500S Manual AKM 3621

REAR AXLE AND FINAL DRIVE

FINAL DRIVE UNIT MOUNTINGS
—Remove and refit 51.25.31

Front mounting 1 to 6, 15 to 17 and 19.

Rear mounting 1 and 7 to 19.

Removing

1. Jack up the car and support on stands.
2. Remove the self-locking nut and plain washer securing the front end of the final drive extension case.
3. Withdraw the bolt and dished washer.
4. Raise and support the front end of the extension case.
5. Remove the two self-locking nuts securing the flexible mounting to the inside of the front mounting bracket.
6. Withdraw the front flexible mounting dished washer and distance piece.

   NOTE: Three designs of rear mountings are in use and these are illustrated in order. The latest design illustrated last is interchangeable with the earlier types if fitted as a complete set.

7. Remove the spare wheel, if fitted in the boot, and peel back the boot trim above the access cover plates.
8. Remove the access cover plates.
9. Position a jack under the final drive unit to support its weight.

   NOTE: It is advisable to remove and refit one mounting at a time to retain the position of the final drive.

10. Remove the self-locking nut and large washer.
11. Withdraw the bolt and plain washer.

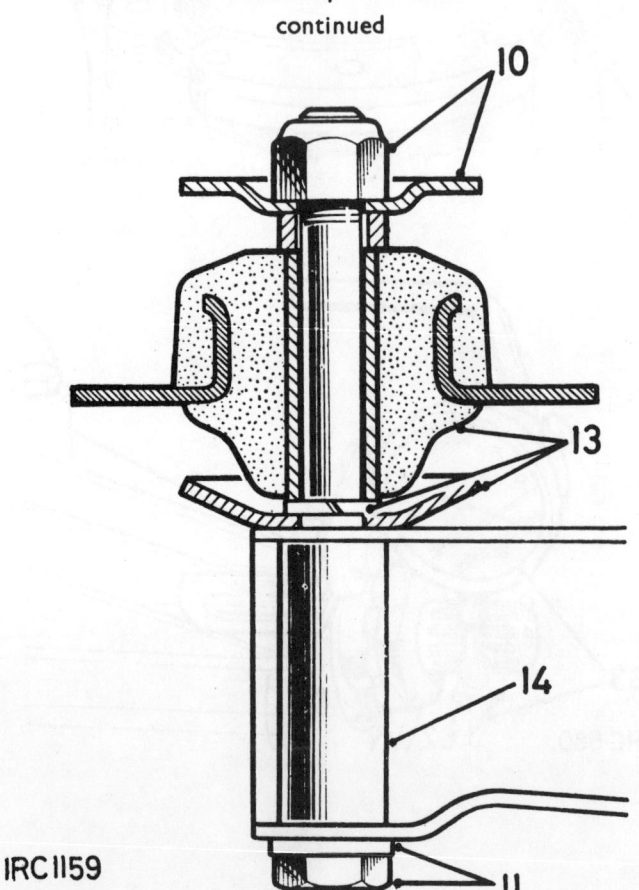

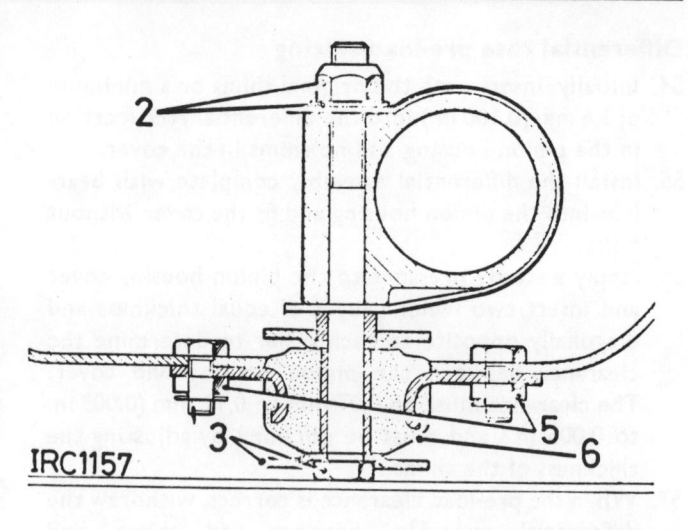

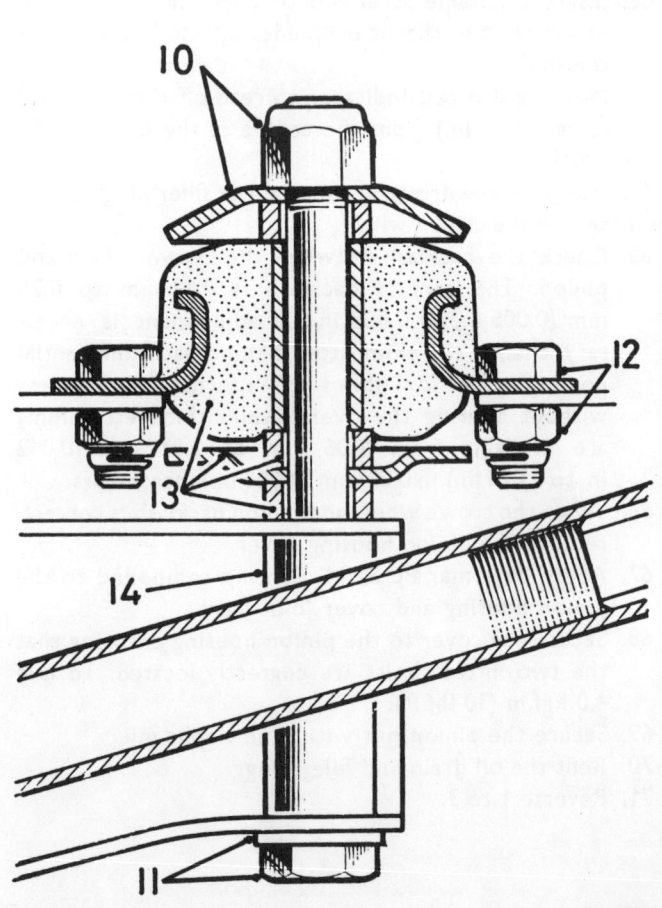

continued

IRC1157

IRC1159

Rover 3500 and 3500S Manual AKM 3621

REAR AXLE AND FINAL DRIVE

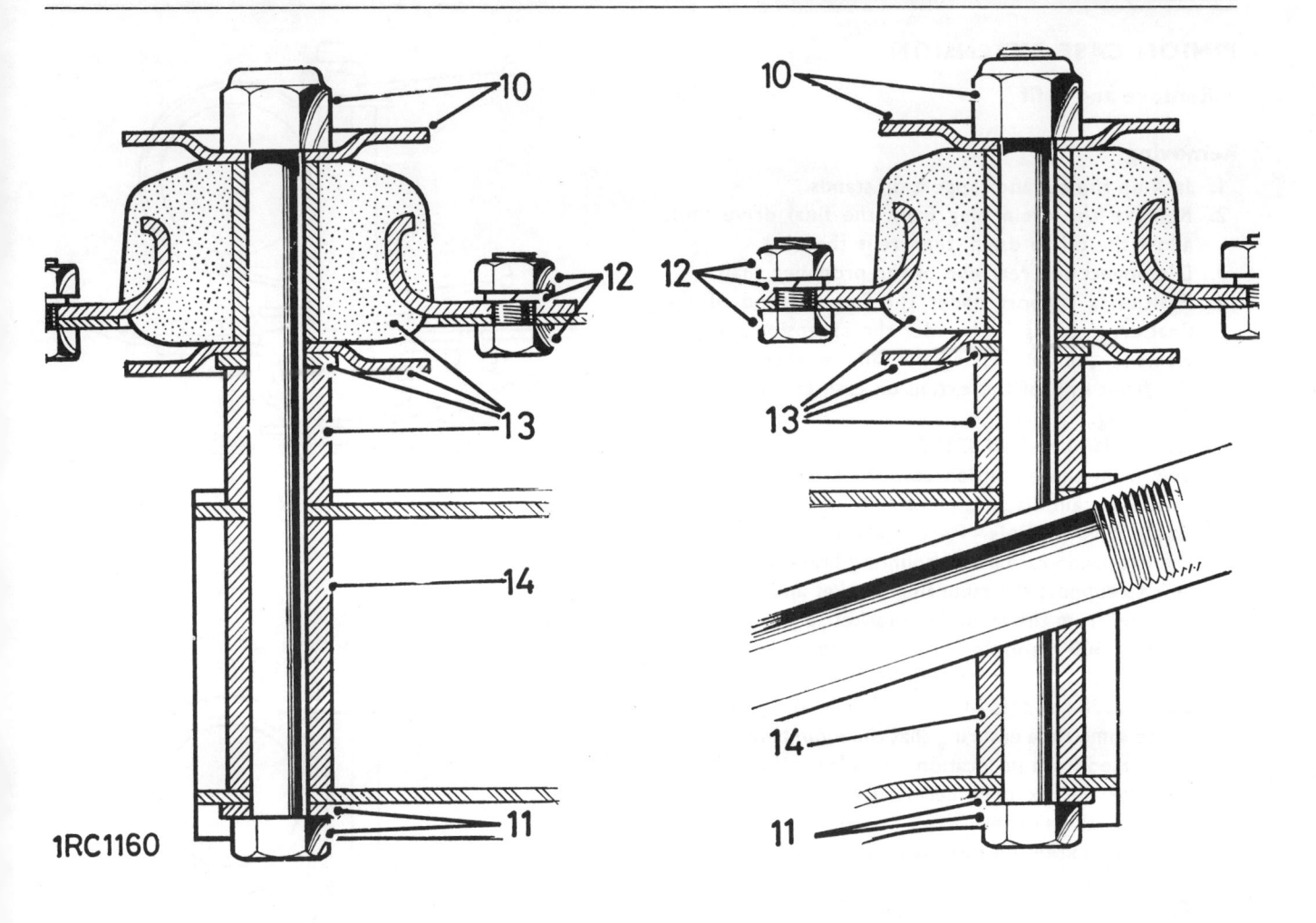

12. Remove the two nuts and bolts securing the flexible mounting to the base unit.
13. Withdraw the flexible mounting distance piece(s) and washers.
14. Withdraw the distance piece from the mounting bracket.

Refitting

15. Assemble the flexible mounting together with the washers and distance pieces as illustrated and retain with the centre bolt and nut, leaving the nut loose.
16. Secure the flexible mounting with the two nuts and bolts. Torque 2,3 kgf.m (17 lbf ft).
17. Secure the centre nut and bolt. Torque: Front mounting 4,8 kgf.m (35 lbf ft). Rear mountings 6,2 kgf.m (45 lbf ft).
18. Reverse 7 to 9.
19. Lower the car to the floor and remove the jack.

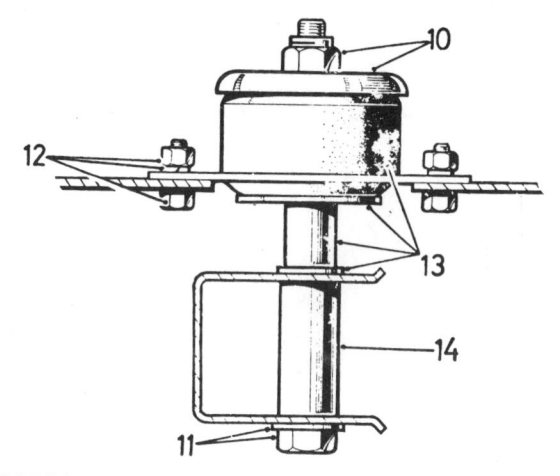

Rover 3500 and 3500S Manual AKM 3621

51.25.31
Sheet 2

REAR AXLE AND FINAL DRIVE

PINION CASE EXTENSION

—Remove and refit   51.25.37

Removing

1. Jack up the car and support on stands.
2. Remove the drain plug from the final drive unit, allow the oil to drain, then refit the plug.
3. Disconnect the rear end of the propeller shaft.
4. Position a support under the forward end of the final drive casing, **not** under the extension case.
5. Remove the self-locking nut and washer securing the front end of the extension case to the flexible mounting.
6. **Latest 3500 and all 3500S models**—Withdraw the bolt from the extension case front mounting, take care to retain distance piece and washers.
7. **Early 3500 models**—Place a jack under one end of the extension case front mounting bracket, raise the jack to support the mounting bracket and to relieve the load from the mounting bracket securing bolt.
8. **Early 3500 models**—Remove one of the bolts securing the front mounting bracket to base unit.
9. **Early 3500 models**—Carefully lower the jack while at the same time ensuring that the mounting bracket is released from its location, use a lever if necessary.
10. Remove the six bolts securing the extension case to the final drive unit.
11. Raise the extension case clear of the front mounting bracket and lift clear.
12. **Early 3500 models**—Remove the bolt and plain washer from the casing front securing lug. Do not lose the distance piece and large plain washer located between the flexible mounting and the extension case securing lug.

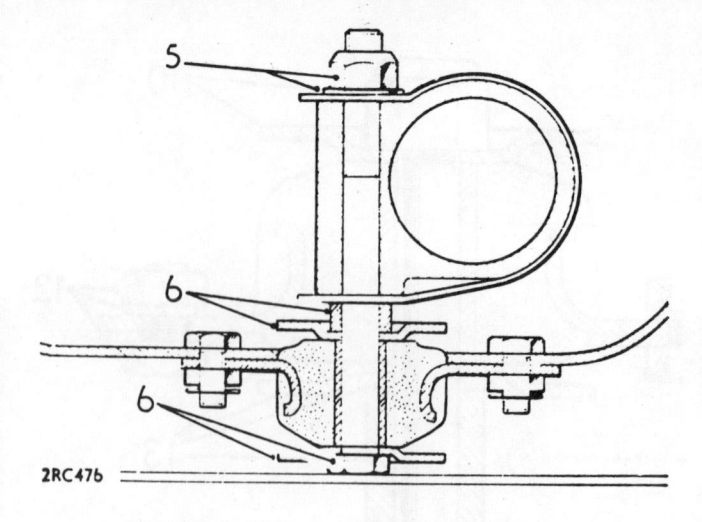

Refitting

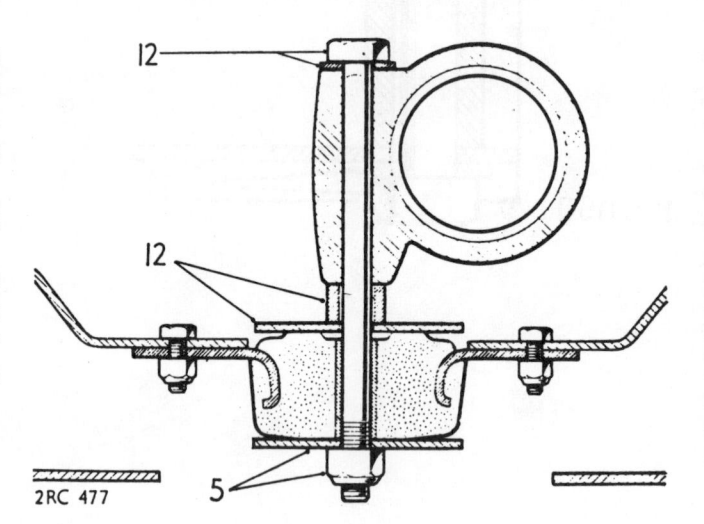

13. Reverse 1 to 12, noting the following.
14. Clean the joint faces and fit a new joint washer.
15. Tightening torque for extension case bolts: 4 kgf.m (30 lbf ft).
16. Tightening torque for nut securing extension case to flexible mounting: 4,8 kgf.m (35 lbf ft).
17. Using the correct grade oil, replenish the final drive unit.

Rover 3500 and 3500S Manual AKM 3621

REAR AXLE AND FINAL DRIVE

PINION CASE EXTENSION

—Overhaul   51.25.49

Service tool: 530105 Coupling flange tool

Dismantling

CAUTION: Do not disturb the harmonic damper fitted to the forward end of the case. This unit is tuned to the case on assembly using special equipment.

1. Remove the pinion case extension 51.25.37
2. Remove the bolt and special washer securing the drive coupling flange. 530105.
3. Withdraw the coupling flange, using a suitable extractor.
4. Remove the oil seal.
5. Remove the circlip.
6. Drift out the extension shaft complete with the bearing from the rear.
7. Remove the bearing from the extension shaft.

   CAUTION: DO NOT remove the coupling sleeve from the extension shaft. New shafts and sleeves are only supplied as assemblies.

Inspecting

8. Obtain a new oil seal and circlip.

   NOTE: The latest type of oil seal has a spiral groove formed on the running diameter. This seal is interchangeable with the earlier type and does not require fitting with jointing compound.

9. Check all other parts for wear or damage and replace as necessary.

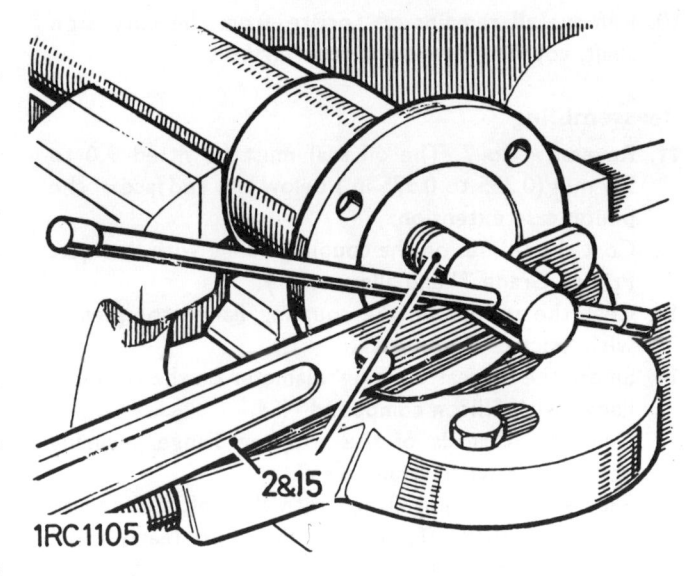

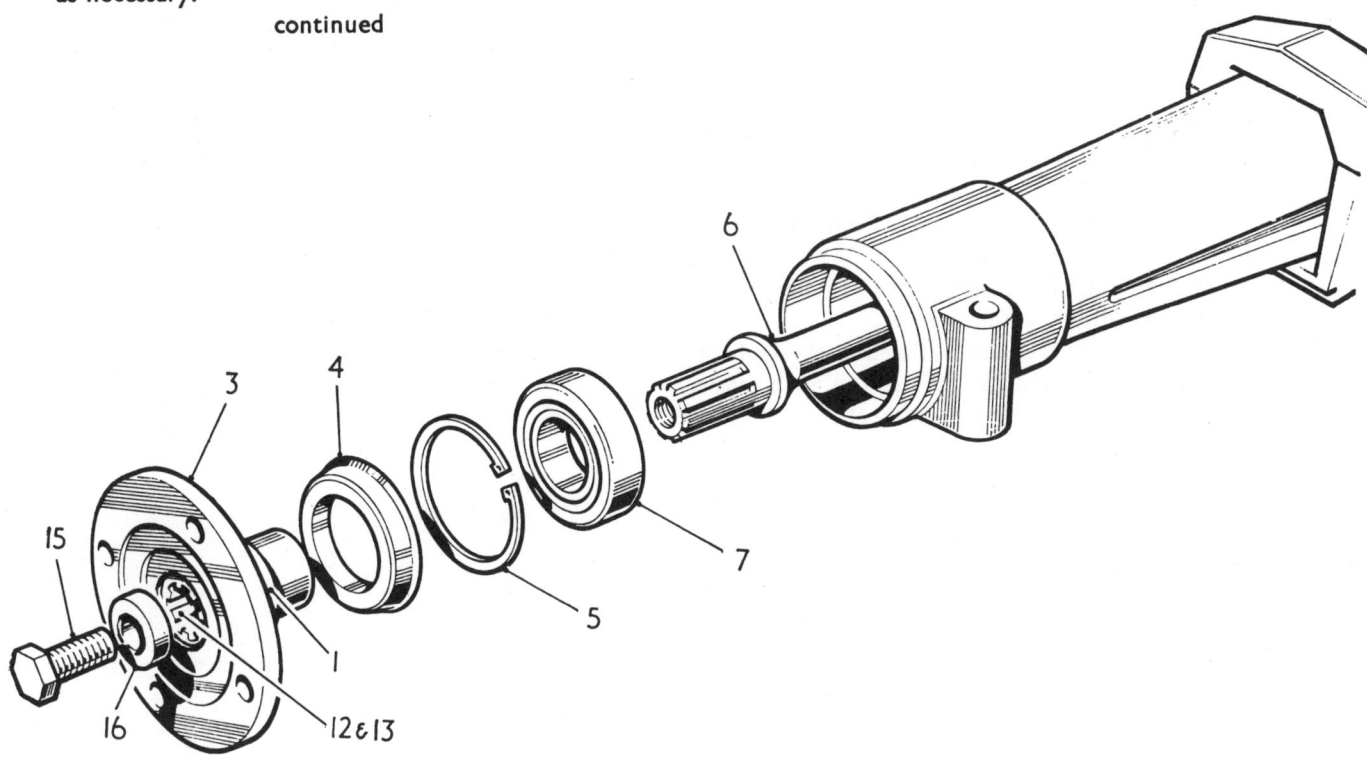

                    continued

Rover 3500 and 3500S Manual AKM 3621

REAR AXLE AND FINAL DRIVE

10. Remove all remains of 'Loctite' from the extension shaft, coupling flange and securing bolt.

Reassembling

11. Reverse 4 to 7. The oil seal must be fitted 9,0 to 9,5 mm (0.355 to 0.375 in.) below the end face of the pinion case extension.
12. Coat the splines on the coupling flange with 'Locquic Primer Grade T' and allow to dry.
13. Coat the splines on the coupling flange completely with 'Loctite Grade AVV'.
14. Smear the oil seal 'running' diameter on the coupling flange with Silicon compound MS4.
15. Coat the threads of the coupling flange securing bolt with 'Loctite Grade AVV'.
16. Secure the coupling flange with the bolt and special washer. 530105. Torque: $\frac{3}{8}$ in. UNF (early type) 4,9 kgf.m (35 lbf ft). $\frac{7}{16}$ in. UNF (latest type) 10,0 kgf.m (75 lbf ft).

NOTE: On latest type units renew the 'O' rings in the double stepped washer before fitting and tightening the $\frac{7}{16}$ in. UNF bolt.

17. Refit the pinion case extension. 51.25.37.

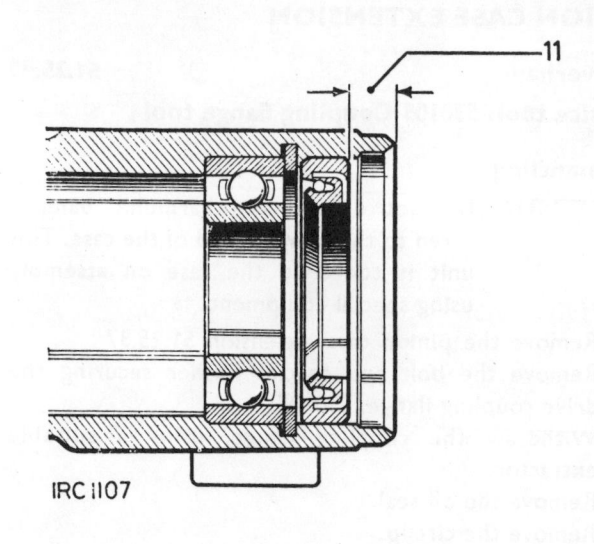

PINION EXTENSION CASE MOUNTING

—Remove and refit   51.25.55

Removing

1. Remove the rear suspension lower links. 64.35.02.
2. Remove the self-locking nut and large washer securing the final drive extension case to the flexible mounting.
3. Support the forward end of the final drive unit, sufficient to remove load from front mounting bracket.
4. Remove the fixings from both ends of the mounting bracket, and withdraw the bracket.
5. If required, new bushes can be fitted at the ends of the front mounting bracket. Locate the bush outer sleeve flush with the mounting bracket.

Refitting

6. Reverse 1 to 5. Tightening torque for the bolt securing the extension case to the flexible mounting: 4,8 kgf.m (35 lbf ft).

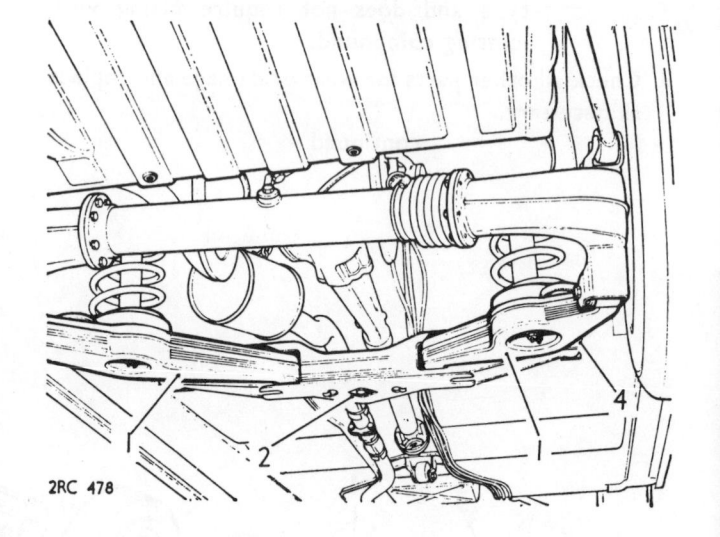

STEERING

LIST OF OPERATIONS

Front wheel alignment—check and adjust	57.65.01
Lock stops—check and adjust	57.65.03
Manual steering box	
—adjust	57.35.01
—remove and refit	57.30.01
—overhaul	57.30.07
Power steering box	
—adjust	57.10.13
—remove and refit	57.10.01
—overhaul	57.10.07
Power steering fluid reservoir—remove and refit	57.15.08
Power steering pump	
—remove and refit	57.20.14
—overhaul	57.20.20
—drive belt—adjust	57.20.01
—drive belt—remove and refit	57.20.02
Power steering system	
—bleed	57.15.02
—test	57.15.01
Steering column assembly	
—remove and refit	57.40.01
—bearings—remove and refit	57.40.18
—lock and switch assembly—remove and refit	57.40.31
Steering relay—remove and refit	57.50.02
Steering side rod—remove and refit	57.55.17
Track rod—remove and refit	57.55.09

Rover 3500 and 3500S Manual AKM 3621

57–1

STEERING

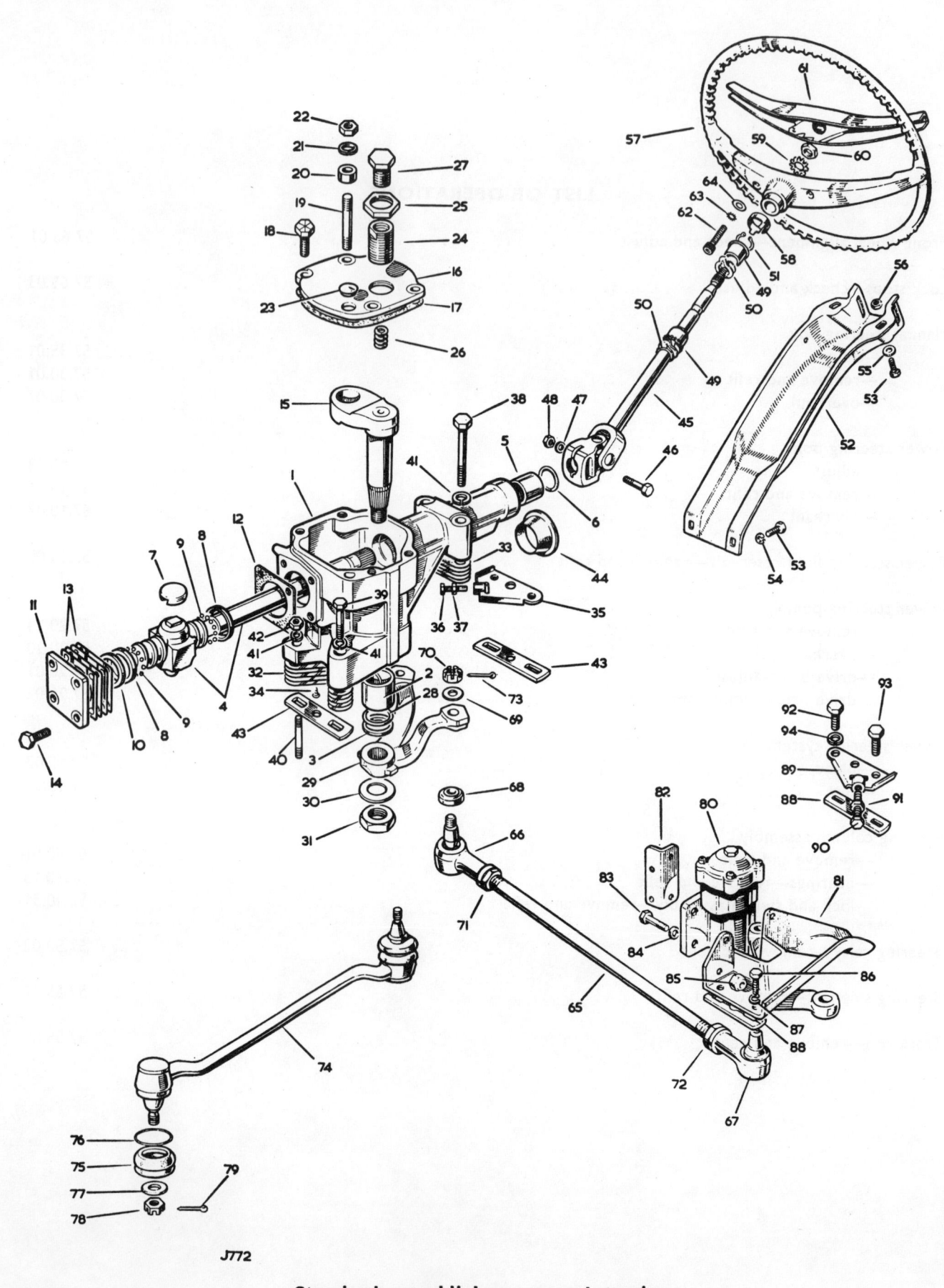

Steering box and linkage, manual steering

STEERING

Key to steering unit and linkage, manual steering

1	Steering box
2	Bush for rocker shaft
3	Washer, oil seal retainer
4	Inner column and main nut
5	Bush for inner column
6	Oil seal for inner column
7	Roller for main nut
8	Adjustable ball race
9	Steel ball (0.281 in diameter) for ball race
10	Spacer washer
11	End cover plate for steering box
12	Joint washer, paper
13	Joint washer, steel
14	Bolt ($\frac{5}{16}$ in UNF x $\frac{5}{8}$ in long) fixing end cover plate
15	Rocker shaft
16	Top cover plate for steering box
17	Joint washer for cover plate
18	Bolt ($\frac{5}{16}$ in UNF x $\frac{3}{4}$ in long) ⎫
19	Stud ⎬ Fixing top cover plate to steering box
20	Distance washer
21	Spring washer
22	Nut ($\frac{5}{16}$ in UNF) ⎭
23	Oil filler plug
24	Adjusting screw for rocker shaft
25	Nut ($\frac{7}{8}$ in UNF) for adjuster screw
26	Spring ⎫ For adjuster screw
27	Special bolt ($\frac{3}{8}$ in UNF x 0.549 in long) ⎭
28	Oil seal for rocker shaft
29	Drop arm
30	Tab washer ⎫ Fixing drop arm to rocker shaft
31	Special nut ⎭
32	Shim, 0.020 in 'V' shape, steering box, front to base unit
33	Shim, 0.020 in, steering box, rear to base unit
34	Drive screw fixing shims to steering box
35	Lock stop bracket, driver's side
36	Bolt $\frac{5}{16}$ in UNF x 1$\frac{1}{8}$ in long ⎫ In lock stop bracket
37	Locknut ($\frac{3}{8}$ in UNF) ⎭
38	Set bolt ($\frac{3}{8}$ in UNF x 3 in long) ⎫
39	Set bolt ($\frac{3}{8}$ in UNF x 2$\frac{1}{4}$ in long) ⎬ Fixing steering box and lock stop bracket to base unit
40	Stud ($\frac{3}{8}$ in UNF)
41	Spring washer
42	Nut ($\frac{3}{8}$ in UNF)
43	Spacer ⎭
44	Grommet, steering box to base unit
45	Steering column shaft
46	Pinch bolt ($\frac{5}{16}$ in UNF x 1$\frac{1}{2}$ in long) ⎫ Fixing shaft yoke to inner column
47	Plain washer
48	Nut ($\frac{5}{16}$ in UNF) ⎭
49	Bearing for steering column shaft
50	Spring washer ⎫ Fixing shaft to bearing
51	Circlip ⎭
52	Support bracket for steering column
53	Bolt ($\frac{1}{4}$ in UNF x $\frac{5}{8}$ in long) ⎫ Fixing support bracket to unit
54	Spring washer
55	Plain washer
56	Nut ($\frac{1}{4}$ in UNF) ⎭
57	Steering wheel
58	Striker for flasher switch
59	Shakeproof washer ⎫ Fixing steering wheel to shaft
60	Special nut ⎭
61	Finisher for steering wheel
62	Special screw ⎫ Fixing finisher to wheel
63	'Starlock' washer ⎬
64	Plain washer ⎭
65	Steering track rod
66	Ball joint, RH thread
67	Ball joint, LH thread
68	Rubber boot for ball joint
69	Plain washer ⎫ Fixing track rod ball joints
70	Special slotted nut ⎭
71	Locknut ($\frac{5}{8}$ in UNF) for ball joint, RH thread
72	Locknut ($\frac{5}{8}$ in UNF) for ball joint, LH thread
73	Split pin for slotted nut
74	Steering side rod
75	Rubber boot ⎫ For side rod ball joints
76	Retaining ring for boot ⎭
77	Plain washer ⎫ Fixing side rod ball joints
78	Special slotted nut ⎭
79	Split pin for slotted nut
80	Steering idler damper complete
81	Mounting bracket for damper
82	Mounting angle for accelerator bracket
83	Bolt ($\frac{3}{8}$ in UNF x 1$\frac{3}{8}$ in long) ⎫ Fixing damper and mounting angle to mounting bracket
84	Plain washer
85	Self-locking nut ($\frac{3}{8}$ in UNF) ⎭
86	Set bolt ($\frac{3}{8}$ in UNF x 1$\frac{1}{4}$ in long) ⎫ Fixing mounting bracket to base unit
87	Spring washer ⎭
88	Spacer
89	Lock stop bracket, passenger's side
90	Bolt ($\frac{5}{16}$ in x 1$\frac{1}{8}$ in long)
91	Locknut ($\frac{5}{16}$ in UNF)
92	Set bolt ($\frac{3}{8}$ in UNF x 1$\frac{1}{8}$ in long) ⎫ Fixing lock stop bracket to base unit
93	Set bolt ($\frac{3}{8}$ in UNF x $\frac{3}{4}$ in long) ⎬
94	Spring washer ⎭

Rover 3500 and 3500S Manual AKM 3621

STEERING

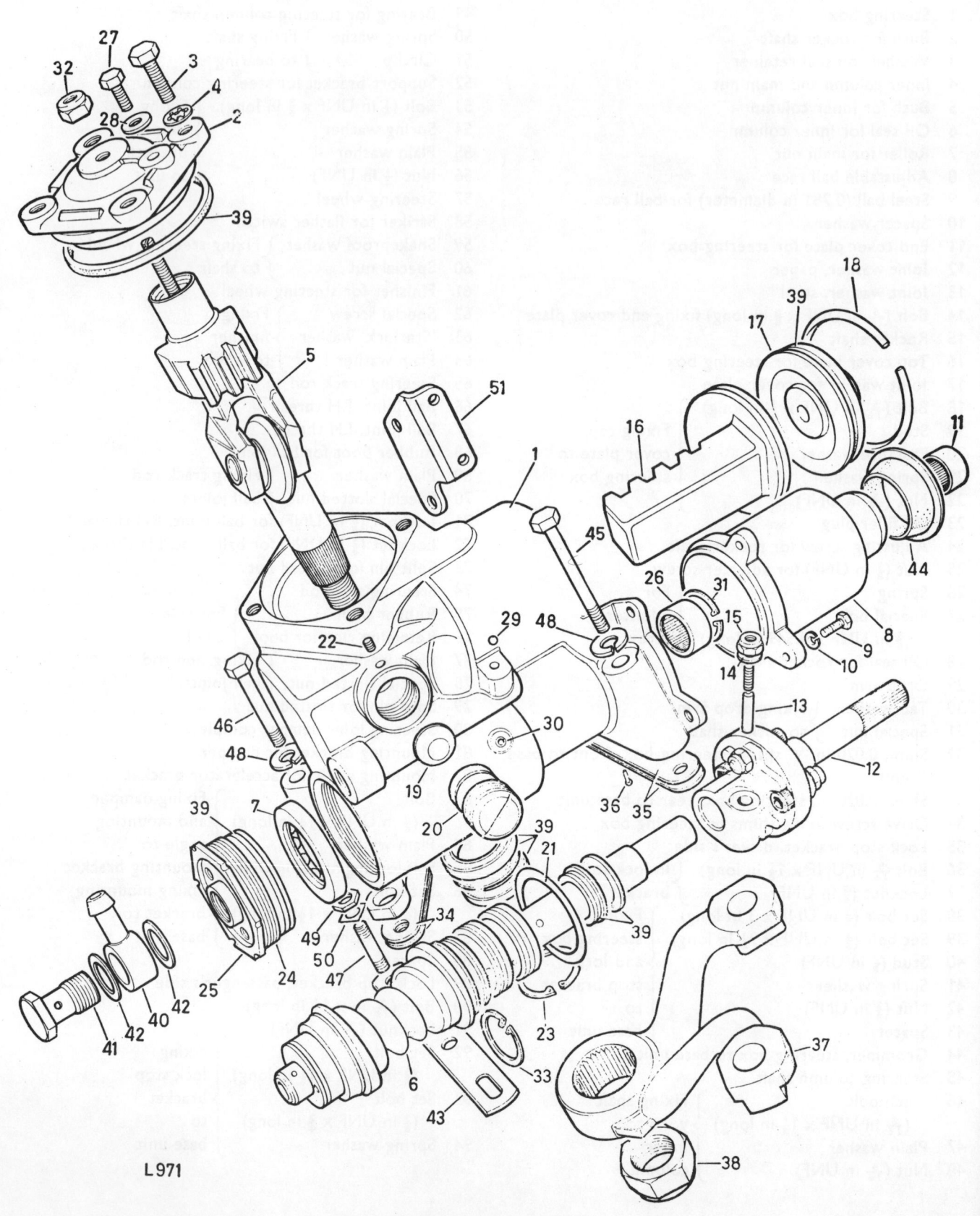

L971

Power steering box

STEERING

Key to power steering unit

1. Housing, bush and seat assembly
2. Cover, bush and seat assembly
3. Set bolt (¾ in UNF) } Fixing top
4. Shakeproof washer } cover plate
5. Sector shaft, follower and nut assembly
6. Valve and worm assembly
7. Bearing complete
8. Coupling housing and bearing
9. Set bolt (0.625 in long) } Fixing coupling housing
10. Spring washer } to main housing
11. Needle bearing for coupling shaft
12. Coupling complete
13. Cotter pin fixing coupling to valve and worm
14. Washer for cotter pin
15. Lock nut for cotter pin
16. Piston and rack
17. Cover plate for cylinder
18. Retainer ring for cylinder cover plate
19. Pad for rack
20. Screw, adjusting rack
21. Shim set for bearing
22. Screw, locking rack adjuster
23. Retaining ring for sector shaft
24. Screw, worm adjusting
25. Lock nut (2.625 in dia) for worm adjusting screw
26. Needle bearing for valve and worm rotor
27. Bleed screw in top cover
28. Special washer for bleed screw
29. Ball (¼ in dia) for external drilling in housing
30. Trim plug
31. Circlip for valve and worm shaft
32. Self-locking nut for sector shaft adjuster
33. Circlip for valve and worm rotor
34. Shim for mounting foot, 'V' shape
35. Shim for mounting foot, straight
36. Grooved pin fixing shims to housing
37. Drop arm
38. Self-locking nut fixing drop arm
39. Kit of seals for power steering unit
40. Banjo connector } For low
41. Banjo bolt } pressure hose
42. Washer for banjo bolt } connection
43. Spacer, steering unit to base unit
44. Grommet, steering unit to base unit
45. Bolt (⅜ in UNF x 3¼ in long) } Fixing
46. Bolt (⅜ in UNF x 2¼ in long) } power
47. Stud } steering
48. Spring washer } unit to
49. Special nut (⅜ in UNF) } base
50. Tab washer } unit
51. Support bracket for accelerator coupling shaft bracket

STEERING

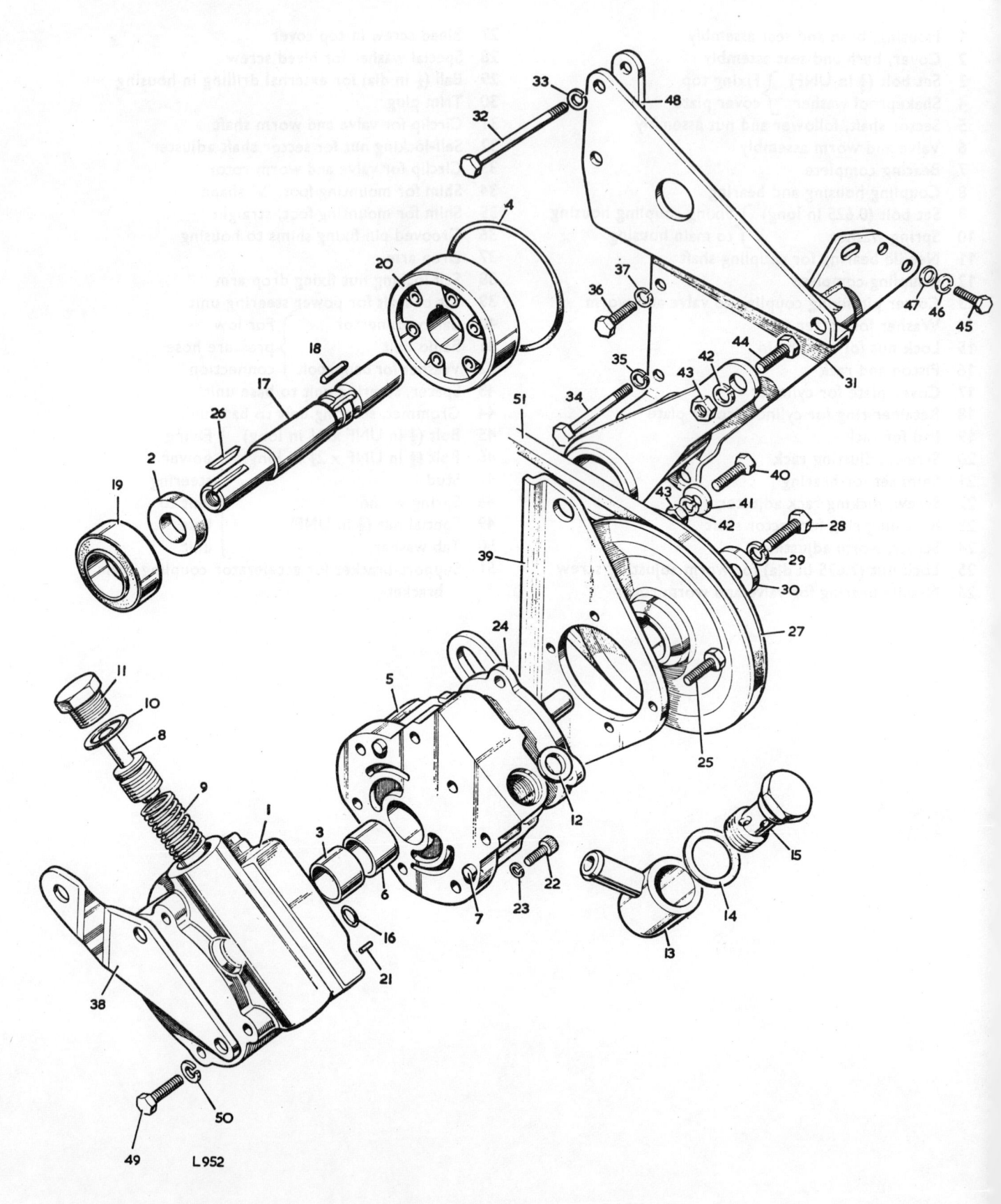

Pump for power steering unit

STEERING

Key to pump for power steering unit

#	Part	#	Part
1	Body assembly	27	Pulley
2	Oil seal } For	28	Bolt ($\frac{5}{16}$ in UNC x $\frac{7}{8}$ in long) } Fixing pulley
3	Bush } body	29	Spring washer } to steering
4	'O' ring, body to cover	30	Plain washer } pump
5	End cover assembly	31	Mounting bracket complete at front cover
6	Bush } For	32	Bolt ($\frac{5}{16}$ in UNC x 5 in long) } Fixing mounting
7	Dowel } end cover	33	Spring washer } bracket to water pump
8	Valve assembly	34	Bolt ($\frac{5}{16}$ in UNC x $3\frac{3}{4}$ in long) } Fixing
9	Spring, flow control	35	Spring washer } mounting
10	'O' ring for valve cap	36	Bolt ($\frac{3}{8}$ in UNC x $1\frac{1}{8}$ in long) } bracket to
11	Valve cap	37	Spring washer } front cover
12	Seal for adaptor	38	Bracket, front, for steering pump
13	Adaptor banjo	39	Bracket, rear, for steering pump
14	Fibre washer for adaptor	40	Bolt ($\frac{3}{8}$ in UNF x $1\frac{1}{8}$ in long) } Fixing front and
15	Banjo bolt	41	Plain washer } rear brackets to
16	'O' ring for by-pass in pump body	42	Spring washer } mounting
17	Shaft	43	Nut ($\frac{3}{8}$ in UNF) } bracket
18	Drive pin, fixing vane carrier	44	Bolt ($\frac{3}{8}$ in UNF x 1 in long) } complete
19	Bearing	45	Bolt ($\frac{5}{16}$ in UNC x $\frac{7}{8}$ in long) } Fixing mounting
20	Vane carrier and roller vane	46	Spring washer } bracket to
21	Cam lock peg	47	Plain washer } cylinder block
22	Special screw } Fixing end cover	48	Distance piece for mounting bracket
23	Lock washer } to body	49	Bolt ($\frac{1}{4}$ in UNF x $\frac{3}{4}$ in long) } Fixing brackets to
24	Bearing retainer plate	50	Spring washer } steering pump
25	Special screw, fixing plate	51	Belt driving steering pump
26	Key for pump shaft		

STEERING

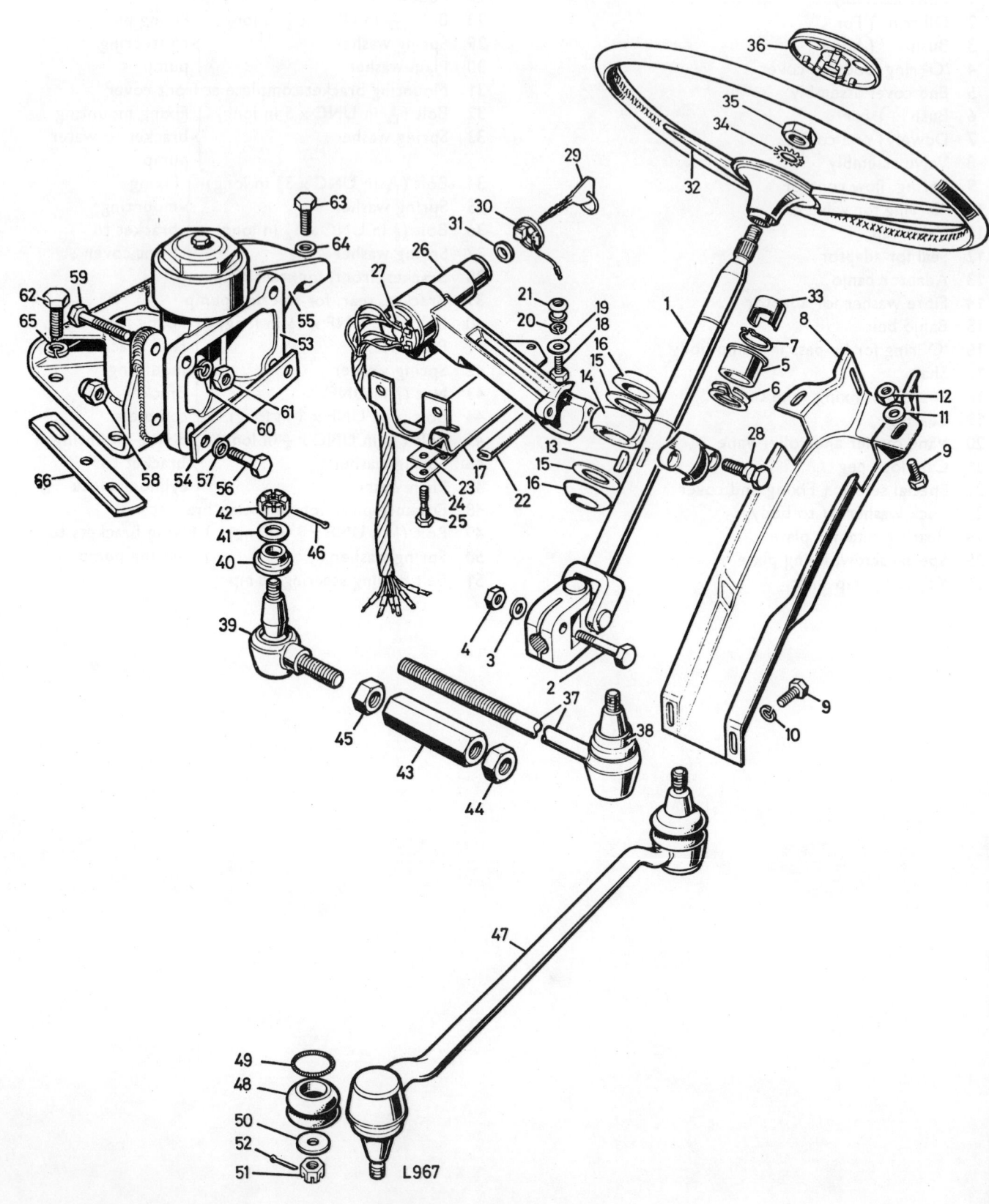

Linkage for power steering

57–8 Rover 3500 and 3500S Manual AKM 3621

STEERING

Key to linkage for power steering

1 Steering column shaft assembly
2 Pinch bolt ($\frac{5}{16}$ in UNF x $1\frac{1}{2}$ in long) ⎫ Fixing shaft
3 Plain washer ⎬ yoke to inner
4 Nut ($\frac{5}{16}$ in UNF) ⎭ column
5 Bearing for steering column shaft
6 Spring washer ⎫ Fixing shaft
7 Circlip ⎭ to bearing
8 Support bracket for steering column
9 Bolt ($\frac{1}{4}$ in UNF x $\frac{5}{8}$ in long) ⎫ Fixing
10 Spring washer ⎬ support
11 Plain washer ⎬ bracket to
12 Nut ($\frac{1}{4}$ in UNF) ⎭ base unit
13 Woodruff key ⎫ For
14 Collar ⎬ steering
15 Thrust washer ⎬ column
16 Bearing ⎭ lock
17 Mounting bracket for steering column lock
18 Bolt ($\frac{1}{4}$ in UNF x $\frac{5}{8}$ in long) ⎫ Fixing
19 Plain washer ⎬ mounting
20 Spring washer ⎬ bracket to
21 Riv-nut ($\frac{1}{4}$ in UNF) ⎭ base unit
22 Plastic protection strip for mounting bracket
23 Mounting plate
24 Spring mounting plate
25 Special bolt ($\frac{1}{4}$ in UNC) fixing lock to mounting bracket
26 Steering column lock assembly
27 Switch for ignition and lock
28 Sheer bolt fixing end cap
29 Special key for steering column lock and ignition switch
30 Switch for warning buzzer
31 Grommet for buzzer switch
32 Steering wheel, leather covered rim
33 Striker for flasher switch
34 Shakeproof washer ⎫ Fixing steering
35 Special nut ⎭ wheel to shaft
36 Finisher for steering wheel
37 Steering track rod and ball joint, RH
38 Rubber boot for ball joint, RH
39 Ball joint assembly, LH thread
40 Rubber boot for ball joint, LH
41 Plain washer ⎫ Fixing track rod
42 Special slotted nut ⎭ ball joints
43 Adjuster sleeve for track rod
44 Locknut ($\frac{5}{8}$ in UNF) for adjuster sleeve, RH thread
45 Locknut ($\frac{5}{8}$ in UNF) for adjuster sleeve, LH thread
46 Split pin for slotted nut
47 Steering side rod assembly
48 Rubber boot ⎫ For side rod
49 Retaining ring for boot ⎭ ball joints
50 Plain washer ⎫ Fixing side rod
51 Special slotted nut ⎭ ball joints
52 Split pin for slotted nut
53 Steering idler damper complete
54 Strap for steering idler damper
55 Mounting bracket for damper
56 Bolt ($\frac{3}{8}$ in UNF x $1\frac{1}{4}$ in long) ⎫
57 Plain washer ⎬ Fixing
58 Self-locking nut ⎬ damper
59 Bolt ($\frac{3}{8}$ in UNF x 2 in long) ⎬ to
60 Spring washer ⎬ bracket
61 Nut ($\frac{3}{8}$ in UNF) ⎭
62 Set bolt ($\frac{3}{8}$ in UNF x $1\frac{1}{8}$ in long) ⎫ Fixing
63 Bolt ($\frac{3}{8}$ in UNF x 1 in long) ⎬ mounting bracket
64 Plain washer ⎬ to base unit
65 Spring washer ⎭
66 Spacer, mounting bracket to base unit

Rover 3500 and 3500S Manual AKM 3621

STEERING

POWER STEERING BOX

—Remove and refit   57.10.01

Service tools: 565446   Ball joint extractor
                  606606   Drop arm ball joint extractor

It is important that whenever any part of the system, including the flexible piping, is removed or disconnected, that the utmost cleanliness is observed.
All ports and hose connections should be suitably sealed off to prevent ingress of dirt etc.
If metallic sediment is found in any part of the system, the complete system should be checked, the cause rectified and the system thoroughly cleaned.
The engine must **NEVER** be started until the reservoir has been filled. Failure to observe this rule will result in damage to the pump.

Procedure

1. Remove the air cleaner. 19.10.01.
2. **RHStg models.** Remove the brake fluid reservoir and mounting bracket.
3. Remove the filler cap from the power steering fluid reservoir. Disconnect the pipes from the pump, drain and discard the fluid.
4. Disconnect the flexible hoses from the steering box.
5. Jack up the front of the car to permit the wheels to turn on lock.
6. Remove the split pins from the two ball joints at the steering box drop arm and the two nuts and washers securing the ball joints.
7. Release the two ball joints using extractor 565446. To remove the track rod ball joint, place steering on lock, loop extractor over drop arm then move the steering back to the straight-ahead position while manoeuvring the extractor over the ball joint. The side rod ball joint is accessible from under the front wing.
8. **Air conditioned models.** If difficulty is experienced in locating extractor 565446 on to the track rod ball joint below the steering box, use instead the extractor 606606. With the extractor edge inserted between the drop arm and the ball joint, as illustrated, drive in to part arm from joint.
9. **LHStg models.** Remove the fixings securing the accelerator bracket to the steering box.
10. Remove the three bolts, lockplate and special nut and spring washers securing the steering box to the base unit.

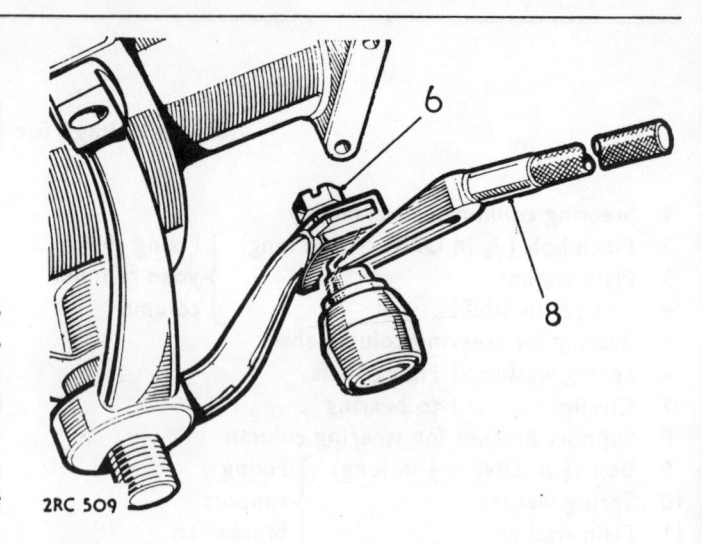

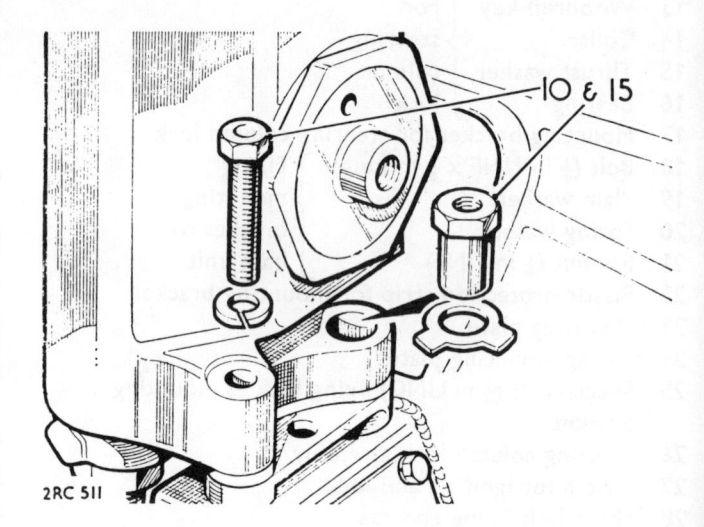

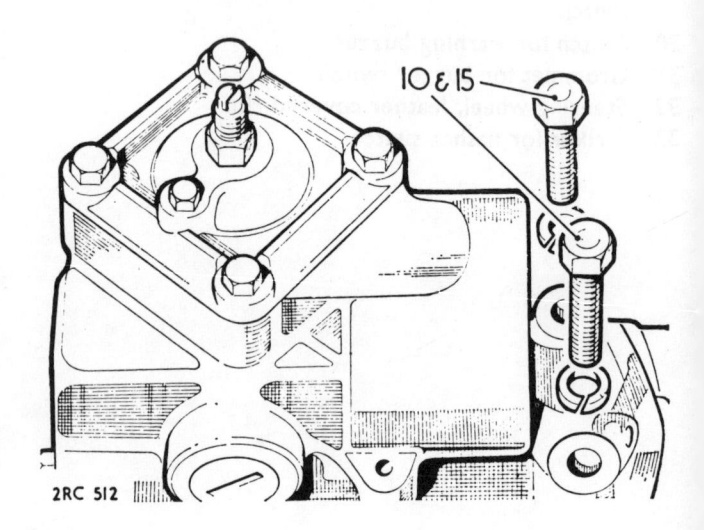

continued

STEERING

11. Remove the pinch bolt, nut and two plain washers from the steering column universal joint, accessible from inside the car, behind glove compartment.
12. Remove the steering box complete. To facilitate removal, vary the position of the drop arm by turning the steering.

    NOTE: There may be spacing plates fitted between the steering box and the base unit, and these must be retained for reassembly.

Refitting

13. If spacer plates were fitted between the steering box and the base unit, ensure that they are in place.
14. Refit the steering box, feeding the shaft into the steering column universal joint. Check to ensure that the spacer plate(s) is not dislodged from the rear mounting position on the base unit.
15. Secure the steering box with three bolts and one special nut together with spring washers. Note that the nut must be fitted first due to its recessed location. A further check should be made to ensure that the bolts are located correctly through the spacer plate(s), front and rear.
16. Reverse 1 to 11, noting the following.
17. Ensure that the steering wheel spokes are horizontal when the wheels are in the straight-ahead position. It may be necessary to remove the steering wheel and reposition on the splines to obtain this condition.
18. Fill the reservoir with one of the recommended fluids (Division 09) and bleed the power steering system. 57.15.02.
19. Check, and if necessary adjust, the front wheel alignment. 57.65.01.
20. Check, and if necessary reset, the steering lock stops. 57.65.03.
21. Check, and if necessary adjust, the steering box. 57.10.13.
22. Test the steering system for leaks, with the engine running, by holding the steering hard on full lock in both directions. **DO NOT** maintain this pressure for more than 30 seconds in any one minute to avoid causing the oil to overheat and possible damage to the seals.

    NOTE: If the steering box is being refitted during the course of repairs to a vehicle that has been damaged in a motoring accident, the steering box height must be checked. 76.10.57.

23. Road test the car.

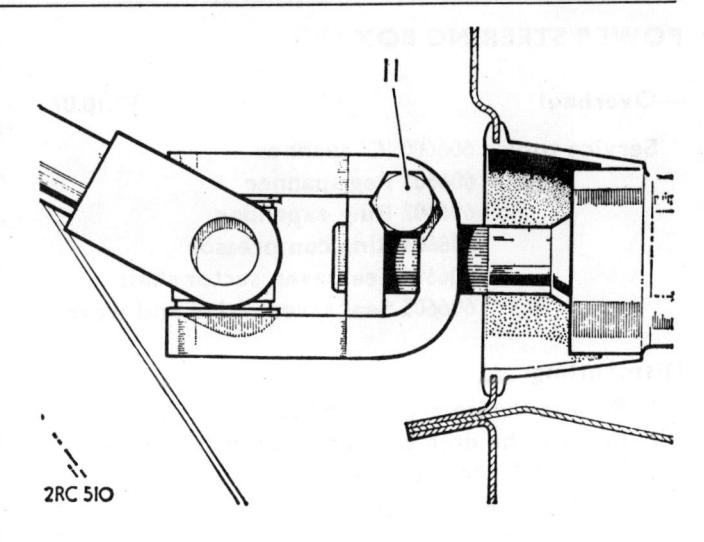

2RC 510

STEERING

POWER STEERING BOX

—Overhaul   57.10.07

Service tools: 606600 'C' spanner
606601 Peg spanner
606602 Ring expander
606603 Ring compressor
606604 Seal saver, sector shaft
606605 Seal saver, valve and worm

Dismantling

1. Remove the steering box. 57.10.01.
2. Remove the drop arm, using a suitable extractor.
3. Remove the bell housing.
4. Lift the cover retaining ring from the groove in the cylinder bore, using a suitable pointed drift applied through the hole provided in the cylinder wall.
5. Complete the retaining ring removal, using a screwdriver.
6. Withdraw the cylinder cover, using molegrips or pliers to grip the centre boss.
7. Rotate the input shaft coupling to position the rack piston 44,5 mm (1.75 in.) approximately from the top of the housing bore.
8. Slacken the retaining grub screw at the rack pad adjuster.
9. Turn back the rack pad adjuster two turns.
10. Remove the sector shaft adjuster locknut.
11. Remove the sector shaft cover fixings.
12. Lift the sector shaft cover free of the housing bore by tapping on the shaft free end, using a soft mallet, and turning in the adjuster screw as necessary.
13. Unscrew the sector shaft cover from the adjuster.
14. Withdraw the sector shaft sufficient only to disengage the sector teeth from the piston and rack teeth.
15. Withdraw the piston and rack, using a suitable ½ in. UNC bolt screwed into the tapped hole in the piston.
16. Withdraw the sector shaft.
17. Lift out the rack adjuster thrust pad, released during the previous operation.
18. Remove the coupling assembly from the input shaft, retained with a cotter pin.
19. Remove the worm adjusting screw locknut, using 'C' spanner 606600.
20. Remove the worm adjusting screw, using peg spanner 606601.
21. Tap in the input shaft to free the outer bearing cup at the other end of the housing.
22. Withdraw the outer bearing cup and the cage and ball bearings assembly.
23. Withdraw the valve and worm assembly. **Do not dismantle this assembly which is calibrated at manufacture.**
24. **Do not disturb the trim screw, otherwise the calibration will be adversely affected.**
25. Withdraw the inner bearing ball race.

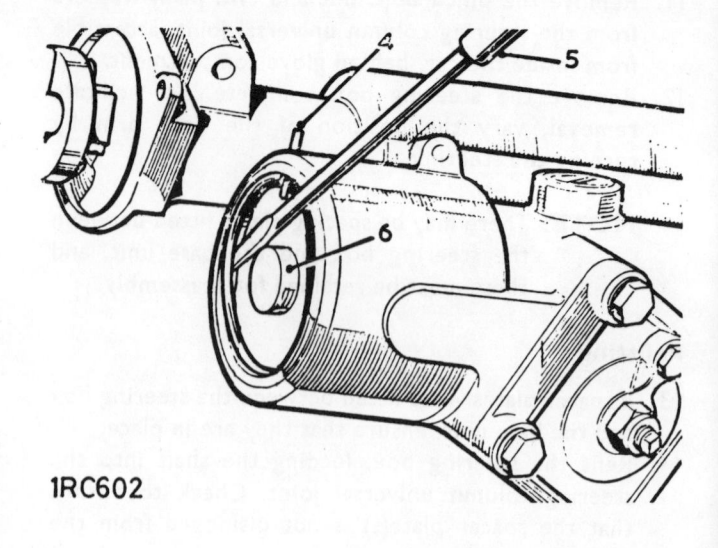

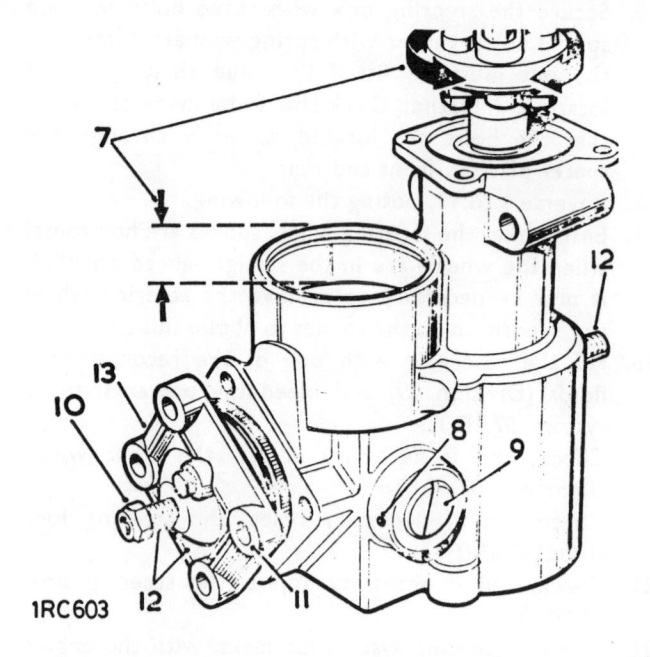

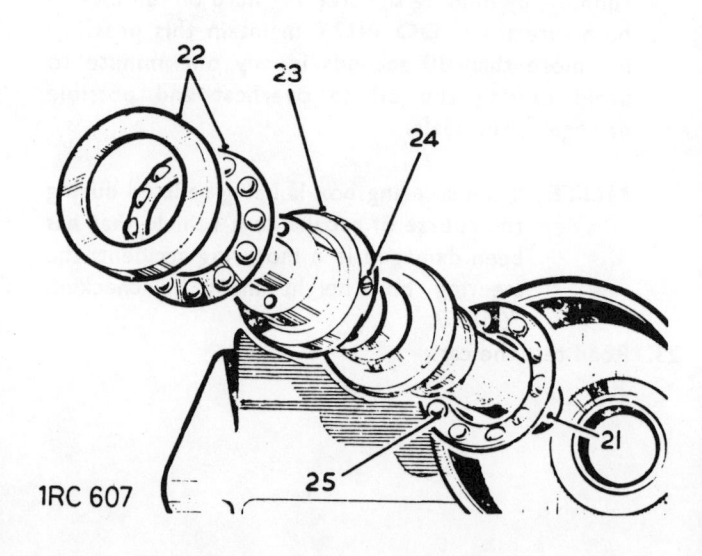

continued

1RC602

1RC603

1RC 607

26. Remove the circlip and seals from the sector shaft housing bore. Do not remove the sector shaft bush unless replacement is required; item 41 refers.
27. Remove the circlip and seals from the input shaft housing bore. Do not remove the input shaft needle bearing unless replacement is required; item 65 refers.
28. Remove the thrust pad adjuster grub screw.
29. Remove the thrust pad adjuster.

Inspecting

30. Discard all rubber seals and provide replacements.

    NOTE: A rubber seal is fitted behind the plastic ring on the rack piston. Discard the seal and the plastic ring and provide replacements.

Coupling and bell housing.

31. If worn or damaged, replace the needle bearing and rubber bush in the bell housing, using Bostik grade 771 on the bush outer diameter.
32. Examine the coupling rubber for wear and damage. When fitting a replacement rubber element, tighten the fixings until the rubber begins to compress.
33. Examine the coupling for wear, damage and corrosion.

Sector shaft assembly

34. Check that there is no side play on the roller.
35. Check that the adjusting screw retainer is securely retained in the sector shaft by staking.
36. The adjusting screw end float must not exceed 0,12 mm (0.005 in.). If necessary, the end float may be decreased by turning in the threaded adjuster retainer then re-staking.

    WARNING: Re-staking of the adjuster retainer must be such that it cannot become loosened during service.

37. Examine the bearing areas on the shaft for excessive wear.
38. Examine the gear teeth for uneven or excessive wear.

Sector shaft cover assembly

39. The cover, bush and seat are supplied as a complete assembly for replacement purposes.

Sector shaft adjuster locknut

40. The locknut functions also as a fluid seal and must be replaced at overhaul.

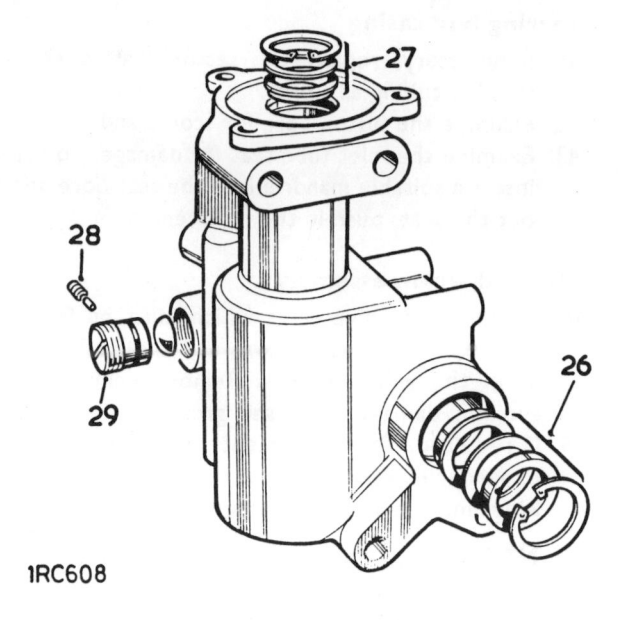

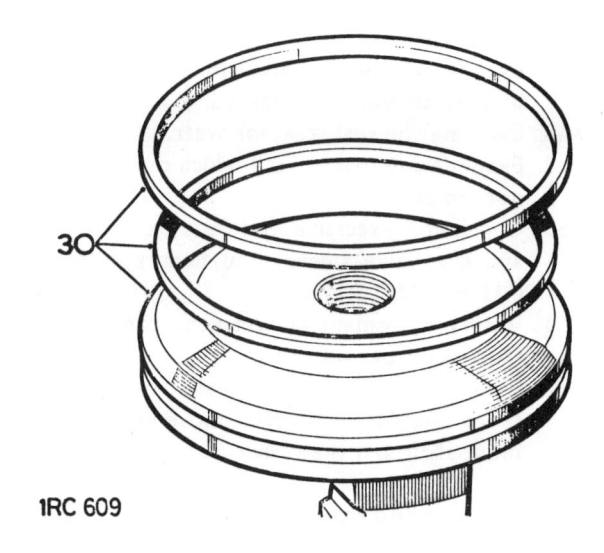

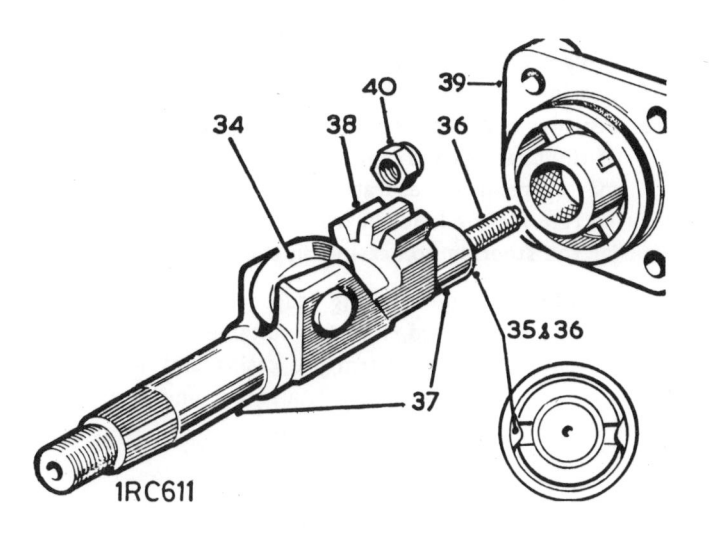

continued

STEERING

Steering box casing

41. If necessary, replace the sector shaft bush, using suitable tubing as a drift.
42. Examine the piston bore for scores and wear.
43. Examine the inlet tube seat for damage. To replace, insert a suitable mandrel into the seat bore and pull out the seat; push in the replacement.

Valve and worm assembly

44. Examine the valve rings which must be free from cuts, scratches and grooves and should be a loose fit in the valve grooves. Remove damaged rings, avoiding damage to the seal grooves.
45. If required, fit replacement rings, using the ring expander 606602.
    Rings may be warmed, if required, using hot water.

    NOTE: The expander will not pass over rings already fitted. These must be discarded to allow access, then replaced.

46. Settle the replaced rings in the grooves, using ring compressor 606603.
47. Examine the bearing areas for wear. The areas must be smooth and not indented.
48. Examine the seal areas for wear.
49. Examine the worm track which must be smooth and not indented.
50. Check for excessive end-float between the worm and the valve sleeve. End float must not exceed 0,12 mm (0.005 in.).
51. Rotary movement between the components at the trim pin is permissible.
52. Check for wear on the torsion bar assembly pins; there should be no free movement between the input shaft and the worm.

Ball bearing and cage assemblies

53. Examine the ball races and cups for wear and general condition.
54. If the ball cage has worn against the bearing cup, fit replacements.
55. Bearing balls must be retained by the cage.
56. Bearings and cage repair is by complete replacement of the bearings and cage assembly. The bearing cup may be replaced separately only.
57. To remove the inner bearing cup and shim washers, jar the steering box casing on the work bench.

    NOTE: If the bearing cup cannot easily be removed, insert a suitable mandrel into the cup, warm the casing and withdraw the cup before it also becomes warmed.

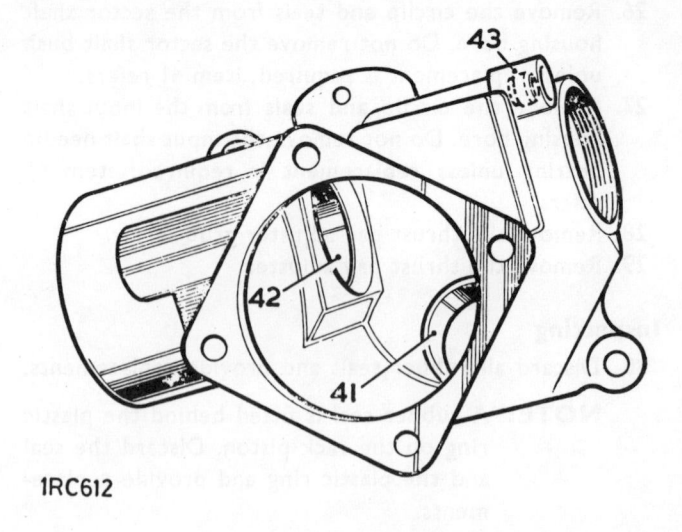

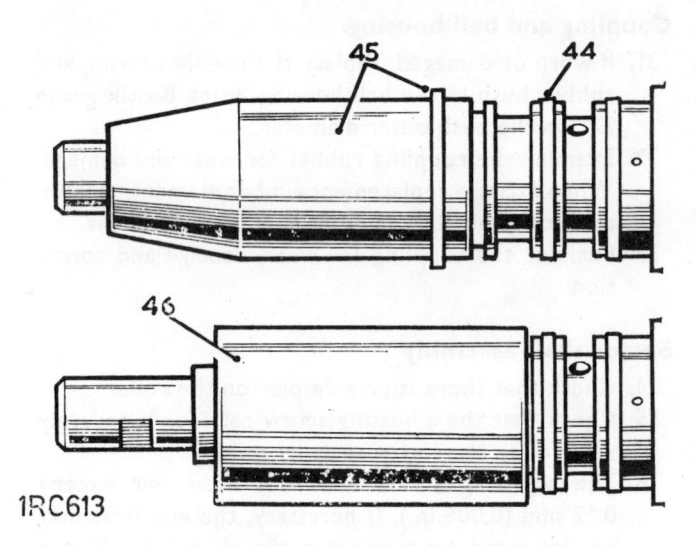

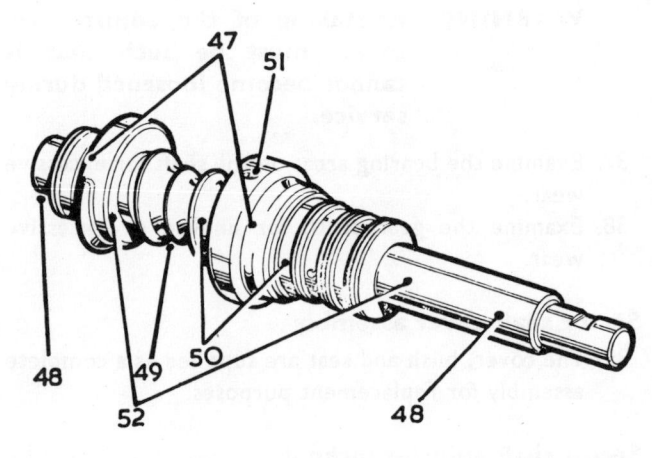

continued

1RC612

1RC613

1RC614

STEERING

Rack thrust pad and adjuster

58. Examine the thrust pad for scores.
59. Examine the adjuster for wear in the pad seat.
60. Examine the nylon pad and adjuster grub screw assembly for wear.

Rack and piston

61. Examine for excessive wear on the rack teeth.
62. Ensure that the thrust pad bearing surface is free of scores and wear.
63. Ensure that the piston outer diameter is free from burrs and damage.
64. Examine the seal and ring groove for scores and damage.

Input shaft needle bearing

65. If necessary, replace the bearing. The replacement must be fitted squarely in the bore (numbered face of bearing uppermost) then carefully pushed in until just flush with the top of the housing bore. Ideally, the bearing will be just clear of the bottom of the housing bore.

Reassembling

NOTE: Oil seals may be fitted dry to their locations. When fitting assemblies to their operating bores, lubricate with the recommended hydraulic operating fluid.

Input shaft oil seal

66. Fit the seal, lipped side first, into the housing. When correctly seated, the seal backing will lie flat on the bore shoulder.
67. Fit the extrusion washer and secure with the circlip.

Sector shaft seal

68. Fit the oil seal, lipped side first.
69. Fit the extrusion washer.
70. Fit the dirt seal, lipped side last.
71. Fit the circlip.

Fitting the valve and worm assembly

72. If removed, refit the original shim washer(s) and the inner bearing cup. If the original shims are mislaid, fit shim(s) of 0,76 mm (0.030 in.) nominal thickness.
73. Fit the inner cage and bearings assembly.
74. Fit the valve and worm assembly, using seal saver 606605, to protect the input shaft seal.
75. Fit the outer cage and bearings assembly.
76. Fit the outer bearing cup.

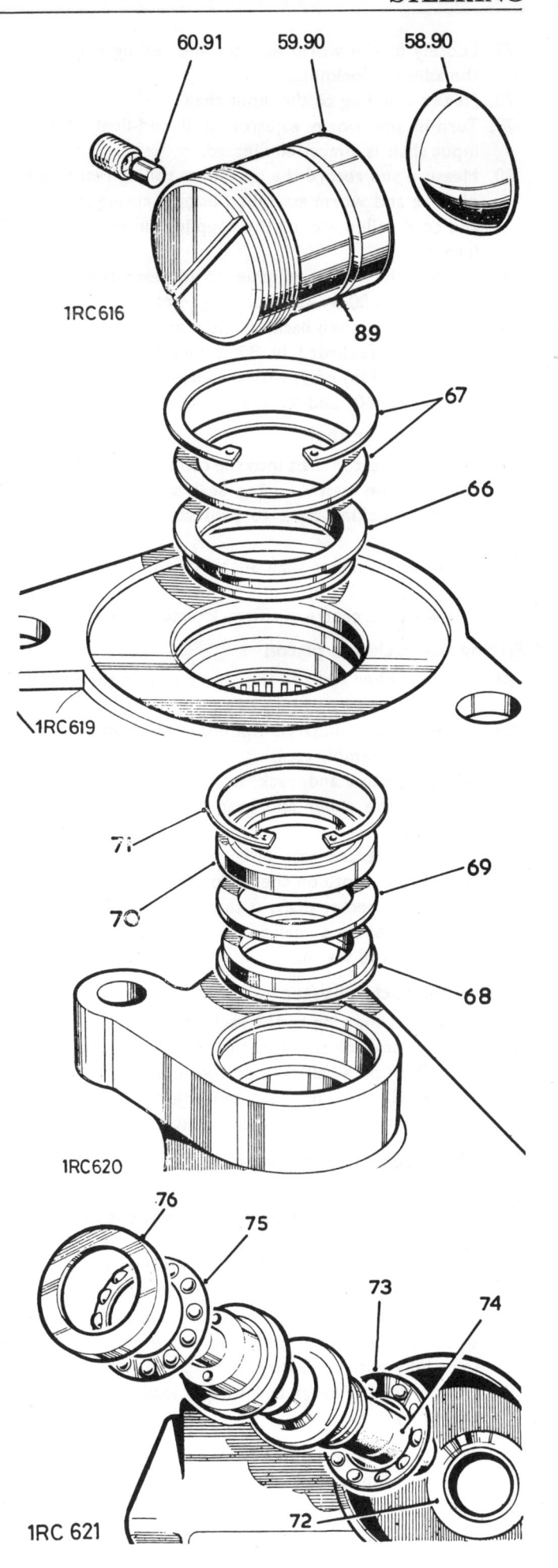

continued

Rover 3500 and 3500S Manual AKM 3621

57.10.07
Sheet 4

STEERING

77. Loosely fit the worm adjuster and sealing ring, and the adjuster locknut.
78. Fit the coupling to the input shaft.
79. Turn-in the worm adjuster until end-float at the input shaft is almost eliminated.
80. Measure and record the maximum rolling resistance of valve and worm assembly, using a spring balance and cord coiled around the coupling rubber circumference.
81. Turn-in the worm adjuster to increase the figure recorded in 80 by 1,8 to 2,2 kg (4 to 5 lb) to settle the bearings, then back-off the worm adjuster until the figure recorded in 80 is increased by 0,9 to 1,3 kg (2 to 3 lb) only, with the locknut tight. Use peg spanner 606601 and 'C' spanner 606600.

Fitting the sector shaft

82. Insert the sector shaft into the casing with the roller towards the worm, using seal saver 606604, or suitable masking tape, to protect the sector shaft seal.
83. Position the sector shaft such that the gap between the roller and gear teeth on the sector shaft is in line with the centre of the rack piston housing bore.

Fitting the rack and piston

84. Fit the rubber seal and the plastic ring to the piston groove.
85. Screw a slave ½ in. UNC bolt into the piston head for use as an assembly tool.
86. Fit the piston and rack so that the rack passes between the roller and gear teeth on the sector shaft.
87. Align the centre gear pitch on the rack with the centre gear tooth on the sector shaft.
88. Push in the sector shaft, at the same time rotate the input shaft about a small arc to allow the sector roller to engage the worm.

Fitting the rack adjuster

89. Fit the sealing ring to the rack adjuster.
90. Fit the rack adjuster and thrust pad to engage the rack. Back off a half-turn on the adjuster.
91. Loosely fit the nylon pad and adjuster grub screw assembly to engage the rack adjuster.

Fitting the sector shaft cover

92. Fit the sealing ring to the cover.
93. Screw the cover assembly fully on to the sector shaft adjuster screw.
94. Position the cover such that the bleed screw will be inboard when the steering box is refitted.
95. Tap home the cover, if necessary back off on the sector shaft adjuster screw to allow the cover to joint fully with the casing.
96. Fit the cover fixings and torque load to 6,0 kgf.m (42 lbf ft).

   NOTE: Before tightening the fixings, rotate the input shaft about a small arc to ensure that the sector roller is free to move in the valve worm.

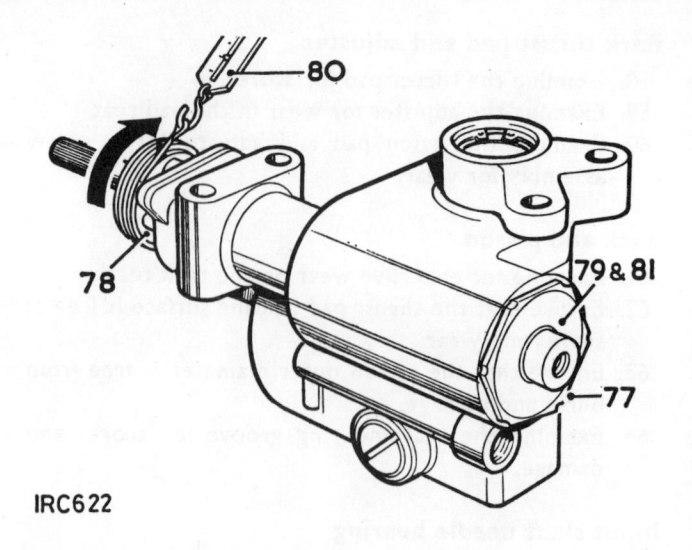

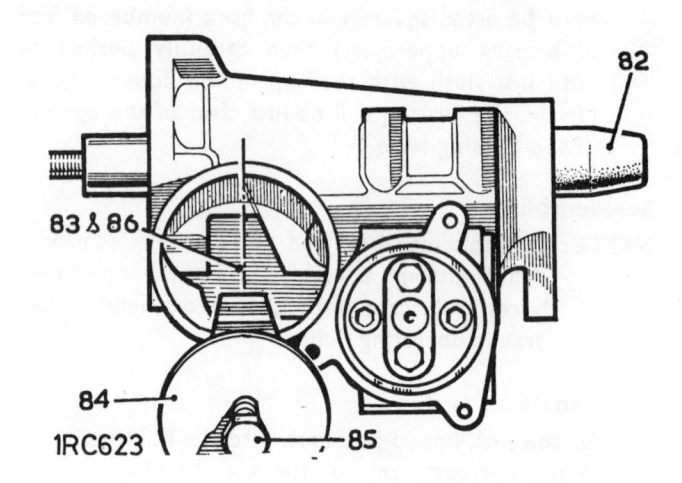

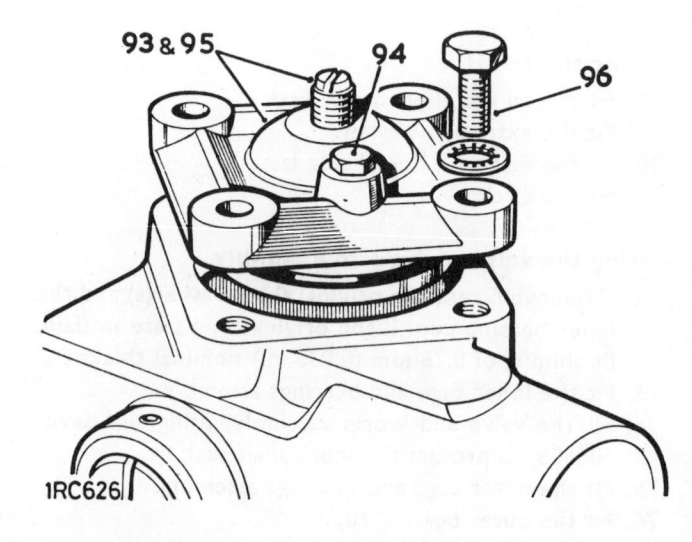

continued

STEERING

Fitting the cylinder cover

97. Fit the square section seal to the cover.
98. Remove the slave bolt and press the cover into the cylinder just sufficient to clear the retainer ring groove.
99. Fit the retainer ring to the groove with one end of the ring positioned 12 mm (0.5 in.) approximately from the extractor hole.

Adjusting the sector shaft

100. Set the worm on centre by rotating the input shaft 1½ turns approximately from either full lock position.
101. Rotate the sector shaft adjusting screw anti-clockwise to obtain backlash between the input shaft and the sector shaft.
102. Rotate the sector shaft adjusting screw clock-wise until the backlash is just eliminated.
103. Measure and record the maximum rolling resistance at the input shaft coupling, using a spring balance and cord.
104. Hold still the sector shaft adjuster screw and loosely fit a new locknut.
105. Turn-in the sector shaft adjuster screw until the figure recorded in 103 is increased by 0,9 to 1,3 kg (2 to 3 lb) with the locknut tightened.
106. Turn-in the rack adjuster to increase the figure recorded in 105 by 0,9 to 1,3 kg (2 to 3 lb). **The final figure may be less than but must not exceed 7,25 kg (16 lb).**
107. Lock the rack adjuster in position with the grub screw.

Torque peak check

With the input shaft rotated from lock-to-lock, the rolling resistance torque figures should be greatest across the centre position (1½ turns approximately from full lock) and equally disposed about the centre position.

This condition depends on the value of shimming fitted between the valve and worm assembly inner bearing cup and the casing. The original shim washer value will give the correct torque peak position unless major components have been replaced.

> **NOTE:** During the following 'Procedure', the stated positioning and direction of rotation of the input shaft applies for both L.H. and R.H. boxes. However, the procedure for shim adjustment where necessary, differs between L.H. and R.H. steering boxes and is described under the applicable L.H. Stg. and R.H. Stg. headings.

Procedure

108. With the input coupling shaft toward the operator, turn the shaft fully anti-clockwise.
109. Check the torque figures obtained from lock-to-lock, using a spring balance and cord coiled around the coupling circumference.

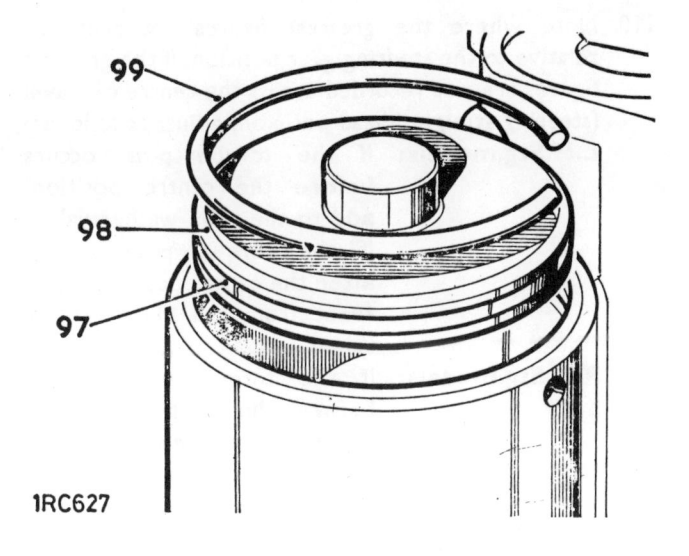

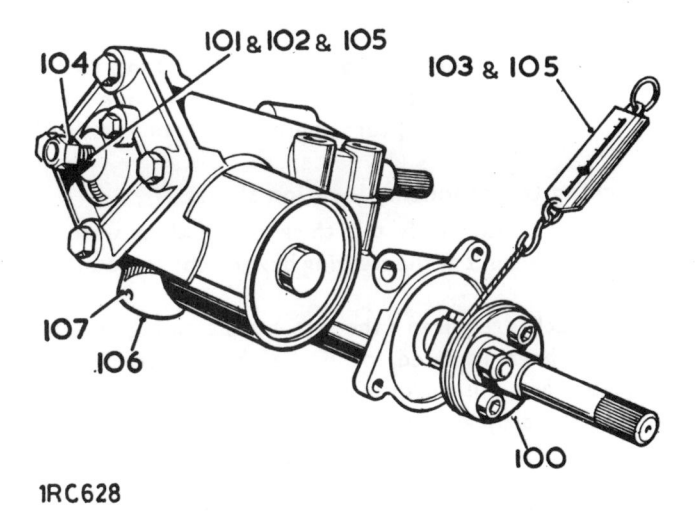

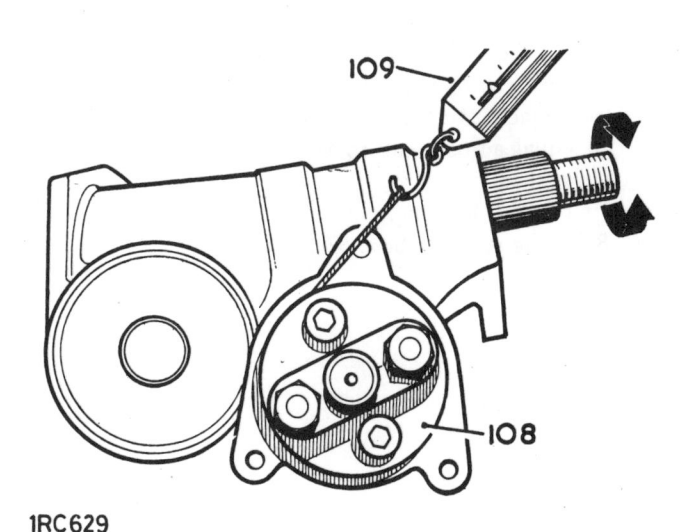

continued

Rover 3500 and 3500S Manual AKM 3621

STEERING

110. Note where the greatest figures are recorded relative to the steering gear position. If the greatest figures are not recorded across the centre of travel (steering straight ahead position) adjust as follows:-

   L.H.Stg.models: If the torque peak occurs **before** the centre position, **add** to the shim washer value; if the torque peak occurs **after** the centre position, **subtract** from the shim washer value.

   R.H.Stg.models: If the torque peak occurs **before** the centre position, **subtract** from the shim washer value; if the torque peak occurs **after** the centre position, **add** to the shim washer value.

   Shim washers are available as follows:
   0,03 mm, 0,07 mm, 0,12 mm and 0,25 mm (0.0015 in. 0.003 in. 0.005 in. and 0.010 in.).

   NOTE: Adjustment of 0,07 mm (0.003 in.) to the shim value will move the torque peak area by $\frac{1}{4}$ turn approximately on the shaft.

Checking the mounting location shims

This check is to verify the thickness of shim washers pegged to the unit at the steering box to vehicle mounting faces.

The check is required if the sector shaft, casing or drop arm has been replaced, or if the shims have become displaced and mislaid or if sector shaft adjustment has altered the drop arm position.

Procedure

111. Fit the steering drop arm.
112. Position the unit as illustrated.
113. Measure the dimension as illustrated between the mounting face and the drop arm face, which should be 22,5 to 24,0 mm (0.885 to 0.945 in).
    Adjust the shimming to suit from the range available.
114. Adjustment to the shim thickness at the mounting face adjacent to the sector shaft must be followed by an equal adjustment to the shims adjacent to the input shaft.
115. Fit the bell housing.
116. Refit the steering box. 57.10.01.

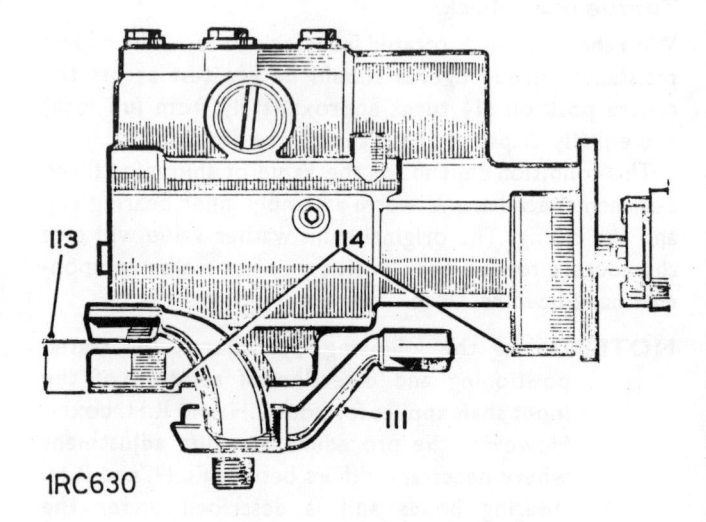

STEERING

POWER STEERING BOX

—Adjust 57.10.13

NOTE: The condition of adjustment which must be achieved is one of minimum backlash without overtightness when the wheels are in the straight-ahead position.

1. Jack the front of the car until the wheels are clear of the ground.
2. Gently rock the steering wheel about the straight-ahead position to obtain 'feel' of backlash present. This should be not more than 9,5 mm (0.375 in) measured at steering wheel rim.
3. Continue the rocking motion whilst an assistant slowly tightens the steering box adjuster screw, after slackening the locknut, until rim movement is reduced to 9,5 mm (0.375 in) maximum.
4. Tighten the locknut, then turn the steering wheel from lock to lock and verify that there is no excessive tightness at any point.
5. Road test the car.

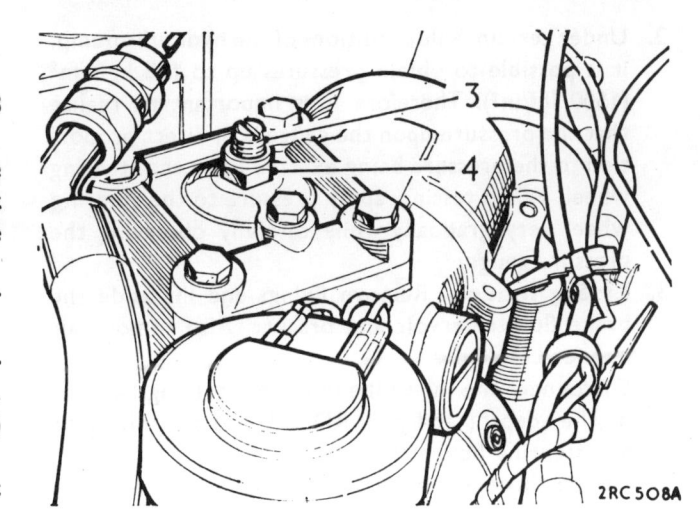

POWER STEERING SYSTEM

—Test 57.15.01

Service tools:
- **RO 1012** Adaptor and pipe assembly for use with JD 10
- **JD 10** Three-way adaptor, hose and pressure gauge for testing power steering

If there is a lack of power assistance evident, it is important that the pressure of the hydraulic pump, which is bolted to the front of the engine, is checked and if necessary corrected or a replacement unit fitted before any action is taken to replace the complete steering unit.

This operation together with the fault finding chart at the end of the operation gives information on how to check this point and also covers other faults on the power steering system which are best diagnosed before attempting removal and overhaul of the system.

Procedure

1. The hydraulic pressure test gauge is used in conjunction with the special adaptor (as illustrated) for testing the power steering system. This gauge is calibrated to read up to 140 kgf/cm^2 (2000 lbf/in^2) and the normal maximum pressure which may be expected in the power steering system is 67 kgf/cm^2 (950 lbf/in^2).

continued

Rover 3500 and 3500S Manual AKM 3621

STEERING

2. Under certain fault conditions of the hydraulic pump, it is possible to obtain pressures up to 105 kgf/cm² (1500 lbf/in²). Therefore it is important to realise that the pressure upon the gauge is in direct proportion to the pressure being exerted upon the steering wheel. When testing, apply pressure to the steering wheel very gradually while carefully observing the pressure gauge.
3. **LHStg models.** Remove fixings and lift aside the brake fluid reservoir and bracket. Also remove air cleaner LH elbow.
4. Disconnect the inlet pipe from the steering box and fit the adaptor and pipe RO 1012, to the inlet pipe and steering box, as illustrated.
5. Fit the three-way adaptor, hose and pressure gauge JD 10, between the previously fitted adaptor and pipe.
6. Bleed the system. 57.15.02. Taking care while doing so, not to overload the pressure gauge.
7. With the system in good condition the pressures should be as follows:
   a Steering wheel held hard on full lock and engine running at 1,000 rev/min, the pressure should be 60 to 67 kgf/cm² (850 to 950 lbf/in²).
   b With engine idling and steering wheel held hard on full lock the pressure should be 28 kgf/cm² (400 lbf/in²) minimum.
   These checks should be carried out first on one lock, then on the other.
   c Release steering wheel and allow engine to idle; pressure should be 7 kgf/cm² (100 lbf/in²) maximum.

Never hold steering wheel on full lock for more than 30 seconds in any one minute to avoid causing the oil to overheat and possible damage to the seals.

8. Failure to build up pressure when items 1 and 2 of the fault finding chart have been eliminated is usually due to a fault in the hydraulic pump relief valve.
9. Should a lack of pressure occur only on one lock, then the cause is undoubtedly a fault in the steering box valve and spool.
10. As it is not possible to distinguish between lack of pressure in the hydraulic pump and a leaking steering box piston seal or worn components in the steering box, it will always be necessary to recheck the pressure after any rectification to the hydraulic pump or after hydraulic pump replacement. If pressure is still low, transfer attention to the steering box and rectify or replace as necessary.

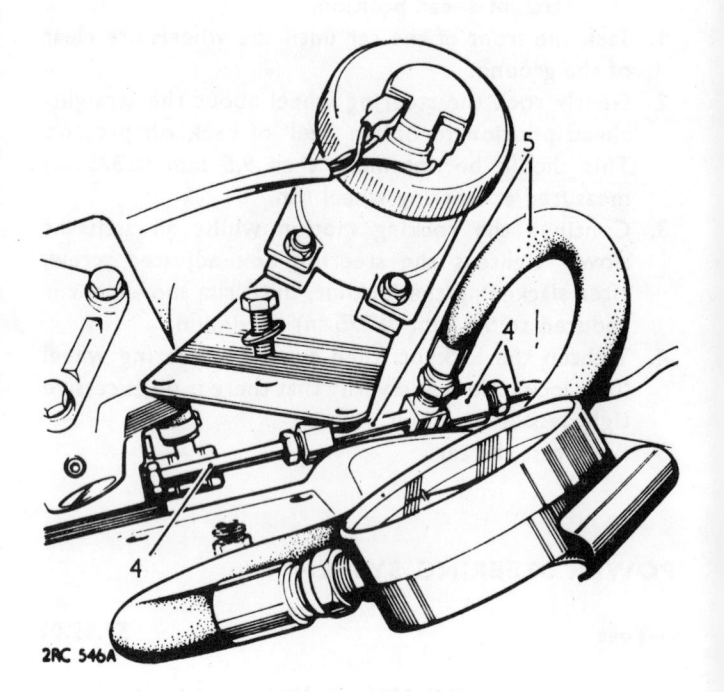

continued

STEERING

FAULT-FINDING CHART—POWER STEERING

SYMPTOM	CAUSE	TEST ACTION	CURE
Insufficient power assistance when parking	(1) Lack of fluid	Check hydraulic fluid tank level	If low, fill and bleed the system. 57.15.02.
	(2) Engine idling speed too low	Try assistance at fast idle	If necessary, reset idle speed.
	(3) Driving belt slipping	Check belt tension	Adjust the driving belt. 57.20.01.
	(4) Defective hydraulic pump and/or pressure relief valve	(a) Fit pressure gauge between high pressure hose and steering box, with steering held hard on full lock, see Note 1 below, and engine running at 1000 rev/min the pressure should be 60 to 67 kgf/cm² (850 to 950 lbf/in²) with engine idling the pressure should be 28 kgf/cm² (400 lbf/in²) minimum	If pressure is outside limits (high or low) after checking items 1 and 3, see Note 2 below.
		(b) Release steering wheel and allow engine to idle, pressure should be 7 kgf/cm² (100 lbf/in²) maximum	If presssure is greater, check steering box for freedom and self-centring action.
Poor handling when car is in motion	Lack of castor action	This is caused by over-tightening the rocker shaft backlash adjusting screw on top of steering box	It is most important that this screw is correctly adjusted. Instructions governing adjustment are given in 57.10.13 and must be strictly adhered to.
	Steering too light and/or over-sensitive	Check for loose torsion bar fixings on steering box valve and worm assembly	Fit new valve and worm assembly
Hydraulic fluid leaks	Damaged pipework, loose connecting unions, etc.	Check by visual inspection; leaks from the high pressure pipe lines are best found while holding the steering on full lock with engine running at fast idle speed (see Note 1 below). Leaks from the steering box tend to show up under low pressure conditions, that is, engine idling and no pressure on steering wheel	Tighten or renew as necessary.
Excessive noise	(1) If the high pressure hose is allowed to come into contact with the body shell, or any component not insulated by the body mounting, noise will be transmitted to the car interior	Check the loose runs of the hoses	Alter hose route or insulate as necessary.
	(2) Noise from hydraulic pump	Check oil level and bleed system. 57.15.02.	If no cure, change hydraulic pump.
Cracked steering box	Excessive pressure due to faulty relief valve in hydraulic pump	Check by visual inspection	Fit new steering box and rectify hydraulic pump or replace as necessary.

Note 1. Never hold the steering wheel on full lock for more than 30 seconds in any one minute, to avoid causing the oil to overheat and possible damage to the seals.

Note 2. High pressure—In general it may be assumed that excessive pressure is due to a faulty relief valve in the hydraulic pump.
Low pressure—Insufficient pressure may be caused by one of the following:
    1 Low fluid level in reservoir ⎫ Most usual cause of
    2 Pump belt slip ⎭ insufficient pressure
    3 Leaks in the power steering system
    4 Faulty relief valve in the hydraulic pump
    5 Fault in steering box valve and spool
    6 Leak at piston sealing in steering box
    7 Worn components in either steering box or hydraulic pump

STEERING

POWER STEERING SYSTEM

—Bleed 57.15.02

Procedure

1. Fill the steering fluid reservoir to the 'F' mark on the dipstick with one of the recommended fluids. Division 09.
2. Start the engine and allow to warm up until normal idling speed is attained. Leave the engine continuing at idle speed.

   NOTE: During the following procedure, ensure that the steering reservoir is kept full. Do not increase the engine speed or move the steering wheel.

3. Slacken the bleed screw. When fluid seepage past the bleed screw is observed, retighten the screw.
4. Ensure that fluid level is up to 'F' mark on reservoir dipstick.
5. Wipe off fluid released during bleeding.
6. Check all hose joints, and pump and steering box for fluid leaks under pressure by holding the steering hard on full lock in both directions. Do not maintain this pressure for more than 30 seconds in any one minute, to avoid causing the oil to overheat and possible damage to the seals. The steering should be smooth lock-to-lock both directions, that is, no heavy or light spots when changing direction when car is stationary.
7. Carry out a short road test. If necessary, repeat the complete foregoing procedure.

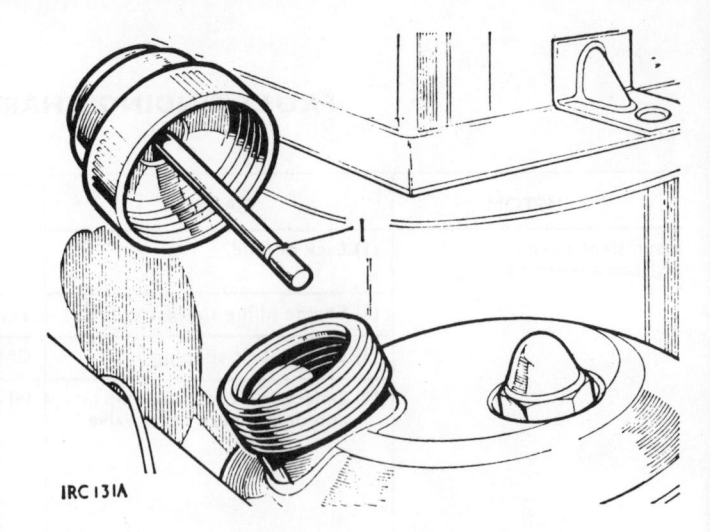

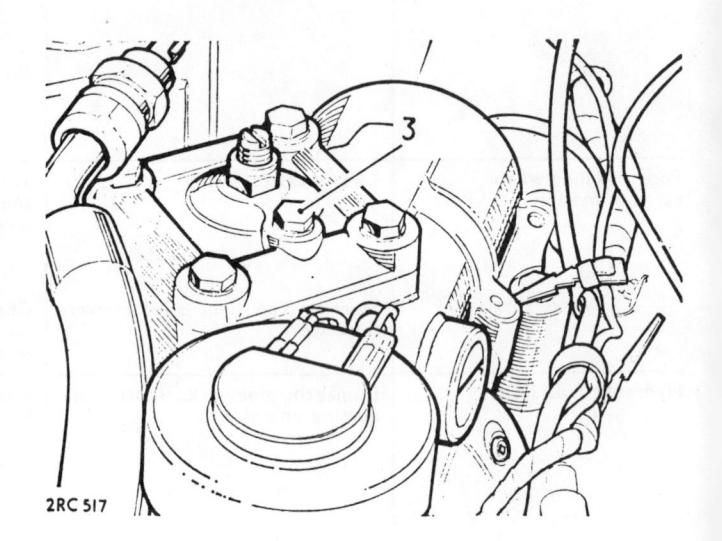

STEERING FLUID RESERVOIR

—Remove and refit 57.15.08

Removing

1. Remove the reservoir filler cap.
2. Disconnect the return pipe from the steering box end cover, drain and discard the fluid.
3. Refit the return pipe to the steering box end cover.
4. Remove the fixings and withdraw the reservoir complete, then disconnect the flexible hoses. If the reservoir is not going to be replaced immediately, the pipes must be sealed to prevent ingress of dirt.
5. Remove the top cover and oil filter element.

Refitting

6. Reverse 1 to 5, fitting a new filter element and ensuring the correct fitting of the ring seal.
7. Fill the reservoir with one of the recommended fluids (Division 09) and bleed the power steering system. 57.15.02.

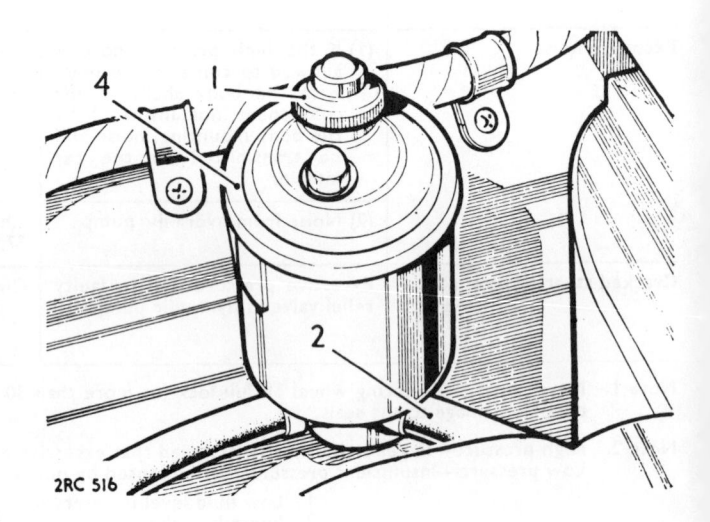

STEERING

POWER STEERING PUMP DRIVE BELT

—Adjust 57.20.01

Procedure

1. Check the belt tension by thumb pressure midway between the crankshaft and pump pulleys, from underneath the vehicle. There should be 6 mm to 9 mm (0.250 in. to 0.375 in.) free movement.
2. If necessary, adjust as follows.
3. Slacken both pivot bolts.
4. Slacken the bolt securing the adjusting link.
5. Pivot the pump inwards or outwards as necessary until the correct belt tension is obtained.
6. Secure the pump adjusting and pivot bolts.

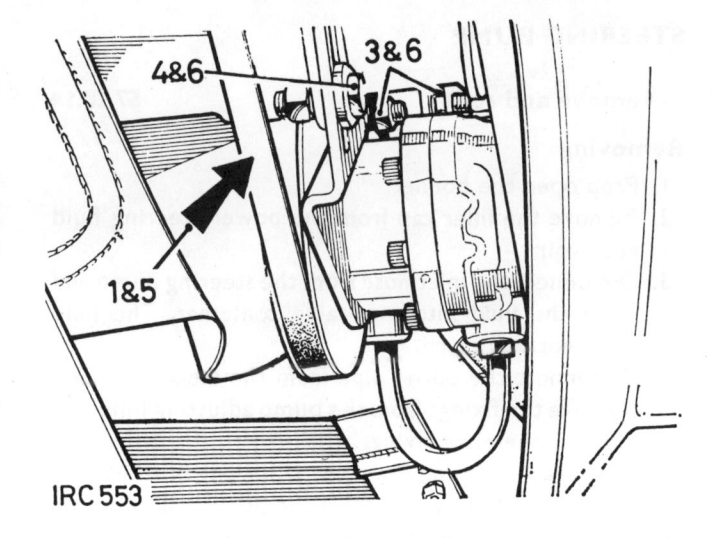

STEERING PUMP DRIVE BELT

—Remove and refit 57.20.02

Removing or preparing for fitting new belt

1. Prop open the bonnet.
2. Slacken the alternator fixings and remove the fan belt.
3. Slacken the timing pointer bolts and move the pointer to one side.
4. Remove the steering pump fixings and move the pump clear of the mounting bracket, then release the driving belt from the pump and crankshaft pulleys.

Refitting

5. Locate the driving belt over the crankshaft and pump pulleys.
6. Fit the pump, leaving the fixings finger tight only.
7. Adjust the position of the pump to give a driving belt tension of 6 to 9 mm (0.250 to 0.375 in) movement when checked by thumb pressure midway between the crankshaft and pump pulleys, then secure the pump adjusting and pivot bolts.
8. Reset the timing pointer, square, and as close as possible to the pulley, and secure the retaining bolts.
9. Fit the fan belt and adjust the tension to give 11 to 14 mm (0.437 to 0.562 in) movement when checked by thumb pressure midway between the crankshaft and alternator pulleys.
10. Close the bonnet.

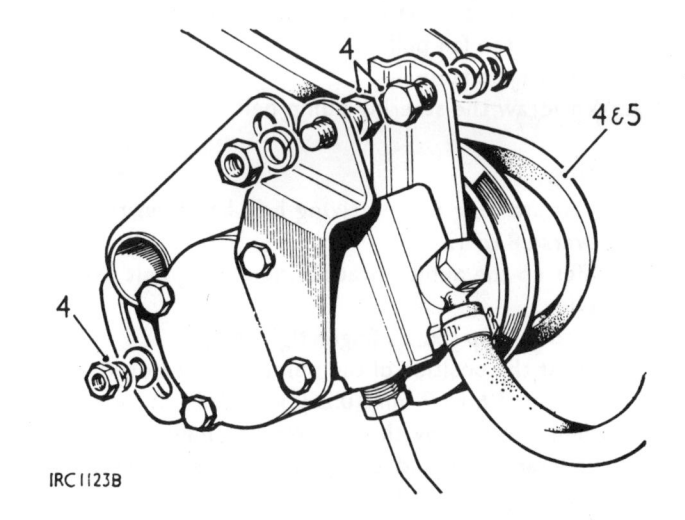

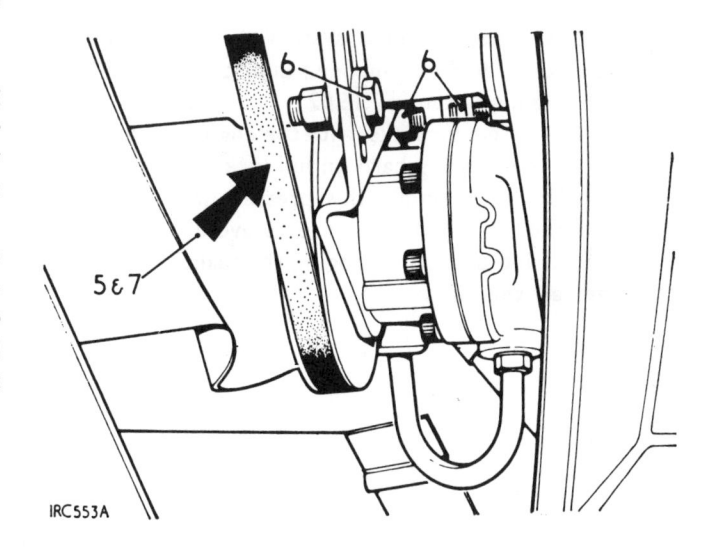

Rover 3500 and 3500S Manual AKM 3621

STEERING

STEERING PUMP

—Remove and refit   57.20.14

Removing

1. Prop open the bonnet.
2. Remove the filler cap from the power steering fluid reservoir.
3. Disconnect the inlet hose from the steering pump and drain the fluid into a suitable container. The fluid must not be reused.
4. Disconnect the outlet pipe from the steering pump.
5. Remove the fixings from the pump adjusting link.
6. Remove the pivot fixings.
7. Withdraw the pump from the mounting brackets and release the driving belt from the pump pulley.

If required, remove the steering pump driving belt. 8 to 12.

8. Remove the fan blades.
9. If air conditioning equipment is fitted, remove the compressor driving belt.
10. Remove the fan pulley.
11. Remove the fan belt.
12. Withdraw the steering pump belt.

Refitting

13. If the steering pump driving belt has been removed, reverse 8 to 12.
14. Offer up the pump, locating the driving belt over the pulley.
15. Fit the pump fixings finger tight only.
16. Adjust the position of the pump to give a driving belt tension of 6 to 9 mm (0.250 to 0.375 in.) movement when checked by thumb pressure midway between the crankshaft and pump pulleys, then secure the pump adjusting and pivot bolts.
17. Reconnect the hydraulic fluid pipes to the steering pump.
18. Replenish the hydraulic system with the correct grade fluid, see Recommended Lubricants—09. Bleed the hydraulic system. 57.15.02
19. Test the system for leaks, with the engine running, by holding the steering hard on full lock in both directions. Do not maintain this pressure for more than 30 seconds in any one minute to avoid overheating the fluid and possibly damaging the seals.
20. Road test the car.

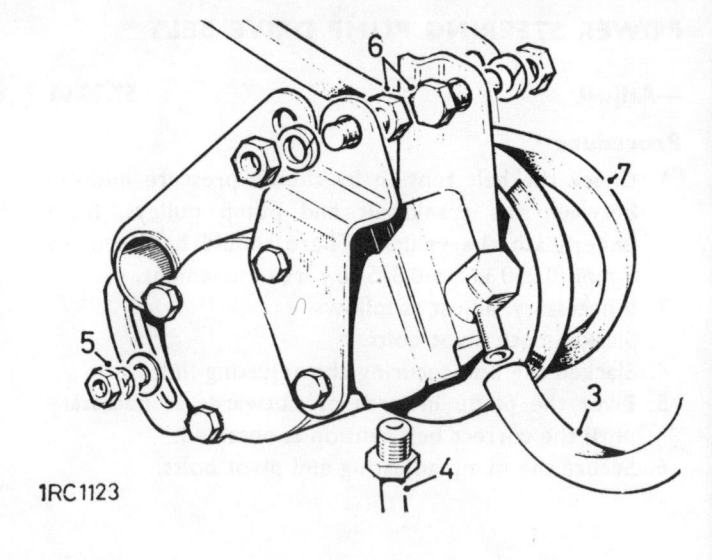

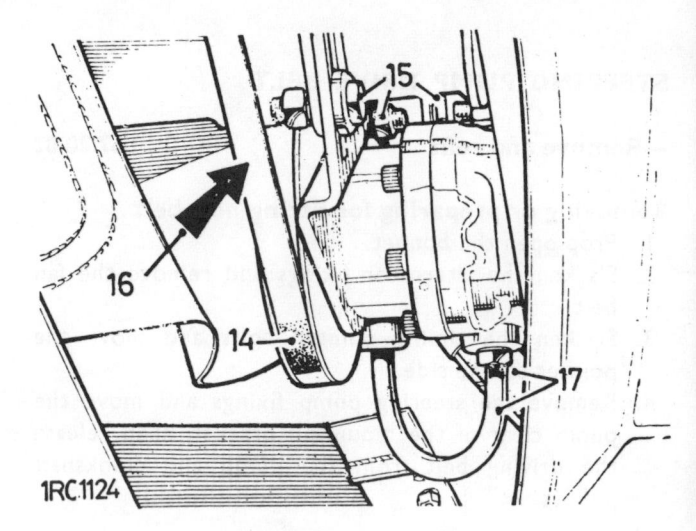

STEERING

STEERING PUMP

—Overhaul   57.20.20

Dismantling

1. Remove the steering pump. 57.20.14.
2. Remove the pump pulley, bolt and special washer.
3. Remove the front mounting bracket.
4. Remove the body end-plate, secured by four Phillips screws.
5. Remove the rear bracket.
6. Clamp the body of the pump in a vice. Remove the adaptor screw, and withdraw the adaptor, fibre washer and rubber seal. The venturi flow director is pressed into the cover and should not be removed.
7. Remove the six Allen screws securing the cover to the pump body, and remove the pump from vice. Separate the cover from the body vertically to prevent parts from falling out.
8. Remove the 'O' ring seals from the groove in the pump body.
9. Carefully tilt the pump body, and remove the six rollers.
10. Draw the carrier off the shaft, and remove the drive pin.
11. Remove the shaft from the body.
12. Remove the cam and the cam lock peg from the pump body.
13. If necessary, remove the key from the shaft, and withdraw the sealed bearings.
14. Remove the shaft seal from the body, ensuring that no damage is caused to the shaft bushing.
15. Remove the valve cap, valve and valve spring from the body. Place all parts where they will not be damaged, or subject to contamination.

Inspection

16. Wash all parts in a suitable solvent, air dry, or wipe clean with a lint-free cloth is air is not available.
17. Check pump body and cover for wear. Replace either part if faces or bushes are worn.

continued

Rover 3500 and 3500S Manual AKM 3621

STEERING

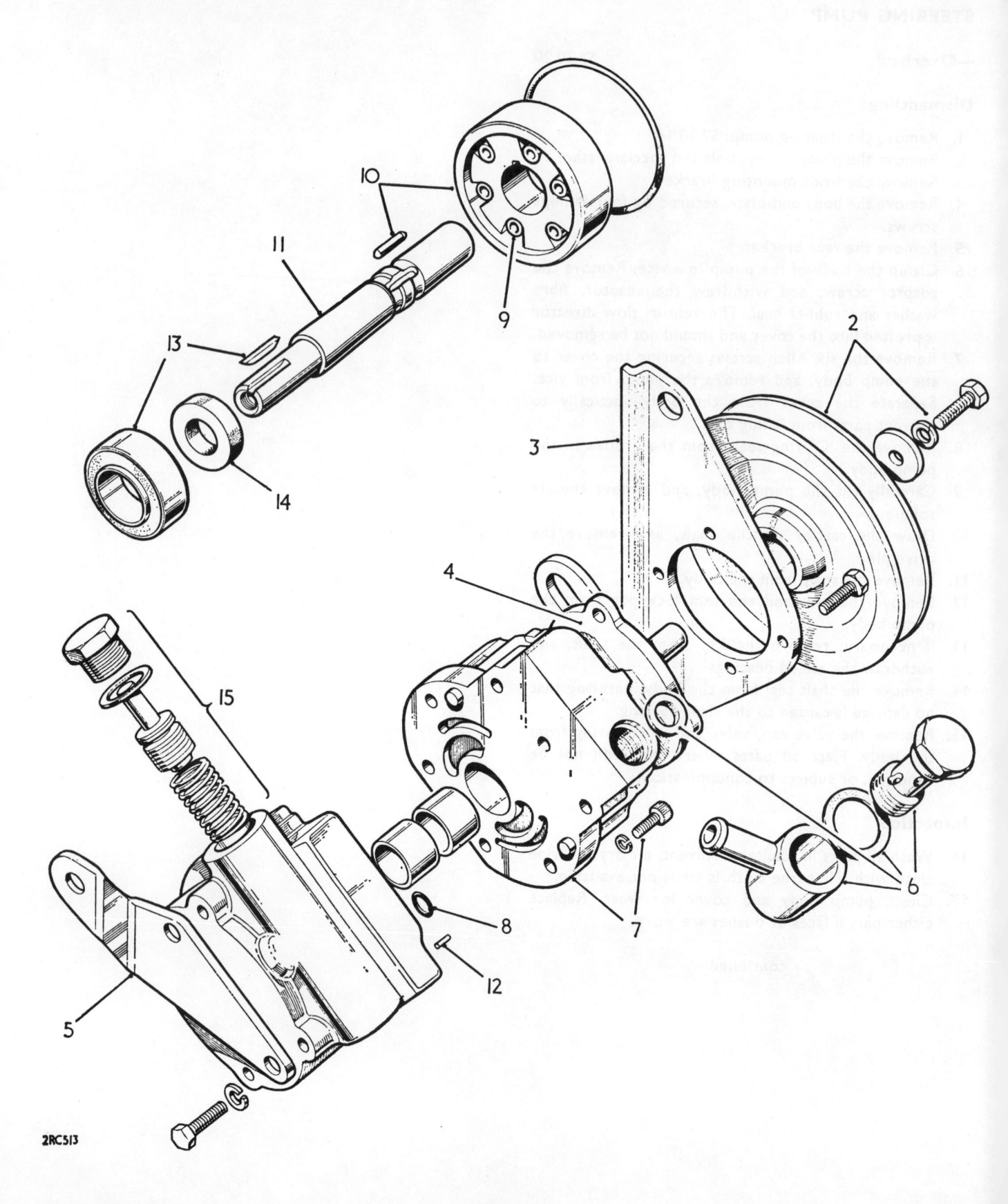

continued

Rover 3500 and 3500S Manual AKM 3621

STEERING

Reassembling

18. Instal a new shaft seal in the pump body with the lip towards the carrier pocket. Care should be taken to ensure that damage is not caused to the lip of the seal.
19. Replace the cam lock peg in the hole in the pocket. Inspect the cam for wear and replace if worn or damaged. Refit the cam in the pocket with the slot over the retaining pin, and ensure that the cam is seated on to the pocket face.
20. If the sealed bearing has been removed, replace it and move it to the limit of its travel on the shaft. Replace the pulley key.
21. Insert the shaft from the seal side of the body, ensuring that there are no sharp edges on the shaft to cut the seal lip.
22. Replace the carrier drive pin in the shaft.
23. Inspect the carrier and replace in position, ensuring that the greater angle on the vein is in the leading position, as illustrated.
24. Inspect rollers, paying particular attention to the finish on the end. Replace if scored, damaged or oval. Refit rollers to carrier.
25. Using a straight edge across the cam surface, and a feeler gauge, check the end clearance of the carrier and rollers in the pump body. If the end clearance is more than 0,05 mm (0.002 in), replace the carrier and rollers.
26. Replace the flow control valve spring in the bore. The spring tension should be 3,6 to 4,8 mkg (8 to 9 lb) at 21 mm (0.820 in). If not, replace with a new spring.
27. Replace the valve in the bore, inserting the valve so that the exposed ball end enters last. Ensure that the valve is not sticking.
28. Replace the 'O' ring on the valve cap and assemble in the pump. Tighten the cap to a torque figure of 4,0 to 4,9 kgf.m (30 to 35 lbf ft).
29. Instal new 'O' rings to the body of the pump.
30. Replace the cover on the pump body and tightly secure with the six Allen screws.
31. Having tightened the Allen screws, check that the shaft rotates freely, and is not binding.
32. Refit the square sectioned oil seal to the groove around the inlet port, and replace the adaptor, fibre washer and adaptor screw.
33. Refit the end plate to the body, ensuring that the larger radius is in line with the valve.
34. Refit the rear bracket.
35. Refit the front bracket, ensuring that the adjustment slot is at the bottom, and that the retaining bolt holes are on the opposite side to the valve.
36. Refit the pulley to the shaft, and secure with the special washer, spring washer and bolt. This should be tightened to a torque figure of 1,4 to 1,6 kgf.m (10 to 12 lbf ft).

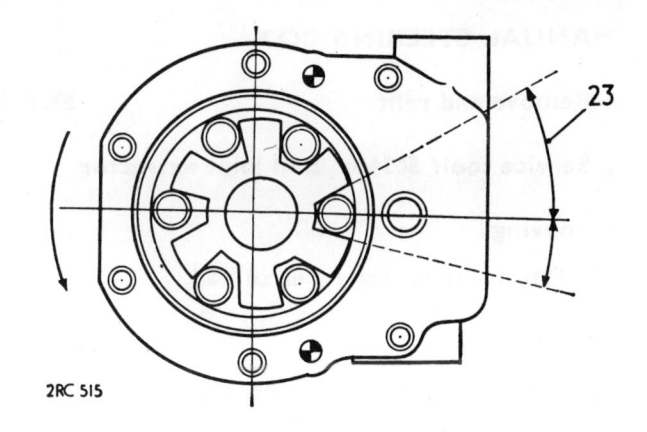

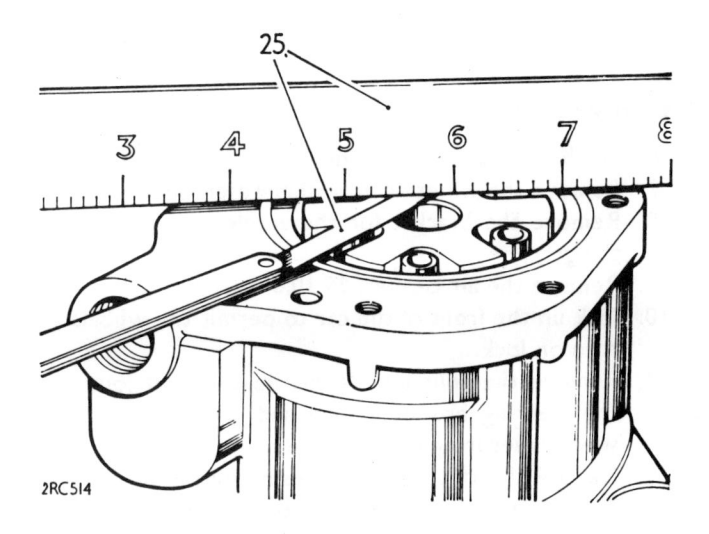

STEERING

MANUAL STEERING BOX

—Remove and refit 57.30.01

Service tool: **565446** Ball joint extractor

Removing

1. Disconnect the battery earth lead.

RHStg 2 to 5

2. Pull off the windscreen wiper blade arms.
3. Lift bonnet and prop open.
4. Remove the complete windscreen wiper motor and link assembly. Division 84.
5. Remove the brake fluid reservoir.

LHStg 6 to 8

6. Lift bonnet and prop open.
7. Remove the windscreen washer reservoir.
8. Remove the bracket for the accelerator shaft from the gearbox.
9. Remove the air cleaner. 19.10.01.
10. Jack up the front of the car to permit the wheels to turn on lock.
11. Remove the split pins from the two ball joints at the steering box drop arm and the two nuts and washers securing the ball joints.
12. Release the two ball joints using extractor 565446. To remove the track rod ball joint, place the steering on lock, loop the extractor over the drop arm then move the steering back to the straight-ahead position while manoeuvring the extractor over the ball joint. The side rod ball joint is accessible from under the front wing, with the front wheel removed.
13. Remove the three bolts, one nut and spring washers securing the steering box to the base unit.
14. Remove the pinch bolt, nut and two plain washers from the steering column universal joint, accessible from inside the car, behind the glove compartment.
15. Remove the steering box complete.

    NOTE: There may be spacing plates fitted between the steering box and the base unit, and these must be retained for reassembly.

Refitting

16. If spacer plates were fitted between the steering box and the base unit, ensure that they are in place.
17. Ensure that the steering stop bracket is in place on the base unit.
18. Refit the steering box, feeding the shaft into the steering column universal joint.
19. Secure the steering box with three bolts and one nut together with spring washers. Note that the nut must be fitted first due to its recessed location.
20. Reverse 1 to 14, noting the following.
21. Ensure that the steering wheel spokes are horizontal when the wheels are in the straight-ahead position. It may be necessary to remove the steering wheel and reposition on the splines to obtain this condition.
22. Check, and if necessary reset, the steering lock stops. 57.65.03.
23. Check, and if necessary replenish, the oil level in the steering box.
24. Check, and if necessary adjust, the front wheel alignment. 57.65.01.
25. Check, and if necessary adjust, the steering box. 57.35.01.

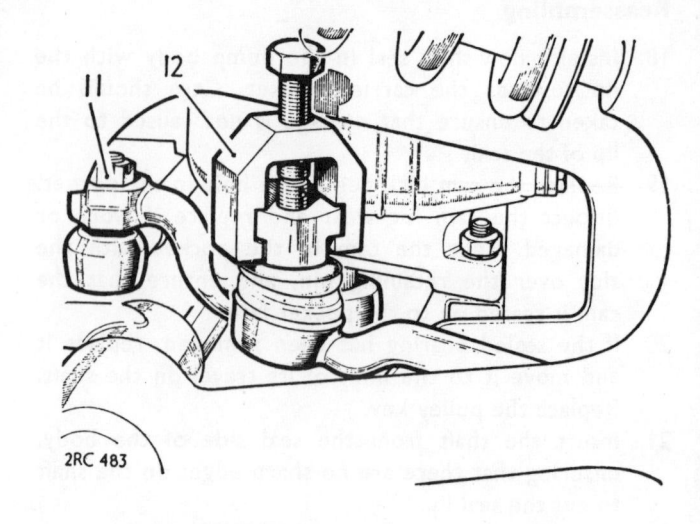

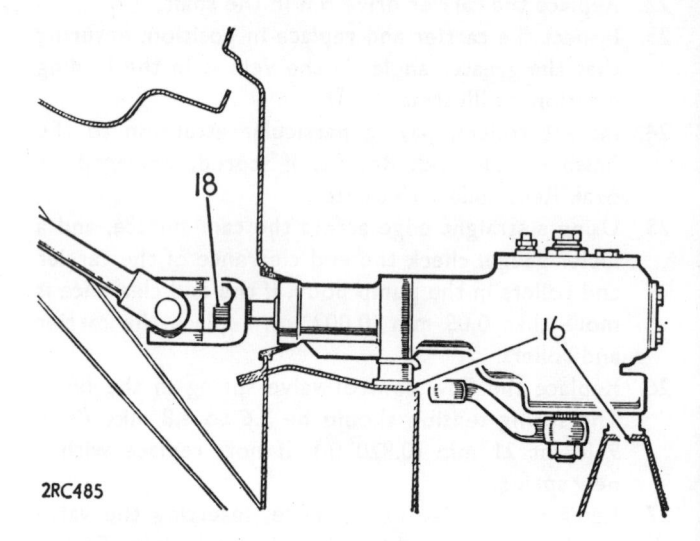

    NOTE: If the steering box is being refitted during the course of repairs to a vehicle that has been damaged in a motoring accident, the steering box height must be checked. 76.10.57.

26. Road test the car.

57.30.01

Rover 3500 and 3500S Manual AKM 3621

STEERING

MANUAL STEERING BOX

—Overhaul   57.30.07

Dismantling

1. Remove the steering box. 57.30.01.
2. Remove the oil filler plug and drain the steering box.
3. Turn the drop arm to full lock either direction, noting the direction chosen, and scribe location marks on the side of the steering box and the drop arm, to ensure correct realignment of the drop arm and rocker shaft on reassembly.
4. Using a vice fitted with soft jaws, clamp the steering box assembly by the drop arm. Bend back the lockwasher and slacken the large nut securing the drop arm to the rocker shaft.
5. Using a suitable two-leg extractor, release the drop arm. Remove the large nut and lift the steering box clear.
6. Clamp the steering box in a vice, using the rib provided; slacken the steering box adjuster and remove the three bolts, nut and spring washers securing the top cover. Lift off cover.
7. Withdraw the rocker shaft and slide the roller off the main nut.
8. Remove the four bolts securing the steering box end cover; lift off cover complete with shims and gaskets.
9. Ensure that the main nut is in the mid-way position, then gently tap the splined end of the shaft to remove the bottom race together with the ten $\frac{9}{32}$ in diameter ball bearings.
10. Wind the worm shaft through the main nut; remove the shaft, nut and ten $\frac{9}{32}$ in diameter ball bearings from the box.
11. Gently tap the bottom end of the steering box with a hide mallet to dislodge the top race.
12. Remove the twenty-six $\frac{5}{16}$ in diameter ball bearings from the main nut and recirculating tube, by tapping the top face of the nut on a wooden block.
13. If necessary press out the rocker shaft bush and oil seal from the box.

Inspecting

14. Check all components for wear or damage.
15. Examine the worm nut ball bearing tracks for signs of indentations or scaling.
16. Examine the worm shaft for similar markings. Slight indentations at the extreme ends of the shaft can be disregarded as this is a normal wear condition, but if indentations have spread to the middle of the shaft, it must be replaced.

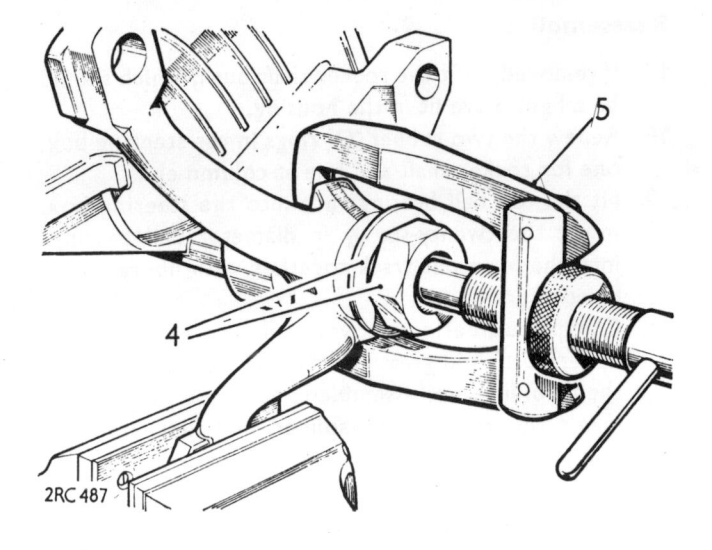

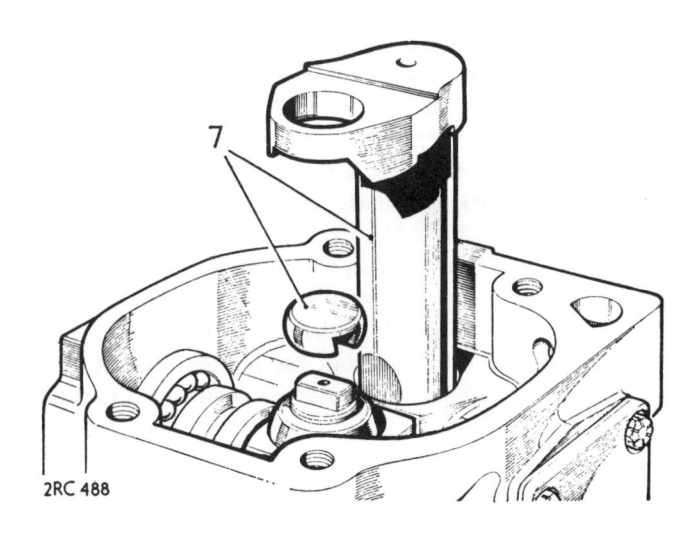

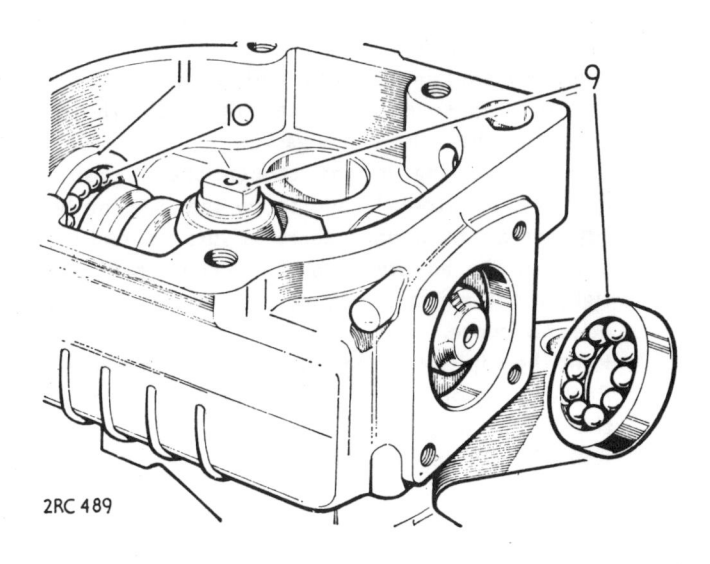

continued

Rover 3500 and 3500S Manual AKM 3621

STEERING

Reassembling

17. If removed, refit the rocker shaft bush, which should be a light drive fit in the housing.
18. Renew the two rubber 'O' rings in the steering box, one for rocker shaft and one at column end.
19. Fit the top ball bearing race into the steering box.
20. Insert the twenty-six $\frac{5}{16}$ in diameter ball bearings into the main nut track together with liberal quantities of grease.
21. Clamp the steering box vertically in a vice, box end uppermost. Add grease to the race fitted in the box, then holding the assembled main nut in position, lower the worm shaft, splined end first, into the box and wind it through the nut, ensuring that all the balls remain in the nut.
22. Raise the worm shaft sufficient to insert the ten $\frac{9}{32}$ in diameter balls into the top race, then lower the shaft again ensuring all balls remain in position and that none have fallen inside the worm shaft case.
23. Using adequate grease, insert the ten $\frac{9}{32}$ in diameter ball bearings and bottom race into position.
24. Apply a non-hardening jointing compound to the tapped holes in the steering box for the end cover securing bolts, and fit the end cover complete with shims and gaskets. Note that there must be a paper gasket fitted each side of the shims.
25. With the steering box and worm shaft still in the vertical position, adjust the worm shaft so that it can be turned by hand but has no end-float. Adjust the shim thickness by adding or removing alternate shims and joint washers under the bottom end cover, to achieve this condition.
26. Move the steering box to horizontal position in vice. Slide the roller on to the main nut and insert the rocker shaft, locating it over the roller.
27. Fill the steering box with the recommended grade of oil. Division 09.
28. Smear a little jointing compound on to the steering box and top cover joint faces, then refit the top cover together with a new gasket.
29. Set the steering in the straight-ahead position (mid-way lock-to-lock) and screw the steering box adjuster by hand until there is just no end-float between the adjuster and the rocker shaft, then tighten the adjuster locknut ensuring the adjuster does not move.
30. Refit the adjuster spring and retainer plug.
31. Recheck the oil level, top-up as necessary and refit the oil filler plug.
32. Refit the steering box. 57.30.01.

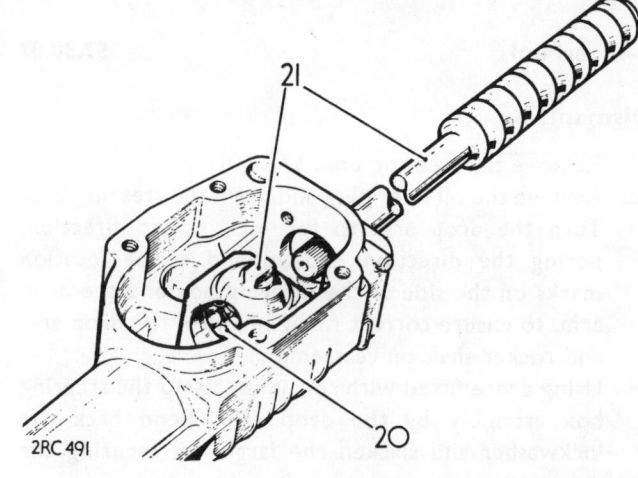

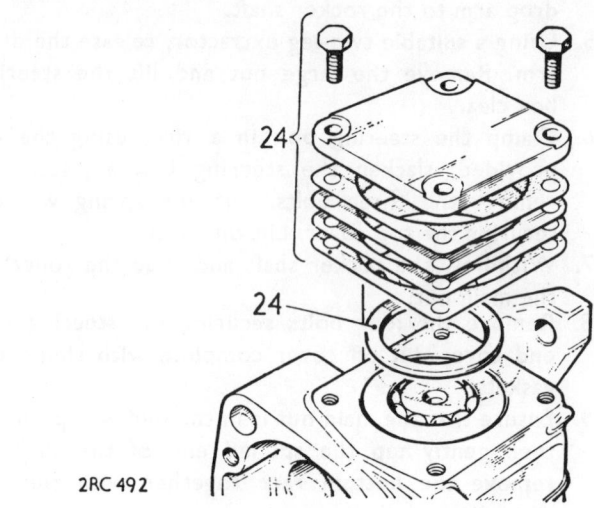

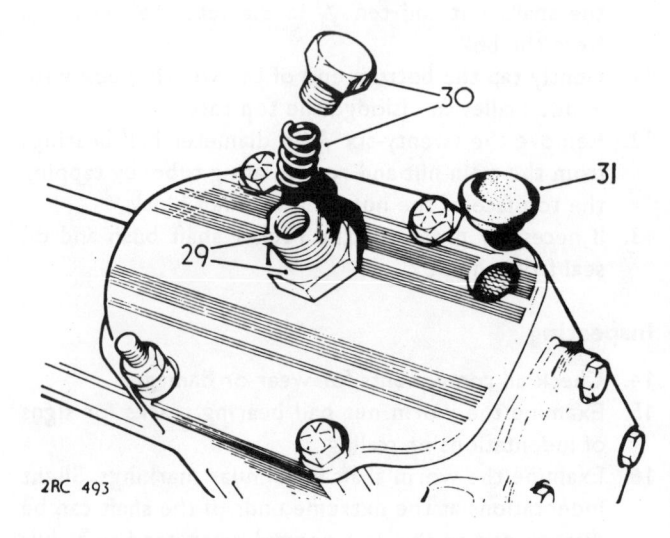

57.30.07
Sheet 2

Rover 3500 and 3500S Manual AKM 3621

STEERING

MANUAL STEERING BOX

—Adjust   57.35.01

Procedure

NOTE: The condition of adjustment which must be achieved is one of zero backlash without over-tightness when the wheels are in the straight-ahead position.

1. Jack the front of the car until the wheels are clear of the ground.
2. Remove the hexagon head plug and spring from the top of the steering box adjuster screw.
3. Gently rock the steering wheel around the straight-ahead position to obtain 'feel' of backlash present.
4. Continue this rocking motion while an assistant slowly tightens the adjuster screw after slackening the locknut.
5. At the instant of zero backlash, tighten the locknut. 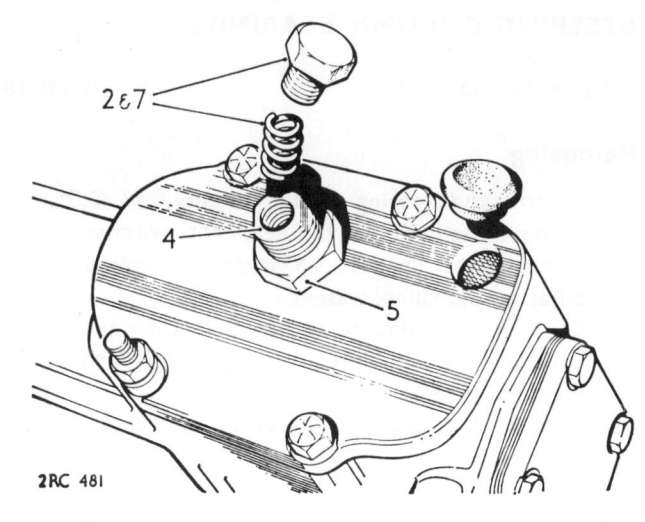
6. Turn the steering wheel from lock to lock and verify that there is no excessive tightness at any point.
7. Finally, replace the adjuster screw spring and plug and road test the car.

STEERING COLUMN ASSEMBLY

—Remove and refit   57.40.01

Removing

1. Remove the driver's side glove box. 76.52.03.
2. Slacken the pinch bolt securing the steering column universal joint, accessible from inside the car, behind the glove box compartment.
3. Remove the finisher from the centre of the steering wheel.
4. Remove the nut and spring washer securing the steering wheel and withdraw it from the steering column.
5. Remove the two bolts, nuts, plain and spring washers securing support bracket to the bulkhead adjacent to the panel.
6. Remove the two bolts securing support bracket to bulkhead.
7. Models equipped with a steering column lock—Remove the fixings securing the steering column lock to the bracket under the dash and remove the bolt securing the bracket to the underside of the dash.
8. Disconnect the wiring harness at the connectors, noting colour coding for reference on refitting. Where applicable disconnect the vacuum pipe for wiper delay switch.
9. Models equipped with a steering column lock—Remove the cable cleat securing the wiring harness to the steering column lock body.
10. Lift out the steering column shaft assembly.
11. Models equipped with a steering column lock—Remove the two Phillips screws from the steering lock switch and detach the switch from the body to allow complete withdrawal of the column and lock from the car. 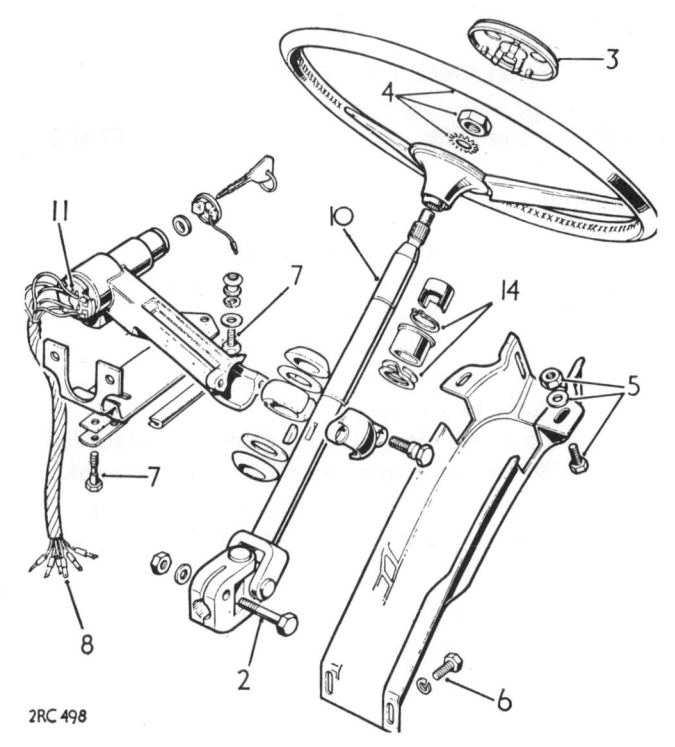
12. Spring off the plastic cover from the steering column nacelle.
13. Remove the horn/indicator switch to gain access to the circlip securing the nacelle to the steering column shaft.
14. Remove the circlip and double spring washer from the top end of the steering column and withdraw. Remove the lower double spring washer and circlip.
15. On early models the steering column journal can be overhauled in exactly the same way as the propeller shafts. 47.15.10. On models equipped with a steering column lock the steering column shaft is complete with journal and cannot be serviced.

Refitting

16. Reverse 1 to 15.

Rover 3500 and 3500S Manual AKM 3621

57.35.01
57.40.01

STEERING

STEERING COLUMN BEARINGS

—Remove and refit   57.40.18

Removing

1. Remove the steering column assembly. 57.40.01.
2. Remove the indicator and headlight switches.
3. Remove the special clip from the column rake adjuster and tufnol washer.
4. Unscrew the adjuster and remove the square nut, plain washer, double coil spring washer and tufnol washer.
5. Remove the bearings (press fit in housing) from steering column nacelle.

Refitting

6. Lubricate the new bearings with one of the recommended greases. Division 09.
7. Reverse 1 to 5.

STEERING COLUMN LOCK AND IGNITION/STARTER SWITCH ASSEMBLY

—Remove and refit   57.40.31

Removing

1. Remove the steering column assembly. 57.40.01, noting that it is not necessary to remove the nacelle assembly from the steering column.
2. Punch drill-guide holes central in the sheared ends of the two shear bolts which secure the end cap to the column lock body.
3. Drill a hole 4,0 mm (0.156 in) diameter in each bolt of sufficient depth to accommodate an extractor.
4. Remove the shear bolts, using a No. 3 'Easy-out' extractor and detach the end cap from the column lock body.

   NOTE: If the bearing arrangement is to be replaced, remove the nacelle assembly from the steering column and proceed as follows.

5. Slide off the plastic bearing and shim washer from the steering column.
6. Slide off the key-located metal collar from steering column and remove the key from the shaft.
7. Slide off the remaining shim washer and plastic bearing.

Refitting

8. Refit the steering column lock to the steering column (ensuring that the column lock shaft engages in the slot in the metal collar), using replacement shear bolts, by reversing the removal procedure.
9. When fitted, tighten the shear bolts using a ½ in AF ring spanner and applying sufficient hand leverage to shear the heads from the bolts.
10. Refit the assembled steering column and lock assembly. 57.40.01.

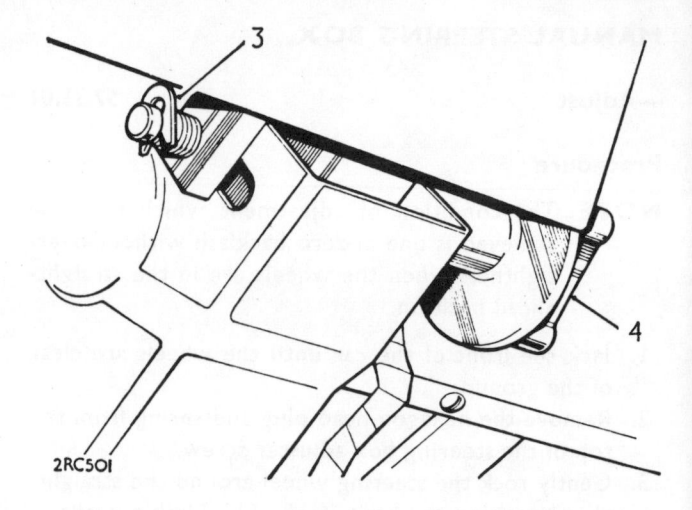

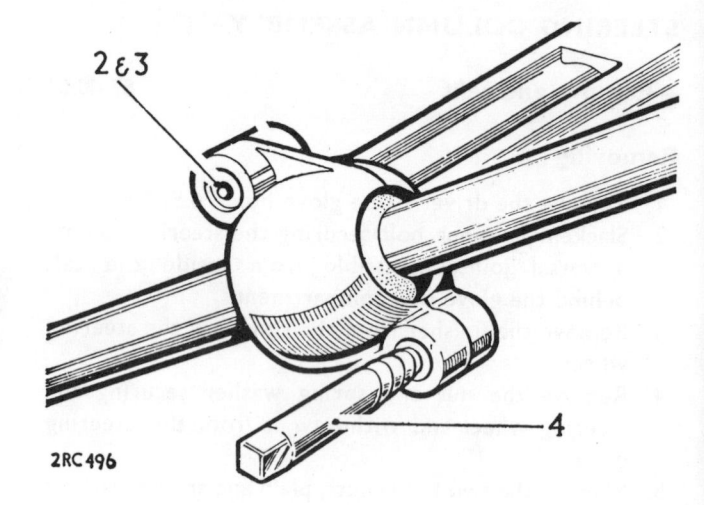

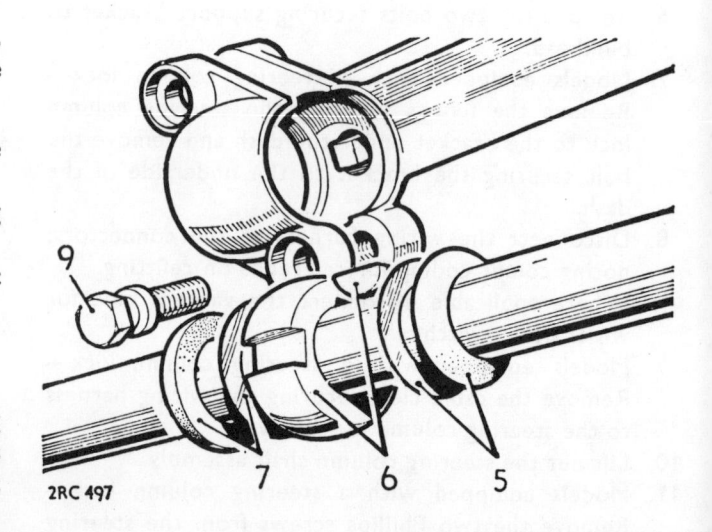

57.40.18
57.40.31

Rover 3500 and 3500S Manual AKM 3621

STEERING

STEERING RELAY

—Remove and refit   **57.50.02**

Service tool: 565446   **Ball joint extractor**

The hydraulically damped steering relay must not be dismantled in any way and requires no maintenance attention during its life.
Should there at any time be signs of oil leaks from the unit, the complete unit must be replaced. The steering relay lever must not be removed since the lever is selectively fitted to the relay shaft and the relationship of the lever position to the internal vanes of the damper is highly critical.

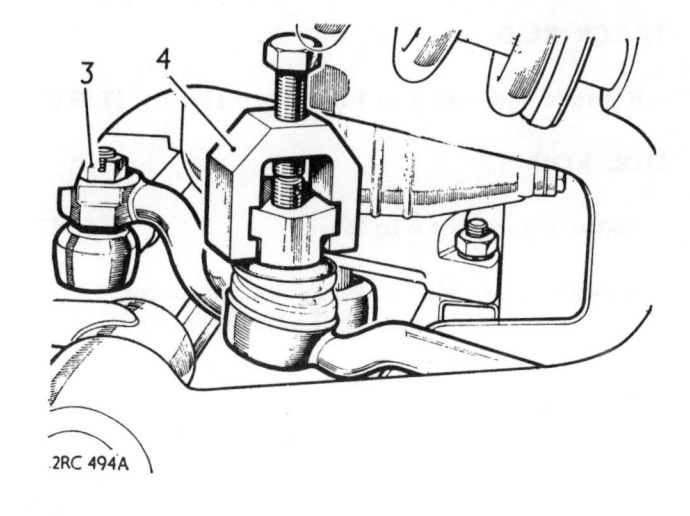

Removing

1. Prop open the bonnet.
2. Jack up the car to permit the wheels to turn on lock.
3. Remove the split pin from the steering side rod and track rod ball joints, and remove the nuts and plain washers.
4. Release the two ball joints, using extractor 565446. To remove the track rod ball joint, place the steering straight-ahead, loop the extractor over the relay lever then move the steering on to lock while manoeuvring the extractor over the ball joint. The side rod ball joint is accessible from under the front wing, with the front wheel removed. Always ensure that the tool locates correctly over the ball joint.
5. **RHStg models.** Remove the bolts securing accelerator bracket to the relay bracket.
6. **LHStg models with power steering.** Release the steering hose clip from the bracket on the steering relay, remove the fixings securing the reservoir to the wing valance and move the reservoir to one side.
7. Remove the three bolts and spring washers securing the mounting bracket to the base unit, and remove the relay unit complete.

    NOTE: There may be spacing plates fitted between the relay mounting plate and the base unit, and these must be retained for reassembly.

8. Remove the mounting and angle brackets from the relay damper.

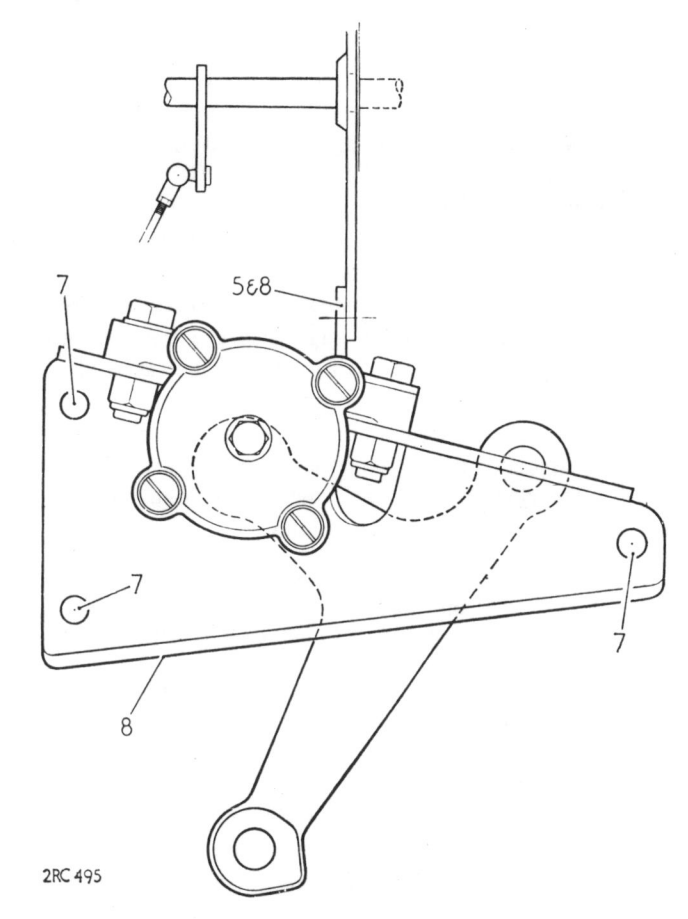

Refitting

9. If spacer plates were fitted between the relay mounting plate and the base unit, ensure that they are in place.
10. Ensure that the steering stop bracket is in place on the base unit.
11. Reverse the removal procedure, ensuring that the ball joints fit into the relay lever before securing brackets.
12. Fit new split pins to the slotted nuts.
13. Check, and if necessary reset, the steering lock stops. 57.65.03.
14. Check, and if necessary adjust, the front wheel alignment. 57.65.01.

    NOTE: If the steering relay is being refitted during the course of repairs to a vehicle that has been damaged in a motoring accident, the steering relay height must be checked. 76.10.57.

Rover 3500 and 3500S Manual AKM 3621

STEERING

TRACK ROD

—Remove and refit 1 to 7 and 12 to 15 57.55.09

SIDE ROD

—Remove and refit 8 to 15 57.55.17

Service tools: 565446 Ball joint extractor
606606 Ball joint extractor

Removing

1. Prop open the bonnet.
2. Remove the air cleaner. 19.10.01.
3. Jack up the front of the car to permit wheels to turn on lock.
4. Remove the split pins retaining the track rod ball joint securing nuts, and remove the nuts.
5. Release the two ball joints using extractor 565446, as follows:
   a. Ball joint, steering box end. Place steering on lock; loop extractor over drop arm, then move the steering back to the straight-ahead position while manoeuvring the extractor over the ball joint.
   b. Power steering or air conditioned models: If difficulty is experienced in locating extractor 565446 on to the track rod ball joint below the steering box, use instead the extractor 606606. With the extractor edge inserted between the drop arm and the ball joint, as illustrated, drive in to part arm from joint.
   c. Ball joint, relay end. Place steering straight-ahead; loop extractor over relay lever, then move the steering on to lock while manoeuvring the extractor over the ball joint.

   NOTE: It is necessary to remove the fixings and lower aside the torsion bar to facilitate removal of the ball joint.

6. Ensure that the extractor locates correctly over the ball joint before extracting the ball joint.
7. Remove the track rod from under the left-hand or right-hand wing valance.
8. For accessibility, jack up the front of the car, place on axle stands and remove the front wheel.
9. Remove the split pin, slotted nut and plain washer securing the side rod ball joint.
10. Using extractor 565446, release the ball joint.
11. Disconnect the side rod from the swivel pillar and withdraw the steering side rod.

Refitting

12. Reverse 1 to 11, noting the following.
13. If the ball joint rubber boots are worn or damaged, replace with new ones.
14. Fit new split pins to the slotted nuts.
15. Lower the car and reset the wheel alignment. 57.65.01.

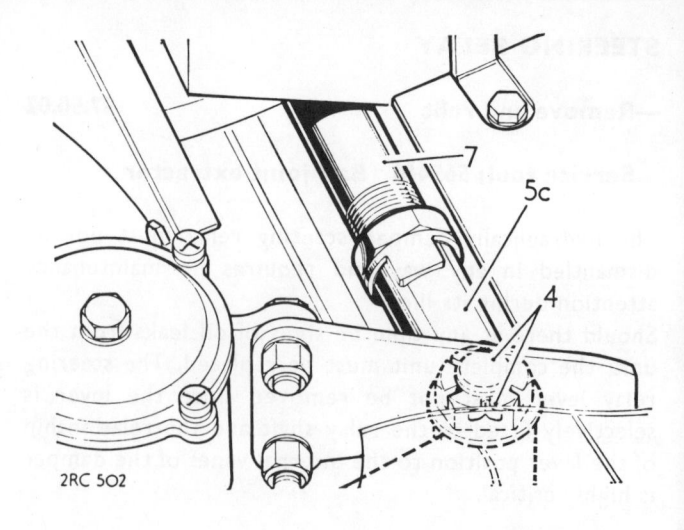

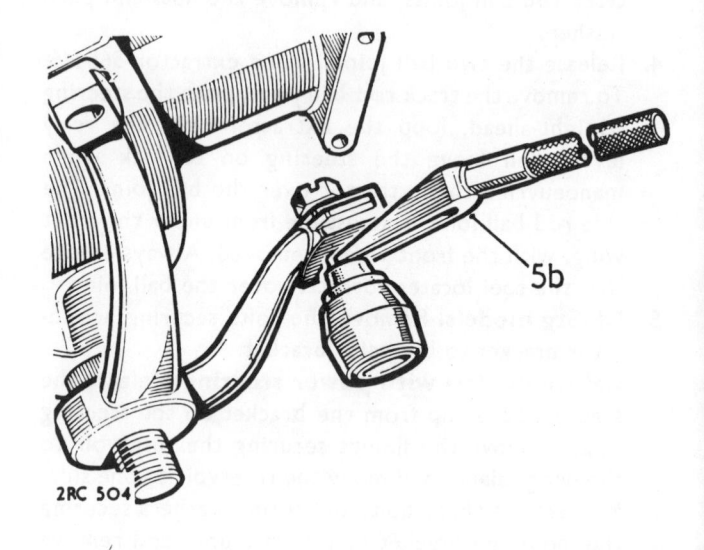

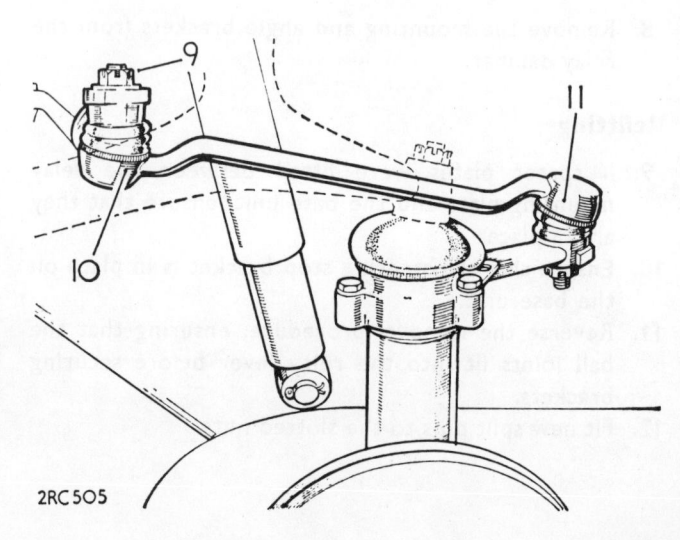

STEERING

FRONT WHEEL ALIGNMENT

—Check and adjust 57.65.01

Procedure

NOTE: The car must be on a flat and level surface for carrying out the Operation.

1. Check, and if necessary adjust, the tyre inflation pressures. Division 10.
2. Before commencing to check the front wheel alignment, first ensure that the wheels are in the straight-ahead position, then push the car forwards for a few feet.
3. Use a wheel alignment gauge and check the front wheel alignment, which should be 3,0 mm (0.125 in) toe in plus or minus 1,5 mm (0.062 in).
4. Air conditioned models: Remove the air cleaner to gain access to the special adjuster.
5. To adjust, slacken the locknuts at each end of the steering track rod at rear of engine and turn rod to obtain the correct alignment.
6. After each adjustment of the track rod, retighten the locknuts and move the car forwards for a short distance, before rechecking. It will be necessary to repeat this check three times to ensure accuracy of readings.
7. Ensure that the track rod ball joints are correctly aligned; that is, the top of the ball joint should be horizontal in the fore and aft position.
8. The camber of the front wheels is 0° with a tolerance of plus or minus 1°. The bottom of the wheels may splay out slightly.

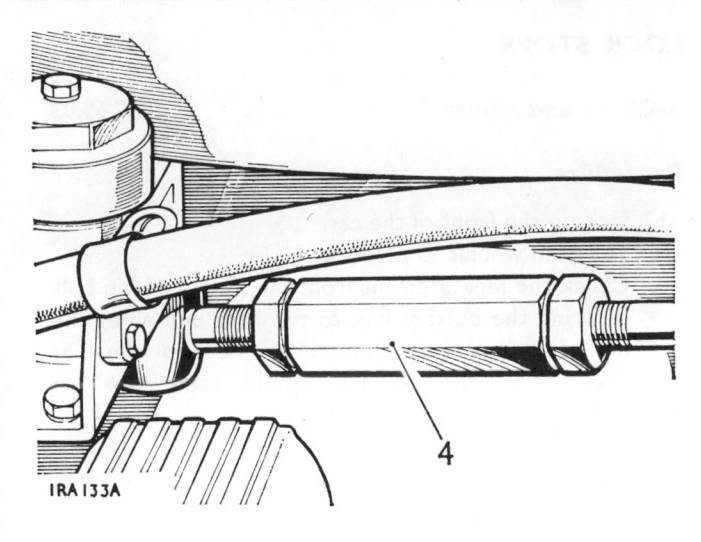

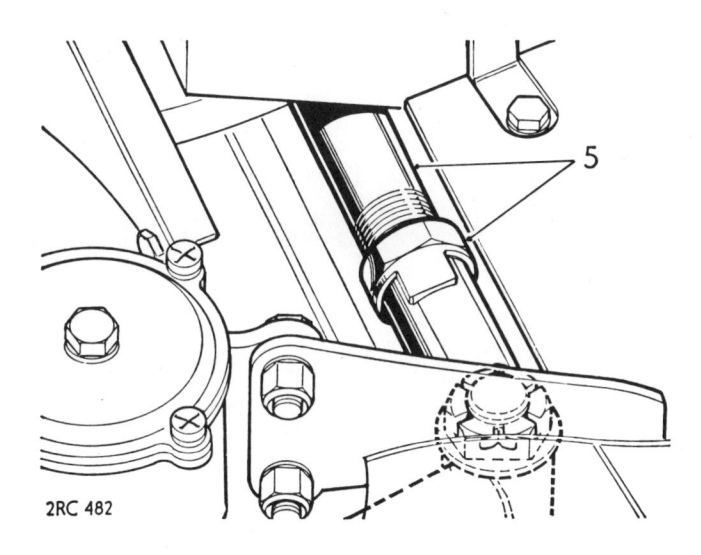

Rover 3500 and 3500S Manual AKM 3621

STEERING

LOCK STOPS

—Check and adjust 57.65.03

Procedure

1. Jack up the front of the car.
2. Turn the wheels to full lock.
3. Check the measurement from the centre of the bolt securing the bottom link to the base unit along the line of the link to the front rim of the road wheel, as illustrated. This distance should read 47,6 cm (18.750 in).
4. Repeat for the opposite lock.
5. Adjust the steering lock stops through the wing valance to obtain this setting.
6. Jack down the car.

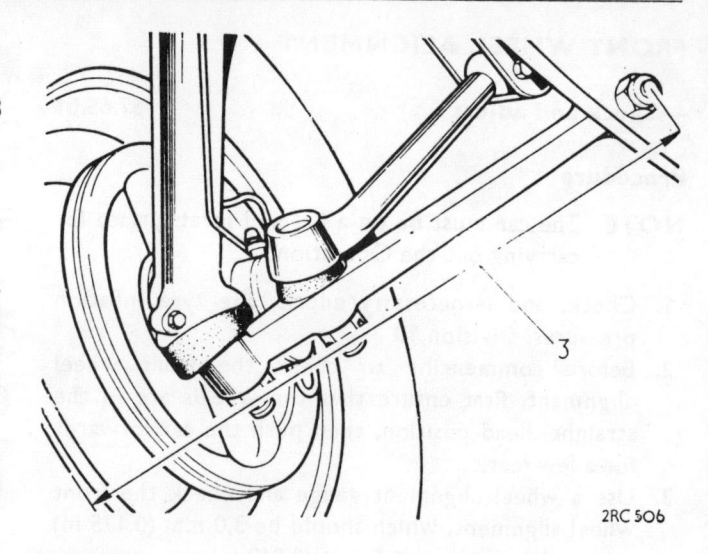

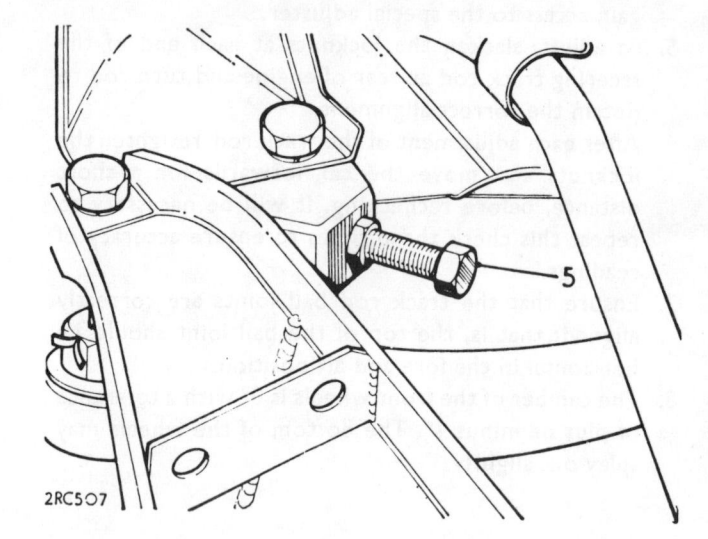

Rover 3500 and 3500S Manual AKM 3621

FRONT SUSPENSION

LIST OF OPERATIONS

Anti-roll bar—remove and refit	60.10.01
Ball joint, lower—remove and refit	60.15.03
Ball joint, upper—remove and refit	60.15.02
Hub assembly	
—remove and refit	60.25.01
—overhaul	60.25.07
Road spring assembly	
—remove and refit	60.20.01
—overhaul	60.20.07
Shock absorber—remove and refit	60.30.02
Suspension bottom link—remove and refit	60.40.02
Suspension top link—remove and refit	60.40.01
Swivel pillar assembly—remove and refit	60.15.30
Tie rod—remove and refit	60.40.09
Trim height—check and adjust	60.45.01

Rover 3500 and 3500S Manual AKM 3621

FRONT SUSPENSION

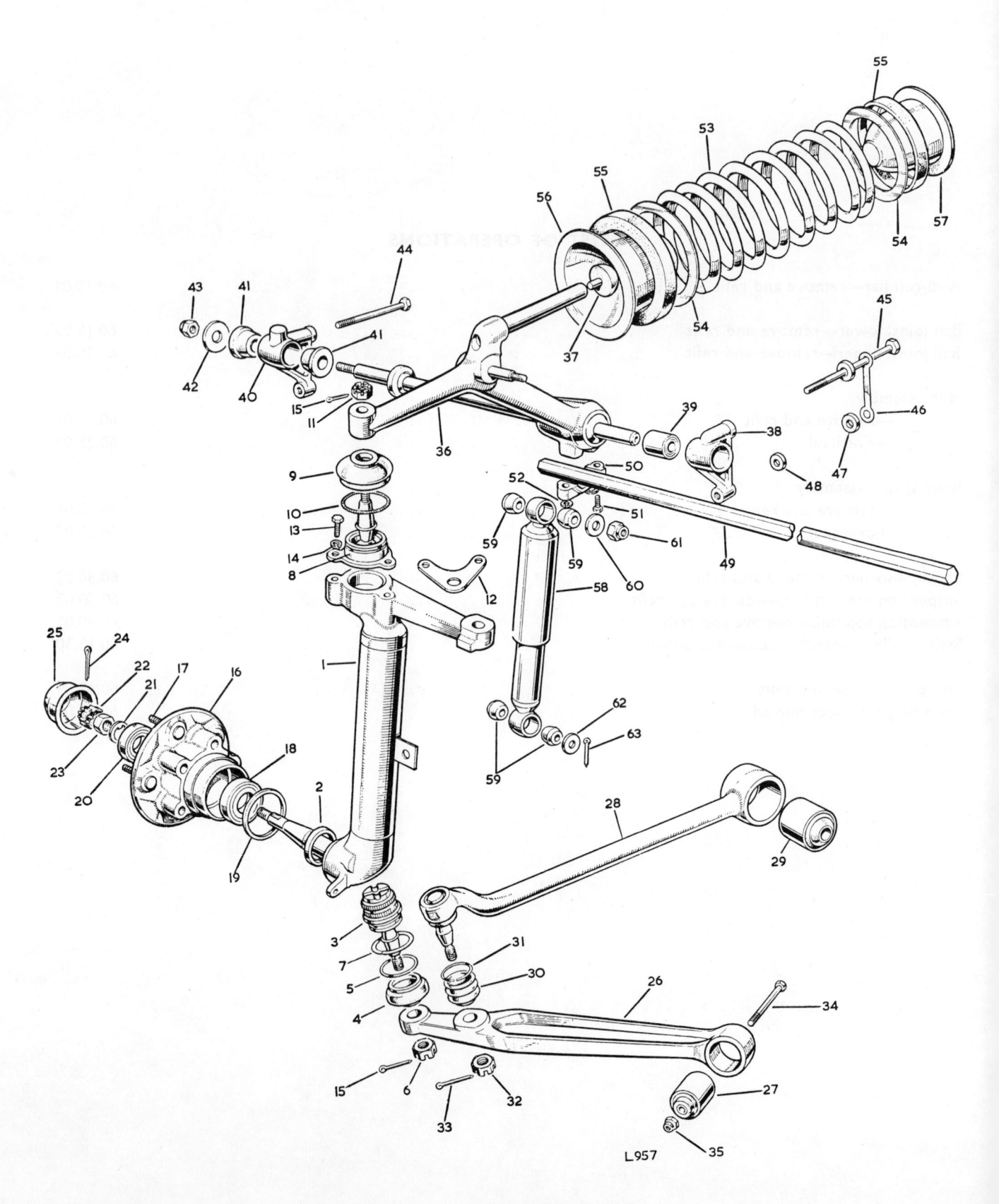

Front suspension and front hubs

Rover 3500 and 3500S Manual AKM 3621

FRONT SUSPENSION

Key to front suspension and front hubs

1 Swivel pillar assembly
2 Distance piece for front hub bearing
3 Bottom ball joint complete
4 Flexible boot for bottom ball joint
5 Retaining ring for flexible boot
6 Special slotted nut
7 Retaining ring for bottom ball joint
8 Top ball joint complete
9 Flexible boot for top ball joint
10 Retaining ring for flexible boot
11 Special slotted nut
12 Mounting plate for front brake hose
13 Set bolt ($\frac{1}{4}$ in UNF x $\frac{3}{4}$ in long) ⎫ Fixing top ball joint to swivel pillar
14 Spring washer ⎭
15 Split pin for top and bottom ball joint nuts
16 Front hub assembly
17 Stud for road wheel
18 Bearing for hub, inner
19 Oil seal for inner bearing
20 Bearing for hub, outer
21 Special washer ⎫
22 Locking cap ⎬ Fixing hub to stub axle
23 Special nut ($\frac{5}{8}$ in UNF) ⎪
24 Split pin ($\frac{5}{32}$ in diameter) ⎭
25 Hub cap
26 Bottom link assembly
27 Bush for bottom link
28 Bottom link strut assembly
29 Bush for bottom link strut
30 Rubber boot for ball joint
31 Retaining ring for boot
32 Special slotted nut fixing strut to link
33 Split pin fixing strut to link
34 Bolt ($\frac{7}{16}$ in UNF x $3\frac{1}{2}$ in long) ⎫ Fixing bottom links and struts to base unit
35 Self-locking nut ($\frac{7}{16}$ in UNF) ⎭
36 Top link assembly
37 Rubber-covered ball end for top link
38 Top link mounting bracket, inner
39 Bush for top link mounting bracket, inner
40 Top link mounting bracket, outer
41 Bush for top link mounting bracket, outer
42 Special washer ⎫ Fixing outer brackets to top links
43 Self-locking nut ($\frac{1}{2}$ in UNF) ⎭
44 Set bolt ($\frac{3}{8}$ in UNF x $5\frac{1}{4}$ in long), outer ⎫ Fixing top link brackets to base unit
45 Set bolt ($\frac{3}{8}$ in UNF x $7\frac{1}{4}$ in long), inner ⎪
46 Locking plate ⎬
47 Packing washer ⎪
48 Spacing washer ⎭
49 Anti-roll bar
50 Cap for anti-roll bar
51 Set bolt ($\frac{5}{16}$ in UNF x $1\frac{3}{8}$ in long) ⎫ Fixing caps to top links
52 Spring washer ⎭
53 Road spring, front
54 Shim for front road spring
55 Cushion for front road spring, front and rear
56 Support cup for front road spring
57 Bump rubber
58 Shock absorber, front
59 Rubber bush for front shock absorber
60 Plain washer ⎫ Fixing shock absorber to top link
61 Self-locking nut ($\frac{7}{16}$ in UNF) ⎭
62 Plain washer ⎫ Fixing shock absorber to base unit
63 Split pin ⎭

FRONT SUSPENSION

ANTI-ROLL BAR

—Remove and refit 60.10.01

Removing

1. Slacken the front road wheel nuts.
2. Jack up the front of the car and support on stands.
3. Remove the front road wheels.
4. Remove the four bolts and spring washers securing the anti-roll bar caps to the top links.
5. Withdraw the anti-roll bar from under the front wing.

Refitting

6. Reverse 1 to 5, ensuring that the contact surfaces of the anti-roll bar and the locating housings are clean. Tightening torque for anti-roll bar securing bolts: 4 kgf.m (30 lbf ft).

BALL JOINT, UPPER

—Remove and refit 60.15.02

Service tools: 511041 Ball joint boot replacer
601476 Ball joint extractor

Removing

1. Slacken the front road wheel nuts.
2. Jack up the front of the car and support on stands.
3. Remove the front road wheel.
4. Remove the split pin and slacken the slotted nut securing the top ball joint to the top link.
5. Remove the three bolts securing the brake hose mounting plate and top ball joint to the swivel pillar.
6. Using a suitable lever on the bottom link, hold the bottom link downwards and, using a drift, tap the top ball joint from swivel pillar.
7. Using 601476, break the taper of the top ball joint at the top link and remove nut and ball joint.

Refitting

8. Ensure that the bolt holes are aligned, then fit the top ball joint, mounting plate for the front brake hose and the three set bolts on to the swivel pillar.
9. Tap the ball joint into position while tightening the securing bolts.
10. Fit a new flexible boot, if required, using 511041.
11. Locate the top link in position and tighten the ball joint nut to 7,5 to 11,5 kgf.m (55 to 85 lbf ft).
12. Fit a new split pin to the slotted nut.
13. Reverse 1 to 3.

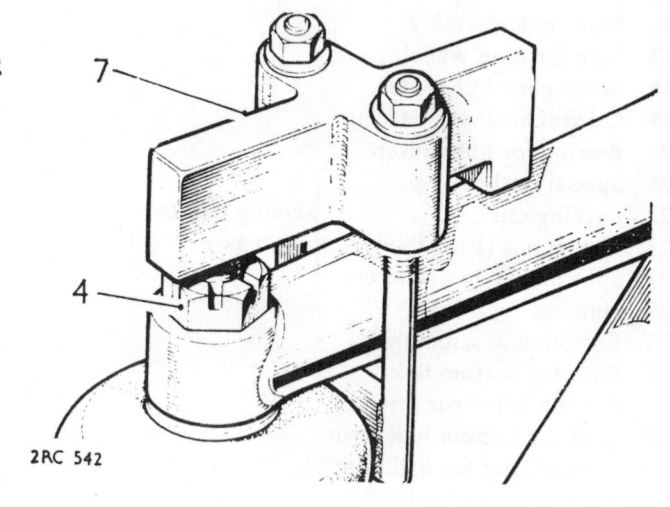

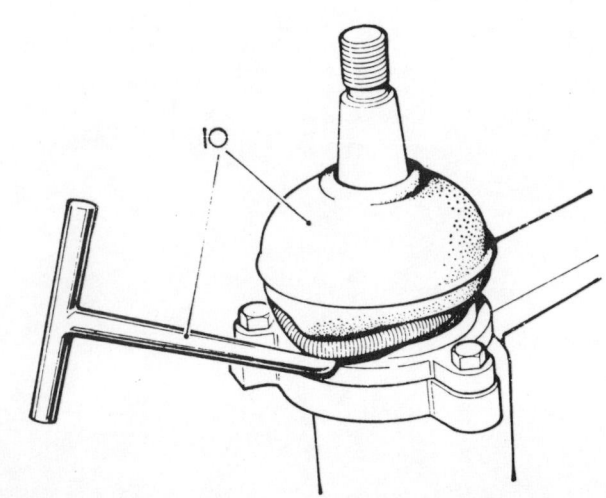

Rover 3500 and 3500S Manual AKM 3621

FRONT SUSPENSION

SWIVEL PILLAR ASSEMBLY

—Remove and refit 60.15.30

LOWER BALL JOINT

—Remove and refit 60.15.03

 Service tools: 511041 Ball joint boot replacer
       600962 Ball joint extractor
       601476 Ball joint extractor

Removing

1. Remove the front hub assembly. 60.25.01.
2. Where applicable, release the brake pipe and pad lead from the swivel pillar.
3. Remove the split pin and slacken the slotted nut from the bottom ball joint of the swivel pillar at the bottom link.
4. Using extractor 601476, break the taper of the ball joint, as shown in the illustration.
5. Remove the slotted nut from the swivel pillar bottom ball joint and separate the swivel pillar from the bottom link.
6. Remove the split pin from the slotted nut of the swivel pillar top ball joint at top link; also the split pin from the slotted nut of the steering side rod ball joint at the swivel pillar end. Slacken both nuts.
7. Using extractor 601476, break the taper of both ball joints and then remove the nuts.
8. Remove the swivel pillar complete.
9. Remove the flexible boot from the lower ball joint, using 511041.
10. Extract the retaining ring from the lower ball joint.
11. Extract the lower ball joint, using 600962.
12. If required, remove the upper ball joint. 60.15.02.

Refitting

13. Drift the bottom ball joint into the swivel pillar housing, using a suitable piece of tubing.
14. Fit the retaining ring for bottom ball joint.
15. Fit the flexible boot, using 511041.
16. Reverse 1 to 8, noting the following.
17. Tightening torque for upper ball joint nut: 7,5 to 11,5 kgf.m (55 to 85 lbf ft).
18. Tightening torque for lower ball joint nut: 8,5 to 10 kgf.m (60 to 75 lbf ft).

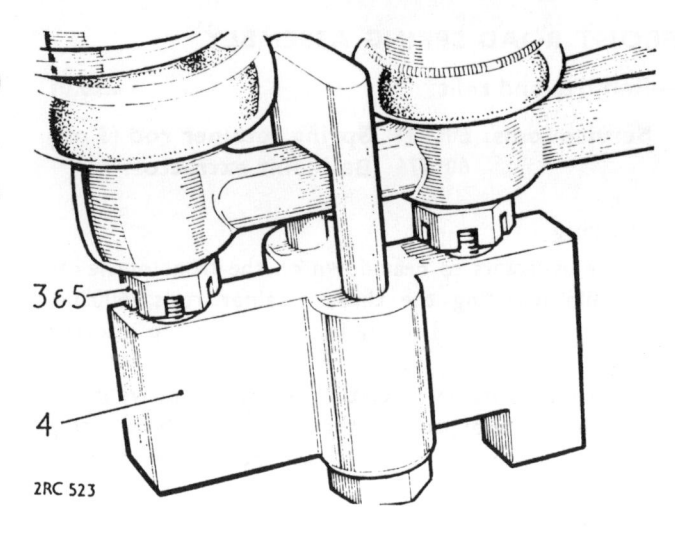

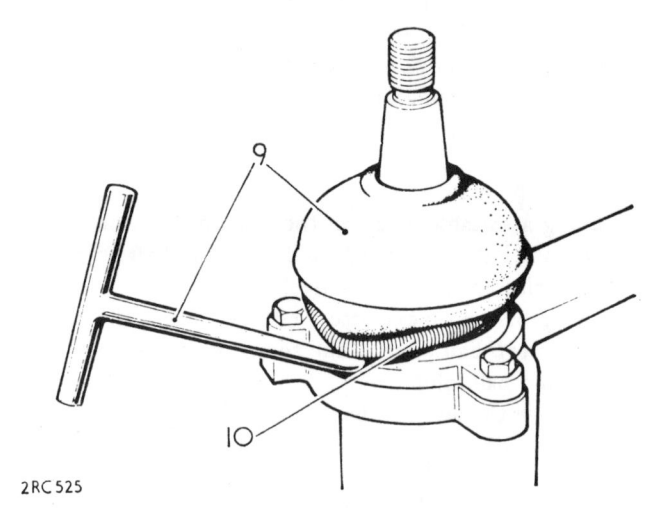

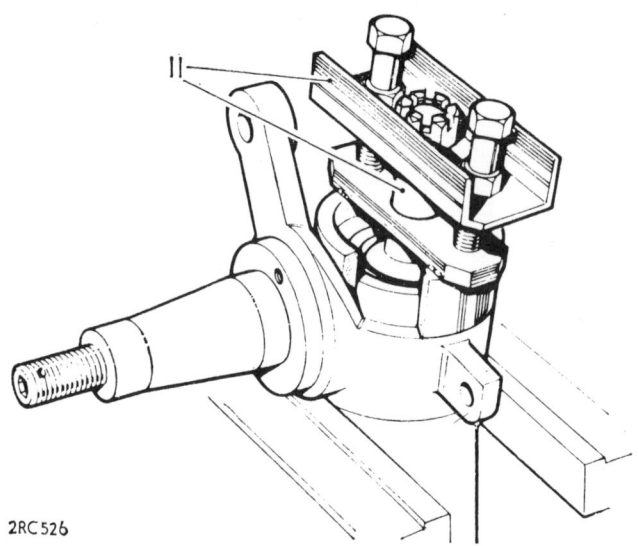

Rover 3500 and 3500S Manual AKM 3621

60.15.03
60.15.30

FRONT SUSPENSION

FRONT ROAD SPRING ASSEMBLY

—Remove and refit 60.20.01

Service tools: 600304 Spring retainer rod (3 off)
601476 Ball joint extractor

Removing

1. Use assistants to bear down on the front of the car while inserting the three retainer rods, 600304, through the road spring front support cup into the slots in the bump rubber rear support.
2. Turn the retainer rods through 90 to prevent the spring from expanding, then turn the whole spring round once to check that the retainer rods have correctly seated.
3. Slacken the front road wheel nuts.
4. Jack up the front of the car and support on stands.
5. Remove the front road wheel.
6. Remove the shock absorber top fixing.
7. Remove the split pin and slacken the slotted nut securing the upper ball joint to the top link.
8. Remove the three bolts securing the brake hose mounting plate and top ball joint to the swivel pillar.
9. Using a suitable lever on the bottom link, hold the bottom link downwards and, using a drift, tap the top ball joint from the swivel column.
10. Using ball joint extractor, 601476, break the taper of the top ball joint at the top link and remove the nut and ball joint.
11. Remove the anti-roll bar.
12. Remove the locker lid and cut and bend the bulkhead insulation panel, as illustrated, from inside the car. (RH side illustrated, LH side symmetrically opposite).

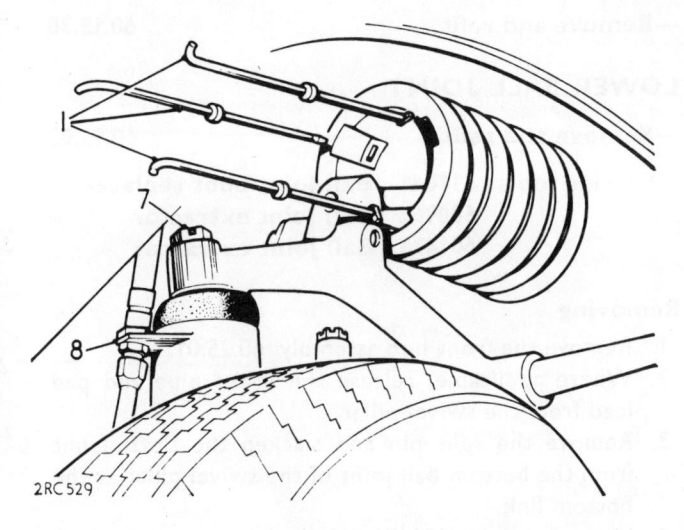

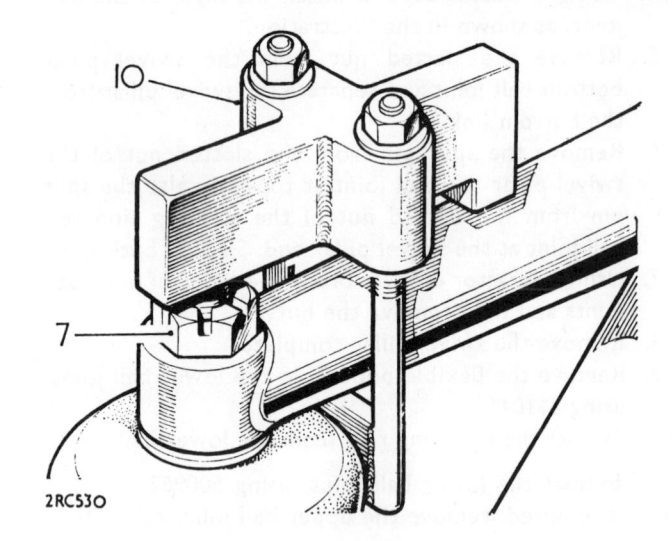

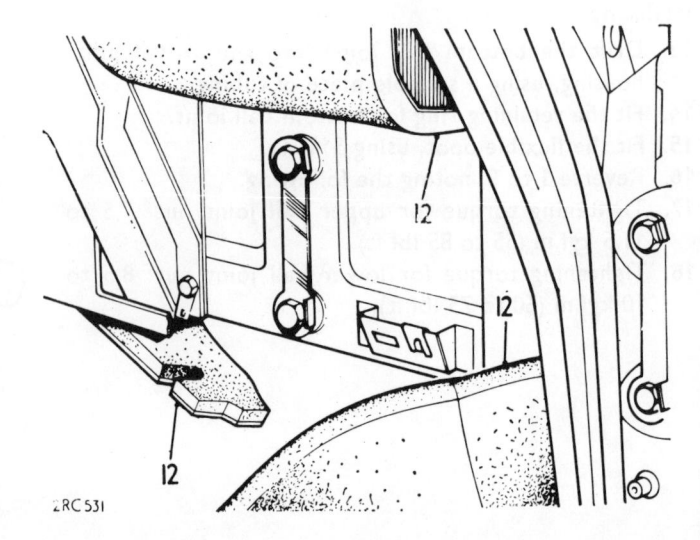

continued

Rover 3500 and 3500S Manual AKM 3621

FRONT SUSPENSION

13. Early models: Release the lockplates and remove the two bolts securing the top link inner mounting bracket.
14. Later models: Remove the locknuts securing the top link inner mounting bracket.
15. Remove the fixings securing the top link outer mounting bracket. These are accessible when the door is open.
16. Withdraw the top link assembly, stiffener, spacing washers and road spring.
17. Withdraw the road spring assembly.

Refitting

18. Reverse 1 to 17, noting the following.
19. The 12,5 mm (0.500 in) wide stripe painted along the full length of the road spring, must be uppermost when the spring is fitted.
20. If required, the later studs and locknuts can be fitted in place of the bolts and lockplates to secure the top link mounting brackets. If bolts are used, fit new lockplates. Tighten torque: 4,0 kgf.m (30 lbf ft).
21. Tightening torque for upper ball joint nut: 7,5 to 11,5 kgf.m (55 to 85 lbf ft) then fit a new split pin.
22. Before fitting the anti-roll bar caps, ensure that the road spring is seating correctly against the bulkhead location and remove the spring retaining rods, then secure the caps. Torque: 4,0 kgf.m (30 lbf ft).

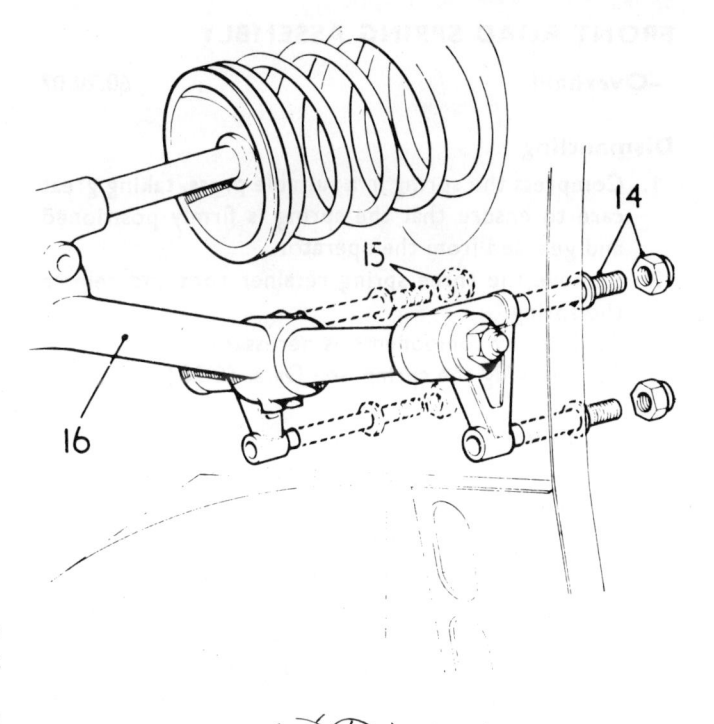

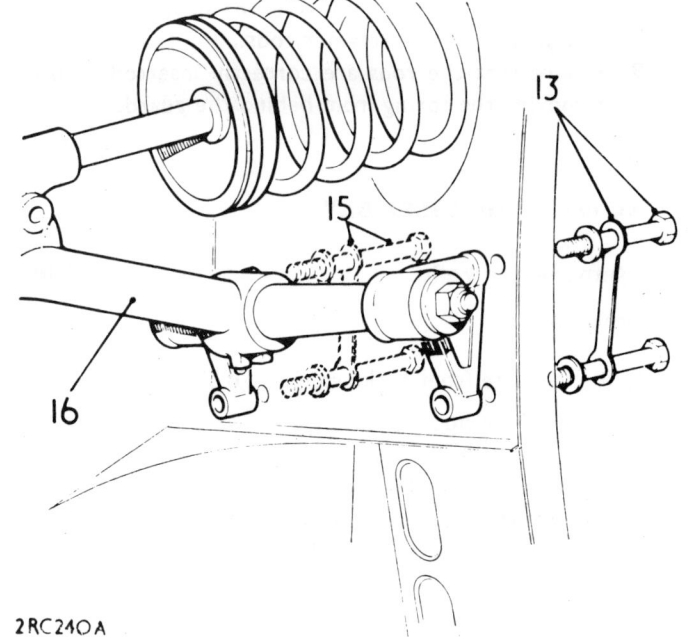

2RC240A

DATA

Front road spring

Manual steering models:
- Number of working coils — $6\frac{3}{8}$
- Free length — 413,5 mm (16.281 in)
- Rate — 30,35 kg/cm (170 lb/in)
- Identification — 12,5 mm (0.500 in) wide white stripe painted on entire length of spring

Power steering:
- Number of working coils — $7\frac{1}{3}$
- Free length — 434,8 mm (17.120 in)
- Rate — 30,35 kg/cm (170 lb/in)
- Identification — 12,5 mm (0.500 in) wide green stripe painted on entire length of spring

Rover 3500 and 3500S Manual AKM 3621

FRONT SUSPENSION

FRONT ROAD SPRING ASSEMBLY

—Overhaul 60.20.07

Dismantling

1. Compress the spring in a suitable press, taking great care to ensure that the spring is firmly positioned and guarded from the operator.
2. Remove the three spring retainer rods and release the spring.
3. Replace any components as necessary.
4. For checking the spring, see Data. 60.20.01.

Reassembling

5. Reassemble the shims, rubber cushions, bump rubber and support cap to the spring.
6. Ensure that the slots in the support cap and bump rubber for the retainer rods are in line.
7. Compress the spring, using a suitable press, taking great care to ensure that the spring is firmly positioned and guarded from the operator.
8. Insert the three retainer rods and turn through just 90°. Release the press, allowing the spring to be held compressed by the retainer rods.
9. Ensure that the rods are correctly inserted before removing the spring from behind the guard.

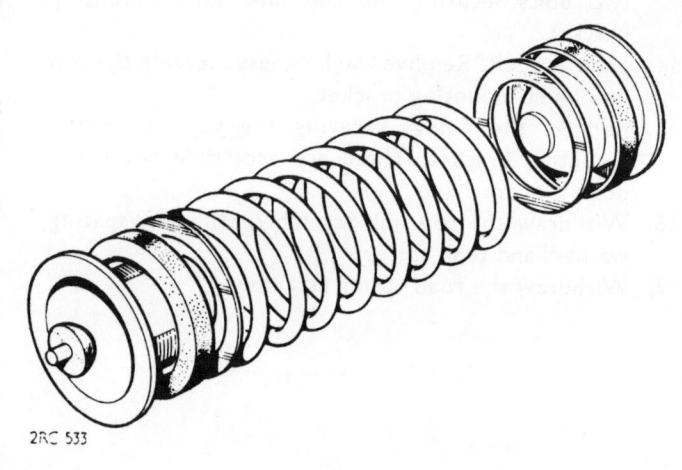

FRONT HUB ASSEMBLY

—Remove and refit. 60.25.01

Removing

1. Remove the front brake caliper. 70.55.02.
2. Prise off the hub cap.
3. Withdraw the split pin.
4. Withdraw the locking cap.
5. Remove the fixings and withdraw the hub.
6. If required remove the brake disc.

Refitting

7. Ensure that the oil seal, distance piece and bearings are in position.
8. Pack the hub bearings with fresh grease of the recommended grade. See Division O9.
9. Fit the hub and adjust the nut to give zero end float.
10. Mount a dial test indicator to read off the brake disc.
11. Check the brake disc run-out at the outer edge, if it exceeds 0,07 mm (0.003 in.), reposition the disc on the hub.
12. Reposition the dial test indicator to read the hub end float.
13. Re-adjust the hub nut to give 0,07 mm to 0,12 mm (0.003 in. to 0.005 in.) end float.
14. Place the locking cap on to the nut so that the split pin can be inserted through the serrations without disturbing the position of the hub nut.
15. Fit a new split pin.
16. Pack the hub cap with grease, as recommended for the hub bearings, and fit the cap in position.
17. Refit the brake caliper. 70.55.02.

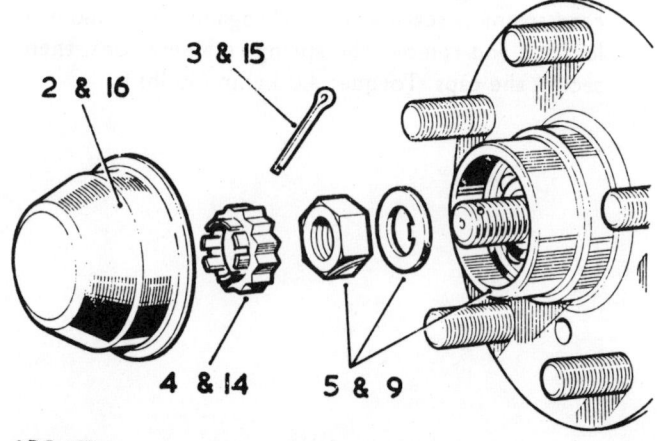

DATA

Brake disc run out limits	0,07 mm (0.003 in.) maximum
Hub end float	0,07 mm to 0,12 mm (0.003 to 0.005 in.)

Rover 3500 and 3500S Manual AKM 3621

FRONT SUSPENSION

FRONT HUB ASSEMBLY

—Overhaul 60.25.07

Dismantling

1. Remove the front hub assembly. 60.25.01.
2. Withdraw the distance piece from the oil seal.
3. Remove the oil seal.
4. Press out the bearings.

Inspecting

5. Clean all components and examine for wear.
6. Obtain new bearings and oil seal.

Reassembling

7. Reverse 1 to 6.

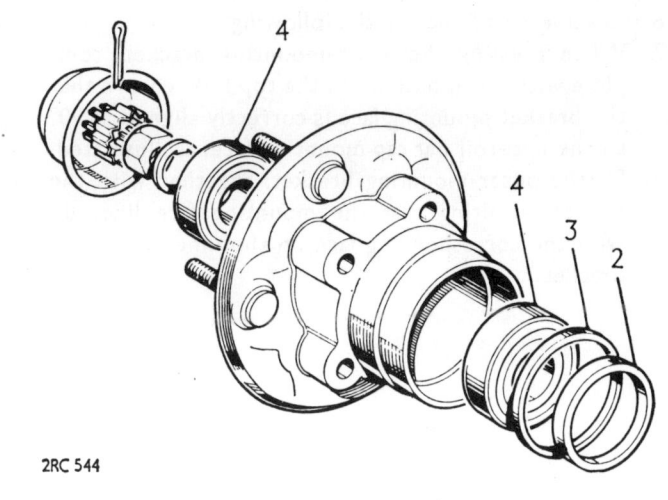

FRONT SHOCK ABSORBER

—Remove and refit 60.30.02

Service tool: 605227 Compressor

Removing

1. Slacken the front road wheel nuts.
2. Jack up the front of the car and support on stands.
3. Remove the front road wheel.
4. Jack up the suspension to take the weight off the shock absorber.
5. Remove the split pin from the shock absorber bottom mounting, using 605227.
6. Remove the washer and outer rubber bushes.
7. Remove the nut washer and outer rubber bush from the top fixing.
8. Remove the shock absorber and inner rubber bush.

Refitting

9. Reverse 1 to 8, using a new split pin. If required, fit new rubber bushes.

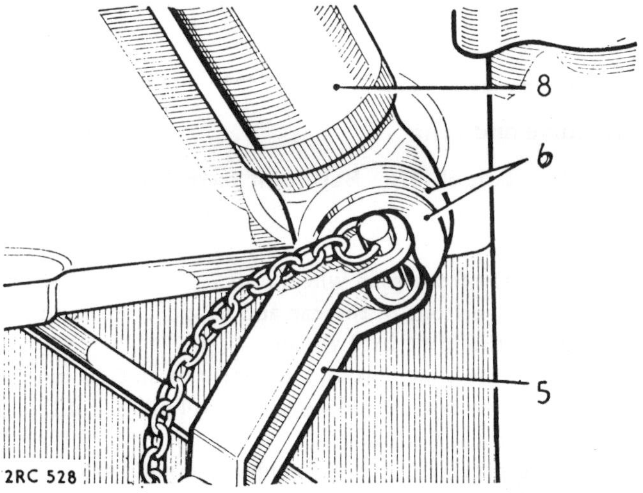

FRONT SUSPENSION TOP LINK

—Remove and refit 60.40.01

Service tools: 600304 Spring retainer rod (3 off)
601476 Ball joint extractor

Removing

1. Remove the front road spring assembly. 60.20.01.
2. Withdraw the top link assembly.
3. Remove the self-locking nut and plain washer securing the top link to the outer mounting bracket.
4. Withdraw the outer mounting bracket together with the two bushes.
5. Press off the inner mounting bracket and bush.

continued

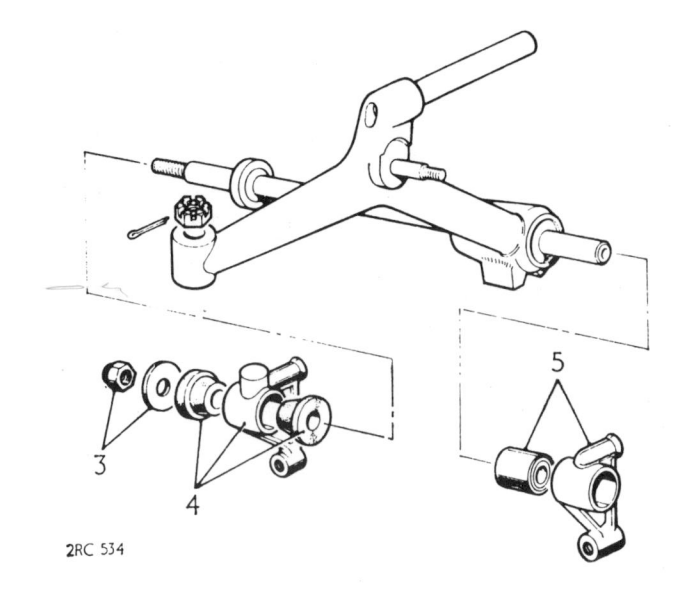

Rover 3500 and 3500S Manual AKM 3621

60.25.07
60.40.01
Sheet 1

FRONT SUSPENSION

Refitting

6. Reverse 1 to 5, noting the following.
7. When pressing the inner mounting bracket, complete with bush, back on to the top link, ensure that the bracket mounting face is correctly aligned at 90° to the anti-roll bar cap mounting face, as illustrated.
8. Fit the outer mounting bracket, together with the bushes, ensuring that the mounting face lines up with the corresponding face on the inner mounting bracket.

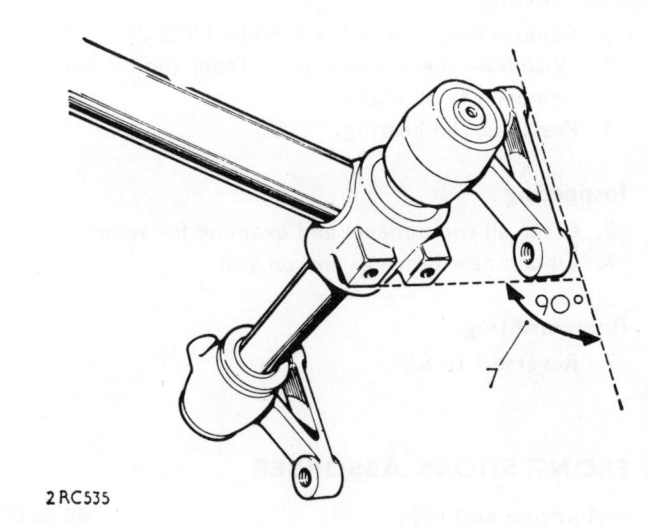

FRONT SUSPENSION BOTTOM LINK

—Remove and refit   60.40.02

Service tool: 601476   Ball joint extractor

Removing

1. Slacken the front road wheel nuts.
2. Jack up the front of the car and support on stands.
3. Remove the front road wheel.
4. Remove the split pins and slacken the slotted nuts securing the bottom link to the swivel pillar and the tie rod to the bottom link.
5. Break the taper of the ball joints, using 601476, and remove the nuts.
6. Remove the self-locking nut and bolt securing the bottom link to the base unit.
7. Withdraw the bottom link.

Refitting

8. Reverse 1 to 7, noting the following.
9. If required, fit a new bush to the bottom link.
10. Do not fully tighten the nut and bolt securing the bottom link to the base unit until the road wheel has been fitted and the car is in the static unladen condition. Then tighten to 7,5 kgf.m (54 lbf ft).
11. Tightening torque for ball joint nuts: 8,5 to 10,0 kgf.m (60 to 75 lbf ft).
12. Fit new split pins to the slotted nuts.

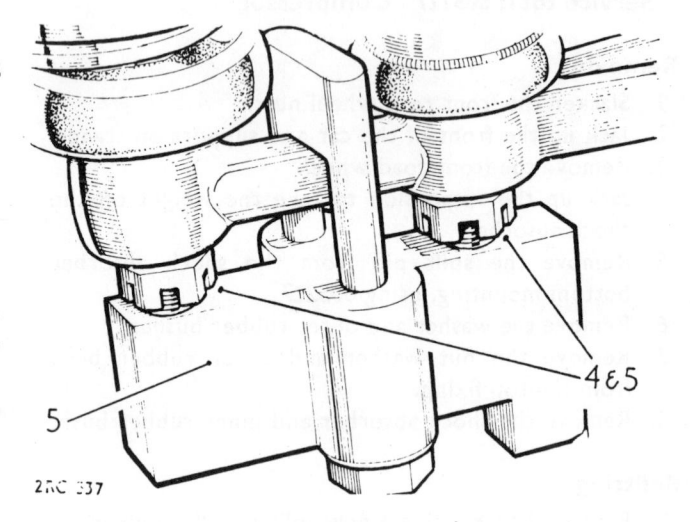

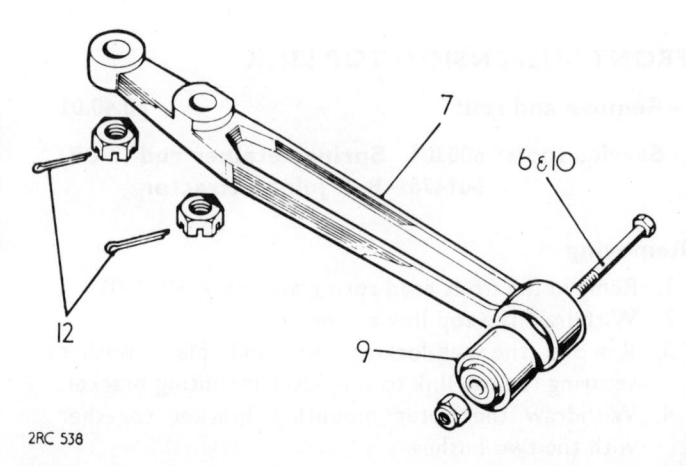

Rover 3500 and 3500S Manual AKM 3621

FRONT SUSPENSION

TIE ROD

—Remove and refit 60.40.09

Service tool: 601476 Ball joint extractor

Removing

1. Slacken the front road wheel nuts.
2. Jack up the front of the car and support on stands.
3. Remove the front road wheel.
4. Remove the split pin and slacken the slotted nut securing the tie rod to the bottom link.
5. Break the taper fit of the ball joint, using 601476, and remove the slotted nut.
6. Remove the nut and bolt securing the tie rod to the base unit.
7. Withdraw the tie rod.

Refitting

8. Reverse 1 to 7, noting the following.
9. If required, fit a new bush to the tie rod.
10. Do not fully tighten the nut and bolt securing the tie rod to the base unit until the road wheel has been fitted and the car is in the static unladen condition, then tighten to 7,5 kgf.m (54 lbf ft).
11. Tightening torque for the ball joint nut: 8,5 to 10,0 kgf.m (60 to 75 lbf ft).
12. Fit a new split pin to the slotted nut.

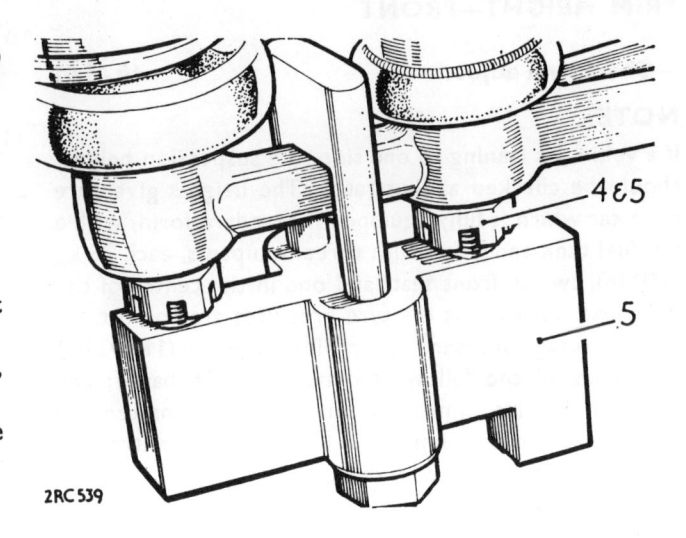

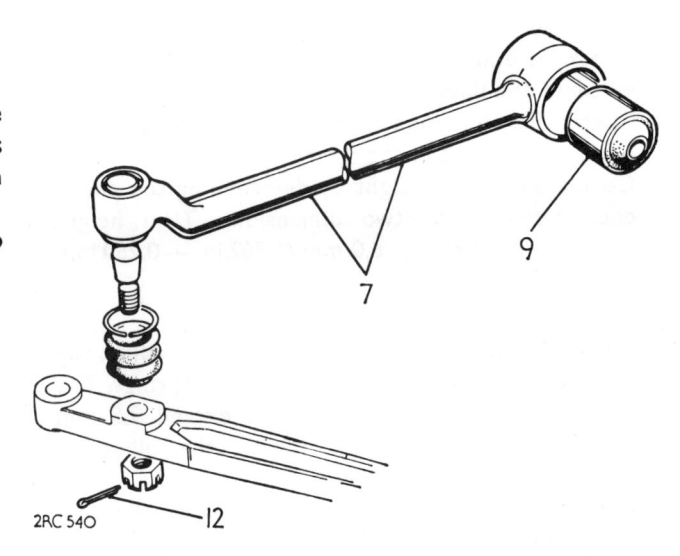

Rover 3500 and 3500S Manual AKM 3621

60.40.09

FRONT SUSPENSION

TRIM HEIGHT—FRONT

—Check and adjust   60.45.01

NOTE:
If a vehicle is leaning to one side, the suspension heights should be checked as illustrated. The heights given are for a car which is fully equipped in standard form, with a full fuel tank and laden with three occupants, each 76 kg (168 lb), two in front seats and one in the centre of the rear seat. Check that the tyre pressures are correct.

The average standard tyre radius is 294 mm (11.593 in.) on which all the following dimensions are based; any variation in this dimension must be compensated for when checking the dimension between the centre line of the bottom link pivot and the floor.

Checking

1. Stand the car on a level floor.
2. Check the dimension from the centre line of the road wheel to the floor.
3. Check the dimension from the centre line of the bottom link pivot to the floor.
4. Calculate the trim height by determining the difference between the two dimensions. Trim height limits are 39,7 mm ± 6,0 mm (1.562 in. ± 0.250 in.).

Adjusting

5. If the trim height is not within the limits, adjust the thickness of shims fitted under the coil springs. The total thickness of shims must not exceed 9,5 mm (0.375 in.), otherwise the location of the coil spring would be adversely affected.
6. If the foregoing procedure fails to rectify the trim height, fit new coil springs.

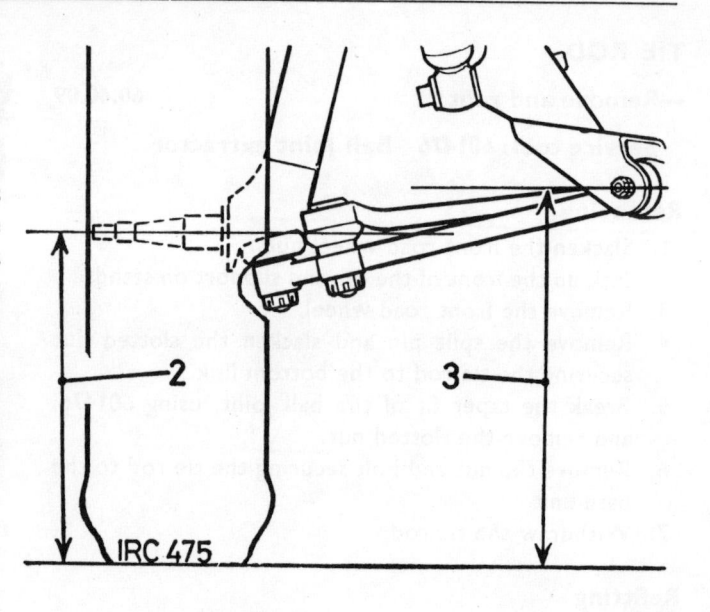

REAR SUSPENSION

LIST OF OPERATIONS

De Dion tube
    —remove and refit 64.10.01
    —overhaul 64.10.07

Hub assembly
    —remove and refit 64.15.10
    —overhaul 64.15.07

Road spring—remove and refit 64.20.01

Shock absorber—remove and refit 64.30.02

Stabiliser rod 64.35.09

Suspension link
    —lower—remove and refit 64.35.02
    —upper—remove and refit 64.35.01

Trim height—check and adjust 64.25.12

Rover 3500 and 3500S Manual AKM 3621

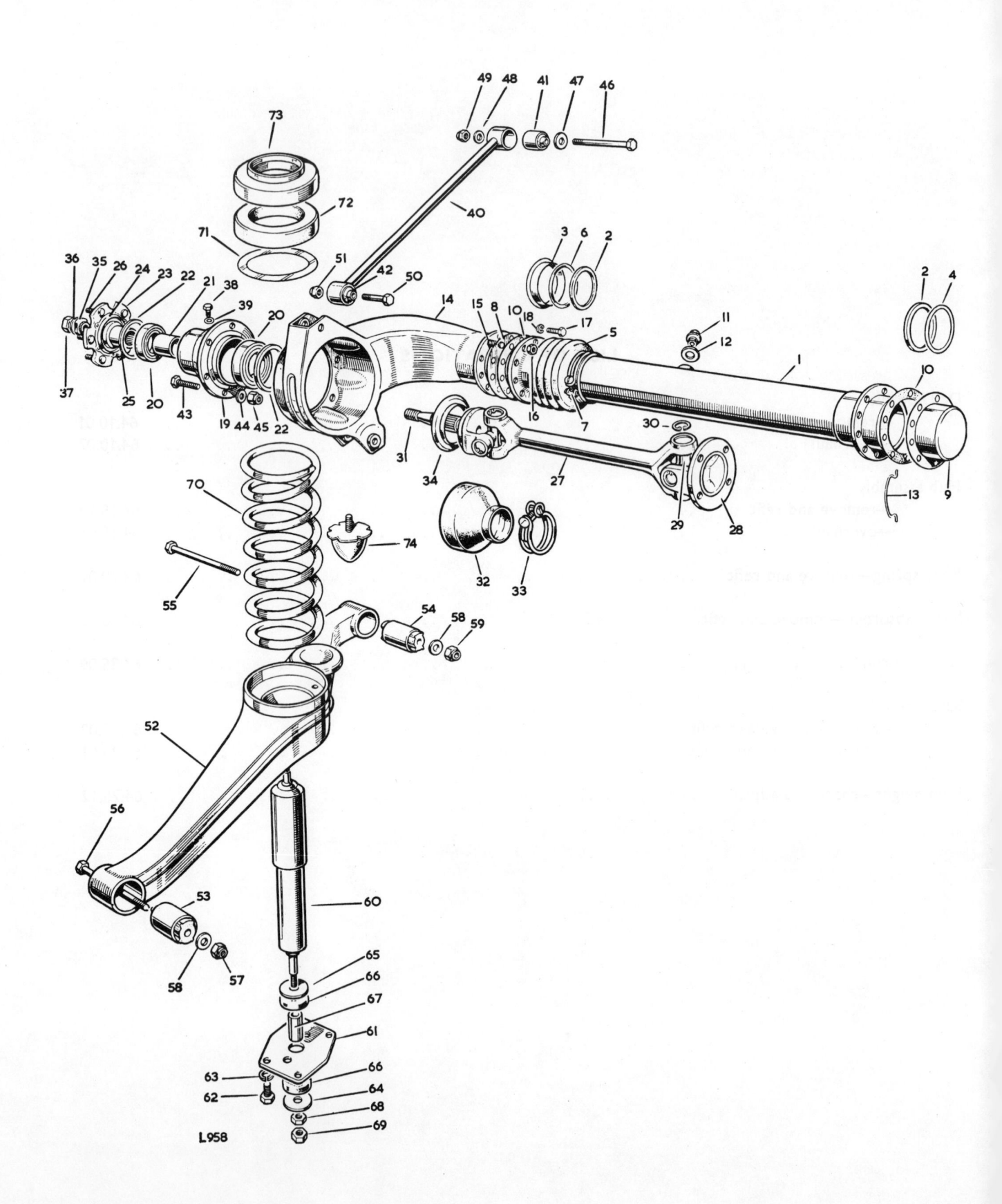

REAR SUSPENSION

64-2   Rover 3500 and 3500S Manual AKM 3621

REAR SUSPENSION

Key to rear suspension, rear hubs and drive shafts

1 De Dion tube assembly
2 Oil seal for de Dion tube
3 Retainer for RH oil seal
4 Retainer for LH oil seal
5 Dust excluder for de Dion tube
6 Garter spring for dust excluder
7 Clip for dust excluder
8 Blanking plate, RH, for de Dion tube
9 Blanking plate, LH, for de Dion tube
10 Joint washer for blanking plates
11 Filler plug for de Dion tube
12 Joint washer for filler plug
13 Spring clip for inner tube
14 De Dion elbow assembly, RH
15 Stud for de Dion elbow
16 Nut ($\frac{1}{4}$ in UNF) ⎫ Fixing elbows
17 Set bolt ($\frac{1}{4}$ in UNF x $\frac{3}{4}$ in long) ⎬ to de Dion
18 Spring washer ⎭ tube
19 Bearing housing for rear hub
20 Bearing for rear hub
21 Collapsible spacer for rear hub bearings
22 Oil seal for rear hub
23 Driving flange assembly
24 Dust excluder, outer
25 Drive screw fixing dust excluder
26 Stud for road wheel
27 Drive shaft assembly for rear hub
28 Flange yoke for drive shaft
29 Journal complete for drive shaft
30 Circlip for journal
31 Yoke shaft for rear hub
32 PVC shield for outer journal
33 Clip fixing shield to drive shaft
34 Dust excluder, inner
35 Special washer ⎫
36 Lockwasher ⎬ For yoke shaft
37 Special nut ($\frac{3}{4}$ in UNF) ⎭
38 Plug for rear hub
39 Joint washer for plug
40 Top link assembly
41 Bush for top link, rear
42 Bush for top link, front
43 Bolt ($\frac{5}{16}$ in UNF x $1\frac{1}{8}$ in long) ⎫ Fixing rear
44 Plain washer ⎬ hubs to
45 Self-locking nut ($\frac{5}{16}$ in UNF) ⎭ de Dion elbows
46 Bolt ($\frac{7}{16}$ in UNF x $3\frac{3}{4}$ in long) ⎫
47 Plain washer, large ⎬ Fixing top link
48 Plain washer, dished ⎬ to base unit
49 Self-locking nut ($\frac{7}{16}$ in UNF) ⎭
50 Bolt ($\frac{7}{16}$ in UNF x $2\frac{3}{4}$ in long) ⎫ Fixing top link
51 Self-locking nut ($\frac{7}{16}$ in UNF) ⎭ to de Dion elbow
52 Bottom link assembly, RH
53 Bush for bottom link, short
54 Bush for bottom link, long
55 Bolt ($\frac{7}{16}$ in UNF x $4\frac{1}{4}$ in long), rear ⎫ Fixing
56 Special bolt, front ⎬ bottom
57 Self-locking nut ⎬ links to base
58 Plain washer ⎬ unit and
59 Self-locking nut ($\frac{7}{16}$ in UNF) ⎭ de Dion elbow
60 Shock absorber, rear
61 Lower mounting for rear shock absorbers
62 Set bolt ⎫ Fixing
 ($\frac{5}{16}$ in UNF x $\frac{3}{4}$ in long) ⎬ mountings to
63 Spring washer ⎭ bottom links
64 Washer, bottom ⎫
65 Washer, top ⎬ Fixing rear shock
66 Rubber cushion ⎬ absorbers to
67 Sleeve ⎬ base unit and
68 Nut ($\frac{3}{8}$ in UNF) ⎬ lower mountings
69 Locknut ($\frac{3}{8}$ in UNF) ⎭
70 Road spring, rear
71 Shim for rear road spring
72 Cushion for rear road spring, top and bottom
73 Top support cup for rear road spring
74 Bump rubber

REAR SUSPENSION

GENERAL SERVICE INFORMATION

MODIFIED REAR SUSPENSION

The rear suspension fitted to early 3500 models is superseded by a modified version, which is fitted to latest 3500 and all 3500S models.

The modification comprises:

a The final drive unit, rear drive flanges, drive shafts and propeller shaft balanced to closer limits.

b Front and rear final drive mountings modified.

c Static laden car height reduced by 6 mm (0.250 in) by fitting shorter springs in conjunction with shims.

Part numbers of modified components:

Component	Qty	Part No.
Final drive assembly extension casing and damper complete, for final drive	1	606255
Mounting bracket, front, for final drive	1	572638
Bolt ($\frac{7}{16}$ in UNF x $5\frac{1}{2}$ in long) — Fixing flexible mounting to extension casing	1	256078
Plain washer	1	570352
Distance piece	1	572615
Support strap	1	572589
Dished washer	2	527394
Self-locking nut ($\frac{7}{16}$ in UNF)	1	252163
Extension casing for final drive	1	576486
Damper plate for extension casing	1	576528
Clip fixing damper plate	2	517585
Mounting bracket, rear, for final drive	1	572953
Propeller shaft assembly	1	576204
Shims for rear road springs	as reqd.	534738
Road spring, rear	2	572967
Dished washer — Fixing final drive to rear flexible mountings	4	527394
Plain washer	2	3680
Rear hub and drive shaft assembly	2	572962
Driving flange assembly	2	572964
Drive shaft assembly for rear hub	2	572966

Commencing numbers:

Suffix letter change: Final drive unit serial number from suffix letter 'B' onwards. Chassis serial number from suffix letter 'C' onwards.

Earlier cars can be modified by fitting all the parts listed. The removal and fitting procedure is as detailed in this Manual, except that the rear road springs must be fitted with the appropriate number of shims to give the rear suspension height. See Operation 64.25.12.

When fitting replacement springs, assemble initially with two shims. Check the suspension height and correct the number of shims as required. The latest rear hub drive shafts, hub driving flanges and rear road springs are completely interchangeable with the previous type and will be supplied for all Service replacements.

Early parts, however, must not be fitted to cars from final drive suffix 'B' and chassis suffix letter 'C' onwards, as this will increase the amount of unbalance beyond the close limits required.

REAR SUSPENSION

GENERAL SERVICE INFORMATION

IMPORTANT: Do not jack under the de Dion tube, use the rear jacking point, as illustrated.

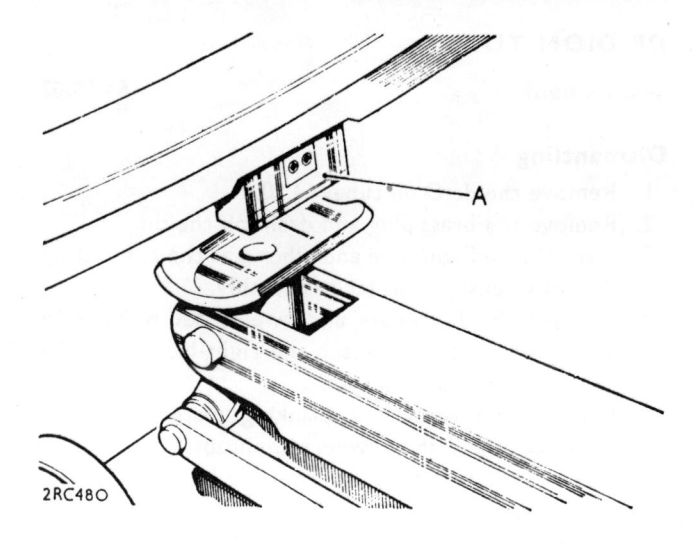

Rear jacking point

A—Jacking bracket

DE DION TUBE
—Remove and refit 64.10.01

Removing
1. Slacken the rear road wheel nuts.
2. Jack up the rear of the car and support with stands under each lower suspension link, below the coil spring.
3. Remove the rear road wheels.
4. Remove the rear hubs and external drive shafts. 64.15.10.
5. Remove the bolts securing the upper links to the de Dion tube, and remove links from elbow.
6. Slacken the bolts securing the lower links to the de Dion tube and allow the de Dion tube to pivot downwards. Remove the bolts.
7. Remove the de Dion tube complete.

Refitting
8. Reverse 1 to 7, but do not fully tighten the fixings for the upper and lower suspension links until the road wheels are fitted and the car is in the static unladen condition. Then tighten to 7,5 kgf.m (54 lbf ft).
9. Tightening torque for wheel hub to de Dion tube bolts: 2,7 kgf.m (20 lbf ft).
10. Tightening torque for drive shaft to brake disc bolts: 11,5 kgf.m (85 lbf ft).

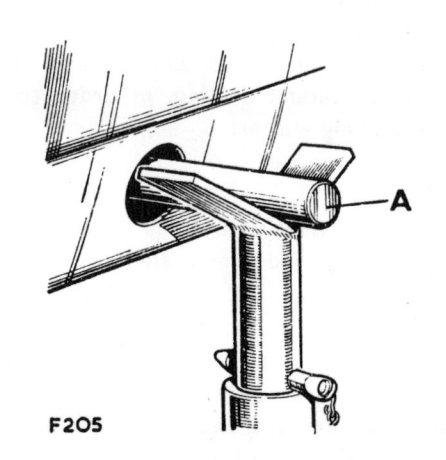

Position of support stands

A—Suitable length of 22 mm ($\frac{7}{8}$ in) diameter rod inserted in jacking tube

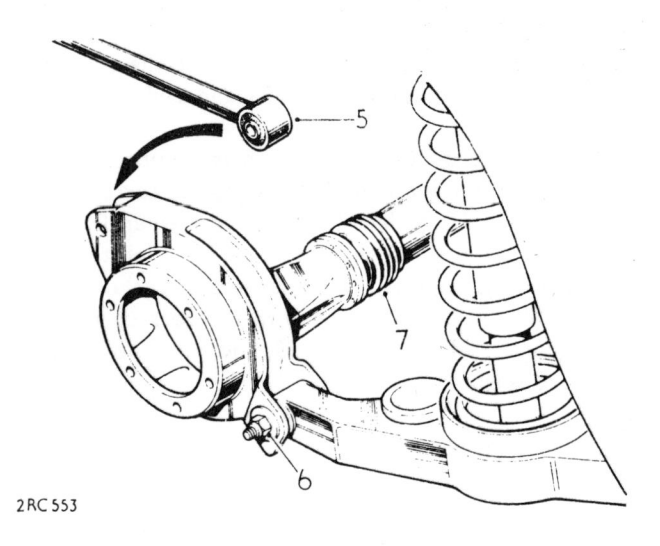

Rover 3500 and 3500S Manual AKM 3621

Rear Suspension

DE DION TUBE

—Overhaul 64.10.07

Dismantling

1. Remove the de Dion tube. 64.10.01.
2. Remove the brass plug and drain off the oil.
3. Mark the de Dion tube and elbows at the connecting flanges to ensure correct reassembly.
4. Remove the four nuts and twelve set bolts with spring washers, and detach the right-hand and left-hand de Dion tube elbows.
5. Remove the respective blanking plates, together with packing washers, where applicable.
6. Remove the retaining clip from the left-hand end of the inner tube.
7. Loosen the dust excluder clip and ease the dust excluder and garter spring from its location.
8. Remove the inner tube assembly from the outer.
9. Remove the dust excluder, garter spring and clip.
10. At this stage the left-hand seal, packing washer, where applicable, and retainer may be removed by prising it out of its housing.
11. Then the right-hand seal and retainer must be removed, by careful drifting, in order to replace seal and packing washer.

Inspecting

12. Examine all parts for wear.
13. Obtain new seals and joint washer.

Reassembling

14. Slide the gaiter, garter spring and clip on to the outer tube, right-hand side.
15. Carefully fit a new seal and flexible packing washer, where applicable, into the retainer and smear with MS4 silicone grease, then fit the seal and retainer to the right-hand end.
16. Fit a seal and retainer to the left-hand end of the outer tube.
17. Refit the respective blanking plate and cup, using a suitable jointing compound.
18. Replace the inner tube assembly into the outer tube.
19. Fit the retaining clip to left-hand end of the inner tube.
20. Refit the de Dion tube elbows, locating the four studs and tightening the bolts to the correct torque of 1,0 kgf.m (8 lbf ft). Ensure that when fitting the right-hand elbow the dust excluder is located before the bolts are fitted; tighten the dust excluder clip.
21. Refill the de Dion tube with the correct grade and quantity of oil. Division 09, and refit the filler plug and washer.

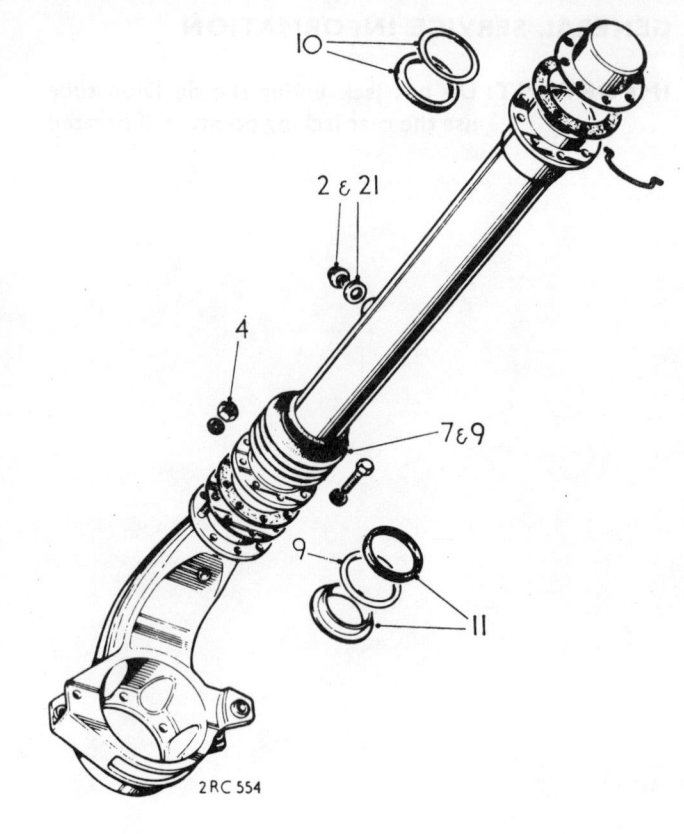

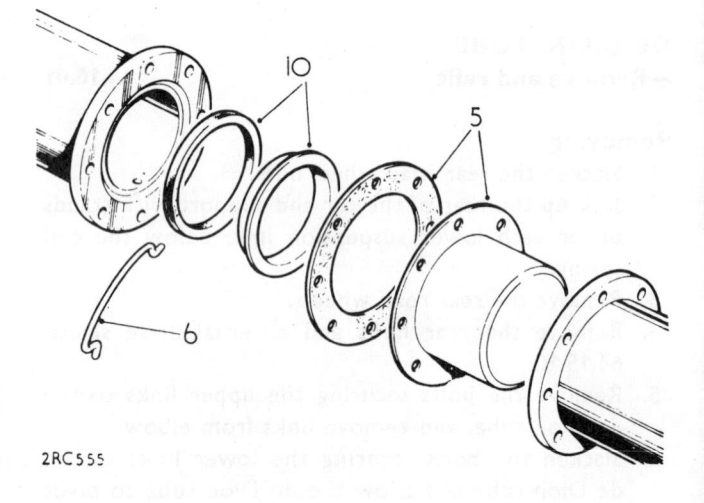

Rear Suspension

REAR HUB AND EXTERNAL DRIVE SHAFT ASSEMBLY

—Remove and refit 64.15.10

Removing

1. Loosen the road wheel nuts.
2. Jack up the car and support on stands at jacking points at each side of car.
3. Remove the road wheel.
4. Remove the four bolts and lock plates securing the drive shaft flange yoke to the differential drive shaft.
5. Remove the six bolts and self-locking nuts securing the hub bearing housing to the de Dion tube assembly.
6. Withdraw the hub and drive shaft assembly.

Refitting

7. Offer the hub and external drive shaft assembly on to the de Dion tube, at the same time locating the drive shaft flange on to the brake disc and differential drive shaft.
8. Secure the hub bearing housing to the de Dion tube with six bolts and nuts, and tighten to 2,7 kgf.m (20 lbf ft).
9. Secure the drive shaft flange to the brake disc and differential drive shaft with four bolts and new locking washers, and tighten to 11,5 kgf.m (85 lbf ft). Ensure that when fitted there is a clearance of at least 0,2 mm (0.010 in) between the bolt ends and the oil catcher.
10. Bend over the locking washers to secure the bolt heads.
11. Fit the road wheel and tighten the securing nuts.
12. Remove the support stands and lower the car, and tighten the road wheel nuts.

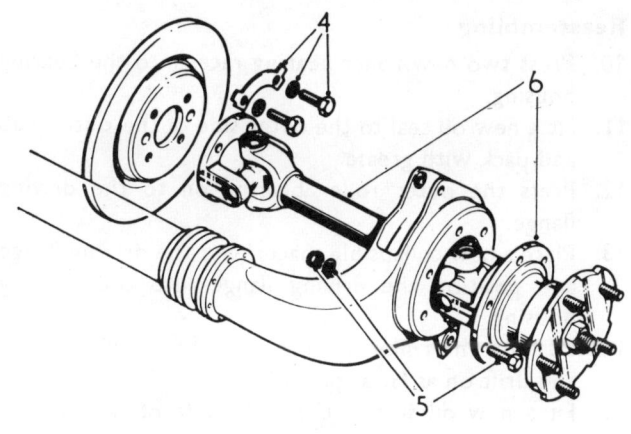

REAR HUB ASSEMBLY

—Overhaul 64.15.07

Service tools: 18G 79 Press
541884 Extractor, hub bearings

Dismantling

1. Remove the rear hub assembly. 64.15.10.
2. Free the lock washer from the yoke shaft nut.
3. Remove the yoke nut and lock washer.
4. Withdraw the shaft from the hub assembly.
5. Drift the driving flange out of the bearing housing.
6. Remove the collapsible spacer, and remove inner bearing, using collets, 541884, and the press, 18G 79.
7. Remove the two oil seals from the bearing housing.
8. Drift out the two roller races from the bearing housing.
9. Remove all traces of 'Loctite' from the splines.

continued

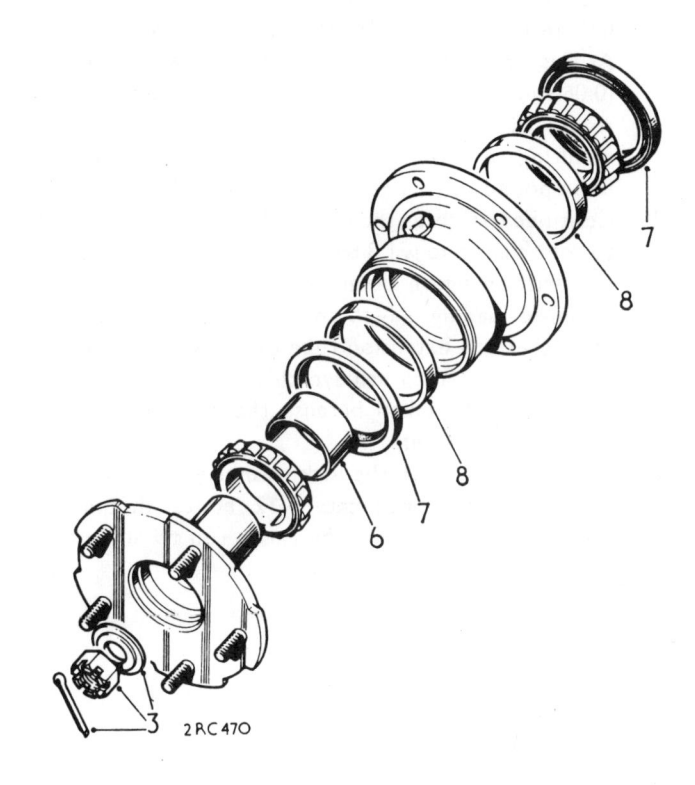

Rover 3500 and 3500S Manual AKM 3621

REAR SUSPENSION

Reassembling

10. Press two new roller bearing races into the bearing housing.
11. Fit a new oil seal to the outer side of the outer hub and pack with grease.
12. Press the outer roller bearing on to the driving flange.
13. Place a new collapsible spacer over the driving flange and position the driving flange into the bearing housing.
14. Position the roller bearing on to the driving flange and drift on as far as possible.
15. Fit a new oil seal to the inner side of the bearing housing.
16. Ensure that the splines of the yoke shaft are free from grease, then apply 'Loctite' sealant grade AVV sparingly. Fit the yoke shaft into the hub assembly.
17. Fit the lock washer and nut.
18. When reassembling the hub with new bearings, hold the assembly in a vice and tighten the nut until the load on the bearings is such that a pull of 2,2 to 4,5 kg (5 to 10 lb) is registered on a spring balance, as illustrated. The required loading is high and it will be found necessary to use an extension tube or bar with a suitable socket in order to achieve this loading.
19. When reassembling the hub with the existing bearings, hold the assembly in a vice and tighten the nut until the new collapsible spacer is just trapped. At this point there should be approximately 1,5 mm (0.060 in) end-float. Check the torque required to rotate the hub by means of a spring balance, the resulting seal friction figure which is obtained will be in pounds or kilogrammes, as applicable.
20. Continue to tighten the nut until the torque required to rotate the hub is 1,3 to 3,6 kg (3 to 8 lb) above the seal friction figure obtained in Item 19. Again the required loading is high and it will be found necessary to use an extension tube or bar with a suitable socket in order to achieve this loading. Particular care should be taken because the required figures are easily exceeded.
21. Secure the nut with the lock washer.
22. A period of approximately 12 hours should be allowed to lapse before the car is run to allow the 'Loctite' sealant to cure fully.

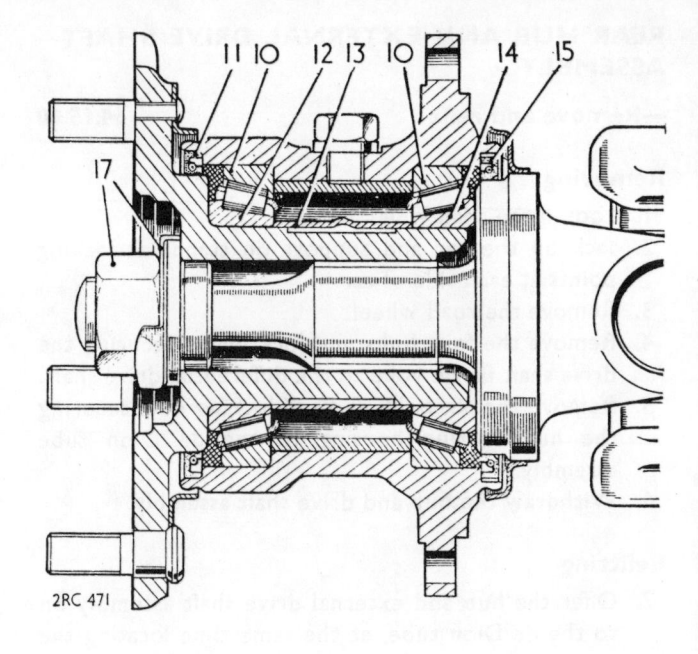

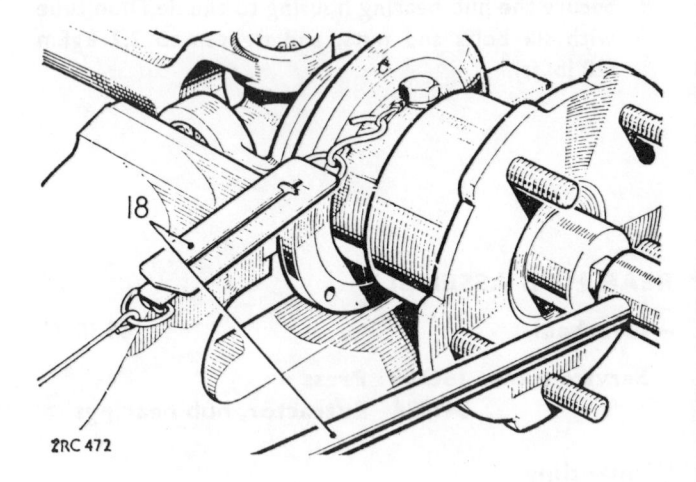

REAR SUSPENSION

REAR ROAD SPRING

—Remove and refit 64.20.01

Removing

1. Slacken the rear road wheel nuts.
2. Jack up the rear of the car until the rear wheels are 230 to 300 mm (9 to 12 in) from the ground and support on stands.
3. Remove the rear road wheel.
4. Place a jack under the bottom link, as shown in the illustration.
5. Raise the bottom link sufficiently to allow the lower end of the shock absorber fixing to be released by removing the two nuts and the mounting rubber.
6. Remove the bolt and nut securing the lower suspension link to the de Dion tube.
7. Lower the bottom link and remove the coil spring, noting the cushions at each end of the spring and the support plate.
8. For coil spring dimensions see Data.

Refitting

9. Reverse 1 to 8, but do not fully tighten the nut and bolt securing the lower suspension link to the de Dion tube until the spring has been compressed by jacking up the bottom link until it is in the approximate static unladen position. Then tighten to 7,5 kgf.m (54 lbf ft).

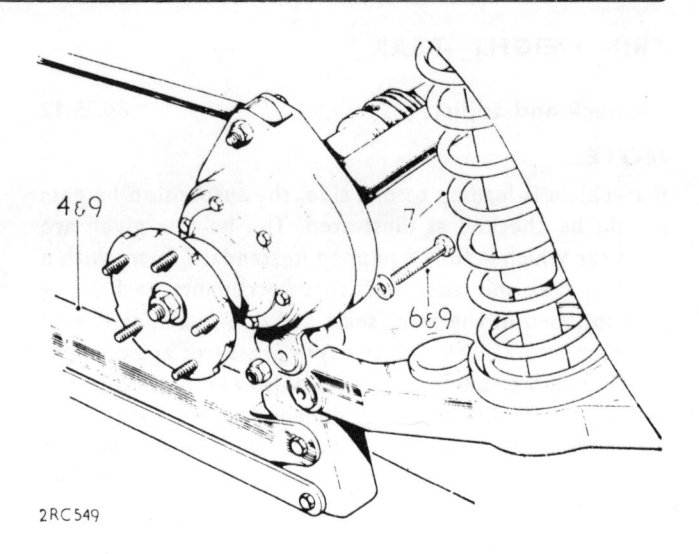

DATA

Rear road spring:
3500 models up to chassis suffix 'A'

　Number of working coils
$5\frac{1}{2}$

　Free length
338 mm (13.312 in)

　Rate
46,44 kg/cm (260 lb/in)

　Identification
12,5 mm (0.500 in) wide green and blue stripes painted on entire length of spring

3500 models from chassis suffix 'B' onwards and all 3500S models:

　Number of working coils
5.4

　Free length
331,1 mm (13.038 in)

　Rate
46,44 kg/cm (260 lb/in)

　Identification
Two 12,5 mm (0.500 in) wide white stripes painted on entire length of spring

Heavy duty road spring:

　Number of working coils
5.27

　Free length
320,8 mm (12.630 in)

　Rate
53,58 kg/cm (300 lb/in)

　Identification
Two 12,5 mm (0.500 in) wide green stripes painted on entire length of spring

REAR SUSPENSION

TRIM HEIGHT—REAR

—Check and adjust.   64.25.12

NOTE:

If a vehicle is leaning to one side, the suspension heights should be checked as illustrated. The heights given are for a car which is fully equipped in standard form, with a full fuel tank and laden with three occupants, each 76 kg (168 lb), two in the front seats and one in the centre of the rear seat. Check that the tyre pressures are correct.

The average standard tyre radius is 294 mm (11.593 in.), on which all the following dimensions are based; any variation in this dimension must be compensated for when checking the dimension between the centre line of the top link pivot and the floor.

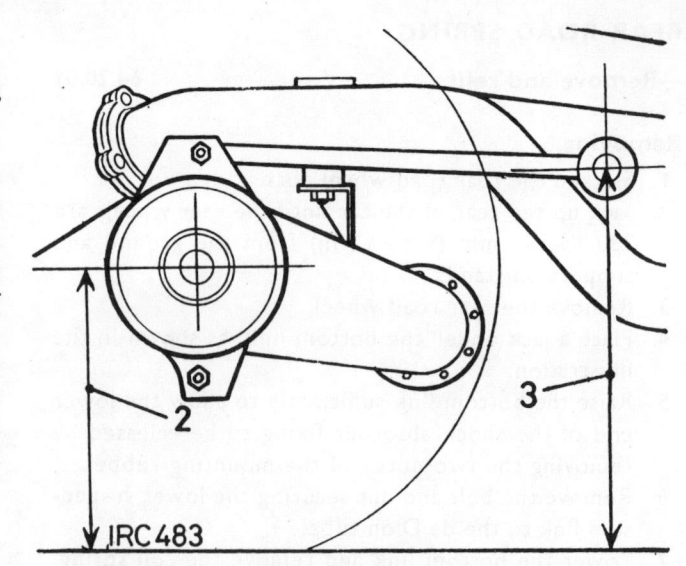

Checking

1. Stand the car on a level floor.
2. Check the dimension from the centre line of the road wheel to the floor.
3. Check the dimension from the centre line of the top link pivot to the floor.
4. Calculate the trim height by determining the difference between the two dimensions. Trim height limits are 95,0 mm ± 6,0 mm (3.750 in. ± 0.250 in.).

Adjusting

5. If the trim height is not within the limits, adjust the thickness of shims fitted under the coil springs. The total thickness of shims must not exceed 9,5 mm (0.375 in.), otherwise the location of the coil spring would be adversely affected.
6. If the foregoing procedure fails to rectify the trim height, fit new coil springs.

REAR SHOCK ABSORBER

—Remove and refit   64.30.02

Removing

1. Remove the rear seat cushion.
2. Remove the two drive screws securing the base of the seat squab and withdraw the squab.
3. Remove the two nuts, retainer and mounting rubber from the top of the shock absorber.
4. Slacken the rear road wheel nuts.
5. Jack up the rear of the car and support on stands.
6. Remove the rear road wheel.
7. Remove the four bolts and spring washers securing the shock absorber lower mounting to the bottom link.
8. Withdraw the shock absorber.
9. Remove the shock absorber lower mounting.

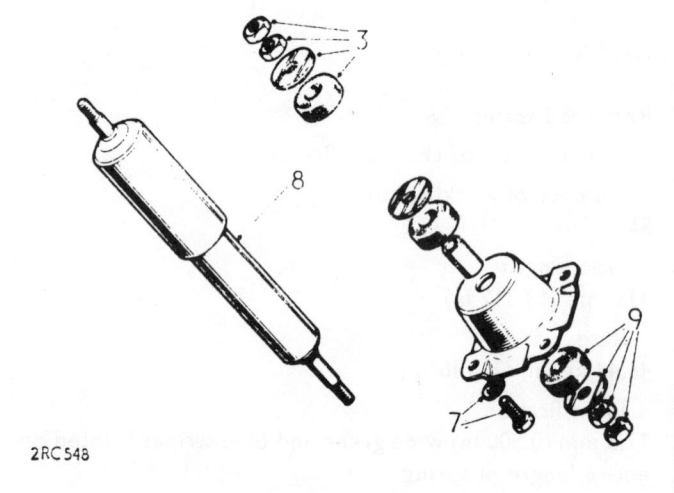

Refitting

10. Refit the shock absorber to the lower mounting.
11. Jack up the bottom link and guide the shock absorber into the base unit mounting; secure the lower mounting to the bottom link, tightening the first nut to 15 lbf.ft (2,0 kgf.m) and the locknut to 23 lbf.ft (3,2 kgf.m).
12. Refit the mounting rubber and nuts to the top of the shock absorber, and lock in position.
13. Refit the road wheels. Jack down.
14. Refit the seat squab and cushion.

REAR SUSPENSION

SUSPENSION LINK, UPPER

—Remove and refit   64.35.01

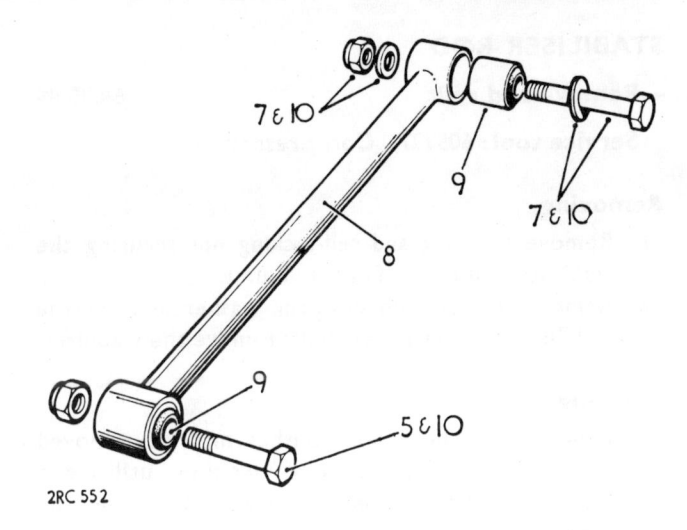

Removing

1. Slacken the rear road wheel nuts.
2. Jack up the rear of the car and support on stands.
3. Remove the rear road wheel.
4. Place a jack under the rear end of the bottom link and raise a few inches to relieve the load on the upper link.
5. Remove the nut and bolt securing the upper suspension link to the de Dion tube.
6. Left-hand upper suspension link—Remove the spare wheel, if fitted in the boot.
7. Lift the boot trim and remove the rear fixings for the upper suspension link.
8. Withdraw the upper suspension link.

Refitting

9. If required, fit new bushes at each end of the upper suspension link.
10. Reverse 1 to 8, but do not fully tighten the nuts and bolts securing the upper suspension link until the road wheel has been fitted and the car is in the static unladen condition. Then tighten to 7,5 kgf.m (54 lbf ft).

SUSPENSION LINK, LOWER

—Remove and refit   64.35.02

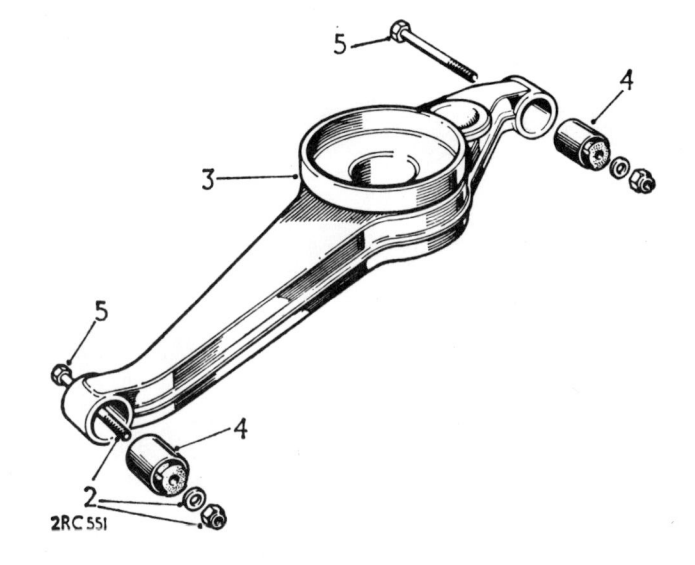

Removing

1. Remove the rear road spring. 64.20.01.
2. Remove the fixings securing the front end of the lower suspension link.
3. Withdraw the lower suspension link.

Refitting

4. If required, fit new bushes at each end of the lower suspension link.
5. Reverse 1 to 3, but do not fully tighten the nuts and bolts securing the lower suspension link until the road wheel has been fitted and the car is in the static unladen condition. Then tighten to 7,5 kgf.m (54 lbf ft).

Rover 3500 and 3500S Manual AKM 3621

64.35.01
64.35.02

REAR SUSPENSION

STABILISER ROD
—Remove and refit   64.35.09

Service tool: 605227   Compressor

Removing

1. Remove the bolt and self-locking nut securing the stabiliser rod to the final drive unit.
2. Remove the split pin and plain washer securing the stabiliser rod to the base unit; remove the stabiliser.

Refitting

3. If necessary, the rubber bushes may be removed and new ones pressed into position, utilising a suitable length of tubing and a workshop vice. The tube should be the same diameter as the outside of the bush.
4. Car to be in normal static unladen condition, that is, with oil, water and five gallons of fuel. Rock car up and down at the front to allow it to take up a static position.
5. Fit the stabiliser rod to the base unit, using tool, 605227, to compress the rubber.
6. Adjust the length so that the bolt fits quite easily at the final drive unit.
7. Secure rod at the final drive unit.

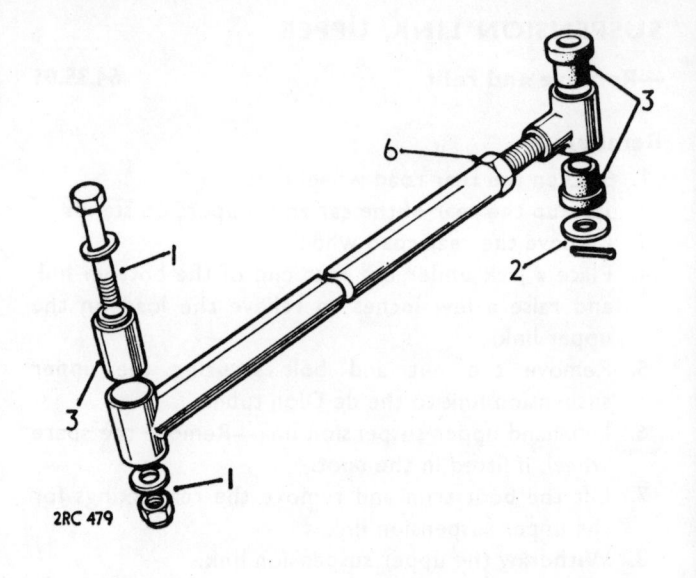

2RC 479

64.35.09

Rover 3500 and 3500S Manual AKM 3621

BRAKES

LIST OF OPERATIONS

Brakes—bleed	70.25.02
Brake failure switch—remove and refit	70.15.36
Calipers	
—front—remove and refit	70.55.02
—rear—remove and refit	70.55.03
—front—overhaul	70.55.13
—rear—overhaul	70.55.14
Discs	
—front—remove and refit	70.10.10
—rear—remove and refit	70.10.11
Fluid reservoir—remove and refit	70.30.15
Hand brake	
—cable and linkage assembly—remove and refit	70.35.17
—lever assembly—remove and refit	70.35.08
Master cylinder	
—remove and refit	70.30.01
—overhaul	70.30.02
Pads	
—front—remove and refit	70.40.02
—rear—remove and refit	70.40.03
Pedal assembly—remove and refit (see Division 33 for manual gearbox models)	70.35.01
Pedal box—remove and refit	70.35.03
Servo assembly	
—remove and refit	70.50.01
—overhaul	70.50.06

Rover 3500 and 3500S Manual AKM 3621

BRAKES

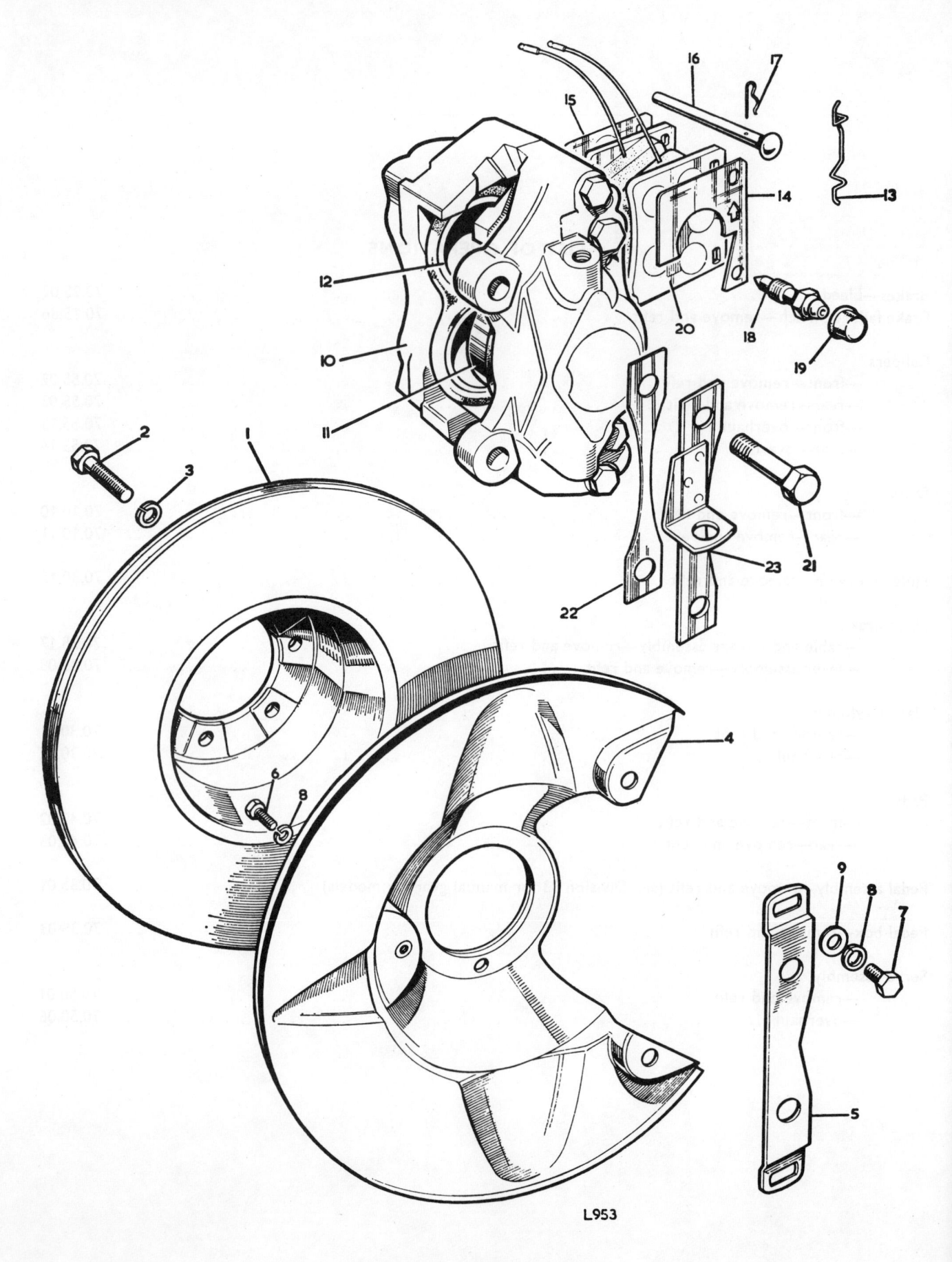

Front brake caliper and disc

BRAKES

Key to illustration of front brake caliper and disc

1 Disc for front brake
2 Set bolt
3 Spring washer
4 Shield and dust cover for disc
5 Strap for shield
6 Set bolt
7 Set bolt
8 Spring washer
9 Plain washer
10 Front brake caliper assembly
11 Piston, inner
12 Piston, outer
13 Damping spring
14 Damping shim, inner
15 Damping shim, outer
16 Retaining pin for pad
17 Clip retaining pin
18 Bleed screw
19 Dust cap for bleed screw
20 Friction pad
21 Special set bolt
22 Tab washer

Rover 3500 and 3500S Manual AKM 3621

BRAKES

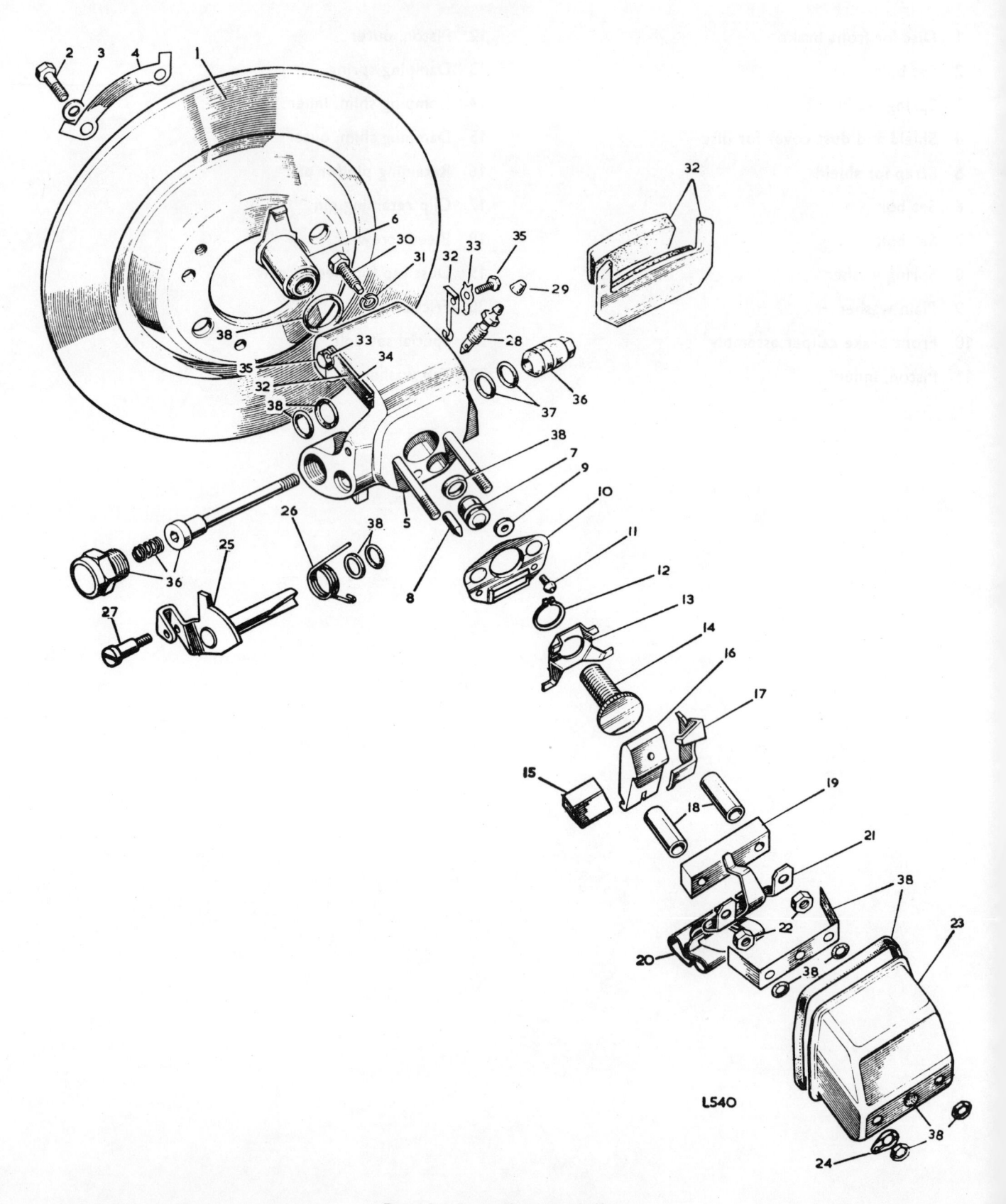

Rear brake caliper and disc

70-4

Rover 3500 and 3500S Manual AKM 3621

BRAKES

Key to illustration of rear brake caliper and disc

1 Disc for rear brake
2 Special bolt ⎫ Fixing brake disc and
3 Shim washer, thick ⎬ rear hub drive shaft to
4 Locking washer ⎭ differential drive shaft
5 Rear brake caliper assembly, RH
6 Piston cup assembly
7 Piston
8 Tappet for hand brake
9 Stop washer for piston
10 Location plate
11 Special screw fixing location plate
12 Circlip for push-rod
13 Collar for push-rod
14 Push-rod
15 Strut
16 Lever for strut
17 Pawl
18 Sleeve
19 Beam
20 'S' spring
21 Main spring
22 Special nut
23 Cover
24 Anchor plate for bias spring
25 Cam lever for hand brake, RH
26 Return spring for cam lever, RH
27 Stop pin for cam lever
28 Bleed screw
29 Dust cap for bleed screw
30 Drag pin
31 Spring washer
32 Friction pads, set of four (includes locking tabs and spring plates)
33 Locking tab for pad retainer bolt
34 Retaining plate for pad
35 Special bolt for pad retainer plate
36 Hinge pin assembly
37 Seal for hinge pin sealing nut
38 Seal repair kit for rear calipers

Rover 3500 and 3500S Manual AKM 3621

BRAKES

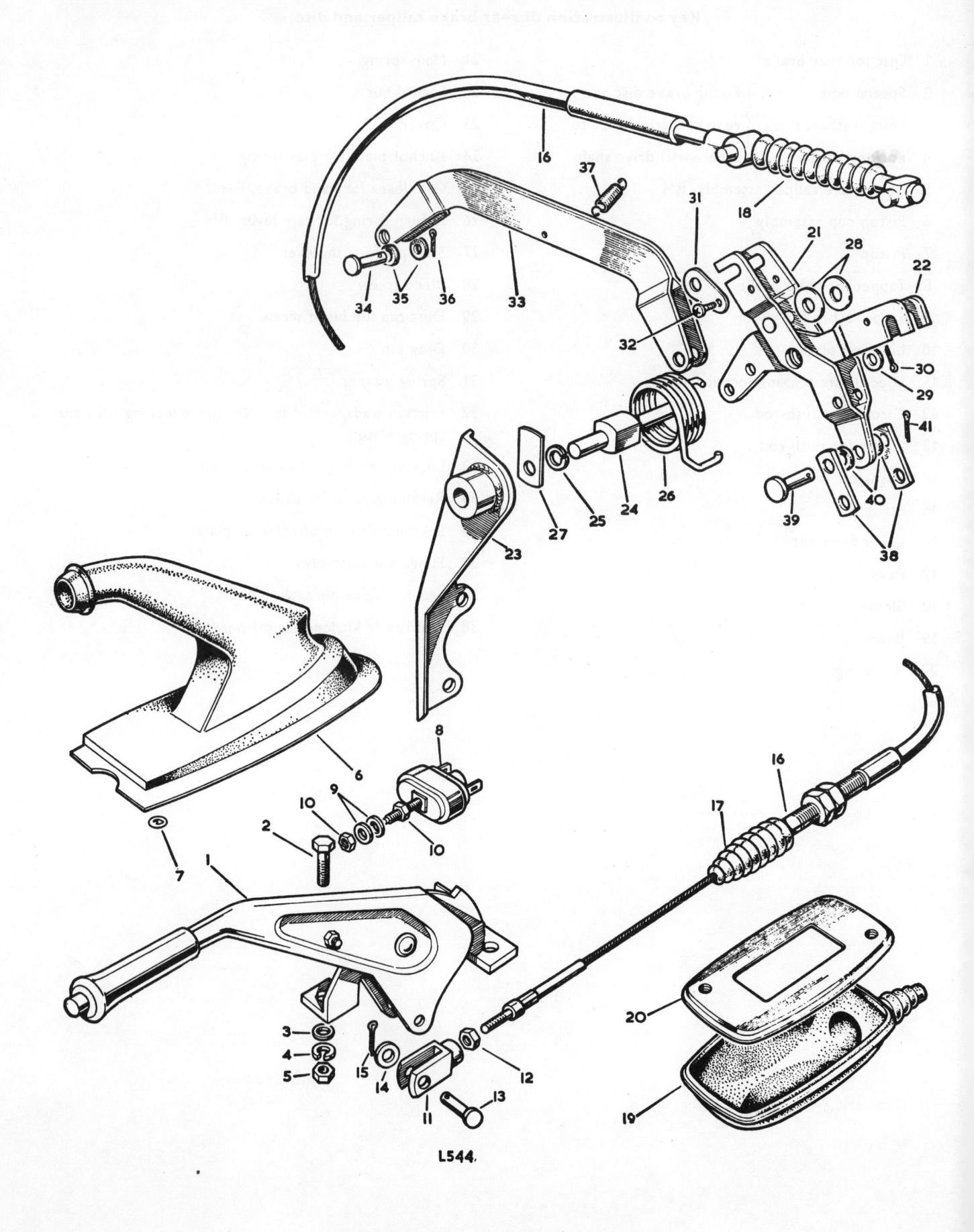

Hand brake lever, cable and linkage

Rover 3500 and 3500S Manual AKM 3621

BRAKES

Key to illustration of hand brake lever, cable and linkage

1	Hand brake lever complete
2	Bolt ($\frac{5}{16}$ in UNF x 1 in long) ⎫ Fixing hand brake lever to base unit
3	Packing washer
4	Spring washer
5	Nut ($\frac{5}{16}$ in UNF)
6	Grommet for hand brake lever
7	'Starlock' washer, fixing grommet to tunnel finisher
8	Switch for warning light
9	Plain washer ⎫ Fixing switch to hand brake lever
10	Nut ($\frac{1}{4}$ in UNF)
11	Fork end for brake cable
12	Nut ($\frac{1}{4}$ in UNF), fork end to cable
13	Clevis pin ⎫ Fixing hand brake lever to cable
14	Plain washer
15	Split pin
16	Hand brake cable complete
17	Grommet, front end ⎫ For hand brake cable
18	Grommet, rear end
19	Grommet for hand brake linkage (under tunnel)
20	Retaining plate for hand brake linkage grommet
21	Bellcrank lever, outer
22	Bellcrank lever, inner
23	Bracket for bellcrank levers
24	Pivot pin for bellcrank levers
25	Spring washer for pivot pin
26	Return spring for bellcrank levers
27	Abutment plate for return spring (earlier type illustrated)
28	Plain washer separating bellcrank levers
29	Plain washer ⎫ Fixing bellcrank levers to pivot pin
30	Split pin
31	Trunnion retaining plate ⎫ Fixing brake cable to bellcrank levers
32	Self-tapping screw
33	Link lever
34	Clevis pin ⎫ Fixing link lever to bellcrank levers and caliper lever
35	Felt washer
36	Split pin
37	Bias spring, RH caliper to link lever
38	Link plate
39	Clevis pin ⎫ Fixing link plates to bellcrank levers and caliper lever
40	Felt washer
41	Split pin

Rover 3500 and 3500S Manual AKM 3621

BRAKES

General service information

IMPORTANT: Do not jack under the de Dion tube, use the rear jacking point, as illustrated.

BRAKES

Preventive maintenance, all models

Preventive maintenance is in addition to routine maintenance, described in Division 10, and consists of the replacement, or overhaul, of hydraulic components incorporated in the braking system at scheduled periods, in order that brake performance is maintained at peak efficiency.

1. **Hydraulic fluid—Every 30.000 km (18,000 miles) or every 18 months, whichever occurs first.**
   All brake fluid absorbs moisture from the air, and as a result its boiling point is lowered with a consequent deterioration in performance. In the sealed brake system, water absorption takes place over a period and can, if not remedied, reduce brake performance to a dangerous level.
   All the fluid in the brake system should be changed at the stated intervals. It should also be changed before touring in mountainous areas if not done in the previous nine months. Use only the correct grade fluid, see Division 09, which complies with Federal Standard 116, from sealed tins. Never use fluid which has been left in an unsealed tin, nor re-use fluid already drained.

2. **Rubber seals in brake system—Every 60.000 km (36,000 miles) or every 36 months, whichever occurs first.**
   Renew all rubber seals in the complete brake system and renew all brake hydraulic hoses. Drain the brake fluid reservoir, flush the system and refill with fresh fluid of the correct grade.

3. **Dual-line brake system**
   As a result of service experience, the brake efficiency has been improved by revising the slave cylinder output pipe connections from the original layout on the Rover 3500 S Automatic special export model. The following diagrams show the revised layout fitted to all current Rover 3500 Automatic and Rover 3500 S Manual models when dual-line brakes are specified. Earlier dual-line systems can be converted to the current layout, and when a new servo assembly is required it is supplied complete with pipes, forming a conversion kit to the current layout.
   A new Operation, 70.50.01, is issued to cover Servo assembly – remove and refit, showing the revised pipe connections.
   A new Operation, 70.25.02, is issued giving a revised brake bleed procedure for both single and dual-line systems.

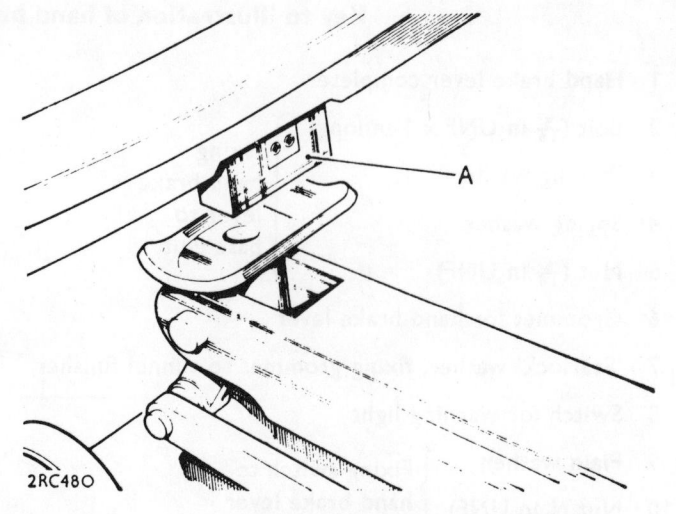

Rear jacking point
A—Jacking bracket

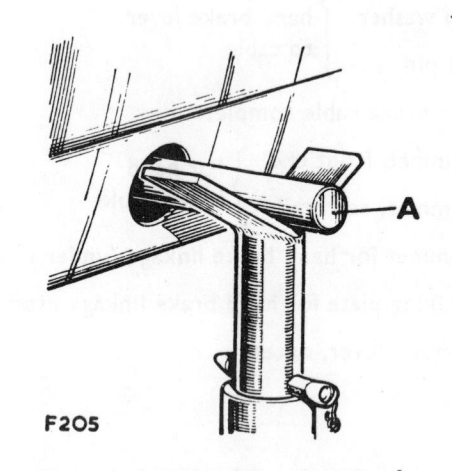

Method of supporting car on stands
A—Insert a suitable length of $\frac{7}{8}$ in. (22 mm) diameter steel rod into jacking tube

BRAKES

Diagram of 'Dual-line' brake system

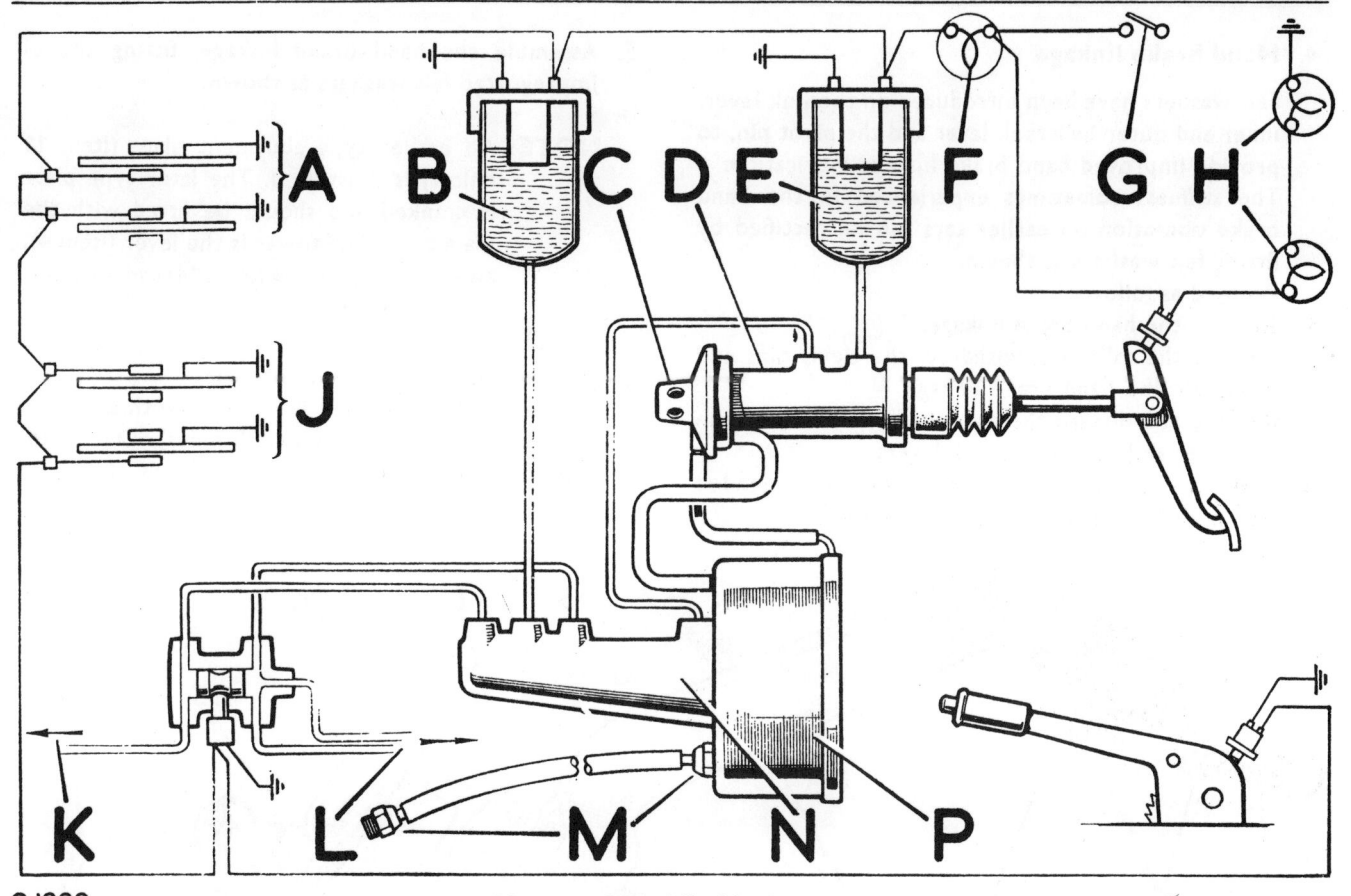

A—Front brake pads
B—Fluid reservoir for front brakes
C—Reaction valve
D—Master cylinder
E—Fluid reservoir for rear brakes
F—Brake warning light
G—Ignition switch
H—Stop lamps
J—Rear brake pads
K—To rear brakes
L—To front brakes
M—Non-return valve in manifold pipe
N—Slave cylinder (tandem)
P—Servo unit

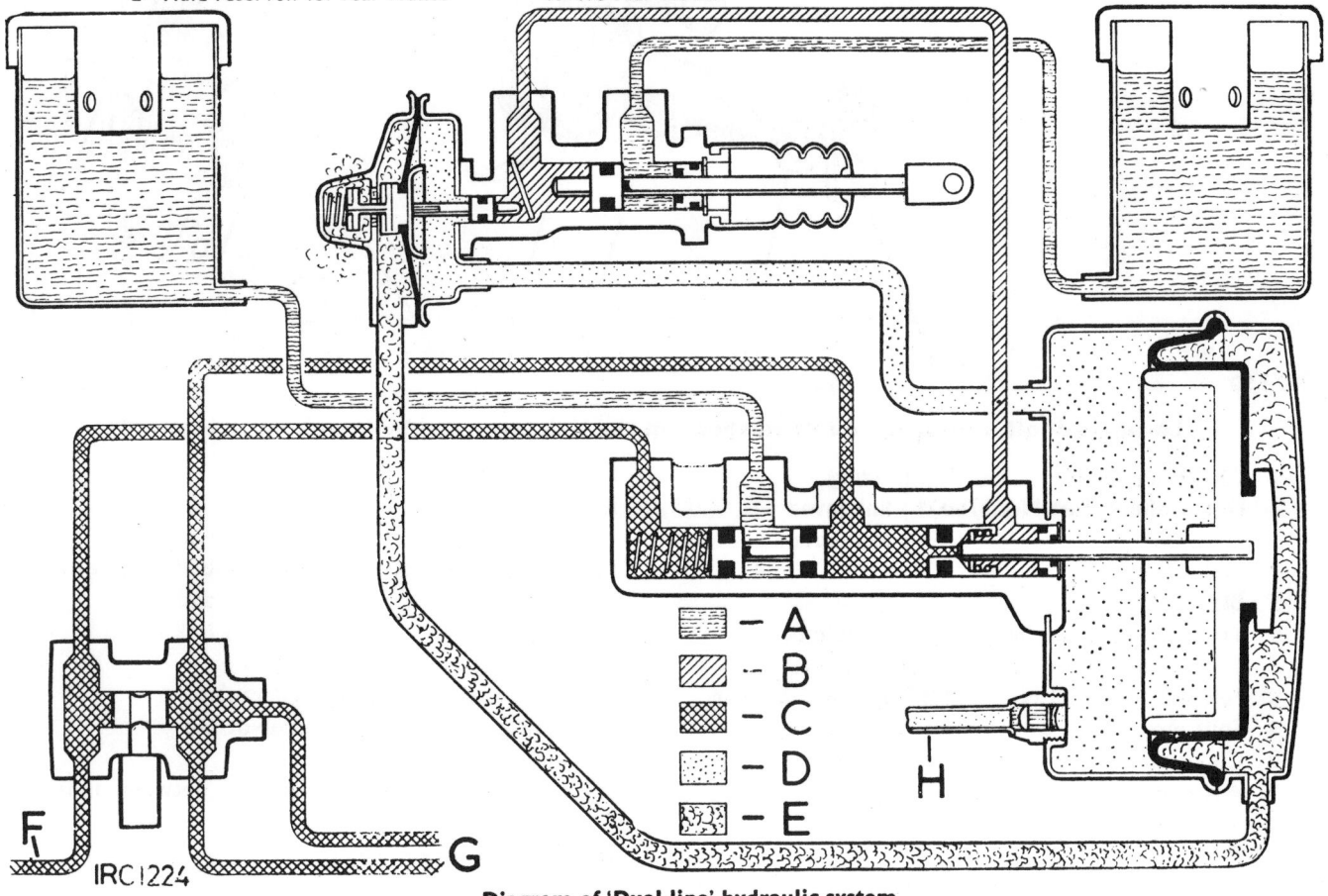

Diagram of 'Dual-line' hydraulic system

A—Reserve fluid
B—Fluid at master cylinder pressure
C—Fluid at operating pressure
D—Vacuum
E—Air at atmospheric pressure
F—To rear brakes
G—To front brakes
H—To induction manifold

Rover 3500 and 3500S Manual AKM 3621

70.00.00
Sheet 2

BRAKES

4. Hand brake linkage

Felt washers have been introduced on the link lever, inner and outer bellcrank lever and the pivot pin, to provide improved hand brake linkage lubrication. The stiffness sometimes experienced in the hand brake operation on earlier cars can be rectified by fitting felt washers as shown.

Proceed as follows:

1. Remove the hand brake linkage.
2. Remove the split pins, withdraw the clevis pins and dismantle the hand brake linkage.
3. Remove and discard the two plain washers between the inner and outer bellcrank levers.
4. Soak the felt washer in one of the recommended engine or final drive oils for 15 minutes.
5. Assemble the hand brake linkage, fitting the oil impregnated felt washers as shown.

   NOTE: An earlier type abutment plate (Item 11 below) is illustrated. The later type plate is cranked and should be fitted with the plain crank half towards the lever (Item 4), away from the spring (Item 14) and inversely to that shown.

6. Refit the hand brake linkage to car.

   The felt washer requires lubrication with a suitable oil can every 10.000 km (6,000 miles). One of the recommended engine or final drive oils should be used for this purpose.

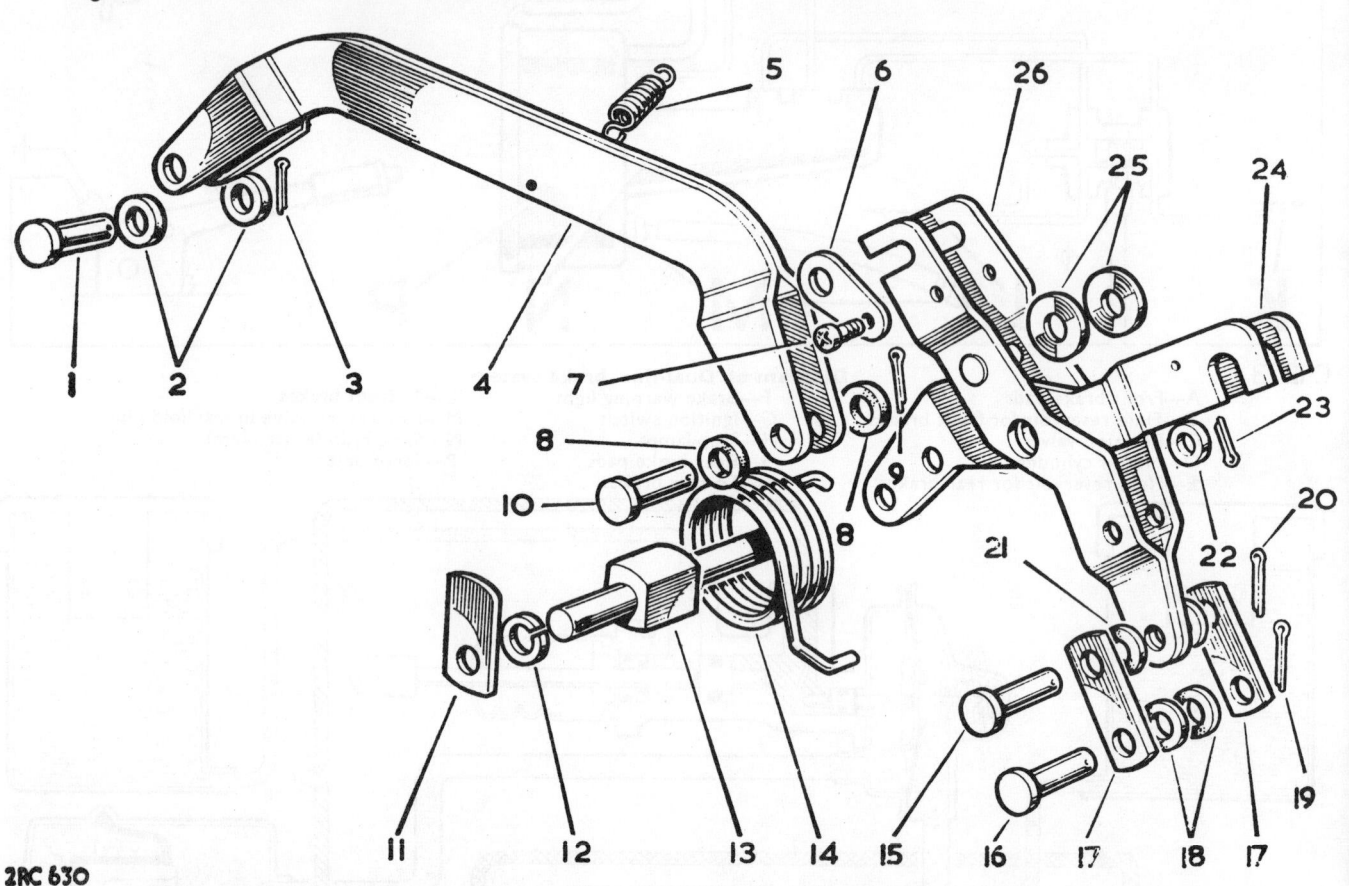

2RC 630

Location of oil impregnated felt washers on the link lever and bellcrank lever assembly

1. Clevis pin, lever ⎫
2. Felt washers, Part No. 578009 ⎬ Fixing link lever to caliper
3. Split pin ⎭
4. Link lever
5. Bias spring
6. Trunnion retaining plate ⎫ Fixing cable to bellcrank levers
7. Self-tapping screw ⎭
8. Felt washers, Part No. 578009 ⎫
9. Split pin ⎬ Fixing link lever to bellcrank levers
10. Clevis pin ⎭
11. Abutment plate for return spring
12. Spring washer for pivot pin
13. Pivot pin
14. Return spring for bellcrank levers
15. Clevis pin
16. Clevis pin
17. Link plates ⎫
18. Felt washers, Part No. 578009 ⎬ Fixing bellcrank lever and link plates
19. Split pin ⎬
20. Split pin ⎭
21. Felt washers, Part No. 578009 ⎭
22. Plain washer ⎫ Fixing bellcrank levers to pivot pin
23. Split pin ⎭
24. Inner bellcrank lever
25. Felt washer, Part No. 578010 between inner and outer bellcrank levers
26. Outer bellcrank lever

70.00.00
Sheet 3

Rover 3500 and 3500S Manual AKM 3621

BRAKES

FRONT DISCS

—Remove and refit 70.10.10

Removing

1. Remove the front hub assembly. 60.25.01.
2. Separate the disc from the front hub flange face by removal of the five securing bolts, using a hide mallet to tap the two units apart.

Refitting

3. Reverse 1 and 2. Torque tighten the disc securing bolts to 6,0 kgf.m (44 lbf ft).
4. Mount a dial test indicator to read off the brake disc.
5. Check the brake disc run-out at the outer edge, if it exceeds 0,07 mm (0.003 in), reposition the disc on the hub.

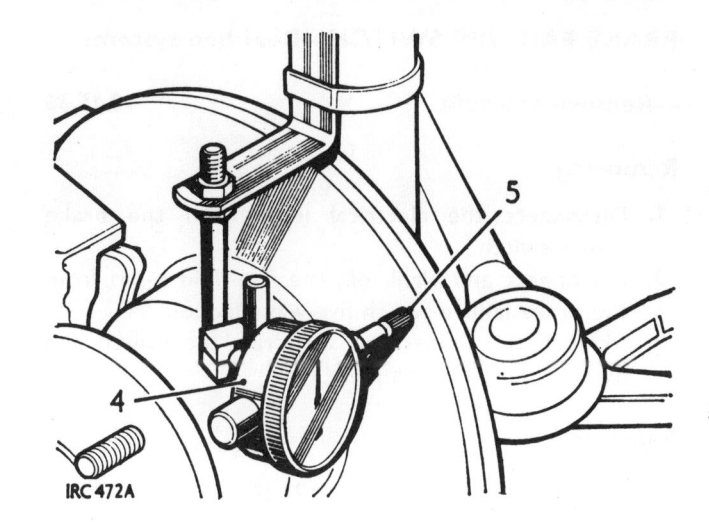

REAR DISCS

—Remove and refit 70.10.11

Removing

1. Slacken the rear road wheel nuts.
2. Jack up the rear of the car and support on stands.
3. Remove the rear road wheel.
4. Remove the rear brake pads. 70.40.03.
5. Remove the four bolts and lock plates securing the drive shaft flange to the final drive output flange, expand the de Dion tube, and allow the shaft to fall clear.
6. Rotate the disc until the dowel holes are parallel with the calipers.
7. Ease the disc off the dowels and withdraw.

Refitting

8. Reverse 1 to 7, including the following.
9. Tightening torque for the drive shaft flange and brake disc securing bolts: 11,5 kgf.m (85 lbf ft).
10. Mount a dial test indicator to read off the brake disc.
11. Check the brake disc run-out at the outer edge. If it exceeds 0,17 mm (0.007 in), reposition the disc on the final drive unit.

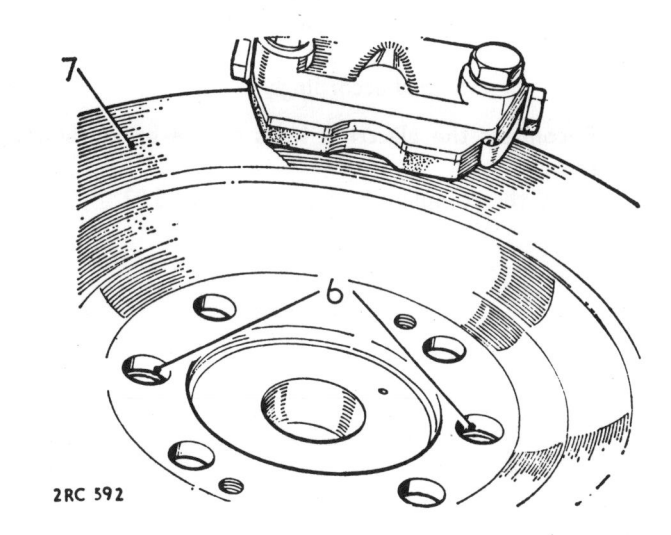

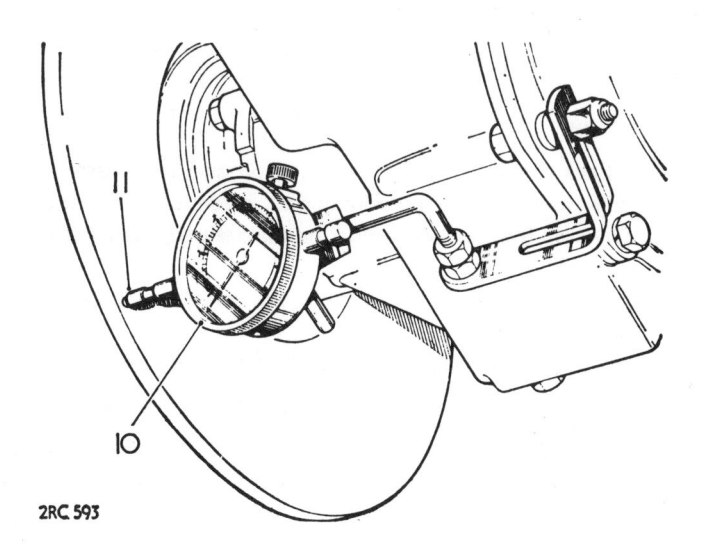

Rover 3500 and 3500S Manual AKM 3621

BRAKES

BRAKE FAILURE SWITCH—Dual line systems

—Remove and refit 70.15.36

Removing

1. Disconnect the electrical leads from the brake failure switch.
2. Disconnect and blank off, the five fluid pipes from the brake failure switch five-way junction.
3. Remove the fixings and withdraw the switch and five-way junction complete.

Refitting

4. Secure the switch and five-way junction assembly in position.
5. Remove the blanks and reconnect the fluid pipes. Ensure correct location, as illustrated.

    NOTE: On early cars the pipe arrangement between the brake failure switch and the servo slave cylinder, differs from that illustrated, and early cars should be modified accordingly.

6. Reconnect the electrical leads to the brake failure switch.
7. Bleed the complete brake system. 70.25.02.

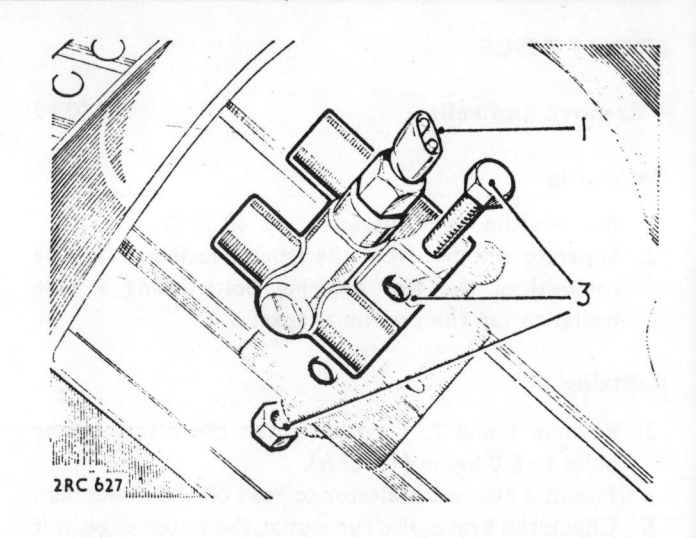

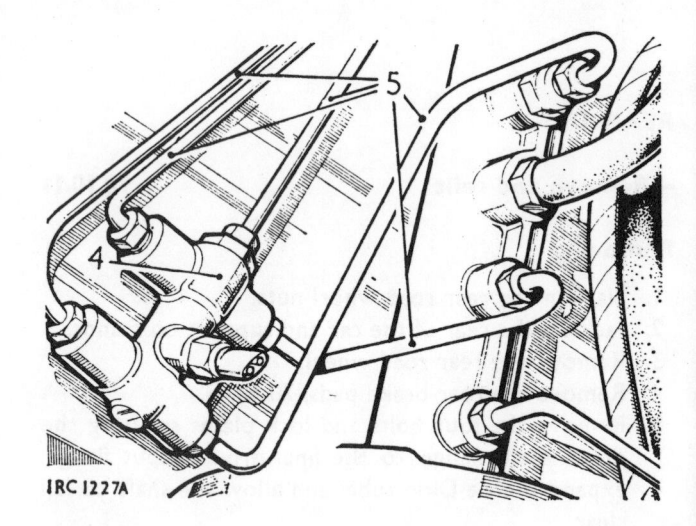

BRAKES

—Bleed 70.25.02

Single-line system 1 to 10
Dual-line system 11 to 23

Bleeding

1. Check the fluid level in the reservoir and, if necessary, top up with the recommended grade of fluid, see Division 09. This level must be maintained during the operation of bleeding.
2. Connect a bleed tube to the bleed screw on the rear left hand caliper.
3. Submerge the free end of the bleed tube in a container of clean brake fluid.
4. Slacken the bleed screw.

continued

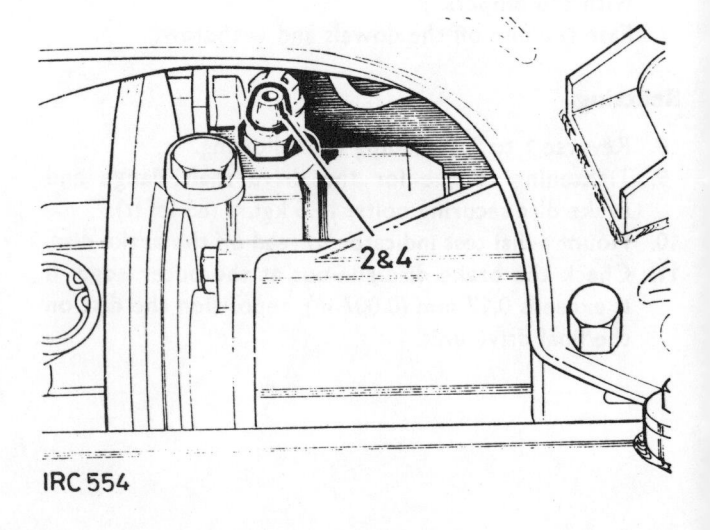

Rover 3500 and 3500S Manual AKM 3621

BRAKES

Latest style grille

5. Remove any floor mat or carpet which might restrict the full travel of the brake pedal.
6. Operate the brake pedal fully and allow to return. **IMPORTANT.** Allow at least five seconds to elapse with the foot right off the pedal to ensure that the system recuperates before operating the pedal again.
7. Repeat 6 until fluid clear of air bubbles appears in the container, then keeping the pedal fully depressed, tighten the bleed screw and remove the bleed tube.

   NOTE: There is only one bleed screw for the rear brakes.

8. Repeat 2 to 7 for the front brakes in turn, finishing with the one nearest the brake pedal.
9. Apply normal working load to the footpedal for three minutes and inspect the system for fluid leaks.
10. Finally, check the fluid level in the reservoir.
11. Check the fluid level in both reservoirs, if necessary, top up with the recommended grade of fluid, see Division 09. This level must be maintained during the operation of bleeding.

    NOTE: The quickest and most efficient method of purging the whole system is by bleeding both primary and secondary circuits at the same time, that is one side, front and rear, of the vehicle together. This necessitates having two bleeder tubes and jars.

12. Connect a bleed tube to the bleed screw on the front, driver's side, caliper, and another bleed tube to the bleed screw on the rear left-hand caliper.

    NOTE: There is only one bleed screw for the rear brakes.

13. Submerge the free ends of the tubes in containers of clean brake fluid.
14. Slacken both bleed screws.
15. Remove any floor mat or carpet which might restrict the full travel of the brake pedal.
16. Operate the brake pedal firmly through its full travel, allowing it to return freely.
17. Allow at least five seconds to elapse with the foot right off the pedal, then repeat item 16 until fluid clear of bubbles is seen to emerge from the front bleed tube. Hold the pedal down and tighten the front brake bleed screw.

    NOTE: The condition of the braking system may affect the ease with which the bleeding operation may be carried out. In cases of difficulty, where repeated pumping does not clear the fluid, allow a much longer interval between strokes of the pedal. This will enable the master cylinder to recuperate fully and ensure a complete charge of fluid on each stroke.

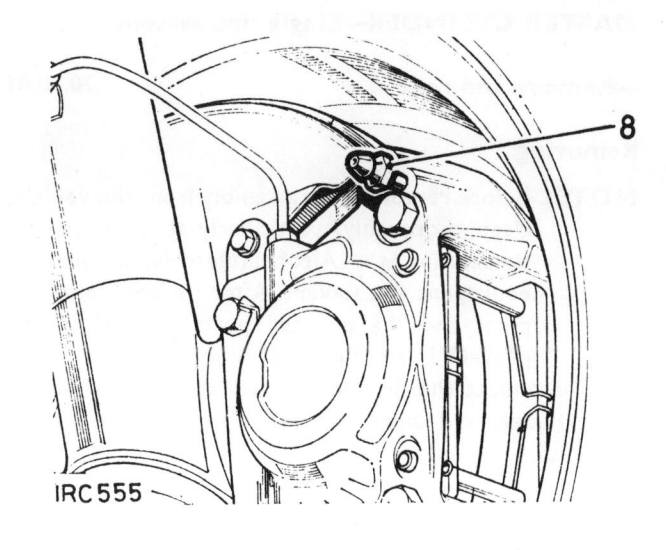

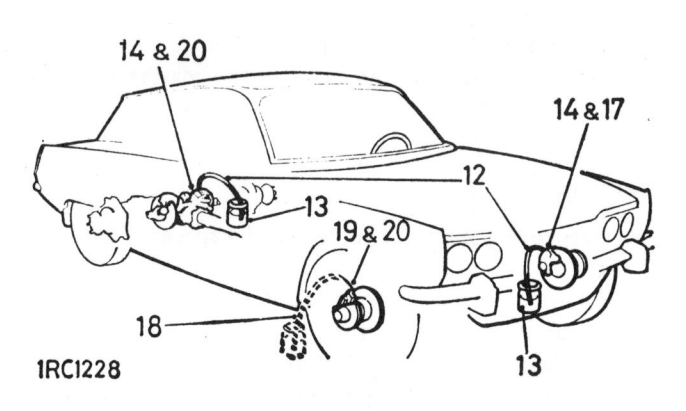

18. Transfer the front bleed tube and jar to the bleed screw on the passenger's side caliper.
19. Slacken the bleed screw on the passenger's side caliper and repeat the pedal pumping procedure until clear fluid emerges.
20. Continue the pedal pumping procedure until clear fluid emerges from the rear brake bleed tube, then hold down the pedal and tighten both front and rear bleed screws.
21. Remove bleed tubes and jars.
22. Apply normal working load to the foot pedal for three minutes and inspect the system for fluid leaks.
23. Finally, check the fluid level in the reservoirs.

Rover 3500 and 3500S Manual AKM 3621

70.25.02
Sheet 2

BRAKES—Single line system

MASTER CYLINDER—Single line system

—Remove and refit 70.30.01

Removing

NOTE: Before removing the assembly from the vehicle, clean thoroughly, particularly at pipe connections, using Ethyl Alcohol (Methylated Spirit) or equivalent as a solvent. After removal, plug the pipes and the exposed ports of the unit to prevent loss of fluid and entry of dirt.

1. Disconnect the fluid feed pipe, reservoir to master cylinder, and allow the reservoir to drain into a suitable container.
2. Remove the reservoir.
3. Disconnect the servo fluid feed pipe at the master cylinder and blank off the connection.
4. Unscrew the three nuts and one bolt securing the heat shield, spacer plate and master cylinder to the base unit, and remove the heat shield only.
5. Release the locknut from the operating rod at the foot pedal and unscrew the rod through the trunnion, at the same time withdrawing the master cylinder complete with spacer plate.
6. Unscrew the bolt and remove the master cylinder from the plate.

Refitting

7. Secure the master cylinder to the spacer plate with the top bolt only.
8. Offer the master cylinder assembly to the base unit, ensuring that the operating rod correctly enters the threaded trunnion in the foot pedal assembly.
9. Screw the rod through the trunnion until approximately 25.4 mm (1 in) protrudes, then secure the cylinder, spacer plate and heat shield and tighten the securing bolts.
10. Ensure that the rod and pedal have free movement when the bolts are tight.
11. Set the foot pedal height to 165 mm (6.500 in) from the floor beneath the carpet to lower edge of the pedal pad.
12. Reconnect the fluid reservoir pipe to the master cylinder.
13. Refit the fluid pipe, master cylinder to servo.
14. Bleed the brake system. 70.25.02.

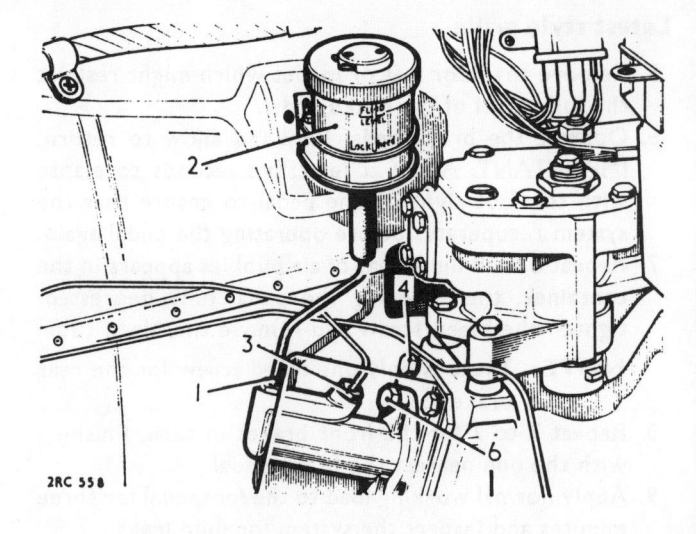

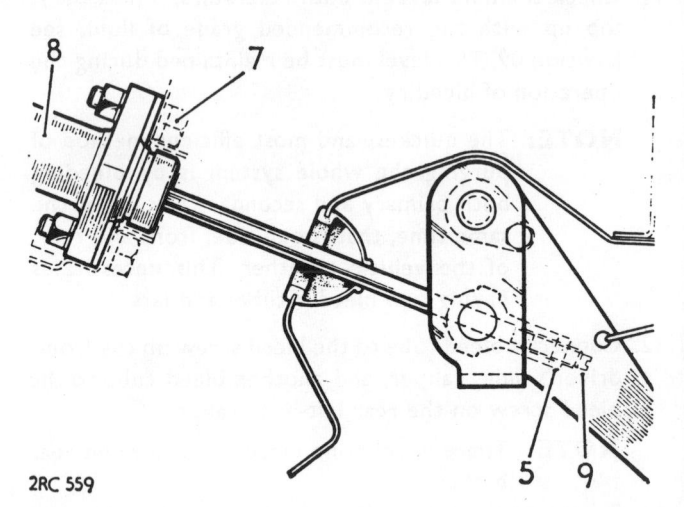

BRAKES—Dual line system

MASTER CYLINDER—Dual-line system

—Remove and refit 70.30.01

Removing

NOTE: Before removing the assembly from the vehicle, clean thoroughly, particularly at pipe connections, using Ethyl Alcohol (Methylated Spirit) or equivalent as a solvent. After removal, plug the pipes and the exposed ports of the unit to prevent loss of fluid and entry of dirt.

1. LHStg. Withdraw the dipstick from the engine sump and remove the left-hand elbow from the air cleaner.
2. Remove the fixings securing the brake fluid reservoir mounting bracket to the wing valance, disconnect the fluid inlet pipe at the master cylinder, and allow the reservoir to drain into a suitable container.
3. Disconnect the servo fluid feed pipe at the master cylinder.
4. Disconnect the air and vacuum pipes from the air control valve on the master cylinder.
5. Remove the two bolts securing the master cylinder to the sleeve and withdraw the master cylinder, leaving the operating rod in position.

Refitting

6. Reverse 1 to 5.
7. Check, and if necessary adjust, the brake pedal setting.
8. Bleed the brake system. 70.25.02.

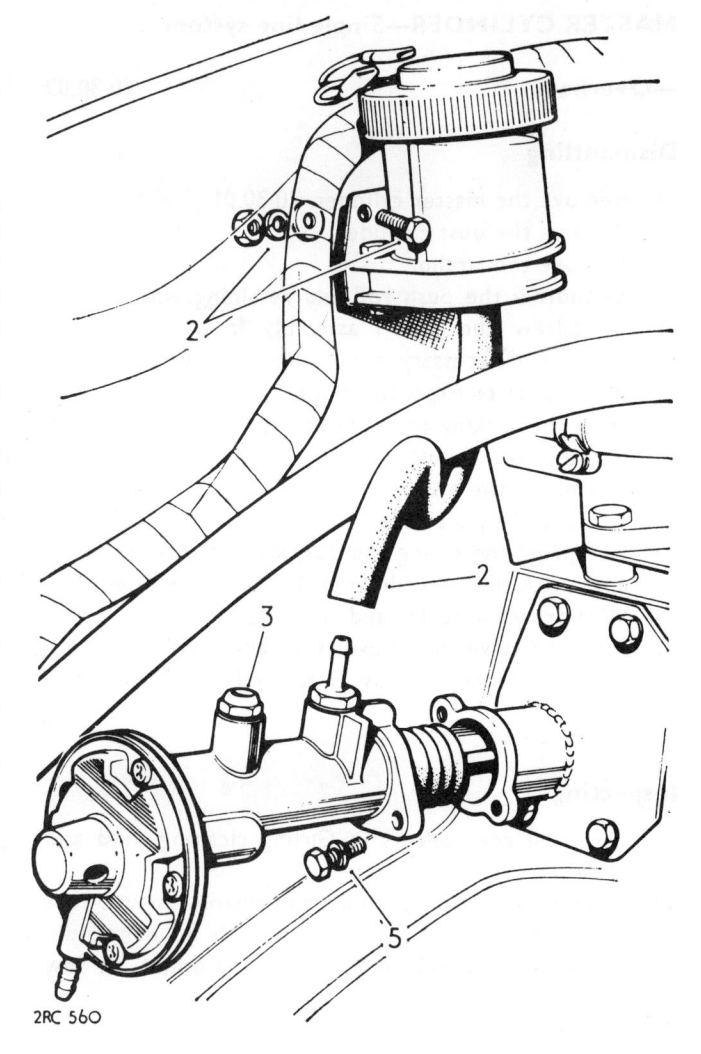

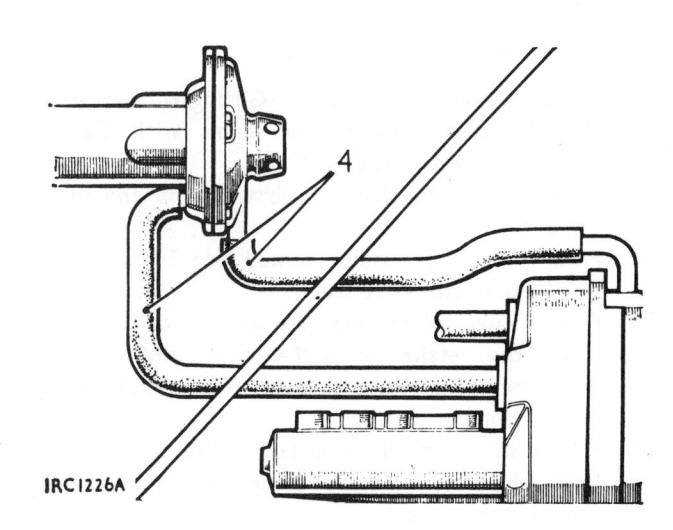

Rover 3500 and 3500S Manual AKM 3621

BRAKES—Single line system

MASTER CYLINDER—Single line system

—Overhaul 70.30.02

Dismantling

1. Remove the master cylinder. 70.30.01.
2. Remove the dust excluder.
3. Remove the circlip.
4. Withdraw the push rod and retaining washer.
5. Withdraw the piston assembly from the master cylinder. If necessary, apply a low air pressure to the outlet port to expel the piston.
6. Prise the locking prong of the spring retainer clear of the piston shoulder.
7. Withdraw the piston.
8. Remove the piston seal.
9. Compress the spring and position the valve stem to align with the larger hole in the spring retainer.
10. Withdraw the spring and retainer.
11. Slide the valve spacer over the valve stem.
12. Remove the spring washer and valve seal from the stem.

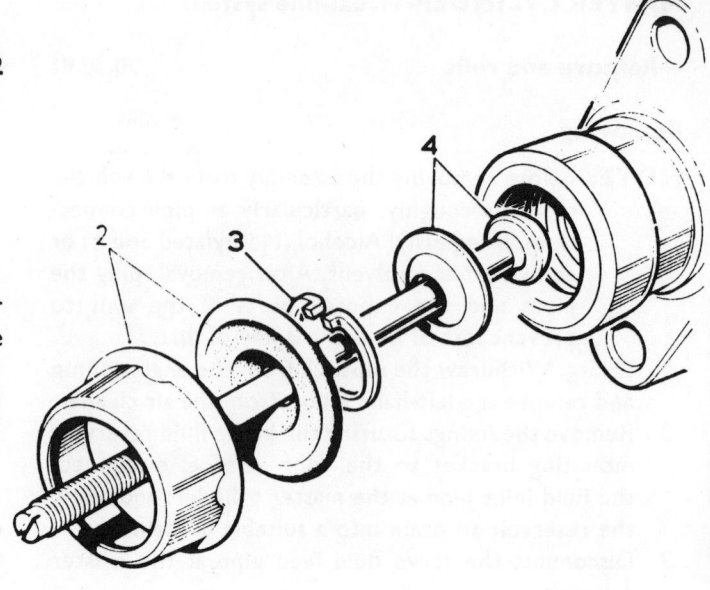

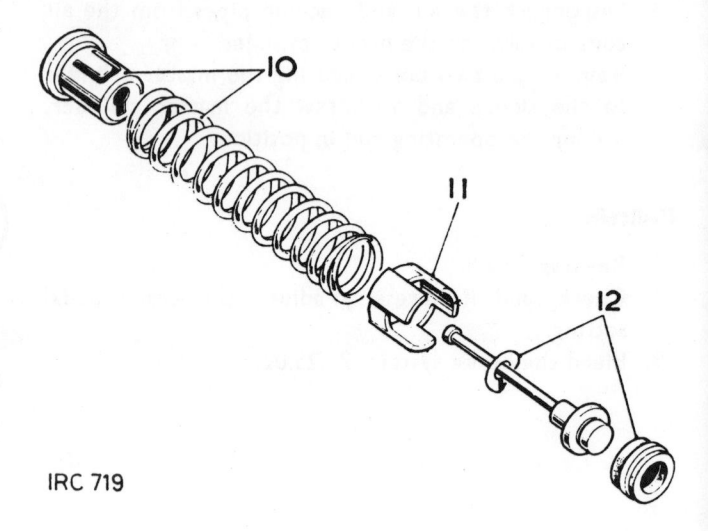

Inspecting

13. Clean all components in Girling cleaning fluid and allow to dry.
14. Examine the cylinder bore and piston, ensure that they are smooth to the touch with no corrosion, score marks or ridges. If there is any doubt, fit new replacements.
15. The seals should be replaced with new components. These items are included in the master cylinder overhaul kit.

Reassembling

16. Smear the seals with Castrol-Girling rubber grease and the remaining internal items with Castrol-Girling Brake and Clutch Fluid.
17. Fit the valve seal, flat side first, to the end of the valve stem.
18. Place the spring washer, domed side first, over the small end of the valve stem.
19. Fit the valve spacer, legs first, and the coil spring.
20. Insert the retainer into the spring and compress until the stem passes through the keyhole and is engaged in the centre.
21. Fit the seal, large diameter last, to the piston.
22. Insert the piston into the spring retainer and engage the locking prong.
23. Smear the piston with Castrol-Girling rubber grease and insert the assembly, valve end first, into the cylinder.
24. Fit the push rod, retaining washer and circlip.
25. Smear liberally the inside of the dust cover with Castrol-Girling rubber grease and fit the cover over the push rod and cylinder.
26. Fit the locknut and washer to the push rod.
27. Refit the master cylinder. 70.30.01.

IRC 719

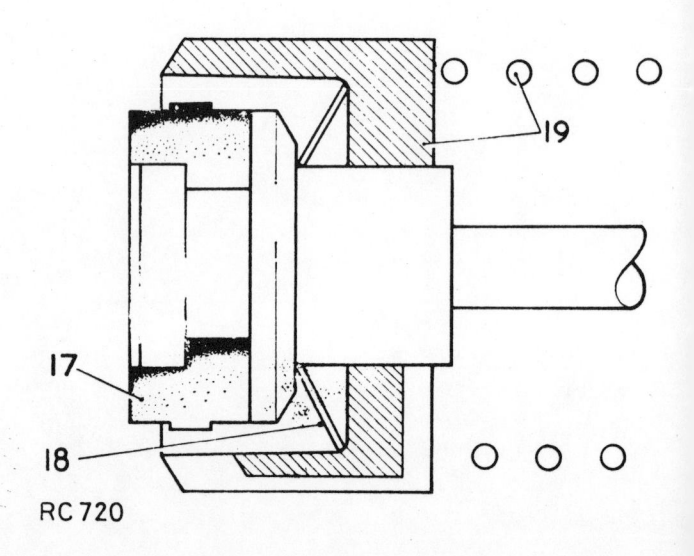

RC 720

BRAKES—Dual line system

MASTER CYLINDER—Dual line system

—Overhaul 70.30.02

Dismantling

1. Remove the master cylinder. 70.30.01.
2. Grip the assembly in a soft-jawed vice with the air control valve uppermost.
3. Extract the five self-tapping screws holding the plastic air valve cover to the valve housing, and remove the cover complete with the air valve sub assembly. Except to gain access to the filter by taking off the snap fitting cap, the air valve should not be further dismantled. Suspect functioning of the air valve must be remedied by fitting a new cover, filter and air valve assembly.
4. Remove the rubber diaphragm and plastic support to expose the two bolts securing the valve housing to the mounting flange on the master cylinder. Extract the bolts and star washers and take off the housing and gasket.
5. Reposition the cylinder assembly in the vice so that the mounting flange is uppermost. Remove the boot and push-rod.
6. Depress the spring retainer to reveal the 'Spirolox' retaining ring which can be eased from the groove at the rear of the piston and then unwound.
7. Press the piston down the bore and remove the circlip retaining the plastic bearings. The piston assembly, complete with plastic bearings, seals, return spring and retainers, may then be withdrawn from the bore.
8. Remove the inlet and outlet adaptors and copper washers, and carefully extract the trap valve and spring from the outlet adaptor.
9. Remove the cylinder body from the vice and tip out the lever in the nose of the cylinder.
10. Remove the valve piston by careful use of a blunt instrument through the cylinder outlet port to ease the piston along the bore until it can be removed by hand.
    Do not attempt to remove the valve piston from the bore by using pliers.
11. The main piston assembly may then be stripped, first removing the plastic bearings complete with 'O' ring and secondary cup. Prise off the plastic spring retainer from the nose of the piston complete with spring. The main seal and piston washer may then be removed.

Inspecting

12. Clean all parts thoroughly with new LOCKHEED Brake Fluid. Dry with a lint free cloth.
13. Carefully inspect the metal components for faults and wear.
14. A replacement cylinder will always be required where the cylinder bore, after cleaning, shows the slightest signs of corrosion or scoring.

continued

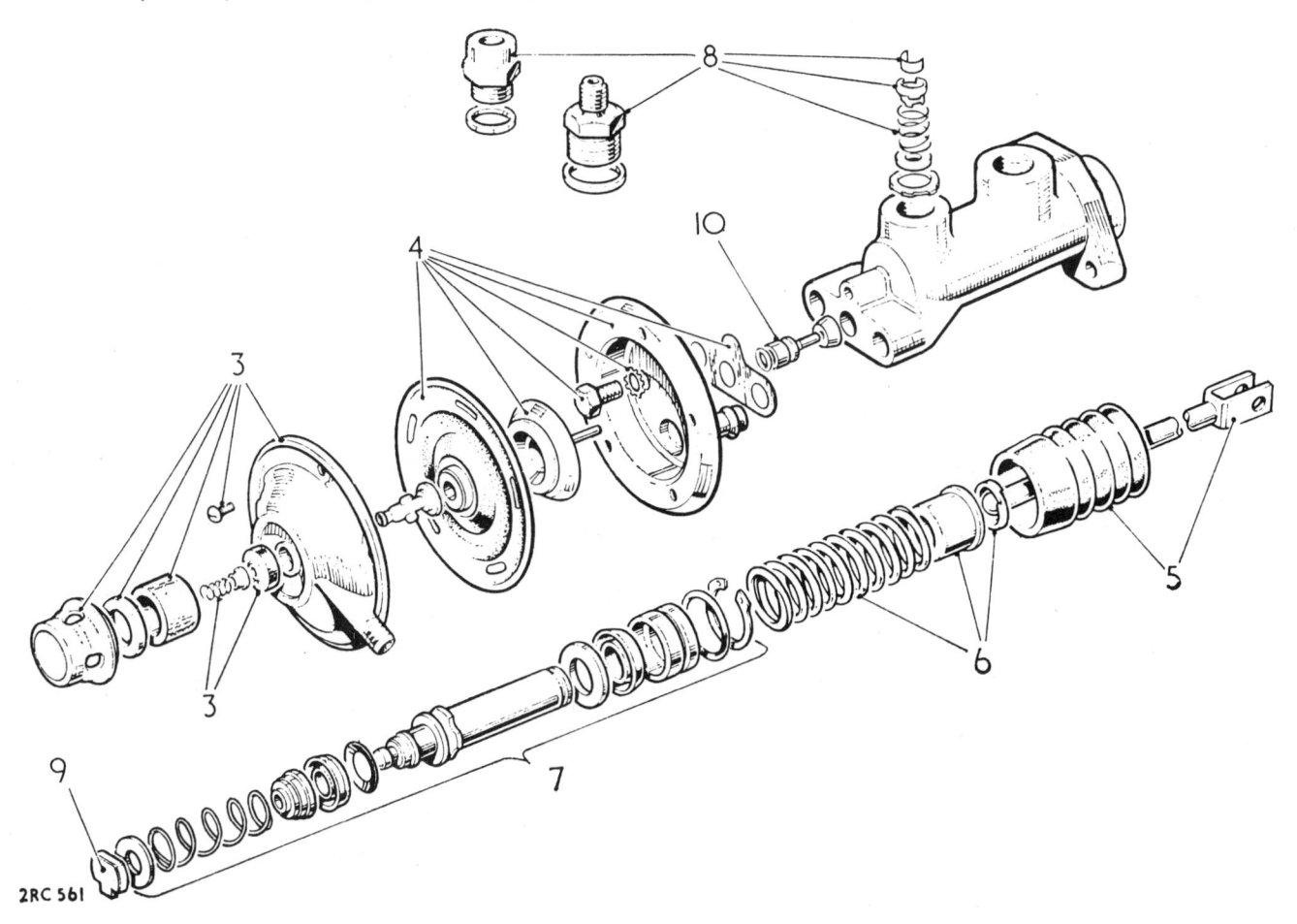

Rover 3500 and 3500S Manual AKM 3621

70.30.02
Sheet 2

BRAKES—Dual line system

15. If the metal parts of the original assembly are found to be in perfect condition, be prepared to fit new rubber parts. These are available in repair kit form. New copper washers will be needed for the inlet and outlet adaptors.
16. Ensure that the by-pass port and feed hole drillings into the cylinder are not obstructed. Likewise the equalising hole in the outlet port trap valve and the fluid feed holes in the head of the main piston. Take care not to enlarge the holes.
17. Check that the curved spring insert in the trap valve body is intact and undamaged.

Reassembling

18. The cylinder bores should be lubricated with new LOCKHEED Brake Fluid before re-assembling.
19. Hold the cylinder at an angle of about 25° to the horizontal, mouth of bore uppermost. Insert the valve piston operating lever into the bore with the tab downwards, ensuring that when in position at the bottom of the bore the tab drops into the recess provided.
20. Position the master cylinder in the vice so that the mouth of the bore is uppermost.
21. Assemble the components on to the main piston as previously illustrated, including the spring and retainers.
22. Coat the seals and plastic bearing with LOCKHEED Disc Brake Lubricant, and insert into the bore, taking care not to bend back the lip of the main seal or damage the 'O' ring seal around the plastic bearing. Insert the circlip and stroke the piston to ensure correct operation. Check security of the circlip.
23. Position the return spring and retainer over the exposed portion of the piston and compress the spring to uncover the groove for the retaining ring. Fit the 'Spirolox' retaining ring and release the pressure on the spring assembly to allow the retainer to seat on the ring.
24. Coat the inside and the beaded edge of the rubber boot with LOCKHEED Rubberlube and slide the push-rod into the boot. Insert the push-rod into the end of the piston and locate the edge of the boot into the groove on the end of the cylinder body.
    Place the trap valve body and spring into the outlet port, spring first, fit a new copper washer on to the adaptor and tighten in the port to a torque of 4,6 kgf.m (33 lbf ft).
25. Similarly, fit a new washer to the inlet adaptor and tighten in the port to a torque of 4,6 kgf.m (33 lbf ft).
26. Reposition the cylinder body in the vice so that the mounting face of the air valve assembly is uppermost. Ensure that the gasket face is clean and undamaged.
27. Using the fingers only, fit a new rubber seal on to the air valve piston with the lip facing away from the recessed head, coat with LOCKHEED Disc Brake Lubricant and insert into the bore, taking care not to bend back the lip of the seal.
28. Locate a new gasket and then the air valve housing on to the mounting face and secure with the two bolts and shakeproof washers, tightening to a torque of 1,9 kgf.m (14 lbf ft). Insert the diaphragm support, spigot first, into the head of the valve piston.
29. Place the valve cover over the diaphragm making sure that the projections on the under surface of the cover engage in the slots in the diaphragm. Insert the five self-tapping screws and tighten them diametrically and evenly. Do not overtighten.
30. Refit the master cylinder. 70.30.01.

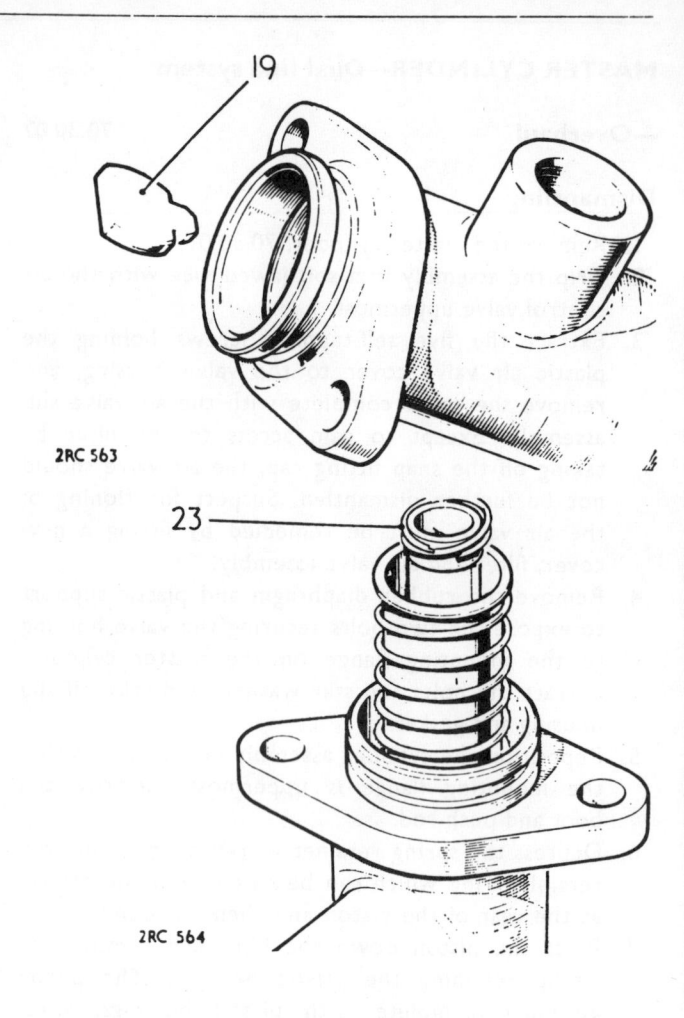

BRAKES

FLUID RESERVOIR

—Remove and refit 70.30.15

Removing

NOTE: Manual gearbox models are fitted with a combined brake and clutch fluid reservoir. Dual-line systems are fitted with two fluid reservoirs, the reservoir nearest the front of the car supplies the front brakes, and the rearmost reservoir supplies the rear brakes.

1. Remove the support bracket from the wing valance.
2. Disconnect the electrical connectors at the reservoir cap.
3. Disconnect, at the lower end, the pipe from the fluid reservoir, drain the reservoir and discard the fluid. Blank off the cylinder union.
4. Remove the reservoir.

Refitting

5. Reverse 1 to 4.
6. Bleed the brakes. 70.25.02.

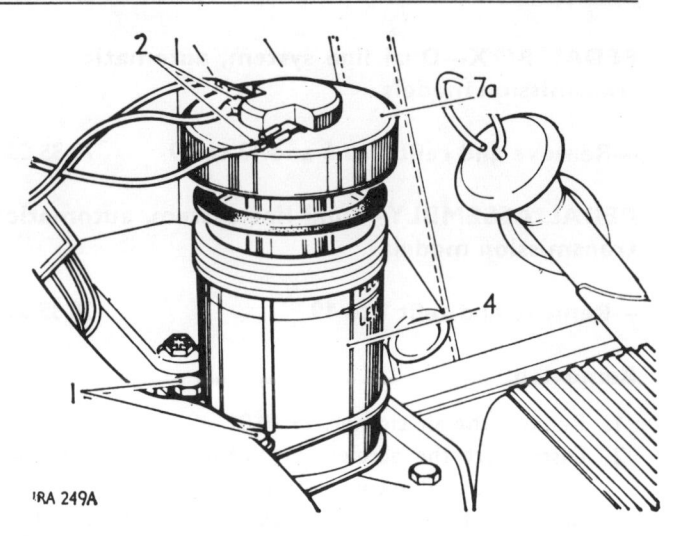

7. Check the operation of the reservoir level indicator switches, as follows:
   a. Switch the ignition 'on', release the hand brake, unscrew the reservoir cap and lift approximately 25 mm (1.000 in), this should cause the warning light to be illuminated.
   b. If the warning light is not illuminated, the operation of the float unit concerned and the wiring connections must be investigated.

PEDAL BOX—Automatic gearbox models with single line system

—Remove and refit 1 to 5 and 10 to 15 70.35.03

PEDAL ASSEMBLY—Automatic gearbox models with single line system

—Remove and refit 1 to 15 70.35.01

Removing

1. Remove the two bolts, nuts and spring washers fixing the accelerator pedal to the body. The nuts are accessible from beneath front wing.
2. Unhook the return spring from the pedal.
3. Remove the four bolts and sealing washers securing the pedal bracket to the body.
4. Unscrew the locknut from the brake operating rod.
5. With a screwdriver, screw the rod through the pedal trunnion while removing the assembly.
6. Remove the trunnion from the pedal.
7. Remove the bolt retaining the pivot spindle to mounting bracket. Withdraw the spindle from the pedal.
8. Remove the pedal from the bracket, together with distance bush and double spring and plain washers.

Refitting

9. Grease the bush and spindle before reassembling the pedal to the bracket, using molybdenum disulphide grease on the trunnion.

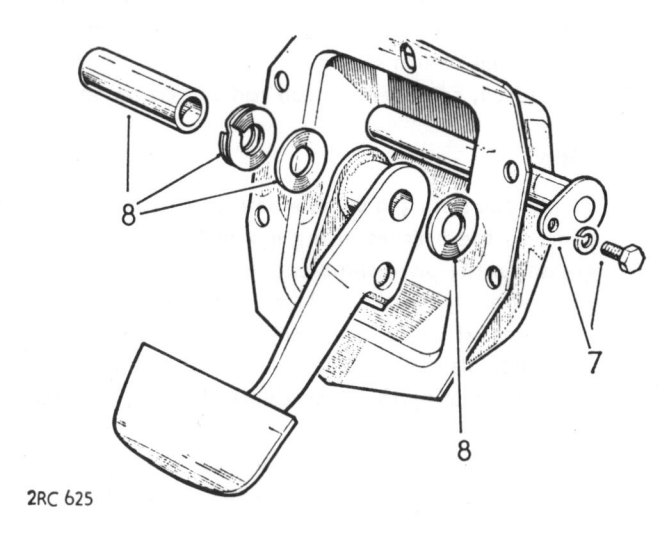

10. Apply Bostik 692 to the pedal box flange and offer the pedal assembly into position.
11. Locate the operating rod into the trunnion with the assistance of a thin screwdriver.
12. Screw the operating rod in until approximately 25,4 mm (1.000 in) of thread is protruding through the trunnion.
13. Locate the pedal bracket with set bolts, spring and plain washers. Locate all the securing bolts before tightening.
14. Re-secure the accelerator linkage to the body with the two bolts, nuts and spring washers.
15. Reconnect the return spring to the pedal. Adjust the brake pedal height to 165 to 170 mm (6.500 to 6.750 in) vertically from the floor beneath the carpet to underside of the pedal rubber.

Rover 3500 and 3500S Manual AKM 3621

BRAKES—Dual line system

PEDAL BOX—Dual line system, automatic transmission models

—Remove and refit 1 to 6 and 13 to 19   70.35.03

PEDAL ASSEMBLY—Dual line system, automatic transmission models

—Remove and refit 1 to 19   70.35.01

Removing

1. Remove the air cleaner. 19.10.01.
2. Disconnect the accelerator linkage at the vertical rod in the engine compartment.
3. Remove the two bolts securing the accelerator pedal to the brake pedal box.
4. Remove the two bolts securing the accelerator pedal to the car floor, and move pedal to one side.
5. Disconnect the leads from the brake stop switch.
6. Remove the remaining bolts from the brake pedal box flange and withdraw the pedal and box assembly.
7. Remove the brake stop switch.
8. Disconnect the brake pedal return spring.
9. Remove the bolt retaining the pivot spindle and withdraw the spindle from the pedal.
10. Remove the pedal from the bracket, together with the distance piece, double spring washer, shim washer and return spring.
11. If required, remove the push rod and trunnion.

Refitting

12. Grease the bush and spindle before reassembling the pedal to the bracket, using molybdenum disulphide grease on the pivot spindle.
13. Reassemble the brake pedal to the box by reversing the removal procedure. Apply Bostik 692 to the pedal box flange and offer the pedal assembly into position, locating the operating rod into the master cylinder.
14. Locate all securing bolts before tightening, leaving the two for the accelerator pedal loose until the pedal is fitted.
15. Refit the accelerator pedal and reconnect the linkage in the engine compartment.
16. Refit the air cleaner. 19.10.01.

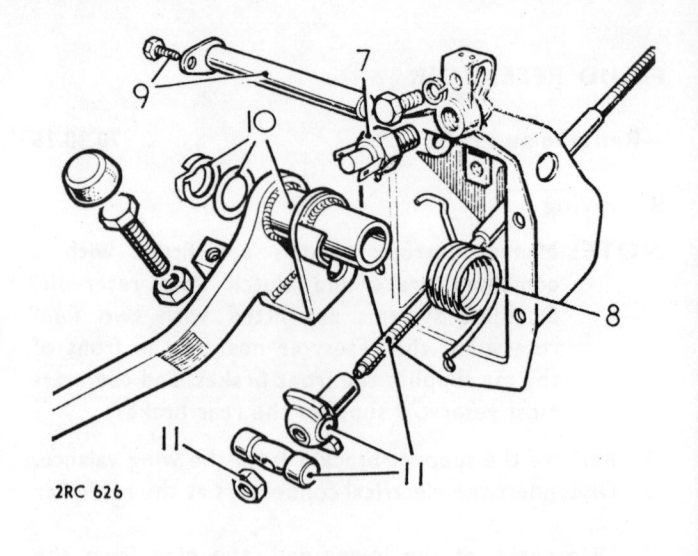

Adjusting

17. Check the brake pedal height, from beneath the carpet to lower edge of pedal pad, this should be 165 to 170 mm (6.500 to 6.750 in). If necessary, adjust the stop pin to give the correct height.
18. Check, by hand operation of the brake pedal, that there is approximately 1,5 mm (0.062 in) clearance between the end of the brake pedal push-rod and the master cylinder piston when the pedal is in the off position. If necessary, slacken the locknut and adjust the push-rod by screwing in or out as required.
19. Connect a test lamp (consisting of a small dry cell battery and bulb) to the terminals on the stop lamp switch. With the brake pedal in the off position, screw the switch in until the test lamp is illuminated, then turn the switch back by five flats of its hexagon. Remove the test lamp and reconnect the stop switch leads.

BRAKES

HAND BRAKE LEVER ASSEMBLY

—Remove and refit   70.35.08

Removing

1. Remove the gearbox tunnel cover. 76.25.07.
2. Remove the two bolts fixing the hand brake lever to the tunnel.
3. Remove the rubber cover from the hand brake lever, at the underside of the car.
4. Disconnect the clevis pin for the hand brake cable.
5. Disconnect the hand brake switch wires, and remove the lever complete.

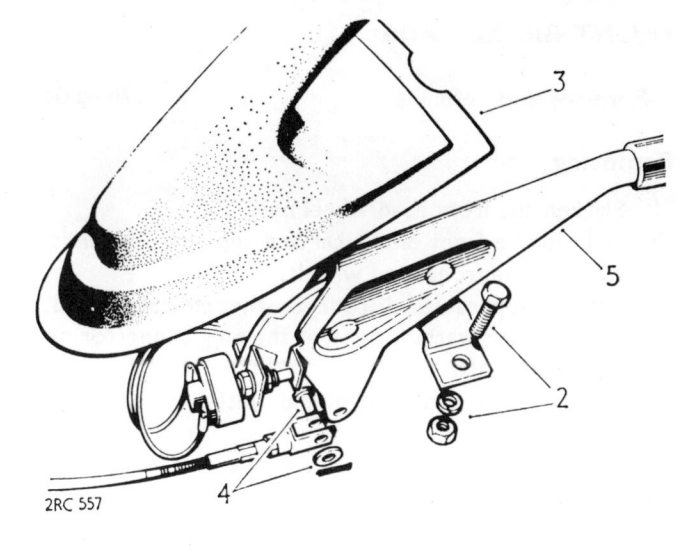

Refitting

6. Reverse 1 to 5.
7. Check the operation of the hand brake and the warning light.

HAND BRAKE CABLE AND LINKAGE ASSEMBLY

—Remove and refit   70.35.17

Removing

1. Remove both trunnion retainers and disconnect hand brake cable from inner and outer bellcrank levers.
2. Slacken the locknut and disconnect the outer cable from the bracket just above the final drive front mounting bracket.
3. Disconnect the link lever and link plates from hand brake lever and bellcrank lever at each side.
4. Remove the split pin and washer securing the bellcrank lever and return spring to the pivot pin. Withdraw the parts as an assembly.
5. If necessary, remove the pivot pin and abutment plate from the bracket and the bracket from the final drive unit.
6. Pull the rubber dust cover clear of the hand brake mechanism to gain access to the front end of the cable. Remove the split pin and clevis pin, noting which hole the clevis pin is fitted, and withdraw the cable.

Reassembling

7. Reverse 1 to 6, noting and including the following.
8. The inner cable clevis must fit to the hand brake lever without pre-loading the cable.
9. The outer cable nuts should be midway on the thread.
10. Ensure correct adjustment and operation, then apply a little Bostik adhesive to the lip of the dust cover, and fit the cover in place.

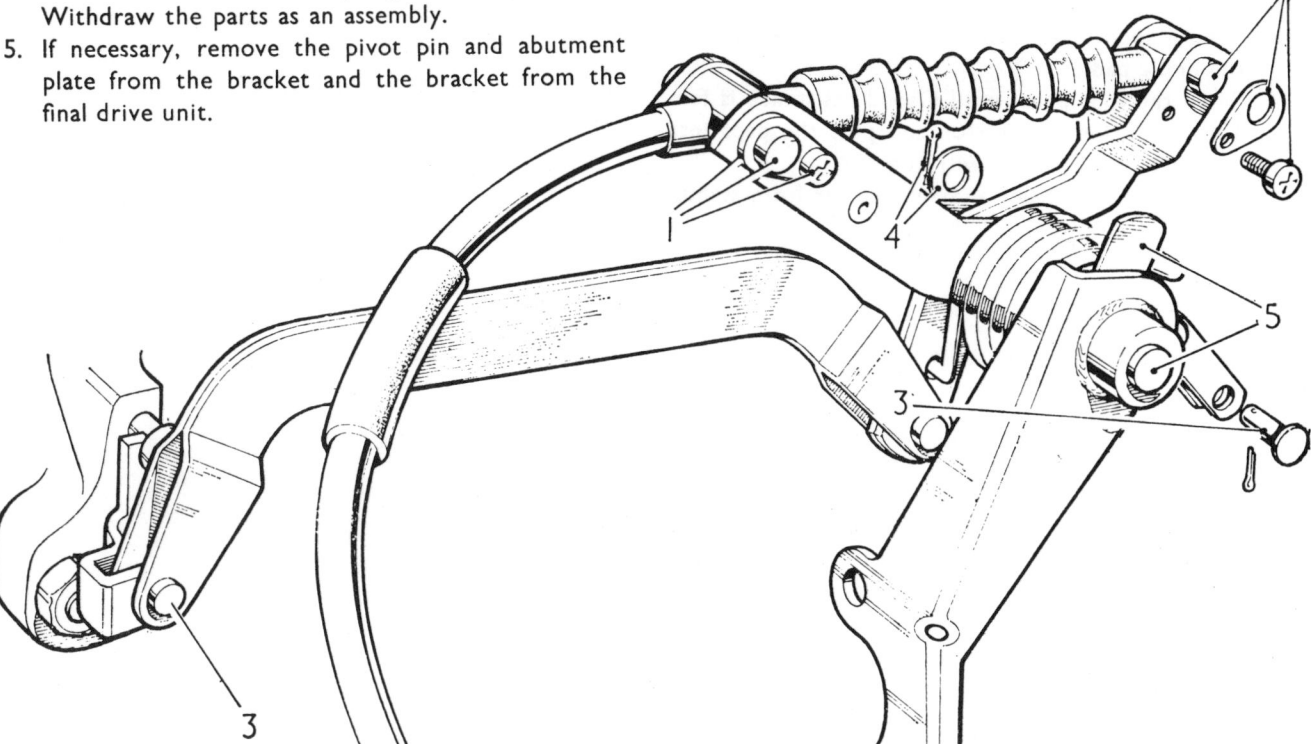

Rover 3500 and 3500S Manual AKM 3621

BRAKES

FRONT BRAKE PADS

—Remove and refit 70.40.02

Removing

1. Slacken the front road wheel nuts.
2. Jack up the front of the car and support on stands
3. Remove the front road wheels.
4. Where applicable, disconnect the electrical lead, brake pad to warning light, at the plug connector on the swivel pillar.
5. Remove the special clip and withdraw the pad retaining pins.
6. Withdraw the pads complete with anti-rattle springs and damping shims.
7. If the linings have worn down to 3 mm (0.125 in) or less, the pads must be replaced. If the pads are not to be replaced, mark each pad in order that it may be refitted in its original position.

Refitting

NOTE: The shim with the largest surface area fits behind the pad on the single piston side of the caliper. Where applicable, the pad with the electrical lead fits on the single piston side of the caliper.

8. Push in the pistons with an even pressure to the bottom of the cylinder bores. Then slide the pads into position, together with the damping shims. Ensure that the arrow cut-out in the shim points in the direction of forward rotation.
9. Refit the anti-rattle springs, one on each pad, then replace the pad retaining pins, ensuring that the anti-rattle springs are clipped under the pins.
10. Fit new special clips.
11. Where applicable, reconnect the electrical leads to the brake pads.
12. Pump the foot pedal until a solid resistance is felt. This repositions the piston and puts the pads into slight frictional contact with the disc.
13. Refit the road wheels and remove car from stands.

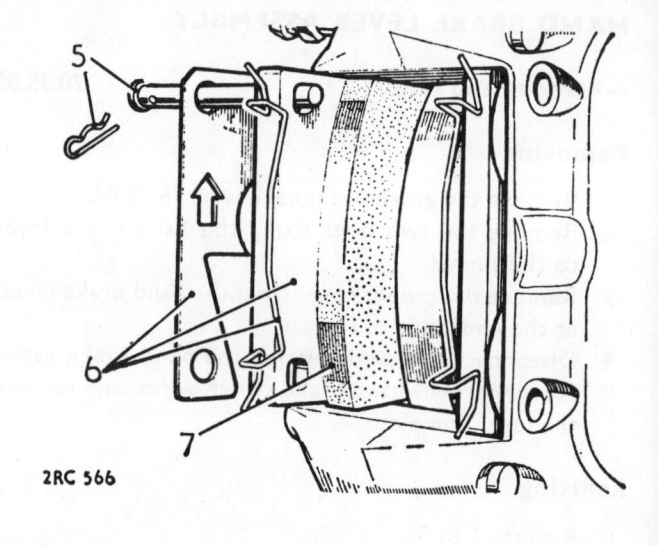

Rover 3500 and 3500S Manual AKM 3621

BRAKES

REAR BRAKE PADS

—Remove and refit   70.40.03

Service tool 601962—Piston setting tool

Removing

1. Disconnect the battery earth lead.
2. Disconnect the electrical leads, rear brake pads to warning light, at the plug connectors above the final drive unit.
3. Remove the clevis pin securing the hand brake link lever to the R.H. caliper.
4. Remove the clevis pin securing the outer bell crank levers and link plates to the L.H. caliper.
5. Disconnect the bias spring between the link lever and R.H. caliper.
6. Remove the fixings securing the bell crank lever bracket to the final drive unit.
7. Withdraw the handbrake linkage and allow to hang clear of the calipers.
8. Disengage the tab washer and remove the securing bolt and anti-rattle spring from the rear of each caliper.
9. Unscrew the drag pins.
10. Withdraw the outer pads in either direction.
11. Withdraw the pad retainer plate from the rear of each caliper.
12. Swing the rear end of the inner pads downward and withdraw them towards the front of the car.

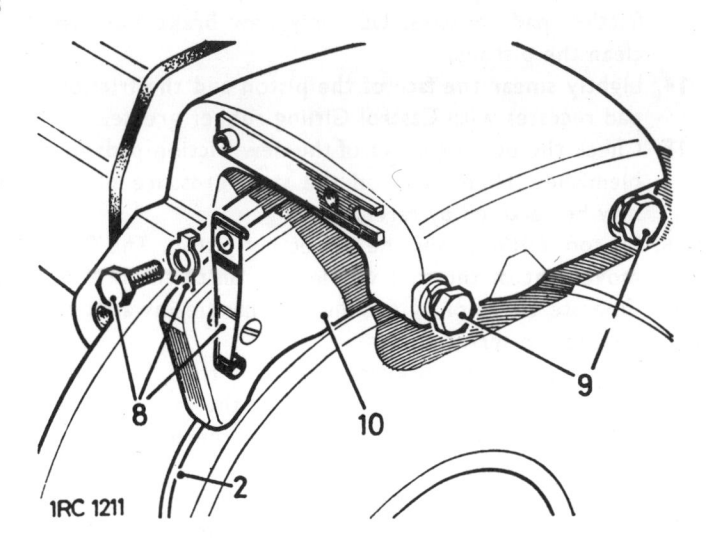

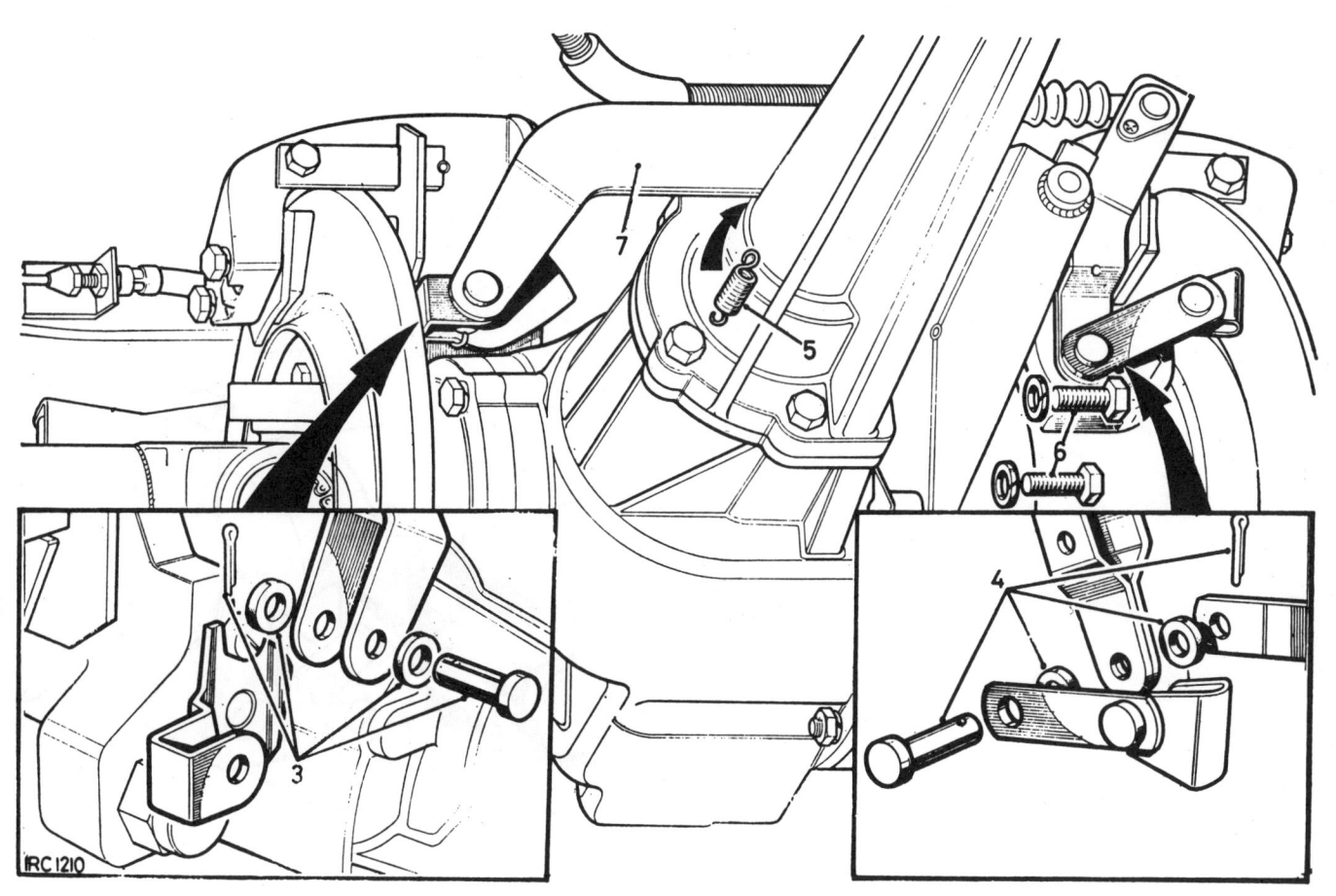

continued

Rover 3500 and 3500S Manual AKM 3621

70.40.03
Sheet 1

BRAKES

Refitting

13. Clean the exposed part of the pistons and the friction pad recesses. Use only new brake fluid to clean the pistons.
14. Lightly smear the face of the piston and the friction pad recesses with Castrol Girling rubber grease.
15. Check the bearing edges of the new friction pads for blemishes. High spots on the steel pressure plates may be rectified by carefully filing.
16. Fit tool 601962 over the projecting piston. The first movement of the tool will lock it on to the piston. Operate the tool by pushing the handle inwards to engage the serrations.
17. Turn the piston clockwise until it is right back and the ratchet is audible. Ensure that the lever on the piston is upright, if necessary turn the piston anti-clockwise. Remove the tool.
18. Locate the inner pad into each caliper and fit the retaining plate, with the off-set cut-out uppermost.
19. Reverse 1 to 10.
20. Pump the brake pedal and operate the driver's handbrake lever until a solid resistance is felt.
21. With the driver's handbrake lever released, check that the handbrake levers on both calipers are abutting the stops.
22. If necessary, adjust the handbrake cable to achieve the the above condition.

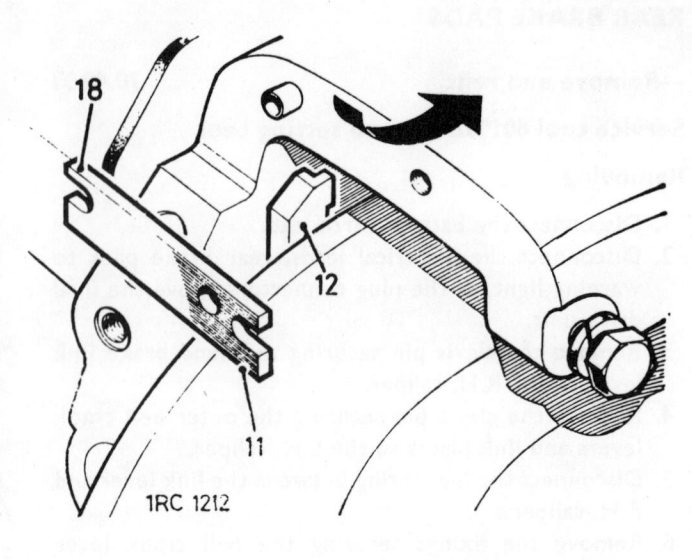

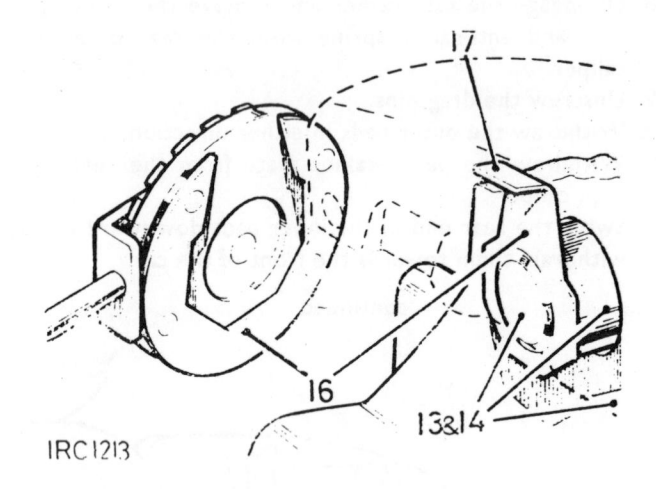

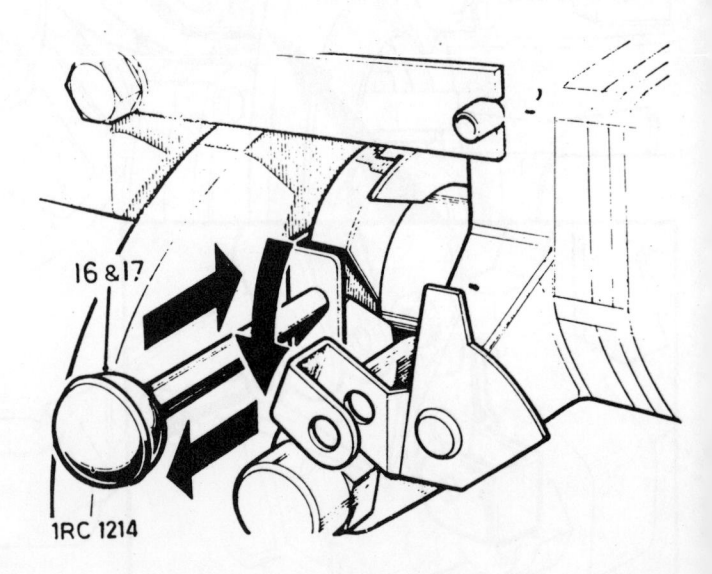

BRAKES

SERVO ASSEMBLY—Single line system

—Remove and refit 70.50.01

Removing

NOTE: Before removing the assembly from the vehicle, clean thoroughly, particularly at pipe connections, using Ethyl Alcohol (Methylated Spirit) or equivalent as a solvent. After removal, plug the pipes and the exposed ports of the unit to prevent loss of fluid and entry of dirt.

1. Remove the servo feed pipe to the five-way junction. and blank off the union on the junction.
2. Disconnect the vacuum pipe at the non-return valve on the servo.
3. Disconnect at the servo, the pipe from the master cylinder to servo.
4. Remove the two nuts and spring washers securing the servo to the base unit, these are accessible from under the front wing.
5. Remove the nut and bolt from the bracket securing the slave cylinder to the base unit.
6. Remove the servo.

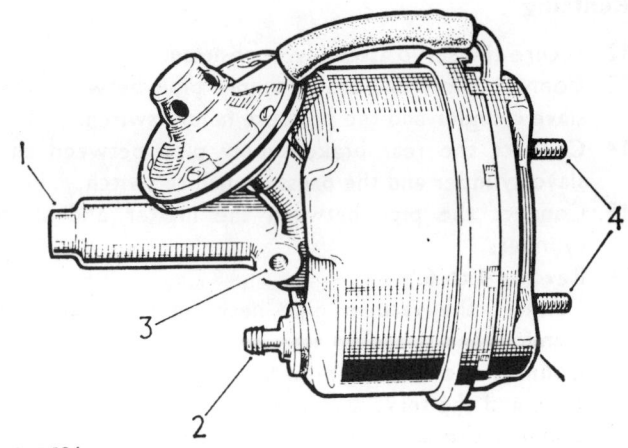

Removing

7. Reverse 1 to 6.
8. Bleed the brakes. 70.25.02.

SERVO ASSEMBLY—Dual-line system

—Remove and refit 70.50.01

Removing

1. Disconnect the battery earth lead.
2. Clean the servo assembly, particularly at pipe connections, using Ethyl Alcohol (methylated spirit) or equivalent as a solvent.
3. Disconnect the three air and vacuum pipes from the vacuum chamber.
4. Disconnect the electrical leads from the R.H. brake fluid reservoir and from the pressure failure switch, move leads to one side.
5. Disconnect, at the servo slave cylinder, the pipe from the R.H. brake fluid reservoir, drain the reservoir and discard the fluid.
6. Remove the R.H. brake fluid reservoir.
7. Disconnect, at the servo slave cylinder, the pipe, master cylinder to servo, and the two feed pipes to the pressure failure switch.
8. Slacken the servo rear mounting bolt. Note that the rear mounting bracket is slotted.
9. Remove the servo fixings, two nuts, plain and spring washers, from under the R.H. front wing valance.
10. Remove servo unit.
11. Blank off the disconnected pipes and ports to prevent loss of fluid and entry of dirt.

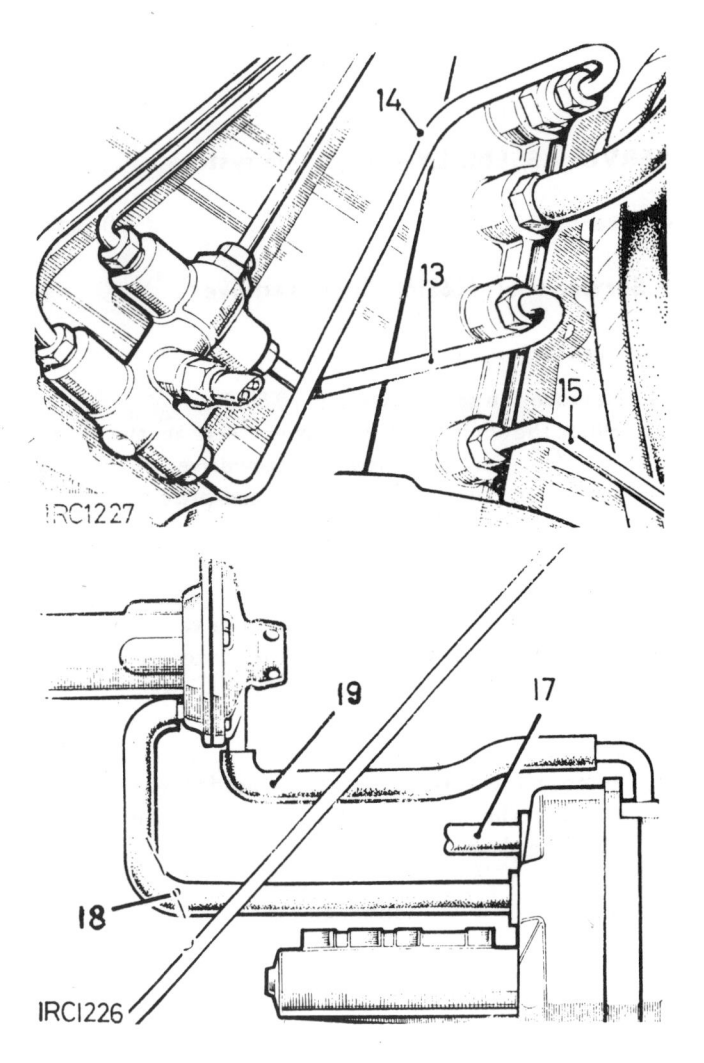

continued

Rover 3500 and 3500S Manual AKM 3621

BRAKES

Refitting

12. Secure the servo assembly in position.
13. Connect the front brake supply pipe between the slave cylinder and the pressure failure switch.
14. Connect the rear brake supply pipe between the slave cylinder and the pressure failure switch.
15. Connect the pipe between the master and slave cylinders.
16. Reverse 4 to 6.
17. Connect the vacuum pipe between the induction manifold and the servo assembly.
18. Connect the vacuum pipe between the air control valve and the servo assembly.
19. Connect the air inlet pipe between the air control valve and the servo assembly.
20. Reconnect the battery earth lead.
21. Bleed the complete brake system. 70.25.02.

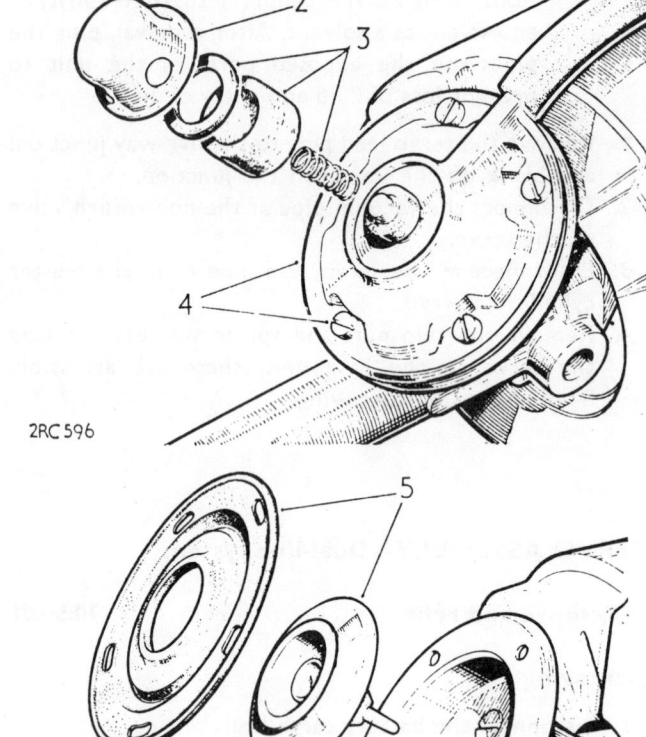

SERVO ASSEMBLY—Single line system

—Overhaul   70.50.06

Service tool: 606516   Cover remover

Dismantling

1. Remove the servo assembly. 70.50.01.
2. Hold the unit in a soft-jawed vice at the slave cylinder. Carefully insert a screwdriver in one of the cover holes of the air filter and prise the cover away.
3. Remove the sorbo washer, filter and spring, as illustrated.
4. Remove the five screws securing the valve cover to the housing and withdraw the cover from the rubber hose.
5. Lift off the reaction valve diaphragm and support, and separate.
6. Remove the three countersunk-headed screws and separate the valve housing, together with gasket from the slave cylinder body.

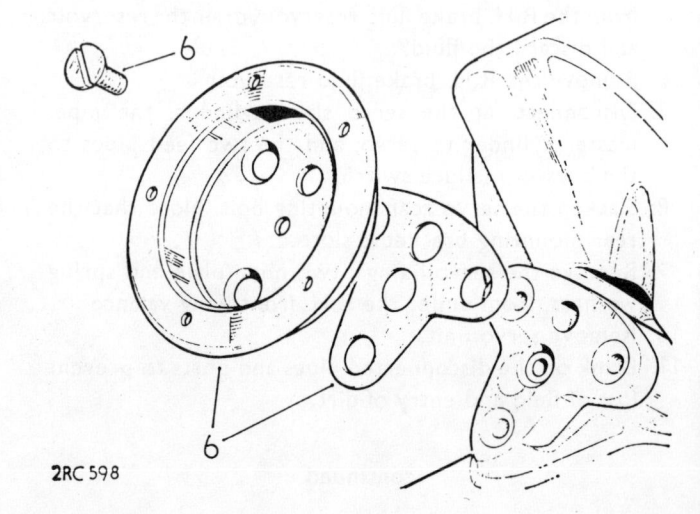

continued

70.50.01
Sheet 2
70.50.06
Sheet 1

Rover 3500 and 3500S Manual AKM 3621

BRAKES—Single line system

7. Remove the air valve piston by applying low air pressure to the fluid inlet connection whilst blanking off the hydraulic outlet.
8. Remove the cup from the valve piston.
9. Prise out the vacuum non-return valve from the vacuum shell.
10. Reposition the unit in the vice so that the vacuum shell is uppermost. Fit 606516, to the end cover fixing with two $\frac{5}{16}$ in UNF nuts, and remove the cover by turning the tool in an anti-clockwise direction.
11. Disengage the diaphragm from the diaphragm support by peeling it out of the rim of the vacuum shell and off the lip of the support.
12. Lightly press the diaphragm support inwards and shake the key from the diaphragm support.
13. Remove the support and the return spring.
14. Bend back the tabs on the locking plate, remove the set bolts, and withdraw locking tab and abutment plate.
15. Separate the vacuum chamber, together with gasket from the slave cylinder body.
16. Withdraw the push-rod and internal parts from the slave cylinder.

continued

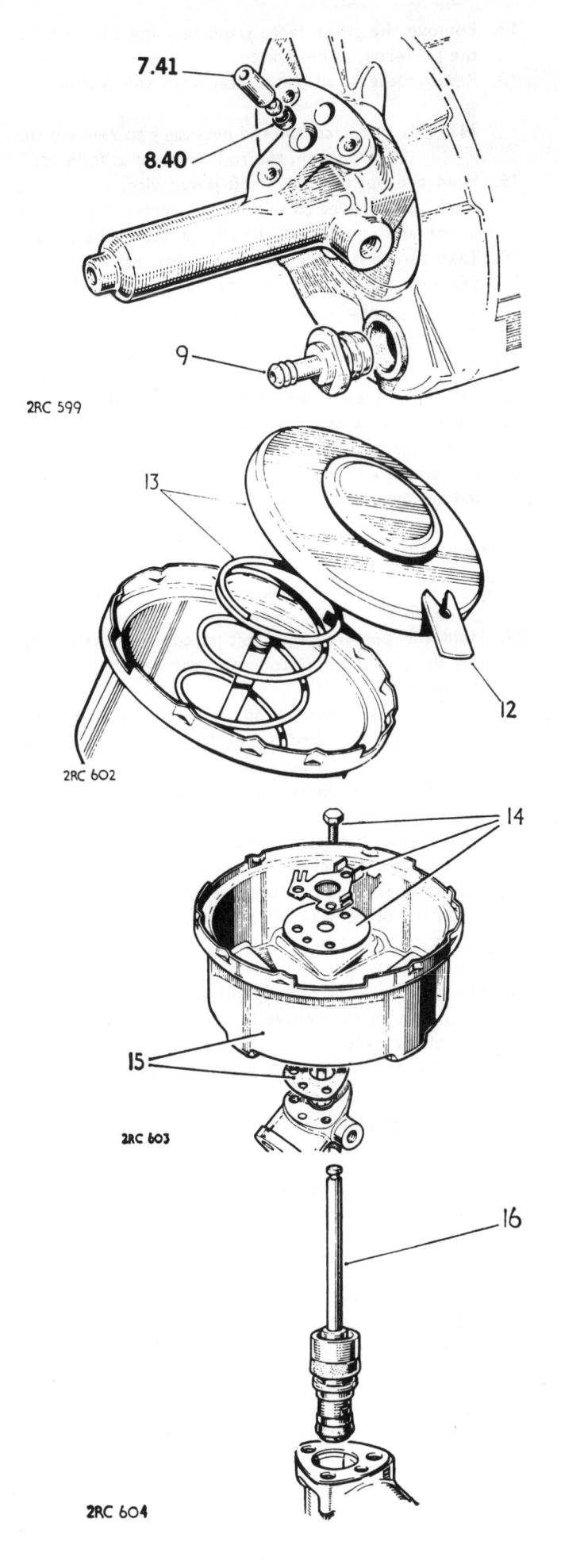

Rover 3500 and 3500S Manual AKM 3621

70.50.06
Sheet 2

BRAKES—Single line system

17. Remove the guide, large gland seal and spacer from the push-rod, as illustrated.
18. Remove the small piston seal from the piston.

   NOTE: If it is considered necessary to remove the piston from the rod, proceed as follows:

19. Hold the push-rod in a soft-jawed vice, then use a small screwdriver to expand the spring clip on the piston and gently ease the clip off along the piston.
20. Take care not to damage the piston surface.
21. Separate the piston from the push-rod by pressing out the small pin, as illustrated.

Inspecting

22. Examine all components for wear or damage and replace any parts necessary. If the bore of the slave cylinder is scored or rusty, this must also be replaced.
23. The piston and all rubber parts should be renewed and are available in a repair kit.

Reassembling

NOTE: If the piston has been removed from the rod, refit as follows: 24 to 27

24. Hold the push-rod in a soft-jawed vice, then slide the slave cylinder piston over the tapered end of the push-rod.
25. Hold the small spring within the piston back towards the diaphragm end of the push-rod, and insert the small pin into the hole in the push-rod, ensuring that the end of the spring rests against the pin and that it does not pass through the coil of the spring.
26. With the pin correctly positioned, expand the spring clip and pass it over the piston to secure the pin. It is most important that the outside diameter of the piston is not damaged in any way, however slight, during this operation.
27. Remove push-rod from vice.
28. Fit a new small 'U' section seal to the slave cylinder piston so that the groove in the seal is towards the head of the piston.
29. Assemble the spacer with the large diameter against the head of the piston.
30. Fit the large gland seal, grooved face leading, on to the push-rod.

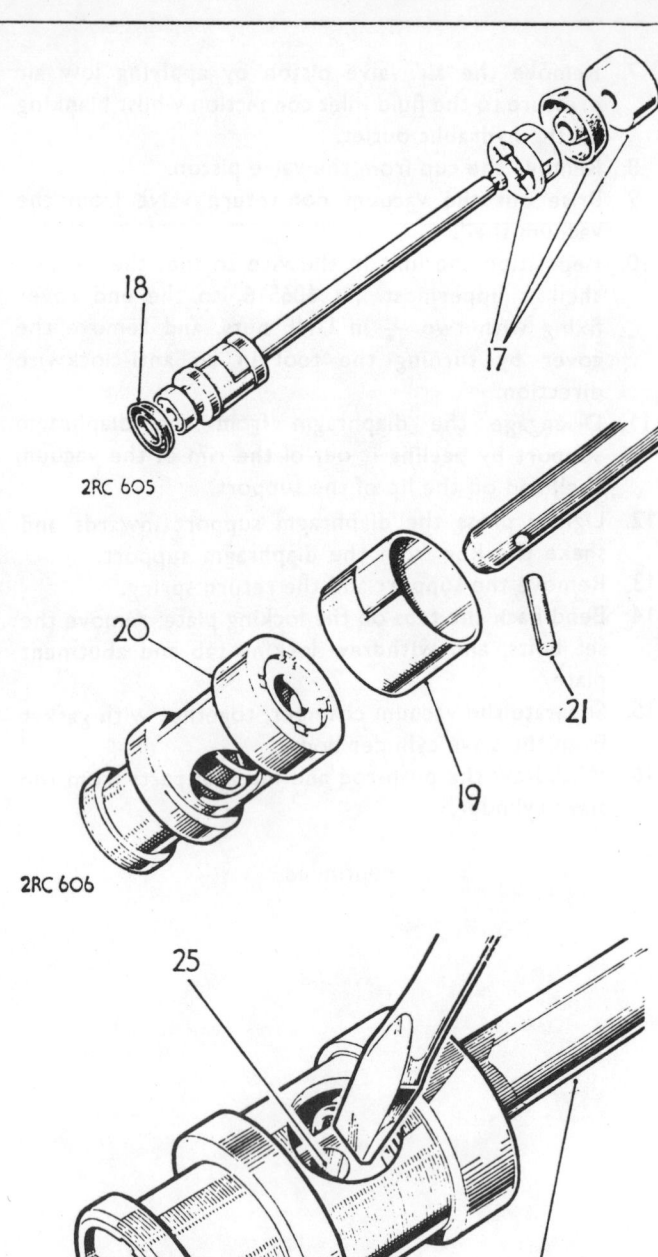

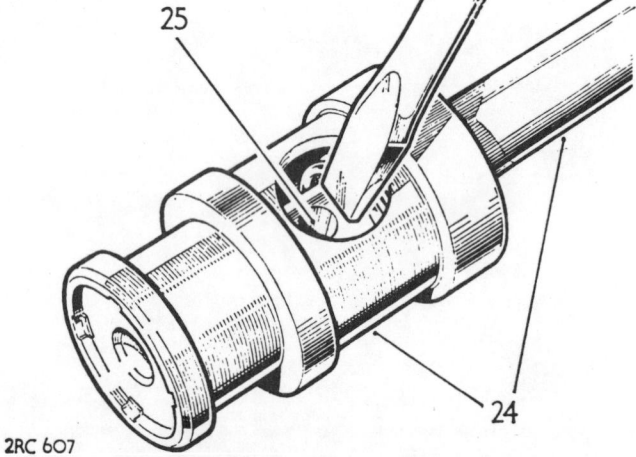

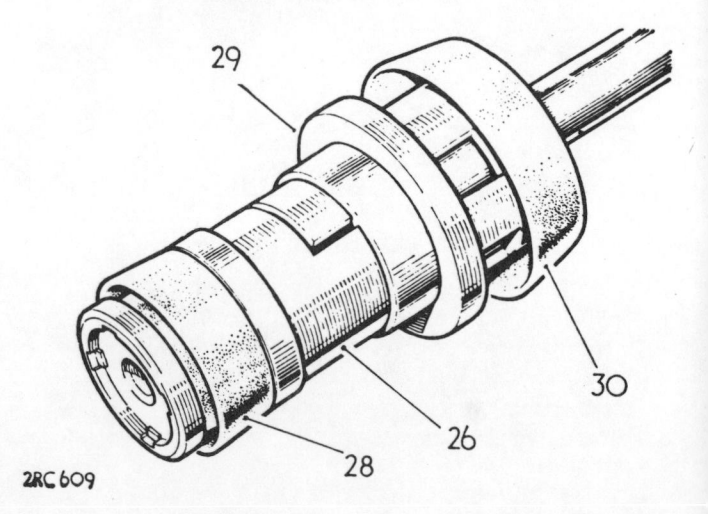

continued

Rover 3500 and 3500S Manual AKM 3621

BRAKES—Single line system

31. Hold the slave cylinder body in a soft-jawed vice, then fit the push-rod assembly into the bore of the slave cylinder, ensuring that the seals are not damaged or distorted.
32. Refit the push-rod guide into the end of the slave cylinder so that its flat face is innermost in the bore.
33. Locate a new gasket over the push-rod guide following up with the vacuum shell, abutment plate and locking plate, line up the holes with the tapped holes in the slave cylinder, and insert the three set bolts, tightening to a torque of 1,6 to 1,9 kgf.m (12 to 14 lbf ft). Bend over the locking tabs.
34. Ensure that the push-rod is withdrawn fully and place the return spring over the push-rod.
35. Locate the diaphragm support on the end of the push rod.
36. Compress support and secure by fitting the key into the slot in the side of the diaphragm support. Ensure that the key engages the groove in the push-rod and that it is pushed right in.
37. Stretch the smaller diameter of the diaphragm into position on the diaphragm support, ensuring that it is correctly bedded into its groove. Locate the outer edge of the diaphragm in the vacuum shell. The diaphragm MUST be fitted dry. DO NOT use any form of lubricant.
38. If the flat face of the diaphragm appears buckled, this indicates that the diaphragm has not been assembled correctly. Ensure correct location in groove of diaphragm support and rim of vacuum shell.
39. With the special tool 606516, attached to the end cover, refit the cover by pressing downwards and turning in a clockwise direction. Ensure that the end cover air pipe is positioned in correct relationship with the face of the air valve. Remove special tool.
40. Reposition the slave cylinder in vice so that the air valve face is uppermost. Ease the cup into the groove in the air valve piston so that the lip of the cup faces the shouldered end of the piston.
41. Insert the piston into the valve bore of the slave cylinder.
42. Position a new gasket on the slave cylinder.
43. Fit the valve housing, line up the holes, and secure by fitting the three countersunk-headed screws. The screws should be tightened to a torque of 0,6 to 0,9 kgf.m (5 to 7 lbf ft).
44. Stretch the reaction valve diaphragm on to the diaphragm support and insert the push-rod portion of the diaphragm support through the hole in the valve housing.
45. Stretch the valve rubber, which is formed with a groove around its inside diameter, on to the valve stem flange with the fingers. Insert the valve stem through the hole in the valve cover, fit the other valve rubber over the valve stem, and secure by fitting the snap-on cap. If necessary, warm cap in hot water for ease of assembly.

continued

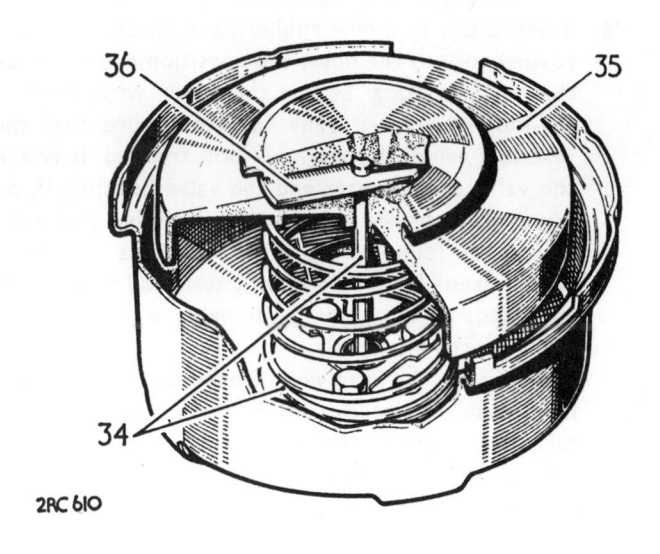

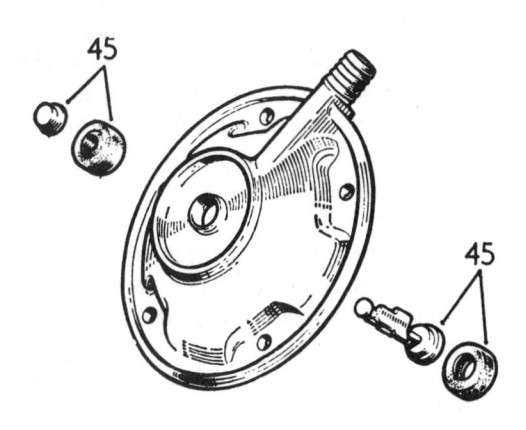

Rover 3500 and 3500S Manual AKM 3621

70.50.06
Sheet 4

BRAKES

46. If necessary, fit a new rubber hose, insert the valve cover end into the hose, then position the cover on the valve housing. Secure the cover with the five self-tapping screws. Take care to ensure that the reaction valve diaphragm is not trapped between the valve cover bosses and the valve housing. If, on removal of the air filter cover, the valve and valve seals were found to be unserviceable, a new filter cover assembly complete with seals and valve must be fitted; this is essential as valve stem and seals are a selective assembly.
47. Examine the air filter and, if necessary, renew; any light dust may be cleared by washing the filter in methylated spirits and drying with an air line. Do not use any other cleaning fluid.
48. Position the air filter over the air valve, then position the small spring over the snap-on cap.
49. Fit the sorbo washer in the air filter cover, then snap the assembly on to the air valve cover.
50. If necessary, replace the sealing rubber for the non-return valve, then push the non-return valve into position.
51. Refit the servo assembly. 70.50.01.

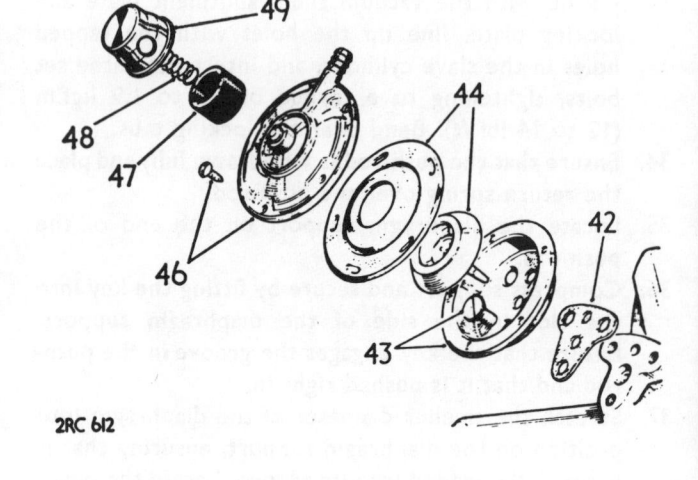

SERVO ASSEMBLY—Dual line system

—Overhaul   70.50.06

Service tool: 606516   Cover remover

Dismantling
1. Remove the servo assembly. 70.50.01.
2. Grip the slave cylinder in a soft-jawed vice, angled at about 45° with the ports uppermost.
3. Fit 606516 to the end cover and turn the cover in an anti-clockwise direction as far as the stops on the cover will allow. Lift off the cover.
4. Free the edge of the rubber servo diaphragm from the rim of the shell, and ease the centre of the diaphragm out of the groove in the plastic diaphragm support.
5. Turn the diaphragm support so that the slot for the push-rod retaining key faces downwards. Light fluctuating pressure on the support into the shell will release the key and permit the support to lift clear of the push-rod under the influence of the main return spring. Extract the spring from the shell.
6. Bend back the tabs of the exposed locking plate. Remove the bolts and the locking plate, abutment plate and servo shell.
7. Withdraw the servo push-rod and attached piston assembly from the slave cylinder bore by pulling gently on the rod.

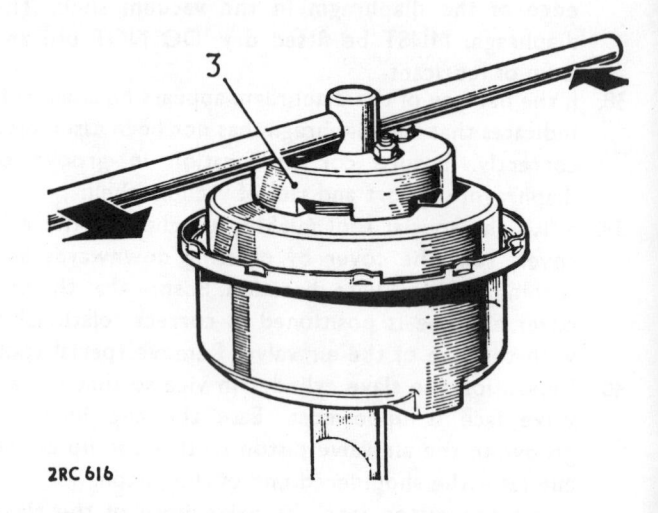

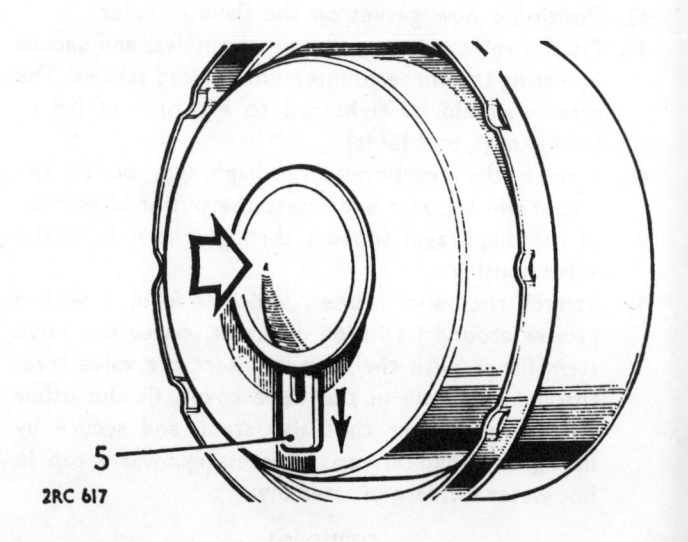

continued

70.50.06
Sheet 5

Rover 3500 and 3500S Manual AKM 3621

BRAKES—Dual line system

8. Slide off the plastic bearing, rubber cup and plastic spacer noting their relative positions for subsequent refitting.
9. The push-rod can be separated from the piston by opening the retaining clip with a small screwdriver, to expose and then drive out the connecting pin.
10. Unscrew the fluid inlet adaptor, or remove the retaining circlip as applicable to release the adaptor.
11. Remove the secondary piston stop pin from the base of the port.
12. Remove the secondary piston and spring from the slave cylinder bore. Ejection of the piston can be assisted by gently tapping the mouth of the cylinder with a rubber or hide hammer.
13. The piston seals, piston washer, and spring and retainer may then be removed from the piston.
14. Finally, unscrew the adaptor in the outlet port at the end of the slave cylinder, and extract the trap valve and spring.

Inspecting

15. Carefully inspect all parts for faults and wear. Be prepared to fit new rubber parts throughout.
16. A replacement assembly will always be required where the cylinder bore, after cleaning, shows the slightest signs of corrosion or scoring.
17. New copper washers will be needed for the adaptors in the slave cylinder.
18. Ensure that the by-pass ports and feed hole drillings into the cylinder are not obstructed. Likewise, the equalising holes in the trap valve body and the fluid feed holes in the head of the secondary piston. Take care not to enlarge the holes.
19. Check that the curved spring insert in the trap valve body is intact and undamaged.

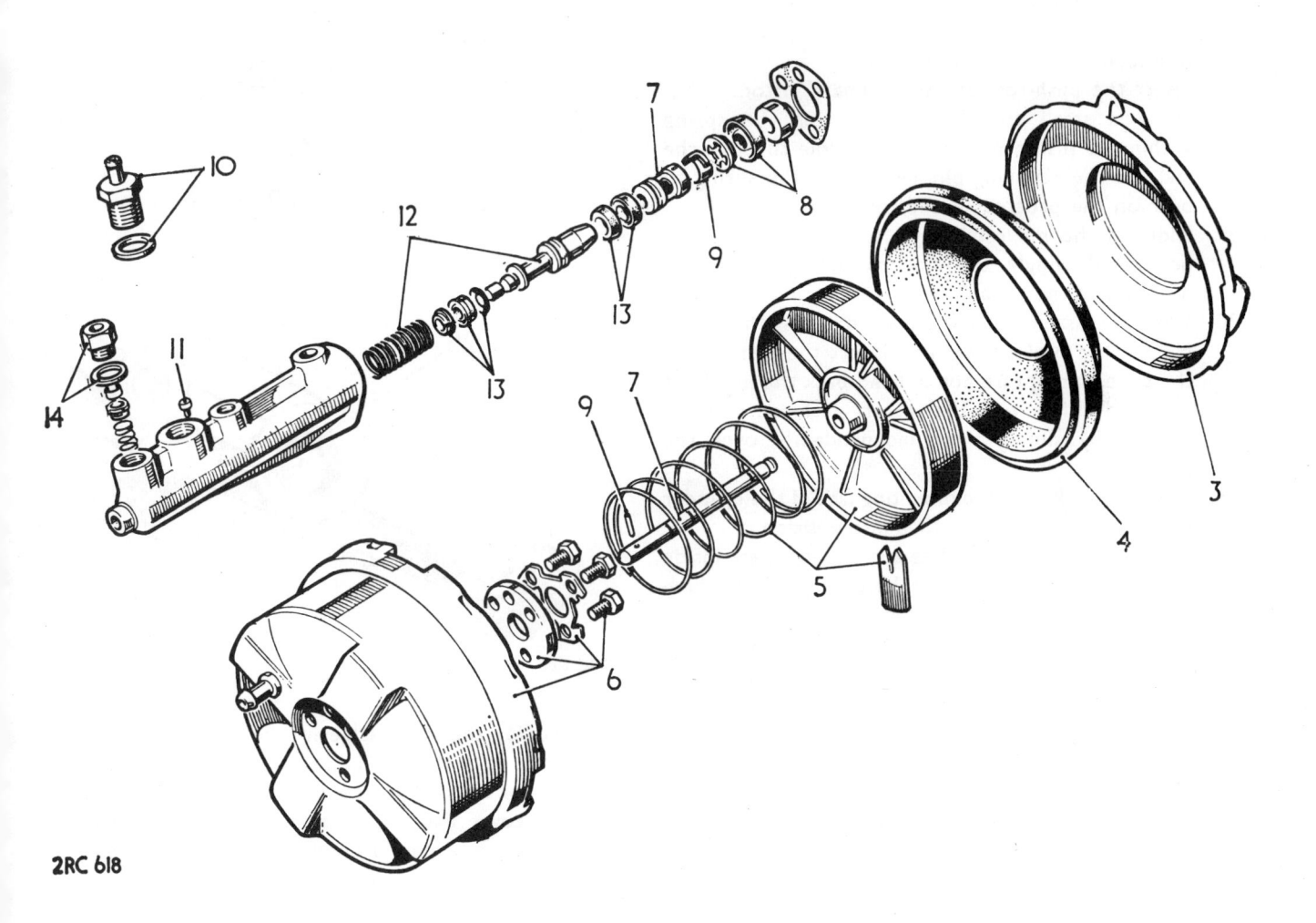

continued

2RC 618

Rover 3500 and 3500S Manual AKM 3621

70.50.06
Sheet 6

BRAKES—Dual line system

Reassembling

20. Use clean brake fluid to lubricate the bore of the slave cylinder before assembly.
21. Locate the piston washer over the secondary piston head extension, convex face towards the piston and using the fingers only, assemble the two seals on to the piston so that their lips face outwards.
22. Press the spring retainer on to the piston head extension and fit the spring.
23. Grip the slave cylinder body in the vice at an angle of approximately 45°, mouth of bore uppermost.
24. Coat the piston seals with LOCKHEED Disc Brake Lubricant and insert the assembly into the slave cylinder bore, spring leading. Ensure that the lip of the leading seal is not bent back on entering the bore.
25. Press down the piston against the action of the return spring, until the drilled piston head passes the piston stop hole in the inlet port. Fit the stop pin and secure by fitting the inlet adaptor. Tighten the adaptor to a torque of 4,6 kgf.m (33 lbf ft).
26. Refit the trap valve assembly in the outlet port at the end of the cylinder body and tighten the adaptor to a torque of 4,6 kgf.m (33 lbf ft).
27. Insert the push-rod into the primary piston.
28. Using a small screwdriver compress the spring within the piston to uncover the hole through the rod. Fit the retaining pin and release the spring to bear on the protruding ends of the pin.
29. Slide on the pin retaining clip ensuring that it is positioned snugly and not exceeding the outer diameter of the piston.
30. Coat the piston seal and push-rod seal with LOCKHEED Disc Brake Lubricant. Insert the piston into the bore, taking care not to bend back the lip of the seal, and follow with the spacer, push-rod seal and bearing by sliding them independently over the push-rod.
31. Place the gasket on the mounting face of the cylinder before putting the servo shell into position.
32. Position the abutment plate and locking plate inside the shell and fit the three bolts, tightening them evenly to a torque of 1,8 kgf.m (13 lbf ft). Bend over the ears of the locking plate to secure the bolts.
33. Insert the main return spring with the first coil spaced around the abutment plate.
34. Fit the diaphragm support, locating this on the end coil of the spring with the key slot uppermost.
35. Gently press the support into the shell against the resistance of the return spring until the groove in the end of the push-rod is visible through the slot. Insert the key.
36. Make sure that the rubber servo diaphragm is completely dry, especially in the vicinity of the centre hole where there must be no trace of lubricant, likewise the diaphragm support particularly at the groove. Fit the diaphragm to the support, carefully bedding the inner edge in the groove. Smear the outer edge of the diaphragm with LOCKHEED Disc Brake Lubricant where it will contact the rim of the end cover and of the shell. Bed the edge of the diaphragm evenly around the rim of the shell.
37. Refit the end cover, taking care not to trap the rubber diaphragm with its edge, and making sure that the vacuum pipe is in its correct position.
38. Take off the end cover removal tool and inspect the servo for correct reassembly.
39. Refit the servo assembly. 70.50.01.

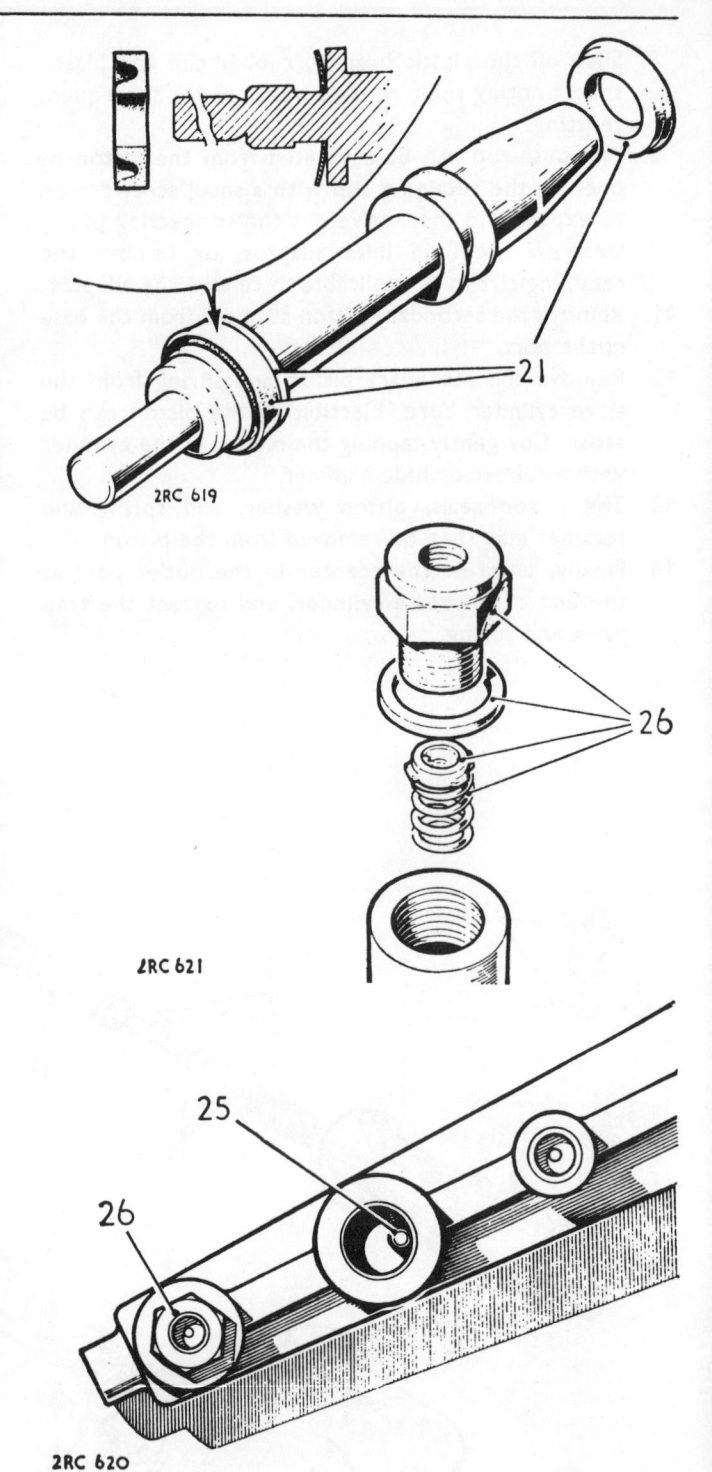

BRAKES

FRONT BRAKE CALIPERS

—Remove and refit 70.55.02

Removing

1. Slacken the front road wheel nuts.
2. Jack up the front of the car and support on stands.
3. Remove the front road wheel.
4. Disconnect the fluid feed pipe between the swivel pillar and the caliper, and withdraw the pipe. Blank off the ends of the pipes and the aperture in the caliper to prevent ingress of dirt.
5. To avoid leakage of brake fluid from the reservoir, depress the foot pedal and secure in that position until the work is completed.
6. Where applicable, disconnect the electrical lead, brake pad to warning light, at the plug connector on the swivel pillar.
7. Remove the front brake pads. 70.40.02.
8. Bend back the tabs on the locking plate, and remove the two bolts, locking plate and the caliper unit.

Refitting

9. Reverse 1 to 8. Tightening torque for the caliper securing bolts: 8,5 kgf.m (60 lbf ft).
10. Bleed the brakes. 70.25.02.

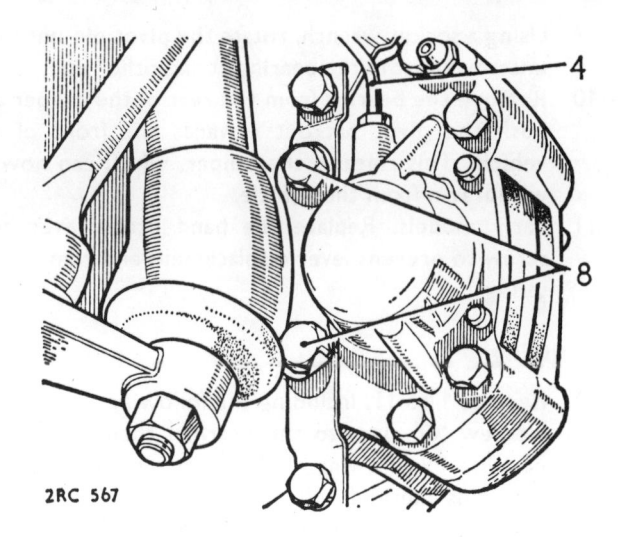

REAR BRAKE CALIPERS

—Remove and refit 70.55.03

Removing

1. Slacken the rear road wheel nuts.
2. Jack up the rear of the car and support on stands.
3. Remove the rear road wheel.
4. Remove the rear brake pads. 70.40.03.
5. RH caliper—Disconnect the fluid feed pipe from the bracket on the base unit.
6. Disconnect interconnecting pipe from the calipers.
7. To provide access to the caliper pivot nut, remove the stop pin for the hand brake lever, rotate the lever inwards and retain it with string to prevent its withdrawal. On some models fitted with a large pivot nut, it is necessary to withdraw the hand brake lever to allow a spanner to be fitted on the nut.
8. Remove the plug and spring.

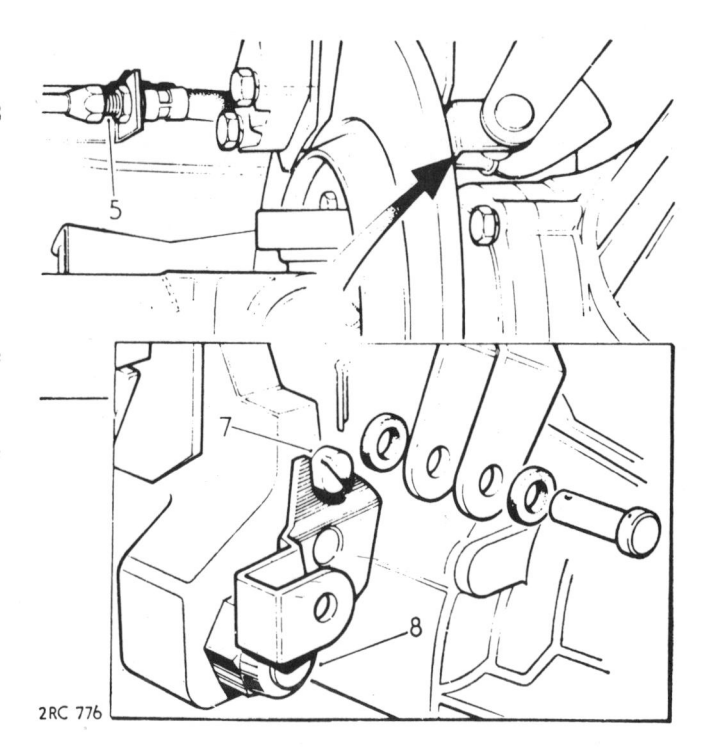

continued

Rover 3500 and 3500S Manual AKM 3621

70.55.02
70.55.03
Sheet 1

BRAKES

9. Using a socket wrench, rotate the pivot pin until it is unscrewed from the bearing at the other end.
10. Remove the bearing from the rear of the caliper and push the pivot pin out towards the front of the vehicle. This releases the caliper, which can now be withdrawn from the vehicle.
11. Early models. Replace the hand brake lever stop screw to prevent lever displacement and remove the string.

Refitting

12. Reverse 1 to 11, including the following.
13. Fit new 'O' rings to the bearing, then pull it into position by screwing in the pivot pin with an Allen key; this method will prevent damage to the two 'O' rings. Then tighten the bearing to a torque figure of 4,9 kgf.m (35 lbf ft).
14. Tightening torque for the spring-loaded plug: 3,8 to 4,9 kgf.m (28 to 35 lbf ft).
15. If it has been necessary to withdraw the hand brake lever, ensure after refitting that there is a resistance when rotating the lever; this indicates that the hand brake tappet is correctly located.
16. If the hand brake lever rotates without resistance, the tappet has dropped out of its bore into the cavity, where it turns sideways and engages in the groove of the cam lever. Reposition the tappet, as illustrated in the previous item.
17. Bleed the brakes. 70.25.02.

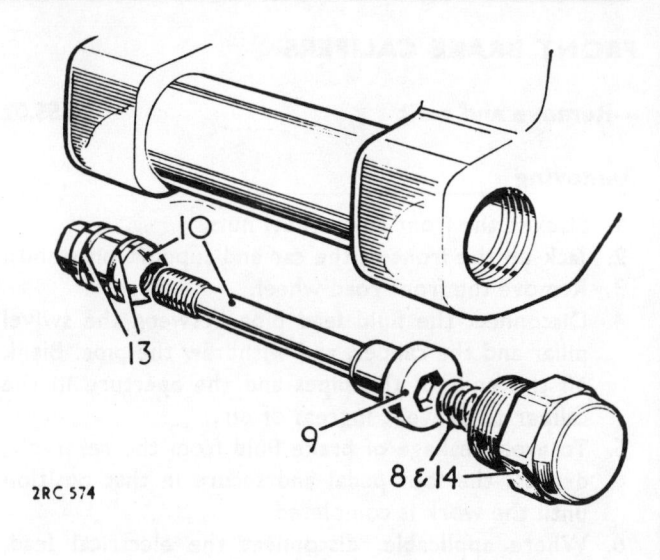

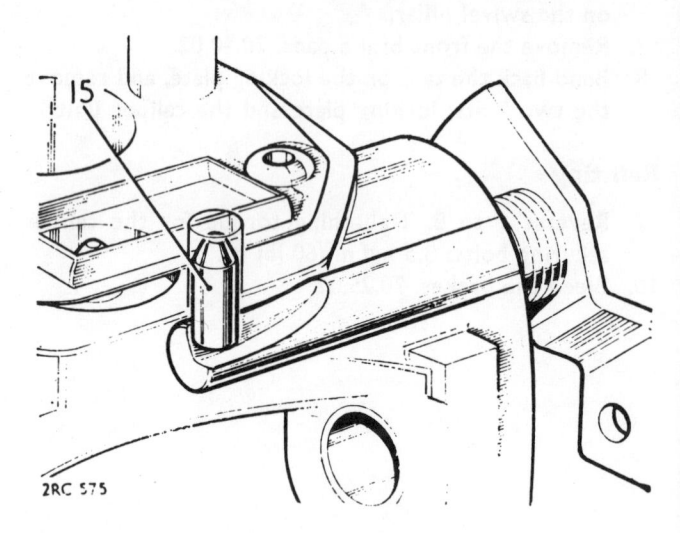

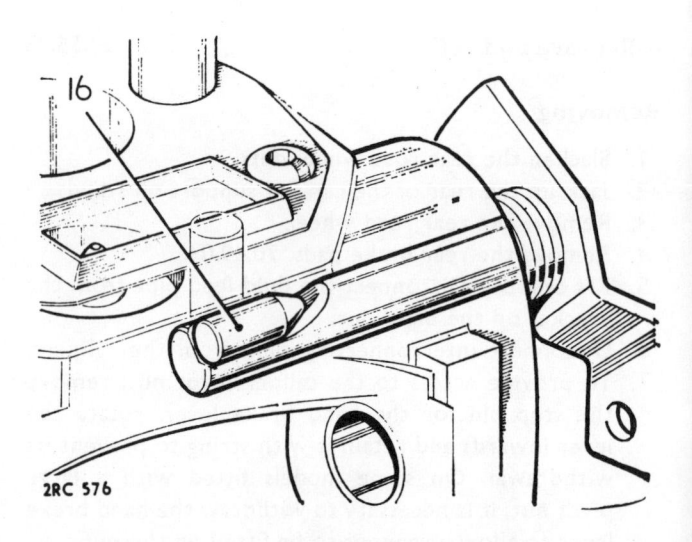

BRAKES

FRONT BRAKE CALIPERS

—Overhaul 70.55.13

Dismantling

NOTE: It is important that the bolts securing the two halves of the brake caliper are not disturbed; the pistons and seals must be removed without separating the caliper.

1. Remove the front brake caliper. 70.55.02.
2. Remove the bleed screw.
3. Remove the rubber boots from the inner and outer pistons and caliper housing.
4. Place a piece of clean rag between the pistons. Apply air pressure to the feed pipe hole to blow out the inner piston, then withdraw the two outer pistons.
5. Remove the inner sealing rings, taking care not to damage the bores or locating groove.

Inspecting

6. Wash all components in Girling cleaning fluid.
7. Have available a front caliper overhaul kit.

Reassembling

8. Clean out the grooves in the caliper housing and fit new sealing rings to the large groove.
9. Apply Castrol Girling Brake Fluid to the piston bore and piston exterior.
10. Fit the rubber boot to the housing and ensure that the lip of the boot is seated in the small groove.
11. Insert the piston squarely into the bore. Press fully in, excessive pressure should not be required. Ensure that the sealing lip is located in the groove in the piston.
12. Refit the bleed screw.
13. Refit the front brake caliper. 70.55.02.

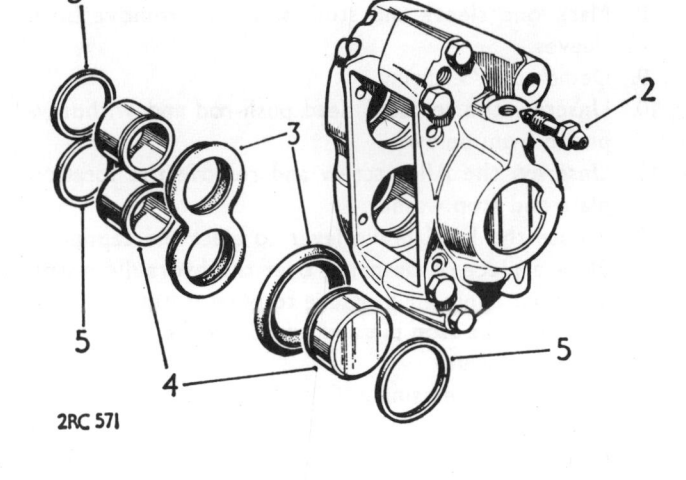

REAR BRAKE CALIPERS

—Overhaul 70.55.14

Dismantling

1. Remove the rear brake caliper. 70.55.03.
2. Remove the cover together with the large rubber seal, two small rubber seals and anti-corrosive paper.
3. Remove the nuts securing the beam.
4. Mark the strut to ensure correct replacement and then remove it by lifting the lever and pulling strut forward.
5. Press down the lever and withdraw it complete with pawl.
6. Push up the piston from below and ease off the main spring over the two studs.
7. Lift off the beam.

continued

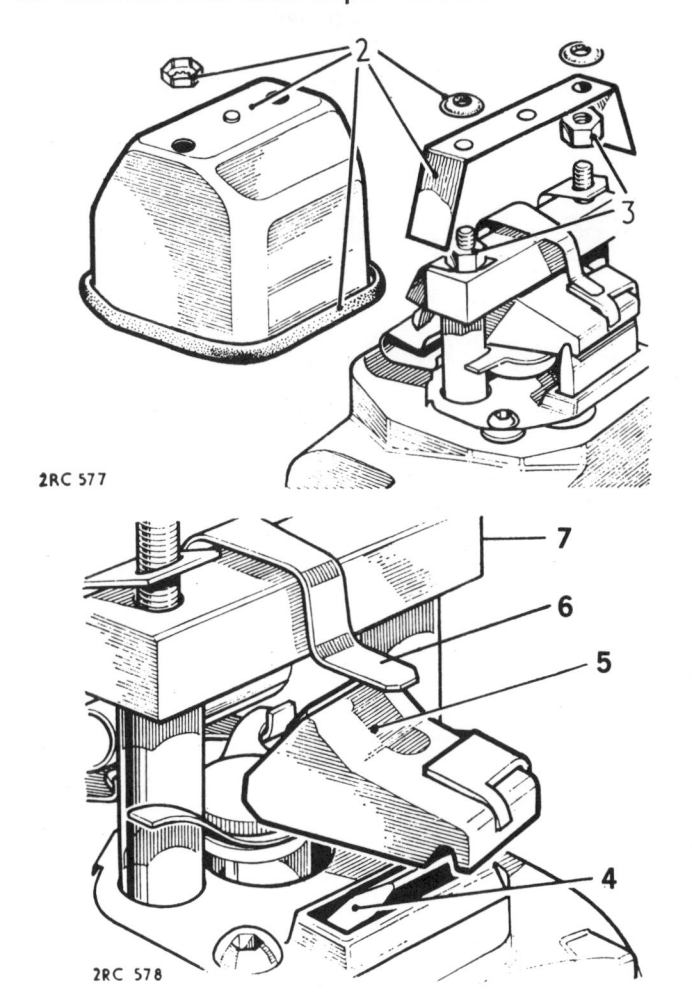

BRAKES

8. Mark one sleeve and stud, and then remove both sleeves.
9. Detach the 'S' spring.
10. Unscrew the serrated head push-rod and withdraw piston from below.
11. Unscrew the Allen screw and remove the location plate and stop washer.
12. Rotate the hand brake lever to eject the tappet.
13. Place a piece of clean rag over the hydraulic piston and then apply air pressure to blow it out.
14. Unscrew the stop pin and withdraw the hand brake lever and shaft.
15. Remove the remaining 'O' rings and retainers.
16. Remove the bleed screw.
17. Remove the drag pins.

Inspecting

18. Obtain new seals, drag pins and cover nuts.
19. Clean all the remaining parts and ensure that they are in good condition. Clean the hydraulic bore and piston with Girling Cleaning Fluid or clean brake fluid.
    NOTE: A caliper overhaul spares kit is available.

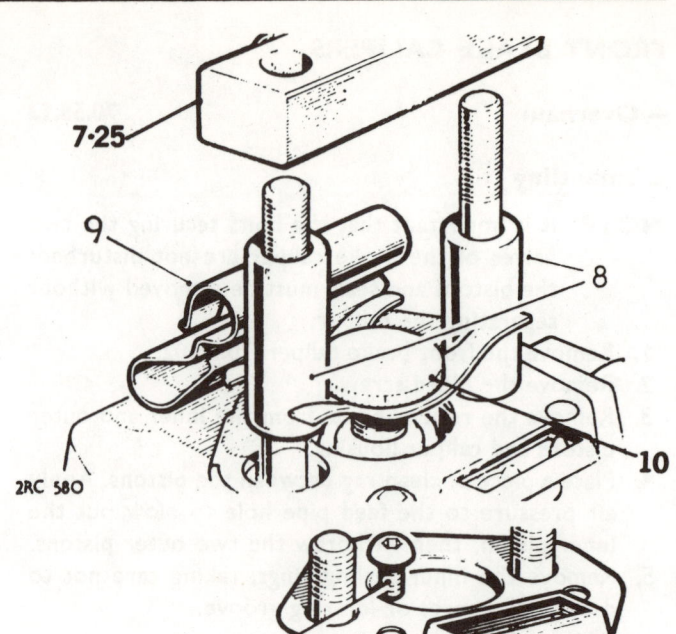

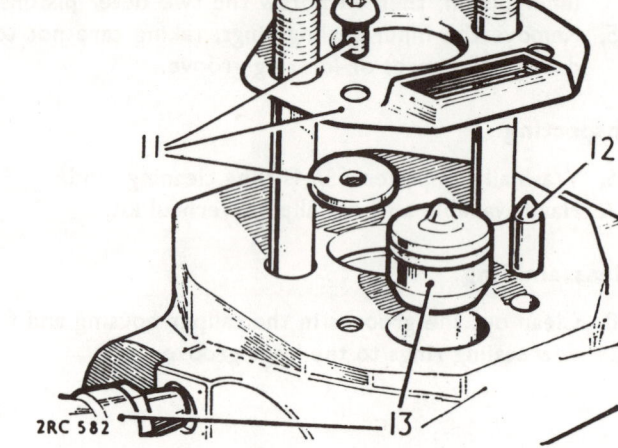

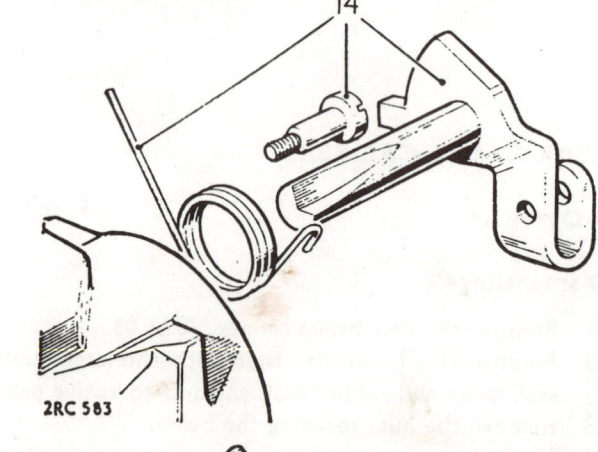

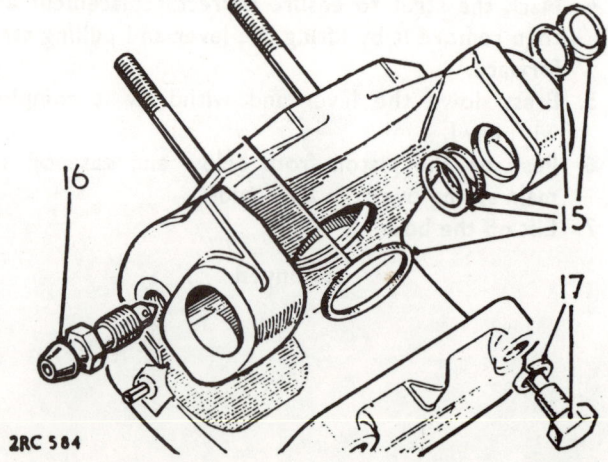

continued

70.55.14
Sheet 2

Rover 3500 and 3500S Manual AKM 3621

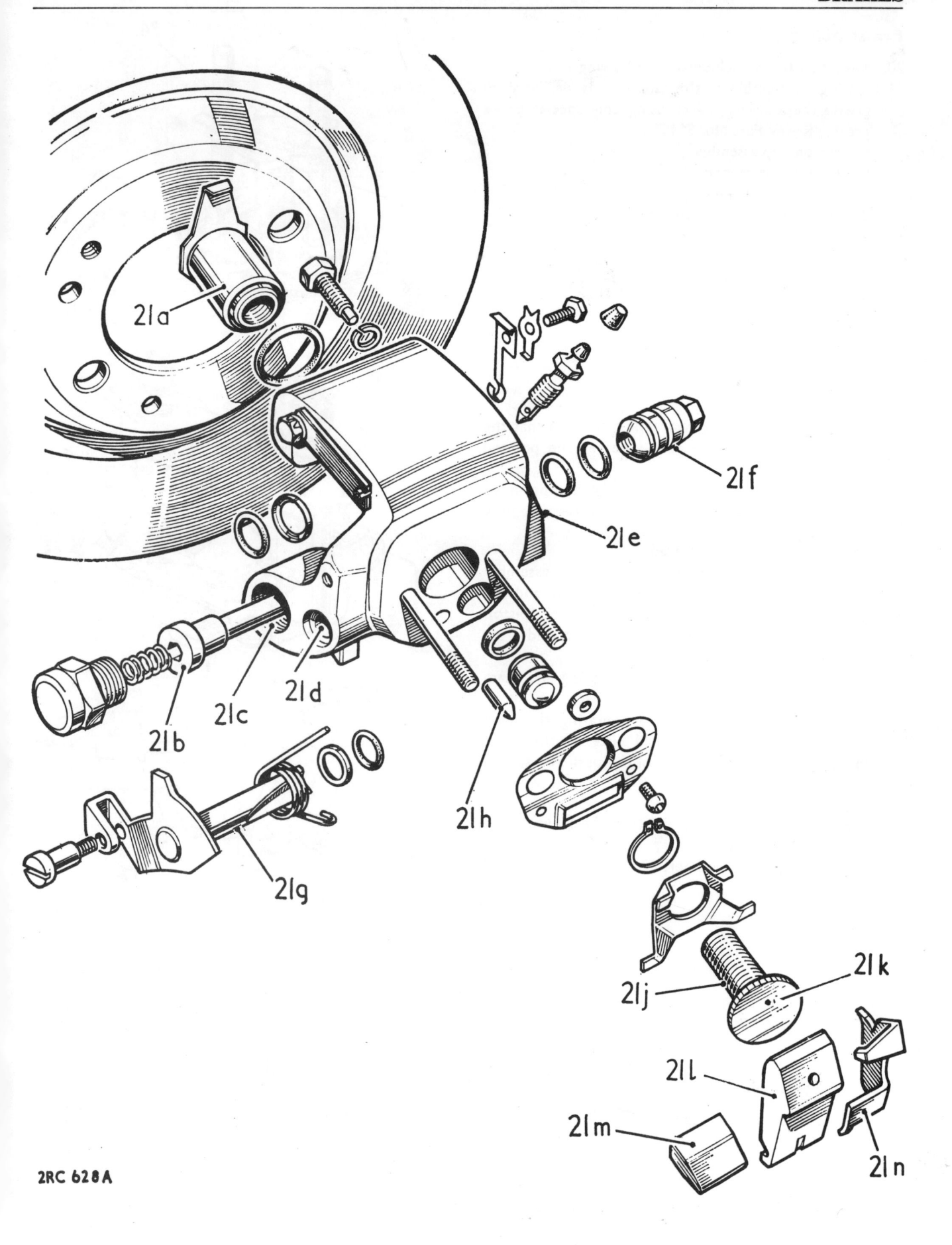

BRAKES

continued

Rover 3500 and 3500S Manual AKM 3621

2RC 628A

70.55.14
Sheet 3

BRAKES

Reassembling

20. Reverse 1 to 16, including the following.
21. During reassembly of the caliper it is essential to grease the following parts, using only special brake grease, Rover Part No. 514577.
    a. Piston cup assembly.
    b. Spindle for caliper.
    c. Aperture for spindle.
    d. Aperture for hand brake lever.
    e. Aperture for hinge pin.
    f. Hinge pin.
    g. Hand brake lever
    h. Tappet for hand brake.
    j. Push rod thread.
    k. Face of push rod.
    l. Lever for strut.
    m. Strut.
    n. Pawl.
22. Fit the new seal to the hydraulic piston, so that the smaller diameter of the seal is nearer the pointed end of the piston.
23. Insert the hand brake tappet, pointed end uppermost.
24. Screw in the serrated-head push rod, three complete turns only.
25. Locate the beam as illustrated.
26. Fit the pawl to the lever and insert the assembly slantwise, as illustrated, between the beam and the head of the push-rod. To assist, pull the piston down from below. Swing the lever square and press into position.
27. Tighten the nuts securing the main spring and beam to a torque figure of 3,8 kgf.m (28 lbf ft).
28. Test the action of the unit by operating the hand brake lever. The pawl should click as the lever moves and the serrated head of the push rod should rotate one tooth on return. If not, the lever with pawl is incorrectly located.
29. Screw in the piston fully and turn the lever to the position illustrated.
30. Refit the rear brake caliper. 70.55.03.

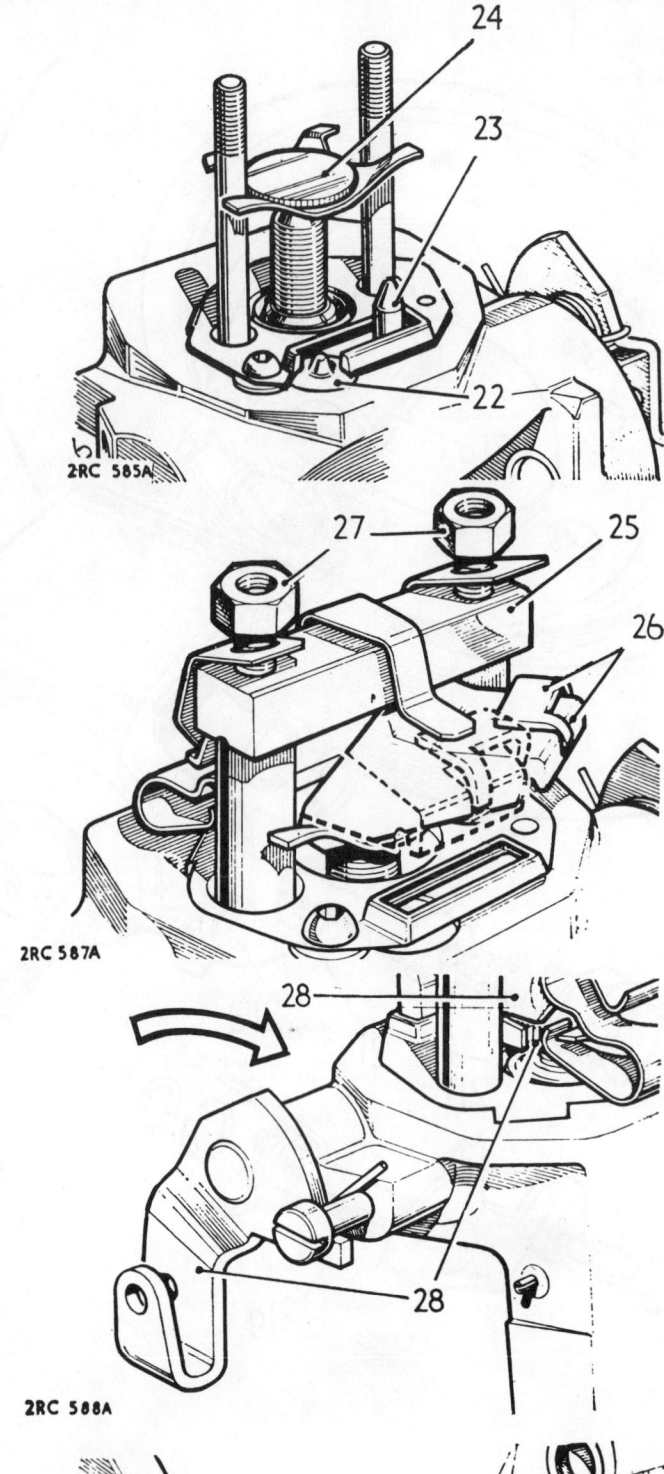

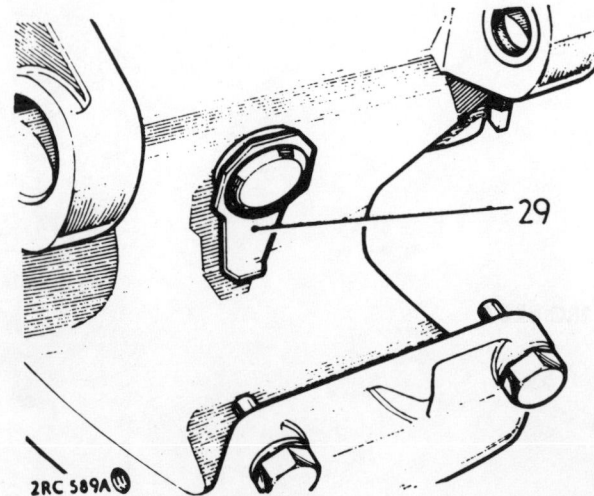

70.55.14
Sheet 4

Rover 3500 and 3500S Manual AKM 3621

WHEELS AND TYRES

LIST OF OPERATIONS

General Service Information	74.00.00
Tyre—remove and refit	74.10.01
Valve unit—remove and refit	74.10.07

Rover 3500 and 3500S Manual AKM 3621

WHEELS AND TYRES

General service information   74.00.00

1. **Tyre repairs**
   Legal requirements. In many countries there are regulations governing the repair of tyres, and any repairs carried out must conform to local laws. In all circumstances, repairs should be vulcanised.

2. **Changing wheel positions**
   With the type of tyre used on Rover 3500 and 3500S models, it is not considered advantageous to change the wheel positions, this in fact can give unpleasant handling characteristics when carried out, particularly if there is considerable difference between the wear pattern of one tyre and the other.

3. **Tyre pressures**
   Recommended tyre inflation pressures for all driving conditions are detailed in Maintenance, Division 10.

4. **Wheel and tyre sizes**
   Wheel and tyre size data is detailed in General Specification Data, Division 04.

5. **Wheel and tyre balancing**
   Ensure that the tyre is fitted correctly with the balance marks aligned. 74.10.01. The wheel and tyre complete should be balanced dynamically, using standard garage equipment.

6. **Dunlop DENOVO**
   Cars fitted with Dunlop DENOVO tyre and wheel assemblies require specialised facilities and knowledge for dealing with repairs and the fitting of new tyres. All repairs and replacements of DENOVO tyres must be carried out strictly in accordance with current Dunlop instructions and, for this reason, a network of Dunlop DENOVO Dealers have been appointed who are able to undertake all necessary work. Therefore, if repairs or new tyres are necessary, the nearest appointed Dealer should be contacted. The list of Dunlop DENOVO Dealers, together with a temporary repair kit and instructions, are supplied with every new car fitted with Dunlop DENOVO tyres.

WHEELS AND TYRES

TYRE

—Remove and refit   74.10.01

Removing

NOTE: The wheel rim incorporates a safety ledge to help prevent the outward facing tyre wall slipping into the well of the rim in conditions of under-inflation and fast cornering. Due to the shape of this design rim it is essential that tyres are removed and fitted from the side of the wheel that faces towards the centre of the car. As inextensible wires are incorporated in the beads of the outer cover, the beads must not be stretched over the wheel rim.

1. Remove the road wheel.
2. Remove the valve cap and core and deflate the tyre.
3. Press each bead in turn off its seating.
4. From the side of the wheel that faces towards the centre of the car, insert a lever at the same radial position as the valve and, while pulling on the lever, press the bead into the well diametrically opposite the valve.
5. Insert a second lever close to the first and prise the bead over the wheel rim. Continue round the bead in small steps until it is completely off the rim.
6. Pull the second bead over the rim.

Refitting

7. Using a damp cloth, wipe clean the tyre beads and the wheel rim.
8. Apply Dunlop Tyre Bead Lubricant to assist tyre fitting. This lubricant is essential, and no other lubricant whatsoever may be used, as it may cause the tyre to rotate on the rim, which upsets the correct balance of wheel and tyre assemblies to a gradually increasing degree.

continued

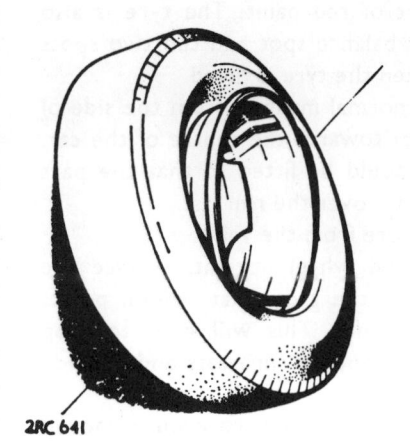

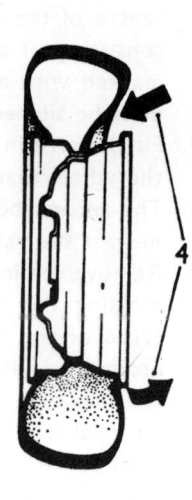

WHEELS AND TYRES

9. Check the side of the wheel that faces towards the centre of the car—there should be a balance mark consisting of a spot of red paint. The tyre is also marked with a red balance spot and the two spots must be aligned when the tyre is fitted.
10. Fit the tyre in the normal manner, from the side of the wheel that faces towards the centre of the car.
11. The second bead should be fitted so that the part nearest the valve goes over the rim last.
12. Remove the inner core from the valve.
13. Holding the tyre and wheel upright, bounce the tread of the tyre on the ground at several points around its circumference. This will help to snap the beads on to the tapered rim seats and provide a partial seal.
14. Connect an air line with the valve core still removed and inflate with the wheel and tyre upright. If the first rush of air does not seal the beads, continue to bounce the tyre with the air line attached until the beads are fully home against the rim flanges.
15. Remove air line and fit valve core, then inflate to 3,5 kgf/cm² (50 lbf/in²).
16. If air continues to escape under the beads after bouncing and the tyre cannot be inflated, use a suitable proprietary tyre tourniquet.
17. Adjust to the correct inflation pressure.
18. Refit the road wheel.

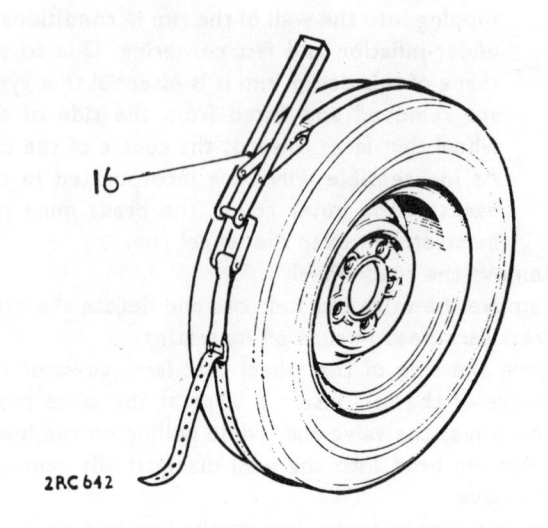

VALVE UNIT

—Remove and refit 74.10.07

Service tool: 268723 Valve extractor/fitter

Removing

1. Remove the tyre. 74.10.01.
2. Fit 268723 to the threaded end of the valve unit.
3. Using the lever of 268723 against the rim of the road wheel, extract the valve unit.
4. Insert a new valve unit from the inside of the rim, and use 268723 to draw it into place.
5. Refit the tyre. 74.10.01.

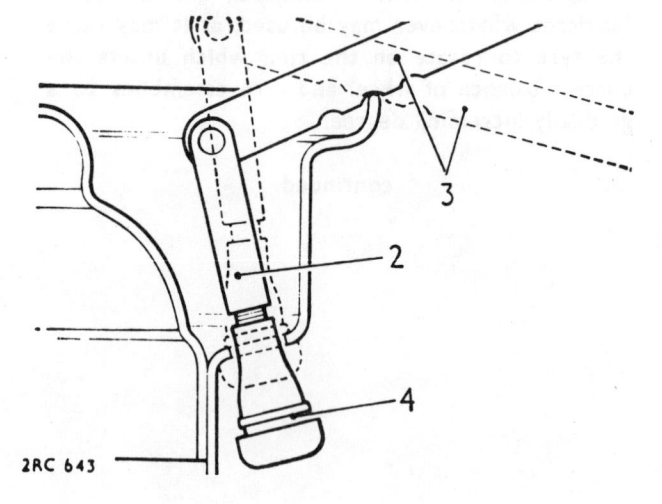

BODY

LIST OF OPERATIONS

Air intake valance—remove and refit	76.79.04
'B' post capping—remove and refit	76.43.34
'B' post trim pad—remove and refit	76.13.08
Backlight glass and sealing rubber—remove and refit	76.81.10

Base unit
—alignment check	76.10.03
—bulkhead—overhaul	76.10.60
—front section—overhaul	76.10.57
—initial check on Churchill fixture	76.10.61
—repair check on Churchill fixture	76.10.62

Body rear quarter panel—remove and refit	76.10.22/23

Bonnet
—remove and refit	76.16.01
—lock control cable—remove and refit	76.16.29
—safety catch—remove and refit	76.16.34

Boot lid
—remove and refit	76.19.01
—hinges—remove and refit	76.19.07
—spring—remove and refit	76.19.08
—seal—remove and refit	76.19.06

Console unit—remove and refit	76.25.01

Doors
—anti-burst device—remove and refit	76.37.46
—check strap—remove and refit	76.40.27
—glass—side front—remove and refit	76.31.01
—glass—side rear—remove and refit	76.31.02
—glass regulator—side—remove and refit	76.31.44
—seal—side front—remove and refit	76.40.01
—seal—side rear—remove and refit	76.40.02
—side—front—adjust	76.28.07
—side—front—remove and refit	76.28.01
—side—rear—adjust	76.28.08
—side—rear—remove and refit	76.28.02
—strut—remove and refit	76.34.18
—trim pad—side front—remove and refit	76.34.01
—trim pad—side rear—remove and refit	76.34.04

Door locks
—check and adjust	76.37.01
—side front—remove and refit	76.37.12
—side rear—remove and refit	76.37.13
—private—side front—remove and refit	76.37.39
—push button—remove and refit	76.58.12
—remote control—side front—remove and refit	76.37.31
—remote control—side rear—remove and refit	76.37.32
—striker plate—remove and refit	76.37.23

BODY

Fascia
    —switch panel—remove and refit 76.46.16
    —top rail—remove and refit 76.46.04
    —veneer strip—remove and refit 76.46.14

Front valance—remove and refit 76.79.03

Gearbox tunnel cover—remove and refit 76.25.07

Glove box—remove and refit 76.52.03

Glove box lock—remove and refit 76.52.08/09

Gutter finisher—backlight—remove and refit 76.43.14

Gutter finisher—cantrail—remove and refit 76.43.11

Headlining—remove and refit 76.64.01

Parcel tray—remove and refit 76.67.02

Quarter light—fixed—side rear—remove and refit 76.31.31

Quarter vent—side front—remove and refit 76.31.28

Quarter vent—side rear—remove and refit 76.31.30

Radiator grille—remove and refit 76.55.03

Rear decker panel—remove and refit 76.10.45

Rear valance—remove and refit 76.79.09

Roof panel—remove and refit 76.10.13

Seats
    —front—remove and refit 76.70.01
    —front—back finisher—remove and refit 76.70.03
    —front reclining mechanism—overhaul 76.70.33
    —front runners—remove and refit 76.70.21
    —front runners and cables—remove and refit 76.70.23
    —rear squab—remove and refit 76.70.38

Sill panel—outer—remove and refit 76.76.04

Waist rail moulding—remove and refit 76.43.02/05

Windscreen finisher—remove and refit 76.43.39/40

Windscreen glass and sealing rubber—remove and refit 76.81.01

Wings
    —front—remove and refit 76.10.26
    —rear—remove and refit 76.10.27/28

BODY

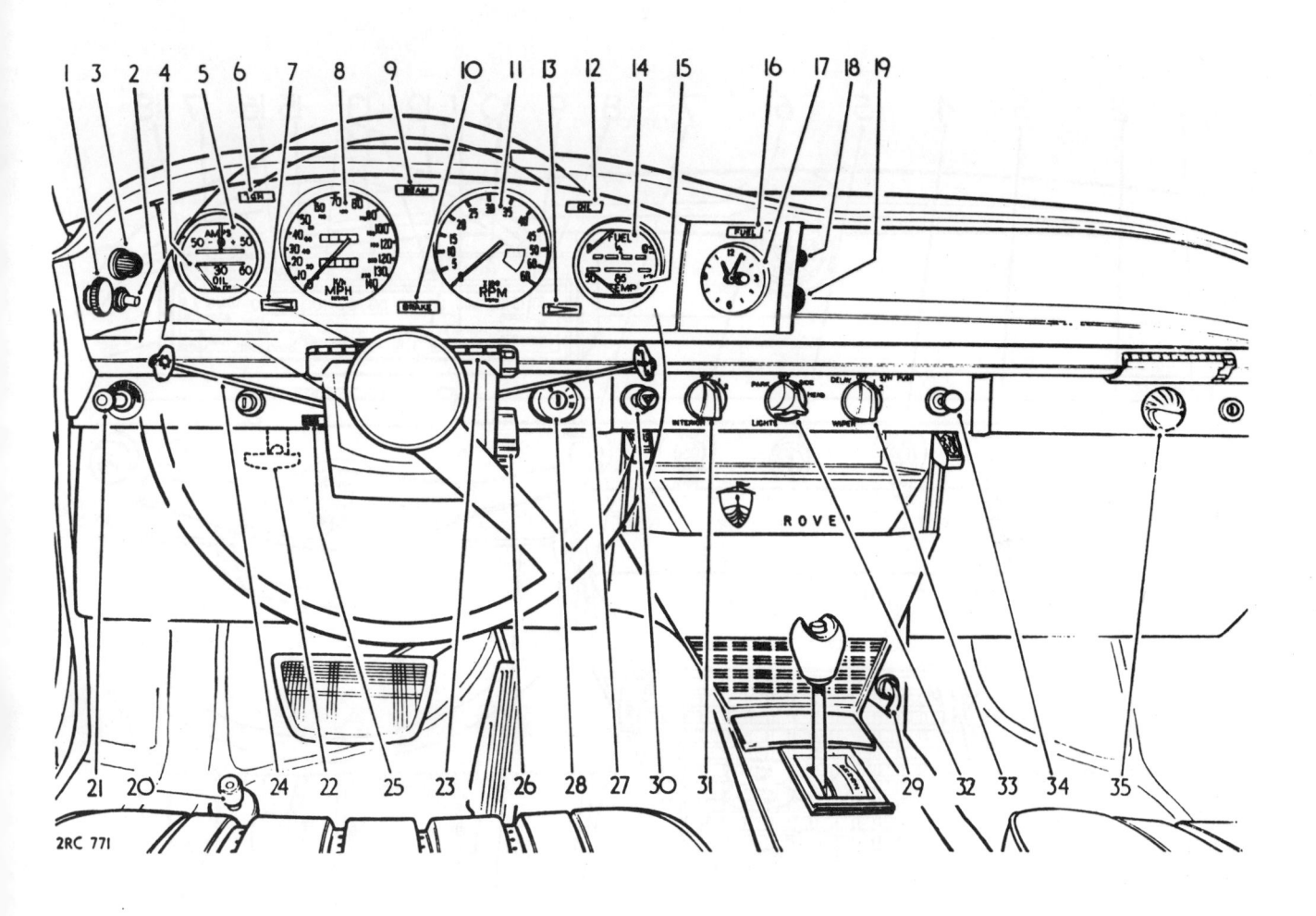

Layout of facia and controls, later 3500 models. Left-hand steering illustrated

NOTE: The layout of 3500S models is similar except for the gear change lever and the addition of the clutch pedal.

1 Rheostat, 'Icelert' warning light, when fitted
2 Test button, 'Icelert' warning light
3 Warning light, 'Icelert'
4 Gauge, oil pressure
5 Ammeter
6 Warning light, ignition
7 Direction indicator arrow, LH
8 Speedometer
9 Warning light, main beam
10 Warning light, brakes
11 Tachometer
12 Warning light, oil pressure
13 Direction indicator arrow, RH
14 Gauge, fuel
15 Gauge, water temperature
16 Warning light, fuel reserve
17 Clock
18 Control, speedometer trip
19 Rheostat, panel lights
20 Switch, headlamp dipper
21 Switch, heated backlight, when fitted
22 Bonnet release, inside driver's glove box
23 Face level vent
24 Switch, headlamp flasher
25 Delay control, wiper
26 Knob, column rake adjuster
27 Switch, direction indicator and horn
28 Switch, ignition and starter with steering column lock
29 Switch, fuel reserve
30 Switch, hazard warning
31 Switch, interior lights
32 Switch, side, park, headlamp and fog lamp, when fitted
33 Switch, wiper
34 Cigar lighter
35 Rotary map light

Rover 3500 and 3500S Manual AKM 3621

BODY

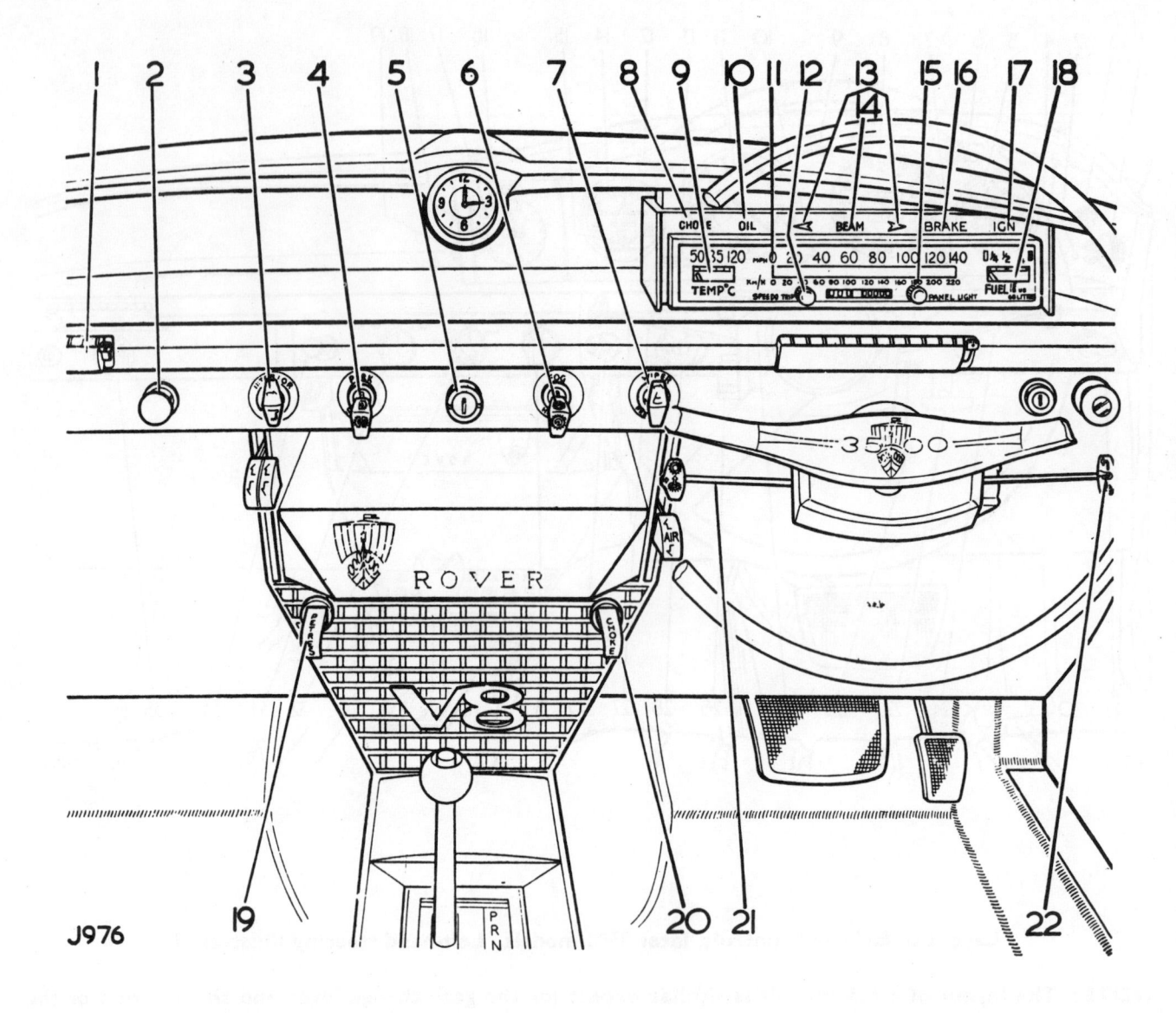

Layout of facia and controls, early 3500 models. Right-hand steering illustrated

1 Face level vent
2 Cigar lighter
3 Switch, interior lights
4 Switch, side and park
5 Switch, ignition
6 Switch, headlamp and foglamp, when fitted
7 Switch, wiper
8 Warning light, choke
9 Gauge, water temperature
10 Warning light, oil pressure
11 Speedometer
12 Speedometer trip
13 Direction indicator arrows
14 Warning light, main beam
15 Switch, panel light
16 Warning light, brake
17 Warning light, ignition
18 Gauge, fuel
19 Control, fuel reserve
20 Control, choke
21 Switch, headlamp dipper and flasher
22 Switch, direction indicator and horn

BODY

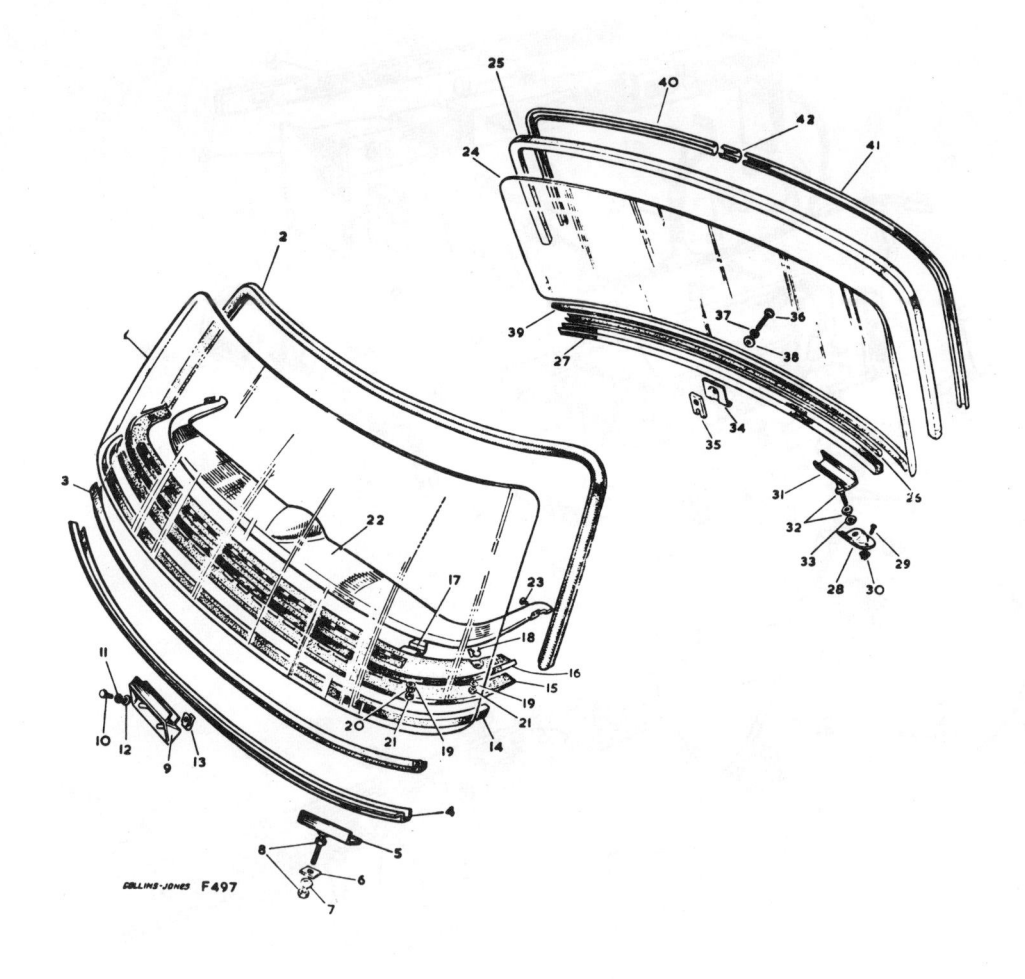

Windscreen and backlight

1. Glass for windscreen
2. Seal for windscreen, top and sides
3. Glazing rubber, for windscreen, lower
4. Support channel for windscreen, lower
5. Jacking bracket complete, for windscreen
6. Plate for jacking bracket
7. Plain washer ⎫ For screw in
8. Nut ($\frac{1}{4}$ in UNF) ⎭ jacking bracket
9. Centre clamp for windscreen
10. Bolt (10 UNF x $\frac{5}{8}$ in long)
11. Spring washer ⎫ Fixing centre clamp
12. Plain washer ⎬ to base unit
13. Spire nut (10 UNF) ⎭
14. Seal, windscreen to base unit
15. Dust seal ⎫ For screen
16. Finisher ⎭ demister
17. Retainer for screen demister finisher
18. Clip, screen finisher end, to dash
19. Plain washer ⎫ Fixing retainer
20. Spring washer ⎬ and finisher to
21. Nut (10 UNF) ⎭ base unit
22. Screen rail complete
23. Drive screw fixing screen rail
24. Glass for backlight
25. Seal for backlight, top and sides
26. Glazing rubber for backlight, lower
27. Support channel for backlight
28. Mounting plate for backlight jacking bracket
29. Drive screw ⎫ Fixing mounting plate
30. Nylon clinch nut ⎭ to base unit
31. Jacking bracket complete for backlight
32. Nut ($\frac{1}{4}$ in UNF) ⎫ For screw in
33. Plain washer ⎭ jacking bracket
34. Centre clamp for backlight
35. Spacer for centre clamp
36. Set bolt ($\frac{1}{4}$ in UNF x $1\frac{1}{4}$ in long) ⎫ Fixing clamp
37. Spring washer ⎬ to
38. Plain washer ⎭ base unit
39. Seal, backlight to base unit
40. Stainless steel finisher, RH, for backlight
41. Stainless steel finisher, LH, for backlight
42. Joint cover clip for backlight finishers

BODY

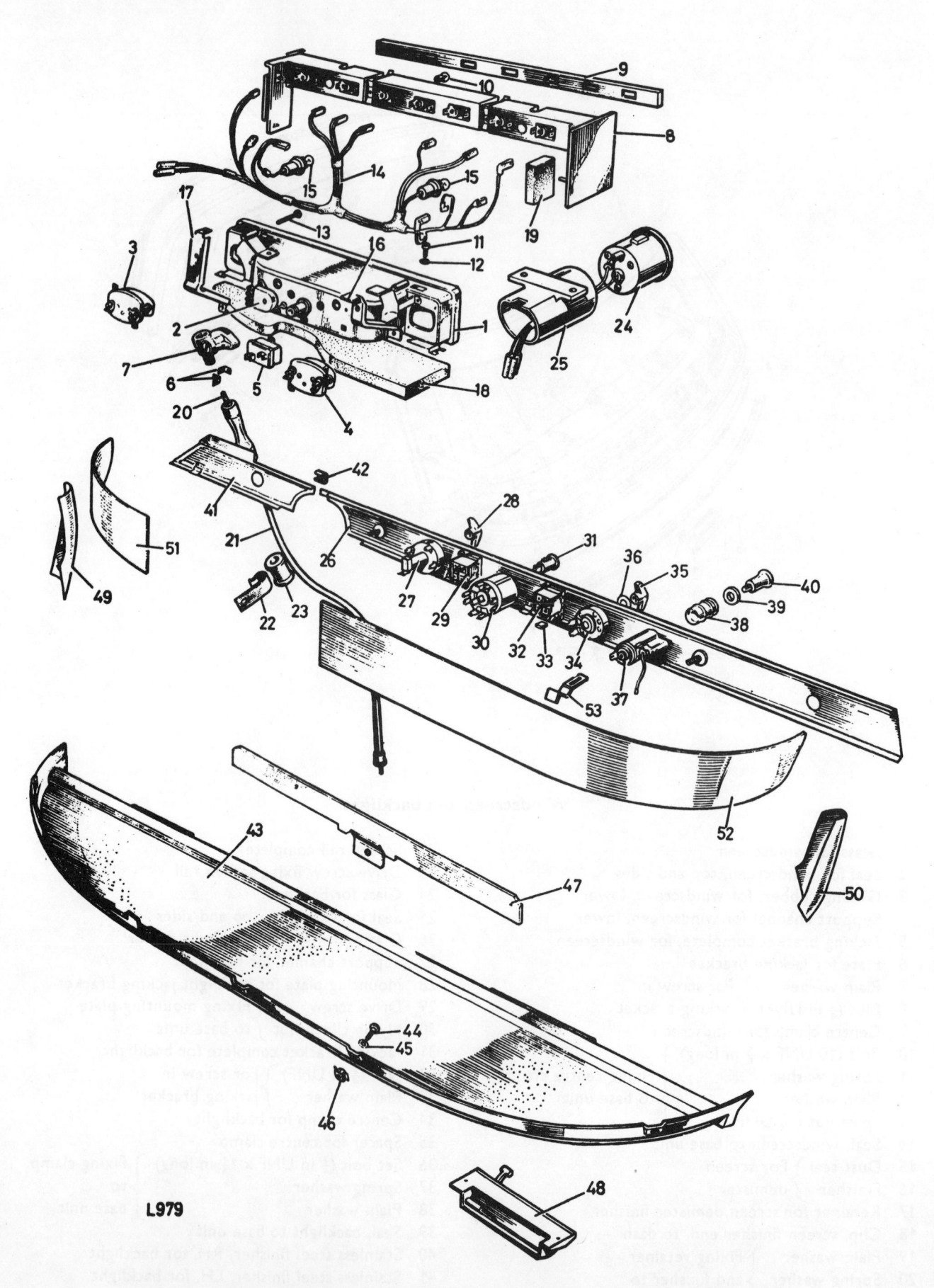

Layout of instrument panel, switch panel and parcel tray, early 3500 models

BODY

Key to illustration of instrument panel, switch panel and parcel tray, early 3500 models

1. Speedometer assembly
2. Panel light rheostat
3. Fuel level indicator
4. Water temperature indicator
5. Voltage regulator
6. Terminal blade for voltage regulator
7. Angle drive complete for speedometer
8. Saddle complete for instrument panel
9. Printed window for warning lights
10. Bulb for warning lights, 2.2 watt
11. Special screw } Fixing saddle
12. Plain washer } to speedometer
13. Special screw fixing saddle to base unit
14. Panel light harness
15. Bulb for panel light, 12 volt 3.6 watt, LU 984
16. Bracket for instrument panel, inner
17. Bracket for instrument panel, outer
18. Foam mounting pad } For instrument
19. Foam side pad } panel
20. Cable, inner
21. Cable, outer
22. Clip for speedometer cable
23. Grommet for clip
24. Clock
25. Moulding for clock
26. Switch panel complete
27. Switch (rheostat) for wiper and washer
28. Knob for wiper and washer switch
29. Switch for head and fog lamps, LU 34631
30. Switch, ignition and starter, LU 39222
31. Barrel lock for ignition switch
32. Switch, side and park, LU 34630
33. Lucar insulating sleeve, LU 54948329
34. Switch for roof lamp, LU 35879
35. Knob for roof lamp switch
36. Escutcheon for roof lamp switch
37. Cigar lighter
38. Heating unit complete }
39. Flange for knob } For cigar lighter
40. Knob }
41. Dash finisher complete, RH
42. Joint piece, dash finisher to switch panel
43. Parcel shelf, front
44. Shouldered screw } Fixing
45. Plain washer } parcel shelf
46. Spire nut (10 UNF) } to base unit
47. Stiffening plate for front parcel shelf
48. Face-level vent
49. Extension piece, RH } For front
50. Extension piece, LH } parcel shelf
51. Finisher, short, for front parcel shelf
52. Finisher, long, for front parcel shelf
53. Retaining clip for parcel shelf finishers

Rover 3500 and 3500S Manual AKM 3621

BODY

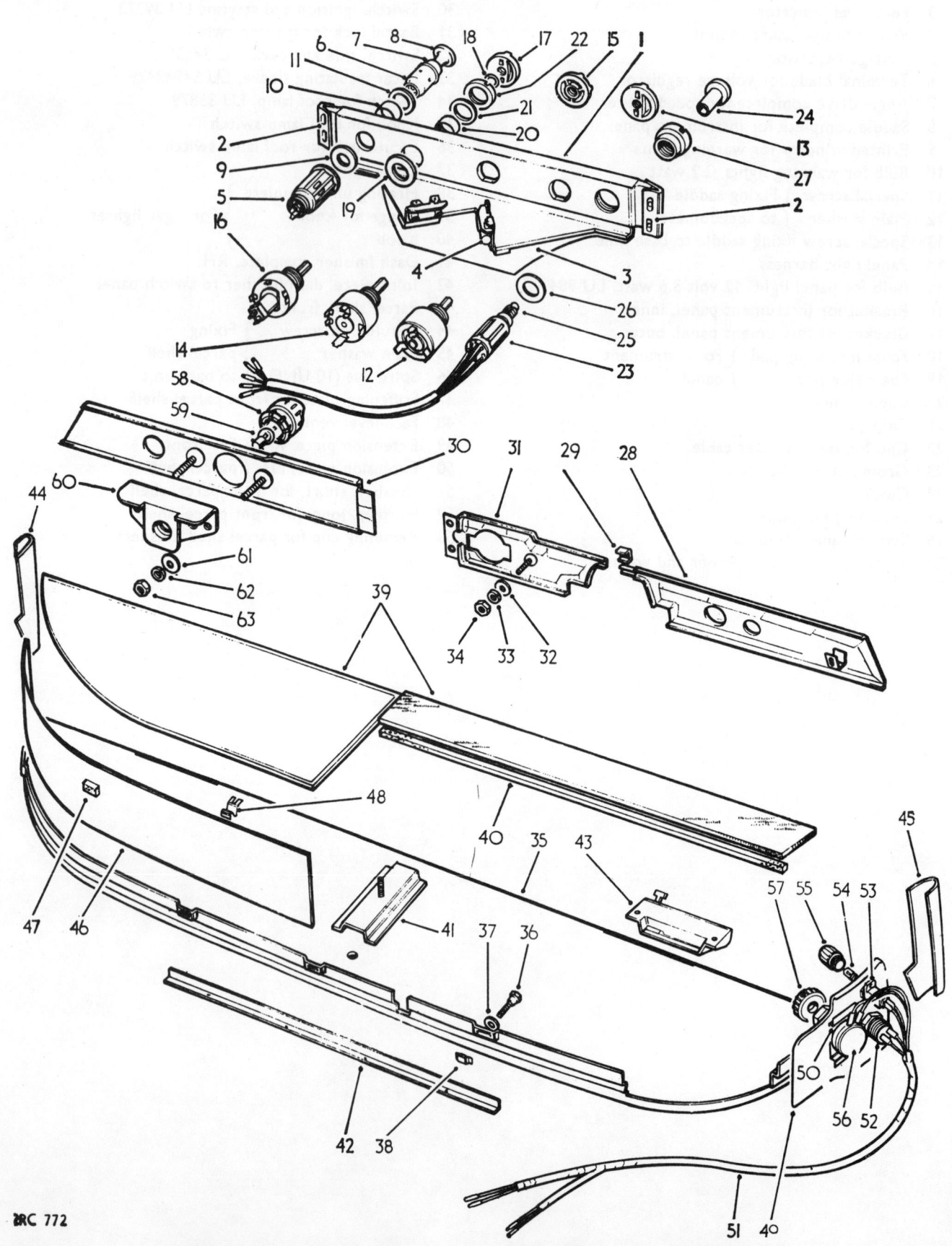

Layout of switch panel and parcel tray, later 3500 and all 3500S models

BODY

Key to illustration of switch panel and parcel tray, later 3500 and all 3500S models

1. Switch panel
2. End cover for switch panel
3. Light shield for switch panel
4. Bulb, 6 watt festoon, switch panel illumination
5. Cigar lighter assembly
6. Heating unit
7. Flange for knob
8. Knob for cigar lighter
9. Rubber washer, cigar lighter to switch panel
10. Spacer for cigar lighter
11. Chrome washer for cigar lighter
12. Switch, interior light
13. Knob for interior light switch
14. Switch, main lighting
15. Knob for main lighting switch
16. Switch, windscreen wiper
17. Knob for windscreen wiper switch
18. Rubber washer for knob
19. Rubber washer, switch to rear of switch panel
20. Spacer for switch
21. Chrome washer for switch
22. Bezel for switch
23. Switch, hazard warning, LU 54033563
24. Knob for hazard warning switch
25. Bulb, LU 281 for switch
26. Washer for rear of hazard warning switch
27. Bezel for hazard warning switch
28. Dash finisher complete, LH
29. Joint piece, dash finisher to switch panel
30. Switch panel, outer, trimmed assembly, RH
31. Switch panel, inner, trimmed assembly, LH
32. Plain washer ⎫ Fixing
33. Spring washer ⎬ inner switch panel
34. Nut (10 UNF) ⎭ to base unit
35. Parcel shelf, front
36. Shouldered screw ⎫ Fixing
37. Plain washer ⎬ parcel shelf
38. Spire nut (10 UNF) ⎭ to base unit
39. Non-slip mat for parcel shelf
40. Packing strip for underside of non-slip mat
41. Reinforcement panel for parcel shelf
42. Stiffening plate for front parcel shelf
43. Face level vent
44. Extension piece, RH
45. Extension piece, LH
46. Finisher, long, for front parcel shelf
47. Foam pad for front parcel shelf finisher
48. Retaining clip for parcel shelf finishers
49. Panel for 'Icelert', when fitted
50. Mounting bracket for 'Icelert' rheostat
51. Cable assembly for 'Icelert'
52. Test switch
53. Warning light for 'Icelert'
54. Bulb for warning light
55. Lens for warning light
56. Rheostat assembly
57. Knob for rheostat
58. Revolving courtesy lamp
59. Bulb, for revolving courtesy lamp
60. Nylon pad for lamp
61. Plain washer ⎫ Fixing
62. Spring washer ⎬ courtesy lamp
63. Nut (10 UNF) ⎭ to dash panel

BODY

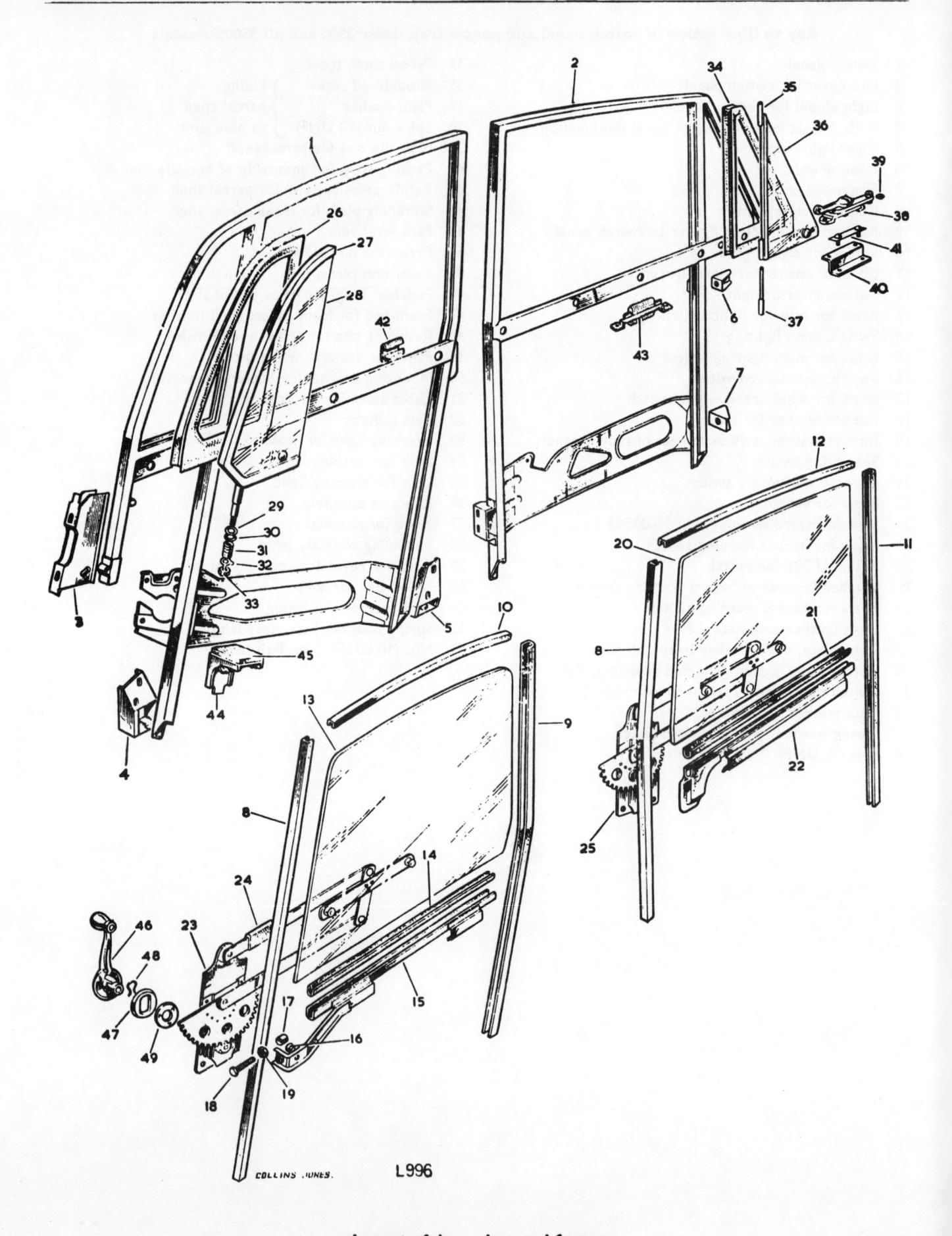

Layout of door glass and frames

Rover 3500 and 3500S Manual AKM 3621

Key to layout of door glass and frames

1. Frame, front door, LH
2. Frame, rear door, LH
3. Bracket for door frame (at waist), front door
4. Bracket for door frame, front lower, front door
5. Bracket for door frame, rear lower, front door
6. Bracket for door frame, rear upper, rear door
7. Bracket for door frame, rear lower, rear door
8. Channel, front and rear door, front ⎫
9. Channel, front door, rear ⎬ For
10. Channel, front door, top ⎬ door
11. Channel, rear door, rear ⎬ glass
12. Channel, rear door, top ⎭
13. Glass assembly for front door, LH
14. Glazing strip
15. Support channel for glass, front door
16. Self-sealing grommet
17. Steady pad for glass lifting channel, front door
18. Bolt ($\frac{1}{4}$ in UNF x $1\frac{1}{8}$ in long) ⎫ Adjusting
19. Locknut ($\frac{1}{4}$ in UNF thin) ⎭ steady pad
20. Glass assembly for rear door
21. Glazing strip
22. Support channel for glass, rear door
23. Window regulator, front door
24. Anti-rattle pad for front window regulator
25. Window regulator, rear door
26. Vent seal, front door
27. Ventilator assembly, front door
28. Glass only for front door ventilator
29. Plain washer, brass ⎫
30. Plain washer, steel ⎬ Retaining
31. Spring ⎬ front door vent
32. Tab washer ⎬ lower pivot
33. Nut ($\frac{1}{4}$ in UNF) ⎭
34. Vent seal, rear door
35. Ventilator assembly, rear door
36. Glass only for rear door ventilator
37. Pivot pin for ventilator
38. Ventilator catch, rear door
39. Fibre washer for catch
40. Support angle for catch
41. Stud plate
42. Window steady pad, outer
43. Window steady pad and bracket, inner
44. Glass stop bracket
45. Pad for glass stop bracket
46. Window regulator handle assembly
47. Escutcheon, black ⎫ For window
48. Spring clip ⎬ regulator
49. Wearing plate ⎭ handle

Rover 3500 and 3500S Manual AKM 3621

BODY

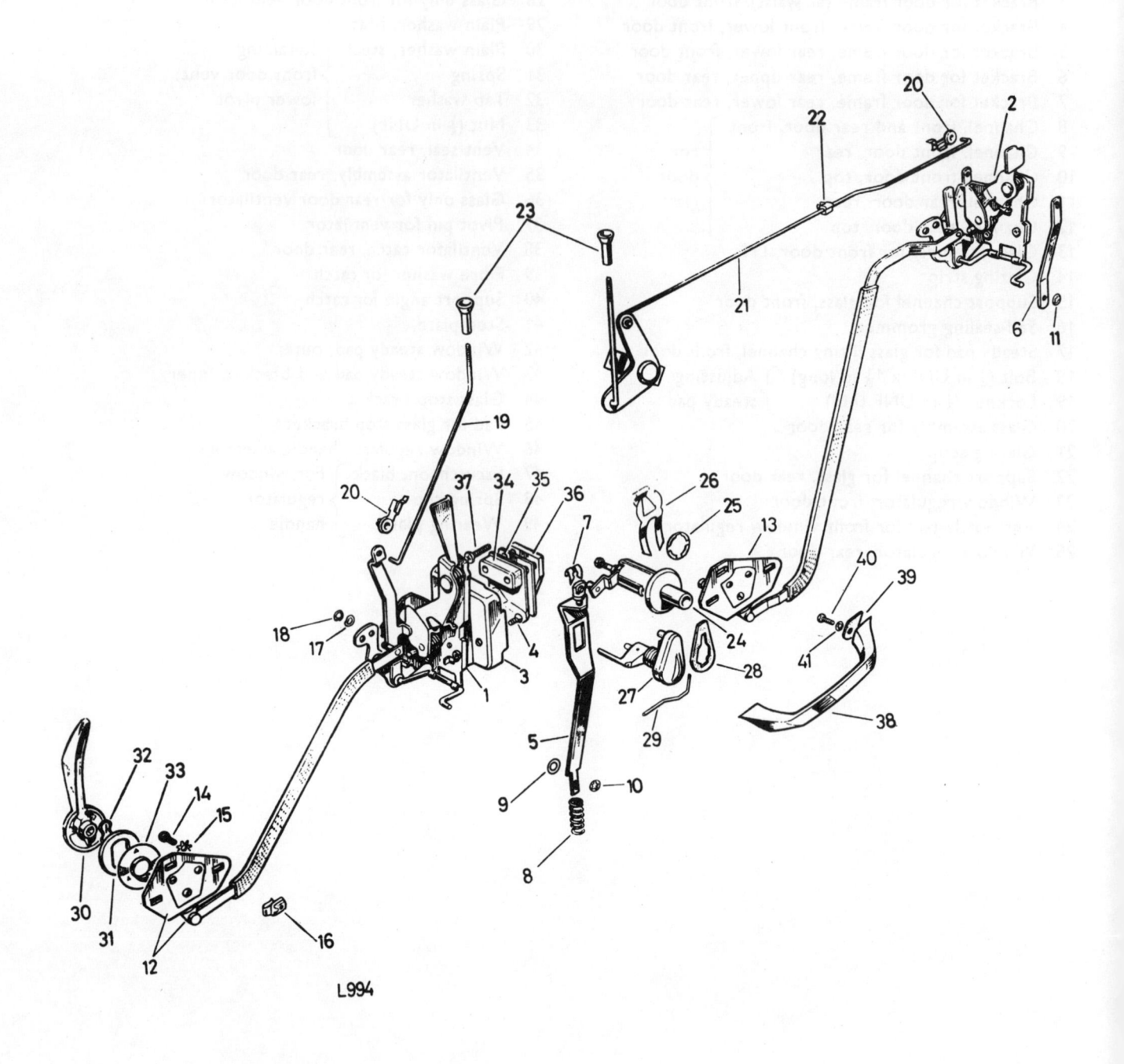

Layout of door locks and handles

76–12

Rover 3500 and 3500S Manual AKM 3621

BODY

Key to layout of door locks and handles

1 Lock complete, front door
2 Lock complete, rear door
3 Dovetail for LH door locks
4 Special screw fixing lock to door
5 Adjustable link for LH front door lock
6 Connecting link for rear door lock
7 Retaining clip, link to push-button
8 Spring for links, front door
9 Plain washer ⎫ Fixing links
10 'Starlock' washer ⎭ to locks
11 'Starlock' washer, link to rear door lock
12 Remote control and link, front door
13 Remote control and link, rear door
14 Screw (10 UNF x ½ in long)
15 Shakeproof washer
16 Spire nut (10 UNF)
17 Waved washer
18 Circlip
19 Locking rod, front door
20 Bush and clip complete, locking rod to front door
21 Bellcrank and link for rear door safety catch
22 Nylon steady bracket for rear door links
23 Interior locking knob, front and rear door
24 Push-button, front and rear door
25 Seating washer for push-button
26 Retaining clip for push-button
27 Private lock, front door
28 Seating washer for private lock
29 Retaining clip for private lock
30 Inside door handle assembly ⎫ To fit
31 Escutcheon for handle, black ⎬ square
32 Spring clip fixing inside handle ⎭ shaft
33 Wearing plate for inside handle
34 Striker for door lock
35 Shim
36 Tapping plate
37 Screw (¼ in UNF x ⅝ in long) fixing strikers to 'B' and 'D' posts
38 Outside door handle
39 Seating washer for outside handle
40 Set bolt (10 UNF x ⅜ in long) ⎫ Fixing outside
41 Shakeproof washer ⎭ handle

Rover 3500 and 3500S Manual AKM 3621

BODY

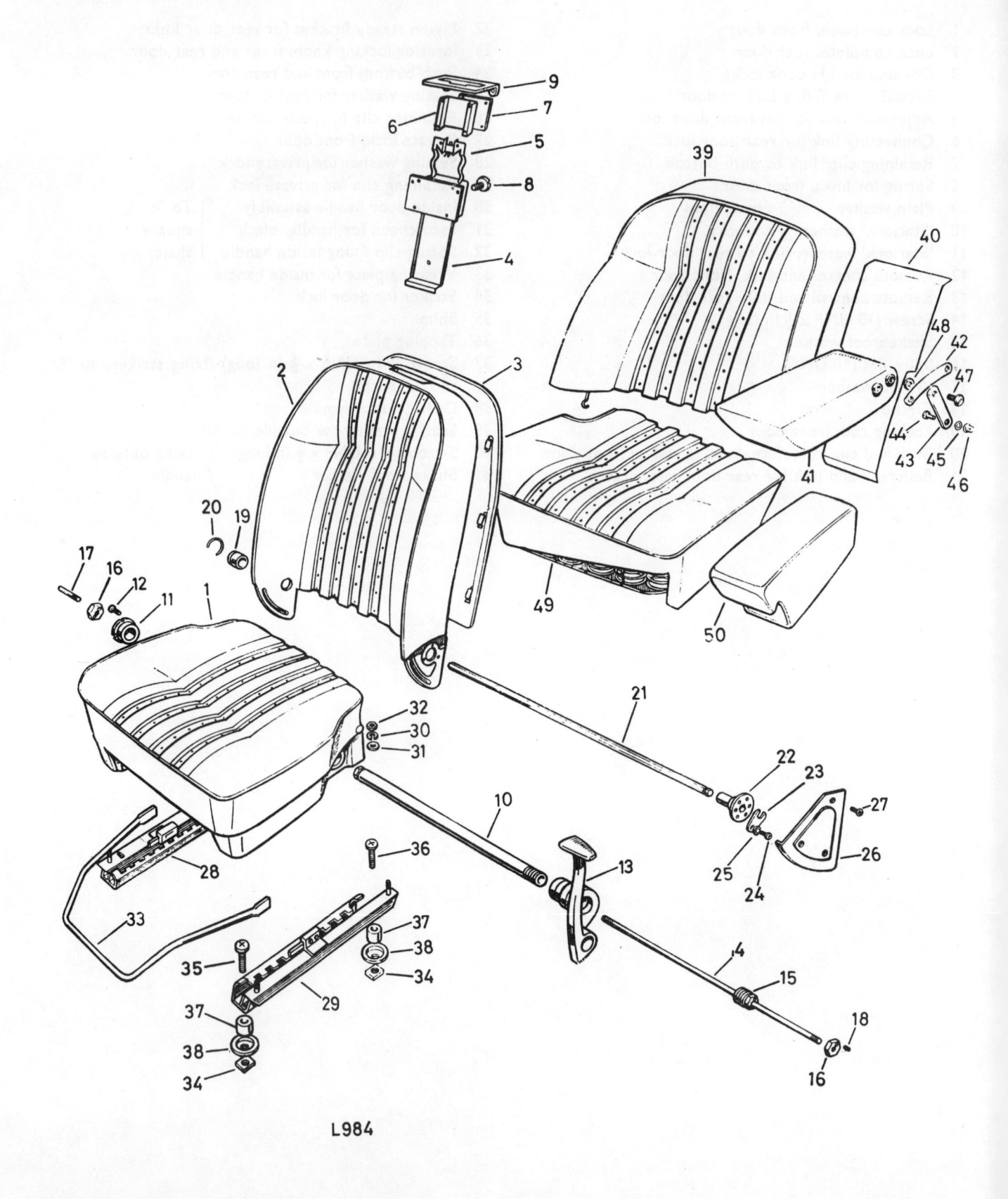

Layout of front and rear seats

L984

Rover 3500 and 3500S Manual AKM 3621

BODY

Key to layout of front and rear seats

1	Front seat cushion complete	26	Cover plate, RH, for pivot leg
2	Front seat squab complete	27	Drive screw fixing cover plate to squab
3	Trimmed back finisher complete for front squab	28	Seat slide, RH
4	Retainer for headrest	29	Seat slide, LH
5	Spring for retainer	30	Spring washer
6	Guide for headrest	31	Plain washer — Fixing slides to seat
7	Compression plate for guide	32	Nut ($\frac{5}{16}$ in UNF)
8	Adjusting screw for headrest	33	Operating bar for seat slides
9	Escutcheon for headrest aperture	34	Tapping plate — Fixing
10	Compression tube for seat lock	35	Special screw ($\frac{1}{4}$ in UNF x $1\frac{7}{8}$ in long) — seat
11	Bearing for compression tube	36	Special screw ($\frac{1}{4}$ in UNF x $1\frac{3}{8}$ in long) — slides to
12	Drive screw fixing bearing to squab frame	37	Distance piece — base
13	Seat locking lever for front seat, RH	38	Retaining washer — unit
14	Tie bar, seat locking	39	Rear squab complete, RH
15	Locking pad, inner, LH thread, for front seat, RH	40	Trimmed finisher complete for centre armrest
16	Locking pad, outer	41	Centre armrest complete
17	Dowel for locking pad, long	42	Link arm, top, for centre armrest
18	Dowel for locking pad, short	43	Link arm, lower, for centre armrest
19	Bearing, fixed pivot to squab	44	Shaft
20	Circlip, squab to bearing	45	Plain washer — Fixing link arms to base unit
21	Torsion bar for squab support	46	Retaining clip
22	Loose pivot, torsion bar to squab	47	Screw ($\frac{1}{4}$ in UNF x $\frac{1}{2}$ in long) — Fixing link arms
23	Retainer for torsion bar	48	Plain washer — to armrest
24	Set screw (10 UNF x $\frac{1}{2}$ in long) — Fixing pivot and	49	Rear cushion complete, RH
25	Spring washer — retainer to squab	50	Spacer complete for rear cushion

Rover 3500 and 3500S Manual AKM 3621

BODY

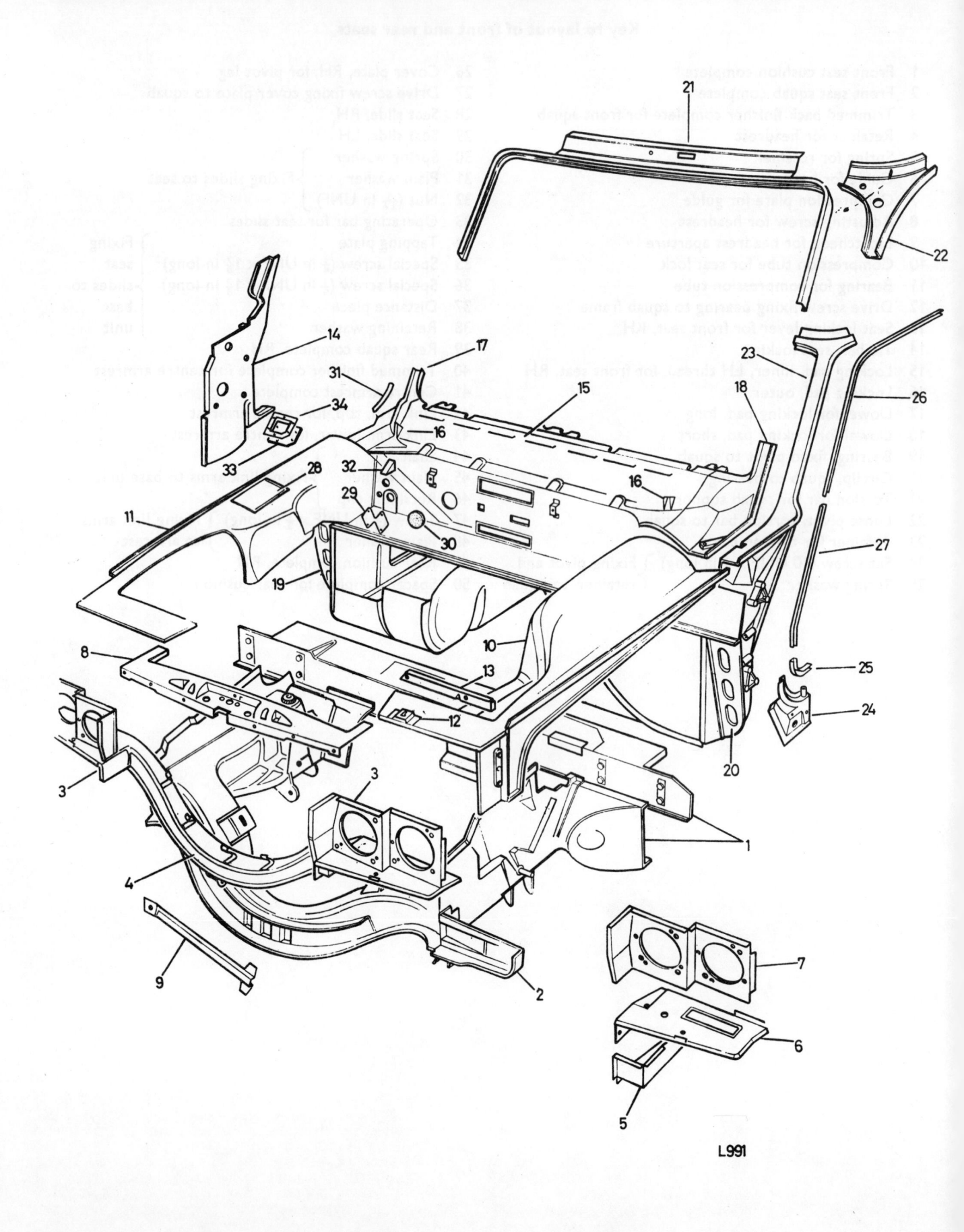

Base unit components, front end

76–16

Rover 3500 and 3500S Manual AKM 3621

BODY

Key to illustration of base unit components, front end

1 Lower valance and dash crossmember assembly
2 Outrigger, side LH
3 Front grille and headlamp mounting panel
4 Closing panel for front crossmember
5 Front outrigger bracket
6 Lower panel support for headlamp mounting panel
7 Headlamp panel
8 Panel for bonnet lock platform
9 Stay support, lock platform
10 Front upper valance assembly
11 Front upper valance panel
12 Retaining bracket for bonnet prop rod on upper valance panel
13 Panel, LH, extension
14 'A' post inner panel, lower
15 Top panel for dash, centre
16 Bracket, windscreen mounting
17 Top panel for dash, outer
18 Top panel for dash, outer
19 Closing (splash) panel for sill, front RH, lower
20 Closing (splash) panel for sill, front LH, lower
21 Windscreen header and glazing channel
22 Top corner gusset, LH, for screen pillar
23 'A' post
24 Lower fairing for 'A' post, LH
25 Retaining channel for door seal, 'A' post lower fairing, LH
26 Retaining channel for door seal, 'A' post, upper LH
27 Retaining channel for door seal, 'A' post, lower LH
28 Service and oil recommendation label
29 Cover plate for redundant master cylinder hole
30 Grommet for redundant steering box hole
31 Seal for dash top panel, outer
32 Seal for front valance to dash, upper
33 Cover plate for redundant pedal box hole
34 Sealing gasket for cover plate

Rover 3500 and 3500S Manual AKM 3621

BODY

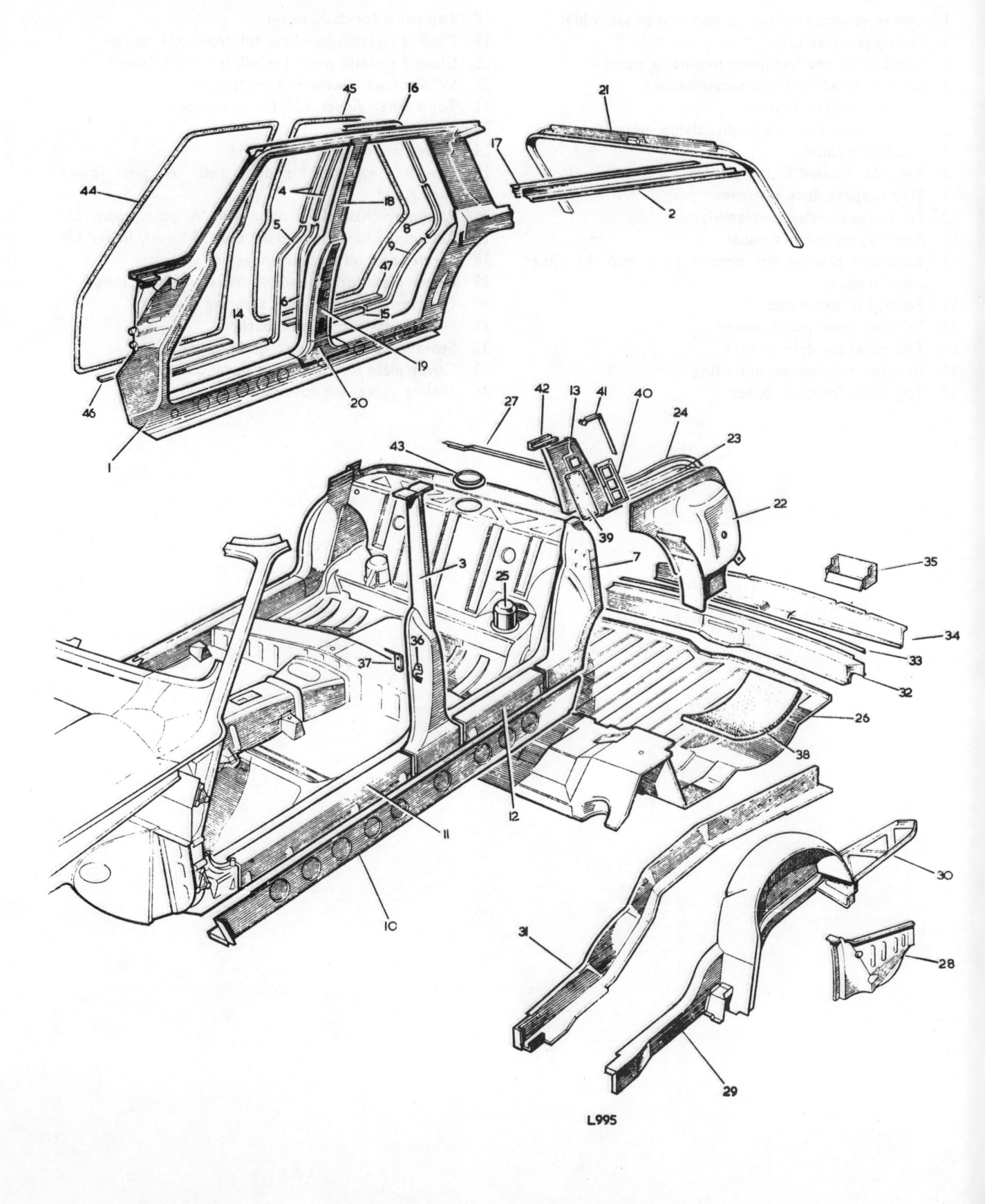

Base unit components, rear end

Rover 3500 and 3500S Manual AKM 3621

BODY

Key to illustration of base unit components, rear end

1. Side frame complete
2. Cant rail outer panel
3. 'BC' post
4. Retaining channel for door seal, 'B' and 'C' post, upper
5. Retaining channel for door seal, 'B' post, lower
6. Retaining channel for door seal, 'C' post, lower
7. 'D' post, lower outer panel
8. Retaining channel for door seal, 'D' post
9. Retaining channel for door seal, wheel arch
10. Sill closing panel
11. Channel, sill mounting, 'A' to 'B' post
12. Channel, sill mounting, 'C' to 'D' post
13. Rear quarter inner panel
14. Retaining channel for door seal, front sill
15. Retaining channel for door seal, rear sill
16. Retaining channel for door seal, 'D' post and top
17. Closing panel for cant rail
18. Closing panel for 'BC' post, upper
19. Closing panel for 'BC' post, centre
20. Closing panel, lower, for 'BC' post
21. Backlight header and glazing channel
22. Rear valance panel
23. Boot aperture drain channel, side
24. Finishing flange for boot aperture drain channel, side
25. Rear shock absorber mounting cup
26. Boot floor panel
27. Finishing flange for boot aperture drain channel, front
28. Vertical side panel for boot floor, LH
29. Rear wheelarch and outer longitudinal member
30. Horizontal side panel for boot floor
31. Inner closing panel for outer longitudinal member
32. Rear lower panel, inner
33. Finishing flange for boot aperture drain channel, rear
34. Reinforcement for rear lower panel
35. Rear jacking pad
36. Anti-burst plate
37. Tapped plate
38. Pad for spare wheel depression in boot floor
39. Blanking plate, inner, for rear quarter
40. Valve, through-flow ventilation
41. Drain channel
42. Seal, base unit to rear quarter panel
43. Cover plate for rear speaker aperture
44. Seal for front door aperture
45. Seal for rear door aperture
46. Retaining strip for front door seal
47. Retaining strip for rear door seal

Rover 3500 and 3500S Manual AKM 3621

BODY

General service information

1. **Paint rectification in low bake ovens**

   'New Look' Rover 3500 and 3500S

   The temperatures in low bake paint ovens are such that they can adversely affect certain synthetic materials used in components of the above models and therefore it is essential that before a car is placed in a low bake oven the following items are removed and certain precautions taken.

   a. Remove all front and rear lamps and flashers except headlamps.
   b. Remove sun visors, glove boxes, radiator grille and radio if fitted.
   c. Mask all badges.
   d. Fit slave wheels.
   e. Ensure that there is not more than 2 galls. of fuel in the tank.

   CAUTION: Do not allow Dunlop DENOVO tyres and wheels to enter a low bake oven, as high temperatures will destroy their special properties. Fit slave wheels in their place.

2. **Base unit**

   The construction is unusual in that it makes use of a base unit which forms a chassis skeleton, to which all exterior panels and mechanical components are bolted.

   Replacement parts are available for the base unit, and special welding and checking fixtures have been developed for the accurate repair of the most vulnerable parts at the front of the vehicle.

   Operation 76.10.57 in this Division covers the use of special checking and welding fixtures which are required when replacing any parts forward of the bulkhead.

   Operation 76.10.61 gives details of the checking fixtures which are for use in conjunction with the Churchill 700 body checking and repair fixture.

   IMPORTANT

   The special tools available for welding and checking the base unit are made to very close tolerances, and care should be taken not to damage them either in use or in storage.

3. **Limitations of repairs**

   A visual inspection should first be made of the damaged vehicle to determine the extent of the damage. To do this it may be necessary to remove interior and exterior panels, mechanical and electrical components and any windows in the vicinity of the damage.

   If there is obvious severe damage to any part of the base unit shown shaded in the three-quarter illustrations and this damage cannot be repaired, the base unit is scrap. With the exception that bulkhead distortion may be rectified within the limitations of Operation 76.10.60.

However, very considerable repairs can be carried out successfully by using the special checking fixtures developed for use with the Churchill 700 body checking and repair fixture. See Operation 76.10.61. The unshaded portions in the illustrations may be obtained as parts or sub-assemblies as listed in the Rover 3500 and 3500S Parts information service. All base unit replacement parts are supplied with the bosses spot-faced and drilled.

BODY

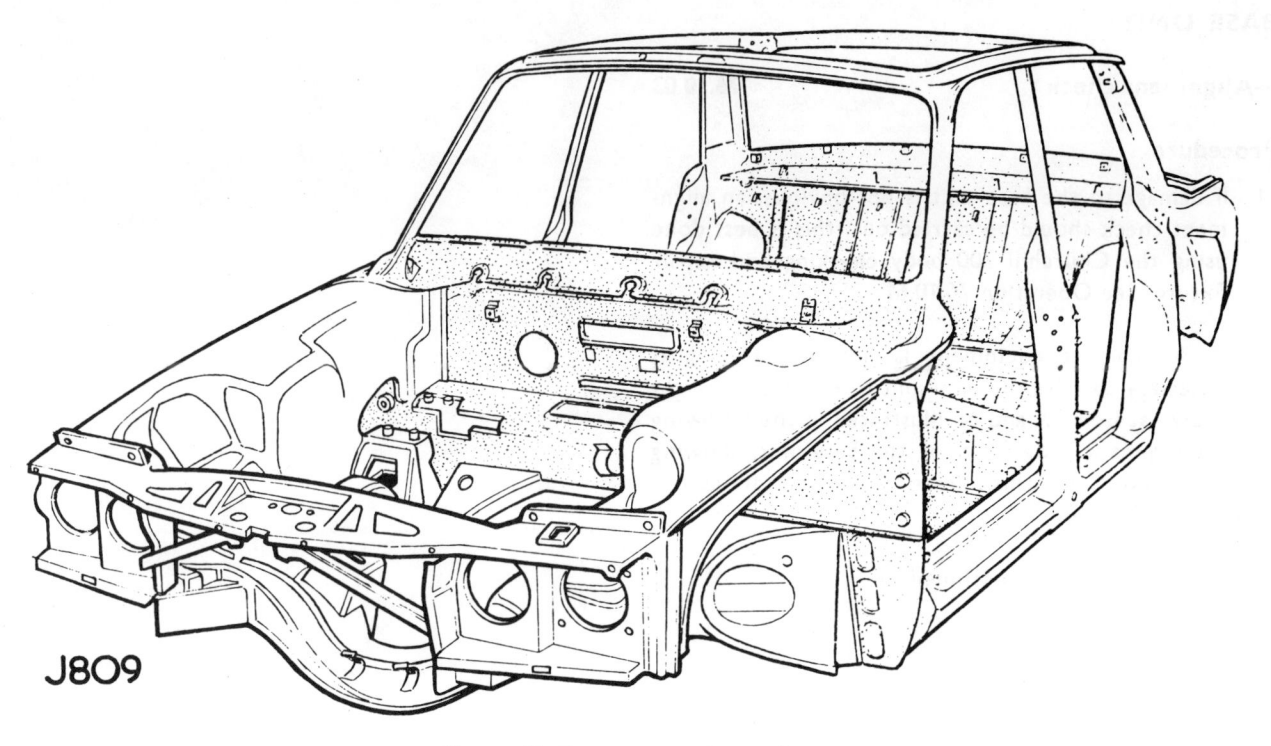

Base unit, three-quarter front view
Shaded portions indicate non-serviceable parts

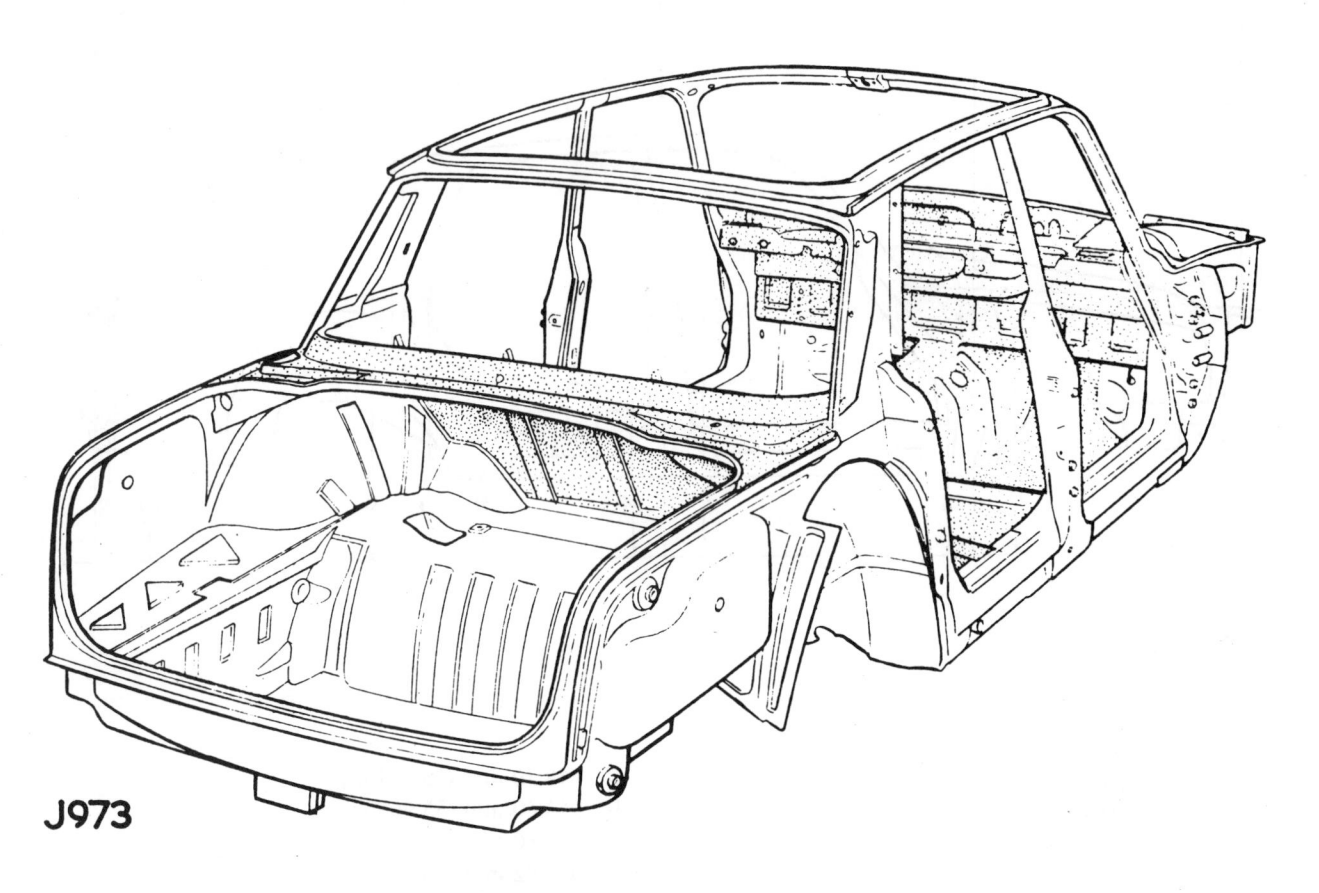

Base unit, three-quarter rear view
Shaded portions indicate non-serviceable parts

BODY

BASE UNIT

—Alignment check 76.10.03

Procedure

1. If damage to the base unit is not obvious, an alignment check should be carried out; this is best done using the Churchill 700 body checking and repair fixture. See Operation 76.10.61.

2. If the correct equipment is not available, it is necessary to raise the base unit to the datum height shown at Items K and N, illustrated in the following diagrammatic side view of the base unit, showing datum positions. This operation must be carried out on a level floor.

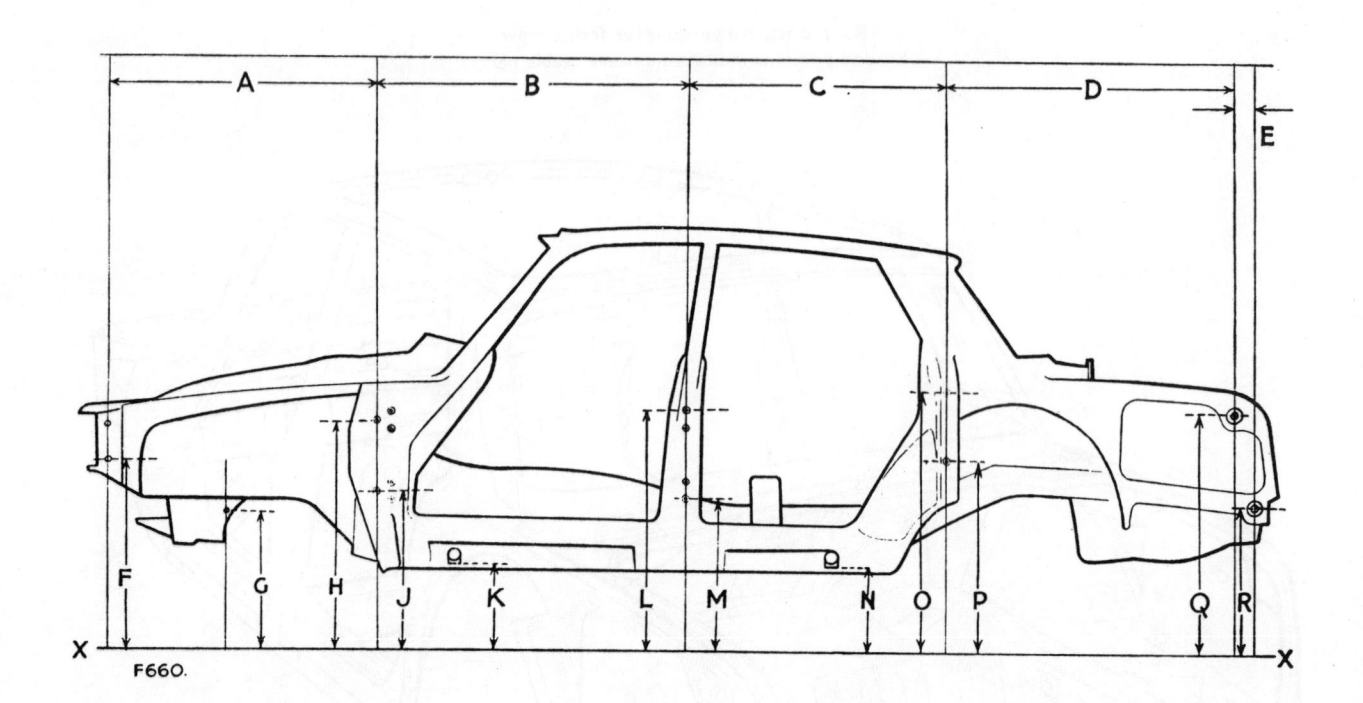

continued

A 937,42 mm (36.906 in)
B 1043,78 mm (41.093 in)
C 860,42 mm (33.875 in)
D 974,72 mm (38.375 in)
E 87,31 mm (3.437 in)
F 736,60 mm (29.000 in)
G 518,32 mm (20.406 in)
H 874,31 mm (34.421 in)
J 639,36 mm (25.171 in)

K 418,70 mm (16.203 in)
L 913,60 mm (35.968 in)
M 621,50 mm (25.468 in)
N 402,82 mm (15.859 in)
O 978,29 mm (38.515 in)
P 740,49 mm (29.546 in)
Q 897,33 mm (35.328 in)
R 583,00 mm (22.953 in)
XX Floor level

Rover 3500 and 3500S Manual AKM 3621

BODY

3. Using a suitable plumb line, project down from the lower bosses on the base unit to the floor, then extend lines to dimensions shown at 'F', 'G' and 'H' and join up these points so that the alignment of the base unit can be checked, as shown in the following diagrammatic plan view of the base unit.

4. If it is impossible to join up the points projected in a straight line from 'X' to 'X', then the base unit is distorted. It will be found that there are usually only one or two points in error, in accordance with the dimensions given in the illustration. For example, if dimensions 'F' will not line up on either one or both sides, this indicates that misalignment is forward of the bulkhead.

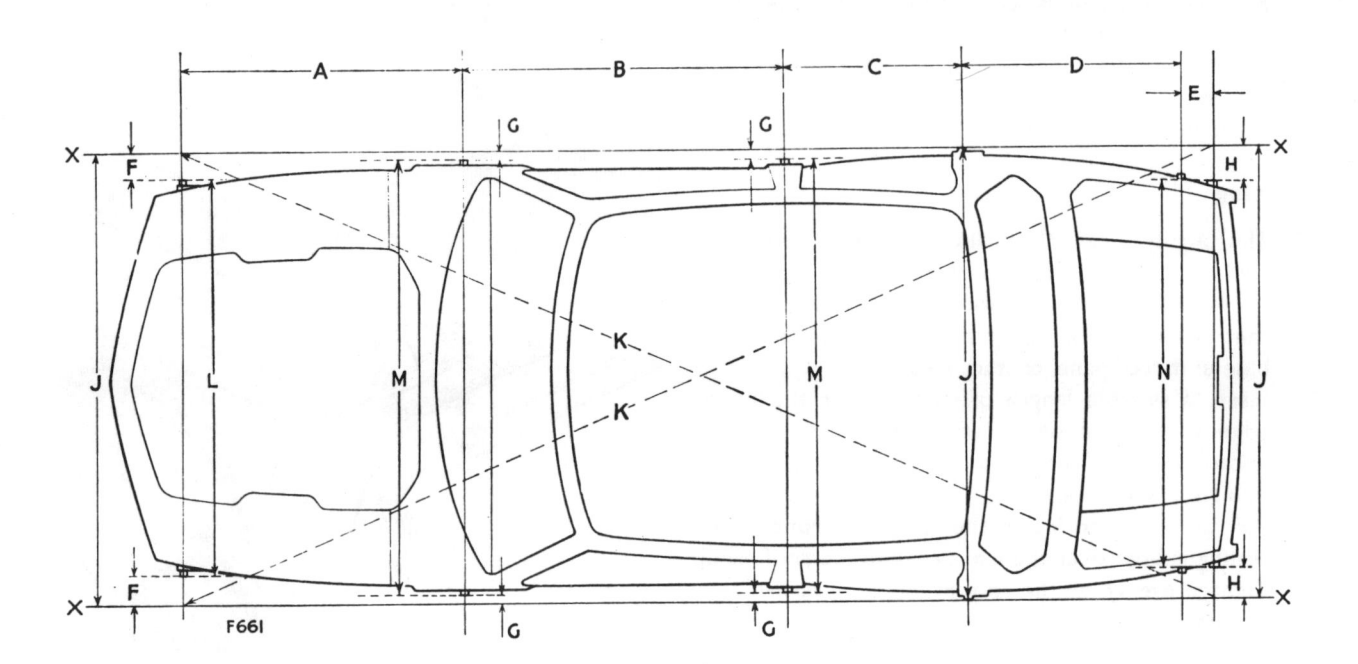

A 937,42 mm (36.906 in)
B 1043,78 mm (41.093 in)
C 860,42 mm (33.875 in)
D 974,72 mm (38.375 in)
E 87,31 mm (3.437 in)
F 19,84 mm (4.781 in)
G 15,87 mm (0.625 in)

H 149,22 mm (5.875 in)
J 1576,38 mm (62.062 in)
K 4210,84 mm (165.781 in)
L 1333,50 mm (52.500 in)
M 1544,63 mm (60.812 in)
N 1277,93 mm (50.312 in)

Rover 3500 and 3500S Manual AKM 3621

BODY

ROOF PANEL

—Remove and refit 76.10.13

Removing

1. Remove the headlining. 76.64.01.
2. Remove the rear decker panel. 76.10.45.
3. Remove the backlight glass and sealing rubber. 76.81.10.
4. Remove the stainless steel finishers from the backlight. 76.43.14.
5. Remove the body rear quarter panels. 76.10.22/23.
6. Remove the stainless steel finishers from the windscreen. 76.43.39/40.
7. Remove all the screws securing the roof panel to the body.
8. Remove the roof panel.

Refitting

9. Thoroughly clean all sealing compound from the roof and the roof panel aperture.
10. Apply a firm sealing compound along the base unit cant rail channels. Apply sealing compound to the base unit roof panel contact areas.
11. Insert a suitable length of string under the upper edge of the front screen sealing rubber.
12. Locate the roof panel in position and fit all the fixing screws loosely.
13. Pull the string to locate the lip of the front screen sealing rubber over the front edge of the roof panel.
14. Tighten evenly, all the roof panel fixing screws.
15. Reverse 1 to 6.

BODY REAR QUARTER PANEL—LH

—Remove and refit 76.10.22

BODY REAR QUARTER PANEL—RH

—Remove and refit 76.10.23

Removing

1. Remove the rear seat squab. 76.70.38.
2. Where applicable, remove the rear seat safety harness outer fixings.
3. Remove the two drive screws retaining the rear quarter trim pad.
4. Ease the trim pad forward to release it from its retaining clips.
5. Remove the Phillips head screws adjacent to the edge of the rear window.
6. Remove the three Phillips head screws located in the body adjacent to the rear edge of the door glass surround.
7. Remove the quarter panel.

Refitting

8. Reverse 1 to 7.

Rover 3500 and 3500S Manual AKM 3621

BODY

FRONT WING

—Remove and refit   76.10.26

Removing

1. Disconnect the battery earth lead.
2. Prop open the bonnet.
3. Remove the bolt, spring washer and plain washer from the top rear edge.
4. Remove the nuts, washers and stud plate securing the wing to the front valance.
5. From underneath the wing, remove the two bolts from the top front fixing.
6. Remove the two screws from the underside of the wing.
7. Disconnect the wiring harness from the side and flasher lamps.
8. Draw the front wing forward clear of the dowel and lift clear.

Refitting

9. Reverse 1 to 8.

REAR WING—LEFT HAND

—Remove and refit   76.10.27

REAR WING—RIGHT HAND

—Remove and refit   76.10.28

Removing

1. Open the luggage boot.
2. Raise the trim from the side panel in the boot to give access to the fixing bolts.
3. Remove the two nuts, spring and plain washers securing the wing to the body side panel, one from inside the boot and the second from the underside of the car at the rear corner.
4. Remove the special bolt and shakeproof washer securing the rear valance to the rear wing.
5. Slacken the two bolts, spring and plain washers securing the forward edge of the wing to the rear door pillar. These are accessible with the rear door opened.
6. Remove the nut and bolt adjacent to the mud flap.
7. Remove the rear wing, withdrawing the cables through the grommet.

Refitting

8. Reverse 1 to 7, ensuring that the rear wing sealing rubber is in place.

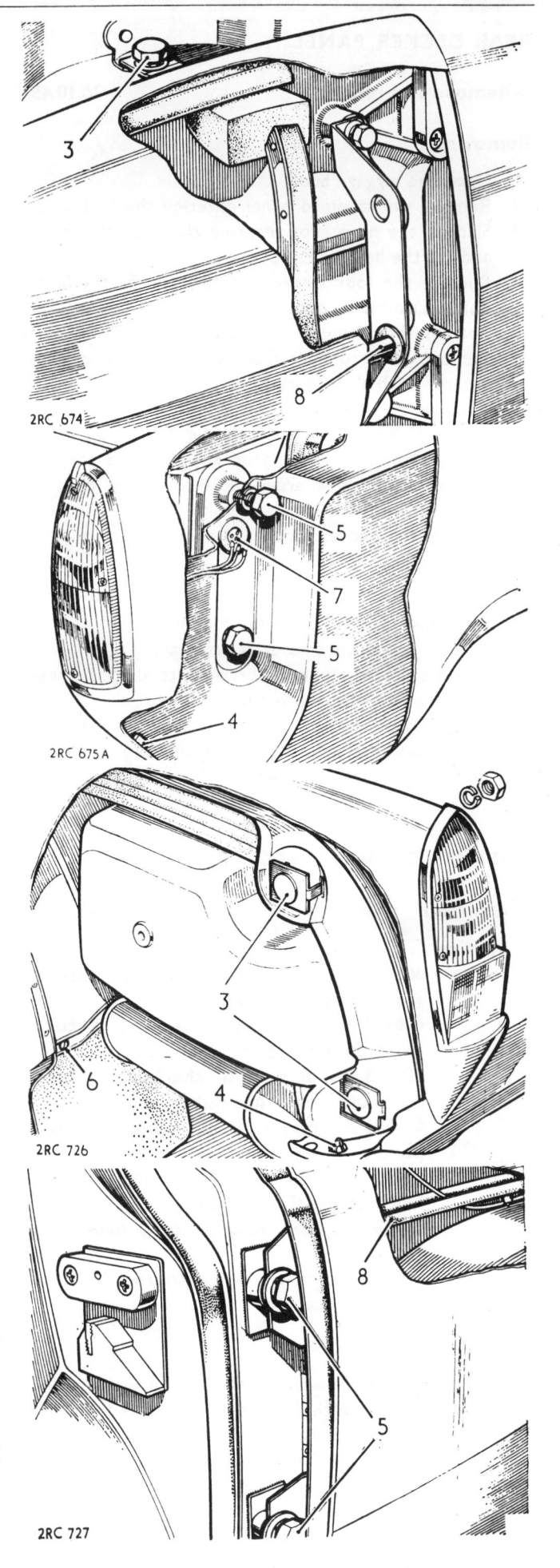

Rover 3500 and 3500S Manual AKM 3621

76.10.26
76.10.28

BODY

REAR DECKER PANEL

—Remove and refit   76.10.45

Removing

1. Open the luggage boot.
2. Remove the trimmed panel covering the fuel tank.
3. Slacken the hose clips securing the top filler hose and cut the hose to remove.
4. Remove the four screws securing the petrol filler unit, accessible by opening the filler cap.
5. Remove the filler unit complete.
6. Remove the grommet from the filler cap orifice, secured with a spring ring.
7. Remove the two nuts, plain and spring washers securing the decker panel to the base unit; these are located at the right-hand and left-hand corners of the decker panel, accessible through the boot.
8. Remove the three nuts and bolts securing the rear edge of the decker panel, and remove the panel.

Refitting

9. Reverse 1 to 8, including the following.
10. Fit new petrol filler hoses as necessary.
11. Use the slotted mounting brackets to align the rear decker panel with the wings and boot.
12. Apply Bostik sealing compound 1753 to the filler end of top hose and Bostik 692 round the mounting bolts.

2RC 729

BASE UNIT, FRONT SECTION

—Overhaul   76.10.57

Service
tools: **605486** Welding and checking fixture, front lower valance
600572 Welding and checking fixture, front upper valance right hand
600573 Welding and checking fixture, front upper valance left hand
600596 Alignment rod, upper valance panels
606457 Welding and checking fixture, front suspension brackets
606456 Support legs (2 off), base unit front end.
605571 Gauge, checking steering box and top link location

NOTE:

This operation gives details of the special welding and checking fixtures which have been developed for the repair of the front end of the base unit.

Assuming the vehicle has sustained a severe frontal impact necessitating the replacement of the 'lower valances', proceed as follows.

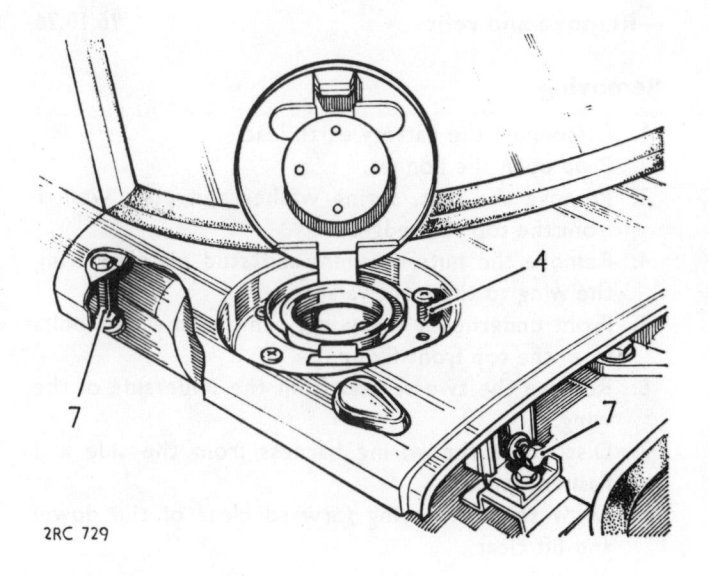

Base unit and engine mountings

Rover 3500.

When carrying out accident repair work on cars up to and including vehicle suffix letter 'D' it may be necessary to fit a base unit intended for cars from suffix 'E' onwards, i.e. Part Nos. 370353 or 370714.

These are designed to accommodate the later type engine mountings and it will be necessary to fit the following items:

Engine mounting plate	2	572854
Mounting rubber	2	572855
Packing plate	2	572869

If repairs are being carried out involving the lower valance and dash crossmember assembly, Part No. 366861, it should be noted that Part No. 389199 is now supplied as a replacement and the above engine mountings and associated parts must be utilised.

Procedure

1. Raise the base unit to the datum position.
2. Remove the damaged portion of the 'lower valance and dash cross-member assembly' from the base unit along the welding seams, illustrated in heavy line.
3. If the 'dash cross-member' is damaged beyond repair, the complete base unit is scrap and a new replacement must be fitted.

continued

Rover 3500 and 3500S Manual AKM 3621

BODY

4. A new 'lower valance and dash cross-member assembly' is only supplied complete with the 'dash cross-member' attached, illustrated in dotted line; however, this section must be removed from the new lower valance before it is offered to the base unit.
5. Secure the welding and checking fixture to the new lower valances. 605486.
6. The lower valance welding seams are illustrated in heavy line.
7. Offer the lower valance assembly to the base unit and secure to the dash cross-member with four bolts.
8. Ensure that the dimension from the datum to the front bottom link location is 518,32 mm (20.406 in.).

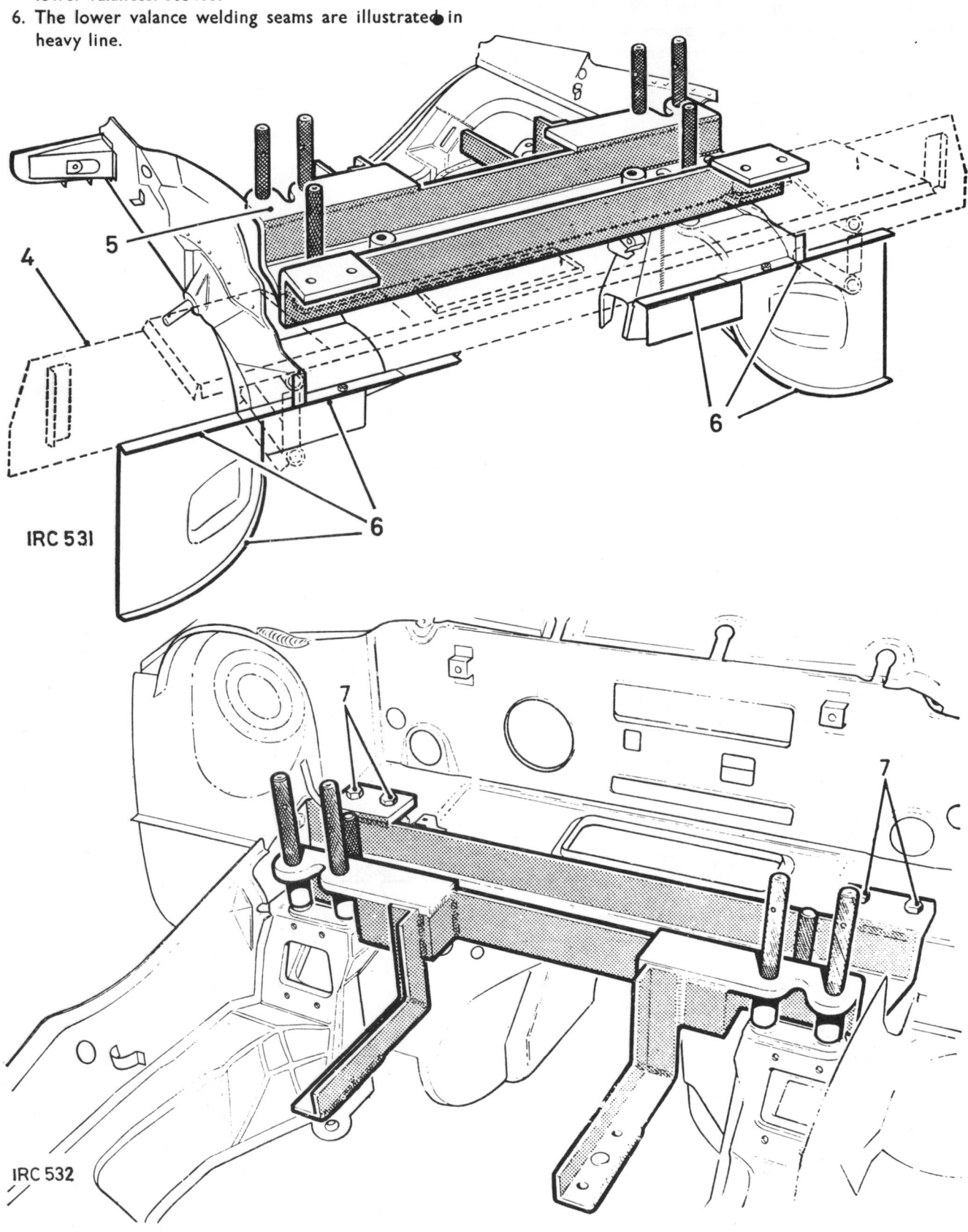

continued

IRC 531

IRC 532

Rover 3500 and 3500S Manual AKM 3621

76.10.57
Sheet 2

BODY

If a Churchill 700 Body Checking and Repair fixture is not available, proceed 9 to 14

9. Mount the base unit on stands on a level floor.
10. Adjust the front datum height to 418,7 mm (16.203 in.).
11. Adjust the rear datum height to 402,8 mm (15.859 in.).
12. Secure a support leg to the front bottom link location bracket each side of the base unit. 606456.
13. Ensure that the support legs are at 90° to the floor.
14. Place a 50 kg (1 cwt) weight on the base unit front end to prevent any height variations during welding and cooling.

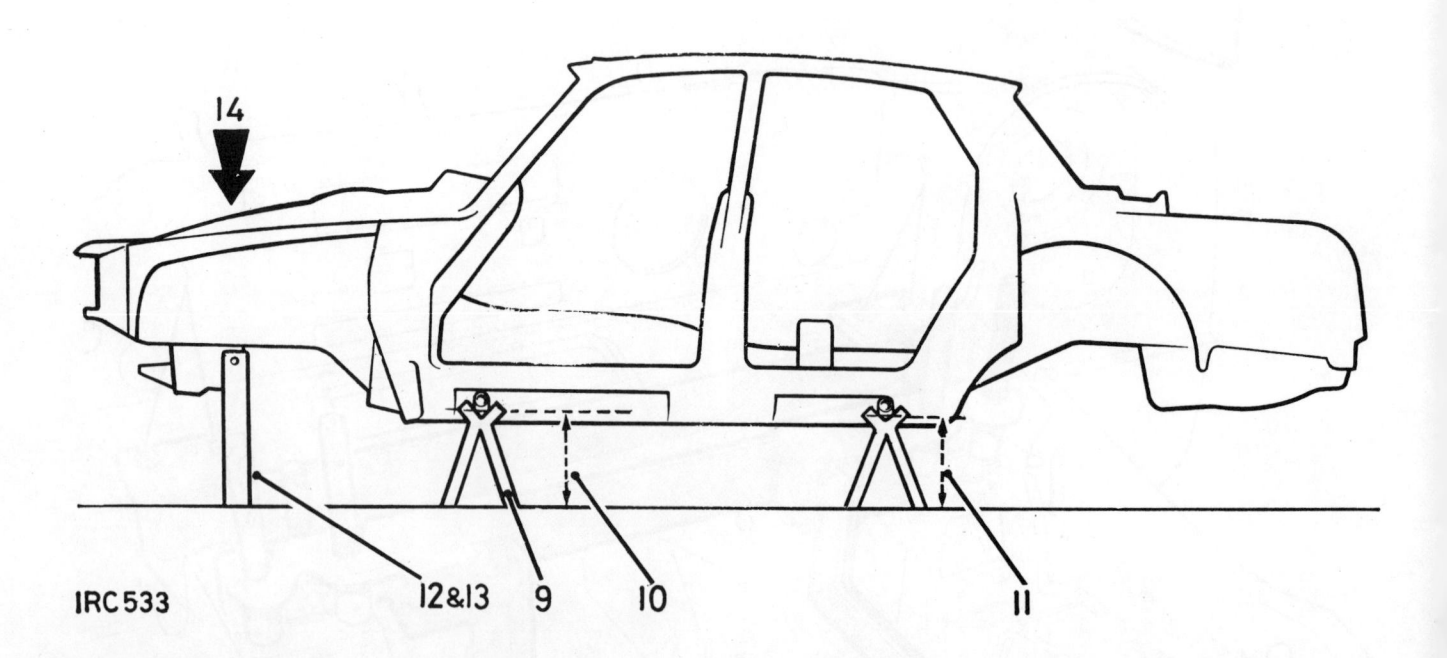

continued

IRC 533

BODY

15. Tack weld the 'lower valance' to the base unit.
16. Remove the panel illustrated in dotted line from the new upper valance panels.
17. Trim any surplus metal from the rear edge of the new upper valances, illustrated in heavy line.

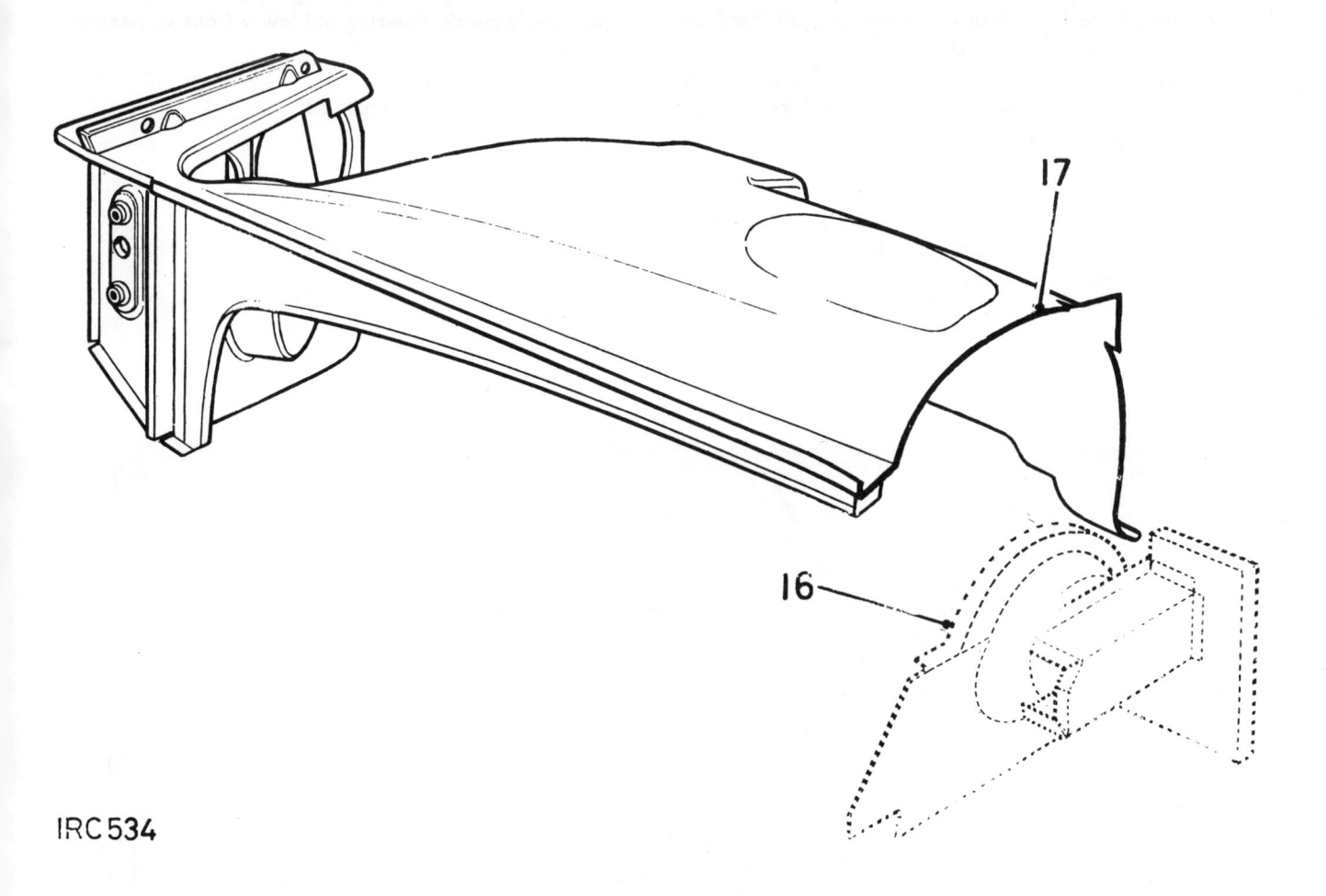

continued

IRC 534

BODY

18. Offer the right hand and left hand upper valance to the base unit and drive screw or pop rivet into position.
19. Bolt the right hand and left hand front upper valance welding and checking fixture to the base unit. 600572 and 600573.
20. Secure the front upper valances to the welding and checking fixtures. 600572 and 600573.
21. Insert the alignment rod through the hole in the centre of the boss plate on either the left hand or right hand front upper valances. 600596.
22. Adjust the length of the alignment rod to give a dimension of 1333,50 mm (52.500 in.) between the upper valance bosses.
23. Tack weld the front upper valances in position.
24. Seam weld the complete assembly into position, ensuring that the dimension in item 8 does not alter.
25. Weld the closing panel for the front cross-member into position.
26. Finally, weld into position the 'front grille and headlamp mounting panel' and 'panel for bonnet lock'.
27. Damage to the front suspension brackets can be checked and rectified using the front suspension welding and checking fixture. 606457.
28. If all the front suspension brackets are damaged, the front suspension welding and checking fixture 606457, should be used in conjunction with the front lower valance welding and checking fixture 605486. This positively checks steering and lower front suspension locating points.

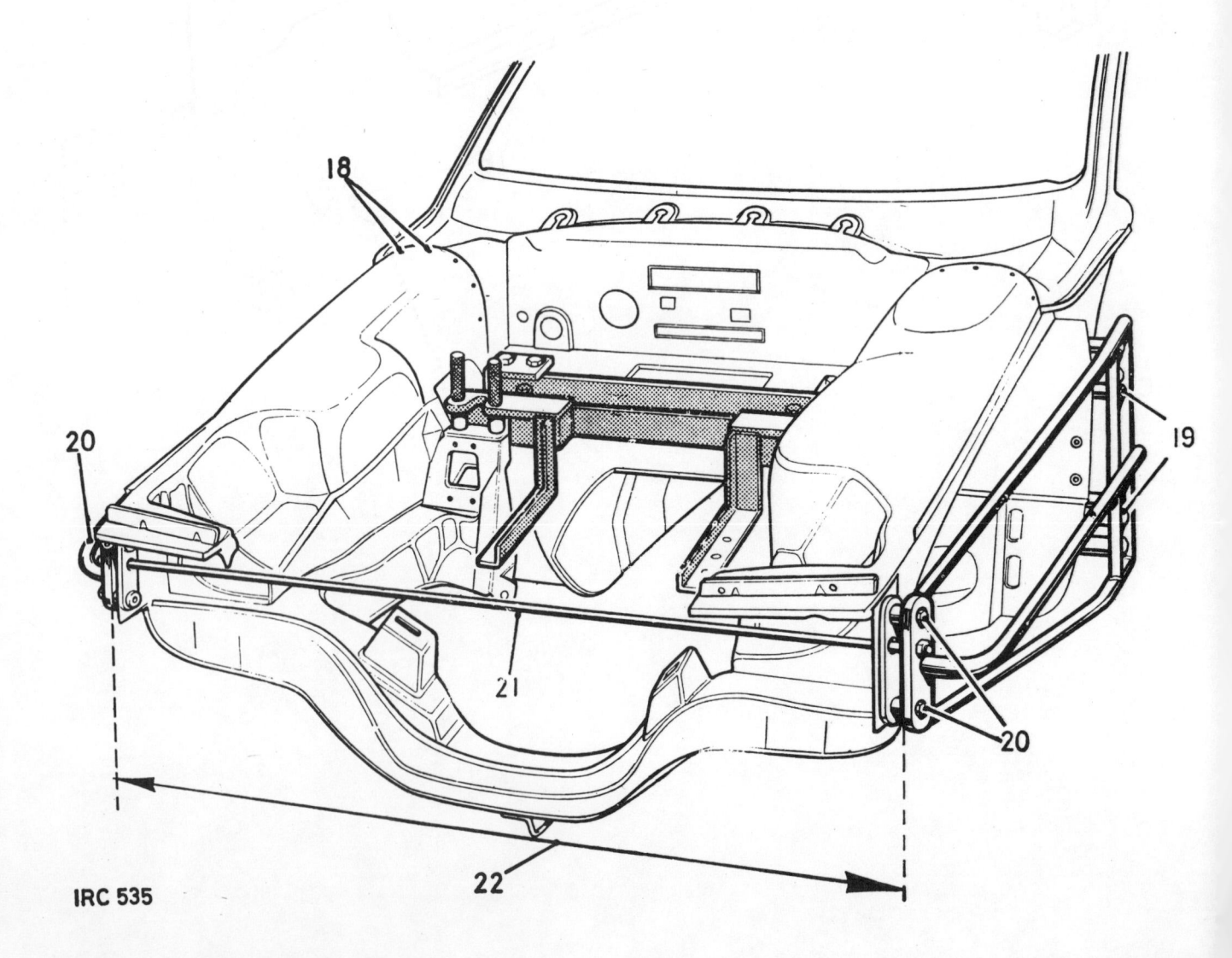

continued

IRC 535

76.10.57
Sheet 5

Rover 3500 and 3500S Manual AKM 3621

BODY

29. On completion of the welding operations, the steering box, steering relay and the front suspension top link mounting locations should be checked for correct alignment, using gauge 605571.

    NOTE: Prior to using gauge 605571, check that it is dimensionally correct.

The overall length A of the pointer must be 43,53 mm + 0,25 mm (1.714 in.+0.010 in.) and if it exceeds this length it must be reduced by grinding at point B.

The length of the plain section C of the gauge must be 37,08 mm ± 0,25 mm (1.460 in. ± 0.010 in.) and if it is found to measure only 33,40 mm (1.315 in.), a spacer 3,68 mm (0.145 in.) thick must be inserted between the end of the tapered post and the plate of the tool at D.

30. Fit the steering box, steering relay and front suspension top links to the base unit.
31. Fit the gauge to the steering drop arm.
32. Adjust the gauge pointer so that it locates in the 'centre' hole of the top link, and lock the pointer in position.
33. Remove the gauge without disturbing the pointer and check the dimension of the post protruding from the gauge plate. The dimension should be 17,07 mm + 3,17 mm (0.675 in. + 0.125 in.).
34. If necessary, the steering box to top link location can be corrected by varying the number of packing plates fitted between the steering box and the base unit. As the tolerance range on the nominal dimension is wide, full use of the tolerances can result in a differing height setting between the steering box and steering relay. When carrying out the above operation, it is beneficial to tyre wear to set the height of the steering box and relay to as near the same nominal dimension as can be achieved within the range of shims available.
35. Repeat 30 to 33 for the steering relay, and adjust as necessary.
36. Damage to the remainder of the base unit front end, should be repaired in the conventional way, by aligning the damaged part with the undamaged.

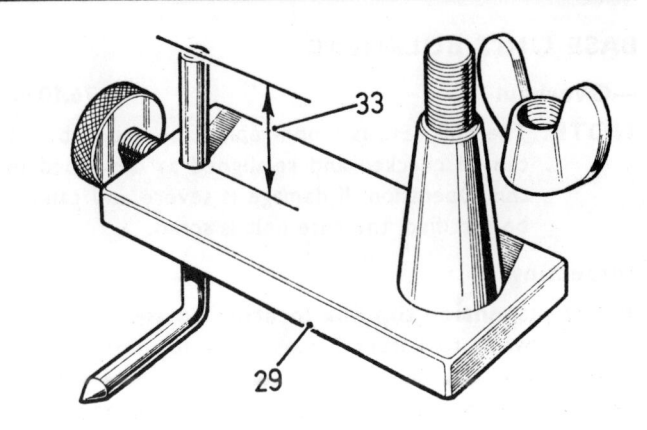

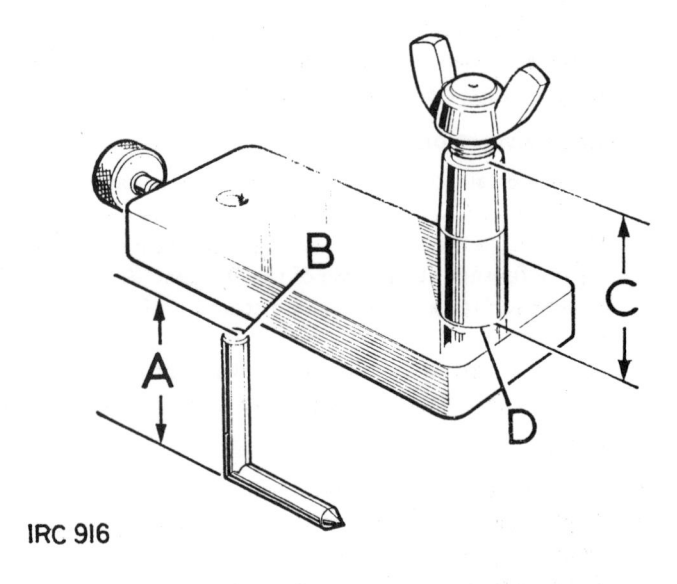

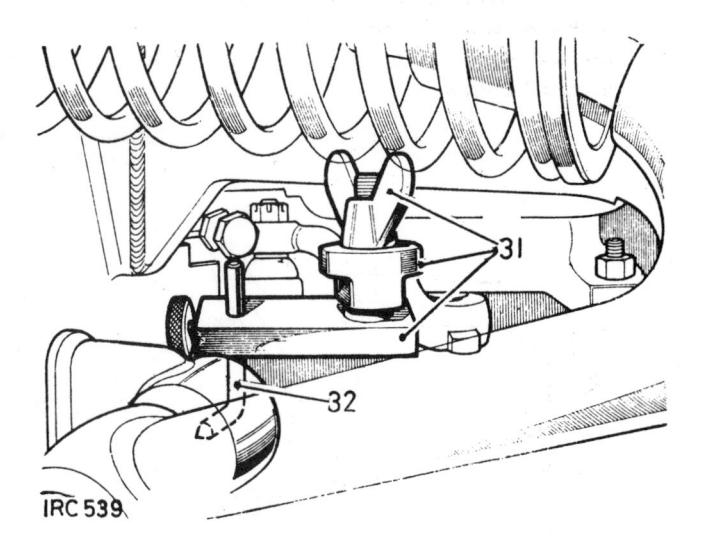

Rover 3500 and 3500S Manual AKM 3621

BODY

BASE UNIT BULKHEAD

—Overhaul 76.10.60

NOTE: The bulkhead is a non-replaceable item, but it can be checked and re-aligned as described in this Operation. If damage is severe and cannot be rectified, the base unit is scrap.

Inspecting

Front suspension top link locating bosses

Where, owing to a frontal impact, the top link locating bosses have been pushed back, on one side of the car, this can be checked without removing the engine, as follows:

1. Remove both top links. 60.40.01.
2. Secure two suitable flat plates across the locating bosses on both sides of the car, to provide a flat surface for checking with a straight edge.
3. Check alignment across both pairs of locating bosses, using a straight edge.

Bulkhead distortion

4. Under certain conditions a frontal impact can cause the bulkhead to distort at the base of the 'A' post. This damage is not always obvious. Check by removing the finisher at the base of the 'A' post. A slightly rippled surface will confirm a distortion of the bulkhead.
5. Where a severe frontal impact has occurred, the windscreen pillars may also have been damaged.

Repairing

6. Position a suitable hydraulic jack in the door aperture.
7. Apply heat at the base of the 'A' post and operate hydraulic jack.
8. If the pressure of the hydraulic jack fails to return bulkhead and 'A' post to their original position, it may be necessary to cut the sill on the inside only, using a hacksaw.
9. Reweld the cut on completion of the straightening operation.

   NOTE: Where pulling equipment is used in conjunction with the hydraulic ram, it may not be necessary to heat and cut the base of the 'A' post.

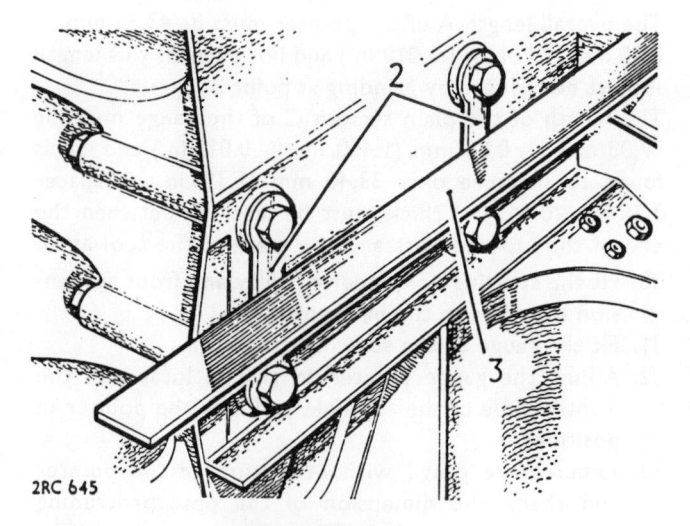

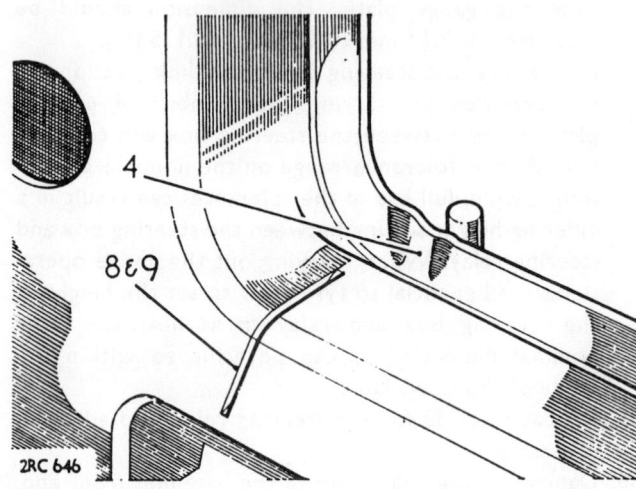

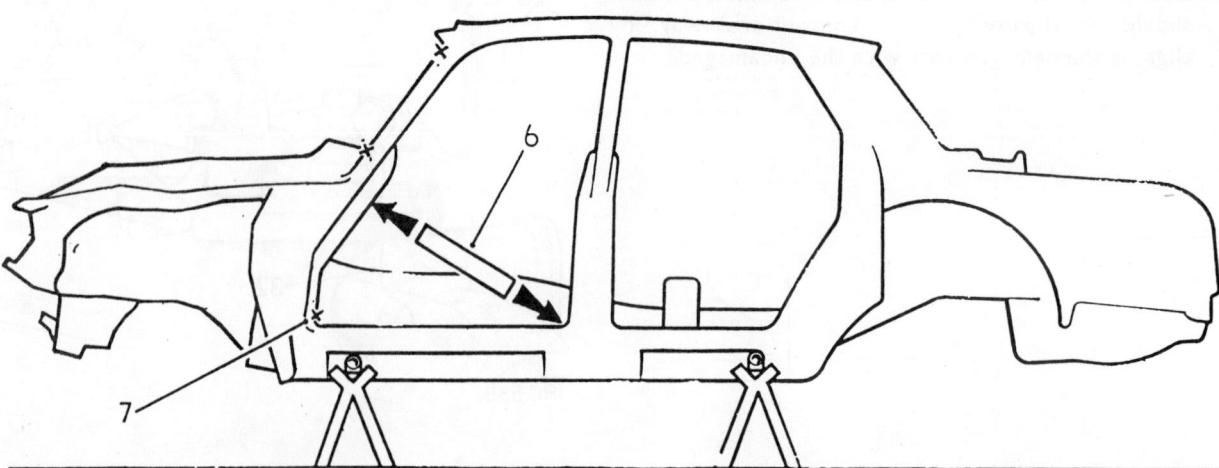

76.10.60

Rover 3500 and 3500S Manual AKM 3621

BODY

BASE UNIT
—Initial check on Churchill fixture
  1 to 10   76.10.61
—Repair check on Churchill fixture
  1 to 38   76.10.62

Service tools: 601462 Checking and repair jig kit
601464 Subsidiary checking and repair jig kit

(Details of the above kits are included in the following general notes.)

NOTE:
The Rover body checking and repair jigs described on the following pages have been designed for use with the Churchill 700 body checking and repair fixture and are an extension of the Rover-designed body repair tools already in use.

The Rover 3500 and 3500S body can be completely checked for body distortion, and front and rear suspension misalignment, with the minimum of dismantling. It is merely necessary to remove the two sill panels, the rear mudflaps and fit gauge blocks at the front and rear suspension points.

The car can then be lowered on to the jacks incorporated in the fixture; this immediately checks basic body alignment, at the same time the position of the gauge blocks within the 'V' of the location brackets will indicate if the front or rear suspension is misaligned or distorted in any way. The car can be positively locked to the fixture to enable any repair work to be carried out. The jigs themselves and the jacks incorporated in the fixture are made to very close limits and, as the jacks locate on the four base unit datum points for all base unit drilling and spotfacing, it follows that the base unit can be accurately repaired, as the damaged area is repaired relative to the undamaged, which is positively locked to the jig. As repair work is carried out it can also be checked to fixed datum points on the jigs.

It is possible, and desirable in many cases, to rectify accident damage by the use of power tools without the need to weld in replacement panels. This applies particularly to the bulkhead area where distortion has taken place at the steering unit, steering relay and front door hinge location points. In cases of this sort, use the jigs mounted on the Churchill 700 to accurately position the 'A' post and Rover front lower valance and dash welding and checking fixture, Part No. 605486, to position the bosses for the steering unit or steering relay. See the detailed instructions which follow.

The checking and welding fixtures are sold as complete kits and can be obtained ex-stock from V. L. Churchill & Co. Ltd.

Those Distributors and Dealers who have the Churchill 700 system already installed, that is the basic floor beams and the four transverse members, should place orders immediately. Distributors and Dealers who carry out extensive body repairs should give earnest consideration to the purchase of the Churchill 700 equipment and the Rover jigs, which are used in conjunction with it, as only with this equipment can the Rover 3500 and 3500S be quickly checked for misalignment and distortion and economically repaired.

We consider that the use of the Churchill 700 checking fixture and the tools designed for use with it will materially reduce labour costs on many body repair jobs as there are positive datum points to work to when the unit is correctly set on the jig and the various location brackets attached to it.

The following illustration details the Rover checking and repair jig kits mounted on the Churchill 700 body checking and repair fixture.

continued

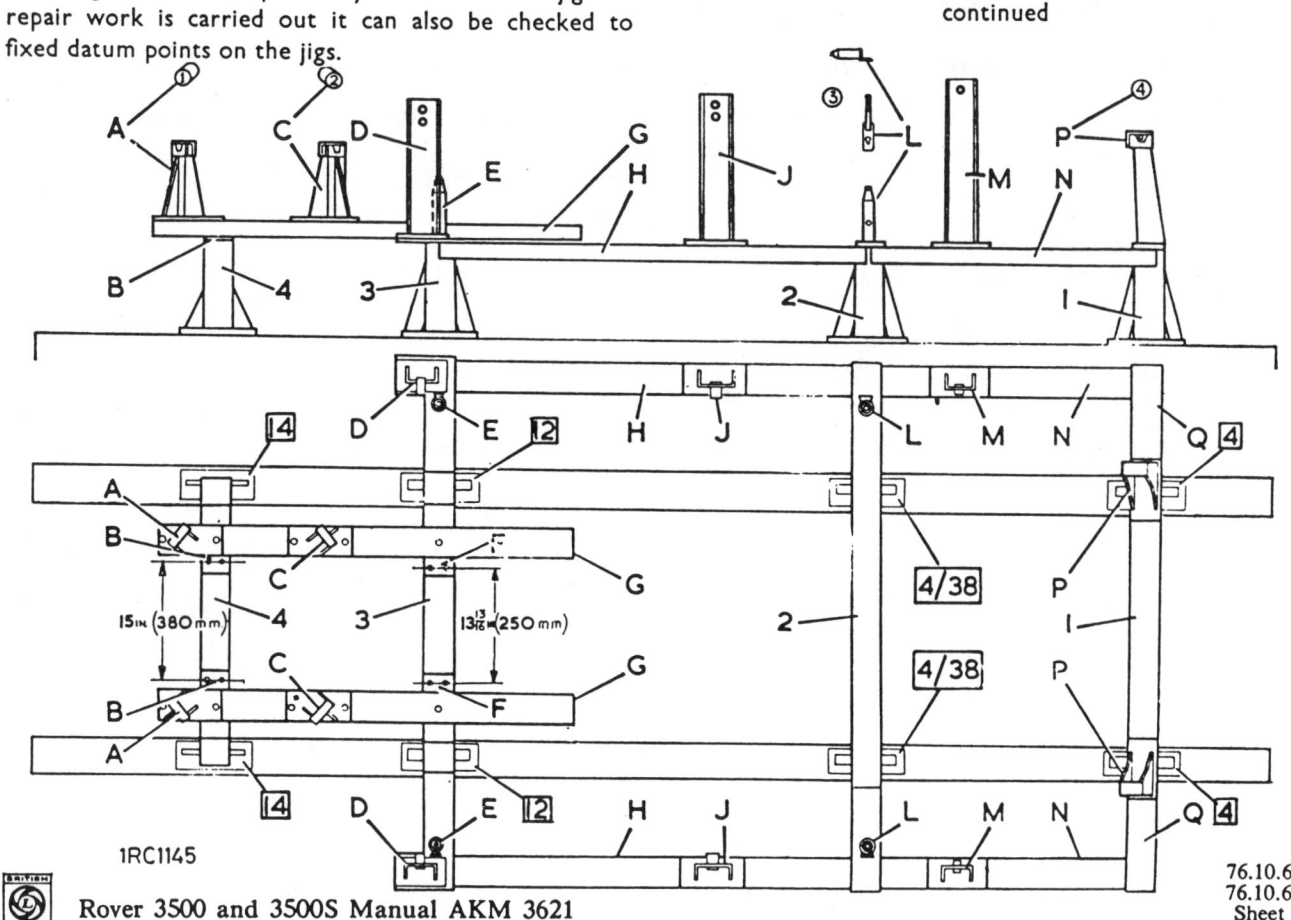

Rover 3500 and 3500S Manual AKM 3621

BODY

601462, Main kit, base unit checking and repair jigs

This kit comprises the items listed below. It should be noted that the Rover part numbers are stamped on each individual item. For identification purposes see illustration.

Description and key to illustration	Qty.	Part No.
A { Location bracket, front suspension, front RH	1	605490
{ Location bracket, front suspension, front LH	1	605491
B { Support bracket, front RH } For front suspension	1	601329
{ Support bracket, front LH } longitudinal members	1	601330
C { Location bracket, front suspension, rear RH	1	605492
{ Location bracket, front suspension, rear LH	1	605493
D —Support bracket, 'A' post, RH or LH	1	600932
E { Location bracket, front jacking tube, RH } Complete with locking	1	600925
{ Location bracket, front jacking tube, LH } bracket and plug	1	600926
F —Support bracket, rear, for front suspension	2	601328
G —Longitudinal members for front suspension location brackets	2	601327
H —Longitudinal members, front, for 'A' and 'BC' post support bracket, RH or LH	1	600937
J —Support bracket, 'BC' post, RH or LH	1	600933
L —Location bracket, rear jacking tubes, complete with locking bracket and plug	2	600927
M —Support bracket, 'D' post, RH or LH	1	600934
N —Longitudinal member, rear, for 'D' post support bracket, RH or LH	1	600936
P { Location bracket, rear end of rear suspension, RH	1	†600930
{ Location bracket, rear end of rear suspension, LH	1	†600931
Q —Adaptor for longitudinal member, rear RH or LH	1	600935

Station No. 14   Station No. 12   Station No. 4/38   Station No. 4

Note: Transverse members are numbered from rear to front.

Transverse member No. 1   Transverse member No. 2   Transverse member No. 3   Transverse member No. 4

†Complete with gauge block and clamping bar.

This kit enables the Rover 3500 and 3500S body to be completely checked and also provides means of holding the body rigidly while repairs are being carried out. Various components of the body can be positioned accurately whilst being welded in position on either the left-hand or right-hand side.

601464, Subsidiary kit, base unit checking and repair jigs

This kit, which can only be used in conjunction with the main kit (Part No. 601462), comprises an additional set of some of the main jig components. For identification see illustration.

Description	Qty.	Part No.
D—Support bracket, 'A' post, RH or LH	1	600932
H—Longitudinal member, front, for 'A' and 'BC' post support bracket, RH or LH	1	600937
J—Support bracket, 'BC' post, RH or LH	1	600933
M—Support bracket, 'D' post, RH or LH	1	600934
N—Longitudinal member, rear, for 'D' post support bracket, RH or LH	1	600936
Q—Adaptor for longitudinal member, rear RH or LH	1	600935

These additional parts enable a complete front or rear end to be accurately positioned and supported at both sides during the welding process.

continued

BODY

Procedure

Checking basic alignment 1 to 10

1. Position the transverse members on the floor beams as follows, when using the Churchill 700 jig.
   a. No. 1 transverse member at Station No. 4.
   b. No. 2 transverse member at Station No. 4/38.
   c. No. 3 transverse member at Station No. 12.
   d. No. 4 transverse member at Station No. 14.

   NOTE: Providing that all the fixing bolts are used, the various location brackets can only be fitted to the transverse members in the correct position.

2. Position the car over and parallel with the floor beams. Make sure car is absolutely parallel with the floor beams, otherwise difficulty will be experienced in locating the four car datum points on the location jacks.

3. Bolt the location brackets for front jacking tube, Part No. 600925/6, at each end of No. 3 transverse member.

4. The car can be jacked up front end first or rear end first, depending on the area of damage. Always try to locate car on two undamaged points first.

5. Jacking front end first now described. Jack up the front of the car, using a standard trolley jack. Slide No. 3 transverse member complete with location brackets into place at Station No. 12. Insert dowel pins and secure transverse member with set bolts.

6. Lower the car until the ends of the location bracket jacks enter into the holes of the datum points on the underside of the car jacking tubes.

7. Bolt location brackets for rear jacking tube, Part No. 600927, two off, to outer end of No. 2 transverse member.

8. Jack up rear of car and position No. 2 transverse member complete with location brackets at Station No. 4/38. Secure location brackets and lower car as detailed for the front end.

9. If one of the jacking tubes on the car has been forced downwards and will obviously foul the jack on the location bracket, then the three remaining jacks should be screwed up to allow the car to locate on these three points.

10. The basic alignment of the car can now be checked. The location pads on the car jacking tubes should fit squarely on to the location bracket jacks; use the clamping brackets to firmly secure the body to the jig at each jacking point. Any misalignment at the jacking points should be rectified as necessary by the use of power tools and/or replacement parts.

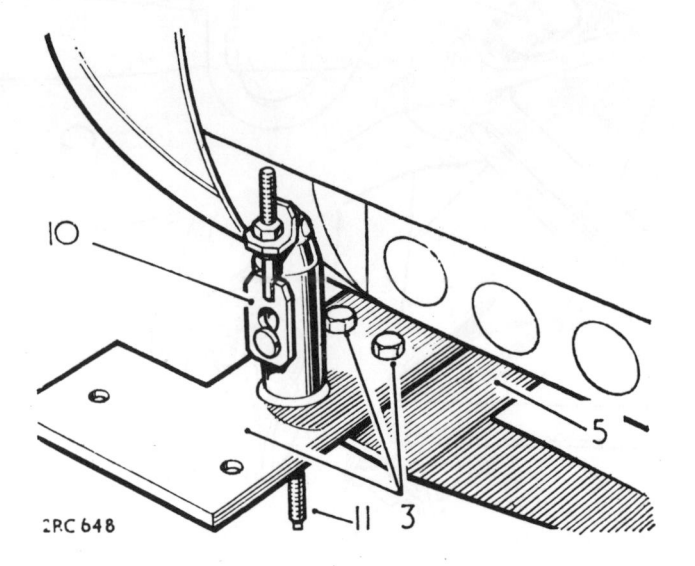

continued

Rover 3500 and 3500S Manual AKM 3621

BODY

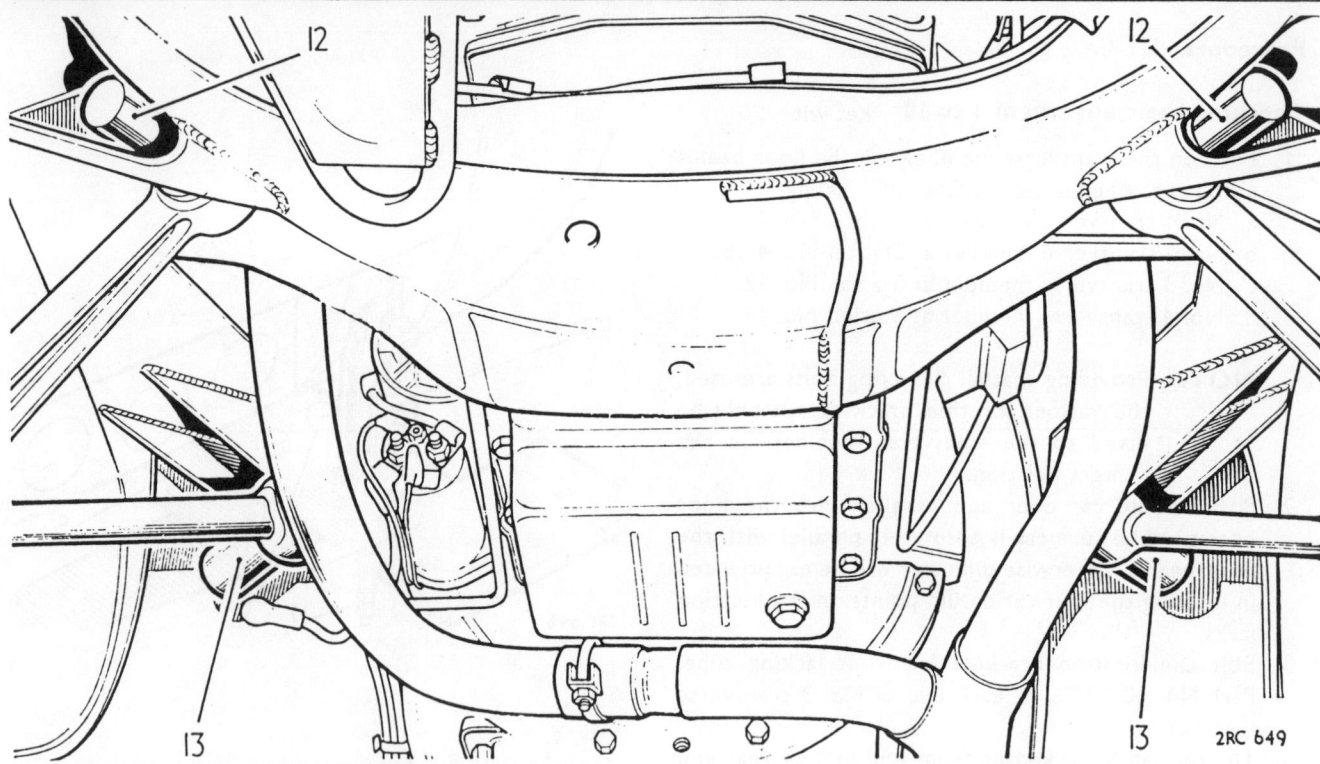

Checking front suspension locations 11 to 22

11. Temporarily remove the securing brackets at the jacking points and, using the jacks incorporated in the fixture, raise the car about 25 mm (1.000 in).
12. Remove the nuts which secure the bottom link to the base unit and screw on the gauge block etched with a figure '1'.
13. Remove the nuts which secure the tie rod to the base unit and screw on the gauge block etched with a figure '2'.
14. Bolt the two support brackets for front suspension longitudinal member, Part No. 601328, to No. 3 transverse member so that the inner fixing holes of the two support brackets are 350 mm (13.812 in) apart, as shown on the illustration of the checking and repair jigs at the beginning of this operation.
15. Bolt the support brackets for front longitudinal member, Part No. 601329/30, to No. 4 transverse member; in this case so that the inner fixing holes are 380 mm (15.000 in) apart.
16. Slide No. 4 transverse member complete with the support bracket into position at Station No. 14. Insert dowel pin and secure with set bolts.
17. Bolt the two longitudinal members for the front location brackets, Part No. 601327, to the support brackets previously fitted to No. 3 and 4 transverse members.
18. Bolt the front location brackets for the front suspension 605490/1 to the longitudinal members, directly underneath the gauge blocks previously fitted at the bottom link attachment point.
19. Bolt the rear location brackets for the front suspension 605492/3 to the longitudinal members, directly underneath the gauge blocks previously fitted at the tie rod attachment point.

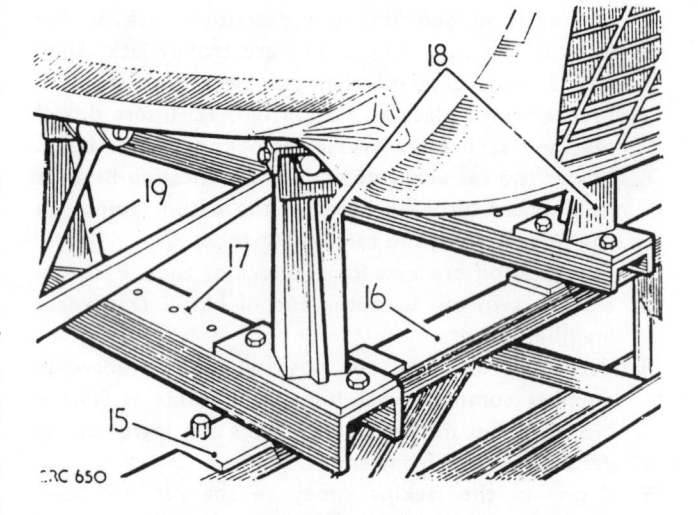

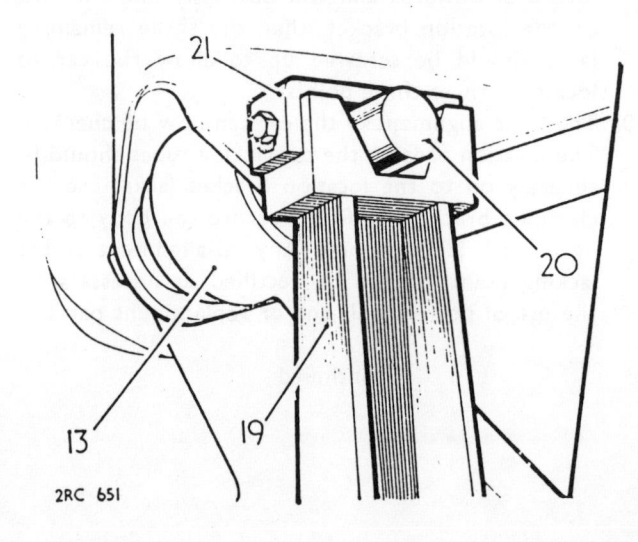

continued

76.10.61
76.10.62
Sheet 4

Rover 3500 and 3500S Manual AKM 3621

BODY

20. Lower the car fully by means of the four jacks at the main location bracket. The gauge blocks should locate in the 'V' of the location bracket with 1,5 mm (0.062 in) side and bottom clearance.
21. The clamping bracket shown should be used when any repair work is being carried out on the body. This will prevent components which are correct being distorted by any welding, etc, which is carried out on adjacent panels.
22. Any misalignment should be rectified as necessary by means of power tools and/or replacement parts.

Checking rear suspension rear location 23 to 25

NOTE: The rear suspension can be checked in a similar manner to the front, except that gauge blocks are only available for the rear ends of the top links. The bottom links must be checked visually. Proceed noting the following.

23. The location of front end of the rear suspension bottom link, can be checked with the removeable crossmember and the close proximity of the rear jacking tubes at location 'L', on the illustration of the checking and repair jigs at the beginning of this operation.
24. When fitting the two gauge blocks at the rear position of top link for rear suspension, it is necessary to first remove the rear mudflaps. The nut retaining the top link is then removed and gauge block etched with figure '4' is screwed into position.
25. Bolt the location brackets for the rear suspension 600930/1 to the transverse members directly underneath the gauge block fitted at the top link rear attachment point.

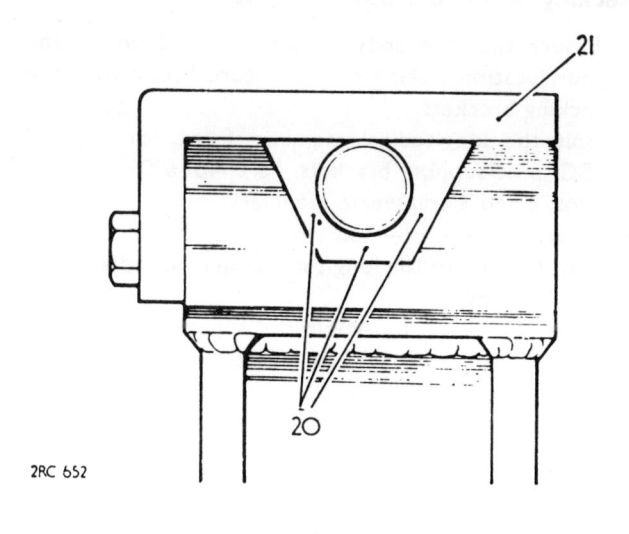

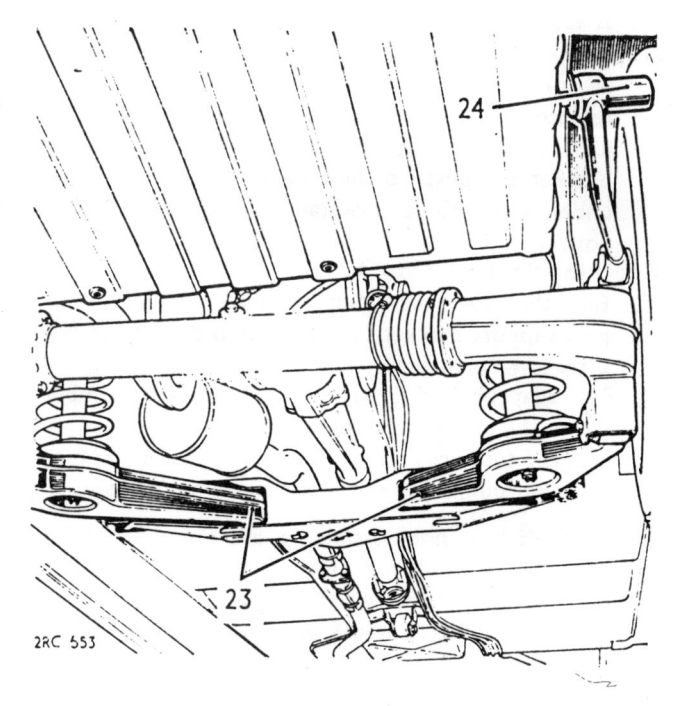

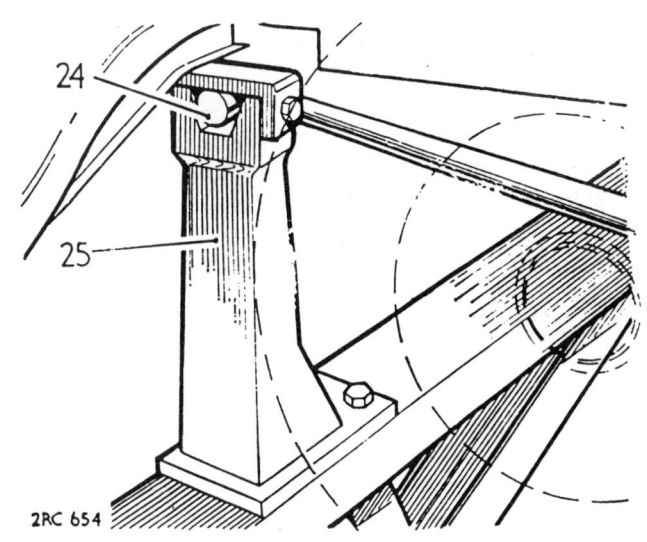

continued

Rover 3500 and 3500S Manual AKM 3621

76.10.61
76.10.62
Sheet 5

BODY

Checking 'A' or 'BC' posts 26 to 30

26. Ensure that the body is jacked fully down at the four location jacking points. Secure firmly with the locking brackets.
27. Bolt the longitudinal member, front, for 'A' and 'BC' post support brackets, Part No. 600937, on to Nos. 2 and 3 transverse members.
28. Checking 'A' post—Bolt the support bracket for 'A' post 600932, to the longitudinal member adjacent to the 'A' post.
29. Checking 'BC' post—Bolt the support bracket for the 'BC' post 600933, to the longitudinal member adjacent to the 'BC' post.
30. The holes in the support brackets should line up with the bottom holes in the base unit with a clearance of 1,5 mm (0.062 in) between the bracket and the 'A' post; and 0,8 mm (0.031 in) between the 'BC' post and bracket or the 'D' post and bracket, as applicable.
31. In addition to checking position, these brackets can be used to positively locate a new panel when a replacement is being welded into position. However, it is necessary to use packing washers between bracket and posts to the dimensions detailed under under item 30 to maintain the correct relative position.

Checking 'D' post 32 to 34

32. Bolt the rear longitudinal member 600936 for 'D' post support bracket to the location bracket on No. 2 transverse member.
33. Bolt adaptor 600935, to No. 1 transverse member and to the longitudinal member.
34. Bolt the support bracket for the 'D' post 600934, to the longitudinal member adjacent to the 'D' post.

   NOTE: It is not possible to use the adaptor for longitudinal member, Part No. 600935, and the rear location brackets, Part No. 600930/1, at the same time.

Fitting a complete front or rear end 35 to 38

When a complete front or rear end is to be fitted, use the additional parts from the subsidiary kit as follows:

35. Bolt longitudinal members for 'A' post, 'BC' post and 'D' post support brackets together with the adaptors to both sides of the jig.
36. Fit gauge blocks at the suspension positions applicable; lower replacement units on to the front or rear location brackets. Use 1,5 mm (0.062 in) packing between gauge block and bottom of 'V' on the location bracket.
37. Bolt the replacement unit to the support brackets at 'A' or 'D' post as applicable; use packing washers between the bracket and the unit as detailed previously.
38. Spot or torch weld the unit into position, then check that the dimensions between 'A' or 'D' post upper fixing points and the 'BC' posts are correct.

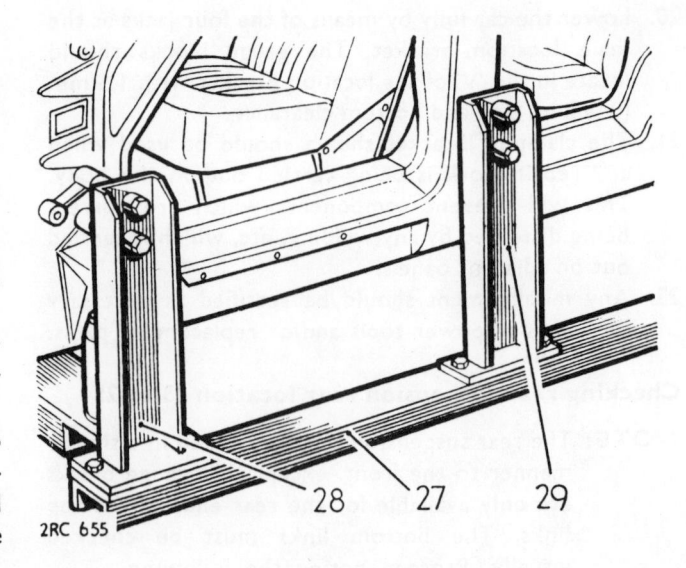

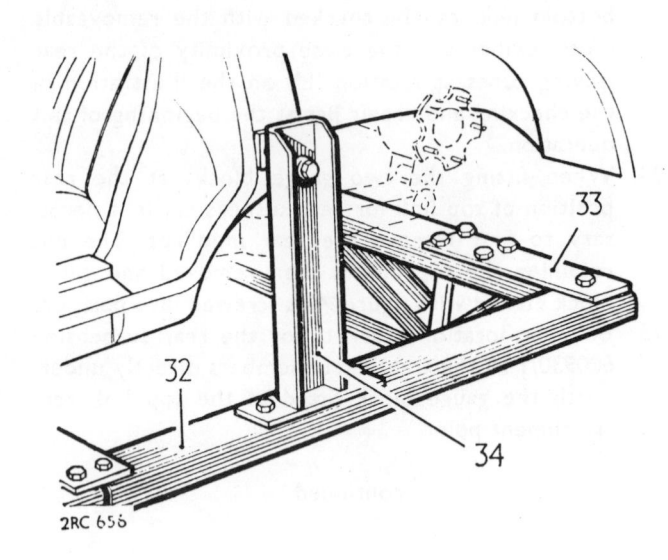

NOTE: It must be emphasised that the tools used in conjunction with the Churchill body checking fixture are an extension of the Rover-designed body tools which are already in use and that the Rover tools are still required when replacing any parts forward of the bulkhead; that is, dash unit bottom half, front upper valance panel, etc, and for positively checking the steering suspension pick-up points in relation to the steering box and steering relay pick-up points on the dash. See Operation 76.10.57.

BODY

'B' POST TRIM PAD

—Remove and refit 76.13.08

Removing

1. Remove the seat belt fixing and stowage hook from the upper 'B' post.
2. Ease the upper finisher from the 'B' post and remove.
3. Remove the door aperture edge finishers from the 'B' post.
4. Remove the lower seat belt fixing from the 'B' post.
5. Remove the clips securing the lower finisher to the 'B' post and remove the finisher.

Refitting

6. Reverse 1 to 5.

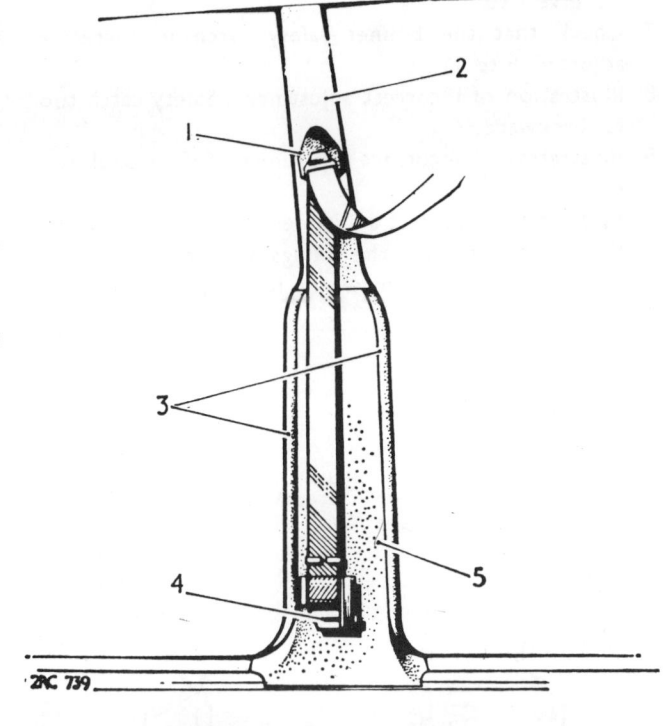

BONNET

—Remove and refit 1 to 4 and 6 to 11 76.16.01

BONNET SAFETY CATCH

—Remove and refit 1, 5 and 6 to 11 76.16.34

Removing

1. Prop open the bonnet.
2. Disconnect the windscreen washer pipe at the T-piece between the jets.
3. Remove the fixings securing the bonnet hinges to the bulkhead.
4. Lift the bonnet clear.
5. Remove the bonnet safety catch.

continued

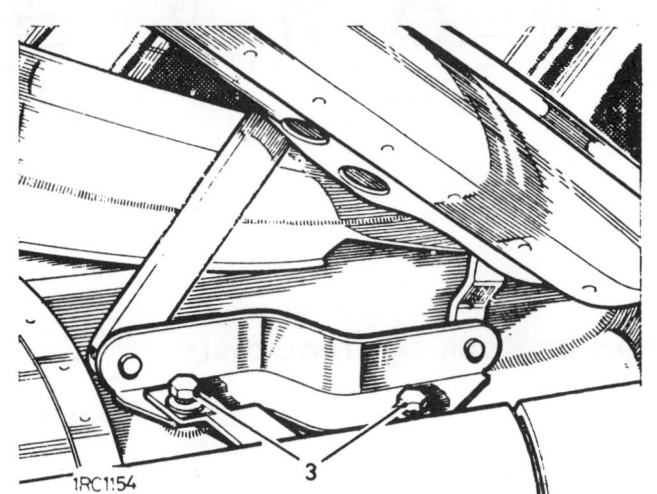

Rover 3500 and 3500S Manual AKM 3621

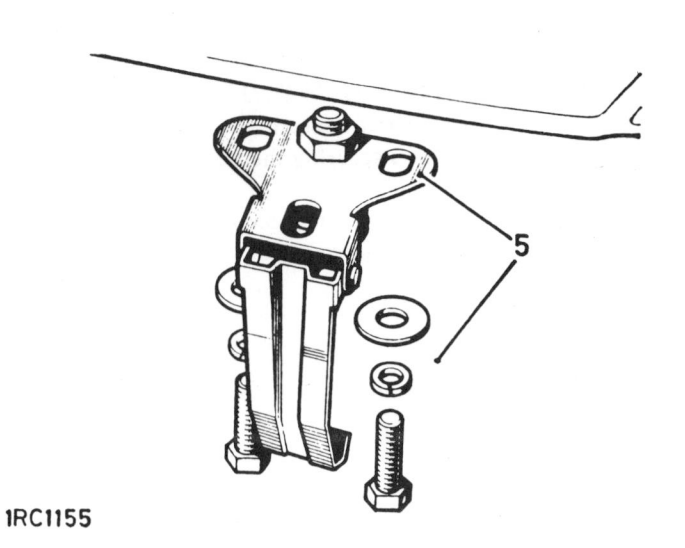

BODY

Refitting

6. Reverse 1 to 5.
7. Check that the bonnet safety catch is correctly adjusted. 8 to 11.
8. Illustration of incorrect adjustment: Safety catch too far backward.
9. Illustration of incorrect adjustment: Safety catch too far forward.
10. Illustration of correct adjustment.
11. If necessary, slacken the fixings and slide the safety catch backwards or forwards to obtain the correct setting.

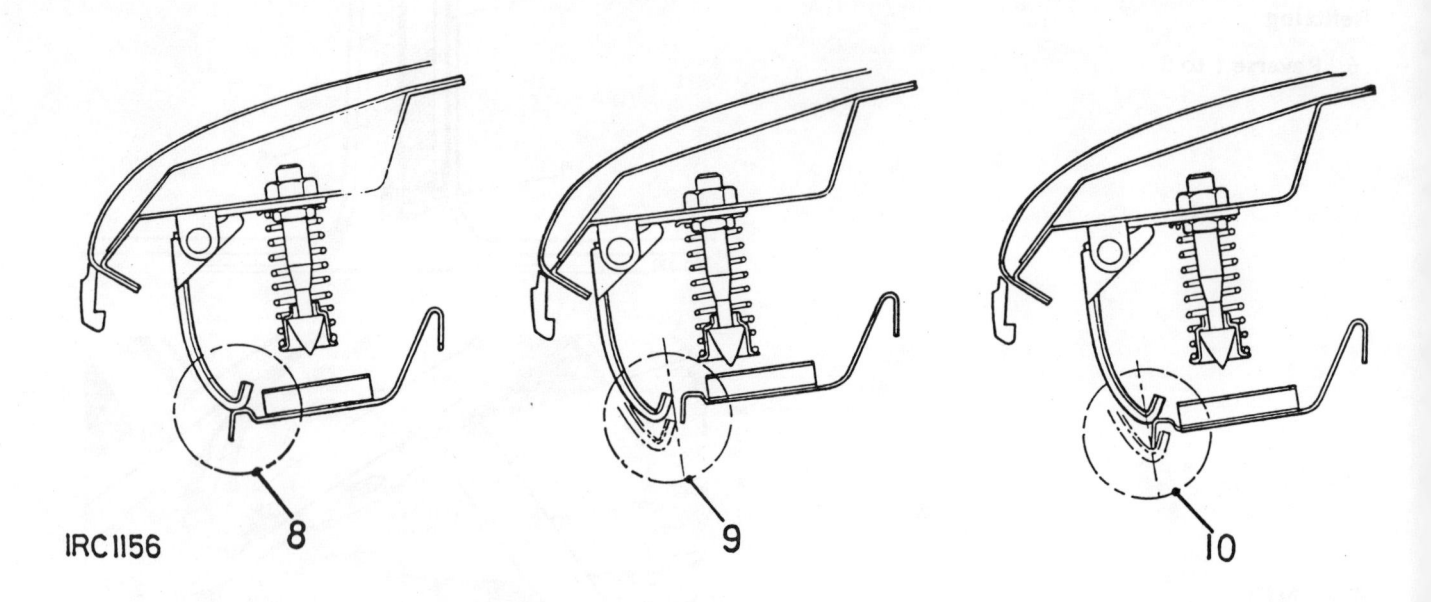

BONNET LOCK CONTROL CABLE

—Remove and refit   76.16.29

Removing

1. Remove the instrument panel. 88.20.01.
2. Early models with ribbon type speedometer— Remove the fascia switch panel. 76.46.16.
3. Remove the fascia top rail. 76.46.04.
4. Remove the parcel tray. 76.67.02.
5. Remove the felt from beneath the parcel tray trim.
6. Remove the screws securing the bonnet lock control to the base unit.
7. Disconnect the cable at the bonnet lock.
8. Slacken the cable securing clips.
9. Withdraw the cable.

Refitting

10. Reverse 1 to 9.

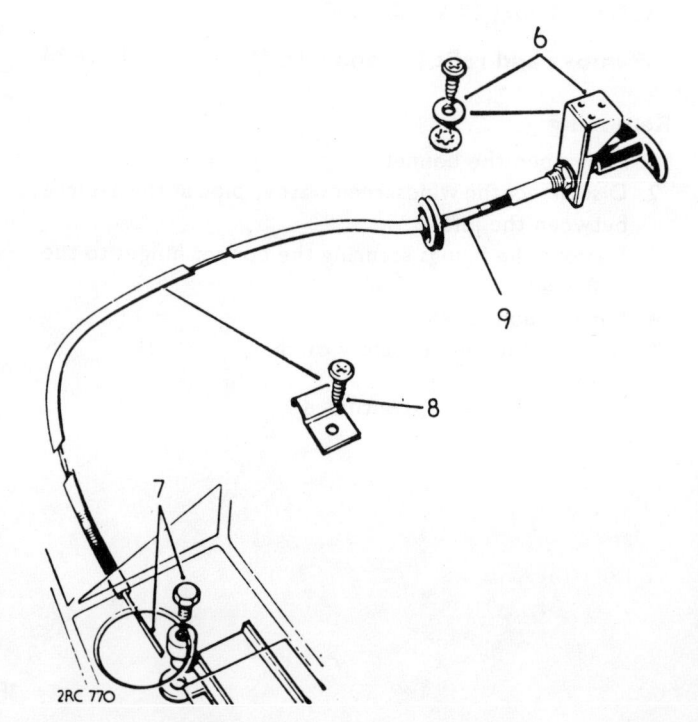

76.16.01/34
Sheet 2
76.16.29

Rover 3500 and 3500S Manual AKM 3621

BODY

BOOT LID

—Remove and refit 1 to 7 and 9 to 11　　76.19.01

BOOT LID HINGES

—Remove and refit 1 to 11　　76.19.07

BOOT LID SPRING

—Remove and refit 1 to 11　　76.19.08

Removing

CAUTION: The boot lid must be supported in the fully open position throughout the following procedure, otherwise, the boot lid may be damaged by the fixing bolts.

1. Open the boot lid.
2. Remove the two hinge covers.
3. Remove the four bolts and plain washers securing the hinges to the boot lid.
4. Remove the boot lid.
5. Remove the three bolts securing the rear edge of the lower decker panel to allow the panel to be eased.
6. Slacken the hinge bracket steady bolts.
7. Remove the four bolts fixing the hinges to the body.
8. Remove the hinges and torsion bars complete.

Refitting

9. Reverse 2 to 8.
10. Adjust the hinge bracket steady bolts to contact the base unit.
11. Close the boot lid and check that it is correctly adjusted.

BOOT LID SEAL

—Remove and refit　　76.19.06

Service tool: 600358　Fitting tool, body sealing rubber

Removing

1. Remove the boot lid. 76.19.01.
2. Withdraw the seal from the boot lid channel.

Refitting

3. Apply silicone grease MS4 to the boot seal.
4. Fit the seal to the boot lid channel using 600358.
5. Refit the boot lid. 76.19.01.

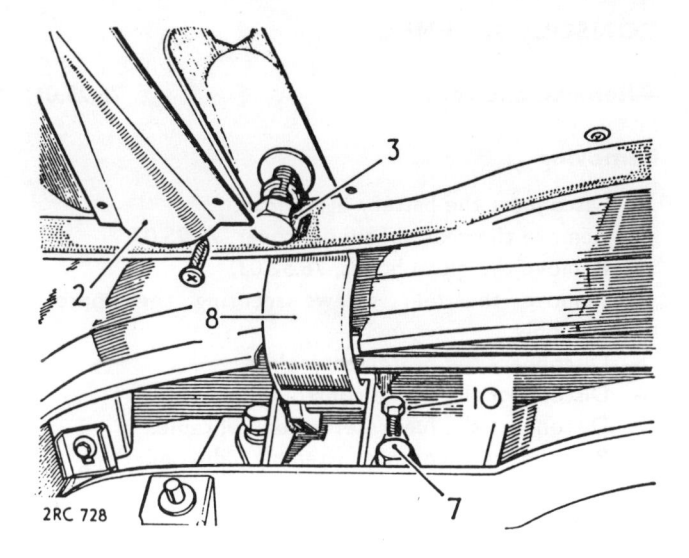

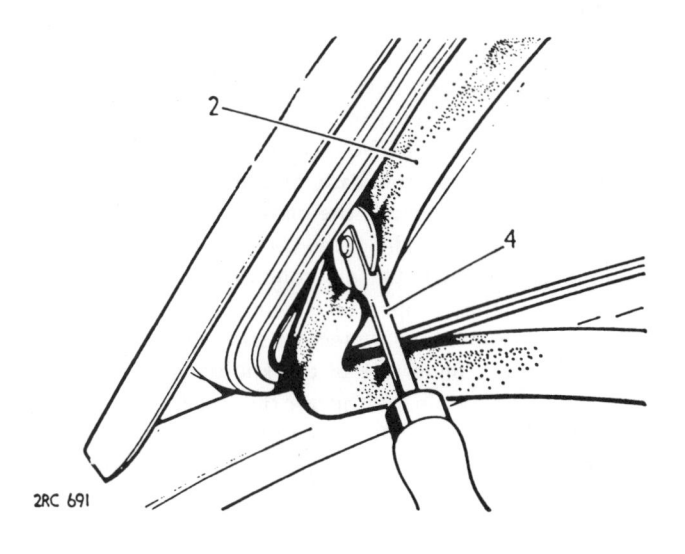

Rover 3500 and 3500S Manual AKM 3621

BODY

CONSOLE ASSEMBLY

—Remove and refit 76.25.01

Removing

1. Disconnect the battery earth lead.
2. Remove the gearbox tunnel cover. 76.25.07.
3. Remove the glove boxes. 76.52.03.
4. Remove the four screws securing the console assembly.
5. Disconnect the heater controls.
6. Disconnect the choke control cable.
7. Disconnect the fuel reserve control cable.
8. Remove the console assembly, disconnecting all wiring.

Refitting

9. Feed the choke and fuel reserve cables through their respective grommets in the base unit.
10. Locate the console assembly in position and reconnect all wiring.
11. Insert the two drive screws adjacent to the left-hand and right-hand glove box hinges.
12. Fit the brackets securing the radio panel to the underside of the switch panel.
13. Refit the heater controls.
14. Refit the gearbox tunnel cover. 76.25.07.
15. Connect the choke and fuel reserve cables and check for correct operation.
16. Refit the glove boxes. 76.52.03.
17. Reconnect the battery earth lead.

GEARBOX TUNNEL COVER

—Remove and refit 76.25.07

Removing

1. Remove the front and rear ashtrays.
2. Remove the two nuts from under the rear ashtray and the single nut from under the front tray.
3. Remove the screws from the front of the cover.
4. Lift the flap at the base of the hand brake lever grommet and remove the drive screw.
5. Remove the grille from the console unit by removing the two screws. On air conditioning models, first remove the temperature selection and air flow knobs by inserting a piece of stiff wire through the hole in the underside of the knob and compressing the retaining pin. Withdraw the knob.
6. Unscrew the knob from the gear lever.
7. Automatic transmission models—Remove the gear selector indicator plate.
8. Slide the front seats to the rearward position and raise the seat locking levers.
9. Manoeuvre the gearbox tunnel cover over the hand brake and gear change levers, and withdraw.

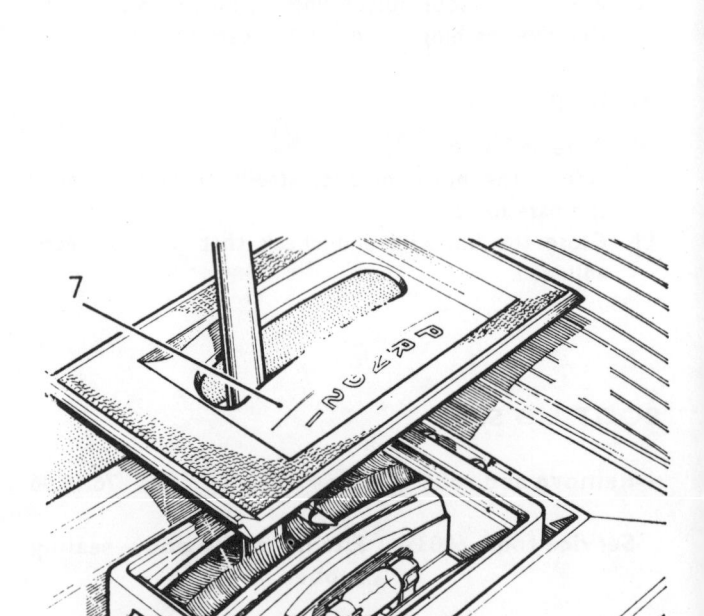

Refitting

10. Reverse 1 to 9.
11. Automatic transmission models—Ensure that the fibre washer is in place beneath knob on the gear-change lever.

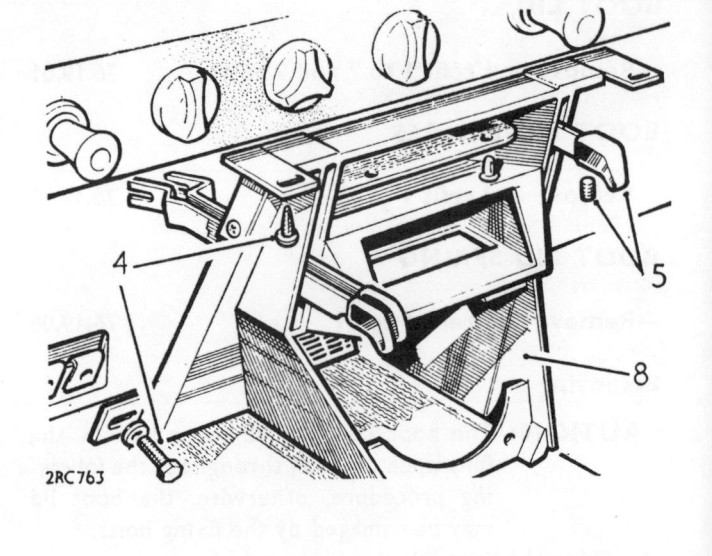

Rover 3500 and 3500S Manual AKM 3621

BODY

SIDE DOOR—FRONT

—Remove and refit   76.28.01

SIDE DOOR—REAR

—Remove and refit   76.28.02

Removing

1. Remove the locking cap from the check-strap locating peg.
2. Slacken the bottom hinge pin nut and screw up the hinge pin far enough to allow the door to be lifted off the top hinge pin.
3. Do not slacken or remove the top hinge pin as this will locate the door when refitting.
4. Lift the door off the top hinge pin.

Refitting

5. Apply Duckham's KG15 grease, or a suitable equivalent, to the door hinge cup.
6. Lift the door on to the top hinge pin and position the check-strap on peg; fit a new locking cap.
7. Screw down the lower hinge pin until all play has been eliminated and check for door closing correctly.
8. If necessary, adjust the door. 76.28.07/08.

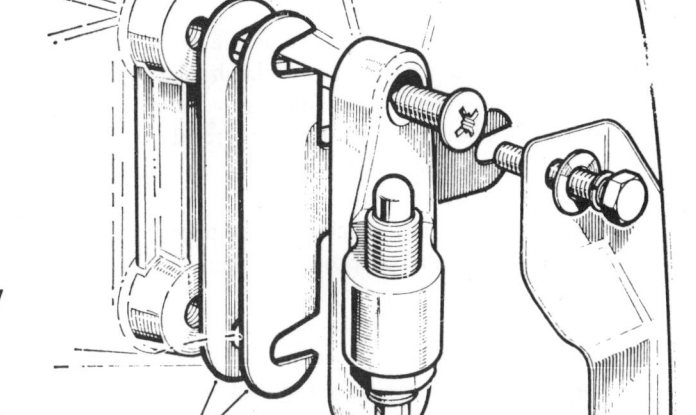

SIDE DOOR—FRONT

—Adjust   76.28.07

SIDE DOOR—REAR

—Adjust   76.28.08

Procedure

Door adjustment can be obtained in three ways, as follows:

1. To adjust the door in or out of the aperture, add or remove shims between the door hinge and the base unit.
2. To adjust the door forward or rearward in the aperture, add or remove shims between the door and the hinge.
3. To adjust the height of the door in the aperture, raise or lower the hinge pin.

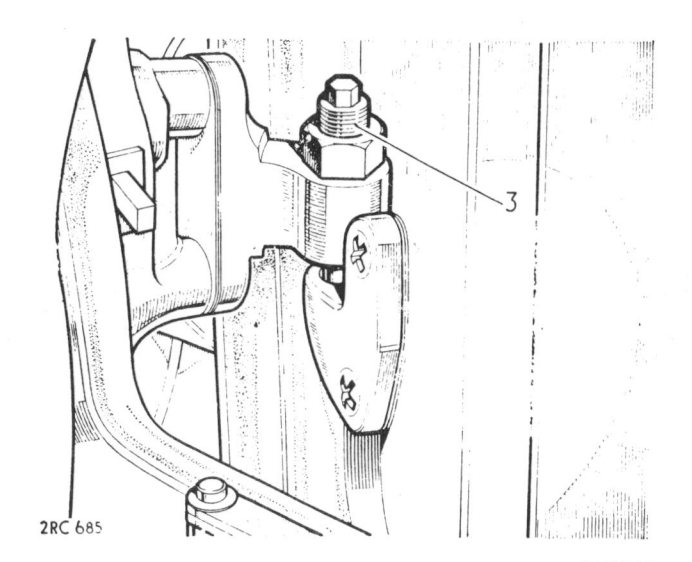

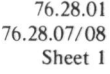

continued

Rover 3500 and 3500S Manual AKM 3621

76.28.01
76.28.07/08
Sheet 1

BODY

4. Ensure that the striker plate is not distorted by placing a straight edge across the two outside faces; if distortion is present, remove the striker plate and level the pillar seating to allow the striker to sit squarely. Distortion at this point can cause the sliding wedge to foul the bottom of the striker.
5. Lightly tighten the fixing screws to support the striker plate in an approximate vertical position on the pillar.
6. Keeping the push-button pressed fully in, gently close the door on to the striker. If necessary, move the striker up or down until the dovetail does not foul either the nylon block or the bottom edge of the sliding wedge. If movement of the striker is required and the striker is at the limit of its adjustment, the door may be raised or lowered by hinge pin adjustment.

   Tighten the screws on the striker plate, close the door fully, then check that over-travel exists; that is, ensure it is possible to push the door inboard a limited amount after the rotary cam has engaged. By adjusting the striker plate inboard or outboard the correct over-travel can be obtained, minimum permissible being 1,5 mm (0.060 in).

   Vertical and horizontal alignment of the lock and striker faces is important, and these should be kept as near parallel as possible.
7. To ensure that the rotary cam of the door lock is fully engaged on to the striker latch plate, check the push-button free play as follows:
   a. With the door open, push the button in, to take up clearance; pencil mark the side of the button adjacent to the escutcheon.
   b. Close the door and check that the pencil mark is in the same relative position when the button clearance is taken up.
   c. If the clearance when the door is closed is more than when the door is open, the rotary cam of the door lock is not fully engaged with the latch of the striker plate. This condition should be rectified by a slight outward adjustment of the striker plate.
   d. Ensure that the lock is fully engaged with the striker plate and that the striker plate does not foul any part of the lock. Adjust by adding or removing shims between striker plate and door pillar.

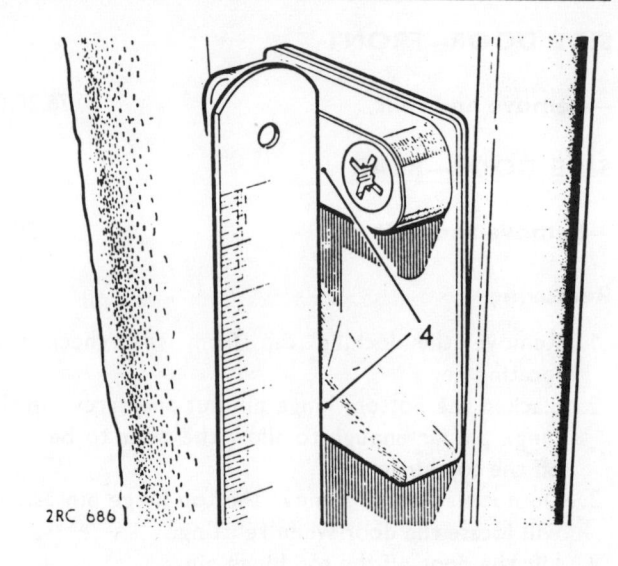

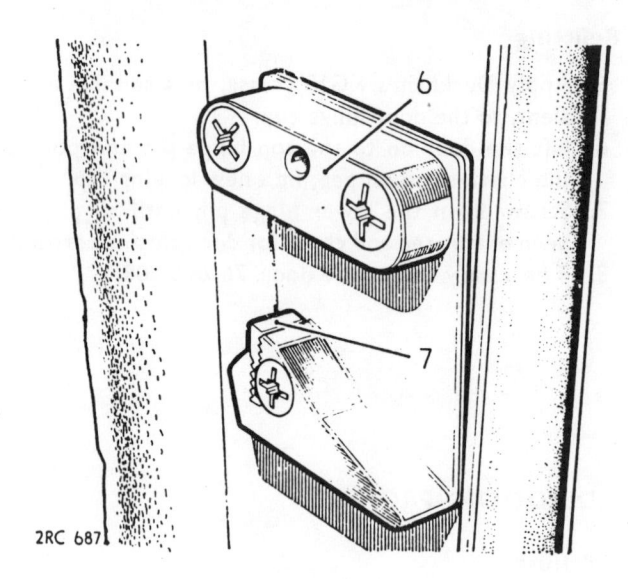

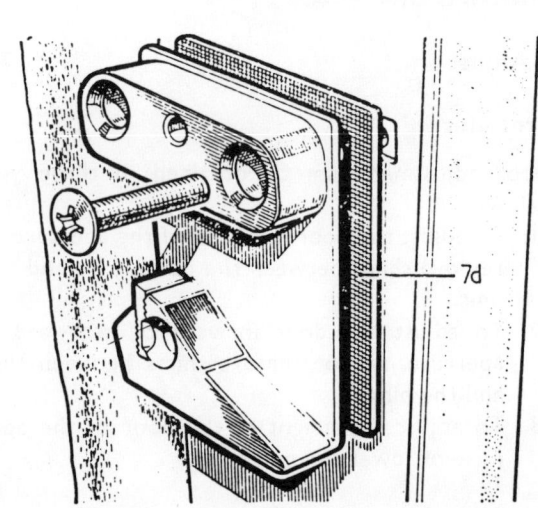

76.28.07/08
Sheet 2

Rover 3500 and 3500S Manual AKM 3621

BODY

DOOR GLASS—SIDE FRONT

—Remove and refit   76.31.01

Removing

1. Remove the door trim pad. 76.34.01.
2. Remove the closing panel from the top of the door.
3. Remove one bolt and washer and slacken the other, securing the glass stop at the bottom of the door.
4. Lower the window and remove the screws securing the window frame cross rail.
5. Remove the fixings retaining the window frame channel at the front edge of the door.
6. Remove the fixings retaining the window frame channel at the rear edge of the door.
7. Remove the four bolts and washers retaining the bottom of the window channel, accessible through the aperture in the bottom of the door.
8. Remove the two long through bolts, flat washers and cage nuts at the front top inside of the door, accessible through the aperture.
9. Remove the door frame bracket.
10. Remove the complete window frame, mechanism and rubber weather seal from the door.
11. Raise the window and remove the bolts securing the window regulator mechanism.
12. Remove the regulator mechanism.
13. Slide the glass down the channel and remove.
14. If required, remove the support channel and glazing strip from the glass.

Refitting

15. Reverse 2 to 14, noting the following.
16. Manoeuvre the complete window frame assembly into position with the weather seal.
17. Wind the glass right down to gain access, and fit to the top, four self-tapping screws and star washers.
18. Fit all other securing bolts loosely.
19. Set the window frame height to a snug fit in the body frame and, when correct, tighten the four set bolts into the two brackets at the bottom of the runners.
20. Set the frame inwards to apply tension to the seal at top of door and tighten all securing bolts.
21. Adjust nylon glass steady pad to minimise door glass movement.
22. Fit the window glass bottom stop bracket and rubber, four set bolts, plain and spring washer.
23. Fit the top closing panel with self-tapping screws.
24. Fit the waist moulding with four self-tapping screws, and slide the veneer strip into position. Refit the locking knob.
25. Stick the plastic sheet to the door.
26. Push the inside trim panel into position on the spring clips.
27. Refit the window winder handle and inside door handle. Refit the arm rest.
28. Refit the door trim pad 76.34.01.

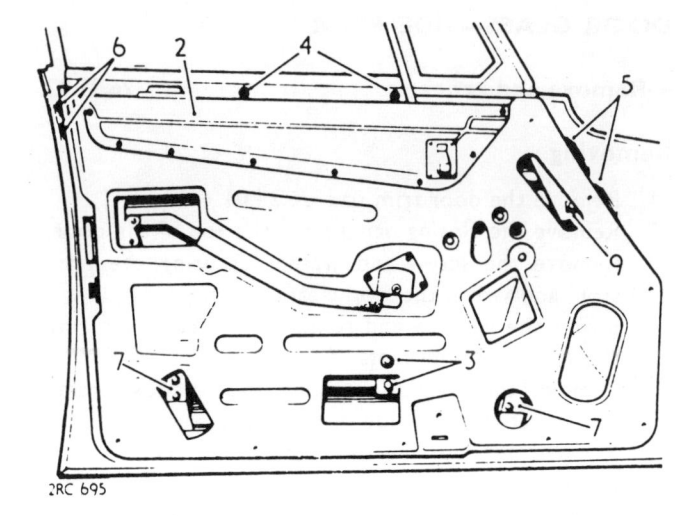

Rover 3500 and 3500S Manual AKM 3621

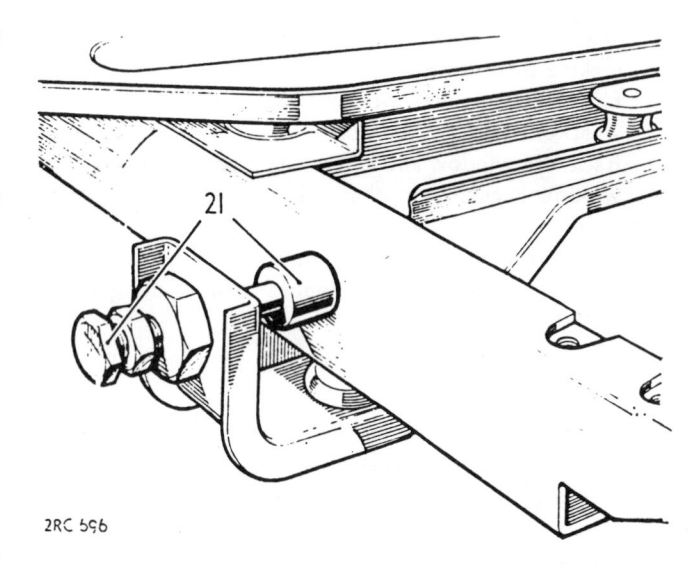

BODY

DOOR GLASS—SIDE REAR

—Remove and refit 76.31.02

Removing

1. Remove the door trim pad. 76.34.04.
2. Remove the closing panel from the top of the door.
3. Remove the screw and washer securing the glass stop, and swivel the stop clear.
4. Lower the window and remove the screws securing the window frame cross rail.
5. Remove the fixings retaining the window frame channel at the rear side of the door.
6. Remove the door check strap. 76.40.27.
7. Remove the fixings retaining the window frame channel, adjacent to the door lower hinge.
8. Remove the plug, nut and washers retaining the window frame, adjacent to the door top hinge.
9. Remove the spring clip securing the linkage for the interior locking knob, and disconnect from the door locking mechanism.
10. Lift the complete window frame, mechanism and weather seal from the door.
11. Raise the window and remove the bolts securing the window regulator mechanism.
12. Remove the regulator mechanism.
13. Slide the glass down the channel and remove.
14. If required, remove the support channel and glazing strip from the glass.

Refitting

15. Reverse 2 to 14, noting the following.
16. Manoeuvre the complete window frame assembly into position with the weather seal.
17. Wind the glass down to gain access to fit the six Phillips head screws in channel cross rail.
18. Fit all bolts, Phillips head screws and nuts, but do not tighten at this stage. Set the window height to a snug fit in the body frame and, when correct, tighten the four set bolts into the two brackets at the bottom of the runners. Set the frame inwards to apply tension to the seal at the top of door and tighten all securing bolts.
19. Check the window for freedom of operation.
20. Fit the window glass stop bracket.
21. Fit the top closing panel retained by the self-tapping screws, and the door locking mechanism secured by Phillips-head screws and star washers to closing panel.
22. Refit door trim pad. 76.34.04.

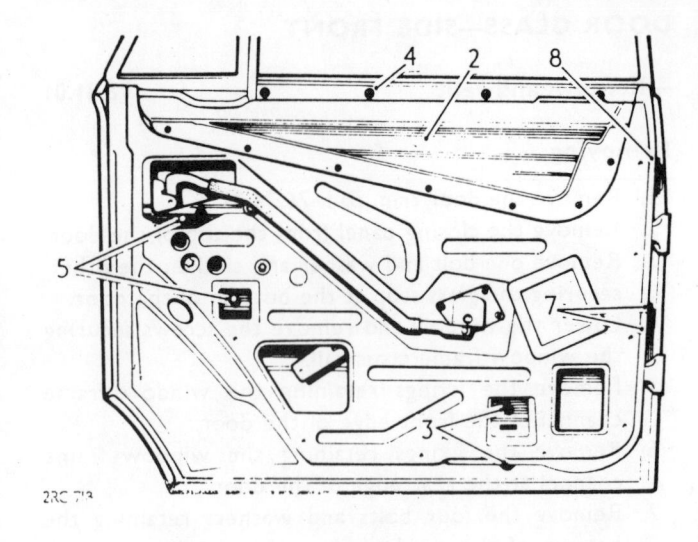

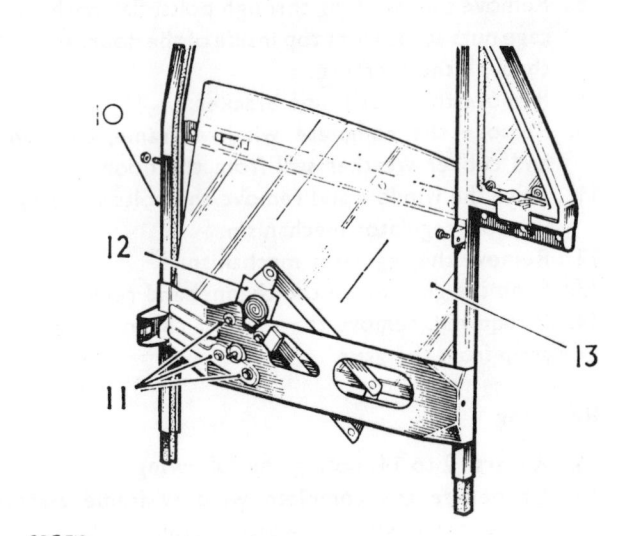

BODY

QUARTER VENT—SIDE FRONT

—Remove and refit　　　　　　　　　76.31.28

　Early models: 1 to 4 and 10
　Latest models: 1 and 5 to 10

Removing

1. Remove the door trim pad. 76.34.01.
2. Remove the closing panel from the top of the door.
3. Remove the lock plate, nut, washers and spring, accessible through the aperture.
4. Lift out the quarter vent.
5. To remove the glass only—Remove the clip retaining the spindle, and pull the glass and frame upwards.
6. To remove the operating mechanism, 7 to 10.
7. Remove the door glass. 76.31.01.
8. Remove the fixings securing the quarter vent mechanism to the door.
9. Withdraw the quarter vent.

Refitting

10. Reverse 1 to 9, as necessary.

QUARTER VENT—SIDE REAR

—Remove and refit 1 to 9, 13 and 14　　　76.31.30

QUARTER LIGHT—FIXED—SIDE REAR

—Remove and refit 10 to 13　　　　　　76.31.31

Removing

1. Remove the door trim pad. 76.34.04.
2. Remove the nine self-tapping screws, two Phillips head screws and star washers securing the closing panel to the door locking mechanism, and remove.
3. Remove the plastic knob retaining the door locking rod.
4. Remove the door frame bracket.
5. Slacken the pinch bolt securing the bottom of the quarter vent spindle.
6. Remove the spring clip securing the remote control lever to the lock.
7. Unscrew the spindle from the bracket in the glass frame.
8. Remove the screw and fibre washers securing the catch to the quarter vent.
9. Remove the quarter vent from the frame.
10. Remove the door glass. 76.31.02.
11. Drill out the pop rivets and remove the closing panel and channel.
12. Withdraw the quarter light together with the rubber surround.

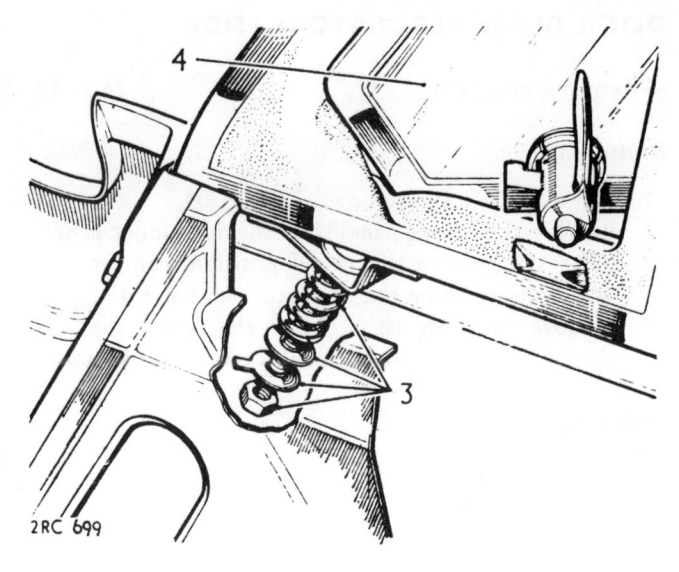

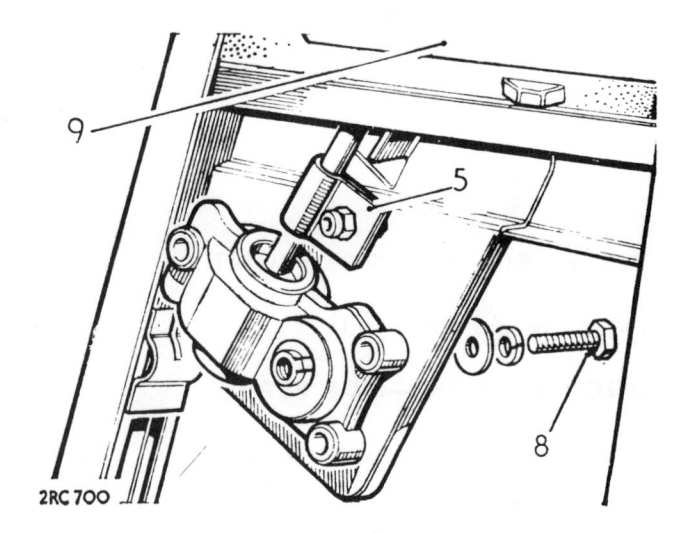

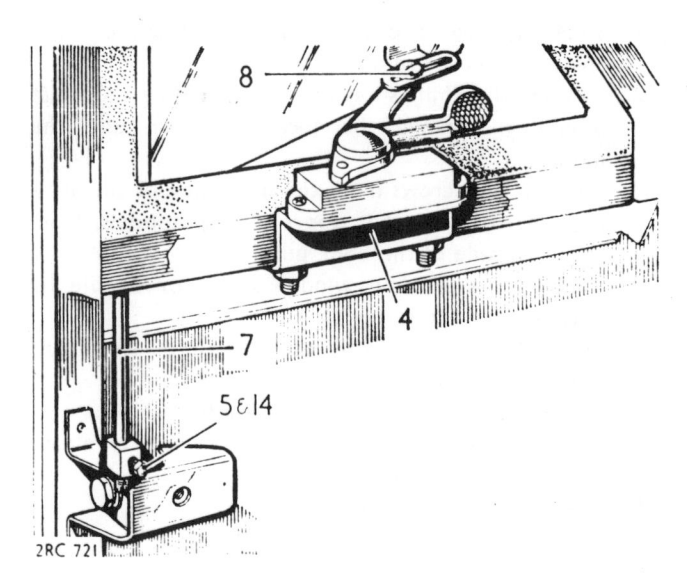

Refitting

13. Reverse 1 to 12, including the following.
14. When refitting the quarter vent, adjust the bottom hinge pin so that there is no up or down movement and lock in position with the bolt.

Rover 3500 and 3500S Manual AKM 3621

76.31.28
76.31.31

BODY

DOOR GLASS REGULATORS—SIDE

—Remove and refit 76.31.44

Removing

1. Remove the door trim pad. 76.34.01/04.
2. Remove the closing panel from the door inner panel.
3. With the glass in the raised position, remove the four bolts securing the regulator mechanism.
4. Remove the regulator from the glass support channel and withdraw through the top of the door.

Refitting

5. Reverse 1 to 4.

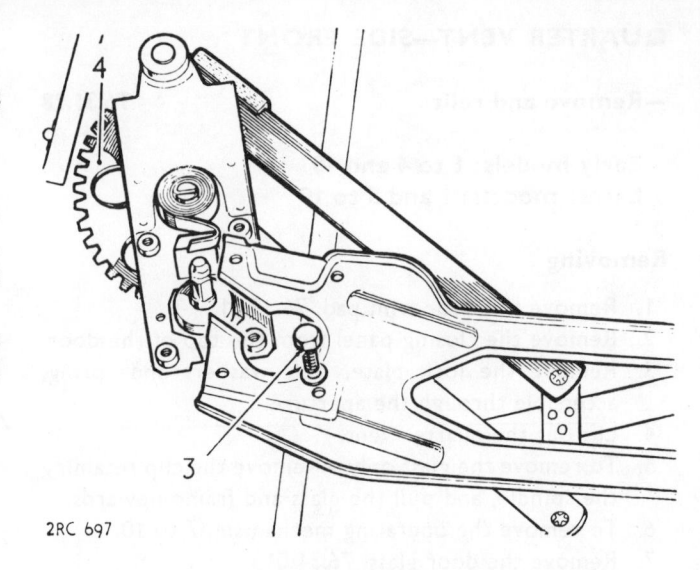

DOOR TRIM PAD—SIDE FRONT

—Remove and refit 1 to 11 76.34.01

DOOR TRIM PAD—SIDE REAR

—Remove and refit 1 to 4 and 7 to 11 76.34.04

Removing

1. Slide the veneer out of the waist moulding and remove the clips retaining the veneer.
2. Remove the Phillips head screws retaining the waist moulding, remove the locking button and the moulding.
3. Remove the screws from the underside of the armrest and, on the front doors of later models, from the underside of the door pull handle.
4. Remove the window winder and door handle by withdrawing the spring clips.
5. Prise the finisher from the centre of the front quarter vent operating knob.
6. Remove the front quarter vent knob.
7. Carefully lever around the door trim pad to release the clips, then withdraw the pad.
8. Remove the plastic sheet.

Refitting

9. Reverse 2 to 8.
10. Position the veneer retaining clips, convex side outwards and with the chamfered sides facing front to rear, over the waist moulding fixing screws.
11. Slide the veneer into position.

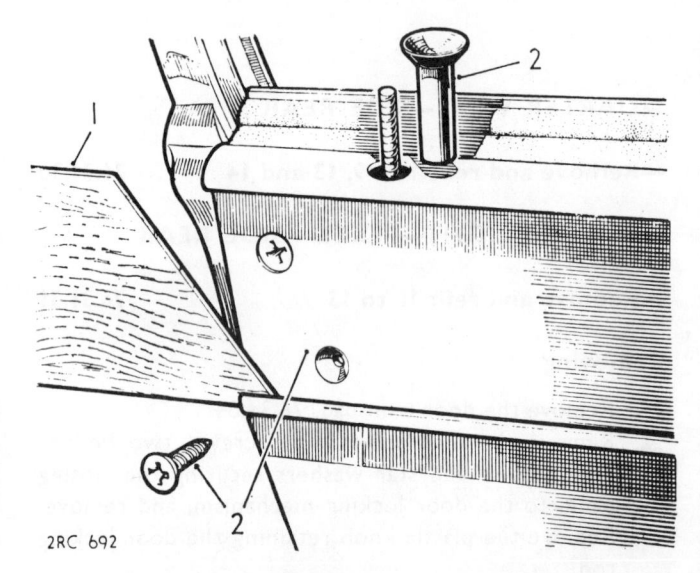

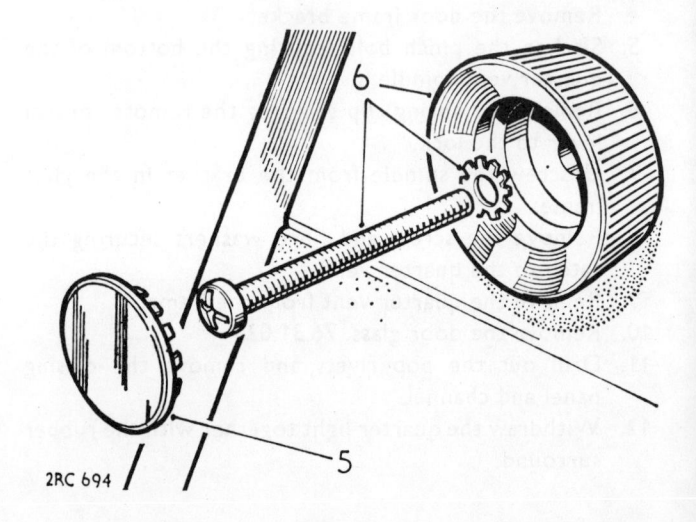

76.31.44
76.34.04

Rover 3500 and 3500S Manual AKM 3621

BODY

DOOR STRUT

—Remove and refit 76.34.18

NOTE: Door struts are normally only fitted to air conditioned models.

Removing

1. Remove the door trim pad. 76.34.01/04.
2. Remove the door glass. 76.31.01/02.
3. Remove the front and rear fixings securing the strut end brackets to the door.
4. Withdraw the strut through the top of the door casing.

Refitting

5. Reverse 1 to 4.

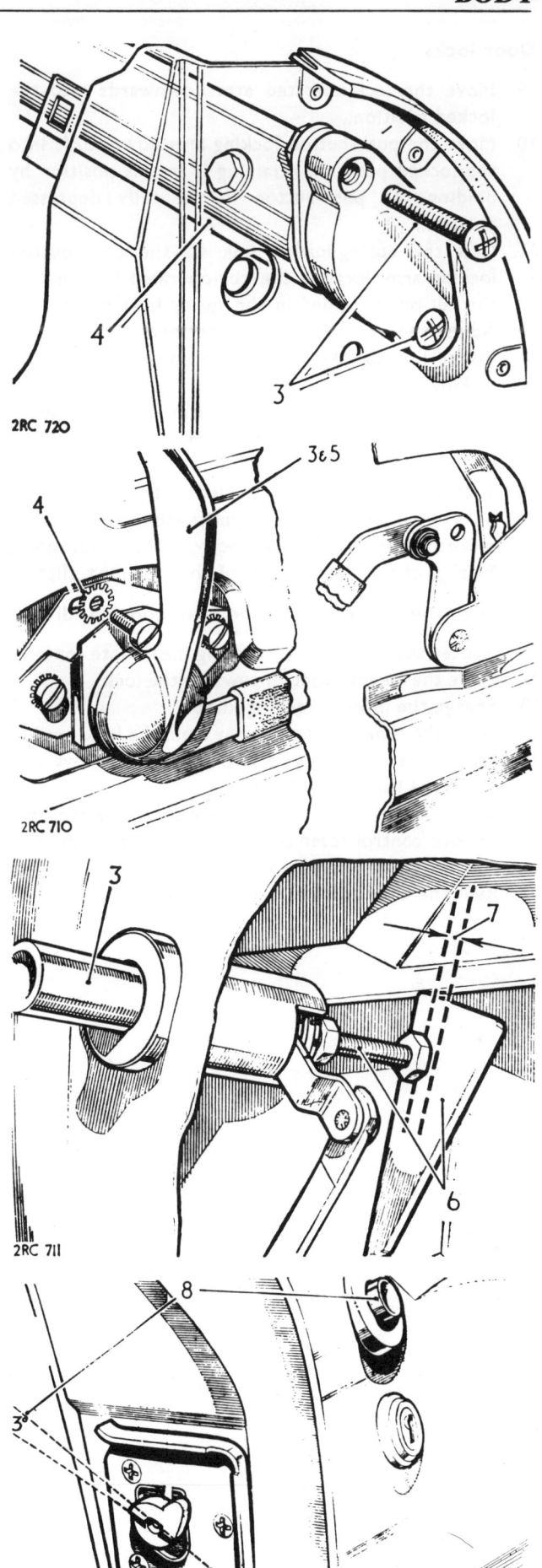

DOOR LOCKS

—Check and adjust 76.37.01

Procedure

1. Remove the door trim pad. 76.34.01/04.
2. If necessary, remove the closing panel from the top of the door.

Remote control and push-button

3. It is important to ensure that both the remote control handle and the push-button have a small amount of free movement before they begin to operate the rotary cam. If free movement is not present at these points, the slightest touch of the push-button or remote control handle will cause the door to open on to the safety catch.

Remote control

4. The adjustment for the remote control is obtained by moving the handle mechanism in the slotted holes to ensure the linkage is not loaded.
5. After adjustment, check that the handle has a slight amount of free movement before the lock begins to operate.

Push button

6. Adjustment for the push-button is obtained by providing clearance between the head of the adjustment bolt and the contact plate.
7. Check the clearance between the adjustment bolt head and the contact plate, and if necessary reset to 1,5 to 2,3 mm (0.062 to 0.093 in).
8. When the clearance is correct, open the door and hold the push-button fully depressed and check that the rotary cam of the lock is fully retracted to within 3°; this check is necessary to ensure clearance of the striker plate latch when opening the door.

continued

Rover 3500 and 3500S Manual AKM 3621

BODY

Door locks

9. Move the lock-operated arm downwards into the locked position.
10. Move the push-button locking arm downwards into the locked position, retaining it in this position by holding the push-button in the fully depressed position.
11. Clip the spring-loaded link to the push-button locking arm, locating the key-operated lock arm in the cut-out provided in the spring-loaded link.
12. Adjust the eye bolt at the upper end of the spring-loaded link to take up any slack in the push-button locking arm.
13. Ensure that the lock-operated arm is not in contact with the frame of the cut-out, either in the locked or unlocked position.
14. If these operations are carried out correctly, the linkage will be completely synchronised and the lock mechanism will operate correctly, without resistance.
15. Finally, check the security of the retaining clips.

Remote control safety lock on rear door handles

16. Slacken the three screws securing the remote control.
17. Move the remote control towards the lock.
18. Engage the interior locking knob.
19. Move the remote control away from the lock until it reaches its stop, then secure the three screws and disengage the locking knob.
20. Check that there is free movement before the remote control lever operates the rotary cam.
21. If removed, refit the door closing panel.
22. Refit the door trim pad. 76.34.01/04.

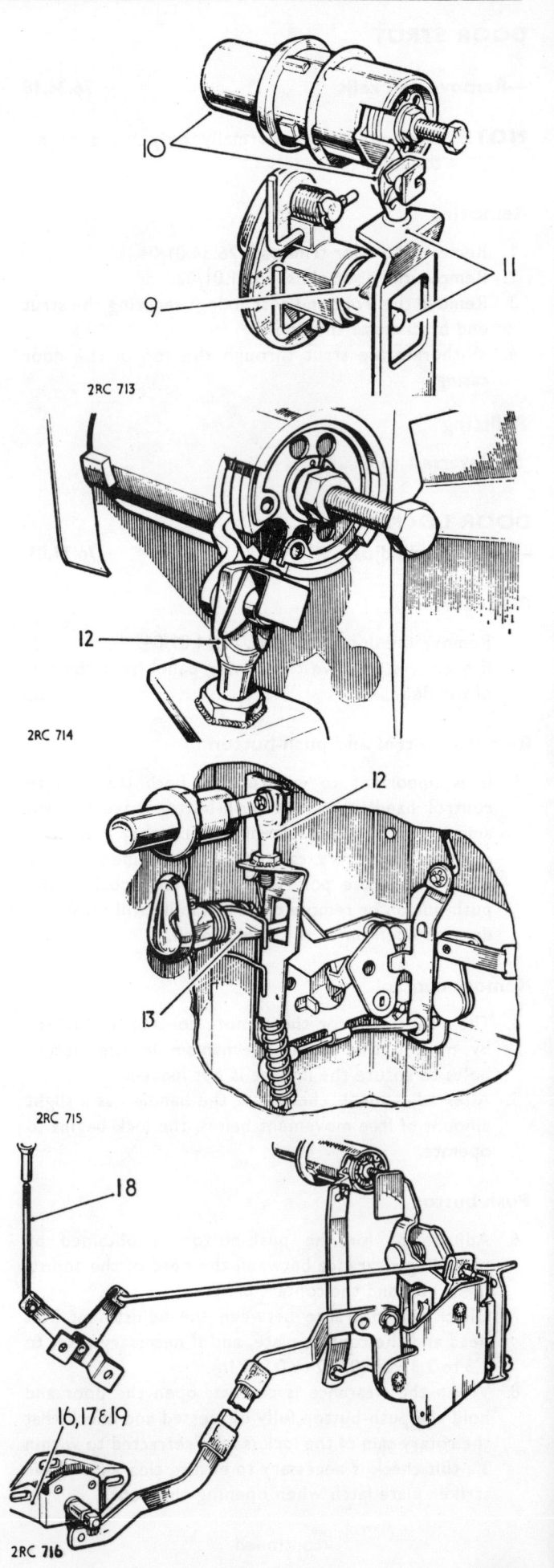

Rover 3500 and 3500S Manual AKM 3621

BODY

DOOR LOCK—SIDE FRONT

—Remove and refit 76.37.12

Removing

1. Remove the door trim pad. 76.34.01.
2. Remove the closing panel from the top of the door.
3. Disconnect the remote control arm from the lock.
4. Disconnect the eye bolt from the push-button locking arm.
5. Remove the dovetail and lock with the lower link.
6. If required, remove the lower link from the lock.

Refitting

7. Reverse 3 to 6.
8. Check, and if necessary adjust, the operation of the front door lock. 76.37.01.
9. Reverse 1 and 2.

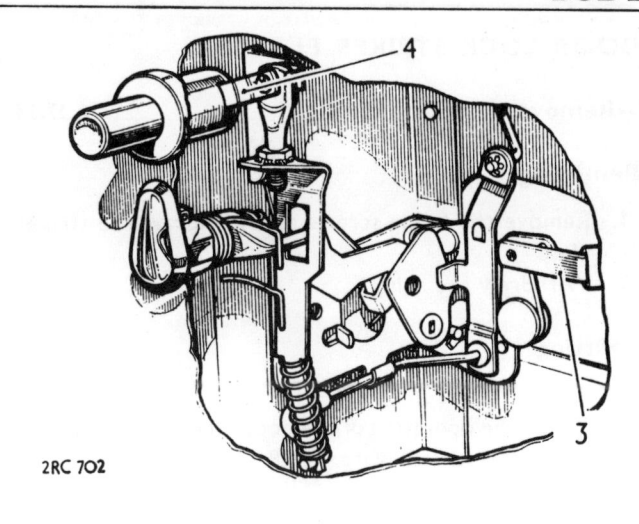

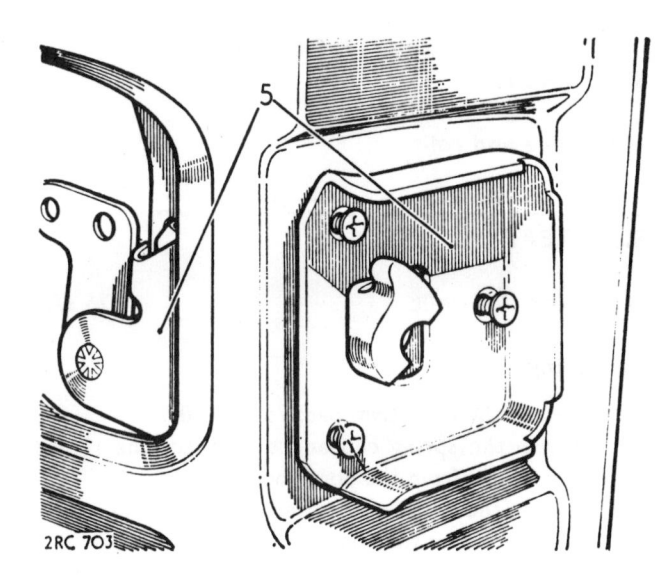

DOOR LOCK—SIDE REAR

—Remove and refit 76.37.13

Removing

1. Remove the door trim pad. 76.34.04.
2. Remove the closing panel from the top of the door.
3. Disconnect the bell crank link from the lock.
4. Disconnect the remote control arm from the lock.
5. Disconnect the link from the push-button lever.
6. Remove the dovetail and lock, with the link.
7. Remove the lower link from the lock.

Refitting

8. Reverse 3 to 7.
9. Check, and if necessary adjust, the operation of the rear door lock. 76.37.01.
10. Reverse 1 and 2.

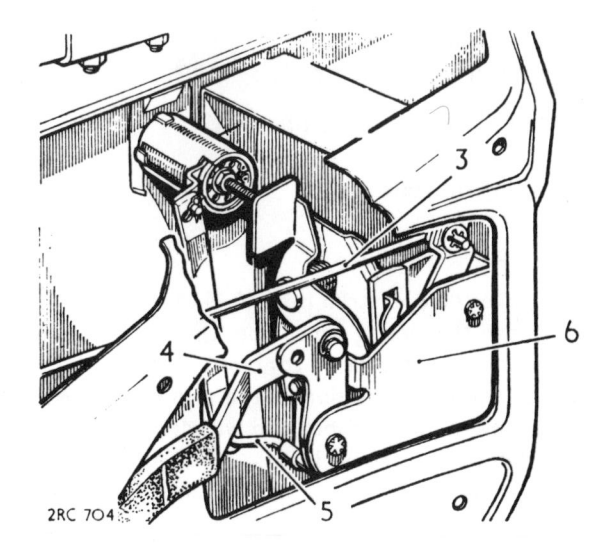

Rover 3500 and 3500S Manual AKM 3621

BODY

DOOR LOCK STRIKER PLATE

—Remove and refit   76.37.23

Removing

1. Remove the fixing screws and withdraw the striker plate.
2. Withdraw the shim.

Refitting

3. Reverse 1 and 2.
4. Check the operation of the door lock and if necessary adjust the striker plate. 76.28.07/08.

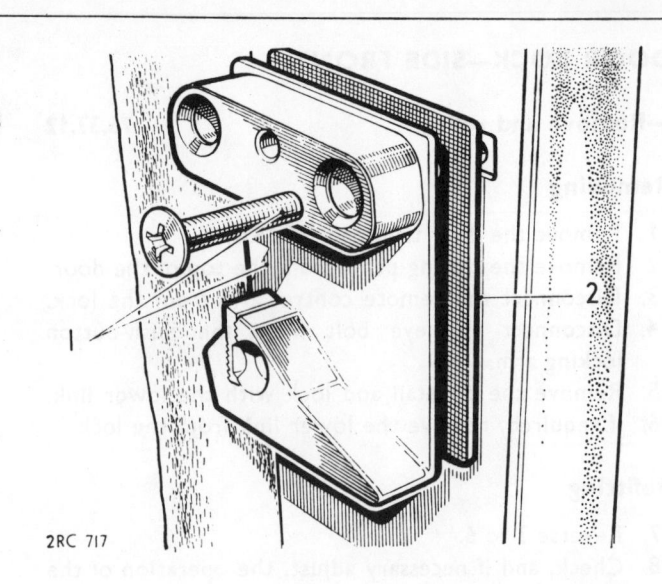

DOOR LOCK REMOTE CONTROL— SIDE FRONT

—Remove and refit   76.37.31

DOOR LOCK REMOTE CONTROL— SIDE REAR

—Remove and refit   76.37.32

Removing

1. Remove the door trim pad. 76.34.01/04.
2. Release the spring clip and disconnect the remote control arm from the door lock.
3. Remove the three screws and withdraw the remote control mechanism from the door.

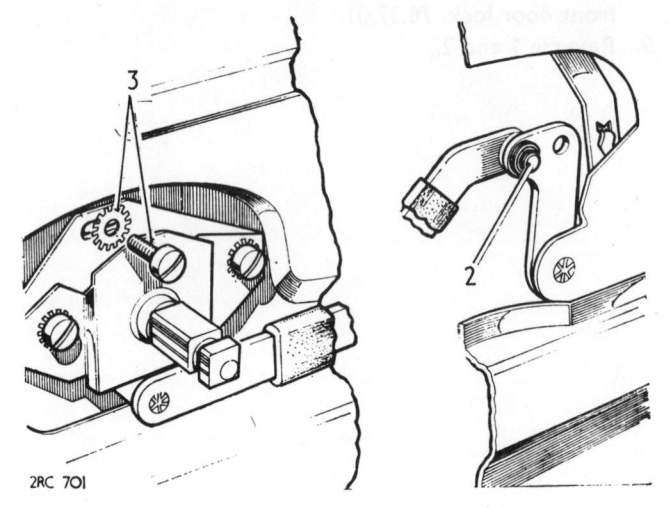

Refitting

4. Reverse 2 and 3.
5. Check, and if necessary adjust, the operation of the remote control. 76.37.01.
6. Refit the door trim pad. 76.34.01/04.

PRIVATE LOCK—SIDE FRONT

—Remove and refit   76.37.39

Removing

1. Remove the door trim pad. 76.34.01.
2. Remove the closing panel from the top of the door.
3. Disconnect the remote control arm from the lock.
4. Raise the angled section of the two retaining wires securing the private lock and withdraw the wires.
5. Withdraw the private lock.

Refitting

6. Reverse 3 to 5.
7. Check, and if necessary adjust, the operation of the front door lock. 76.37.01.
8. Reverse 1 and 2.

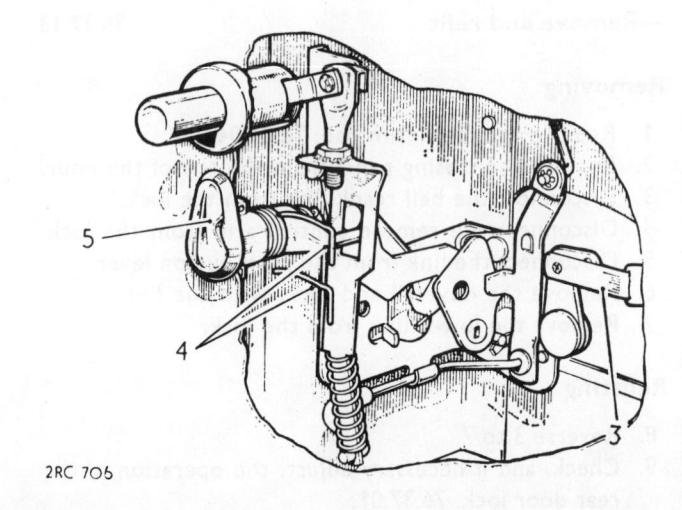

Rover 3500 and 3500S Manual AKM 3621

BODY

ANTI-BURST DEVICE

—Remove and refit　　　　　　　　　　76.37.46

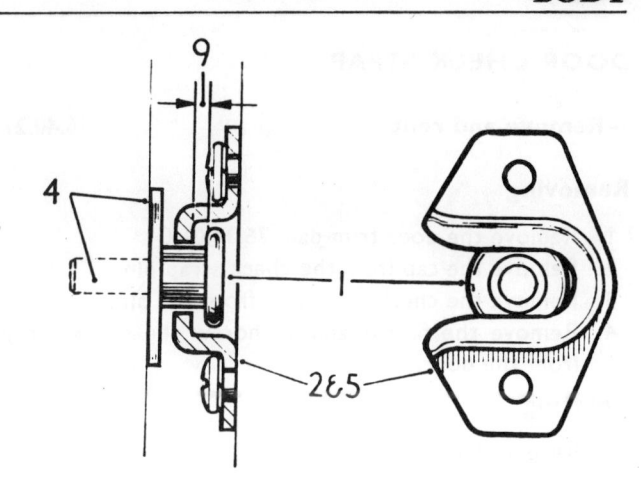

Removing

1. Remove the anti-burst pin and large plain washer from the door casing.

   NOTE: Some anti-burst pins may have two washers fitted.

2. Remove the anti-burst striker plate from the door pillar.

Refitting

3. Check that the door closes correctly. If not, adjust the door lock striker plate as necessary. 76.28.07/08.

   NOTE: Adjustment of the door lock striker plate must be carried out with the anti-burst pin removed from the door casing, to ensure that the anti-burst device is not interfering with the closing of the door.

4. Fit the anti-burst pin and one large washer to the door casing edge and tighten the pin fully.
5. Fit the anti-burst striker plate to the door pillar, leaving the screws not fully tightened at this stage.
6. Carefully close the door, observing the entry of the pin into the striker plate.

   CAUTION: If the lower face of the pin fouls the striker plate jaws, do not close the door fully, otherwise it may jam.

7. If the pin does foul the striker plate, remove the pin and fit a second plain washer.
8. Refit the pin and close the door, checking the entry of the pin into the striker plate.
9. There must be clearance between the head of the pin and the jaws of the striker plate. If necessary, a third plain washer can be fitted.
10. When the door closes fully without the anti-burst pin fouling the jaws of the striker plate, the screws securing the striker plate can be tightened securely.

2RC 689

DOOR SEAL—SIDE FRONT

—Remove and refit　　　　　　　　　　76.40.01

DOOR SEAL—SIDE REAR

—Remove and refit　　　　　　　　　　76.40.02

   Service tool: 600358　Fitting tool, body sealing rubber

Removing

1. Disconnect the applicable door check strap. 76.40.27. For front doors, protect the rear edge of the front wing. For rear doors, also open the front door.
2. Remove the sill plate.
3. Remove the strip fixing the door aperture seal to the sill.
4. Remove the door seal by pulling out.

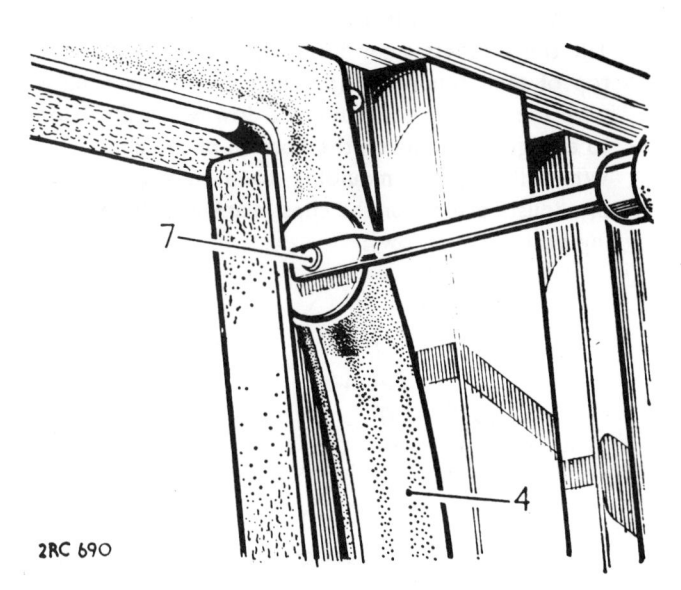

2RC 690

Refitting

5. Apply silicone grease MS4 to the door seal.
6. Locate the seal into the door sill channel and secure with the metal strip and screws.
7. Fit the remainder of the seal, using 600358.
8. Refit the sill plate.
9. Refit the door check strap. 76.40.27.

Rover 3500 and 3500S Manual AKM 3621

BODY

DOOR CHECK STRAP

—Remove and refit 76.40.27

Removing

1. Remove the door trim pad. 76.34.01/04.
2. Remove the cap from the check strap pin.
3. Remove the check strap arm from the pin.
4. Remove the fixings and withdraw the check strap from the door.

Refitting

5. Reverse 1 to 4.

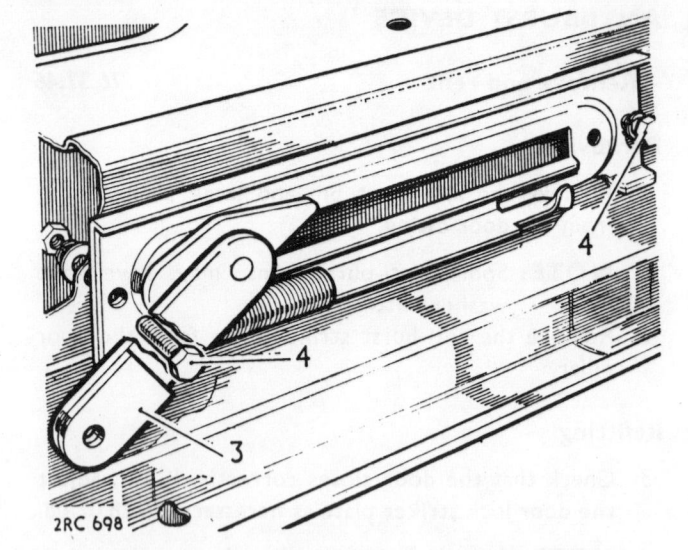

WAIST RAIL MOULDING

—Remove and refit

—Front wings	76.43.02
—Rear wings	76.43.03
—Side door—front	76.43.04
—Side door—rear	76.43.05

Removing

1. Lever the moulding gently off the clips.
2. If necessary, remove the clips.

Refitting

3. Insert the shank of the nylon clip in the hole and place the clip horizontally.
4. Tap the pin through the hollow shank with a light hammer until it is level with the clip surface.
5. Tilt the stainless steel mouldings to engage on the top of each retaining clip. Then give a sharp blow with the palm of the hand in a downward direction to engage moulding on the bottom of the clip.
   When fitting the moulding to a new panel not previously drilled for the clips, mark out the moulding position as follows:
6. Mark a line from the lower position of the front side lamp to lower position of the rear side lamp, as illustrated.
7. Drill 4,7 mm (0.187 in) diameter holes in the wings and doors, three in each front wing, seven in each front door, six in each rear door and eight in each rear wing. Space out as shown in illustration.
8. Fit the nylon clips and stainless steel mouldings as already detailed.

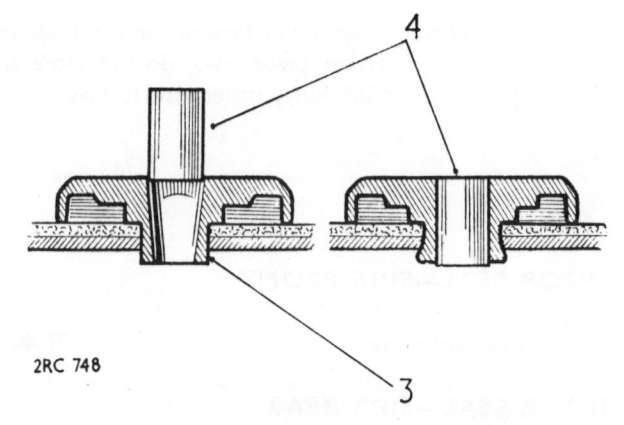

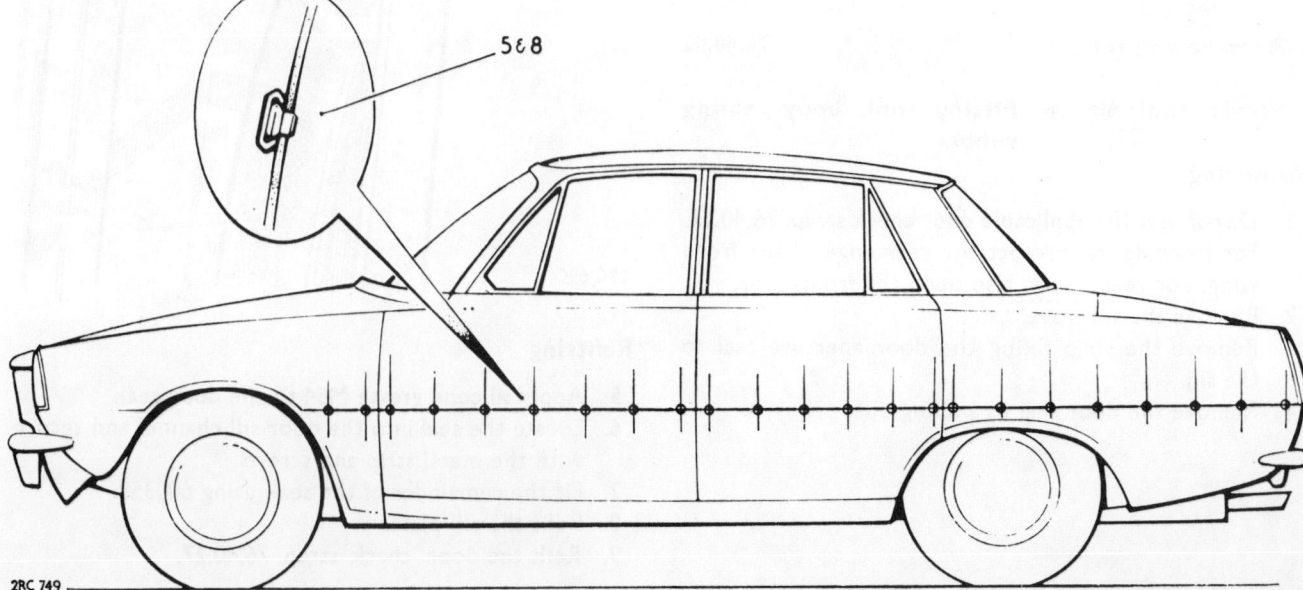

BODY

GUTTER FINISHER—CANTRAIL

—Remove and refit 76.43.11

Removing

1. Remove the windscreen finishers from the 'A' posts. 76.43.39/40.
2. If fitted, tap the rear corner finishers off the end of the cantrails.
3. Lever off the gutter finishers.

Refitting

4. Fit the gutter finishers to the cantrails, using a hide mallet.
5. If applicable, refit the rear corner finishers, bonding to the gutter finishers with Bostik 1776.
6. Refit the windscreen finishers. 76.43.39/40.

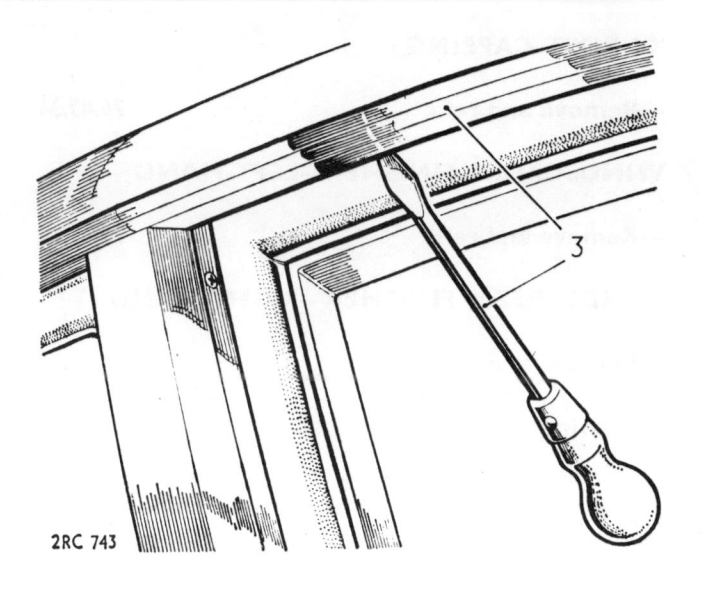

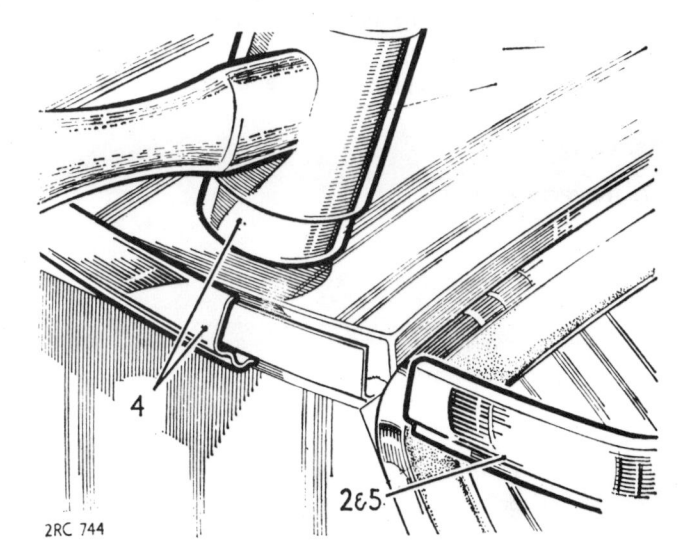

GUTTER FINISHER—BACKLIGHT

—Remove and refit 76.43.14

Removing

1. Remove the backlight glass and sealing rubber. 76.81.10.
2. Remove the gutter finisher one side at a time by gently prising away from the base unit.
3. Remove the joint clip.

Refitting

4. Insert sealing compound into the channel of each half of the finishers.
5. Apply sealing compound to the screen aperture corners in the base unit.
6. Fit the finishers by tapping with a mallet and ensure satisfactory corner seating. The joint clip should be fitted with first half and finally positioned to cover the joint.
7. Apply a small amount of sealing compound between the finisher and the panels.
8. Refit the backlight glass and sealing rubber. 76.81.10.

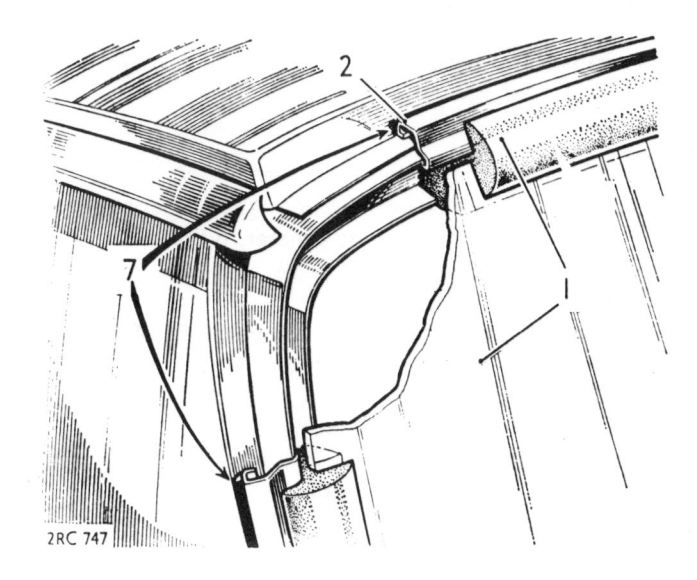

Rover 3500 and 3500S Manual AKM 3621

BODY

'B' POST CAPPING

—Remove and refit   76.43.34

WINDSCREEN FINISHER—LEFT-HAND

—Remove and refit   76.43.39

WINDSCREEN FINISHER—RIGHT-HAND

—Remove and refit   76.43.40

Removing

1. Open the front doors and drill out the rivets securing the windscreen top corner and side finishers.
2. Slide off the side finishers, downwards.
3. Withdraw the corner finishers.
4. If required, slide off the finisher from the top of the screen.
5. Open the rear doors and drill out the rivets securing the 'B' post capping.
6. Withdraw the capping from the 'B' post.

Refitting

7. Reverse 1 to 6.

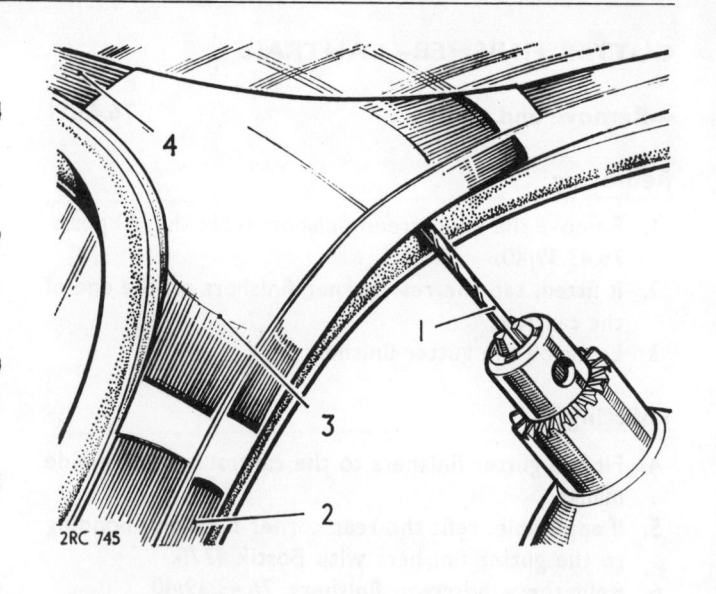

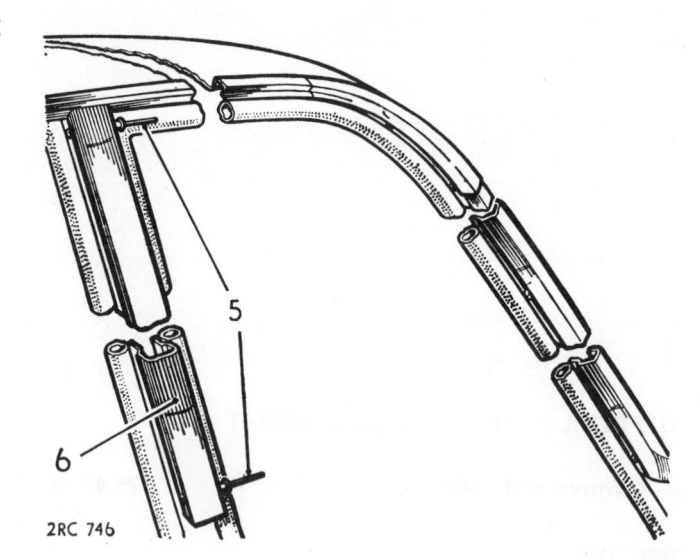

FASCIA—TOP RAIL

—Remove and refit   76.46.04

Removing

1. Disconnect the battery earth lead.
2. Early 3500 models with ribbon type speedometer—Remove the clock and tachometer, if fitted. Division 88.
3. Later 3500 and all 3500S models with circular speedometer—Remove the instrument panel. 88,20.01.
4. Remove the drive screws securing each end of the fascia top rail.
5. Slacken the four nuts securing the fascia top rail to the screen rail.
6. Withdraw the fascia top rail.

Refitting

7. Reverse 1 to 6. Ensure that the fascia top rail is correctly located on the clips at the screen rail.

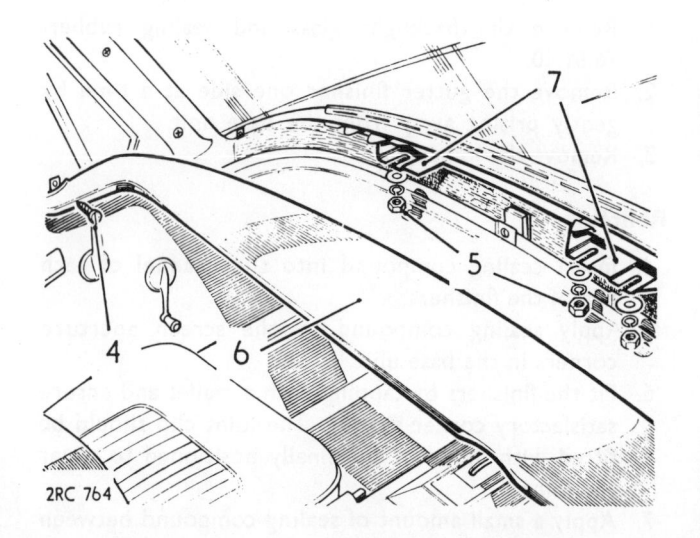

76.43.34
76.46.04

Rover 3500 and 3500S Manual AKM 3621

BODY

FASCIA—VENEER STRIP

—Remove and refit 76.46.14

Early models with ribbon type speedometer
1 to 5 and 7

Later models with circular type speedometer
1, 3, 6 and 7

Removing

1. Remove the fascia top rail. 76.46.04.
2. Remove the instrument panel. 88.20.01.
3. Spring off the moulded finishers from either end of the parcel tray.
4. Remove the three clips securing the left-hand veneered strip to the screen rail clip and carefully ease the strip out, taking care not to damage veneer.
5. Slide the right-hand veneered strip clear of the clip.
6. Remove the clips securing the top edge of the veneer strip to the screen rail and ease the strip out.

Refitting

7. Reverse 1 to 6.

FASCIA—SWITCH PANEL

—Remove and refit 76.46.16

Early models with ribbon type speedometer
1 to 6 and 24

Later models with circular type speedometer
1 and 7 to 24

Service tool: 601942 Lock ring spanner
601952 Lock ring spanner

Removing

1. Disconnect the battery earth lead.
2. Remove the locking rings from the RH and LH glove box locks, using 601952.
3. Remove the cigar lighter by unscrewing at the rear of the panel.
4. Slacken the steering column clamping nut, lower the steering column and remove the two nuts, spring and plain washers securing the switch panel to base unit.
   These are located one to the left of the steering column and one to the left of the cigar lighter, and are accessible from behind the panel.
5. Remove the two drive screws, one at each end of panel, accessible from under the panel.
6. Withdraw the switch panel in two parts, disconnect the wiring and remove switches if necessary.

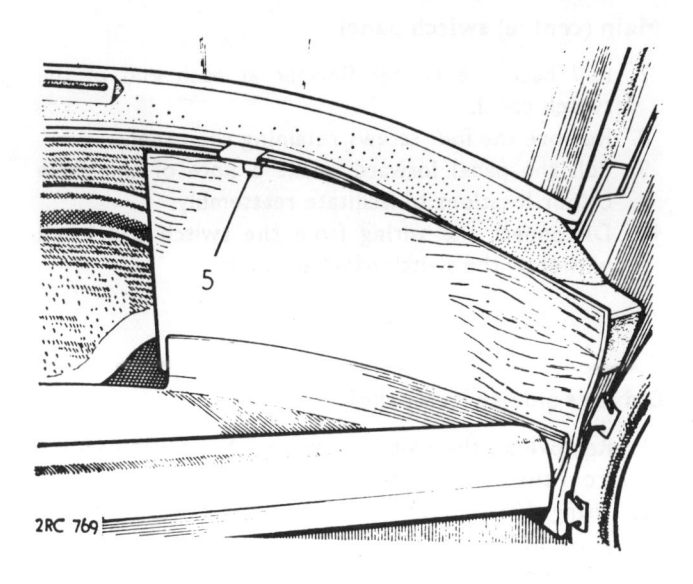

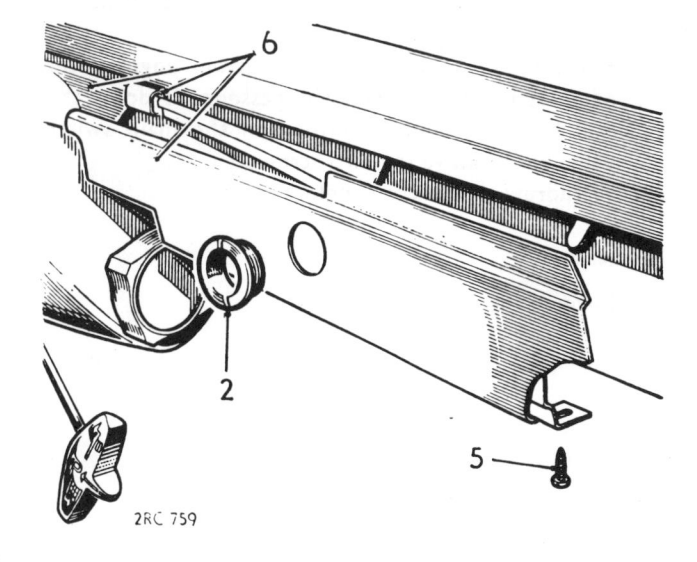

continued

Rover 3500 and 3500S Manual AKM 3621

76.46.14
76.46.16
Sheet 1

BODY

Main (centre) switch panel

7. Pull back the rubber finisher at each end of the switch panel.
8. Remove the four screws retaining the panel.
9. Pull the panel forward, make a note of the cable colour locations to facilitate reassembly.
10. Disconnect the wiring from the switch panel and withdraw the switch panel assembly.

Driver's side switch panel

11. Remove all the switches, except the ignition switch from the panel. Division 86.
12. Using 601952, remove the locking ring from the glove box lock.
13. Slacken the steering column clamping nut and lower the steering column.
14. Remove the screw from the bottom edge of the panel at the end nearest the driver's door.
15. Pull back the rubber finisher on the main switch panel and remove the two screws.
16. Disconnect the lead at the snap connector and carefully prise the RH side of the ignition switch finisher away from the panel.
17. Remove the nut from the back of the ignition switch panel.
18. Withdraw the panel, passing the ignition switch finisher through the hole. Note that the panel is in two halves, joined by a small, loose joint piece.

Passenger's side switch panel

19. Using 601952, remove the locking ring from the glove box lock.
20. Open the glove box and withdraw the bulb holder from the map light.
21. Remove the screw from the bottom edge of the panel at the end nearest the passenger's door.
22. Pull back the rubber finisher on the main switch panel and remove the two screws.
23. Withdraw the switch panel.

Refitting

24. Reverse the removal procedure.

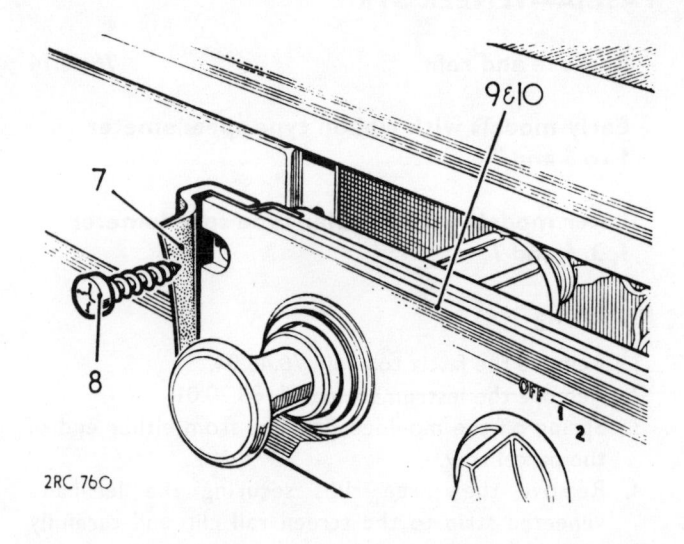

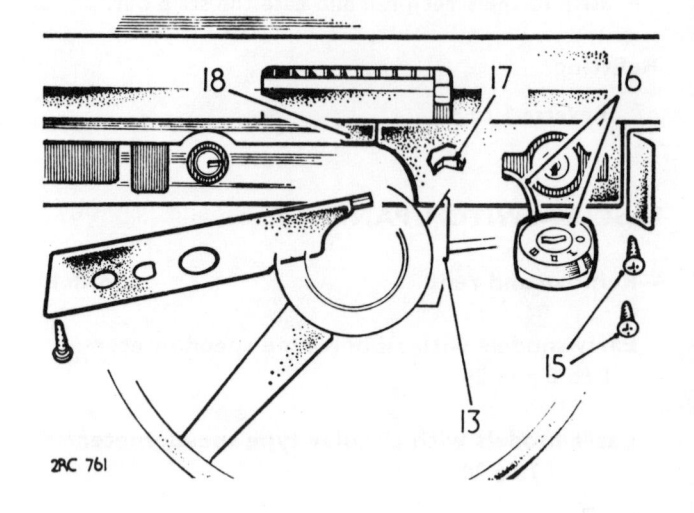

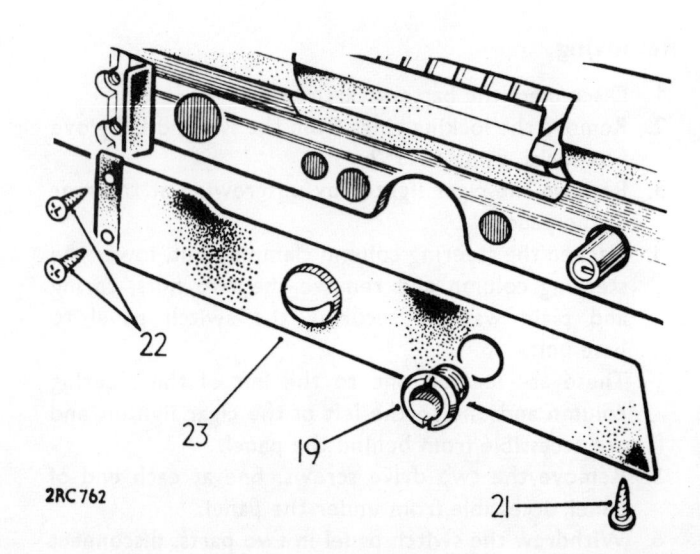

76.46.16
Sheet 2

Rover 3500 and 3500S Manual AKM 3621

BODY

GLOVE BOX

—Remove and refit   76.52.03

Removing

1. Open the glove box and detach one strap end for right-hand box and two strap ends for left-hand box from the spring clips, as illustrated.
2. Remove the two bolts retaining the hinges on both the left-hand and the right-hand glove boxes.
3. Remove the glove boxes.

Refitting

4. Reverse 1 to 3.

NOTE:
When refitting a replacement glove box to the driver's side of a LHStg car, ensure that the new glove box has a larger radius at the upper end of the large compartment to provide clearance for the speedometer cable.

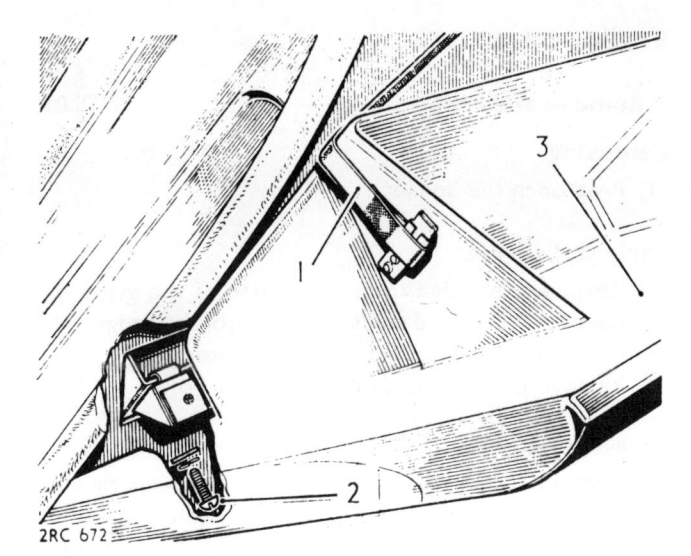

GLOVE BOX LOCK

—Remove and refit
  left-hand   76.52.08
  right-hand   76.52.09

Service tool: 601952   Lock ring spanner

Removing

1. Remove the fascia switch panel. 76.46.16.
2. Remove the two screws securing the lock to the base unit.
3. Withdraw the lock complete.

Refitting

4. Reverse 1 to 3.

Rover 3500 and 3500S Manual AKM 3621

BODY

RADIATOR GRILLE

—Remove and refit 76.55.03

Removing

1. Prop open the bonnet.

Early style grille

2. Remove the badge from the centre of the grille by tapping the knurled studs at the rear of the badge.
3. Prise the plastic caps from the screw heads.
4. Remove both sections of the radiator grille.

Latest style grille

5. Remove the fixings from both headlamp bezels and the radiator grille.
6. Withdraw the headlamp bezels and the radiator grille.

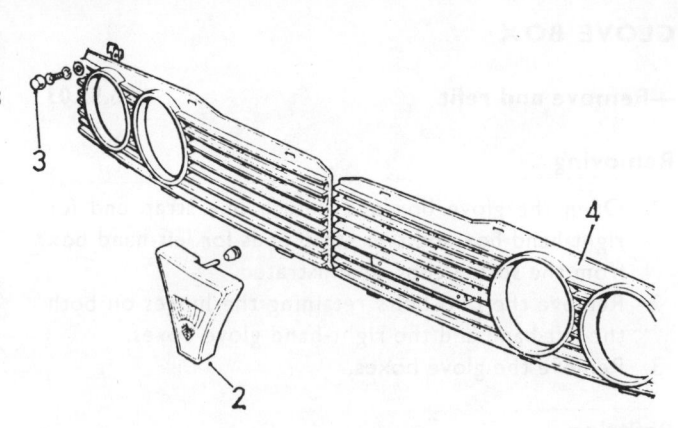

Refitting

Early style grille

7. Reverse 1 to 4.

Latest style grille

8. Fit the radiator grille, engaging the location tab into the lower grille panel.
9. Fit both headlamp bezels, engaging the location tabs into the lower grille panel and the radiator grille.
10. Secure the assembly with screws and washers.

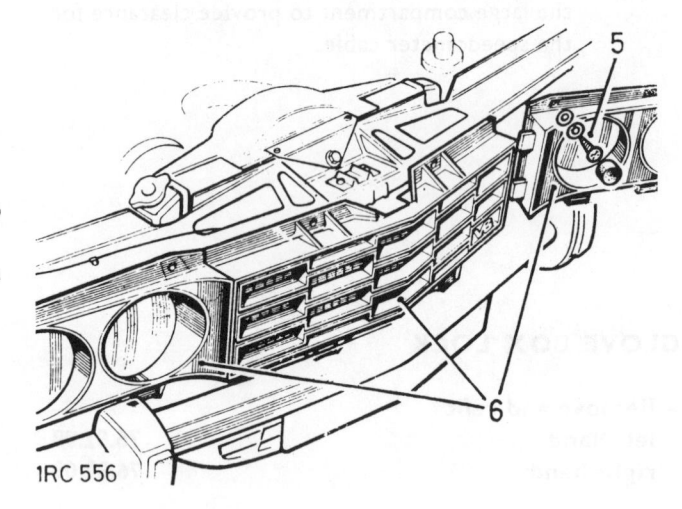

DOOR LOCK PUSH-BUTTON

—Remove and refit 76.58.12

Removing

1. Remove the door trim pad. 76.34.01/04.
2. Remove the closing panel from the top of the door.
3. Disconnect the upper link from the push-button lever.
4. Remove the leaf spring retaining the push-button.
5. Withdraw the push-button from the door.

Refitting

6. Reverse 3 to 5.
7. Check, and if necessary adjust, the operation of the push-button. 76.37.01.
8. Reverse 1 and 2.

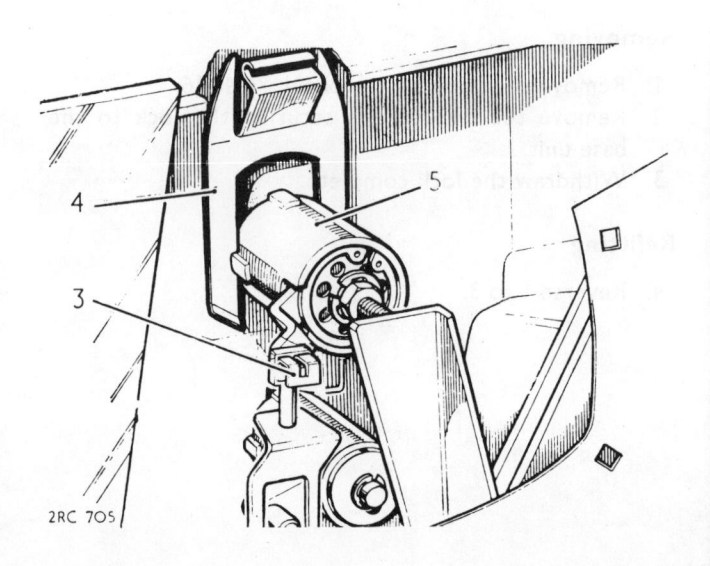

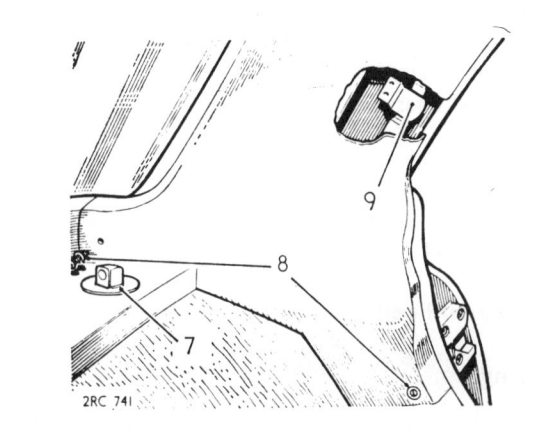

BODY

HEADLINING

—Remove and refit 76.64.01

Removing

1. Remove the sun visors.
2. Remove the trim pads from the windscreen pillars, secured by two drive screws.
3. Remove the rear view mirror and interior lamp.
4. Remove the header panel, secured by two drive screws.
5. Remove 'B' post trim pads. 76.13.08.
6. Remove the rear seat squabs. 76.70.38.
7. Where applicable, remove the rear seat safety harness outer fixing.
8. Remove the two drive screws retaining the rear quarter trim pad.
9. Ease the trim pad forward to release it from its retaining clips.
10. Lay the front seat squabs flat.
11. Remove the drive screws securing the headlining, and withdraw the headlining together with the underlining through the front passenger door.

Refitting

12. Reverse 1 to 11.

PARCEL TRAY—FRONT

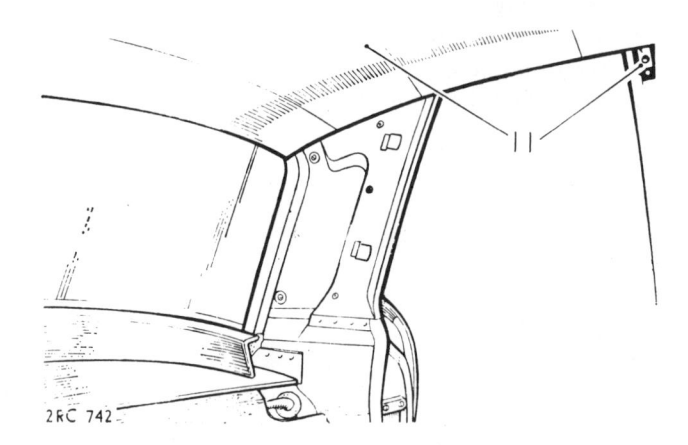

—Remove and refit 76.67.02

Early models with ribbon type speedometer
1 to 9 and 14

Later models with circular type speedometer
2, 5 and 10 to 16

Removing

1. Disconnect the battery earth lead.
2. Remove the fascia top rail. 76.46.04.
3. Remove the instrument panel. 88.20.01.
4. Remove the fascia switch panel. 76.46.16.
5. Spring off the moulded finishers from either end of the parcel tray.
6. Remove the three clips securing the left-hand veneered strip to the screen rail clip and carefully ease the strip out, taking care not to damage veneer.
7. Slide the right-hand veneered strip clear of the clip.
8. Remove the five set screws securing the forward edge of the parcel tray to the base unit.
9. Lift out the parcel tray complete.

continued

Rover 3500 and 3500S Manual AKM 3621

76.64.01
76.67.02
Sheet 1

BODY

10. Remove the two clips retaining the top edge of the veneer strip and lift the strip clear of the bottom groove.
11. Remove the instrument panel support bracket from the centre of the parcel tray.
12. Remove the three screws and plain washers from the forward edge of the parcel tray.
13. Ease the front edge of the parcel tray up clear of the face level vent controls, then lift the tray clear.

Refitting

14. Reverse 1 to 13, as applicable, including the following on later models with circular type speedometers.
15. Slacken the fixings at the fascia switch panels sufficient to provide access for the front edge of the parcel tray which fits behind the panels.
16. Secure the fascia switch panels.

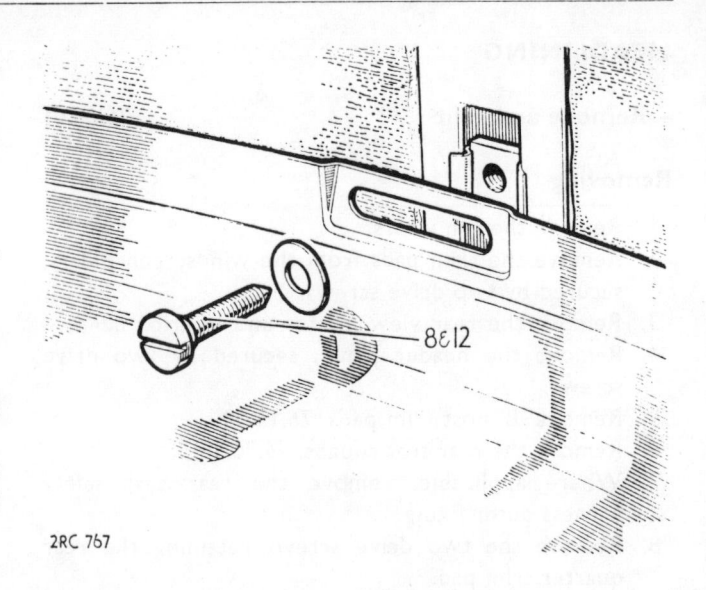

FRONT SEAT

—Remove and refit 76.70.01

Removing

1. Slide the seat runners fully forward and remove the rear fixing.
2. Slide the seat runners fully rearward and remove the front fixing.
3. Withdraw the front seat.

Refitting

4. Reverse 1 to 3.

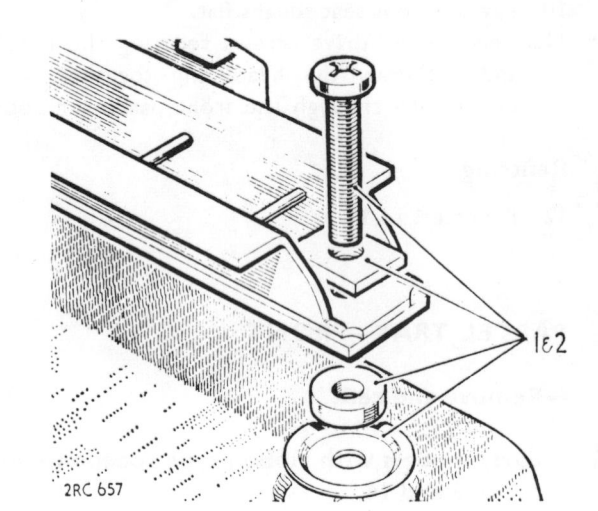

FRONT SEAT BACK FINISHER

—Remove and refit 76.70.03

Removing

1. Remove the front seat. 76.70.01.
2. Remove the cover plates from both sides of the seat squab.
3. Where applicable, slacken the screw at the rear of the seat back and withdraw the head restraint.
4. Bend back the two steel clips under the bottom edge of the back finisher.
5. Lift the back finisher upwards to remove. It is located by six tongued clips and three screws.

Refitting

6. Reverse 1 to 5.

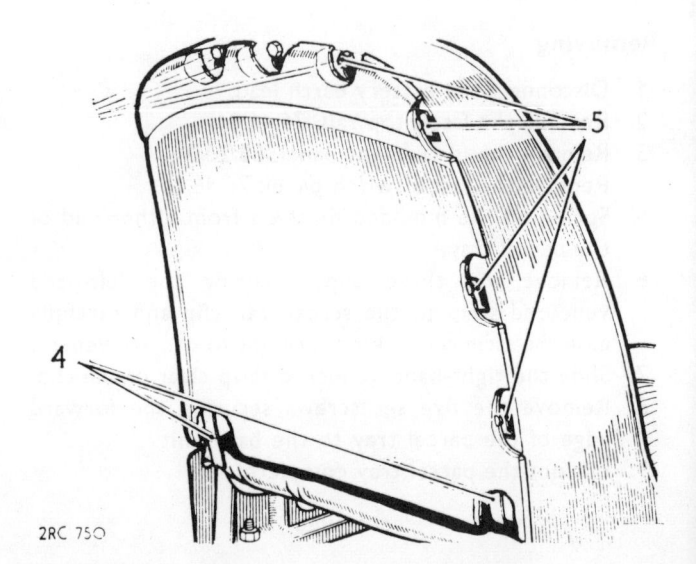

76.67.02
Sheet 2
76.70.03

Rover 3500 and 3500S Manual AKM 3621

BODY

FRONT SEAT RUNNERS

—**Remove and refit**   76.70.21

Removing

1. Remove the front seat from the runners.
2. Slide the seat runners fully forward and remove the rear fixings.
3. Slide the seat runners fully rearward and remove the front fixings.
4. Withdraw the seat runner assembly.
5. Lift the retaining springs and withdraw the operating bar.

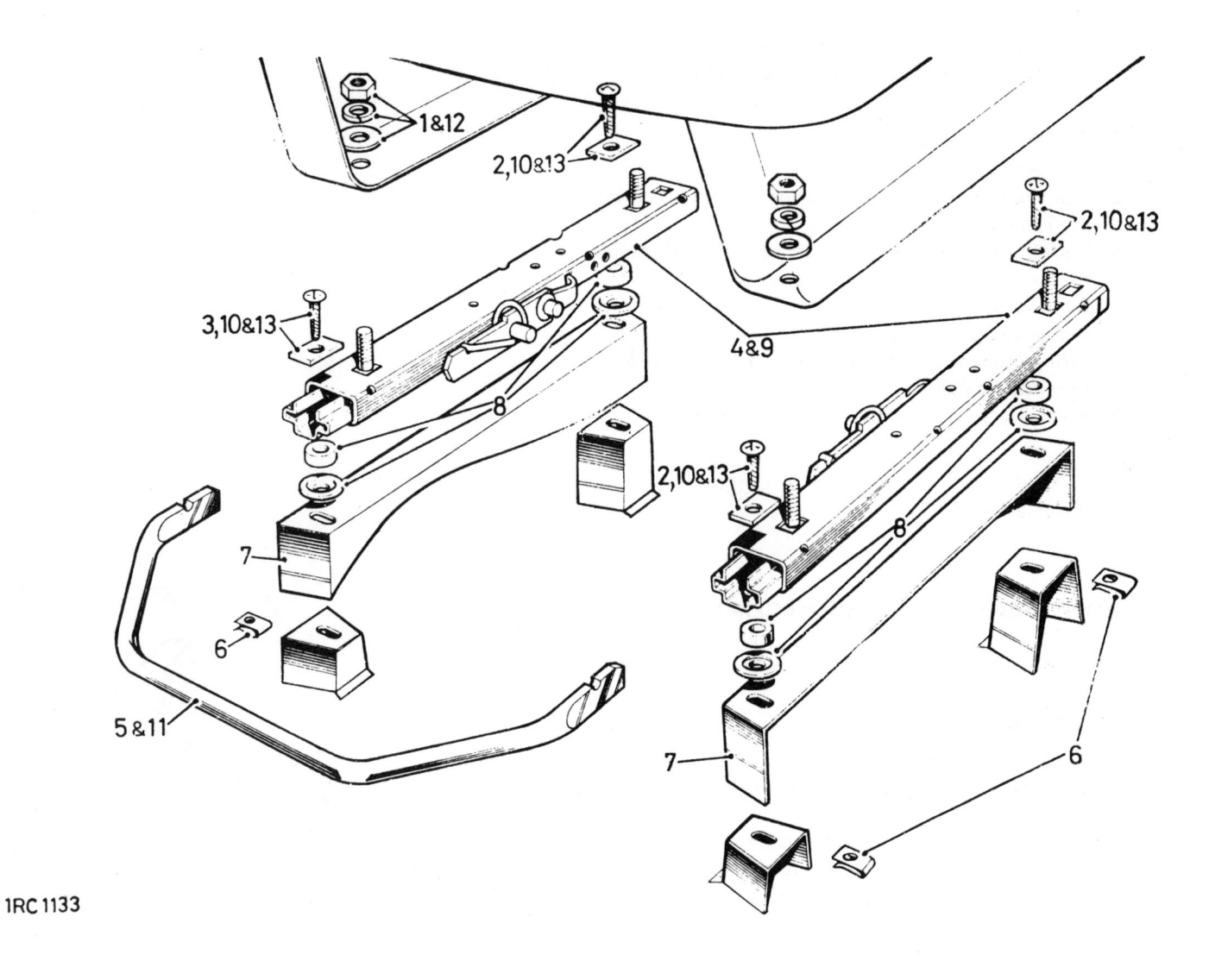

continued

1RC 1133

Rover 3500 and 3500S Manual AKM 3621

76.70.21
Sheet 1

BODY

Refitting

6. Align the tapping plates centrally with the slot in the seat support brackets.
7. Locate the covers over the brackets, aligning the holes with the tapping plates.
8. Place a retaining washer and distance piece into each hole in the top face of the covers.
9. Locate the seat runners in position, aligning the holes with the items previously fitted.
10. Fit the washers and screws, but do not fully tighten at this stage.
11. Push the operating bar, notched side uppermost, over the trigger unit on each seat runner, engaging the retaining spring in the notches.
12. Secure the front seat to the runners.
13. Secure the seat runner fixings.

FRONT SEAT RUNNERS AND CABLES

—Remove and refit   76.70.23

NOTE: Cable operated front seat runners are fitted to early 3500 models. Later 3500 and all 3500S models have bar operation, 76.70.21 refers.

Removing

1. Remove the front seat. 76.70.01.
2. Remove the four nuts and spring washers securing the runners to the seat frame.
3. Slacken the Phillips head screws retaining the cables to operating lever, and detach cables from lever.
4. Remove the split pins retaining the locking plungers and withdraw plungers, slotted washers and coil springs complete with cables (and retaining clip on long cable only).
5. If necessary, remove the clip retaining lever pivot.

Refitting

6. Reverse 1 to 5.

FRONT SEAT RECLINING MECHANISM

—Overhaul   76.70.33

Dismantling

1. Remove the front seat. 76.70.01.
2. Remove the cover plates from both sides of the seat squab.
3. Remove the retainer for the torsion bar.
4. Drift the torsion bar out approximately 25 mm (1.000 in) to release the bar tension.
5. Withdraw the torsion bar.
6. Remove the adjuster plate.

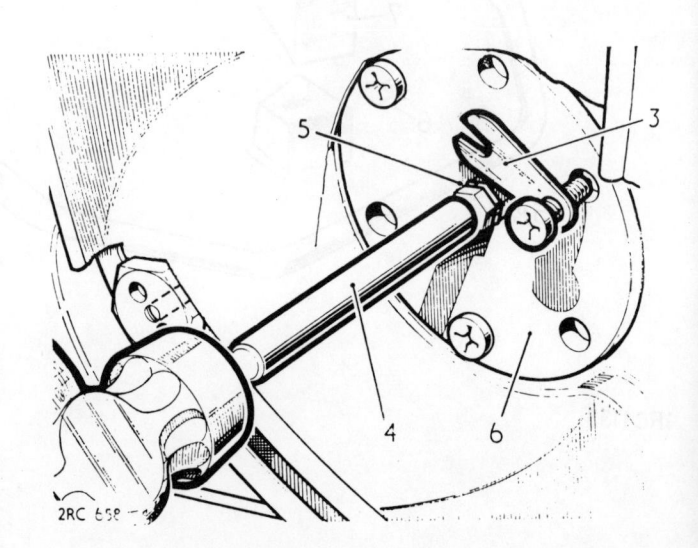

continued

76.70.21
Sheet 2
76.70.33
Sheet 1

Rover 3500 and 3500S Manual AKM 3621

BODY

7. Remove the circlip from the spigot bush on the squab frame.
8. Withdraw the spigot bush.

   NOTE: The spigot bush must be removed to prevent distorting the squab frame during the following procedure.

9. Remove the long locking screw from the locking pad.
10. Remove the short locking screw from the opposite side.
11. Unscrew the locking pad from the brake rod.
12. Withdraw the brake rod complete with the locking pad from the other side.
13. Ease the seat squab from the cushion frame.
14. Withdraw the compression tube complete with the locking lever.
15. Unscrew the coarse adjusting pad from the locking lever.
16. Unscrew the fine adjusting pad from the compression tube.
17. Unscrew the compression tube from the lever.
18. If required, remove the bush from the seat cushion frame.

Reassembling

19. Smear the inside of the locking lever with grease and screw it on to the compression tube as far as possible, then make alignment marks on the tube and lever boss.
20. Unscrew the lever approximately one complete turn to give a dimension of 475 mm (18.687 in) as illustrated.
21. Lightly smear the coarse locking pad with grease, then screw it into the lever until the pad face protrudes 2,5 mm (0.093 in) from the lever boss face.
22. Grease the thread of the fine adjustment locking pad and screw it into the compression tube, ensuring that the previously made adjustments are not disturbed.
23. Offer the lever and tube assembly to the squab frame and adjust the fine adjustment locking pad as necessary to the inside width of the squab frame. Ensure that the locking pad holes align with the screw holes in the compression tube.
24. If removed, grease the inner diameter of the bush for the compression tube and fit it to the seat cushion frame.
25. Grease the boss of the locking lever and insert the lever and tube complete through the seat cushion frame in such a way that the lever is at right angles to the seat as shown. Also ensure that the marks made previously on tube and lever boss are aligned, and that the fine adjustment locking pad in the opposite end of the tube does not turn.
26. Offer the squab frame to the seat cushion frame ensuring that the position of the locking lever and locking pads are not disturbed.

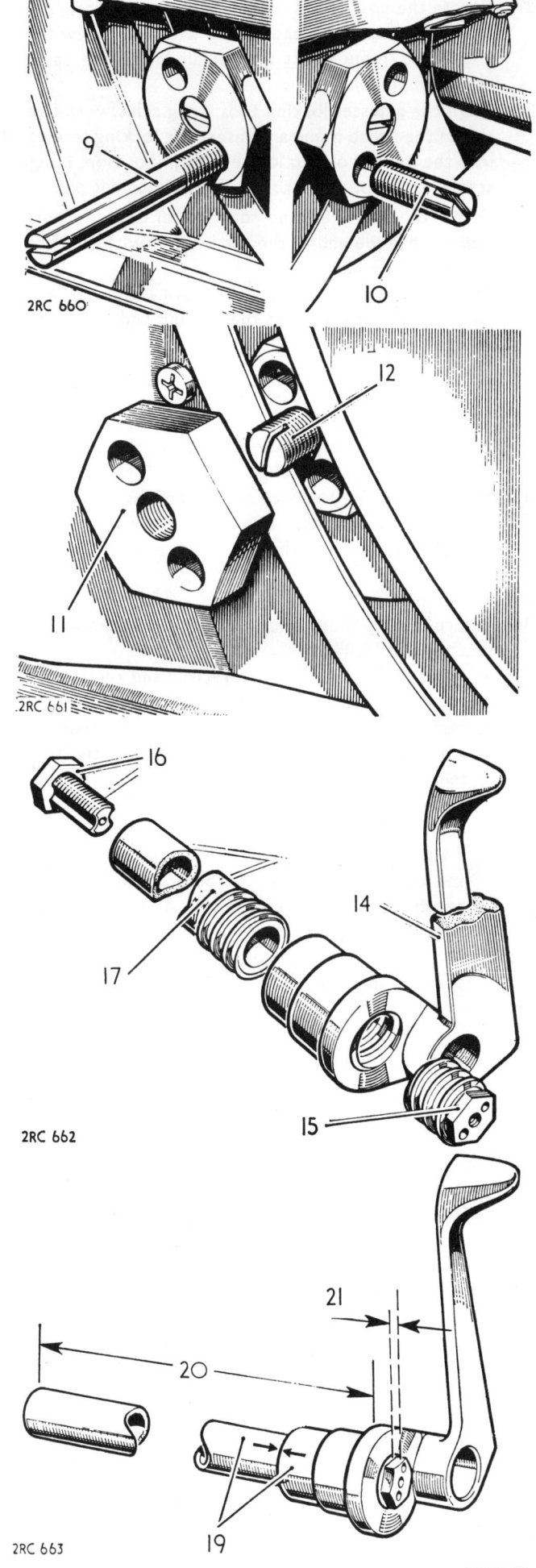

continued

Rover 3500 and 3500S Manual AKM 3621

76.70.33
Sheet 2

BODY

27. Grease the boss of the torsion bar adjuster, insert it into the squab frame and secure with one screw, to locate the frame whilst the remainder of the parts are assembled.
28. Align the adjuster locking pads, both sides, with the slot in the squab frame and insert the locking screws into the lower holes to locate pads in position. Long screw should be fitted to the fine adjustment pad.
29. Screw the round locking rod into an outer locking pad so that the end of the rod is flush with the face of the locking pad.
30. Insert the locking rod through both the squab and seat frames ensuring that the locking pad is located over locking screw.
31. Remove the other locking screw from the opposite side.
32. With the locking lever in the vertical position, fit the remaining outer locking pad to the rod and tighten until the holes align.
33. Refit the locking screw to the lower hole, to secure the outer locking pad.
34. Grease the bush for the squab pivot and insert it into the squab frame with the clip groove innermost.
35. Fit the circlip.
36. Before fitting the torsion bar, check operation of seat squab as follows:
    a. Move the locking lever rearwards and recline the squab; the action should be smooth and free.
    b. Move the lever forward to the locked position which is with the lever end just above the seat trim and check that the squab is now locked. Repeat this check through all the operating angles of the squab, from fully reclining to fully upright.
    c. If adjustment is required, remove locking screws and tighten or slacken the outer locking pad as necessary to give the condition described under a and b.
37. Remove the screw previously fitted to secure torsion bar adjuster.
38. With the seat squab in the upright position insert the torsion bar from the right-hand side, the grooved retainer end going in first.
39. Push the bar through in direction of arrow and enter into the adjuster. Allow the grooved end to protrude through the adjuster by not more than 9,5 mm (0.375 in).
40. Using a suitable ring spanner on the projecting end of the bar, wind up the torsion bar in a clockwise direction to align three of the holes in the adjuster with the three threaded holes in the squab frame. Insert and tighten two screws. Maintain tension on the spanner to allow easy entry of the screws.
41. Tap the torsion bar back into the adjuster to allow the retainer plate to be fitted. Lock plate in position with the third screw.
42. Recheck the operation of the seat squab, ensuring that the squab will come upright from the reclining position without assistance.
43. Refit the cover plates to both sides of the seat squab.
44. Refit the front seat. 76.70.01.

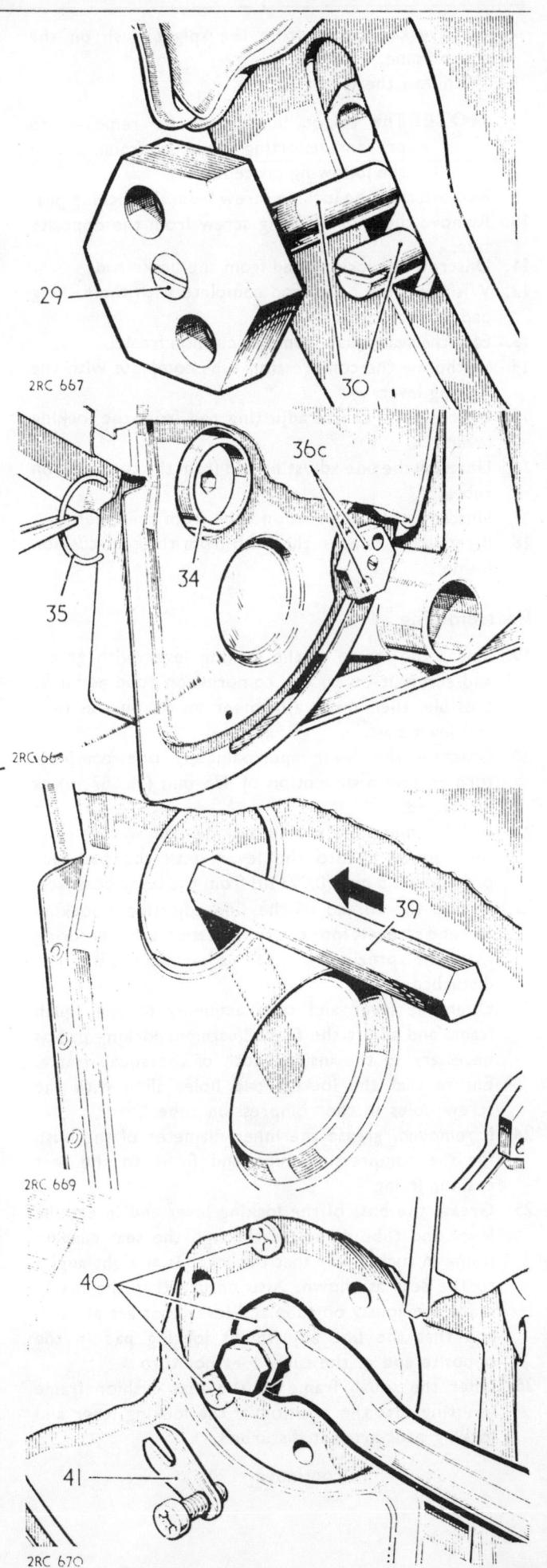

76.70.33
Sheet 3

Rover 3500 and 3500S Manual AKM 3621

BODY

REAR SEAT SQUAB

—Remove and refit 76.70.38

Removing

1. Withdraw the rear seat cushion.
2. Remove the drive screw at the base of the squab.
3. Peel back the wheel arch trim and remove the securing screw at the wheel arch.
4. Lift the squab clear of the retaining clips and remove.

Refitting

5. Reverse 1 to 4.

SILL PANEL—OUTER

—Remove and refit 76.76.04

Removing

1. Remove the fixings securing the sill panel to the base unit.
2. Withdraw the sill panel.

Refitting

3. Reverse 1 and 2.

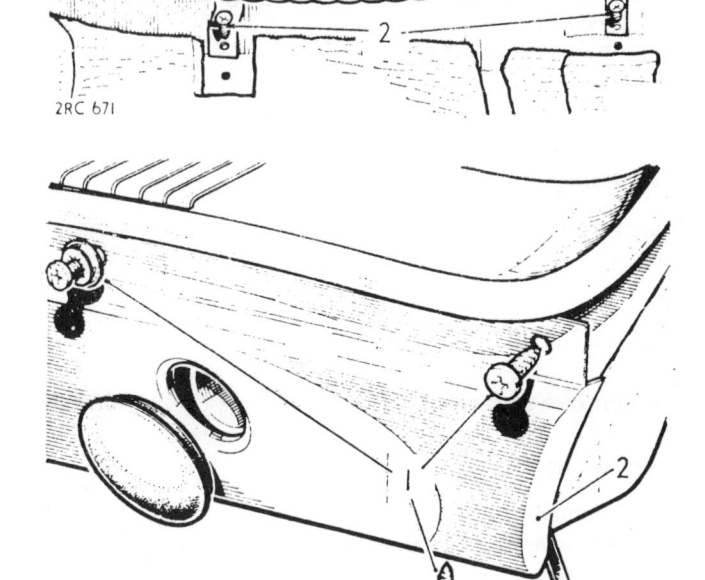

FRONT VALANCE

—Remove and refit 76.79.03

Removing

1. Remove the front bumper.
2. Remove the two outer brackets from the body, secured by two bolts, spring and plain washers, accessible from under front wings.
3. Where applicable, remove the fixings securing the towing eye bracket and withdraw the bracket.
4. Remove the radiator grille. 76.55.03.
5. Remove the drive screw securing the badge mounting bracket to the front valance, if necessary remove the bracket.
6. Remove the two screws, plain and spring washers securing the valance to the front member of the base unit.
7. Remove the four nuts, plain washers and spring washers securing the wing to the valance, and withdraw the stud plates.
8. Remove the valance.

Refitting

9. Reverse 1 to 8.

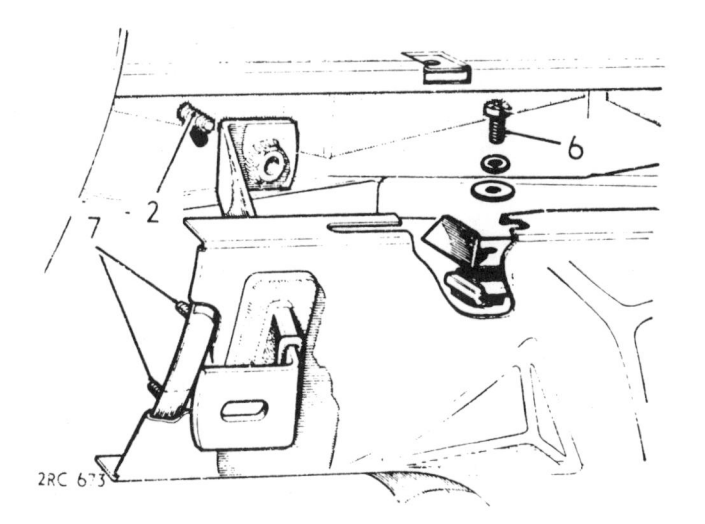

Rover 3500 and 3500S Manual AKM 3621

BODY

AIR INTAKE VALANCE

—Remove and refit   76.79.04

Removing

1. Prop open the bonnet.
2. Withdraw the windscreen wiper arms and blades.
3. Remove the fixings from the windscreen wiper spindles.
4. Where applicable, withdraw the windscreen washer tube from the clip on the valance.
5. Remove the drive screws securing each side of the valance.
6. Remove the fixings securing the valance to the heater.
7. Withdraw the air intake valance.

Refitting

8. Reverse 1 to 7.

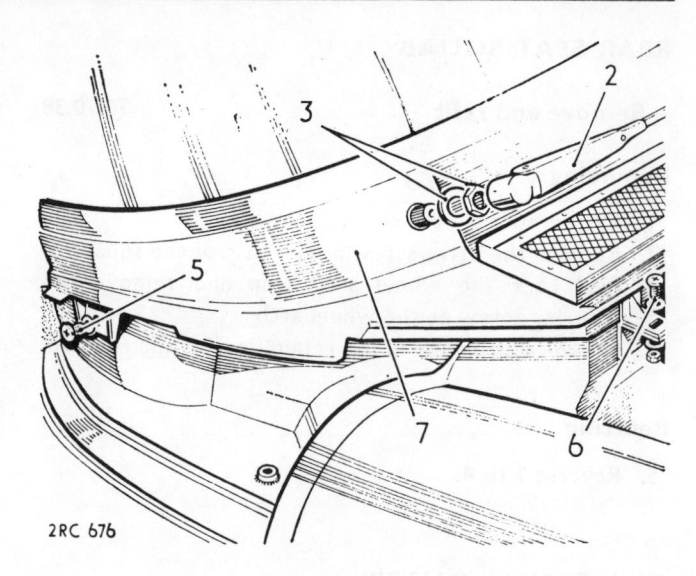

REAR VALANCE

—Remove and refit   76.79.09

Removing

1. Remove the rear bumper.
2. Remove the six screws securing the rear valance to the base unit.
3. Remove the bolts securing the rear valance to the rear wing.
4. Remove the two drive screws securing the rear valance to the rear jacking point.
5. Disconnect the electrical leads and withdraw the rear valance.

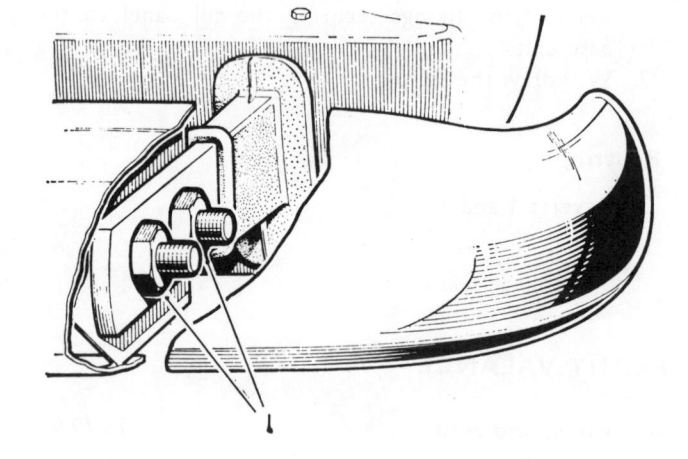

Refitting

6. Reverse 1 to 5.

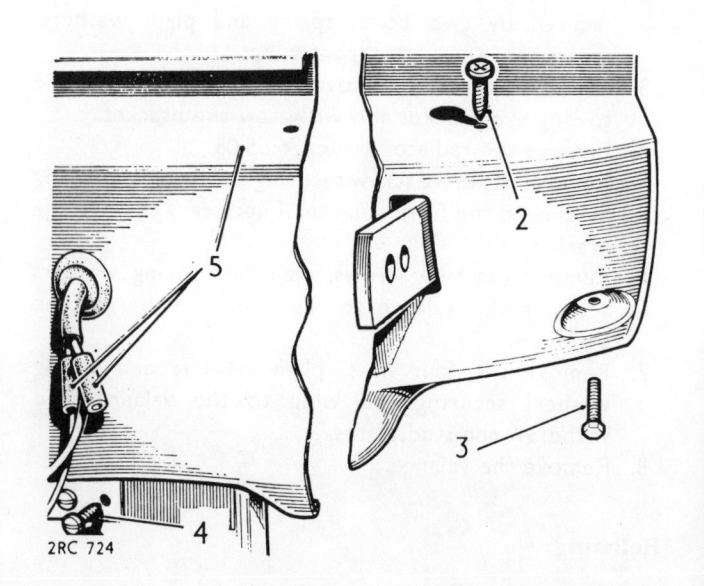

76.79.04
76.79.09

Rover 3500 and 3500S Manual AKM 3621

BODY

WINDSCREEN GLASS AND SEALING RUBBER

—Remove and refit   76.81.01

Removing

1. Remove the bonnet. 76.16.01.
2. Remove the air intake valance. 76.79.04.
3. Open the front doors and drill out the rivets securing the top corner stainless steel finishers and the side finishers.
4. Slide off the side finishers, downwards and withdraw the corner finishers.
5. Slide off the finisher from the top of the screen.
6. Remove the screen centre clamp.
7. Slacken the nuts, then remove the two screen retaining brackets.
8. Push the screen outwards.
9. Withdraw the rubber seal.

Refitting

10. Clean off the old sealing compound from the screen and channel and apply 'Sealastik' to the inner corner of the outer edge of the screen recess.
11. Insert sealing compound into the glass channel of the rubber seal and fit the seal to the glass.
12. Fit a length of string under the lip of the seal as shown.
13. Lubricate the exterior of the seal with soapy water and push the glass and seal upwards into the screen aperture of the base unit. Support the screen with the retainer brackets at the lower edge and adjust the brackets to retain the glass in position.
14. Ensure that the glass and rubber are positioned correctly in the base unit aperture. Then pull out the string from the seal, to peel the rubber lip out of the channel and over the finisher.
15. Carefully adjust the retainer brackets so that the glass and rubber are firmly in position. Do not overstress the glass.
16. Then tighten the locknuts.
17. Fit the centre clamp.
18. Apply 'Sealastik' between the glass and seal, around the top and sides.
19. Fit the screen top finisher.
20. Fit the corner finishers.
21. Slide the side finishers into place and secure with 'pop' rivets.
22. Refit the air intake valance. 76.79.04.
23. Refit the bonnet. 76.16.01.

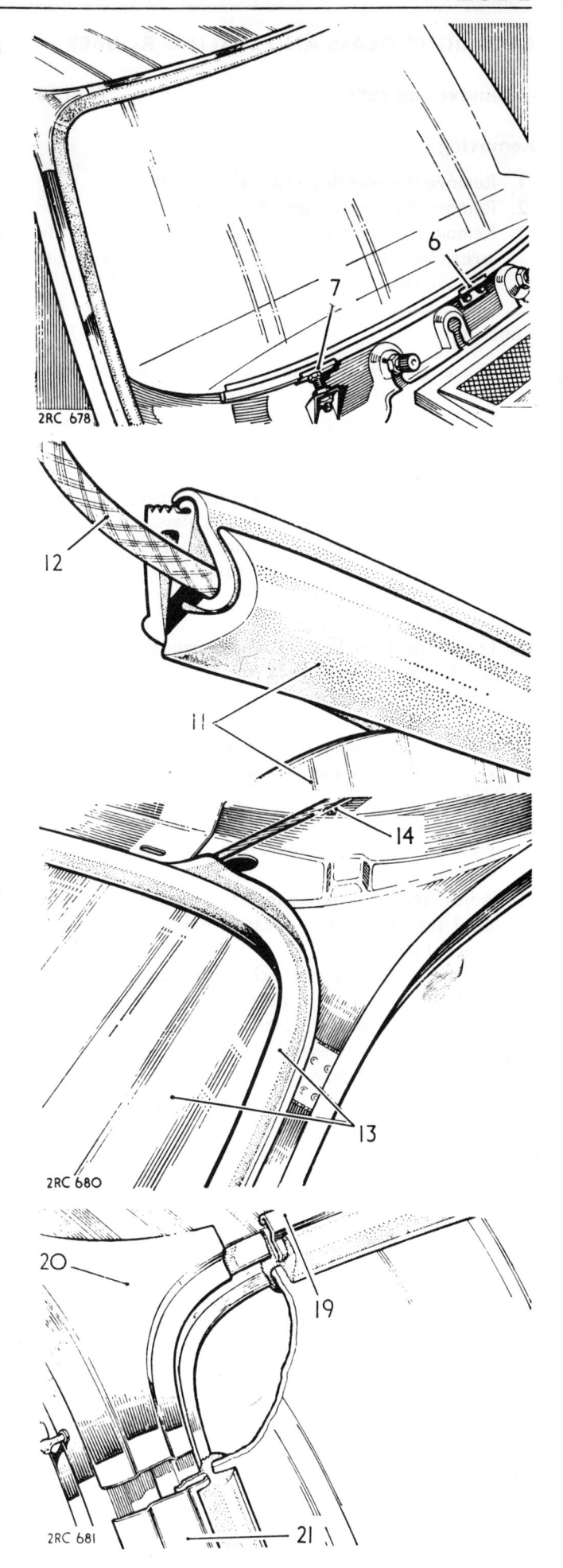

Rover 3500 and 3500S Manual AKM 3621

BODY

BACKLIGHT GLASS AND SEALING RUBBER

—Remove and refit 76.81.10

Removing

1. Remove the rear decker panel. 76.10.45.
2. Remove the centre clamp bracket.
3. Remove the two retaining brackets and push the screen from the inside of the car downwards and out.
4. Remove the rubber seal.

Refitting

5. Thoroughly clean off all traces of old sealing compound from the screen and channel.
6. Insert sealing compound into the new rubber seal and fit the seal to the glass.
7. Fit a length of string to the rubber seal to ease fitting of the seal over the stainless steel rim.
8. Fit the bottom glazing seal to the glass.
9. Lubricate the exterior of the seal and insert the assembled glass into the aperture of the base unit.
10. Position the angle support channel centrally on the base of the glass.
11. Loosely fit the right-hand and left-hand side retaining brackets to the base unit.
12. Adjust the brackets up to the angle support of the glass by hand.
13. Fit the centre clamp bracket to the base unit with the nylon spacer and tighten just enough to nip the glass.
14. Ensure that the rubber seal is in position round the glass and lift the lip over the stainless steel trim with the string.
15. Adjust the right-hand and left-hand brackets evenly until the glass is fully located in the seal. Do not overtighten.
16. Lightly tighten the centre clamp bracket to the glass.
17. Lock the bracket in position.
18. Refit the rear decker panel. 76.10.45.
19. Remove excess sealing compound from around rubber seal and clean glass.

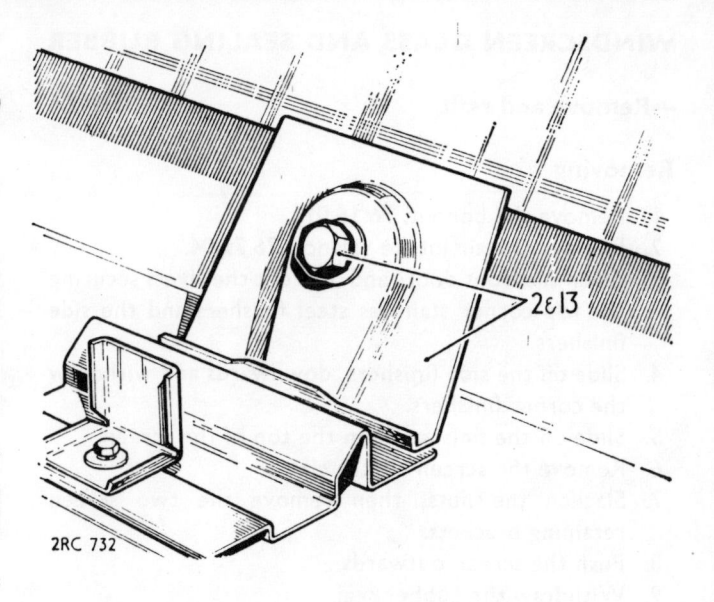

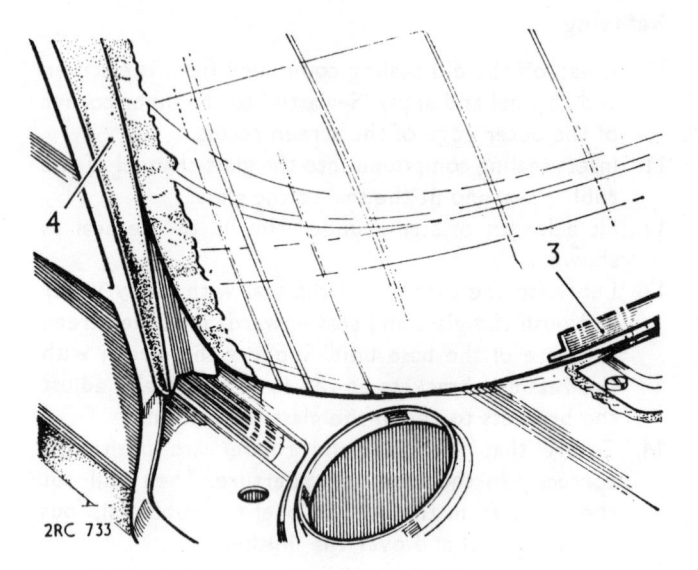

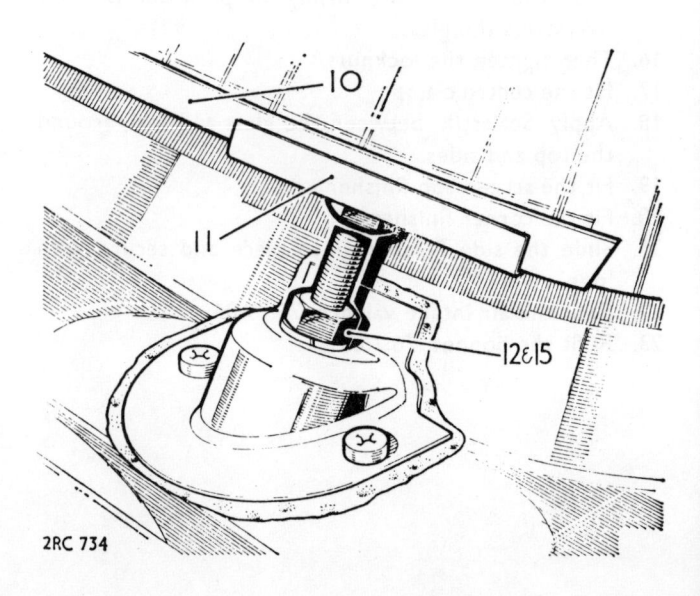

76.81.10

Rover 3500 and 3500S Manual AKM 3621

HEATING AND VENTILATION

LIST OF OPERATIONS

Heater

    —controls—adjust 80.10.03
    —radiator—remove and refit 80.20.29
    —unit—remove and refit 80.20.01

Heater fan

    —motor—remove and refit 80.20.15
    —resistance unit—remove and refit 80.20.17
    —switch—remove and refit 80.10.22

Ventilators—remove and refit 80.15.22/23

Ventilator outlets—remove and refit 80.15.09

Rover 3500 and 3500S Manual AKM 3621

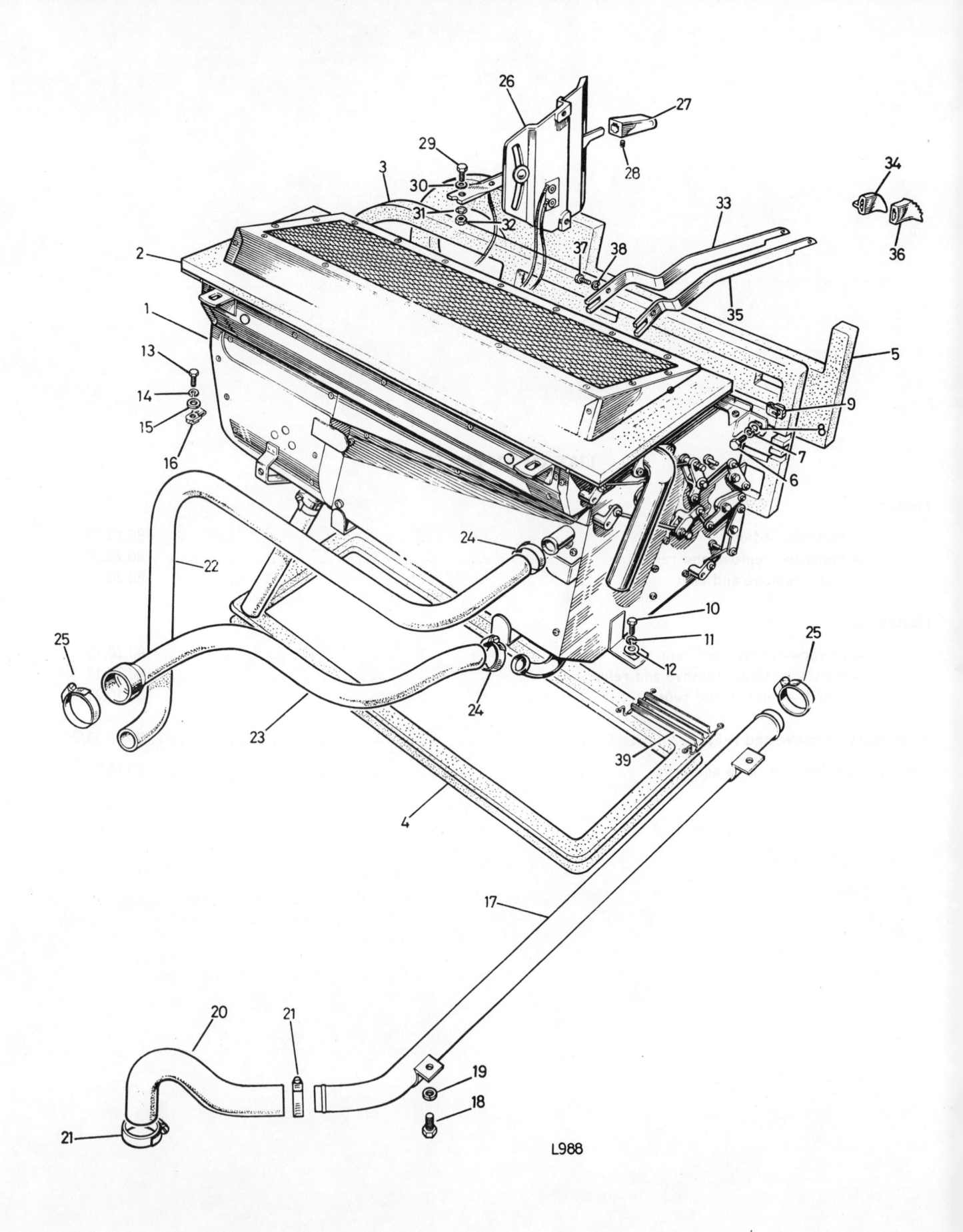

HEATING AND VENTILATION

80-2

Rover 3500 and 3500S Manual AKM 3621

HEATING AND VENTILATION

1 Heater and demister complete, SM
2 Rubber seal, heater to air intake panel
3 Rear seal ⎫
4 Bottom seal ⎬ Heater and demister unit to base unit
5 Pad for rear seal ⎭
6 Bolt (¼ in UNF x ⅝ in long) ⎫
7 Spring washer ⎬ Fixing heater to base unit, upper
8 Plain washer ⎪
9 Spire nut (¼ in UNF) ⎭
10 Set bolt (¼ in UNF x ⅝ in long) ⎫
11 Spring washer ⎬ Fixing heater to base unit, LH side
12 Plain washer ⎭
13 Bolt (¼ in UNF x ⅝ in long) ⎫
14 Spring washer ⎬ Fixing heater to base unit, RH side
15 Plain washer ⎪
16 Spire nut (¼ in UNF) ⎭
17 Outlet pipe beneath inlet manifold
18 Bolt (¼ in UNC x ½ in long) ⎫ Fixing outlet pipe to inlet manifold
19 Spring washer ⎭
20 Hose, outlet pipe to engine
21 Clip fixing hose
22 Water outlet hose for heater
23 Water inlet hose for heater
24 Clip for hose, small
25 Clip for hose, large
26 Main control lever for heater
27 Knob for main control lever
28 Grub screw fixing knob
29 Bolt (10 UNF x ½ in long) ⎫
30 Plain washer ⎬ Fixing main control lever to heater arm
31 Shakeproof washer ⎪
32 Nut (10 UNF) ⎭
33 Temperature control lever
34 Knob for temperature control lever
35 Distribution control lever
36 Knob for distribution control lever
37 Bolt (10 UNF x ⅜ in long) ⎫ Fixing levers to heater
38 Spring washer ⎭
39 Air deflector

Rover 3500 and 3500S Manual AKM 3621

HEATING AND VENTILATION

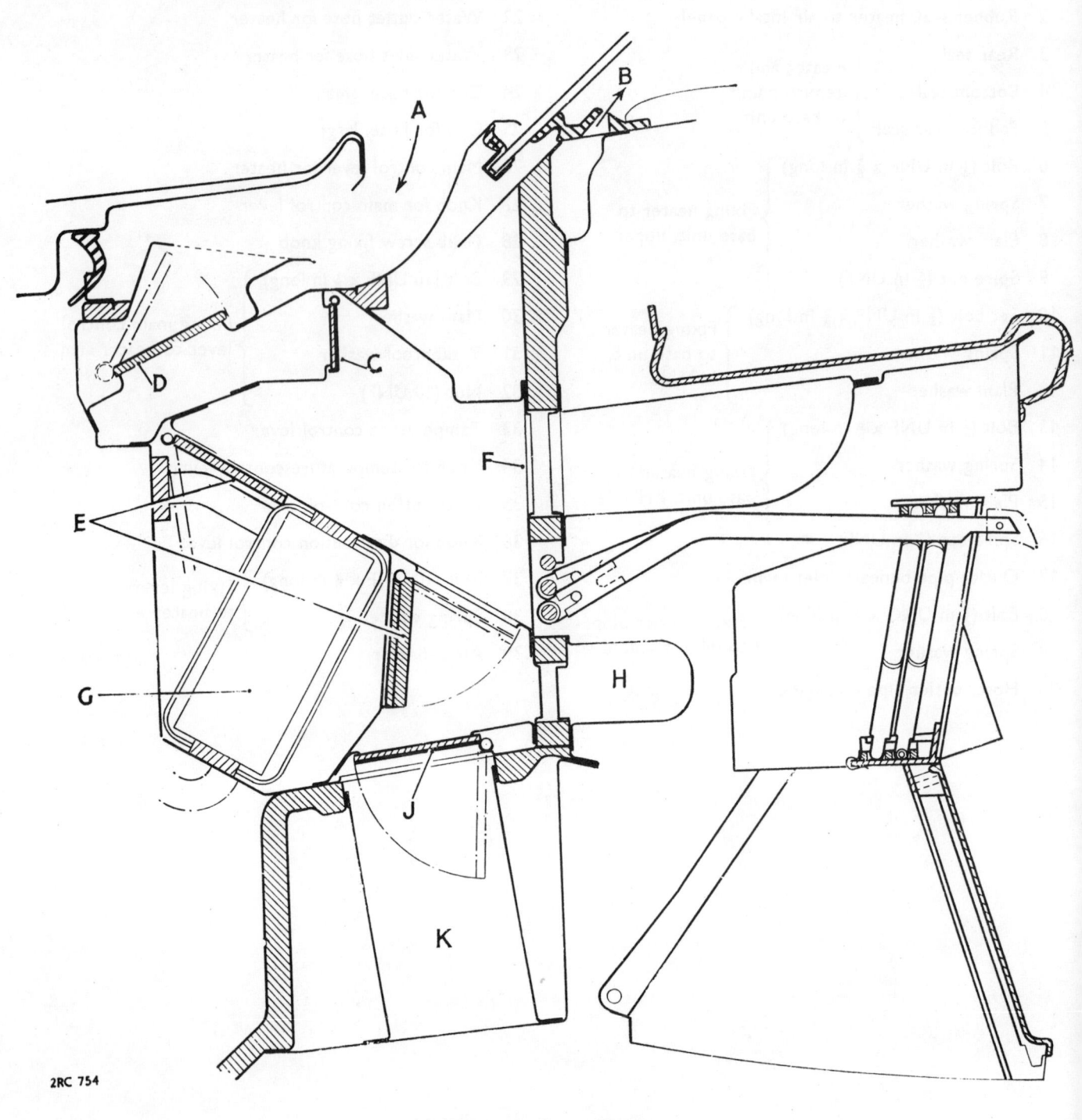

Cross-section of heater

A—Fresh air inlet
B—Outlet from demister duct
C—Windblown valve
D—Main valve, shown in closed position, open position shown dotted
E—Mixing valves, shown in 'Cold' position, full heat position shown dotted
F—Inlet to face-level vents
G—Heater matrix
H—Inlet to demister duct
J—Distribution valve, shown in screen only position, screen and car position shown dotted
K—Inlet to car heater

HEATING AND VENTILATION

HEATER CONTROLS

—Adjust 80.10.03

NOTE: If the heater control linkage has been disturbed, or if the heater should fail to function correctly, the controls should be checked, and if necessary adjusted, in the following sequence. All adjustments are made to the linkage outside the heater unit.

Procedure

Main control

1. Slacken the locking screw on the upper lever.
2. Set the main control lever on the console approximately 6,3 mm (0.250 in) below the upper detent position.
3. Apply sufficient pressure to the upper link on the heater unit, in a clockwise direction, to ensure that the main flap valve is sealed, then maintaining this pressure, tighten the locking screw.
4. Move the control knob to 'ram air only' detent (about halfway down) and look through the intake grille to see whether the metal back of the inlet valve is touching or almost touching the small piece of sponge pad which acts as a stop.
5. If this valve position is not correct, slacken the adjusting screw and slide the end of the spring link into the trunnion so as to increase the effective length of the link by about 1,5 to 3,0 mm (0.062 to 0.125 in). Retighten the adjusting screw firmly.
6. Repeat items 1 to 4. Care must be taken not to increase the length of the spring link more than necessary, otherwise there will be insufficient adjustment at the upper lever locking screw to ensure that the flap can be shut when control is 'off'.

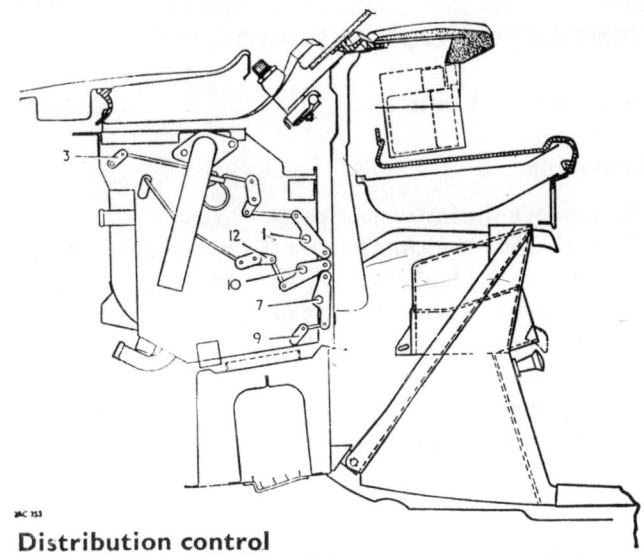

Distribution control

7. Slacken the locking screw on the lower control lever.
8. Set the distribution lever on the console approximately 6,3 mm (0.250 in) below the upper detent position.
9. Apply sufficient pressure to the lower link on the heater unit, in a clockwise direction, to ensure that the distribution flap valve is sealed, then maintaining this pressure, tighten the locking screw.

Temperature control

10. Slacken the locking screw on the centre control lever.
11. Set the temperature control lever on the console approximately 6,3 mm (0.250 in) below the upper detent position.
12. Apply sufficient pressure to the centre link on the heater unit, in a clockwise direction, to seal the temperature flap valves, then maintaining this pressure, tighten the locking screw.

HEATER FAN SWITCH

—Remove and refit 80.10.22

Removing

1. Disconnect the battery earth lead.
2. Remove the 'Rover' motif from the front of the console unit.
3. Open the RH glove box and remove the bolt, nut and spring washer connecting the switch lever to the heater control rod.
4. Disconnect the two feed and earth wires at the snap connectors and note the colour coding.
5. Remove the heater switch knob by removing the small grub screw on the underside of the knob.
6. From the underside of the switch panel, clip out the cover panel.
7. Remove the two $\frac{3}{8}$ in AF bolts and shakeproof washers securing the switch to the console unit and lift out the switch complete.

Refitting

8. Reverse 1 to 7.
9. Test the switch function.

Rover 3500 and 3500S Manual AKM 3621

80.10.03
80.10.22

HEATING AND VENTILATION

VENTILATOR OUTLETS—FRONT

—Remove and refit 80.15.09

Removing

1. Remove the instrument panel. 88.20.01.
2. Early 3500 models—Remove the switch panel.
3. Remove the fascia top rail. 76.46.04.
4. Remove the parcel tray. 76.67.02.
5. Remove the insulation pads.
6. Remove the drive screws securing the cover plates and vent outlets.
7. Withdraw the ventilator outlets.

Refitting

8. Reverse 1 to 7, renewing the ventilator to base unit sealing.

VENTILATORS

—Remove and refit
  left-hand   80.15.22
  right-hand 80.15.23

NOTE: Through-flow ventilators are only fitted to cars that are also fitted with fixed rear quarter lights.

Through-flow ventilation is achieved completely automatically by means of three fibre-glass cloth one-way flap valves at each side of the car under the rear quarter trim, without the use of internal or external grilles. Air flows round and under the rear quarter trim, out through the flaps and is dispersed under the exterior rear quarter panel.

Removing

1. Remove the rear seat squab. 76.70.38.
2. Remove the body rear quarter panel. 76.10.22/23.
3. Drill out the pop rivets retaining the flap valve and drain channel.

Refitting

4. Apply a thin coating of Suppra Seal C66 or similar sealing compound around the inner face of the flap valve and drain channel, ensure that the sealing compound does not come into contact with the fibre-glass cloth valves.
5. Position the flap valve and drain channel on to the base unit and secure with six pop rivets.
6. Brush sealing compound round the edge of the flap valve and drain channel.

Fitting new seals, as required, 7 to 12

7. Remove the original seals and all traces of old adhesive.
8. Apply Bostik 1GA 167 or a similar adhesive to the gutter seal and press it into position against the inner panel, behind the guttering.

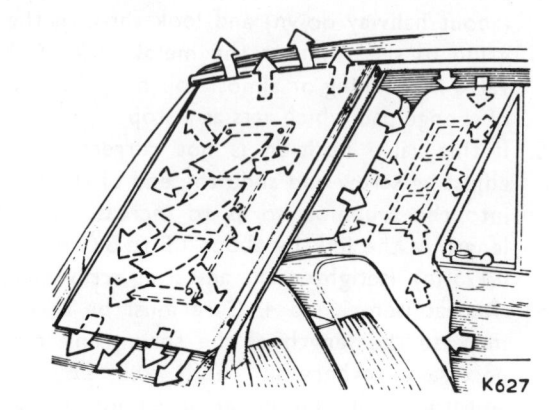

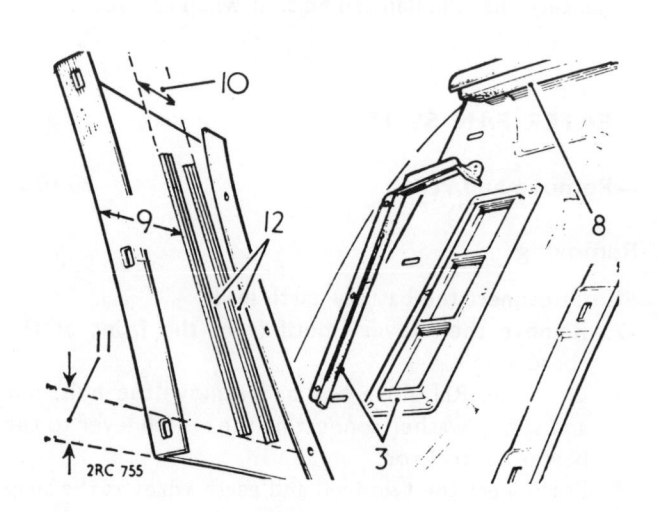

9. Mark off the inside of the body rear quarter panel 47,5 mm (1.875 in) from the front edge.
10. Mark off a second line 37,5 mm (1.500 in) from the first.
11. Mark off 108,0 mm (4.250 in) from the panel bottom edge.
12. Secure the sealing strips to the marked positions on the inside of the body rear quarter panel, using Bostik 1GA 167, or a similar adhesive.
13. Reverse 1 and 2.

Rover 3500 and 3500S Manual AKM 3621

HEATING AND VENTILATION

HEATER UNIT

—Remove and refit 80.20.01

Removing

1. Disconnect the battery earth lead.
2. Remove the glove box. 76.52.03
3. Disconnect the heater controls from the adjustment fork.
4. Remove the bonnet. 76.16.01.
5. Remove the air cleaner. 19.10.01.
6. Drain the cooling system. 26.10.01.
7. Remove the windscreen wiper arms.
8. Remove the air intake valance. 76.79.04.
9. Remove the accelerator coupling shaft bracket from the relay.
10. Disconnect the electrical leads from the heater.
11. Disconnect the heater hoses.
12. Remove the fixings securing the heater to the bulkhead.
13. Hold the air intake flap on the heater in the closed position and lift out the heater unit complete.
14. If required, remove the insulation pad.

Refitting

15. If removed, refit the insulation pad.
16. Close the air intake flap and place the heater unit in position on the bulkhead. Ensure correct seating on the sealing rubbers and that the heater wiring does not become trapped.
17. Reverse 1 to 12.
18. Run the engine and check the heater for correct operation, also check for coolant leaks.

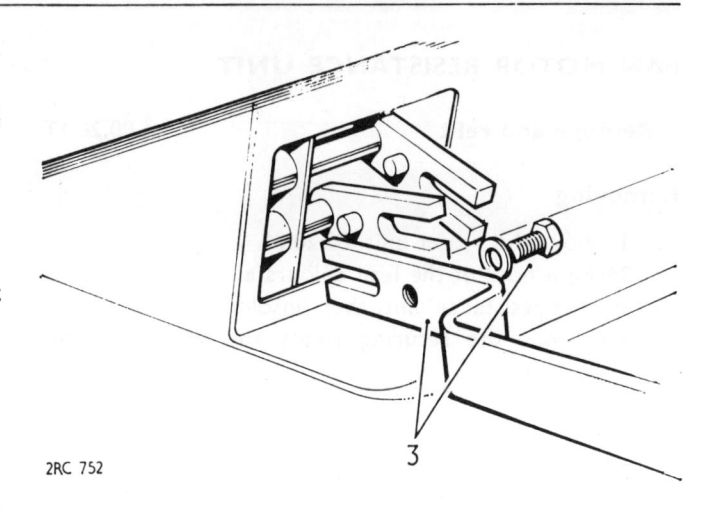

HEATER FAN MOTOR

—Remove and refit 80.20.15

Removing

1. Remove the heater unit. 80.20.01.
2. Remove the domed cover plate.
3. Make a note of the lead colours at the connections to the heater fan motor, then disconnect the leads.
4. Remove the three small bolts securing the heater fan motor.
5. Withdraw the heater fan motor.

Refitting

6. Reverse 1 to 5.

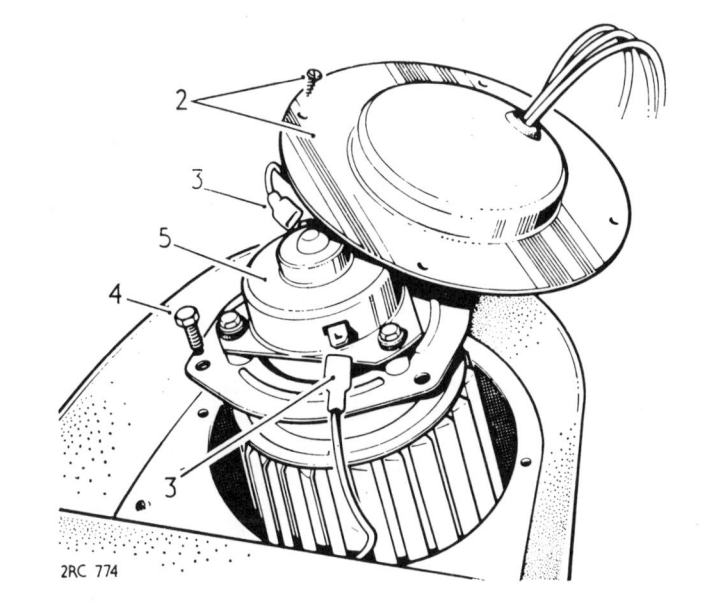

Rover 3500 and 3500S Manual AKM 3621

HEATING AND VENTILATION

FAN MOTOR RESISTANCE UNIT

—Remove and refit 80.20.17

Removing

1. Remove the heater unit. 80.20.01.
2. Make a note of the lead colours at the connections on the resistance unit, then unsolder the leads.
3. Drill out the securing rivets and withdraw the resistance unit.

Refitting

4. Reverse 1 to 3.

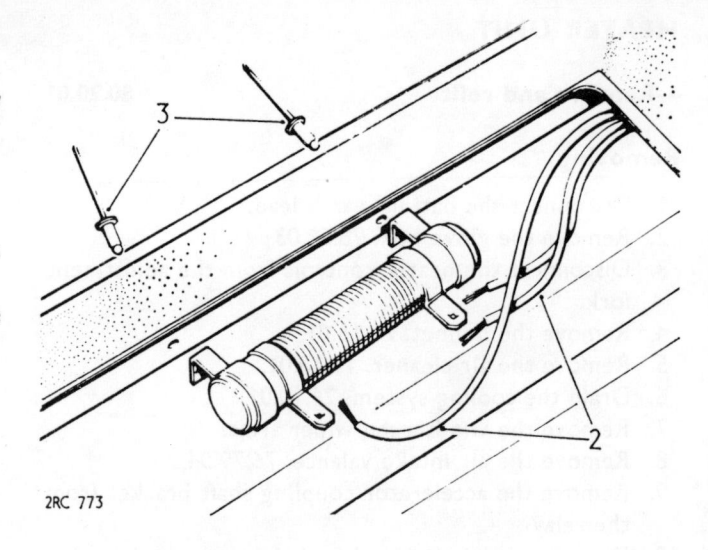

HEATER RADIATOR

—Remove and refit 80.20.29

Removing

1. Remove the heater unit. 80.20.01.
2. Remove the radiator cover.
3. Release the fixings from the end cover.
4. Manoeuvre the radiator from the heater case.

Refitting

5. Reverse 1 to 4.

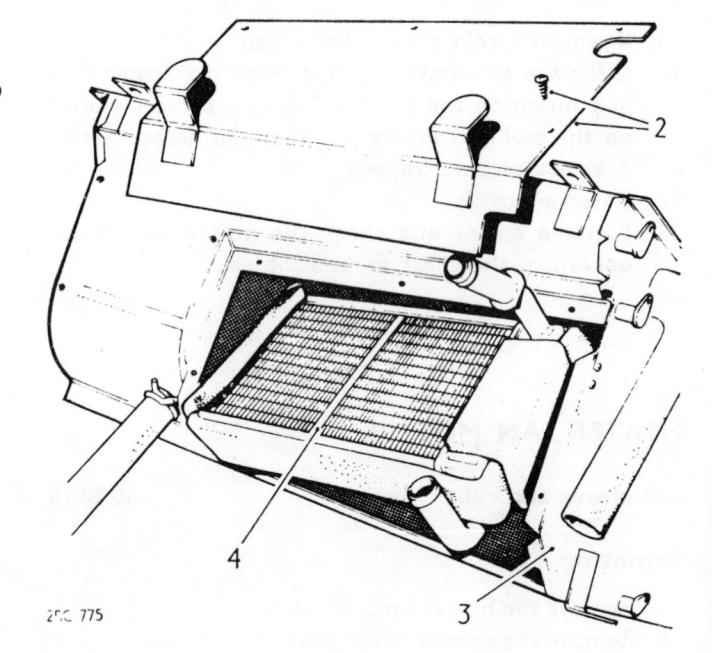

AIR CONDITIONING

LIST OF OPERATIONS

Air conditioning system	
—charge	82.30.08
—depressurise	82.30.05
—evacuate	82.30.06
—leak test	82.30.09
—pressure test	82.30.10
—sweep	82.30.07
—test	82.30.16
Blower assembly—remove and refit	82.25.13
Blower motor	
—brushes—remove and refit	82.25.31
—resistance unit—remove and refit	82.20.26
Charging and testing equipment—fit and remove	82.30.01
Compressor	
—remove and refit	82.10.20
—overhaul	82.10.26
—drive belt—adjust	82.10.01
—drive belt—remove and refit	82.10.02
—drive clutch—remove and refit	82.10.08
—oil level—check	82.10.14
Condenser—remove and refit	82.15.07
Console unit—remove and refit	82.25.07
Control switches	
—air flow switch—remove and refit	82.20.11
—compressor isolator switch—remove and refit	82.20.12
—temperature control switch—remove and refit	82.20.10
Evaporator—remove and refit	82.25.20
Expansion valve—remove and refit	82.25.01
High pressure cut-out—remove and refit	82.20.20
Heater/cooler unit—remove and refit	82.25.21
Liquid receiver/drier—remove and refit	82.17.01
Main relay—remove and refit	82.20.08
Water hoses—heater/cooler unit—remove and refit	82.25.27
Water valve and radiator—remove and refit	82.20.32
Vacuum motors—remove and refit	82.20.14
Vacuum reservoir—remove and refit	82.20.13

Rover 3500 and 3500S Manual AKM 3621

AIR CONDITIONING

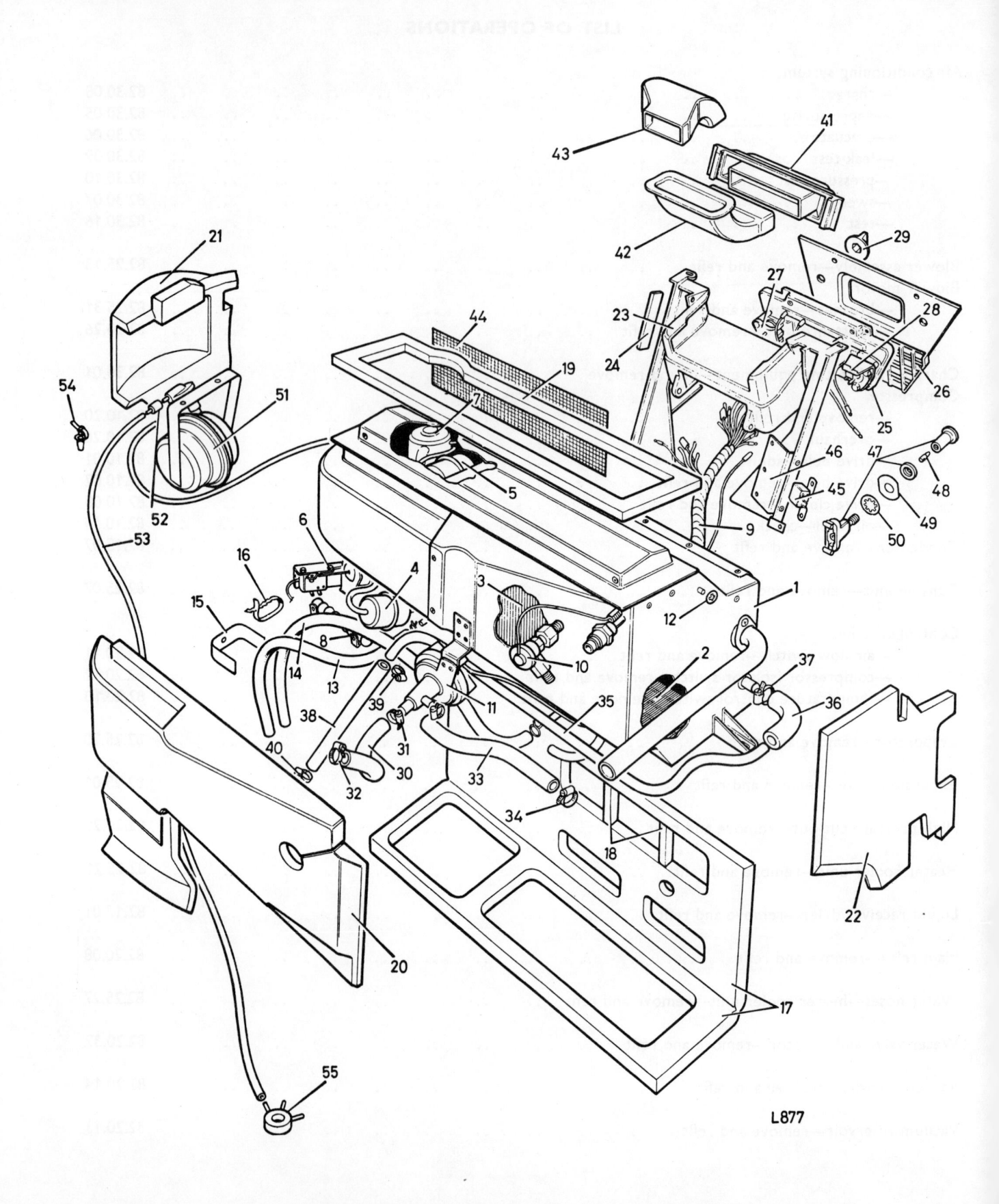

Illustration of the heater/cooler unit and ducting

L877

82-2

Rover 3500 and 3500S Manual AKM 3621

AIR CONDITIONING

Key to illustration of heater-cooler unit and ducting

1. Heater/cooler unit
2. Radiator for heater
3. Evaporator and filter
4. Actuator (vacuum motor) for defrost valve
5. Actuator (vacuum motor) fresh air/recirculating valve
6. Thermostat and actuator (vacuum motor) for compressor
7. Blower motor
8. Resistor for blower motor
9. Harness, electric cables and vacuum pipes
10. Expansion valve
11. Water valve for heater
12. Clip retaining spindle
13. Drain tube, RH ⎫ For heater/cooler unit
14. Drain tube, LH ⎬ casing
15. Hose clip, double for drain tubes
16. Cleat for drain tubes
17. PVC seal for heater/cooler unit
18. Sealing strip, intermediate duct to heater/cooler unit
19. Top seal for heater/cooler unit
20. Insulation cover, front ⎫
21. Insulation cover, RH ⎬ For heater/cooler unit
22. Insulation cover, LH ⎭
23. Console unit assembly
24. Filler piece blanking off heater lever slots
25. Housing for window lift switch panel
26. Switch panel and texture plate
27. Switch, air flow control
28. Switch, temperature control
29. Knob for switch
30. Hose, inlet, heater/cooler unit
31. Clip fixing inlet hose to heater/cooler unit
32. Clip fixing inlet hose to adaptor
33. Hose, valve to heater/cooler unit
34. Clip fixing hose
35. Pipe outlet, heater/cooler unit
36. Elbow for pipe
37. Clip fixing elbow hose
38. Connecting hose, outlet pipe to outlet pipe on engine
39. Clip fixing hose to heater pipe
40. Clip fixing hose to engine pipe
41. Air outlet nozzle for console unit
42. Air duct, outlet nozzle to intermediate duct
43. Intermediate duct at parcel shelf
44. Mesh cover for re-circulating hose in dash
45. Relay
46. Mounting bracket for relay
47. Switch magnetic clutch isolator
48. Bulb for switch
49. Escutcheon for switch
50. Earth washer
51. Vacuum reservoir tank
52. Adaptor for vacuum pipe
53. Vacuum hose, engine to tank
54. Clip fixing hose
55. Banjo connection at inlet manifold

Rover 3500 and 3500S Manual AKM 3621

AIR CONDITIONING

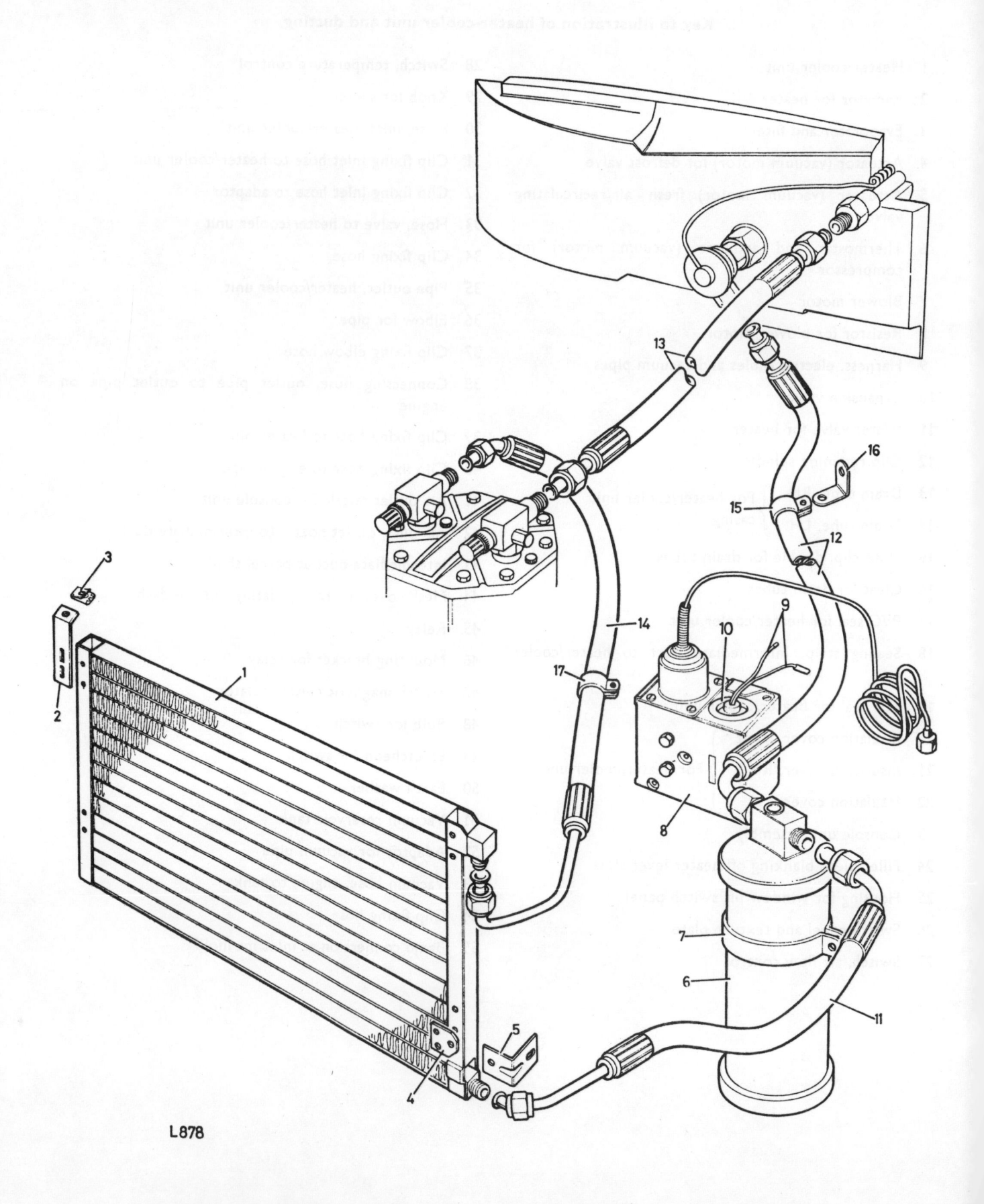

Layout of condenser, receiver drier, high pressure cut-out switch and pipes

82-4   Rover 3500 and 3500S Manual AKM 3621

Key to illustration of condenser, receiver drier, high pressure cut-out switch and pipes

1. Condenser
2. Mounting bracket, top
3. Spire nut
4. Fixing plate
5. Mounting bracket, condenser, bottom
6. Receiver drier bottle
7. Clip for receiver drier bottle
8. High pressure cut-out switch
9. Lead for cut-out switch
10. Grommet
11. Hose, condenser to drier bottle
12. Hose, drier bottle to evaporator
13. Hose, evaporator to compressor
14. Hose compressor to condenser
15. Hose clip for drier bottle to evaporator hose
16. Bracket for clip
17. Hose clip for compressor to condenser hose

AIR CONDITIONING

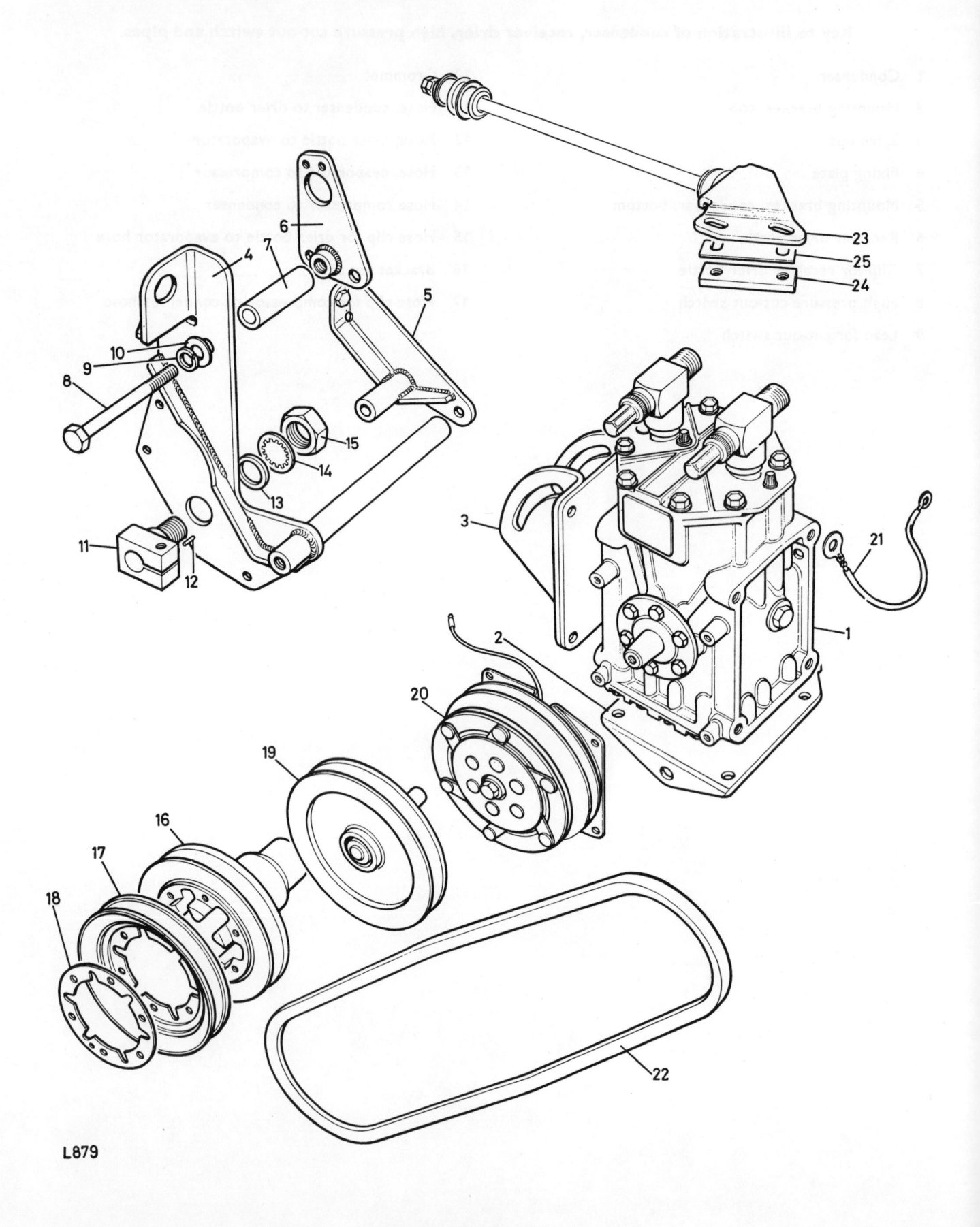

Layout of compressor and drive

Rover 3500 and 3500S Manual AKM 3621

AIR CONDITIONING

Key to illustration of compressor and drive

1 Compressor complete
2 Mounting bracket at base of compressor
3 Adjusting bracket for compressor
4 Mounting bracket, front, for compressor
5 Mounting bracket, rear, bottom for compressor
6 Mounting bracket, top rear, for compressor
7 Spacer, mounting brackets to adjusting brackets
8 Bolt ($\frac{7}{16}$ in. UNF x 4 in. long) ⎫
9 Spring washer ⎬ Fixing front mounting bracket, adjusting bracket and rear upper mounting bracket together
10 Plain washer ⎭
11 Bearing support assembly
12 Roll pin
13 Backing washer
14 Shakeproof washer
15 Nut ($\frac{3}{4}$ in. UNF)
16 Vibration damper
17 Pulley driving compressor
18 Reinforcing plate
19 Idler pulley complete
20 Electro-magnetic clutch
21 Earth lead for electro-magnetic clutch
22 Driving belt for compressor
23 Bracket, tie bar to valance
24 Tapping plate ⎫ Tie-bar bracket to valance
25 Packing plate ⎭

AIR CONDITIONING

General service information

1. Introduction

This Division provides the instruction necessary for the adjustment of the compressor driving belt, the checking of the oil level in the compressor, and for the removal and refitting of the components of the system. Removal of components involves depressurising the system, and instructions for this operation are included, together with those for the subsequent evacuation and charging.

Before any component of the air conditioning system is removed the system must be depressurised. When the component is replaced the system must be evacuated to remove all traces of old refrigerant and moisture. Then the system must be recharged with new refrigerant.

Any service work that requires loosening of a refrigerant line connection should be performed only by qualified service personnel. Refrigerant and/or oil will escape whenever a hose or pipe is disconnected.

All work involving the handling of refrigerants requires special equipment, a knowledge of its proper use, and attention to safety measures.

2. Servicing equipment

The following equipment is required for full servicing of air conditioning.
Charging trolley.
Leak detector.
Tachometer (Part Number 601284).
Lock ring spanner (Part Number 601942).
Valve key (Britool D71 or D72).
Safety goggles.
Refrigerant charging line gaskets.
Compressor dip stick.
$\frac{5}{8}$ in. UNC bolt or Union nut (Part Number 534127) for extraction of the compressor pulley.
Thermometer—20°C to—60°C (0°F to —120°F).

The charging trolley (as illustrated in the applicable operations) is manufactured by the Service Tool Division of Kent Moore Organisation Inc, 28635 Mound Road, Warren, Michigan, U.S.A. It is available through their Overseas Subsidiaries and comprises vacuum pump, manifold gauge set with flexible charging lines and a measuring cylinder. An alternative refrigeration equipment is manufactured by Perfection Parts Limited, 59 Union Street, London S.E.1., their particular trolley being intended specifically for vehicle systems.

3. Servicing materials

Refrigerant R12.
Refrigerant oil.
Refer to Division 09 for specific details.

4. Precautions in handling refrigerant 12

Refrigerant 12 is transparent and colourless in both the gaseous and liquid state. It has a boiling point of —30°C (—22°F) and at all normal pressures and temperatures it is a vapour. The vapour is heavier than air, non-flammable and non-explosive. It is non-poisonous except when in contact with an open flame, and non-corrosive until it comes into contact with water.

The following precautions in handling refrigerant 12 should be observed at all times:

a. Do not leave a drum of refrigerant without its heavy metal cap fitted.
b. Do not carry a drum in the passenger compartment of a car.
c. Do not subject drums to high temperature.
d. Do not weld or steam clean near an air conditioning system.
e. Do not discharge refrigerant vapour into an area with an exposed flame, or into the engine air intake. Heavy concentrations of refrigerant in contact with a live flame will produce a toxic gas that will also attack metal.
f. Do not expose the eyes to liquid refrigerant. ALWAYS wear safety goggles.

5. Precautions in handling refrigerant lines

WARNING: Always wear safety goggles when opening refrigerant connections

a. When disconnecting any pipe or flexible connection the system must be discharged of all pressure. Proceed cautiously, regardless of gauge readings. Open connections slowly, keeping hands and face well clear, so that no injury occurs if there is liquid in the line. If pressure is noticed allow it to bleed off slowly.
b. Lines, flexible end connections and components must be capped immediately they are opened to prevent the entrance of moisture and dirt.
c. Any dirt or grease on fittings must be wiped off with a clean alcohol dampened cloth. Do not use chlorinated solvents such as trichloroethylene. If dirt, grease or moisture cannot be removed from inside pipes, they must be replaced with new.
d. All replacement components and flexible end connections are sealed, and should only be opened immediately prior to making the connection. (They must be at room temperature before uncapping to prevent condensation of moisture from the air that enters.)
e. Components must not remain uncapped longer than 15 minutes. In the event of delay the caps must be replaced.

continued

AIR CONDITIONING

f. Receiver driers must never be left uncapped as they contain Silica Gel which will absorb moisture from the atmosphere. A receiver drier left uncapped must be replaced, and not used.
g. A new compressor contains an initial charge of 0,28 litre (0.5 UK pint) of refrigerant oil when received, part of which is distributed throughout the system when it has been run. The compressor contains a holding charge of gas when received which should be retained until the hoses are connected.
h. The compressor shaft must not be rotated until the system is entirely assembled and contains a charge of refrigerant.
j. The receiver-dryer should be the last component connected to the system to ensure optimum dehydration and maximum moisture protection of the system.
k. All precautions must be taken to prevent damage to fittings and connections. Minute damage could cause a leak with the high pressures used in the system.
l. Always use two spanners of the correct size, one on each hexagon, when releasing and tightening refrigeration unions.
m. Joints should be coated with refrigeration oil to aid correct seating.
n. All lines must be free of kinks. The efficiency of the system is reduced by a single kink or restriction.
o. Flexible hoses should not be bent to a radius less than ten times the diameter of the hoses.
p. Flexible connections should not be within 50 mm (2.000 in.) of the exhaust manifold.
q. Completed assemblies must be checked for refrigerant lines touching sheet metal panels. Any direct contact of lines and sheet metal transmits noise and must be eliminated.

6. Periodic maintenance

The design of the system is such that routine servicing, apart from visual checks is not necessary.
These visual inspections are listed as follows and should be carried out at bi-monthly intervals:

a. **Condenser.** With a hose pipe or air line, clean the face of the condenser to remove flies, leaves, etc. Check pipe connections for signs of oil leakage.
b. **Compressor.** Check hose connections for signs of oil leakage. Check flexible hoses for swellings. Examine the compressor belt for tightness and condition. 82.10.01. Checking compressor oil level and topping-up is only necessary after:
   i Charging the system.
   ii Any mal-function of the equipment. See 82.10.14.
c. **Liquid receiver.** Examine the sight glass with the system running for bubbles. Check connections for leakage.
d. **Heater/cooler unit.** Examine refrigeration connections at the unit also water hose connections at the unit and engine for leakage.

If the system should develop any fault, or if erratic operation is noticed, refer to the fault diagnosis charts in this Division.

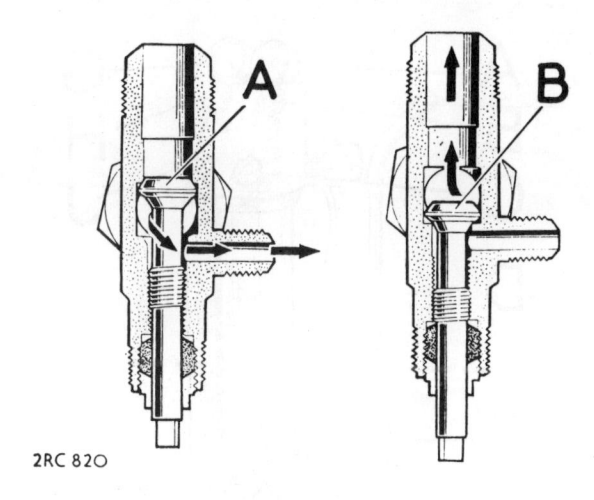

Service valve operation
A—Valve fully front-seated, compressor isolated
B—Valve fully back-seated, normal operation

e. **Service valves.** These are secured to the head of the compressor, and the suction and discharge flexible end connections are secured to them by unions.
The service valves are identified as suction or low pressure, and discharge or high pressure. Whilst they are identical in operation they are not interchangeable, as the connections are of different sizes. When in position the valve can be identified by the letters 'DD DISCH' and 'SS SUCTION' on the compressor head. The valve with the largest connection fits the suction side.
As the name suggests, these valves are for service purposes, providing connections to external pressure/vacuum gauges for test purposes, and isolating the compressor to check the oil level. In combination with charging and testing equipment they are used to charge the system with refrigerant.
The illustration shows the left-hand service valve fully front-seated, compressor isolated. The right-hand valve is shown fully back-seated, normal operating position.
It is important that the compressor is not operated with the discharge service valve in its front-seated, clockwise, position. Severe damage to the compressor could result.

Rover 3500 and 3500S Manual AKM 3621

AIR CONDITIONING

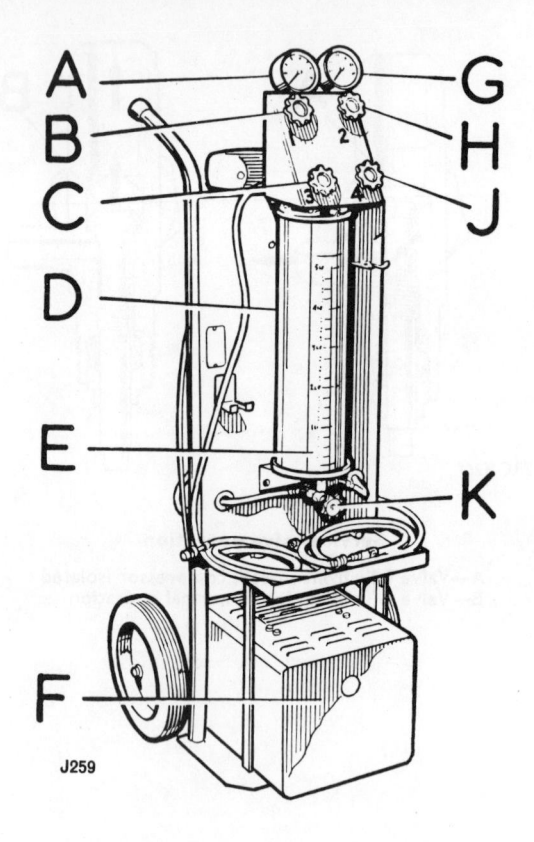

Refrigerant charging station, front view

A—Compound gauge
B—Low pressure control valve No. 1
C—Vacuum control valve No. 3
D—Charging cylinder
E—Sight glass
F—Vacuum pump case
G—High pressure gauge
H—High pressure control valve No. 2
J—Refrigerant control valve No. 4
K—Cylinder base valve

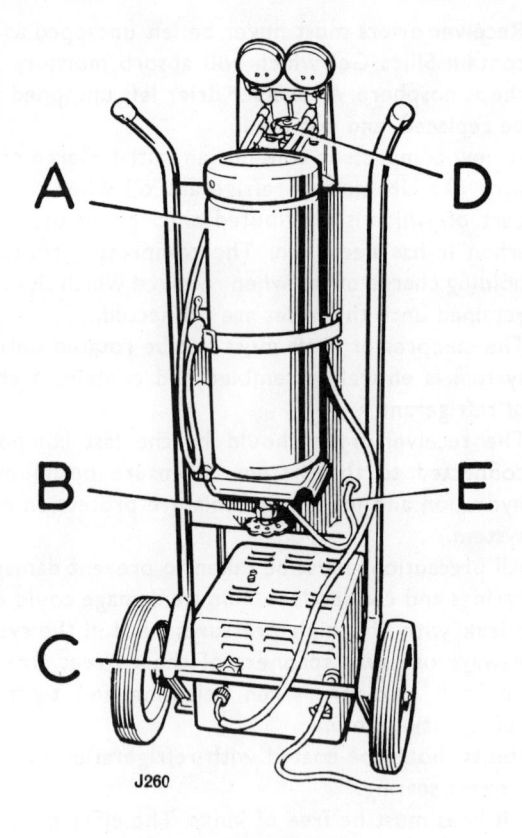

Refrigerant charging station, rear view

A—Refrigerant drum
B—Refrigerant drum valve
C—Vacuum pump valve
D—Top cylinder valve
E—Charging cylinder hose

7. Test procedure

Efficient testing of the air conditioning system requires the use of charging and testing equipment and an accurate thermometer, in addition to normal workshop equipment.

8. Charging and testing equipment.

This is standard equipment for the servicing of automotive air conditioners, and is used for testing, trouble shooting, evacuating and charging the system.

All evacuating and charging equipment is assembled into a compact portable unit, as shown in the illustration. The use of this equipment reduces air conditioner servicing to a matter of connecting two hoses and manipulating clearly labelled valves. An instruction plate indicates the operations required for evacuating and charging the car air conditioner system.

The compound or low pressure gauge is connected to the suction service valve to check pressure and vacuum on the low pressure side of the system. It is graduated in inches Hg of vacuum 0 to 30, and in pounds per square inch from 0 to 60.

The high pressure gauge is connected to the discharge service valve to check the pressures on the high pressure side of the system. It is graduated in pounds per square inch from 0 to 600.

The connections beneath the gauges are for attaching the hoses to the appropriate service valve.

The valves are opened and closed by means of handwheels at each end of the control panel. At no time is it possible to close off the gauges. The valves close off the centre section from each end connection and from one another. When the valves are turned fully clockwise, the centre section is closed to both sides of the system. As each valve is turned counter-clockwise, the appropriate service valve is put into communication with the centre section, gas flow being according to the pressure difference.

For further information and pipe connection diagrams, refer to 82.30.01.

AIR CONDITIONING

9. Electrical supply, switches and fuses

The electrical supply to the compressor magnetic clutch is in series with the blower control switch, compressor isolator switch, thermostat control switch contacts, a 10 amp in line fuse and the high pressure cut-out switch. The supply to the blower control is taken from the car main circuit, through a 35 amp fuse, and also through the contacts of a relay that is operated by the first movement of the 'air conditioning' control. Current to the relay is monitored by the evaporator thermostat which has the effect of cutting the supply to the compressor when reduced cooling is required.

Fuses

35 amp. Location: Main fuse box.

10 amp. Location: Engine compartment, in line between the high pressure cut-out switch and the thermostat control.

The 10 amp fuse is in a line holder, retained between two tubular halves with a bayonet fixing. If either fuse should blow, investigate the cause before renewing.

Blower motor resistance unit

Location: inside the heater/cooler unit. For removal and refitting instructions. 82.20.26 refers.

Relay unit

Location: Behind console unit. For removal and refitting instructions, 82.20.08 refers.

Compressor isolator switch

Location: Driver's side switch panel, 82.20.07 refers.

10. Model variations

Over a period, slight variations in design have been introduced, and the following summarises the affect on servicing.

a. **Bonnet**

   On early 3500 models, the bonnet incorporates a centrally mounted air intake nacelle which directs an air supply straight to the air cleaner intake. Later 3500 models and all 3500S models are fitted with the standard bonnet.
   Operation 76.16.01 is applicable to all models.

b. **Radiator**

   A fan cowl is fitted to the rear face of the radiator assembly to improve the efficiency of the fan in drawing cooling air through the air conditioning condenser and the engine coolant radiator.
   Operation 26.40.04 is applicable after releasing the fan cowl fixings.

c. **Air cleaner**

   On early 3500 models an automatically controlled valve in the air cleaner intake duct permits intake air to be drawn either from an exhaust manifold 'hot' box or through the bonnet intake nacelle. The system is designed to achieve an air intake temperature of approximately 38°C (100°F) as quickly as possible and maintain this temperature whilst ambient conditions are below 38°C (100°F).
   Later 3500 models and all 3500S models are fitted with the standard air cleaner.
   Operation 19.10.01 is applicable to all models.

d. **Carburetters**

   Early 3500 models were fitted with twin SU HS 6 carburetters which varied from the standard specification as follows:
   Needle   BBC
   Engine idle speed 700 to 750 rev/min.
   Operation 19.15.02 and 19.15.18 modified by the above specification, are applicable.
   Later 3500 models and all 3500S models are fitted with twin SU HIF 6 carburetters with standard specification except that the engine idle speed setting must be 700 to 750 rev/min.

e. **Fuel filter**

   The cartridge type fuel filter is mounted above the engine right-hand rocker cover.
   Operation 19.25.01 applies to all models.

f. **Charcoal canister**

   On early 3500 models the design of the engine air intake system requires the carburetters to be pressure balanced by venting the float chambers into the carburetter intakes.
   To prevent hot starting problems which could be caused by a build-up of vaporised fuel in the venting system, the vent pipe is fed via a canister filled with activated charcoal. During hot standing conditions, vapour is adsorbed by the charcoal, then drawn into the engine during running and burned in the combustion chamber.
   Operation 19.25.10 applies.

g. **Fan, engine cooling system**

   To ensure adequate cooling of the air conditioning condenser and the engine coolant radiator, a high performance fan is used. However, to prevent the fan causing excessive engine load at high speed a viscous coupling is incorporated which limits the fan speed to approximately 2,500 rev/min.
   Operation 26.25.01 applies.

continued

Rover 3500 and 3500S Manual AKM 3621

AIR CONDITIONING

h. Water pump

The design of the water pump has been modified in order to accept the viscous coupling for the fan drive.
Operation 26.50.01 applies.

j. Fuel pump

The AC mechanical fuel pump, which is normally mounted on the front left-hand side of the engine, is replaced by a Bendix electrical fuel pump, mounted beneath the luggage boot floor. This obviates the possibility of fuel vaporisation which could result from the slightly higher engine running temperatures, imposed by the additional load of driving the air conditioning compressor.
Operation 19.45.05 and 19.45.08 are applicable.

k. Fuel reserve switch

The fuel reserve switch is relocated alongside the console unit. Operation 19.20.20 applies after allowing for the relocation.

l. Electrical circuit

The basic electrical circuit for the car is the same as standard models, and is included in Division 86. An additional circuit diagram covering the air conditioning equipment in included in this division.

m. Body

In order to comply with certain safety regulations, all four doors incorporate strengthening struts.
Operation 76.34.18 applies.

AIR CONDITIONING

11. FAULT DIAGNOSIS

Introduction
The following charts cover general fault diagnosis from symptoms encountered in the passenger compartment, and continue with pressure check diagnosis for suspected faults in the refrigeration system.
If a fault is suspected in the refrigeration system, it is recommended that the test procedure, 82.30.16, is carried out. Symptoms that are apparent during this test can be located in this section, and the appropriate additional tests and remedies performed.

i. General fault diagnosis

Control position	SYMPTOM	CAUSE	CURE
RH control 'OFF' LH control 'MAX COOLING'	1. Cold air leak to foot level	Fresh air flap open: (a) Control not venting (b) Linkage between vacuum motor and flap valve jammed (c) Vacuum pipes crossed at control	Change control Remove unit and rectify Remove control, check pipes
	2. Blower motor operates	(a) Switch broken	Change RH control
RH control 'LOW' LH control 'MAX COOLING'	3. Blower motor does not run	(a) Fuse blown (b) Wire off switch terminal (c) Resistors failed (d) Lead not connected to motor inside unit (e) Switch failed	Find reason and change fuse Remove switch, replace wire Remove unit and replace resistors Remove unit and replace lead Replace RH control
	4. Blower motor runs at high speed	(a) Switch wiring incorrect (b) Check harness connections all round for crossing leads	Check and rectify
	5. Unit runs but rattles and/or delivers no air	(a) Motor faulty (b) Runner has come off motor spindle (c) Motor running backwards	Remove unit and replace motor Remove unit, replace and tighten runner Reverse motor leads
	6. Unit delivers uncooled air	Compressor not engaged: (a) Clutch defective (b) Wiring fault in switch or clutch (c) Thermostat contacts failed (d) Vee-belt broken, or slack (e) High pressure switch failed (f) Restriction in pipe (g) Compressor faulty (h) Condenser dirty or blocked (HP switch cuts out) (i) Blocked expansion valve (HP switch cuts out) (j) Expansion valve capillary damaged (k) Capillary not touching evaporated outlet pipe Compressor engaged: (l) Complete loss of charge (m) Low charge due to leak (look for bubbles in sight glass) (n) Liquid receiver connected wrong way round	Replace Trace and rectify Remove unit and replace Replace or tighten Replace Locate restriction and change pipe, re-charge Change and re-charge Clean or replace Change valve and re-charge Change valve and re-charge Tighten spring clip Find leak and re-charge Find leak and re-charge Reverse
RH control 'MED'	7. Blower motor runs at high or low speed	(a) Switch faulty (b) Incorrect connections	Change RH control Check and rectify
	8. Unit delivers uncooled air	As 6 (a) to (m)	
RH control 'HIGH'	9. Blower motor runs at low or medium speed	As 7 (a)/(b)	
	10. Unit delivers un-cooled air	As 6 (a) to (m)	

Rover 3500 and 3500S Manual AKM 3621

AIR CONDITIONING

Control position	SYMPTOM	CAUSE	CURE
LH control 'MAX' cooling RH control 'HIGH' for rest of checks	11. Unit delivers un-cooled air	As 6 (a) to (m)	
	12. Cooled air is not discharged at centre face outlet, but is discharged in quantity at feet level	(a) No vacuum to mode change valve (b) No vacuum, pipe failed	Check piping to control and valve (remove unit) Replace pipe
	13. Air not cooled to maximum, also fumes enter car	Fresh air valve is open: (a) Piping incorrect	Check piping
Rotate LH control through cooling mode	14. Air remains at maximum coldness	(a) Thermostat not working (b) Thermostat vacuum motor leaking (c) Thermostat vacuum not connected at control or motor (d) Vacuum pipes crossed at control (e) Capillary has become detached from evaporator fins	Replace Remove unit Check and replace Check and rectify Remove unit, replace
Rotate LH control into heating mode	15. Very cold air comes out of face level centre outlet	Mode valve not working: (a) Crossed vacuum pipes (b) Flap stuck in 'cold' position	Remove control, rectify Remove unit and rectify
	16. Very cold air comes out of foot level outlets	Water valve not working: (a) Vacuum pipe off valve (b) Water valve broken (c) Vacuum pipe off control	Replace Remove unit and replace Remove control and replace
	17. Air is diverted to screen before 'DEFROST' position is reached	(a) Defrost valve motor leaking (b) Vacuum pipes crossed	Replace motor Remove control and rectify
	18. Air does not increase in temperature as control is rotated clockwise	Water valve not opening: (a) Vacuum pipe off valve or control (b) Water valve faulty (c) Control has vacuum leak. Mode valve closing (d) Mode valve motor leaking (e) Water valve capillary not in position (f) Water hose leaking	Trace and rectify Remove unit and change valve Change control Remove unit and change motor Remove unit, assemble capillary correctly Trace leak. Tighten clip or replace hose
	19. Air temperature increases to maximum instantly as control is rotated slightly clockwise	Water valve opening fully and not metering: (a) Control applying full vacuum (b) Water valve broken	Change control Remove unit and replace valve
	20. During normal heating, windows mist up	Compressor has disengaged: (a) Compressor clutch faulty (b) Wiring fault, switch or clutch (c) Vee-belt broken or slipping (d) High pressure switch failed (e) Condenser blocked	Replace clutch Trace and rectify Replace or adjust Replace switch Clean
Rotate LH control into 'DEFROST' position	21. Airflow not diverted to screen	Defrost valve not closing: (a) Linkage sticking (b) Vacuum pipes crossed (c) Vacuum motor faulty	Remove unit and adjust flap valve Remove control and rectify Replace motor
	22. Air temperature gradually decreases	Leaking water valve: (a) Control venting (b) Water valve faulty (c) Vacuum supply fails (d) Leaking hose	Change control Remove unit and change water valve Replace vacuum tank or pipe as necessary As 18 (f)

AIR CONDITIONING

ii. Refrigeration system fault diagnosis

SYMPTOM	CAUSE	TEST AND CURE

A—COMPRESSOR DISCHARGE PRESSURE TOO HIGH	1. Engine overheated
	2. Condenser air flow blocked
	
	3. Discharge service valve partially closed
	4. Restrictions in high pressure flexible connections, pipes or components
	5. Too much lubricating oil in system
	6. Overcharge of refrigerant
	
	7. Air in system
	
	8. Restriction at expansion valve
	
	9. Restriction in the condenser
	10. Restriction in receiver-drier
	11. Compressor suction pressure too high
		1. Check engine cooling system
		2. Thoroughly clean the condenser and radiator surfaces, including the space between them
		3. Fully backseat the discharge service valve
		4. Evident to the touch by cold spot at the restriction. Remove, inspect, clean or replace as necessary. See items 8, 9 and 10
		5. Check compressor oil level. 82.10.14
		6. Discharge some refrigerant by opening the high pressure gauge valve
		  a. If discharge pressure drops slowly it indicates excessive refrigerant. Depressurise with system working until bubbles appear in sight glass, then add 4 fluid ounces of refrigerant
		  b. If discharge pressure drops rapidly it indicates air and possibly moisture in the system. See item 7
		  c. If discharge pressure remains high there is a restriction in the high pressure side of the system. See items 3, 8, 9 and 10
		7. Pour cold water on the condenser. If there is excessive refrigerant the discharge pressure will momentarily fall. If the pressure remains high suspect air in the system. Evacuate the system. 82.30.06 and recharge with refrigerant
		8. Feel the valve body and pipes with the hand. The inlet pipe to the evaporator should feel warm, whereas a restriction will cause a cold line. Depressurise the system, remove and flush the expansion valve
		9. Remove, inspect, clean or replace condenser
		10. Replace receiver-drier
		11. See 'Suction Pressure too high', Symptom E

B—FLUCTUATION OF DISCHARGE PRESSURE	1. Moisture in expansion valve	1. Alternate high and low readings of the high pressure gauge, together with a low reading of the compound gauge, indicate moisture in the expansion valve, causing internal icing. Depressurise the system, remove, flush and dry out the expansion valve. Replace the receiver-drier. Evacuate the system for at least 20 minutes. Recharge with refrigerant and re-test
C—COMPRESSOR DISCHARGE PRESSURE TOO LOW	1. Insufficient refrigerant because of low initial charge or leak in system 2. Compressor defective 3. Compressor suction pressure too low	1. Check the sight glass for bubbles or foam after five minutes running. If bubbles persist check the system for leaks. 82.30.09 If no leaks are evident, add refrigerant until sight glass clears then add additional 4 fluid ounces 2. Slowly front seat the suction valve—fully clockwise. If vacuum is applied too quickly the residual oil may be drawn out. The compound gauge should quickly record a high vacuum. If not, the compressor valves, head plate, pistons, etc. are suspect. Remove and examine compressor. 82.10.20 and 82.10.26. Repair or replace 3. See 'Suction pressure too low', Symptom E

Rover 3500 and 3500S Manual AKM 3621

AIR CONDITIONING

SYMPTOM	CAUSE	TEST AND CURE
D—COMPRESSOR CYCLES TOO FREQUENTLY	1. High heat load system overloaded 2. Intermittent electrical fault in compressor supply 3. High pressure switch faulty	1. High pressure cut-out switch operating as normal, causing the compressor to cut out. Reduce the heat load and/or improve condenser cooling 2. Check electrical wiring, relay, switches, etc. 3. Operate the system under high heat load condition with the charging and testing equipment fitted, temperature control and blower at maximum, and ventilation to the condenser restricted. Discharge pressure should rise to $19{,}3 \pm 0{,}7$ kgf/cm² (275 ± 10 lbf/in²) before the high pressure switch operates. Cut-in pressure should be 3,5 kg (50 lbs) below cut-out pressure. Replace switch if operating below these figures
E—SUCTION PRESSURE TOO HIGH	1. Thermal bulb of expansion valve loose on evaporator outlet, or capillary tube broken 2. Expansion valve faulty 3. Overcharge of refrigerant	1. The expansion valve will show signs of frosting. Clean the contact areas and refit the capillary tube or replace the expansion valve if the capillary is broken 2. If this is combined with a low reading of the high pressure gauge and the sight glass is clear, replace the expansion valve 3. See Symptom A, item 6
F—SUCTION PRESSURE TOO LOW	1. Restriction in high pressure lines, flexible connections or components 2. Suction service valve partially closed 3. Leak in system 4. Capillary tube of expansion valve broken or faulty 5. Insufficient refrigerant	1. Evident to the touch by cold spot at point of restriction. See Symptom A, items 3, 4, 8, 9 and 10 2. Fully backseat the suction service valve 3. See Symptom C, item 1, 82.30.09 4. Wrap the capillary tube with a warm hand. This should cause the expansion valve to flood and give a high suction pressure. If pressure does not rise, replace the expansion valve 5. See Symptom C, item 1

12. Service Depots

The air conditioning compressor is manufactured by York Division of Borg-Warner Limited, and they and their Distributors provide a range of servicing facilities for refrigeration equipment. Therefore, if additional advice or service is required, it may be obtained by contacting.

A. United Kingdom and Eire

YORK DIVISION OF BORG-WARNER LIMITED
North Circular Road, London NW2 7AU, England
Tel: 01-452 5411

Factory
Gardiners Lane South, Basildon, Essex
Tel: Basildon 22231

continued

AIR CONDITIONING

B. Overseas

ARABIAN GULF
Bahrain .. .. Y. K. Almoayyed & Sons
P.O. Box 143
Cable address: ALMOAYYED
Dubai .. .. General Enterprises Co.
P.O. Box 363
Cable address: GECO

ARGENTINA
Buenos Aires .. Industrias Tecnicas Aire S.A.I.C.
Calle Peru 1472
Cable address: PETRAR
Buenos Aires .. Est. Metalurgicos Crespo, S.A.
Crespo 2924/26
Cable address: CRESPO 2924/26

AUSTRALIA
Gladesville, N.S.W. York Division, Borg-Warner (Australia) Ltd.
P.O. Box 103, Gladesville, NSW 2111
Cable address: YORKAIR-SYDNEY

AUSTRIA
Vienna 1 .. .. Climaco Ges. M.B.H.
Postgasse 16
Telex No. 07-4789

BANGLADESH
Dacca, 3 .. .. Shahnawaz (Bangladesh) Ltd.
9/C Motijheel

BELGIUM
Brussels 5 .. Van Den Bosch S.A.
92 Rue Lesbroussart
Cable address: BOSCHAUFFAGE
Brussels 7 .. Dehaes S.A.
146 Rue Heyvaert

BERMUDA
Hamilton .. Appliance Services Ltd.
Church & King Streets
P.O. Box 962
Cable address: APSERCO

BRAZIL
Sao Paulo .. Tecfril S/A
Industria e Comercio
Rue Corrientes 130
Lapa
Cable address: TECFRIL

BURMA
Rangoon .. .. Rangoon Refrigeration Service
P.O. Box 367
Cable address: DEEPFREEZE

CAMBODIA
Phnom-Penh .. Comin Khmere
P.O. Box 625
Cable address: COMINK

CANADA
Rexdale, Ontario.. York Division of Borg-Warner (Canada) Ltd.
326 Rexdale Boulevard

CEYLON
Colombo .. .. Colombo Commercial Co. (Engineers) Ltd.
P.O. Box 1191
Cable address: EXTRACTOR

COLOMBIA
Bogota 1 .. .. Gamko Ingenieros
Apartado Aereo 29629

CONGO
Democratic
Republic of
Kinshasa .. .. Erce
P.O. Box 80-23

COSTA RICA, C.A.
San Jose .. .. Construcciones Industriales Ltda.
P.O. Box 433
Cable address: COINSA

CURACAO, N.A.
Willemstad .. United Agencies Inc.
P.O. Box 779
Cable address: UNIGENCIES

CYPRUS
Nicosia .. .. Vahan Karaian
P.O. Box 1189
Cable address: VAHAN KARAIAN

CZECHOSLOVAKIA
Calp Establishment
Hauptstrasse 26
Vaduz, F Liechtenstein
Cable address: CALP
Address all correspondence concerning Czechoslovakia to Calp in Liechtenstein.

DENMARK
Glostrup 2600 .. IWO A/S
Roskildevej 527
Cable address: SKANFRYS

DOMINICAN REPUBLIC
Santo Domingo .. Santos Dalmau, S.A.
Apartado 210
Cable address: SADASA

ECUADOR
Guayaquil .. Teague Associates Del Ecuador
P.O. Box 6516
Cable address: TEAGUE

EL SALVADOR, C.A.
San Salvador .. Refrigeracion e Industria, S.A.
Apartado Postal 811
Cable address: REISA

FINLAND
Helsinki 30 .. Oy Morus AB
P.O. Box 30017
Cable address: MORUS

FRANCE
44 Carquefou
(L-A) .. .. L. F. I. Brissonneau-York, S.A.
Zone Industrielle de Nantes
Carquefou, Boite Postale 10
Cable address: FIBY
Paris XVII .. 11 Rue Laugier
Cable address: FIBY
La Corneuve .. Froid SATAM Neve
(Seine) Boite Postale 96
Cable address: FRIDUSTRIE

continued

Rover 3500 and 3500S Manual AKM 3621

AIR CONDITIONING

GHANA
Accra .. .. Caramafra Ltd.
P.O. Box 2400
Cable address: CARAMAFRA

Accra .. .. Hoeks (Ghana) Ltd.
F800/1 Cantonments Road
P.O. Box 1888
Cable address: HOEKS

GREECE
Athens 301 .. Klimapsyktiki Ltd.
Iera Odos 79
Cable address: KLIMAPSY

GUATEMALA, C.A.
Guatemala City .. Marwilco
Apartado 1524
Cable address: CAMWI

HONDURAS
San Pedro Sula .. Gases Industriales, S.A.
P.O. Box No. 214
Cable address: GISA

HONG KONG
The Jardine Engineering Corporation Ltd.
P.O. Box 517
Cable address: JARDENG

HUNGARY
Calp Establishment
Haupstrasse 26
Vaduz, F. Liechtenstein
Cable address: CALP

Address all correspondence concerning Hungary to Calp in Liechtenstein.

ICELAND
Reykjavik .. Samband Isl. Samvinnufelaga
P.O. Box 180
Cable address: SIS/50

INDIA
Faridabad .. York India Limited
(Haryana) Mathura Road
Cable address: YORKINDIA

INDONESIA
Djakarta .. .. P.T. Jaya Teknik Indonesia
Dje. K. H. Wahid Hasjim 114A
Cable address: TEKINDPT

IRAN
Teheran .. .. Techno Frigo Corporation
280 Iranshahr Avenue
Cable address: TECFRI

ISRAEL
Tel-Aviv .. .. Mashav
91 University Street
Ramat-Aviv

Tel-Aviv .. .. A.S.K.-Kor Ltd.
P.O. Box 25046

ITALY
Milan .. .. Officine di Seveso, Soc.p.Az.
P.O. Box 1308
Cable address: SEVESINT

JAMAICA, W.I.
Kingston 11 .. Geddes Refrigeration Ltd.
P.O. Box 49
Cable address: GEDREFRIG

JAPAN
Tokyo .. .. Mitsubishi-York Ltd.
Mitsubishi-Juko Building
5-1 Marunouchi, 2-Chome,
Chiyoda-ku
Cable address: HISHIYORK

JORDAN
Amman .. .. Supplies & Contracts Co. Ltd.
P.O. Box 748
Cable address: SAC

KOREA (South)
Seoul .. .. Hanjin Transportation Co. Ltd.
Samsung Building
Room 1001-10
P.O. Box 389 Central
Cable address: HJTRANSCO

KUWAIT .. .. A. H. Alsagar & Brothers
P.O. Box 244
Cable address: ALSAGAR

Modern Lights Co.
P.O. Box 4648
Cable address: MEZERLIGHT

LEBANON
Beirut .. .. International Traders, S.A.L.
P'O' Box 473
Cable address: INTRACAR

Beirut .. .. Ranec Engineers & Contractors
P.O. Box 4185
Cable address: RANEC

Beirut .. .. Borg-Warner International Corp.
P.O. Box 5246
Cable address: YORKAIR

LIBERIA
Monrovia .. .. Technico-Auriole Engineering Co.
P.O. Box 1336
Cable address: TECHNICO

LIBYA
Tripoli .. .. Libya Desert Operations Co.
P.O. Box 491
Cable address: LIDO

MALAYSIA (States of Malaya)
BRUNEI
Brunei Town .. Guthrie-Waugh (Malaysia) Sdn. Bhd.
P.O. Box 77
Kota Kinabalu
Cable address: GAWAIR

MALAYA
Petaling Jaya .. Guthrie-Waugh (Malaysia) Sdn. Bhd.
West Malaysia 19 Jalan Semangat
P.O. Box 30
Cable address: GAWAIR PETALING JAYA

SABAH .. Guthrie-Waugh (Malaysia) Sdn. Bhd.
P.O. Box 77
Kota Kinabalu
Cable address: GAWAIR

SARAWAK .. Guthrie-Waugh (Malaysia) Sdn. Bhd.
P.O. Box 401
Cable address: GAWAIR

MARTINIQUE .. Ets. Maurice Larrouy
Boite Postal 296
Fort de France

MEXICO
Mexico City 4, D.F. Recold de Mexico S.A. de C.V.
Apartado Postal 32-930

Monterey, N.L. .. York Aire, S.A.
Apartado 1900

NETHERLANDS
Amersfoort .. I.B.K. — Nederland N.V.
Postbus 138
Cable address: IBEKA

continued

AIR CONDITIONING

NEW ZEALAND
Auckland, C.1 .. James J. Niven & Co. Ltd.
P.O. Box 5543
Wellington, C.1 .. P.O. Box 2096
Cable address: NIVENCO

NIGERIA
Apapa .. .. Holt Engineering
P.O. Box 217
Cable address: HOLTENG
Ibaden .. .. Holt Engineering
Technical Sales Dept.
P.O. Box 47
Warri .. .. John Holt Mid-West Ltd.
P.O. Box 1
Cable address: HOLTS MIDWEST

NIGERIA, East
Port Harcourt .. John Holt Limited
P.O. Box 8
Cable address: EBANI

NORWAY
Oslo 2 .. .. Lehmkuhl Kjøling (A/S Lehmkuhl)
P.O. Box 145
Cable address: LEHMCOLD

PAKISTAN
Karachi 2 .. .. Shahnawaz Limited
West Wharf Road
P.O. Box 4766
Cable address: AUTOSHEZ

PANAMA, Republic of
Panama 9A .. Aircond, S.A.
P.O. Box B2
Cable address: HOJALATERIA

PERU
Lima .. .. W. R. Teague Associates
Av. Bolivar 720
Orrantia
Cable address: TEAGUE

PHILIPPINES
Manila .. .. Atlantic, Gulf & Pacific Co. of Manila Inc.
P.O. Box 626
Cable address: DREDGING

PORTUGAL
Lisbon 6 .. .. Metalurgica Luso-Italiana S.A.R.L.
Av. Marechal Gomes da Costa, N15-15A
Cable address: LUSITALIANA

PUERTO RICO
00910 San Juan .. Abarca Warehouses Corporation
P.O. Box 2352
Cable address: ABARCA
00923 Rio Piedras York International
Borg-Warner Corporation
Condominium Green Village
Apt. A-1
Calle Jose de Diego 472
Cable address: YORKAIR

SAUDI ARABIA
Jeddah .. .. E. A. Juffali & Bros.
P.O. Box 1049
Cable address: JUFFALICENT

SINGAPORE, Republic of
Singapore 10 .. Guthrie-Waugh (Singapore) Pte. Ltd.
P.O. Box 495
Cable address: GAWAIR
Singapore 9 .. York International
P.O. Box 179
Killiney Road Post Office
Cable address: YORKAIR

SOUTH AFRICA, Republic of
Johannesburg .. Mining & Industrial Air
Conditioning (Pty) Ltd.
P.O. Box 9977
Cable address: HAVERCROFT

SPAIN
Hospitalet del Refracsa
Llobregat .. Carretera del Medio 48
Barcelona Cable address: REFRIGERACION
Gava .. .. Clima-Roca
Prov. Barcelona Rambla Lluch, 2
Cable address: CLIMAROCA

SWEDEN
Goteborg .. AB Lehmkuhl
P.O. Box 90
433 01 Partille 1
Cable address: LEHMGOT

SWITZERLAND
Chatelaine, Geneva Technicair S.A.
Avenue de L'Etang 53
Cable address: TECHNICAIR
Geneva .. .. Borg-Warner International
York Division
21 Chemin Francois Lehmann
1218 Grand-Saconnex

TAIWAN
Taipei .. .. Silo Corporation
P.O. Box 3301
Cable address: SILOCO

THAILAND
Bangkok .. .. The Ua Withya Air Conditioning &
Refrigeration Co. Ltd.
484 Rama 1 Road
Prathumwan
Siam Square SOI 6
Cable address: UAAIR
Bangkok .. .. Jardine Waugh Ltd.
P.O. Box 40
Cable address: WAUGHCO

TUNISIA
Tunis .. .. Le Confort
13 Avenue de Carthage
Cable address: CONFORTU

TURKEY
Istanbul .. .. Ekrem Akomer ve Ortaklari Tic Ltd., Sti.
Buyukdere Cad. Otobus Garaji yani
Hukukcular Sitesi, Mecidiyekoy
Cable address: EKREM

UNITED ARAB REPUBLIC
Cairo .. .. The Tractor & Engineering Co. S.A.E.
P.O. Box 366
Cable address: CROCODILE

UNITED STATES OF AMERICA
York, Pa. 17405 .. York International
P.O. Box 1592

URUGUAY
Montevideo .. Benech Industrial y Comercial S.A.
Casilla de Correo 858
Cable address: BENECHILL

VENEZUELA
Caracas .. .. York Venezuela C.A.
P.O. Box 70010 Los Ruices
Cable address: YORKVEN

WEST GERMANY
Mannheim .. Brown Boveri-York, G.m.b.H.
Postfach 346
6800 Mannheim 1

AIR CONDITIONING

Air conditioning system, description and operation

Air conditioning is available as optional equipment on Rover 3500 and 3500S models, but because it forms a major basic installation it can only be specified during the manufacture of the car.

Air conditioning is provided from an integral heating and cooling unit designed to give a comfortable interior temperature under all climatic conditions with the simplicity of driver control.

The system incorporates a mechanical compression refrigeration circuit, with thermostatic control for air cooling; and for heating, a thermostatically controlled heat exchanger utilising hot water from the engine cooling system. Both circuits are vacuum operated utilising one rotary control which will select any desired temperature, from maximum cooling to maximum heating. A second, electrical/vacuum control, provides an 'off' position and selects any one of three fan speeds. Thus permitting control over air flow through the car interior.

Circuit description—Unit

The heater/cooler units fits into the space normally occupied by the standard heater, i.e. on the engine side of the bulkhead. The unit itself contains the following parts: an electrically driven blower which draws its air from both a recirculatory intake under one foot-well, and a fresh air intake open to the plenum chamber. Under all conditions except maximum cooling which is recirculation only, the blower delivers a mixture of fresh and recirculatory air. All air from the blower passes through the evaporator, where it is cooled and de-humidified, when heating is required it is then directed through the heater radiator. Evaporator temperature is regulated by a variable thermostat also mounted within the unit, having an electrical link with the compressor clutch. The mode change valve immediately in front of the evaporator, diverts cool air, either in the cooling mode through three airways, to the sides and centre

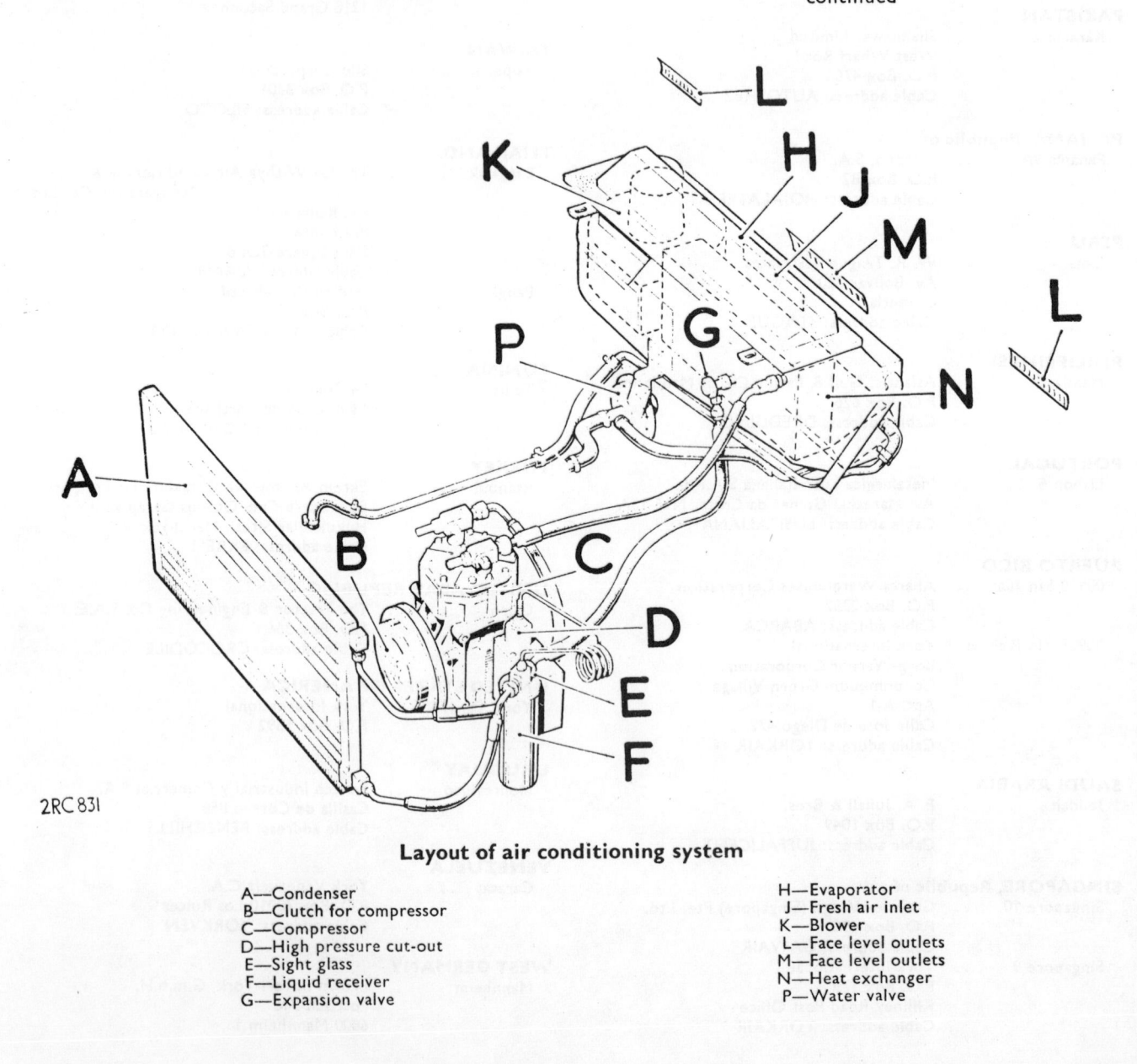

continued

Layout of air conditioning system

A—Condenser
B—Clutch for compressor
C—Compressor
D—High pressure cut-out
E—Sight glass
F—Liquid receiver
G—Expansion valve
H—Evaporator
J—Fresh air inlet
K—Blower
L—Face level outlets
M—Face level outlets
N—Heat exchanger
P—Water valve

82.00.01
Sheet 1

Rover 3500 and 3500S Manual AKM 3621

AIR CONDITIONING

face level outlets; or, in the heating mode through the heat exchanger, in order that hot air may be directed to foot level outlets. This mode change valve has two perforations in its outer edge such as that in the cooling mode, a small quantity of cool air is metered to the foot level outlets, and conversely in the heating mode, a small quantity of air is directed to the side face level outlets. This air remains cool when the air to foot level is warmed; thus a differential between foot and head level temperatures is available. If this condition is not desired, the side outlets may be closed manually.

The heater radiator output is controlled by a metering water valve, mounted on the underside of the unit case. This valve varies the flow of hot coolant and hence the heat transfer according to the position selected on the control, the heat output is then kept at a constant rate by the valve sensing capillary which senses the air temperature leaving the heater radiator. This warmed air is directed to foot level and screen until 'defrost' is selected, in which case the defrost valve closes and forces all air at maximum temperature to the screen.

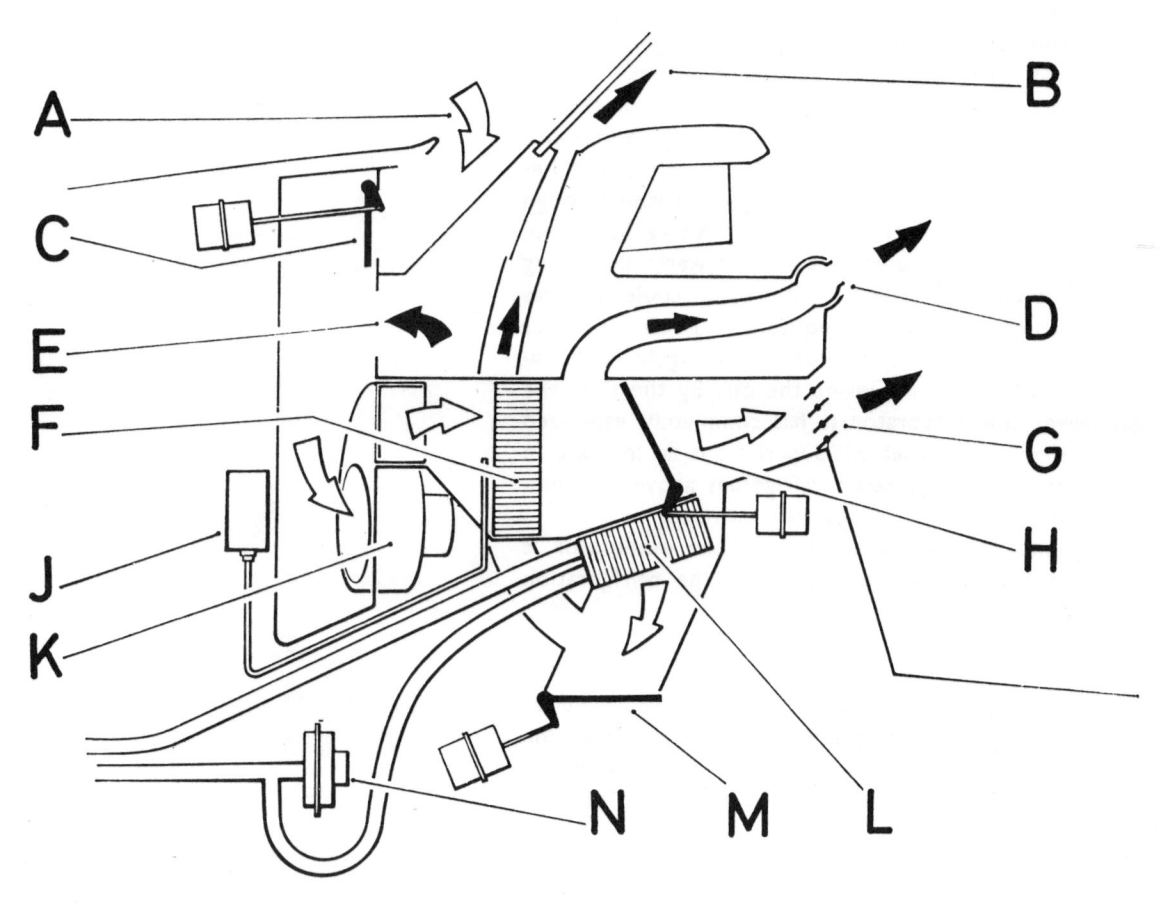

Heater cooler unit general arrangement

A—Fresh air inlet
B—Outlets to windscreen
C—Fresh/recirculation air valve
D—Outlets to side face level
E—Recirculated air
F—Evaporator
G—Outlets to centre face level
H—Mode change valve
J—Thermostat
K—Blower
L—Heater radiator
M—Defrost valve
N—Water valve

Rover 3500 and 3500S Manual AKM 3621

AIR CONDITIONING

Refrigeration cycle

Cool refrigeration gas is drawn into the compressor from the evaporator and pumped from the compressor to the condenser under high pressure. This high-pressure gas will also have a high temperature as a result of being subjected to compression. As it passes through the condenser, the high-pressure high-temperature gas rejects its heat to the outside air passing over the cooling surfaces of the condenser. The cooling of the gas causes it to condense into liquid refrigerant. The liquid refrigerant, still under high pressure, passes from the bottom of the condenser into the receiver-drier. The receiver acts as a reservoir for the liquid.

The liquid refrigerant flows from receiver-dryer to the expansion valve. The valve meters the high-pressure refrigerant flow into the evaporator. Since the pressure in the evaporator is relatively low, the refrigerant immediately begins to boil. As the refrigerant passes through the evaporator it continues to boil, drawing heat from the surface of the evaporator core. The warm air passing over the evaporator rejects its heat to the cooler surfaces of the evaporator core. Any moisture in the air condenses on the cool surface of the core, resulting in cool, dehumidified air entering inside the car. By the time the gas leaves the evaporator it has completely vaporised and is slightly superheated. Superheat is an increase in temperature of the gaseous refrigerant above the temperature at which the refrigerant vaporised.

Refrigerant vapour passing through the evaporator is returned to the compressor, where the refrigeration cycle is repeated.

Heating cycle

Heating follows conventional car heater practice in that hot coolant is tapped from the cylinder head galleries before the engine thermostat and impelled by the engine circulation pump into the thermostatic water valve which in turn meters it into the heater radiator, after which it is returned to the engine block.

Operation

Controls

The system is controlled by two switches, an 'air flow' switch marked 'off', 'LO', 'MED' and 'HI', and a temperature switch marked 'Cooling', 'Heating' and 'Defrost'. The air flow control operates the compressor clutch and blower motor electrically, and has a vacuum override function which ensures that the fresh air flap is closed in the 'off' position when no airflow is desired, regardless of the position of the temperature control.

The temperature control is vacuum only, and diverts vacuum to give closure of the mode valve and defrost valve at the appropriate positions. It also modulates vacuum depression over a range 100 mm to 300 mm (4 in to 12 in) Hg to produce a progressive movement of the thermostat and the water valve diaphragm giving fine control of cooling and heating.

The following chart gives an indication of the functions performed by the valve actuators related to angular movements of the control.

AIR CONDITIONING

LH control (temperature) from 'cold' (clockwise rotation)	RH control (air flow) from 'off' (clockwise rotation)	Function
	0°	No electric current, compressor clutch disengaged, fresh air valve closed against incoming ram air.
0°		Fresh air valve closed—no vacuum supply to RH control.
	45° (low)	Compressor engaged, blower motor running at lowest speed. System is now delivering recirculatory air to face level at lowest possible temperature. Fresh air valve 'unlocked' by RH control, but held closed by LH control.
	90° (med)	Blower motor runs at medium speed, airflow increased.
	135° (high)	Blower motor runs at high speed, airflow at maximum.
At 7½°		Fresh air valve opens under full vacuum depression, system now operates on mixture of fresh and recirculatory air, at low temperature. Airflow mainly to face level, but small flow to feet.
From 7½° To 127°	At (low) (med) or (high) depending on airflow desired	Increasing modulated vacuum applied to evaporator thermostat through actuator, thus progressively decreasing degree of air cooling up to an air temperature of approximately 60°F.
At 132° At 135° At 138° (These functions take place simultaneously)		Modulated vacuum applied to water valve to admit coolant to heat exchanger, in readiness for change to heating mode. Vacuum is released from actuator, thus lifting mode change valve to uncover heat exchanger. Airflow is now mainly to foot level, but perforations in valve allow reduced flow of cold air to side face level vents only. Vacuum is released from compressor thermostat. Evaporator temperature stabilises at maximum coldness for de-humidification. Air passage through heater unit however produces overall effect of mild heating.
143° to 270°		Increasing modulated vacuum is applied to water valve flow of coolant through heat exchanger increases, and air temperature rises progressively. Water valve capillary monitors flow to maintain constant unit outlet temperature at any selected position.
270°		Water valve is fully open, unit is delivering air at maximum heat to foot level.
280° to 290°	Select 135° for max	Vacuum is released from actuator to defrost valve which then closes foot outlets and diverts all hot air output to screen for de-icing.

Rover 3500 and 3500S Manual AKM 3621

AIR CONDITIONING

Control circuits—vacuum

The illustration shows a line diagram of the vacuum circuit. The vacuum reservoir provides sufficient capacity to ensure adequate vacuum for all operating conditions, and is evacuated by induction manifold depression through a non-return valve. All vacuum tubes have lettered sleeves at each end which correspond to letters cast on to the back plate of the controls.

The four vacuum actuators (motors) are mounted within the unit which must be removed from the vehicle if replacement is required; the water valve, which is mounted outside the unit has its actuator built in; but unit removal is still necessary due to the capillary clamped to the heater radiator.

Control circuits—electrical

The function of the electrical circuit is to switch the compressor 'on' or 'off' and to provide three fan speeds for variation of air flow. The supply to the switch is through a 35 amp fuse on 3500 models, and a 15 amp fuse on 3500S models. Switching from 'off' to 'low' energises the compressor clutch (this current may be interrupted by (a) the compressor thermostat according to control setting and evaporator temperature, or (b) the high pressure switch, which will stop current supply to the clutch should compressor pressure rise above 125 kgf/cm^2 (275 lbf/in^2), or (c) the compressor isolating switch), it also energises the blower motor.

Current to the blower motor passes two 0.65 Ω resistors in 'low' position one 0.65 Ω resistor in 'med' position and is unmodified in 'high' position giving full motor speed.

Note:

An independent isolating switch is fitted between the control and compressor in order that the driver may switch off the compressor when neither cooling nor de-humidification is desired to reduce load on car engine but the compressor must be run at least once a month for lubrication purposes.

AIR CONDITIONING

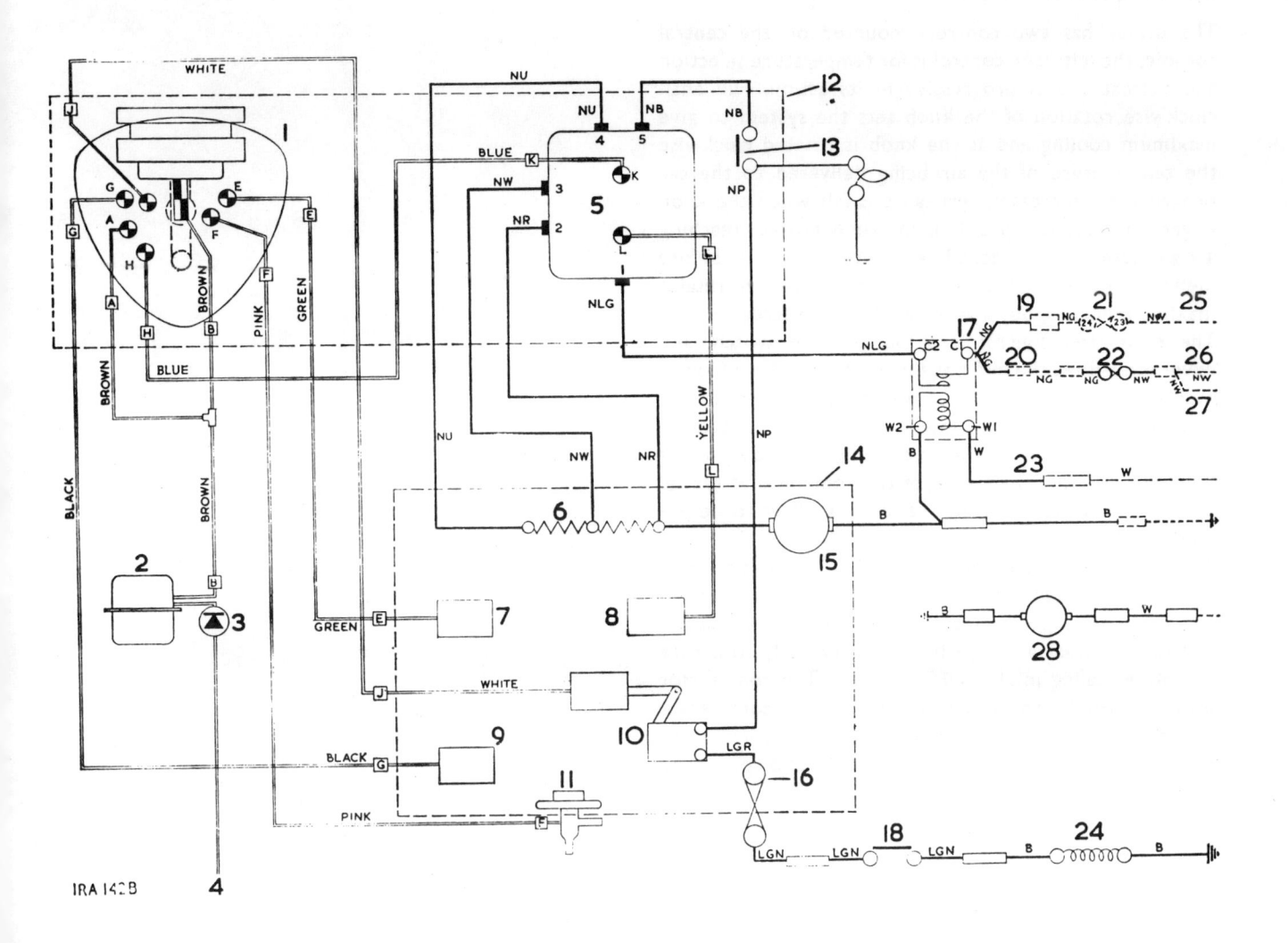

Key to circuit diagram, air conditioning

Double light lines represent vacuum pipe connections   Single heavy lines represent electrical circuit

Dotted lines denote existing parts

1. Temperature switch
2. Vacuum reserve tank
3. Non-return valve
4. To inlet manifold
5. Air-flow switch
6. Resistor for blower motor
7. Mode valve
8. Fresh air circulation
9. Defrost circulation
10. Variable thermostat
11. Water valve
12. Switch box on console unit
13. Switch and warning light, air conditioning isolation
14. Heater-cooler unit
15. Blower motor
16. In-line fuse, 10 amp
17. Relay 6 RA
18. Switch, high pressure cut-out
19. Pick-up point on main circuit diagram, 3500S
20. Pick-up point on main circuit diagram 3500
21. Existing fuse 24–23, 15 amp at fuse box 3500S
22. In-line fuse, 35 amp, 3500
23. Pick-up point on main circuit diagram
24. Electro magnetic clutch
25. To the ammeter, 3500S
26. To the ammeter, 3500
27. To the lighting switch, 3500
28. Electric fuel pump

Key to cable colours

B—Black   G—Green   NG—Brown/Green   Y—Yellow   BW—Black/White   P—Purple   PN—Purple/Brown
PW—Purple/White   PY—Purple/Yellow   RP—Red/Purple   RY—Red/Yellow   UB—Blue/Black   UY—Blue/Yellow
W—White   NW—Brown White

Rover 3500 and 3500S Manual AKM 3621

AIR CONDITIONING

Driver operation

The driver has two controls mounted on the central console, the left-hand control is for temperature selection and defrost and is progressive in its action. Full anti-clockwise rotation of the knob sets the system to give maximum cooling and as the knob is rotated clockwise the temperature of the air being delivered to the car progressively increases, resistance is felt when the knob is vertical. Past this position, the air is heated, reaching a maximum when a second resistance is felt just before 'defrost'. Rotation into the 'defrost' position retains maximum heating but diverts all air to the screen.

The right-hand control is a four-position switch, and governs air-flow, the 'off' position switches off the blower motor and closes all flaps, sealing the system against incoming air; positions 'low', 'med', 'high' represent air flow, 'high' being maximum air delivery. The following illustrations show the layout of the controls and typical settings for various climatic conditions likely to be encountered.

The side face level outlets are manually operated and, from these, cooled air is available under all conditions of control setting. The central face level outlet is adjustable in two directions and may be closed manually to ensure complete sealing in the 'OFF' position. The compressor isolating switch can be used to cut out the compressor under cool dry climatic conditions, but the compressor must be run at least once a month for lubrication purposes.

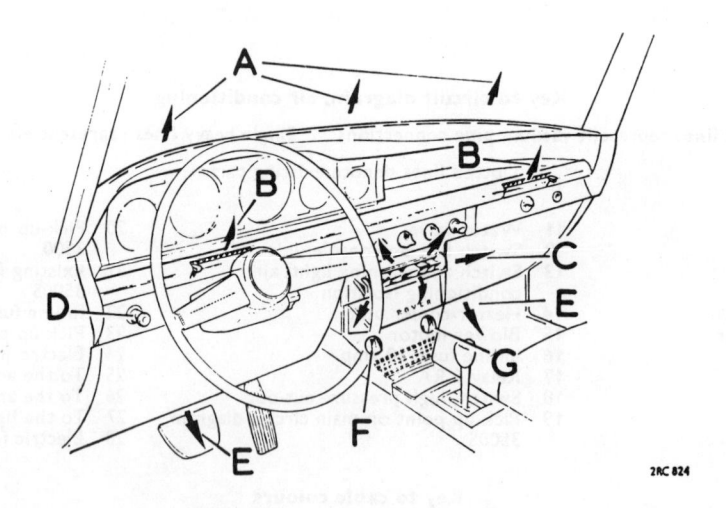

Layout of air conditioning controls

A—Windscreen outlets
B—Side face level outlets
C—Central face level outlet
D—Compressor isolating switch
E—Fixed foot level outlets
F—Temperature control
G—Air-flow control

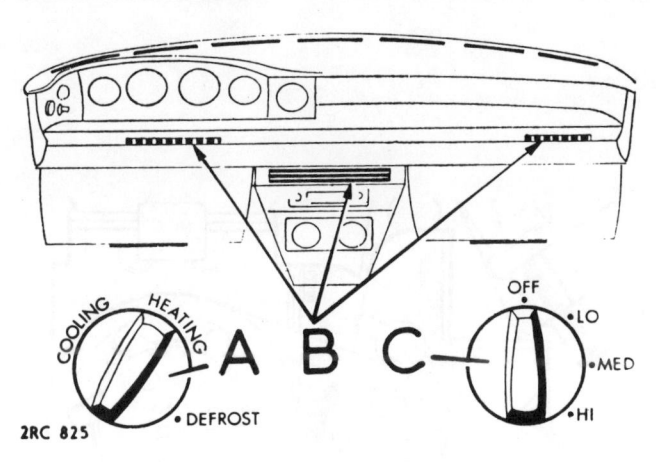

AIR CONDITIONING

Complete system OFF—when no air flow is required: or when driving in a polluted atmosphere

A—Left-hand control as shown, or may be pre-set for temperature in any position
B—Close outlets manually to ensure sealing
C—Right-hand control set to 'OFF'

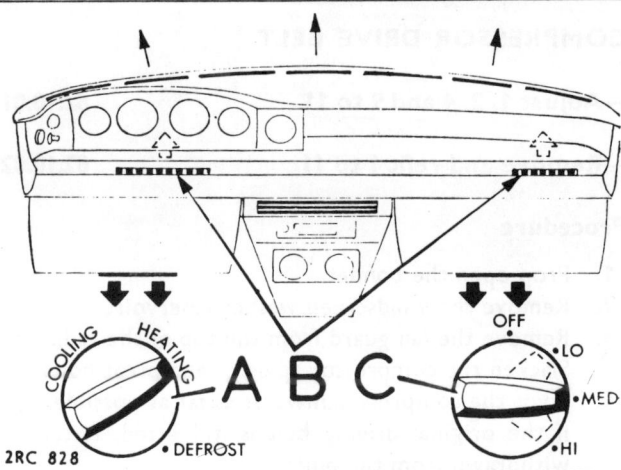

General heating—cool weather, 5° to 15°C (40° to 60°F)

A—Left-hand control, adjust within heating range for desired warmth
B—Open face level outlets manually if a flow of cooled air is required
C—Right-hand control, select airflow, low, medium or high, as required

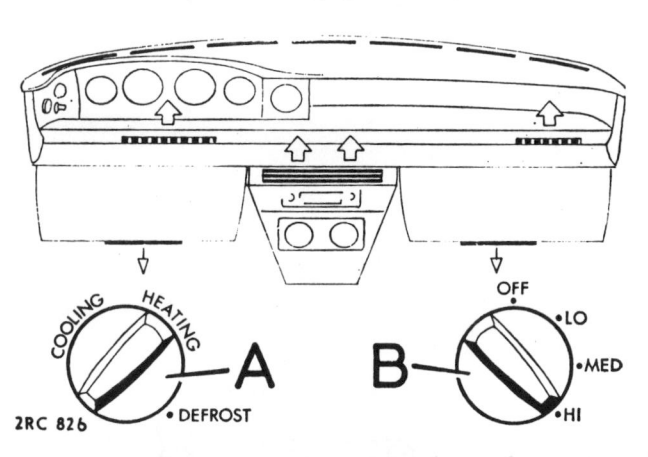

Maximum cooling—very hot weather, 27°C (80°F) and above

A—Left-hand control set in maximum anti-clockwise position
B—Right-hand control as shown, or move to medium or low to decrease air flow

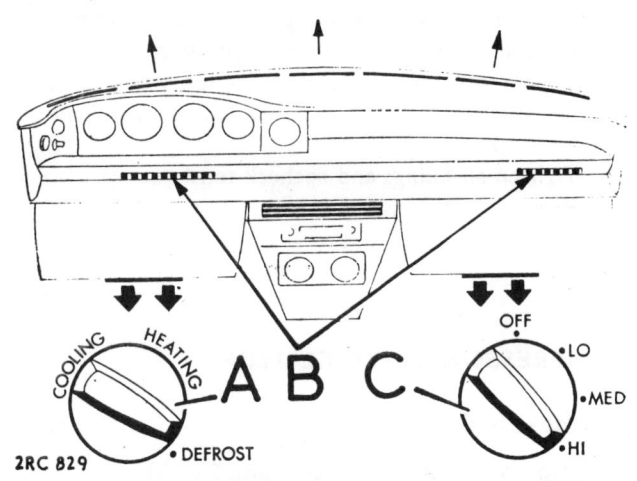

Maximum heating—cold/freezing weather 5°C (40°F) and below

A—Left-hand control, rotate clockwise until resistance is felt
B—Close face level outlets
C—Right-hand control, set to high

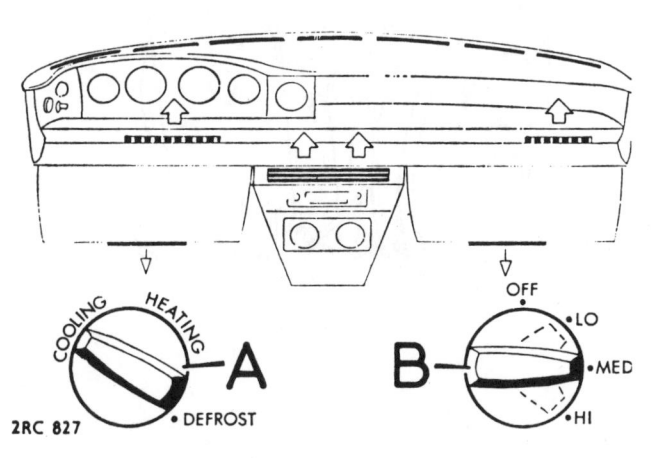

General cooling—Hot/warm weather, 15° to 27°C (60° to 80°F)

A—Left-hand control, adjust within cooling range for desired degree of coldness
B—Right-hand control, select airflow, low, medium or high, as required

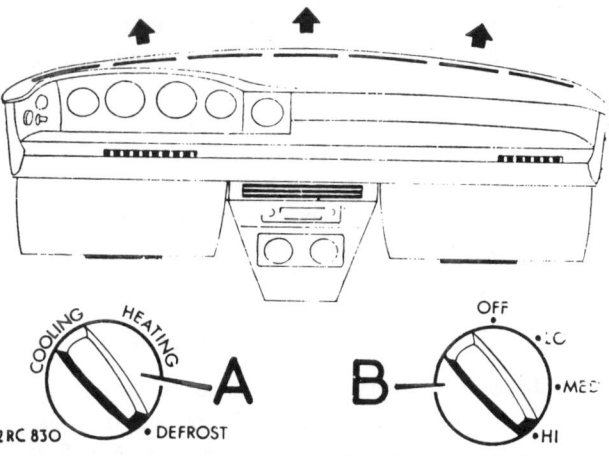

Maximum screen clearance—windscreen coated with snow or ice

A—Left-hand control, rotate fully clockwise
B—Right-hand control, set to high

Rover 3500 and 3500S Manual AKM 3621

AIR CONDITIONING

COMPRESSOR DRIVE BELT

—Adjust 1, 2, 4 and 9 to 11　　　　　　　82.10.01

—Remove and refit 1 to 11　　　　　　　82.10.02

Procedure

1. Prop open the bonnet.
2. Remove the windscreen washer reservoir.
3. Remove the fan guard from the top of the radiator.
4. Slacken the compressor adjuster and pivot bolts.
5. Pivot the compressor inwards as far as possible.
6. If the original driving belt is still fitted, it can be withdrawn from the pulleys.
7. Fit the new driving belt from underneath the car locating it on to the idler and compressor pulleys first, and finally on to the crankshaft pulley.
8. Fit the fan guard to the top of the radiator.
9. Adjust the position of the compressor by means of its pivot and slotted fixing, to give the correct belt tension. The belt must be tight with 4 to 6 mm (0.187 to 0.250 in.) total deflection when checked midway between the compressor and idler pulleys, by hand.
10. Tighten all fixings and recheck tension.
11. Reverse 1 to 3, as applicable.

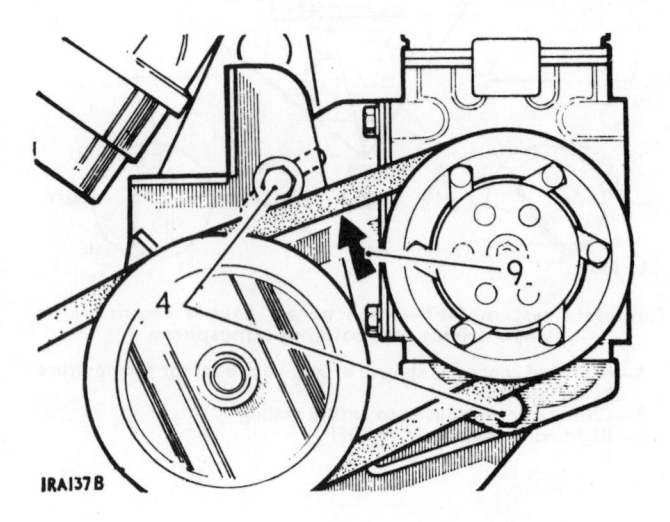

COMPRESSOR DRIVE CLUTCH

—Remove and refit　　　　　　　　　　　82.10.08

　Compressor installed 1 to 3 and 5 to 22

　Compressor removed 4 to 10 and 12 to 18

Removing

1. Remove the radiator block. 26.40.04.
2. Remove the compressor drive belt. 82.10 02.
3. Switch on the ignition and the air conditioning air flow control.
4. Connect the clutch lead to the positive terminal of a twelve volt battery, and connect the compressor crankcase to the battery negative terminal.
5. Remove the bolt and washer securing the compressor pulley to the crankshaft.
6. Screw a $\frac{5}{8}$ in. UNC bolt into the thread provided in the pulley bore to extract the pulley from the crankshaft and remove the extractor bolt.

　　Note: If difficulty is experienced in removing the pulley, an alternative method is available, using a union nut, Rover Part No. 534127, and a 6mm (0.250 in.) diameter bolt by 32 mm (1.250 in.) long. The bolt must be a clearance fit in the bore of the union nut. Proceed items 7 to 10.

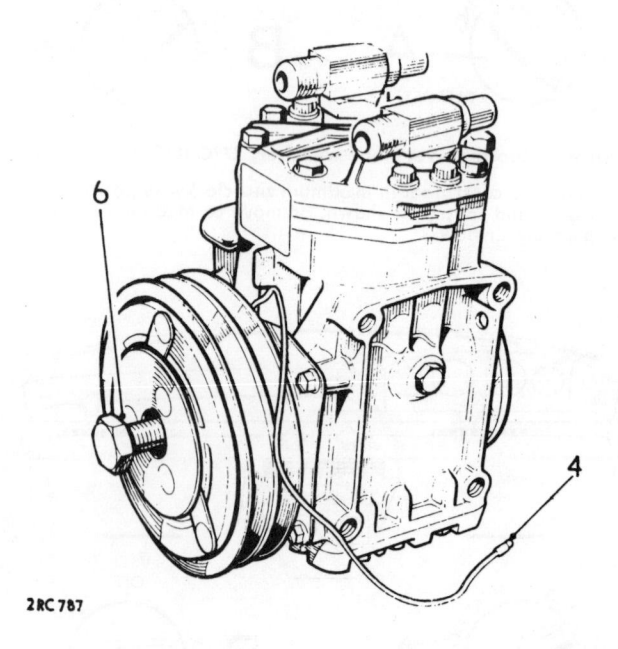

continued

AIR CONDITIONING

7. Assemble the bolt and union nut as illustrated.
8. Screw the assembly into the tapping in the pulley centre.
9. Tighten the union nut to load the bolt head on to the compressor crankshaft end face.
10. Alternately tap with a hammer on the bolt end, screw in the union nut to free the pulley from the tapered seating.
11. Switch off the ignition and air flow control or disconnect the battery from the compressor, as applicable.
12. Disconnect the electrical lead from the compressor clutch at the snap connector.
13. Remove the four bolts and washers securing the drive clutch to the compressor crankcase.
14. Withdraw the compressor drive clutch and base plate assembly.

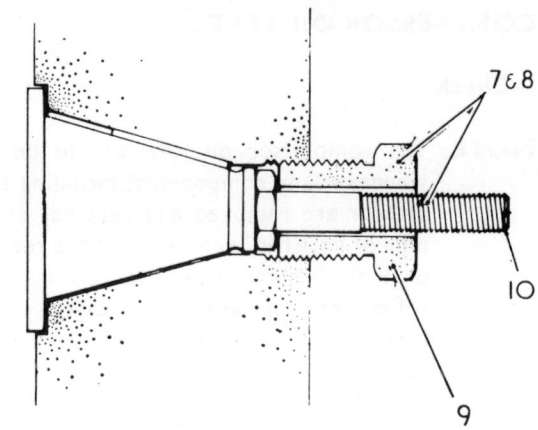

Refitting

15. Secure the clutch and base plate assembly to the compressor crankcase.
16. Connect the clutch lead to the positive terminal of a 12 volt battery, and connect the compressor crankcase to the battery negative terminal. If the compressor is installed in the car switch on the ignition and air conditioning air flow control.
17. With the clutch energised, fit the pulley and tighten the securing bolt to a torque of 2,2 kgf.m (16 lbf ft).
18. Disconnect the battery, or switch off the ignition and air-flow control, as applicable.
19. Reconnect the electrical lead for the compressor clutch at the snap connector on the left-hand wing valance.
20. Refit the compressor drive belt. 82.10.02
21. Refit the radiator block. 26.40.04.
22. Refill the engine cooling system, using the correct mixture. 26.10.01.

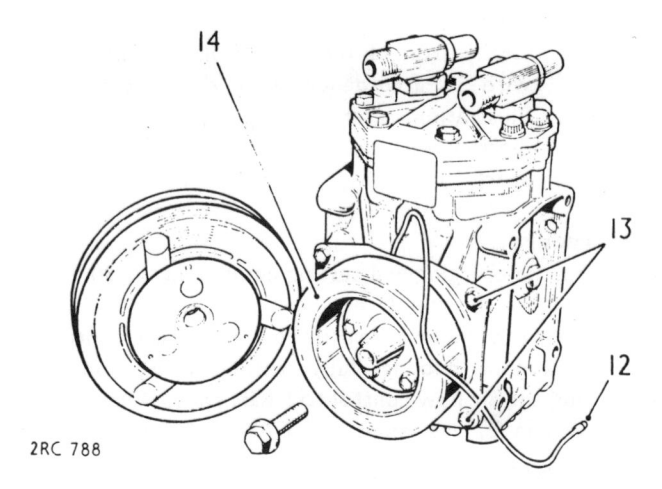

Rover 3500 and 3500S Manual AKM 3621

AIR CONDITIONING

COMPRESSOR OIL LEVEL

— Check   82.10.14

NOTE: The compressor oil level should be checked whenever any components, including the compressor are removed and refitted, or when a pipe or hose has been removed and reconnected or, if a refrigerant leak is suspected.

All compressors are factory charged with 0,28 litre (0.5 UK pint) of oil. When the air conditioning equipment is operated some of the oil circulates throughout the system with the refrigerant, the amount varying with engine speed. When the system is switched off the oil remains in the pipe lines and components, so the level of oil in the compressor is reduced.

The compressor oil level must finally be checked after the system has been fully charged with refrigerant and operated to obtain a refrigerated temperature of the car interior. This ensures the correct oil balance throughout the system.

The compressor is not fitted with an oil level dipstick, and a suitable dipstick must be made locally from 3 mm (0.125 in.) diameter soft wire in accordance with the accompanying illustration. After shaping, mark the end of the dipstick with twelve notches, 3 mm (0.125 in.) apart.

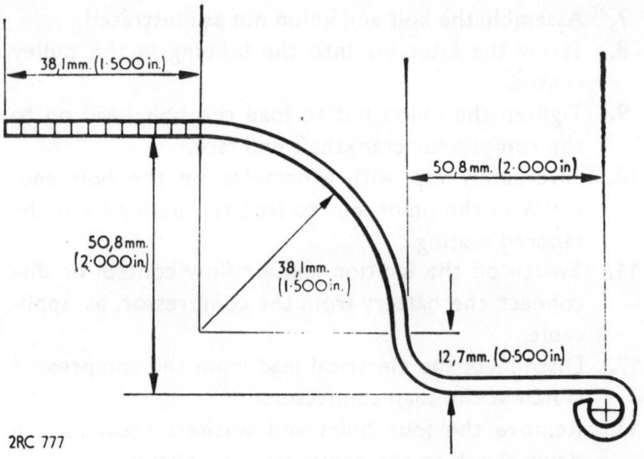

Procedure

1. Fit the charging and testing equipment. 82.30.01.
2. Remove the windscreen washer reservoir.
3. Start the engine and turn the temperature control to maximum cooling—fully anti-clockwise—position, and the air flow control to HIGH speed. Operate the system for five minutes at 1,200—1,500 rev./min.

   NOTE: It is important to close the suction service valve slowly during the following item to avoid a sudden pressure reduction in the compressor crankcase that could cause a large amount of oil to leave the compressor.

4. Reduce the engine speed to idling, and SLOWLY forward-seat—turn clockwise—the suction service valve on the compressor until the compound gauge reads 0 or a little below.
5. Stop the engine at this point and quickly forward-seat the suction service valve and discharge service valve.
6. Loosen the oil level plug and unscrew it slowly by 5 turns to bleed off crankcase pressure.
7. Remove the oil level plug, wipe the dipstick and insert it as near vertical as possible and to the lowest point of the crankcase. It may be necessary to turn the compressor crank to obtain clearance.

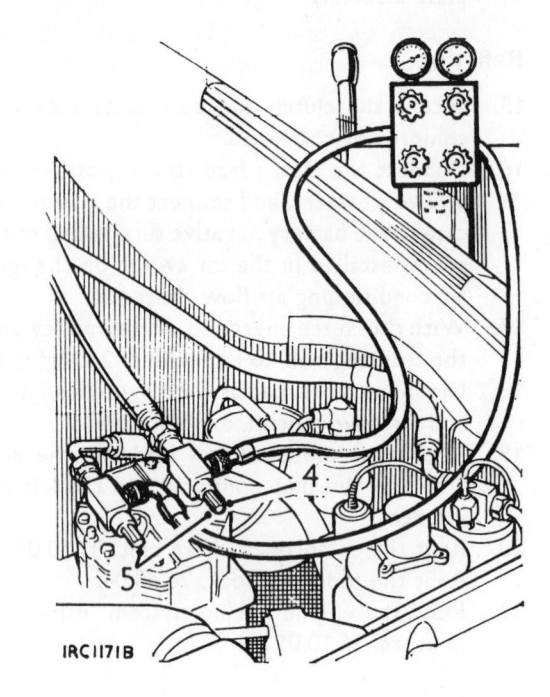

continued

82.10.14
Sheet 1

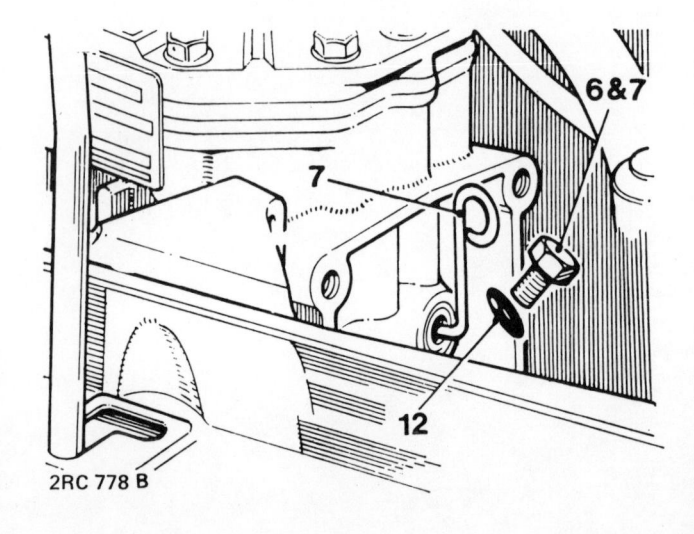

Rover 3500 and 3500S Manual AKM 3621

AIR CONDITIONING

8. Withdraw the dipstick and determine the depth of oil. The depth should be 25 mm. (1.00 in.) maximum, with a minimum of 22 mm. (0.875 in.)

   NOTE: 25mm (1.00 in.) depth corresponds to 0,23 litre (0.4 UK pint).
   22 mm (0.875 in.) depth corresponds to 0,17 litre (0.3 UK pint).

9. If required, top up the compressor crankcase to the correct level, using special refrigerant oil, Division 09.
10. If the system has been cleaned out, or the compressor has been overhauled, the oil should be initially filled to a depth of 28 mm (1.125 in.), corresponding to 0,28 litre (0.5 UK pint).
11. When the system has been charged with refrigerant it must be operated and checked as described in items 1 to 8, and then the oil level rechecked, to ensure correct oil balance.
12. Lubricate a new 'O' ring with refrigerant oil, fit it over the threads of the level plug without twisting, and install the level plug loosely.
13. Evacuate the air from the compressor crankcase, using the vacuum pump on the charging and testing equipment, as follows:
    a. Open valve numbers 1 and 2.
    b. Ensure that valve number 4 is closed.
    c. Start the vacuum pump and open the vacuum pump valve.
    d. Slowly open valve number 3.
    e. Tighten the crankcase level plug securely.
    f. Close valve number 3.
    g. Switch off the vacuum pump.
14. Back-seat (fully anti-clockwise) both service valves.
15. Start and run the engine at 1,200 rev/min and check for leak at the compressor level plug. Do not overtighten to correct a leak. In the event of a leak isolate the compressor as previously described in items 4 to 6, and check the 'O' ring seats for dirt, etc.
16. Stop the engine.
17. Close both service valves (turn fully anti-clockwise)
18. Close all valves on the charging and testing equipment.
19. Disconnect the charging lines from the compressor.
20. Refit the blanking caps to the compressor valve stems and gauge connections, and to the charging lines.
21. Close the bonnet.

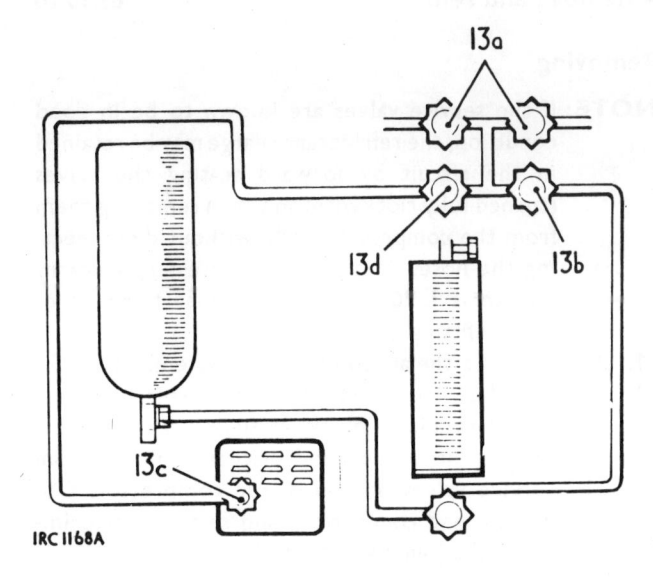

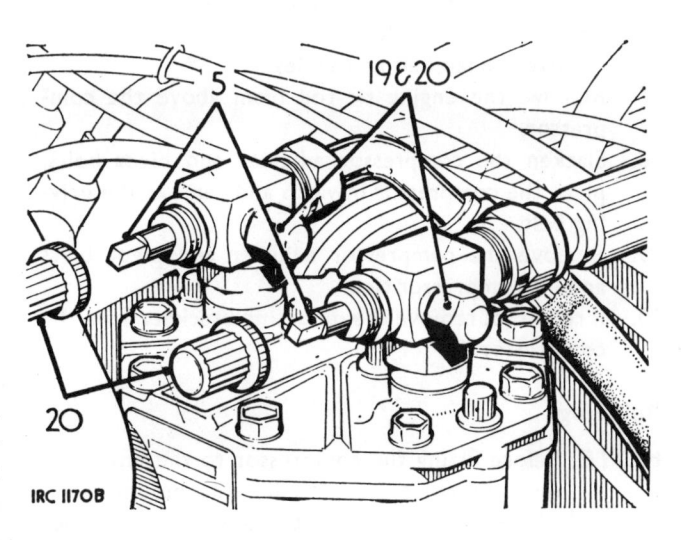

Rover 3500 and 3500S Manual AKM 3621

82.10.14
Sheet 2

AIR CONDITIONING

COMPRESSOR

—Remove and refit 82.10.20

Removing

NOTE: If the service valves are known to be in good condition, the refrigerant charge may be retained in the circuit by forward seating the valves (turned fully clockwise) and then removing them from the compressor body without disconnecting the hoses. It is not then necessary to carry out items 1, 20, 21 and 22. Back-seat the valves after refitting.

1. Depressurise the air conditioning system. 82.30.05
2. Remove the windscreen washer reservoir.
3. Remove the radiator block. 26.40.04.
4. Disconnect the suction and discharge unions from the service valves on top of the compressor. Cap the flexible end connections and service valves immediately the joints are opened.
5. Disconnect the lead to the compressor magnetic clutch at the snap connector on the wing valance.
6. Remove the fan blades and viscous coupling. Do not remove the fan belt and pulley.
7. Remove the engine tie-rod from above the compressor.
8. Slacken the compressor adjuster and pivot bolts, pivot the compressor inwards and remove the driving belt.
9. Remove the compressor adjuster and pivot bolts and lift compressor clear.
10. If required, remove the mounting brackets from the compressor.

Refitting

NOTE: Before fitting the compressor to the engine, it must be complete with its side mounting/adjuster bracket and its bottom mounting/pivot bracket.

11. Locate the compressor in position and fit the pivot and adjuster bolts, finger tight only.
12. Fit the compressor driving belt and adjust the compressor by means of its pivot and slotted fixing, to give the correct belt tension. The belt must be tight with 4 to 6 mm (0.187 to 0.250 in.) total deflection when checked midway between the compressor and idler pulleys, by hand.
13. Tighten the adjuster and pivot bolts and recheck the driving belt tension.
14. Fit the viscous coupling and fan blades, with the larger diameter fan fixing bosses to the front.
15. Fit the radiator. 26.40.04.
16. Refill the engine cooling system, using the correct mixture. 26.10.01
17. Refit the suction and discharge flexible end connections to the service valves, lubricating the flares and threads of the unions with refrigeration oil.
18. Fit the engine tie-rod.
19. Connect the lead to the compressor magnetic clutch at the snap connector.
20. Evacuate the air conditioning system. 82.30.06. Maintain the vacuum for five minutes.
21. Charge the air conditioning system. 82.30.08.
22. Check, and if necessary replenish, the compressor oil level. 82.10.14.
23. Start and run the engine. Check for oil and water leaks.
24. Switch on the air conditioning and check for correct operation.

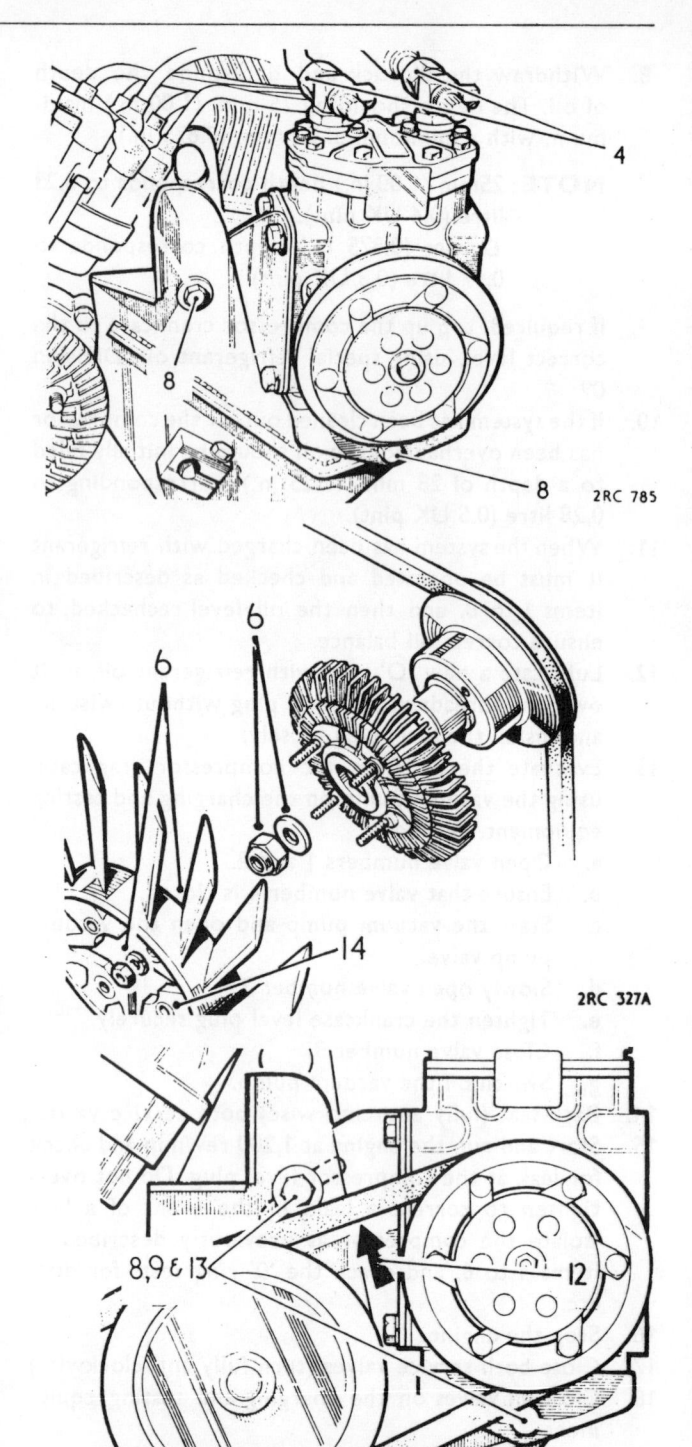

AIR CONDITIONING

COMPRESSOR

—Overhaul   82.10.26

NOTE: The compressor crankcase and covers are made of aluminium alloys and care must be taken to avoid scratches, nicks and burrs on the machined surfaces. Even tightening of the securing bolts to the specified torque is important. All gaskets should be replaced.

IMPORTANT: Cleaning of parts should be done with petroleum or alcohol-based solvents. Chlorinated solvents such as trichloroethylene must not be used.

Dismantling

1. Remove the compressor. 82.10.20
2. Remove the protective caps from the valve stems and open both service valves to free any gas pressure that may be in the compressor.
3. Scribe a line down both sides of each service valve to ensure correct angle of valve when refitted.
4. Undo the hexagon coupling on the 'Rotalock' valves and remove the valves complete with O-ring.
5. Remove the eight bolts securing the head and valve plate to the cylinder block.
6. Remove the valve plate and head from the block by tapping under the ears which extend from the valve plate. Do not prise.
7. If the head and valve plate adhere together, hold the head and tap the valve plate ears away from the head. Do not strike the head, or prise between the head and the valve plate.
8. Remove the nuts and washers securing the valve assemblies.
9. Withdraw the reed restrainers.
10. Withdraw the discharge valve reeds.
11. Remove the bolts and suction reeds.
12. Remove the base plate from the crankcase.
13. Withdraw the gasket.
14. Mark the connecting rods and caps for correct reassembly, then remove the caps.
15. Withdraw the connecting rod and piston assembly.

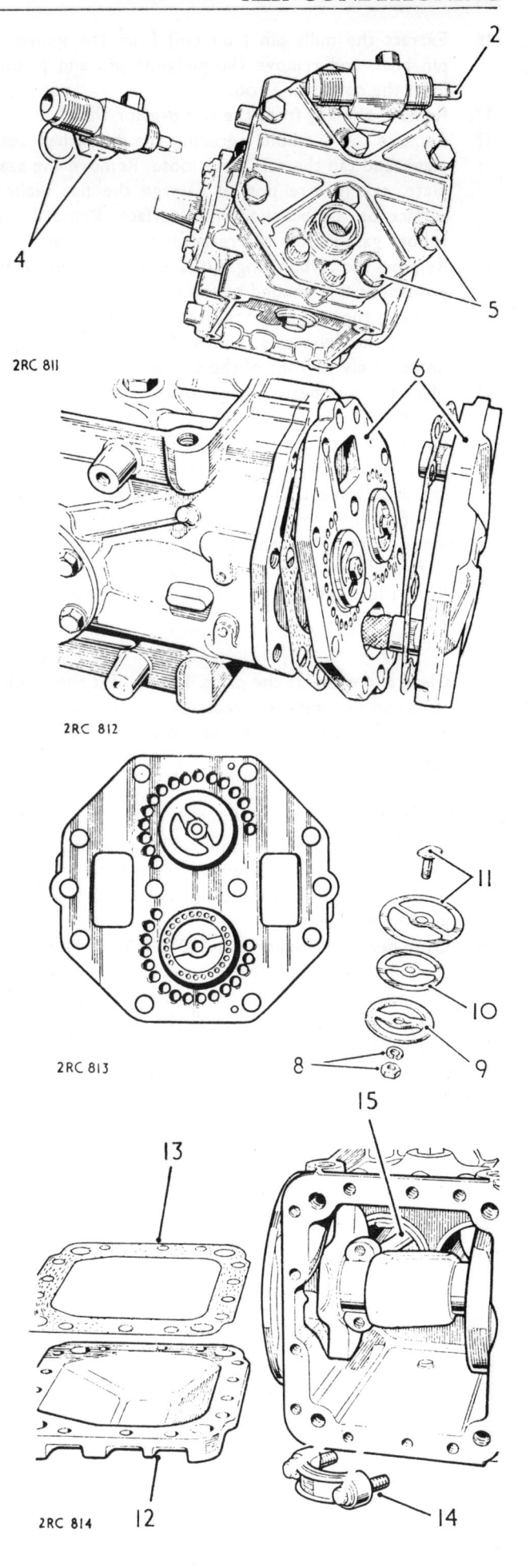

continued

Rover 3500 and 3500S Manual AKM 3621

82.10.26
Sheet 1

AIR CONDITIONING

16. Extract the mills pin (roll pin) from the gudgeon pin boss, and remove the gudgeon pin and piston from the connecting rod.
17. Remove the key from the compressor shaft.
18. Remove the six bolts securing the crankshaft seal plate, and tap the seal plate loose. Remove the seal plate, taking care not to damage the flat sealing surface or the polished shaft surface. Remove the rubber gasket from the recess in the crankcase.
19. Remove the carbon ring if it is loose, but leave it in position it if is bonded to the retainer.
20. Remove the seal assembly from the shaft by prying behind the drive ring, which is the part of the seal immediately in front of the ball bearing.
21. Mark the rear bearing cover plate to ensure correct assembly. Remove the four countersunk screws and withdraw the cover plate and 'O' ring.
22. Clean the crankcase and crankshaft assembly in an approved solvent. Oven heat the assembly to 150°C (300°F) when the crankshaft and bearing can be removed from the crankcase with very little pressure. (Localised heating may crack the crankcase.)
23. Press the crankshaft out of the ball bearing.

Inspecting

24. Clean and examine the parts. Ensure that the crankshaft and the crankcase recesses are free from burrs and nicks. Check that the ports in the suction reservoir are clear.

Reassembling

25. Lubricate all bearings and rubbing surfaces with clean refrigerant oil when assembling. Gaskets should be coated with refrigerant oil. Do not use jointing compound.
26. Oven heat the crankcase to 150°C (300°F). Note that the inner and outer races of the bearing are flush on only one side. Insert the bearing in the heated crankcase so that the flush side of the bearing is at the front of the crankcase, facing outwards.
27. Allow the crankcase to cool. Support the inner race of the bearing and press the crankshaft into the bearing.

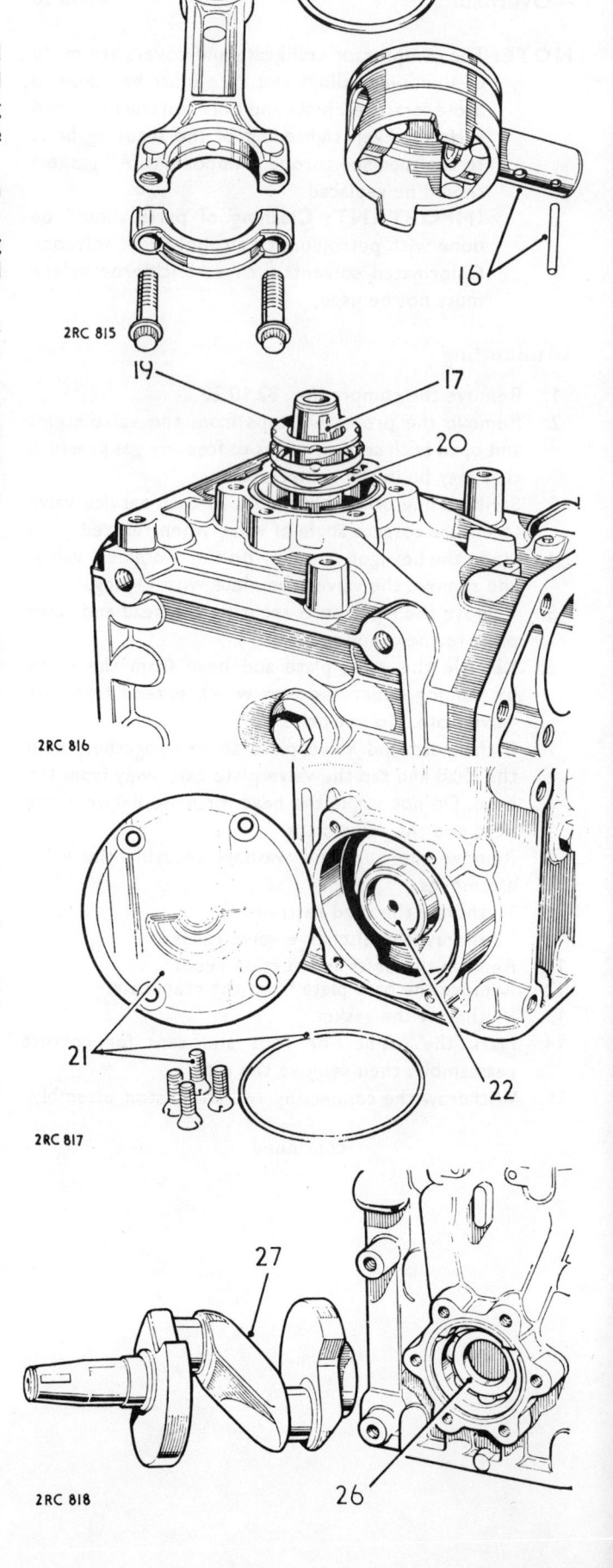

continued

82.10.26
Sheet 2

Rover 3500 and 3500S Manual AKM 3621

AIR CONDITIONING

28. Fit the rear bearing cover plate with a new 'O' ring. Check the mating marks, ensuring the oil holes in the cover are towards the top. Tighten the screws evenly to 1,2 to 2,3 kgf.m (9 to 17 lbf.ft).
29. Place the rubber gasket of the front seal plate in the crankcase recess with a coating of refrigerant oil. Note that seal and gasket packs for replacement purposes are suitable for all compressors, Series 60 onwards. Select only the gasket which matches the one removed and discard the others.
30. Push the seal assembly, less the carbon ring if it is free, over the end of the crankshaft, with the carbon ring retainer outwards. Place the carbon ring in the retainer with the flat surface outward and the indentations in the rim engaging the lugs of the retainer.
31. Use the seal cover plate to push the seal into position on the crankshaft. Fit the bolts and rotate the crankshaft whilst tightening them evenly. Check that there is even clearance between the crankshaft and the hole in the seal plate. Tap the seal plate into position if necessary.
32. Fully tighten the seal plate bolts to 0,9 to 1,7 kgf.m (7 to 13 lbf ft.).
33. Fit the gudgeon pins to the pistons and connecting rods and tap the mills pins (roll pins) securely into position.
34. Fit the piston rings with the chamfered face upward.
35. Fit the connecting rods and pistons from the top of the crankcase, with the mills pins (roll pins) towards the centre of the compressor. Note the mating marks, and fit the connecting rod caps. Tighten the bolts evenly to 1,7 to 2,2 kgf.m (13 to 16 lbf. ft) torque.
36. Fit the base plate with a new gasket and tighten the bolts to 1,9 to 3,0 kgf.m (14 to 22 lbf. ft.) torque.
37. Fit the suction valve reeds (larger) over the bolts, and insert the bolts from the underside of the valve plate.
38. Fit the discharge valve reeds over the bolts from the top of the valve plate.
39. Fit the valve retainer with the concave surface uppermost, and the arms covering the arms of the discharge reed. Fit the washers and nuts. Tighten the nuts to the shoulder of the bolts, and stake them in position.
40. Fit the valve plate to the crankcase with a new gasket. The discharge valves and retainers should be uppermost and the valve plate located on the dowel pins.
41. Fit the head with a new gasket, locating it on the dowel pins.
42. Tighten the head bolts evenly to 2,0 to 3,0 kgf.m (15 to 23 lbf. ft.).
43. Fit the suction valve—$\frac{5}{8}$ in. outlet—on the suction side, and the discharge valve—$\frac{1}{2}$ in. outlet— on the discharge side, using new 'O' rings.
44. Tighten the 'Rotalock' valve hexagon couplings ensuring that the valves are in line with the scribed marks made on removal. Tighten 'Rotalock' valve hexagon to 4,0 to 4,9 kgf.m (30 to 35 lbf. ft.).
45. Backseat the service valves and fit the protective caps.
46. Replenish the compressor crankcase with 0,23 litre (0.5 UK pint) of refrigerant oil. See Division 09 for recommendations. Tighten the oil filler plug to 0,3 to 0,8 kgf.m (2 to 6 lbf. ft.).
47. Refit the compressor. 82.10.20.

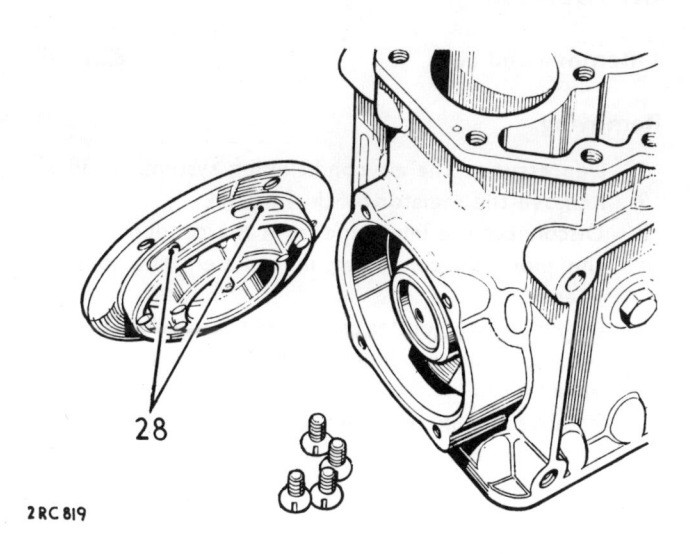

Rover 3500 and 3500S Manual AKM 3621

AIR CONDITIONING

CONDENSER

—Remove and refit　　　　　　　　82.15.07

Removing

1. Depressurise the air conditioning system. 82.30.05
2. Remove the radiator block. 26.40.04.
3. Disconnect the inlet pipe from the condenser. Cap the pipe and condenser immediately the joint is opened.
4. Disconnect the outlet pipe from the condenser. Cap the pipe and condenser immediately the joint is opened.
5. Remove the fixings securing the condenser.
6. Withdraw the condenser.

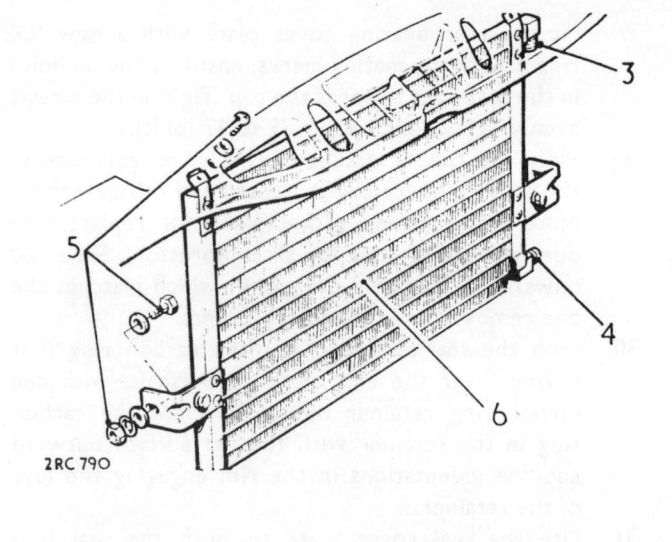

Refitting

7. Reverse 2 to 6.
8. Evacuate the air conditioning system. 82.30.06. Maintain the vacuum for five minutes.
9. Charge the air conditioning system. 82.30.08.
10. Check, and if necessary replenish, the compressor oil level. 82.10.14.

LIQUID RECEIVER/DRIER

—Remove and refit　　　　　　　　82.17.01

Removing

1. Depressurise the air conditioning system. 82.30.05.
2. Remove the windscreen washer reservoir.
    CAUTION: Fit blanking caps to the liquid receiver/drier and the pipe unions immediately each pipe is disconnected during the following procedure, otherwise the liquid receiver/drier will be scrap.
3. Disconnect the inlet pipe.
4. Disconnect the outlet pipe.
5. Disconnect the pipe for the high pressure cut-out.
6. Remove the fixing and withdraw the liquid receiver/drier complete with the mounting clip.
7. If required, remove the clip from the liquid receiver/drier.

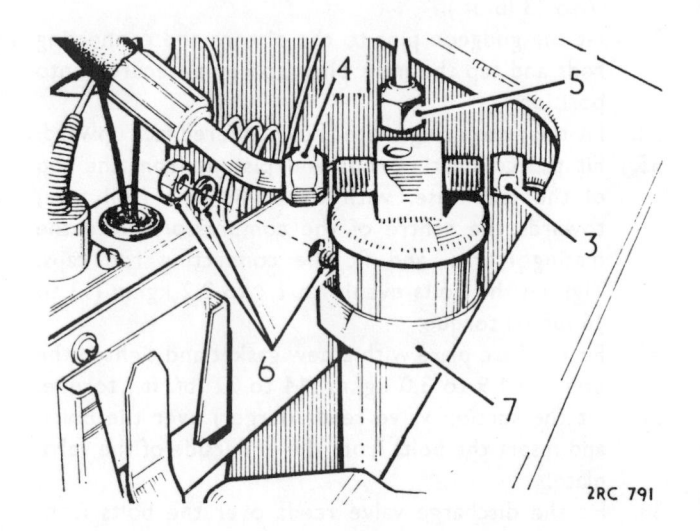

Refitting

8. If the retaining clip has been removed from the liquid receiver/drier, ensure that the fixing bolt is in place and fit the clip to the liquid receiver/drier body.
9. Offer up the liquid receiver/drier to its mounting bracket. If necessary, adjust the retaining clip position to align the pipe connections, then secure the liquid receiver/drier in place.
    NOTE: It is essential to ensure that the receiver/drier is connected up correctly as the pipe union sizes are identical. The union end marked 'IN' must be connected to the condenser.
10. Remove the blanks and connect the inlet, outlet and high pressure cut-out pipes to the liquid receiver/drier, lubricating the flares and threads of the unions with refrigerant oil. Use a spanner on both hexagons to prevent distortion.
11. Fit the windscreen washer reservoir.
12. Evacuate the air conditioning system. 82.30.06. Maintain the vacuum for five minutes.
13. Charge the air conditioning system. 82.30.08.
14. Check, and if necessary replenish, the compressor oil level. 82.10.14.

Rover 3500 and 3500S Manual AKM 3621

AIR CONDITIONING

MAIN RELAY

—Remove and refit   82.20.08

Removing

1. Disconnect the battery earth lead.
2. Remove the left-hand glove box.76.52.03.
3. Disconnect the leads from the relay, remove the fixings and lift the relay clear.

Refitting

4. Secure the relay in position and connect the electrical leads in accordance with the circuit diagram. Page 82.00.01.
5. Refit the glove box.
6. Reconnect the battery earth lead.

CONTROL SWITCHES

—Remove and refit

Temperature control switch 2 to 6, 13 and 15 to 17   82.20.10

Air-flow control switch 1 to 4, 7, 8 and 14 to 17   82.20.11

Compressor isolator switch 1, 9 to 12, 18 and 19   82.20.12

Service tool: 601942, Lock ring spanner

Removing

1. Disconnect the battery earth lead.
2. Push in the plungers retaining the temperature control and air-flow control switch knobs to their spindles and withdraw both knobs.
3. Remove the two screws and lift out the front cover from the console unit.
4. Remove the four screws from the switch panel and ease the panel forward.
5. Remove the locknut and washer from the temperature control switch spindle and withdraw the switch from the panel.
6. Carefully pull the vacuum piping from the switch tubes, noting the letter coding, and lift the switch clear. Take precautions against loss of identification sleeves on the vacuum pipes.
7. Remove the locknut and washer from the air flow control switch spindle and withdraw the switch from the panel.
8. Carefully pull the vacuum piping from the switch and disconnect the electrical leads, noting the colour coding, and lift the switch clear. Take precautions against loss of identification sleeves on the vacuum pipes.
9. Open the glove box lid.
10. Unscrew the switch knob.
11. Unscrew the locking ring from the switch, 601942, and withdraw the switch towards the front of the car.
12. Disconnect the wiring and remove the switch.

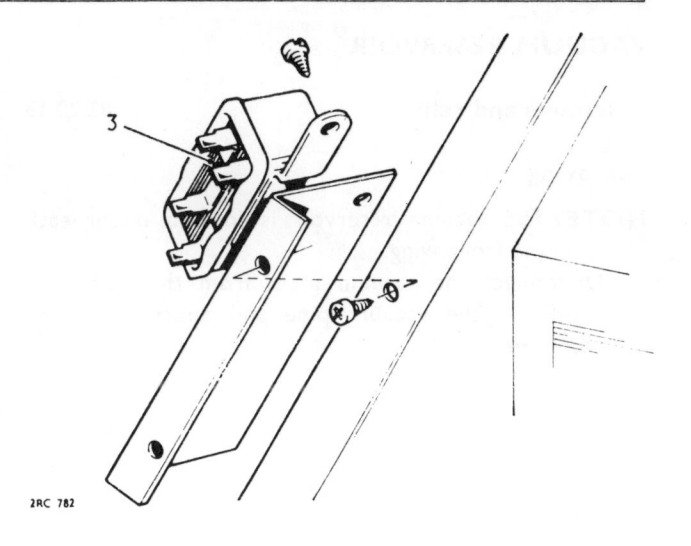

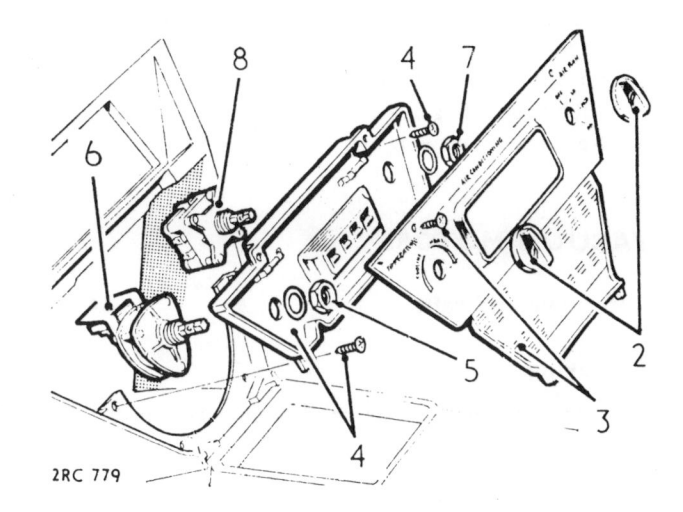

Refitting

13. Locate the temperature control switch into the console unit and connect the vacuum pipes in accordance with the identification letters on the pipes and switch.
14. Locate the air-flow control switch into the console unit and connect the vacuum pipes in accordance with the identification letters on the pipes and switch. Connect the electrical leads in accordance with the circuit diagram. Page 82.00.01.
15. Fit the switch in position on the switch panel and secure with a locknut and washer.
16. Secure the switch panel and front cover in position on the console unit.
17. Fit the temperature control and air-flow switch knobs to their spindles.
18. Reverse 9 to 12
19. Reconnect the battery earth lead.

Rover 3500 and 3500S Manual AKM 3621

AIR CONDITIONING

VACUUM RESERVOIR

—Remove and refit   82.20.13

Removing

NOTE: The vacuum reservoir is located underneath the front wing.

1. Disconnect the vacuum pipes from the reservoir. Blank off the vacuum pipe and aperture in the reservoir.
2. Remove the fixings and lift the reservoir clear.

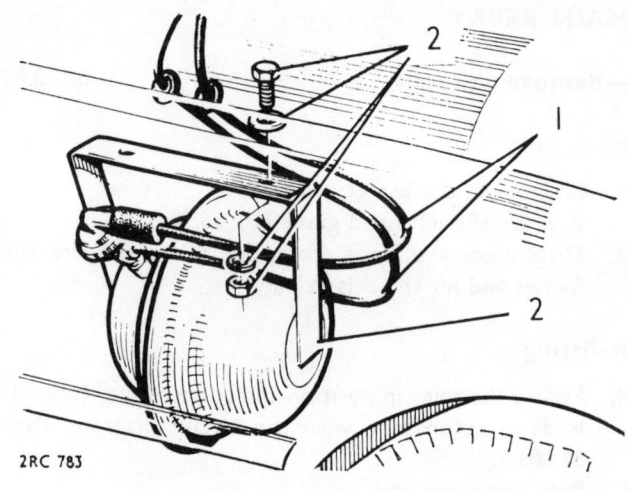

Refitting

3. Remove the RH front wing. 76.10.26.
4. Secure the reservoir in position.
5. Remove the blanks and connect the vacuum pipes to the reservoir. The connections are different diameter to prevent incorrect fitting, the pipe from the engine manifold fits the large diameter connection on the reservoir, through the medium of a short rubber coupling.
6. Refit the front wing.

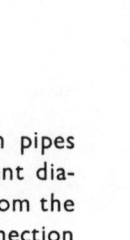

VACUUM MOTORS

—Remove and refit   82.20.14

Removing

1. Depressurise the air conditioning system 82.30.05.
2. Disconnect the battery earth lead.
3. Drain the engine coolant. 26.10.01.
4. Remove the heater/cooler unit. 82.25.21.
5. Remove the fixings from the joint flanges between the top cover, blower case projection and the blower outlet, leaving the blower motor in position.
6. Withdraw the end shell as far as the capillary will allow.
   NOTE: The lower vacuum motors are accessible through the recirculation intake, using a socket spanner. The upper vacuum motor is accessible, using an open end spanner.
7. Remove the fixings from the fresh/recirculation valve motor, slacken the trunnion and unscrew the link rod.
8. Remove the fixing from the defrost valve motor, slacken the trunnion and unscrew the link rod.
9. Remove the fixings from the mode change valve motor, slacken the trunnion and unscrew the link rod.
10. Remove the fixings from the evaporator thermostat and motor assembly, slacken the trunnion and unscrew the link rod.

Refitting

11. Assemble the motor to the case and tighten the screws. 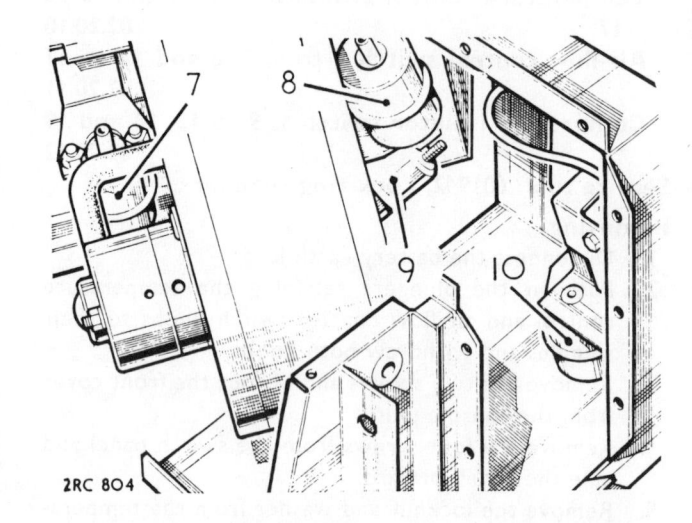
12. Push the relevant lever to its extreme position away from the motor, assemble the trunnion and screw to the lever, thread the motor shaft through the trunnion, depress the motor diaphragm plate 3 mm (0.125 in.) to give pre-loading, and tighten the trunnion screw.
13. Re-assemble the heater/cooler unit.
14. Fit heater/cooler unit. 82.25.21.
15. Reconnect the battery earth lead.
16. Refill the engine cooling system, using the correct mixture. 26.10.01.
17. Evacuate the air conditioning system. 82.30.06. Maintain the vacuum for five minutes.
18. Charge the air conditioning system. 82.30.08.
19. Check, and if necessary replenish, the compressor oil level. 82.10.14.
20. Start and run the engine. Check for leaks from the water and refrigerant hose connections.
21. Switch on the air conditioning and check for correct operation.

82.20.13
82.20.14

Rover 3500 and 3500S Manual AKM 3621

AIR CONDITIONING

HIGH PRESSURE CUT-OUT

—Remove and refit 82.20.20

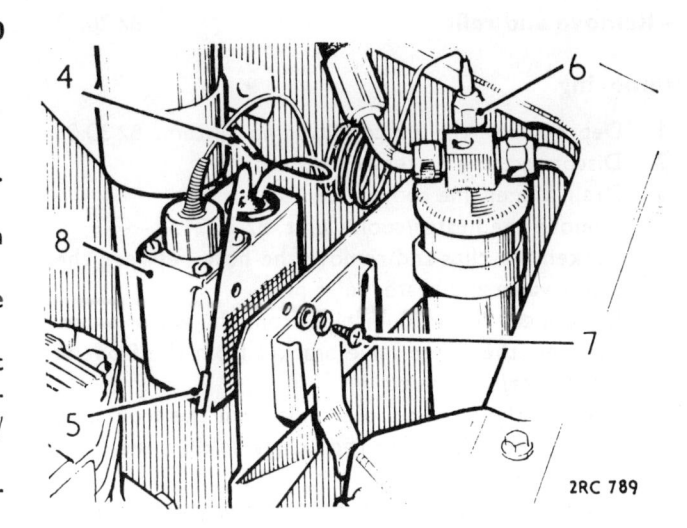

Removing

1. Disconnect the battery earth lead.
2. Depressurise the air conditioning system. 82.30.05.
3. Remove the windscreen washer reservoir.
4. Disconnect the electrical feed lead for the high pressure cut-out, at the snap connector.
5. Disconnect the electrical lead from the high pressure cut-out to the compressor clutch.
6. Disconnect the pipe for the high pressure cut-out at the liquid receiver/drier. Blank off the disconnected pipe and the aperture in the liquid receiver/drier.
7. Remove the fixings securing the high pressure cut-out to the mounting bracket.
8. Withdraw the high pressure cut-out.

Refitting

9. Reverse 3 to 8.
10. Reconnect the battery earth lead.
11. Evacuate the air conditioning system. 82.30.06. Maintain the vacuum for five minutes.
12. Charge the air conditioning system. 82.30.08.
13. Check, and if necessary replenish, the compressor oil level. 82.10.14.

BLOWER MOTOR RESISTANCE UNIT

—Remove and refit 82.20.26

Removing

1. Depressurise the air conditioning system. 82.30.05.
2. Disconnect the battery earth lead.
3. Drain the engine coolant. 26.10.01.
4. Remove the heater/cooler unit. 82.25.21.
5. Drill out the two 'pop' rivets securing the resistance unit to the case.
6. Withdraw the resistance unit through the recirculation inlet.
7. Unsolder the leads and remove the resistance unit.

Refitting

8. Draw the resistance unit leads through the recirculation inlet of the heater/cooler unit. Solder the brown/blue lead to the resistance low speed terminal, the brown/white lead to the mid terminal and the brown/red and green leads to the high speed terminal.
9. Using 'pop' rivets, secure the resistance unit in position inside the heater/cooler unit.
10. Refit the heater/cooler unit. 82.25.21.
11. Reconnect the battery earth lead.
12. Refill the engine cooling system, using the correct mixture. 26.10.01. 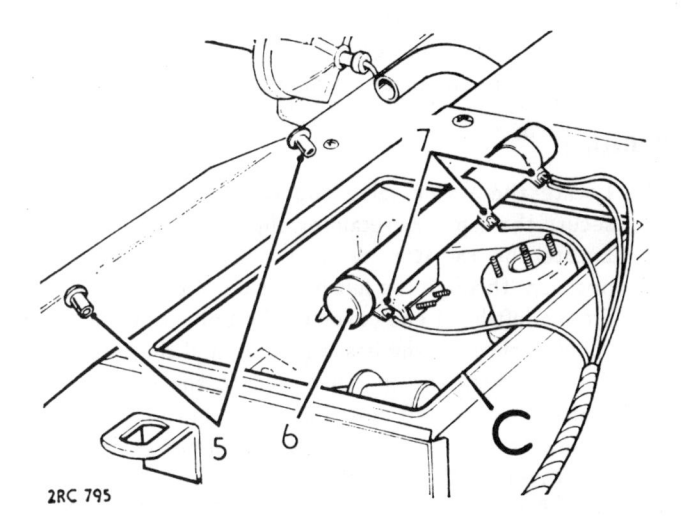
13. Evacuate the air conditioning system. 82.30.06. Maintain the vacuum for five minutes.
14. Charge the air conditioning system. 82.30.08.
15. Check, and if necessary replenish, the compressor oil level. 82.10.14.
16. Start and run the engine. Check for leaks from the water and refrigerant hose connections.
17. Switch on the air conditioning and check for correct operation.

Rover 3500 and 3500S Manual AKM 3621

82.20.20
82.20.26

AIR CONDITIONING

WATER VALVE AND RADIATOR

—Remove and refit 82.20.32

Removing

1. Depressurise the air conditioning system. 82.30.05
2. Disconnect the battery earth lead.
3. Drain the engine coolant. 26.10.01.
4. Remove the heater/cooler unit. 82.25.21.
5. Slacken the clips and remove the hose between the water valve and heater inlet pipe.
6. Disconnect the vacuum pipe from the water valve.
7. Unscrew the water valve bracket from the front of the heater/cooler unit.
8. Position the heater/cooler unit with the front face uppermost and remove the fixings holding the front fan tail cover.
9. Remove the four screws holding the patch plate at the base of expansion valve on the fan tail cover.
10. Withdraw the patch plate and fan tail cover.
11. Remove the clips holding the spindles in the bearing holes in the left-hand end cover. These clips must be broken to effect removal.
12. Remove the two screws holding the drain tube flange to the end cover. Remove the flange and drain tube.
13. Push the internal drain back through the aperture into the unit.
14. Remove the fixings and withdraw the end cover from the heater/cooler unit.
15. Withdraw the radiator, seals and patch grommet endways from the unit.
16. Remove the water valve, carefully easing the looped capillary out of the retaining clips and withdrawing the capillary complete towards the end of the unit.

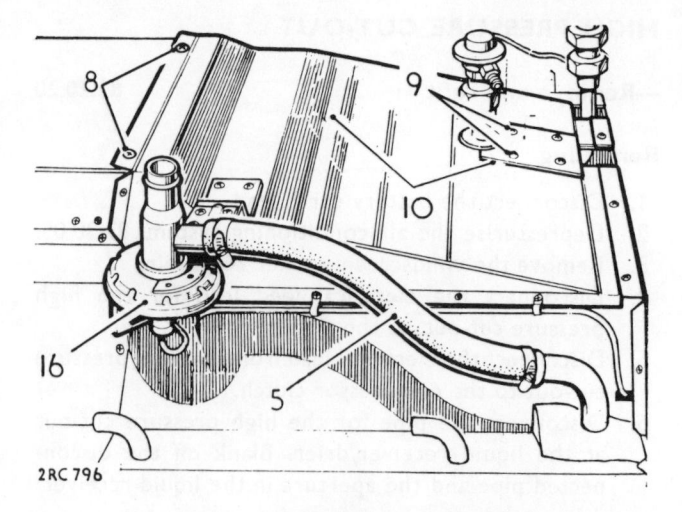

Refitting

17. Locate the water valve and capillary in position, secure the capillary retaining clips.
18. Locate the radiator, seals and patch grommet in position, through the left-hand side of the unit.
19. Fit the left-hand end cover to the unit, locating the spindles through the bearing holes, and the internal drain through the aperture. Secure the end cover fixing screws and fit new clips to retain the spindles.
20. Secure the external drain tube and flange to the end cover.
21. Fit the fan tail cover in position on the front face of the unit. Secure the patch plate to the fan tail cover and secure the cover to the unit.
22. Screw the water valve and bracket into the front face of the unit.
23. Connect the hose between the water valve and heater inlet pipe, secure the hose clips. Connect the vacuum pipe to the water valve.
24. Fit the heater/cooler unit. 82.25.21.
25. Reconnect the battery earth lead.
26. Refill the engine cooling system, using the correct mixture. 26.10.01.
27. Evacuate the air conditioning system. 82.30.06. Maintain the vacuum for five minutes.
28. Charge the air conditioning system. 82.30.08.
29. Check, and if necessary replenish, the compressor oil level. 82.10.14.
30. Start and run the engine. Check for leaks from the water and refrigerant hose connections.
31. Switch on the air conditioning and check for correct operation.

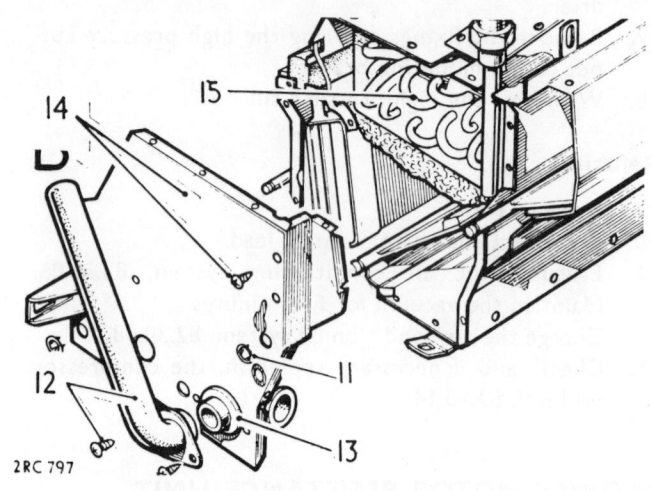

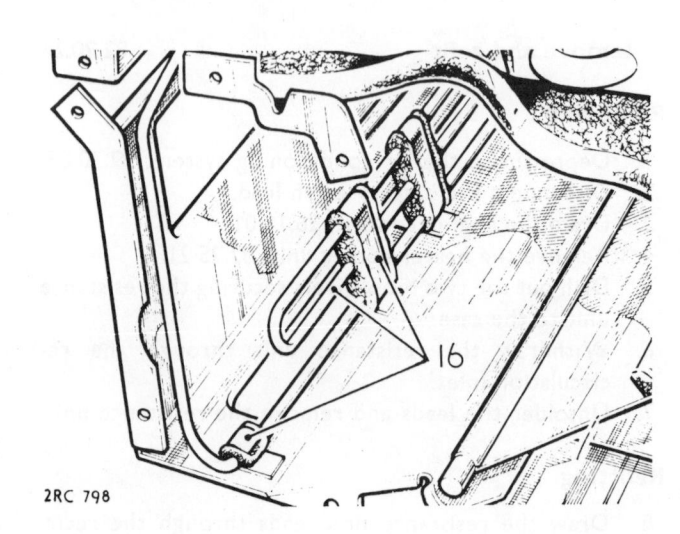

82.20.32

Rover 3500 and 3500S Manual AKM 3621

AIR CONDITIONING

EXPANSION VALVE

—Remove and refit   82.25.01

Removing

1. Depressurise the air conditioning system. 82.30.05.
2. Remove the air cleaner. 19.10.01
3. Carefully prise off the clip holding the valve capillary coil to the evaporator outlet pipe.
4. Hold the expansion valve body with a ¾ in. AF spanner and disconnect the refrigeration hose from the valve. Blank off the disconnected hose and aperture in the valve.
5. Unscrew the expansion valve from the evaporator inlet. Blank off apertures in the expansion valve and evaporator.

Refitting

6. Remove the blanks and fit the expansion valve to the evaporator inlet, aligning the inlet hose connection and capillary coil as illustrated.
7. Remove the blanks and connect the refrigeration hose to the expansion valve. Use two spanners to avoid distortion.
8. Clip the capillary coil to the evaporator outlet.
9. Fit the air cleaner. 19.10.01.
10. Evacuate the air conditioning system. 82.30.06. Maintain the vacuum for five minutes.
11. Charge the air conditioning system. 82.30.08.
12. Check, and if necessary replenish, the compressor oil level 82.10.14.
13. Run the engine and check the air conditioning system for leaks.

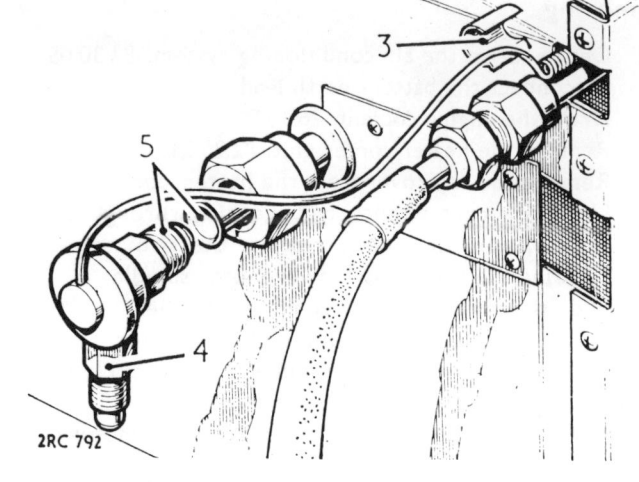

CONSOLE UNIT

—Remove and refit   82.25.07

Removing

1. Disconnect the battery earth lead.
2. Remove the gearbox tunnel cover. 76.25.07.
3. Remove both glove boxes. 76.52.03.
4. Remove the four screws securing the panel for the air conditioning control switches.
5. Turn the switch panel and pass it into the console unit.
6. From the RH side of the console unit, release the bracket carrying the fuel reserve switch.
7. From the LH side of the console unit release the bracket carrying the relay for the air conditioning system.
8. Remove the four drive screws securing the console unit to the bulkhead and withdraw the unit.

Refitting

9. Reverse 1 to 8.

Rover 3500 and 3500S Manual AKM 3621

AIR CONDITIONING

BLOWER ASSEMBLY

—Remove and refit   **82.25.13**

Removing

1. Depressurise the air conditioning system. 82.30.05.
2. Disconnect the battery earth lead.
3. Drain the engine coolant. 26.10.01.
4. Remove the heater/cooler unit. 82.25.21.
5. Remove the top cover from the heater/cooler unit.
6. Using a socket wrench and working through the recirculation air intake, remove the two screws securing the fan to the blower motor spindle.
7. Slide the fan off the blower motor spindle.
8. Disconnect the leads from the blower motor.
9. Withdraw the ventilator tube from the blower motor top cover.
10. Drill out the three 'pop' rivets holding the motor to the blower case.
11. Lift out the motor, noting the position of the identification label.

Refitting

12. Locate the blower motor in place, ensuring that the label is in the same relative position noted during removal, and secure, using 5 mm. (0.200 in.) diameter rivets.
13. Push the fan as far as possible on to the motor spindle, then withdraw it 1,5 mm. (0.062 in.) to provide adequate clearance and tighten the securing screws.
14. Connect the ventilator tube to the blower motor top cover.
15. Connect the electrical feed lead to the blower motor.
16. Connect the black earth lead to the blower motor as illustrated.

    NOTE: If the leads are incorrectly fitted, the blower motor will run backwards, when in use.

17. Secure the top cover to the heater/cooler unit.
18. Fit the heater/cooler unit. 82.25.21.
19. Reconnect the battery earth lead.
20. Refill the engine cooling system, using the correct mixture. 26.10.01.
21. Evacuate the air conditioning system. 82.30.06. Maintain the vacuum for five minutes.
22. Charge the air conditioning system. 82.30.08.
23. Check, and if necessary replenish, the compressor oil level, 82.10.14.
24. Start and run the engine. Check for leaks from the water and refrigerant hose connections.
25. Switch on the air conditioning and check for correct operation.

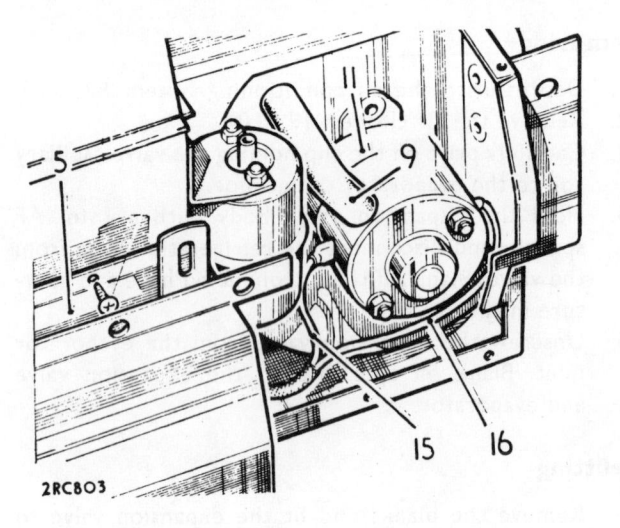

AIR CONDITIONING

EVAPORATOR

—Remove and refit   82.25.20

Removing

1. Depressurise the air conditioning system. 82.30.05.
2. Disconnect the battery earth lead.
3. Drain the engine coolant. 26.10.01.
4. Remove the heater/cooler unit. 82.25.21.
5. Release the water valve from the front fan tail cover.
6. Remove the fixings holding front fan tail cover.
7. Remove the four screws holding the patch plate at the base of the expansion valve on the fan tail cover.
8. Withdraw the patch plate and fan tail cover.
9. Remove the two screws from each side of the blower outlet duct and withdraw the curved deflector plate.
10. Carefully withdraw the capillary loop from the evaporator fins.
11. Remove the fixings and lift out the evaporator end plate. If necessary remove the unit end cover.
12. Withdraw the evaporator and filter assembly from the end of the unit.

Refitting

13. Locate the evaporator and filter assembly into the heater/cooler unit.
14. Fit and secure the evaporator end plate.
15. Fit the unit end cover, if removed.
16. Carefully feed the capillary loop through the evaporator fins.
17. Secure the curved deflector plate inside the end of the blower outlet duct, with the main portion of the plate outside the duct.
18. Fit the fan tail cover in position on the front face of the unit. Secure the patch plate to fan tail cover and secure the cover to the unit.
19. Fit the water valve.
20. Fit the heater/cooler unit 82.25.21.
21. Reconnect the battery earth lead.
22. Refill the engine cooling system, using the correct mixture. 26.10.01.
23. Evacuate the air conditioning system. 82.30.06. Maintain the vacuum for five minutes.
24. Charge the air conditioning system. 82.30.08
25. Check, and if necessary replenish, the compressor oil level. 82.10.14.
26. Start and run the engine. Check for leaks from the water and refrigerant hose connections.
27. Switch on the air conditioning and check for correct operation.

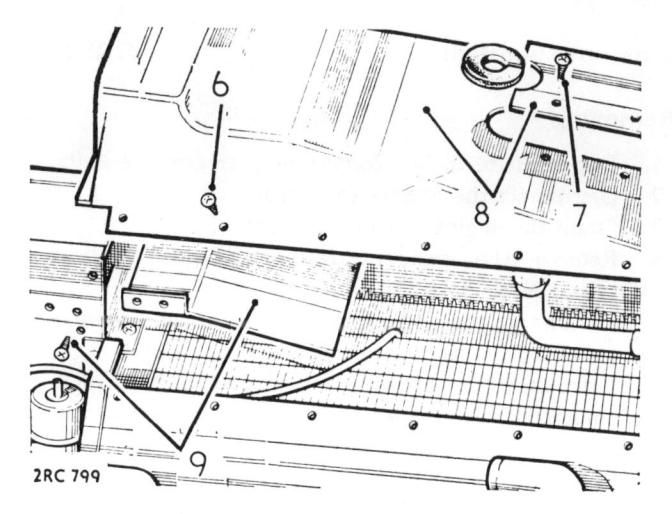

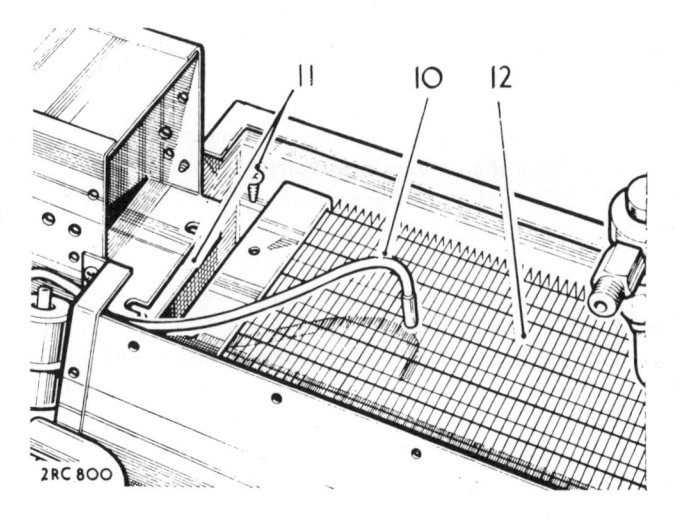

Rover 3500 and 3500S Manual AKM 3621

AIR CONDITIONING

HEATER/COOLER UNIT

—Remove and refit   82.25.21

Removing

1. Depressurise the air conditioning system. 82.30.05.
2. Disconnect the battery earth lead.
3. Drain the engine coolant. 26.10.01.
4. Remove the temperature control and air-flow switches. 82.20.10/11.
5. Disconnect the single black lead from the two-way connector, accessible behind the console unit.
6. Disconnect the brown/green lead from the single connector, also the white lead from the single connector, behind the console unit.
7. Disconnect the Brown/light green, Brown/dark green, White and Black leads from the relay unit located the LH side of console unit, accessible through the LH glove compartment.
8. Disconnect the Brown/black and Brown/purple isolating switch leads from the two single connectors behind the console unit.
9. Remove the bonnet. 76.16.01.
10. From under the RH front wing, disconnect the vacuum pipe from the vacuum reservoir.
11. Disconnect the Light green/red electrical lead from the high pressure cut-out.
12. Remove the windscreen wiper arms.
13. Remove the air intake valance. 76.79.04.
14. Remove the air cleaner. 19.10.01.
15. Using a spanner on both hexagons, disconnect the refrigeration pipe from the evaporator outlet. Blank off the disconnected pipe and the aperture in the evaporator.
16. Disconnect the refrigeration pipe from the expansion valve. Blank off the disconnected pipe and the aperture in the expansion valve.
17. Slacken the hose clip and disconnect the hose, engine to water valve, at water valve.
18. Slacken the hose clip and disconnect the heater outlet pipe from the LH side of the unit.
19. Remove the four fixings, securing the heater/cooler unit to the bulkhead.
20. **LHStg. models**—Remove the fixings securing the accelerator cross-shaft bracket to the steering box and move the accelerator linkage to one side.
21. Lift the heater/cooler unit from the vehicle, breaking the seals between the bulkhead wall and the bulkhead base. As the unit is removed, carefully pull the vacuum and wiring loom through the control aperture in the vertical bulkhead.

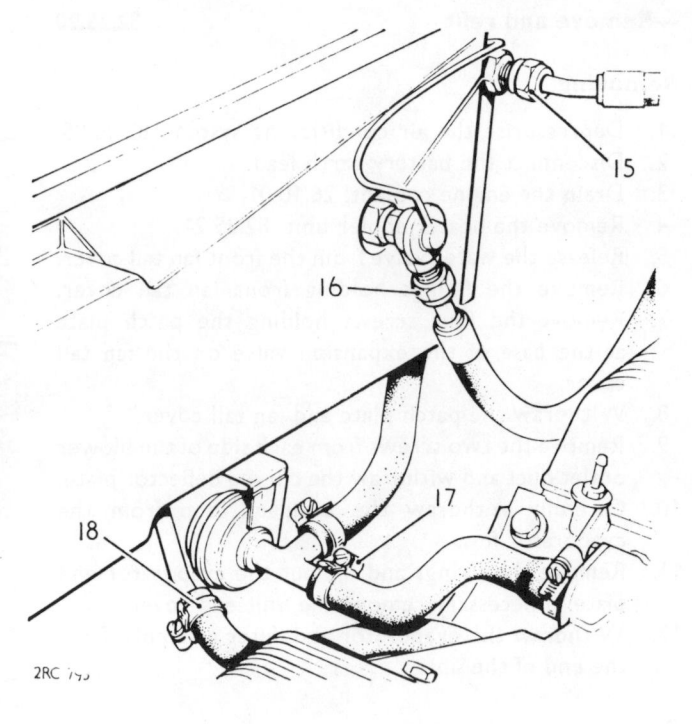

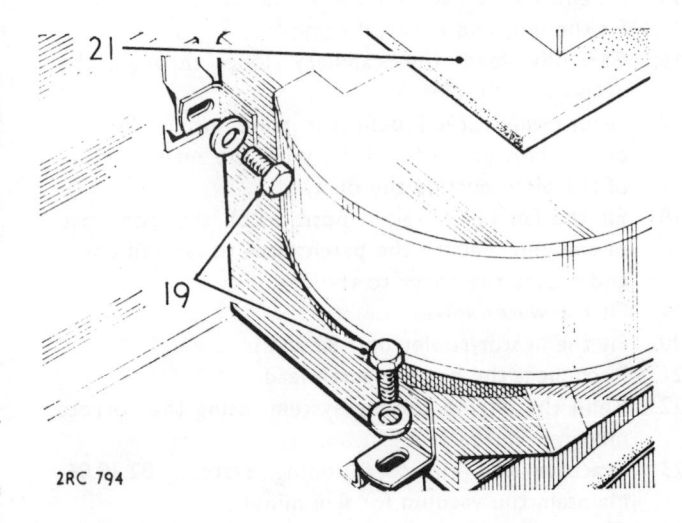

continued

Rover 3500 and 3500S Manual AKM 3621

AIR CONDITIONING

Refitting

22. Remove all remains of the old seals from the rear and underside of the heater/cooler unit.
23. Using Bostik 'C' or a similar adhesive, fit new seals to the rear and underside of the unit.
24. When the seals are firmly stuck to unit, coat the seal outer mating faces with the adhesive, in order to make an air-tight seal around the unit inlet and outlet apertures.
25. Locate the heater/cooler unit in position on the vehicle, feeding the vacuum and wiring loom through the control aperture in the vertical bulkhead. Secure the upper and lower fixings.
26. **On LHStg. models**—Secure the accelerator cross-shaft to the steering box.
27. Fit the heater outlet pipe to the left-hand side of the unit and secure the hose clip.
28. Connect the hose between the engine and the water valve, and secure the hose clip.
29. Remove the blanks, lubricate the flares and threads of the unions with refrigerant oil and connect the refrigeration pipe to the expansion valve.
30. Remove the blanks, lubricate the flares and threads of the unions with refrigerant oil and connect the refrigeration pipe to the evaporator outlet. Use a spanner on both hexagons to avoid distortion.
31. Fit the air cleaner. 19.10.01.
32. Fit the air intake valance. 76.79.04.
33. Fit the windscreen wiper arms.
34. Connect the Light green/red electrical lead to the high pressure cut-out.
35. Push the vacuum pipe on to the vacuum reservoir.
36. Fit the bonnet. 76.16.01.
37. Connect the electrical leads to the compressor isolating switch, in accordance with the circuit diagram. Page 82.00.01.
38. Connect the electrical leads to the relay on the console unit.
39. Connect the White lead at the single connector, the Brown/green lead at the single connector, and the Black lead to the two-way connector, all located behind the console unit.
40. Fit the temperature control and air-flow switches. 82.20.10/11.
41. Connect the Light green/red electrical lead to the high pressure cut-out.
42. Reconnect the battery earth lead.
43. Refill the engine cooling system, using the correct mixture. 26.10.01.
44. Evacuate the air conditioning system. 82.30.06. Maintain the vacuum for five minutes.
45. Charge the air conditioning system. 82.30.08.
46. Check, and if necessary replenish, the compressor oil level. 82.10.14.
47. Start and run the engine. Check for leaks from the water and refrigerant hose connections.
48. Switch on the air conditioning and check for correct operation.

WATER HOSES-HEATER/COOLER UNIT

—Remove and refit 82.25.27

WARNING: Do not disconnect the refrigeration hoses fitted between the compressor and the heater/cooler unit unless the air conditioning system has been de-pressurised. 82.30.05.

Removing

1. Drain the engine coolant. 26.10.01.
2. Remove the air cleaner. 19.10.01.
3. Slacken the clips and remove the hose between the inlet manifold and the water valve.
4. Slacken the clips and remove the heater inlet hose.
5. Slacken the clips and remove the heater outlet hose.

Refitting

6. Reverse 2 to 5.
7. Refill the engine cooling system, using the correct mixture 26.10.01.
8. Check the water hoses for leaks.

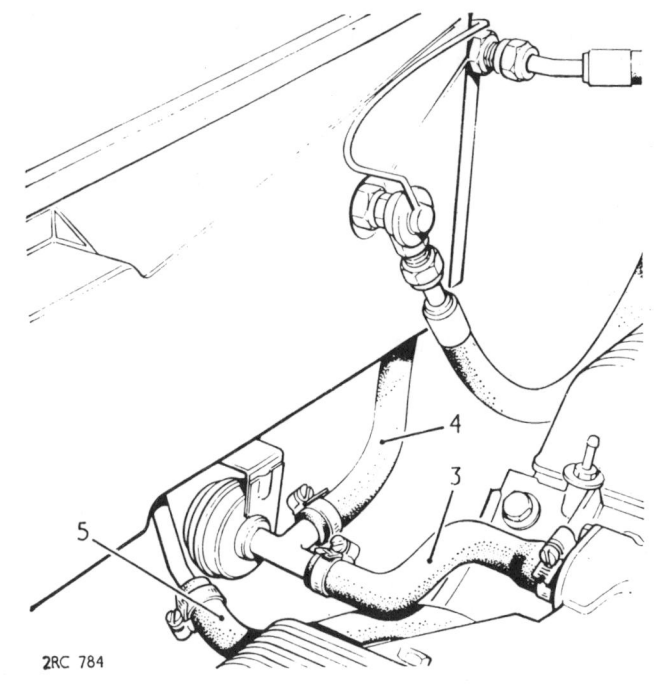

Rover 3500 and 3500S Manual AKM 3621

82.25.21
Sheet 2
82.25.27

AIR CONDITIONING

BLOWER MOTOR BRUSHES

—Remove and refit 82.25.31

Removing

1. Depressurise the air conditioning system. 82.30.05.
2. Disconnect the battery earth lead.
3. Drain the engine coolant. 26.10.01.
4. Remove the heater/cooler unit. 82.25.21.
5. Remove the top cover from the heater/cooler unit.
6. Disconnect the leads from the blower motor.
7. Withdraw the ventilator tube from the blower motor top cover.
8. Remove the fixings and withdraw the top cover from the blower motor.
9. Remove the shims from the motor spindle.
10. Remove the springs from the studs.
11. Withdraw the brush plate complete.

Refitting

12. Locate the brush plate in position on the blower motor.
13. Place the shims on the motor spindle.
14. Fit a spring to each end cover fixings stud.
15. Take care not to displace the motor fixing studs and fit the top cover. Secure with washers and nuts.
16. Connect the electrical feed lead to the blower motor.
17. Connect the black earth lead to the blower motor, as illustrated.

    NOTE: If the leads are incorrectly fitted, the blower motor will run backwards, when in use.

18. Connect the ventilator tube to the blower motor top cover.
19. Secure the top cover to the heater/cooler unit.
20. Fit the heater/cooler unit. 82.25.21.
21. Reconnect the battery earth lead.
22. Refill the engine cooling system, using the correct mixture. 26.10.01.
23. Evacuate the air conditioning system. 82.30.06. Maintain the vacuum for five minutes.
24. Charge the air conditioning system. 82.30.08.
25. Check, and if necessary replenish, the compressor oil level. 82.10.14
26. Start and run the engine. Check for leaks from the water and refrigerant hose connections.
27. Switch on the air conditioning and check for correct operation.

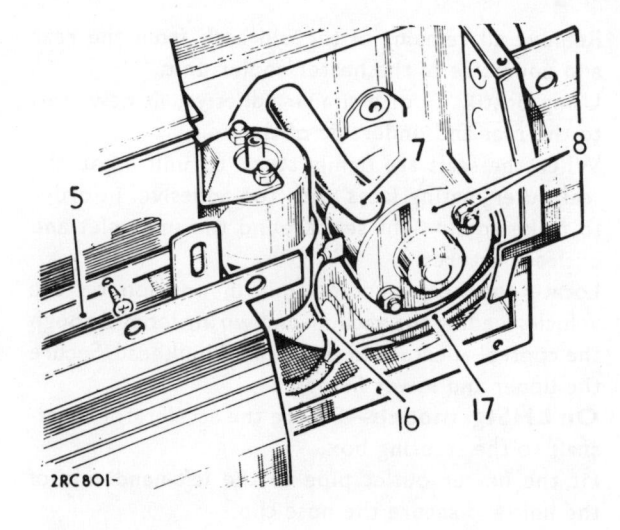

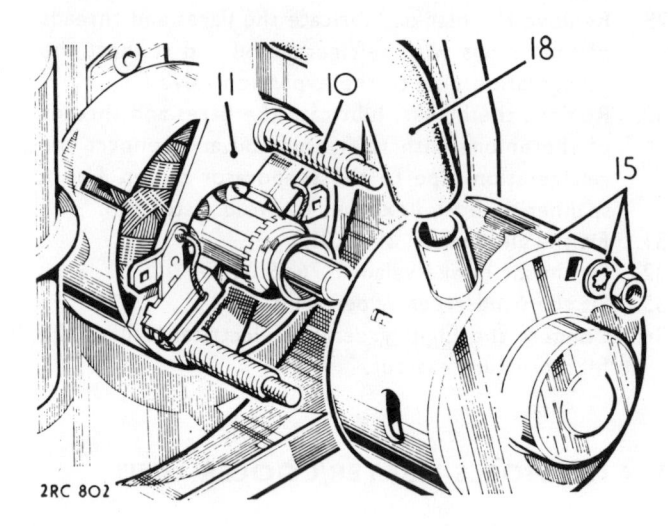

AIR CONDITIONING

CHARGING AND TESTING EQUIPMENT

—Fit and remove   82.30.01

For evacuating or charging with liquid refrigerant

1 to 5 and 7 to 23.

For sweeping or charging with gaseous refrigerant

1 to 4, 6 to 13 and 18 to 23.

NOTE:

There are two methods of connecting the charging and testing equipment, depending on the Operation to be carried out. The method described for 'evacuating or charging with liquid refrigerant' also applies to 'Pressure test' and 'Compressor oil level check' operations.

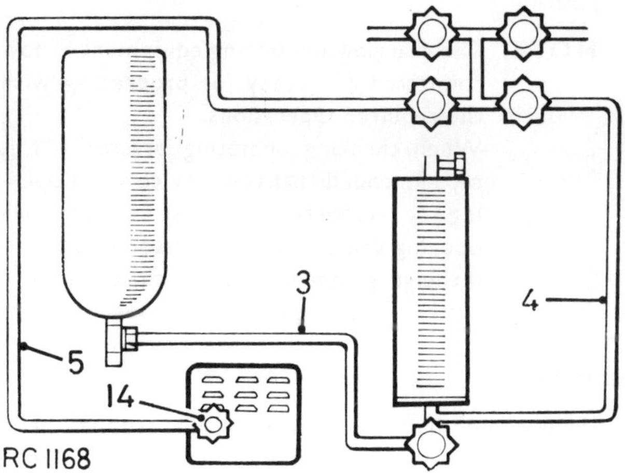

Fitting

1. Ensure that all the valves on the charging and testing equipment are closed.
2. Mount a 11,3 kg (25 lb) drum of refrigerant upside-down on the support at the rear of the charging equipment, and secure with the web strap.
3. Connect the hose from the bottom of the charging cylinder to the refrigerant drum valve.
4. Connect the hose between the bottom of the charging cylinder and the refrigerant control valve (No. 4).
5. **For evacuating or charging with refrigerant in a liquid state**—Connect the hose between the vacuum pump valve and the vacuum control valve (No. 3).
6. **For sweeping or charging with refrigerant in gaseous state**—Connect the hose between the top of the charging cylinder and the vacuum control valve (No. 3).
7. Prop open the car bonnet.
8. Remove the caps from the compressor service valves.
9. Check that both service valves are fully back seated (turned fully anti-clockwise).
10. Remove the caps from the gauge connections on both service valves.
11. Remove the blanking caps from the equipment charging lines and coat the threads and flares with refrigerant oil.
12. Connect the low pressure charging line (blue) from valve No. 1, to the compressor suction service valve.
13. Connect the high pressure charging line (red) from valve No. 2, to the compressor discharge service valve.
14. Start the vacuum pump and open the vacuum pump valve.
15. Open valve numbers 1, 2 and 3, to evacuate the charging lines.
16. Close valve numbers 1, 2 and 3 and stop the vacuum pump.

continued

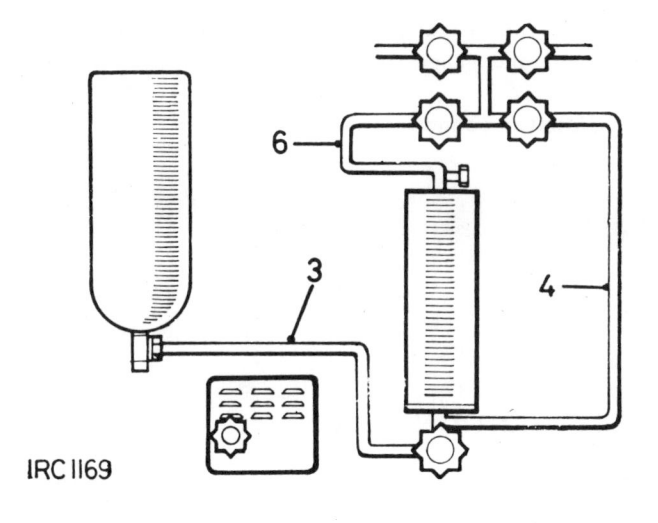

Rover 3500 and 3500S Manual AKM 3621

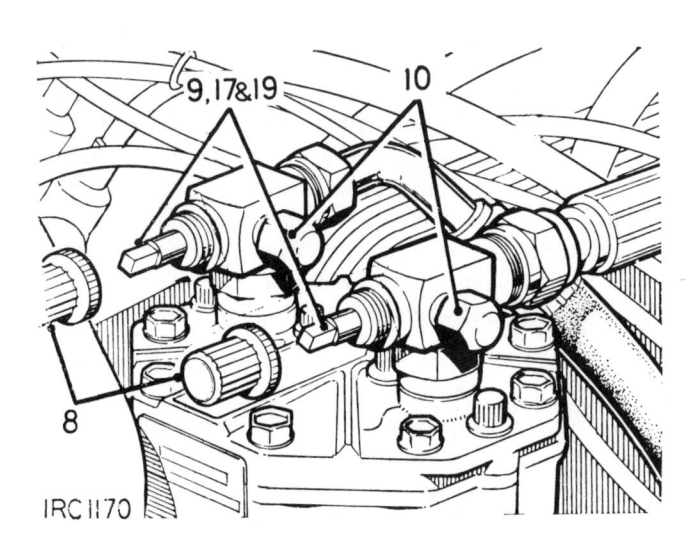

AIR CONDITIONING

17. Open the service valves three turns to the midway position.

    NOTE: The charging and testing equipment is now connected and ready for proceeding with the required Operations.
    When checking operating pressures it is recommended that the service valve opening be restricted to ¼-turn. This small opening damps the compressor pulsations. If the gauge needle flutters, close the valve further until the fluttering stops.

Removing

18. If the engine has been run, it must be stopped prior to disconnecting the charging and testing equipment.
19. Close both service valves (turn fully anti-clockwise).
20. Close all valves on the charging and testing equipment.
21. Disconnect the charging lines from the compressor.
22. Refit the blanking caps to the compressor valve stems and gauge connections, and to the charging lines.
23. Close the bonnet.

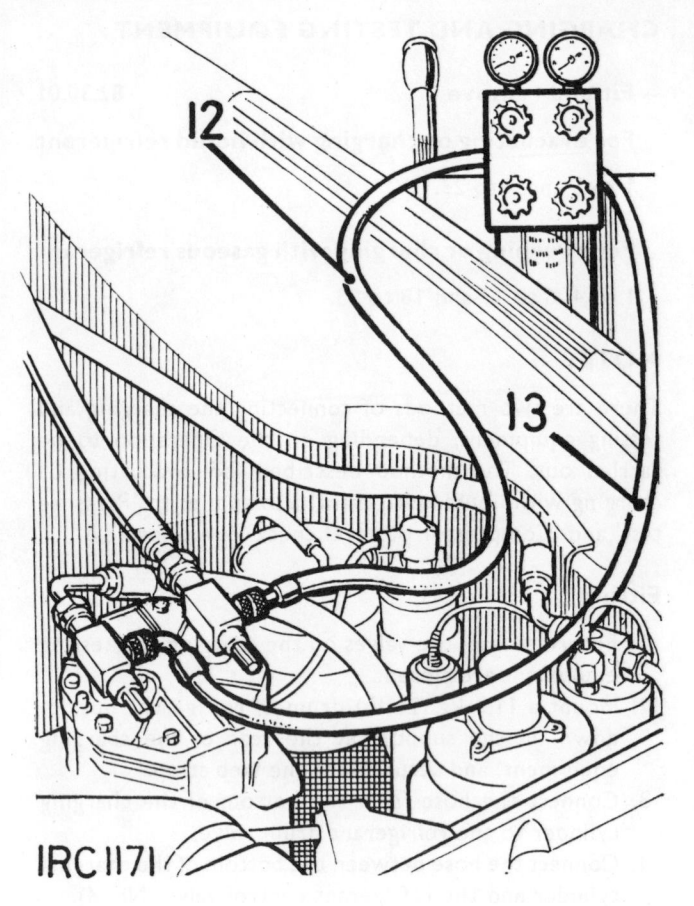

AIR CONDITIONING SYSTEM

—Depressurise   82.30.05

NOTE: The air conditioning refrigeration system contains 'Refrigerant 12' under pressure, and before any component is disconnected or removed, the system must be discharged of all pressure.

Refrigerant 12 is transparent and colourless in both the gaseous and liquid state. It has a boiling point of —30°C (—21.7°F) and at all normal pressures and temperatures it is a vapour. The vapour is heavier than air, non-flammable and non-explosive. It is non-poisonous except when in contact with an open flame, and non-corrosive until it comes into contact with water.

Proceed cautiously, regardless of gauge readings.

WARNING: Open connections slowly, keeping the hands and face well clear, so that no injury occurs if there is liquid in the line. If pressure is noticed allow it to bleed off slowly.
Always wear safety goggles when opening refrigerant connections.

continued

82.30.01
Sheet 2
82.30.05
Sheet 1

Rover 3500 and 3500S Manual AKM 3621

AIR CONDITIONING

Depressurising

1. Place the car in a ventilated area away from open flames and heat sources.
2. Stop the engine.
3. Prop open the bonnet.
4. Remove the caps from the compressor service valves.
5. Check that both service valves are fully back seated (turned fully anti-clockwise).
6. Close all valves on the charging and testing equipment.
7. Put on safety goggles.
8. Connect the high pressure charging line (red) from valve No. 2, to the compressor discharge service valve.
9. Open the compressor discharge service valve (turn clockwise) a quarter of a turn.
10. Slowly open valve No. 2, one turn.
11. Hold the end of the low pressure charging line (blue) in an absorbent rag.
12. Slowly open valve No. 1, and discharge the refrigerant vapour into the rag. If oil is discharged, reduce the valve opening.
13. When the pressure has been reduced, and the hissing sound ceases, close the valves Nos. 1 and 2 on the charging and testing equipment.
14. Close the compressor discharge service valve (turn fully anti-clockwise).
15. Disconnect the high pressure charging line from the compressor service valve.

    NOTE: If it is necessary to disconnect the compressor hoses, the compressor should be sealed by front seating the relevant service valve (turn fully clockwise). It is essential to ensure that both service valves are back seated before operating the compressor. Similarly, any other component of the refrigerant system should be capped immediately when disconnected.

16. Open the refrigeration drum valve.
17. Open the valve at the base of the charging cylinder and allow approximately 0,25 kg (½ lb) of refrigerant to enter the cylinder.
18. Close the refrigeration drum valve and the valve at the base of the charging cylinder.
19. Open valve No. 4 (refrigerant control) and flush out the high and low pressure lines by opening valve numbers 1 and 2 momentarily until a white stream of refrigerant is observed.
20. Close all valves on the charging and testing equipment, and fit the blanking caps.
21. Remove safety goggles.
22. The air conditioning system is now depressurised.

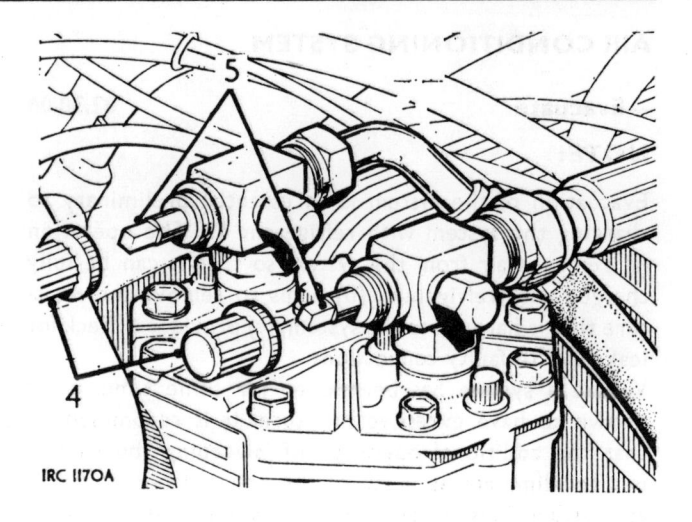

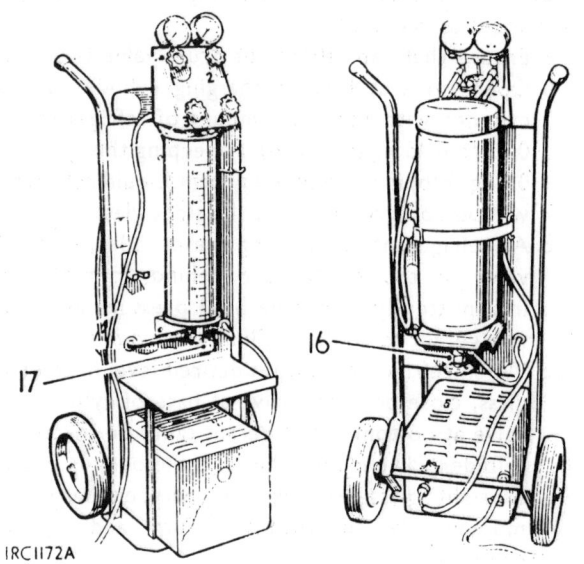

Rover 3500 and 3500S Manual AKM 3621

AIR CONDITIONING

AIR CONDITIONING SYSTEM

—Evacuate 82.30.06

NOTE:

Evacuation of the system is an essential preliminary to charging the system with refrigerant 12. The operation (a) removes air from the system so that it can be fully charged with refrigerant, (b) helps to remove the moisture that is harmful to the system, (c) provides a check for leaks due to faulty connections.

Where a system has been open for some time, or is known to have excessive moisture, it is recommended that the additional operation of 'sweeping' be carried out to eliminate as much of the accumulated moisture as possible. This is done after evacuation, and is detailed in Operation 82.30.07. If sweeping is intended, the receiver-dryer unit must be replaced before commencing evacuation. It is recommended that vacuum of the system is maintained for 20 minutes, following sweeping of the system.

Evacuating

1. Depressurise the air conditioning system 82.30.05.
2. Connect the charging and testing equipment as for evacuating 82.30.01.
3. Open the low pressure valve (No. 1).
4. Open the high pressure valve (No. 2).
5. Start the vacuum pump and check that the vacuum pump valve is open.
6. Slowly open the vacuum control valve (No. 3). If vacuum is applied to the system too quickly the residual oil may be drawn out.
7. A vacuum of 711 mm (28 in.) Hg at or near sea level should be reached. Allow 25 mm (1 in.) Hg reduced vacuum for each 300 metres (1,000 feet) of elevation.
8. While the system is evacuating fill the charging cylinder as required:
   a Ensure that the refrigerant drum valve is opened.
   b Open valve at base of charging cylinder and fill cylinder with required amount of refrigerant, i.e. 0,25 to 0,45 kg (½ to 1 lb) if sweeping the system or 0,9 kg (2 lb) if charging the system. Liquid refrigerant will be observed rising in the sight glass.
   c As refrigerant stops filling the cylinder, open the valve at top of cylinder behind control panel intermittently to relieve head pressure and allow refrigerant to continue filling the cylinder.
   d When refrigerant reaches desired level in the sight glass, close both the valve at base of cylinder and valve at bottom of refrigerant tank. Be certain top cylinder valve is fully closed. If bubbling is present in sight glass, reopen the cylinder base valve momentarily to equalise drum and cylinder pressures.
9. If 711 mm (28 in.) Hg of vacuum cannot be obtained, close the vacuum control valve (No. 3), stop the vacuum pump and check the system for leaks.
10. Close the vacuum control valve (No. 3).
11. Close the vacuum pump valve, switch off the pump and allow the vacuum to hold for 15 minutes, then check that no pressure rise—loss of vacuum—is evident on the compound gauge. Any pressure rise denotes a leak that must be rectified before proceeding further. See leak detection, Operation 82.30.09 It is possible for residual liquid refrigerant in the compressor oil to vapourise and create a slight pressure rise. This can be eliminated by starting the engine and energising the magnetic clutch to rotate the compressor for about 30 seconds, then evacuate the system again.
12. With the system satisfactorily evacuated, the system is ready for sweeping or charging with refrigerant.

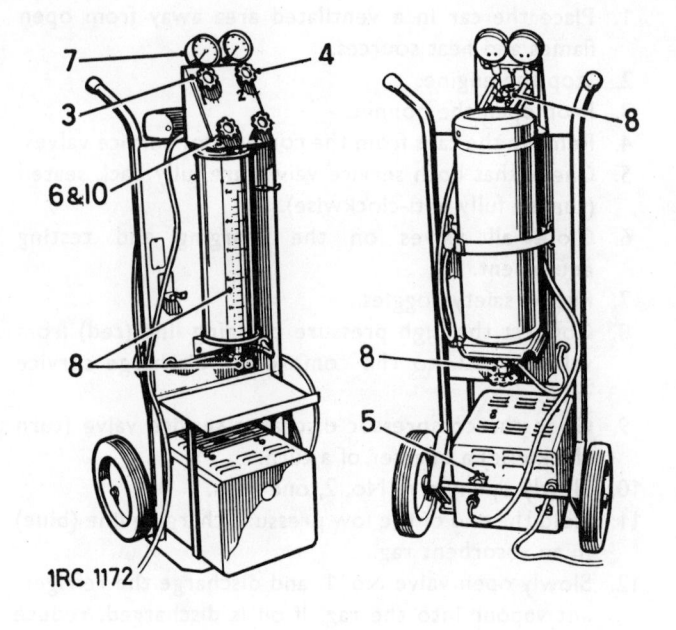

Rover 3500 and 3500S Manual AKM 3621

AIR CONDITIONING

AIR CONDITIONING SYSTEM

—Sweep 82.30.07

NOTE: This operation is in addition to evacuating, and is to remove moisture from systems that have been open to atmosphere for a long period, or that are known to contain excessive moisture.

Sweeping

1. Fit a new liquid receiver/drier. 82.17.01.
2. Ensure that a full drum of refrigerant is fitted on the charging and testing equipment.
3. Fit the charging and testing equipment, 82.30.01, as described for evacuating.
4. Evacuate the air conditioning system, 82.30.06, allowing 0,25 to 0,45 kg ($\frac{1}{2}$ to 1 lb) of refrigerant to enter the charging cylinder.
5. Close all valves on the charging and testing equipment.
6. Disconnect the intake hose from the vacuum pump.
7. Connect the intake hose to the valve at the top of the charging cylinder.
8. Open the valve at the top of the charging cylinder.
9. Put on safety goggles.
10. Crack open the hose connection at valve No. 3, and allow some refrigerant to purge the hose, then close the connection.
11. Open the high pressure valve (No. 2).
12. Slowly open valve No. 3, which is now connected to the top valve of the charging cylinder, and allow gas to flow into the system until the reading on the compound gauge remains steady. Between 0,25 and 0,45 kg ($\frac{1}{2}$ and 1 lb) of refrigerant will enter the system.
13. Allow the dry refrigerant introduced into the system to remain for 10 minutes.
14. Crack the suction valve charging line at the connection on the compressor to allow an escape of refrigerant, at the same time observing the sight glass in the charging cylinder. A slight drop in the level should be allowed before closing the connection at the compressor.
15. Close the high pressure valve (No. 2).
16. Close valve No. 3.
17. Close the valve at the top of the charging cylinder.
18. Reconnect the charging and testing equipment, 82.30.01, as described for evacuating.
19. Evacuate the air conditioning system, 82.30.06. Maintain the vacuum for twenty minutes.
20. The air conditioning system is now ready for charging with refrigerant. 82.30.08.

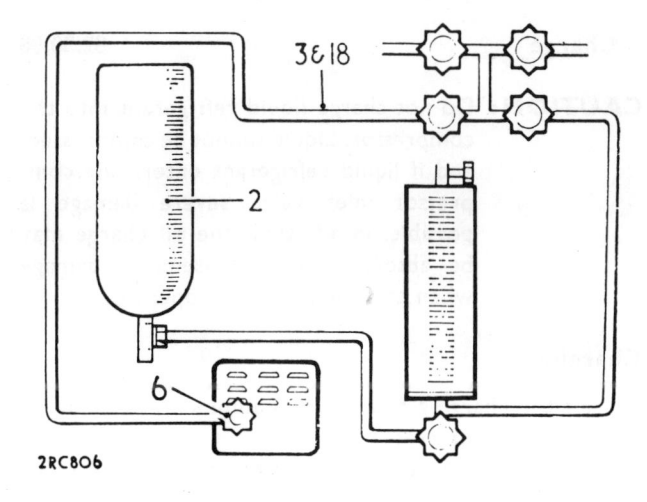

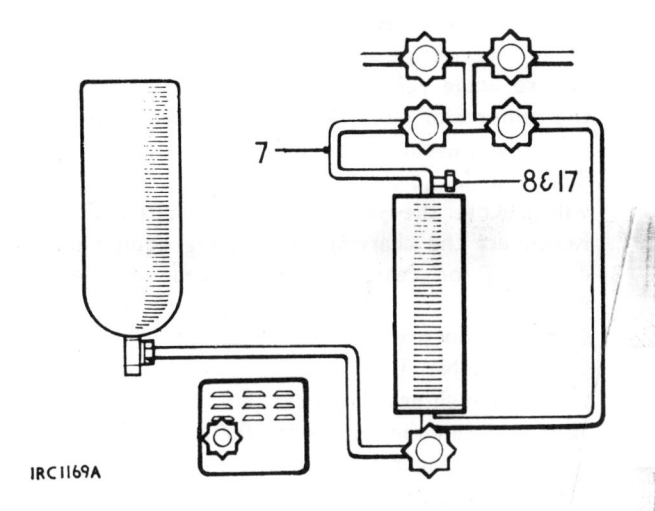

AIR CONDITIONING

AIR CONDITIONING SYSTEM

—Charge 82.30.08

CAUTION: Do not charge liquid refrigerant into the compressor. Liquid cannot be compressed, and if liquid refrigerant enters the compressor inlet valve severe damage is possible. In addition, the oil charge may be absorbed, with consequent damage when the compressor is operated.

Charging

1. Fit the charging and testing equipment, 82.30.01, as described for evacuating.
2. Evacuate the air conditioning system, 82.30.06, allowing 0,9 kg (2 lb) of refrigerant to enter the charging cylinder.
3. Put on safety goggles.
4. Close the low pressure valve (No. 1).
5. Open the refrigerant control valve (No. 4) and release liquid refrigerant into the system through the compressor discharge valve. The pressure in the system will eventually balance.
6. If the full charge of 0,9 kg (2 lb) of liquid refrigerant will not enter the system, proceed with items 7 to 12.
7. Reconnect the charging and testing equipment as described for charging with gaseous refrigerant. 82.30.01.
8. Open the low pressure valve (No. 1).
9. Open valve No. 3.
10. Close the high pressure valve (No. 2).
11. Start and run the engine at 1000 to 1500 rev/min and allow refrigerant to be drawn through the low pressure valve (No. 1) until the full charge of 0,9 kg (2 lb) has been drawn into the system.
12. Close valve numbers 1 and 3.
13. Close valve No. 4.
14. Check that the air conditioning system is operating satisfactorily by carrying out a pressure test. 82.30.10.

CAUTION: Do not overcharge the air conditioning system as this will cause excessive head pressure.

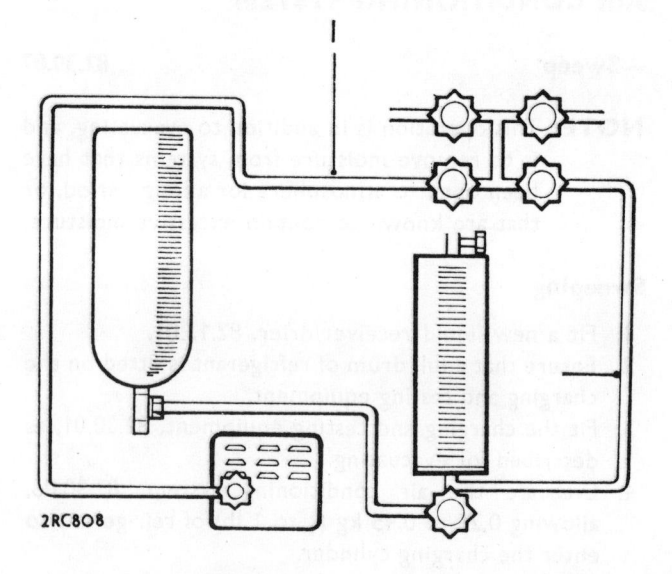

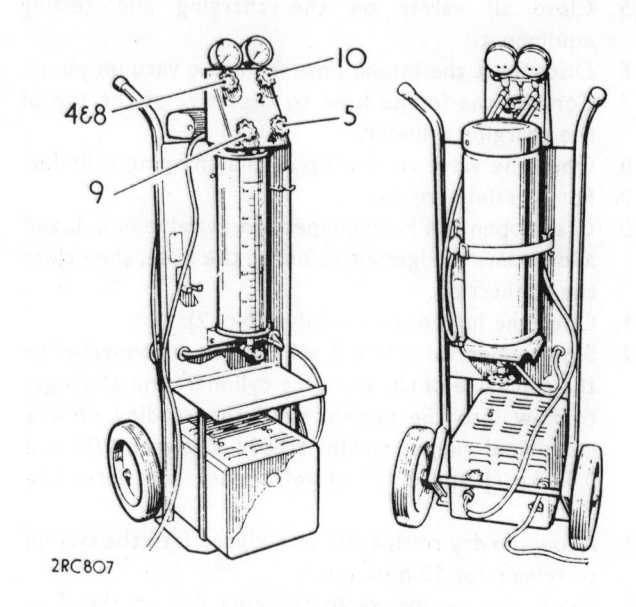

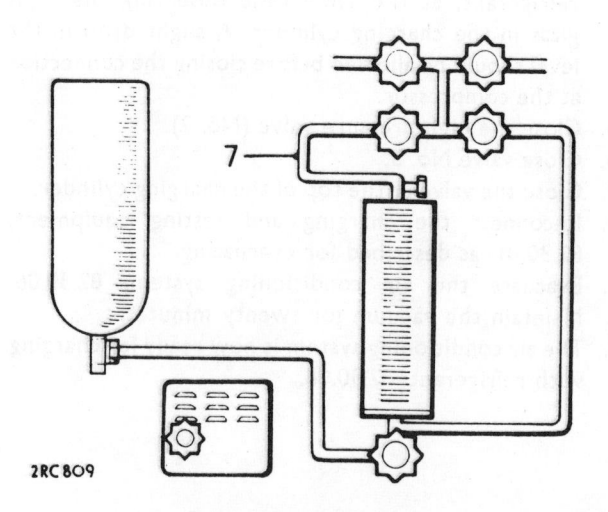

Rover 3500 and 3500S Manual AKM 3621

AIR CONDITIONING

AIR CONDITIONING SYSTEM

—Leak test 82.30.09

NOTE: Loss of vacuum during the evacuation operation indicates a leak in the system that must be rectified.

The leak detector pulls the refrigerant gas through the search hose into the burner of a hydro-carbon gas. The burner includes a copper element that causes the refrigerant to burn with a green or purple flame.

Every precaution must be taken to ensure that there is no possible danger of petrol or other inflammable material coming into contact with the naked flame of the burner. The leak detector is intentionally provided with a long search hose so that the operator can check for leaks whilst the burner is a safe distance from the car.

WARNING: The product of the refrigerant and burning hydro-carbon is Chlorine Gas, and must not be inhaled.

Leak testing

1. Place the car in a well ventilated area or refrigerant may persist in the vicinity and give misleading results. Strong draughts are to be avoided, as a seepage from a leak could be dissipated without detection.
2. Charge the system with refrigerant to a pressure of about 2,8 kgf/cm^2 (40 lbf/in^2).
3. Open the shut-off valve on the gas cylinder.
4. Open the burner shut-off valve and light the burner.
5. Pass the search hose of the lighted leak detector very slowly around the circumference of each refrigerant connection, with particular attention to the underside, as the refrigerant gas is heavier than air.
6. Insert the search hose into an air outlet of the evaporator. Switch the air conditioner blower on and off at intervals of 10 seconds. Leaking refrigerant will be gathered in by the blower and discharged into the search hose. With the blower switched off insert the search hose into one of the evaporator drain tubes. The heavy refrigerant vapour will sink through the evaporator to the drains.
7. Insert the search hose between the magnetic clutch and the compressor to check the shaft seal for leaks. Check all service valve connections, valve plate, head and base plate joints, and back seal plate.
8. Check the condenser at the pipe connections.
9. If any leaks are found the system must be depressurised before attempting rectification. If repairs by brazing are to be carried out the component must be removed from the car and all traces of refrigerant expelled before heat is applied.
10. After repairs the system should be retested for leaks and evacuated prior to charging.

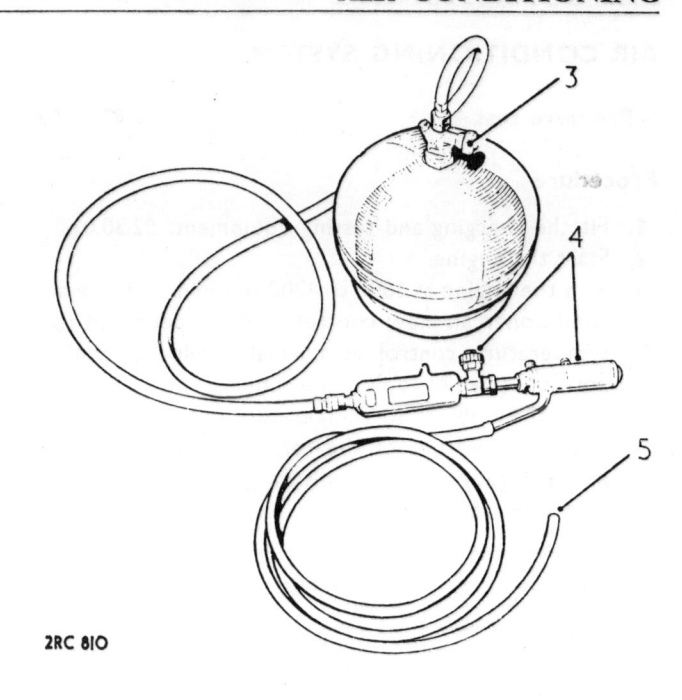

    NOTE: An electronic leak test unit can also be used; however, these units are much more sensitive than the burner type described above.

Rover 3500 and 3500S Manual AKM 3621

AIR CONDITIONING

AIR CONDITIONING SYSTEM

—Pressure test 82.30.10

Procedure

1. Fit the charging and testing equipment. 82.30.01.
2. Start the engine.
3. Run the engine at 1000 to 1200 rev/min with the air conditioner air flow control at high speed and the temperature control at normal cooling operating position. (Half of cooling movement.)
4. Note the ambient air temperature in the immediate test area in front of the car, and check the high pressure gauge reading—discharge side—against the following table.

Ambient Temperature		Compound Gauge Readings		High Pressure Gauge Readings	
°C	°F	kgf/cm²	lbf/in²	kgf/cm²	lbf/in²
16	60	1,05–1,4	15–20	7,0–10,2	100–150
26,7	80	1,4–1,75	20–25	9,8–13,3	140–190
38	100	1,75–2,1	25–30	11,6–15,8	180–225
43,5	110	2,1–2,45	30–35	15,1–17,5	215–250

The pressure gauge readings will vary within the range quoted with the rate of flow of air over the condenser, the higher readings resulting from a low air flow. It is advisable to place a fan for additional air flow over the condenser if the system is to be operated for a long time. Always use a fan if temperatures are over 26,7°C (80°F) so that consistent analysis can be made of readings.

5. If the pressure readings are outside the limits quoted, refer to the fault diagnosis chart at the beginning of this Division.
6. Stop the engine.
7. Close both service valves (turn fully anti-clockwise).
8. Close all valves on the charging and testing equipment.
9. Disconnect the charging lines from the compressor.
10. Refit the blanking caps to the compressor valve stems and gauge connections, and to the charging lines.
11. Close the bonnet.

AIR CONDITIONING EQUIPMENT

—Test 82.30.16

Procedure

1. Place the car in a ventilated, shaded area free from excessive draught, with the car doors and windows open.
2. Check that the surface of the condenser is not restricted with dirt, leaves, flies, etc. Do not neglect to check the surface between the condenser and the car radiator. Clean as necessary.
3. Switch on the ignition and the air conditioner air flow control. Check that the blower is operating efficiently at low, medium and high speeds. Switch off the blower and the ignition.
4. Check that the evaporator condensate drains are open and clear.
5. Check the tension of the compressor driving belt, and adjust if necessary. 82.10.01.
6. Inspect all connections for the presence of refrigerant oil. If oil is evident, check for leaks, 82.30.09, and rectify as necessary.

   NOTE: The compressor oil is soluble in Refrigerant 12 and is deposited when the refrigerant evaporates from a leak.

7. Start the engine.
8. Set the temperature control switch to maximum cooling (fully anti-clockwise) and switch the air conditioner blower control on and off several times, checking that the magnetic clutch on the compressor engages and releases each time.
9. With the temperature control at maximum cooling and the blower control at high speed, warm up the engine and fast idle at 1000 rev/min. Check the sight glass in the top of the receiver-dryer for bubbles or foam. The sight glass should be generally clear after five minutes running, occasional bubbles being acceptable. Continuous bubbles may appear in a serviceable system on a cool day, or if there is insufficient air flow over the condenser at a high ambient temperature.
10. Repeat at 1800 rev/min.
11. Gradually increase the engine speed to the high range, and check the sight glass at intervals.
12. Check for frosting on the service valves and evaporator fins.
13. Check the high pressure pipes and connections by hand for varying temperature. Low temperature indicates a restriction or blockage at that point.
14. Switch off the air conditioning blower and stop the engine.
15. If the air conditioning equipment is still not satisfactory, proceed with the pressure test. 82.30.10.

WINDSCREEN WIPERS AND WASHERS

LIST OF OPERATIONS

Wiper arms—remove and refit	84.15.01
Wiper motor and cable drive—remove and refit	84.15.09
Wiper motor and linkage—remove and refit	84.15.11
Wiper motor—remove and refit	84.15.12
Wiper motor—overhaul	84.15.18
Wiper wheel boxes—remove and refit	84.15.28/29
Wiper wheel boxes and drive cable tubes—remove and refit	84.15.27

WINDSCREEN WIPERS AND WASHERS

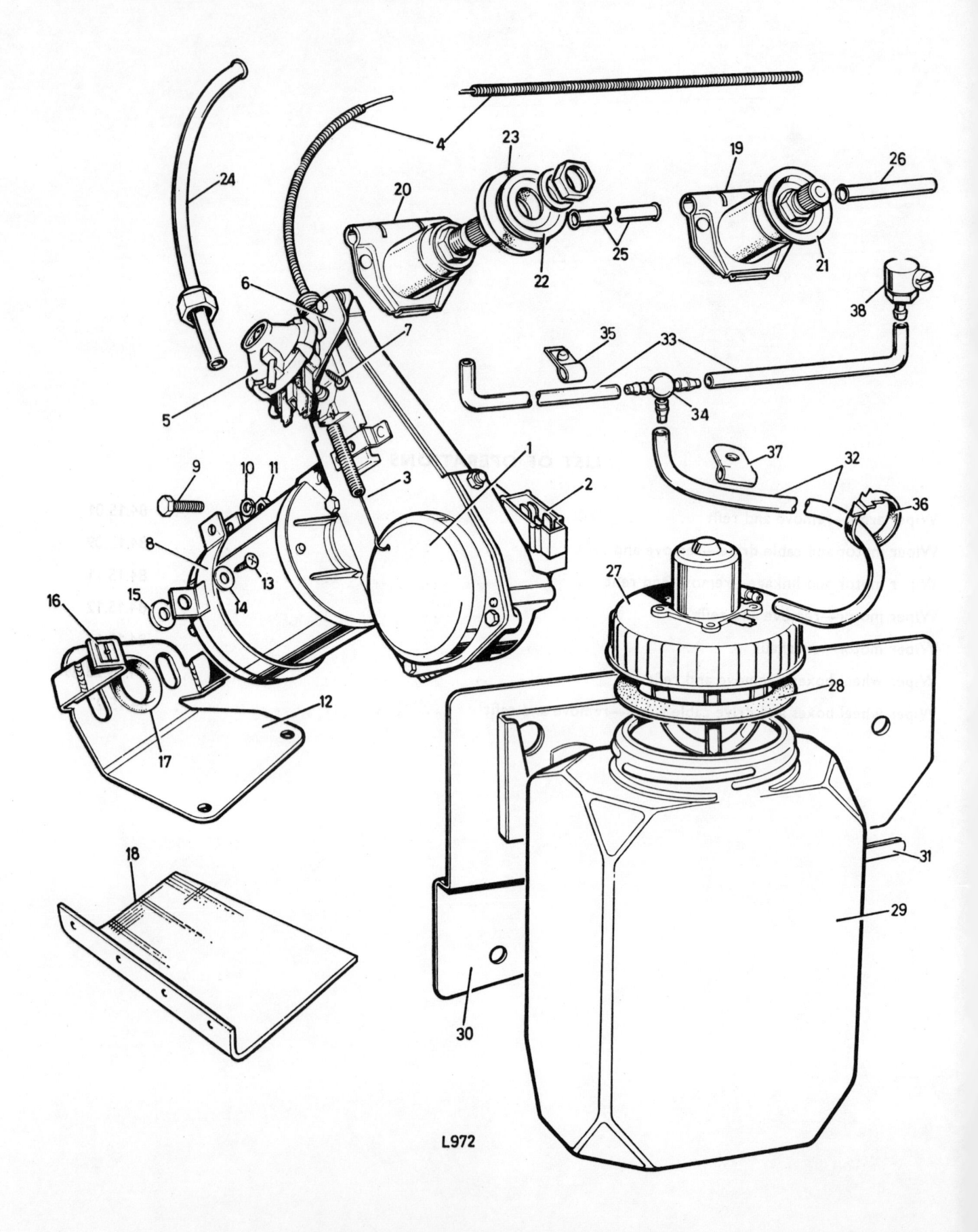

Illustration of windscreen wipers and washers

Rover 3500 and 3500S Manual AKM 3621

WINDSCREEN WIPERS AND WASHERS

Key to illustration of windscreen wiper and washer

1	Screen wiper motor assembly	20	Wheelbox assembly, passenger's side
2	Park switch	21	Cover washer, driver's side
3	Return spring for breech block	22	Cover washer, passenger's side
4	Drive cable for windscreen wiper	23	Grommet under cover washer
5	Governor for wiper motor delay	24	Outer casing, motor to wheelbox
6	Bracket for governor	25	Outer casing, wheelbox to wheelbox
7	Screw, taptite fixing governor to bracket	26	Outer casing, wheelbox end
8	Clamp bracket for wiper motor	27	Motor, pump and cap assembly
9	Screw (10 UNF x 5/8 in long) ⎫ Fixing clamp bracket to motor	28	Seal for cap
10	Spring washer	29	Reservoir only
11	Nut (10 UNF) ⎭	30	Mounting bracket for reservoir
12	Mounting bracket for screen wiper motor	31	Protection strip for mounting bracket
13	Drive screw ⎫ Fixing clamp bracket to mounting bracket	32	Plastic tubing, reservoir to tee piece
14	Plain washer	33	Plastic tubing, tee piece to nozzle
15	Grommet	34	'T' piece connector for tubing
16	Locknut ⎭	35	Clip for tubing
17	Grommet for mounting bracket	36	Cleat for tubing at LH wing valance
18	Protection flap for wiper motor	37	Clip for tubing at heater and drain channel
19	Wheelbox assembly, driver's side	38	Windscreen washer nozzle complete

Rover 3500 and 3500S Manual AKM 3621

WINDSCREEN WIPERS AND WASHERS

WIPER ARMS—SET

—Remove and refit　　　　　　　　84.15.01

Removing

1. Hold back the small spring clip that retains the wiper arm on the spindle boss, by means of a suitable tool.
2. Gently prise off the wiper arm from the spindle boss.

Refitting

3. Allow the wiper motor to move to the 'park' position.
4. Push the arm on to the boss, locating it on the splines so that the wiper blades are just clear of the screen rail.
5. Ensure that the spring retaining clip is located in the retaining groove on the spindle boss.

Cable drive type windscreen wiper

6. Final fine adjustment is carried out by rotating the knurled nut located near the cable outlet, half a turn at a time, until satisfactory.

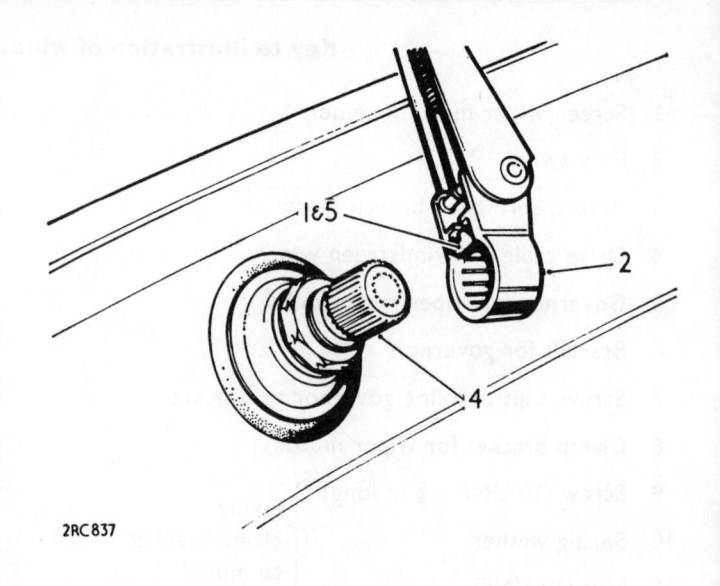

WIPER MOTOR AND DRIVE

—Remove and refit　　　　　　　　84.15.09

NOTE: The cable drive type of windscreen wiper is fitted to later 3500 models and all 3500S models. For early 3500 models, refer to 84.15.11.

Removing

1. Disconnect the battery earth lead.
2. Detach the wiper arms from the wheelboxes.
3. Remove the wheelbox locknuts.
4. Open and prop the bonnet.
5. Remove the two screws and nuts securing the air intake valance to heater box.
6. Remove the two drive screws securing the air intake valance outer ends to base unit.
7. Withdraw air intake valance.
8. Air conditioned models—withdraw the section of the heater/cooler insulator cover nearest to the windscreen wiper motor.
9. Disconnect the vacuum pipe from the wiper motor delay governor.
10. Note the colour coding and detach the electrical leads from the delay governor.
11. Unscrew the hexagon nut securing the drive cable tubing to the wiper motor threaded ferrule.
12. Remove the two drive screws securing the wiper motor to the mounting bracket.
13. Turn and lift the wiper motor and remove the electrical leads plug from the motor casing.
14. Remove the wiper motor and at the same time withdraw the attached drive cable from the wheelboxes and drive cable tubing.

If required, remove the drive cable, 15 to 18

15. Remove the fixings and withdraw delay governor and cover from wiper motor casing.
16. Remove the circlip and plain washer from the cross-shaft pivot pin.
17. Lift off cross-shaft from cable and pivot pin.
18. Ease out the threaded ferrule from its location in the wiper motor body and withdraw the drive cable.

Refitting

19. Apply Ragosine Listate 225 grease to the wheelboxes and the interior of the protective tubing.
20. Reverse 1 to 18.

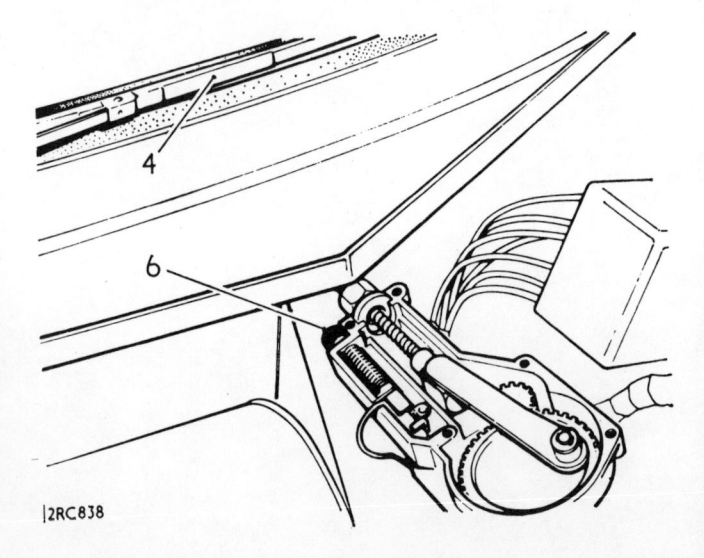

WINDSCREEN WIPERS AND WASHERS

WIPER MOTOR AND LINKAGE

—Remove and refit, 1 to 6 and 9   84.15.11

WIPER MOTOR

—Remove and refit, 1 to 9   84.15.12

NOTE: The link drive type of windscreen wiper is fitted to early 3500 models only. For later 3500 and all 3500S models, refer to 84.15.09.

Removing

1. Disconnect the battery earth lead.
2. Remove the air intake valance. 76.79.04.
3. Slacken the locknut clamping the link drive and spacer to the base unit.
4. Remove the locknut from the wiper motor bracket, held in position on the steering box cover plate.
5. Disconnect the electrical wiring from the motor.
6. Carefully manoeuvre the link drive bracket complete with motor and link drive, between the heater box and base unit, until the complete assembly can be lifted clear.
7. Release the spring circlip from the link to driver's side, at the motor end, and lift up clear of the bush.
8. Remove the three screws fixing the wiper motor to the bracket and remove the motor.

Refitting

9. Reverse 1 to 8.

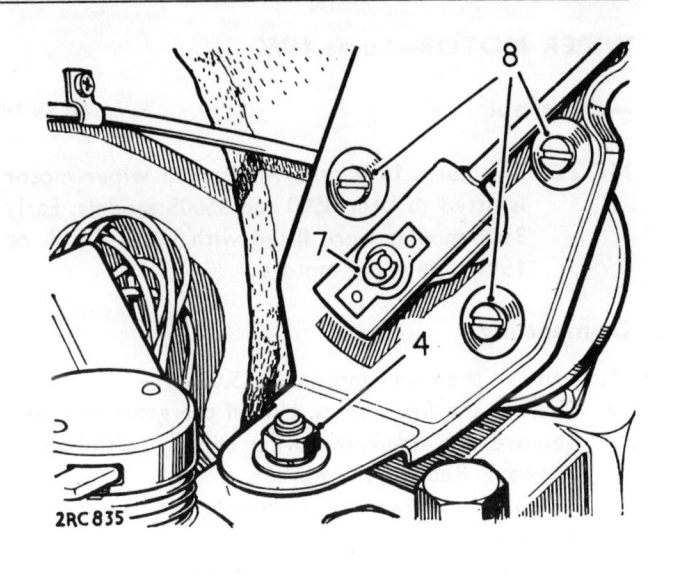

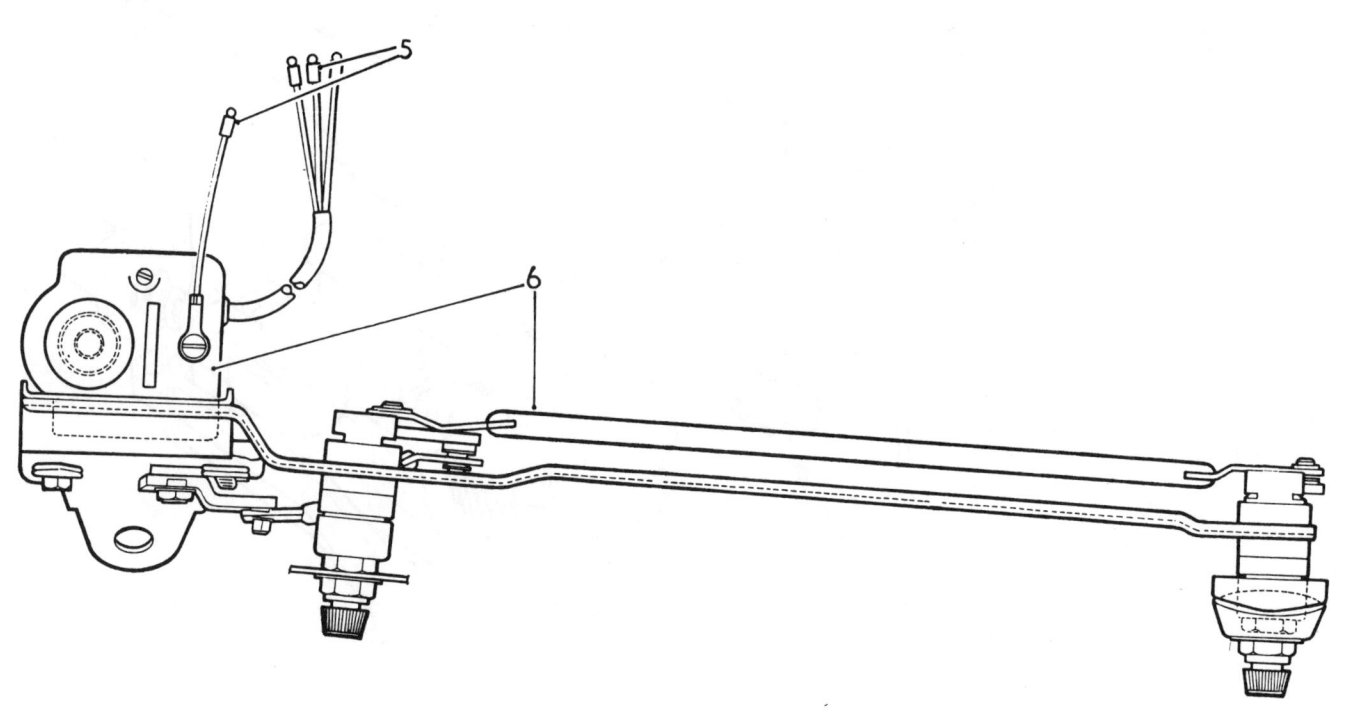

Rover 3500 and 3500S Manual AKM 3621

84.15.11
84.15.12

WINDSCREEN WIPERS AND WASHERS

WIPER MOTOR—Lucas 16W

—Overhaul 84.15.18

NOTE: The Lucas 16W type windscreen wiper motor is fitted to later 3500 and 3500S models. Early 3500 models were fitted with Lucas DL3A or 15W type wiper motors.

Dismantling

1. Remove the wiper motor. 84.15.09.
2. Remove the five screws. Lift off the gearbox cover.
3. Remove the crankpin spring clip by withdrawing sideways. Remove the washer.
4. Carefully withdraw the connecting rod. Remove the washer.
5. If necessary, remove the cross-head.
6. Remove the slider block. Note direction of cam slope.
7. Remove the final gearshaft spring clip by withdrawing sideways. Remove the washer.
8. Ensure that the shaft is burr-free and withdraw. Remove the dished washer.
9. Remove the thrust screw and locknut.
10. Remove the through-bolts.
11. Carefully withdraw the cover and armature about 5 mm (0.200 in). Continue withdrawal, allowing the brushes to drop clear of the commutator. Ensure that the brushes are not contaminated with grease.
12. Pull the armature from the cover against the action of permanent magnet.
13. Note position of the limit switch on the gearbox by scribing a pencil line.
14. Remove the five screws to release the brush assembly and limit switch. Remove both units joined together by wires.
15. Remove the plate.

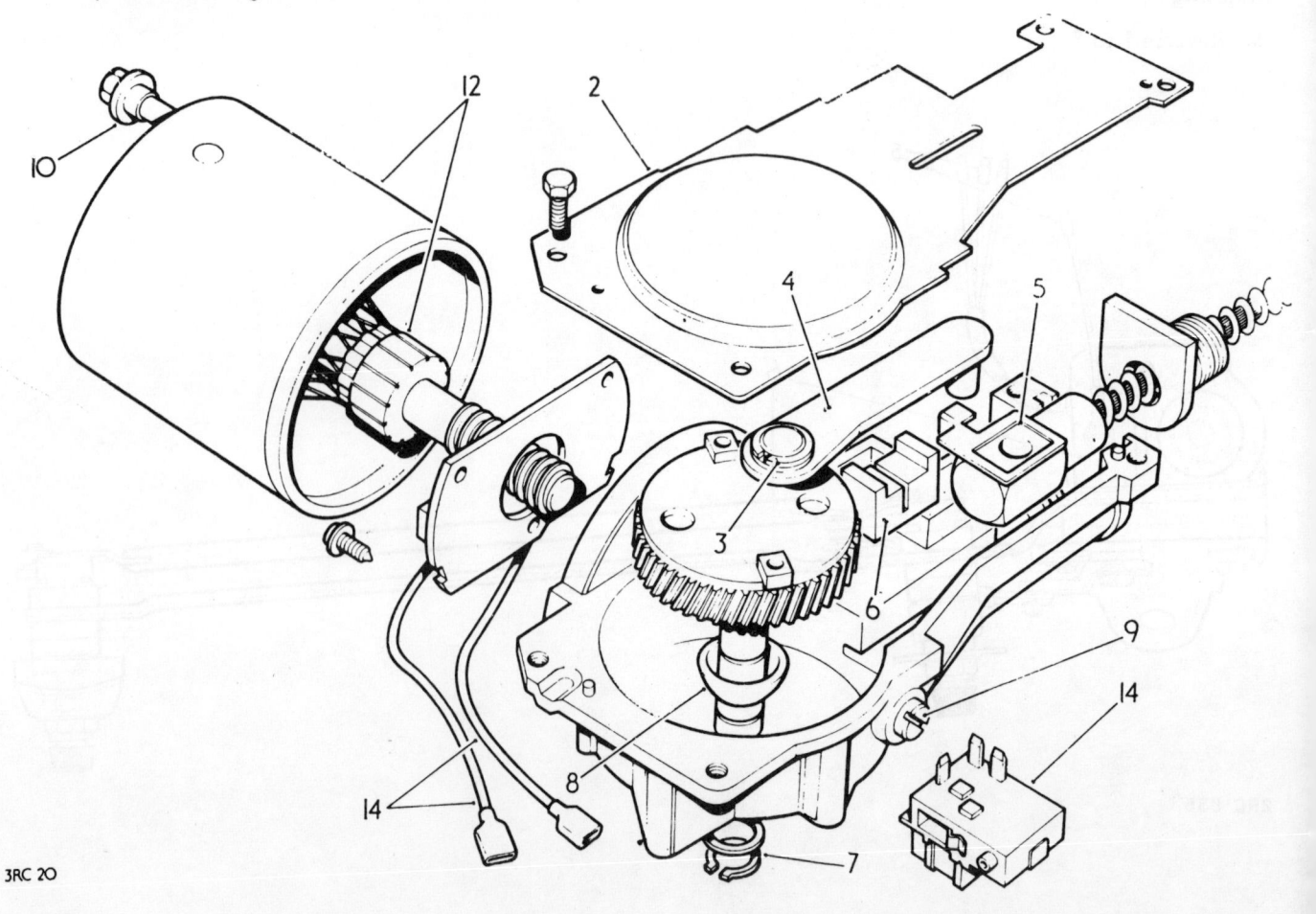

continued

3RC 20

84.15.18
Sheet 1

Rover 3500 and 3500S Manual AKM 3621

WINDSCREEN WIPERS AND WASHERS

Reassembling

16. Position the plate so that the round hole will accommodate the plunger.
17. Position the brush assembly and limit switch joined together by wires. Secure with five screws.
18. Slacken the two limit switch screws and align pencil lines. Tighten the screws.
19. Lubricate the cover bearing and saturate the cover bearing felt washer with Shell 41 Turbo oil. Position the armature to the cover against the action of the permanent magnet.
20. Lubricate the self-aligning bearing with Shell Turbo 41 oil. Carefully insert the armature shaft through the bearing. Ensure that the brushes are not contaminated with lubricant. Push the three brushes back to clear the commutator.
21. Seat the cover against the gearbox. Turn the cover to align the marks shown. Fit the through-bolts.

    NOTE: If the marks are not aligned, the motor will run in the reverse direction.

22. Fit the thrust screw and locknut.
23. Adjust the armature end-float as follows. Slacken the locknut. Screw the thrust screw in until resistance is felt. Screw the thrust screw out a quarter of a turn, maintain in this position and tighten the locknut.
24. Lubricate the final gear bushes with Shell Turbo 41 oil.
25. Fit the dished washer with concave surface facing the final gear. Insert the shaft.
26. Fit the washer. Fit the spring clip by inserting sideways.
27. Lubricate the slider block cam slope, block sides and the guide channel with Ragosine Listate grease.
28. Position the slider block with direction of cam slope as shown.
29. Pack Ragosine Listate grease around the worm gear, final gear and into the cross-head guide channel.
30. If necessary, fit the cross-head locating projection in the slider block slot.
31. Fit the washer. Lubricate the final gear crankpin with Shell Turbo 41 oil. Lubricate the cross-head end of the connecting rod, including pin, with Ragosine Listate grease. Carefully insert the connecting rod.
32. Fit the washer. Fit the spring clip by inserting sideways.
33. Position the gearbox cover. Secure with five screws.
34. Refit the wiper motor. 84.15.09.

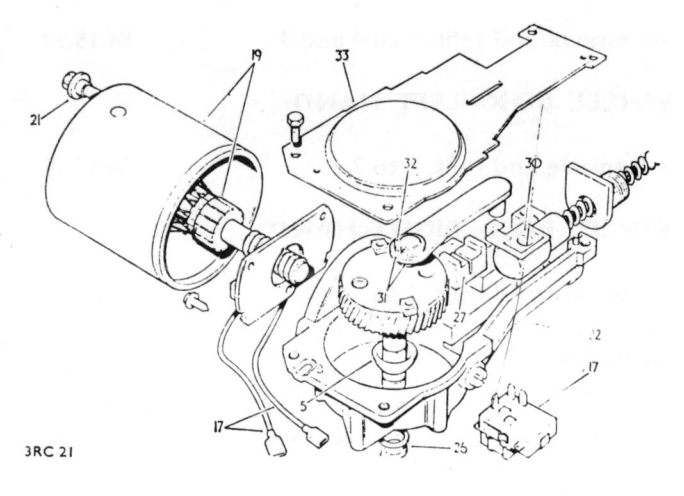

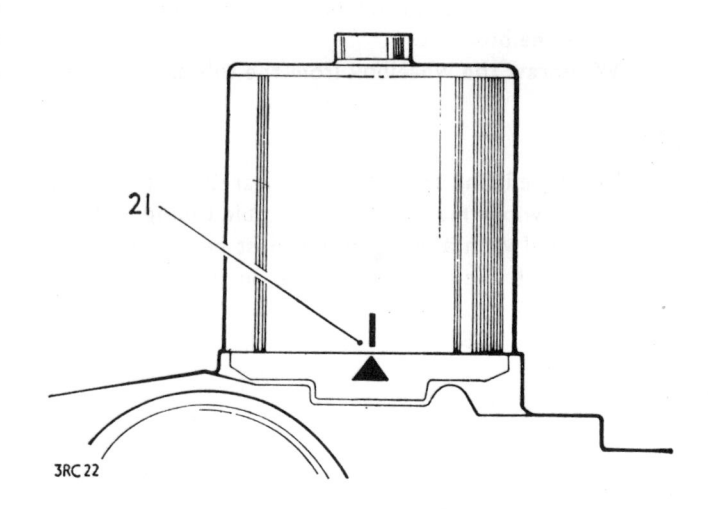

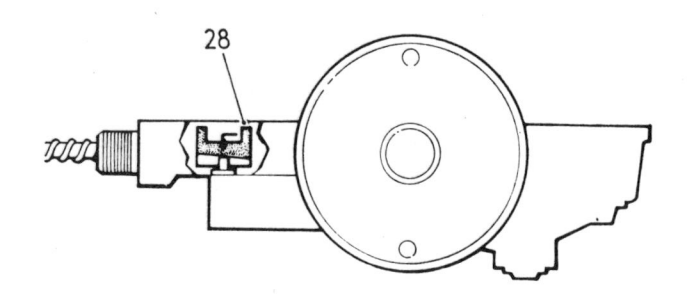

Rover 3500 and 3500S Manual AKM 3621

84.15.18
Sheet 2

WINDSCREEN WIPERS AND WASHERS

WIPER WHEEL BOXES AND DRIVE CABLE TUBES

—Remove and refit, 1 to 4 and 7   84.15.27

WHEEL BOX—LEFT HAND

—Remove and refit, 1 to 7   84.15.28

WHEEL BOX—RIGHT HAND

—Remove and refit, 1 to 7   84.15.29

Removing

1. Remove the wiper motor and drive. 84.15.09.
2. Remove the locknuts retaining the wheelboxes to the base unit.
3. Withdraw the rubber grommets and distance tubes from the wheelboxes.
4. Lower the wheelboxes until they are free of the body panel slots and withdraw complete with the attached drive cable tubing.
5. Slacken the fixings retaining the drive cable tubes in the wheelbox.
   Withdraw the wheelbox from the tubes.

Refitting

6. When refitting the wheelbox nearest to the windscreen wiper motor, the drive cable tubing which is provided with a hexagon nut must be positioned to align with the threaded ferrule on the motor when fitted.
7. Reverse 1 to 5.

ELECTRICAL

LIST OF OPERATIONS

Alternator	
—test (in position)	86.10.01
—remove and refit	86.10.02
—overhaul	86.10.08
—control system relay—check	86.10.18
—control system relay—adjust	86.10.19
—control box—test	86.10.25
—control box—adjust	86.10.27
—warning light control—test	86.10.33
Bulbs—remove and refit	
—automatic transmission selector lamp	86.45.40
—boot lamp	86.45.15
—headlamp	86.40.09
—heated backlight switch	86.45.82
—'Icelert' warning lamp	86.45.80
—map lamp	86.45.09
—number plate lamp	86.40.85
—panel lamp	86.45.31
—roof lamp	86.45.01
—warning lights	86.45.61
Cigar lighter—remove and refit	86.65.60
Distributor	
—remove and refit	86.35.20
—overhaul	86.35.26
Flasher unit—remove and refit	86.55.11
Fuse—remove and refit	86.70.02
Fuse box—remove and refit	86.70.01
Hazard flasher unit—remove and refit	86.55.12
Horn—remove and refit	86.30.10
Icelert warning unit	86.55.16
Lamps—remove and refit	
—automatic transmission selector	86.45.39
—boot	86.45.16
—front side/flasher	86.40.26
—head	86.40.02
—'Icelert' warning	86.45.81
—number plate	86.40.86
—tail, stop, flasher and reverse	86.40.74
Headlamps—align beams	86.40.18
Starter motor	
—remove and refit	86.60.01
—overhaul	86.60.13
—relay—remove and refit	86.60.10
Switches—remove and refit	
—boot light	86.65.22
—choke warning light	86.65.53
—combined direction indicator/headlight/horn	86.65.55
—door pillar	86.65.15
—handbrake warning	86.65.45
—hazard warning	86.65.50
—'Icelert' test	86.65.62
—'Icelert' rheostat	86.65.63
—ignition	86.65.02
—panel	86.65.06
—rheostat	86.65.07
—stop light	86.65.51
Wiring diagrams	86.00.01

Rover 3500 and 3500S Manual AKM 3621

ELECTRICAL

General service information

1. Engine earth lead

Rover 3500 and 3500S with brake pad wear indicator.

Ensure that the engine earth lead is securely connected, as it should be noted that in the event of the engine earth lead not being connected, or left loose on the above models, the shortest path to earth is through the rear caliper earth lead, which has been provided in connection with the brake pad wear indicator system.

In the case of an engine earthing through the rear brake caliper lead, it is possible that the body harness may burn out.

2. Distributor contact points

If burning of the contact points or melting of the actuating heel is experienced, items should be checked in the following order.

a. Regulator 4TR — Earlier models

i   Check condition of wiring and rectify as necessary.
ii Initially adjust the existing contact points by means of a dwell meter sufficiently to start and run the engine.
iii Warm the engine and regulator by running at 3000 rev/min for at least 8 minutes to ensure the system voltage has stabilised.
iv Reduce the engine speed to 2000 rev/min.
v   Place a voltmeter across the battery terminals; if reading is in excess of 14.5 volts, the regulator must be considered unserviceable and replaced.
vi If an excessive charge is allowed to pass through the system, burning of the contact points will occur.

b. Ballast resistance

i   Stop the engine. Switch on the ignition again.
ii Ensure that contact points are closed.
iii Introduce a voltmeter between the positive terminal on the coil and earth.
iv If full battery voltage is recorded, then the ballast resistance wire must be suspect and replaced. On early cars this was an independent wire with a fixed resistance but on later cars it is incorporated in the main dash harness which would have to be replaced.
v   The correct reading should be 6–7 volts.
vi Should full battery voltage be allowed to pass through the contact points, burning will result.

c. Condenser

i   With the engine stopped, examine the condenser connections.
ii If connections are dirty and loose, arcing of the contact points will occur.
iii Rectification in this case, after checking regulator and ballast resistance, will be to renew the contact points, clean and tighten condenser connections and reset the distributor.

If the foregoing items are found to be in good order, then the contact points themselves must be suspect. Rectification should be affected by replacing the contact set and condenser. Ensure that there is no copper weld splash on the new contact.

It is essential that condenser Part No. 606594 is fitted and this may be identified by a spot of blue paint on the body.

3. Rheostat switch for instrument panel illumination

Rover 3500 (New Look).

Incidents have occurred where the rheostat switch for the instrument panel illumination has burnt out owing to a short circuit caused by the illumination wiring for the gear selector indicator being trapped between the gear selector lever housing and the transmission tunnel.

When replacing a burnt-out rheostat switch, the wiring to the gear selector indicator should be checked and repaired as necessary. Additionally, care should be taken not to overtighten the retaining nut for the rheostat switch. This should only be nipped up. If it is overtightened, it will cut into the rubber washer and earth the rheostat.

4. Lucas 11 AC alternator

a. Description

Battery charging is maintained by a 43-ampere, three-phase alternator. The eight-pole rotor carries the field winding, which is connected to two face-type slip rings. The rotor is supported by a ball bearing in the drive end bracket, and a needle roller bearing in the slip-ring end cover.

The slip-ring end cover also carries six silicon diodes connected in a three-phase bridge circuit to give rectification of the generated AC output.

continued

b. Output control

The alternator terminal voltage is controlled by an electronic voltage regulator that utilises transistors as switching devices in place of moving contacts.

At zero or very low rotor speeds the alternator diodes prevent battery current flowing through the output windings, so that a cut-out unit is unnecessary.

c. Precaution when using a fast charger

Before using a fast charger, either to boost the battery or to start the engine, first withdraw the three-way connector from the 4TR control unit terminals. Do not reconnect the terminals until the charger has been disconnected, and in the case of assisted starting the engine speed is reduced to idling.

Failure to observe this precaution may result in irreparable damage to the semi-conductors in the control unit.

d. Field isolation

The voltage regulator and alternator field windings are isolated from the battery when the ignition is switched off by the normally open contacts of a model 6RA relay on early 3500 models or by isolating contacts incorporated in the ignition switch on 3500S models and later 3500 models. The operating coil of the relay is supplied by the ignition switch.

e. Alternator warning light

This is similar in operation to the ignition warning light or 'no charge' warning light used with generator charging systems. It is operated by a thermal relay, model 3AW, connected to one pair of the alternator diodes.

f. Battery polarity

The Rover 3500 and 3500S models to which the alternator is fitted incorporate a **negative earth** electrical system.

CAUTION:

The diodes and transistors in the alternator and voltage regulator, and any other transistorised equipment that may be fitted, will be irreparably damaged if the battery connections are reversed. Care must be taken to check the polarity of all connections in the electrical system.

Damage to the diodes can also be caused by a short circuit to earth of a live lead of the charging system. Battery voltage is applied to the alternator output cable even when the ignition is switched off, so it is recommended that the battery be disconnected when any work is being done in the vicinity of the alternator or voltage regulator.

The battery should also be disconnected when repairs to the body structure are being done by arc-welding.

ELECTRICAL

5. **Replacement headlamp sealed beam units and bulbs**
   a. **3500 models with ribbon type speedometer**

	Inner headlamp			Outer headlamp		
	Sealed beam unit	Bulb	Wattage	Sealed beam unit	Bulb	Wattage
Home market, RHStg	LU 54522973	—	37.5	LU 54521806	—	50–37.5
Export:						
Except France, Austria, Italy, Sweden and North America	LU 54521805	—	37.5	—	LU 410	40–45
France	LU 54520931	—	37.5	—	LU 411	40–45
Austria	LU 54521615	—	37.5	—	LU 410	40–45
Italy	—	LU 410	40–45	—	LU 410	40–45
Sweden	LU 54521805	—	37.5	—	LU 410	40–45

Bulb location	Make and type	Value	
Headlamp, inner	Lucas No. 59641 or GE 59272A	12v, 37½w	
Headlamp, outer	Lucas No. 59647 or GE 59273A	12v, 50/37½w	
Side (park) lamps	Lucas No. 989	12v, 6w	Exterior lamps
Stop/tail lamps	Lucas No. 380	12v, 21/6w	
Reverse lamp	Lucas No. 382	12v, 21w	
Direction indicator lamps	Lucas No. 382	12v, 21w	
Licence plate lamp	Lucas No. 989	12v, 6w	
Instrument panel light	Lucas No. 987	12v, 2.2w	
Warning lights for oil, brake and ignition		12v, 3.6w	
Clock	Lucas No. 281	12v, 2w	
Hazard warning, heated rear window and compressor switch warning lights	Lucas No. 281	12v, 2w	
'Icelert' warning light	Lucas MES cap	6v, .04 amp	Interior lights
Selector illumination, automatic transmission	Lucas No. 256	12v, 3w Festoon bulb	
Interior lights	Lucas No. 272	12v, 6w Festoon bulb	
Warning lights for direction indicator and main beam	Lucas No. 643	12v, 2.2w	
Boot lamp	Lucas No. 209	12v, 6w	

b. **3500 (with circular speedometer) and all 3500S models**

	Inner headlamp		Outer headlamp	
	Sealed beam unit	Bulb	Sealed beam unit	Bulb
UK all RHStg and Japan	Lucas 12v 75w	—	Lucas 12v 37½/50w	—
LHStg, Europe except UK and France	—	Lucas SP 410 12v 45w	—	Lucas SP 410 12v 40/45w
France	—	Lucas SP 411 12v 45w Yellow	—	Lucas SP 411 12v 40/45w Yellow

86.00.00
Sheet 3

Rover 3500 and 3500S Manual AKM 3621

ELECTRICAL

Bulb location	Make and type	Value	
Side lamps	Lucas No. 989	12v 6w	⎫
Stop/tail lamps	Lucas No. 380	12v 21/6w	
Direction indicator lamps	Lucas No. 382	12v 21w	⎬ Exterior lamps
Rear number plate lamp	Lucas No. 989	12v 6w	
Reverse lamps	Lucas Mo. 382	12v 21w	⎭
Instrument panel lights		12v 2.2w capless	
Warning lights		12v 3w capless	
Interior light, festoon bulb	Lucas No. 272	12v 10w	⎫
Rotary map lamp	Lucas No. 989	12v 6w	
Rear luggage boot lamp	Lucas No. 209	12v 5w	
Selector illumination, automatic transmission, festoon bulb	Lucas No. 256	12v 3w	⎬ Interior lights
Switch panel, festoon bulb	Lucas No. 254	12v 6w	
Hazard warning switch light	Lucas No. 281	12v 2w	⎭

6. Fuses

a. 3500 models with ribbon type speedometer

Equipment controlled by fuse	Amperage of fuse	Position marked
Horns, cigar lighter illumination and interior lights	35	1–2
Windscreen washer, stop lights and flasher lights	35	3–4
Cigar lighter, clock and panel illumination	2	5–6
Heater and wiper motor	15	7–8

b. Early 3500 with circular speedometer, and 3500S models

Equipment controlled by fuse	Amperage of fuse	Position marked
Horns, clock, cigar lighter, interior lights	25	1–2
Hazard warning	25	3–4
Side light RH, tail light RH	5	5–6 ⎫ Reversed on
Side light LH, tail light LH and number plate illumination	5	7–8 ⎭ LHStg models
Instrument panel, switch panel and gear change selection illumination (automatic transmission)	8	9–10
Headlamp main, inner, and main beam warning light	25	11–12
Headlamp, main, outer	15	13–14
Headlamp, RH, dip	10	15–16
Headlamp, LH, dip	10	17–18
Screen wiper switch, delay governor and screen wiper	25	19–20
Screenwasher, stop lamps, reverse lamps, gauges and flashers	25	21–22
Heater	15	23–24
Heated backlight		
In-line fuse adjacent to fuse box. Use 25 amp fuse		

c. Later 3500 and 3500S models

Equipment controlled by fuse	Amperage of fuse	Position marked
Horns, clock, cigar lighter, interior lights	25	1–2
Heated backlight	25	3–4
Side light RH, tail light RH	5	5–6 ⎫ Reversed on
Side light LH, tail light LH and number plate illumination	5	7–8 ⎭ LHStg models
Instrument panel, switch panel and gear change selection illumination (automatic transmission)	8	9–10
Headlamp main, inner, and main beam warning light	25	11–12
Headlamp, main, outer	15	13–14
Headlamp, RH, dip	15	15–16
Headlamp, LH dip	15	17–18
Instrument gauges, stop lamps, reverse lamps, flashers, hazard warning and screen washer	25	19–20
Screen wiper switch, delay governor and screen wiper motor	25	21–22
Heater motor	15	23–24

Rover 3500 and 3500S Manual AKM 3621

ELECTRICAL

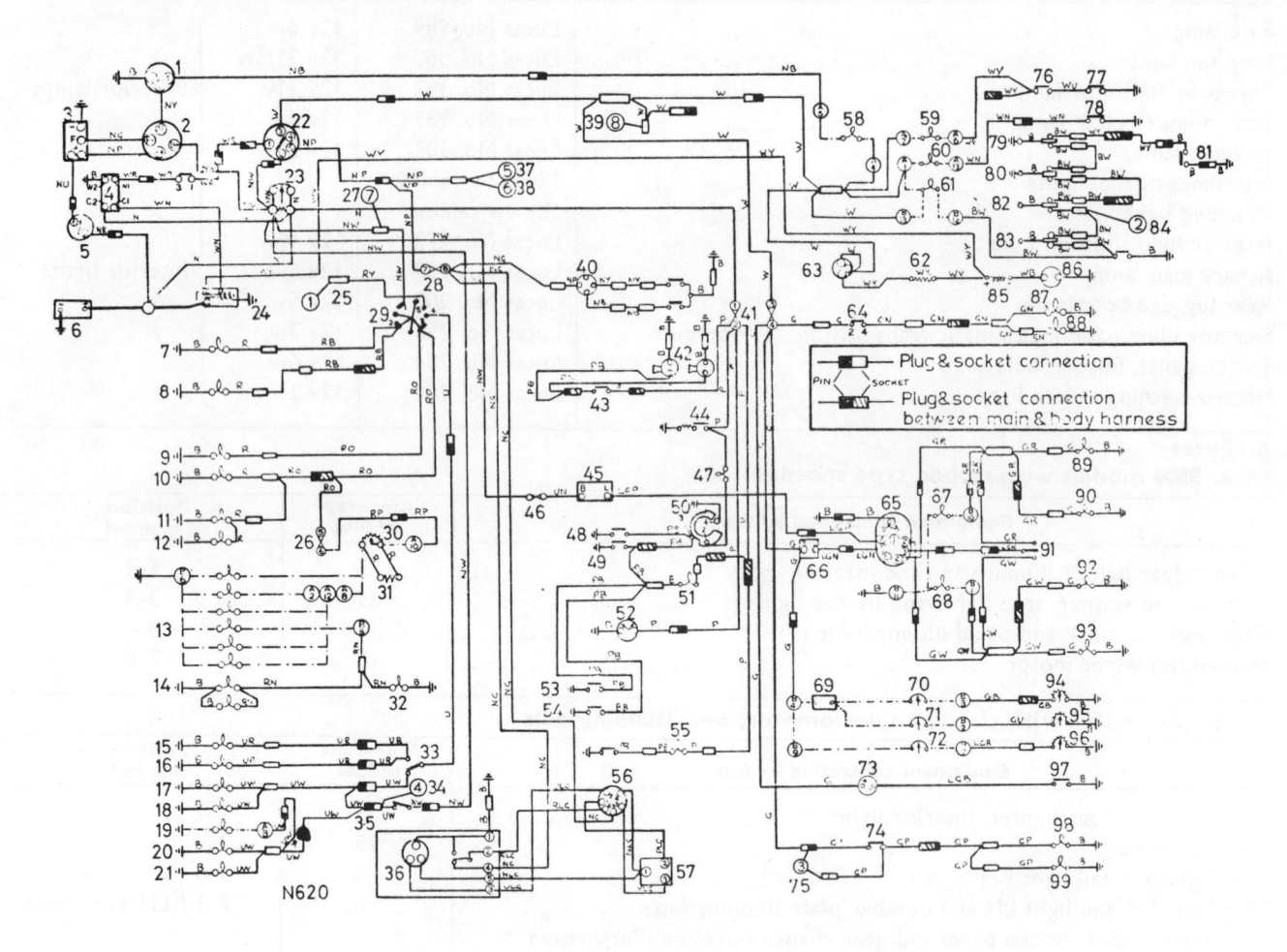

Key to circuit diagram
Rover 3500 Automatic 'New Look' models Chassis Suffix 'A'

Encircled figures denote pick-up points for optional extra equipment
Encircled letters and figures denote printed circuit and plug pick-up points on printed circuit board. Figure denotes pin number, letter denotes plug

1 Relay for ignition warning light, type 3AW	34 Pick-up point for headlamp flash, special markets	66 Flasher unit, type 8FL
2 Alternator, Lucas, type 11AC	35 Switch for headlamp flash	67 Warning light, indicator, LH
3 Control box, Lucas, type 4TR	36 Screenwiper motor	68 Warning light, indicator, RH
4 Relay for starter motor	37 Pick-up point for heated backlight	69 Bi-metal voltage regulator for instruments
5 Alternator contacts, ignition switch	38 Pick-up point for radio	70 Fuel contents gauge
6 Battery	39 Pick-up point for air conditioning circuit	71 Coolant temperature gauge
7 Side and park lamp	40 Heater blower unit, two-speed	72 Oil pressure gauge
8 Tail and park lamp	41 Fuses, 1, 2, 3 and 4, 35 amp	73 Screenwasher motor
9 Side lamp	42 Twin horns	74 Switch for stop lights, hydraulic
10 Tail lamp	43 Switch for horns	75 Pick-up point, stop light switch, mechanical
11 Number plate illumination	44 Cigar lighter and switch	76 Switch for choke control
12 Number plate illumination	45 Hazard warning flasher unit	77 Thermostat switch for choke control
13 Instrument panel illumination	46 In-line fuse	78 Switch for oil pressure warning light
14 Switch panel illumination	47 Rotary map light	79 Brake pad wear warning electrode, rear LH
15 Headlamp dip beam, LH	48 Switch for front door, LH	80 Brake pad wear warning electrode, rear RH
16 Headlamp dip beam, RH	49 Switch for front door, RH	81 Switch for hand brake warning light
17 Headlamp main beam, RH	50 Switch for interior light	82 Brake pad wear warning electrode, front RH
18 Headlamp main beam, RH	51 Interior light	83 Brake pad wear warning electrode, front LH
19 Warning light for headlamp main beam	52 Clock	84 Pick-up point for dual braking system
20 Headlamp main beam	53 Switch for rear door, RH	85 Ignition coil
21 Headlamp main beam, LH	54 Switch for rear door, LH	86 Distributor
22 Steering column lock switch	55 Boot lamp	87 Reverse lamp, RH
23 Ammeter shunt	56 Switch for windscreen wiper	88 Reverse lamp, LH
24 Starter motor, pre-engaged	57 Delay switch for windscreen wiper	89 Direction indicator lamp, front LH
25 Pick-up point for auxiliary lamps	58 Warning light for ignition	90 Direction indicator lamp, rear LH
26 Fuses 5 and 6, 2 amp	59 Warning light for choke	91 Direction indicator switch
27 Pick-up point for air conditioning circuit	60 Warning light for oil pressure	92 Direction indicator lamp, rear RH
28 Fuses 7 and 8, 35 amp	61 Warning light for brake fluid reservoir, hand brake and pad wear indicators	93 Direction indicator lamp, front RH
29 Main lighting switch	62 Ballast resistor cable	94 Tank unit, fuel contents gauge
30 Rheostat for panel lights	63 Tachometer	95 Transmitter, coolant temperature
31 Ballast resistor	64 Switch for reverse lamps	96 Transmitter, oil pressure
32 Illumination for gear selector	65 Switch and warning light for hazard warning	97 Switch for screenwasher
33 Switch for headlamp dip		98 Stop lamp, RH
		99 Stop lamp, LH

Key to cable colours

B—Black   G—Green   LG—Light green   N—Brown   P—Purple   R—Red   W—White   U—Blue   Y—Yellow

The last letter of a colour code denotes the tracer colour

Rover 3500 and 3500S Manual AKM 3621

ELECTRICAL

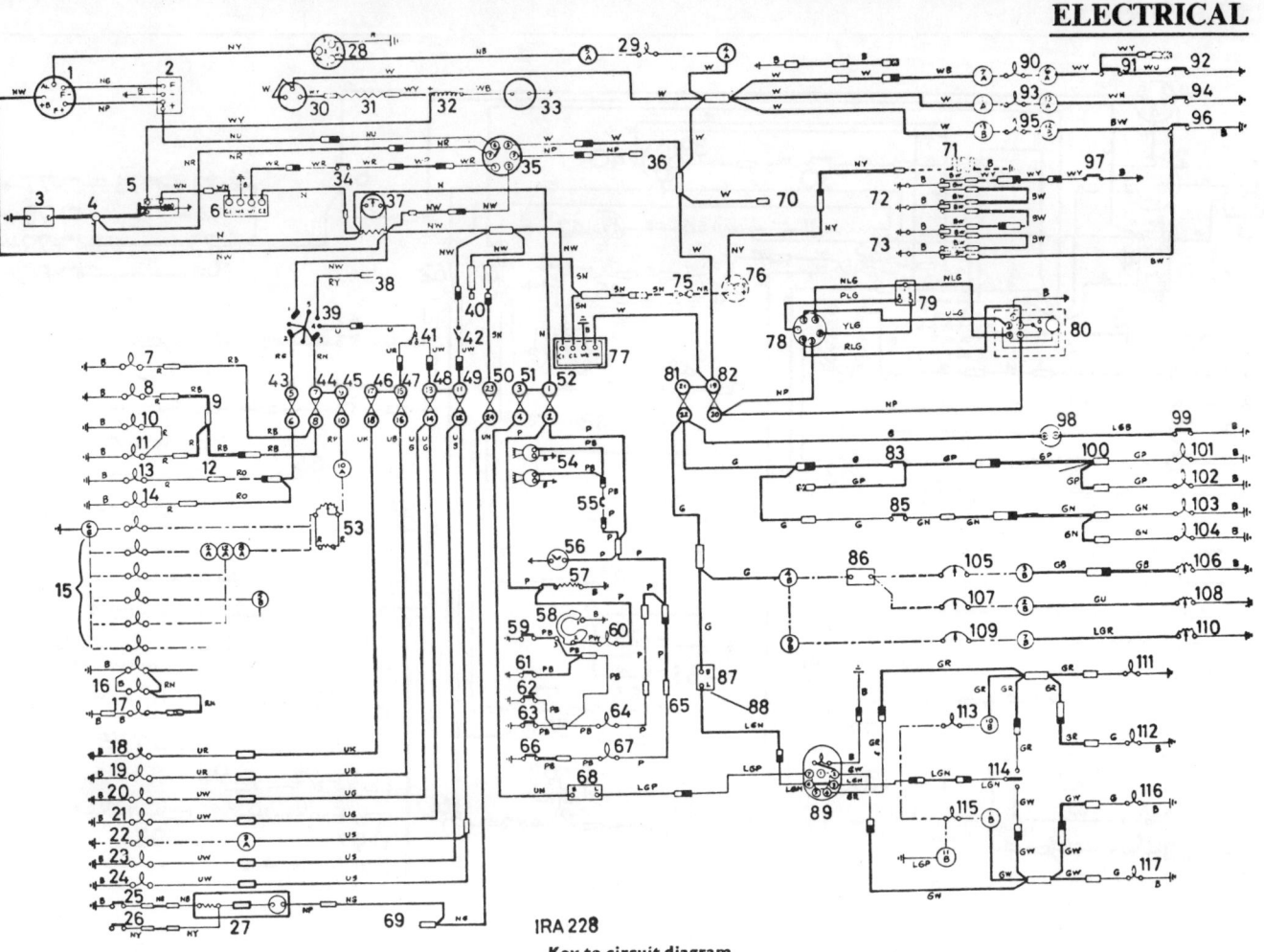

Key to circuit diagram
Rover 3500 Automatic 'New Look' models Chassis Suffix 'B' and Rover 3500S Manual Chassis suffix 'A'
Encircled figures and letters denote printed circuit and plug pick-up points on printed circuit board—figure for pin number, letter for plug

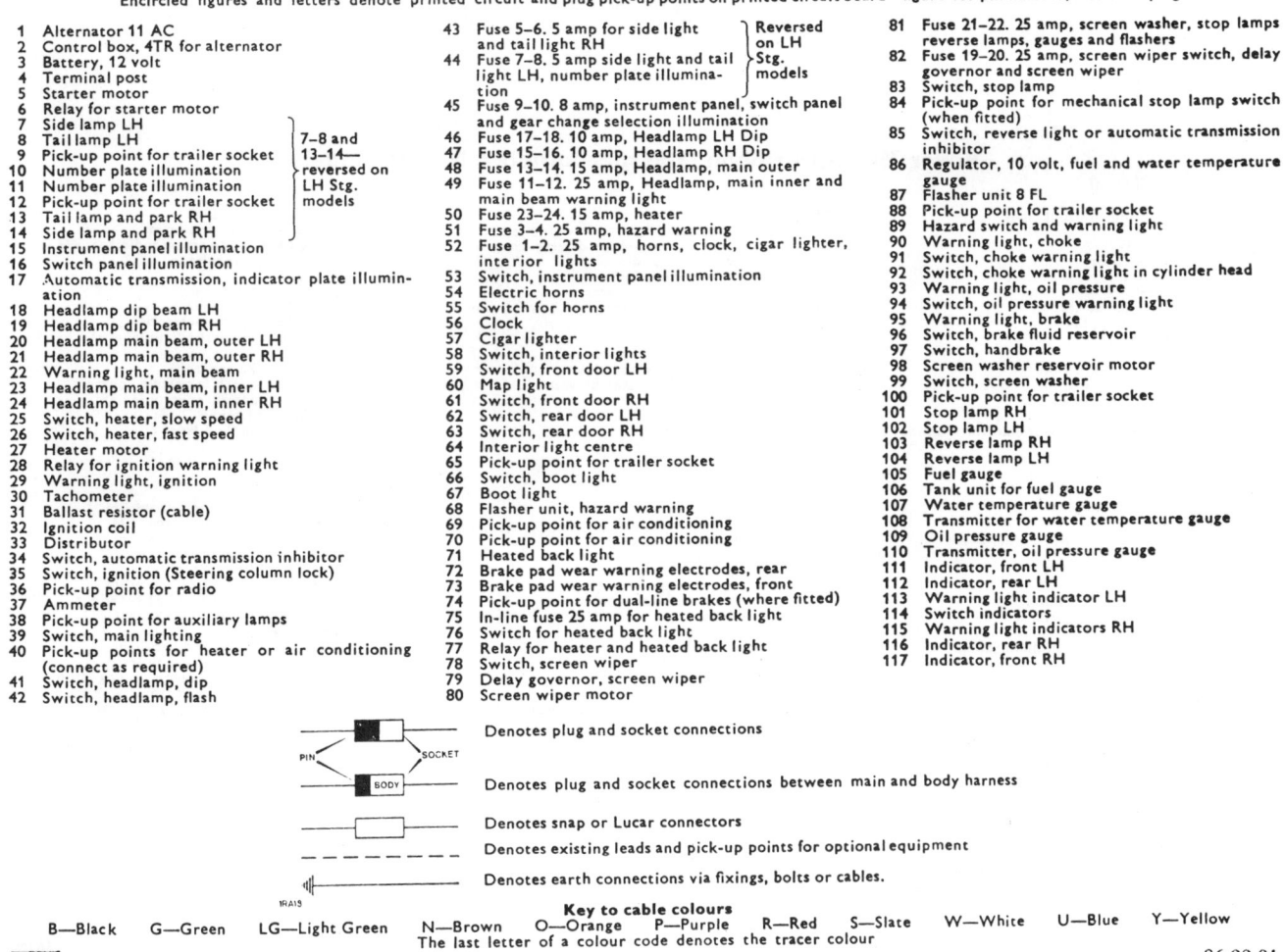

1. Alternator 11 AC
2. Control box, 4TR for alternator
3. Battery, 12 volt
4. Terminal post
5. Starter motor
6. Relay for starter motor
7. Side lamp LH
8. Tail lamp LH
9. Pick-up point for trailer socket
10. Number plate illumination
11. Number plate illumination
12. Pick-up point for trailer socket
13. Tail lamp and park LH
14. Side lamp and park RH
15. Instrument panel illumination
16. Switch panel illumination
17. Automatic transmission, indicator plate illumination
18. Headlamp dip beam LH
19. Headlamp dip beam RH
20. Headlamp main beam, outer LH
21. Headlamp main beam, outer RH
22. Warning light, main beam
23. Headlamp main beam, inner LH
24. Headlamp main beam, inner RH
25. Switch, heater, slow speed
26. Switch, heater, fast speed
27. Heater motor
28. Relay for ignition warning light
29. Warning light, ignition
30. Tachometer
31. Ballast resistor (cable)
32. Ignition coil
33. Distributor
34. Switch, automatic transmission inhibitor
35. Switch, ignition (Steering column lock)
36. Pick-up point for radio
37. Ammeter
38. Pick-up point for auxiliary lamps
39. Switch, main lighting
40. Pick-up points for heater or air conditioning (connect as required)
41. Switch, headlamp, dip
42. Switch, headlamp, flash

{7-8 and 13-14—reversed on LH Stg. models}

43. Fuse 5–6. 5 amp for side light and tail light RH
44. Fuse 7–8. 5 amp side light and tail light LH, number plate illumination

{Reversed on LH Stg. models}

45. Fuse 9–10. 8 amp, instrument panel, switch panel and gear change selection illumination
46. Fuse 17–18. 10 amp, Headlamp LH Dip
47. Fuse 15–16. 10 amp, Headlamp RH Dip
48. Fuse 13–14. 15 amp, Headlamp, main outer
49. Fuse 11–12. 25 amp, Headlamp, main inner and main beam warning light
50. Fuse 23–24. 15 amp, heater
51. Fuse 3–4. 25 amp, hazard warning
52. Fuse 1–2. 25 amp, horns, clock, cigar lighter, interior lights
53. Switch, instrument panel illumination
54. Electric horns
55. Switch for horns
56. Clock
57. Cigar lighter
58. Switch, interior lights
59. Switch, front door LH
60. Map light
61. Switch, front door RH
62. Switch, rear door LH
63. Switch, rear door RH
64. Interior light centre
65. Pick-up point for trailer socket
66. Switch, boot light
67. Boot light
68. Flasher unit, hazard warning
69. Pick-up point for air conditioning
70. Pick-up point for air conditioning
71. Heated back light
72. Brake pad wear warning electrodes, rear
73. Brake pad wear warning electrodes, front
74. Pick-up point for dual-line brakes (where fitted)
75. In-line fuse 25 amp for heated back light
76. Switch for heated back light
77. Relay for heater and heated back light
78. Switch, screen wiper
79. Delay governor, screen wiper
80. Screen wiper motor

81. Fuse 21–22. 25 amp, screen washer, stop lamps reverse lamps, gauges and flashers
82. Fuse 19–20. 25 amp, screen wiper switch, delay governor and screen wiper
83. Switch, stop lamp
84. Pick-up point for mechanical stop lamp switch (when fitted)
85. Switch, reverse light or automatic transmission inhibitor
86. Regulator, 10 volt, fuel and water temperature gauge
87. Flasher unit 8 FL
88. Pick-up point for trailer socket
89. Hazard switch and warning light
90. Warning light, choke
91. Switch, choke warning light
92. Switch, choke warning light in cylinder head
93. Warning light, oil pressure
94. Switch, oil pressure warning light
95. Warning light, brake
96. Switch, brake fluid reservoir
97. Switch, handbrake
98. Screen washer reservoir motor
99. Switch, screen washer
100. Pick-up point for trailer socket
101. Stop lamp RH
102. Stop lamp LH
103. Reverse lamp RH
104. Reverse lamp LH
105. Fuel gauge
106. Tank unit for fuel gauge
107. Water temperature gauge
108. Transmitter for water temperature gauge
109. Oil pressure gauge
110. Transmitter, oil pressure gauge
111. Indicator, front LH
112. Indicator, rear LH
113. Warning light indicator LH
114. Switch indicators
115. Warning light indicators RH
116. Indicator, rear RH
117. Indicator, front RH

Denotes plug and socket connections

Denotes plug and socket connections between main and body harness

Denotes snap or Lucar connectors

Denotes existing leads and pick-up points for optional equipment

Denotes earth connections via fixings, bolts or cables.

Key to cable colours

B—Black G—Green LG—Light Green N—Brown O—Orange P—Purple R—Red S—Slate W—White U—Blue Y—Yellow
The last letter of a colour code denotes the tracer colour

Rover 3500 and 3500S Manual AKM 3621

ELECTRICAL

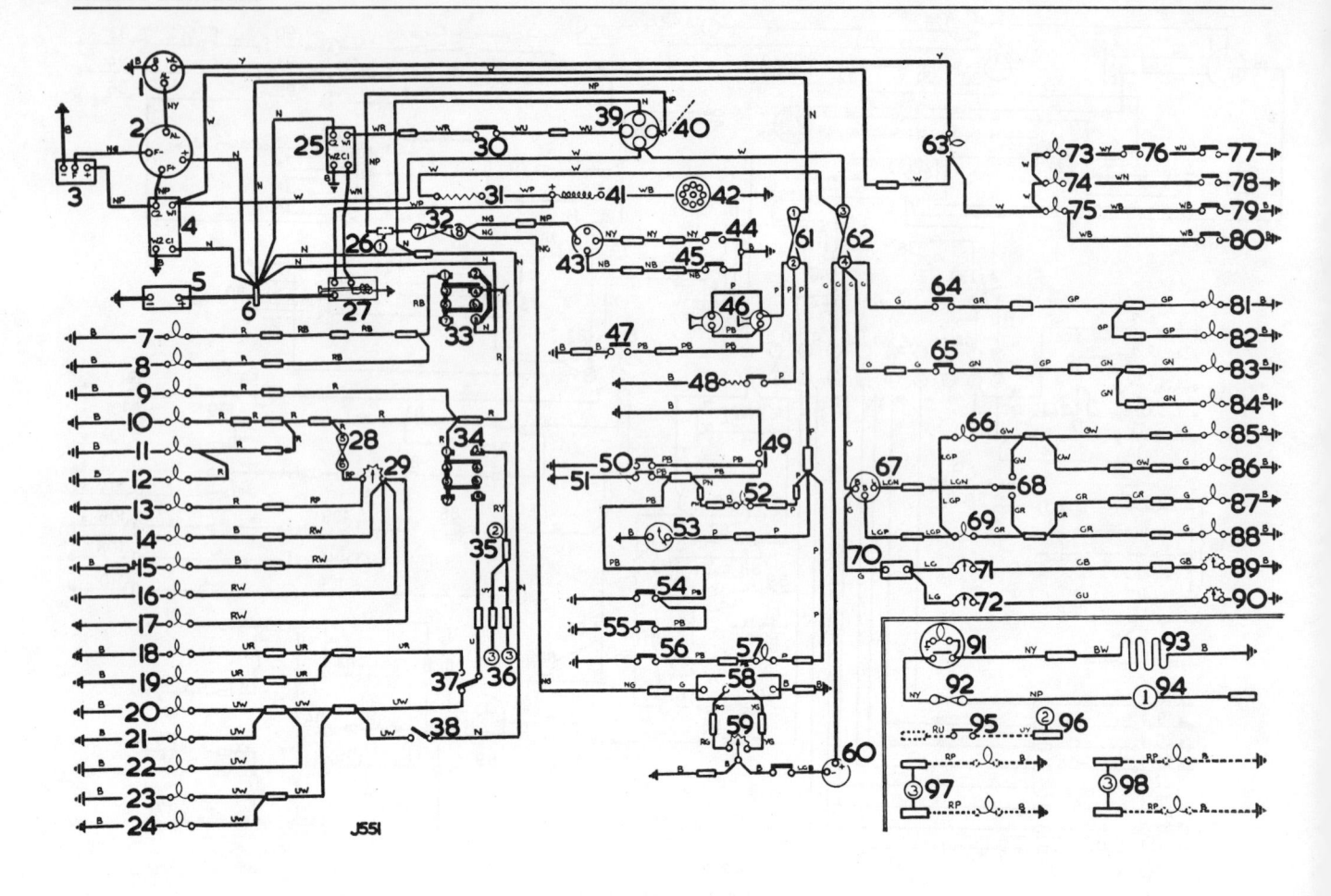

Rover 3500 with ribbon type speedometer, RHStg models

1. Relay for ignition warning light
2. Alternator
3. Control box, 4TR
4. Relay for alternator
5. Battery, 60 AH
6. Terminal post
7. Side and park lamp, LH
8. Tail and park lamp, LH
9. Side lamp, RH
10. Tail lamp, RH
11. Number plate illumination
12. Number plate illumination
13. Cigar lighter illumination
14. Clock illumination
15. Automatic transmission gear selector illumination
16. Panel illumination
17. Panel illumination
18. Headlamp dip beam, LH
19. Headlamp dip beam, RH
20. Headlamp main beam, RH
21. Headlamp main beam, RH
22. Warning light for headlamp main beam
23. Headlamp main beam, LH
24. Headlamp main beam, LH
25. Relay for starter motor
26. Pick-up point for heated backlight
27. Starter motor, pre-engaged type
28. Fuse, 5–6, 2 amp
29. Rheostat switch for panel illumination
30. Inhibitor switch for automatic transmission
31. Ballast resistor
32. Fuse, 7–8, 15 amp
33. Switch for side and park lamp
34. Switch for headlamps
35. Pick-up point for long-range driving lamp switch
36. Pick-up point for long-range driving lamps
37. Switch for headlamp, dip
38. Switch for headlamp, flash
39. Switch for ignition and starter
40. Pick-up point for radio feed and illumination
41. Ignition coil
42. Distributor
43. Heater blower unit, 2-speed
44. Switch, fast speed, heater blower
45. Switch, slow speed, heater blower
46. Clock
47. Twin horns
48. Switch for horns
49. Cigar lighter and switch
50. Switch for interior light
51. Switch for front door, LH
52. Switch for front door, RH
53. Switch for rear door, RH
54. Switch for rear door, LH
55. Interior light
56. Switch for boot light
57. Boot lamp
58. Screen wiper motor
59. Switch for screen wiper and screen washer
60. Screen washer
61. Fuse, 1–2, 35 amp
62. Fuse, 3–4, 35 amp
63. Warning light for ignition
64. Switch for stop lamp
65. Switch for reverse lamp
66. Warning light for direction indicator, LH
67. Direction indicator unit
68. Switch for direction indicator
69. Warning light for direction indicator, RH
70. Bi-metal voltage regulator
71. Fuel gauge
72. Temperature gauge
73. Warning light for choke
74. Warning light for oil pressure
75. Warning light for brake fluid level and hand brake
76. Switch for choke control
77. Thermostatic switch for choke control
78. Switch for oil pressure warning light
79. Switch for hand brake
80. Switch for brake fluid level
81. Stop lamp, RH
82. Stop lamp, LH
83. Reverse lamp, RH
84. Reverse lamp, LH
85. Direction indicator lamp, front RH
86. Direction indicator lamp, rear RH
87. Direction indicator lamp, rear LH
88. Direction indicator lamp, front LH
89. Fuel level indicator in fuel tank
90. Transmitter for water temperature
91. Illuminated switch for heated backlight
92. In-line fuse for heated backlight
93. Heated backlight
94. Pick-up point for heated backlight
95. Switch for long-range driving lamp
96. Pick-up point for long-range driving lamp
97. Pick-up point and long-range driving lamp
98. Pick-up point and long-range driving lamp

Key to cable colours

| B—Black | G—Green | LG—Light Green | N—Brown | R—Red | W—White | U—Blue | Y—Yellow | P—Purple |

Rover 3500 and 3500S Manual AKM 3621

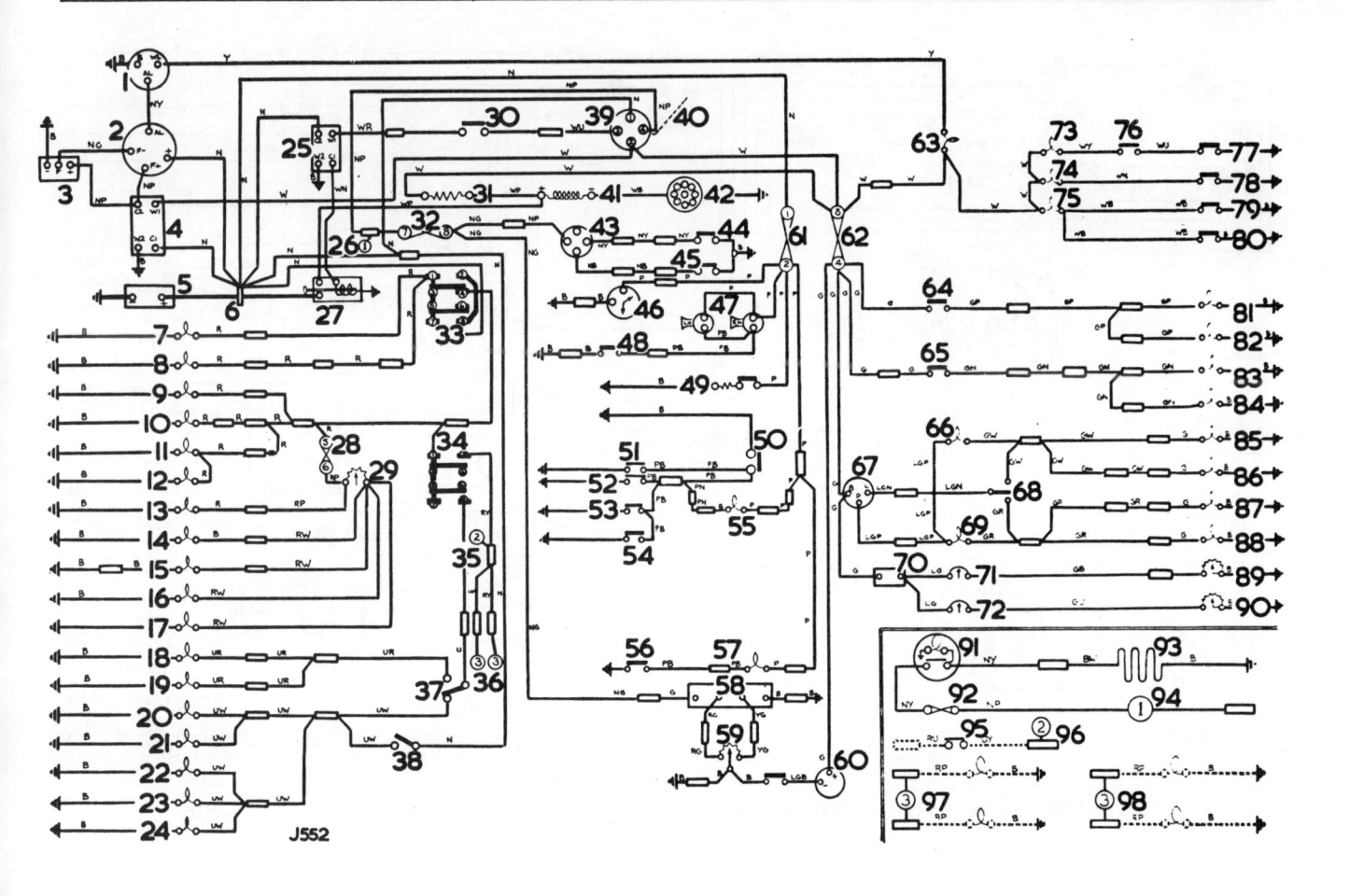

ELECTRICAL

Rover 3500 with ribbon type speedometer, LHStg models

1. Relay for ignition warning light
2. Alternator
3. Control box, 4TR
4. Relay for alternator
5. Battery, 60 AH
6. Terminal post
7. Side and park lamp, LH
8. Tail and park lamp, LH
9. Side lamp, RH
10. Tail lamp, RH
11. Number plate illumination
12. Number plate illumination
13. Cigar lighter illumination
14. Clock illumination
15. Automatic transmission gear selector illumination
16. Panel illumination
17. Panel illumination
18. Headlamp dip beam, LH
19. Headlamp dip beam, RH
20. Headlamp main beam, RH
21. Headlamp main beam, RH
22. Warning light for headlamp main beam
23. Headlamp main beam, LH
24. Headlamp main beam, LH
25. Relay for starter motor
26. Pick-up point for heated backlight
27. Starter motor, pre-engaged type
28. Fuse, 5–6, 2 amp
29. Rheostat switch for panel illumination
30. Inhibitor switch for automatic transmission
31. Ballast resistor
32. Fuse, 7–8, 15 amp
33. Switch for side and park lamp
34. Switch for headlamps
35. Pick-up point for long-range driving lamp switch
36. Pick-up point for long-range driving lamps
37. Switch for headlamp, dip
38. Switch for headlamp, flash
39. Switch for ignition and starter
40. Pick-up point for radio feed and illumination
41. Ignition coil
42. Distributor
43. Heater blower unit, 2-speed
44. Switch, fast speed, heater blower
45. Switch, slow speed, heater blower
46. Clock
47. Twin horns
48. Switch for horns
49. Cigar lighter and switch
50. Switch for interior light
51. Switch for front door, LH
52. Switch for front door, RH
53. Switch for rear door, RH
54. Switch for rear door, LH
55. Interior light
56. Switch for boot light
57. Boot lamp
58. Screen wiper motor
59. Switch for screen wiper and screen washer
60. Screen washer
61. Fuse, 1–2, 35 amp
62. Fuse, 3–4, 35 amp
63. Warning light for ignition
64. Switch for stop lamp
65. Switch for reverse lamp
66. Warning light for direction indicator, LH
67. Direction indicator unit
68. Switch for direction indicator
69. Warning light for direction indicator, RH
70. Bi-metal voltage regulator
71. Fuel gauge
72. Temperature gauge
73. Warning light for choke
74. Warning light for oil pressure
75. Warning light for brake fluid level and hand brake
76. Switch for choke control
77. Thermostatic switch for choke control
78. Switch for oil pressure warning light
79. Switch for hand brake
80. Switch for brake fluid level
81. Stop lamp, RH
82. Stop lamp, LH
83. Reverse lamp, RH
84. Reverse lamp, LH
85. Direction indicator lamp, front RH
86. Direction indicator lamp, rear RH
87. Direction indicator lamp, rear LH
88. Direction indicator lamp, front LH
89. Fuel level indicator in fuel tank
90. Transmitter for water temperature
91. Illuminated switch for heated backlight
92. In-line fuse for heated backlight
93. Heated backlight
94. Pick-up point for heated backlight
95. Switch for long-range driving lamp
96. Pick-up point for long-range driving lamp switch
97. Pick-up point and long-range driving lamp
98. Pick-up point and long-range driving lamp

Key to cable colours

B—Black   G—Green   LG—Light/Green   N—Brown   R—Red   W—White   U—Blue   Y—Yellow   P—Purple

Rover 3500 and 3500S Manual AKM 3621

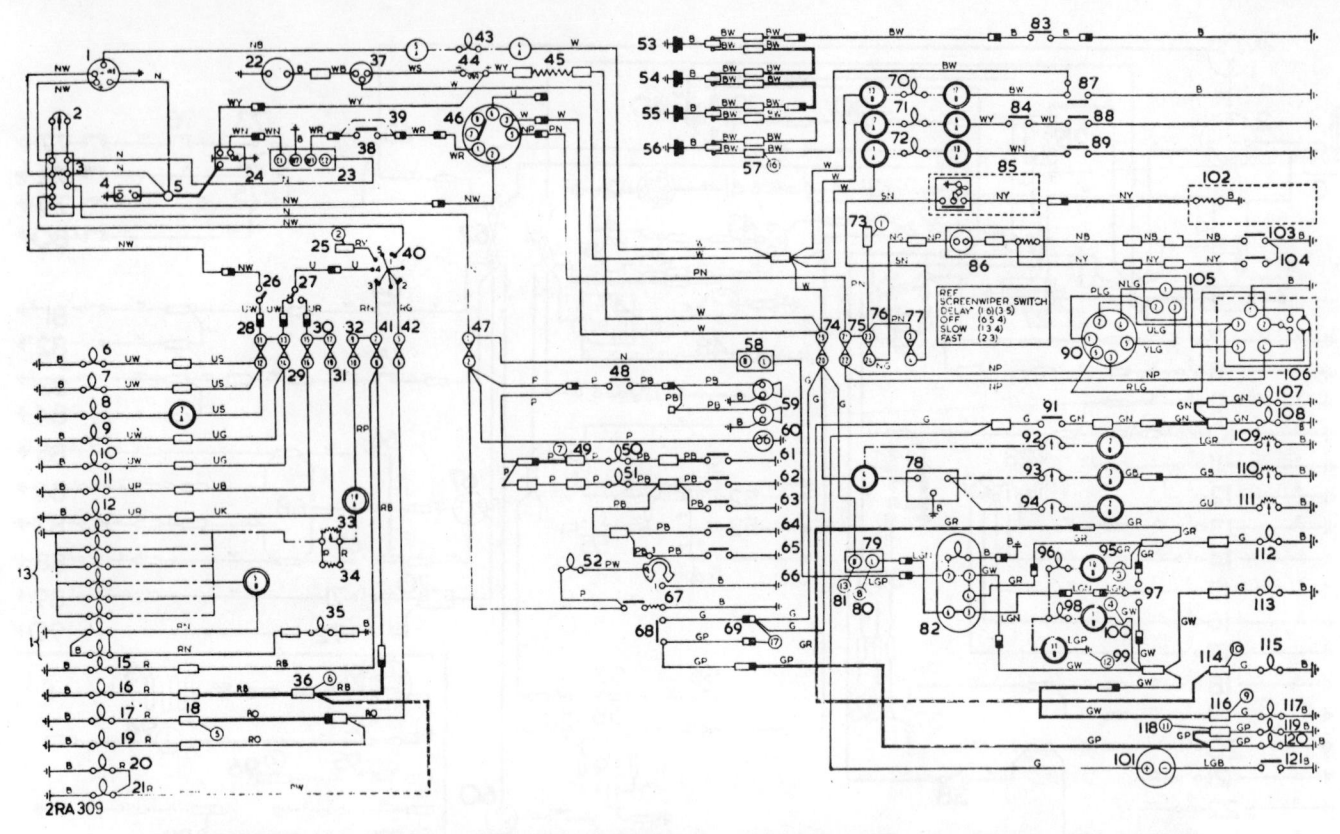

ELECTRICAL

Rover 3500 from chassis suffix 'D' and 3500S from chassis suffix 'C', RHStg 'New Look' models with circular type speedometer

Encircled figures and letters denote printed circuit and plug pick-up points on printed circuit board- figure for pin number, letter for plug

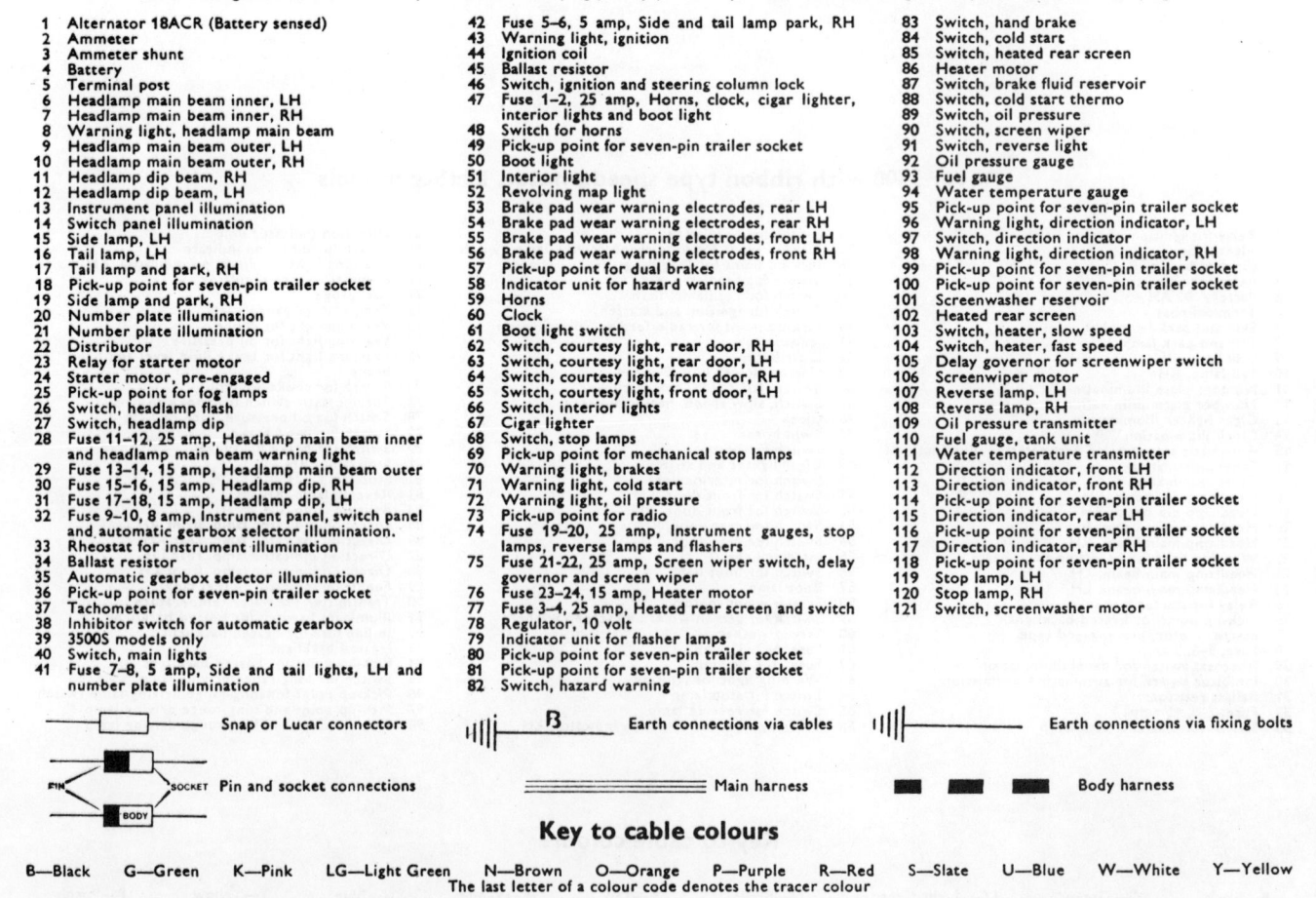

1 Alternator 18ACR (Battery sensed)
2 Ammeter
3 Ammeter shunt
4 Battery
5 Terminal post
6 Headlamp main beam inner, LH
7 Headlamp main beam inner, RH
8 Warning light, headlamp main beam
9 Headlamp main beam outer, LH
10 Headlamp main beam outer, RH
11 Headlamp dip beam, RH
12 Headlamp dip beam, LH
13 Instrument panel illumination
14 Switch panel illumination
15 Side lamp, LH
16 Tail lamp, LH
17 Tail lamp and park, RH
18 Pick-up point for seven-pin trailer socket
19 Side lamp and park, RH
20 Number plate illumination
21 Number plate illumination
22 Distributor
23 Relay for starter motor
24 Starter motor, pre-engaged
25 Pick-up point for fog lamps
26 Switch, headlamp flash
27 Switch, headlamp dip
28 Fuse 11–12, 25 amp, Headlamp main beam inner and headlamp main beam warning light
29 Fuse 13–14, 15 amp, Headlamp main beam outer
30 Fuse 15–16, 15 amp, Headlamp dip, RH
31 Fuse 17–18, 15 amp, Headlamp dip, LH
32 Fuse 9–10, 8 amp, Instrument panel, switch panel and automatic gearbox selector illumination.
33 Rheostat for instrument illumination
34 Ballast resistor
35 Automatic gearbox selector illumination
36 Pick-up point for seven-pin trailer socket
37 Tachometer
38 Inhibitor switch for automatic gearbox
39 3500S models only
40 Switch, main lights
41 Fuse 7–8, 5 amp, Side and tail lights, LH and number plate illumination

42 Fuse 5–6, 5 amp, Side and tail lamp park, RH
43 Warning light, ignition
44 Ignition coil
45 Ballast resistor
46 Switch, ignition and steering column lock
47 Fuse 1–2, 25 amp, Horns, clock, cigar lighter, interior lights and boot light
48 Switch for horns
49 Pick-up point for seven-pin trailer socket
50 Boot light
51 Interior light
52 Revolving map light
53 Brake pad wear warning electrodes, rear LH
54 Brake pad wear warning electrodes, rear RH
55 Brake pad wear warning electrodes, front LH
56 Brake pad wear warning electrodes, front RH
57 Pick-up point for dual brakes
58 Indicator unit for hazard warning
59 Horns
60 Clock
61 Boot light switch
62 Switch, courtesy light, rear door, RH
63 Switch, courtesy light, rear door, LH
64 Switch, courtesy light, front door, RH
65 Switch, courtesy light, front door, LH
66 Switch, interior lights
67 Cigar lighter
68 Switch, stop lamps
69 Pick-up point for mechanical stop lamps
70 Warning light, brakes
71 Warning light, cold start
72 Warning light, oil pressure
73 Pick-up point for radio
74 Fuse 19–20, 25 amp, Instrument gauges, stop lamps, reverse lamps and flashers
75 Fuse 21–22, 25 amp, Screen wiper switch, delay governor and screen wiper
76 Fuse 23–24, 15 amp, Heater motor
77 Fuse 3–4, 25 amp, Heated rear screen and switch
78 Regulator, 10 volt
79 Indicator unit for flasher lamps
80 Pick-up point for seven-pin trailer socket
81 Pick-up point for seven-pin trailer socket
82 Switch, hazard warning

83 Switch, hand brake
84 Switch, cold start
85 Switch, heated rear screen
86 Heater motor
87 Switch, brake fluid reservoir
88 Switch, cold start thermo
89 Switch, oil pressure
90 Switch, screen wiper
91 Switch, reverse light
92 Oil pressure gauge
93 Fuel gauge
94 Water temperature gauge
95 Pick-up point for seven-pin trailer socket
96 Warning light, direction indicator, LH
97 Switch, direction indicator
98 Warning light, direction indicator, RH
99 Pick-up point for seven-pin trailer socket
100 Pick-up point for seven-pin trailer socket
101 Screenwasher reservoir
102 Heated rear screen
103 Switch, heater, slow speed
104 Switch, heater, fast speed
105 Delay governor for screenwiper switch
106 Screenwiper motor
107 Reverse lamp, LH
108 Reverse lamp, RH
109 Oil pressure transmitter
110 Fuel gauge, tank unit
111 Water temperature transmitter
112 Direction indicator, front LH
113 Direction indicator, front RH
114 Pick-up point for seven-pin trailer socket
115 Direction indicator, rear LH
116 Pick-up point for seven-pin trailer socket
117 Direction indicator, rear RH
118 Pick-up point for seven-pin trailer socket
119 Stop lamp, LH
120 Stop lamp, RH
121 Switch, screenwasher motor

Snap or Lucar connectors
Pin and socket connections
Earth connections via cables
Main harness
Earth connections via fixing bolts
Body harness

Key to cable colours

B—Black G—Green K—Pink LG—Light Green N—Brown O—Orange P—Purple R—Red S—Slate U—Blue W—White Y—Yellow

The last letter of a colour code denotes the tracer colour

Rover 3500 and 3500S Manual AKM 3621

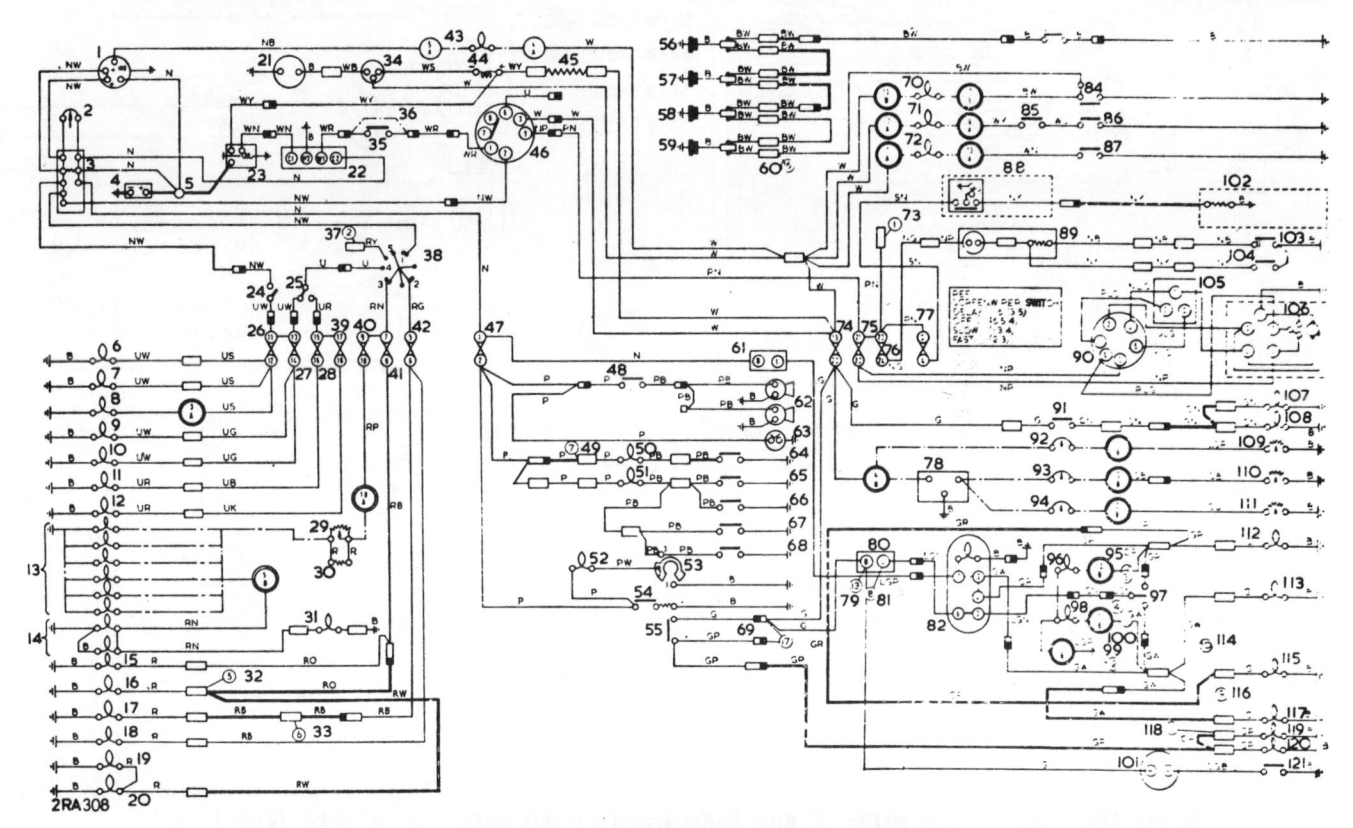

ELECTRICAL

Rover 3500 from chassis suffix 'D' and 3500S from chassis suffix 'C', LHStg 'New Look' models with circular type speedometer

Encircled figures and letters denote printed circuit and plug pick-up points on printed circuit board—figure for pin number, letter for plug

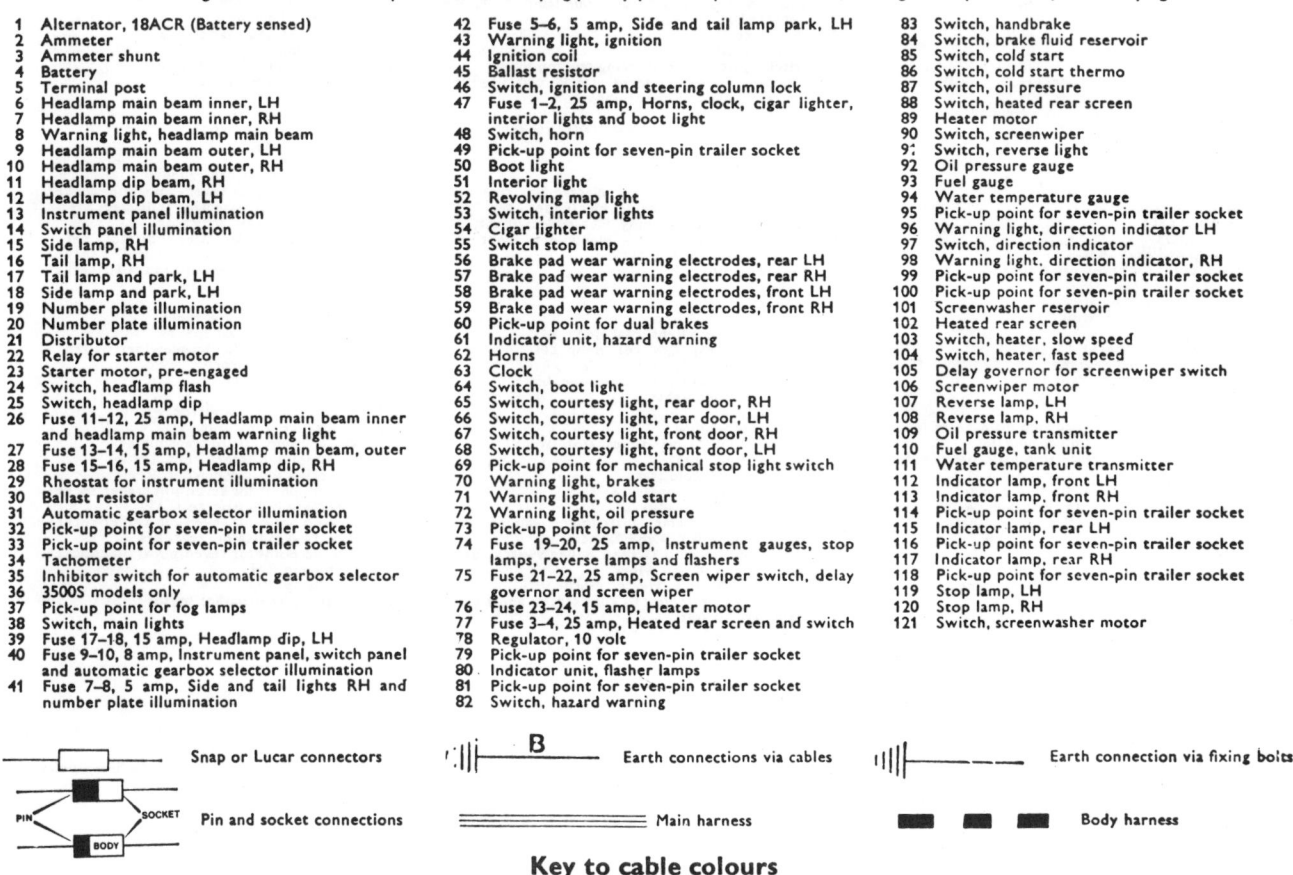

1	Alternator, 18ACR (Battery sensed)	42	Fuse 5–6, 5 amp, Side and tail lamp park, LH	83	Switch, handbrake
2	Ammeter	43	Warning light, ignition	84	Switch, brake fluid reservoir
3	Ammeter shunt	44	Ignition coil	85	Switch, cold start
4	Battery	45	Ballast resistor	86	Switch, cold start thermo
5	Terminal post	46	Switch, ignition and steering column lock	87	Switch, oil pressure
6	Headlamp main beam inner, LH	47	Fuse 1–2, 25 amp, Horns, clock, cigar lighter, interior lights and boot light	88	Switch, heated rear screen
7	Headlamp main beam inner, RH	48	Switch, horn	89	Heater motor
8	Warning light, headlamp main beam	49	Pick-up point for seven-pin trailer socket	90	Switch, screenwiper
9	Headlamp main beam outer, LH	50	Boot light	91	Switch, reverse light
10	Headlamp main beam outer, RH	51	Interior light	92	Oil pressure gauge
11	Headlamp dip beam, RH	52	Revolving map light	93	Fuel gauge
12	Headlamp dip beam, LH	53	Switch, interior lights	94	Water temperature gauge
13	Instrument panel illumination	54	Cigar lighter	95	Pick-up point for seven-pin trailer socket
14	Switch panel illumination	55	Switch stop lamp	96	Warning light, direction indicator LH
15	Side lamp, RH	56	Brake pad wear warning electrodes, rear LH	97	Switch, direction indicator
16	Tail lamp, RH	57	Brake pad wear warning electrodes, rear RH	98	Warning light, direction indicator, RH
17	Tail lamp and park, LH	58	Brake pad wear warning electrodes, front LH	99	Pick-up point for seven-pin trailer socket
18	Side lamp and park, LH	59	Brake pad wear warning electrodes, front RH	100	Pick-up point for seven-pin trailer socket
19	Number plate illumination	60	Pick-up point for dual brakes	101	Screenwasher reservoir
20	Number plate illumination	61	Indicator unit, hazard warning	102	Heated rear screen
21	Distributor	62	Horns	103	Switch, heater, slow speed
22	Relay for starter motor	63	Clock	104	Switch, heater, fast speed
23	Starter motor, pre-engaged	64	Switch, boot light	105	Delay governor for screenwiper switch
24	Switch, headlamp flash	65	Switch, courtesy light, rear door, RH	106	Screenwiper motor
25	Switch, headlamp dip	66	Switch, courtesy light, rear door, LH	107	Reverse lamp, LH
26	Fuse 11–12, 15 amp, Headlamp main beam inner and headlamp main beam warning light	67	Switch, courtesy light, front door, RH	108	Reverse lamp, RH
27	Fuse 13–14, 15 amp, Headlamp main beam, outer	68	Switch, courtesy light, front door, LH	109	Oil pressure transmitter
28	Fuse 15–16, 15 amp, Headlamp dip, RH	69	Pick-up point for mechanical stop light switch	110	Fuel gauge, tank unit
29	Rheostat for instrument illumination	70	Warning light, brakes	111	Water temperature transmitter
30	Ballast resistor	71	Warning light, cold start	112	Indicator lamp, front LH
31	Automatic gearbox selector illumination	72	Warning light, oil pressure	113	Indicator lamp, front RH
32	Pick-up point for seven-pin trailer socket	73	Pick-up point for radio	114	Pick-up point for seven-pin trailer socket
33	Pick-up point for seven-pin trailer socket	74	Fuse 19–20, 25 amp, Instrument gauges, stop lamps, reverse lamps and flashers	115	Indicator lamp, rear LH
34	Tachometer	75	Fuse 21–22, 25 amp, Screen wiper switch, delay governor and screen wiper	116	Indicator lamp, rear RH
35	Inhibitor switch for automatic gearbox selector	76	Fuse 23–24, 15 amp, Heater motor	117	Indicator lamp, rear RH
36	3500S models only	77	Fuse 3–4, 25 amp, Heated rear screen and switch	118	Pick-up point for seven-pin trailer socket
37	Pick-up point for fog lamps	78	Regulator, 10 volt	119	Stop lamp, LH
38	Switch, main lights	79	Pick-up point for seven-pin trailer socket	120	Stop lamp, RH
39	Fuse 17–18, 15 amp, Headlamp dip, LH	80	Indicator unit, flasher lamps	121	Switch, screenwasher motor
40	Fuse 9–10, 8 amp, Instrument panel, switch panel and automatic gearbox selector illumination	81	Pick-up point for seven-pin trailer socket		
41	Fuse 7–8, 5 amp, Side and tail lights RH and number plate illumination	82	Switch, hazard warning		

Snap or Lucar connectors

Pin and socket connections

Earth connections via cables

Earth connection via fixing bolts

Main harness

Body harness

Key to cable colours

B—Black G—Green K—Pink LG—Light Green N—Brown O—Orange P—Purple R—Red S—Slate U—Blue W—White Y—Yellow

The last letter of a colour code denotes the tracer colour

Rover 3500 and 3500S Manual AKM 3621

ELECTRICAL

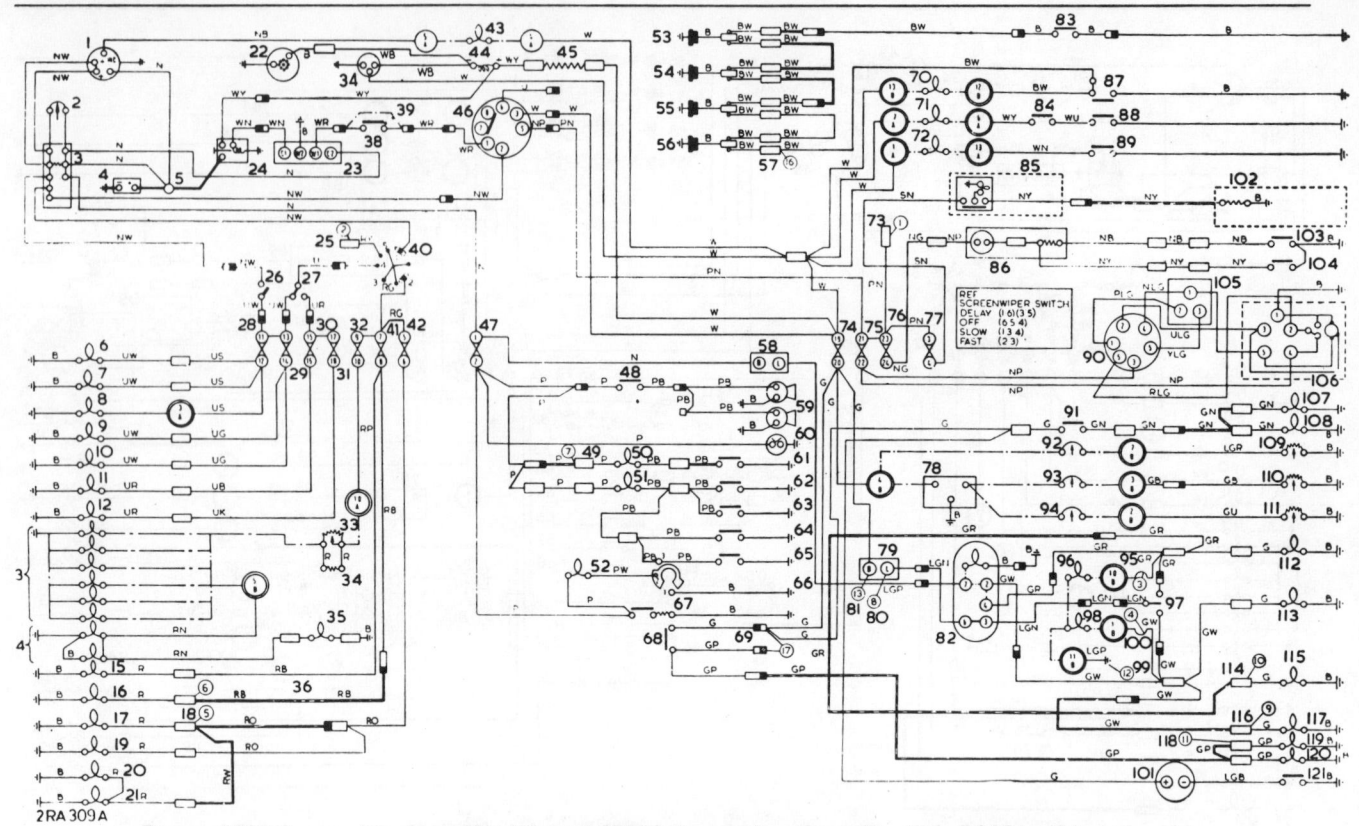

Rover 3500 from chassis suffix 'E' and 3500S from chassis suffix 'D', RHStg 'New Look' models with circular type speedometer

Encircled figures and letters denote printed circuit and plug pick-up points on printed circuit board—figure for pin number, letter for plug

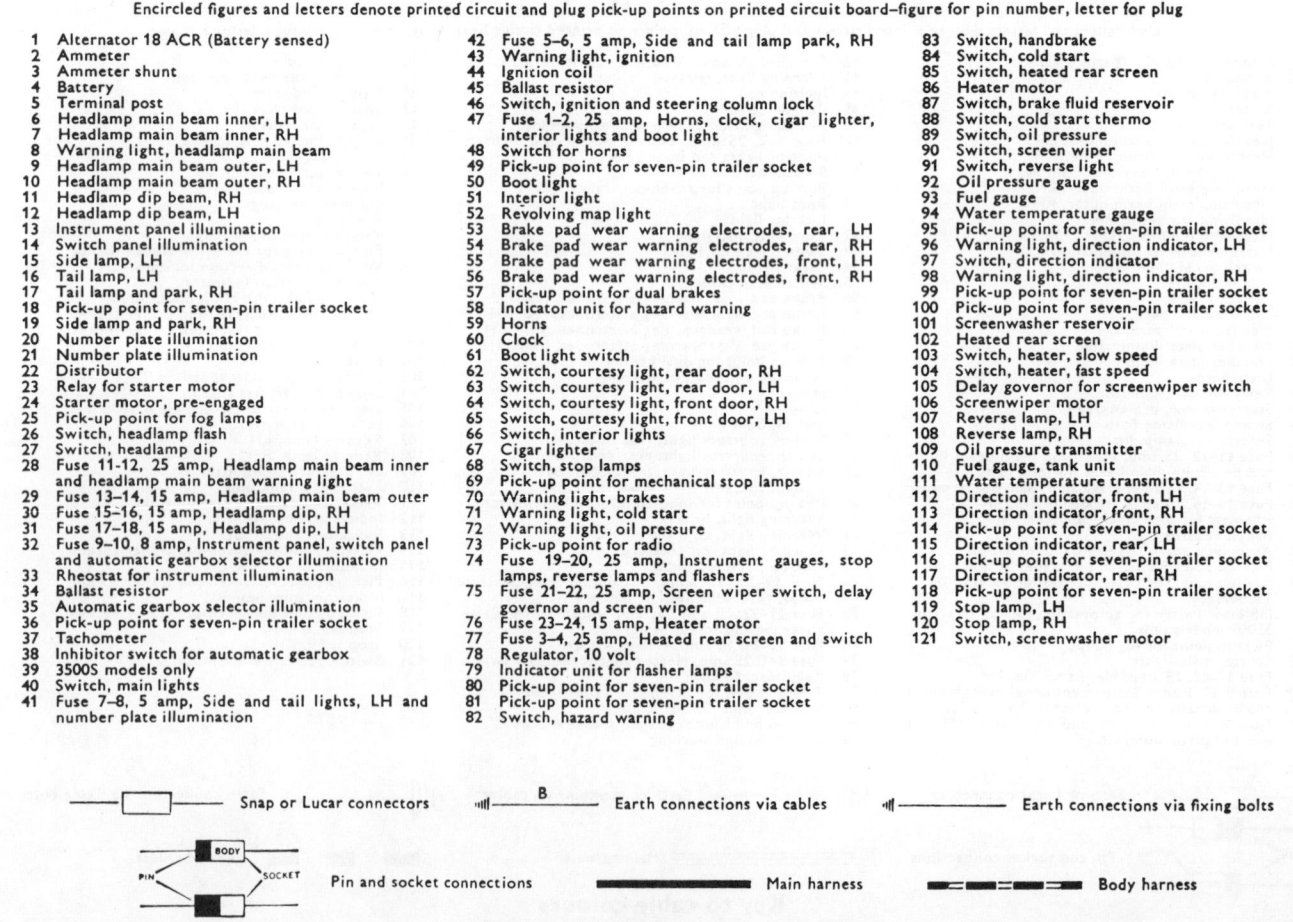

1. Alternator 18 ACR (Battery sensed)
2. Ammeter
3. Ammeter shunt
4. Battery
5. Terminal post
6. Headlamp main beam inner, LH
7. Headlamp main beam inner, RH
8. Warning light, headlamp main beam
9. Headlamp main beam outer, LH
10. Headlamp main beam outer, RH
11. Headlamp dip beam, RH
12. Headlamp dip beam, LH
13. Instrument panel illumination
14. Switch panel illumination
15. Side lamp, LH
16. Tail lamp, LH
17. Tail lamp and park, RH
18. Pick-up point for seven-pin trailer socket
19. Side lamp and park, RH
20. Number plate illumination
21. Number plate illumination
22. Distributor
23. Relay for starter motor
24. Starter motor, pre-engaged
25. Pick-up point for fog lamps
26. Switch, headlamp flash
27. Switch, headlamp dip
28. Fuse 11-12, 25 amp, Headlamp main beam inner and headlamp main beam warning light
29. Fuse 13-14, 15 amp, Headlamp main beam outer
30. Fuse 15-16, 15 amp, Headlamp dip, RH
31. Fuse 17-18, 15 amp, Headlamp dip, LH
32. Fuse 9-10, 8 amp, Instrument panel, switch panel and automatic gearbox selector illumination
33. Rheostat for instrument illumination
34. Ballast resistor
35. Automatic gearbox selector illumination
36. Pick-up point for seven-pin trailer socket
37. Tachometer
38. Inhibitor switch for automatic gearbox
39. 3500S models only
40. Switch, main lights
41. Fuse 7-8, 5 amp, Side and tail lights, LH and number plate illumination
42. Fuse 5-6, 5 amp, Side and tail lamp park, RH
43. Warning light, ignition
44. Ignition coil
45. Ballast resistor
46. Switch, ignition and steering column lock
47. Fuse 1-2, 25 amp, Horns, clock, cigar lighter, interior lights and boot light
48. Switch for horns
49. Pick-up point for seven-pin trailer socket
50. Boot light
51. Interior light
52. Revolving map light
53. Brake pad wear warning electrodes, rear, LH
54. Brake pad wear warning electrodes, rear, RH
55. Brake pad wear warning electrodes, front, LH
56. Brake pad wear warning electrodes, front, RH
57. Pick-up point for dual brakes
58. Indicator unit for hazard warning
59. Horns
60. Clock
61. Boot light switch
62. Switch, courtesy light, rear door, RH
63. Switch, courtesy light, rear door, LH
64. Switch, courtesy light, front door, RH
65. Switch, courtesy light, front door, LH
66. Switch, interior lights
67. Cigar lighter
68. Switch, stop lamps
69. Pick-up point for mechanical stop lamps
70. Warning light, brakes
71. Warning light, cold start
72. Warning light, oil pressure
73. Pick-up point for radio
74. Fuse 19-20, 25 amp, Instrument gauges, stop lamps, reverse lamps and flashers
75. Fuse 21-22, 25 amp, Screen wiper switch, delay governor and screen wiper
76. Fuse 23-24, 15 amp, Heater motor
77. Fuse 3-4, 25 amp, Heated rear screen and switch
78. Regulator, 10 volt
79. Indicator unit for flasher lamps
80. Pick-up point for seven-pin trailer socket
81. Pick-up point for seven-pin trailer socket
82. Switch, hazard warning
83. Switch, handbrake
84. Switch, cold start
85. Switch, heated rear screen
86. Heater motor
87. Switch, brake fluid reservoir
88. Switch, cold start thermo
89. Switch, oil pressure
90. Switch, screen wiper
91. Switch, reverse light
92. Oil pressure gauge
93. Fuel gauge
94. Water temperature gauge
95. Pick-up point for seven-pin trailer socket
96. Warning light, direction indicator, LH
97. Switch, direction indicator
98. Warning light, direction indicator, RH
99. Pick-up point for seven-pin trailer socket
100. Pick-up point for seven-pin trailer socket
101. Screenwasher reservoir
102. Heated rear screen
103. Switch, heater, slow speed
104. Switch, heater, fast speed
105. Delay governor for screenwiper switch
106. Screenwiper motor
107. Reverse lamp, LH
108. Reverse lamp, RH
109. Oil pressure transmitter
110. Fuel gauge, tank unit
111. Water temperature transmitter
112. Direction indicator, front, LH
113. Direction indicator, front, RH
114. Pick-up point for seven-pin trailer socket
115. Direction indicator, rear, LH
116. Pick-up point for seven-pin trailer socket
117. Direction indicator, rear, RH
118. Pick-up point for seven-pin trailer socket
119. Stop lamp, LH
120. Stop lamp, RH
121. Switch, screenwasher motor

Key to cable colours

B—Black G—Green K—Pink LG—Light green N—Brown O—Orange P—Purple R—Red S—Slate U—Blue W—White Y—Yellow

The last letter of a colour code denotes the tracer colour

86.00.01
Sheet 7

Rover 3500 and 3500S Manual AKM 3621

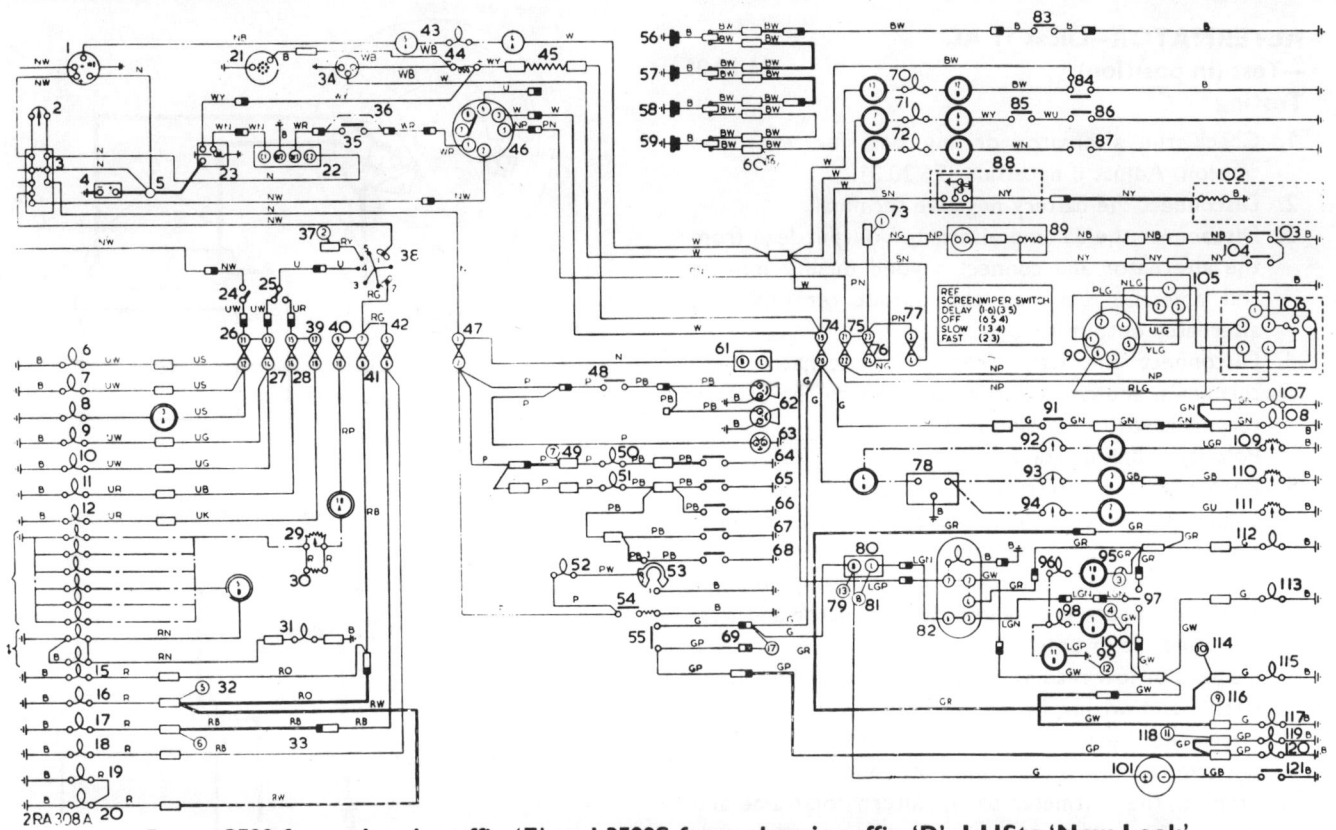

ELECTRICAL

Rover 3500 from chassis suffix 'E' and 3500S from chassis suffix 'D', LHStg 'New Look' models with circular type speedometer

Encircled figures and letters denote printed circuit and plug pick-up points on printed circuit board—figure for pin number, letter for plug

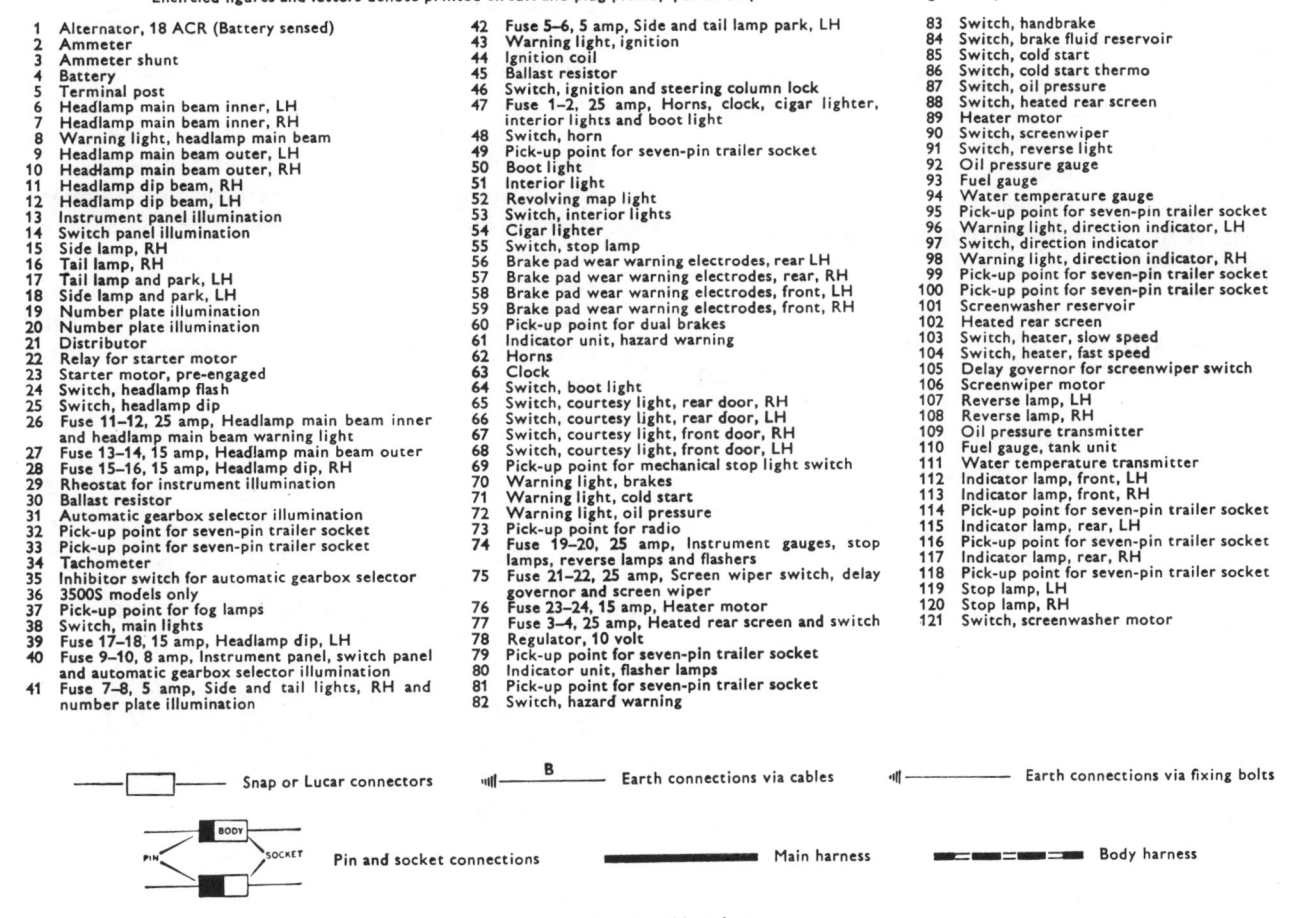

1. Alternator, 18 ACR (Battery sensed)
2. Ammeter
3. Ammeter shunt
4. Battery
5. Terminal post
6. Headlamp main beam inner, LH
7. Headlamp main beam inner, RH
8. Warning light, headlamp main beam
9. Headlamp main beam outer, LH
10. Headlamp main beam outer, RH
11. Headlamp dip beam, RH
12. Headlamp dip beam, LH
13. Instrument panel illumination
14. Switch panel illumination
15. Side lamp, RH
16. Tail lamp, RH
17. Tail lamp and park, LH
18. Side lamp and park, LH
19. Number plate illumination
20. Number plate illumination
21. Distributor
22. Relay for starter motor
23. Starter motor, pre-engaged
24. Switch, headlamp flash
25. Switch, headlamp dip
26. Fuse 11–12, 25 amp, Headlamp main beam inner and headlamp main beam warning light
27. Fuse 13–14, 15 amp, Headlamp main beam outer
28. Fuse 15–16, 15 amp, Headlamp dip, RH
29. Rheostat for instrument illumination
30. Ballast resistor
31. Automatic gearbox selector illumination
32. Pick-up point for seven-pin trailer socket
33. Pick-up point for seven-pin trailer socket
34. Tachometer
35. Inhibitor switch for automatic gearbox selector
36. 3500S models only
37. Pick-up point for fog lamps
38. Switch, main lights
39. Fuse 17–18, 15 amp, Headlamp dip, LH
40. Fuse 9–10, 8 amp, Instrument panel, switch panel and automatic gearbox selector illumination
41. Fuse 7–8, 5 amp, Side and tail lights, RH and number plate illumination
42. Fuse 5–6, 5 amp, Side and tail lamp park, LH
43. Warning light, ignition
44. Ignition coil
45. Ballast resistor
46. Switch, ignition and steering column lock
47. Fuse 1–2, 25 amp, Horns, clock, cigar lighter, interior lights and boot light
48. Switch, horn
49. Pick-up point for seven-pin trailer socket
50. Boot light
51. Interior light
52. Revolving map light
53. Switch, interior lights
54. Cigar lighter
55. Switch, stop lamp
56. Brake pad wear warning electrodes, rear LH
57. Brake pad wear warning electrodes, rear, RH
58. Brake pad wear warning electrodes, front, LH
59. Brake pad wear warning electrodes, front, RH
60. Pick-up point for dual brakes
61. Indicator unit, hazard warning
62. Horns
63. Clock
64. Switch, boot light
65. Switch, courtesy light, rear door, RH
66. Switch, courtesy light, rear door, LH
67. Switch, courtesy light, front door, RH
68. Switch, courtesy light, front door, LH
69. Pick-up point for mechanical stop light switch
70. Warning light, brakes
71. Warning light, cold start
72. Warning light, oil pressure
73. Pick-up point for radio
74. Fuse 19–20, 25 amp, Instrument gauges, stop lamps, reverse lamps and flashers
75. Fuse 21–22, 25 amp, Screen wiper switch, delay governor and screen wiper
76. Fuse 23–24, 15 amp, Heater motor
77. Fuse 3–4, 25 amp, Heated rear screen and switch
78. Regulator, 10 volt
79. Pick-up point for seven-pin trailer socket
80. Indicator unit, flasher lamps
81. Pick-up point for seven-pin trailer socket
82. Switch, hazard warning
83. Switch, handbrake
84. Switch, brake fluid reservoir
85. Switch, cold start
86. Switch, cold start thermo
87. Switch, oil pressure
88. Switch, heated rear screen
89. Heater motor
90. Switch, screenwiper
91. Switch, reverse light
92. Oil pressure gauge
93. Fuel gauge
94. Water temperature gauge
95. Pick-up point for seven-pin trailer socket
96. Warning light, direction indicator, LH
97. Switch, direction indicator
98. Warning light, direction indicator, RH
99. Pick-up point for seven-pin trailer socket
100. Pick-up point for seven-pin trailer socket
101. Screenwasher reservoir
102. Heated rear screen
103. Switch, heater, slow speed
104. Switch, heater, fast speed
105. Delay governor for screenwiper switch
106. Screenwiper motor
107. Reverse lamp, LH
108. Reverse lamp, RH
109. Fuel gauge transmitter
110. Fuel gauge, tank unit
111. Water temperature transmitter
112. Indicator lamp, front, LH
113. Indicator lamp, front, RH
114. Pick-up point for seven-pin trailer socket
115. Indicator lamp, rear, LH
116. Pick-up point for seven-pin trailer socket
117. Indicator lamp, rear, RH
118. Pick-up point for seven-pin trailer socket
119. Stop lamp, LH
120. Stop lamp, RH
121. Switch, screenwasher motor

Key to cable colours

B—Black G—Green K—Pink LG—Light green N—Brown O—Orange P—Purple R—Red S—Slate U—Blue W—White Y—Yellow
The last letter of a colour code denotes the tracer colour

Rover 3500 and 3500S Manual AKM 3621

ELECTRICAL

ALTERNATOR—Lucas 11 AC
—Test (in position) 86.10.01
Testing

1. Check the alternator driving belt for wear and tension. Adjust if necessary. 26.20.01.
2. Disconnect the battery negative terminal.
3. Disconnect the 35 amp alternator output lead from the alternator and connect a good quality moving coil ammeter between the output terminal and alternator output lead.
4. Disconnect the leads from the alternator field terminals, and connect the terminals to the battery terminals with extension leads.
   Polarity is unimportant.
5. Reconnect the battery negative terminal. Start the engine and increase the speed until the alternator is running at approximately 4,000 rev/min. The ammeter reading should be approximately 40 amps and if this figure is obtained the alternator and output cable are in order.
6. If a zero or low reading is obtained check the output circuit and wiring by connecting a voltmeter between the alternator output terminal and the battery positive terminal and noting the reading.
7. Transfer the voltmeter to the alternator frame and the battery negative terminal and noting the reading. If either reading exceeds 0.5 volt there is a high resistance in the circuit that must be traced and remedied.
8. If the test in item 7 does not reveal high resistance, and alternator output is low, remove and examine the alternator brush gear. Retest, and if low output persists, replace or overhaul the alternator.
9. If the alternator output is in order, disconnect the battery negative terminal, remove the ammeter and connect the alternator output cable to its terminal. Remove the extension leads from the alternator field cables.

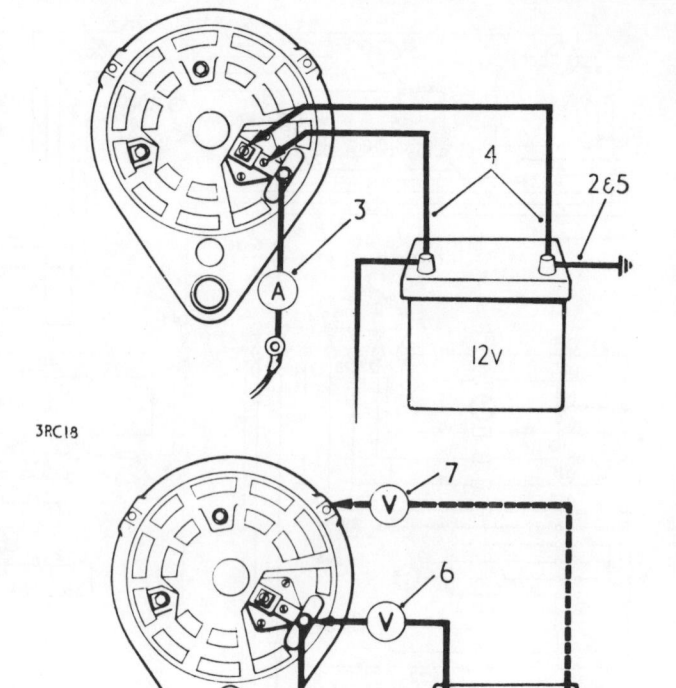

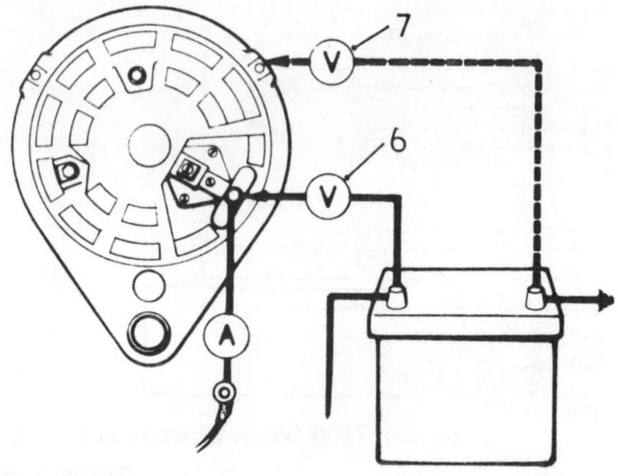

ALTERNATOR
—Remove and refit 86.10.02
Removing

1. Disconnect the battery earth lead.
2. Disconnect the leads from the alternator.
3. Slacken the alternator fixings, pivot the alternator inwards and remove the fan belt.
4. Remove the alternator.

Refitting

5. Connect the leads prior to fitting the alternator.
6. Fit the alternator, locating the fan belt over the pulleys, and the fan guard under the heads of the bolts at the adjusting link and the front of the mounting bracket.
7. Adjust the fan belt to the correct tension of 11 to 14 mm (0.437 to 0.562 in) free movement when checked midway between the alternator and crankshaft pulleys, by hand.
8. Secure the alternator fixings.
9. Reconnect the battery earth lead.

10. Connect the battery negative terminal. Switch on the ignition, and check that battery voltage is applied to the cable ends normally attached to the alternator field terminals by connecting a voltmeter across them. A reading of battery voltage proves that the field isolating relay circuit and wiring is in order. Conversely, a low or zero reading indicates a fault in the field isolating relay or the isolating contacts in the ignition switch (as applicable) associated wiring, or the alternator output control unit. See 86.10.18 for testing of field isolating relay. Refer to circuit diagram for continuity testing of wiring in the charging system.

For 18 ACR alternator test in position details see 86.10.08

86.10.01
86.10.02

Rover 3500 and 3500S Manual AKM 3621

ELECTRICAL—11 AC Alternator

ALTERNATOR—Lucas 11 AC

—Overhaul   86.10.08

Dismantling

1. Remove the alternator. 86.10.02.
2. Remove the nut and spring washer from the rotor shaft.
3. Withdraw the pulley and fan.
4. Scribe the drive end bracket, lamination pack and slip-ring end bracket so that they can be reassembled in the correct angular relation to each other.
5. Remove the three through bolts and washers.
6. Withdraw the drive end bracket and rotor as an assembly from the stator.
7. Remove the Woodruff key and bearing collar.
8. Press out the rotor shaft from the drive end bracket.
9. Remove the circlip from the drive end bracket.
10. Remove the retaining plate.
11. Press out the ball bearing.
12. Remove the 'O' ring and retaining washer.
13. Remove the two securing screws and washers and withdraw the brushbox.
14. Withdraw the brush-spring-and-terminal assemblies from the brushbox.

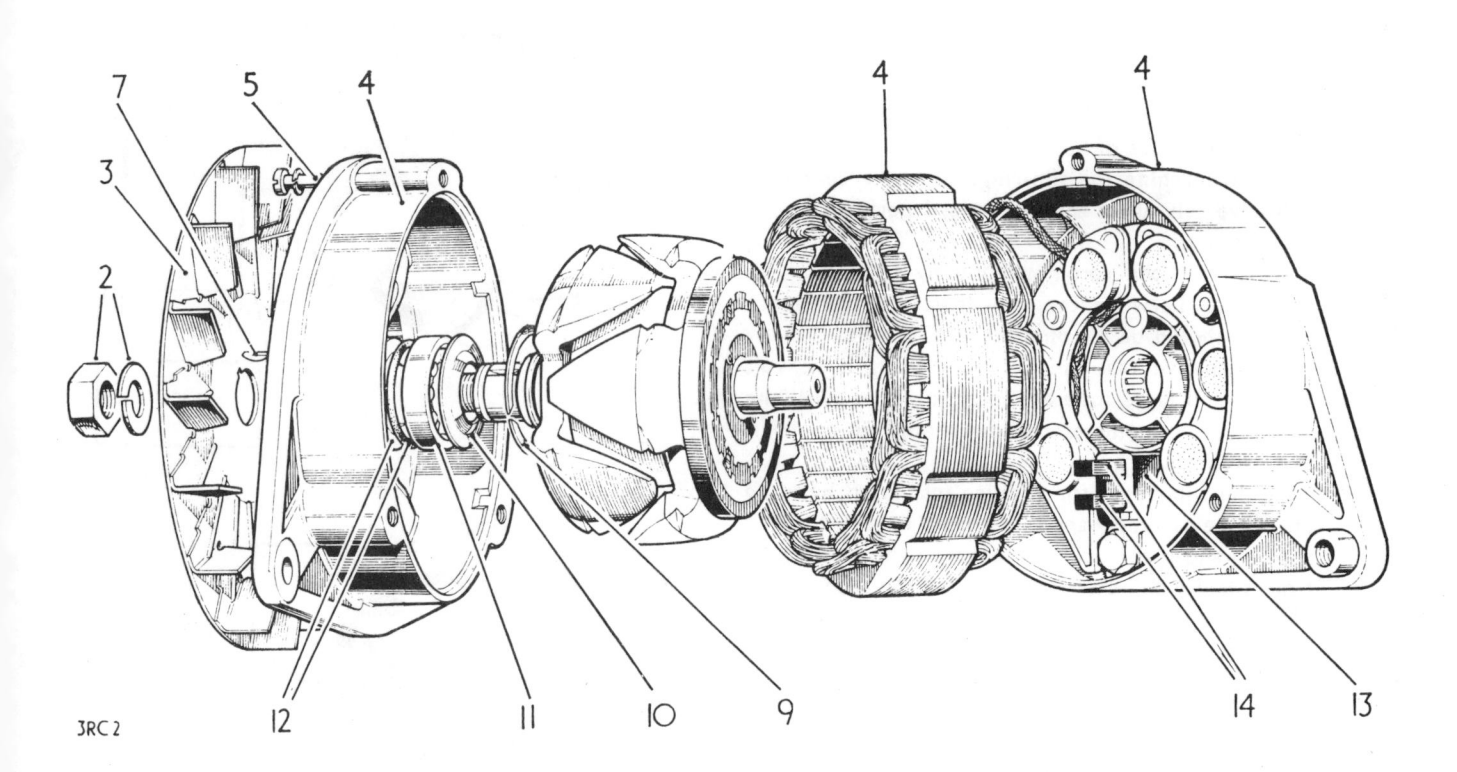

continued

Rover 3500 and 3500S Manual AKM 3621

ELECTRICAL—11 AC Alternator

15. Remove the nut and washer from the warning light 'AL' terminal.
16. Lift off the 17½ amp Lucar blade and the insulation bush.
17. Remove the bolt and washer securing the slip-ring end bracket to the heat sinks.
18. Withdraw the stator and heat sinks from the end bracket, noting the insulating and plain washers on the 'AL' and output terminal posts.
19. Note the connections of the wires to the diodes, and unsolder the wires, using a pair of long-nosed pliers as a thermal shunt as in the illustration. Great care must be taken to avoid overheating the diodes, or bending the diode pins and the operation should be done as quickly as possible.
20. Remove the nut securing the heat sinks together, and separate the heat sinks, noting the insulation washers.

Inspecting and testing

21. Measure the brush length. A new brush is 15,9 mm (0.625 in) long; a fully worn brush is 4 mm (0.156 in) long, and must be replaced at, or approaching this length.
22. Check the brush spring pressures against the General Specification Data.
23. Check that the brushes move freely in their holders. Clean the brushes with a petrol-moistened cloth, and lightly polish the brush sides with a smooth file if necessary.
24. **Slip-rings.** The surfaces of the slip-rings should be smooth and free from oil. Clean the surfaces with a petrol-moistened cloth, or very fine glass paper if there is evidence of burning. No attempt should be made to machine the slip-rings, as any eccentricity will adversely affect the high speed performance of the alternator.
25. **Rotor.** Test the rotor winding by connecting an ohmmeter between the slip-rings. The resistance reading should be 3.8 ohms at 20°C (68°F).
26. An alternative test can be made using a 12 volt battery and large scale ammeter in series with the winding. The ammeter reading should be approximately 3.2 amperes.
27. Test the slip-ring/rotor winding insulation by using an AC mains supply and a test lamp in series with a slipring and rotor pole. If the lamp lights the insulation is faulty and the rotor must be replaced.
No attempt should be made to machine the rotor poles or true a distorted shaft.

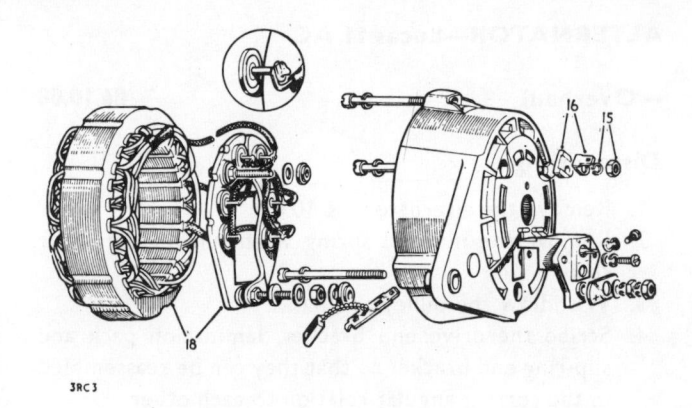

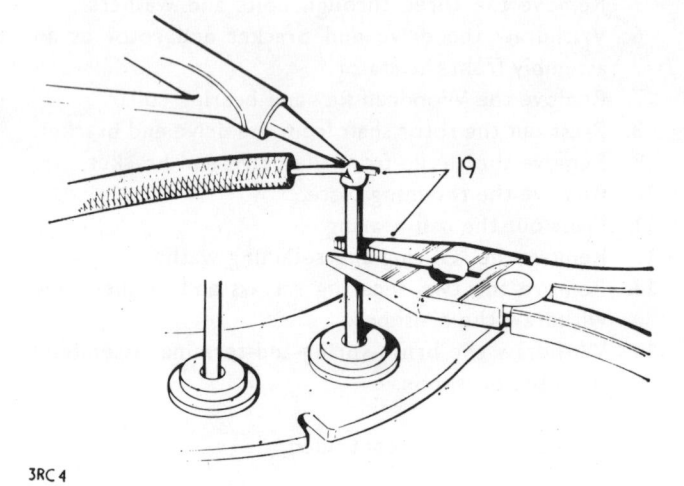

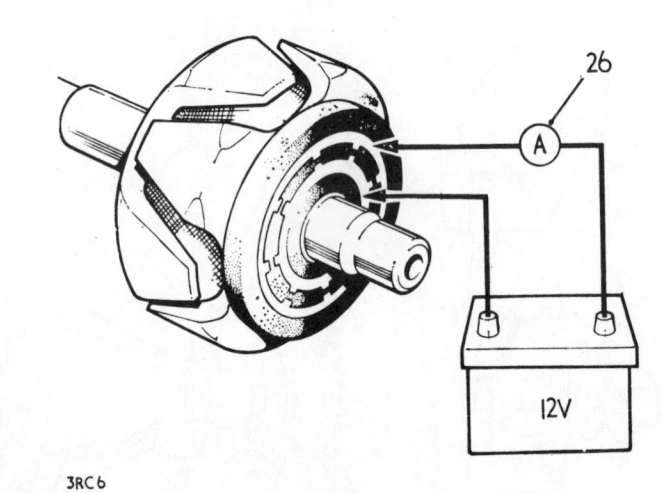

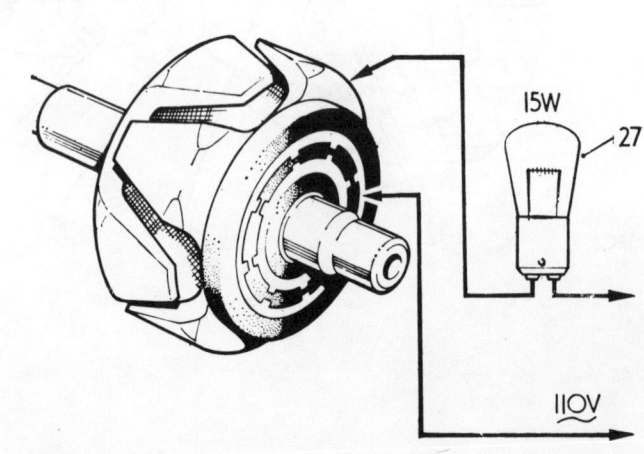

continued

86.10.08
Sheet 2

Rover 3500 and 3500S Manual AKM 3621

ELECTRICAL—11 AC Alternator

28. **Stator.** Check the continuity of the stator windings with the stator cables separated from the heat sinks. Connect any two of the three stator cables in series with a 1.5 watt test lamp and a 12 volt battery. Repeat the test, replacing one of the two cables by the third cable. Failure of the test lamp to light on either occasion means that part of the stator winding is open-circuit, and the stator must be replaced.

29. Test the insulation between the stator coils and lamination pack with an AC mains supply and a bulb in series with any one of the three cable ends and the lamination pack. If the bulb lights the stator coils are earthing and the stator must be replaced.

30. **Heat sinks and diodes.** There are two heat sink assemblies, one of positive polarity and the other negative. Each carries three diodes that are pressed into position. The diodes are not individually replaceable, and if a diode is found defective the heat sink must be replaced.

31. The diodes can be tested for service purposes with the stator wires separated from the heat sinks. Connect a 12 volt battery and 1.5 watt bulb in series with each diode, and reverse the connections. Current should flow, and the bulb light up, in one direction only. If the bulb lights up in both directions, or does not light up in either, the diode is defective and the heat sink assembly must be replaced.

32. Accurate measurement of diode resistance requires factory equipment. Since the forward resistance of a diode varies with the voltage applied, no realistic readings can be obtained with battery powered ohmmeters. If a battery-ohmmeter is used, a good diode will yield infinity in one direction, and some indefinite but much lower reading in the other.

    CAUTION: Ohmmeters of the type incorporating a hand-driven generator must never be used for checking diodes.

33. **Bearings.** Bearings that are worn to the extent that they allow excessive side movement of the rotor shaft must be renewed. The needle-roller bearing in the slip-ring end bracket is not serviced separately, and in the unlikely event that it becomes unserviceable the end bracket must be replaced.

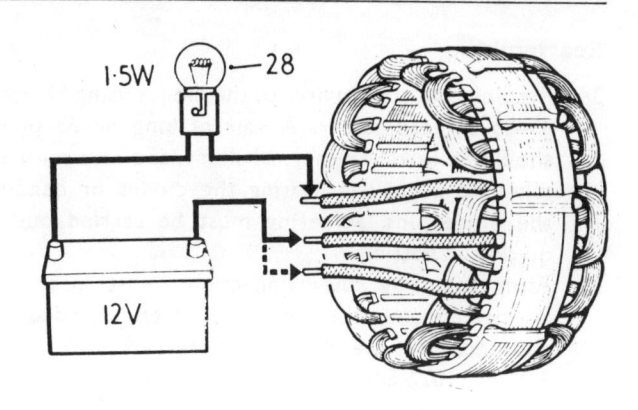

continued

3RC 8

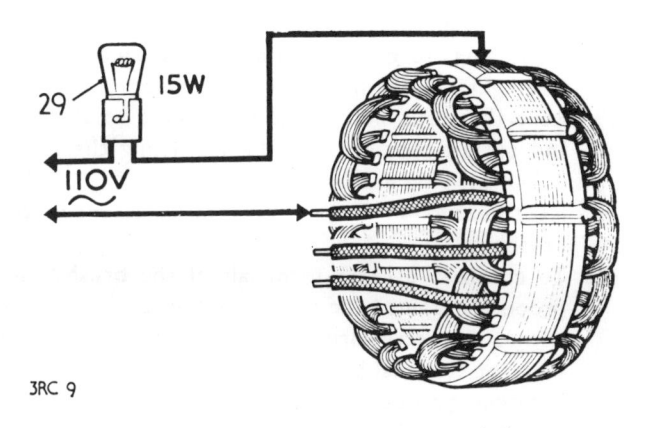

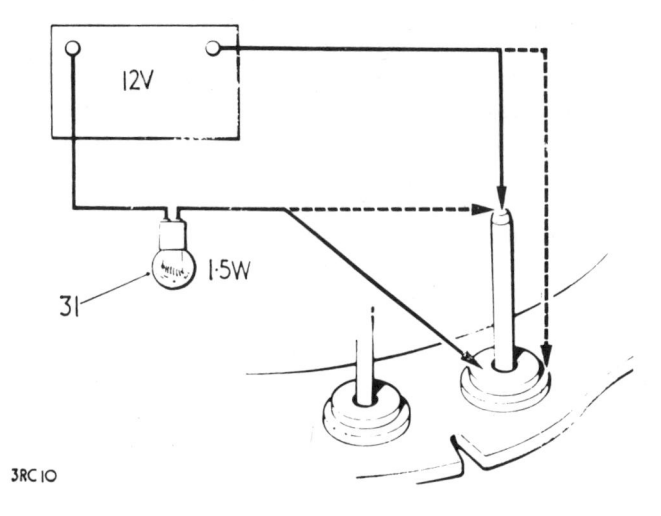

3RC 9

3RC 10

Rover 3500 and 3500S Manual AKM 3621

ELECTRICAL—11 AC Alternator

Reassembling

34. Resolder the stator wires to the diodes using M grade 45–55 tin/lead solder. A pair of long nosed pliers should be used as a thermal shunt, and care must be taken to avoid overheating the diodes or bending the diode pins. Soldering must be carried out as quickly as possible.
35. After soldering, the connections must be neatly arranged around the heat sinks to ensure adequate clearance for the rotor, and tacked down with 3 M EC 1022 adhesive where indicated in the illustration. The stator connections must pass through the appropriate notches at the edge of the heat sink.
36. Reverse 1 to 20, including the following.
37. Assemble the insulation washers between the heat sinks in the following order: thick plain washer, thin plain washer and small plain washer.
38. Tightening torque for diode heat sink fixings: 0,28 kgf.m (25 lbf. in).
39. Refit the plain and insulating washers to the 'AL' and output terminal posts before fitting the stator to the slip-ring end bracket.
40. Fit the two small insulating washers under the brushbox.
41. To ensure the Lucar terminals of the brushes are properly retained the tongue should be levered up with a fine screwdriver to make an angle of 30° with the terminal blade.
42. Tightening torque for the brush box fixing screws: 0,11 kgf.m (10 lbf. in).
43. Support the inner journal of the drive end bearing when pressing in the rotor shaft. Do not use the bracket as a support for the bearing while fitting the rotor shaft.
44. Align the drive end bracket, lamination pack and slip-ring bracket, and tighten the three bolts evenly. Check that the rotor is quite free in its bearing.
45. Tightening torque for alternator through bolts: 0,51 to 0,57 kgf.m (45 to 50 lbf. in).
46. Refit the alternator. 86.10.02.

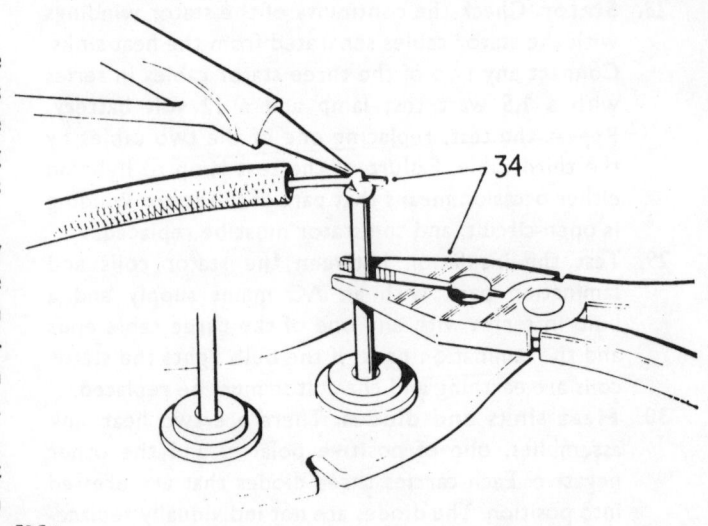

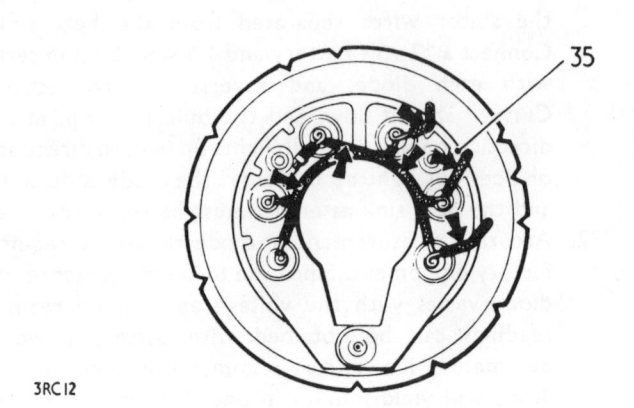

continued

ELECTRICAL—11 AC Alternator

DATA

Alternator:

Type	Lucas 11 AC with transistorised current-voltage regulator
Rating	45 amps
Nominal voltage	12 volts
Nominal direct current output	43 amperes
Resistance of field coil at 20°C (68°F)	3.8 ohms
Maximum rotor speed	12,500 rev/min
Stator phases	3
Stator connection	Star
Number of rotor poles	8
Number of field coils	1
Slip-ring brushes length: new	15,9 mm (0.625 in)
replace at	4 mm (0.156 in)
Brush spring tension:	
load at 19,9 mm ($\tfrac{25}{32}$ in)	113 to 142 grms (4 to 5 ozs)
load at 10,3 mm ($\tfrac{13}{32}$ in)	212 to 241 grms (7.5 to 8.5 ozs)
Assembly torques, maximum:	
Brushbox fixing screws	0,115 kgf. m (10 lbf in)
Diode heat sink fixings	0,288 kgf. m (25 lbf in)
Through bolts	0,518 to 0,576 kgf. m (45 to 50 lbf in)
Alternator control box	Lucas 4TR
To test control box	Switch on side and tail lamps to give an electrical load of approximately 2 amps.
	Start engine and run alternator at 3,000 rev/min for at least 8 minutes, to ensure that the system voltage has been stabilised, and that the charging current is not more than 10 amps. The unit can be adjusted to give 13.9 to 14.3 volts, but it is recommended that adjustment is effected by a Lucas agent.

Rover 3500 and 3500S Manual AKM 3621

ELECTRICAL—Alternator type 18ACR

ALTERNATOR—Lucas 18ACR

—Overhaul
86.10.08

Note. Alternator charging circuit
The ignition warning light is connected in series with the alternator field circuit. Bulb failure would prevent the alternator charging, except at very high engine speeds, therefore, the bulb should be checked before suspecting an alternator fault.

Precautions
Battery polarity is NEGATIVE EARTH, which must be maintained at all times.
No separate control unit is fitted; instead a voltage regulator of micro-circuit construction is incorporated on the slip-ring end bracket, inside the alternator cover.
Battery voltage is applied to the alternator output cable even when the ignition is switched off, the battery must be disconnected before commencing any work on the alternator. The battery must also be disconnected when repairs to the body structure are being done by arc welding.

Surge protection device
Some protection of the alternator is provided by a surge protection device, connected across the 'IND' terminal, to absorb high transient voltages.

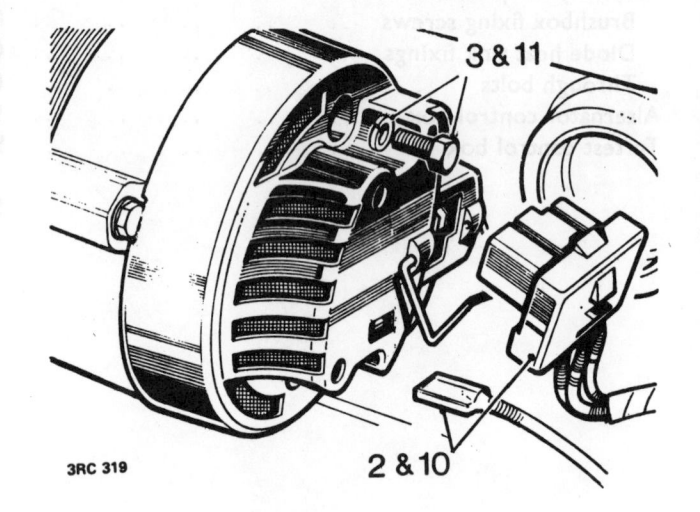

Testing in position
Surge protection device
If the alternator output falls to zero, the fault may be caused by the surge protection device failing safe short-circuit, or a fault in the alternator circuit. Check the surge protection device as follows:

1. Check that the fan belt is correctly tensioned.
2. Withdraw the connectors from the alternator.
3. Remove the alternator rear cover.
4. Remove the surge protection device from the 'IND' terminal.
5. Refit the alternator cover, and connectors. Ensure that all circuit connections are clean and tight.
6. Start and run the engine. If the alternator output is now normal, fit a new surge protection device.

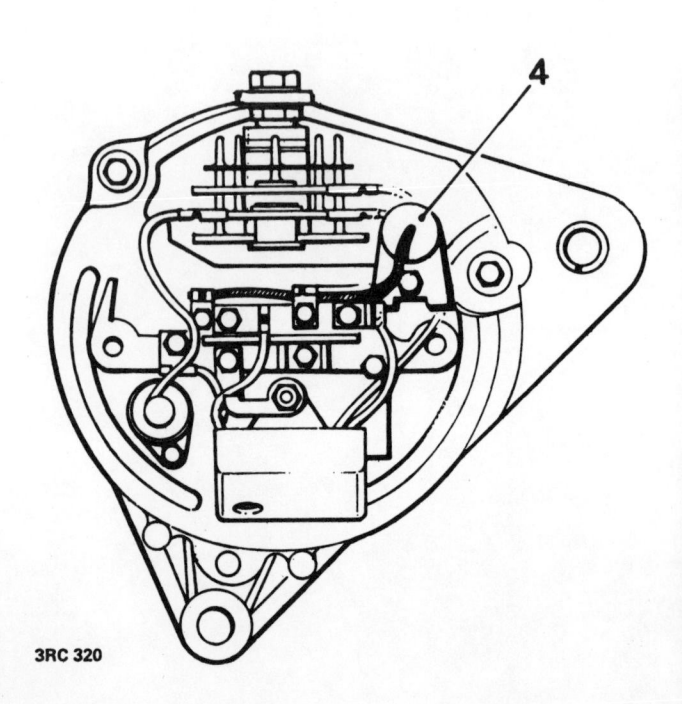

continued

Rover 3500 and 3500S Manual AKM 3621

ELECTRICAL—Alternator type 18ACR

Output test

7. Check that the fan belt is correctly tensioned and that all charging circuit connections are secure.
8. Run the engine at fast idle until normal operating temperature is attained.
9. Stop the engine.
10. Withdraw the connectors from the alternator.
11. Remove the alternator rear cover.
12. Connect the regulator case to the alternator frame.
13. Connect a 0–60 ammeter between the alternator and the battery.
14. Connect a 0–20 voltmeter across the battery terminals.
15. Connect a 15 ohm 35 amp, variable resistor across the battery terminals.

    CAUTION: Do not leave the variable resistor connected across the battery terminals for longer than is necessary to carry out the following test, items 10 and 11.

16. Start the engine and run at 750 rev/min. The warning light bulb should be extinguished.
17. Increase the engine speed to 3000 rev/min, and adjust the variable resistance until the voltmeter reads 13.6 volts. The ammeter reading should then be approximately 45 amps. Any appreciable deviation from this figure will necessitate removing and dismantling the alternator. If the output test is satisfactory, proceed with the regulator test.

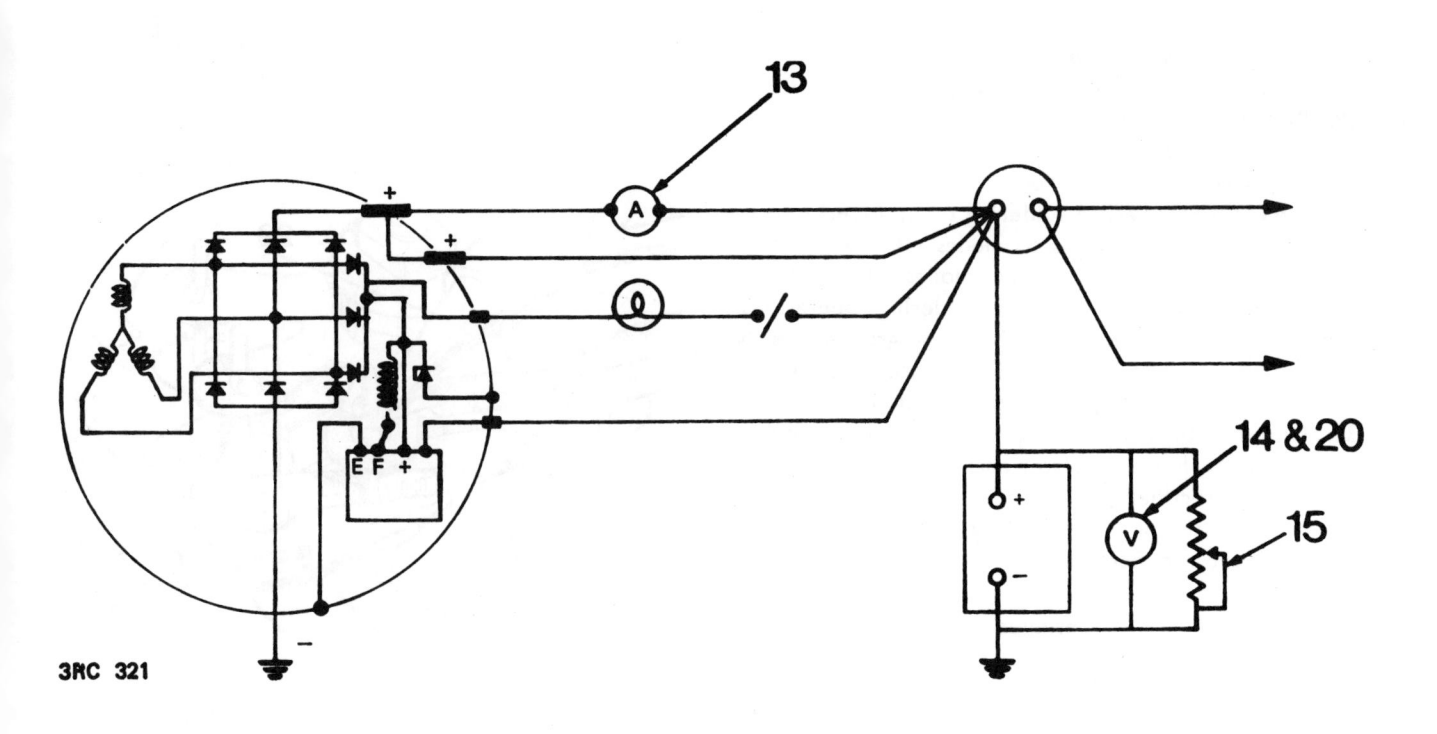

continued

Rover 3500 and 3500S Manual AKM 3621

86.10.08
Sheet 7

ELECTRICAL—Alternator type 18ACR

Regulator test

18. Disconnect the variable resistor and remove the connection between the regulator and the alternator frame.
19. With the remainder of the circuit connected as for the alternator output test, start the engine and run at 3,000 rev/min, until the ammeter shows an output current of less than 10 amperes.
20. The voltmeter should now give a reading of 13.6 to 14.4 volts. Any appreciable deviation from this (regulating) voltage indicates a faulty regulator which must be replaced.
21. If the foregoing output and regulator tests show the alternator and regulator to be performing satisfactorily, disconnect the test circuit, reconnect the alternator terminal connector and proceed with the charging circuit resistance test.

Charging circuit resistance test

22. Connect a low-range voltmeter between either of the alternator terminals marked + and the positive terminal of the battery.
23. Switch on the headlamps.
24. Start the engine and run at approximately 3,000 rev/min. Note the voltmeter reading.
25. Transfer the voltmeter connections to the frame of the alternator and the negative terminal of the battery, and again note the voltmeter reading.
26. If the reading exceeds 0.5 volt on the positive side or 0.25 volt on the negative side, there is a high resistance in the charging circuit which must be traced and remedied.

Testing—alternator removed

27. Remove the alternator. 86.10.02.
28. Remove the alternator rear cover.
29. Unsolder stator connections from rectifier pack, noting connection positions.
    CAUTION: When soldering or unsoldering connections to diodes take care not to overheat the diodes or bend the pins. During soldering operations, diode pins should be gripped lightly with a pair of long nosed pliers which will act as a thermal shunt.
30. Unscrew brush moulding securing screws and if necessary, lower regulator pack securing screw.
31. Slacken rectifier pack retaining nuts and withdraw both brush moulding, with or without regulator pack, and rectifier pack.

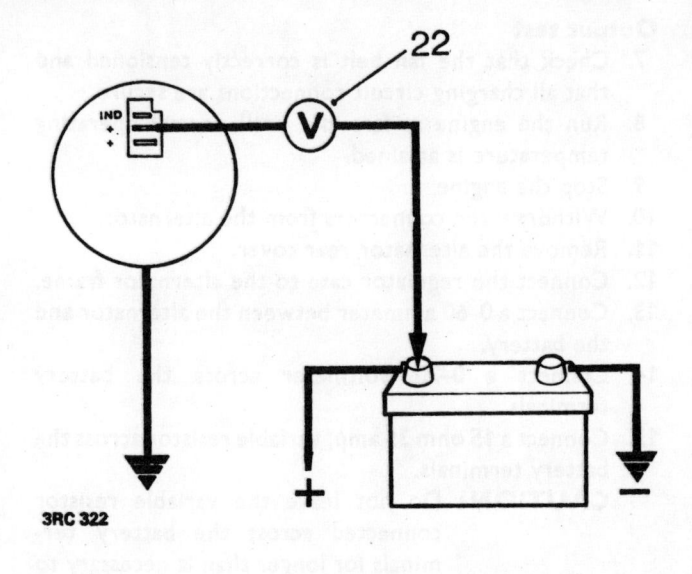

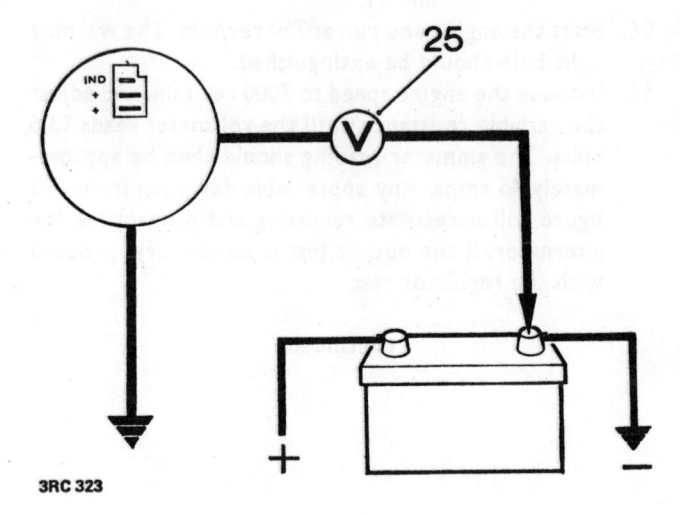

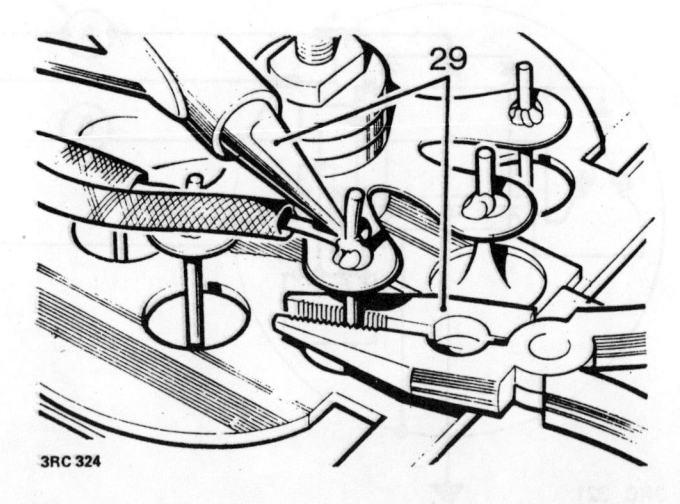

continued

86.10.08
Sheet 8

Rover 3500 and 3500S Manual AKM 3621

ELECTRICAL—Alternator type 18ACR

Brushes

32. Check brushes for wear by measuring length of brush protruding beyond brush box moulding. If length is 8 mm (0.3 in.) or less, brush must be renewed.
33. Check that brushes move freely in holders. If brush is sticking, clean with petrol moistened cloth or polish sides of brush with fine file.
34. Check brush spring pressure using push-type spring gauge. Gauge should register 255 to 368g (9 to 13 oz) when brush is pushed back until face is flush with housing. If reading is outside these limits, renew brush assembly.

Slip-rings

35. Clean surfaces of slip-rings using petrol moistened cloth.
36. Inspect slip-ring surfaces for signs of burning; remove burn marks using very fine sandpaper. On no account should emery cloth or similar abrasives be used, or any attempt made to machine the slip-rings.

Rotor

37. Connect an ohmmeter or a 12-volt battery and an ammeter to the slip-rings. An ohmmeter reading of 3.2 ohms or an ammeter reading of 4 amps should be recorded.
38. Using a 110-volt a.c. supply and a 15-watt test lamp, test for insulation between one of the slip-rings and one of the rotor poles. If the test lamp lights, the rotor must be renewed.

Stator

39. Connect a 12-volt battery and a 36-watt test lamp to two of the stator connections. Repeat the test replacing one of the two stator connections with the third. If test lamp fails to light in either test, stator must be renewed.
40. Using a 110-volt a.c. supply and a 15-watt test lamp, test for insulation between any one of the three stator connections and stator laminations. If test lamp lights, stator must be renewed.

Diodes

41. Connect a 12-volt battery and a 1.5-watt test lamp in turn to each of the nine diode pins and corresponding heat sink on the rectifier pack, then reverse the connections. Lamp should light with current flowing in one direction only. If lamp lights in both directions or fails to light in either, rectifier pack must be renewed.

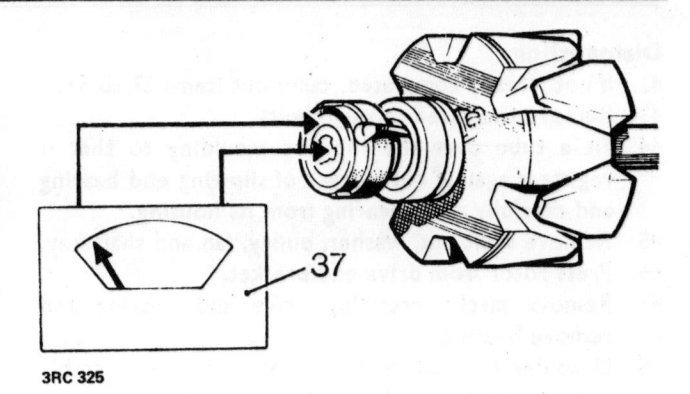

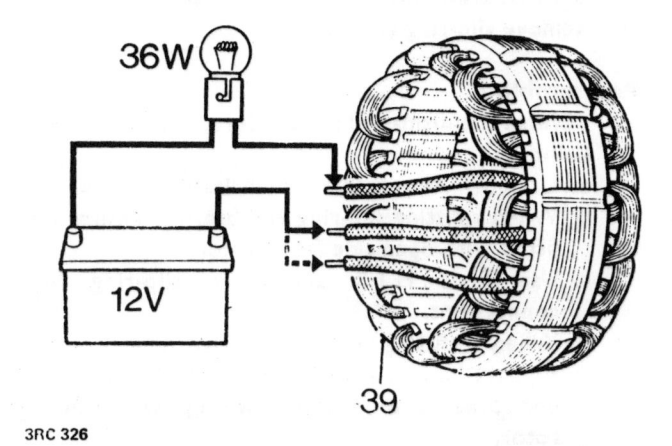

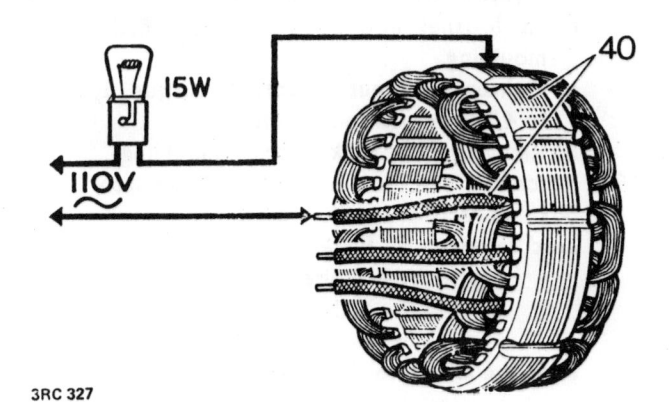

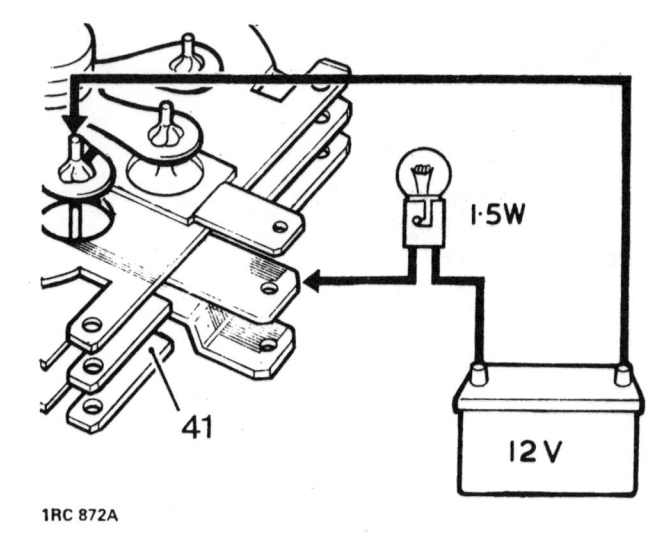

continued

Rover 3500 and 3500S Manual AKM 3621

ELECTRICAL—Alternator type 18ACR

Dismantling

42. If not already completed, carry out items 27 to 31.
43. Remove the three through bolts.
44. Fit a tube over the slip-ring moulding so that it registers against outer track of slip-ring end bearing and carefully drive bearing from its housing.
45. Remove shaft nut, washer, pulley, fan and shaft key.
46. Press rotor from drive end bracket.
47. Remove circlip retaining drive end bearing and remove bearing.
48. Unsolder field connections from slip-ring assembly and withdraw assembly from rotor shaft.
49. Remove slip-ring end bearing.

Reassembling

50. Reverse the dismantling procedure, noting following points.
    a. Use Shell Alvania 'RA' to lubricate bearings.
    b. When refitting slip-ring end bearing, ensure it is fitted with open side facing rotor.
    c. Use Fry's H.T.3. solder on slip-ring field connections.
    d. When refitting rotor to drive end bracket, support inner track of bearing. Do not use drive end bracket to support bearing when fitting rotor.
    e. Tighten through-bolts evenly.
    f. Fit brushes into housings before fitting brush moulding.
    g. Tighten shaft nut to correct torque figure 3,5 to 4,2 kgf.m (25 to 30 lbf ft).
    h. Refit regulator pack to brush moulding.
51. Reconnect the leads between the regulator; brush box and rectifier, as illustrated.
    Lead colours   B—Black
                   W—White
                   Y—Yellow
                   S—Sensing terminal
52. Refit the alternator 86.10.02.

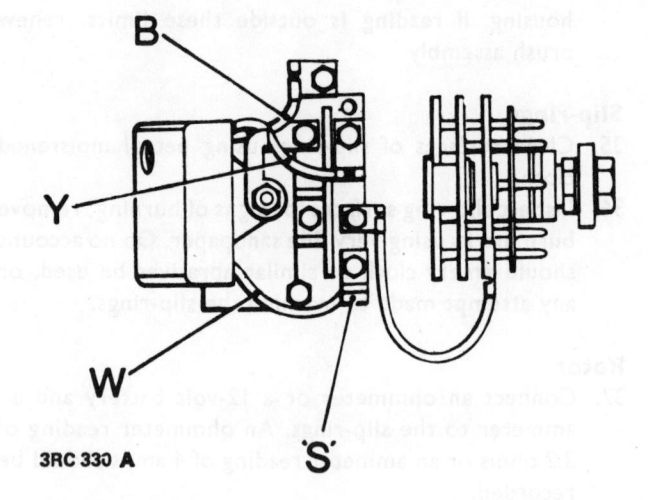

DATA

Alternator	Lucas 18ACR battery sensed
Nominal output	45 amps at 6000 alternator rev/min
Field resistance	3.2 ohms
Brush spring pressure	255 to 368g (9 to 13 oz)
Brush minimum length	8 mm (0.312 in)
Regulating voltage	13.6 to 14.4 volts

ELECTRICAL—Alternator type 18ACR

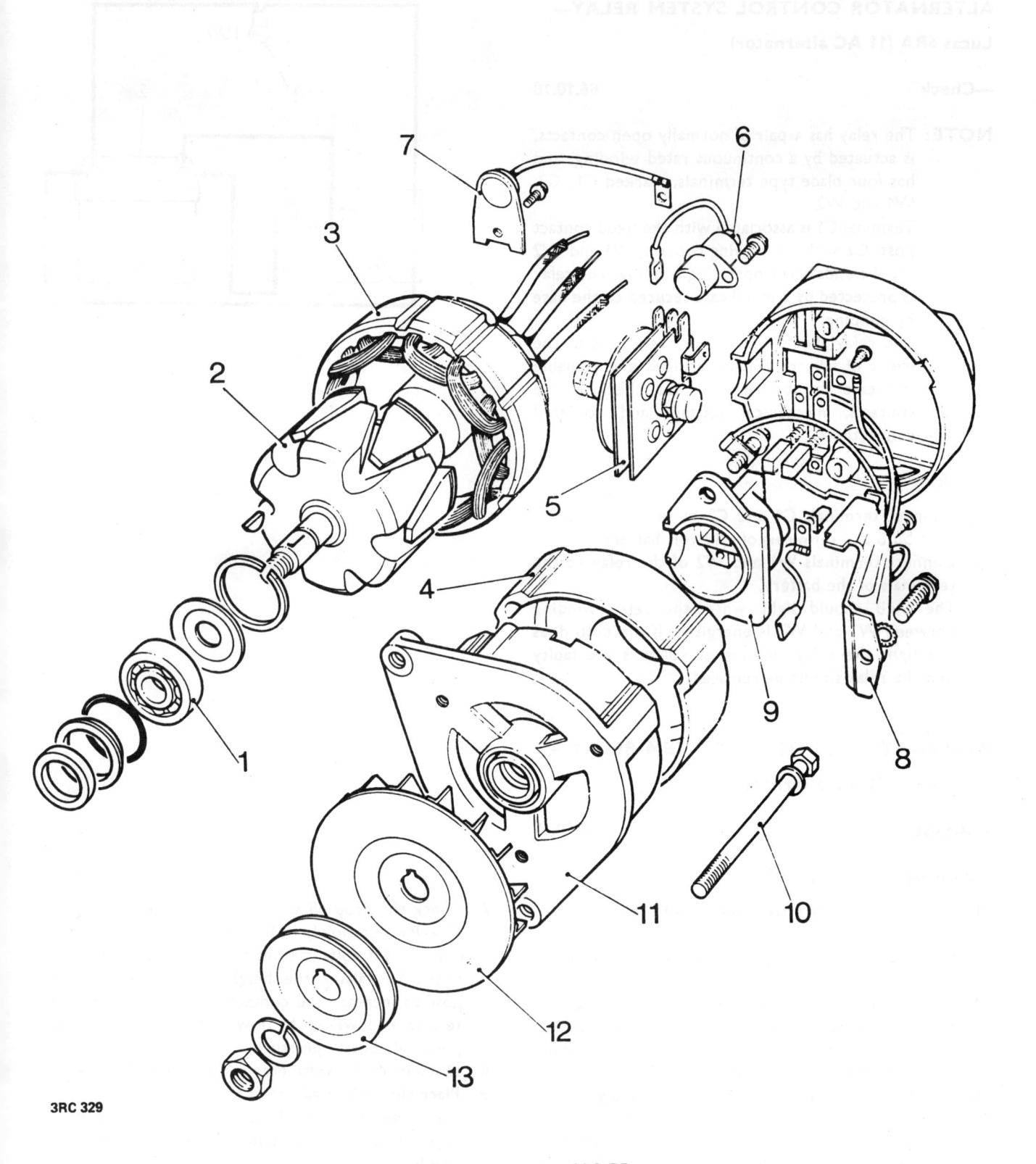

Alternator, type 18ACR

1. Drive end bearing
2. Rotor and slip ring
3. Stater
4. Slip ring bracket
5. Rectifier
6. Suppressor
7. Surge protection device
8. Regulator unit
9. Brush Box
10. Through bolt
11. Drive end bracket
12. Fan
13. Pulley

ELECTRICAL—11 AC Alternator

ALTERNATOR CONTROL SYSTEM RELAY—

Lucas 6RA (11 AC alternator)

—Check 86.10.18

NOTE: The relay has a pair of normally open contacts, is actuated by a continuous rated winding, and has four blade type terminals, marked C1, C2, W1 and W2.

Terminal C1 is associated with the fixed contact post, C2 with the moving contact; W1 and W2 are the ends of the operating winding. The relay is protected by a metal case secured to the base by gimping.

The relay should be replaced as a unit in the event of failure. Where facilities are available the case can be removed for checking the contacts, mechanical settings and soldered connections.

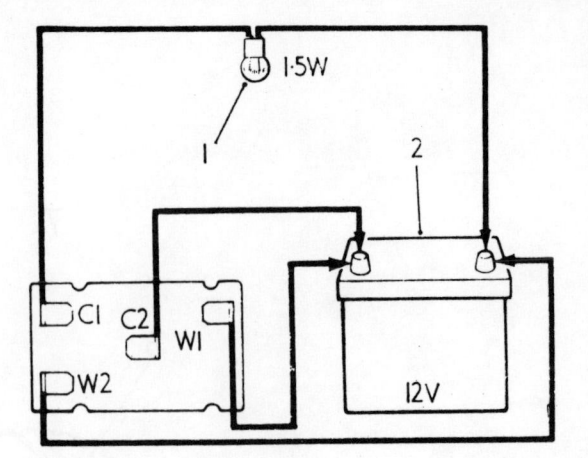

Checking

1. Connect terminals C1 and C2 in series with a 1.5 watt bulb and terminals of a 12 volt battery.
2. Connect terminals W1 and W2 of the relay to the terminals of the battery.
3. The bulb should light when the relay winding between W1 and W2 is energised. If the bulb does not light the relay winding or contacts are faulty and the relay should be replaced.

ALTERNATOR CONTROL SYSTEM RELAY—

Lucas 6RA (11 AC alternator)

—Adjust 86.10.19

Adjusting

1. Check the control system relay. 86.10.18.
2. Remove the relay.
3. Prise open the gimping securing the cover to the base.
4. Connect a first grade moving coil voltmeter across the relay operating winding, terminals W1 and W2.
5. Connect a variable direct current supply 0–15 volts to the winding terminals, W1 and W2.
6. Check the cut-in setting by raising the supply voltage from zero. The contacts should close when the voltage is 6.0 to 7.5 volts. The setting is raised by increasing the air gap between the underside of the armature and the bobbin core, and lowered by decreasing the air gap.
7. Check the drop-off setting by raising the voltage to 15 volts then slowly reducing it. The contacts should open at 4.0 volts minimum. The drop-off setting is raised by raising the height of the fixed contact post to increase the contact pressure. The drop-off setting is lowered by lowering the fixed contact post and contact pressure.
8. Disconnect the voltmeter and voltmeter supply.
9. Place the sealing gasket in the case flange and insert the relay assembly. Press the components firmly together and secure at four points by gimping the case lip.
10. Recheck the cut-in and drop-off settings.
11. Refit the relay.

DATA

Nominal voltage	12 volts	Early 3500 models.
Cut-in voltage	6.0 to 7.5 volts	On later models and 3500S, the alternator is isolated by contacts in the ignition switch
Drop-off voltage	4.0 volts minimum	
Resistance of operating winding	76 ohms	

ELECTRICAL—11 AC Alternator

CONTROL BOX—Lucas 4TR (11 AC alternator)

—Test   86.10.25

NOTE: Before checking and adjusting the control unit it must be established that the alternator and charging circuit wiring are in good order. See 86.10.01. The battery to control unit wiring, (including the 6RA field isolating relay on early 3500 models or the isolating contacts in the ignition switch on later models and 3500S) must also be in good order. The resistance of this circuit, including the relay, must not exceed 1 ohm. Higher resistance must be traced and remedied.

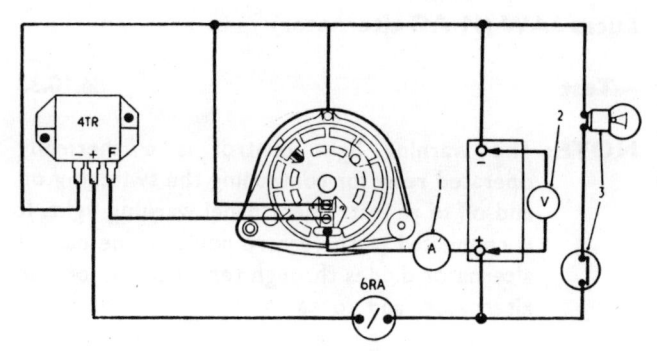

Testing

1. Disconnect the alternator main output lead from the alternator or starter solenoid, and connect an ammeter of 50 amp range in series with the lead and its connection.
2. Connect a voltmeter of 1 per cent accuracy or better between the battery terminals and note the reading with all electrical equipment switched off. If available, use a voltmeter of the suppressed-zero type, reading 12–15 volts.
3. Switch on the side and tail lights to give an electrical load of approximately 2 amperes.
4. Start the engine and run the alternator at approximately 3,000 rev/min for at least 8 minutes to ensure the system voltage has stabilised. If the charging current is greater than 10 amperes, continue to run the engine until this figure is reached. The voltmeter should then read 13.9 to 14.3 volts.
5. If the reading obtained is stable, but outside these limits, the unit can be adjusted to obtain the correct voltage. See 86.10.27.
6. If the voltmeter reading remains unchanged at open circuit battery voltage, or increases in an uncontrolled manner, the control unit is faulty and must be replaced.

CONTROL BOX—Lucas 4TR (11 AC alternator)

—Adjust   86.10.27

Adjusting

1. Test the control box. 86.10.25.
2. Stop the engine and remove the control unit securing screws, invert the unit and carefully scrape away the sealing compound that conceals the potentiometer adjuster.
3. Start the engine and run the alternator at approximately 3,000 rev/min with the voltmeter firmly connected to the battery terminals. Turn the potentiometer adjuster slot clockwise to increase the voltage setting, or counter-clockwise to decrease it. A small movement of the adjuster causes an appreciable difference in the voltage reading.
4. Recheck the setting by stopping the engine, then restarting and running the alternator at 3,000 rev/min.
5. Disconnect the voltmeter and ammeter. Reconnect the alternator output lead.
6. Refit the control unit.

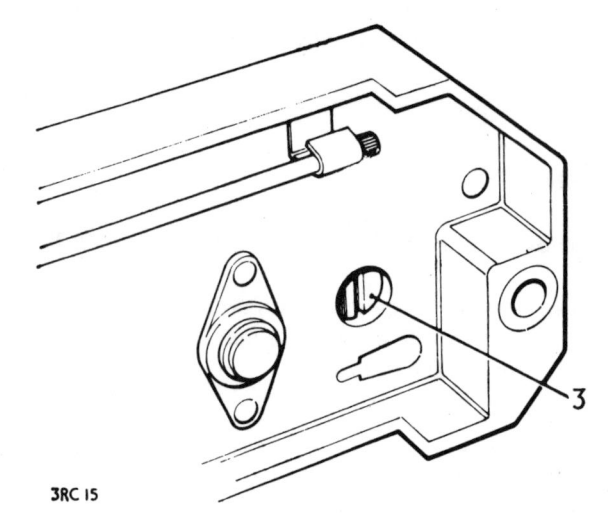

Rover 3500 and 3500S Manual AKM 3621

ELECTRICAL

ALTERNATOR WARNING LIGHT CONTROL—

Lucas 3AW (11 AC alternator)

—Test   86.10.33

NOTE: The warning light control is a thermally operated relay for controlling the switching on and off of an instrument panel warning light. It is connected to the centre point of one pair of alternator diodes through terminal 'AL' on the alternator, and to earth. The indication given by the warning light is similar to that provided by the ignition ('No Charge') warning light used with dynamo charging systems.

Due to the external similarity of the alternator warning light control model 3AW to flasher unit model FL5, a distinctive green label is applied to the aluminium case of model 3AW. Care must be taken to avoid connecting either of these units into a circuit designed for the other.

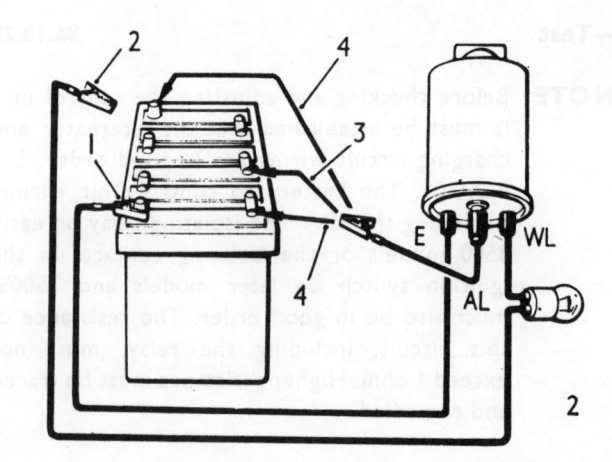

Testing

1. Connect terminal 'E' of the warning light control to the battery negative terminal.
2. Connect a 2.2 watt bulb in series with the terminal 'WL' of the warning light control and the positive battery terminal. The bulb should light up immediately. If the bulb does not light up the warning light control is faulty and must be replaced.
3. With the terminals connected as in 1 and 2 above, connect terminal 'AL' of the warning light control to the 6 volt tapping of the battery. The bulb should go out within five seconds.
4. Transfer the battery connection of terminal 'AL' to the positive battery terminal for ten seconds only. Then quickly transfer it to the battery 2 volt tapping. The bulb should light up within five seconds.
5. If the performance in items 3 and 4 differs appreciably from the test requirements the unit is faulty and must be replaced.

DATA

Resistance of actuator wire and internal ballast resistor (terminals 'AL' and 'E') .. .. .. 14 to 16 ohm

HORN

—Remove and refit   86.30.10

NOTE: The car is equipped with twin horns, one fitted at each side of the base unit, below and to the rear of the headlamps.

Removing

1. Disconnect wiring from horn.
2. Remove the bolts, nuts and spring washers fixing horn to base unit.
3. Withdraw horn.

Refitting

4. Reverse 1 to 3.

86.10.33
86.30.10

Rover 3500 and 3500S Manual AKM 3621

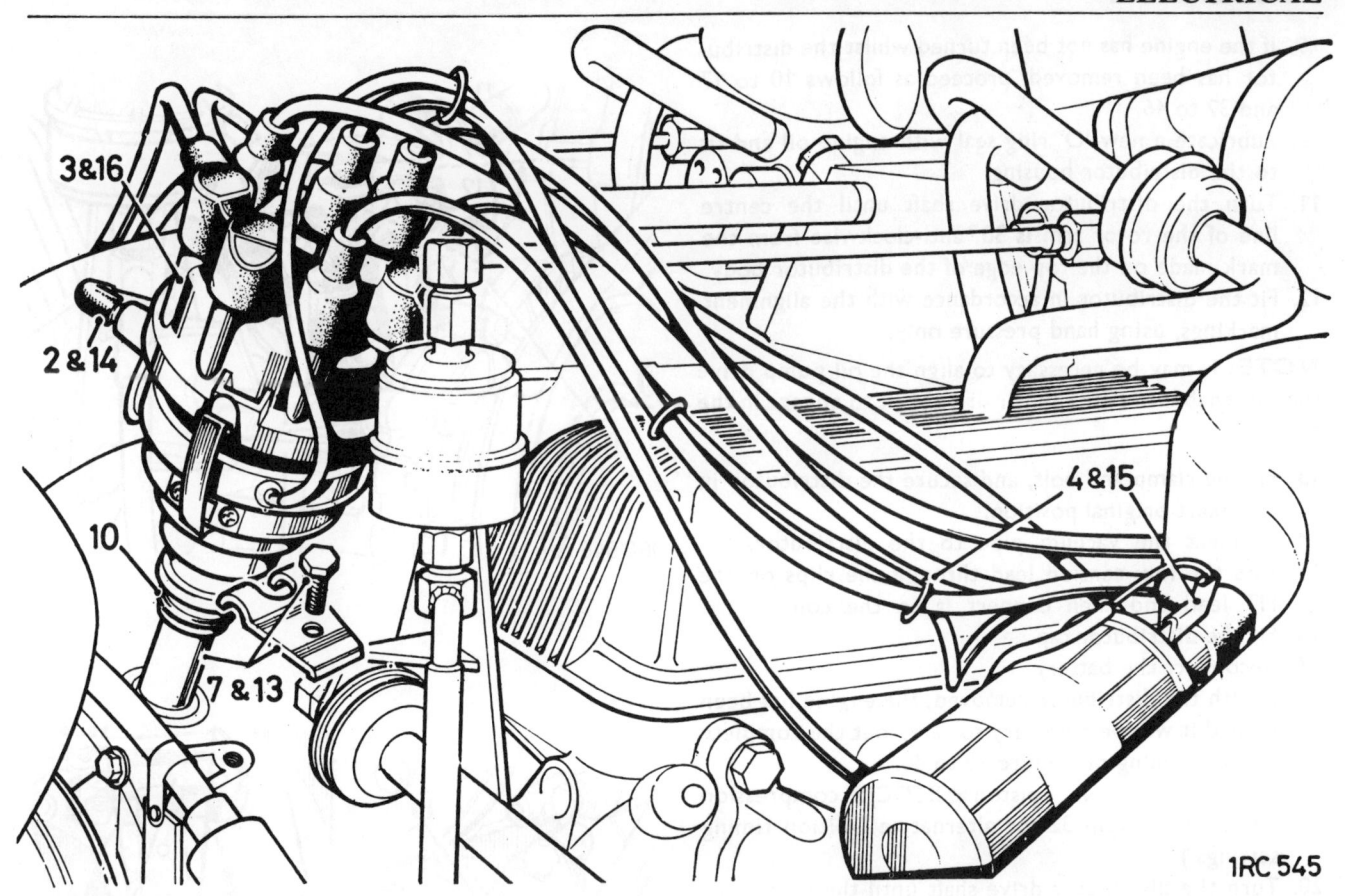

ELECTRICAL

DISTRIBUTOR

—Remove and refit 86.35.20

Removing

1. Disconnect the battery earth lead.
2. Disconnect the vacuum pipe.
3. Remove the distributor cap.
4. Disconnect the low tension lead from the coil and release from clips.
5. Mark the distributor body in relation to the centre line of the rotor arm.
6. Add alignment marks between the distributor and the timing gear cover.

    NOTE: Marking the distributor enables refitting in the exact original position, but if the engine is turned while the distributor is removed, the complete ignition timing procedure must be followed.

7. Remove the distributor.

Refitting

NOTE: If a new distributor is being fitted, mark the body in the same relative position as the distributor removed.

8. The leads for the distributor cap should be connected as illustrated. Figures 1 to 8 inclusive indicate the plug lead numbers.
    RH—Right hand side of engine when viewed from the rear.
    LH—Left hand side of engine when viewed from the rear.

continued

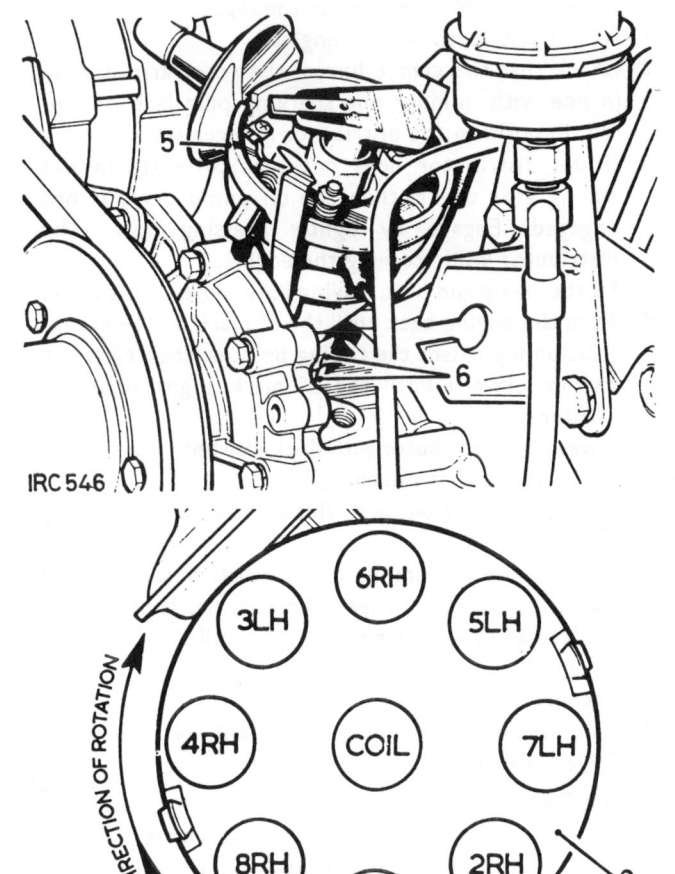

Rover 3500 and 3500S Manual AKM 3621

86.35.20
Sheet 1

ELECTRICAL

9. If the engine has not been turned whilst the distributor has been removed, proceed as follows 10 to 17 and 32 to 46.
10. Lubricate a new 'O' ring seal with engine oil and fit to the distributor housing.
11. Turn the distributor drive shaft until the centre line of the rotor arm is 30° anti-clockwise from the mark made on the top edge of the distributor body.
12. Fit the distributor in accordance with the alignment markings, using hand pressure only.

NOTE: It may be necessary to align the oil pump drive shaft to enable the distributor drive shaft to engage in the slot.

13. Fit the clamp and bolt, and secure the distributor in the exact original position.
14. Connect the vacuum pipe to the distributor.
15. Pass the low tension lead through the clips on the HT lead and then connect it to the coil.
16. Fit the distributor cap.
17. Reconnect the battery.
18. If with the distributor removed, the engine has been turned it will be necessary to carry out the complete ignition timing procedure 19 to 46.
19. Set the engine—No. 1 piston 6° BTDC on compression stroke. (See item 32 for alternative ignition timing settings.)
20. Turn the distributor drive shaft until the rotor arm is approximately 30° anti-clockwise from number one sparking plug lead position on the cap.
21. Fit the distributor to the engine.
22. Check that the centre line of the rotor arm is now in line with number one sparking plug lead on the cap. Reposition the distributor if necessary.
23. If the distributor does not seat correctly in the timing gear cover, then the oil pump drive is not engaged. Engage by lightly pressing down the distributor while turning the engine.
24. Fit the clamp and bolt, leaving both loose at this stage.
25. Turn the engine back until the 6° mark on the crankshaft pulley passes the timing pointer, then turn the engine forward until the 6° mark aligns with the pointer.
26. Rotate the distributor anti-clockwise until the contact points just start to open.
27. Secure the distributor in this position by tightening the clamp bolt.
28. Connect the vacuum pipe to the distributor.
29. Pass the low tension lead through the clips on the HT lead and then connect it to the coil.
30. Fit the distributor cap.
31. Reconnect the battery.
32. Using suitable electronic equipment, set the dwell angle and ignition timing.—See Data

   NOTE: It is essential that the following procedure is adhered to. Inaccurate timing can lead to serious engine damage. If the car is fitted with air conditioning equipment, isolate the compressor.

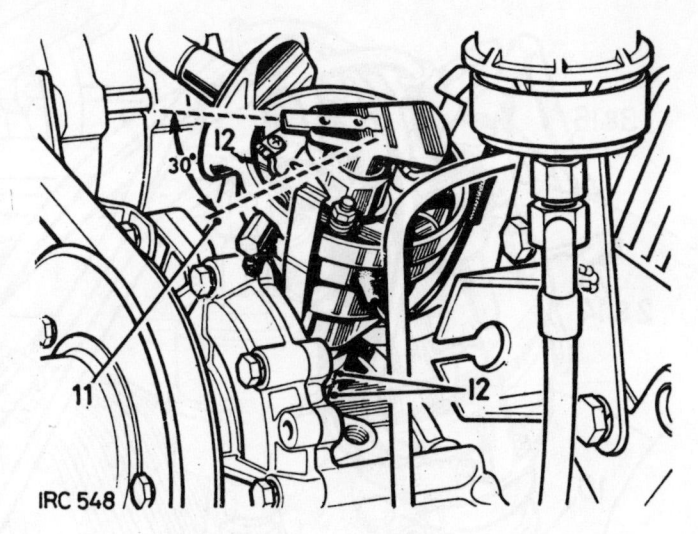

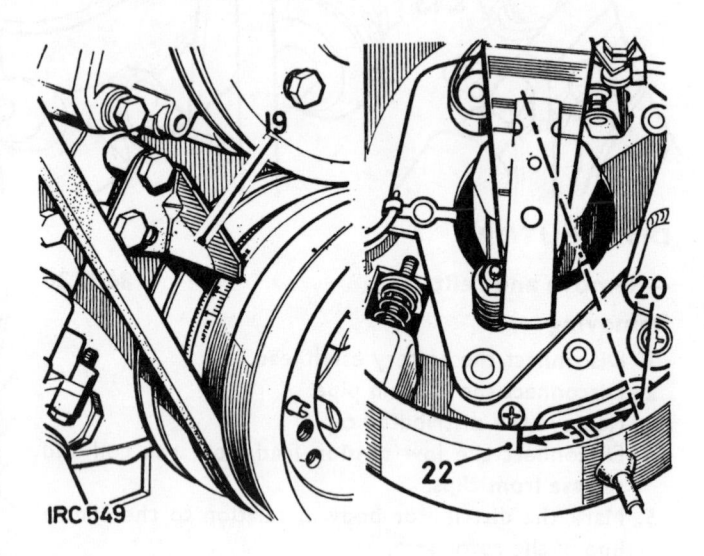

continued

86.35.20
Sheet 2

Rover 3500 and 3500S Manual AKM 3621

ELECTRICAL

33. Set the ignition timing statically, prior to the engine being run, by the basic lamp timing method (this sequence is to give only an approximation in order that the engine may be run. On no account should the engine be started before this check is carried out).
34. Start the engine and set to the correct idling speed as follows:

10.5:1 and 8.5:1 compression ratios: 600 to 650 rev/min (700 to 750 rev/min. for emission controlled and air conditioned vehicles).

9.25:1 compression ratio: 700 to 750 rev/min.

Set the dwell angle as follows (using a Tach-dwell meter).

35. Set the Tach-dwell meter switches to DWELL and CALIBRATE positions.
    Adjust the calibration with the test leads disconnected from the engine until the meter pointer reads on the set line.
36. Couple the Tach-dwell meter to the engine following the manufacturer's instructions.

    NOTE: When adjusting the dwell angle, it is essential that the correct setting of 26° to 28° is arrived at by a reducing adjustment. Therefore, if the dwell angle is less than 26° initially, adjust it to 30° and then reduce to 26° to 28°.

37. Set the selector knob to the 8 cylinder position and the Tach-dwell selector knob to 'dwell'. Adjust the distributor dwell angle by turning the hexagon headed adjustment screw on the distributor body until the meter reading is reduced to 26° to 28°. If the meter used does not have an 8 cylinder position, set the selector knob to the four cylinder position and adjust at the distributor until the meter reads 52° to 56°.
38. Uncouple the Tach-dwell meter.

Set the ignition timing as follows:

39. Couple a stroboscopic timing lamp to the engine following the manufacturer's instructions, with the high tension lead attached into No. 1 cylinder plug lead.
40. Disconnect the vacuum pipe from the distributor and block the pipe by some suitable means to prevent an air leak into the manifold.
41. Check that the distributor clamping bolt is slack and that the engine idle speed is 600 to 650 rev/min. (700 to 750 rev/min for emission controlled or air conditioned vehicles).

    NOTE: The ignition timing must be set at engine speed of 600 rev/min on all models; this can be achieved by lifting the piston slightly, on one carburetter only.

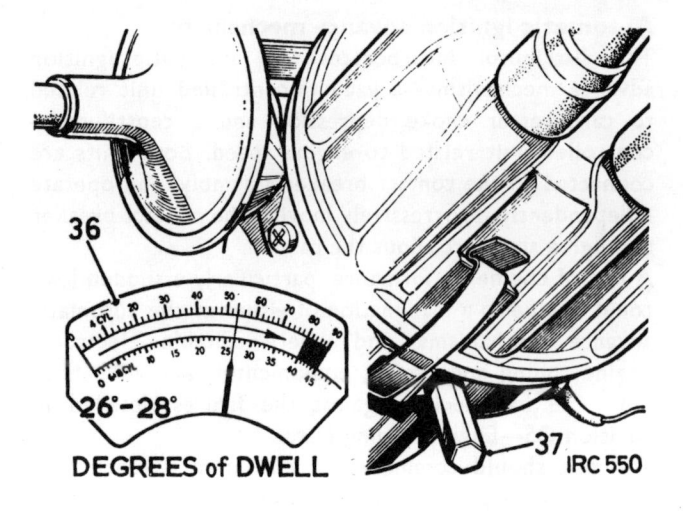

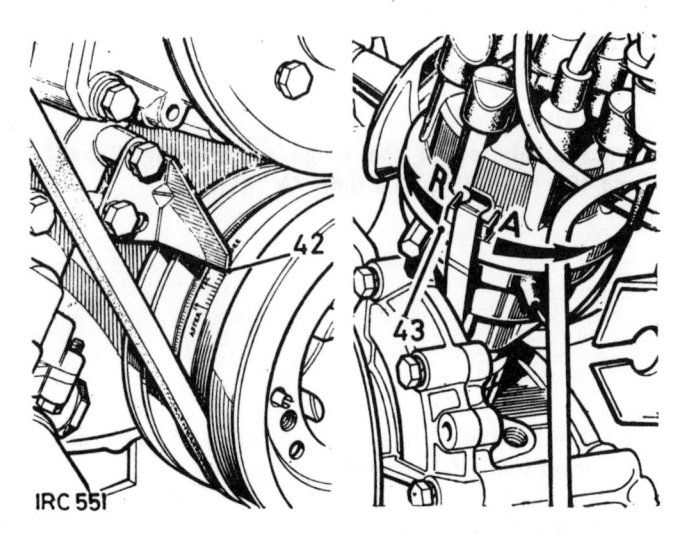

42. Check the ignition timing. The stroboscopic lamp must synchronise the timing pointer and the timing mark at 6° BTDC on the crankshaft pulley (or TDC for 10.5:1 compression ratio engines if using 96 octane fuel or if emission controlled.)
43. If necessary, adjust the timing. Turn the distributor clockwise to retard or anti-clockwise to advance.
44. Tighten the distributor clamping bolt.
45. Unplug the vacuum pipe and reconnect it to the distributor.
46. Disconnect the stroboscopic timing lamp.

    NOTE: Engine speed accuracy during ignition timing is of paramount importance. Any variation from the required 600 rev/min. particularly in an upwards direction will lead to wrongly set ignition timing.

47. If the car is fitted with air conditioning switch to select the compressor in circuit, as required.

continued

Rover 3500 and 3500S Manual AKM 3621

ELECTRICAL

Automatic ignition advance mechanism
The distributor incorporates two automatic ignition advance mechanisms—a vacuum-controlled unit related to carburetter choke depression and a centrifugally-controlled unit related to engine speed. Both units are connected to the contact breaker assembly, and operate independently, progressively moving the contact breaker through a small arc about the cam.

A loss of engine performance, particularly a sudden loss, could be due to a malfunction of either of the automatic advance mechanisms, and where suitable electronic engine tuning and testing equipment is available, both units can be checked against the figures detailed in Division O5—Engine Tuning Data.

The test should commence at maximum advance conditions and be checked during deceleration.

DATA

Dwell angle — 26° to 28°

Ignition timing:
Engines numbered in the range commencing 425, 427 and 451, 453, 455 with a suffix 'A' 'B' or 'C'

10.5:1 compression ratio — 6° BTDC static and dynamic when using fuel of 100 octane, minimum
TDC static and dynamic when using fuel of 96 octane, minimum and emission controlled engines

8.5:1 compression ratio — 6° BTDC static and dynamic when using fuel of 90 octane minimum

Engines numbered in the range commencing 451, 453 and 455 from suffix 'D' onwards

9.25:1 compression — 6° BTDC static and dynamic when using fuel of 96 octane minimum

Engine idle speed for setting dwell angle and ignition timing

600 rev/min

Emission control
3500 and 3500S models supplied to Europe: TDC static and dynamic for use with fuel of 96 octane, minimum.

8.5:1 compression ratio: — 6° BTDC static and dynamic when using fuel of 90 octane, minimum.

9.25:1 compression ratio: — These engines are all emission controlled. 6° BTDC static and dynamic for use with fuel of 96 octane, minimum.

86.35.20
Sheet 4

Rover 3500 and 3500S Manual AKM 3621

ELECTRICAL

DISTRIBUTOR

—Overhaul 86.35.26

NOTE: Two slightly varying designs of distributor are in use, the first two illustrations show the early type, the third illustration shows the latest, concentric base plate, type.

Dismantling

1. Remove the distributor. 86.35.20.
2. Unclip and remove distributor cap.
3. Withdraw rotor arm and felt lubricating pad.
4. Remove contact spring and lift off the insulating bush together with the low tension and capacitor leads.
5. Lift off the moving contact point and the insulating washer from the contact pivot and spring post.
6. Remove the capacitor.
7. Remove the fixed contact.
8. Early type distributor—Remove the nut, plain washer and spring from the contact breaker base plate pivot pin.
9. Remove the dwell angle adjuster screw and spring.
10. Remove the earth lead from the centrifugal advance cover plate.
11. Remove the contact breaker base plate.
12. Remove the vacuum unit and grommet.
13. Remove the centrifugal advance cover plate.
14. Carefully withdraw the two springs from the centrifugal advance unit.
15. Remove the screw from inside the cam and lift off the cam and cam foot.
16. Remove the two weights.
17. Drive out the pin securing the driving gear and remove the gear and tab washer.

Inspecting

18. Check all parts for wear or damage and replace as necessary.
19. Obtain a new contact points set.

Reassembling

20. Reverse 1 to 17, including the following.
21. When fitting the centrifugal governor springs, take care not to stretch them.
22. When fully assembled the points can be set to a clearance of 0 35 to 0,40 mm (0.014 to 0.016 in) as an initial guide before refitting to the engine.
23. Refit the distributor. 86.35.26.

    NOTE: It is most important that the dwell angle is adjusted to 26° to 28° using specialised equipment, when the distributor has been refitted.

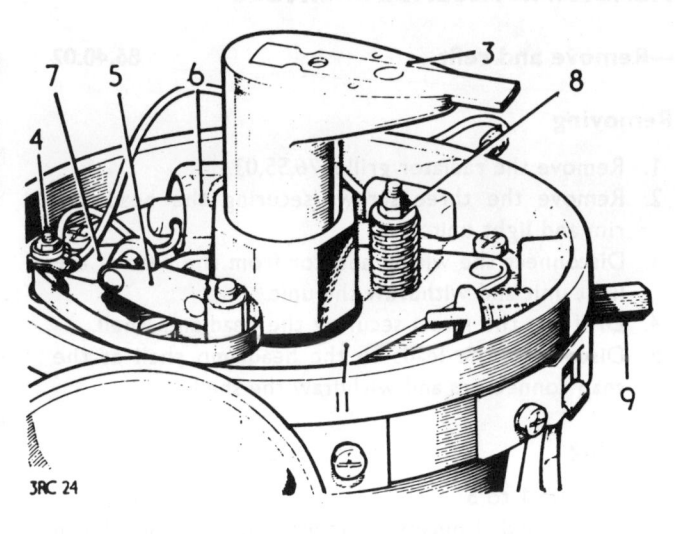

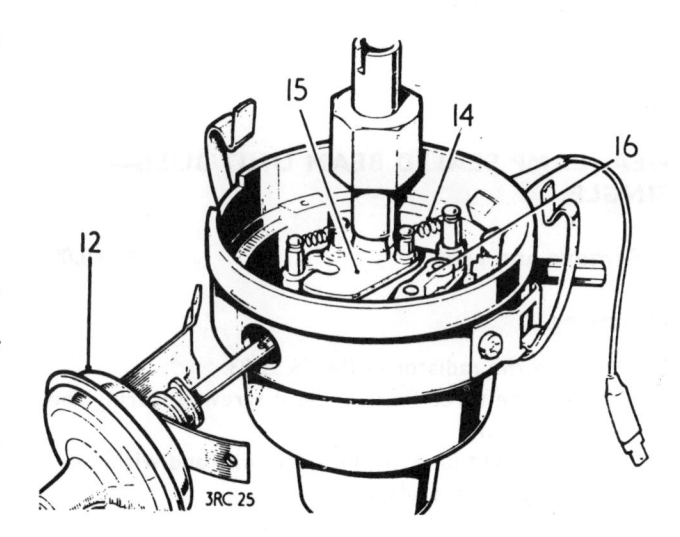

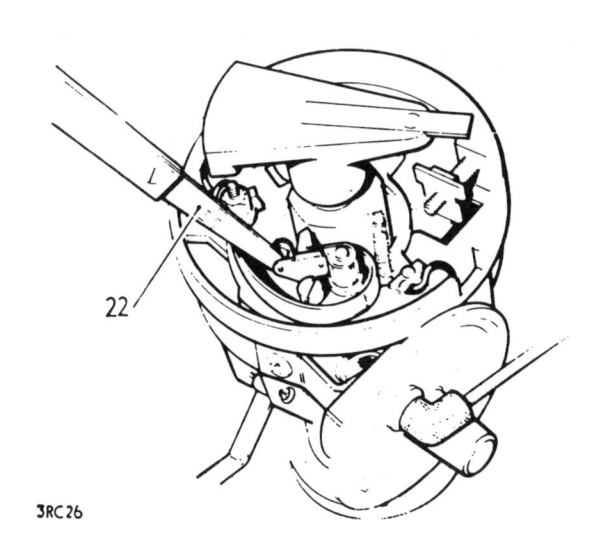

Rover 3500 and 3500S Manual AKM 3621

ELECTRICAL

HEADLAMP ASSEMBLY—SINGLE

—Remove and refit 86.40.02

Removing

1. Remove the radiator grille. 76.55.03.
2. Remove the three screws securing the headlamp rim and light unit.
3. Disconnect the wiring adaptor from the rear of the light unit and withdraw the unit.
4. Drill out the rivets securing the headlamp shell.
5. Disconnect the leads to the headlamp shell at the snap connectors and withdraw the shell.

Refitting

6. Reverse 1 to 5.
7. Check, and if necessary adjust, the headlamp beam alignment. 86.40.18.

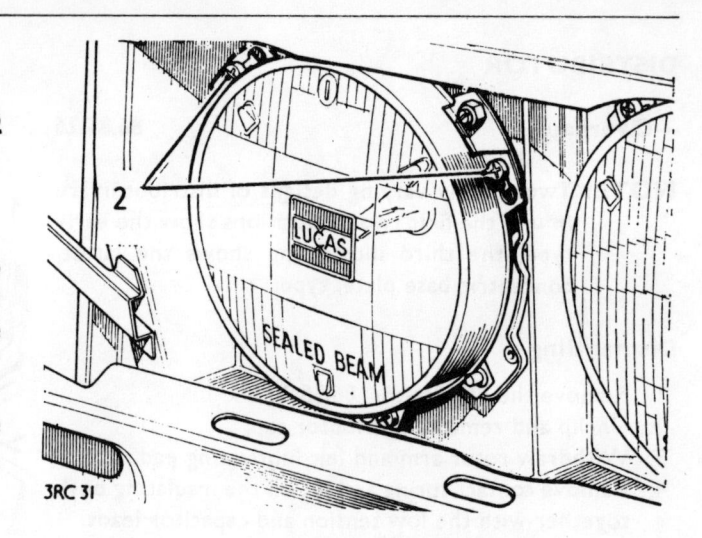

HEADLAMP SEALED BEAM UNIT/BULB—SINGLE

—Remove and refit 86.40.09

Removing

1. Remove the radiator grille. 76.55.03.
2. Slacken the three Pozidriv head screws securing the headlamp rim.
3. Turn the light unit rim anti-clockwise and withdraw.
4. Remove the light adaptor.

Refitting

5. Fit a new light unit or bulb to the adaptor.
6. Reverse 1 to 3.
7. Check, and if necessary adjust, the headlamp beam alignment. 86.40.18.

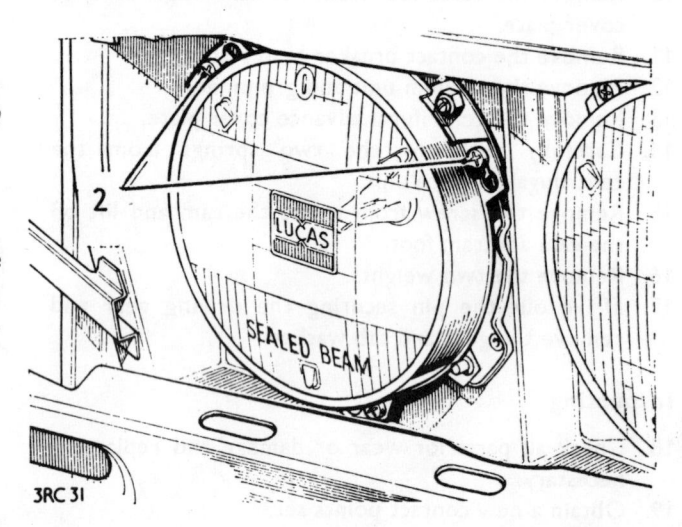

ELECTRICAL

HEADLAMPS—FOUR

—Align beams 86.40.18

NOTE: The setting of the pair of twin-beam head lights necessitates the use of a meter such as the Lucas Lev-L-Lite mechanical aimer for curved glass lenses, or Lucas 'Beamsetter' for flat glass lenses. The method used for setting single beam headlamps is unsatisfactory. When using the Lucas Lev-L-Lite beam setting device, it is very important that the device is set to zero before adjustment is carried out on the headlamps. Instructions for setting are supplied with the setting equipment.

With the device set correctly the headlamps will be set as follows:

Aligning

1. Check that the car is level and in the static unladen condition (no driver or passengers).
2. Check the main beam and dip beam which should be two divisions down from the vertical.
3. If necessary, align the headlamps by means of the headlamp horizontal adjusting screw.
4. If necessary, align the headlamps by means of the headlamp vertical adjusting screw.

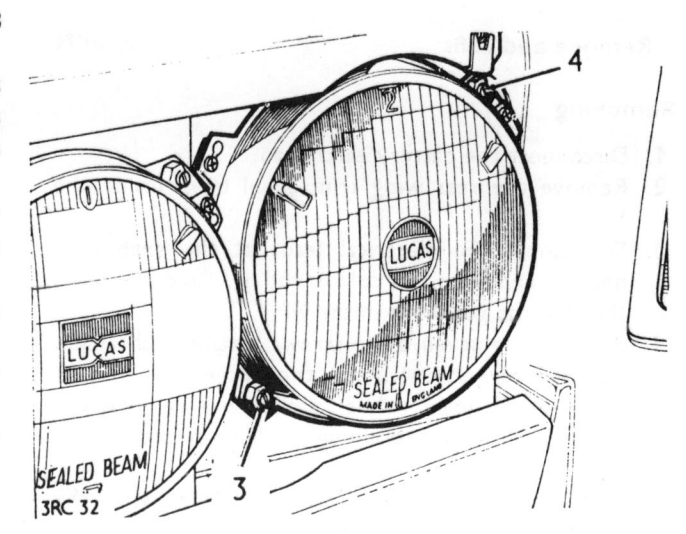

FRONT SIDE/FLASHER LAMP ASSEMBLY

—Remove and refit 86.40.26

Removing

1. Disconnect the battery earth lead.
2. Remove the four Phillips head screws securing the lamp glass to the lamp body.
3. Remove the foam rubber seal.
4. Remove the three screws securing the lamp body to the front wing.
5. Withdraw the lamp and disconnect the wiring from the three-way connector.

Refitting

6. Reverse 1 to 5.

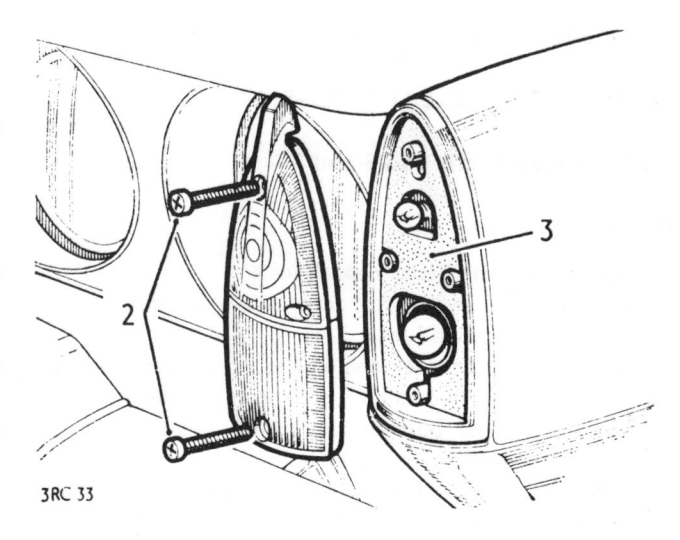

Rover 3500 and 3500S Manual AKM 3621

ELECTRICAL

TAIL, STOP, FLASHER AND REVERSE LAMP ASSEMBLY

—Remove and refit   86.40.74

Removing

1. Disconnect the battery earth lead.
2. Remove the rear wing trim panel from inside the boot.
3. Disconnect the electrical leads at the snap connectors and note the colour coding. Feed the leads through the grommet in the wing.
4. Remove the seven drive screws retaining the lamp cluster lenses and withdraw the lenses.
5. Remove the foam rubber gasket and remove the three drive screws retaining the lamp cluster to the wing.
6. Withdraw the lamp cluster together with the electrical leads.

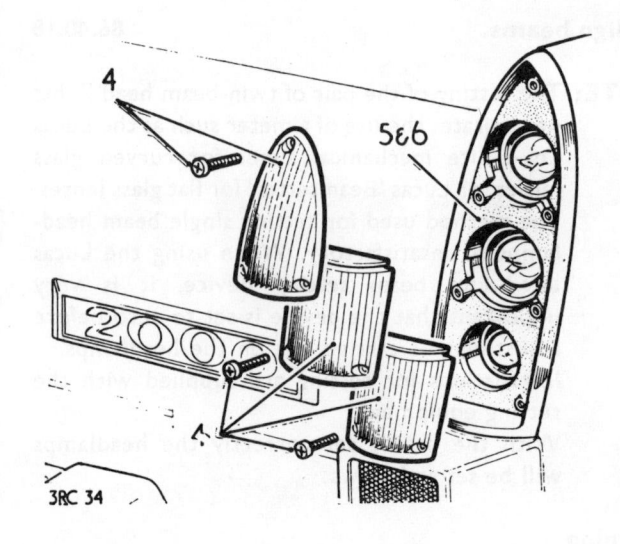

Refitting

7. Reverse 1 to 6.

NUMBER PLATE LAMP BULB

—Remove and refit 2 and 6   86.40.85

NUMBER PLATE LAMP ASSEMBLY

—Remove and refit 1 to 5 and 7   86.40.86

Removing

1. Disconnect the battery earth lead.
2. Remove the two fixing screws and withdraw the lamp surround and lens.
3. Detach the electrical feed lead at the lamp terminal.
4. Remove the two nuts and spring washers from inside the rear bumper and detach the earth lead for the lamp from its fixing.
5. Withdraw the lamp base complete with foam rubber joint washer and lamp bulbs from the bumper.

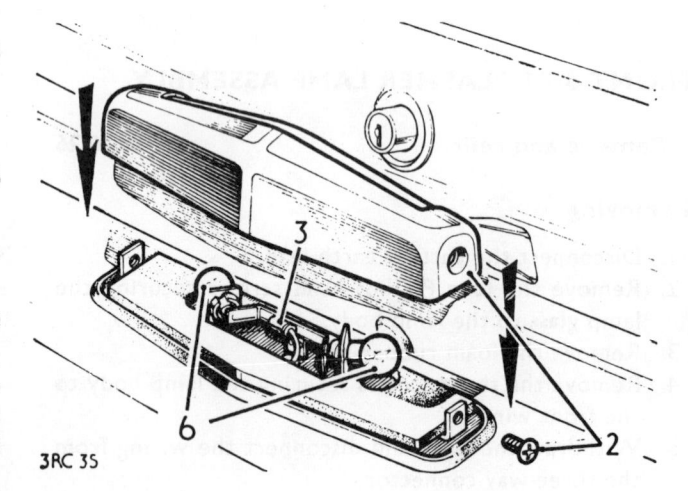

Refitting

6. Fit a new bulb as required.
7. Reverse 1 to 5.

Rover 3500 and 3500S Manual AKM 3621

ELECTRICAL

ROOF LAMP BULB

—Remove and refit 86.45.01

Removing

1. Remove the lens from the courtesy lamp by pressing it upward and turning it anti-clockwise.
2. Withdraw the bulb from its spring clip holder.

Refitting

3. Reverse 1 and 2.

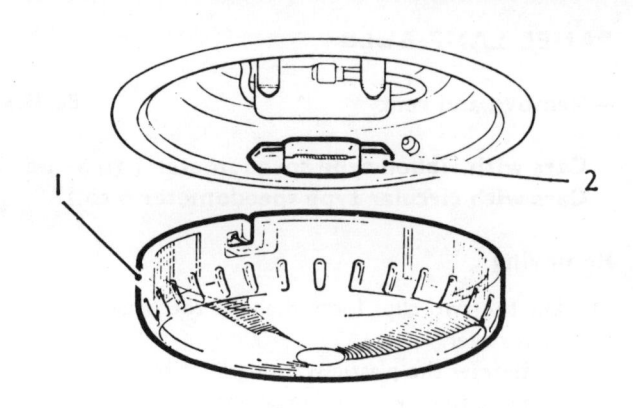

MAP LAMP BULB

—Remove and refit 86.45.09

Removing

1. Open the glove box.
2. Withdraw the bulb holder and bulb towards the front of the car.
3. Remove the bulb from the holder.

Refitting

4. Reverse 1 to 3.

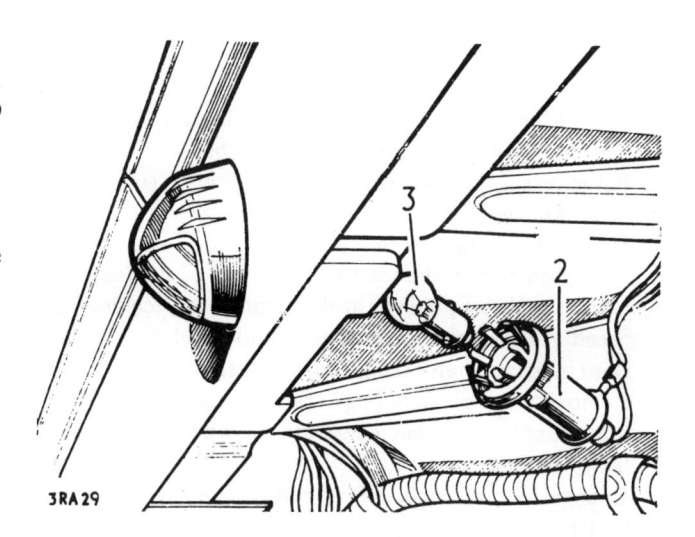

BOOT LAMP BULB

—Remove and refit 2 and 6 86.45.15

BOOT LAMP ASSEMBLY

—Remove and refit 1 to 5 and 7 86.45.16

Removing

1. Disconnect the battery earth lead.
2. Remove the two screws retaining the lamp glass cover.
3. Remove the rear boot trim panel.
4. Disconnect the lamp electrical leads at the snap connectors and note the colour coding.
5. Withdraw the lamp complete.

Refitting

6. Replace the bulb.
7. Reverse 1 to 5.

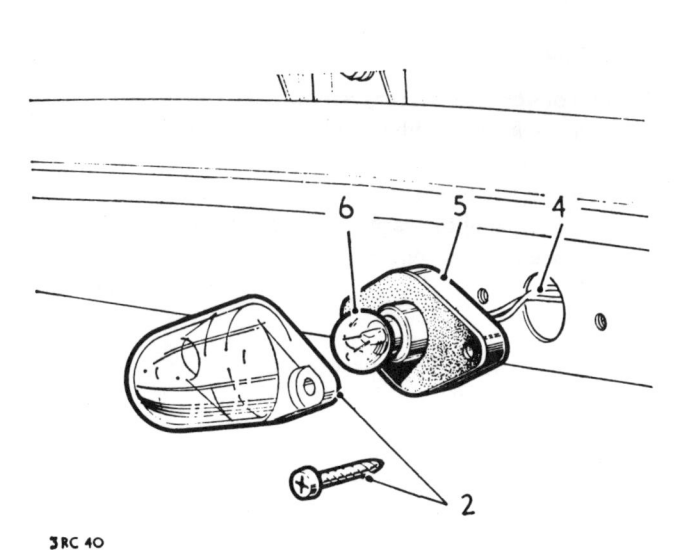

Rover 3500 and 3500S Manual AKM 3621

86.45.01
86.45.16

ELECTRICAL

PANEL LAMP BULB

—Remove and refit 86.45.31

Cars with ribbon type speedometer 1 to 5 and 17
Cars with circular type speedometer 6 to 17

Removing

1. Lift the two clips located above the plastic warning light cover.
2. Withdraw the plastic warning light cover.
3. Remove the two panel retaining screws.
4. Disconnect the speedometer drive cable, this can be done from inside the glove box.
5. Withdraw the panel, then remove the bulb holder and replace bulb as required.

Instrument panel

6. Pull off the knob for the speedometer trip control and the knob for the panel light switch.
7. Remove the locking rings retaining the instrument panel finisher.
8. Slide the finisher to one side and withdraw.
9. Spring off the finisher at the other end of the instrument panel, by pulling the knob or, if an 'Icelert' is fitted remove the grub screw, knob and nut from the rheostat, then spring off the finisher.
10. Remove the two retaining screws at each end of the instrument. The front panel can now be eased away from the instruments.
11. Change bulb as required by pulling the capless bulb straight out of the holder.

Switch panel

12. Disconnect the battery earth lead.
13. Pull back the rubber finisher at each end of the switch panel.
14. Remove the four screws retaining the panel.
15. Pull the panel forward to give access to the bulbs.
16. Replace bulb as required.

Refitting

17. Reverse the removal procedure.

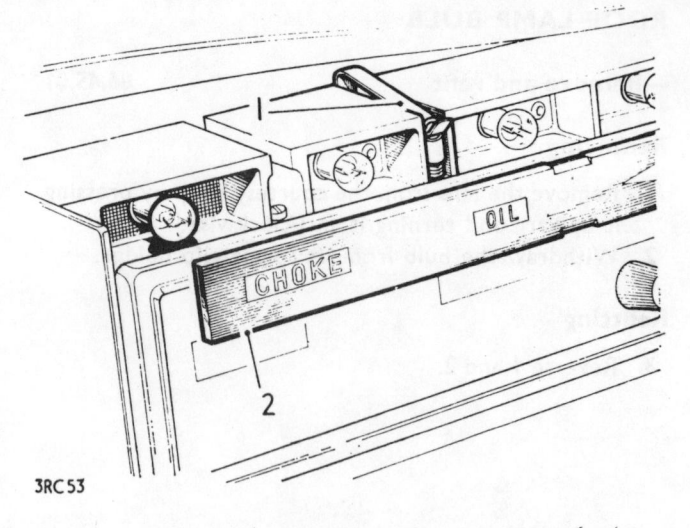

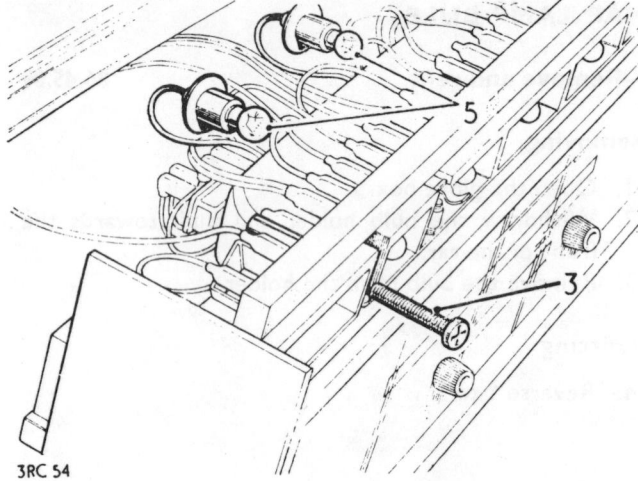

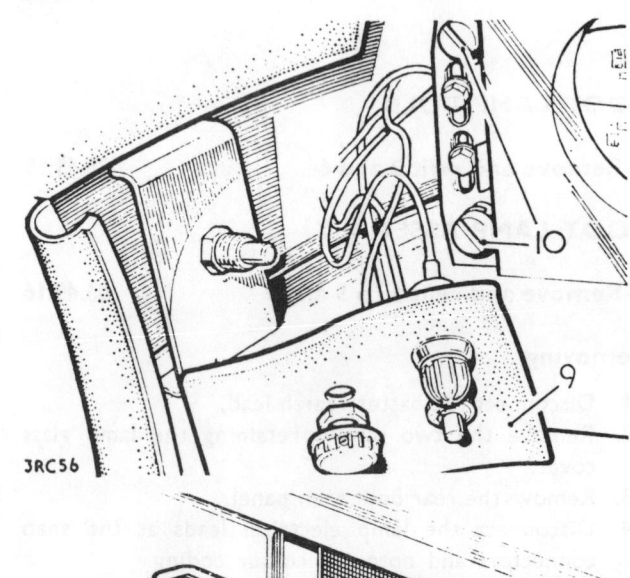

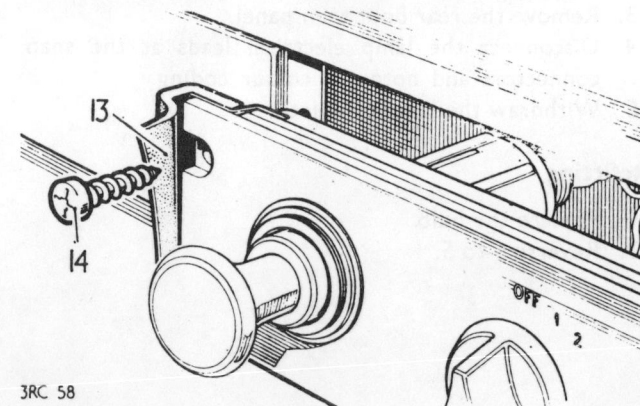

Rover 3500 and 3500S Manual AKM 3621

ELECTRICAL

AUTOMATIC GEARBOX SELECTOR LAMP BULB

—Remove and refit 2, 3 and 6   86.45.40

AUTOMATIC GEARBOX SELECTOR LAMP

—Remove and refit 1 to 6   86.45.39

Removing

1. Disconnect the battery earth lead.
2. Spring out the indicator plate from the selector lever housing.
3. Remove the illumination bulb from the spring clips.
4. Remove drive screw at the centre of lamp body.
5. Withdraw the lamp body and disconnect the electrical leads at the snap connectors under the transmission tunnel carpet. Note the colour code to facilitate refitting.

Refitting

6. Reverse the removal procedure.

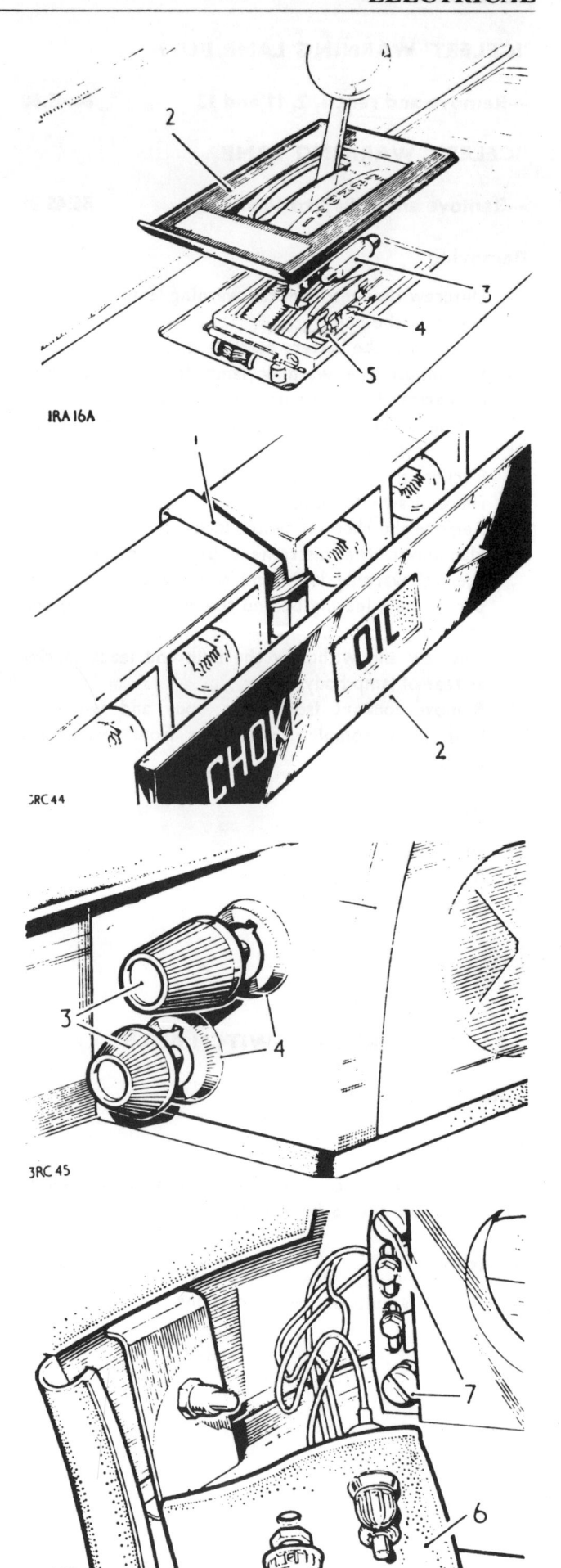

WARNING LIGHT BULB

—Remove and refit   86.45.61

Cars with ribbon type speedometer 1, 2 and 9
Cars with circular speedometer 3 to 9

Removing

1. Lift the two clips located above the plastic warning light cover.
2. Remove the plastic warning light cover and renew bulb or bulbs as necessary.
3. Pull off the knob for the speedometer trip control and the knob for the panel light switch.
4. Remove the locking rings retaining the instrument panel finisher.
5. Slide the finisher to one side and withdraw.
6. Spring off the finisher, at the other end of the instrument panel, by pulling the knob or, if an 'Icelert' is fitted remove the grub-screw, knob and nut from the rheostat then spring off the finisher.
7. Remove the two retaining screws at each end of the instrument panel. Front panel can now be eased away from the instruments.
8. Change bulb as required by pulling the capless bulb straight out of the holder.

Refitting

9. Reverse the removal procedure.

Rover 3500 and 3500S Manual AKM 3621

86.45.40
86.45.61

ELECTRICAL

'ICELERT' WARNING LAMP BULB

—Remove and refit 1, 2, 11 and 12   86.45.80

'ICELERT' WARNING LAMP

—Remove and refit 3 to 10 and 13   86.45.81

Removing

1. Unscrew the lens from the warning lamp.
2. Unscrew the bulb.
3. Disconnect the battery earth lead.
4. Disconnect the warning lamp leads at the snap connectors inside the driver's side glove box.
5. Spring off the parcel shelf moulded finisher at the driver's side door 'A' post.
6. Slacken the grub screw and remove the 'Icelert' rheostat switch knob.
7. Remove the rheostat switch locknut.
8. Manoeuvre the instrument panel complete with 'Icelert' warning lamp and test switch away from parcel shelf, leaving behind the rheostat switch and mounting bracket.
9. Unscrew and withdraw the bulb and leads carrier at rear of lamp body.
10. Remove locknut from lamp body and withdraw lamp body complete with lens from instrument panel.

Refitting

11. Fit a new bulb.
12. Refit the warning lamp lens.
13. Reverse 3 to 10.

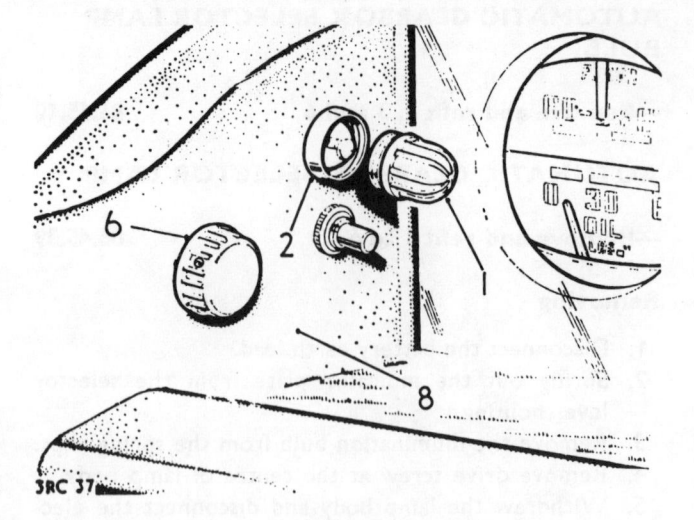

HEATED BACKLIGHT SWITCH BULB

—Remove and refit   86.45.82

Removing

1. Unscrew the knob from the heated backlight switch.
2. Withdraw the spring and bulb.

Refitting

3. Fit a new bulb and refit the spring and switch knob.

86.45.80
86.45.82

Rover 3500 and 3500S Manual AKM 3621

ELECTRICAL

FLASHER UNIT

—Remove and refit   86.55.11

HAZARD FLASHER UNIT

—Remove and refit   86.55.12

Removing

NOTE: On some early 3500 cars, the flasher unit is fitted behind the instrument panel. Therefore, if the car is an early model and the flasher unit is not visible when the driver's glove box is opened, remove the instrument panel. 88.20.01.

1. Disconnect the battery earth lead.
2. Open driver's side glove box.
3. Withdraw flasher unit from its carrier.

ICELERT WARNING UNIT

—Remove and refit   86.55.16

Removing

1. Remove the fixings securing the warning unit to the radiator grille.
2. Disconnect the electrical leads at the snap connectors.
3. Withdraw the 'Icelert' warning unit.

Refitting

4. Remove the 'Rover' motif badge from the front grille.
5. Remove the RH side front grille and place aside the fixing screws and washers, and the three plastic boots which house the grille locating lugs.
6. Offer up the sensor unit and attached mounting bracket to the grille front, positioning the mounting bracket between the first two full-length vertical spars inboard from the inner head lamp aperture, and across the second and third grille louvres.
7. Fit the $\frac{1}{4}$ in UNF bolt to the sensor bracket fixing hole, with the bolt head toward the sensor unit.
8. Locate the flanged small bracket on the fixing bolt threads, position the spring washer and nut on the bolt and tighten to secure the sensor unit and bracket to the grille.
9. Pass the sensor unit cable ends through the grille to the rear.
10. Connect the sensor unit cables to the harness cables clipped to the cross member, connecting colour to like colour and using the cable connectors provided.
11. Refit the plastic boots to the grille lower lugs and fit the grille and attached sensor unit to the grille mounting panel. Secure with the existing fixings. Refit the motif badge and close the bonnet.
    Check the 'Icelert' warning lamp operation, using the test switch provided on the 'Icelert' instrument panel.

4. Note the colour coding and position of the electrical leads and disconnect the leads from the flasher unit terminals.

Refitting

5. Reverse 1 to 4. If required, refer to the circuit diagram at the beginning of this Division.

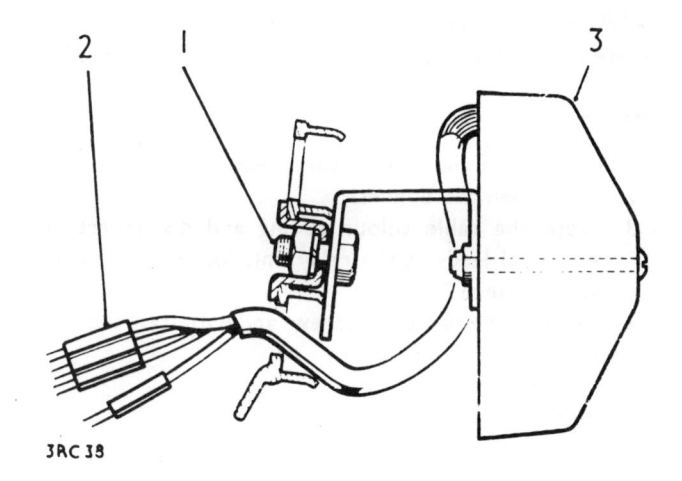

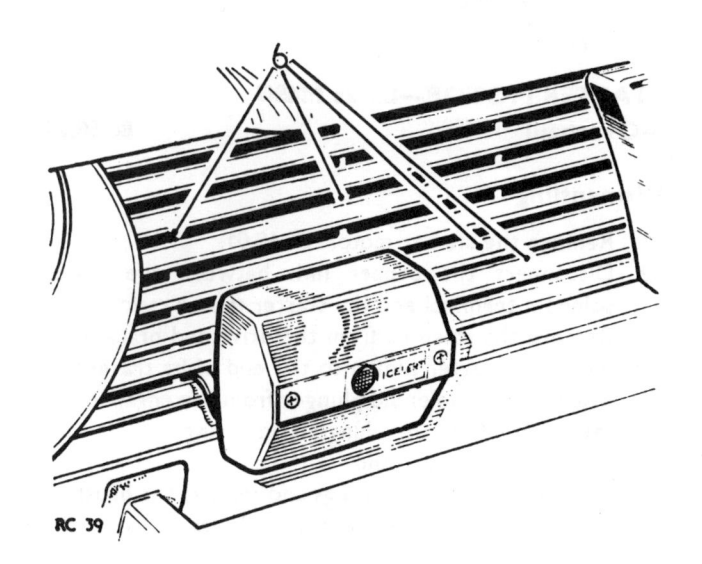

Rover 3500 and 3500S Manual AKM 3621

86.55.11
86.55.16

ELECTRICAL

STARTER MOTOR
—Remove and refit 86.60.01

Removing

1. Place car on a suitable ramp.
2. Disconnect the battery.
3. Disconnect the leads from the solenoid and starter motor.
4. Remove the two bolts securing the starter motor to the flywheel housing.
5. Remove starter motor from underneath the vehicle.

Refitting

6. Reverse 1 to 5. Torque tighten the starter motor securing bolts to 4,0 to 4,9 kgf.m (30 to 35 lbf ft).

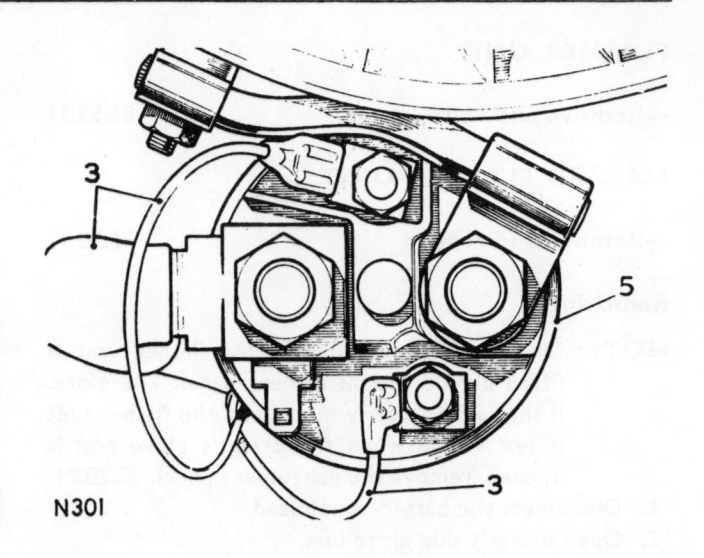

STARTER RELAY
—Remove and refit 86.60.10

Removing

1. Disconnect the battery earth lead.
2. Prop open the bonnet.
3. Note the cable colour coding and disconnect the electrical leads at the relay unit located on the RH wing valance.
4. Remove the two drive screws and withdraw the relay unit.

Refitting

5. Reverse 1 to 4. Ensure that the two earth lead terminals are located between the rear drive screw and the relay flange.

STARTER MOTOR—Lucas M45G
—Overhaul 86.60.13

Dismantling

1. Remove the starter motor. 86.60.01.
2. Disconnect the copper link between the lower solenoid terminal and the starter motor casing.
3. Remove the solenoid from the drive end bracket.
4. Move the starter pinion to the end of its travel and disengage the solenoid plunger from the engagement lever; withdraw the plunger and spring.
5. Remove the cover band.
6. Hold back the brush springs and remove the brushes.
7. Remove the commutator end bracket from the starter yoke.
8. Separate the yoke from the drive end bracket.
9. Remove the eccentric pin and engagement lever from the drive end bracket.
10. Withdraw the armature, drive gear and intermediate bracket.

continued

Rover 3500 and 3500S Manual AKM 3621

ELECTRICAL—M45G Starter

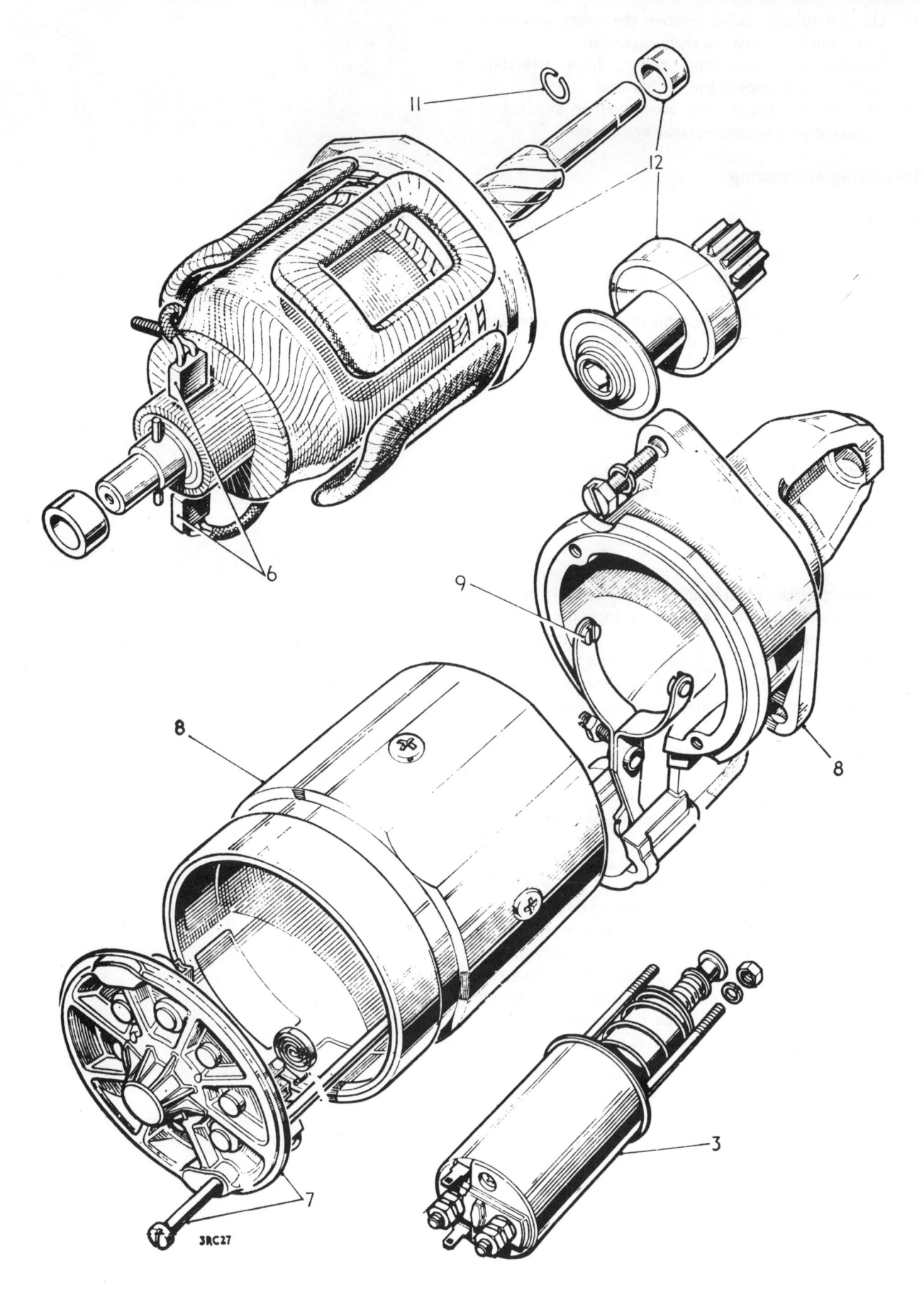

continued

Rover 3500 and 3500S Manual AKM 3621

86.60.13
Sheet 2

ELECTRICAL—M45G Starter

11. Using a suitable tube, remove the collar and jump ring from the armature shaft extension.
12. Remove the jump ring, collar, drive assembly, intermediate bracket and rubber seal.
13. Remove the brake ring, steel washer and tufnol washer from the commutator end bracket.

Inspecting and testing

Clutch

14. Check that the clutch will:
    (a) Give instantaneous take up of the drive in one direction.
    (b) Rotate easily and smoothly in the other direction.
    (c) Be free to move round and along the shaft splines without any tendency to bind. All moving parts should be smeared with Shell S3 2628 grease for cold climates or Shell Retinax 'A' grease for hot climates.

    NOTE: The roller clutch drive is sealed in a rolled steel outer cover and must not be dismantled.

Brushes

15. Check that the brushes move freely in their holders while holding back the brush springs. If a brush is damaged or worn so that it does not make good contact with the commutator, all the brushes must be renewed. The new brushes are pre-formed; 'bedding' to the commutator is therefore unnecessary.
16. The flexible connectors are soldered to the terminal tabs, two to the field coils and two to the brush boxes. The flexible connectors must be unsoldered and the new brushes secured in position by re-soldering.
17. Using a spring balance, check that the tension of the brush springs is between 0.85 Kg. to 1.13 Kg. (30 oz. to 40 oz.). If tension is low, fit new springs.
18. Check the commutator, having first cleaned with a petrol (gasoline) moistened cloth. If necessary, rotate the armature and remove pits and burrs using fine glass paper. If the commutator is badly worn, mount in a lathe and with a very sharp tool take a light cut, removing no more metal than is necessary.

    NOTE: The commutator segments must not be undercut.

continued

ELECTRICAL—M45G Starter

Auxiliary contacts, to check

19. Disconnect all cables from the solenoid terminals and connectors.
20. Connect a test lamp between the connector marked 'IGN' and a good earth.
21. Connect the battery to the small unmarked connector.
22. Momentarily connect the large battery terminal to earth. The solenoid contacts should close fully and remain closed; the test lamp should emit a steady light.
23. Disconnect the battery from the small unmarked connector. The contacts should open, the solenoid will release and the test lamp extinguish. The period of energising should be as brief as possible to avoid overheating end windings.

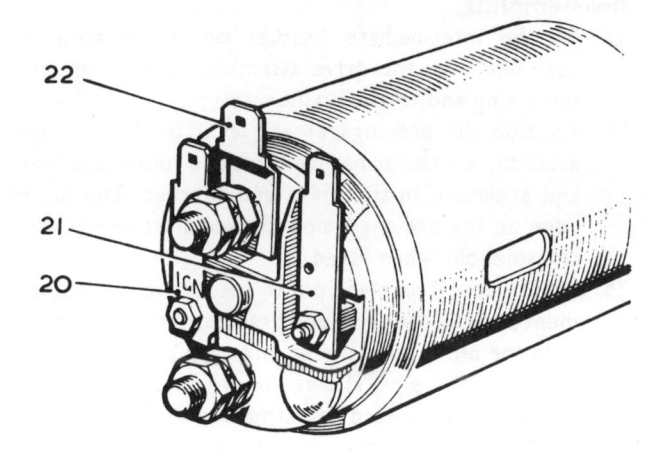

N362

Bushes

24. Examine the bushes in the drive end bracket, intermediate and commutator end brackets; renew as necessary.

Armature insulation

25. Connect a 110 V a.c 15W test lamp between any one of the commutator segments and the shaft.
26. The lamp should not light, if it does light, fit a new armature.

Field coil insulation

27. Connect a 110V a.c 15W test lamp between the yoke terminal and the yoke.
28. Ensure that the brushes are not touching the yoke during the test.
29. The lamp should not light, if it does light, fit a new field coil assembly.

IRC 855A

Field coil continuity

30. Connect a 110V a.c. 15W test lamp between the two field coil brushes.
31. The lamp should light, if it does not light, fit a new field coil assembly.

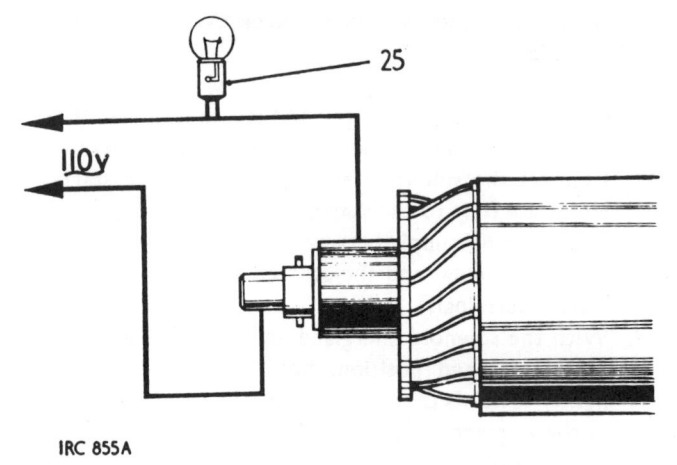

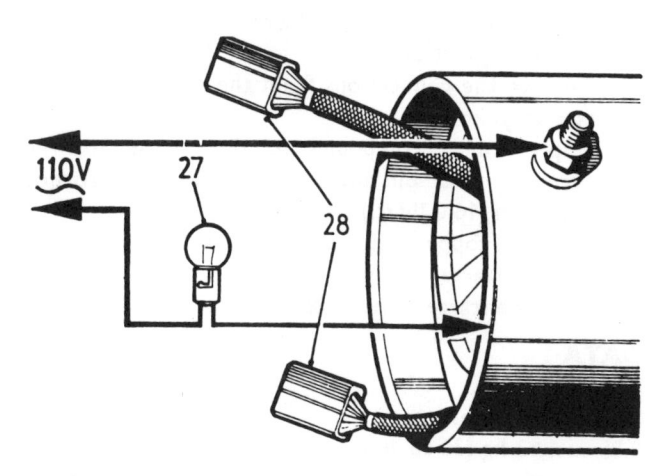

continued

IRC 856A

Rover 3500 and 3500S Manual AKM 3621

ELECTRICAL—M45G Starter

Reassembling

32. Fit the intermediate bracket on to the armature together with the drive assembly, stop collar, new jump ring and thrust washer.
33. Position the pre-engagement lever on to the drive assembly on the armature shaft and locate the lever and armature in the drive end bracket. The flatter edge on the pre-engagement lever must face toward the solenoid when fitted.
34. Fit the eccentric pivot pin to retain the pre-engagement lever in the drive end bracket, then temporarily turn the pin in the reverse direction until the eccentric cam on the pin allows the pre-engagement lever to move fully toward the armature shaft.
35. Fit the yoke over the armature and locate on to the end bracket dowel.
36. Fit the tufnol washer, thrust washer and brake ring into the commutator end bracket, with the brake ring angled slots uppermost.
37. With the brushes positioned clear of the commutator, fit the bracket on to the yoke, locating the dowel and the drive pin into the brake ring.
38. Replace the brushes.
39. Replace the cover band.
40. With the drive assembly held in the forward position, locate the solenoid plunger over the engagement lever.
41. Fit the return spring over the plunger. Fit the solenoid on to the plunger and locate it in the drive end bracket. Secure the nuts and spring washers.
42. Connect the solenoid terminal 'STA' to the starter yoke case (not to the starter terminals).
43. Connect a 6V supply between the solenoid-operating 'Lucar' terminal and the starter yoke case (not the starter terminals).
44. With the solenoid energised and the drive assembly in the engaged position, hold the pinion pressed lightly towards the armature to take up any free play in the engagement linkage.
45. Measure the clearance between the pinion and the thrust collar on the armature shaft. The correct clearance is 0,12 mm to 0,38 mm (0.005 in to 0.015 in).
46. If necessary, adjust the clearance by rotating the eccentric pivot pin.
47. Remove the connections from the solenoid and the starter yoke case.
48. Apply sealing compound to the threads of the eccentric pivot pin and secure the locknut. Torque 2,2 kgf.m (16.0 lbf.ft.).
49. Refit the copper link between the solenoid terminal and the starter casing.
50. Refit the starter motor. 86.60.01.

DATA
Starter motor
Brush spring pressure
0,85 to 1,13 kg (30 to 40 oz)
Brush minimum length
8,0 mm (0.312 in)

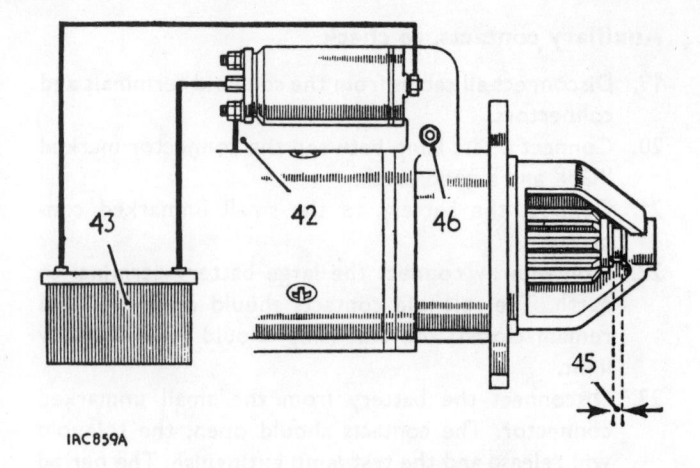

ELECTRICAL—3M100PE Starter

STARTER MOTOR—Lucas 3M100PE

—Overhaul 86.60.13

Dismantling

1. Remove the starter motor. 86.60.01.
2. Remove the connecting link between the starter and the solenoid terminal 'STA'.
3. Remove the solenoid from the drive end bracket.
4. Grasp the solenoid plunger and lift the front end to release it from the top of the drive engagement lever.
5. Remove the end cap seal.
6. Using an engineer's chisel, cut through a number of the retaining ring claws until the grip on the armature shaft is sufficiently relieved to allow the retaining ring to be removed.
7. Remove the two through bolts.
8. Partially withdraw the commutator end cover and disengage the two field coil brushes from the brush box.
9. Remove the commutator end cover.
10. Withdraw the yoke and field coil assembly.
11. Remove the retaining ring from the drive engagement lever pivot-pin, using the method previously described.
12. Withdraw the pivot pin.
13. Withdraw the armature.
14. Using a suitable tube, remove the collar and jump ring from the armature shaft.
15. Slide the thrust collar and the roller clutch drive and lever assembly off the shaft.

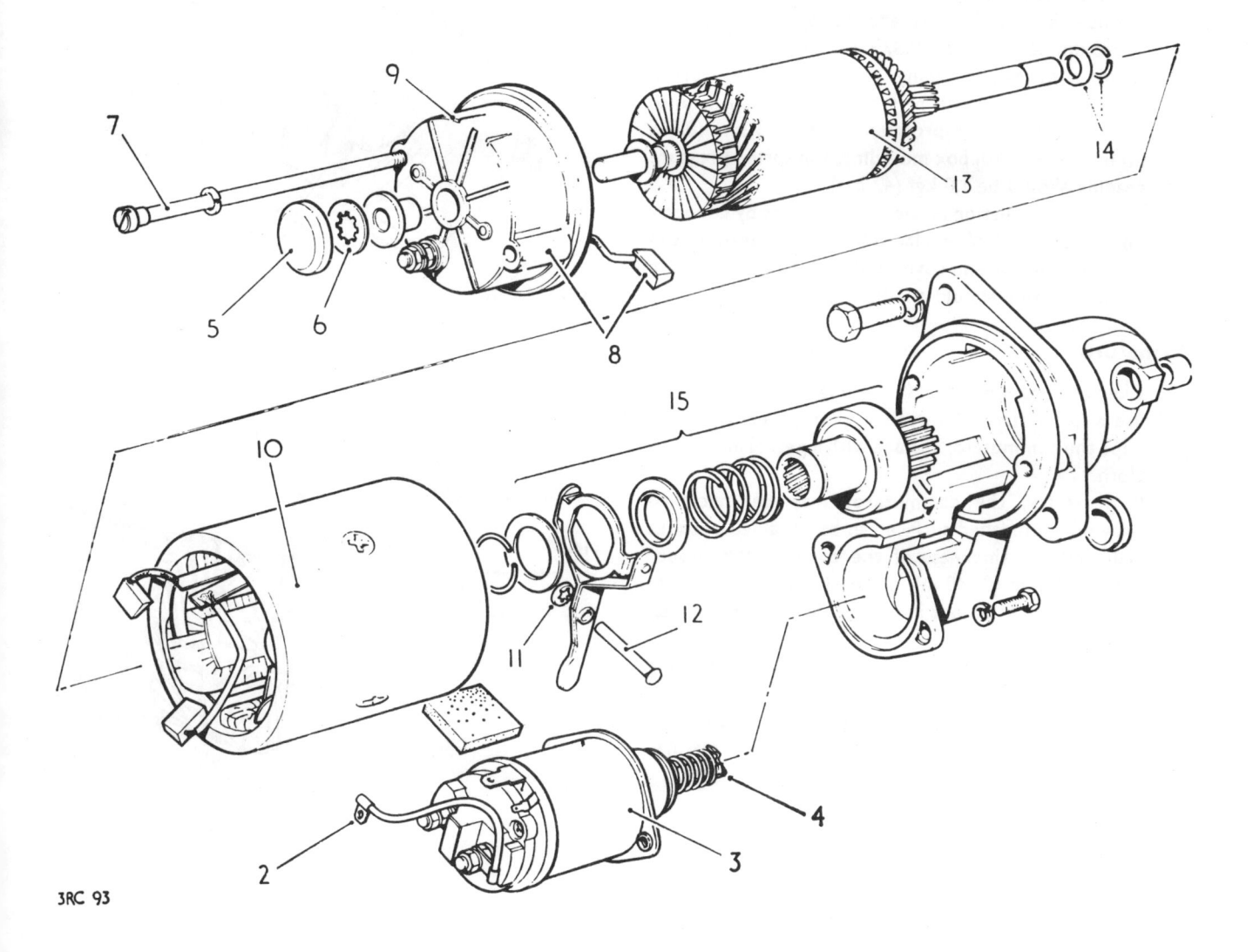

continued

continued

Rover 3500 and 3500S Manual AKM 3621

86.60.13
Sheet 6

ELECTRICAL—3M100PE Starter

Inspecting

Clutch

16. Check that the clutch gives instantaneous take-up of the drive in one direction and rotates easily and smoothly in the other direction.
17. Ensure that the clutch is free to move round and along the shaft splines without any tendency to bind.

   NOTE: The roller clutch drive is sealed in a rolled steel cover and cannot be dismantled.

18. Lubricate all clutch moving parts with Shell SB 2628 grease for cold and temperate climates or Shell Retinax 'A' for hot climates.

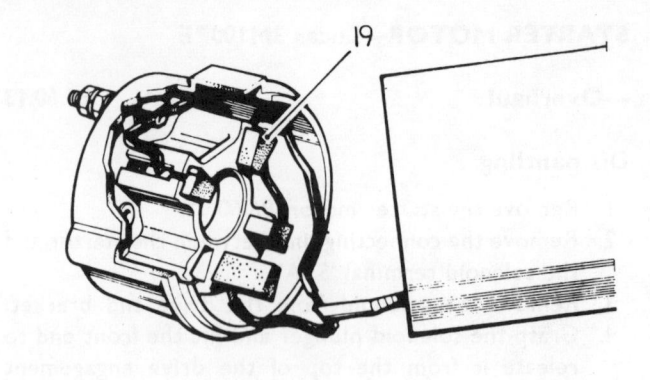

Brushes

19. Check that the brushes move freely in the brush box moulding. Rectify sticking brushes by wiping with a petrol moistened cloth.
20. Fit new brushes if they are damaged or worn to approximately 9,5 mm (0.375 in).
21. Using a push-type spring gauge, check the brush spring pressure. With new brushes pushed in until the top of the brush protrudes about 1,5 mm (0.065 in) from the brush box moulding, the spring pressure reading should be 1,0 kgf (42 ozf).
22. Check the insulation of the brush springs by connecting a 110V a.c 15W test lamp between a clean part of the commutator end cover and each of the springs in turn. The lamp should not light.

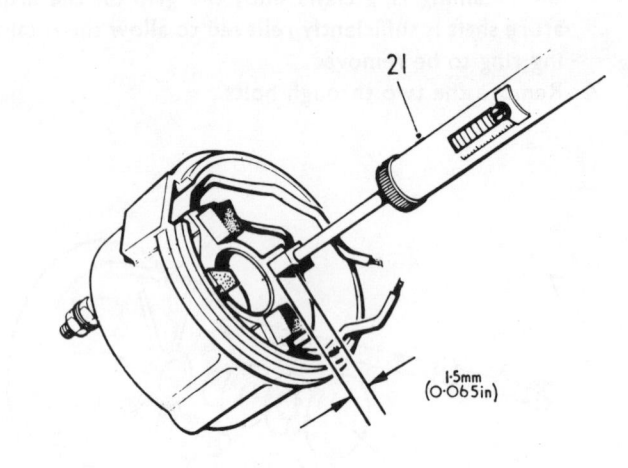

Armature

23. Check the commutator. If cleaning only is necessary, use a flat surface of very fine glass paper, and then wipe the commutator surface with a petrol moistened cloth.
24. If necessary, the commutator may be machined providing a finished surface can be obtained without reducing the thickness of the commutator copper below 3,5 mm (0.140 in), otherwise a new armature must be fitted. Do not undercut the insulation slots.
25. Check the armature insulation by connecting 110V a.c. 15W test lamp between any one of the commutator segments and the shaft. The lamp should not light, if it does light fit a new armature.

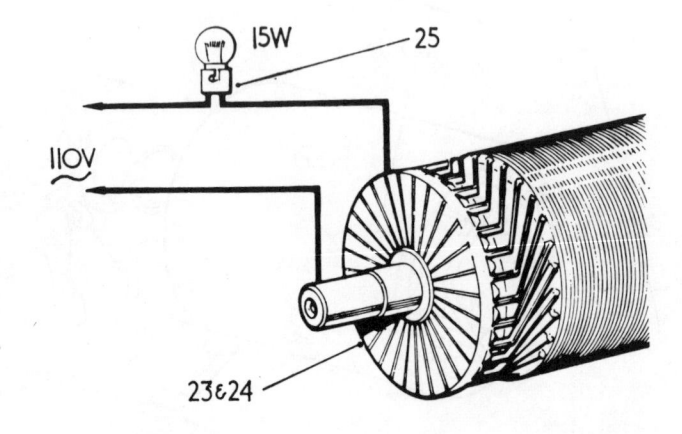

continued

86.60.13
Sheet 7

Rover 3500 and 3500S Manual AKM 3621

ELECTRICAL—3M100PE Starter

Field coil insulation

26. Disconnect the end of the field winding where it is riveted to the yoke, by filing away the riveted over end of the connecting-eyelet securing rivet, sufficient to enable the rivet to be tapped out of the yoke.
27. Connect a 110V a.c. 15W test lamp between the disconnected end of the winding and a clean part of the yoke.
28. Ensure that the brushes or bare parts of their flexibles are not touching the yoke during the test.
29. The lamp should not light, if it does light, fit a new field coil assembly.
30. Resecure the end of the field winding to the yoke.

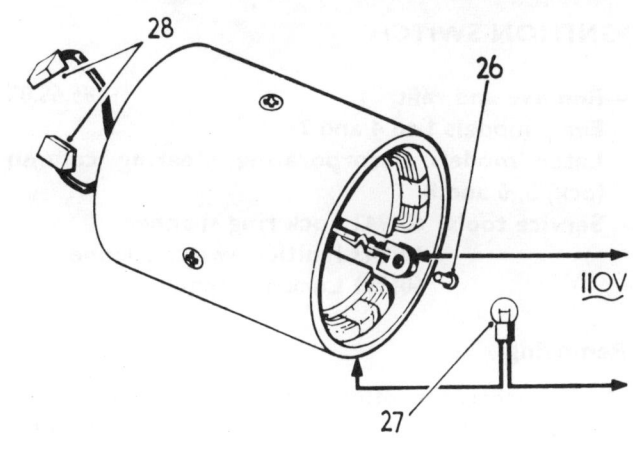

Field coil continuity

31. Connect a 12V battery operated test lamp between each of the brushes in turn and a clean part of the yoke.
32. The lamp should light, if it does not light, fit a new field coil assembly.

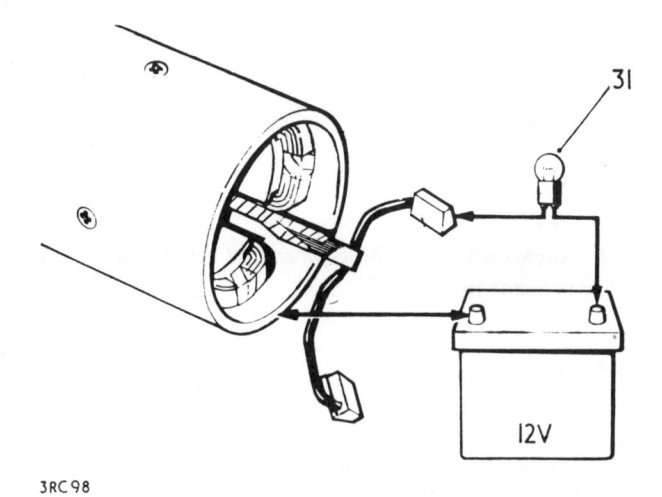

Solenoid

33. Disconnect all cables from the solenoid terminals and connectors.
34. Connect a 12V battery and a 12V 60W test lamp between the solenoid main terminals. The lamp should not light, if it does light, fit new solenoid contacts or a new solenoid complete.
35. Leave the test lamp connected and, using the same 12V battery supply, energise the solenoid by connecting 12V between the small solenoid operating 'Lucar' terminal blade and a good earth point on the solenoid body.
36. The solenoid should be heard to operate and the test lamp should light with full brilliance, otherwise fit new solenoid contacts or a new solenoid complete.

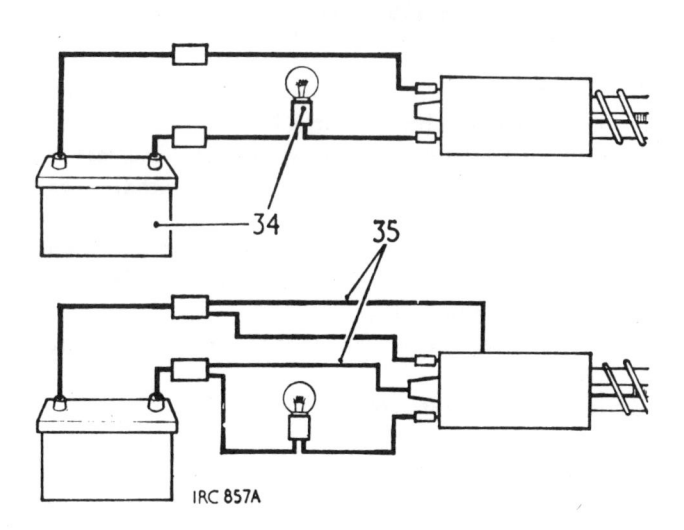

Reassembling

37. Reverse 1 to 15, including the following:.
38. Fit the commutator end cover before refitting the solenoid to facilitate assembly of the block shaped grommet which, when assembled, is compressed between the yoke, solenoid and fixing bracket.
39. Ensure that the internal thrust washer is fitted to the commutator end of the armature shaft.
40. Tightening torques:
Through bolts 1,1 kgf.m (8.0 lbf ft).
Solenoid fixing stud nuts 0,6 kgf.m (4.5 lbf ft).
Solenoid upper terminal nuts 0,4 kgf.m (3.0 lbf ft).
41. Set the armature end float by driving the retaining ring on the armature shaft into a position that provides a maximum of 0,25 mm (0.010 in) clearance between the retaining ring and the bearing bush shoulder.

Rover 3500 and 3500S Manual AKM 3621

ELECTRICAL

IGNITION SWITCH

—Remove and refit   86.65.02
Early models 1 to 4 and 7
Later models (incorporating steering column lock) 5, 6 and 8
Service tools: 601942 Lock ring spanner
601943 Ignition switch spanner
601952 Lockring spanner

Removing

1. Disconnect the battery earth lead.
2. Locate 601943 on to the ignition switch, behind the panel.
3. Using 601952, unscrew the lock ring from the ignition switch.
4. Withdraw the ignition switch from the rear of the panel and disconnect the wiring.
5. Remove the steering column assembly. 57.40.01.
6. Remove the small retaining screws and withdraw the ignition switch.

Refitting

7. Reverse 1 to 4.
8. Reverse 5 and 6. Note that the switch has a key to prevent incorrect fitting.

PANEL SWITCHES

—Remove and refit   86.65.06
Cars with ribbon type speedometer 1 to 4 and 11 to 15
Cars with circular speedometer 1 and 5 to 15
Service tools: 601942 Lock ring spanner
601952 Lock ring spanner

Removing

1. Disconnect the battery earth lead.
2. Interior light switch and wiper switch—Depress the plunger at the side of the knob and withdraw the knob.
3. Using 601942, unscrew the lock ring from the switch.
4. Withdraw the switch from the rear of the panel and disconnect the wiring.
5. Pull back the rubber finisher at each end of the switch panel.
6. Remove the four screws retaining the panel.
7. Pull the panel forward, make a note of cable colour locations to facilitate re-assembly.
8. Disconnect the wiring from the switch panel and withdraw the switch panel assembly.

Rotary switches 9 and 10

9. Depress the plunger through the bottom of the switch knob and withdraw the knob.
10. Unscrew the locking rings and withdraw the switches. 601942, 601952.

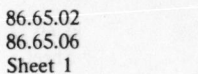

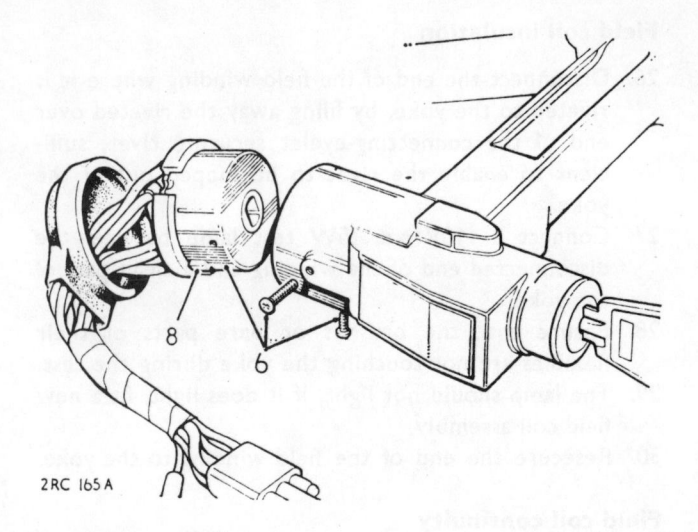

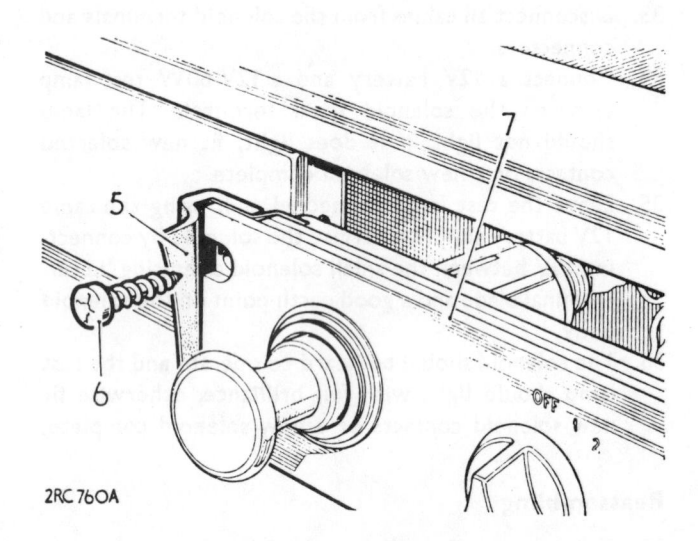

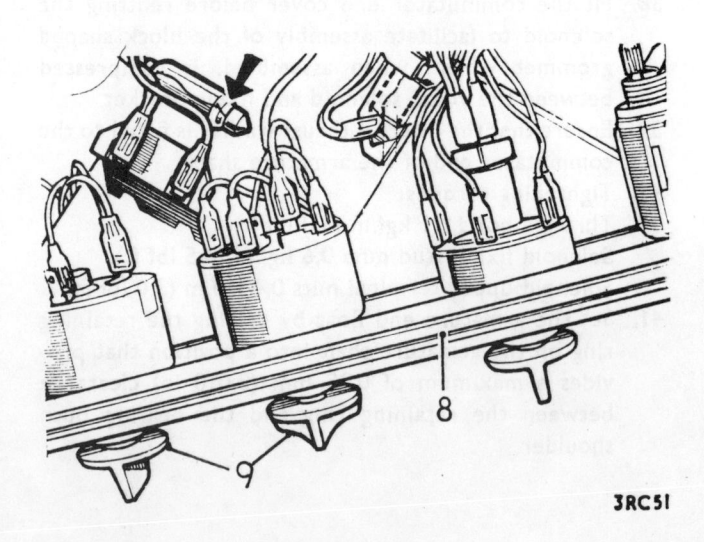

continued

86.65.02
86.65.06
Sheet 1

ELECTRICAL

11. Unscrew the knob from the heated backlight switch.
12. Withdraw the bulb and spring.
13. Unscrew the locking ring from the switch.
14. Withdraw the switch from the rear of the panel and disconnect the wiring.

Refitting

15. Reverse the removal procedure. If necessary, refer to the applicable circuit diagram at the beginning of this Division.

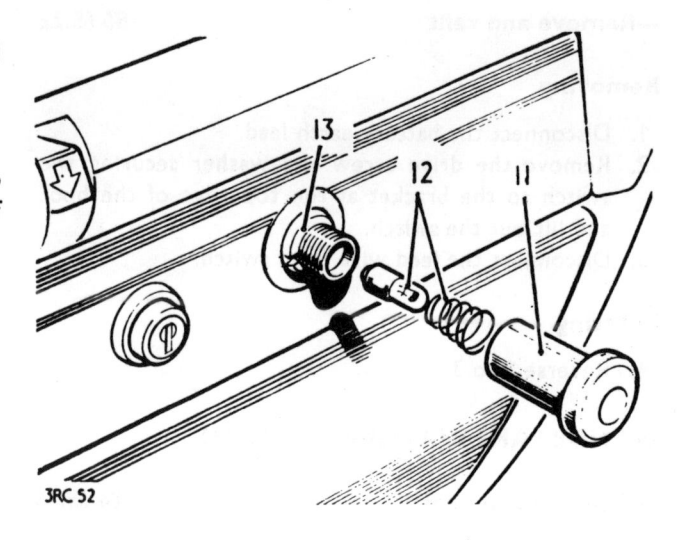

RHEOSTAT

—Remove and refit   86.65.07
Cars with ribbon type speedometer 1 to 3 and 5.
Cars with circular type speedometer 1, 4 and 5.

Removing

1. Remove the instrument panel. 88.20.01.
2. Remove the knob from the panel light switch.
3. Remove the rheostat from the rear of the instrument panel.
4. Withdraw the switch from the housing slot and disconnect the lead from the panel circuit terminals.

Refitting

5. Reverse the removal procedure.

DOOR PILLAR SWITCH

—Remove and refit   86.65.15

Removing

1. Disconnect the battery earth lead.
2. Open the car door.
3. Remove the switch, secured by a small drive screw.
4. Disconnect the wiring.

Refitting

5. Reverse 1 to 4.

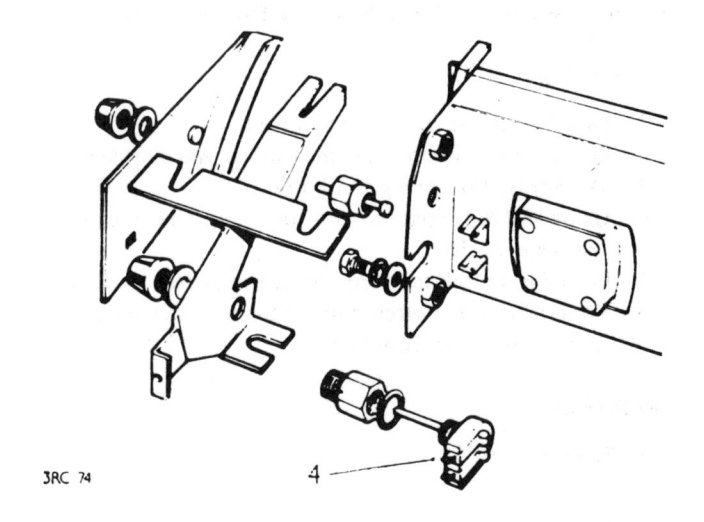

Rover 3500 and 3500S Manual AKM 3621

ELECTRICAL

BOOT LIGHT SWITCH

—Remove and refit   86.65.22

Removing

1. Disconnect the battery earth lead.
2. Remove the drive screw and washer securing the switch to the bracket at the top edge of the boot and lift out the switch.
3. Disconnect the feed wire from switch.

Refitting

4. Reverse 1 to 3.

HANDBRAKE WARNING SWITCH

—Remove and refit   86.65.45

Removing

1. Remove the gearbox tunnel cover. 76.25.07.
2. Remove the locknut and plain washer from the hand brake switch mounting stud and withdraw the switch, adjusting nut and mounting stud complete from the mounting bracket on the hand brake quadrant.
3. Disconnect the electrical leads.

Refitting

4. Reverse 2 and 3.
5. Adjust the position of the switch, by moving the adjuster nut, until the warning lamp will operate with the hand brake applied one or two notches only. Tighten the locknut.
6. Refit the gearbox tunnel cover. 76.25.07.

HAZARD WARNING SWITCH

—Remove and refit   86.65.50

Service tools: 601952 Spanner for lock ring

Removing

1. Disconnect the battery earth lead.
2. Unscrew the knob from the hazard warning switch.
3. Withdraw the spring and bulb.
4. Using 601952, unscrew the lock ring and withdraw the plain washer.
5. Push the switch forward out of the switch panel.
6. Withdraw the switch and disconnect the electrical leads at the snap connectors.

Refitting

7. Reverse 1 to 6.

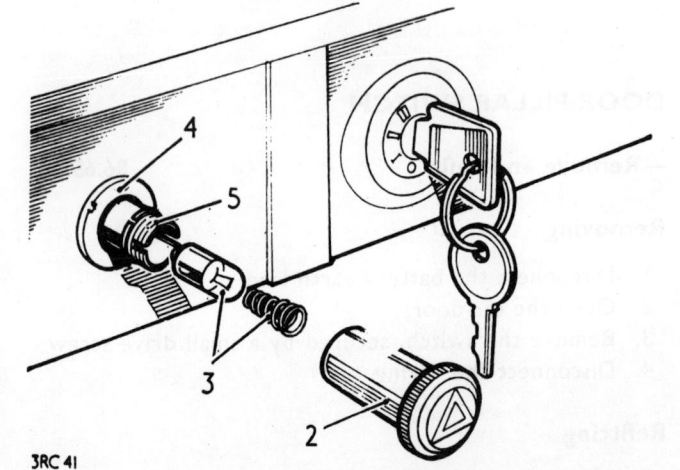

ELECTRICAL

STOP LIGHT SWITCH

—Remove and refit 86.65.51

Removing

1. Prop open the bonnet.
2. Disconnect the electrical leads from the stop light switch, located in the brake pipe connector below the servo.
3. Unscrew the stop light switch. Take precautions against brake fluid spillage.
4. Withdraw the sealing washer.

Refitting

5. Reverse 2 to 4.
6. Bleed the brakes. 70.25.02.

NOTE: On some dual braking circuit cars the stop light switch is located in the foot pedal mounting box assembly to the left of the brake pedal. It is mechanically operated by movement of the brake pedal and may be adjusted, removed and refitted by screwing the switch body into or out of its mounting bracket.

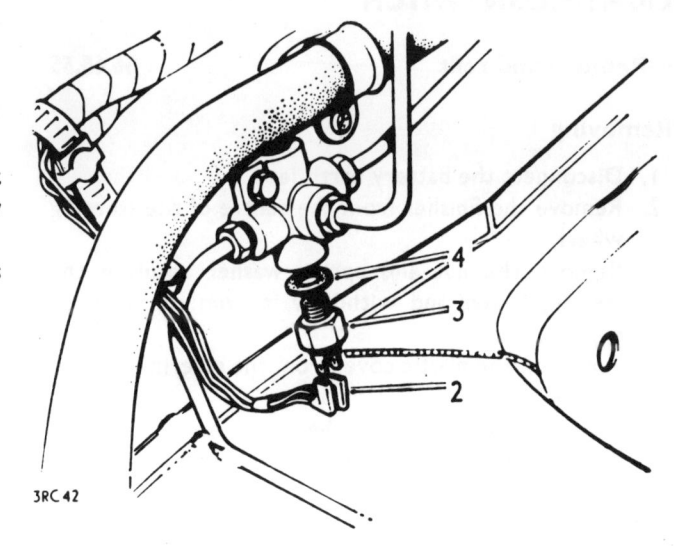

CHOKE WARNING LIGHT SWITCH

—Remove and refit 86.65.53

NOTE: Two switches are incorporated in the cold-start warning light system, a manually-operated switch and a thermo or bi-metal switch, no adjustment being required for either switch. The manual switch is located behind the facia panel on the choke control cable and is covered by this Operation. The bi-metal switch is mounted on the top face of the induction manifold, and is retained by the three screws.

Removing

1. Disconnect the battery earth lead.
2. Remove the radio speaker panel.
3. Disconnect the electrical leads from the choke warning light switch.
4. Release the cable clamping bolt and withdraw the switch.

Refitting

5. Reverse 1 to 4. Ensure that the small projection, incorporated in the switch moulding, is located in the centre hole in the choke cable.

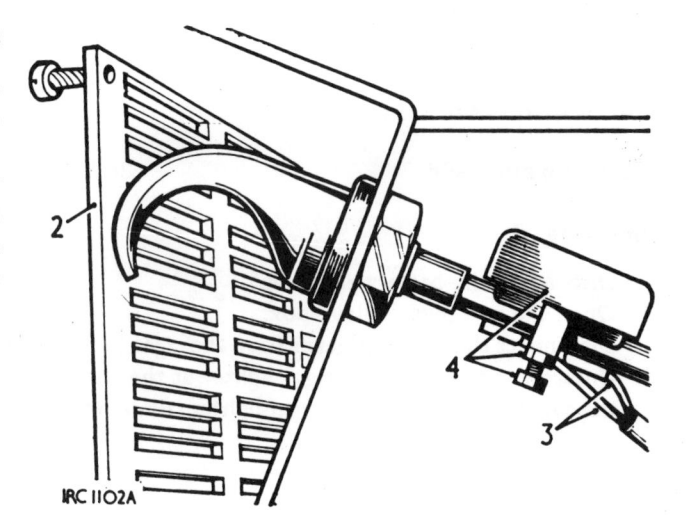

Rover 3500 and 3500S Manual AKM 3621

ELECTRICAL

COMBINED DIRECTION INDICATOR/HEAD-LIGHT/HORN SWITCH

—Remove and refit   86.65.55

Removing

1. Disconnect the battery earth lead.
2. Remove the finisher from the centre of the steering wheel.
3. Remove the nut and spring washer securing the steering wheel and withdraw it from the steering column.
4. Spring off the plastic cover from the steering column nacelle.
5. Disconnect the wiring to the switches.
6. Remove the switches, fixed by drive screws.

Refitting

7. Reverse 1 to 6.
8. Check the operation of the switches.

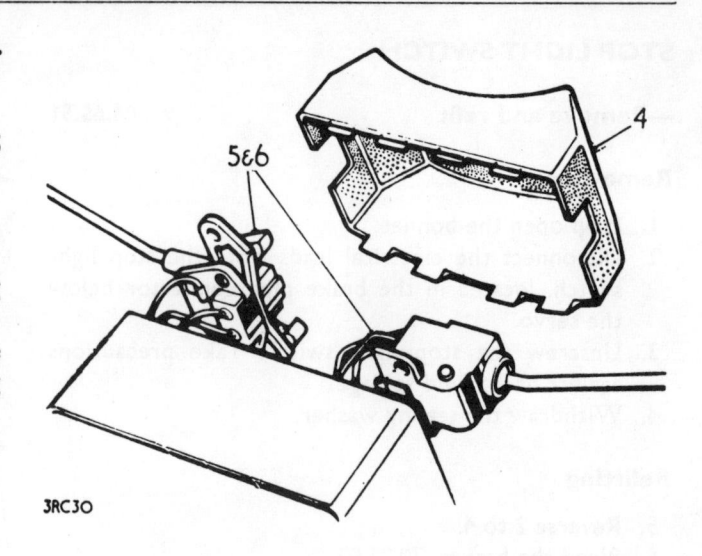

CIGAR LIGHTER

—Remove and refit   86.65.60

Removing

1. Disconnect the battery earth lead.
2. Open the passenger's glove box.
3. Disconnect the electrical leads from the cigar lighter.
4. Unscrew the body of the cigar lighter at the rear of the panel and withdraw the cigar lighter components.

Refitting

5. Reverse 1 to 4.

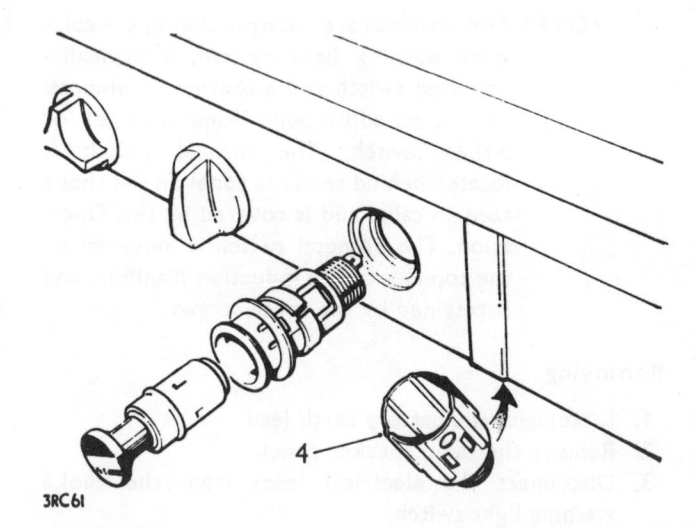

Rover 3500 and 3500S Manual AKM 3621

ELECTRICAL

ICELERT TEST SWITCH
—Remove and refit 1 to 8 and 11   86.65.62

ICELERT RHEOSTAT SWITCH
—Remove and refit 1 to 7 and 9 to 11   86.65.63

Removing

1. Disconnect the battery earth lead.
2. Disconnect the 'Icelert' switch leads at the snap connectors inside the driver's side glove box.
3. Spring off the parcel shelf moulded finisher at the driver's side door 'A' post.
4. Slacken the grub screw and remove the 'Icelert' rheostat switch knob.
5. Remove the locknut from the rheostat switch.
6. Manoeuvre the instrument panel complete with the 'Icelert' test switch and warning lamp, away from the parcel shelf.
7. Note the fitted position of the rheostat switch mounting bracket, to facilitate re-assembly, then ease out the bracket and switch from between the parcel shelf and padded screen rail.
8. Remove the locknut and withdraw the test switch from the instrument panel.
9. Remove the locknut and withdraw the rheostat switch from the bracket.

Refitting

10. Offer the rheostat switch to the bracket with the lead terminals towards the front edge of the bracket, when fitted, then fit the locknut.
11. Reverse 1 to 8 as appropriate.

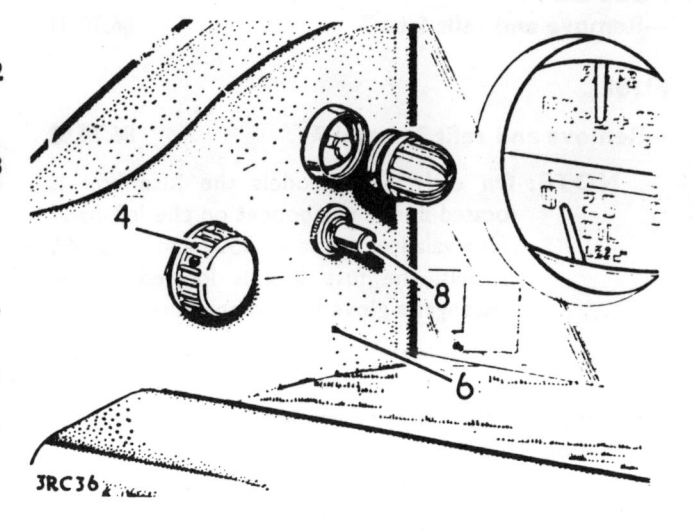

Rover 3500 and 3500S Manual AKM 3621

86.65.62
86.65.63

ELECTRICAL

FUSE BOX
—Remove and refit 1 to 7   86.70.01

FUSE
—Remove and refit 2, 6 and 7   86.70.02

NOTE: On early 3500 models the fuse box is located under the bonnet on the left-hand wing valance. On later 3500 and all 3500S models, the fuse box is located in the passenger's glove box compartment.

Removing

1. Disconnect the battery earth lead.
2. Pull off the cover from the fuse box.
3. Remove the fixings securing the fuse box.
4. Withdraw the fuse(s) box and disconnect the wiring.

Refitting

5. Reverse 3 and 4. Refer to the circuit diagram at the beginning of this Division.
6. Replace fuse as required.
7. Reverse 1 and 2.

NOTE: The fuse position and ratings are marked on the fuse box and specified at the beginning of this Division. See also the applicable circuit diagram at the beginning of this Division.

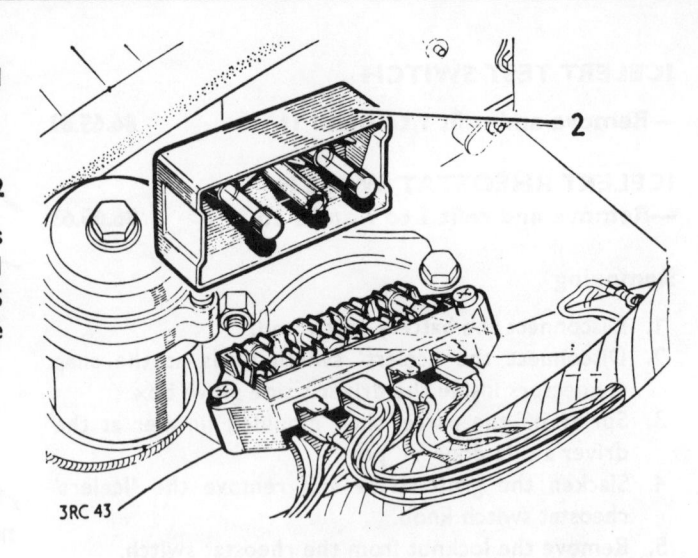

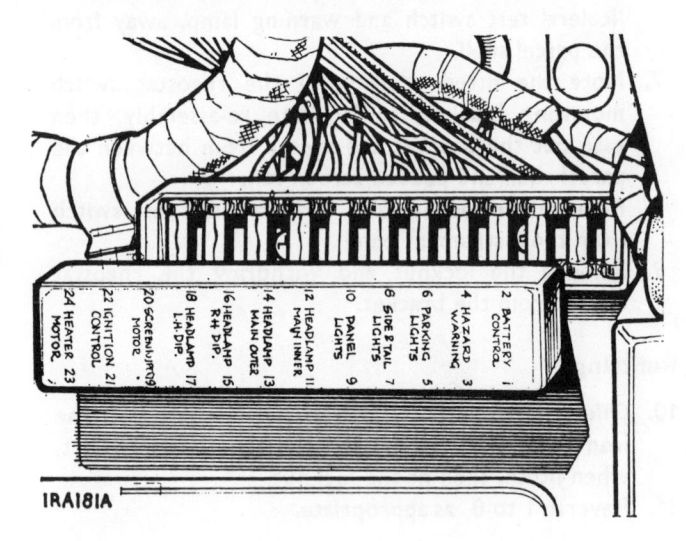

INSTRUMENTS

LIST OF OPERATIONS

Clock	
—adjust	88.15.04
—remove and refit	88.15.07
Combined ammeter and oil gauge—remove and refit	88.25.03
Combined coolant temperature and fuel gauge—remove and refit	88.25.16
Coolant temperature gauge—remove and refit	88.25.14
Coolant temperature transmitter—remove and refit	88.25.20
Fuel gauge—remove and refit	88.25.26
Fuel gauge tank unit—remove and refit	88.25.32
Instrument panel—remove and refit	88.20.01
Oil pressure switch—remove and refit	88.25.08
Oil pressure transmitter—remove and refit	88.25.07
Speedometer—remove and refit	88.30.01
Speedometer cable assembly	88.30.06
Tachometer	88.30.21
Voltage stabilizer	88.20.26

Rover 3500 and 3500S Manual AKM 3621

88-1

INSTRUMENTS

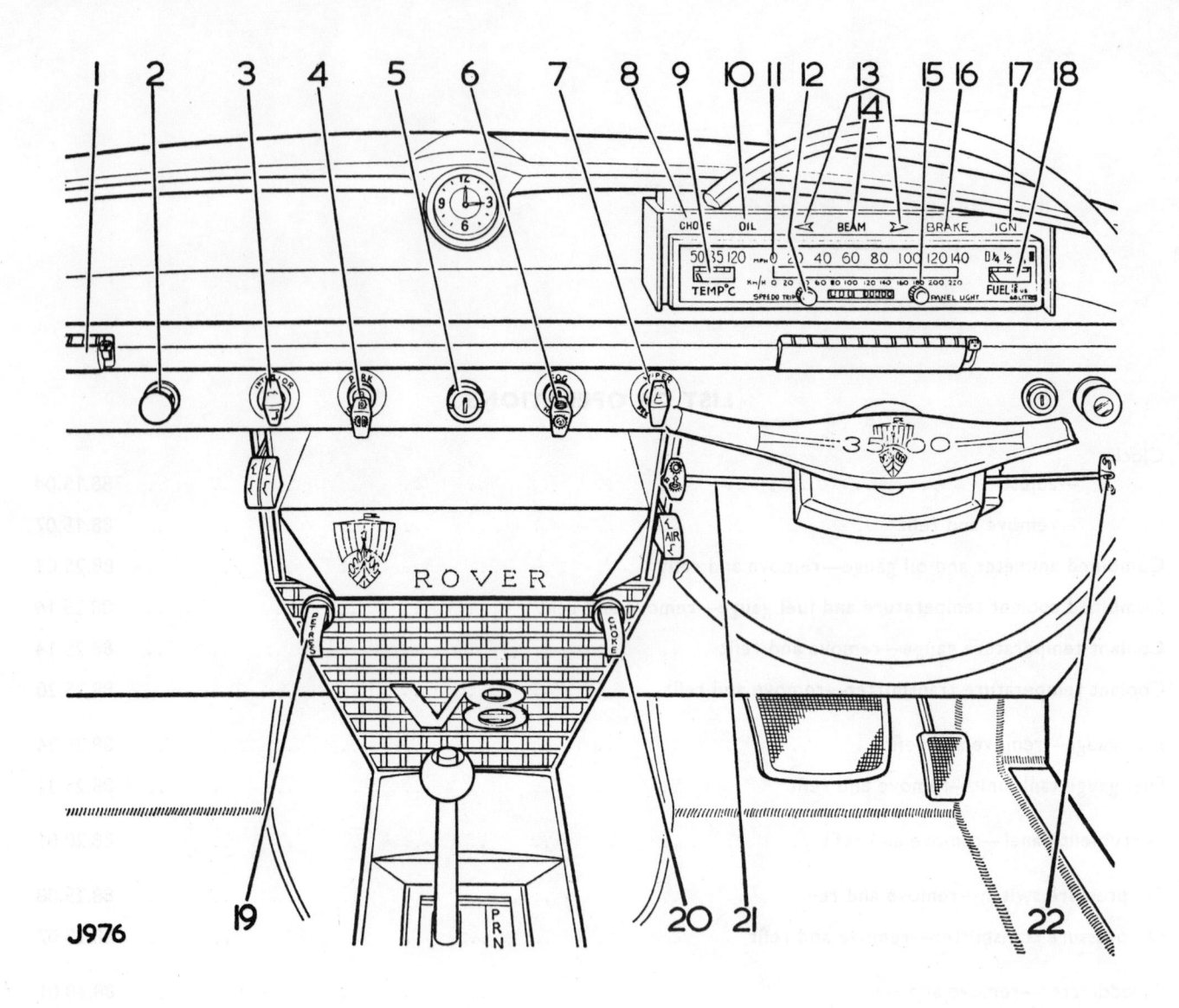

Layout of facia and controls, early 3500 models. Right-hand steering illustrated

1 Face level vent
2 Cigar lighter
3 Switch, interior lights
4 Switch, side and park
5 Switch, ignition
6 Switch, headlamp and foglamp, when fitted
7 Switch, wiper
8 Warning light, choke
9 Gauge, water temperature
10 Warning light, oil pressure
11 Speedometer
12 Speedometer trip
13 Direction indicator arrows
14 Warning light, main beam
15 Switch, panel light
16 Warning light, brake
17 Warning light, ignition
18 Gauge, fuel
19 Control, fuel reserve
20 Control, choke
21 Switch, headlamp dipper and flasher
22 Switch, direction indicator and horn

88-2

Rover 3500 and 3500S Manual AKM 3621

INSTRUMENTS

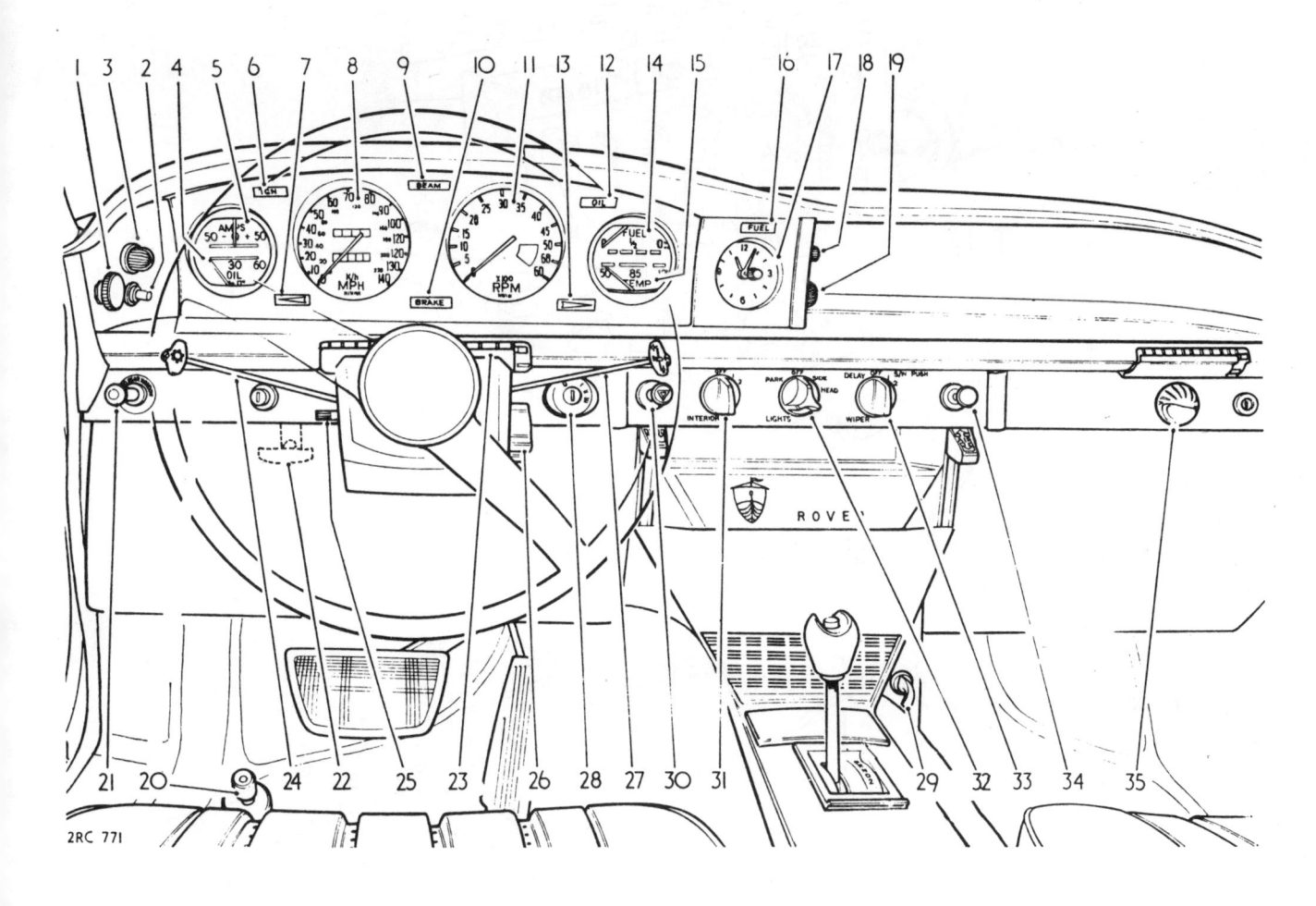

Layout of facia and controls, later 3500 models. Left-hand steering illustrated

NOTE: The layout of 3500S models is similar except for the gear change lever and the addition of the clutch pedal.

1. Rheostat, 'Icelert' warning light, when fitted
2. Test button, 'Icelert' warning light
3. Warning light, 'Icelert'
4. Gauge, oil pressure
5. Ammeter
6. Warning light, ignition
7. Direction indicator arrow, LH
8. Speedometer
9. Warning light, main beam
10. Warning light, brakes
11. Tachometer
12. Warning light, oil pressure
13. Direction indicator arrow, RH
14. Gauge, fuel
15. Gauge, water temperature
16. Warning light, fuel reserve
17. Clock
18. Control, speedometer trip
19. Rheostat, panel lights
20. Switch, headlamp dipper
21. Switch, heated backlight, when fitted
22. Bonnet release, inside driver's glove box
23. Face level vent
24. Switch, headlamp flasher
25. Delay control, wiper
26. Knob, column rake adjuster
27. Switch, direction indicator and horn
28. Switch, ignition and starter with steering column lock
29. Switch, fuel reserve
30. Switch, hazard warning
31. Switch, interior lights
32. Switch, side, park, headlamp and fog lamp, when fitted
33. Switch, wiper
34. Cigar lighter
35. Rotary map light

Rover 3500 and 3500S Manual AKM 3621

INSTRUMENTS

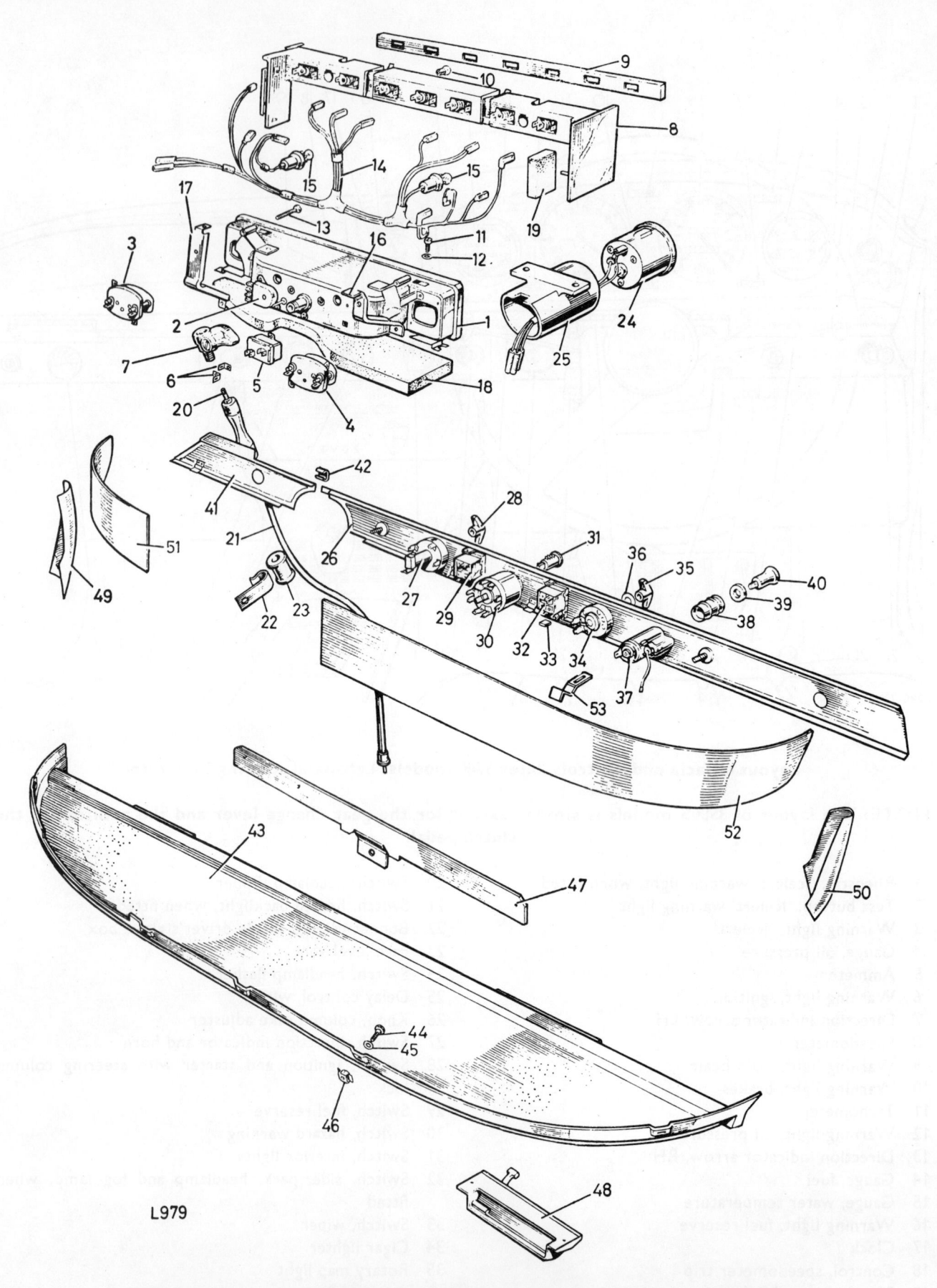

L979

Layout of instrument panel, switch panel and parcel tray, early 3500 models, with ribbon type speedometer

Rover 3500 and 3500S Manual AKM 3621

INSTRUMENTS

Key to illustration of instrument panel, switch panel and parcel tray, early 3500 models, with ribbon type speedometer

1 Speedometer assembly
2 Panel light rheostat
3 Fuel level indicator
4 Water temperature indicator
5 Voltage regulator
6 Terminal blade for voltage regulator
7 Angle drive complete for speedometer
8 Saddle complete for instrument panel
9 Printed window for warning lights
10 Bulb for warning lights, 2.2 watt
11 Special screw ⎫ Fixing saddle
12 Plain washer ⎭ to speedometer
13 Special screw fixing saddle to base unit
14 Panel light harness
15 Bulb for panel light, 12 volt, 3.6 watt, LU 984
16 Bracket for instrument panel, inner
17 Bracket for instrument panel, outer
18 Foam mounting pad ⎫ For instrument
19 Foam side pad ⎭ panel
20 Cable, inner
21 Cable, outer
22 Clip for speedometer cable
23 Grommet for clip
24 Clock
25 Moulding for clock
26 Switch panel complete
27 Switch (rheostat) for wiper and washer
28 Knob for wiper and washer switch
29 Switch for head and fog lamps, LU 34631
30 Switch, ignition and starter, LU 39222
31 Barrel lock for ignition switch
32 Switch, side and park, LU 34630
33 Lucar insulating sleeve, LU 54948329
34 Switch for roof lamp, LU 35879
35 Knob for roof lamp switch
36 Escutcheon for roof lamp switch
37 Cigar lighter
38 Heating unit complete ⎫
39 Flange for knob ⎬ For cigar lighter
40 Knob ⎭
41 Dash finisher complete, RH
42 Joint piece, dash finisher to switch panel
43 Parcel shelf, front
44 Shouldered screw ⎫ Fixing
45 Plain washer ⎬ parcel shelf
46 Spire nut (10 UNF) ⎭ to base unit
47 Stiffening plate for front parcel shelf
48 Face-level vent
49 Extension piece, RH ⎫ For front
50 Extension piece, LH ⎭ parcel shelf
51 Finisher, short, for front parcel shelf
52 Finisher, long, for front parcel shelf
53 Retaining clip for parcel shelf finishers

Rover 3500 and 3500S Manual AKM 3621

INSTRUMENTS

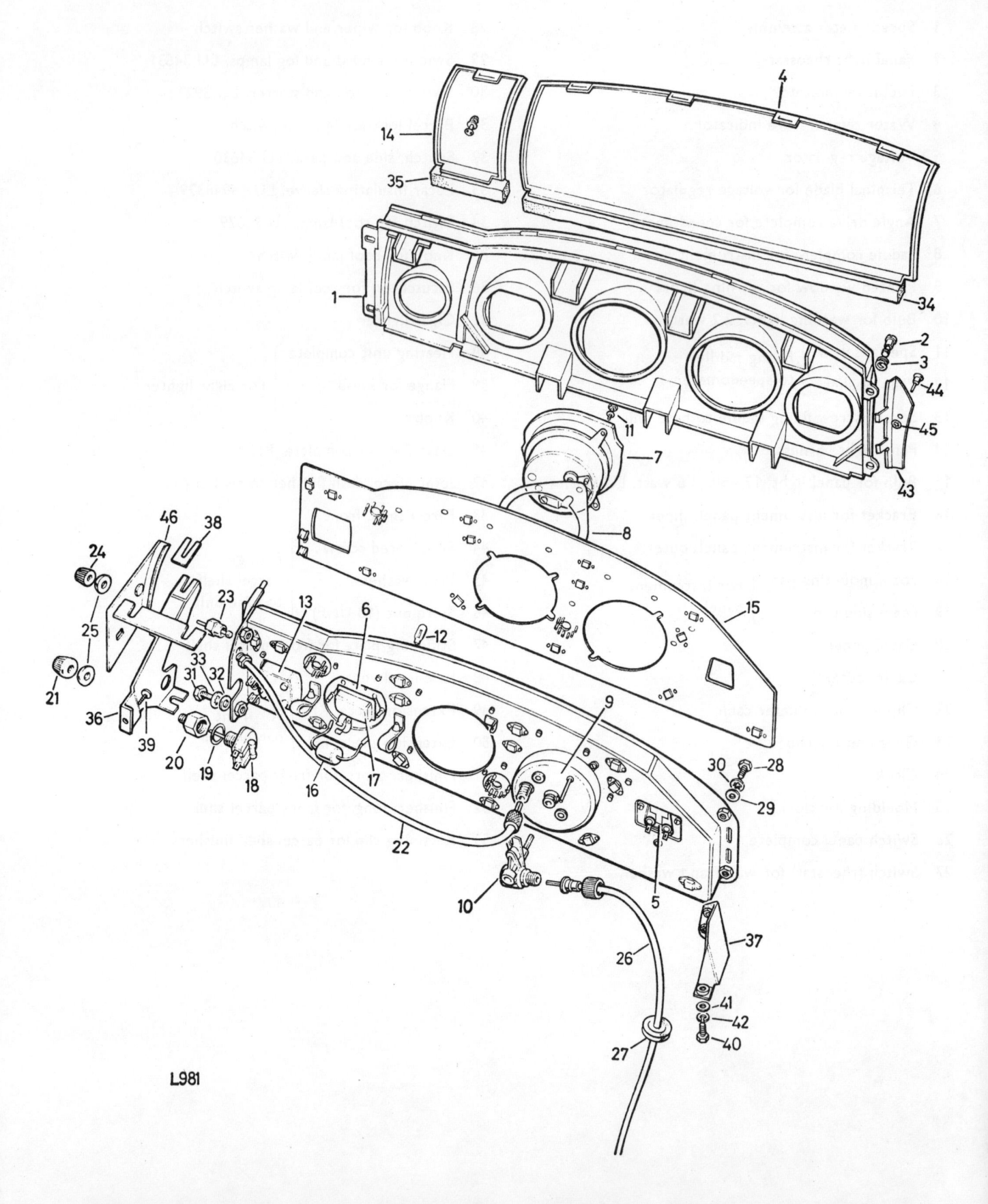

General layout of instrument panel for later 3500 and 3500S models, with circular type speedometer

Rover 3500 and 3500S Manual AKM 3621

INSTRUMENTS

Key to layout of instrument panel for later 3500 and 3500S models, with circular type speedometer

1. Front moulding complete for instrument unit
2. Special screw fixing front moulding to casing
3. Special spring washer for screw
4. Window for instrument unit
5. Ammeter and oil pressure gauge
6. Fuel and water temperature gauge
7. Tachometer
8. Insulating washer for tachometer
9. Speedometer
10. Angled drive for speedometer
11. Screw fixing instruments to case
12. Bulb, capless 2.2 watt for instrument panel and warning lights
13. Clock complete
14. Window and reset knob for clock
15. Printed circuit board
16. Radio suppressor (when fitted)
17. Voltage stabilizer
18. Rheostat for instrument lighting
19. Rubber washer for rheostat
20. Adaptor for rheostat
21. Knob complete for rheostat
22. Trip reset cable for speedometer
23. Adaptor for trip reset
24. Knob for trip reset
25. Escutcheon for rheostat and trip reset knob
26. Speedo drive cable complete
27. Grommet for speedo cable
28. Bolt (10 UNF x $\frac{3}{8}$ in long) ⎫ Fixing instrument unit, outer
29. Plain washer
30. Shakeproof washer ⎭
31. Bolt ($\frac{1}{4}$ in UNF x $\frac{3}{8}$ in long) ⎫ Fixing instrument unit, inner
32. Plain washer
33. Shakeproof washer ⎭
34. Pad, long ⎫ Instrument unit to parcel shelf
35. Pad, short ⎭
36. Mounting bracket, centre ⎫ For instrument unit
37. Mounting bracket, outer ⎭
38. Shim for centre mounting bracket
39. Drive screw fixing bracket to dash
40. Bolt ($\frac{1}{4}$ in UNF x $\frac{5}{8}$ in long) ⎫ Fixing outer bracket to base unit
41. Plain washer
42. Spring washer ⎭
43. End finisher assembly, small for instrument unit
44. Knob for end finisher
45. Starlock washer fixing knob to end finisher
46. End finisher, large for instrument unit

INSTRUMENTS

CLOCK

—Adjust   88.15.04

NOTE: Adjustment should be carried out only if the clock gains or loses more than two or three minutes per week. The accompanying illustrations show the clock fitted to cars with the ribbon type speedometer and the circular type speedometer.

Adjusting

1. Remove the clock. 88.15.07.
2. Adjust by means of the screw at the rear of the clock:
   clock gaining, turn to — mark
   clock losing, turn to + mark
3. Refit the clock. 88.15.07.

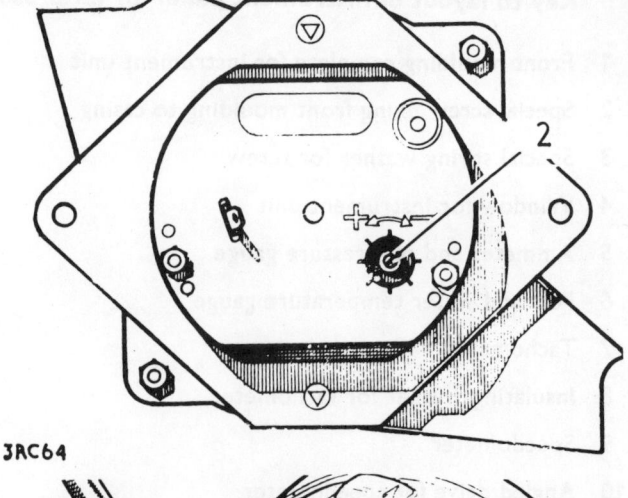

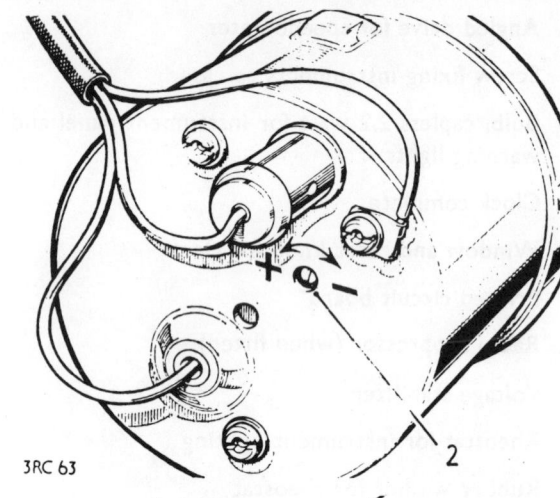

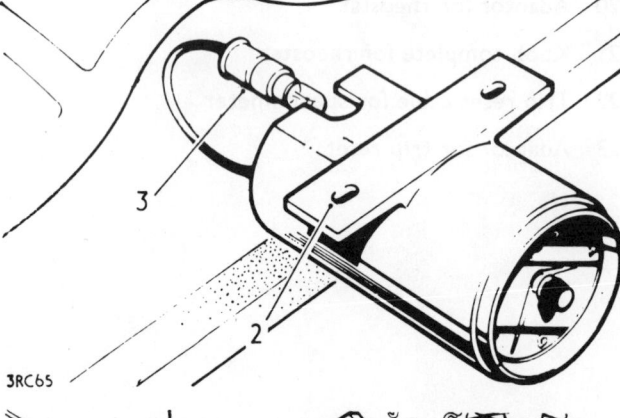

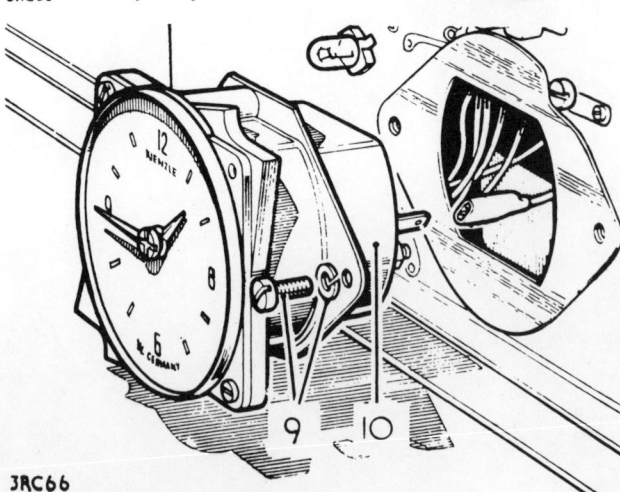

CLOCK

—Remove and refit   88.15.07

Cars with ribbon type speedometer 1 to 5 and 11

Cars with circular type speedometer 1 and 6 to 11

Removing

1. Disconnect the battery earth lead.
2. Remove the two nuts, together with the washers, securing the clock shroud to the facia top rail.
3. Disconnect the wiring from the connectors at the rear of the panel.
4. Withdraw the clock and shroud complete.
5. Slacken the brass set screw and remove the clock from the shroud.
6. Remove the instrument panel. 88.20.01.
7. Remove the two retaining screws from each end of the instrument panel.
8. Ease the front of the panel away from the instruments.
9. Remove the clock fixings.
10. Withdraw the clock and insulation strip.

Refitting

11. Reverse the removal procedure.

88.15.04
88.15.07

Rover 3500 and 3500S Manual AKM 3621

INSTRUMENTS

INSTRUMENT PANEL

—Remove and refit   88.20.01

Cars with ribbon type speedometer 1 to 7, 16 to 22, 24 and 25

Cars with circular type speedometer 1, 8 to 15 and 23 to 25

Removing

1. Disconnect the battery earth lead.
2. Lift the two securing clips located above the plastic warning light covers and remove cover.
3. Remove the top facia rail. 76.46.04.
4. Disconnect the speedometer cable.
5. Remove the two screws securing the instrument panel and pull the panel forward.
6. Disconnect the wiring from the rear of the instruments.
7. Remove the instrument panel.
8. Pull off the knob for the speedometer trip control and the knob for the panel light switch.
9. Remove the locking rings retaining the instrument panel finisher.
10. Slide the finisher to one side and withdraw.
11. Spring off the finisher at the other end of the instrument panel, by pulling the knob, or, if an 'Icelert' is fitted remove the grub-screw, knob and nut from the rheostat then spring off the finisher.
12. Remove the two retaining hexagon head bolts at each end of the instrument panel, and pull the panel forward.
13. Disconnect the speedometer cable, leaving the angled drive on the back of the panel.
14. Make a note of cable colour locations to facilitate reassembly, then disconnect the wiring from the instrument panel, consisting of two multi-pin plugs and five separate leads.
15. Remove the instrument panel.

continued

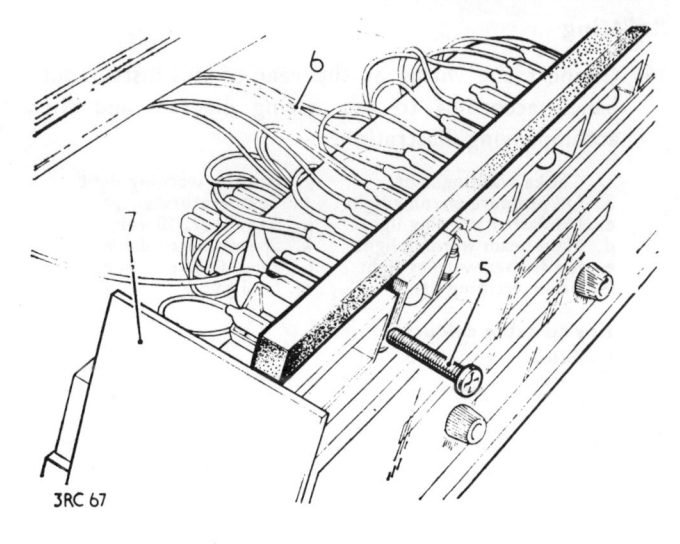

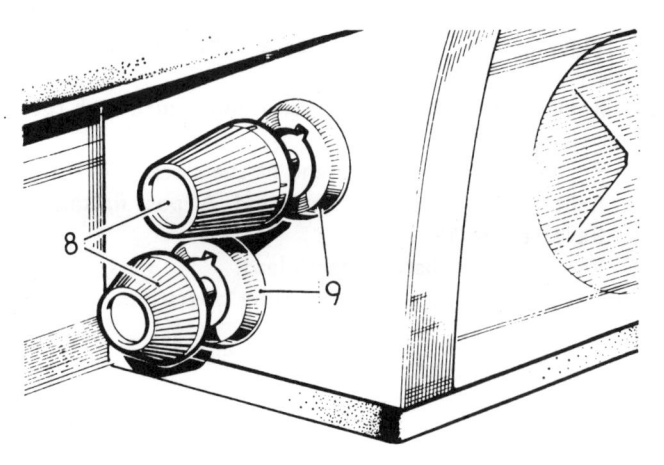

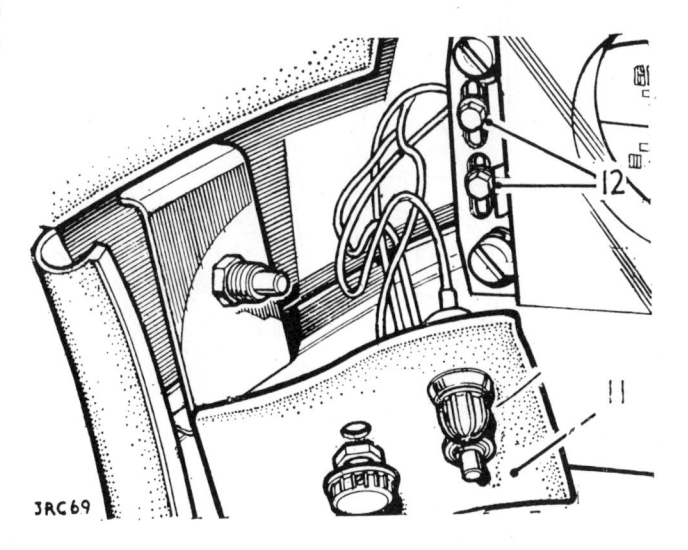

Rover 3500 and 3500S Manual AKM 3621

88.20.01
Sheet 1

INSTRUMENTS

Refitting

16. Connect the wiring at the rear of the instrument panel, according to the wiring diagram and the accompanying illustration.

    a. Choke warning light
    b. Oil pressure warning light
    c. LH flasher warning light
    d. Main-beam warning light
    e. RH flasher warning light
    f. Brake warning light
    g. Ignition warning light
    h. Temperature gauge
    j. Regulator, 10 volt
    k. Speedometer drive
    l. Panel light rheostat
    m. Fuel gauge

    Key to cable colours:
    B—Black   N—Brown   U—Blue
    G—Green   P—Purple   W—White
    LG—Light green R—Red Y—Yellow

17. Remove the angled drive from the speedometer head, and screw it on to the cable.
18. Connect the cable and angled drive to the speedometer head.
19. Ensure that the instrument panel is correctly located and that the securing screw holes are aligned, then refit and tighten the two securing screws.
20. Replace the warning light cover and secure with the two spring clips.
21. Replace the top facia rail and fit the clock.
22. Refit the right-hand and left-hand 'A' post finishers.
23. Reverse 8 to 15.
24. Reconnect the battery earth lead.
25. Check all instruments for correct operation.

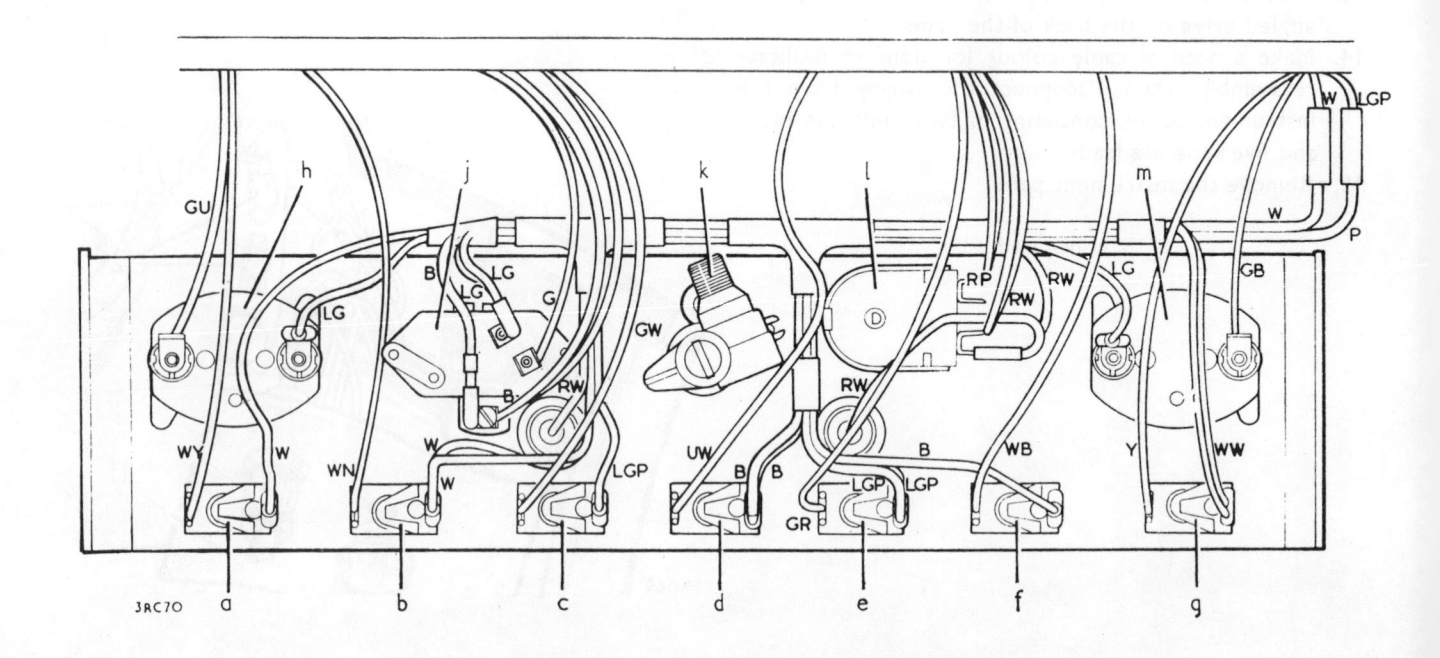

88.20.01
Sheet 2

Rover 3500 and 3500S Manual AKM 3621

INSTRUMENTS

VOLTAGE STABILIZER

—Remove and refit　　　　　　　　　88.20.26

Cars with ribbon type speedometer 1, 2 and 5

Cars with circular type speedometer 1 and 3 to 5

Removing

1. Remove the instrument panel. 88.20.01.
2. Remove the voltage stabilizer from the rear of the instrument panel.
3. Withdraw the voltage stabilizer from the terminal sockets on the rear of the combined fuel contents and coolant temperature gauge.
4. Remove the insulation strip.

Refitting

5. Reverse 1 to 4, as applicable.

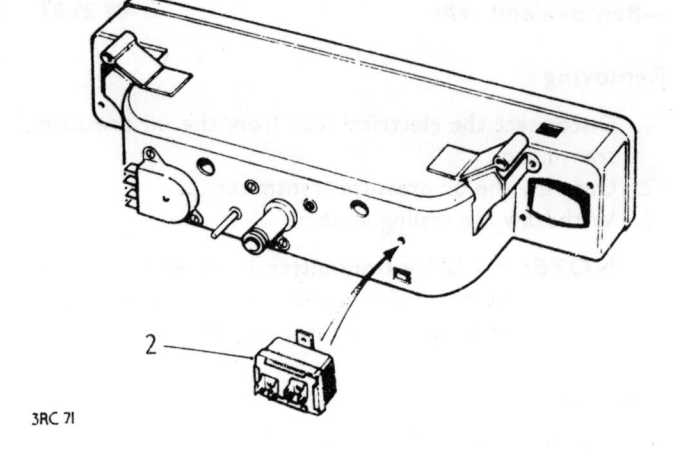

COMBINED AMMETER AND OIL GAUGE

—Remove and refit 1 to 3 and 5　　　　88.25.03

COMBINED COOLANT TEMPERATURE AND FUEL GAUGE

—Remove and refit 1, 2, 4 and 6　　　　88.25.16

NOTE: This Operation refers to cars fitted with circular instruments.

Removing

1. Remove the instrument panel. 88.20.01.
2. Remove the two retaining screws at each end of the instrument panel, and ease the front of the panel away from the instruments.
3. Remove the three fixing screws and washers, and withdraw the combined ammeter and oil gauge.
4. Remove the fixing screws and washers, and withdraw the combined coolant temperature and fuel gauge.

Refitting

5. Reverse 1 to 3, noting that the spring washer and plain washer are fitted to the screw positioned at the lower right-hand fixing hole (adjacent to electrical circuit pick-up point).
6. Reverse 1, 2 and 4, noting that the plain washers are fitted to the two lower fixing screws (adjacent to the two electrical pick-up points).

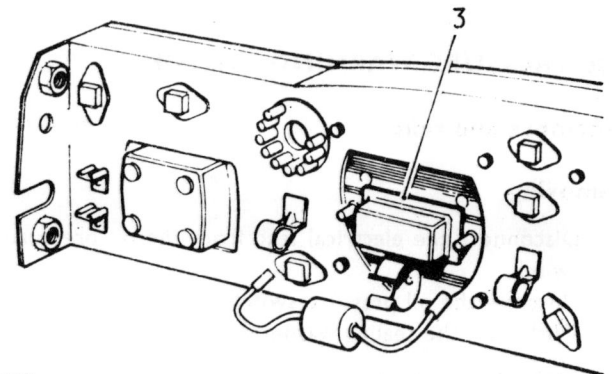

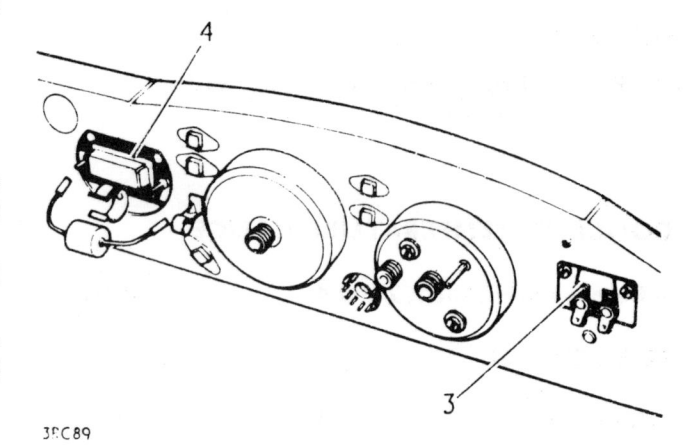

Rover 3500 and 3500S Manual AKM 3621

INSTRUMENTS

OIL PRESSURE TRANSMITTER

—Remove and refit 88.25.07

Removing

1. Disconnect the electrical lead from the oil pressure transmitter.
2. Unscrew the oil pressure transmitter.
3. Withdraw the sealing washer.

    NOTE: If a faulty transmitter is suspected, it can be checked by substitution or by temporarily fitting a pressure gauge in its place.

Refitting

4. Reverse 1 to 3, using a new sealing washer.

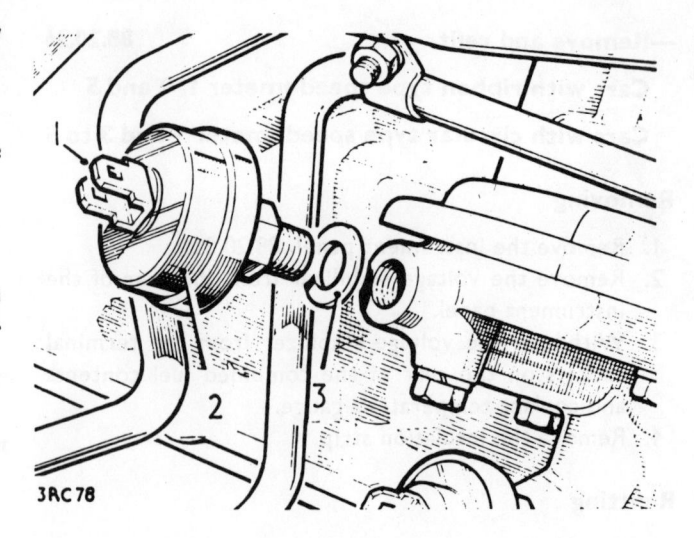

OIL PRESSURE WARNING SWITCH

—Remove and refit 88.25.08

Removing

1. Disconnect the electrical lead from the oil pressure switch.
2. Unscrew the oil pressure switch.
3. Withdraw the sealing washer.

    NOTE: If a faulty switch is suspected it can be checked by substitution or by temporarily fitting a pressure gauge in its place.

Refitting

4. Reverse 1 to 3, using a new sealing washer.

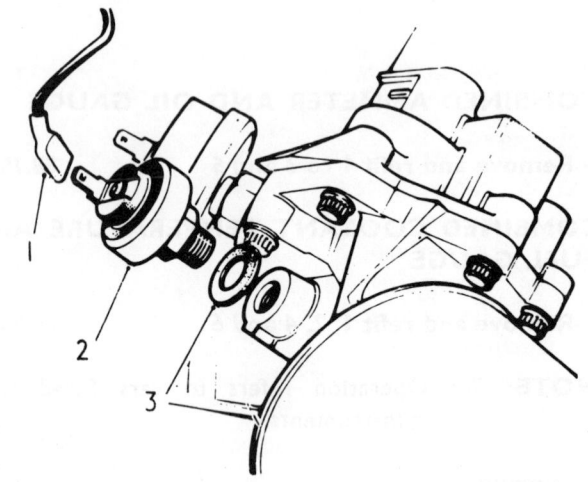

COOLANT TEMPERATURE GAUGE

—Remove and refit 1, 2 and 4 88.25.14

FUEL GAUGE

—Remove and refit 1, 3 and 4 88.25.26

NOTE: This Operation refers to cars fitted with a ribbon type speedometer.

Removing

1. Remove the instrument panel. 88.20.01.
2. Remove the coolant temperature gauge.
3. Remove the fuel gauge.

Refitting

4. Reverse the removal procedure.

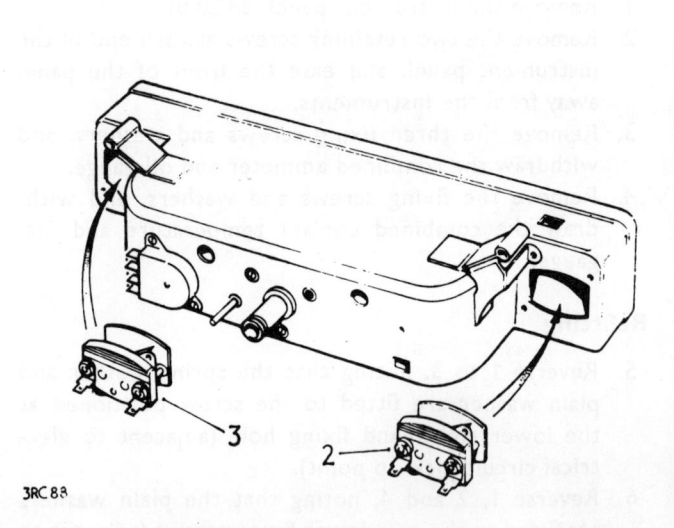

88.25.07
88.25.26

Rover 3500 and 3500S Manual AKM 3621

INSTRUMENTS

COOLANT TEMPERATURE TRANSMITTER

—Remove and refit   88.25.20

Removing

1. Disconnect the electrical lead from the coolant temperature transmitter.
2. Unscrew the coolant temperature transmitter.
3. Withdraw the sealing washer.

Refitting

4. Reverse 1 to 3, using a new sealing washer.

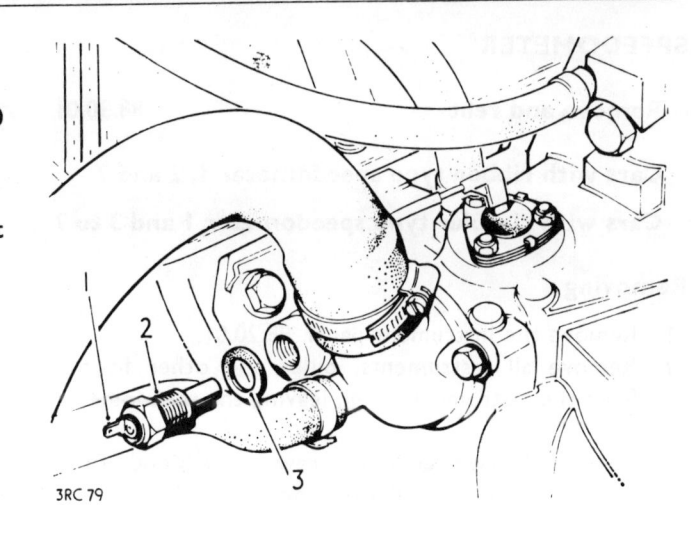

FUEL GAUGE TANK UNIT

—Remove and refit   88.25.32

Service tool: 600964

Removing

NOTE: Two differing fuel pipe arrangements are illustrated, the first refers to cars with the fuel reserve tap in the engine compartment, the second shows cars with the reserve tap beneath the fuel tank.

1. Disconnect the battery earth lead.
2. Drain the fuel tank by disconnecting the fuel pipes from the tank unit.
3. Note the colour and positioning of the electrical leads to facilitate reassembly, then disconnect the leads from the tank unit.
4. Using 600964, turn the tank unit locking ring in an anti-clockwise direction.
5. Withdraw the tank unit and remove the sealing ring.
6. Release the retaining spring and withdraw and clean the filter.

Refitting

7. Locate the filter over the reserve outlet of the tank unit and secure with the spring.
8. Position the large rubber seal on to the tank unit and insert the unit into the tank.
9. Fit the locking ring and, using 600964, secure by turning the ring in a clockwise direction.
10. Connect the fuel pipes to the tank unit, the main feed to the long union.
11. Reconnect the electrical leads to the tank unit.
12. Refill the fuel tank and check for leaks.
13. Reconnect the battery earth lead.
14. Check the operation of the fuel contents gauge.

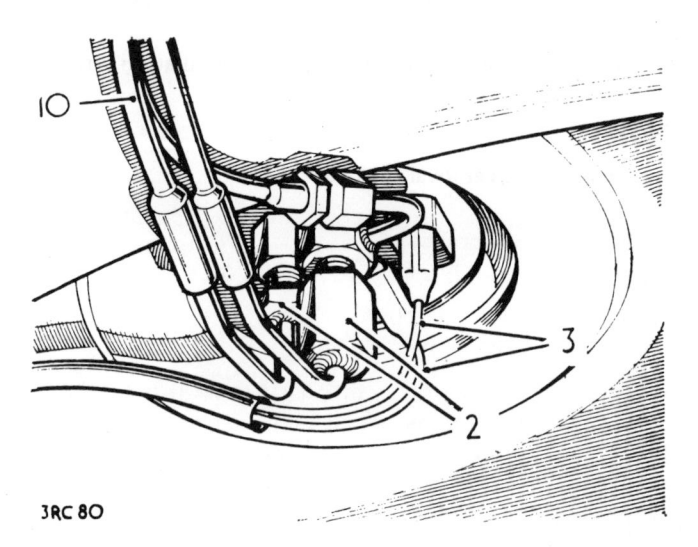

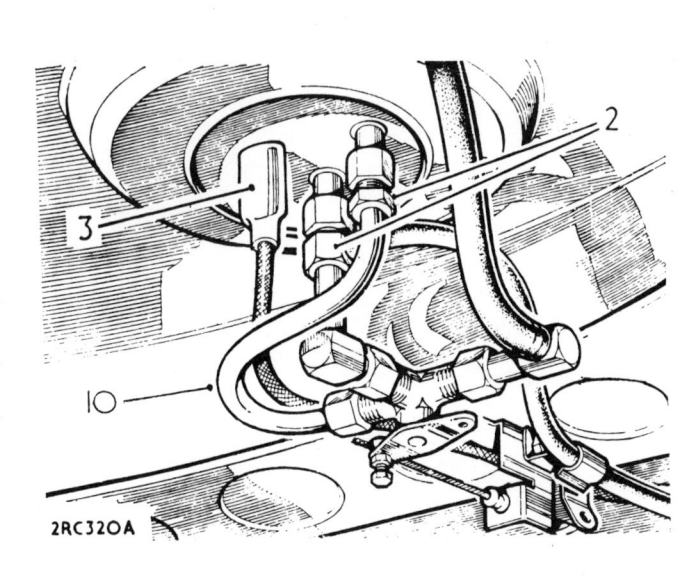

Rover 3500 and 3500S Manual AKM 3621

INSTRUMENTS

SPEEDOMETER

—Remove and refit   88.30.01

Cars with ribbon type speedometer 1, 2 and 7

Cars with circular type speedometer 1 and 3 to 7

Removing

1. Remove the instrument panel. 88.20.01.
2. Remove all instruments, bulbs and other fittings from the instrument panel, leaving the speedometer only.
3. Remove the two retaining screws at each end of the instrument panel, and ease the front of the panel away from the instruments.
4. Disconnect the speedometer trip cable at the knurled nut on the rear of the unit.
5. Remove the three fixing screws and shakeproof washers.
6. Withdraw the speedometer unit and insulation strip.

Refitting

7. Reverse the removal procedure.

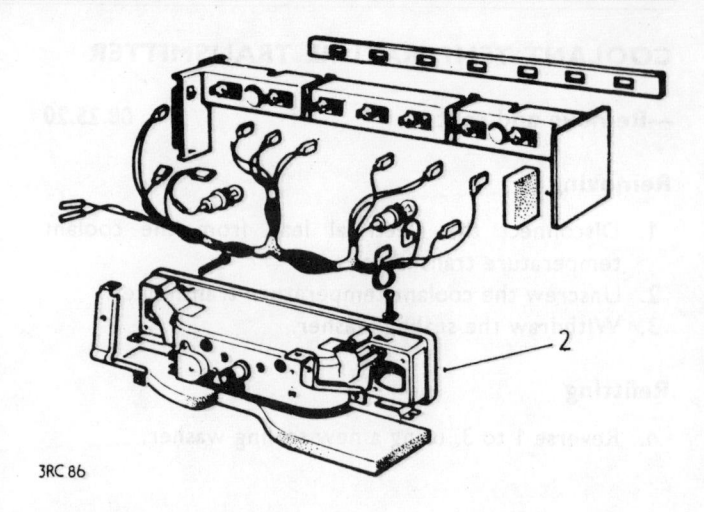

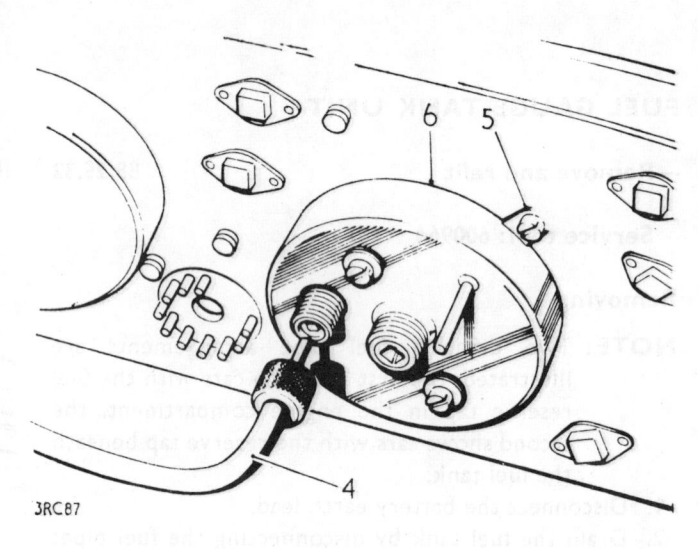

88.30.01

Rover 3500 and 3500S Manual AKM 3621

INSTRUMENTS

SPEEDOMETER CABLE ASSEMBLY

—Remove and refit　　　　　　　　　　88.30.06

　Cars with ribbon type speedometer 2 to 11, 13 and 14

　Cars with circular type speedometer 1, 3 to 10 and 12 to 14

Removing

1. Remove the instrument panel. 88.20.01.
2. Disconnect the speedometer cable from the instrument.
3. Remove the radio speaker panel.
4. Lift the carpet at the left-hand side of the gearbox tunnel and remove the cover plate.
5. Disconnect the speedometer cable from the gearbox.
6. Withdraw the speedometer cable and grommet through the base unit, releasing it from retaining clips as applicable.

Refitting

7. Connect the cable to the speedometer.

    NOTE: Models with ribbon type speedometer—It may be necessary to slacken or remove the instrument panel securing screws to facilitate reconnecting the cable to the speedometer.

8. Feed the speedometer cable through the grommet in the base unit and connect it to the gearbox, ensuring correct entry of the inner cable into the speedometer drive.
9. Secure with the special retainer and self-locking nut.
10. Refit the cover to the gearbox tunnel.
11. Refit the instrument securing screws or retighten if not actually removed.
12. Refit the instrument panel. 88.20.01.
13. Refit the radio speaker panel.
14. Check the speedometer for correct operation.

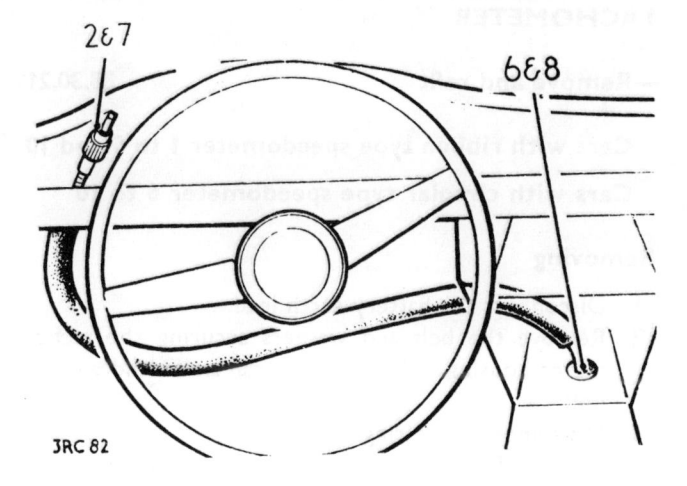

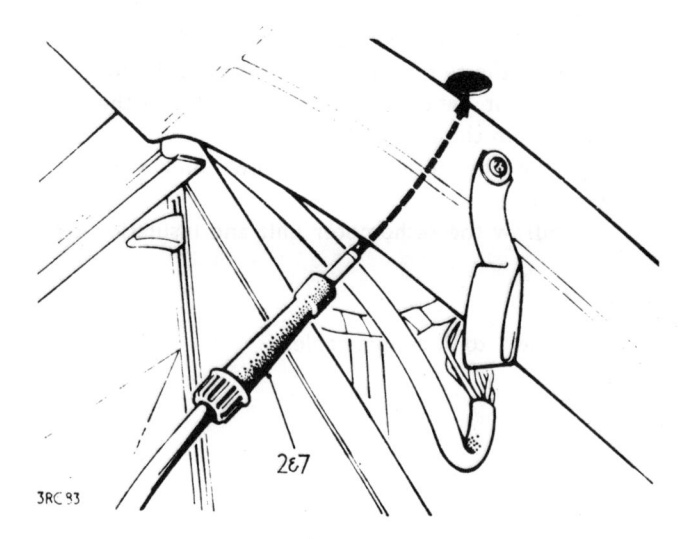

Rover 3500 and 3500S Manual AKM 3621　　　　　88.30.06

INSTRUMENTS

TACHOMETER

—Remove and refit   **88.30.21**

Cars with ribbon type speedometer 1 to 5 and 10

Cars with circular type speedometer 6 to 10

Removing

1. Disconnect the battery earth lead.
2. Remove the bolt and washers securing the tachometer housing to the facia top rail and withdraw the unit.
3. Disconnect the wiring.
4. Slacken the brass set screw retaining the clock and remove the clock.
5. Remove the tachometer retaining strap and remove the tachometer.
6. Remove the instrument panel. 88.20.01.
7. Remove the two retaining screws at each end of the instrument panel, and ease the front of the panel away from the instruments.
8. Remove the three fixing screws and shakeproof washers.
9. Withdraw the tachometer unit and insulation strip.

Refitting

10. Reverse 1 to 9, as applicable.

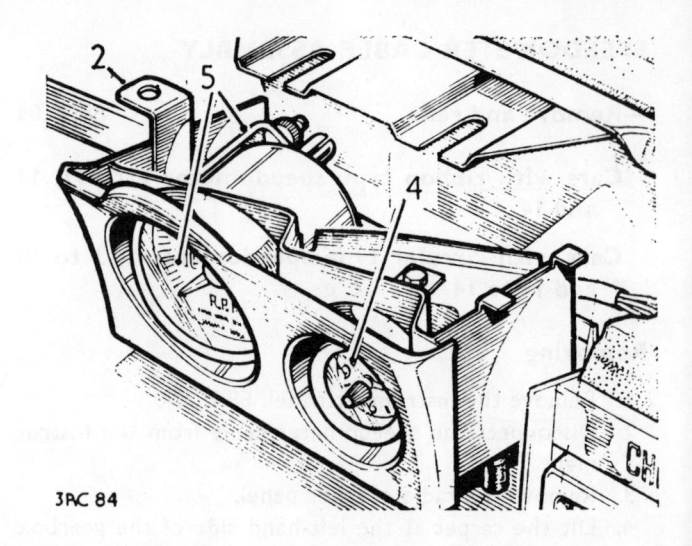

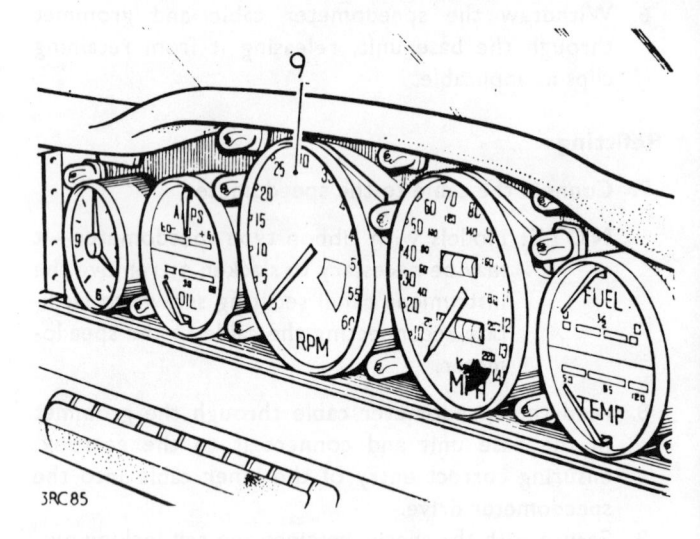

Rover 3500 and 3500S Manual AKM 3621

88.30.21

SERVICE TOOLS

JD 10	Power steering test set
18G2 (6312A)	Two legged puller
18G126 (606606)	Ball joint extractor
18G134 (550)	Bearing and oil seals replacer
18G1001 (600964)	Lock ring spanner
18G1004 (7066)	Circlip pliers
18G1205	Propeller shaft flange wrench
JD 34 (606604)	Selector shaft spline seal protector
47	Bearing remover
RO 1009	Upper rear main bearing oil seal remover/replacer
RO 1012	Power steering hydraulic adapter
RO 1014	Rear crankshaft oil seal protection sleeve
262757A	Differential pinion bearing extractor
262758	Pinion bearing press block
274401	Exhaust valve guide remover
275870	Axle shaft bearing remover
511041	Ball swivel rubber boot replacer
530102	Differential pinion nut and dog nut spanner
530105	Differential overhaul spanner
530106	Dial indicator gauge bracket
541884	Hub bearing remover adapter
541885	Pinion bearing remover adapter
565446	Steering ball joints remover
600192	Differential bearing remover adapter
600287	Inlet valve guide remover
600304	Front road springs retainer rods
600358	Door sealing rubbers fitting tool
600447	Pinion splines torque wrench adapter
600504	Carburettor adjusting spanner
600571	Front suspension brackets checking fixture
600572	Front upper valance R.H. checking fixture
600573	Front upper valance checking fixture
600596	Front upper valance alignment rod
600962	Bottom ball joint remover
600963	Engine sling
600967	Choke and fuel reverse cables spanner
600968	Pinion preload gauge
601476	Ball swivels extractor
601942	Switch spanner
601943	Switch spanner
601952	Glove box lock spanner
605004	Differential pinion setting gauge
605008	Differential drive collar shaft remover
605227	Shock absorber and stabiliser rod compressor
605330	Carburettor balancing kit
605350	Gudgeon pin remover/replacer
605351	Connecting rod bolt guide tool
605486	Front lower valance and dash checking fixture
605571	Steering box location checking tool
605641	Final drive disc brake shaft remover adapter
605774	Valve guide distance piece
605927	Carburettor adjustment spanner
606456	Front suspension support leg
606457	Front lower valance checking fixture
606600	Steering box lock screw spanner
606601	Steering box adjustment screw spanner
606602 (JD 32)	Steering box worm seals ring expander
606603 (JD 33)	Steering box worm seals ring compressor
606604 (JD 34)	Steering box selector shaft seal saver
606605 (JD 35)	Steering box input shaft seal saver

AUTOMATIC TRANSMISSION

18G 537	Torque wrench
18G 537A (APES)	Torque wrench adapter
18G 1004H (7066H)	Circlip plier points
18G 1004J (7066J)	Circlip plier points
CBW 1C	Pressure test set
CBW 31A	Tool kit
CBW 35B (18G673)	Gearbox cradle
CBW 37A	Clutch spring compressor
CBW 60	Bench cradle
CBW 61	Servo adjuster wrench adapters
CBW 62	Throttle cable seal remover
CBW 87	End float checking gauge
CBW 547B-75	Torque wrench
CBW 547A-50-1B	Multi disc clutch rear and front pump take off adapters
CBW 547A-50-2A (18G537B)	Rear servo adjuster adapter
CBW 547A-50-3	Inhibitor switch locknut wrench
CBW 547A-50-4	Take off plug adapter
CBW 547A-50-5	Screwdriver bit adapter
CBW 548 (18G681)	Rotary torque wrench
CBW 548-1	Screwdriver bit adapter
CBW 548-2A	Front servo adjuster adapter
CBW 548-3 (18G678)	Front servo adjuster adapter
606328	Front brake band adjuster

All service tools mentioned in this manual must be obtained direct from the manufacturers:
Messrs. V. L. Churchill and Co. Ltd., P.O. Box 3, London Road, Daventry.

SERVICE TOOLS

JD 10

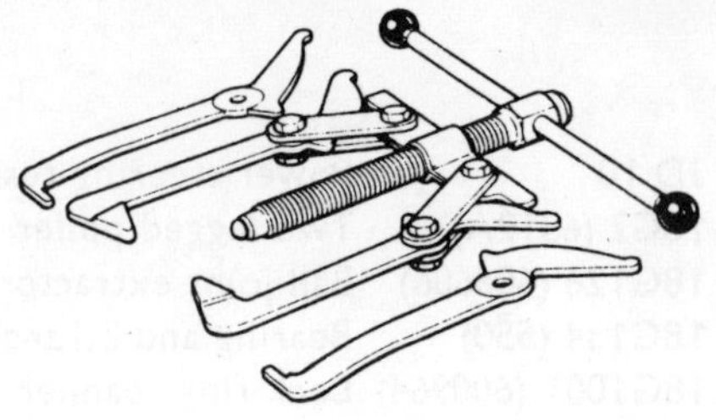

18G 2

18G 126

18G 134

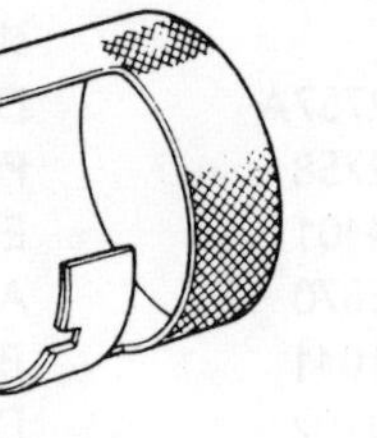

18G 1001

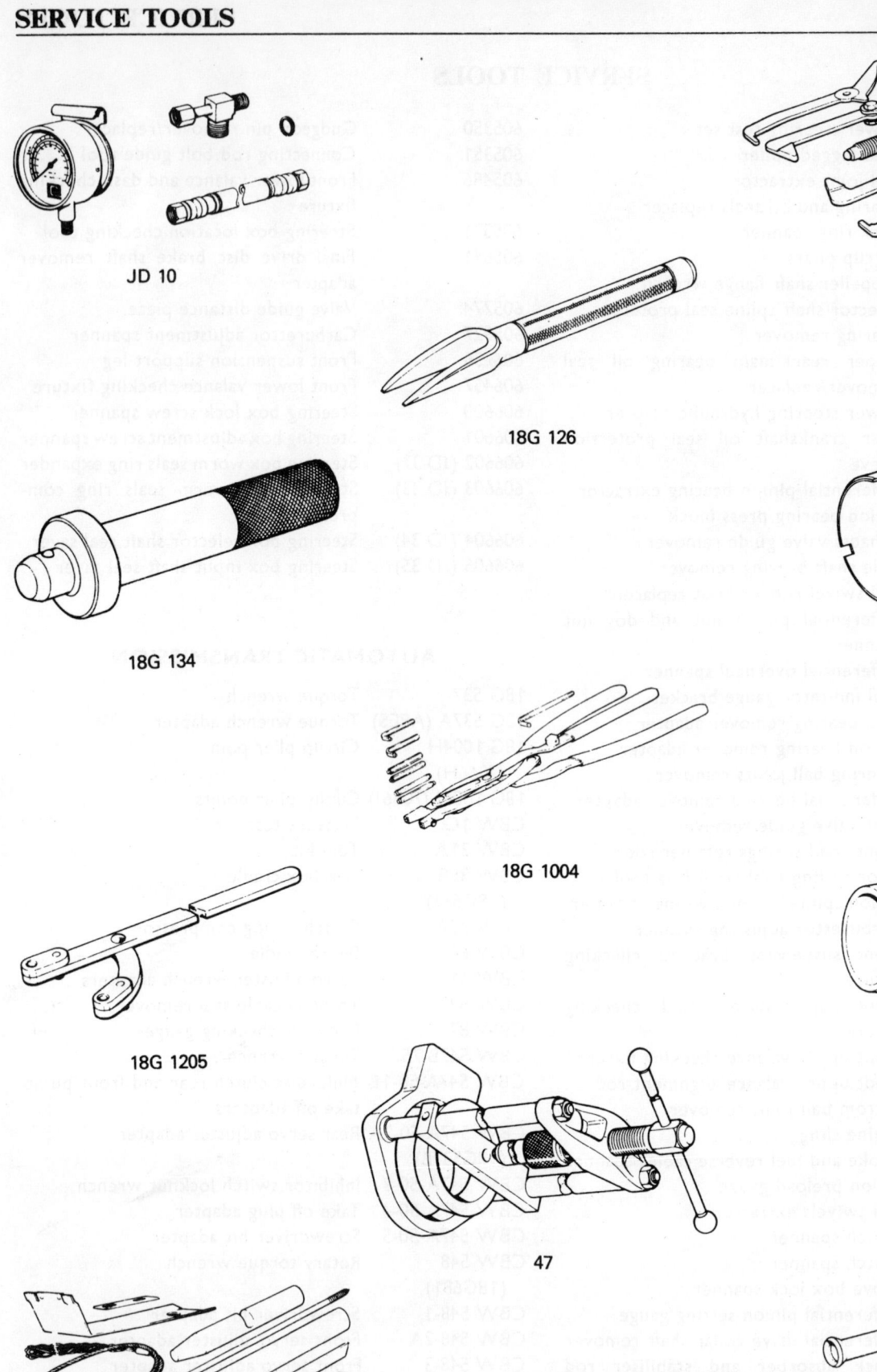

18G 1004

18G 1205

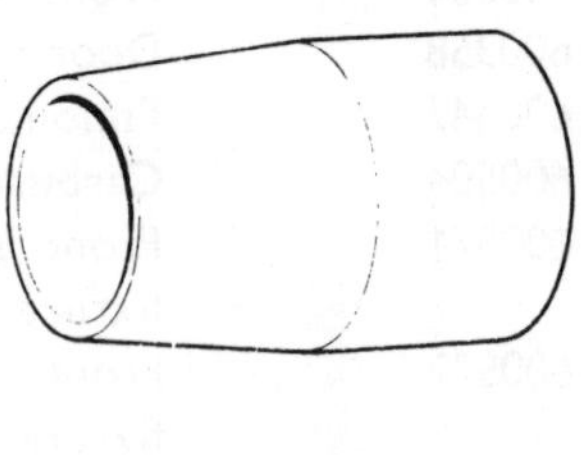

JD 34

47

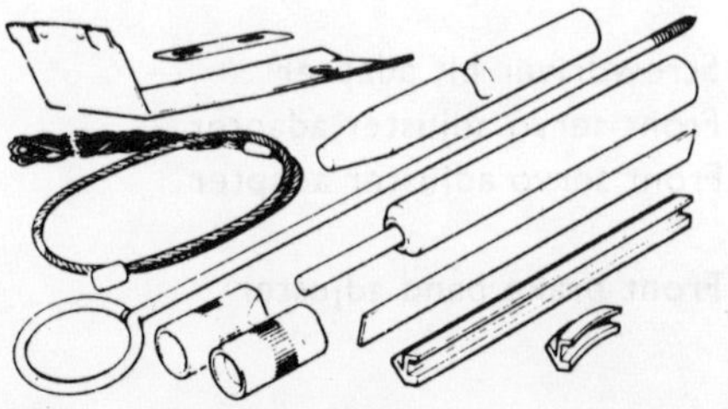

RO 1009

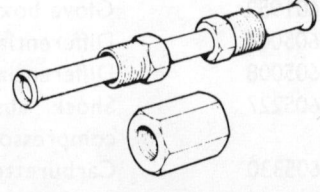

RO 1014

RO 1012

99.00.01

Rover 3500 and 3500S Manual AKM 3621

SERVICE TOOLS

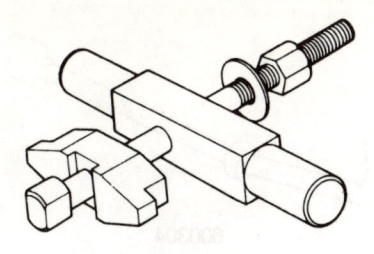

262757A

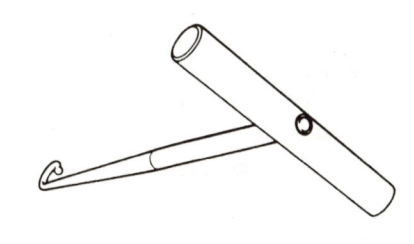

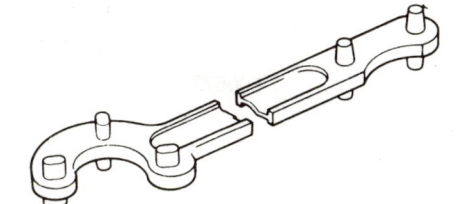

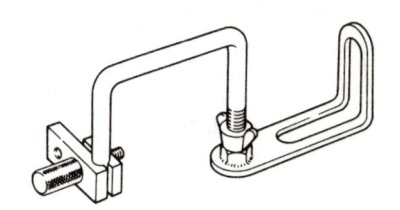

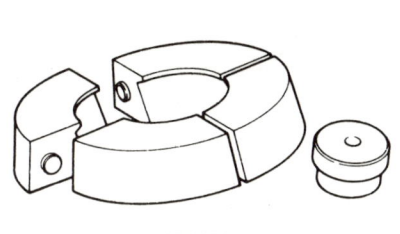

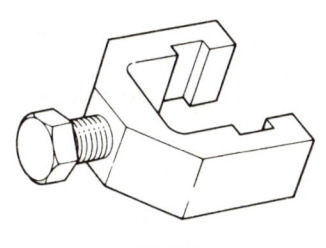

262758

274401

275870

511041

530102

530105

530106

541884

541885

565446

600192

Rover 3500 and 3500S Manual AKM 3621

99.00.02

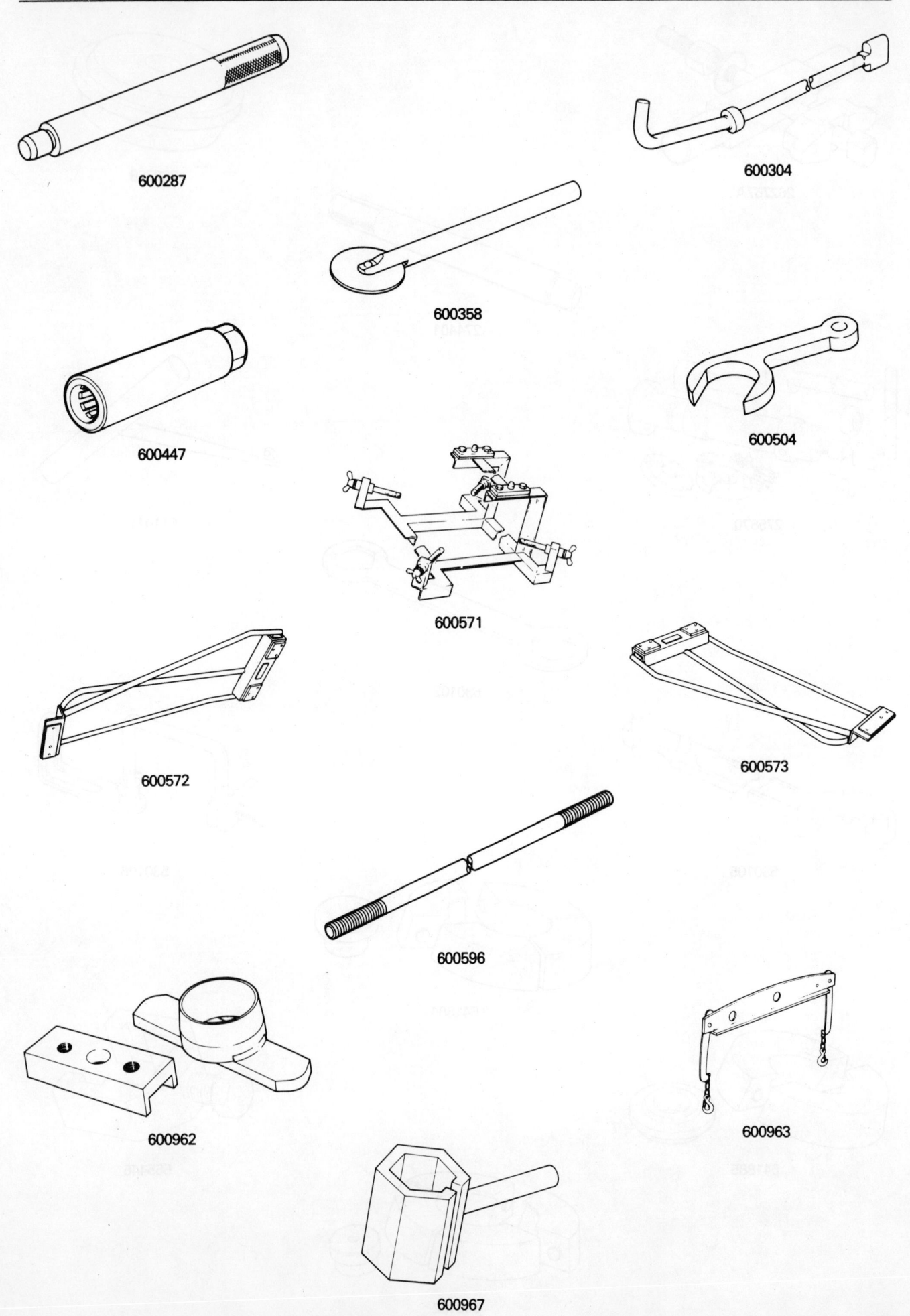

SERVICE TOOLS

600287

600304

600358

600447

600504

600571

600572

600573

600596

600962

600963

600967

99.00.03

Rover 3500 and 3500S Manual AKM 3621

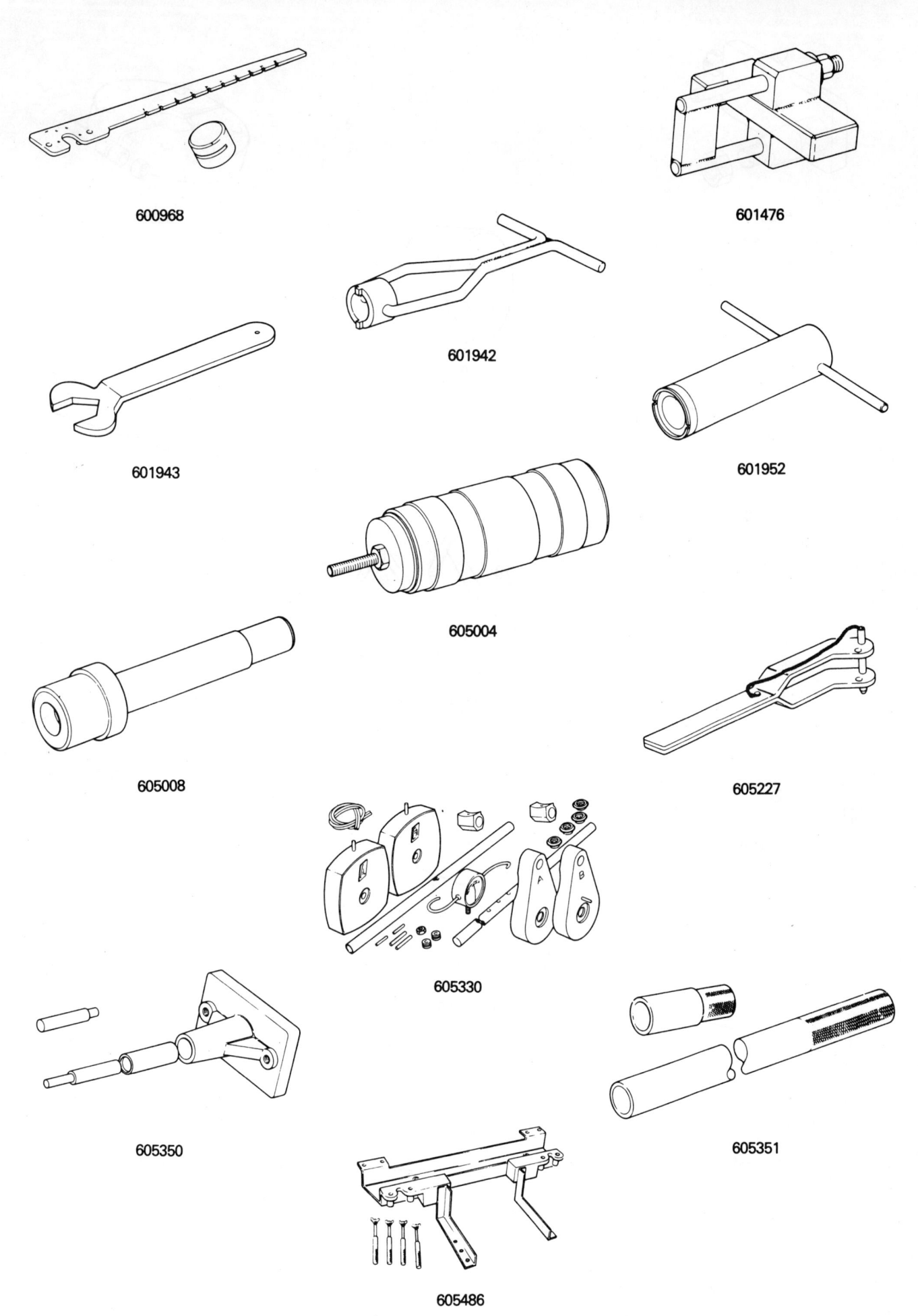

SERVICE TOOLS

600968

601476

601942

601943

601952

605004

605008

605227

605330

605350

605351

605486

Rover 3500 and 3500S Manual AKM 3621

99.00.04

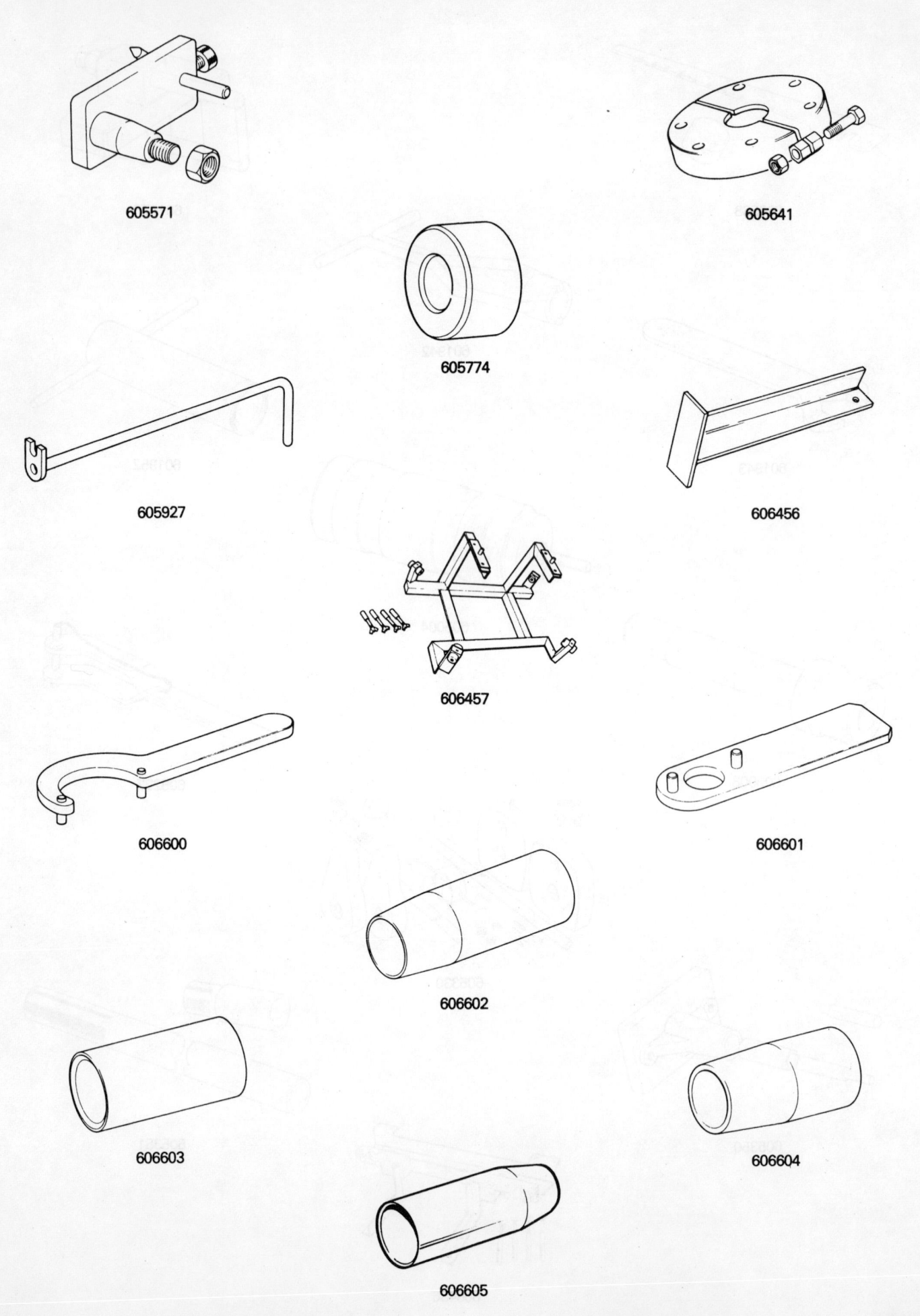

SERVICE TOOLS

605571

605641

605774

605927

606456

606457

606600

606601

606602

606603

606604

606605

99.00.05

Rover 3500 and 3500S Manual AKM 3621

SERVICE TOOLS

AUTOMATIC TRANSMISSION

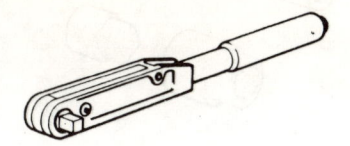

18G 537

18G 537A

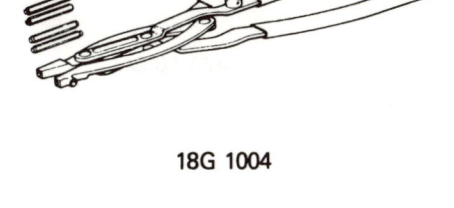

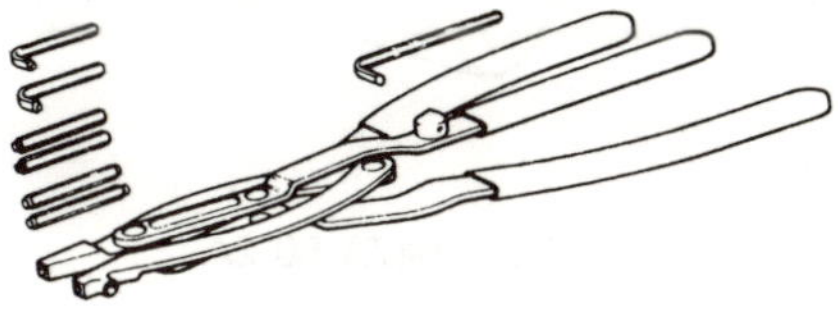

18G 1004

18G 1004H

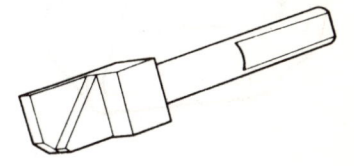

18G 1004J

CBW 1C

CBW 31A

CBW 35B

CBW 37A

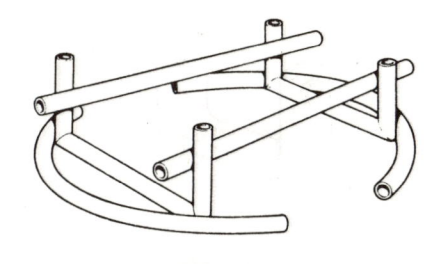

CBW 60

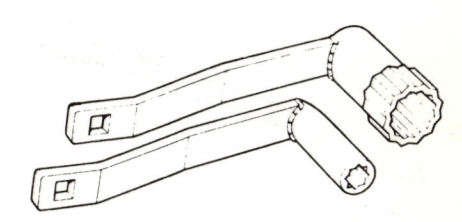

CBW 61

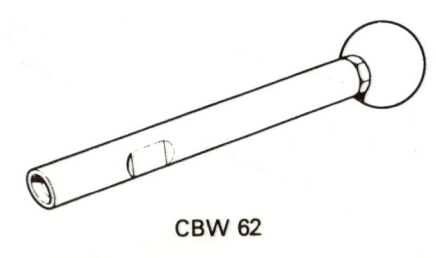

CBW 62

Rover 3500 and 3500S Manual AKM 3621

SERVICE TOOLS

CBW 87

CBW 547A-50-2A

CBW 547A-50-1B

CBW 547A-50-3

CBW 547A-50-4

CBW 547A-50-5

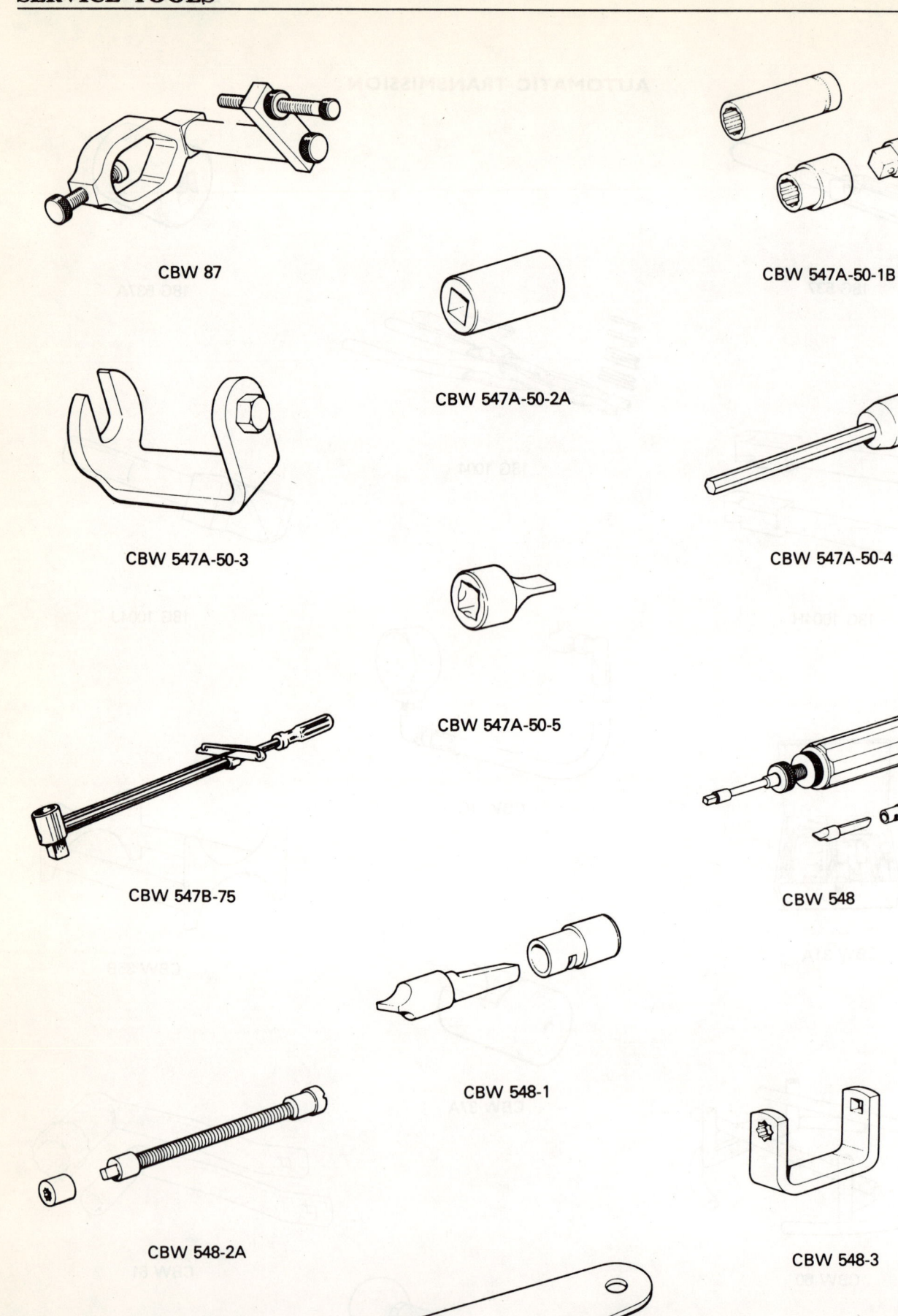

CBW 547B-75

CBW 548

CBW 548-1

CBW 548-2A

CBW 548-3

606328

99.00.07

Rover 3500 and 3500S Manual AKM 3621

PRINTED AND DISTRIBUTED BY BROOKLANDS BOOKS LTD
PO BOX 146, COBHAM, SURREY, KT11 1LG, ENGLAND